AR

ANNUAL REVIEW OF NEUROSCIENCE

Production Editor: JENNIFER E. MANN
Bibliographic Quality Control: MARY A. GLASS
Electronic Content Coordinator: SUZANNE K. MOSES
Subject Indexer: SUZANNE COPENHAGEN

ANNUAL REVIEW OF NEUROSCIENCE

VOLUME 27, 2004

STEVEN E. HYMAN, *Editor*
Harvard University

THOMAS M. JESSELL, *Associate Editor*
Columbia University

CARLA J. SHATZ, *Associate Editor*
Harvard Medical School

CHARLES F. STEVENS, *Associate Editor*
Salk Institute for Biological Studies

www.annualreviews.org   science@annualreviews.org   650-493-4400

ANNUAL REVIEWS
4139 El Camino Way • P.O. Box 10139 • Palo Alto, California 94303-0139

ANNUAL REVIEWS
Palo Alto, California, USA

International Standard Serial Number: 0147-006X
International Standard Book Number: 0-8243-2427-7

∞ The paper used in this publication meets the minimum requirements of American National Standards for Information Sciences—Permanence of Paper for Printed Library Materials, ANSI Z39.48-1992.

TYPESET BY TECHBOOKS, FAIRFAX, VA
PRINTED AND BOUND BY MALLOY INCORPORATED, ANN ARBOR, MI

Annual Review of Neuroscience
Volume 27, 2004

CONTENTS

ERRATA

An online log of corrections to *Annual Review of Neuroscience* chapters may be found at http://neuro.annualreviews.org/

Related Articles

From the ***Annual Review of Biochemistry***, Volume 73 (2004)

Cytochrome C-Mediated Apoptosis, Xuejun Jiang and Xiaodong Wang

Crawling Toward a Unified Model of Cell Motility: Spatial and Temporal Regulation of Actin Dynamics, Susanne M. Rafelski and Julie A. Theriot

Inositol 1,4,5-Trisphosphate Receptors as Signal Integrators, Randen L. Patterson, Darren Boehning, and Solomon H. Snyder

Emerging Principles of Conformation-Based Prion Inheritance, Peter Chien, Jonathan S. Weissman, and Angela H. DePace

Opioid Receptors, Maria Waldhoer, Selena E. Bartlett, and Jennifer L. Whistler

From the ***Annual Review of Biomedical Engineering***, Volume 6 (2004)

Advances in High-Field Magnetic Resonance Imaging, Xiaoping P. Hu and David G. Norris

Advances in Quantitative Electroencephalogram Analysis Methods, Nitish V. Thakor and Shanbao Tong

Robotics, Motor Learning, and Neurologic Recovery, David J. Reinkensmeyer, Jeremy L. Emken, and Steven C. Cramer

From the ***Annual Review of Cell and Developmental Biology***, Volume 19 (2003)

Adult Stem Cell Plasticity: Fact or Artifact, Martin Raff

Cyclic Nucleotide-Gated Ion Channels, Kimberly Matulef and William N. Zagotta

Regulation of MAP Kinase Signaling Modules by Scaffold Proteins in Mammals, Deborah Morrison and Roger J. Davis

From the ***Annual Review of Genomics and Human Genetics***, Volume 4 (2003)

Nonsyndromic Seizure Disorders: Epilepsy and the Use of the Internet to Advance Research, Mark F. Leppert and Nanda A. Singh

The Genetics of Narcolepsy, Dorothée Chabas, Shahrad Taheri, Corinne Renier, and Emmanuel Mignot

From the ***Annual Review of Medicine***, Volume 55 (2004)

The Impact of the Completed Human Genome Sequence on the Development of Novel Therapeutics for Human Disease, Christopher P. Austin

Toward Alzheimer Therapies Based on Genetic Knowledge, John Hardy

The Scientific Basis for the Current Treatment of Parkinson's Disease, C. Warren Olanow

Molecular Neurobiology of Drug Addiction, Jennifer Chao and Eric J. Nestler

From the ***Annual Review of Physiology***, Volume 66 (2004)

Mechanism of Rectification in Inward-Rectifier K^+ *Channels*, Zhe Lu

Metabolic Regulation of Potassium Channels, Xiang Dong Tang, Lindsey Ciali Santarelli, Stefan H. Heinemann, and Toshinori Hoshi

Structure and Function of Glutamate Receptor Ion Channels, Mark L. Mayer and Neali Armstrong

Metabolic Rate and Body Temperature Reduction During Hibernation and Daily Torpor, Fritz Geiser

Sleep and Circadian Rhythms in Mammalian Torpor, H. Craig Heller and Norman F. Ruby

Estrogens in the Nervous System: Mechanisms and Nonreproductive Functions, Adriana Maggi, Paolo Ciana, Silvia Belcredito, and Elisabetta Vegeto

Learning Mechanisms in Addiction: Synaptic Plasticity in the Ventral Tegmental Area as a Result of Exposure to Drugs of Abuse, Julie A. Kauer

Localization of Voltage-Gated Ion Channels in Mammalian Brain, James S. Trimmer and Kenneth J. Rhodes

Myosin-1c, the Hair Cell's Adaptation Motor, Peter G. Gillespie and Janet L. Cyr

Interpreting the BOLD Signal, Nikos K. Logothetis and Brian A. Wandell

Live Optical Imaging of Nervous System Development, Cristopher M. Niell and Stephen J Smith

From the ***Annual Review of Psychology***, Volume 54 (2003)

Addiction, Terry E. Robinson and Kent C. Berridge

Language Processing: Functional Organization and Neuroanatomical Basis, Randi C. Martin

Neuroimaging Studies of Language Production and Comprehension, Morton Ann Gernsbacher and Michael Kaschak

ANNUAL REVIEWS is a nonprofit scientific publisher established to promote the advancement of the sciences. Beginning in 1932 with the *Annual Review of Biochemistry*, the Company has pursued as its principal function the publication of high-quality, reasonably priced *Annual Review* volumes. The volumes are organized by Editors and Editorial Committees who invite qualified authors to contribute critical articles reviewing significant developments within each major discipline. The Editor-in-Chief invites those interested in serving as future Editorial Committee members to communicate directly with him. Annual Reviews is administered by a Board of Directors, whose members serve without compensation.

Annual Reviews of

Anthropology
Astronomy and Astrophysics
Biochemistry
Biomedical Engineering
Biophysics and Biomolecular Structure
Cell and Developmental Biology
Earth and Planetary Sciences
Ecology, Evolution, and Systematics
Entomology
Environment and Resources
Fluid Mechanics
Genetics
Genomics and Human Genetics
Immunology
Materials Research
Medicine
Microbiology
Neuroscience
Nuclear and Particle Science
Nutrition
Pharmacology and Toxicology
Physical Chemistry
Physiology
Phytopathology
Plant Biology
Political Science
Psychology
Public Health
Sociology

SPECIAL PUBLICATIONS
Excitement and Fascination of Science, Vols. 1, 2, 3, and 4

Annu. Rev. Neurosci. 2004. 27:1–28
doi: 10.1146/annurev.neuro.27.070203.144157

THE AMYGDALA MODULATES THE CONSOLIDATION OF MEMORIES OF EMOTIONALLY AROUSING EXPERIENCES

James L. McGaugh
Center for the Neurobiology of Learning and Memory and Department of Neurobiology and Behavior, University of California, Irvine 92697–3800; email; jlmcgaug@uci.edu

Key Words basolateral amygdala, emotional arousal, memory consolidation, hippocampus, cortex

■ **Abstract** Converging findings of animal and human studies provide compelling evidence that the amygdala is critically involved in enabling us to acquire and retain lasting memories of emotional experiences. This review focuses primarily on the findings of research investigating the role of the amygdala in modulating the consolidation of long-term memories. Considerable evidence from animal studies investigating the effects of posttraining systemic or intra-amygdala infusions of hormones and drugs, as well as selective lesions of specific amygdala nuclei, indicates that (*a*) the amygdala mediates the memory-modulating effects of adrenal stress hormones and several classes of neurotransmitters; (*b*) the effects are selectively mediated by the basolateral complex of the amygdala (BLA); (*c*) the influences involve interactions of several neuromodulatory systems within the BLA that converge in influencing noradrenergic and muscarinic cholinergic activation; (*d*) the BLA modulates memory consolidation via efferents to other brain regions, including the caudate nucleus, nucleus accumbens, and cortex; and (*e*) the BLA modulates the consolidation of memory of many different kinds of information. The findings of human brain imaging studies are consistent with those of animal studies in suggesting that activation of the amygdala influences the consolidation of long-term memory; the degree of activation of the amygdala by emotional arousal during encoding of emotionally arousing material (either pleasant or unpleasant) correlates highly with subsequent recall. The activation of neuromodulatory systems affecting the BLA and its projections to other brain regions involved in processing different kinds of information plays a key role in enabling emotionally significant experiences to be well remembered.

INTRODUCTION

Emotionally arousing experiences tend to be well remembered. Studies of the past several decades have provided considerable evidence suggesting that stress hormones released by emotional experiences play an important role in mediating the effects of emotional arousal on lasting memory. Moreover, there is also substantial evidence that neuromodulatory influences occurring selectively within the basolateral amygdala (BLA) regulate the consolidation of memory for various kinds of experiences through BLA projections to many other brain regions involved in storing newly acquired information. This review focuses explicitly on evidence from studies investigating the role of the amygdala in modulating memory consolidation. The role(s) of the amygdala in other aspects of learning, memory, and behavior are considered in other recent reviews (Davis et al. 2003, Everitt et al. 2003, Gallagher 2000, LeDoux 2000, Sah et al. 2003, Schafe et al. 2001, See et al. 2003, Stork & Pape 2002).

THE BASOLATERAL AMYGDALA MODULATES MEMORY CONSOLIDATION

Although studies published over six decades ago of the effects of brain lesions suggested a possible involvement of the amygdala in learning and memory (e.g., Klüver & Bucy 1937, Weiskrantz 1956), studies of the effects of electrical stimulation of the amygdala on learning and memory (Goddard 1964) were the first to suggest that the amygdala plays a role in influencing memory consolidation. Electrical stimulation of the amygdala administered shortly after rats were trained on an aversively motivated task impaired their memory of the training. The conclusion that the amygdala stimulation disrupted the consolidation of the memory of the training was confirmed by many subsequent findings from other laboratories (Kesner & Wilburn 1974, McGaugh & Gold 1976). One possible interpretation of these findings is that the stimulation disrupted the consolidation of memory processes occurring within the amygdala. However, several subsequent findings suggested that amygdala stimulation modulates memory consolidation and does so through influences mediated by amygdala efferents. The finding that posttraining electrical stimulation of the amygdala can either enhance or impair memory for aversive training (inhibitory avoidance), depending on the stimulation intensity and the training conditions (Gold et al. 1975), clearly indicates that the effects are modulatory, not simply memory impairing. Further, the finding that lesions of the *stria terminalis* block the impairing effects of posttraining electrical stimulation of the amygdala on memory for inhibitory avoidance training (Liang & McGaugh 1983) strongly suggested that the modulation involves projections from the amygdala to other brain regions.

Experiments by Ellis & Kesner (1983) and Gallagher et al. (1981) were the first to use posttraining drug infusions to investigate the involvement of the amygdala in memory consolidation. β-adrenoceptor antagonists infused into the amygdala impaired rats' retention of inhibitory avoidance, and concurrent infusion of norepinephrine (NE) blocked the memory impairment (Gallagher et al. 1981). These investigators also found that posttraining intra-amygdala infusions of opioid peptidergic agonists impaired memory and that antagonists enhanced memory. The use of posttraining drug treatments, which were introduced in early studies of drug influences on memory (Breen & McGaugh 1961, McGaugh & Petrinovich 1965), excluded direct influences of the drugs on acquisition or retrieval. Thus, the findings provided compelling evidence suggesting that the amygdala infusions affected memory by influencing memory consolidation (McGaugh 1989, 2000).

Subsequent research has provided extensive confirming evidence that posttraining treatments affecting amygdala functioning influence memory consolidation. The finding that posttraining intra-amygdala infusions of drugs influence retention performance tested 24 h or longer after the training, but do not affect performance tested within a few hours after training, indicates that the treatments selectively affect the consolidation of long-term memory (Barros et al. 2002, Bianchin et al. 1999, Schafe & LeDoux 2000). Moreover, as discussed below, the findings of many recent studies indicate that the BLA is selectively involved in the memory-modulatory influences. The adjacent central nucleus does not appear to play a significant, if any, role in modulating memory consolidation (DaCunha et al. 1999, McGaugh et al. 2000, Parent & McGaugh 1994, Tomaz et al. 1992). Thus, the effects of relatively large intra-amygdala drug infusions (i.e., $>0.2\ \mu l$) that may spread to several amygdala regions, including the infusion volumes (e.g., $1.0\ \mu l$) typically used in many early studies, are most likely due to selective influences on the BLA.

Although many of the experiments investigating BLA involvement in memory consolidation have used inhibitory avoidance training and testing (Izquierdo et al. 1997, McGaugh et al. 2000, McGaugh & Izquierdo 2000, Parent & McGaugh 1994, Wilensky et al. 2000), comparable effects of posttraining amygdala treatments have been obtained in experiments using a wide variety of training tasks, including contextual fear conditioning (LaLumiere et al. 2003, Sacchetti et al. 1999, Vazdarjanova & McGaugh 1999), cued fear conditioning (Sacchetti et al. 1999, Schafe & LeDoux 2000), Y-maze discrimination training (McGaugh et al. 1988), change in reward magnitude (Salinas et al. 1997), conditioned place preference (Hsu et al. 2002, Schroeder & Packard 2000), radial maze appetitive training (Packard & Chen 1999), water-maze spatial and cued training (Packard et al. 1994), conditioned taste aversion (Miranda et al. 2003), olfactory training (Kilpatrick & Cahill 2003a), and extinction of conditioned reward (Schroeder & Packard 2003). Additionally, Walker et al. (2002) reported that intra-amygdala drug infusions (D-cycloserine) administered before extinction training enhance the extinction of conditioned fear. Studies to date have not examined the effects of amygdala infusions administered after fear extinction trials. Although there is abundant evidence that the BLA is involved in modulating memory of aversively motivated training, such as footshock training used in inhibitory avoidance and Pavlovian fear conditioning, the evidence also clearly indicates that the BLA is quite promiscuous in influencing the consolidation of memory for many different kinds of motivationally arousing training experiences (McGaugh 2002b, Packard & Cahill 2001).

As memory for different types of training is known to involve different brain regions (Eichenbaum & Cohen 2001, Packard & Knowlton 2002, Poldrack & Packard 2003), the extensive evidence that posttraining intra-amygdala drug infusions affect memory for many kinds of training is consistent with the hypothesis that the amygdala regulates memory consolidation occurring at other brain sites. In view of the evidence suggesting that the amygdala may be a locus of plasticity mediating fear-based learning (Davis 2000, LeDoux 2000, Shumyatsky et al. 2002, Stork & Pape 2002), posttraining intra-amygdala infusions may also directly influence consolidation occurring within the amygdala (Schafe et al. 2001). This hypothesis is supported by considerable evidence that intra-amygdala infusions of the NMDA receptor antagonist AP-5 typically impair memory for fear-based learning (Fanselow & Kim 1994, Kim & McGaugh 1992, Maren et al. 1996, Stork & Pape 2002, Walker & Davis 2000). However, the evidence that posttraining intra-BLA infusions of AP-5 can either enhance or impair memory consolidation, depending on the training conditions (LaLumiere et al. 2004), suggests that AP-5 effects on memory are not solely due to blocking of neuroplasticity within the BLA. Moreover, memory consolidation is modulated by activating or blocking either NMDA receptors or non-NMDA glutamate receptors in the BLA posttraining (Bonini et al. 2003, Rubin et al. 2001). Additionally, there is considerable evidence that an intact and functioning amygdala is not required for many types of fear-based learning, including inhibitory avoidance and contextual fear conditioning (Amorapanth et al. 2000; Berlau & McGaugh 2003; Cahill et al. 2000; Killcross et al. 1997; Lehmann et al. 2000, 2003; Vazdarjanova & McGaugh 1998). There is also much evidence that memory for these kinds of training is modulated by posttraining drug infusions administered to many other brain regions (e.g., Sacchetti et al. 1999, 2002; Walz et al. 2000). Such evidence clearly suggests that if the BLA is a locus of fear-based memory it is not a unique locus; brain regions other than the amygdala are involved in the consolidation of memory of fear-based training.

STRESS HORMONES INFLUENCE MEMORY CONSOLIDATION VIA NEUROMODULATORY INTERACTIONS WITHIN THE BASOLATERAL AMYGDALA

The early evidence that drugs administered systemically after training can enhance memory consolidation in rats (McGaugh & Petrinovich 1965) suggested the possibility that adrenal stress hormones, epinephrine and corticosterone, released by the emotionally arousing training typically used in such studies may play a role in aiding the consolidation of memory of the training experiences (Gold & McGaugh 1975). There is now extensive evidence supporting this hypothesis. Posttraining systemic injections of epinephrine or corticosterone, as well as drugs affecting adrenergic and glucocorticoid receptors, produce dose-dependent and time-dependent enhancement of memory (Bohus 1994, de Kloet 1991, Gold &

van Buskirk 1975, Izquierdo & Diaz 1983, Lupien & McEwen 1997, McEwen & Sapolsky 1995, Sandi & Rose 1994). Furthermore, these two hormones interact in influencing memory consolidation (Borrell et al. 1983). Administration of metyrapone, a drug that attenuates the increase in corticosterone induced by aversive stimulation, blocks the memory-enhancing effects of posttraining systemic injections of epinephrine (Roozendaal et al. 1996). Moreover, there is considerable evidence that the BLA is critically involved in mediating the memory-modulating effects of epinephrine and corticosterone (McGaugh et al. 2000; McGaugh & Roozendaal 2002; Roozendaal 2000, 2002).

Noradrenergic Influences

The finding that *stria terminalis* lesions block the memory-enhancing effects of systemically administered epinephrine (Liang & McGaugh 1983) was the first to suggest that epinephrine effects on memory might involve the amygdala (see also Torras-Garcia et al. 1998). Additional evidence supporting this possibility was provided by the finding that epinephrine effects on memory are also blocked by lesions of the amygdala (Cahill & McGaugh 1991) as well as by intra-amygdala infusions of the β-adrenoceptor antagonist propranolol (Liang et al. 1986). An important implication of these findings, as well as the earlier findings of Gallagher et al. (1981), was that memory consolidation is influenced by NE release in the amygdala. In strong support of this implication, many subsequent studies have reported that posttraining intra-amygdala (or intra-BLA) infusions of NE or the β-adrenoceptor agonist clenbuterol enhance memory in rats (e.g., Ferry & McGaugh 1999; Hatfield & McGaugh 1999; Introini-Collison et al. 1991, 1996; Izquierdo et al. 1992; Liang et al. 1990, 1995). Moreover, intra-amygdala infusions of the noradrenergic toxin DSP-4 impair inhibitory avoidance memory, and posttraining intra-amygdala infusions of NE block the impairment (Liang 1998). The finding of Liang et al. (1990) that *stria terminalis* lesions block the memory-enhancing effects of posttraining intra-amygdala infusions of NE is consistent with the hypothesis that the effects involve projections to other brain regions mediated, at least in part, by that pathway.

In addition to β-adrenoceptor influences, α-adrenoceptor influences in the BLA also modulate memory consolidation. Posttraining intra-BLA infusions of the α_1-adrenoceptor antagonist prazosin impair inhibitory avoidance memory, whereas infusions of the α_1 agonist phenylephrine, together with yohimbine, a presynaptic α_2-adrenoceptor antagonist, enhance retention (Ferry et al. 1999b). The α_1-induced enhancement is likely due to an interaction involving β-adrenoceptors because posttraining intra-BLA infusions of the β-adrenoceptor antagonist atenolol block the memory enhancement produced by activation of α_1 receptors. The evidence that retention is enhanced by posttraining intra-amygdala infusions of the synthetic cAMP analog 8-bromo-cAMP (Liang et al. 1995) suggests that activation of β-adrenoceptors modulates memory via a direct coupling to adenylate cyclase. Thus, the finding that blocking α_1-adrenoceptors with intra-amygdala infusions of prazosin does not prevent the memory enhancement induced by concurrently infused 8-bromo-cAMP is consistent with the hypothesis that the memory-enhancing effects of α-adrenoceptor activation are mediated by an interaction with β-adrenoceptors (Ferry et al. 1999a).

There is also considerable evidence that the amygdala is involved in mediating the memory-modulating effects of systemically administered stress hormones and drugs. As noted above, *stria terminalis* or amygdala lesions or infusions of β-adrenoceptor antagonists block the memory-enhancing effects of epinephrine. Findings of several studies suggest that epinephrine effects involve activation of β-adrenoreceptors on the ascending vagus projecting to the nucleus of the solitary tract (NTS), which provides direct NE activation of the amygdala as well as indirect activation via projections to the locus coeruleus (Liang 2001, Williams & Clayton 2001). The finding that intra-NTS infusions of the β-adrenoceptor antagonist propranolol (Clayton & Williams 2000b) or inactivation of the NTS with lidocaine (Williams & McGaugh 1992) prevent epinephrine enhancement of memory provides evidence that the NTS is part of a brain stem system that, together with the locus coeruleus, enables epinephrine-induced memory enhancement. Amygdala or *stria terminalis* lesions also block the effects of systemically administered drugs affecting GABAergic, opioid peptidergic, and muscarinic cholinergic receptors (Ammassari-Teule et al. 1991, Introini-Collison et al. 1989, McGaugh et al. 1986, Tomaz et al. 1992). Moreover, the evidence that β-adrenoceptor antagonists infused into the amygdala (and, more specifically, the BLA) block the memory-modulating effects of these systemically administered drugs affecting GABAergic and opiate peptidergic receptors indicates that their effects, like those produced by intra-amygdala infusions, involve a common action on NE activation within the amygdala (McGaugh et al. 1988, 2000). Ragozzino & Gold (1994) reported that posttraining intra-amygdala infusions of glucose block the memory-impairing effects of the opiate drug morphine. However, as such glucose infusions do not attenuate the memory impairment induced by propranolol (Lennartz et al. 1996, McNay & Gold 1998), glucose effects appear to act prior to adrenergic activation within the amygdala, consistent with evidence that intra-amygdala propranolol infusions block opiate effects on memory consolidation (McGaugh et al. 1988). The finding that glutamate infused concurrently with propranolol prevented propranolol-induced memory impairment (Lennartz et al. 1996) suggests that glutamate influences in the amygdala occur at a step beyond noradrenergic activation.

Evidence from studies using in vivo microdialysis and high-performance liquid chromatography (HPLC) to assess NE release in the amygdala supports the view that NE release in the amygdala plays an important, perhaps even critical, role in amygdala influences on memory. Footshock stimulation like that used in inhibitory avoidance training induces NE release in the amygdala; the NE levels vary directly with the footshock intensity (Galvez et al. 1996, Quirarte et al. 1998). Systemic administration of memory-enhancing doses of picrotoxin or naloxone (GABA and opioid receptor antagonists, respectively) increases NE release, whereas injections of memory-impairing doses of muscimol or β-endorphin (GABA and opioid receptor agonists, respectively) decrease NE release (Hatfield et al. 1999, Quirarte et al. 1998). Moreover, systemic administration of memory-enhancing doses of epinephrine, electrical stimulation of the vagus nerve projecting to the NTS, or infusions of glutamate or the β-adrenoceptor clenbuterol into the NTS also increase NE release and enhance memory (Clayton & Williams 2000a, Hassert 2004, Miyashita & Williams 2002, Williams et al. 1998).

Recent experiments also examined the effects of inhibitory avoidance training on amygdala NE release (McIntyre et al. 2003). As is shown in Figure 1*A*, the training induced an increase in NE release that was sustained for at least 2 h. Additionally, the NE release induced by the single footshock training trial was greater than that induced in animals given a similar footshock in a holding cage (Galvez et al. 1996, Quirarte et al. 1998). Figure 1*B* shows the NE levels of the individual rats given inhibitory avoidance training as well as the retention score (latency to reenter the shock compartment) on a 24-h test. Animals with higher NE levels after training displayed better retention performance than those with lower levels. The NE levels during the first five sample periods following the training correlated highly with subsequent retention performance. These in vivo microdialysis findings fit well with the evidence from pharmacological studies suggesting that noradrenergic activation of the amygdala plays an important role in amygdala modulation of memory consolidation.

Cholinergic Influences

The finding that ST lesions block the memory-enhancing effect of systemically administered cholinergic drugs (Introini-Collison et al. 1989) was the first to suggest that cholinergic drugs may affect memory consolidation, at least in part, through an influence involving the amygdala. Subsequent studies provided extensive evidence that, in rats, posttraining intra-amygdala infusions of muscarinic cholinergic agonists and antagonists enhance and impair, respectively, memory for many kinds of training, including inhibitory avoidance, Pavlovian fear conditioning, conditioned place preference, and change in reward magnitude (Introini-Collison et al. 1996, Passani et al. 2001, Power et al. 2003, Power & McGaugh 2002, Salinas et al. 1997, Schroeder & Packard 2002, Vazdarjanova & McGaugh 1999). Findings of experiments using posttraining infusions of the muscarinic cholinergic agonist oxotremorine administered together with selective M1 and M2 antagonists indicate that both receptor types are involved in the memory-enhancing effects of cholinergic activation (Power et al. 2003). Lesions of the nucleus basalis, the major source of cholinergic activation of the BLA, impair the learning and retention of inhibitory avoidance, and posttraining intra-BLA infusions of either oxotremorine or the acetylcholinesterase inhibitor physostigmine attenuate the memory impairment (Power & McGaugh 2002). In contrast to the effects of GABAergic and opioid peptidergic drugs, cholinergic effects are not mediated by adrenergic activation. Intra-amygdala infusions of β-adrenoceptor antagonists do not block the

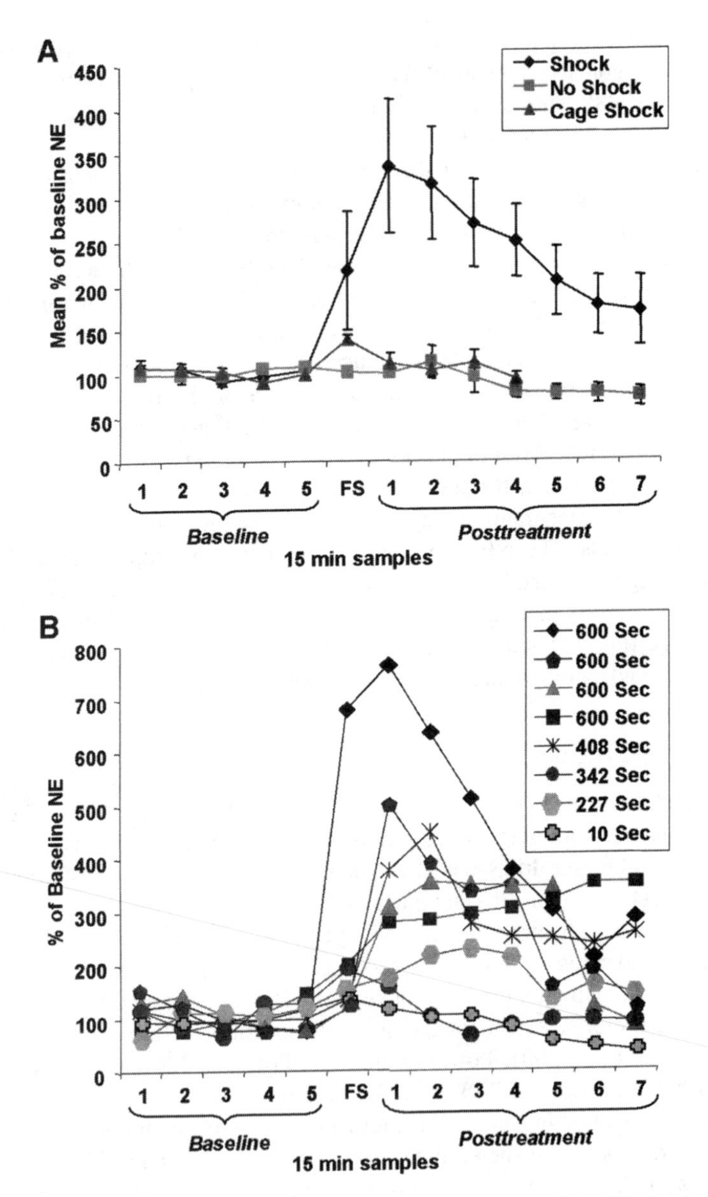

Figure 1 (*A*) Inhibitory avoidance training increases amygdala NE levels as assessed by in vivo microdialysis and high-performance liquid chromatography (HPLC). The No Shock animals were treated as the Shock (i.e., trained) animals but received no footshock. The Cage Shock animals received footshock in a different apparatus. (*B*) Amygdala NE release assessed by in vivo microdialysis and HPLC and inhibitory avoidance retention latencies of individual rats given footshock training. The correlation between NE levels assessed on the first 5 posttraining samples and 24-h retention latencies varied from +0.75 to +0.92. From McIntyre et al. 2002.

memory-enhancing effects of concurrently infused oxotremorine. Moreover, the muscarinic cholinergic antagonist atropine, at a subeffective dose when infused alone, blocks the memory-enhancing effects of intra-amygdala infusions of clenbuterol (Dalmaz et al. 1993, Salinas et al. 1997). Thus, cholinergic activation in the BLA, like that of glutamate effects discussed above, appears to provide modulatory influences on memory storage downstream from adrenergic activation.

Studies of the effects of histamine antagonists and agonists infused into the BLA provide additional evidence of a role of acetylcholine in the BLA in modulating consolidation (Cangioli et al. 2002, Passani et al. 2001). H_3 antagonists (ciproxifan, clobenprobit, or thioperamide) infused into the BLA decreased acetylcholine release, as assessed by in vivo microdialysis, and concurrent infusions of the H_2 agonist cimetidine blocked the decreased acetylcholine release. Moreover, posttraining intra-BLA infusions of H_3 antagonists, administered in doses found to decrease acetylcholine release, impaired memory for contextual fear conditioning (Passani et al. 2001). There is also evidence that acetylcholine is released within the amygdala during training. Studies using in vivo microdialysis have shown that acetylcholine levels in the amygdala increase while rats perform a spontaneous alternation task and that the increase is correlated with good performance on the task (Gold 2003, McIntyre et al. 2003). Because an intact amygdala is not required for the learning or retention of this task it is highly unlikely that the acetylcholine affects performance by influencing neuroplasticity within the amygdala. Rather, because this task is known to involve hippocampal functioning, these findings are consistent with other evidence, discussed below, suggesting that the amygdala influences memory processing that involves the hippocampus.

Glucocorticoid Influences

The glucocorticoid corticosterone (cortisol in humans) secreted from the adrenal cortex following arousing or stressful stimulation readily enters the brain. As noted above, extensive evidence indicates that posttraining systemically administered corticosterone and the synthetic glucocorticoid dexamethasone enhance memory consolidation and that the enhancement is due to selective activation of glucocorticoid receptors (Bohus 1994; Cordero & Sandi 1998; de Kloet 1991; Hui et al. 2004; Lupien & McEwen 1997; McGaugh & Roozendaal 2002; Oitzl & de Kloet 1992; Pugh et al. 1997; Roozendaal 2000, 2002; Sandi et al. 1997; Sandi & Rose 1994; Zorawski & Killcross 2002). Furthermore, the effects of corticosterone, like those of epinephrine, are mediated by the BLA. Selective lesions of the BLA block the memory-enhancing effects of dexamethasone (Roozendaal & McGaugh 1996a), and selective posttraining intra-BLA infusions of the glucocorticoid receptor agonist RU28362 enhance memory for inhibitory avoidance training (Roozendaal & McGaugh 1997b).

Additionally, glucocorticoid influences on memory, like that of epinephrine, involve β-adrenoceptor and cholinergic activation within the BLA. Intra-BLA infusions of a β-adrenoceptor antagonist block the memory-enhancing effects of systemic injections of dexamethasone as well as the effects of the glucocorticoid receptor agonist RU28362 infused into the BLA concurrently (Quirarte et al. 1997). The enhancement of memory for inhibitory avoidance training induced by intra-BLA infusions of RU28362 is also blocked by concurrent infusion of Rp-cAMPS, a drug that inhibits PKA activity and thus blocks the norepinephrine signal cascade. Moreover, intra-BLA infusions of the glucocorticoid receptor antagonist RU38486 attenuate the memory-enhancing effects of clenbuterol infused concurrently such that a much higher dose of the β-adrenoceptor agonist clenbuterol (100 ng versus 1 ng) is required to induce memory enhancement (Roozendaal et al. 2002a). Such findings suggest that activation of glucocorticoid receptors in the BLA may facilitate memory consolidation by potentiating the norepinephrine signal cascade through an interaction with G protein–mediated effects.

Glucocorticoid effects on memory also appear to involve activation of brain stem nuclei, including the NTS, that send noradrenergic projections to the amygdala. The finding that infusions of RU28362 into the NTS enhance inhibitory avoidance retention and that intra-BLA infusions of a β-adrenoceptor antagonist block the enhancement (Roozendaal et al. 1999a) provides additional evidence that the NTS influences on memory consolidation involve noradrenergic influences within the amygdala (Clayton & Williams 2000a,b; Miyashita & Williams 2002; Williams et al. 1998, 2001). Other findings indicate that cholinergic activation within the BLA is also critical for enabling glucocorticoid enhancement of memory consolidation. Atropine infused into the BLA blocks the memory-enhancing effects of RU28362 infused concurrently posttraining, as well as the effects of systemically administered dexamethasone (Power et al. 2000). Finally, memory-modulating effects like those found with glucocorticoids are also found with corticotropin-releasing hormone (CRH). Posttraining intra-amygdala infusions of CRH enhance inhibitory avoidance retention (Liang & Lee 1988), and posttraining intra-BLA infusions of a specific CRH receptor antagonist impair retention (Roozendaal et al. 2002b).

Neuromodulatory Interactions Within the Amygdala Modulate Memory Consolidation

Figure 2 summarizes schematically the interactions of posttraining neuromodulatory influences within the amygdala affecting memory consolidation, as discussed in the preceding sections. The influences of epinephrine, opioid peptides, and GABA (gamma amino butyric acid) converge in regulating norepinephrine release in the amygdala. The effects of activation of α-adrenoceptors involve an interaction with β-adrenoceptor activation. Acetylcholine and glutamate influences occur at steps beyond the activation of β-adrenoceptors. Glucocorticoid effects on memory consolidation involve activation of brain stem nuclei (NTS and locus coeruleus) that send noradrenergic projections to the amygdala as well as potentiating the noradrenergic signal cascade within the amygdala. As discussed below, extensive evidence suggests that projections from the amygdala mediated by the ST, as

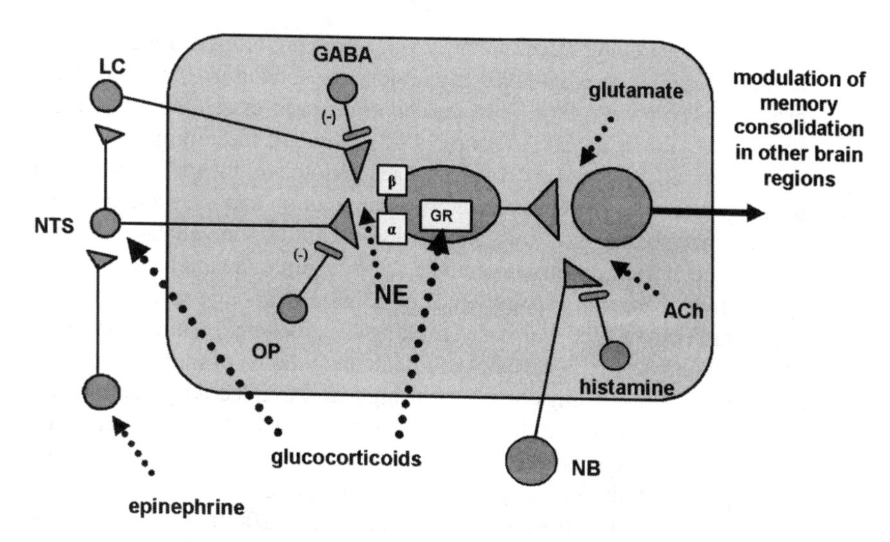

Figure 2 Schematic representation of neuromodulatory interactions within the basolateral amygdala affecting memory consolidation suggested by the experimental findings. NE release within the amygdala is critical for memory-modulatory influences. Epinephrine released from the adrenal medulla activates receptors on the ascending vagus projecting to the nucleus of the solitary tract (NTS), which sends noradrenergic projections to the amygdala, as well as the locus coeruleus (LC), which also sends noradrenergic projections to the amygdala. Opioid peptides (OP) and GABA inhibit norepinephrine (NE) release. Corticosterone released from the adrenal cortex activates glucocorticoid receptors (GR) in the NTS and BLA and elsewhere in the brain. In the BLA, corticosterone interacts with β-adrenergic activation. Glutamatergic and cholinergic activation (ACh) from the nucleus basalis (NB) occur at a step beyond noradrenergic activation. Activation of histamine receptors regulates ACh release. These modulatory influences converge in activating amygdala projections to other brain regions involved in memory consolidation.

well as other pathways, modulate memory consolidation occurring in other brain regions.

THE BASOLATERAL AMYGDALA INTERACTS WITH OTHER BRAIN SYSTEMS IN MODULATING MEMORY CONSOLIDATION

There is now abundant evidence that neuromodulatory interactions occurring within the amygdala influence memory consolidation. Although it remains possible that such influences may be due, at least in part, to influences on neuroplasticity within the amygdala, several kinds of evidence suggest that alterations in amygdala functioning affect memory consolidation through amygdala influences on other brain regions involved in memory consolidation. First, the BLA sends projections to many other brain regions (Petrovich et al. 2001, Pitkänen 2000, Price 2003, Sah et al. 2003, Young 1993). Some of these projections are mediated by the ST. Thus, the evidence, discussed above, that ST lesions block the memory-modulating effects of electrical stimulation and intra-amygdala drug infusions strongly suggests that modulation within the amygdala is not sufficient to affect memory; efferent projections seem required. Second, posttraining treatments affect memory for many kinds of training. Although many (perhaps most) of the studies have used fear-based training (i.e., footshock), the evidence summarized above clearly indicates that amygdala modulation of memory consolidation is not restricted to findings of studies using fear-based learning tasks. Third, there is evidence that training known to involve the amygdala (Pavlovian fear conditioning) induces the expression of several transcriptionally regulated genes implicated in synaptic plasticity in many brain areas, including the hippocampus, striatum, and cortex, as well as the amygdala (Ressler et al. 2002). As these effects were seen only when the stimuli used in the training induced behavioral learning, they were not simply due to nonspecific effects of stress or arousal but appeared to be involved in memory consolidation. Fourth, as discussed below, extensive evidence from many types of studies indicates that the amygdala interacts with other brain regions in modulating the consolidation of memory for different kinds of training (McGaugh 2002a).

Caudate Nucleus, Hippocampus, and Nucleus Accumbens

The amygdala sends direct projections to the caudate nucleus (via the ST) and both direct and indirect projections to the hippocampus (Petrovich et al. 2001, Pitkänen 2000). The finding that ST lesions block the memory-enhancing effects of oxotremorine infused posttraining into the caudate nucleus suggests that efferents from the amygdala influence memory processing involving the caudate nucleus (Packard et al. 1996). Much evidence indicates that the caudate and hippocampus are involved in different kinds of learning (e.g., McDonald & White 1993, Packard & McGaugh 1992). In a study of rats given water-maze training (swimming to a platform submerged below the water surface), Packard and colleagues (Packard & Teather 1998, Packard et al. 1994) found that amphetamine infused posttraining into the caudate selectively enhanced memory of visually cued training, whereas infusions administered into the dorsal hippocampus selectively enhanced memory of spatial training. In contrast, amphetamine infused into the amygdala posttraining enhanced memory for both types of training. Additional findings indicated that inactivation of the hippocampus (with lidocaine) prior to testing blocked retention of spatial training and inactivation of the caudate blocked retention of visually cued training. However, inactivation of the amygdala prior to retention testing did not block memory of either kind of training. These findings provide strong evidence that the amygdala modulates the consolidation of both caudate-dependent and hippocampus-dependent tasks and is not a locus of memory for either type of training. In other experiments, Packard and his colleagues found that glutamate infused into the caudate or hippocampus posttraining enhanced response learning or spatial learning, respectively (Packard 1999). Additionally, in rats trained in a radial-maze task, lidocaine infused into the amygdala blocked the memory enhancement induced by posttraining intrahippocampal infusions of glutamate (Packard & Chen 1999).

Other recent findings of studies of the effects of posttraining intra-amygdala infusions of a glucocorticoid agonist provide additional evidence of amygdala-hippocampus interactions in memory consolidation. Unilateral posttraining intrahippocampal infusions of RU28362 enhanced rats' retention of inhibitory avoidance training. The glucocorticoid-induced retention enhancement was blocked selectively by ipsilateral infusions of a β-adrenoceptor antagonist into the BLA or by lesions of the BLA (Roozendaal & McGaugh 1997a; Roozendaal et al. 1999b). These findings indicate that the hippocampus is involved in memory for inhibitory avoidance and that glucocorticoids influence memory by activating hippocampal glucocorticoid receptors as well as receptors located in the NTS and BLA, as discussed above. Other evidence indicates that glucocorticoids affect memory by activating the nucleus accumbens. The BLA projects to the nucleus accumbens primarily via the ST (Kelley et al. 1982, Wright et al. 1996). The possible importance of the BLA-ST-nucleus accumbens pathway in memory consolidation is suggested by the finding that lesions of the nucleus accumbens, like lesions of the BLA, block the memory-enhancing effects of systemically administered dexamethasone (Roozendaal & McGaugh 1996b, Setlow et al. 2000). Furthermore, the finding that unilateral lesions of the BLA combined with contralateral (unilateral) lesions of the nucleus accumbens also blocked dexamethasone effects strongly indicates that these two structures interact via the ST in influencing memory consolidation (Setlow et al. 2000).

Hippocampal and BLA glucocorticoid influences on memory for inhibitory avoidance training also involve the nucleus accumbens as well as the ST. As is shown in Figures 3*A* and 3*B*, bilateral lesions of either the ST or nucleus accumbens block the memory-enhancing effects of posttraining intra-BLA or intrahippocampal infusions of RU28362 (Roozendaal et al. 2001). Because the BLA does not project to the hippocampus via the ST, indirect projections to the hippocampus via the entorhinal cortex seem likely to enable the BLA-hippocampus interaction in memory modulation. Because the hippocampus is known to project to the nucleus accumbens, that region may be a critical locus of converging BLA and hippocampal modulatory influences on memory consolidation. The finding that inactivation of the nucleus accumbens with infusions of bupivacaine prior to training blocks the acquisition of contextual fear conditioning provides evidence consistent with this hypothesis (Haralambous & Westbrook 1999). Understanding the roles of the BLA-nucleus accumbens and hippocampus-nucleus accumbens pathways in modulating glucocorticoid effects on memory consolidation will require further investigation. It also remains to be determined whether these pathways are

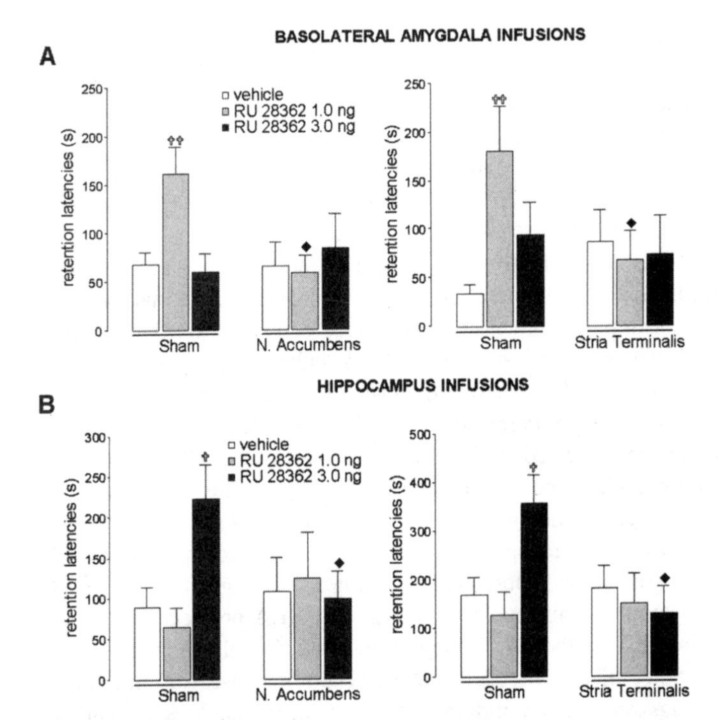

Figure 3 (*A*) Bilateral lesions of the nucleus accumbens or *stria terminalis* block the enhancement of inhibitory avoidance retention induced by posttraining intra-BLA infusions of the glucocorticoid receptor agonist RU28362. (*B*) Bilateral lesions of the nucleus accumbens or *stria terminalis* block the enhancement of inhibitory avoidance retention induced by posttraining intrahippocampal infusions of the glucocorticoid receptor agonist RU28362. From Roozendaal et al. 2001.

also involved in mediating other BLA neuromodulatory influences on memory consolidation.

In most of the studies summarized above, the animals given footshocks during fear-based training received only a few footshocks; only one footshock was given for inhibitory avoidance training. Other studies have reported that more extensive stressful experiences, such as those induced by prolonged restraint, many footshocks, or confinement on an elevated platform prior to training can either enhance or impair acquisition, depending on the type of training task used (Kim et al. 2001, Shors 2001). Additionally, consistent with findings discussed above, amygdala lesions block the stress-induced modulation of learning (Kim et al. 2001) as well as stress-induced enhancement of conditioning (trace conditioning) that is known to depend on the hippocampus and to involve the actions of glucocorticoids (Beylin & Shors 2003). The evidence that stress induced in rats shortly after the induction of long-term potentiation in the dentate gyrus of the hippocampus disrupts the maintenance of LTP (long-term potentiation) (Korz & Frey 2003) is consistent with the evidence suggesting that stress effects on memory involve influences on hippocampal functioning.

Studies of BLA influences on hippocampal neuroplasticity provide additional important evidence of amygdala-hippocampal interactions (Abe 2001). Electrical stimulation of the BLA enhances the induction of LTP in the dentate gyrus of the hippocampus (Akirav & Richter-Levin 1999, Frey et al. 2001, Ikegaya et al. 1995a). Also, selective lesions of the BLA or infusions of a β-adrenoceptor antagonist into the BLA block the induction of LTP in the dentate gyrus (Ikegaya et al. 1994, 1995b, 1997). Also, consistent with the findings of BLA modulation of memory, NE and corticosterone both influence the effects of BLA stimulation on dentate gyrus LTP (Akirav & Richter-Levin 2002). Thus, although it is not known what specific pathway(s) connect the BLA with the hippocampus, it is clear from these findings that alterations in BLA functioning regulate hippocampal neuroplasticity. It is also not yet known whether the BLA influences the induction of LTP in other regions of the hippocampus or in other areas of the brain. The recent findings that Pavlovian fear conditioning induces an increase in synchronization of theta-frequency activity in the lateral amygdala and CA1 region of the hippocampus strongly suggest that activation of an amygdala-hippocampus circuit is involved in fear-based learning (Seidenbecher et al. 2003). More generally, studies of synchronized oscillatory activity within the BLA suggest that such activity may facilitate the temporal lobe as well as neocortical processes involved in consolidating explicit or declarative memory (Paré 2003, Pelletier & Paré 2004).

BLA-Cortical Interactions in Memory Consolidation

It is now well established that various regions of the cortex are involved in memory consolidation. Posttraining infusions of drugs into cortical regions can impair or enhance the consolidation of memory for several kinds of training (Ardenghi et al. 1997, Baldi et al. 1999, Izquierdo et al. 1997, Sacchetti et al. 1999). Moreover, the findings of several recent studies indicate that the BLA modulates cortical functioning involved in memory consolidation (Paré 2003). Neurons within the BLA project directly to the entorhinal cortex (Paré & Gaudreau 1996, Paré et al. 1995, Petrovich et al. 2001, Pikkarainen et al. 1999). Posttraining drug infusions administered into the entorhinal cortex modulate consolidation of memory for inhibitory avoidance training (Izquierdo & Medina 1997). Recent findings have shown that such modulation requires a functioning BLA (Roesler et al. 2002). As is shown in Figure 4*A*, unilateral posttraining infusions of 8-Bromo-cAMP produce dose-dependent memory enhancement. The enhancement is blocked in animals with ipsilateral lesions of the BLA. Further, as is shown in Figure 4*B*, lesions of the BLA contralateral to the entorhinal cortex infusions do not block the memory enhancement induced by 8-Bromo-cAMP. These findings clearly suggest that the BLA influence is mediated by direct connections with the ipsilateral entorhinal cortex and is not due to other possible effects of a BLA lesion. Other recent

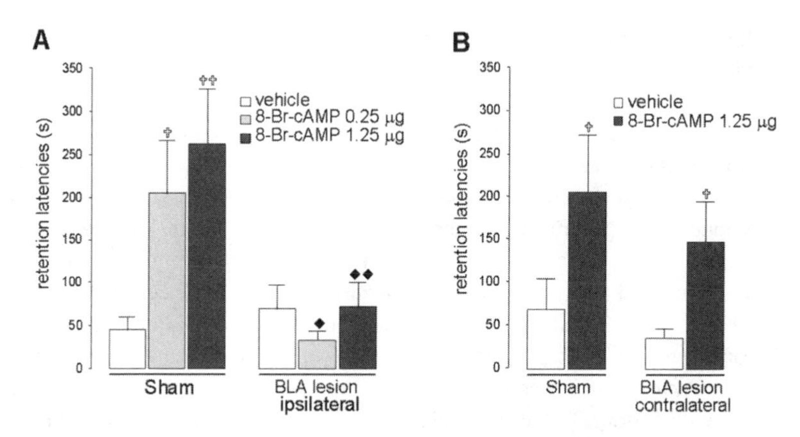

Figure 4 (*A*) Unilateral BLA lesions block the enhancement of inhibitory avoidance retention induced by posttraining ipsilateral infusions of 8-Bromo-cAMP administered into the entorhinal cortex. (*B*) Unilateral BLA lesions do not block the memory-enhancing effects of 8-Bromo-cAMP infused posttraining into the contralateral entorhinal cortex. From Roesler et al. 2002.

studies have reported that BLA lesions also block the memory-enhancing effects of oxotremorine infused posttraining into the anterior cingulate cortex (Malin & McGaugh 2003) and that a β-adrenoceptor antagonist infused into the BLA blocks the memory enhancement induced by 8-Br-cAMP infused into the insular cortex posttraining (M.I. Miranda and J.L. McGaugh, submitted manuscript).

It is likely that the BLA also influences cortical functioning via its projection, largely through the ST (Price 1981), to the nucleus basalis (NB), which provides cholinergic activation of the cortex. The NB-cortical projections are essential for learning-induced cortical plasticity (Miasnikov et al. 2001). Stimulation of the BLA activates the cortex, as indicated by EEG desynchronization, and potentiates NB influences on cortical activation. Moreover, inactivation of the NB with lidocaine blocks the BLA effects on cortical activation (Dringenberg & Vanderwolf 1996, Dringenberg et al. 2001). These findings suggest that the BLA may influence cortical functioning in memory consolidation, at least in part, through its effects on the NB and consequent cholinergic activation of the cortex. Recent findings support this suggestion (Power et al. 2002). Selective lesions of cortical NB corticopetal cholinergic projections induced by 192-IgG saporin blocked the dose-dependent enhancement of inhibitory avoidance induced by posttraining intra-BLA infusions of NE. Although these findings indicate that memory-modulatory influences within the BLA require cholinergic activation of the cortex, it is not clear from these findings whether BLA-induced cholinergic activation is required for such modulation or, alternatively, whether cortical cholinergic activity is required for BLA influences on memory consolidation. In any case, it is clear that a functioning cortex is required for BLA influences on memory consolidation.

EMOTIONAL AROUSAL, STRESS HORMONES, AND AMYGDALA ACTIVATION MODULATE HUMAN MEMORY CONSOLIDATION

The extensive findings of animal experiments discussed above provide considerable evidence that stress hormones released by emotional experiences influence memory consolidation and that the influence is mediated by activation of the amygdala. The findings of many studies of effects of emotional arousal, stress hormones, and amygdala activation on human memory are consistent with those of animal studies (Buchanan & Adolphs 2003, Cahill 2000, Cahill & McGaugh 1998, Dolan 2000, McGaugh 2000). The stress hormone cortisol administered prior to presentations of words or pictures enhanced subsequent recall (Abercrombie et al. 2003, Buchanan & Lovallo 2001). Amphetamine administered to human subjects, either before or after they learned lists of words, also enhanced long-term memory (Soetens et al. 1993, 1995). Administration of the β-adrenoceptor antagonist propranolol to subjects prior to their viewing an emotionally arousing slide presentation blocked the enhancing effects of emotional arousal on long-term memory (Cahill et al. 1994). Epinephrine or cold pressor stress (known to induce the release of epinephrine) administered after subjects viewed emotionally arousing slides enhanced the subjects' long-term memory of the slides (Cahill & Alkire 2003, Cahill et al. 2003). Similar effects were produced by administration of the α_2 adrenoceptor antagonist yohimbine, which acts to stimulate NE release (O'Carroll et al. 1999, Southwick et al. 2002). The findings of these human studies are consistent with those of animal experiments in indicating a critical role of emotional activation and stress hormones in memory consolidation.

There is also considerable evidence from human studies indicating that the amygdala is involved in enabling the enhanced memory induced by emotional arousal. In human subjects with selective bilateral lesions of the amygdala, memory for emotionally arousing material is not enhanced, as it is in normal subjects (Adolphs et al. 1997, Cahill et al. 1995). Unilateral lesions, involving the temporal lobe, that damage the amygdala produce similar effects (Frank & Tomaz 2003, LaBar & Phelps 1998). Studies using positron emission tomography (PET) and fMRI brain imaging have provided additional evidence that the influence of emotional arousal on memory involves activation of the amygdala. Cahill et al. (1996) reported that amygdala activity assessed by PET imaging of subjects as they viewed emotionally arousing films correlated highly (+0.93) with subjects' recall of the films three weeks later. The valence of the emotional influence appears not to be critical for activation of the amygdala (Anderson et al. 2003, Small et al. 2003). In subsequent studies using PET imaging Hamann et al. (1999, 2002) reported that amygdala activity induced by viewing either pleasant or unpleasant slides correlated highly with memory for the slides assessed one month later. Studies using fMRI have obtained highly similar findings. Canli et al. (2000) found that subjects' memory for a series of scenes tested three weeks after brain scanning correlated highly with amygdala activity induced by viewing the scenes. Furthermore, and importantly, the relationship between amygdala activity during encoding and memory was greatest for the scenes rated as most emotionally intense. An additional finding of these studies was that, with both PET and fMRI experiments, activity of the right amygdala was related to enhanced memory in men, whereas activity of the left amygdala correlated with enhanced memory in women (Cahill et al. 2001, Canli et al. 2002). Such findings indicate that it will be essential to consider possible sex differences in future studies of emotionally influenced memory and that understanding the bases of such differences will provide further insight into mechanisms of emotional arousal underlying influences on memory consolidation. There is evidence from human imaging studies that the amygdala is also activated during the retrieval of previously learned emotionally arousing material and that the effect is independent of the valence of the emotional material (Dolan 2000).

Other findings based on an analysis of PET scans provide evidence, consistent with that from many animal studies, indicating that amygdala activation influences memory processing in other brain regions. Hamann et al. (1999) and Dolcos et al. (2003) found that the activity levels of amygdala and hippocampal/parahippocampal regions were correlated during the encoding of emotionally arousing material. The findings of a "path analysis" (structural equation modeling) study (Kilpatrick & Cahill 2003b) of amygdala activity scanned while subjects viewed neutral or emotionally arousing films (Cahill et al. 1996) suggest that emotional arousal increased amygdala influences on activity of the ipsilateral parahippocampal gyrus and ventrolateral prefrontal cortex. These findings provide additional evidence that amygdala influences on other brain regions are critical in creating lasting memories.

CONCLUSIONS

It has long been known, or at least commonly believed, that emotionally arousing experiences tend to be well remembered. The findings of animal and human studies reviewed here provide extensive evidence that supports this conclusion. Additionally, the findings have revealed some of the neurobiological processes that enable emotional experiences to create strong, long-lasting memories. Emotionally significant experiences, whether pleasant or unpleasant, activate hormonal and brain systems that regulate the consolidation of newly acquired memories. Many neuromodulatory systems play a role, and critical interactions among them occur in the basolateral region of the amygdala. These effects are integrated through common actions on noradrenergic and cholinergic activation within the basolateral amygdala, which, in turn, regulate memory consolidation through the amygdala's projections to many other brain regions involved in processing memories of different kinds. Through the activation of these interacting systems, our emotionally exciting experiences become well remembered.

ACKNOWLEDGMENTS

Supported by NIMH Grant MH12526. I thank Nancy Collett and Dan Berlau for assistance in preparation of the manuscript.

The *Annual Review of Neuroscience* is online at http://neuro.annualreviews.org

LITERATURE CITED

Abe K. 2001. Modulation of hippocampal long-term potentiation by the amygdala: a synaptic mechanism linking emotion and memory. *Jpn. J. Pharmacol.* 86:18–22

Abercrombie HC, Kalin NH, Thurow ME, Rosenkranz MA, Davidson RJ. 2003. Cortisol variation in humans affects memory for emotionally laden and neutral information. *Behav. Neurosci.* 117:506–16

Adolphs R, Cahill L, Schul R, Babinsky R. 1997. Impaired declarative memory for emotional material following bilateral amygdala damage in humans. *Learn. Mem.* 4:51–54

Aggleton JP, ed. 2000. *The Amygdala*. London: Oxford Univ. Press. 690 pp.

Akirav I, Richter-Levin G. 1999. Biphasic modulation of hippocampal plasticity by behavioral stress and basolateral amygdala stimulation in the rat. *J. Neurosci.* 19:10530–35

Akirav I, Richter-Levin G. 2002. Mechanisms of amygdala modulation of hippocampal plasticity. *J. Neurosci.* 22:9912–21

Ammassari-Teule M, Pavone F, Castellano C, McGaugh JL. 1991. Amygdala and dorsal hippocampus lesions block the effects of GABAergic drugs on memory storage. *Brain Res.* 551:104–9

Amorapanth P, LeDoux JE, Nader K. 2000. Different lateral amygdala outputs mediate reactions and actions elicited by a fear-arousing stimulus. *Nat. Neurosci.* 3:74–79

Anderson AK, Christoff K, Stappen I, Panitz D, Ghahremani DG, et al. 2003. Dissociated neural representations of intensity and valence in human olfaction. *Nat. Neurosci.* 6:196–202

Ardenghi P, Barros D, Izquierdo LA, Bevilaqua L, Schroder N, et al. 1997. Late and prolonged post-training memory modulation in entorhinal and parietal cortex by drugs acting on the cAMP/protein kinase a signalling pathway. *Behav. Pharmacol.* 8:745–51

Baldi E, Ambrogi Lorenzini C, Sacchetti B, Tassoni G, Bucherelli C. 1999. Effects of combined medial septal area, fimbria-fornix and entorhinal cortex tetrodotoxin inactivations on passive avoidance response consolidation in the rat. *Brain Res.* 821:503–10

Barros DM, Pereira P, Medina JH, Izquierdo I. 2002. Modulation of working memory and of long- but not short-term memory by cholinergic mechanisms in the basolateral amygdala. *Behav. Pharmacol.* 13:163–67

Ben-Ari Y, ed. 1981. *The Amygdaloid Complex*. North Holland: Elsevier

Berlau DJ, McGaugh JL. 2003. Basolateral amygdala lesions do not prevent memory of context-footshock training. *Learn Mem.* 10:495–502

Beylin AV, Shors TJ. 2003. Glucocorticoids are necessary for enhancing the acquisition of associative memories after acute stressful experience. *Horm. Behav.* 43:124–31

Bianchin M, Mello e Souza T, Medina JH, Izquierdo I. 1999. The amygdala is involved in the modulation of long-term memory, but not in working or short-term memory. *Neurobiol. Learn. Mem.* 71:127–31

Bohus B. 1994. Humoral modulation of memory processes. Physiological significance of brain and peripheral mechanisms. In *The Memory System of the Brain*, ed. J Delacour, pp. 337–64. River Edge, NJ: World Sci., Adv. Ser. Neurosci. Vol. 4

Bonini JS, Rodrigues L, Kerr DS, Bevilaqua LR, Cammarota M, Izquierdo I. 2003. AMPA/kainate and group-I metabotropic receptor antagonists infused into different brain areas impair memory formation of inhibitory avoidance in rats. *Behav. Pharmacol.* 14:161–66

Borrell J, De Kloet ER, Versteeg DH, Bohus B. 1983. Inhibitory avoidance deficit following short-term adrenalectomy in the rat: the role of adrenal catecholamines. *Behav. Neural Biol.* 39:241–58

Breen RA, McGaugh JL. 1961. Facilitation of maze learning with posttrial injections of picrotoxin. *J. Comp. Physiol. Psych.* 54:498–501

Buchanan TW, Adolphs R. 2003. The neuroanatomy of emotional memory in humans. In *Memory and Emotion*, ed. D Reisberg, P Hertel, pp. 42–75. New York: Oxford Univ. Press. In press

Buchanan TW, Lovallo WR. 2001. Enhanced memory for emotional material following stress-level cortisol treatment in humans. *Psychoneuroendocrinology* 26:307–17

Cahill L. 2000. Modulation of long-term memory in humans by emotional arousal: adrenergic activation and the amygdala. See Aggleton 2000, pp. 425–46

Cahill L, Alkire M. 2003. Epinephrine enhancement of human memory consolidation: interaction with arousal at encoding. *Neurobiol. Learn. Mem.* 79:194–98

Cahill L, Babinsky R, Markowitsch HJ, McGaugh JL. 1995. The amygdala and emotional memory. *Nature* 377:295–96

Cahill L, Gorski L, Le K. 2003. Enhanced human memory consolidation with post-learning stress: interaction with the degree of arousal at encoding. *Learn. Mem.* 10:270–74

Cahill L, Haier RJ, Fallon J, Alkire M, Tang C, et al. 1996. Amygdala activity at encoding correlated with long-term, free recall of emotional information. *Proc. Natl. Acad. Sci. USA* 93:8016–21

Cahill L, Haier RJ, White NS, Fallon J, Kilpatrick L, et al. 2001. Sex-related difference in amygdala activity during emotionally influenced memory storage. *Neurobiol. Learn. Mem.* 75:1–9

Cahill L, McGaugh JL. 1991. NMDA-induced lesions of the amygdaloid complex block the retention enhancing effect of posttraining epinephrine. *Psychobiology* 19:206–10

Cahill L, McGaugh JL. 1998. Mechanisms of emotional arousal and lasting declarative memory. *TINS* 21:294–99

Cahill L, Prins B, Weber M, McGaugh JL. 1994. β-adrenergic activation and memory for emotional events. *Nature* 371:702–4

Cahill L, Vazdarjanova A, Setlow B. 2000. The basolateral amygdala complex is involved with, but is not necessary for, rapid acquisition of Pavlovian 'fear' conditioning. *Eur. J. Neurosci.* 12:3044–50

Cangioli I, Baldi E, Mannaioni PF, Bucherelli C, Blandina P, Passani MB. 2002. Activation of histamine H_3 receptors in the rat basolateral amygdala improves expression of fear memory and enhances acetylcholine release. *Eur. J. Neurosci.* 16:521–28

Canli T, Desmond JE, Zhao Z, Gabrieli JD. 2002. Sex differences in the neural basis of emotional memories. *Proc. Natl. Acad Sci. USA* 99:10789–94

Canli T, Zhao Z, Brewer J, Gabrieli JD, Cahill L. 2000. Event-related activation in the human amygdala associates with later memory for individual emotional experience. *J. Neurosci.* 20:RC99, 1–5

Clayton EC, Williams CL. 2000a. Adrenergic activation of the nucleus tractus solitarius potentiates amygdala norepinephrine release and enhances retention performance in emotionally arousing and spatial tasks. *Behav. Brain Res.* 112:151–58

Clayton EC, Williams CL. 2000b. Noradrenergic receptor blockade of the NTS attenuates the mnemonic effects of epinephrine in an appetitive light-dark discrimination learning task. *Neurobiol. Learn. Mem.* 74:135–45

Cordero MI, Sandi C. 1998. A role for brain glucocorticoid receptors in contextual fear conditioning: dependence upon training intensity. *Brain Res.* 786:11–17

DaCunha C, Roozendaal B, Vazdarjanova A, McGaugh JL. 1999. Microinfusions of flumazenil into the basolateral but not the central nucleus of the amygdala enhance memory consolidation in rats. *Neurobiol. Learn. Mem.* 72:1–7

Dalmaz C, Introini-Collison IB, McGaugh JL. 1993. Noradrenergic and cholinergic interactions in the amygdala and the modulation of memory storage. *Behav. Brain Res.* 58:167–74

Davis M. 2000. The role of the amygdala in conditioned and unconditioned fear and anxiety. See Aggleton 2000, pp. 213–88

Davis M, Walker DL, Myeres KM. 2003. Role of the amygdala in fear extinction measured with potentiated startle. *Ann. NY Acad. Sci.* 985:218–32

de Kloet ER. 1991. Brain corticosteroid receptor balance and homeostatic control. *Front. Neuroendocrinol.* 12:95–164

Dolan RJ. 2000. Functional neuroimaging of the amygdala during emotional processing and learning. See Aggleton 2000, pp. 631–54

Dolcos F, Graham R, Labar K, Cabeza R. 2003. Coactivation of the amygdala and hippocampus predicts better recall for emotional than for neutral pictures. *Brain Cogn.* 51:221–23

Dringenberg H, Saber AJ, Cahill L. 2001. Enhanced frontal cortex activation in rats by convergent amygdaloid and noxious sensory signals. *NeuroReport* 12:1295–98

Dringenberg H, Vanderwolf C. 1996. Cholinergic activation of the electrocorticogram: an amygdaloid activating system. *Exp. Brain Res.* 108:285–96

Eichenbaum H, Cohen NJ. 2001. *From Conditioning to Conscious Recollection.* New York: Oxford. 583 pp.

Ellis ME, Kesner RP. 1983. The noradrenergic system of the amygdala and aversive memory processing. *Behav. Neurosci.* 97:399–415

Everitt BJ, Cardinal RN, Parkinson JA, Robbins TW. 2003. Appetitive behavior impact of amygdala-dependent mechanisms of emotional learning. *Ann. NY Acad. Sci.* 985:233–50

Fanselow MS, Kim JJ. 1994. Acquisition of contextual Pavlovian fear conditioning is blocked by application of an NMDA receptor antagonist D,L-2-amino-5-phosphonovaleric acid to the basolateral amygdala. *Behav. Neurosci.* 108:210–12

Ferry B, McGaugh JL. 1999. Clenbuterol administration into the basolateral amygdala post-training enhances retention in an inhibitory avoidance task. *Neurobiol. Learn. Mem.* 72:8–12

Ferry B, Roozendaal B, McGaugh JL. 1999a. Basolateral amygdala noradrenergic influences on memory storage are mediated by an interaction between beta- and alpha$_1$-receptors. *J. Neurosci.* 19:5119–23

Ferry B, Roozendaal B, McGaugh JL. 1999b. Involvement of alpha$_1$-adrenergic receptors in the basolateral amygdala in modulation of memory storage. *Eur. J. Pharmacol.* 372:9–16

Frank JE, Tomaz C. 2003. Lateralized impairment of the emotional enhancement of verbal memory in patients with amygdala-hippocampal lesion. *Brain Cogn.* 52:223–30

Frey S, Bergado-Rosado J, Seidenbecher T, Pape HC, Frey JU. 2001. Reinforcement of early long-term potentiation (early-LTP) in dentate gyrus by stimulation of the basolateral amygdala: heterosynaptic induction mechanisms of late-LTP. *J. Neurosci.* 21:3697–703

Gallagher M. 2000. The amygdala and associative learning. See Aggleton 2000, pp. 391–423

Gallagher M, Kapp BS, Pascoe JP, Rapp PR. 1981. A neuropharmacology of amygdaloid systems which contribute to learning and memory. See Ben-Ari 1981, pp. 311–30

Galvez R, Mesches M, McGaugh JL. 1996. Norepinephrine release in the amygdala in response to footshock stimulation. *Neurobiol. Learn. Mem.* 66:253–57

Goddard GV. 1964. Amygdaloid stimulation and learning in the rat. *J. Comp. Physiol. Psychol.* 58:23–30

Gold PE. 2003. Acetylcholine modulation of neural systems involved in learning and memory. *Neurobiol. Learn. Mem.* 80:194–210

Gold PE, Hankins L, Edwards RM, Chester J, McGaugh JL. 1975. Memory interference and facilitation with posttrial amygdala stimulation: effect on memory varies with footshock level. *Brain Res.* 86:509–13

Gold PE, McGaugh JL. 1975. A single-trace, two-process view of memory storage processes. In *Short Term Memory*, ed. D Deutsch, JA Deutsch, pp. 355–78. New York: Academic

Gold PE, van Buskirk R. 1975. Facilitation of time-dependent memory processes with posttrial epinephrine injections. *Behav. Biol.* 13:145–53

Greenough WT, Gold PE, eds. 2001. *Memory Consolidation: Essays in Honor of James L. McGaugh*. Washington, DC: Am. Psychol. Assoc. 402 pp.

Hamann SB, Eli TD, Grafton ST, Kilts CD. 1999. Amygdala activity related to enhanced memory for pleasant and aversive stimuli. *Nat. Neurosci.* 2:289–303

Hamann SB, Eli TD, Hoffman JM, Kilts CD. 2002. Ecstacy and agony: activation of the human amygdala in positive and negative emotions. *Psychol. Sci.* 13:135–41

Haralambous T, Westbrook RF. 1999. An infusion of bupivacaine into the nucleus accumbens disrupts the acquisition but not the expression of contextual fear conditioning. *Behav. Neurosci.* 113:925–40

Hassert DL, Miyashita T, Williams CL. 2004. The effects of peripheral vagal nerve stimulation at a memory modulating intensity on norepinephrine output in the basolateral amygdala. *Behav. Neurosci.* 118:In press

Hatfield T, McGaugh JL. 1999. Norepinephrine infused into the basolateral amygdala posttraining enhances retention in a spatial water maze task. *Neurobiol. Learn. Mem.* 71:232–39

Hatfield T, Spanis C, McGaugh JL. 1999. Response of amygdalar norepinephrine to footshock and GABAergic drugs using *in vivo* microdialysis and HPLC. *Brain Res.* 835:340–45

Hsu EH, Schroeder JP, Packard M. 2002. The amygdala mediates memory consolidation for an amphetamine conditioned place preference. *Behav. Brain Res.* 129:93–100

Hui GK, Davila IR, Poytress BS, Roozendaal B, McGaugh JL, Weinberger NM. 2004. Memory enhancement of classical fear conditioning by post-training injections of corticosterone in rats. *Neurobiol. Learn Mem.* 81: 67–94

Ikegaya Y, Saito H, Abe K. 1994. Attenuated hippocampal long-term potentiation in basolateral amygdala-lesioned rats. *Brain Res.* 656:157–64

Ikegaya Y, Saito H, Abe K. 1995a. High-frequency stimulation of the basolateral amygdala facilitates the induction of long-term potentiation in the dentate gyrus in vivo. *Neurosci. Res.* 22:203–7

Ikegaya Y, Saito H, Abe K. 1995b. Requirement of basolateral amygdala neuron activity for the induction of long-term potentiation in the dentate gyrus in vivo. *Brain Res.* 671:351–54

Ikegaya Y, Saito H, Abe K, Nakanishi K. 1997. Amygdala beta-noradrenergic influence on hippocampal long-term potentiation in vivo. *NeuroReport* 8:3143–46

Introini-Collison IB, Arai Y, McGaugh JL. 1989. Stria terminalis lesions attenuate the effects of posttraining oxotremorine and atropine on retention. *Psychobiology* 17:397–401

Introini-Collison IB, Dalmaz C, McGaugh JL. 1996. Amygdala β-noradrenergic influences on memory storage involve cholinergic activation. *Neurobiol. Learn. Mem.* 65:57–64

Introini-Collison IB, Miyazaki B, McGaugh JL. 1991. Involvement of the amygdala in the memory-enhancing effects of clenbuterol. *Psychopharmacology* 104:541–44

Izquierdo I, DaCunha C, Rosat R, Jerusalinsky D, Ferreira MBC, Medina JH. 1992. Neurotransmitter receptors involved in posttraining memory processing by the amygdala, medial septum and hippocampus of the rat. *Behav. Neur. Biol.* 58:16–26

Izquierdo I, Dias RD. 1983. Effect of ACTH, epinephrine, β-endorphin, naloxone, and of the combination of naloxone or β-endorphine with ACTH or epinephrine on memory consolidation. *Psychoneuroendocrinology* 8:81–87

Izquierdo I, Medina JH. 1997. Memory formation: the sequence of biochemical events in the hippocampus and its connection to activity in other brain structures. *Neurobiol. Learn. Mem.* 68:285–316

Izquierdo I, Quillfeldt JA, Zanatta MS, Quevedo J, Schaeffer E, et al. 1997. Sequential role of hippocampus and amygdala, entorhinal cortex and parietal cortex in formation and retrieval of memory for inhibitory avoidance in rats. *Eur. J. Neurosci.* 9:786–93

Kelley AE, Domesick VB, Nauta WJH. 1982. The amygdalostriatal projection in the rat: an anatomical study by anterograde and retrograde tracing methods. *Neuroscience* 7:615–30

Kesner RP, Wilburn MW. 1974. A review of electrical stimulation of the brain in context of learning and retention. *Behav. Biol.* 10:259–93

Killcross S, Robbins TW, Everitt BJ. 1997. Different types of fear-conditioned behaviour mediated by separate nuclei within amygdala. *Nature* 388:377–80

Kilpatrick L, Cahill L. 2003a. Modulation of memory consolidation for olfactory learning by reversible inactivation of the basolateral amygdala. *Behav. Neurosci.* 117:184–88

Kilpatrick L, Cahill L. 2003b. Amygdala modulation of parahippocampal and frontal regions during emotionally influenced memory storage. *NeuroImage* 20:2092–100

Kim JJ, Lee HJ, Han J-S, Packard MG. 2001. Amygdala is critical for stress-induced modulation of hippocampal long-term potentiation and learning. *J. Neurosci.* 21:5222–28

Kim M, McGaugh JL. 1992. Effects of intra-amygdala injections of NMDA receptor antagonists on acquisition and retention of inhibitory avoidance. *Brain Res.* 585:35–48

Klüver H, Bucy PC. 1937. "Psychic blindness" and other symptoms following bilateral temporal lobectomy in rhesus monkeys. *Am. J. Physiol.* 119:352–53

Korz V, Frey JU. 2003. Stress-related modulation of hippocampal long-term potentiation in rats: involvement of adrenal steroid receptors. *J. Neurosci.* 23:7281–87

LaBar KS, Phelps EA. 1998. Arousal mediated memory consolidation: role of the medial temporal lobe in humans. *Psychol. Sci.* 9:490–93

LaLumiere RT, Buen T-V, McGaugh JL. 2003. Posttraining intra-basolateral amygdala infusions of norepinephrine enhance consolidation of memory for contextual fear conditioning. *J. Neurosci.* 23:6754–58

LaLumiere RT, Pizano E, McGaugh JL. 2004. Intra-basolateral amygdala infusions of AP-5 impair or enhance retention of inhibitory avoidance depending on training conditions. *Neurobiol. Learn. Mem.* 81:60–66

LeDoux JE. 2000. Emotion circuits in the brain. *Annu. Rev. Neurosci.* 23:155–84

Lehmann H, Treit D, Parent MB. 2000. Amygdala lesions do not impair shock-probe avoidance retention performance. *Behav. Neurosci.* 114:107–16

Lehmann H, Treit D, Parent MB. 2003. Spared anterograde memory for shock-probe fear conditioning after inactivation of the amygdala. *Learn. Mem.* 10:261–69

Lennartz RC, Hellems KL, Mook ER, Gold PE. 1996. Inhibitory avoidance impairments induced by intra-amygdala propranolol are reversed by glutamate but not glucose. *Behav. Neurosci.* 110:1033–39

Liang K, Chen L, Huang T-E. 1995. The role of amygdala norepinephrine in memory formation, involvement in the memory enhancing effect of peripheral epinephrine. *Chin. J. Physiol.* 38:81–91

Liang KC. 1998. Pretraining infusion of DSP-4 into the amygdala impaired retention in the inhibitory avoidance task: involvement of norepinephrine but not serotonin in memory facilitation. *Chin. J. Physiol.* 41:223–33

Liang KC. 2001. Epinephrine modulation of memory: amygdala activation and regulation of long-term memory storage. See Greenough & Gold 2001, pp. 165–84

Liang KC, Juler RG, McGaugh JL. 1986. Modulating effects of post-training epinephrine on memory: involvement of the amygdala noradrenergic system. *Brain Res.* 368:125–33

Liang KC, Lee EH. 1988. Intra-amygdala injections of corticotropin releasing factor facilitate inhibitory avoidance learning and reduce exploratory behavior in rats. *Psychopharmacology* (Berl.) 96:232–36

Liang KC, McGaugh JL. 1983. Lesions of the stria terminalis attenuate the enhancing effect of post-training epinephrine on retention of an inhibitory avoidance response. *Behav. Brain Res.* 9:49–58

Liang KC, McGaugh JL, Yao H-Y. 1990. Involvement of amygdala pathways in the influence of posttraining amygdala norepinephrine and peripheral epinephrine on memory storage. *Brain Res.* 508:225–33

Lupien SJ, McEwen BS. 1997. The acute effects of corticosteroids on cognition: integration of animal and human model studies. *Brain Res. Rev.* 24:1–27

Malin E, McGaugh JL. 2003. Basolateral amygdala lesions block the memory enhancing effect of oxotremorine infused into the rostral anterior cingulated cortex after inhibitory avoidance training. *Soc. Neurosci. Abstr.* 290.11

Maren S, Aharonov G, Stote DL, Fanselow MS. 1996. N-methyl-D-aspartate receptors in the basolateral amygdala are required for both acquisition and expression of conditional fear in rats. *Behav. Neurosci.* 110:1365–74

McDonald RJ, White NM. 1993. A triple dissociation of memory systems: hippocampus, amygdala and dorsal striatum. *Behav. Neurosci.* 107:3–22

McEwen BS, Sapolsky RM. 1995. Stress and cognitive function. *Curr. Opin. Neurobiol.* 5:205–16

McIntyre CK, Hatfield T, McGaugh JL. 2002. Amygdala norepinephrine levels after training predict inhibitory avoidance retention performance in rats. *Eur. J. Neurosci.* 16: 1223–26

McIntyre CK, Marriot LK, Gold PE. 2003. Cooperation between memory systems: acetylcholine release in the amygdala correlates positively with performance on a hippocampus-dependent task. *Behav. Neurosci.* 117:320–26

McGaugh JL. 1989. Involvement of hormonal and neuromodulatory systems in the regulation of memory storage. *Annu. Rev. Neurosci.* 12:255–87

McGaugh JL. 2000. Memory: a century of consolidation. *Science* 287:248–51

McGaugh JL. 2002a. Memory consolidation and the amygdala: a systems perspective. *TINS* 25:456–61

McGaugh JL. 2002b. The amygdala regulates memory consolidation. In *Neuropsychology of Memory*, eds. LR Squire, D Schacter, pp. 437–49. New York: Guilford. 3rd ed.

McGaugh JL, Ferry B, Vazdarjanova A, Roozendaal B. 2000. Role in modulation of memory storage. See Aggleton 2000, pp. 391–423

McGaugh JL, Gold PE. 1976. Modulation of memory by electrical stimulation of the brain. In *Neural Mechanisms of Learning and Memory*, ed. MR Rosenzweig, EL Bennett pp. 549–60. Cambridge, MA: MIT Press

McGaugh JL, Introini-Collison IB, Juler RG, Izquierdo I. 1986. Stria terminalis lesions attenuate the effects of posttraining naloxone and β-endorphin on retention. *Behav. Neurosci.* 100:839–44

McGaugh JL, Introini-Collison IB, Nagahara AH. 1988. Memory-enhancing effects of posttraining naloxone: involvement of β-noradrenergic influences in the amygdaloid complex. *Brain Res.* 446:37–49

McGaugh JL, Izquierdo I. 2000. The contribution of pharmacology to research on the mechanisms of memory formation. *TIPS* 21:208–10

McGaugh JL, Petrinovich LF. 1965. Effects of drugs on learning and memory. *Int. Rev. Neurobiol.* 8:139–96

McGaugh JL, Roozendaal B. 2002. Role of adrenal stress hormones in forming lasting memories in the brain. *Curr. Opin. Neurobiol.* 12:205–10

McNay EC, Gold PE. 1998. Memory modulation across neural systems: intra-amygdala glucose reverses deficits caused by intraseptal morphine on a spatial task but not on an aversive task. *J. Neurosci.* 18:3853–58

Miasnikov AA, McLin D 3rd, Weinberger NM. 2001. Muscarinic dependence of nucleus basalis induced conditioned receptive field plasticity. *NeuroReport* 12:1537–42

Miranda MI, LaLumiere RT, Buen T-V, Bermudez-Rattoni F, McGaugh JL. 2003. Blockade of noradrenergic receptors in the basolateral amygdala impairs taste memory. *Eur. J. Neurosci.* 18:2605–10

Miyashita T, Williams CL. 2002. Glutamatergic transmission in the nucleus of the solitary tract modulates memory through influences on amygdala noradrenergic systems. *Behav. Neurosci.* 116:13–21

O'Carroll RE, Drysdale E, Cahill L, Shajahan P, Ebmeier KP. 1999. Stimulation of the noradrenergic system enhances and blockade reduces memory for emotional material in man. *Psychol. Med.* 29:1083–88

Oitzl MS, de Kloet ER. 1992. Selective corticosteroid antagonists modulate specific aspects of spatial orientation learning. *Behav. Neurosci.* 108:62–71

Packard MG. 1999. Glutamate infused posttraining into the hippocampus or caudate putamen differentially strengthens place and response learning. *Proc. Natl. Acad. Sci. USA* 96:12881–86

Packard MG, Cahill L. 2001. Affective modulation of multiple memory systems. *Curr. Opin. Neurobiol.* 11:752–56

Packard MG, Cahill L, McGaugh JL. 1994. Amygdala modulation of hippocampal-dependent and caudate nucleus-dependent memory processes. *Proc. Natl. Acad. Sci. USA* 91:8477–81

Packard MG, Chen SA. 1999. The basolateral amygdala is a cofactor in memory enhancement produced by intrahippocampal glutamate injections. *Psychobiology* 27:377–85

Packard MG, Introini-Collison I, McGaugh JL. 1996. Stria terminalis lesions attenuate memory enhancement produced by intra-caudate nucleus injections of oxotremorine. *Neurobiol. Learn. Mem.* 65:278–82

Packard MG, Knowlton BJ. 2002. Learning and memory functions of the basal ganglia. *Annu. Rev. Neurosci.* 25:563–93

Packard MG, McGaugh JL. 1992. Double dissociation of fornix and caudate nucleus lesions on acquisition of two water maze tasks: further evidence for multiple memory systems. *Behav. Neurosci.* 106:439–46

Packard MG, Teather L. 1998. Amygdala modulation of multiple memory systems: hippocampus and caudate-putamen. *Neurobiol. Learn. Mem.* 69:163–203

Paré D. 2003. Role of the basolateral amygdala in memory consolidation. *Prog. Neurobiol.* 70:409–20

Paré D, Dong J, Gaudreau H. 1995. Amygdalo-entorhinal relations and their reflection in the hippocampal formation: generation of sharp potentials. *J. Neurosci.* 15:2482–503

Paré D, Gaudreau H. 1996. Projection cells and interneurons of the lateral and basolateral amygdala: distinct firing patterns and differential relation to theta and delta rhythms in conscious cats. *J. Neurosci.* 16:3334–50

Parent MB, McGaugh JL. 1994. Posttraining infusion of lidocaine into the amygdala basolateral complex impairs retention of inhibitory avoidance training. *Brain Res.* 661:97–103

Passani MB, Cangioli I, Baldi E, Bucherelli C, Mannaioni PF, Blandina P. 2001. Histamine H_3 receptor-mediated impairment of contextual fear conditioning and *in vivo* inhibition of cholinergic transmission in the rat basolateral amygdala. *Eur. J. Neurosci.* 14:1522–32

Pelletier JG, Paré D. 2004. Role of amygdala oscillations in the consolidation of emotional memories. *Neurosci. Perspec.* In press

Petrovich GD, Canteras NS, Swanson LW. 2001. Combinatorial amygdalar inputs to hippocampal domains and hypothalamic behavior systems. *Brain Res. Rev.* 38:247–89

Pikkarainen M, Ronko S, Savander V, Insausti R, Pitkänen A. 1999. Projections from the lateral, basal, and accessory basal nuclei of the amygdala to the hippocampal formation in rat. *J. Comp. Neurol.* 403:229–60

Pitkänen A. 2000. Connectivity of the rat amygdaloid complex. See Aggleton 2000, pp. 31–115

Poldrack RA, Packard MG. 2003. Competition among multiple memory systems: converging evidence from animal and human studies. *Neuropsychologia* 41:245–51

Power AE, McGaugh JL. 2002. Phthalic acid amygdalopetal lesion of the nucleum basalis magnocellularis induces reversible memory deficits in rats. *Neurobiol. Learn. Mem.* 77:372–88

Power AE, McIntyre CK, Litmanovich A, McGaugh JL. 2003. Cholinergic modulation of memory in the basolateral amygdala involves activation of both m1 and m2 receptors. *Behav. Pharmacol.* 14:207–13

Power AE, Roozendaal B, McGaugh JL. 2000. Glucocorticoid enhancement of memory consolidation in the rat is blocked by muscarinic receptor antagonism in the basolateral amygdala. *Eur. J. Neurosci.* 12:3481–87

Power AE, Thal LJ, McGaugh JL. 2002. Lesions of the nucleus basalis magnocellularis induced by 192 IgG-saporin block memory enhancement with posttraining norepinephrine in the basolateral amygdala. *Proc. Natl. Acad. Sci. USA* 99:2315–19

Price JL. 1981. Toward a consistent terminology for the amygdaloid complex. See Ben-Ari 1981, pp. 13–18

Price JL. 2003. Comparative aspects of amygdala connectivity. *Ann. NY Acad. Sci.* 985:50–58

Pugh CR, Tremblay D, Fleshner M, Rudy JW. 1997. A selective role for corticosterone in contextual-fear conditioning. *Behav. Neurosci.* 111:503–11

Quirarte GL, Galvez R, Roozendaal B, McGaugh JL. 1998. Norepinephrine release in the amygdala in response to footshock and opioid peptidergic drugs. *Brain Res.* 808: 134–40

Quirarte GL, Roozendaal B, McGaugh JL. 1997. Glucocorticoid enhancement of memory storage involves noradrenergic activation in the basolateral amygdala. *Proc. Natl. Acad. Sci. USA* 94:14048–53

Ragozzino ME, Gold PE. 1994. Task-dependent effects of intra-amygdala morphine injections: attenuation by intra-amygdala glucose injections. *J. Neurosci.* 14: 7478–85

Ressler KJ, Paschall G, Zhou X-L, Davis M. 2002. Regulation of synaptic plasticity genes during consolidation of fear conditioning. *J. Neurosci.* 22:7892–902

Roesler R, Roozendaal B, McGaugh JL. 2002. Basolateral amygdala lesions block the memory-enhancing effect of 8-Br-cAMP infused into the entorhinal cortex of rats after training. *Eur. J. Neurosci.* 15:905–10

Roozendaal B. 2000. Glucocorticoids and the regulation of memory consolidation. *Psychoneuroendocrinology* 25:213–38

Roozendaal B. 2002. Stress and memory: opposing effects of glucocorticoids on memory consolidation and memory retrieval. *Neurobiol. Learn. Mem.* 78:578–95

Roozendaal B, Carmi O, McGaugh JL. 1996. Adrenocortical suppression blocks the memory-enhancing effects of amphetamine and epinephrine. *Proc. Natl. Acad. Sci. USA* 93:1429–33

Roozendaal B, de Quervain J-F, Ferry B, Setlow B, McGaugh JL. 2001. Basolateral amygdala-nucleus interactions in mediating glucocorticoid effects on memory consolidation. *J. Neurosci.* 21:2518–25

Roozendaal B, Holloway BL, Brunson KL, Baram TZ, McGaugh JL. 2002b. Involvement of stress-released corticotropin-releasing hormone in the basolateral amygdala in regulating memory consolidation. *Proc. Natl. Acad Sci. USA* 99:13908–913

Roozendaal B, McGaugh JL. 1996a. Amygdaloid nuclei lesions differentially affect glucocorticoid-induced memory enhancement in an inhibitory avoidance task. *Neurobiol. Learn. Mem.* 65:1–8

Roozendaal B, McGaugh JL. 1996b. The memory-modulatory effects of glucocorticoids depend on an intact stria terminalis. *Brain Res.* 709:243–50

Roozendaal B, McGaugh JL. 1997a. Basolateral amygdala lesions block the memory-enhancing effect of glucocorticoid administration in the dorsal hippocampus of rats. *Eur. J. Neurosci.* 9:76–83

Roozendaal B, McGaugh JL. 1997b. Glucocorticoid receptor agonist and antagonist administration into the basolateral but not central amygdala modulates memory storage. *Neurobiol. Learn. Mem.* 67:176–79

Roozendaal B, Nguyen BT, Power A, McGaugh JL. 1999b. Basolateral amygdala noradrenergic influence enables enhancement of memory consolidation induced by hippocampal glucocorticoid receptor activation. *Proc. Natl. Acad. Sci. USA* 96:11642–47

Roozendaal B, Quirarte GL, McGaugh JL. 2002a. Glucocorticoids interact with the basolateral amygdala β-adrenoceptor-cAMP/PKA system in influencing memory consolidation. *Eur. J. Neurosci.* 15:553–60

Roozendaal B, Williams CL, McGaugh JL. 1999a. Glucocorticoid receptor activation in the rat nucleus of the solitary tract facilitates memory consolidation: involvement of the basolateral amygdala. *Eur. J. Neurosci.* 11:1317–23

Rubin MA, Stiegemeier JA, Volkweis MA, Oliveira DM, Fenili AC, et al. 2001. Intra-amygdala spermidine administration improves inhibitory avoidance performance in rats. *Eur. J. Pharmacol.* 423:35–39

Sacchetti B, Baldi E, Lorenzini CA, Bucherelli C. 2002. Cerebellar role in fear conditioning consolidation. *Proc. Natl. Acad. Sci. USA* 99:8406–11

Sacchetti B, Lorenzini CA, Baldi E, Tassoni G, Bucherelli C. 1999. Auditory thalamus, dorsal hippocampus, basolateral amygdala, and perihinal cortex role in the consolidation of conditioned freezing to context and to acoustic conditioned stimulus in the rat. *J. Neurosci.* 19:9570–78

Sah P, Faber ES, Lopez De Armentia M, Power J. 2003. The amygdaloid complex: anatomy and physiology. *Physiol. Rev.* 83:803–34

Salinas JA, Introini-Collison IB, Dalmaz C, McGaugh JL. 1997. Posttraining intra-amygdala infusion of oxotremorine and propranolol modulate storage of memory for reductions in reward magnitude. *Neurobiol. Learn. Mem.* 68:51–59

Sandi C, Loscertales M, Guaza C. 1997. Experience-dependent facilitating effect of corticosterone on spatial memory formation in the water maze. *Eur. J. Neurosci.* 9:637–42

Sandi C, Rose SPR. 1994. Corticosterone enhances long-term potentiation in one-day-old chicks trained in a weak passive avoidance learning paradigm. *Brain Res.* 647:106–12

Schafe GE, LeDoux JE. 2000. Memory consolidation of auditory Pavlovian fear conditioning requires protein synthesis and protein kinase A in the amygdala. *J. Neurosci.* 20:RC96, 1–5

Schafe GE, Nader K, Blair HT, LeDoux JE. 2001. Memory consolidation of Pavlovian fear conditioning: a cellular and molecular perspective. *TINS* 24:540–46

Schroeder JP, Packard MG. 2000. Differential effects of intra-amygdala lidocaine infusion on memory consolidation and expression of a food conditioned place preference. *Psychobiology* 28:486–91

Schroeder JP, Packard MG. 2002. Posttraining intra-basolateral amygdala scopolamine impairs food- and amphetamine-induced conditioned place preferences. *Behav. Neurosci.* 116:922–27

Schroeder JP, Packard MG. 2003. Systemic or intra-amygdala injections of glucose facilitate memory consolidation for extinction of drug-induced conditioned reward. *Eur. J. Neurosci.* 17:1482–88

See RE, Fuches RA, Ledford CC, McLaughlin J. 2003. Drug addiction, relapse, and the amygdala. *Ann. NY Acad. Sci.* 985:294–307

Seidenbecher T, Laxmi TR, Stork O, Pape H-C. 2003. Synchronization of amygdalar and hippocampal theta oscillations during retrieval of Pavlovian fear memory. *Science* 301:846–50

Setlow B, Roozendaal B, McGaugh JL. 2000. Involvement of a basolateral amygdala complex–nucleus accumbens pathway in glucocorticoid-induced modulation of memory storage. *Eur. J. Neurosci.* 12:367–75

Shors TJ. 2001. Acute stress rapidly and persistently enhances memory formation in the male rat. *Neurobiol. Learn. Mem.* 75:10–29

Shumyatsky GP, Tsvetkov E, Malleret G, Vronskaya S, Hatton M, et al. 2002. Identification of a signaling network in lateral nucleus of amygdala important for inhibiting memory specifically related to learned fear. *Cell* 111:905–18

Small DM, Gregory MD, Mak YE, Gitelman D, Mesulam MM, Parrish T. 2003. Dissociation of neural representation of intensity and affective valuation in human gustation. *Neuron* 39:701–11

Soetens E, Casaer S, D'Hooge R, Hueting JE. 1995. Effect of amphetamine on long-term retention of verbal material. *Psychopharmacology* (Berl.) 119:155–62

Soetens E, D'Hooge R, Hueting JE. 1993. Amphetamine enhances human-memory consolidation. *Neurosci. Lett.* 161:9–12

Southwick S, Davis M, Horner B, Cahill L, Morgan D, et al. 2002. Relationship of enhanced norepinephrine activity during memory consolidation to enhanced long-term memory in humans. *Am. J. Psychiatry* 159: 1420–22

Stork O, Pape H-C. 2002. Fear memory and the amygdala: insights from a molecular perspective. *Cell Tissue Res.* 310:271–77

Tomaz C, Dickinson-Anson H, McGaugh JL. 1992. Basolateral amygdala lesions block diazepam-induced anterograde amnesia in an inhibitory avoidance task. *Proc. Natl. Acad. Sci. USA* 89:3615–19

Torras-Garcia M, Costa-Miserachs D, Portell-Cortés I, Morgado-Bernal I. 1998. Posttraining epinephrine and memory consolidation in rats with different basic learning capacities. The role of the stria terminalis. *Exp. Brain Res.* 121:20–28

Vazdarjanova A, McGaugh JL. 1998. Basolateral amygdala is not critical for cognitive memory of contextual fear conditioning. *Proc. Natl. Acad. Sci. USA* 95:15003–7

Vazdarjanova A, McGaugh JL. 1999. Basolateral amygdala is involved in modulating consolidation of memory for classical fear conditioning. *J. Neurosci.* 19:6615–22

Walker DL, Davis M. 2000. Involvement of NMDA receptors within the amygdala in short- versus long-term memory for fear conditioning as assessed with fear-potentiated startle. *Behav. Neurosci.* 114:1019–33

Walker DL, Ressler KJ, Lu KT, Davis M. 2002. Facilitation of conditioned fear extinction by systemic administration or intra-amygdala infusions of D-cycloserine as assessed with fear-potentiated startle in rats. *J. Neurosci.* 22:2343–51

Walz R, Roesler R, Quevedo J, Sant'Anna MK, Madruga M, et al. 2000. Time-dependent impairment of inhibitory avoidance retention in rats by posttraining infusion of a mitogen-activated protein kinase inhibitor into cortical and limbic structures. *Neurobiol. Learn. Mem.* 73:11–20

Weiskrantz L. 1956. Behavioral changes associated with ablation of the amygdaloid complex in monkeys. *J. Comp. Physiol. Psychol.* 49:381–91

Wilensky AE, Schafe GE, LeDoux JE. 2000. The amygdala modulates memory consolidation of fear-motivated inhibitory avoidance learning but not classical conditioning. *J. Neurosci.* 20:759–66

Williams CL, Clayton EC. 2001. Contribution of brainstem structures in modulating memory storage processes. See Greenough & Gold 2001, pp. 141–64

Williams CL, McGaugh JL. 1992. Reversible inactivation of the nucleus of the solitary tract impairs retention performance in an inhibitory avoidance task. *Behav. Neur. Biol.* 58:204–10

Williams CL, Men D, Clayton EC, Gold PE. 1998. Norepinephrine release in the amygdala after systemic injections of epinephrine or inescapable footshock: contribution of the nucleus of the solitary tract. *Behav. Neurosci.* 112:1414–22

Wright CI, Beijer AVJ, Groenewegen HF. 1996. Basal amygdaloid afferents to the rat nucleus accumbens are compartmentally organized. *J. Neurosci.* 16:1877–93

Young MP. 1993. The organization of neural systems on the primate cerebral cortex. *Proc. R. Soc.* 252:13–18

Zorawski M, Killcross S. 2002. Posttraining glucocorticoid receptor agonist enhances memory in appetitive and aversive Pavlovian discrete-cue conditioning paradigms. *Neurobiol. Learn. Mem.* 78:458–64

Annu. Rev. Neurosci. 2004. 27:29–51
doi: 10.1146/annurev.neuro.27.070203.144143

First published online as a Review in Advance on February 23, 2004

CONTROL OF CENTRAL SYNAPTIC SPECIFICITY IN INSECT SENSORY NEURONS

Jonathan M. Blagburn[1] and Jonathan P. Bacon[2]
[1]Institute of Neurobiology and Department of Physiology, Medical Sciences Campus, University of Puerto Rico, San Juan, Puerto Rico 00901-1123; email: jmblagbu@neurobio.upr.clu.edu
[2]School of Life Sciences, University of Sussex, Falmer, Brighton BN1 9QG, United Kingdom; email: J.P.Bacon@sussex.ac.uk

Key Words *Drosophila*, cell-adhesion molecule, optic lobe, olfactory, Dscam

■ **Abstract** Synaptic specificity is the culmination of several processes, beginning with the establishment of neuronal subtype identity, followed by navigation of the axon to the correct subdivision of neuropil, and finally, the cell-cell recognition of appropriate synaptic partners. In this review we summarize the work on sensory neurons in crickets, cockroaches, moths, and fruit flies that establishes some of the principles and molecular mechanisms involved in the control of synaptic specificity. The identity of a sensory neuron is controlled by combinatorial expression of transcription factors, the products of patterning and proneural genes. In the nervous system, sensory axon projections are anatomically segregated according to modality, stimulus quality, and cell-body position. A variety of cell-surface and intracellular signaling molecules are used to achieve this. Synaptic target recognition is also controlled by transcription factors such as Engrailed and may be, in part, mediated by cadherin-like molecules.

INTRODUCTION

The brain is often likened to a computer. One of the many flaws in this comparison is that the components of a brain, unlike those of a computer, wire themselves together. The accuracy with which neurons select their synaptic partners during development is critical to proper functioning of any nervous system, but how neurons achieve this synaptic specificity is still not understood at the molecular level. In general, it is clear that synaptic specificity is the culmination of a series of developmental processes. First of all, the type and subtype of neuronal identity must be established. According to this identity, the axon must then navigate to the appropriate region of the nervous system and form a terminal arborization in a suitable subdivision of the neuropil. Then the correct synaptic partners must be selected from the cell types available in that local environment. Following synaptic

0147-006X/04/0721-0029$14.00

assembly, in many systems, particularly those of vertebrates, activity-dependent refinement of synaptic connections takes place.

For the last three decades, insects have been favorable model systems for addressing this problem. There are several reasons for this. Insects exhibit complex behaviors, yet their nervous systems comprise much fewer cells than those of vertebrates. Many of these neurons are individually identifiable; this is an obvious advantage for addressing questions of neuronal selectivity. These identifiable neurons are often relatively large in size, making them amenable to electrophysiological or anatomical assays of connectivity. More recently, the genetic tractability of the fruit fly, *Drosophila melanogaster*, has made possible several major advances in our understanding of the biochemical mechanisms underlying specificity. Sensory neurons of insects offer particular experimental advantages because they are located in the periphery. They are accessible to transplantation and are often present in stereotyped arrays. They are also accessible to RNA interference, allowing knock out of genes in insect species that are refractory to conventional molecular genetic intervention.

The purpose of this review is to summarize our understanding of the control of synaptic specificity by insect sensory neurons. Some major advances have been made in the fields of neuronal determination, axon guidance, and synaptic target recognition by studying insect CNS neurons or motor neurons—it is not within our purview to cover these areas. We make no claim that this will be a complete or exhaustive review of all the literature, but simply an overview that reflects our particular research interests.

DIFFERENT TYPES OF INSECT SENSORY NEURONS

Type I sensory organs have a specialized sensory structure associated with them, often clonally related to the sensory neurons themselves. They can be divided into (*a*) external sense (es) organs, which include the mechanosensory bristles and hairs that are distributed all over the body, strain-sensitive (mechanosensory) campaniform sensilla at the bases of appendages, olfactory sensilla that are concentrated on the antennae, and gustatory bristles present on mouthparts, legs, and even wings; and (*b*) internally located stretch receptors or chordotonal (ch) organs. Type II sensory organs have no associated structures and include the internal touch-sensitive and nociceptive multidendritic (md) neurons that branch extensively under the epidermis. Lastly, light-sensitive photoreceptors are found in the compound eyes and ocelli. These cells are distinguished by a microvillar stack of membranes, the rhabdomere, which contains the rhodopsin photopigments.

How are the identities of these different types of sensory neurons determined so that they can form appropriate synaptic connections in the CNS? We consider several sensory systems in turn, focusing on the genetic mechanisms that distinguish sensillum type and position, since these are most relevant to synaptic specificity.

Neuronal Specification of Type I Sense Organs of the *Drosophila* PNS

Type I mechanosensory neurons develop from a single precursor, the sensory organ precursor (SOP), otherwise known as a sensory mother cell (SMC) (see reviews by Ghysen & Dambly-Chaudiere 2000, Jan & Jan 1993). The SOP divides asymmetrically, giving rise to a family of support cells and neurons, the number of which varies according to organ type. Mechanosensory bristles comprise a clone of five cells: the neuron, the sheath cell that enwraps it (termed the thecogen), the support cells that secrete the bristle shaft (trichogen) and cuticular socket (tormogen), and a glial cell that migrates away (Gho et al. 1999). The Delta-Notch signaling pathway and segregation of Numb protein are required to control the fates of these different cell types—these complex mechanisms have been studied extensively and are reviewed elsewhere (Jan & Jan 2000, Posakony 1994).

The location at which a sensory organ develops is determined by the expression of proneural genes. These genes, encoding bHLH transcription factors, are initially expressed by a cluster of ectodermal cells, known as an equivalence group, making them all competent to form neural precursors. Subsequently, interactions between these cells restrict the expression of the proneural protein to only one of them, which becomes the SOP. The involvement of Notch and the *Enhancer of split* gene complex in these interactions has been reviewed previously (Bray 2000, Simpson 1997). Proneural proteins form functional dimers with the Daughterless protein (Murre et al. 1989).

The type of sensory organ (e.g., external bristle or internal chordotonal organ) that develops at a given location is determined by the particular proneural gene that the cluster expresses. External sensory organs are specified by proneural genes of the *achaete-scute* complex (AS-C), which induces expression of the homeobox selector gene *cut* (Blochlinger et al. 1991, Campuzano & Modolell 1992). Internal chordotonal organs, on the other hand, are specified by the proneural gene *atonal*, which represses *cut* (Jarman & Ahmed 1998, Jarman et al. 1993). Other genes acting at later stages in the SOP lineage affect the type of sense organ produced: SOPs that express *pox-neuro* (*poxn*) form an open-shafted hair innervated by several gustatory neurons (Dambly-Chaudiere et al. 1992), one of which is distinguished by expression of the bHLH gene *tap* (Gautier et al. 1997). The *Bar* genes determine the formation of campaniform sensilla, rather than mechanosensory bristles (Higashijima et al. 1992).

The positioning of proneural clusters, and ultimately of the sensory organs themselves, appears to be controlled by domains of transcription factor expression in the ectoderm. Pre-pattern factors that control bristle position on the adult *Drosophila* thorax and wing include Araucan and Caupolican (Gomez-Skarmeta et al. 1996), Pannier (Heitzler et al. 1996), and the products of the *BarH* genes (Sato et al. 1999). The expression of these factors is, in turn, influenced by global positioning signals such as Wingless, Engrailed, Hedgehog, and Decapentaplegic (Gomez-Skarmeta & Modolell 1996) and by the homeotic genes that control body regional identity, although this aspect of sensory organ positioning is not fully understood (Ghysen & Dambly-Chaudiere 2000).

Specification of Olfactory Neuron Identity in *Drosophila*

Flies have two organs that bear olfactory sensilla: the antenna and the maxillary palp. There are several types of sensilla on the antenna, each one innervated by up to four sensory neurons. As with mechanosensory organs, differential expression of proneural genes in a cluster of ectodermal cells determines the type of sensillum that forms. The bHLH transcription factor Atonal, required for chordotonal organ formation, also specifies the formation of the coeloconic type of olfactory sensillum (Jhaveri et al. 2000), whereas Amos is required for the other types (Goulding et al. 2000). Lozenge (Lz), a Runt domain-containing transcription factor, regulates the amount of Amos; high levels of Lz result in basiconic sensilla, low levels in trichoid sensilla (Gupta et al. 1998).

Within antennal basiconic sensilla, there are 16 functional classes of olfactory receptor neuron based on physiological responses to odors; pairs of neurons are combined within 7 classes of sensillum, each restricted to a defined region (de Bruyne et al. 2001). The response characteristics of an individual neuron arise from the expression of 1 of at least 60 different types of odor receptor (Or) protein in its dendrite (Dobritsa et al. 2003). In one class of maxillary palp olfactory neuron it is known that the POU-domain transcription factor Acj6 controls these characteristics (Clyne et al. 1999), but otherwise the genetic mechanisms that determine the different subtypes and functional classes of olfactory organ are unknown.

Specification of Type II Sensory Neurons

Type II multidendritic (md) neurons have mixed lineages, many arising as siblings of external sensory or chordotonal neurons, others arising from their own precursor (Brewster & Bodmer 1995). Some md precursors, whether producing a mixed lineage or not, require AS-C and Cut (Vervoort et al. 1997); others require Atonal (Jarman et al. 1993). The remainder are specified by another proneural gene, *amos* (Huang et al. 2000). In mixed lineages, expression of Hamlet, a Zn-finger transcription factor, ensures that a neuron adopts the external sensory fate, forming a single dendrite (Moore et al. 2002). It was shown recently that the size and complexity of the somatic dendritic tree formed by the dendritic arborization class of md neuron are regulated by the amount of Cut that the neurons express (Grueber et al. 2003).

Specification of Photoreceptor Identity

The *Drosophila* compound eye comprises approximately 750 facets or ommatidia (Wolff & Ready 1993). Beneath the lens of an ommatidium lies a bundle of 8 photoreceptors or retinula cells, R1–R8, and 12 auxiliary cells. The UV-sensitive

R7 is located inside the outer ring of six wide-spectrum UV- to green-sensitive photoreceptors (R1–R6), and underneath that is located a blue- or green-sensitive R8 (Figure 1*C*, see color insert) (Hardie 1986, Harris et al. 1976, Salcedo et al. 1999). In *Musca*, R1–R6 code achromatic contrast, whereas R7 and R8 together code wavelength distribution and polarization plane (Anderson & Laughlin 2000).

The adult eye develops in the late third instar larva and the pupal stage. The eye-antennal imaginal disc enlarges by proliferative cell division until the third instar, when the morphogenetic furrow sweeps across it from posterior to anterior, followed by the differentiation of new rows of ommatidia (reviewed by Frankfort & Mardon 2002). R8 is always the first cell to appear; as with a chordotonal organ, its formation depends on a progressive refinement of Atonal expression (Jarman et al. 1994). R8 then orchestrates the development of the other photoreceptors via the Spitz-EGF receptor and, in the case of R7, the Boss-Sevenless and Delta-Notch signaling pathways (Frankfort & Mardon 2002, Tomlinson & Struhl 2001). R8 differentiation requires continued expression of the zinc-finger transcription factor, Senseless (Frankfort et al. 2001). Conversely, Rough expression specifies the R2/R5 fate (Kimmel et al. 1990).

Subsequently, ommatidia in the dorsal and ventral halves of the eye rotate 90° in opposite directions so that R3 is always facing away from the equator of the eye. This is the end result of the R3/R4 precursors sensing their relative dorso-ventral positions in two opposing gradients of membrane proteins, Dachsous and Four-jointed, initially set up by Wingless activity at the poles of the wing disc (Yang et al. 2002). This positional information is interpreted using a complex pathway involving Fat, a cadherin superfamily member; Frizzled, the Wingless receptor; and then Delta-Notch signaling (Tomlinson & Struhl 1999, Yang et al. 2002).

Later in pupal development the *spalt* gene complex is required for expression of different opsins in R1–R6 versus R7 and R8 (Mollereau et al. 2001), with the former expressing *Rhodopsin 1* (*Rh1*). R7 morphology and rhodopsin expression is distinguished from that of R8 by its expression of the transcription factor Prospero (Cook et al. 2003). There are two types of R7: one expressing *Rh3* and the other, *Rh4*. The first induces its neighbor R8 to express the blue-sensitive *Rh5* instead of the default, green-sensitive *Rh6* (Chou et al. 1999, Salcedo et al. 1999).

AXON GUIDANCE TO THE CNS

Once the sensory organ has differentiated, the neuron or neurons must send out their axons toward the CNS. The mechanisms involved at this stage are probably those used to guide all axonal growth cones; in general, growth cones are guided in steps toward their target region, integrating at least four simultaneous types of signal: contact attraction, contact repulsion, long-range chemoattraction and long-range chemorepulsion (Tessier-Lavigne & Goodman 1996). These signals are mediated through highly conserved families of guidance molecules, which include the netrins, Slits, semaphorins, and ephrins (Huber et al. 2003).

For reasons of space, we do not consider this aspect further; there are no indications that the specificity of connections made by sensory afferents within the CNS is affected by the pathways they take to it.

SEGREGATION OF SENSORY AXON PROJECTIONS IN THE CNS

Segregation of Afferents According to Stimulus Modality, Stimulus Quality, and Sensillum Position

It has long been established that, as in other organisms, insect sensory neurons encoding different modalities project to different regions of the CNS (Murphey et al. 1985). In segmental ganglia, axons of touch-sensitive bristles project into a ventral layer of neuropil, while those of chordotonal neurons and neurons that innervate hair plates project into a more dorsal, proprioceptive layer. Wind-sensitive filiform hairs and gravity-sensitive clavate hairs of orthopteroid insects project to an intermediate neuropil region. Transplantation experiments have demonstrated that during postembryonic development, these modality-specific projection areas appear to restrict the availability of potential targets to those neurons that form dendrites in the same region (Killian et al. 1993, Murphey et al. 1983). This principle of modality-specific projection areas has also been established for adult flies (Murphey et al. 1989) and for the embryonic and larval *Drosophila* peripheral nervous system (Merritt & Whitington 1995, Schrader & Merritt 2000); this allows the molecular basis of modality-specific targeting within the CNS to be addressed (see below).

In addition to being segregated according to modality, sensory afferents may form different patterns of synaptic outputs according to the quality of the stimulus. In the cricket cercal system, some giant interneurons preferentially receive inputs from short filiform hairs, which are more sensitive to acceleration, whereas other interneurons receive inputs from long hairs, which are sensitive to wind velocity. As the insect grows, and short hairs become longer, the synapses undergo rearrangement to maintain this pattern of inputs (Chiba et al. 1988). In this system there is no anatomical segregation of afferents responding to different frequencies (Paydar et al. 1999), so a process of local cell-cell recognition must ensure appropriate sensory characteristics of interneurons. However, in other sensory systems, anatomical segregation of afferents based on stimulus quality also contributes to synaptic specificity. In the olfactory system, afferents arborize in characteristic globular regions of neuropil (Figure 2*D*, see color insert); each so-called glomerulus contains the branches of neurons that respond to a particular odor even though the cell bodies of these neurons may be scattered in different locations on the antenna (Bhalerao et al. 2003, Oland & Tolbert 1996). In the visual system of the fly, photoreceptors responding to blue and UV light do not terminate in the lamina but form their own topographic projections in the medulla (Figure 1*A*,*B*) (Meinertzhagen & Hanson 1993).

Finally, many sensory neurons form topographically ordered projections so that the position of the sense organ in the periphery is reflected in the position of the afferent's terminal arborization within the CNS. Often the somatotopic ordering is nested within a modality-specific region of neuropil; thus on the cricket cercus, wind-sensitive filiform hair sensory neurons form a highly ordered projection in the cercal glomerulus that encodes wind direction (Bacon & Murphey 1984, Jacobs & Theunissen 1996), whereas the interspersed touch-sensitive bristles form their own separate, topographically ordered projection in the bristle neuropil (Figure 2*A*) (Murphey 1985). In rarer cases, a somatotopic projection is present with no anatomical separation of two different modalities, as in the projection of the basiconic sensilla, mixed mechano- and chemosensory sense organs on the leg of the locust (Newland et al. 2000).

Anatomical segregation of afferent projections in this way serves as a preliminary determinant of synaptic specificity, reducing the range of possible postsynaptic targets that are available (Bacon & Murphey 1984). However, as studies of the simpler first-instar cockroach cercal system indicate, accurate patterns of synaptic connections can still form even when axonal projections overlap (Figure 2*B*), so clearly additional local cell-cell recognition mechanisms exist, ensuring synaptic specificity (Blagburn & Thompson 1990). We now examine a few sensory systems (mostly in *Drosophila*), in which progress has been made in unraveling the molecular and genetic control of afferent segregation.

Adult Sensory Neurons Follow Embryonic Pioneers

The guidance cues and steering mechanisms used by embryonic sensory axons to form anatomically segregated arborizations within the CNS may not necessarily be employed by postembryonically developing neurons. In *Drosophila*, a subset of larval sensory neurons persists through metamorphosis into the adult stage. Most are internal sensory neurons, but a few are external. Their axons prefigure specific adult sensory axon pathways (Williams & Shepherd 1999). Ablation of these larval sensory neurons shows they are required by adult sensory neurons for peripheral pathfinding, entry into the CNS, and, most significantly, growth guidance within the CNS. Existing dorsal neurons are required for subsequent axon guidance across the midline, and lateral neurons are required for posterior growth (Williams & Shepherd 2002).

Genetic Control of Modality-Specific Projections in the *Drosophila* PNS

Ectopic expression of *poxn* transforms leg mechanosensory bristles of *Drosophila* into chemosensory bristles; the neurons innervating these morphologically transformed bristles follow pathways and establish connections within the thoracic ganglion that are appropriate for their new modality (Nottebohm et al. 1992). Poxn expression after the neuron is born can still affect its fate, showing that this aspect of sensory neuron phenotype is not determined at the neural precursor stage (Nottebohm et al. 1994).

Sensory neurons of the peripheral nervous system (PNS) of the *Drosophila* larva exhibit modality-specific axonal projections in the CNS of the embryo. The axons of mechanosensory es neurons project to the ventral region of neuropil, within which they are also somatotopically mapped. The axons of stretch-receptive (ch) organs project onto an intermediately situated longitudinal fascicle, whereas those of most md neurons terminate on medial fascicles (Figure 2*C*). One particular md neuron, dbd, terminates on a more dorsal medial fascicle (Merritt & Whitington 1995).

Loss of the proneural gene *cut* transforms AS-C-dependent SOPs from es to ch precursors (Jarman & Ahmed 1998, Merritt 1997). Loss of *cut* also influences the projection pattern of the axons within the CNS, although its effects are variable, which suggests that other genes are also involved (Merritt et al. 1993). The combinations of proneural genes expressed by sensory neurons correlate closely with the positions and shapes of their axonal arbors, and it was proposed that these combinations of transcription factors serve to determine the modality-specific projection patterns (Merritt & Whitington 1995).

A recent study provides direct support for this proposal (Zlatic et al. 2003). Chordotonal neurons must express the Robo3 receptor for the midline-secreted, repellent glycoprotein Slit (Kidd et al. 1999) in order to form their terminal arbor on the appropriate, intermediately situated, longitudinal fascicle instead of more medial fascicles (Figure 2*C*). Ectopic expression of Robo3 in the normally medially arborizing dbd neurons forces the neurons to terminate more laterally, although their dorso-ventral positioning is not affected. These effects are completely independent of the positions of the target fascicles themselves. Ectopic expression of the chordotonal proneural gene *atonal* in md neurons showed that, although too late to change the cell-body morphology, Atonal protein is sufficient to induce Robo3 expression, thereby also altering the position of the dbd arbor. In this case, the dorso-ventral position of dbd is also changed and its arbor no longer undergoes its characteristic late remodeling phase, which suggests that Atonal affects other guidance cues in addition to Robo3. Thus, rather than a sensory axon being guided by its targets, the position of the terminal arbor within the neuropil is regulated by factors that specify its position in two dimensions (Zlatic et al. 2003). Also, Robo is expressed by both ch and md neurons—it serves to prevent crossing of the midline (Zlatic et al. 2003).

Homeotic and Pre-Pattern Genes Control Position-Specific Projections in the *Drosophila* PNS

Less is known about how neuronal position specifies connectivity within the CNS, but the available evidence suggests that patterning genes and prepattern genes that have early roles in setting up the body plan play persistent or recurring roles at later stages of neuronal development.

It was demonstrated 20 years ago that homeotic genes of the Bithorax complex that control the identity of the third thoracic segment of *Drosophila* also control the morphology of axonal projections of thoracic mechanosensory bristles (Ghysen et al. 1983). The prepattern genes *araucan* and *caupolican* control the position-specific axonal projections of bristle sensory neurons on the adult *Drosophila* notum. Medially born neurons form a branch that crosses the midline of the CNS; lateral ones do not. Araucan and Caupolican are expressed in lateral neurons and prevent this branch formation (Grillenzoni et al. 1998). In recent studies we showed that Engrailed, a transcription factor involved in specifying segmental identity, controls the axonal projection of an identified sensory neuron of the cockroach cercal system (this work is discussed in more detail below, in the Synaptic Connections section).

Less information is available about downstream effector genes that may be involved in the guidance of sensory axons forming somatotopic projections. The family of fasciclins, cell-surface-adhesion molecules, are essential for proper axon guidance of *Drosophila* wing mechanosensory neurons within the CNS (Whitlock 1993). The cytoplasmic dynein light chain also has a role in axon pathfinding of both stretch receptor and mechanosensory neurons; mutations disrupt sensory axon trajectories in the imaginal nervous system but do not prevent these axons from reaching their normal termination regions (Phillis et al. 1996).

Odor-Specific Projections in the Olfactory System

As in vertebrates, the olfactory neuropil of insects is subdivided into structures called glomeruli. Each glomerulus consists of the highly branched arbors of olfactory receptor axons and the dendrites of antennal-lobe neurons, bounded by an envelope of glial cells (Figure 2*D*). Another similarity with vertebrates is that olfactory receptor neurons that express a particular type of odorant receptor protein, and therefore respond to a particular odor, project to small subsets of glomeruli, thus forming a spatial representation of the olfactory world (Vosshall et al. 2000, Wang et al. 2003).

In the tobacco hornworm moth, *Manduca sexta*, it is clear that this central functional representation is not simply a recapitulation of order on the antenna because each antennal segment bears many olfactory receptor types (reviewed by Hildebrand & Shepherd 1997). Prior to entering the antennal lobe, olfactory receptor axons undergo extensive reorganization in a zone composed of CNS glia. Here, they are sorted into bundles of axons that each project to a common glomerulus (Oland et al. 1998, Rossler et al. 1999). It seems clear that differential axonal adhesion plays a role in this reorganization because a subset of Fasciclin II-positive axons segregate from the others in this zone—on entering the olfactory lobe these fasciclin-expressing axons target certain glomeruli and avoid others, such as the sexually dimorphic glomerulus (Higgins et al. 2002).

Work in *Drosophila* has begun to elucidate some of the molecular mechanisms involved in establishing glomerular projections. The axons of one group of sensory neurons, those of the Atonal-dependent coeloconic sensilla, arrive first in the olfactory lobe; in their absence the formation of glomeruli is prevented (Jhaveri & Rodrigues 2002). However, these axons are required to pioneer all glomeruli, which suggests that Atonal does not directly control odor-specific guidance mechanisms.

A couple of recent studies have begun to shed light on the signaling pathways that guide olfactory receptor axons to the appropriate target glomerulus. The axons grow directly around the periphery of the antennal lobe to their target; this takes place before odorant receptor expression, which implies that the latter have no role in targeting, in contrast to the developing mouse olfactory lobe (Mombaerts et al. 1996, Wang et al. 1998). Axons mutant for either of the growth cone guidance genes, *dreadlocks* (*dock*) and *PAK-kinase* (*Pak*), navigate through the antennal nerve but, on arrival at the antennal lobe, take chaotic routes across it and fail to form glomeruli (Ang et al. 2003). Axons with more distant targets appear to suffer greater perturbation. The SH2/SH3 adaptor protein Dock recruits the serine/threonine kinase Pak to the cell membrane; together they appear to function in a signaling pathway that mediates the guidance of growth cones to specific glomeruli.

The cell-surface Down syndrome cell-adhesion molecule (Dscam) was also found to affect guidance of olfactory receptor axons (Hummel et al. 2003). Axons mutant for *Dscam* can form glomeruli, but for some receptor subtypes the selectivity of axon guidance to the target is perturbed. As with *dock* and *Pak* mutants, more distant, contralateral targets are rarely innervated. In addition, olfactory axons from the maxillary palps show a propensity to form glomeruli in any neuropil they encounter en route to the antennal lobe. Other Dscams, as well as the cadherin superfamily members N-cadherin and Flamingo, may be used for guidance by other groups of olfactory receptor neurons (Hummel et al. 2003).

Retinotopic Projections of *Drosophila* Photoreceptors

The development of the visual ganglia of *Drosophila* is dependent upon the arrival of axons from the compound eye (Meyerowitz & Kankel 1978). In fact, removal of the retinal axons leads to lamina cell death (Selleck et al. 1992). As the retinal axons arrive in the lamina, they release Hedgehog, which triggers the final division of lamina precursor cells (Huang & Kunes 1996), and Spitz, which activates the epidermal growth factor receptor (EGFR) and induces the differentiation of the lamina neurons (Huang et al. 1998). This interplay of signals ensures a precise correspondence in the numbers and timing of development of the pre- and postsynaptic cells.

Photoreceptor axons form two separate retinotopic projections within the optic lobe: R1–R6 axons in the lamina, R7 and R8 in the medulla (Figure 1*A*) (Meinertzhagen & Hanson 1993). The sequential way in which photoreceptor axons trigger the growth of the lamina suggests a simple mechanical explanation for the formation of that retinotopic map, in which newly arrived axons come to occupy the next available space in the developing target fields. However, for both the lamina projection of R1–R6 (Kunes et al. 1993) and the medulla projection

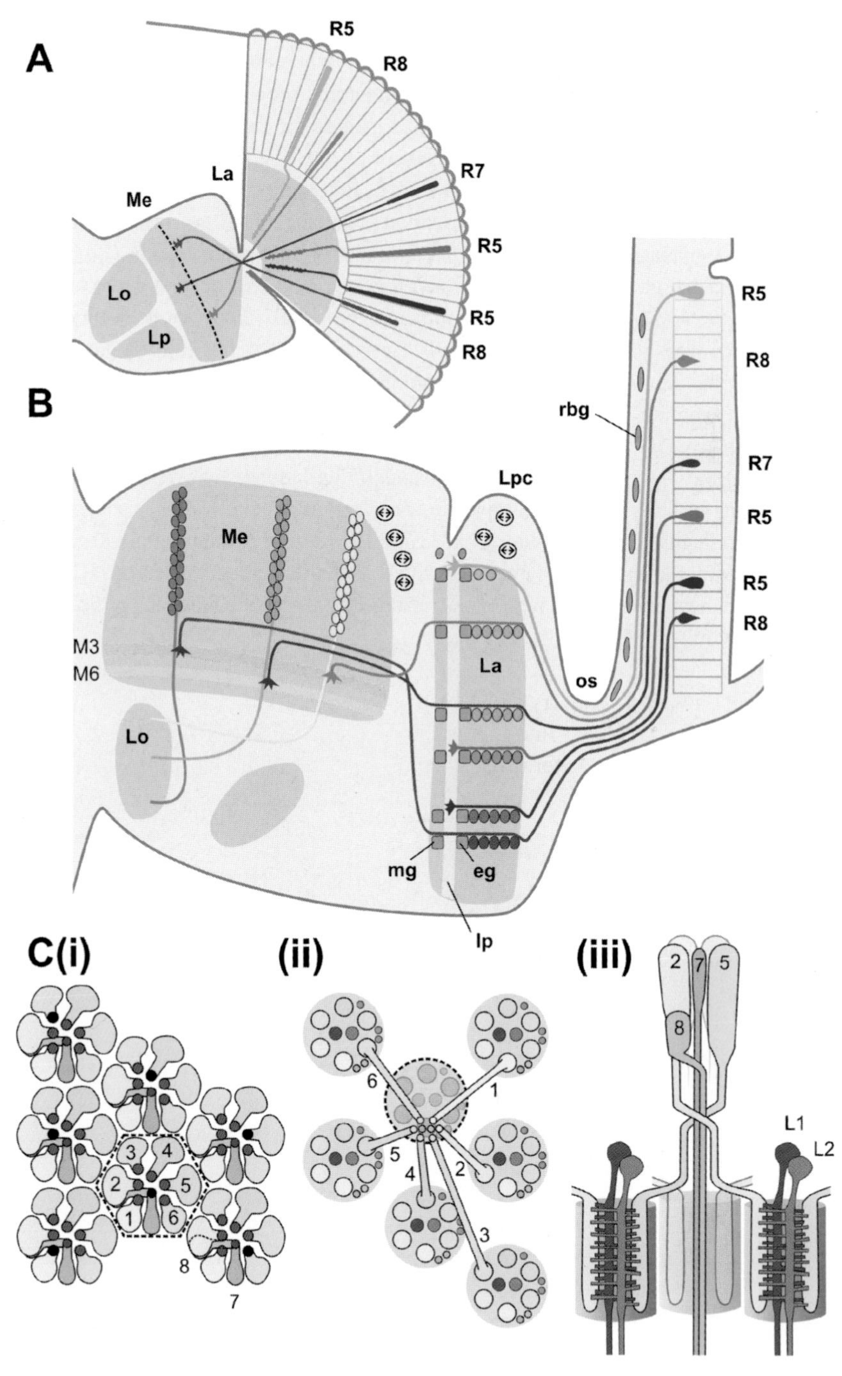

See legend on next page

Figure 1 The *Drosophila* eye and optic lobe. (*A*) In the adult fly, R1–R6 photoreceptor neurons from each ommatidium (only R5 is shown, in only a few ommatidia) project retinotopically to the lamina (La); R7 and R8 form retinotopic projections in different layers of the medulla (Me). Other regions of the optic lobe include the lobula (Lo) and lobula plate (Lp). (*B*) At the end of the larval stage, photoreceptor axons have projected from the eye imaginal disc through the optic stalk (os) to the optic lobe, following the retinal basal glia (rbg). Axons of R1–R6 (*green*) induce the division of lamina precursor cells (Lpc) and stop in the lamina at the marginal glia layer (mg), their growth cones forming the lamina plexus (lp) between these and the epithelial glia (eg). Axons of R7 and R8 (*purple*, *blue*) continue to the medulla, forming a retinotopic projection. They then project centripetally along axons of medulla neurons (*yellow*); R8 stops in layer 3 (M3) while R7 projects to layer 6 (M6). (*C*) Neural superposition. [*C*(*i*)] Surface view of seven retinal ommatidia, in the dorsal half of the left eye, showing photoreceptors R1–R6 (*yellow-green)* surrounding R7 (*purple*) and R8 (*blue*). The rhabdomeres that receive light from the same point in space are shown filled in black. [*C*(*ii*)] Profiles of the same photoreceptors in the underlying lamina, also viewed from the surface. Axons from the central ommatidium (*dashed outline*) project outward in a stereotypic array to the surrounding lamina cartridges. [*C*(*iii*)] Side view of the photoreceptors in one ommatidium, along with the underlying lamina cartridges. Dendrites of lamina neurons L1 and L2 form synaptic connections with axon terminals of R1–R6; axons of R7 and R8 project through the cartridge.

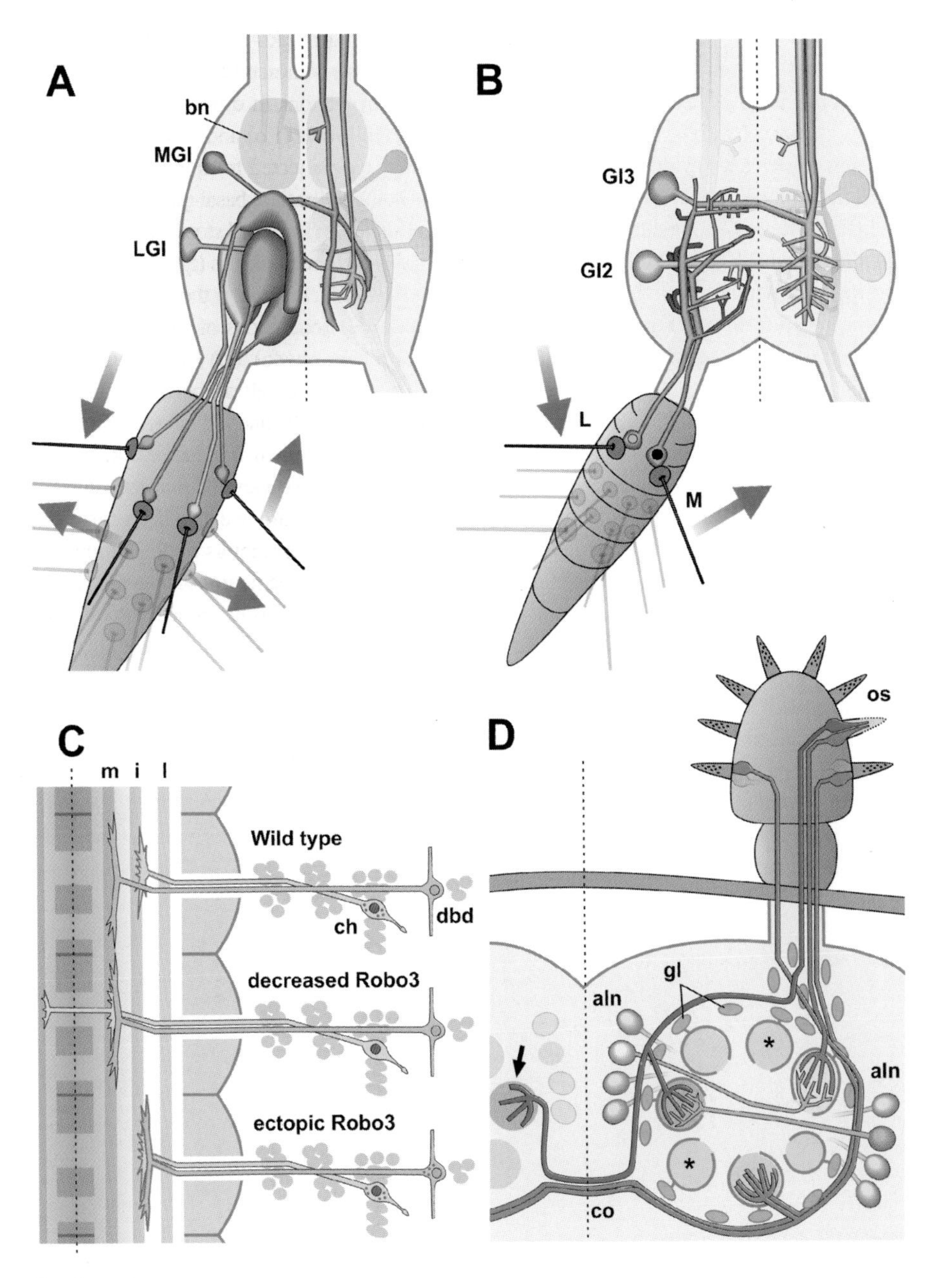

See legend on next page

Figure 2 Segregation of sensory afferents according to stimulus modality and quality. Cartoons of different insect sensory systems; in all cases the cells shown are duplicated about the midline (*dashed line*). Anterior is upward. (*A*) Cricket cercal sensory system. Filiform hair sensory neurons with sensitivities to wind from different directions (*arrows*) form nonoverlapping projections in different regions of the cercal glomerulus in the terminal ganglion (*green, blue, yellow and purple*). Afferents of bristle hairs on the cercus (*not shown*) project to a different region, the bristle neuropil (bn). The dendrites of giant interneurons, two of which are illustrated (MGI and LGI), selectively project into certain regions of neuropil, establishing the basis for neural selectivity. (*B*) Cercal sensory system of the second instar cockroach. Two of the 39 filiform hair sensilla are shown, one lateral (L) and the other medial (M). They respond to wind from different directions (*arrows*) and have anatomically distinct, but overlapping, arbors in the terminal ganglion. However, they form specific patterns of synaptic connections to giant interneurons: GI2 receives input only from M (*purple afferent*), and GI3 receives input only from L (*green afferent*). The axonal projection and synaptic connectivity of M is controlled by its expression of Engrailed (*black nucleus*). (*C*) Slit-Robo signaling controls the termination site of *Drosophila* chordotonal sensory axons. In the wild-type embryo, axons of chordotonal neurons (ch) express Robo3 (*red outline*) and terminate on an intermediate longitudinal fascicle (i); those of dbd neurons (dbd) do not and terminate on a medial fascicle (m). Reduction of Robo3 allows ch axons to terminate medially, whereas ectopic expression of Robo3 in dbd forces them to terminate on the intermediate fascicle. (*D*) Olfactory receptors on the *Drosophila* antenna. Some olfactory sensilla (os) have two neurons (*green, purple*), which respond to different odors. Their axons project to different glomeruli (*two are asterisked*) in the antennal lobe of the brain, where they synapse with antennal lobe neurons (aln). Axons from sensilla in different locations that respond to the same odor (*blue*) project to the same glomerulus. Many olfactory receptor axons also project to the contralateral glomerulus (*arrow*) via the olfactory commissure (co). Glial cells (gl) surround the antennal lobe neuropil and form a sheath around each glomerulus.

of R7 (Ashley & Katz 1994), photoreceptor axons can find the appropriate positions in their target fields, even if their neighbors are absent or aberrant. These studies suggest that the simple mechanical explanation is not adequate and that a chemoaffinity mechanism, along with local interactions between photoreceptor axons, is required for formation of the retinotopic projections.

It is not known how positional information in the eye, perhaps that encoded by the Dachsous and Four-jointed gradients (Yang et al. 2002), influences the formation of the retinotopic projections in the optic lobe. However, some of the cell-surface molecules involved have been identified. Expression of the netrin receptor, Frazzled, is required in the lamina for the orderly projection of retinal axons, although netrins themselves do not appear to be required (Gong et al. 1999). Recently, investigators showed, as in the vertebrate visual system, the Eph receptor tyrosine kinase has a role in directing *Drosophila* photoreceptor axons to topographically appropriate dorso-ventral sites in the medulla (Dearborn et al. 2002).

Other cell-surface molecules may be involved in interactions between the growth cones. *N-cadherin* mutant axons exhibit local disruptions of the topographic mapping in both lamina and medulla, with axon terminals clumping together instead of being spaced regularly (Lee et al. 2001). Dynamic expression of the seven-transmembrane protocadherin, Flamingo (Fmi), also has a similar role in establishing the regular array of nonoverlapping projections of R8 and R7 axons within the medulla (Lee et al. 2003, Senti et al. 2003). In addition, removal of the synaptic vesicle protein N-synaptobrevin by expression of tetanus toxin in photoreceptors perturbs the orderly array of R7 terminals (Hiesinger et al. 1999).

Arresting the Growth of Retinula Axons in the *Drosophila* Lamina

As retinal axons enter the optic lobe during late larval development, the pioneer R8 axon passes through the first optic ganglion, the lamina, to terminate in the second ganglion, the medulla. R1–R6 axons follow the R8 axon only as far as the marginal glial layer near the base of the lamina (Perez & Steller 1996). Here their growth cones cease extension and form the lamina plexus (Figure 1*B*). Synapses are not formed until later, in the pupal stage (see below). The late-growing axon of R7 also does not stop in the lamina plexus but follows R8 through the lamina to the medulla (Fischbach & Dittrich 1989) (Figure 1*B*). Axons of the five lamina neurons in each cartridge also project, along with R7 and R8, to different medulla layers (Fischbach & Dittrich 1989). Two separate retinal projections are thus formed, the one in the lamina consisting of axons from photoreceptors responsive to a wide spectrum, the medulla projection consisting of UV-responsive (R7) and blue- or green-responsive (R8) axons.

Many recent studies have focused upon the molecular mechanisms controlling these aspects of axonal targeting (reviewed by Clandinin & Zipursky 2002, Tayler & Garrity 2003). Postsynaptic lamina neurons themselves are not actually required for R1–R6 to stop in the lamina plexus (Huang & Kunes 1996). However, the lamina glial cells are critical for this stop signal; when their numbers are reduced the R1–R6 projection is severely disrupted, with many axons projecting through the lamina to the medulla (Poeck et al. 2001, Suh et al. 2002). Interpretation of the stop signal, the nature of which is still unknown, requires that the photoreceptors express the nuclear protein Brakeless (Bks), which is necessary but not sufficient for R1–R6 lamina targeting (Rao et al. 2000, Senti et al. 2000). In R2 and R5, Bks represses the expression of a second transcription factor, Runt, which is normally expressed only by R7 and R8 (Kaminker et al. 2002). Although *runt* mutants show no targeting defects, misexpression of Runt in R2 and R5 alone is sufficient to direct all R1–R6 through the lamina to target incorrectly in the medulla, without otherwise transforming the identity of the neurons (Kaminker et al. 2002). This also shows that interactions between the photoreceptor axons themselves, as well as between the axons and glia, play an important role in targeting to the lamina plexus.

The links between the genes involved in the establishment of photoreceptor identity and the regulation of Runt expression are not yet clear. Although the *spalt* genes are required for expression of the appropriate rhodopsins in R7 and R8, they have no effects on their axons' decision to continue growth through the lamina (Mollereau et al. 2001). No direct connection has yet been made between Runt and the expression of growth cone signaling molecules that respond to the stop signal. However, several of the latter have been identified. Two cell-surface molecules have permissive roles in the decision of R1–R6 to stop in the lamina: the receptor tyrosine phosphatases, PTP69D (Garrity et al. 1999, Newsome et al. 2000a) and Leukogen-antigen-related-like (LAR) (Clandinin et al. 2001). Mutations in these genes also only affect small numbers of retinal axons, which suggests that these receptor phosphatases have auxiliary or overlapping roles (Tayler & Garrity 2003), perhaps in decreasing the adhesion of R1–R6 axons to that of R8 (Newsome et al. 2000a).

At an intermediate level in the pathway, cytoplasmic molecules have been identified that contribute to transducing the lamina stop signal. Mutations of *dock* cause some axons to project through to the medulla (Garrity et al. 1996); this protein is required in growth cones for signal transduction and interacts with Misshapen (Msn), a serine/threonine kinase (Ruan et al. 1999, Su et al. 2000). Overexpression of the latter leads to premature arrest of the photoreceptor growth cones before they reach the lamina (Ruan et al. 1999). A downstream component of the stop signal pathway is Bifocal (Bif), a cytoskeletal regulator protein that associates with, and is phosphorylated by, Msn (Ruan et al. 2002). Other molecules, Pak, Trio and Kette, have been identified that interact with Dock and are required for photoreceptor axon growth cone guidance in general (Hing et al. 1999, Hummel et al. 2000, Newsome et al. 2000b), their mutations having complex effects on the retinal projection. Recently investigators showed that the *Drosophila* insulin receptor (DInR) tyrosine kinase interacts with Dock in affecting axon guidance, and, to some extent, photoreceptor layer targeting (Song et al. 2003).

Photoreceptor Axon Targeting to Lamina Cartridges

In the pupal stage of *Drosophila* and larger flies, the axons of R1–R6 resume growth laterally within the lamina plexus, interweaving to form the adult pattern of neural superposition (Meinertzhagen & Hanson 1993). The curvature of the adult eye, along with the separate, stereotypic positions of the rhabdomere of each photoreceptor within a single ommatidium, results in the central R7 and R8 photoreceptors receiving light from the same point in space as R1–R6 in neighboring ommatidia (Figure 1*C*). Neural superposition ensures that these spatially coincident inputs all converge on a single column of lamina neurons, forming a unit of neuropil, the lamina cartridge (Figure 1*C*). Photoreceptor growth cones in the lamina undergo this highly reproducible feat of navigation between 12.5% and 50% of pupal development (Meinertzhagen & Hanson 1993); subsequently, the receptor axons turn and extend parallel to the axons of lamina neurons, forming the scaffold of the lamina cartridge neuropil.

Anatomical observations (Meinertzhagen & Hanson 1993) and subsequent genetic ablation studies (Clandinin & Zipursky 2000) showed that contact-mediated interactions between photoreceptor growth cones in, and perhaps between, ommatidial bundles are necessary to ensure that the axons target to the appropriate lamina cartridge. Genetic removal of R3 and R4 perturbs the targeting of the other axons, whereas removal of R1 and R6 affects R2 and R5 but has no effect on R3 and R4 (Clandinin & Zipursky 2000). The direction of extension of R3 and R4 depends primarily on the orientation of the ommatidium with respect to the equator of the eye, and perhaps on weaker directional cues in the lamina (Clandinin & Zipursky 2000), although the evidence for the latter is not unequivocal (Meinertzhagen 2000).

Recent work has begun to elucidate the molecular cues involved in the interactions within R1–R6 as they interweave in the lamina plexus. So far, these cues appear to be proteins used at earlier stages in layer-specific targeting. In *LAR* mutants, the axons fail to extend outward from the ommatidial bundle (Clandinin et al. 2001). LAR is required cell-autonomously for R1 and R6 (and probably the others) to leave the bundle. However, mutant axons can still form synapses in inappropriate cartridges (Clandinin et al. 2001). N-cadherin is also required for this process: *N-cadherin* mutant photoreceptor axons behave similarly to *LAR* mutants (Lee et al. 2001). N-cadherin may also be required for the formation of morphologically normal synapses (Iwai et al. 2002). Mutations in the protocadherin, Flamingo, also disrupt cartridge selection, although they do not prevent axons from leaving the ommatidial bundle as do the mutations described above (Lee et al. 2003). Synapse formation in the aberrant cartridges is otherwise normal.

During the period between the photoreceptor growth cones arriving at the appropriate cartridge and the formation of synapses, the photoreceptors express a soluble guanylate cyclase that responds to NO produced by nitric oxide synthase in the optic lobe (Gibbs et al. 2001, Gibbs & Truman 1998). Blocking this retrograde signaling system leads to disorganization of the retinal projection and the elongation of axons through the lamina and out past the medulla, which suggests that NO normally prevents the further extension of the axons, thus allowing synaptic stabilization to take place.

Layer Targeting of R7 and R8 in the *Drosophila* Medulla

R7 and R8, the UV- and blue- or green-sensitive photoreceptors that project to the medulla neuropil, also exhibit layer-specific targeting within this region, with R8 stopping in the M3 layer and R7 projecting approximately 10 microns beyond it to M6 (Fischbach & Dittrich 1989) (Figure 1*A*,*B*). This separation of photoreceptor axon terminals is established within the first 24 h of pupal development (Clandinin et al. 2001). At this early stage the cellular composition of these neuropil layers is not known but could include branches of the precocious medulla tangential neurons (Meinertzhagen & Hanson 1993).

No transcription factors have yet been found that control this layer targeting; Prospero in R7 controls rhodopsin expression but not axon targeting (Cook et al. 2003). However, several downstream molecules have been identified; these include the receptor phosphatases also used for R1–R6 lamina targeting, LAR (Clandinin et al. 2001, Maurel-Zaffran et al. 2001) and PTP69D (Newsome et al. 2000a). In animals with *LAR* mutant photoreceptors, the R7 axon transiently projects to the correct M6 layer but then retracts to M3 or an intermediate position (Clandinin et al. 2001, Maurel-Zaffran et al. 2001). The phosphatase domains of LAR are required for its activity in R7, but not in R8, which suggests that it can act both as a receptor in R7 and as a ligand provided by R8. Genetic interactions suggest that both Enabled and Trio are involved in LAR signal transduction (Maurel-Zaffran et al. 2001).

Another important cell-surface molecule involved in R7/R8 layer targeting is N-cadherin (Lee et al. 2001). N-cadherin in R8 is essential for layer formation in the medulla, whereas selective removal of the protein from some R7 axons prevents them from projecting into the appropriate layer and significantly disrupts the fly's phototactic response to UV light (Lee et al. 2001). Similar to LAR, it appears that *N-cadherin* mutant R7 axons initially do project to M6 but, perhaps failing to form stable contacts, retract to M3 (Clandinin & Zipursky 2002). This finding has led to the proposal that N-cadherin and LAR work together or in parallel to stabilize photoreceptor growth cone contacts in the appropriate target neuropil layer. Coupled with the role of Flamingo in stabilizing R8 terminals (Lee et al. 2003, Senti et al. 2003), these results suggest there is a cadherin-based system for target layer specificity in the *Drosophila* visual system.

Relatively late in development of the eye, well after these axonal projections have formed, R8 in some ommatidia expresses the blue-sensitive rhodopsin, Rh5; in other ommatidia it expresses the green-sensitive Rh6 (Salcedo et al. 1999). It is possible that the different color sensitivities of the two subtypes of R8 neuron are functionally encoded in the medulla by different patterns of synaptic output. However, these detailed aspects of local cell-cell synaptic recognition within optic lobe layers have not yet been addressed.

SYNAPTIC RECOGNITION AT THE CELLULAR LEVEL

Photoreceptor Synapses in the Fly Lamina

After the axons of R1–R6 have selected the appropriate cartridges for neural superposition, filopodia of lamina neurons form dendrites that grow to the photoreceptor terminals and start to form synapses from about 60% of pupal development (Meinertzhagen & Hanson 1993, Meinertzhagen et al. 2000). Synapse formation takes place in a highly stereotypic sequence so that in the adult each receptor synapse contacts four postsynaptic elements (a tetrad); this comprises one dendrite of each of the L1 and L2 neurons, usually one amacrine cell process, a dendrite of L3, and sometimes a glial process (Meinertzhagen & Hanson 1993). L4 and L5 do not receive synaptic input from the photoreceptors, but L4 is presynaptic to them (Meinertzhagen & O'Neil 1991). This complex, specific pattern of synaptic inputs and outputs presumably requires that active, mutual recognition takes place between pre- and postsynaptic partners, but little is known about the molecular and genetic mechanisms that control this process.

Engrailed Controls Synaptic Recognition in the Cercal Afferent System of the Cockroach

The cockroach, *Periplaneta americana*, has a very ancient escape system that dates back to the first flightless insects (Edwards & Reddy 1986). The cerci, conical appendages at the rear of the animal, bear highly mobile filiform hairs, each innervated by a single sensory neuron. In the newly hatched (first instar) animal there are only two pairs of these sensory neurons, which form synapses with restricted subsets of identified giant interneurons (GIs) in the CNS (Figure 2*B*) (Blagburn 1989). Lateral sensory neurons respond mainly to air from the front of the animal and form synaptic connections with GI3 and GI6; medial neurons respond to air from the rear and connect to GIs 1, 2, and 5. This pattern of inputs determines the GIs' directional sensitivity; the GIs' outputs to interneurons in the thoracic ganglia mediate the animal's directed escape response away from a predator.

Within the cercal neuropil of the terminal abdominal ganglion, lateral and medial filiform hair afferents form anatomically distinguishable, yet overlapping, arborizations. Unlike cricket cercal afferents, position-specific differences between the projections play little part in determining synaptic connectivity with the GIs, since the presynaptic axons and postsynaptic dendrites all arborize in the same local region of neuropil (Blagburn & Thompson 1990). Despite this close anatomical apposition, synapses are specific from the time they form in the embryo, before spontaneous activity begins in the sensory afferents (Blagburn et al. 1996). As in the fly lamina, it is the postsynaptic neurons that grow toward the presynaptic axons. Dendrites of GIs extend filopodia that selectively contact the appropriate sensory axon; only 1% of the anatomical synapses formed are "incorrect." This specificity of synaptic choice at such close range must be brought about by local cell-cell recognition events.

The transcription factor Engrailed (En) plays a vital role early in development in establishing the polarity of body segments (Kornberg 1981, Patel et al. 1989). In the cockroach, En is expressed in the medial half of the cercal epidermis and in the medial sensory neuron. This expression persists in medially born cercal neurons of the next larval stage (Blagburn et al. 1995). A recent study focused on one of these identified second instar neurons, 6m, and used dsRNA interference to show that En controls its arbor shape and, more significantly, the specificity of its synaptic connections to GIs (Marie et al. 2000). In the absence of En, the neuron adopts an arbor morphology and pattern of synaptic connections similar to those of a neighboring, normally En-negative, neuron (Marie et al. 2000).

Rather surprisingly, En knockout at different stages during the development of 6m showed that persistent expression of the transcription factor is required in the postmitotic neuron, at each stage of the formation of its axonal arborization and synaptic connections (Marie et al. 2002). Removal of En, after 6m has entered the CNS, comes too late to change the main axonal trajectory but does alter the growth of secondary axonal branches and causes the formation of synaptic connections typical of an En-negative cell, which suggests that En controls target recognition molecules independently from those guiding the axon. In a further study, we showed that the two En paralogues, Pa-En1 and Pa-En2, have different effects on the control of axonal arborization and synaptic connections (Marie & Blagburn 2003). Axon guidance decisions are affected only by the total amount of En activity in 6m; the two paralogues are present in different amounts so their individual removal has different effects. In contrast, synaptic connections are only affected by removal of Pa-En1, which suggests that it alone controls the expression of synaptic recognition molecules.

So far, the effector molecules downstream of En have not been identified. However, in *Drosophila*, En downregulates the cell-adhesion molecules Connectin and Neuroglian (Siegler & Jia 1999), and it is probable that this pathway has been conserved, at least in insects. Whether these molecules have roles in determining the specificity of synaptic connections in the cockroach remains to be determined.

Role of Postsynaptic Neurons

We cannot leave the subject of synaptic specificity without at least touching on the role of the postsynaptic neuron. Many insect sensory synapses are dyadic, with two postsynaptic processes; those in the fly lamina are tetrads. In these configurations sheer logic dictates that the postsynaptic cells must play an active role in seeking out the appropriate presynaptic partner. Certainly in two widely disparate insect sensory systems, the fly eye and the cockroach cercus, this is the case (Blagburn et al. 1996, Meinertzhagen et al. 2000). Even at neuromuscular synapses, once the epitome of an active presynaptic axon seeking out a passive postsynaptic target, it is now apparent that muscle cells actively extend filopodia or "myopodia" that play a dynamic role in recognition between synaptic partners (Ritzenthaler & Chiba 2003, Ritzenthaler et al. 2000).

Many of the molecular mechanisms used to guide the growth of axons are also used during dendritic growth. In a similar fashion to the positioning of sensory arbors, dendrites of embryonic (Furrer et al. 2003) and adult motorneurons (Godenschwege et al. 2002) also use Robo receptors to sense the Slit gradient. Motorneuron dendrites are also regulated by Netrin-Frazzled signaling (Furrer et al. 2003). There is also evidence for specific dendritic targeting in the olfactory system of *Drosophila*. The location of the glomerulus, to which a given population of projection neurons sends its dendrites, is determined in part by expression of one of the POU domain transcription factors, Acj6 or Drifter (Komiyama et al. 2003). Interestingly, Acj6 also determines the odor sensitivity, and presumably the glomerular projection, of a subset of olfactory afferents (Clyne et al. 1999), which raises the possibility that, as in the vertebrate spinal cord (Arber et al. 2000), expression of the same transcription factor in sensory neurons and their targets may be one strategy for coordinating synaptic specificity.

The *Annual Review of Neuroscience* is online at http://neuro.annualreviews.org

LITERATURE CITED

Anderson JC, Laughlin SB. 2000. Photoreceptor performance and the co-ordination of achromatic and chromatic inputs in the fly visual system. *Vision Res.* 40:13–31

Ang LH, Kim J, Stepensky V, Hing H. 2003. Dock and Pak regulate olfactory axon pathfinding in *Drosophila. Development* 130:1307–16

Arber S, Ladle DR, Lin JH, Frank E, Jessell TM. 2000. ETS gene *Er81* controls the formation of functional connections between group Ia sensory afferents and motor neurons. *Cell* 101:485–98

Ashley JA, Katz FN. 1994. Competition and position-dependent targeting in the development of the *Drosophila* R7 visual projections. *Development* 120:1537–47

Bacon JP, Murphey RK. 1984. Receptive fields of cricket giant interneurons are related to their dendritic structure. *J. Physiol.* 352:601–23

Bhalerao S, Sen A, Stocker R, Rodrigues V. 2003. Olfactory neurons expressing identified receptor genes project to subsets of glomeruli within the antennal lobe of *Drosophila melanogaster*. *J. Neurobiol.* 54:577–92

Blagburn JM. 1989. Synaptic specificity in the first instar cockroach: patterns of monosynaptic input from filiform hair afferents to giant interneurons. *J. Comp. Physiol. A* 166:133–42

Blagburn JM, Gibbon CR, Bacon JP. 1995. Expression of *engrailed* in an array of identified sensory neurons: comparison with position, axonal arborization, and synaptic connectivity. *J. Neurobiol.* 28:493–505

Blagburn JM, Sosa MA, Blanco RE. 1996. Specificity of identified central synapses in the embryonic cockroach: appropriate connections form before the onset of spontaneous afferent activity. *J. Comp. Neurol.* 373:511–28

Blagburn JM, Thompson KS. 1990. Specificity of filiform hair afferent synapses onto giant interneurons in *Periplaneta americana*: anatomy is not a sufficient determinant. *J. Comp. Neurol.* 302:255–71

Blochlinger K, Jan LY, Jan YN. 1991. Transformation of sensory organ identity by ectopic expression of Cut in *Drosophila. Genes Dev.* 5:1124–35

Bray S. 2000. Notch. *Curr. Biol.* 10: R433–5

Brewster R, Bodmer R. 1995. Origin and

specification of type II sensory neurons in *Drosophila. Development* 121:2923–36

Campuzano S, Modolell J. 1992. Patterning of the *Drosophila* nervous system: the *achaete-scute* gene complex. *Trends Genet.* 8:202–8

Chiba A, Shepherd D, Murphey RK. 1988. Synaptic rearrangement during postembryonic development in the cricket. *Science* 240:901–5

Chou WH, Huber A, Bentrop J, Schulz S, Schwab K, et al. 1999. Patterning of the R7 and R8 photoreceptor cells of *Drosophila*: evidence for induced and default cell-fate specification. *Development* 126:607–16

Clandinin TR, Lee CH, Herman T, Lee RC, Yang AY, et al. 2001. *Drosophila* LAR regulates R1–R6 and R7 target specificity in the visual system. *Neuron* 32:237–48

Clandinin TR, Zipursky SL. 2000. Afferent growth cone interactions control synaptic specificity in the *Drosophila* visual system. *Neuron* 28:427–36

Clandinin TR, Zipursky SL. 2002. Making connections in the fly visual system. *Neuron* 35:827–41

Clyne PJ, Certel SJ, de Bruyne M, Zaslavsky L, Johnson WA, Carlson JR. 1999. The odor specificities of a subset of olfactory receptor neurons are governed by Acj6, a POU-domain transcription factor. *Neuron* 22:339–47

Cook T, Pichaud F, Sonneville R, Papatsenko D, Desplan C. 2003. Distinction between color photoreceptor cell fates is controlled by Prospero in *Drosophila. Dev. Cell* 4:853–64

Dambly-Chaudiere C, Jamet E, Burri M, Bopp D, Basler K, et al. 1992. The paired box gene *pox neuro*: a determinant of poly-innervated sense organs in *Drosophila. Cell* 69:159–72

Dearborn R Jr, He Q, Kunes S, Dai Y. 2002. Eph receptor tyrosine kinase-mediated formation of a topographic map in the *Drosophila* visual system. *J. Neurosci.* 22:1338–49

de Bruyne M, Foster K, Carlson JR. 2001. Odor coding in the *Drosophila* antenna. *Neuron* 30:537–52

Dobritsa AA, van der Goes van Naters W, Warr CG, Steinbrecht RA, Carlson JR. 2003. Integrating the molecular and cellular basis of odor coding in the *Drosophila* antenna. *Neuron* 37:827–41

Edwards JS, Reddy GR. 1986. Mechanosensory appendages and giant interneurons in the firebrat (*Thermobia domestica*, Thysanura): a prototype system for terrestrial predator evasion. *J. Comp. Neurol.* 243:535–46

Fischbach K-F, Dittrich APM. 1989. The optic lobe of *Drosophila melanogaster*. I. A Golgi analysis of wild-type structure. *Cell Tissue Res.* 258:441–75

Frankfort BJ, Mardon G. 2002. R8 development in the *Drosophila* eye: a paradigm for neural selection and differentiation. *Development* 129:1295–306

Frankfort BJ, Nolo R, Zhang Z, Bellen H, Mardon G. 2001. *senseless* repression of *rough* is required for R8 photoreceptor differentiation in the developing *Drosophila* eye. *Neuron* 32:403–14

Furrer MP, Kim S, Wolf B, Chiba A. 2003. Robo and Frazzled/DCC mediate dendritic guidance at the CNS midline. *Nat. Neurosci.* 6:223–30

Garrity PA, Lee CH, Salecker I, Robertson HC, Desai CJ, et al. 1999. Retinal axon target selection in *Drosophila* is regulated by a receptor protein tyrosine phosphatase. *Neuron* 22:707–17

Garrity PA, Rao Y, Salecker I, McGlade J, Pawson T, Zipursky SL. 1996. *Drosophila* photoreceptor axon guidance and targeting requires the Dreadlocks SH2/SH3 adapter protein. *Cell* 85:639–50

Gautier P, Ledent V, Massaer M, Dambly-Chaudiere C, Ghysen A. 1997. *tap*, a *Drosophila* bHLH gene expressed in chemosensory organs. *Gene* 191:15–21

Gho M, Bellaiche Y, Schweisguth F. 1999. Revisiting the *Drosophila* microchaete lineage: a novel intrinsically asymmetric cell division generates a glial cell. *Development* 126:3573–84

Ghysen A, Dambly-Chaudiere C. 2000. A genetic programme for neuronal connectivity. *Trends Genet.* 16:221–26

Ghysen A, Janson R, Santamaria P. 1983. Segmental determination of sensory neurons in *Drosophila*. *Dev. Biol.* 99:7–26

Gibbs SM, Becker A, Hardy RW, Truman JW. 2001. Soluble guanylate cyclase is required during development for visual system function in *Drosophila*. *J. Neurosci.* 21:7705–14

Gibbs SM, Truman JW. 1998. Nitric oxide and cyclic GMP regulate retinal patterning in the optic lobe of *Drosophila*. *Neuron* 20:83–93

Godenschwege TA, Simpson JH, Shan X, Bashaw GJ, Goodman CS, Murphey RK. 2002. Ectopic expression in the giant fiber system of *Drosophila* reveals distinct roles for roundabout (Robo), Robo2, and Robo3 in dendritic guidance and synaptic connectivity. *J. Neurosci.* 22:3117–29

Gomez-Skarmeta JL, Diez del Corral R, de la Calle-Mustienes E, Ferre-Marco D, Modolell J. 1996. *araucan* and *caupolican*, two members of the novel *iroquois* complex, encode homeoproteins that control proneural and vein-forming genes. *Cell* 85:95–105

Gomez-Skarmeta JL, Modolell J. 1996. *araucan* and *caupolican* provide a link between compartment subdivisions and patterning of sensory organs and veins in the *Drosophila* wing. *Genes Dev.* 10:2935–45

Gong Q, Rangarajan R, Seeger M, Gaul U. 1999. The netrin receptor Frazzled is required in the target for establishment of retinal projections in the *Drosophila* visual system. *Development* 126:1451–56

Goulding SE, zur Lage P, Jarman AP. 2000. *amos*, a proneural gene for *Drosophila* olfactory sense organs that is regulated by *lozenge*. *Neuron* 25:69–78

Grillenzoni N, van Helden J, Dambly-Chaudiere C, Ghysen A. 1998. The *iroquois* complex controls the somatotopy of *Drosophila* notum mechanosensory projections. *Development* 125:3563–69

Grueber WB, Jan LY, Jan YN. 2003. Different levels of the homeodomain protein Cut regulate distinct dendrite branching patterns of *Drosophila* multidendritic neurons. *Cell* 112:805–18

Gupta BP, Flores GV, Banerjee U, Rodrigues V. 1998. Patterning an epidermal field: *Drosophila lozenge*, a member of the AML-1/Runt family of transcription factors, specifies olfactory sense organ type in a dose-dependent manner. *Dev. Biol.* 203:400–11

Hardie RC. 1986. The photoreceptor array of the dipteran retina. *Trends Neurosci.* 9:419–23

Harris WA, Stark WS, Walker JA. 1976. Genetic dissection of the photoreceptor system in the compound eye of *Drosophila melanogaster*. *J. Physiol.* 256:415–39

Heitzler P, Haenlin M, Ramain P, Calleja M, Simpson P. 1996. A genetic analysis of *pannier*, a gene necessary for viability of dorsal tissues and bristle positioning in *Drosophila*. *Genetics* 143:1271–86

Hiesinger PR, Reiter C, Schau H, Fischbach KF. 1999. Neuropil pattern formation and regulation of cell adhesion molecules in *Drosophila* optic lobe development depend on synaptobrevin. *J. Neurosci.* 19:7548–56

Higashijima S, Michiue T, Emori Y, Saigo K. 1992. Subtype determination of *Drosophila* embryonic external sensory organs by redundant homeo box genes *BarH1* and *BarH2*. *Genes Dev.* 6:1005–18

Higgins MR, Gibson NJ, Eckholdt PA, Nighorn A, Copenhaver PF, et al. 2002. Different isoforms of Fasciclin II are expressed by a subset of developing olfactory receptor neurons and by olfactory-nerve glial cells during formation of glomeruli in the moth *Manduca sexta*. *Dev. Biol.* 244:134–54

Hildebrand JG, Shepherd GM. 1997. Mechanisms of olfactory discrimination: converging evidence for common principles across phyla. *Annu. Rev. Neurosci.* 20:595–631

Hing H, Xiao J, Harden N, Lim L, Zipursky SL. 1999. Pak functions downstream of Dock to regulate photoreceptor axon guidance in *Drosophila*. *Cell* 97:853–63

Huang ML, Hsu CH, Chien CT. 2000. The proneural gene *amos* promotes multiple dendritic neuron formation in the *Drosophila* peripheral nervous system. *Neuron* 25:57–67

Huang Z, Kunes S. 1996. Hedgehog, transmitted along retinal axons, triggers neurogenesis in the developing visual centers of the *Drosophila* brain. *Cell* 86:411–22

Huang Z, Shilo BZ, Kunes S. 1998. A retinal axon fascicle uses Spitz, an EGF receptor ligand, to construct a synaptic cartridge in the brain of *Drosophila*. *Cell* 95:693–703

Huber AB, Kolodkin AL, Ginty DD, Cloutier JF. 2003. Signaling at the growth cone: ligand-receptor complexes and the control of axon growth and guidance. *Annu. Rev. Neurosci.* 26:509–63

Hummel T, Leifker K, Klambt C. 2000. The *Drosophila* HEM-2/NAP1 homolog KETTE controls axonal pathfinding and cytoskeletal organization. *Genes Dev.* 14:863–73

Hummel T, Vasconcelos ML, Clemens JC, Fishilevich Y, Vosshall LB, Zipursky SL. 2003. Axonal targeting of olfactory receptor neurons in *Drosophila* is controlled by Dscam. *Neuron* 37:221–31

Iwai Y, Hirota Y, Ozaki K, Okano H, Takeichi M, Uemura T. 2002. DN-cadherin is required for spatial arrangement of nerve terminals and ultrastructural organization of synapses. *Mol. Cell. Neurosci.* 19:375–88

Jacobs GA, Theunissen FE. 1996. Functional organization of a neural map in the cricket cercal sensory system. *J. Neurosci.* 16:769–84

Jan YN, Jan LY. 1993. The peripheral nervous system. In *The Development of Drosophila melanogaster*, ed. M Bate, A Martinez Arias, pp. 1207–44. Plainview, NY: Cold Spring Harbor Lab.

Jan YN, Jan LY. 2000. Polarity in cell division: What frames thy fearful asymmetry? *Cell* 100:599–602

Jarman AP, Ahmed I. 1998. The specificity of proneural genes in determining *Drosophila* sense organ identity. *Mech. Dev.* 76:117–25

Jarman AP, Grau Y, Jan LY, Jan YN. 1993. *atonal* is a proneural gene that directs chordotonal organ formation in the *Drosophila* peripheral nervous system. *Cell* 73:1307–21

Jarman AP, Grell EH, Ackerman L, Jan LY, Jan YN. 1994. *atonal* is the proneural gene for *Drosophila* photoreceptors. *Nature* 369:398–400

Jhaveri D, Rodrigues V. 2002. Sensory neurons of the Atonal lineage pioneer the formation of glomeruli within the adult *Drosophila* olfactory lobe. *Development* 129:1251–60

Jhaveri D, Sen A, Reddy GV, Rodrigues V. 2000. Sense organ identity in the *Drosophila* antenna is specified by the expression of the proneural gene *atonal*. *Mech. Dev.* 99:101–11

Kaminker JS, Canon J, Salecker I, Banerjee U. 2002. Control of photoreceptor axon target choice by transcriptional repression of Runt. *Nat. Neurosci.* 5:746–50

Kidd T, Bland KS, Goodman CS. 1999. Slit is the midline repellent for the Robo receptor in *Drosophila*. *Cell* 96:785–94

Killian KA, Merritt DJ, Murphey RK. 1993. Transplantation of neurons reveals processing areas and rules for synaptic connectivity in the cricket nervous system. *J. Neurobiol.* 24:1187–206

Kimmel BE, Heberlein U, Rubin GM. 1990. The homeo domain protein Rough is expressed in a subset of cells in the developing *Drosophila* eye where it can specify photoreceptor cell subtype. *Genes Dev.* 4:712–27

Komiyama T, Johnson WA, Luo L, Jefferis GS. 2003. From lineage to wiring specificity. POU domain transcription factors control precise connections of *Drosophila* olfactory projection neurons. *Cell* 112:157–67

Kornberg T. 1981. *Engrailed*: a gene controlling compartment and segment formation in *Drosophila*. *Proc. Natl. Acad. Sci. USA* 78:1095–99

Kunes S, Wilson C, Steller H. 1993. Independent guidance of retinal axons in the developing visual system of *Drosophila*. *J. Neurosci.* 13:752–67

Lee CH, Herman T, Clandinin TR, Lee R, Zipursky SL. 2001. N-cadherin regulates target specificity in the *Drosophila* visual system. *Neuron* 30:437–50

Lee RC, Clandinin TR, Lee CH, Chen PL, Meinertzhagen IA, Zipursky SL. 2003. The protocadherin Flamingo is required for axon target selection in the *Drosophila* visual system. *Nat. Neurosci.* 6:557–63

Marie B, Bacon JP, Blagburn JM. 2000. Double-stranded RNA interference shows that Engrailed controls the synaptic specificity of identified sensory neurons. *Curr. Biol.* 10:289–92

Marie B, Blagburn JM. 2003. Differential roles of Engrailed paralogs in determining sensory axon guidance and synaptic target recognition. *J. Neurosci.* 23:7854–62

Marie B, Cruz-Orengo L, Blagburn JM. 2002. Persistent *engrailed* expression is required to determine sensory axon trajectory, branching, and target choice. *J. Neurosci.* 22:832–41

Maurel-Zaffran C, Suzuki T, Gahmon G, Treisman JE, Dickson BJ. 2001. Cell-autonomous and -nonautonomous functions of LAR in R7 photoreceptor axon targeting. *Neuron* 32:225–35

Meinertzhagen IA. 2000. Wiring the fly's eye. *Neuron* 28:310–13

Meinertzhagen IA, Hanson TE. 1993. The development of the optic lobe. In *The Development of Drosophila melanogaster*, ed. M Bate, A Martinez Arias, pp. 1363–491. Plainview, NY: Cold Spring Harbor Laboratory Press

Meinertzhagen IA, O'Neil SD. 1991. Synaptic organization of columnar elements in the lamina of the wild type in *Drosophila melanogaster*. *J. Comp. Neurol.* 305:232–63

Meinertzhagen IA, Piper ST, Sun XJ, Frohlich A. 2000. Neurite morphogenesis of identified visual interneurons and its relationship to photoreceptor synaptogenesis in the flies, *Musca domestica* and *Drosophila melanogaster*. *Eur. J. Neurosci.* 12:1342–56

Merritt DJ. 1997. Transformation of external sensilla to chordotonal sensilla in the *cut* mutant of *Drosophila* assessed by single-cell marking in the embryo and larva. *Microsc. Res. Tech.* 39:492–505

Merritt DJ, Hawken A, Whitington PM. 1993. The role of the *cut* gene in the specification of central projections by sensory axons in *Drosophila*. *Neuron* 10:741–52

Merritt DJ, Whitington PM. 1995. Central projections of sensory neurons in the *Drosophila* embryo correlate with sensory modality, soma position, and proneural gene function. *J. Neurosci.* 15:1755–67

Meyerowitz EM, Kankel D. 1978. A genetic analysis of visual system development in *Drosophila melanogaster*. *Dev. Biol.* 62:112–42

Mollereau B, Dominguez M, Webel R, Colley NJ, Keung B, et al. 2001. Two-step process for photoreceptor formation in *Drosophila*. *Nature* 412:911–13

Mombaerts P, Wang F, Dulac C, Chao SK, Nemes A, et al. 1996. Visualizing an olfactory sensory map. *Cell* 87:675–86

Moore AW, Jan LY, Jan YN. 2002. *hamlet*, a binary genetic switch between single- and multiple-dendrite neuron morphology. *Science* 297:1355–58

Murphey RK. 1985. A second cricket cercal sensory system: bristle hairs and the interneurons they activate. *J. Comp. Physiol. A* 156:357–67

Murphey RK, Bacon JP, Johnson SE. 1985. Ectopic neurons and the organization of insect sensory systems. *J. Comp. Physiol. A* 156:381–89

Murphey RK, Bacon JP, Sakaguchi DS, Johnson SE. 1983. Transplantation of cricket sensory neurons to ectopic locations: arborizations and synaptic connections. *J. Neurosci.* 3:659–72

Murphey RK, Possidente D, Pollack G, Merritt DJ. 1989. Modality-specific axonal projections in the CNS of the flies *Phormia* and *Drosophila*. *J. Comp. Neurol.* 290:185–200

Murre C, McCaw PS, Vaessin H, Caudy M, Jan LY, et al. 1989. Interactions between heterologous helix-loop-helix proteins generate complexes that bind specifically to a common DNA sequence. *Cell* 58:537–44

Newland PL, Rogers SM, Gaaboub I, Matheson T. 2000. Parallel somatotopic maps of gustatory and mechanosensory neurons in the central nervous system of an insect. *J. Comp. Neurol.* 425:82–96

Newsome TP, Asling B, Dickson BJ. 2000a. Analysis of *Drosophila* photoreceptor axon guidance in eye-specific mosaics. *Development* 127:851–60

Newsome TP, Schmidt S, Dietzl G, Keleman K, Asling B, et al. 2000b. Trio combines with Dock to regulate Pak activity during photoreceptor axon pathfinding in *Drosophila. Cell* 101:283–94

Nottebohm E, Dambly-Chaudiere C, Ghysen A. 1992. Connectivity of chemosensory neurons is controlled by the gene *poxn* in *Drosophila. Nature* 359:829–32

Nottebohm E, Usui A, Therianos S, Kimura K, Dambly-Chaudiere C, Ghysen A. 1994. The gene *poxn* controls different steps of the formation of chemosensory organs in *Drosophila. Neuron* 12:25–34

Oland LA, Pott WM, Higgins MR, Tolbert LP. 1998. Targeted ingrowth and glial relationships of olfactory receptor axons in the primary olfactory pathway of an insect. *J. Comp. Neurol.* 398:119–38

Oland LA, Tolbert LP. 1996. Multiple factors shape development of olfactory glomeruli: insights from an insect model system. *J. Neurobiol.* 30:92–109

Patel NH, Martin-Blanco E, Coleman KG, Poole SJ, Ellis MC, et al. 1989. Expression of Engrailed proteins in arthropods, annelids, and chordates. *Cell* 58:955–68

Paydar S, Doan CA, Jacobs GA. 1999. Neural mapping of direction and frequency in the cricket cercal sensory system. *J. Neurosci.* 19:1771–81

Perez SE, Steller H. 1996. Migration of glial cells into retinal axon target field in *Drosophila melanogaster. J. Neurobiol.* 30: 359–73

Phillis R, Statton D, Caruccio P, Murphey RK. 1996. Mutations in the 8 kDa dynein light chain gene disrupt sensory axon projections in the *Drosophila* imaginal CNS. *Development* 122:2955–63

Poeck B, Fischer S, Gunning D, Zipursky SL, Salecker I. 2001. Glial cells mediate target layer selection of retinal axons in the developing visual system of *Drosophila. Neuron* 29:99–113

Posakony JW. 1994. Nature versus nurture: asymmetric cell divisions in *Drosophila* bristle development. *Cell* 76:415–18

Rao Y, Pang P, Ruan W, Gunning D, Zipursky SL. 2000. *brakeless* is required for photoreceptor growth-cone targeting in *Drosophila. Proc. Natl. Acad. Sci. USA* 97:5966–71

Ritzenthaler S, Chiba A. 2003. Myopodia (postsynaptic filopodia) participate in synaptic target recognition. *J. Neurobiol.* 55:31–40

Ritzenthaler S, Suzuki E, Chiba A. 2000. Postsynaptic filopodia in muscle cells interact with innervating motoneuron axons. *Nat. Neurosci.* 3:1012–17

Rossler W, Oland LA, Higgins MR, Hildebrand JG, Tolbert LP. 1999. Development of a glia-rich axon-sorting zone in the olfactory pathway of the moth *Manduca sexta. J. Neurosci.* 19:9865–77

Ruan W, Long H, Vuong DH, Rao Y. 2002. Bifocal is a downstream target of the Ste20-like serine/threonine kinase misshapen in regulating photoreceptor growth cone targeting in *Drosophila. Neuron* 36:831–42

Ruan W, Pang P, Rao Y. 1999. The SH2/SH3 adaptor protein dock interacts with the Ste20-like kinase misshapen in controlling growth cone motility. *Neuron* 24:595–605

Salcedo E, Huber A, Henrich S, Chadwell LV, Chou WH, et al. 1999. Blue- and green-absorbing visual pigments of *Drosophila*: ectopic expression and physiological characterization of the R8 photoreceptor cell-specific Rh5 and Rh6 rhodopsins. *J. Neurosci.* 19:10716–26

Sato M, Kojima T, Michiue T, Saigo K. 1999. *Bar* homeobox genes are latitudinal prepattern genes in the developing Drosophila notum whose expression is regulated by the concerted functions of *decapentaplegic* and *wingless. Development* 126:1457–66

Schrader S, Merritt DJ. 2000. Central projections of *Drosophila* sensory neurons in the transition from embryo to larva. *J. Comp. Neurol.* 425:34–44

Selleck SB, Gonzales C, Glover DM, White K. 1992. Regulation of the G1-S transition in postembryonic neuronal precursors by axonal ingrowth. *Nature* 355:253–55

Senti K, Keleman K, Eisenhaber F, Dickson BJ. 2000. *brakeless* is required for lamina targeting of R1-R6 axons in the *Drosophila* visual system. *Development* 127:2291–301

Senti KA, Usui T, Boucke K, Greber U, Uemura T, Dickson BJ. 2003. Flamingo regulates R8 axon-axon and axon-target interactions in the *Drosophila* visual system. *Curr. Biol.* 13:828–32

Siegler MV, Jia XX. 1999. Engrailed negatively regulates the expression of cell adhesion molecules Connectin and Neuroglian in embryonic *Drosophila* nervous system. *Neuron* 22:265–76

Simpson P. 1997. Notch signaling in development: on equivalence groups and asymmetric developmental potential. *Curr. Opin. Genet. Dev.* 7:537–42

Song J, Wu L, Chen Z, Kohanski RA, Pick L. 2003. Axons guided by insulin receptor in *Drosophila* visual system. *Science* 300:502–5

Su YC, Maurel-Zaffran C, Treisman JE, Skolnik EY. 2000. The Ste20 kinase misshapen regulates both photoreceptor axon targeting and dorsal closure, acting downstream of distinct signals. *Mol. Cell Biol.* 20:4736–44

Suh GS, Poeck B, Chouard T, Oron E, Segal D, et al. 2002. *Drosophila* JAB1/CSN5 acts in photoreceptor cells to induce glial cells. *Neuron* 33:35–46

Tayler TD, Garrity PA. 2003. Axon targeting in the *Drosophila* visual system. *Curr. Opin. Neurobiol.* 13:90–95

Tessier-Lavigne M, Goodman CS. 1996. The molecular biology of axon guidance. *Science* 274:1123–33

Tomlinson A, Struhl G. 1999. Decoding vectorial information from a gradient: sequential roles of the receptors Frizzled and Notch in establishing planar polarity in the *Drosophila* eye. *Development* 126:5725–38

Tomlinson A, Struhl G. 2001. Delta/Notch and Boss/Sevenless signals act combinatorially to specify the *Drosophila* R7 photoreceptor. *Mol. Cell* 7:487–95

Vervoort M, Merritt DJ, Ghysen A, Dambly-Chaudiere C. 1997. Genetic basis of the formation and identity of type I and type II neurons in *Drosophila* embryos. *Development* 124:2819–28

Vosshall LB, Wong AM, Axel R. 2000. An olfactory sensory map in the fly brain. *Cell* 102:147–59

Wang F, Nemes A, Mendelsohn M, Axel R. 1998. Odorant receptors govern the formation of a precise topographic map. *Cell* 93: 47–60

Wang JW, Wong AM, Flores J, Vosshall LB, Axel R. 2003. Two-photon calcium imaging reveals an odor-evoked map of activity in the fly brain. *Cell* 112:271–82

Whitlock KE. 1993. Development of *Drosophila* wing sensory neurons in mutants with missing or modified cell surface molecules. *Development* 117:1251–60

Williams DW, Shepherd D. 1999. Persistent larval sensory neurons in adult *Drosophila melanogaster*. *J. Neurobiol.* 39:275–86

Williams DW, Shepherd D. 2002. Persistent larval sensory neurones are required for the normal development of the adult sensory afferent projections in *Drosophila*. *Development* 129:617–24

Wolff T, Ready DF. 1993. Pattern formation in the *Drosophila* retina. In *The Development of Drosophila melanogaster*, ed. M Bate, A Martinez-Arias, pp. 1277–325. Plainview, NY: Cold Spring Harbor Lab.

Yang CH, Axelrod JD, Simon MA. 2002. Regulation of Frizzled by Fat-like cadherins during planar polarity signaling in the *Drosophila* compound eye. *Cell* 108:675–88

Zlatic M, Landgraf M, Bate M. 2003. Genetic specification of axonal arbors: *atonal* regulates *robo3* to position terminal branches in the *Drosophila* nervous system. *Neuron* 37:41–51

Annu. Rev. Neurosci. 2004. 27:53–77
doi: 10.1146/annurev.neuro.26.041002.131032

First published online as a Review in Advance on April 2, 2004

SENSORY SIGNALS IN NEURAL POPULATIONS UNDERLYING TACTILE PERCEPTION AND MANIPULATION

Antony W. Goodwin and Heather E. Wheat
Department of Anatomy and Cell Biology, University of Melbourne, Parkville, Victoria 3010, Australia; email: a.goodwin@unimelb.edu.au, h.wheat@unimelb.edu.au

Key Words somatosensory, cutaneous, mechanoreceptor, grasp, primary afferent

■ **Abstract** For humans to manipulate an object successfully, the motor control system must have accurate information about parameters such as the shape of the stimulus, its position of contact on the skin, and the magnitude and direction of contact force. The same information is required for perception during haptic exploration of an object. Much of these data are relayed by the mechanoreceptive afferents innervating the glabrous skin of the digits. Single afferent responses are modulated by all the relevant stimulus parameters. Thus, only in complete population reconstructions is it clear how each of the parameters can be signaled to the brain independently when many are changing simultaneously, as occurs in most normal movements or haptic exploration. Modeling population responses reveals how resolution is affected by neural noise and intrinsic properties of the population such as the pattern and density of innervation and the covariance of response variability.

INTRODUCTION

Primates, an order which includes humans, apes and monkeys, are the only animals in which hands have evolved, and this has played a significant part in their evolutionary success (Napier 1980). In our daily lives we explore and manipulate a wide range of objects, and there are myriad examples of our dependence on manual dexterity and of our reliance on tactile perception of these objects. As illustrations, we can palpate an avocado to determine if it is ripe, scan our fingers over a surface to assess its roughness, lift an object to estimate its weight, do up our shoe laces, separate a peanut from its shell, pick a cherry, or write a message.

Hand function is a vast field that we do not attempt to cover in breadth; rather in this review we focus on a crucial component, namely the cutaneous neural signals from the digits, which are vital to tactile perception and manipulation. We place particular emphasis on the importance of neural population analyses.

0147-006X/04/0721-0053$14.00

THE SPECTRUM OF HAND FUNCTION

The repertoire of hand movements used by humans in their everyday lives is extensive. In the majority of cases the movement comprises several components commencing with a reaching phase and culminating in manipulation or haptic exploration of the object (Jeannerod et al. 1998).

There is a continuous spectrum of hand function. At the one extreme, the object is grasped and manipulated and the main objective is to accomplish the task; a simple example is lifting a cup of coffee off a table. Here the sensory signals from the digits are used by the motor control system to effect the task. However, even at this extreme, sensory perception plays a role and the person is certainly aware of whether the cup is light or heavy and whether the handle is highly curved or not. At the other extreme, the object is explored with the fingertips and the main objective is sensory perception; a simple example is determining how smooth the surface of the object is. Here the sensory signals from the digits are the source for conscious sensory perception and cognition, but they are also important for the motor control system to enable effective scanning of the surface, particularly if its shape is complex.

Although, in general, vision plays a significant part in manipulation, there is abundant evidence that somatosensory feedback is essential. Mott & Sherrington (1895) sectioned the dorsal roots in monkeys and dramatically illustrated the dependence of all phases of upper limb movements on somatosensory feedback despite the fact that vision was unaffected. There are now several reports in the literature of patients with large fiber sensory neuropathies, and in these patients hand function is severely compromised (Sanes et al. 1985). Even in the presence of vision, the case report of Rothwell et al. (1982) states that "...he could no longer execute certain manual tasks; thus he was unable to do up his shirt buttons or manipulate a pen properly." More specifically, when feedback from the glabrous skin of the digits is eliminated by blocking the digital nerves in humans, manipulation is impaired (Johansson & Westling 1984, Monzee et al. 2003). In the reverse direction, precise manipulation is possible without vision. For instance, many tasks like doing up a button or switching on a light in the dark are done without vision; also blind people are dexterous.

Grasp and Manipulation

We cover only those aspects of grasp and manipulation that relate directly to tactile feedback—for a broad review see Johansson (1996). For experimental convenience and tractable analysis, the "manipulation" in most studies is highly simplified. Two-digit (thumb and forefinger) grasp and lift tasks are used extensively (Gordon & Duff 1999, Westling & Johansson 1984), but there have also been studies of multi-digit grasp and lift tasks (Reilmann et al. 2001) and a variety of tasks involving movements more complex than simple lifts (Flanagan & Wing 1997, Gysin et al. 2003).

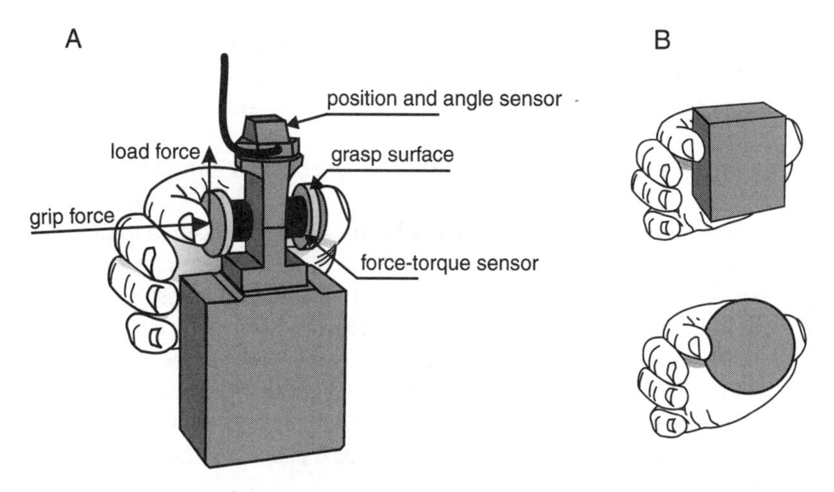

Figure 1 A typical two-digit grip and lift task. (*A*) Apparatus with two sensors registering the three-dimensional forces and torques on the thumb and the index finger. From Jenmalm et al. (1998). (*B*) Grasping the block with flat surfaces or the sphere results in the same joint angles even though the shapes of the objects are different.

When an object is gripped between the thumb and index finger and lifted off a table (Figure 1*A*), there are several key stimulus parameters, as detailed below. Relaying information about these parameters to the CNS is crucial, whether it be for feedback correction during the task itself or to update memory for future use in planning and feedforward control (Johansson 1996).

The load force applied to lift the object must be tailored to the weight of the object. The reaction to the load force is a force on each digit that acts in a direction tangential to the skin surface. In principle, load force could be signaled by receptors in the muscles generating the forces or even by efference copy of the motor commands (Gandevia et al. 1992). Important for this review, low-threshold mechanoreceptors in the glabrous skin of the finger tips are highly sensitive to forces tangential to the skin (Birznieks et al. 2001).

Maintaining a stable grasp on the object requires a grip force of sufficient magnitude to prevent slip, but the grip force should not be excessive so that unnecessary energy expenditure and muscle fatigue do not occur and delicate objects are not damaged. Humans select a grip force with a "reasonable" safety margin that varies with the task and varies between subjects, but usually grip force is of the order of 1.2 to 2 times the minimum force required to prevent slip (Gysin et al. 2003, Westling & Johansson 1984). The reaction of the grip force acts on the skin of the digits in a direction normal to the skin surface and is equivalent to the indenting forces commonly used in somatosensory experiments. Like load force, grip force could be signaled by muscle receptors or efference copy. Indentation normal to the skin excites low-threshold cutaneous mechanoreceptors (Darian-Smith 1984,

Vallbo 1995). There are two equivalent ways of treating contact force. Either it can be resolved into the two orthogonal component magnitudes, normal and tangential to the skin, or it can be treated as a single force vector with a magnitude and a direction.

The shape of the object has a substantial effect on grip and lift tasks (Jenmalm & Johansson 1997, Jenmalm et al. 1998), particularly if some tilt is involved (Goodwin et al. 1998). More precisely, this should be referred to as the local shape of the object because it is only the shape in the region of contact with the skin that is relevant. The distinction between local and global shape is critical when considering sources of feedback in a grip and lift task. It is natural to associate object shape with joint angles. However, Figure 1*B* shows that when gripping a block with flat parallel sides or a sphere of equivalent size, the joint angles are the same. The local shapes of the two objects contacting the skin are vastly different, and the only signals from the hand for local shape are those from the cutaneous afferents.

The position of contact on the digits effects the balance of forces between the thumb and the index finger and impacts on grasp stability (Montana 1988). Excluding vision, position on the skin can only be gauged from cutaneous responses, in particular those from mechanoreceptors.

Limits on the relationship between the grip and load forces are set by the effective coefficient of friction between the object and the digits. The nature of the coefficient of friction between skin and a flat surface is complex; it depends on factors like skin hydration, related to sweating, and does not obey Amonton's law, which assumes a constant coefficient of friction, independent of the grip force magnitude (Koudine et al. 2000). The concept of an effective coefficient of friction between the skin and a curved surface that partially indents into the skin is even more challenging (Pawluk & Howe 1999). Humans adjust their grip and lift strategies appropriately when the frictional characteristics of flat grip surfaces are changed (Johansson & Westling 1984, Forssberg et al. 1995). How frictional information is calculated by the brain is not known, but it is possible to calculate it from the force ratios at instances of slip or microslip (Tremblay & Cutkoksy 1993). Cutaneous afferents are responsive to slips (Johansson & Westling 1987, Srinivasan et al. 1990).

Although the grip and lift tasks, and variants of them, are relatively simple manipulations, they have provided a wealth of information, largely because it is possible to control the tasks so well and to define and measure the component events of the manipulation with precision. Everyday manipulations like peeling a potato or eating a grape are far more complex and demanding, but the principles of sensory feedback are the same. An instructive example for our purposes is the task, familiar to all of us, of doing up a button in the absence of vision. It is clear that the motor control system requires information about the local shape of the button, its position on the digital skin, and the normal and tangential force components. Much of this information is provided by the cutaneous receptors, as is shown below.

Haptic Exploration

Tactile perception of an object requires manual exploration of the object. This may be relatively simple, like scanning the fingertips over a flat textured surface, or more complex, like exploring an intricate three-dimensional carving. This form of hand movement is often covered under the rubric of haptics—for reviews see Craig & Rollman (1999) and Klatzky & Lederman (2003). It is self-evident that signals from the cutaneous mechanoreceptive afferents are essential for tactile perception and cognition. Perception of surface texture, which encompasses roughness, is a quintessential tactile task. If you wish to determine the fine texture of a surface, it is invariably tactile perception that you resort to, not vision (Heller 1989). Because of space constraints, we deliberately exclude the large field of texture; it has been reviewed by Johnson & Hsiao (1992) and is covered in Craig & Rollman (1999).

CUTANEOUS MECHANORECEPTORS AND AFFERENTS

Glabrous skin covering the volar surface of the digits, with its characteristic fingerprint pattern, is highly specialized. Its structure and innervation have been reviewed previously (Darian-Smith 1984, Vallbo 1995) and is not dealt with extensively here. There are four classes of low-threshold mechanoreceptors in human glabrous skin innervated by four classes of peripheral afferent nerve fibers. The more superficial Merkel complexes and Meissner corpuscles are innervated by slowly adapting type I afferents (SAI afferents) and so-called fast-adapting type I afferents (FAI afferents), respectively. The deeper Ruffini organs and Pacinian corpuscles are innervated by slowly adapting type II (SAII) afferents and fast-adapting type II (FAII) afferents, respectively. This nomenclature, developed for human microneurography studies, is invariably used for classifying human afferents.

Glabrous skin in the monkey is similar to that in humans except that there are no reports of Ruffini organs or SAII afferents. It is not known what is the significance of this difference. Nor is it known whether glabrous skin in the apes, which fall between humans and monkeys on the evolutionary scale, is innervated by SAII afferents. The nomenclature used for the monkey is more variable than that used for humans; Merkel afferents are usually called SA (sometimes SAI) afferents, Meissner afferents are usually called RA (sometimes FAI, and in the older literature, QA) afferents, and Pacinian afferents are usually called PC or (sometimes FAII) afferents. In this review we use the human nomenclature for both human afferents and the three afferents present in the monkey.

The relationship between receptors and afferents is complex; single afferents innervate multiple receptors and single receptors are innervated by multiple afferents, but the details of the anatomical convergence and divergence are not known (Cauna 1956, Pare et al. 2002b). The distinction between afferents and receptors is important in this review because we concentrate on resolution and relating it to the afferent innervation density.

Skin Mechanics and Models of the Finger

In the simplest models of glabrous skin mechanics, which have provided valuable insights into afferent response characteristics, it was assumed that the skin was flat and/or uniform in structure, and no account was taken of its layered or ridged nature (Phillips & Johnson 1981b, Srinivasan 1989). However, the majority of experiments are performed on the distal segment of the finger, which is the region that has most contact during manipulation and haptic exploration (Christel 1993). The distal segment has the following mechanical complications: (*a*) Part of the volar surface is relatively flat (or gently curved), but the radial and ulnar sides are highly and nonuniformly curved, and the same is true for the distal tip; (*b*) the distal phalanx, an incompressible bone with a distinctive shape, is at the center of the segment; (*c*) at the proximal end of the segment, the skin on the volar surface is partially anchored by the interphalangeal crease; (*d*) on the dorsal surface, the nail is a stiff structure that is tightly anchored to the distal phalanx. Some models of skin mechanics have begun addressing these issues by modeling the finger in two dimensions and including bone at the center (Maeno & Kobayashi 1998, Srinivasan & Dandekar 1996). As yet, no model comes close to including all the factors that influence real fingertip skin mechanics. This is a significant gap in our knowledge because skin mechanics plays a major role in the transduction of tactile stimuli into afferent responses; without a realistic model it is not possible to predict the responses of an afferent from any location on the fingertip to an arbitrary stimulus.

SINGLE AFFERENT RESPONSES TO LOCAL SHAPE

A convenient way of defining the local shape of an object is in terms of its curvatures. In fact the shape of any three-dimensional object can be defined in terms of two orthogonal curvatures at each point on its surface (Koenderink & van Doorn 1992). In many studies of shape, simple objects such as spheres, cylinders, or torroids are used. These are particularly amenable to experimentation and analysis because of the ease of quantifying the shape.

Vierck (1979) observed that slowly adapting afferents in the cat's hairy skin were sensitive to the edges of indenting stimuli. Phillips & Johnson (1981a) characterized the edge sensitivity of SAI afferents in the monkey's finger (Figure 2*A*), and Johansson et al. (1982) showed that in human glabrous skin SAI afferents, and to a lesser extent FAI afferents, were edge sensitive. Although the authors did not discuss their findings explicitly in the context of shape, these were the first indications that cutaneous afferents may be signaling the local shape of objects contacting the skin.

To investigate tactile shape more directly, LaMotte & Srinivasan (LaMotte & Srinivasan 1987a,b; Srinivasan & LaMotte 1987) applied sinusoidally shaped steps to the monkey's fingerpad. In some experiments the stimuli were indented into the skin (analogous to the situation when grasping and lifting a glass of water), and in

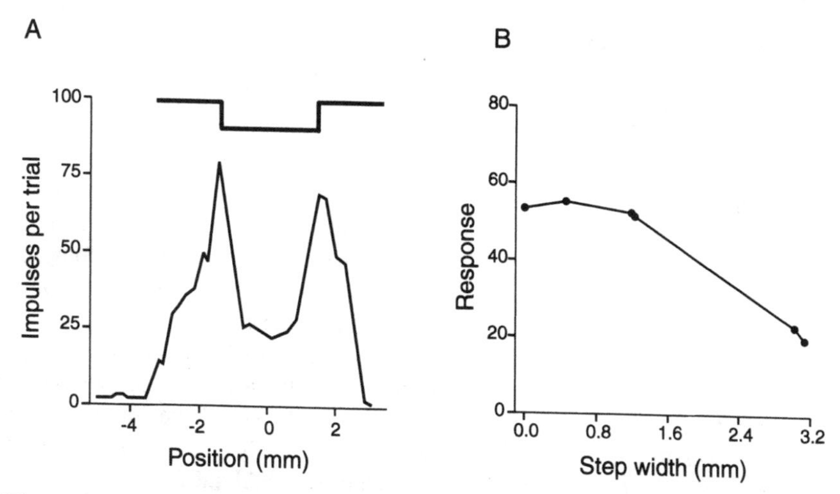

Figure 2 Responses of single SAI afferents in the monkey's fingerpad. (*A*) Responses to a 3-mm bar stepped across the receptive field; the origin for position in the receptive field is arbitrary. Redrawn from Phillips & Johnson (1981a). (*B*) Responses to shaped steps. Each step is a half sine wave (from peak to trough) characterized by the step width (half the period). Redrawn from Srinivasan & LaMotte (1987).

others the stimuli were scanned across the skin (analogous to haptic exploration). Both the SAI and FAI afferent responses were modulated by the shapes of the stimuli, but the effect was more pronounced and more reliable for the SAI afferents (Figure 2*B*). Although the sinusoidal steps could be graded and specified precisely, the local shape over such a step is not constant, and this compromised quantifying the relationship between afferent responses and object shape.

This limitation was not present in the experiments of Goodwin and colleagues (Goodwin et al. 1995), who chose spherical stimuli; the shape of a sphere is completely specified by a single number, its curvature, which is constant over the whole sphere. For this simple shape the curvature is the reciprocal of the radius of the sphere. When spheres were indented into a monkey's fingerpad, a clear monotonic relationship existed between the curvature of the stimulus and the responses of SAI afferents. Moreover, all the SAI afferents had the same curvature-response function except for a multiplicative factor. This factor was a reflection of the variation of sensitivity among SAI afferents, which had been described previously for indenting probes (Knibestol 1975). The curvature of the spheres had only a minor effect on FAI responses.

An important feature in both of the above series of experiments is that the shapes were applied at constant force. In the majority of experiments on cutaneous mechanoreceptive afferents, the stimulus has been applied with a controlled depth of indentation, which was entirely appropriate to the aims of those experiments. However, when relating sensory neural events to manipulation or haptic exploration, force (which has both a tangential and a normal component) is a more appropriate parameter than skin displacement. Normal force and depth of indentation are, of course, related but in a complex nonlinear manner with considerable depth saturation (Pawluk & Howe 1999).

For any shape more complex than a sphere, more than one parameter is needed to define it, and most objects have an orientation. Two relatively simple objects with an orientation that are used to study tactile shape are cylinders and torroids. A cylinder has a curvature of zero in the direction of the cylinder's long axis and a constant curvature (the reciprocal of the radius) in the orthogonal direction, and it is defined by the orientation of the major axis and the curvature along the minor axis. SAI afferent responses are monotonically related to the curvature of an indenting cylinder in much the same way as for spheres, but they are also modulated by the orientation of the cylinder (Dodson et al. 1998). The shapes of "wavy surfaces" (a succession of convex and concave cylinders with decrementing radii) scanned over the monkey's fingerpad affect responses in SAI and FAI afferents (LaMotte & Srinivasan 1996). A torroid is defined by three parameters, two orthogonal curvatures and an orientation. Both SAI and FAI afferent responses are modulated by torroids scanned over the monkey's finger with more information evident in the SAI afferent responses (LaMotte et al. 1998).

Only two studies have addressed the responses of human cutaneous afferents to shape. For spheres and cylinders, the results are comparable to those described above for the monkey (Goodwin et al. 1997). In a more recent study, Jenmalm et al. (2003) employed forces similar to those used by humans in everyday manipulations. Compared to the monkey experiments, the forces were larger in magnitude (4 N), more rapid (125-ms duration for the rising phase), and applied in 5 different directions. They used a flat surface and two spheres and found that curvature significantly modulated responses in SAI, SAII, and FAI afferents.

The experiments described in this section show that responses of mechanoreceptive afferent fibers are affected by the local shape of a stimulus contacting the receptive field. The effect is most pronounced for SAI afferents but is also present in FAI afferents and in SAII afferents in humans.

POPULATION RESPONSES

Rudimentary Population Responses

Demonstrating that the curvature or shape of a stimulus modulates single afferent responses is the crucial first step, but in itself it does not allow us to appreciate how the brain might extract details about the local shape of an object. Such information can only be obtained by combining responses from a number of afferents or, in general, by examining the entire population of active afferents.

Ideally the responses from each afferent in the entire active population (many hundreds of fibers) should be recorded simultaneously, but currently this is not remotely possible. Virtually all isolated-unit peripheral nerve recording is from single fibers; occasionally, with microneurography, two or three fibers are recorded from together (Wessberg et al. 2003). Thus, the only way to analyze whole-population responses is to model them in some way. There are many examples of population models in a diverse range of neural systems; a small sample includes sound localization (Fitzpatrick et al. 1997), visual motion perception (Shadlen et al. 1996), determination of visual shape (Pasupathy & Connor 2002), motor control of upper limb trajectories (Lukashin & Georgopoulos 1993), and cockroach escape turns (Levi & Camhi 2000).

In the somatosensory system, a simple method of approximating rudimentary population responses has been used extensively. By plotting the receptive field profile of a single cell (the response of the cell as a function of the position of the stimulus in the receptive field), that profile can be viewed as a population response, described by Mountcastle & Powell (1959) as "reciprocal interpretation." Despite the limitations of this approach (see later sections) it does provide a visual appreciation of the basic pattern of responses across the population. For example, the profiles in Figure 3*A*, derived essentially by the above method, clearly depict how the orientation and shape of a cylinder is represented in the SAI afferent population.

For stimuli scanned across the skin, the equivalent approach is termed a spatial event plot (Johnson & Lamb 1981). As the stimulus is scanned over the two-dimensional receptive field, a dot marks the occurrence of each action potential and, by reciprocal interpretation, depicts the pattern of impulses in a population of afferents. For example, in Figure 3*B* spatial event plots are shown for a torroid scanned over the receptive field of an SAI afferent, and it is evident how the shape and orientation of the torroid could be represented.

Multiple-Parameter Population Reconstruction

An obvious deficiency of the rudimentary population responses described above is that, in general, all stimulus parameters except one are held constant. However, the primary afferent responses modulated by the shape of the stimulus are also affected by other parameters of the stimulus, such as force magnitude and direction (Birznieks et al. 2001), the position of the stimulus in the receptive field (Wheat et al. 1995), the orientation of nonspherical surfaces (Dodson et al. 1998), and the speed and direction of the scanning motion (LaMotte & Srinivasan 1987b). During real manipulation and haptic exploration, many of the stimulus parameters change concurrently, and a useful population reconstruction should reveal how each of these parameters might be signaled to the brain independently of the others. It is obvious that a change in a single afferent's response cannot convey which parameter or parameters have changed; this can only be determined unambiguously from the population response.

Approaches to Reconstruction

For stimuli contacting the finger, two approaches have been used for acquiring the single fiber responses that form the basis of the reconstruction. In the more

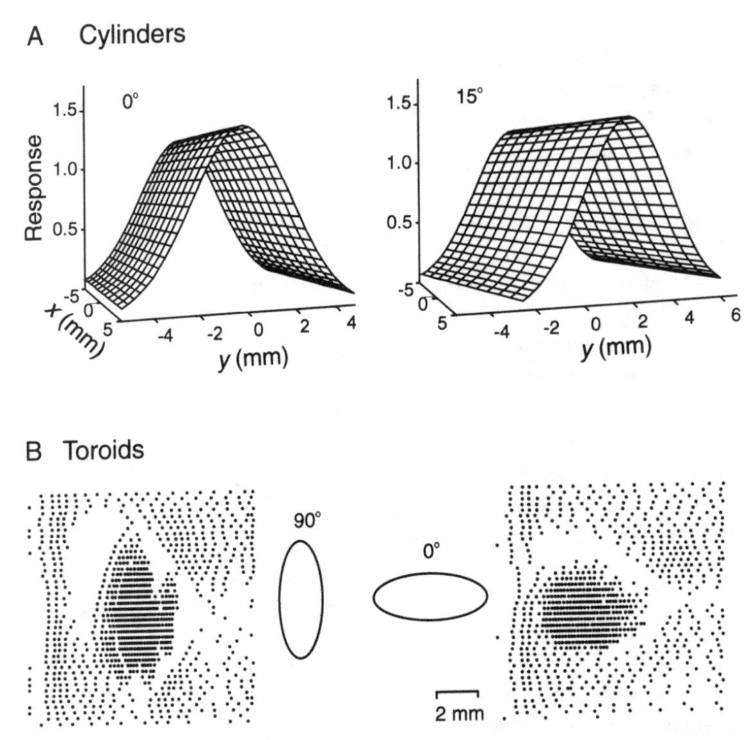

Figure 3 Rudimentary population reconstructions from responses of SAI afferents in the monkey's fingerpad. (*A*) Two-dimensional receptive field profiles, interpreted as population responses, depict the representation of the orientation of a cylinder indented into the fingerpad. The cylinder, with a curvature of 172 m^{-1}, was oriented at 0° or 15°. Modified from Dodson et al. (1998). (*B*) Spatial event plots of a torroid scanned over the fingerpad show the representation of its orientation. The 1 × 5–mm torroid was oriented at 90° or 0°. Redrawn from LaMotte et al. (1998).

common approach, the stimulus is positioned relative to the receptive field center, regardless of the location of that receptive field on the digits (Johnson 1974, Srinivasan & LaMotte 1987). The advantage of this method is that it allows a precise characterization of the receptive field properties of the fibers, and it is usually simple to factor out the effects of differing fiber sensitivities. The disadvantage is that it can only be used for populations (or subpopulations) in which the skin mechanics can be expected to be uniform for all fibers, for example, the relatively flat portion of the fingerpad. In the alternative approach, the stimulus is always located at the same nominal position on the digit (Birznieks et al. 2001, Khalsa et al. 1998). The advantage of this method is that it allows responses to be characterized for afferents innervating parts of the digits with vastly different skin mechanics, including the highly curved sides and distal tip of the digit as well as the relatively flat pad. The disadvantage of this approach is that there is more experimental error because of the variation in geometry from finger to finger, the difficulty of accurately locating a common contact point on different fingers, and the uncertainty in characterizing the position of receptive fields relative to the contact point; it is also more difficult to factor out the differences in fiber sensitivities.

Using the receptive field approach (first approach described above) for spheres indented into the relatively flat portion of the monkey's fingerpad, Goodwin and colleagues (Goodwin et al. 1995, Wheat et al. 1995) performed a detailed parametric analysis of the three key parameters, namely the curvature of the stimulus, the position of contact on the finger, and the contact force normal to the skin. All SAI afferents could be characterized by a single multiparametric stimulus-response function. The only difference between fibers was a difference in sensitivity, which could be accounted for by a multiplicative factor; the statistical distribution of sensitivities was measured.

The mathematical description of single afferent responses enabled a population reconstruction for any arbitrary combination of curvature, position, and force. Figure 4*A* shows a reconstruction for a sphere of curvature 694 m^{-1} (radius of sphere 1.44 mm) contacting the center of the fingerpad. The effect of changing either the curvature, position, or contact force is depicted in Figure 4*B*, *C*, and *D*, respectively. At a qualitative level it is evident that changing the curvature changes the shape of the population response profile, changing the position of contact on the skin shifts the population response correspondingly, and changing the contact force scales the whole profile. This depiction also suggests that the brain could extract each of the three parameters independently even when they are changing concurrently. In this series of studies only SAI afferent responses were reconstructed because, for these stimuli, the FAI afferents did not contain shape information, although they did contain some information about position (Goodwin et al. 1995, Wheat et al. 1995).

For their torroidal stimuli, Khalsa et al. (1998) adopted the alternative approach to population reconstruction. They indented the torroids into the monkey's fingers at a constant nominal position and harvested a population of single afferents at a variety of two-dimensional positions on the fingertip. Their measured population response included the variation in sensitivity among afferents. For their SAI afferent population, the results were analogous to those in Figure 4. For the torroids, the radially symmetric Gaussian-like profiles in Figure 4 became Gaussian-like profiles with different major and minor axes reflecting the major and minor axes of the torroids. Thus, both the shape and orientation of the torroids are evident in the population responses. They also reconstructed FAI afferent population responses and found no reliable representation of the shape or orientation of the torroid in that population.

The power of this second approach to population analysis is evident in experiments from which afferents innervating all regions of glabrous skin on the distal segment of the digit are recorded. Indenting spheres at a constant nominal point on the monkey's fingerpad, Bisley et al. (2000) showed that SAI afferents

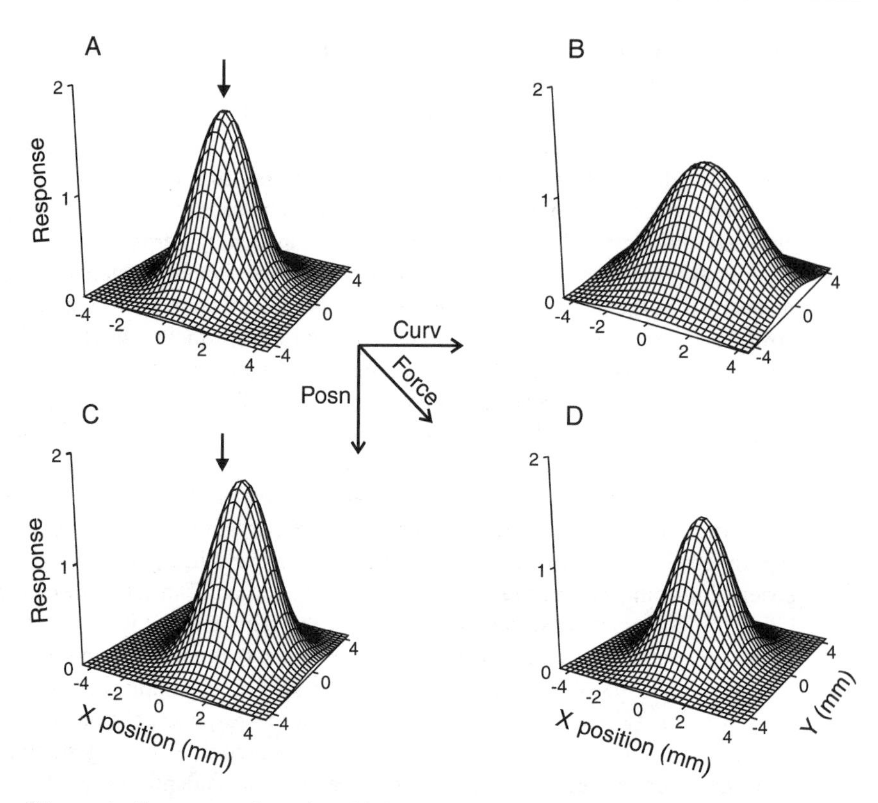

Figure 4 Representation of multiple parameters in the population response of SAI afferents in the monkey's fingerpad. (*A*) A sphere of curvature 694 m^{-1} indenting the fingerpad. (*B*) Changing the curvature (shape) of the sphere to 256 m^{-1} changes the shape of the population response. (*C*) Changing the position of the sphere on the skin by 1 mm shifts the profile a corresponding amount. (*D*) Changing the contact force scales the population response. Modified from Goodwin et al. (1995).

on the highly curved sides and end of the distal segment respond significantly. In a study of human primary afferents, a major goal of Birznieks et al. (2001) and Jenmalm et al. (2003) was to obtain data applicable to real-life human manipulation. Thus they applied spherical stimuli to a constant nominal point on the fingertip and recorded from afferents terminating all over the glabrous skin of the distal segment of the digit using relatively larger and faster contact forces than used previously (rising to 4 N in 125 ms). They also varied the direction of the contact force. Under these conditions, significant information about the force direction and the stimulus curvature was present in the SAI, SAII, and FAI afferent populations. Although a specific population reconstruction was not performed, their data for the effect of curvature on SAI afferent responses is consistent with the data from monkeys. The marked effect of curvature on their FAI afferent responses is different from that in the monkey, presumably because of the larger and faster contact forces.

This second approach to population reconstruction can also be used for stimuli scanned over the skin. Friedman et al. (2002) scanned torroids over the monkey's fingerpad along one of eight nominally standard trajectories and recorded from SAI and FAI afferents over the volar surface of the distal segment. Analysis of these population responses showed that the location and direction of the object moving over the skin was represented in both the SAI and FAI afferent populations, but with greater fidelity in the SAI afferent population.

The reconstructions illustrated in this section indicate how multiple stimulus parameters can be represented, simultaneously, in a single population of afferents, allowing each of the parameters to be determined independently.

PSYCHOPHYSICS

The primacy of sensorimotor integration in hand function poses a number of difficulties for psychophysical studies. We have argued above that sensory input about stimulus parameters is critical for both manipulation and tactile perception. The initial pathway for processing the sensory signals is the same in both cases, but the final processing is different (Figure 5). Nor is it clear where the branch point is, and it may vary from task to task. The advantage of this parallel pathway is that information about sensory input can be obtained in two ways: either from behavioral experiments in manipulation or from psychophysical experiments in sensory perception. The disadvantage is that caution is required in using data from

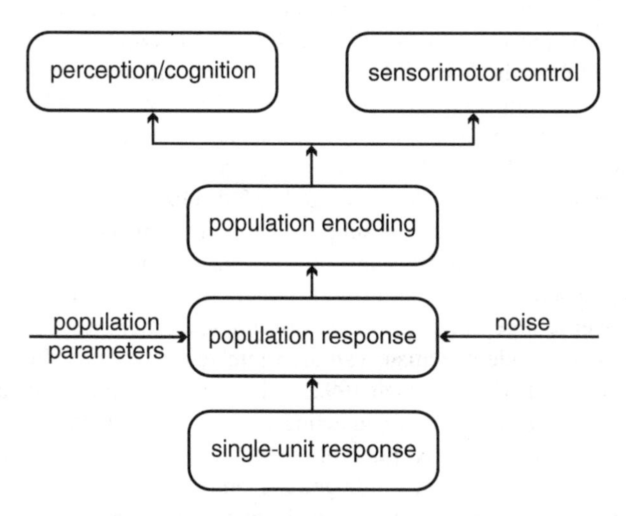

Figure 5 Processing pathways leading to perception and cognition or to sensorimotor control of manipulation. In both cases sensory information is encoded in population responses formed from single-unit responses. The population responses are subject to noise and depend on intrinsic characteristics of the population.

manipulation experiments to draw inferences about sensory perception and in using data from perceptual psychophysics to draw conclusions about manipulation.

Behavioral experiments embracing a number of grip and lift tasks, and variants thereof, have clearly demonstrated the critical role of tactile sensory input (Johansson 1996). However, the design of most studies is such that definitive separation of cutaneous input from proprioceptive input is not possible, and in many cases visual input is also a potential confounding factor.

Similarly, in most experiments in the large field of haptic perception, multisensory information is present (Klatzky & Lederman 2003). Even in those experiments where vision is occluded, it is usually not possible to determine whether the sensory information arises from cutaneous sources, proprioceptive sources, or a combination of both. For example, Davidson (1972) and Kappers et al. (1994) measured the subject's ability to determine curvature using touch, but because haptic exploration involves movement of the fingers over the curved surface it was not possible to isolate cutaneous from proprioceptive contributions. In a subset of haptics experiments, the subject contacts the stimulus without scanning the skin over the surface. In such experiments proprioceptive information may play a role in enabling the subject to execute the task, but the stimulus attributes being tested can only be conveyed by cutaneous cues. For example, in the experiments of Loomis (1979) a finger guide prevented lateral motion of the subject's finger; the digit could only be flexed to contact a tack, and subjects could judge its position with a resolution of about 0.2 mm.

In some psychophysical experiments, the subject's finger is restrained or immobilized in some way, and the stimulus is applied to the finger by a stimulator. With this design, the only information available to the subjects is that arising from cutaneous receptors in the contacted digit. Applying spheres at different positions on the skin, Wheat et al. (1995) found a difference limen of 0.38 mm and 0.55 mm for spheres of curvature 521 and 172 m^{-1}, respectively. It is significant that the resolution for position is markedly superior to figures obtained from simplistic application of the sampling theorem using the estimated receptor spacing of a millimeter or more. When Srinivasan & LaMotte (1987) indented sinusoidally shaped steps into the fingerpad, the subjects could discriminate different steps; however, because of their complex shape, it is not easy to interpret the data quantitatively. Using spheres, Goodwin et al. (1991) showed that humans could scale curvature over a large range and could discriminate differences in curvature of about 10%. For cylinders, humans can discriminate orientation differences of about 5° (Dodson et al. 1998, Lechelt 1992). It has been shown that humans can scale the depth of indentation of a contacting probe (Mountcastle 1967), but there have been few studies of perception of contact force (Cohen & Vierck 1993b). When spheres are indented into the finger, humans can scale the contact force, and they can perceive both the curvature and the contact force of the spheres when both are changed concurrently (Goodwin et al. 1991, Goodwin & Wheat 1992).

These psychophysical data, with stimuli applied to an immobilized finger, demonstrate that precise information about the local shape of handled objects, their position on the skin, and the contact force is relayed by the mechanoreceptive afferents from the digital glabrous skin. Although there may be differences in the use of sensory information by the motor control system and by the processes leading to perception and cognition (Figure 5), the results of the manipulation experiments described at the beginning of this section make it likely that most, if not all, of this sensory information is accessed by the motor control system.

MODELING POPULATION RESPONSES

One reason for reconstructing population responses is to reveal how multiple stimulus parameters can be represented unambiguously even though the responses of individual single units are completely ambiguous (Figure 4). There are two additional compelling reasons (Figure 5). First, neural responses and neural computations are variable or noisy. Second, there are critical parameters intrinsic to the populations themselves, such as the pattern and density of innervation. These factors have a major impact on signal processing in the population, and unless they are addressed, it is not possible to fully appreciate the nature of the neural mechanisms and not possible to make meaningful comparisons with human behavior. Realistic analysis of population responses requires a step beyond the rudimentary reconstructions described in a previous section and is dependent on mathematical models or computer simulations.

Neural Codes

In order to assess the accuracy with which individual stimulus parameters are represented in the population response, and to relate this to psychophysical measures of perception, it is necessary to extract a measure for that parameter from the population response. This metric is often called the neural code for the parameter. The metric quantifies the representation and indicates the potential for the brain to extract information about the parameter, but it is usually not meant to imply that the brain uses that particular computational process. As an example, the pattern of the spatial event plot in Figure 6*A* resembles the pattern of dots in the stimulus scanned over the monkey's finger, but to relate the response to human perception of the roughness of the dots requires that some measure of roughness be extracted from the spatial event plot. Blake et al. (1997) extracted the spatial variation in the SAI afferent impulse pattern and showed a close correlation with perceived roughness (Figure 6*B*) over a wide range of dot heights and spacing.

As a metric for the position of the stimulus on the skin, Goodwin & Wheat (1999) used the centroid of the population response profile. The centroid was closely correlated with the position of the stimulus, and it was insensitive to changes in the curvature or contact force of the stimulus. In the presence of considerable noise, and even at innervation densities lower than that estimated for the human (Johansson & Vallbo 1979), the reliability of the centroid was sufficient to explain the human difference limen of about 0.5 mm. Their analysis also explained why

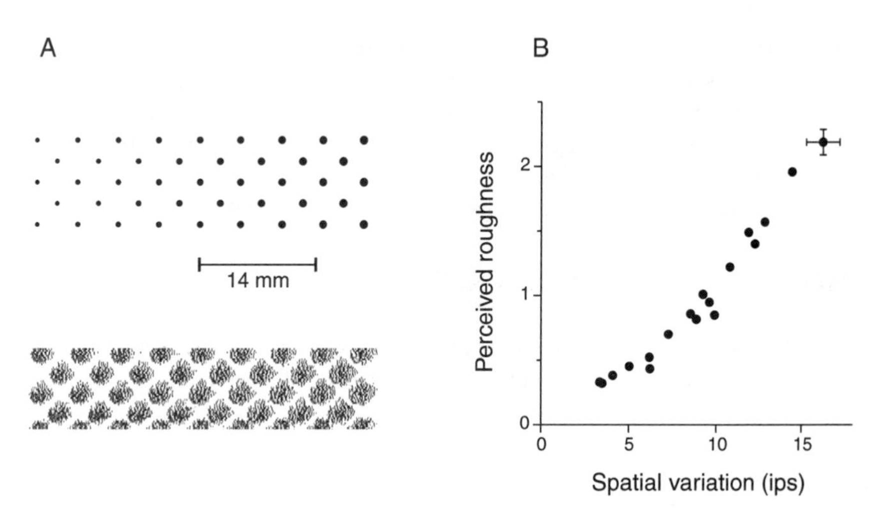

Figure 6 Roughness of a textured surface. (*A*) Spatial event plot (*below*) for an SAI afferent when the dot pattern (*above*) is scanned over the monkey's finger. (*B*) Correlation between spatial variation within the SAI afferent responses and human perception of roughness. Redrawn from Blake et al. (1997).

the difference limen is less than that predicted from simplistic application of the sampling theorem, a phenomenon termed hyperacuity by Westheimer (1975). It is the shift of the whole population profile, determined by many afferents, that provides the positional information. Resolution in populations comprising neurons with overlapping broadly tuned response profiles, which is equivalent to the case here, has been analyzed in a number of systems including electro-reception in fish (Heiligenberg 1987) and has been addressed theoretically (Pouget et al. 1999, Snippe 1996). Center-of-gravity codes have been proposed in a range of sensory and motor systems (Georgopoulos et al. 1988, Lee et al. 1988, Zohary 1992).

Modeling the population response for a stimulus contacting the skin also highlights the dangers of using overly simplistic indicators of neural signals or overly simplistic codes. The variation in sensitivity from afferent to afferent is considerable. This, plus the presence of noise, distorts the response profiles and the position of the stimulus on the skin does not correspond to the point of maximum response in the population, but the centroid is insensitive to these variations (Goodwin & Wheat 1999). This is nontrivial because, within a single human finger, the variations in afferent sensitivity are large so that a robust signal is needed for the position of an object contacting different points on the finger.

If contact force is entirely normal to the skin, then the total response of the population (simply the summed response) is a viable code for this parameter. In several studies, the total population response has been used as a metric for the intensity of a probe indenting the skin (Cohen & Vierck 1993a, Johnson 1974, Ray & Doetsch 1990). Total responses have also been used as population codes for other intensive sensory stimuli such as skin temperature (Johnson et al. 1979) and sound intensity (Relkin & Doucet 1997). In most manipulations or haptic explorations there are also components of force tangential to the skin, or, equivalently, the direction of the contact force varies and is not always normal to the skin. This complicates the extraction of force from the population response. It has been shown that humans can perceive tangential force components (Pare et al. 2002a, Wheat et al. 2004) and Birznieks et al. (2001) showed that the peripheral afferent responses are modulated by force direction; however, how force direction is represented in the population independently of force magnitude and other parameters of the stimulus is still not clear.

The local shape of the stimulus in contact with the skin is perhaps the most challenging stimulus parameter for which to formulate population codes. Most analyses of shape have been in relation to visual perception, but the principles are the same (Hegde & Van Essen 2000). The metric for curvature used by Goodwin & Wheat (Goodwin & Wheat 1999) demonstrates how curvature representation in the population can explain the human psychophysical performance and how humans can perceive curvature independently of the other stimulus parameters.

Noise

All neurons have some variability or noise associated with their responses. Noise is apparent in single-unit studies, but there is also covariance of variability between neurons, which is only manifest when analyzing groups or whole populations of neurons.

In populations responding to a single stimulus parameter, with all the constitutive neurons having the same characteristics, the fundamental question is, how does the size of the population influence the signal-to-noise ratio or the resolution? In such cases the metric used is usually the total or mean population response, and a statistical analysis can be used to show that resolution increases as the number of units in the population increases. Covariance between neurons decreases resolution and limits the benefit of increasing pool size (Johnson et al. 1979, Zohary et al. 1994).

For more complex population responses or more complex noise, analytical solutions are difficult and a computer simulation of some sort is usually used (Shadlen et al. 1996, Vogels 1990). Different types of noise have been observed in different structures in the central nervous system. Noise independent of the mean discharge rate of the neuron has been observed in the retina (Croner et al. 1993), but more commonly, variance proportional to the mean response has been reported (for review see Lestienne 2001). It is not known what types and levels of variance and covariance occur along the entire pathway for tactile perception and manipulation.

The noise in tactile peripheral afferents themselves is not significant (Jenmalm et al. 2003, Phillips & Johnson 1981a, Wheat et al. 1995). However, as the information from the peripheral afferent population is processed further in the CNS, leading to perception and/or manipulation, substantial variability will be introduced, and this variability will interact with the peripherally determined population parameters (like the pattern and density of innervation). Goodwin & Wheat (Goodwin & Wheat 1999; Wheat & Goodwin 2000, 2001) used a single-hybrid population simulation in which the primary afferents were assigned noise attributes reflecting further processing along the CNS pathway. Their population metrics were superior to human performance even in the presence of considerable variance, whether dependent on or independent of the neurons' mean responses. What is striking in their models is that while covariance has a deleterious effect on resolution for contact force (a total response code), it markedly improves resolution for position (a pattern code). This difference has been discussed in a theoretical context (Abbott & Dayan 1999, Johnson 1980), and it has been shown experimentally that resolution improves with increasing covariance for a measure based on differences in neural responses (a form of pattern code) (Romo et al. 2003).

A paradoxical property of noise is its capacity to enhance weak signals in nonlinear systems, known as stochastic resonance. In the tactile system, noise can enhance the detection of subthreshold indentation (Collins et al. 1997), and this enhancement is reflected in neural responses (Collins et al. 1996).

Intrinsic Population Characteristics

Establishing the geometry of cutaneous mechanoreceptive afferent innervation of the finger is not straightforward. Unlike the elegant studies of rods and cones in the retina where the whole receptor mosaic can be visualized (Curico et al. 1987), it is not currently possible to visualize the mosaic of afferent innervation. Thus, indirect approaches combining single-unit recording with histological fiber counts are necessary. Only one study exists in the human, which is that of Johannson & Vallbo (1979). Their data show that, even in the distal segment of the finger, density is not uniform and reaches a maximum toward the tip of the segment of 0.7 and 1.4 mm^{-2} for the SAI and FAI afferents, respectively. In the monkey there is also only one study, which is that of Darian-Smith & Kenins (1980). They did not subdivide the distal segment of the digit, and they obtained an overall estimate for this segment of 1.34 and 1.78 mm^{-2} for the SAI and FAI afferents, respectively. These estimates are invaluable, but they are based on a number of uncertainties and simplifying assumptions as readily acknowledged by the authors.

When comparing human behavior with neural responses it is prudent to allow for this uncertainty in innervation density. Nor do the existing data give any indication of whether these numbers vary between individual humans or change with age. As our baseline we, and most others, use the peak figures from Johannson & Vallbo (1979), which, on the basis of uniform innervation, correspond to a separation between adjacent receptive field centers of 1.2 and 0.85 mm for the SAI and FAI afferents, respectively. But innervation is not uniform; at the very least there is a proximo-distal gradient within the distal segment of the digit, but the exact pattern of innervation is not known. Aging (Cauna 1965, Swallow 1966) and a number of neurological diseases (Dyck & Winkelmann 1966) result in a decrease in the number of peripheral nerve fibers and/or mechanoreceptors. Moreover, it has been shown that in diabetic neuropathy the cutaneous afferents are affected before the onset of symptoms (Mackel 1989). Thus in many humans (perhaps most adults) there is some fiber or receptor dropout leading to a patchy pattern of innervation.

Rudimentary population reconstructions, such as those in Figure 4, should be viewed with caution. Those patterns are not what the brain "sees"; the patterns are sampled, probably nonuniformly, and they are distorted by noise and variation in sensitivity among fibers. Only through population simulations that allow arbitrary variations in the innervation pattern is it possible to appreciate the large impact on representation and resolution (Goodwin & Wheat 1999; Wheat & Goodwin 2000, 2001). These simulations also expose the high degree of interaction between the intrinsic population parameters and noise in limiting resolution.

FUTURE DIRECTIONS

When the fingertip contacts an object, the visco-elastic skin surrounding the entire distal phalanx is distorted. Thus an afferent innervating skin on any part of the segment may be activated even though its classical receptive field (delineated by von Frey hairs) is remote from the region of contact. Thus, to obtain a complete picture of sensory signals when objects are handled or explored, existing population models need to be extended to cover afferents over the entire distal segment. Moreover, these models should include all four cutaneous mechanoreceptive afferent types in a comprehensive interactive manner. In some current models the stimulus has been deliberately chosen to preferentially activate one subtype of afferent (Wheat & Goodwin 2001). In others, more than one subtype is reconstructed, but the populations are handled separately and no interactions between subtypes are considered (Khalsa et al. 1998). Even for relatively simple stimuli, it has been shown that the responses of SAI, FAI, and SAII afferents are modulated by the direction of contact force and the curvature of the stimulus (Jenmalm et al. 2003), and in most realistic tasks all four afferent types would respond.

The stimuli used to date have been relatively simple, partly to facilitate tractable analysis and partly because of the limitation of the stimulators. Hand movements generating everyday manipulations and haptic explorations are rich and varied; ultimately, stimulators will need to simulate more of the essential components of these elaborate movements, and population models will need to cope with the more complex information.

There is an obvious need to incorporate somatosensory signals other than tactile in models of manipulation and haptic perception. When we handle objects, a potential wealth of information could be signaled by joint afferents from the hand as well as from muscle spindles and Golgi tendon organs in the extrinsic and intrinsic hand muscles. There is a great deal of information about muscle and joint receptors, but most of it was gathered in experiments on structures such as the cat's hind limb (Banks 1994). There is only scant information about responses of these receptors in situations equivalent to normal hand function and not nearly enough to form the basis of a quantitative population reconstruction (Wessberg & Vallbo 1995). Movements of the hand are invariably accompanied by movements of the rest of the upper limb to varying degrees, and eventually sensory signals from the whole limb will need to be included. Some experimental paradigms used in somatosensory cortex research include many of these features, but as of yet we do not have a sound theoretical framework to drive the analysis of the neural responses (Gardner et al. 1999).

A major challenge for the future is to develop a global model of manipulation and haptic exploration encompassing all the senses. Under normal circumstances, vision plays a significant role in hand movements (Desmurget et al. 1998), and this is starting to be addressed at a number of levels including hand-eye coordination (Johansson et al. 2001). In related research, experiments in the posterior parietal cortex are beginning to unravel the complex interactions between visual sensory input, somatosensory input, and other sensory input in planning movements (Andersen et al. 1997).

The *Annual Review of Neuroscience* is online at http://neuro.annualreviews.org

LITERATURE CITED

Abbott LF, Dayan P. 1999. The effect of correlated variability on the accuracy of a population code. *Neural Comput.* 11:91–101

Andersen RA, Snyder LH, Bradley DC, Xing J. 1997. Multimodal representation of space in the posterior parietal cortex and its use in planning movements. *Annu. Rev. Neurosci.* 20:303–30

Banks RW. 1994. The motor innervation of mammalian muscle spindles. *Prog. Neurobiol.* 43:323–62

Birznieks I, Jenmalm P, Goodwin AW, Johansson RS. 2001. Encoding of direction of fingertip forces by human tactile afferents. *J. Neurosci.* 21:8222–37

Bisley JW, Goodwin AW, Wheat HE. 2000. Slowly adapting type I afferents from the sides and end of the finger respond to stimuli on the center of the fingerpad. *J. Neurophysiol.* 84:57–64

Blake DT, Hsiao SS, Johnson KO. 1997. Neural coding mechanisms in tactile pattern recognition: the relative contributions of slowly and rapidly adapting mechanoreceptors to perceived roughness. *J. Neurosci.* 17:7480–89

Cauna N. 1956. Nerve supply and nerve endings in Meissner's corpuscles. *Am. J. Anat.* 99:315–50

Cauna N. 1965. The effects of aging on the receptor organs of the human dermis. *Adv. Biol. Skin* 6:63–96

Christel M. 1993. Grasping techniques and hand preferences in Hominoidea. In *Hands of Primates*, ed. H Preuschoft, DJ Chivers, pp. 91–108. Vienna: Springer-Verlag

Cohen RH, Vierck CJ Jr. 1993a. Population estimates for responses of cutaneous mechanoreceptors to a vertically indenting probe on the glabrous skin of monkeys. *Exp. Brain Res.* 94:105–19

Cohen RH, Vierck CJ Jr. 1993b. Relationships between touch sensations and estimated population responses of peripheral afferent mechanoreceptors. *Exp. Brain Res.* 94:120–30

Collins JJ, Imhoff TT, Grigg P. 1996. Noise-enhanced information transmission in rat

SA1 cutaneous mechanoreceptors via aperiodic stochastic resonance. *J. Neurophysiol.* 76:642–45

Collins JJ, Imhoff TT, Grigg P. 1997. Noise-mediated enhancements and decrements in human tactile sensation. *Phys. Rev. E* 56: 923–26

Craig JC, Rollman GB. 1999. Somesthesis. *Annu. Rev. Psychol.* 50:305–31

Croner LJ, Purpura K, Kaplan E. 1993. Response variability in retinal ganglion cells of primates. *Proc. Natl. Acad. Sci. USA* 90: 8128–30

Curico CA, Sloan KR Jr, Packer O, Hendrickson AE, Kalina RE. 1987. Distribution of cones in human and monkey retina: individual variability and radial asymmetry. *Science* 236:579–82

Darian-Smith I. 1984. The sense of touch: performance and peripheral neural processes. In *Handbook of Physiology—The Nervous System III*, ed. JM Brookhart, VB Mountcastle, I Darian-Smith, SR Geiger, pp. 739–88. Bethesda, MD: Am. Physiol. Soc.

Darian-Smith I, Kenins P. 1980. Innervation density of mechanoreceptive fibres supplying glabrous skin of the monkey's index finger. *J. Physiol. (Lond.)* 309:147–55

Davidson PW. 1972. Haptic judgments of curvature by blind and sighted humans. *J. Exp. Psychol.* 93:43–55

Desmurget M, Pelisson D, Rossetti Y, Prablanc C. 1998. From eye to hand: planning goal-directed movements. *Neurosci. Biobehav. Rev.* 22:761–88

Dodson MJ, Goodwin AW, Browning AS, Gehring HM. 1998. Peripheral neural mechanisms determining the orientation of cylinders grasped by the digits. *J. Neurosci.* 18: 521–30

Dyck PJ, Winkelmann RK. 1966. Quantitation of Meissner's corpuscles in hereditary neurological disorders. *Neurology* 16:10–17

Fitzpatrick DC, Batra R, Stanford TR, Kuwada S. 1997. A neuronal population code for sound localization. *Nature* 388:871–74

Flanagan JR, Wing AM. 1997. The role of internal models in motion planning and control: evidence from grip force adjustments during movements of hand-held loads. *J. Neurosci.* 17:1519–28

Forssberg H, Eliasson AC, Kinoshita H, Westling G, Johansson RS. 1995. Development of human precision grip. IV. Tactile adaptation of isometric finger forces to the frictional condition. *Exp. Brain Res.* 104:323–30

Friedman RM, Khalsa PS, Greenquist KW, LaMotte RH. 2002. Neural coding of the location and direction of a moving object by a spatially distributed population of mechanoreceptors. *J. Neurosci.* 22:9556–66

Gandevia SC, McCloskey DI, Burke D. 1992. Kinaesthetic signals and muscle contraction. *Trends Neurosci.* 15:62–65

Gardner EP, Ro JY, Debowy D, Ghosh S. 1999. Facilitation of neuronal activity in somatosensory and posterior parietal cortex during prehension. *Exp. Brain Res.* 127:329–54

Georgopoulos AP, Kettner RE, Schwartz AB. 1988. Primate motor cortex and free arm movements to visual targets in three-dimensional space. II. Coding of the direction of movement by a neuronal population. *J. Neurosci.* 8:2928–37

Goodwin AW, Browning AS, Wheat HE. 1995. Representation of curved surfaces in responses of mechanoreceptive afferent fibers innervating the monkey's fingerpad. *J. Neurosci.* 15:798–810

Goodwin AW, Jenmalm P, Johansson RS. 1998. Control of grip force when tilting objects: effect of curvature of grasped surfaces and applied tangential torque. *J. Neurosci.* 18:10724–34

Goodwin AW, John KT, Marceglia AH. 1991. Tactile discrimination of curvature by humans using only cutaneous information from the fingerpads. *Exp. Brain Res.* 86:663–72

Goodwin AW, Macefield VG, Bisley JW. 1997. Encoding of object curvature by tactile afferents from human fingers. *J. Neurophysiol.* 78:2881–88

Goodwin AW, Wheat HE. 1992. Magnitude estimation of force when objects with different shapes are applied passively to the fingerpad. *Somatosens. Mot. Res.* 9:339–44

Goodwin AW, Wheat HE. 1999. Effects of nonuniform fiber sensitivity, innervation geometry, and noise on information relayed by a population of slowly adapting type I primary afferents from the fingerpad. *J. Neurosci.* 19:8057–70

Gordon AM, Duff SV. 1999. Fingertip forces during object manipulation in children with hemiplegic cerebral palsy. I: anticipatory scaling. *Dev. Med. Child Neurol.* 41:166–75

Gysin P, Kaminski TR, Gordon AM. 2003. Coordination of fingertip forces in object transport during locomotion. *Exp. Brain Res.* 149:371–79

Hegde J, Van Essen DC. 2000. Selectivity for complex shapes in primate visual area V2. *J. Neurosci.* 20:RC61–66

Heiligenberg W. 1987. Central processing of sensory information in electric fish. *J. Comp. Physiol. A* 161:621–31

Heller MA. 1989. Texture perception in sighted and blind observers. *Percept. Psychophys.* 45:49–54

Jeannerod M, Paulignan Y, Weiss P. 1998. Grasping an object: one movement, several components. In *Sensory Guidance of Movement*, ed. GR Bock, JA Goode, pp. 5–20. New York: Wiley

Jenmalm P, Birznieks I, Goodwin AW, Johansson RS. 2003. Influence of object shape on responses of human tactile afferents under conditions characteristic of manipulation. *Eur. J. Neurosci.* 18:164–76

Jenmalm P, Goodwin AW, Johansson RS. 1998. Control of grasp stability when humans lift objects with different surface curvatures. *J. Neurophysiol.* 79:1643–52

Jenmalm P, Johansson RS. 1997. Visual and somatosensory information about object shape control manipulative fingertip forces. *J. Neurosci.* 17:4486–99

Johansson RS. 1996. Sensory and memory information in the control of dextrous manipulation. In *Neural Bases of Motor Behaviour*, ed. F Lacquaniti, P Viviani, pp. 205–60. Dordrecht: Kluwer

Johansson RS, Landstrom U, Lundstrom R. 1982. Sensitivity to edges of mechanoreceptive afferent units innervating the glabrous skin of the human hand. *Brain Res.* 244:27–32

Johansson RS, Vallbo AB. 1979. Tactile sensibility in the human hand: relative and absolute densities of four types of mechanoreceptive units in glabrous skin. *J. Physiol. (Lond.)* 286:283–300

Johansson RS, Westling G. 1984. Roles of glabrous skin receptors and sensorimotor memory in automatic control of precision grip when lifting rougher or more slippery objects. *Exp. Brain Res.* 56:550–64

Johansson RS, Westling G. 1987. Signals in tactile afferents from the fingers eliciting adaptive motor responses during precision grip. *Exp. Brain Res.* 66:141–54

Johansson RS, Westling G, Backstrom A, Flanagan JR. 2001. Eye-hand coordination in object manipulation. *J. Neurosci.* 21:6917–32

Johnson KO. 1974. Reconstruction of population response to a vibratory stimulus in quickly adapting mechanoreceptive afferent fiber population innervating glabrous skin of the monkey. *J. Neurophysiol.* 37:48–72

Johnson KO. 1980. Sensory discrimination: neural processes preceding discrimination decision. *J. Neurophysiol.* 43:1793–815

Johnson KO, Darian-Smith I, LaMotte C, Johnson B, Oldfield S. 1979. Coding of incremental changes in skin temperature by a population of warm fibers in the monkey: correlation with intensity discrimination in man. *J. Neurophysiol.* 42:1332–53

Johnson KO, Hsiao SS. 1992. Neural mechanisms of tactual form and texture perception. *Annu. Rev. Neurosci.* 15:227–50

Johnson KO, Lamb GD. 1981. Neural mechanisms of spatial tactile discrimination: neural patterns evoked by Braille-like dot patterns in the monkey. *J. Physiol. (Lond.)* 310:117–44

Kappers AML, Koenderink JJ, Lichtenegger I. 1994. Haptic identification of curved surfaces. *Percept. Psychophys.* 56:53–61

Khalsa PS, Friedman RM, Srinivasan MA, LaMotte RH. 1998. Encoding of shape and orientation of objects indented into the monkey fingerpad by populations of slowly and rapidly adapting mechanoreceptors. *J. Neurophysiol.* 79:3238–51

Klatzky RL, Lederman SJ. 2003. Touch. In *Experimental Psychology. Volume 4 in I. B. Weiner (Editor-in-Chief) Handbook of Psychology*, ed. AF Healy, RW Proctor, pp. 147–76. New York: Wiley

Knibestol M. 1975. Stimulus-response functions of slowly adapting mechanoreceptors in the human glabrous skin area. *J. Physiol. (Lond.)* 245:63–80

Koenderink JJ, van Doorn AJ. 1992. Surface shape and curvature scales. *Image Vision Comput.* 10:557–64

Koudine AA, Barquins M, Anthoine PH, Aubert L, Leveque J-L. 2000. Frictional properties of skin: proposal of a new approach. *Int. J. Cosmet. Sci.* 22:11–20

LaMotte RH, Friedman RM, Lu C, Khalsa PS, Srinivasan MA. 1998. Raised object on a planar surface stroked across the fingerpad: responses of cutaneous mechanoreceptors to shape and orientation. *J. Neurophysiol.* 80:2446–66

LaMotte RH, Srinivasan MA. 1987a. Tactile discrimination of shape: responses of rapidly adapting mechanoreceptive afferents to a step stroked across the monkey fingerpad. *J. Neurosci.* 7:1672–81

LaMotte RH, Srinivasan MA. 1987b. Tactile discrimination of shape: responses of slowly adapting mechanoreceptive afferents to a step stroked across the monkey fingerpad. *J. Neurosci.* 7:1655–71

LaMotte RH, Srinivasan MA. 1996. Neural encoding of shape: responses of cutaneous mechanoreceptors to a wavy surface stroked across the monkey fingerpad. *J. Neurophysiol.* 76:3787–97

Lechelt EC. 1992. Tactile spatial anisotropy with static stimulation. *Bull. Psychon. Soc.* 30:140–42

Lee C, Rohrer WH, Sparks DL. 1988. Population coding of saccadic eye movements by neurons in the superior colliculus. *Nature* 332:357–60

Lestienne R. 2001. Spike timing, synchronization and information processing on the sensory side of the central nervous system. *Prog. Neurobiol.* 65:545–91

Levi R, Camhi JM. 2000. Population vector coding by the giant interneurons of the cockroach. *J. Neurosci.* 20:3822–29

Loomis JM. 1979. An investigation of tactile hyperacuity. *Sens. Process* 3:289–302

Lukashin AV, Georgopoulos AP. 1993. A dynamical neural network model for motor cortical activity during movement: population coding of movement trajectories. *Biol. Cybern.* 69:517–24

Mackel R. 1989. Properties of cutaneous afferents in diabetic neuropathy. *Brain* 112:1359–76

Maeno T, Kobayashi K. 1998. FE analysis of the dynamic characteristics of the human finger pad in contact with objects with/without surface roughness. *Proc. ASME Int. Mech. Eng. Congr. Expo.* 64:279–86

Montana DJ. 1988. The kinematics of contact and grasp. *Int. J. Robot Res.* 7:17–32

Monzee J, Lamarre Y, Smith AM. 2003. The effects of digital anesthesia on force control using a precision grip. *J. Neurophysiol.* 89:672–83

Mott FW, Sherrington CS. 1895. Experiments upon the influence of sensory nerves upon movement and nutrition of the limbs. Preliminary communication. *Proc. R. Soc.* 57:481–88

Mountcastle VB. 1967. The problem of sensing and the neural coding of sensory events. In *The Neurosciences: A Study Program*, ed. FO Schmitt, G Quarton, T Melnuchuk, pp. 393–408. New York: Rockefeller Univ. Press

Mountcastle VB, Powell TPS. 1959. Neural mechanisms subserving cutaneous sensibility, with special reference to the role of afferent inhibition in sensory perception and discrimination. *Bull. Johns Hopkins Hosp.* 105:201–32

Napier JR. 1980. *Hands*. London: Allen & Unwin

Pare M, Carnahan H, Smith AM. 2002a. Magnitude estimation of tangential force applied to the fingerpad. *Exp. Brain Res.* 142:342–48

Pare M, Smith AM, Rice FL. 2002b. Distribution and terminal arborizations of cutaneous mechanoreceptors in the glabrous finger pads of the monkey. *J. Comp. Neurol.* 445:347–59

Pasupathy A, Connor CE. 2002. Population coding of shape in area V4. *Nat. Neurosci.* 5:1332–38

Pawluk DTV, Howe RD. 1999. Dynamic contact of the human fingerpad against a flat surface. *J. Biomech. Eng.* 121:605–11

Phillips JR, Johnson KO. 1981a. Tactile spatial resolution. II. Neural representation of bars, edges, and gratings in monkey primary afferents. *J. Neurophysiol.* 46:1192–203

Phillips JR, Johnson KO. 1981b. Tactile spatial resolution. III. A continuum mechanics model of skin predicting mechanoreceptor responses to bars, edges, and gratings. *J. Neurophysiol.* 46:1204–25

Pouget A, Deneve S, Ducom JC, Latham PE. 1999. Narrow versus wide tuning curves: What's best for a population code? *Neural Comput.* 11:85–90

Ray RH, Doetsch GS. 1990. Coding of stimulus location and intensity in populations of mechanosensitive nerve fibers of the raccoon: II. Across-fiber response patterns. *Brain Res. Bull.* 25:533–50

Reilmann R, Gordon AM, Henningsen H. 2001. Initiation and development of fingertip forces during whole-hand grasping. *Exp. Brain Res.* 140:443–52

Relkin EM, Doucet JR. 1997. Is loudness simply proportional to the auditory nerve spike count? *J. Acoust. Soc. Am.* 101:2735–40

Romo R, Hernandez A, Zainos A, Salinas E. 2003. Correlated neuronal discharges that increase coding efficiency during perceptual discrimination. *Neuron* 38:649–57

Rothwell JC, Traub MM, Day BL, Obeso JA, Thomas PK, Marsden CD. 1982. Manual motor performance in a deafferented man. *Brain* 105:515–42

Sanes JN, Mauritz KH, Dalakas MC, Evarts EV. 1985. Motor control in humans with large-fiber sensory neuropathy. *Hum. Neurobiol.* 4:101–14

Shadlen MN, Britten KH, Newsome WT, Movshon JA. 1996. A computational analysis of the relationship between neuronal and behavioral responses to visual motion. *J. Neurosci.* 16:1486–510

Snippe HP. 1996. Parameter extraction from population codes: a critical assessment. *Neural Comput.* 8:511–29

Srinivasan M, Dandekar K. 1996. An investigation of the mechanics of tactile sense using two-dimensional models of the primate fingertip. *J. Biomech. Eng. Trans. ASME* 118:48–55

Srinivasan MA. 1989. Surface deflection of primate fingertip under line load. *J. Biomech.* 22:343–49

Srinivasan MA, LaMotte RH. 1987. Tactile discrimination of shape: responses of slowly and rapidly adapting mechanoreceptive afferents to a step indented into the monkey fingerpad. *J. Neurosci.* 7:1682–97

Srinivasan MA, Whitehouse JM, LaMotte RH. 1990. Tactile detection of slip: surface microgeometry and peripheral neural codes. *J. Neurophysiol.* 63:1323–32

Swallow M. 1966. Fibre size and content of the anterior tibial nerve of the foot. *J. Neurol. Neurosurg. Psychiatry* 29:205–13

Tremblay MR, Cutkoksy MR. 1993. Estimating friction using incipient slip sensing during a manipulation task. *Proc. IEEE Conf. Rob. Autom.* 1:429–34

Vallbo AB. 1995. Single-afferent neurons and somatic sensation in humans. In *The Cognitive Neurosciences*, ed. MS Gazzaniga, pp. 237–52. Cambridge: MIT Press

Vierck CJ Jr. 1979. Comparison of punctate, edge and surface stimulation of peripheral, slowly adapting, cutaneous, afferent units of cat. *Brain Res.* 175:155–59

Vogels R. 1990. Population coding of stimulus orientation by striate cortical cells. *Biol. Cybern.* 64:25–31

Wessberg J, Olausson H, Fernstrom KW, Vallbo

AB. 2003. Receptive field properties of unmyelinated tactile afferents in the human skin. *J. Neurophysiol.* 89:1567–75

Wessberg J, Vallbo AB. 1995. Human muscle spindle afferent activity in relation to visual control in precision finger movements. *J. Physiol. (Lond.)* 482:225–33

Westheimer G. 1975. Editorial: Visual acuity and hyperacuity. *Invest. Ophthalmol.* 14: 570–72

Westling G, Johansson RS. 1984. Factors influencing the force control during precision grip. *Exp. Brain Res.* 53:277–84

Wheat HE, Goodwin AW. 2000. Tactile discrimination of gaps by slowly adapting afferents: effects of population parameters and anisotropy in the fingerpad. *J. Neurophysiol.* 84:1430–44

Wheat HE, Goodwin AW. 2001. Tactile discrimination of edge shape: limits on spatial resolution imposed by parameters of the peripheral neural population. *J. Neurosci.* 21:7751–63

Wheat HE, Goodwin AW, Browning AS. 1995. Tactile resolution: peripheral neural mechanisms underlying the human capacity to determine positions of objects contacting the fingerpad. *J. Neurosci.* 15:5582–95

Wheat HE, Salo LM, Goodwin AW. 2004. Human ability to scale and discriminate forces typical of those occurring during grasp and manipulation. *J. Neurosci.* 24:3394–401

Zohary E. 1992. Population coding of visual stimuli by cortical neurons tuned to more than one dimension. *Biol. Cybern.* 66:265–72

Zohary E, Shadlen MN, Newsome WT. 1994. Correlated neuronal discharge rate and its implications for psychophysical performance. *Nature* 370:140–43

Annu. Rev. Neurosci. 2004. 27:79–105
doi: 10.1146/annurev.neuro.27.070203.144323

First published online as a Review in Advance on February 23, 2004

E PLURIBUS UNUM, EX UNO PLURA[1]: Quantitative and Single-Gene Perspectives on the Study of Behavior

Ralph J. Greenspan
The Neurosciences Institute, San Diego, California; email: greenspan@nsi.edu

Key Words genome, natural variation, mutant, pleiotropy, epistasis

■ **Abstract** Genetic studies of behavior have traditionally come in two flavors: quantitative genetic studies of natural variants and single-gene studies of induced mutants. Each employed different techniques and methods of analysis toward the common, ultimate goal of understanding how genes influence behavior. With the advent of new genomic technologies, and also the realization that mechanisms underlying behavior involve a considerable degree of complex gene interaction, the traditionally separate strands of behavior genetics are merging into a single, synthetic strategy.

INTRODUCTION

The contrast between quantitative genetic and single-gene perspectives on heredity is as old as the science of genetics itself (Provine 1971). Francis Galton, the original quantitative geneticist, published his foundational paper in the same year as did Gregor Mendel, originator of the concept of single, stable hereditary factors (later called genes). Galton maintained that heredity is controlled by a multitude of factors, each of small effect, that sum to give the overall trait. His first efforts were directed at demonstrating the hereditary nature of "genius" in men (Galton 1865). The Mendelian perspective, although originating in the study of garden peas, was applied to human behavior as soon as it became well known to the general scientific community at the beginning of the twentieth century, largely through the efforts of Charles B. Davenport (1911), founder of the Eugenics Records Office at the Cold Spring Harbor Laboratory in 1910. During those years, an academic battle raged between biometricians and Mendelians over the issue of whether Darwinian evolution occurred by selection on small continuous variation or by discontinuous leaps. Although the proximate conflict was resolved by the development of theoretical population genetics (Provine 1971), the ultimate conflict has persisted in the form

[1]Out of many, one; out of one, many.

of two contrasting approaches: quantitative versus single-gene. These perspectives have influenced studies of the genetic contributions to behavior as much as any other area of genetic research, and they represent the classic distinction between a systems approach and a reductionist approach.

Quantitative genetics has generally taken for its raw material the naturally occurring genetic variation that affects behavior, whereas single-gene mutant analysis often creates its own raw material by inducing and isolating mutations affecting behavior and neural function. The two approaches began from different premises, and despite the fact that they had the same ultimate goal—an understanding of the genetic underpinnings of behavior—they necessarily had different proximate goals that depended on what was feasible. Quantitative genetics concerned itself with naturally occurring variation, either in the wild or in laboratory selection experiments. The focus was on the composite genotype, variance due to genetics, and evolutionary concerns; much of it is based on the theoretical work of R.A. Fisher (1930). Throughout most of this field's history, the genetic factors responsible for such naturally occurring, continuously varying traits were unknown, and the analysis consisted mainly of general statistical inferences based on phentoypic comparisons—e.g., rough esimates of the number of contributing loci and of the independence (additivity) or nonindependence (epistasis) of their interactions. Often, these results followed from artificial selection experiments in the laboratory (Ehrman & Parsons 1981, Greenspan 2004).

In its contemporary incarnations, the most commonly practiced form of quantitative genetics is quantitative trait locus (QTL) analysis. A QTL is a polymorphic locus that contains alleles with differential effects on the expression of a continuously distributed phenotypic trait. Usually it is detected by means of a DNA polymorphism, often not actually part of the gene in question, that shows association with quantitative variation in a particular phenotypic trait. The full extent of variation for the phenotype is assumed to be determined by the cumulative action of alleles at many loci, as well as by nongenetic factors. In QTL analysis, a phenotypic difference between two strains is mapped against an extensive set of distributed, gratuitous genetic markers that also differ between them, chromosomal regions mediating significant effects are mapped, and candidate genes in those segments are identified with the ulitmate goal of identifying DNA polymorphisms (QTNs or quantitative trait nucleotides) that mediate the phenotypic effects (e.g., Belknap et al. 2001, Bucan & Abel 2002, Mackay 2002). Many of the applications of this approach have been carried out in the mouse because of the existence of many well-characterized inbred strains whose inbreeding has minimized any intrastrain genetic variation (Nguyen & Gerlai 2002, Williams 2000).

In addition to inbred strains, mouse geneticists also have at their disposable a series of recombinant inbred (RI) lines. These are sets of inbred lines made from F1 hybrids between two strains, such that random mixtures of their genomes are fixed in nearly homozygous condition after repeated generations of brother-sister mating (Bailey 1971). Known polymorphisms between the two starting strains are then scored in each RI line, and a rough map is made of the chromosomal segments originating from each parental strain. The characterization of these RI lines over the decades has allowed accumulation of a detailed set of markers, whose resolution has been vastly improved with the advent of sequence length polymorphisms (Dietrich et al. 1992) and now single-nucleotide polymorphisms (Lindblad-Toh et al. 2000). The upshot of this work is a standard set of well-characterized lines for QTL analysis of any phenotype.

Single-gene mutant analysis as a means of defining mechanisms traces its origins to microbial genetics, which concentrated on cellular mechanisms to the virtual exclusion of all evolutionary or environmental questions. It consisted of the induction of mutants to identify relevant genes and direct demonstration of biochemically definable roles and interactions among them. A particular strength of the approach is the theoretical ability to perform saturation mutagenesis for a particular trait and thereby presumably identify all of the genes that are mutable to produce alterations in the selected phenotype. Initially worked out on such prokaryotic phenomena as phage assembly (Wood et al. 1968), saturation-scale mutagenesis was later applied to more complex multicellular phenotypes such as embryonic pattern formation (Nusslein-Volhard & Wieschaus 1980). [Recent, genome-level analysis suggests that saturation mutagenesis is actually quite difficult to achieve in practice (cf. Giaever et al. 2002).]

As first applied to behavior by Seymour Benzer (1967), this approach represented a distinct departure from traditional "behavior genetics," a largely quantitative genetic discipline focused on the behavioral influences of natural genetic variation, most of which seemed to result from relatively mild effects of multiple genes, or on spontaneous mutations that occasionally appeared in laboratory strains. The single-gene, induced-mutation school was not interested in whether a gene would exhibit natural variation. Instead, the interest lay in which genes contributed to the behavior and which neural or biochemical components were selectively altered.

The majority of mutations uncovered in these searches caused disruptions more severe than allelic variants found in the wild. The single-gene practitioners' concern is with whether a mutant is "normal," not with subtle variations from the mean. The severity of a mutant phenotype is a complex function of many factors. It depends, among other things, on the nature and extent of inactivation or alteration of the gene in question, on the gene's role in development and behavior, and on the genetic background in which it is expressed (deBelle & Heisenberg 1996, Hall 1994, Gerlai 1996). Many of the recent knock-out mutations in the mouse have been notable for their lack of an obvious abnormal phenotype, sometimes referred to inappropriately as having no phenotype (cf. Hall 1994 for discussion). (Knock-out refers to the engineering of a mutation that completely eliminates a gene's product, thus producing a complete loss of function. Traditionally, these are known in the genetics literature as null mutations. Mutations producing partial loss of function are called hypomorphs and generally result from lower levels of expression, or from a change in amino acid sequence that reduces the activity, broadly defined, of the gene product.)

In flies and nematodes, most of the mutations were obtained from screening procedures deliberately designed to reveal distinctive changes in phenotype ("forward" genetics), as opposed to the mouse where most knockouts have been generated from cloned genes in an effort to ask if mutations in the particular gene will produce a distinctive phenotype ("reverse" genetics). Forward genetics is now possible in the mouse as well (Bucan & Abel 2002).

Despite the nature of Benzer's mutant hunting strategy in flies, aggressive relative to the collection of natural variants, some of the resulting mutations were considerably milder in phenotype, and often more informative, than null mutations of the same genes. The milder phenotypes associated with these mutations are still more drastic than those commonly found in nature, but they begin to approach the subtler influence of those occurring naturally. At the same time, recent molecular studies of natural variation in single genes influencing behavior have begun to close the gap between the two schools.

The purpose of this article is not to review this extensive literature, which has been done repeatedly, and more focally (e.g., Bucan & Abel 2002, Greenspan & Ferveur 2000, Hall 2003, Mackay 2002, Sokolowski, 2002, Waddell & Quinn 2001). Nor is the purpose to decide which is "better," a task analogous to asking if eating with a fork is better than eating with a spoon. They are independent ways of getting to a common goal: understanding the genetic foundations and mechanisms of behavior. Environmental effects, another critical set of parameters in behavioral genetic studies (Sokolowski & Wahlsten 2001), and interspecies differences are also not considered here. Instead, the goal is to consider the characteristics of the two major genetic approaches and the synthesis between them that is now emerging (cf. Tully 1996).

Sources of Genetic Variation

Studies of the genetic aspects of behavior have drawn on naturally occurring variants and on induced mutations. Natural variants are defined simply as spontaneously occurring, heritable variations of a gene. As such, they can be found in the wild and also in laboratory populations, the main difference residing in which kinds of new mutations will survive and be tolerated in the different environments. Artificial selection experiments represent a special case of variation within a species, in which the experimenter applies external constraints on the population by selecting for a particular phenotype. This procedure biases the polymorphisms and combinations of variants passed on to the next generation. A corollary concern of this approach is genetic background—the naturally occurring variation in laboratory strains—and its ability to influence phenotype.

Induced mutations are the mainstay of single-gene analysis. Originating in Muller's X-ray-induced mutations in the fruit fly in the 1920s, generated in an effort to ask what the gene itself is, they began to be used as an analytical tool in microbial genetics for elaborating metabolic pathways in the 1940s and 1950s, and contributed to the blossoming of eukaryotic developmental genetics in the 1980s. In contrast to quantitative genetics, less attention has been paid to genetic background.

Although genetic studies of behavior have drawn on both sources of genetic variation, each has contributed a dimension that could not have come from the other.

THE CLASSICAL APPROACHES

Genetic Architecture: Geotaxis in *Drosophila*

Artificial selection in the laboratory has been the conventional method for detecting the presence of naturally occurring genetic variation in behavior (Ehrman & Parsons 1981). Such variation has been clearly demonstrable in virtually every behavioral selection attempted (Greenspan 2004), as had already been shown for nonbehavioral phenotypes (Lewontin 1974). Behavioral differences are generally due to the contribution of many genes, usually of rather small effect, and thus refractory to conventional mapping techniques. Only their approximate number and putative interactions could be estimated from statistical analyses of selected lines and of F1, F2, and backcross progeny.

Bidirectional selection for positive versus negative geotaxis constitutes one of the classic experiments in behavioral genetics. Inaugurated by Hirsch in the 1950s, it was inspired by the selectional experiments on maze learning in rats carried out by his mentor, Robert Tryon (Hirsch & Tryon 1956). Hirsch chose *Drosophila* as his experimental material because of its genetic prowess and performed his selection on flies collected from various spots in the New York area. (One of these was a farm in Syosset, Long Island, which shows just how long ago these experiments were undertaken.) Using a choice maze that required the flies to choose 16 times between going down (positively geotaxic) or up (negatively geotaxic), he began a multigenerational selection.

Over the course of the next thirty years, he and his students selected, reselected, reverse selected, performed F1, F2, and backcrosses, tested for the relative contributions of each chromosome, and tested for correlated effects on other behaviors (reviewed by Greenspan 2004). Chromosomal analysis studies, in which strains are constructed containing various combinations of chromosomes from each of the original selected strains, were undertaken in large measure to demonstrate that the behavior had a genetic component. The strategy was originally developed for morphological studies in *Drosophila* (Robertson 1954) and has recently been rediscovered in the mouse (Nadeau et al. 2000). Because these studies compared scores with and without particular chromosomes from each selected line, they provided a very rough map of relevant loci. Moreover, since they could be put back together in various combinations after being tested individually, the presence of interactions between genes on separate chromosomes could also be inferred. The analyses revealed contributions from all chromosomes (Hirsch & Erlenmayer-Kimling 1962, Hirsch & Ksander 1969, Ricker & Hirsch 1988). A statistical (biometrical) analysis of these data subsequently revealed extensive interactions among all three chromsomes (McGuire 1992).

Hirsch's experiments demonstrated clearly that behavioral phenotypes respond to selection and have heritable components. They also hint at the complexity of the genetics, both in the number of loci likely to be involved and in the intricacy of their interactions. The studies could not say anything, however, about the identities of any of the genes.

Genetic Dissection: Circadian Rhythms in *Drosophila*

The clearest contribution single-gene mutants have made is in the realm of identifying individual genes that are central to behavioral mechanisms, pointing the way to the unraveling of the cellular mechanisms. Perhaps the best example of this strategy can be seen in studies of the circadian clock. These discoveries began with the isolation of long-day, short-day, and arrhythmic mutants in *Drosophila*, all of which proved to be alleles of the same gene, dubbed *period* (Konopka & Benzer 1971). The subsequent isolation of additional mutants in flies, fungi, and mice and the cloning of these genes revealed the cellular mechanism of the circadian clock to be nearly universal in the biological world (Hall 1995, 2003; Harmer et al. 2001; Takahashi 2004). Many of the actual genes involved, such as *period*, are conserved between flies and mammals.

The screen for these mutants was begun by Benzer's student Ron Konopka, who placed flies in the lab's spectrophotometer and left them there for days to track their daily activity rhythms. From measuring eclosion (emergence of the adult after metamorphosis) and locomotor activity came the first rhythm mutants of the *period* gene (Konopka & Benzer 1971). Konopka's *period* mutants exhibit normal activity cycles as a function of the light:dark cycle, but they cannot maintain that rhythm in constant darkness. The arrhythmic mutant (*pero*) has no discernible cycle, the short-day mutant (*pers*) has a 19-h cycle, and the long-day mutant (*perl*) has a 29-h cycle in constant darkness.

Following the discovery of the *period* gene, the quest progressed (after a hiatus of ~15 years, see Greenspan 2003) with the isolation and cloning of additional mutants: *timeless*, *doubletime*, *Clock* (neé *Jerk*), *cycle*, and *cryptochrome* (Hall 2003). What has emerged is a picture of the circadian clock as a transcriptional regulatory sequence in which each cell counts out its own 24-h period by means of an oscillating cycle of transcription and translation. A key element is that transcription of *period* and *timeless* (in *Drosophila*) is negatively regulated by their own protein products, with the consequence that as the proteins accumulate, they begin to dampen their own synthesis. Interjection of a time delay between transcription, translation, and import back into the nucleus sets the period of the events to the circadian cycle. This serves as the central time-keeping mechanism for all circadian rhythmic activities.

The first few clock genes isolated in *Drosophila* came from forward genetic screens, testing behaviorally for alterations either in adult eclosion (*per*) or in locomotor activity (*timeless*, Sehgal et al. 1994; *doubletime*, Price et al. 1998; *Clock*, Allada et al. 1998; *cycle*, Rutila et al. 1998). These make up much of the basic cycling mechanism, *per* and *tim* repressing their own transcription, which requires activation by the transcription factors *Clock* and *cycle*, with the *doubletime* kinase regulating the degradation of *period* and *timeless* proteins.

Subsequent mutant screens played variations on the original paradigm. *shaggy*, a kinase well known for its role in embryonic development, showed up as a clock gene in a screen for dominant effects of overexpressed genes (Martinek et al. 2001), and it phosphorylates the product of *timeless*. Another kinase, casein kinase 2α, was isolated as a dominant mutation (*Timekeeper*) that suppresses the short-day phenotype of *per*^{s} (Lin et al. 2002), and it phosphorylates the products of both *period* and *timeless*. A screen based on a molecular phenotype, monitoring alterations in the circadian cycling of a reporter gene (luciferase) fused to the *per* promoter, identified *cryptochrome*, a flavin-containing protein involved in extraocular light-reception for entrainment (Stanewsky et al. 1998). The discovery of the neuropeptide gene that mediates the implementation of circadian locomotor rhythms, *Pdf* (*Pigment-dispersing-factor*), was discovered accidentally while staining immunocytochemically with antisera to the PDF neuropeptide as part of a control experiment (Renn et al. 1999). These represent only the starting members of an ever-expanding set of genes, many of which were isolated by forward genetic screens and isolated in mutant screens for totally different behaviors or for developmental defects, and subsequently shown to affect rhythms (Hall 2003).

The picture that has emerged of the cellular clock is a tour de force of the Benzerian, single-gene mutant approach. In many respects, the analysis of circadian rhythms at the level of the cellular clock resembles the pathway dissections of microbial genetics. By themselves, however, these studies do not address evolutionary issues of natural variation in these genes (see below).

Reverse Genetic Dissection: Pheromone Response in Mice

Reverse genetics has become the predominant mode of analysis in the mouse world, where the mutants are called knockouts in reference to the technology of gene replacement used to create null alleles. It falls squarely into the single-gene camp, but unlike the foregoing examples of random mutagenesis or natural variant analysis, reverse genetics starts from knowledge of a particular gene and then works backward to create a knockout (or some other specific and deliberate alteration) of that gene. It has the virtue (when it works properly, cf. Maldonado et al. 1996) of being well defined from the outset and of providing a clear test of the requirement for that gene. It suffers from the drawback that it relies on our preexisting knowledge of the relevant universe of genes, and our imagination as to which genes will play important roles. Often, the results confirm our expectations based on previous, nongenetic experiments (e.g., the effect of a serotonin receptor on aggression, Sadou et al. 1994, or of an NMDA receptor on learning, Tang et al. 1999). There are occasional surprises in these experiments, but usually the surprise is in the lack of an expected effect of the mutation on a particular phenotype.

A case illustrating all of these features concerns the vomeronasal organ (VNO) in the mouse, the part of the rodent olfactory system involved in pheromone perception. A phospholipase-activated ion channel, TRP2, that shows highly restricted expression to the VNO (Liman et al. 1999) was knocked out as a test of its requirement in pheromone discrimination. VNO neurons in the resulting mice did indeed show a failure to respond to urine-derived pheromones, as expected, and they displayed decreased aggression, also as expected. But the mutant male mice did not live up to expectations in failing to initiate sexual advances toward females. Instead, these males indiscriminately tried to mount both sexes (Leypold et al. 2002, Stowers et al. 2002). On the other hand, knockouts of a specific set of VNO receptor genes (of the *V1ra* and *V1rb* clusters) did reduce the likelihood that a male will mount a female (Del Punta et al. 2002). The failure to achieve this result in the TRP2 knockout may be due to incomplete elimination of VNO function in this mutation.

By making cloned genes accessible to targeted mutagenesis, reverse genetics has opened up a new major avenue of genetic analysis, especially in the mouse where forward genetics is more difficult. It does not, however, offer the same range of alleles that emerge from forward genetic screens, such as the long-day and short-day alleles of the *period* gene, or from natural variants (see below). This is because our imaginations pale in comparison to the myriad variations that emerge from the generation of random variants and selection for specific phenotypes, as occurs in both forward genetics and in natural selection.

MUTUAL CRITICISMS

A measure of the separation between single-gene and quantitative geneticists can be seen in their criticisms of each other. The genetic architecture camp criticizes its single-gene counterpart for the lack of relevance of the extreme phenotypes of its mutants to anything that could contribute to species evolution, as well as for its failure to take into account the contributions of genetic background to phenotype. On the other hand, the mono-geneticists disparaged the architecturists for their failure to say anything about which genes might be involved, as well as for the lack of relevance of small-effect genes to underlying mechanism. As it turns out, induced mutants do have natural counterparts, small-effect genes are relevant to mechanism, and now the genes for complex natural variation can be identified.

Natural Variants of a Single-Gene Mutant: *period*

In contrast to the severe consequences of the mutant alleles used to define the circadian clock, the *per* gene also exhibits milder, natural, polymorphic variation. This, in itself, is not surprising because virtually all genes show some degree of DNA sequence variation. The key question is whether there are detectable, functional consequences to the polymorphisms.

Two alleles of the *per* gene predominate in the wild: one containing 17 repeats of the pair of amino acids threonine and glycine $(Thr\text{-}Gly)_{17}$ in the middle of the coding reigion, and one containing 20 repeats $(Thr\text{-}Gly)_{20}$ (Costa et al. 1992). These two alleles are nonrandomly distributed in the wild, following a north-south gradient (known as a latitudinal cline) from northern Europe down to North Africa, with $(Thr\text{-}Gly)_{20}$ predominating in the north and $(Thr\text{-}Gly)_{17}$ predominating in the south. Distributions of this sort are traditionally explained as an adaptive response to climatic variation. In this case, the strongest correlation is with temperature range.

If these variants are undergoing selection, as the existence of a latitudinal cline implies, then the functional differences between them become relevant. A clue to such a functional difference was seen in tests of temperature compensation—one of the distinctive features of circadian rhythms that allow cold-blooded organisms to maintain their 24-h cycle at different temperatures (Pittendrigh 1954). Flies carrying the northerly $(Thr\text{-}Gly)_{20}$ allele exhibit more robust temperature compensation than those with the more southerly $(Thr\text{-}Gly)_{17}$ allele. The temperature range is much greater in the north and presumably imposes greater demands on the temperature-compensating ability of the clock (Sawyer et al. 1997). The size of the effect is small, perhaps reflecting a trade-off between temperature compensation and conservation of the overall 24-h rhythm.

By showing that naturally occurring variation with selectional and functional implications could be found for one of the canonical rhythm genes, Kyriacou and colleagues effectively answered the objection that induced mutations do not contribute to our understanding of natural processes. On the other hand, the *per* effects could not account for all of the detailed aspects of the variation between natural isolates, underlining the (inevitable) genetic complexity of the strains isolated from northern versus southern locales.

Reverse Genetic Natural Variation: The 5-HT Transporter in Humans

Human genetic studies must necessarily rely on analysis of the existing natural variation in the population, as no one (not even the Raelians) is proposing that we deliberately induce new mutations in ourselves. In one incarnation, the analysis is no different from QTL analysis in laboratory animals, with the exception that the available materials for study (identified subjects and recombination events for mapping purposes) are finite. The challenges to the identification and verification of individual genes from such studies, as compared to the analogous studies in laboratory animals, are correspondingly greater. Yet, these studies constitute much of the contemporary work on mapping disease susceptibility and pharmacogenomics, the genetic influences of drug response in disease treatment. In a variation on this theme, the ability to take a known gene, look for polymorphisms in it, and test for the association of a given allele with some phenotype is entirely feasible and has yielded a series of interesting insights.

The serotonin transporter (5-HTT) occupies a key position in the physiology of mood and anxiety, most notably because of its role as a target for the most popular and effective antidepressants and antianxiolytics. A polymorphism in the length of the transcriptional control region of this gene affects its transcriptional efficiency (Lesch et al. 1996). In a survey of some 500 individuals, an association was found between the short form of the polymorphism (low efficiency of transcription) and various personality measures of anxiety and depression. The short allele behaved dominantly in that there was no difference between homozygotes and heterozygotes. The effect, though statistically significant, was nonetheless small because the allelic association accounted for only 7%–9% of the variance for those measures. This kind of result—statistically significant but small in effect—is typical of such human association studies. It also exemplifies the difficulties and multigenic nature of genetic analysis in humans.

Subsequent attempts to find an overall association between 5-HTT and depression have been inconsistent, a common problem in human association studies (Merikangas et al. 2002). When the subjects' life histories are also taken into account, however, the strength of the association goes up considerably. In a study of nearly 1000 young adults, the correlation of the 5-HTT short allele with depression increased if the individual had also experienced either childhood maltreatment or several major stressful life events, according to a set of standardized measures (Caspi et al. 2003). In this study, there was a graded effect when homozygotes and heterozygotes were compared: Homozygotes for the short allele were more likely to experience depression after stressful life events or childhood maltreatment than heterozygotes, who were more likely to experience depression than those homozygous for the long allele. The subjects in this study had been followed since birth, and their traumatic life events were documented long before the diagnosis of depression, thus strengthening the association between cause and effect.

Thus, nongenetic (i.e., environmental) factors must be added to genetic factors as components of the inherent complexity of these phenotypes (Kendler et al. 2003). It is possible, nonetheless, to assign functional significance to identified, natural variants even if their contribution is only as one among a plethora of others. Moreover, it is easy to imagine that if fruit flies were capable of depression, a severe 5-HTT mutation might produce a strain of consistently and totally "bummed out" flies.

Exceptions to Natural Multi-Genicity

Not all naturally occurring variation is necessarily complex and polygenic. A few notable exceptions illustrate that a single-gene mode of inheritance can occasionally account for the majority of natural variation for a trait.

Natural and laboratory populations of *D. melanogaster* harbor two behavioral types with respect to food foraging behavior: Rovers, which search widely, and sitters, which do not (Sokolowski 1980). The effect is not merely a difference in locomotor activity because it is expressed only in the presence of food. Analysis of F1 and F2 generation progeny showed that the vast majority of the behavioral variance could be accounted for by a single gene for which the sitter phenotype is recessive to Rover (Sokolowski 1980). Utilizing some of the more sophisticated tools of *Drosophila* chromosome manipulation, the locus was mapped and X-ray mutagenesis used to produce new alleles of the locus that was dubbed *foraging* (de Belle et al. 1989). Molecular analysis showed it to be one of two structural loci for cGMP-dependent protein kinase (*dg2*) and that natural Rovers have ~12% more activity of the enzyme than sitters, with correspondingly higher levels of *dg2* mRNA and protein (Osborne et al. 1997). Severe mutations of the *foraging* gene are lethal (de Belle et al. 1993), emphasizing once again the mild character of natural variants as compared to null alleles (Greenspan 1997).

An analogous set of findings has identified a locus in the nematode *Caenorhabditis elegans* similar to vertebrate neuropeptide receptors that accounts for natural variation in aggregation, feeding, and foraging behavior (de Bono & Bargmann 1998, de Bono 2003). These nematodes exist as "solitary" or "social" strains that feed either individually or in clumps. In all natural isolates examined, a single amino acid substitution in the *npr-1* (neuropeptide receptor resemblance) gene distinguishes the 5 solitary cases from the 12 social cases. Solitariness is dominant, null mutants are social, and social animals can be made solitary by expression of the solitary allele or by overexpression of the naturally occurring social allele. The *npr-1* mutants are not affected exclusively in aggregation behavior but also differ with respect to hyperactivity on food, burrowing into agar, and accumulating on the border of a bacterial lawn. Sequence comparisons among natural isolates of three other *Caenorhabditis* species as well as *C. elegans* indicate that the solitary allele is only found in *C. elegans*, which suggests that the social allele is ancestral (Rogers et al. 2003).

An emerging story in the human genetics literature attributes handedness to a single gene difference, with the unusual feature that the recessive phenotype is stochastic. Handedness has generally been assumed to exhibit a complex mode of inheritance (e.g., McManus 1985), if it is heritable at all (e.g., Bishop 2001). By postulating that individuals homozygous for the Non-Right-Handed (*NRH*) allele have a 50:50 chance of being non-right-handed (where non-right-handed is defined as anyone who favors their left hand for any task), the population data fit very well (Klar 1996, 1999, 2003). Heterozygosity or homozygosity for the Right-Handed (*RH*) allele ensures consistent right-handedness. This model accounts well for the population frequency of nonright-handedness if one assumes a 40% frequency of the *NRH* allele. It also accounts very well for the observation of discordance in handedness among indentical (MZ) twins, as well as for the frequency of such discordant twins and for the 50:50 proportion of non-right-handed:right-handed children of the discordant twins. A stochastic, recessive phenotype of this sort finds a precedent in another developmental asymmetry: the *situs inversus* mutation of the mouse, in which half of the homozygous mutant individuals have reversed asymmetry of their internal organs (Layton 1976). One salient feature of this mode of inheritance is the challenge it presents to mapping the pertinent locus, especially if one is not consciously looking for it. This may account for lack of progress in identifying the putative *NRH* locus in humans. If the hypothesis is confirmed, it may eventually break the long-standing logjam in the identification of genetic factors in schizophrenia (Harrison & Owen 2003), if non-right-handedness (*NRH*/*NRH*) proves to be a predisposing factor in the disease (Klar 1999).

The foregoing examples, however, are single-gene exceptions to the quantitative complexity of most polymorphic traits.

Genetic Background: Bane of the Single-Gene Approach

Behavioral mutants are notoriously sensitive to variations in genetic background, the natural, genetic heterogeneity in laboratory stocks (also known in the classical genetics literature as modifiers). Whereas quantitative geneticists have long been aware of this ever-present variable, single-gene practitioners have not, in general, sometimes to their embarassment. A case in point is the erroneous identification of the temperature-sensitive paralytic mutation *shibire* as the voltage-sensitive sodium channel (Kelly 1974), based on the greater resistance to tetrodotoxin of the mutant relative to a control strain. The control, however, was not of the same genetic background as the mutant. As it turned out, the "control" strain was exceptionally sensitive to tetrodotoxin (Gitschier et al. 1980), whereas the *shibire* mutant was as resistant to it as most other fly strains (mutant or normal). The genetic basis of resistance in this control strain has never been properly determined.

Not only do experiments need to be conducted such that mutants and controls are on the same background (Dubnau & Tully 1998), but also mutant phenotypes will often fade over time as the result of unintended selection for such modifiers. The dependence of mutant phenotype on strain background has been well documented for learning mutants in the mouse (Gerlai 1996). The spontaneous disappearance of mutant phenotypes in the laboratory over time, well known at the level of folklore, has been documented for mutants affecting learning and brain development in the fly (de Belle & Heisenberg 1996). The range of phenotypic effects of several anatomical brain mutants is narrowed or widened in different genetic backgrounds (de Belle & Heisenberg 1996).

The ubiquity of the problem has bedeviled mutant studies all along. In a legendary incident at a *Drosophila* meeting in the early 1970s, one investigator began his talk by saying, "I would like to announce that *Hyperkinetic* is now a recessive" (J.C. Hall, personal communication, but not the speaker). That is, the strain no longer showed a dominant mode of inheritance for the mutation, as reported originally. Such phenomena are presumed to be the result of spontaneous selection for modifying alleles that are present in the population—a distinctly quantitative genetic problem! Moreover, the potency with which a given background can mask or exacerbate the phenotype of a mutation underlines its relevance to the issue of genetic mechanism. A graphic example of the range of these effects was shown in a study of modifiers of the *sevenless* mutation in *Drosophila*, a mutant originally isolated as part of a genetic dissection of phototaxis behavior (Harris et al.

1976) and subsequently studied in great depth for its role in cell fate determination in photoreceptors (e.g., Brennan & Moses 2000). When a moderate allele of *sevenless* (roughly midway between the most severe and wild-type) was placed on a range of different genetic backgrounds, phenotypes were found that ranged from fully wild-type to more severe than the most effective enhancer mutations previously isolated (Polaczyk et al. 1998). Clearly, genetic mechanisms cannot be properly understood without paying attention to such background effects. That is, single-gene effects fade into quantitative genetics at the margins.

MUTUAL BENEFITS

At this point in the discussion, it would probably seem to the single-gene school that quantitative genetics would have a lot to gain from the single-gene studies, in the form of information on the function of individual genes and on cellular mechanisms underlying behavior. Similarly, it would probably seem to the quantitative school that single-gene studies would have a lot to gain from quantitative genetics, in the form of a wider net for capturing relevant genes, and a conceptual framework for confronting the problems of gene interaction at the systems level. Both schools, on the other hand, would certainly expect everyone to gain from the new genomic technologies.

Gene Interactions: A Common Currency

Gene interactions have been an explicit part of the quantitative and single-gene programs from the outset. For the quantitative approach, the issue of interaction arises as soon as a trait is seen to be polygenic. If many genes are involved, their interactions are either additive (independent) or nonadditive (epistatic). For the single-gene approach, interactions at the molecular level are inseparable from the issue of mechanism. They constitute one of the main strategies for identifying, through screens for enhancer and suppressor mutations, components that interact molecularly. These sometimes lead to demonstration of direct, physical interactions between gene products. It is fitting, therefore, that gene interactions should emerge as one of the crossover points at which the two genetic approaches meet.

The first major proponent of gene interaction as an important element in evolutionary genetics was Sewall Wright, one of the architects of the modern synthesis of Darwinian theory with Mendelian genetics. His view contrasted with that of his coarchitects, J.B.S. Haldane (1932), who believed that alterations in a few, major individual genes were critical, or R.A. Fisher (1930), who emphasized the importance of small contributions from multiple genes acting in an additive manner. Beginning with his detailed studies of gene interaction in the formation of coat color in guinea pigs (Wright 1916) and continuing through his work on evolution by alterations in gene frequencies (Wright 1932), Wright saw a vast universe of potential gene interaction (1963). With the development of more sophisticated analyses and statistical models for quantitative trait analysis, epistasis could be inferred from the phenotypic scores of the various classes of progeny from test crosses of wild strains or of selected lines (e.g., McGuire 1992, Zeng et al. 1999).

The problem of epistasis emerged in some of the earliest studies by Mendelians (Bateson 1909) but became of major conceptual importance when metabolic pathways began to be analyzed by means of mutants. In this context, epistasis took on a biochemical meaning when a mutation in an upstream enzyme masked the presence of a mutation in one that was downstream (Beadle & Tatum 1941). The analogy would later be applied to developmental pathways and become the paradigm for demonstrating gene interactions in molecular genetic analyses (Avery & Wasserman 1992, Greenspan 2001).

One immediately obvious difference between the use of the term epistasis by the two schools (cf. Phillips 1998) is the insistence in single-gene analysis on a strong interfering effect—e.g., blockage of a pathway's output by a knockout mutation and rescue of that blockage by an activating mutation downstream. That is, an epistatic relationship represents either a direct physical interaction between gene products, or else an interaction only one or two steps away from being directly physical, as in a metabolic or signal transduction pathway. Directness, in turn, is valued as evidence of specificity, and specificity is the shibboleth of modern molecular biology (Greenspan 2001). No such requirement is imposed by the quantitative geneticists. For that school, epistasis occurs whenever two genes interact nonadditively, regardless of how far removed physically or temporally their sphere of activity may be. The corresponding shibboleth for this view is that anything that affects the final phenotype is fair game for natural selection, and is thus relevant.

Are these two views of epistasis reconcilable? More specifically, because the quantitative version of epistasis encompasses the single-gene version, the question is whether less direct varieties of interaction are relevant for understanding core, functional mechanisms. One step in favor of satisfying this demand is the increasing evidence that genes influencing behavior are pleiotropic (Greenspan 1997, Hall 1994, Pflugfelder 1998, Sokolowski 2002). That is, the genes that have mutated to produce behavioral variants are, almost universally, genes that also play other, often vital, roles in the organism's biology. This imposes a stringent selection on behavioral mutants. They must pass through the eye of a needle in order to retain viability and relative normality in most respects, while exhibiting defects in the behavior in question.

A case in point is the *optomotor-blind* mutation in a T-box domain transcription factor, which affects the optomotor response to horizontal motion in *Drosophila* (Heisenberg et al. 1978). It is an unusual allele of *bifid*, a vital gene required for many aspects of fly development, selectively altering transcription in a restricted part of the fly's optic lobes (Heisenberg 1997). As a consequence, these mutants are missing critical, motion-detecting neurons. Similarly, the learning mutant *Volado* is a special allele of the vital *scab* locus, which encodes an α-integrin required in many aspects of fly development (Grotwiel et al. 1998). That the same can be said of the *foraging* gene, i.e., null alleles are lethal (de Belle et al. 1989; and see above), illustrates a fundamental similarity between quantitative and single-gene traits: Behavioral variants are special or mild mutations of genes that have much broader roles.

If pleiotropy is widespread (perhaps even ubiquitous) among genes that affect behavior, then the potential for interactions between genetic variants is vastly expanded. The wider the network of contacts a gene product makes, the more chances there are for an alteration in another gene to influence it. This suggests the basis for a rapprochment between single-gene and quantitative analyses. If the traits studied in single-gene mutant studies were measured more sensitively (i.e., quantitatively) so that less extreme phenotypes were also examined, then mutant screens would yield a wider range of genes. As a result, many more of these wide-ranging interactions would be revealed, many more elements of the core mechanisms would likely emerge, and our concept of a core set of genes would be correspondingly enlarged. Such an approach has already shown its value in the application of a large set of mild, quantitative variants in bristle number to the study of peripheral nervous system development in *Drosophila* (Norga et al. 2003).

Identifying "All" of the Genes: A Second Crossover Point

A prominent feature of our current genomic era is the drive for encyclopedic coverage, to identify "all" of the genes subserving a given process. There were precedents in the saturation mutant screens of the past, carried out in microorganisms (e.g., Wood et al. 1968, for phage morphogenesis), and in a few instances in *Drosophila* (e.g., Nusslein-Volhard & Wieschaus 1980, for embryonic cuticle patterning) and *C. elegans* (Ferguson et al. 1987, for vulva development). For the quantitative geneticists, this has been the quest all along: to know all of the players and not to assume that there are only a few major ones. For the single-gene practitioners, it has motivated the extensive mutant hunts.

Both approaches are currently benefiting from the capabilities offered by genomic technologies. The availability of whole-genome sequences has vastly accelerated the process of identifying a new mutation, high-resolution DNA polymorphism maps have given a major boost to QTL studies, and genome-wide expression analysis with DNA microarrays has provided a new avenue into the range of gene activities that underlie a phenotype. These technologies have enabled a new synthesis between the quantitative and single-gene strategies, not owing to any conceptual breakthrough but simply to the newfound ability to identify individual genetic factors.

CLOSING THE GAP

Geotaxis in *Drosophila* Revisited

The ability to monitor the expression level of every gene in the genome under different conditions and in different genotypes is a technique whose promise has been touted widely (e.g., Lander 1999, White 2001). Use of the technology has already complemented and extended quantitative genetic analysis. One of the first applications of the new technology to quantitative genetic problems utilized microarrays to identify gene expression differences between Hirsch's bidrectionally selected geotaxis lines (described above). Although not all such mRNA expression differences would be due to actual genetic polymorphisms in the affected genes, some would, and others could be indicative of downstream effects of polymorphic genes. With these caveats in mind, RNA from adult heads of the positively geotaxic (Lo) and negatively geotaxic (Hi5) lines were analyzed on an array representing roughly two thirds of the fly genome (Toma et al. 2002). Many genes were differentially expressed ($\sim$250), representing a wide range of functions (e.g., transcription, signal transduction, cytoskeleton, and metabolism).

To test for the functional significance of the differences, mutants in some of the affected genes were tested in the geotaxis maze, after first being standardized with respect to genetic background (Toma et al. 2002). Several of the mutants were found to deflect the geotaxis response in the predicted direction from the geoneutral response of the background strain. The predicted direction for a mutant was based on the expression level of that gene in the selected lines; a null or hypomorphic mutant should behave like the line with the lower expression level. Thus, for several of the loci, a severe, single-gene lesion could mimic a selected phenotype. In most cases, however, no single-gene effect was as strong as that of the aggregate effect in a selected line.

The genes found to affect geotaxis are pleiotropic. (Actually, this had to be true, simply because they were existing mutants that had been isolated on the basis of some other phenotype.) One is relatively restricted in the nervous system: the neuropeptide gene *Pigment-dispersing-factor* (*Pdf*), which is involved in mediating the locomotor output of the fly's circadian rhythms (Park et al. 2000, Renn et al. 1999). The others are rather widely distributed, including *Pendulin*, encoding the nuclear import protein importin-α (Torok et al. 1995), and *cryptochrome*, encoding a flavin-binding protein that serves as an extraocular photoreceptor in circadian rhythms (Stanewsky et al. 1998).

The foregoing example bridges quantitative and single-gene studies in two ways: first, the use of selected strains to identify the many contributions of individual genes, and second, the use of preexisting single-gene mutants to test and validate the functional relevance of genes differing between the selected strains (cf. Long et al. 1996). The idea of using laboratory selection as an avenue toward identifying genetic mechanisms, unthinkable in the traditional single-gene world, has now become the method of choice in certain instances (e.g., Dierick & Greenspan 2003).

A QTL for Sleep EEG in Mice

Many aspects of sleep behavior and physiology vary among inbred strains of the mouse (Tafti & Franken 2002). EEG measurements revealed a particularly prominent difference in theta oscillations (4–12 Hz) during both paradoxical (REM) and slow-wave sleep among mouse strains (Franken et al. 1998). Theta oscillations are thought to modulate REM sleep insofar as treatments that suppress theta also suppress REM sleep (M. Tafti, personal communication). The theta-peak frequency (TPF) was found to vary from 6.12 to 7.61 Hz in the lowest (A/J) and highest (C58BR/cdJ) strains.

To apply genetic analysis to theta oscillations during paradoxical sleep, Tafti et al. (2003) first determined that slow TPF was recessive by making F1 hybrids between a slow TPF strain (BALB/cByJ) and a fast TPF strain (C57BL/6J). By producing F2 progeny from these two strains and scoring TPF phenotype and chromosomal markers, they found a major QTL on chromosome 5. Further crosses were performed, designed to subdivide the region further, and a 2.4 cM chromosomal segment was identified—small but likely to contain upwards of 20 genes.

At this point, the analysis would have stalled had it not been for the existence of relatively severe mutations in two of the promising candidate genes in that region: a knockout in *Nos1* (neuronal nitric oxide synthase), and a naturally occurring mutation of *Acads* (short-chain acyl-coenzyme A dehydrogenase), the first enzymatic step in fatty acid beta-oxidation. *Nos1* was a likely candidate because nitric oxide is known to affect theta oscillations (Datta et al. 1997), but the mutant's TPF score was similar to its own background C57BL/6J score. The other well-defined variant in that chromosomal segment is the *Acads* mutation, a spontaneously arising deletion of several hundred base pairs in the gene in Balb/cByJ that produces a null phenotype (Reue & Cohen 1996).

The *Acads* mutation in Balb/cByJ arose spontaneously after it had been separated in the 1980s from its parental Balb/cBy strain. Thus, these two strains differ exclusively (or nearly so) in the *Acads* gene, and when TPF was compared between Balb/cByJ and Balb/cBy, it differed. Further tests of recombinant inbred lines between the original C57BL/6J and Balb/cBy strains, and of additional recombinant progeny generated from them, confirmed that there were no other loci influencing TPF segregating in these strains. Additional evidence supporting *Acads* involvement came from DNA microarray studies and metabolite administration, strengthening the correlation between TPF and fatty acid beta-oxidation in mitochondria. As an enzyme involved in energy metabolism, the *Acads* gene has potential significance for sleep insofar as one of the active hypotheses for understanding the underlying function of sleep postulates a central role for energy metabolism (Benington & Heller 1995, Kong et al. 2002).

This study draws on nearly all aspects of the genetic arsenal available in the mouse: strain differences, DNA markers, genome sequence information, spontaneous and induced mutants, recombinant inbred strains, and microarrays. In making a successful gene identification, all were needed.

Long-Term Memory in *Drosophila*—Combining Forward and Reverse Genetics

In a forward genetic, single-gene mutagenesis attempt to get at all of the genes subserving a behavior, Dubnau et al. (2003) carried out a large-scale screen for new mutants ($N = 60$) defective in long-term memory, and in parallel, they performed a DNA microarray analysis to identify genes ($N = 42$) expressed in the brains of flies under conditions that produce long-term memory. These genes (both sets) run the gamut of biological functions: transcription, translation, signal transduction, cytoskeleton, and metabolism. The investigators then determined the overlap between the two sets of genes and tested mutants identified on the arrays that fell into the overlap. The results suggested a requirement for the machinery of mRNA localization and translational regulation in the consolidation of long-term memory. The approach relies on training protocols that had previously produced 3-h-versus-1-d memory of an odor made aversive when coupled to electric shock in the fly (Tully et al. 1994). The distinction in memory duration depends on whether these trials are administered all at once (massed) for short-term memory or with 15-min intervals (spaced) for long-term memory. This distinction provided an internal control for judging mutants, e.g., are they normal for immediate memory but abnormal for long-term memory? The same criterion—massed versus spaced training—was the differential applied to the gene profiling results.

The microarrays pointed to the mRNA localization genes *staufen* and *moesin* and the translational regulation genes *pumilio*, *orb*, and *eIF2G* as upregulated selectively after spaced training. The mutant screen isolated new alleles of the related genes ($oskar^{norka}$, $pumilio^{milord}$, and $eIF5C^{krasavietz}$) as showing defective long-term memory. Extant mutants in *staufen*, identified on the arrays, were tested and also found to have abnormal long-term memory. Further confirmation came from a temperature-sensitive genotype of *staufen*, which permitted a demonstration that the gene product is needed in the period soon after training to be effective. The issue of a critical period for these effects is especially relevant, given that all of these genes are capable of affecting development and viability: *staufen*, *oskar*, and *pumilio* were isolated originally as maternal effect genes, certain alleles of which produce grossly abnormal embryos when the mother is mutant (Palacios & St. Johnston 2001), and the one preexisting allele of *eIF5C* has severely reduced viability (Spradling et al. 1999).

The foregoing is an example of a single-gene study that begins to look quantitative based on having cast such a wide net for "all" (or at least as many as possible) of the genes mutable to that phenotype. The spectrum of biological functions revealed in the mutant screen alone should dispel any notions of the unimportance of pleiotropic genes and their far-ranging activities for a core mechanism.

Odor-Guided Behavior in *Drosophila*

A study of odor-guided behavior in the fruit fly bridges quantitative with single-gene analysis in a different way. A series of insertional mutations was identified in which flies fail to jump in response to benzaldehyde (Anholt et al. 1996). Fourteen *smell impaired* (*smi*) mutant lines were recovered from several hundred strains in which a *P* element, a transposable DNA sequence routinely used for insertional mutagenesis in *Drosophila*, had been inserted at random on a chromosome. Any lines that were not viable and healthy as homozygotes were discarded, as were those that showed any locomotor defects. Because they are insertional mutants, the resulting 14 variants could be easily mapped to their exact chromosomal location using genomic information, and the genes cloned and sequenced (e.g., Ganguly et al. 2003, Kulkarni et al. 2002).

A classical quantitative genetic analysis was then performed: interactions among the genes were assessed by constructing pairwise combinations of the mutants and testing the resulting double heterozygotes (i.e., *mutant1*/+, *mutant2*/+) for their olfactory responsiveness. Many of the pairwise combinations showed epistatic interactions: a more potent effect of the combination than predicted from the average effect of each mutant by itself (Federowicz et al. 1998).

Had the study stopped there, the conclusion would have been that mutants isolated for a common phenotype can show interactions—this is not particularly surprising or informative. Fortunately, it did not, and the investigators went on to measure the genome-wide transcriptional profiles of five of these *smell impaired* (*smi*) mutants with DNA microarrays (Anholt et al. 2003). The results showed, once again, a wide range of genes whose transcription levels are altered in each mutant. More importantly, there was an overlapping set of gene expression effects among the interacting *smi* mutants. Behavioral tests of extant mutants in several of the genes identified on the microarrays (e.g., the ion channels *Shaker* and *trp1*, the component of synaptic release *Syntaxin1A*, *calmodulin*, and the GABA receptor *Rdl*) further demonstrated their functional relevance to the *smi* phenotype.

BEHAVIOR-GENETIC ANALYSIS: THE NEW SYNTHESIS

> Almost never can a complex system of any kind be understood as a simple extrapolation from the properties of its elementary components.
>
> D. Marr (1982)

Reductionist schools of thought usually define themselves in opposition to systems schools. This certainly describes the relationship between the single-gene analysts of behavior and the behavior-genetic architecturalists. The disagreement usually revolves around each side's view that the other is missing some important point. The behavior geneticists felt that the single-gene approach ignored the complexity of interactions of genes and the inherent variability of genes in each population. For their part, the single-gene analysts objected to the lack of identification and mechanistic explanation of the entities being described by the behavior geneticists. Neither was wrong because neither had the whole answer.

The passage of time and the elaboration of findings in opposing schools sometimes allow a reconciliation. As a result, there is now a basis for synthesizing the viewpoints and principles from each school.

Single Genes in Genetic Architecture Terms

From its inception, the concept of genetic architecture implicitly acknowledged the existence of single genes influencing behavior and of the interactions of several or many such genes determining a particular phenotype. The expansion in the modes of analysis described above enlarges our view of the varieties of gene action and sensitizes us to the network aspects of the system.

The recognition of the ubiquity of pleiotropy in gene action (Greenspan 1997, Hall 1994, Pflugfelder 1998, Sokolowski 2002) means that each gene has, in effect, its own architecture—a distributed pattern of action through the various stages and tissues of the organism. In this sense, the summated action of the genes is not so much a jigsaw puzzle in which each piece fits together with its immediate neighbors in one spot, but is rather a flexible, multilayered network (cf. Greenspan 2001)—a viewpoint that was implicit in quantitative genetics and that single-gene genetics has been slowly approaching.

Genetic Architecture in Single-Gene Terms

In populations, the pleiotropic, network attributes of genes have consequences for how genetic variation can produce behavioral variation. Each allele of a gene can potentially contribute in several ways to phenotype. These contributions, in turn, depend on the partners with which a gene interacts. Variation can thus occur in a restricted portion of a gene's range of activities if its interacting partners are more sensitive to perturbation in one place than in another. If its interacting partners also come in allelic variants, a further dimension is added.

Phenotypic variation in a population, which is what one measures, is thus not a monotonic function of allelic variation. Instead, it may well represent a more complex fabric than the distribution of alleles alone might suggest. This may seem to present an even-more-bewildering picture than the traditional view. Its saving grace, however, is knowing that the network nature of a gene's interactions ultimately makes its contributions to phenotype more comprehensible. Further study of the interacting nature of one gene's variation with that of another, in turn, brings its population genetic architecture within the realm of comprehensibility.

The two perspectives can be distilled into one: many genes for each behavior (e pluribus unum), many behaviors from each gene (ex uno plura).

The Relative Contributions of Genes Differ

The concept of genetic architecture has always assumed that where there are multiple genes, they do not necessarily contribute equally to the behavioral phenotype. Some are strong effect, some weak, and every stripe in between, but all are subject to changes in the strength of their effect in different genetic backgrounds. This view finds support in both classical and molecular studies.

In contrast, the idea that the various contributing genes sum to produce the phenotype, R.A. Fisher's concept of additive, independent factors in quantitative genetics (Fisher 1930), has found less support (e.g., Weber et al. 1999, 2001) than its converse, nonadditive interactions (e.g., Clark & Wang 1997, Federowicz et al. 1998, Mackay 2002). The recent findings described above provide further support for nonadditivity by suggesting a molecular basis for it in the wide-ranging effects seen on expression across the genome.

Synergism and network flexibility make it easier to conceive of how new properties in behavior can emerge: Tune an allele up here, tune another one down there, combine them with some other preexisting variants, allow it all to ripple through the networks, and boom! you have a new behavior. Although no one is yet at the point of demonstrating this in the lab, the threshold effects frequently seen in selection experiments, in which the phenotype does not move at all for many generations and then diverges rapidly (e.g., Erlenmeyer-Kimling et al. 1962), or in which the phenotype fluctuates dramatically before diverging consistently (e.g., Manning 1961), suggest that such effects can occur in the laboratory, where they can be studied in the ways exemplified above.

At the same time, it is also easy to imagine that the number of ways for genes to influence behavior will be manifold. It will depend on the context of other alleles present (i.e., genetic background), as well as on the actual role(s) a given gene plays in that behavior. The impact of one level, the individual gene, on the other, the gene system, is reciprocal: individual genes influence the network, and the network properties, in turn, influence the action of individual genes.

At the beginning of the single-gene era of behavioral studies, Sydney Brenner (1973) remarked, "Understanding the genetic foundations of behavior may well require solving all of the outstanding questions of biology." The thirty years that have passed since then suggest that this may not be quite true. But to the extent that we must understand the nature and principles of how gene networks influence complex phenotypes, the synthesis of quantitative and single-gene approaches currently underway would seem to be a prerequisite.

ACKNOWLEDGMENTS

Helpful discussions and comments on the manuscript were provided by Rozi Andretic, Kathy Crossin, Herman Dierick, Joe Gally, Indrani Ganguly, Mehdi Tafti, and Dan Toma. R.J.G. is the Dorothy and Lewis B. Cullman Fellow at The Neurosciences Institute, which is supported through the Neurosciences Research Foundation.

The *Annual Review of Neuroscience* is online at http://neuro.annualreviews.org

LITERATURE CITED

Allada R, White NE, So WV, Hall JC, Rosbash M. 1998. A mutant *Drosophila* homolog of mammalian *Clock* disrupts circadian rhythms and transcription of *period* and *timeless*. *Cell* 93:791–804

Anholt RRH, Dilda CL, Chang S, Fanara

J-J, Kulkarni NH, et al. 2003. The genetic architecture of odor-guided behavior in Drosophila: epistasis and the transcriptome. *Nat. Genet.* 34: doi:10.1038/ng1240

Anholt RRH, Lyman RF, Mackay TFC. 1996. Effects of single P-element insertions on olfactory behavior in *Drosophila melanogaster*. *Genetics* 143:293–301

Avery L, Wasserman S. 1992. Ordering gene function: the interpretation of epistasis in regulatory hierarchies. *Trends Genet.* 8:312–16

Bailey DW. 1971. Recombinant-inbred strains. An aid to finding identity, linkage, and function of histocompatibility and other genes. *Transplantation* 11:325–27

Bateson W. 1909. *Mendel's Principles of Heredity*. Cambridge: Cambridge Univ. Press.

Beadle GW, Tatum EL. 1941. Genetic control of biochemical reactions in Neurospora. *Proc. Natl. Acad. Sci. USA* 27:499–506

Belknap JK, Hitzemann R, Crabbe JK, Phillips TJ, Buck KJ, Williams RW. 2001. QTL analysis and genome-wide mutagenesis in mice: complementary genetic approaches to the dissection of complex traits. *Behav. Genet.* 31:5–15

Benington JH, Heller HC. 1995. Restoration of brain energy metabolism as the function of sleep. *Prog. Neurobiol.* 45:347–60

Benzer S. 1967. Behavioral mutants of *Drosophila* isolated by countercurrent distribution. *Proc. Natl. Acad. Sci. USA* 58:1112–19

Bishop DV. 2001. Individual differences in handedness and specific speech and language impairment: evidence against a genetic link. *Behav. Genet.* 31:339–51

Brennan CA, Moses K. 2000. Determination of Drosophila photoreceptors: Timing is everything. *Cell Mol. Life Sci.* 57:195–214

Brenner S. 1973. The genetics of behaviour. *Br. Med. Bull.* 29:269–71

Bucan M, Abel T. 2002. The mouse: Genetics meets behaviour. *Nat. Rev. Genet.* 3:114–23

Caspi A, Sugden K, Moffitt TE, Taylor A, Craig IW, et al. 2003. Influence of life stress on depression: moderation by a polymorphism in the 5-HTT gene. *Science* 301:386–89

Clark AG, Wang L. 1997. Epistasis in measured genotypes: *Drosophila* P-element insertions. *Genetics* 147:157–63

Costa R, Peixoto AA, Barbujani G, Kyriacou CP. 1992. A latitudinal cline in a Drosophila clock gene. *Proc. R. Soc. Biol. Sci.* 250:43–49

Datta S, Patterson EH, Siwek DF. 1997. Endogenous and exogenous nitric oxide in the pedunculopontine tegmentum induces sleep. *Synapse* 27:69–78

Davenport CB. 1911. *Heredity in Relation to Eugenics*. New York: Holt. 298 pp.

de Belle JS, Heisenberg M. 1996. Expression of *Drosophila* mushroom body mutations in alternative genetic backgrounds: a case study of the mushroom *body miniature gene (mbm)*. *Proc. Natl. Acad. Sci. USA* 93:9875–80

de Belle JS, Hilliker AJ, Sokolowski MB. 1989. Genetic localization of *foraging* (*for*): a major gene for larval behavior in *Drosophila melanogaster*. *Genetics* 123:157–63

de Belle JS, Sokolowski MB, Hilliker AJ. 1993. Genetic analysis of the *foraging* microregion of *Drosophila melanogaster*. *Genome* 36:94–101

de Bono M. 2003. Molecular approaches to aggregation behavior and social attachment. *J. Neurobiol.* 54:78–92

de Bono M, Bargmann CI. 1998. Natural variation in a neuropeptide Y receptor homolog modifies social behavior and food response in *C. elegans*. *Cell* 94:679–89

Del Punta K, Leinders-Zufall T, Rodriguez I, Jukam D, Wysocki CJ, et al. 2002. Deficient pheromone responses in mice lacking a cluster of vomeronasal receptor genes. *Nature.* 419:70–74

Dierick HA, Greenspan RJ. 2003. Selection for aggressive behavior in *Drosophila melanogaster*. *Cold Spring Harbor Neurobiol. Drosoph.* 2003:2

Dietrich W, Katz H, Lincoln SE, Shin HS, Friedman J, et al. 1992. A genetic map of the mouse suitable for typing intraspecific crosses. *Genetics* 131:423–47

Dubnau J, Chiang AS, Grady L, Barditch J, Gossweiler S, et al. 2003. The *staufen/pumilio* Pathway is involved in Drosophila long-term memory. *Curr. Biol.* 13:286–96

Dubnau J, Tully T. 1998. Gene discovery in *Drosophila*: new insights for learning and memory. *Annu. Rev. Neurosci.* 21:407–44

Ehrman L, Parsons PA. 1981. *Behavior Genetics and Evolution.* New York: McGraw Hill. 450 pp.

Erlenmeyer-Kimling LF, Hirsch J, Weiss JM. 1962. Studies in experimental behavior genetics. *J. Comp. Physiol. Psychol.* 55:722–31

Fedorowicz GM, Fry JD, Anholt RR, Mackay TF. 1998. Epistatic interactions between smell-impaired loci in *Drosophila melanogaster*. *Genetics* 148:1885–91

Ferguson EL, Sternberg PW, Horvitz HR. 1987. A genetic pathway for the the specification of vulval cell lineages of *Caenorhabditis elegans*. *Nature* 326:259–67

Fisher RA. 1930. *The Genetical Theory of Natural Selection.* Oxford, UK: Clarendon. 291 pp.

Franken P, Malafosse A, Tafti M. 1998. Genetic variation in EEG activity during sleep in inbred mice. *Am. J. Physiol.* 275:R1127–37

Galton F. 1865. Hereditary talent and character. *MacMillan's Mag.* 11:157–66, 318–27

Ganguly I, Mackay TFC, Anholt RRH. 2003. Scribble is essential for olfactory behavior in *Drosophila melanogaster*. *Genetics* 164:1447–57

Gerlai R. 1996. Gene-targeting studies of mammalian behavior: Is it the mutation or the background genotype? *Trends Neurosci.* 19:177–81

Giaever G, Chu AM, Ni L, Connelly C, Riles L, et al. 2002. Functional profiling of the *Saccharomyces cerevisiae* genome. *Nature* 418:387–91

Gitschier J, Strichartz GR, Hall LM. 1980. Saxitoxin binding to sodium channels in head extracts from wild-type and tetrodotoxin-sensitive strains of *Drosophila melanogaster.* *Biochim. Biophys. Acta.* 595:291–303

Greenspan RJ. 1997. A kinder, gentler genetic analysis of behavior: Dissection gives way to modulation. *Curr. Opin. Neurobiol.* 7:805–11

Greenspan RJ. 2001. The flexible genome. *Nat. Rev. Genet.* 2:383–87

Greenspan RJ. 2003. The 2003 GSA Medal: Jeffey C. Hall. *Genetics* 164:1446–47

Greenspan RJ. 2004. The varieties of selectional experience in behavioral genetics. *J. Neurogenet.* 17:241–70

Greenspan RJ, Ferveur JF. 2000. Courtship in *Drosophila*. *Annu. Rev. Genet.* 34:205–32

Grotewiel MS, Beck CD, Wu KH, Zhu XR, Davis RL. 1998. Integrin-mediated short-term memory in *Drosophila*. *Nature* 391:455–60

Haldane JBS. 1932. *The Causes of Evolution.* London: Longmans, Green & Co. 222 pp.

Hall JC. 1994. Pleiotropy of behavioral genes. In *Flexibility and Constraint in Behavioral Systems.* ed. RJ Greenspan, CP Kyriacou. pp. 15–28. Berlin: Dahlem Konferenzen. 313 pp.

Hall JC. 1995. Tripping along the trail to the molecular mechanisms of biological clocks. *Trends Neurosci.* 18:230–40

Hall JC. 2003. Genetics and molecular biology of rhythms in *Drosophila* and other insects. *Adv. Genet.* 48:1–280

Harmer SL, Panda S, Kay SA. 2001. Molecular bases of circadian rhythms. *Annu. Rev. Cell Dev. Biol.* 17:215–53

Harris WA, Stark WS, Walker JA. 1976. Genetic dissection of the photoreceptor system in the compound eye of *Drosophila melanogaster*. *J. Physiol.* 256:415–39

Harrison PJ, Owen MJ. 2003. Genes for schizophrenia? Recent findings and their pathophysiological implications. *Lancet* 361:417–19

Heisenberg M. 1997. Genetic approaches to neuroethology. *BioEssays* 19:1065–73

Heisenberg M, Wonneberger R, Wolf R. 1978. *optomotor-blind*^{H31}—a *Drosophila* mutant of the lobula plate giant neurons. *J. Comp. Physiol. A* 124:287–96

Hirsch J, Erlenmeyer-Kimling LF. 1962. Studies in experimental behavior genetics: IV. Chromosome analyses for geotaxis. *J. Comp. Physiol. Psychol.* 55:732–39

Hirsch J, Ksander G. 1969. Studies in experimental behavior genetics. *J. Comp. Physiol. Psychol.* 67:118–22

Hirsch J, Tryon RC. 1956. Mass screening and reliable individual measurement in the experimental behavior genetics of lower organisms. *Psychol. Bull.* 53:402–10

Kelly LE. 1974. Temperature-sensitive mutations affecting the regenerative sodium channel in *Drosophila melanogaster*. *Nature* 248:166–68

Kendler KS, Prescott CA, Myers J, Neale MC. 2003. The structure of genetic and environmental risk factors for common psychiatric and substance use disorders in men and women. *Arch. Gen. Psychiatry* 60:929–37

Klar AJS. 1996. A single locus, *RGHT*, specifies preference for hand utilization in humans. *Cold Spring Harb. Symp. Quant. Biol.* 61:59–65

Klar AJS. 1999. Genetic models for handedness, brain lateralization, schizophrenia, and manic-depression. *Schizophrenia Res.* 39:207–18

Klar AJS. 2003. Human handedness and scalp hair-whorl direction develop from a common genetic mechanism. *Genetics* 165:269–76

Kong J, Shepel PN, Holden CP, Mackiewicz M, Pack AI, Geiger JD. 2002. Brain glycogen decreases with increased periods of wakefulness: implications for homeostatic drive to sleep. *J. Neurosci.* 22:5581–87

Konopka RJ, Benzer S 1971. Clock mutants of *Drosophila melanogaster*. *Proc. Natl. Acad. Sci. USA* 68:2112–16

Kulkarni NH, Yamamoto AH, Robinson KO, Mackay TF, Anholt RR. 2002. The DSC1 channel, encoded by the smi60E locus, contributes to odor-guided behavior in Drosophila melanogaster. *Genetics* 161: 1507–16

Lander ES. 1999. Array of hope. *Nat. Genet.* 21(1 Suppl.):3–4

Layton WM Jr. 1976. Random determination of a developmental process: reversal of normal visceral asymmetry in the mouse. *J. Hered.* 67:336–38

Lesch KP, Bengel D, Heils A, Sabol SZ, Greenberg BD, et al. 1996. Association of anxiety-related traits with a polymorphism in the serotonin transporter gene regulatory region. *Science* 274:1527–31

Lewontin R. 1974. *The Genetic Basis of Evolutionary Change*. New York: Columbia Univ. Press. 346 pp.

Leypold BG, Yu CR, Leinders-Zufall T, Kim MM, Zufall F, et al. 2002. Altered sexual and social behaviors in trp2 mutant mice. *Proc. Natl. Acad. Sci. USA* 99:6376–81

Liman ER, Corey DP, Dulac C. 1999. TRP2: a candidate transduction channel for mammalian pheromone sensory signaling. *Proc. Natl. Acad. Sci. USA* 96:5791–96

Lin JM, Kilman VL, Keegan K, Paddock B, Emery-Le M, et al. 2002. A role for casein kinase 2alpha in the *Drosophila* circadian clock. *Nature* 420:816–20

Lindblad-Toh K, Winchester E, Daly MJ, Wang DG, Hirschhorn JN, et al. 2000. Large-scale discovery and genotyping of single-nucleotide polymorphisms in the mouse. *Nat. Genet.* 24:381–86

Long AD, Mullaney SL, Mackay TF, Langley CH. 1996. Genetic interactions between naturally occurring alleles at quantitative trait loci and mutant alleles at candidate loci affecting bristle number in *Drosophila melanogaster. Genetics* 144:1497–510

Mackay TF. 2002. Quantitative trait loci in *Drosophila*. *Nat. Rev. Genet.* 2:11–20

Maldonado R, Blendy JA, Tzavara E, Gass P, Roques BP, et al. 1996. Reduction of morphine abstinence in mice with a mutation in the gene encoding CREB. *Science* 273:657–59

Manning A. 1961. The effects of artificial selection for mating speed in *Drosophila melanogaster*. *Anim. Behav.* 9:82–92

Marr D. 1982. *Vision*. Cambridge, MA: MIT Press. 397 pp.

Martinek S, Inonog S, Manoukian AS, Young MW. 2001. A role for the segment polarity gene shaggy/GSK-3 in the Drosophila circadian clock. *Cell* 105:769–79

McGuire TR. 1992. A biometrical genetic approach to chromosome analysis in *Drosophila*: detection of epistatic interactions in geotaxis. *Behav. Genet.* 22:453–67

McManus IC. 1985. Handedness, language dominance and aphasia: a genetic model. *Psychol. Med. Monogr.* 8(Suppl.):1–40

Merikangas KR, Chakravarti A, Moldin SO, Araj H, Blangero JC, et al. 2002. Future of genetics of mood disorders research. *Biol. Psychiatry* 52:457–77

Nadeau JH, Singer JB, Matin A, Lander ES. 2000. Analysing complex genetic traits with chromosome substitution strains. *Nat. Genet.* 24:221–25

Nguyen PV, Gerlai R. 2002. Behavioural and physiological characterization of inbred mouse strains: prospects for elucidating the molecular mechanisms of mammalian learning and memory. *Genes Brain Behav.* 1:72–81

Norga KK, Gurganus MC, Dilda CL, Yamamoto A, Lyman RF, et al. 2003. Quantitative analysis of bristle number in *Drosophila* mutants identifies genes involved in neural development. *Curr. Biol.* 13:1388–97

Nusslein-Volhard C, Wieschaus E. 1980. Mutations affecting segment number and polarity in *Drosophila. Nature* 287:795–801

Osborne KA, Robichon A, Burgess E, Butland S, Shaw RA, et al. 1997. Natural behavior polymorphism due to a cGMP-dependent protein kinase of *Drosophila. Science* 277:834–36

Palacios IM, St Johnston D. 2001. Getting the message across: the intracellular localization of mRNAs in higher eukaryotes. *Annu. Rev. Cell Dev. Biol.* 17:569–614

Park JH, Helfrich-Forster C, Lee G, Liu L, Rosbash M, Hall JC. 2000. Differential regulation of circadian pacemaker output by separate clock genes in *Drosophila. Proc. Natl. Acad. Sci. USA* 97:3608–13

Pflugfelder GO. 1998. Genetic lesions in *Drosophila* behavioural mutants. *Behav. Brain Res.* 95:3–15

Phillips PC. 1998. The language of gene interaction. *Genetics* 149:1167–71

Pittendrigh CS. 1954. On temperature independence in the clock system controlling emergence time in *Drosophila. Proc. Natl. Acad. Sci. USA* 40:1018–29

Polaczyk PJ, Gasperini R, Gibson G. 1998. Naturally occurring genetic variation affects *Drosophila* photoreceptor determination. *Dev. Genes Evol.* 207:462–70

Price JL, Blau J, Rothenfluh A, Abodeely M, Kloss B, Young MW. 1998. *double-time* is a novel *Drosophila* clock gene that regulates PERIOD protein accumulation. *Cell* 94:83–95

Provine WB. 1971. *The Origins of Theoretical Population Genetics.* Chicago: Univ. Chicago Press. 211 pp.

Renn SC, Park JH, Rosbash M, Hall JC, Taghert PH. 1999. A *pdf* neuropeptide gene mutation and ablation of PDF neurons each cause severe abnormalities of behavioral circadian rhythms in *Drosophila. Cell.* 99:791–802

Reue K, Cohen RD. 1996. *Acads* gene deletion in BALB/cByJ mouse strain occurred after 1981 and is not present in BALB/cByJ-fld mutant mice. *Mamm. Genome* 7:694–95

Ricker JP, Hirsch J. 1988. Genetic changes occurring over 500 generations in lines of *Drosophila melanogaster* selected divergently for geotaxis. *Behav. Genet.* 18:13–25

Robertson FW. 1954. Studies in quantitative inheritance. V. Chromosome analyses of crosses between selected and unselected lines of different body size in *Drosophila melanogaster. J. Genet.* 52:494–20

Rogers C, Reale V, Kim K, Chatwin H, Li C, et al. 2003. Inhibition of *Caenorhabditis elegans* social feeding by FMRFamide-related peptide activation of NPR-1. *Nat. Neurosci.* 6:1178–85

Rutila JE, Suri V, Le M, So WV, Rosbash M, Hall JC. 1998. CYCLE is a second bHLH-PAS clock protein essential for circadian rhythmicity and transcription of *Drosophila period* and *timeless. Cell* 93:805–14

Saudou F, Amara DA, Dierich A, LeMeur M,

Ramboz S, et al. 1994. Enhanced aggressive behavior in mice lacking 5-HT1B receptor. *Science* 265:1875–78

Sawyer LA, Hennessy JM, Peixoto AA, Rosato E, Parkinson H, et al. 1997. Natural variation in a *Drosophila* clock gene and temperature compensation. *Science* 278:2117–20

Sehgal A, Price JL, Man B, Young MW. 1994. Loss of circadian behavioral rhythms and per RNA oscillations in the *Drosophila* mutant *timeless*. *Science* 263:1603–6

Sokolowski MB. 1980. Foraging strategies of *Drosophila melanogaster*: a chromosomal analysis. *Behav. Genet.* 10:291–302

Sokolowski MB. 2002. *Drosophila*: Genetics meets behaviour. *Nat. Rev. Genet.* 2:879–90

Sokolowski MB, Wahlsten D. 2001. Gene-environment interaction and complex behavior. In *Methods in Genomic Neuroscience*, ed. HR Chin, SO Moldin, pp. 3–27. Boca Raton, FL: CRC Press. 344 pp.

Spradling AC, Stern D, Beaton A, Rhem EJ, Laverty T, et al. 1999. The Berkeley *Drosophila* genome project gene disruption project. Single P-element insertions mutating 25% of vital *Drosophila* genes. *Genetics* 153:135–77

Stanewsky R, Kaneko M, Emery P, Beretta B, Wager-Smith K, et al. 1998. The *cry*^{b} mutation identifies cryptochrome as a circadian photoreceptor in *Drosophila*. *Cell* 95:681–92

Stowers L, Holy TE, Meister M, Dulac C, Koentges G. 2002. Loss of sex discrimination and male-male aggression in mice deficient for TRP2. *Science* 295:1493–500

Tafti M, Franken P. 2002. Invited review: genetic dissection of sleep. *J. Appl. Physiol.* 92:1339–47

Tafti M, Petit B, Chollet D, Neidhart E, de Bilbao F, et al. 2003. Deficiency in short-chain fatty acid beta-oxidation affects theta oscillations during sleep. *Nat. Genet.* 34:320–25

Takahashi JS. 2004. Circadian rhythms. *Annu. Rev. Genom. Human Genet.* 5:doi:10.1146/annurev.genom.5.06 1903. 175925. In press

Tang YP, Shimizu E, Dube GR, Rampon C, Kerchner GA, et al. 1999. Genetic enhancement of learning and memory in mice. *Nature* 401:63–69

Toma DP, White KP, Hirsch J, Greenspan RJ. 2002. Identification of genes involved in *Drosophila melanogaster* geotaxis, a complex behavioral trait. *Nat. Genet.* 31:349–53

Torok I, Strand D, Schmitt R, Tick G, Torok T, et al. 1995. The overgrown hematopoietic organs-31 tumor suppressor gene of *Drosophila* encodes an Importin-like protein accumulating in the nucleus at the onset of mitosis. *J. Cell Biol.* 129:1473–89

Tully T. 1996. Discovery of genes involved with learning and memory: an experimental synthesis of Hirschian and Benzerian perspectives. *Proc. Natl. Acad. Sci. USA* 93:13460–67

Tully T, Preat T, Boynton SC, Del Vecchio M. 1994. Genetic dissection of consolidated memory in *Drosophila*. *Cell* 79:35–47

Waddell S, Quinn WG. 2001. Flies, genes, and learning. *Annu. Rev. Neurosci.* 24:1283–309

Weber K, Eisman R, Higgins S, Morey L, Patty A, et al. 2001. An analysis of polygenes affecting wing shape on chromosome 2 in *Drosophila melanogaster*. *Genetics* 159:1045–57

Weber K, Eisman R, Morey L, Patty A, Sparks J, et al. 1999. An analysis of polygenes affecting wing shape on chromosome 3 in *Drosophila melanogaster*. *Genetics* 153:773–86

White KP. 2001. Functional genomics and the study of development, variation, and evolution. *Nat. Rev. Genet.* 2:528–37

Williams RW. 2000. Mapping genes that modulate mouse brain development: a quantitative genetic approach. *Results Probl. Cell Differ.* 30:21–49

Wood WB, Edgar RS, King J, Lielausis I, Henninger M. 1968. Bacteriophage assembly. *Fed. Proc.* 27:1160–66

Wright S. 1916. An intensive study of the inheritance of color and of other coat characters in guinea pigs with especial reference to graded variation. *Carnegie Inst. Wash.* Publ. No. 241:59–160

Wright S. 1932. The roles of mutation, inbreeding, crossbreeding, and selection in evolution. *Proc. Sixth Int. Congr. Genet.* 1:356–66

Wright S. 1963. Genic interaction. In *Methods in Mammalian Genetics*, ed. WJ Burdette, pp. 159–92. San Francisco: Holden-Day

Zeng ZB, Kao CH, Basten CJ. 1999. Estimating the genetic architecture of quantitative traits. *Genet. Res.* 74:279–89

Annu. Rev. Neurosci. 2004. 27:107–44
doi: 10.1146/annurev.neuro.27.070203.144206

DESENSITIZATION OF G PROTEIN–COUPLED RECEPTORS AND NEURONAL FUNCTIONS

Raul R. Gainetdinov,[1] Richard T. Premont,[3]
Laura M. Bohn,[1,4] Robert J. Lefkowitz,[2,3]
and Marc G. Caron[1,3]

Howard Hughes Medical Institute Laboratories, Departments of Cell Biology[1], Biochemistry[2], and Medicine[3], Duke University Medical Center, Durham, North Carolina 27710; email: r.gainetdinov@cellbio.duke.edu, richard.premont@duke.edu, lefko001@receptor-biol.duke.edu, caron002@mc.duke.edu
[4]Present Address: Department of Pharmacology, The Ohio State University College of Medicine and Public Health, Columbus, Ohio 43210; email: bohn.24@osu.edu

Key Words sensitization, tolerance, opiates, antinociception, psychostimulants

■ **Abstract** G protein–coupled receptors (GPCRs) have proven to be the most highly favorable class of drug targets in modern pharmacology. Over 90% of nonsensory GPCRs are expressed in the brain, where they play important roles in numerous neuronal functions. GPCRs can be desensitized following activation by agonists by becoming phosphorylated by members of the family of G protein–coupled receptor kinases (GRKs). Phosphorylated receptors are then bound by arrestins, which prevent further stimulation of G proteins and downstream signaling pathways. Discussed in this review are recent progress in understanding basics of GPCR desensitization, novel functional roles, patterns of brain expression, and receptor specificity of GRKs and βarrestins in major brain functions. In particular, screening of genetically modified mice lacking individual GRKs or βarrestins for alterations in behavioral and biochemical responses to cocaine and morphine has revealed a functional specificity in dopamine and μ-opioid receptor regulation of locomotion and analgesia. An important and specific role of GRKs and βarrestins in regulating physiological responsiveness to psychostimulants and morphine suggests potential involvement of these molecules in certain brain disorders, such as addiction, Parkinson's disease, mood disorders, and schizophrenia. Furthermore, the utility of a pharmacological strategy aimed at targeting this GPCR desensitization machinery to regulate brain functions can be envisaged.

INTRODUCTION

The cell-surface receptors for most neuromodulators are members of the large superfamily of G protein–coupled receptors (GPCRs). These receptors share similar primary amino acid sequences, a common seven–transmembrane-spanning domain architecture, and the ability to modulate intracellular metabolism through the activation of heterotrimeric GTP-binding proteins (G proteins) (Hamm & Gilchrist 1996, Watson & Arkinstall 1994). GPCRs exist for many biologically active molecules such as amines (dopamine, noradrenaline, serotonin, histamine), amino acid transmitters (glutamate, GABA), peptides (opioids, tachykinins, neurotensin, somatostatin, cholecystokinin, gut-brain peptides such as GLP-1 and VIP, and most endocrine-releasing factors), and lipid-derived products (lysophosphatidic acid, sphingosine-1-phosphate, eicosinoids). GPCRs thus mediate a large variety of physiological events throughout the body, from chemosensory recognition (vision, olfaction, taste) to endocrine regulation to complex behavioral events. Indeed, over 360 nonsensory GPCRs, which are activated by about 200 endogenous substances, have been characterized, and over 160 orphan GPCRs remain whose natural ligands are still unknown (Wise et al. 2004). Of these nonsensory GPCRs, over 90% are expressed in the brain (Vassilatis et al. 2003).

In the CNS, GPCRs function primarily, but not exclusively, as mediators of slow neuromodulators rather than fast neurotransmitters, and their role is critical to normal brain function. Under- or overactivity of many individual GPCR systems in the brain may contribute to pathological conditions, ranging from hypodopaminergic movement disorders to mania and depression. Thus these receptors are primary or downstream targets for a variety of useful therapeutic agents and continue to be the focus of intense pharmaceutical development (Wise et al. 2004).

GPCR Signaling

In the absence of the appropriate activating ligand or agonist, both receptors and G proteins are generally inactive. GPCRs respond to the presence of their activating ligands or agonists by activating coupled G proteins. Over a range of agonist concentrations, G protein activation is proportionate with receptor binding by the activating ligand.

Each receptor subtype can couple to and activate only certain G protein types, each leading to distinct downstream signals. G proteins consist of three associated protein subunits, called α, β, and γ (Hamm & Gilchrist 1996, Watson & Arkinstall 1994). G proteins are classified based on their α-subunits, and there are 15 known α-subunits that have been categorized into four subfamilies (G_s, G_i, G_q, and G_{12}) based on sequence and functional similarities. There are also five β and fourteen γ proteins. The α-subunit contains the guanine nucleotide binding site, whereas β and γ form a tightly associated $\beta\gamma$-complex. When inactive, the α-subunit is bound to GDP and to $\beta\gamma$-complex to form a trimeric protein complex. Agonist binding to the cell-surface GPCR activates the receptor, which then serves to both facilitate GDP release from and stimulate GTP binding to the α-subunit of coupled G proteins; that is, receptors are guanine nucleotide exchange factors for heterotrimeric G proteins. This GTP binding activates the α-subunit, leading to its dissociation from the $\beta\gamma$-complex. Both α·GTP and $\beta\gamma$ can then bind to and activate intracellular effectors, such as second messenger–generating enzymes as well as specific ion channels (Dickey & Birnbaumer 1993, Hall et al. 1999, Wickman & Clapham 1995). For example, the activated G_s-α proteins stimulate adenylyl cyclases, activated G_i-α proteins inhibit adenylyl cyclases, activated G_q-α proteins turn on phospholipase C-β, and activated G_{12}-α proteins stimulate guanine nucleotide exchange factors for the small GTP-binding protein Rho. The freed $\beta\gamma$-subunits can activate or inhibit various adenylyl cyclases and activate phospholipase C-β and inward rectifying potassium channels (GIRK), among other effectors (Dickey & Birnbaumer 1993, Hall et al. 1999). Upon GTP hydrolysis, the GDP-bound α-subunit and the $\beta\gamma$-subunits reassociate into the inactive G protein and cease activating the effector enzymes. GTP hydrolysis may occur through the intrinsic GTPase activity of the α-subunit or may be enhanced by the action of specific GTPase-activating proteins of the Regulators of G protein Signaling (RGS) family (Berman & Gilman 1998, Dohlman & Thorner 1997, Neubig & Siderovski 2002) or by effectors themselves. Receptors vary in their specificity for activating or coupling to distinct G protein types, and thus activating downstream signaling pathways, with some receptor types activating only a single class of G protein to generate one class of intracellular signal, whereas other receptors more promiscuously couple to many G protein classes to generate multiple intracellular signals. Further, GPCRs may form homo- or heterodimers that could result in a complex variety of signaling events (Angers et al. 2002).

Mechanisms of GPCR Desensitization

One important feature of G protein signaling systems is that they are not constant but exhibit a memory of prior activation or signaling tone (Hausdorff et al. 1990). Thus, high activation of a receptor leads to a reduced ability to be stimulated in the future (desensitization), whereas low activation leads to an increased ability to be stimulated (sensitization). A given dose of agonist or drug thus may give distinctly different responses depending on the prior activation state of the system. This is an important regulatory feature that prevents overstimulation and allows for the linear response range to vary near the ambient stimulation level; in the visual system, such adaptation allows the G protein–coupled "light receptor" rhodopsin to adjust to both dark and light within moments.

GPCRs respond to activating ligands in a dose-dependent manner so that the concentration of agonist is the primary control point for signaling downstream of any given receptor. Receptors also differ in their basal or constitutive (that is, agonist-independent) activity, and in the extent of stimulation that a maximal dose of agonist can achieve. The ability of receptors to signal is regulated at the level of the receptor itself in two main ways: by controlling the number of receptors present on the cell surface and by regulating the signaling efficiency of receptors that are on the cell surface. Receptors are not static but are in equilibrium between cell-surface and endosomal pools and between synthesis and degradation. Receptor activation often leads to the removal of receptors from the cell surface by internalization, and less often, to recruitment of new receptors to the cell surface. Internalized receptors can be recycled to the cell surface (resensitization) for further duty or targeted for degradation in lysosomes (downregulation). Prolonged stimulation generally leads to a profound receptor loss from the cell surface (Bohm et al. 1997).

One major mechanism controlling GPCR responsiveness is the activation-dependent regulation of receptors, also called homologous desensitization (Claing et al. 2002, Ferguson et al. 1998, Hausdorff et al. 1990, Lefkowitz 1998, Perry & Lefkowitz 2002, Sterne-Marr & Benovic 1995). This is discussed in detail below. Other mechanisms also contribute to intrinsic regulation of GPCR signaling (Bohm et al. 1997, Hamm & Gilchrist 1996, Watson & Arkinstall 1994). These include receptor activation-independent regulation of receptors, or heterologous desensitization, as well as mechanisms that act after the receptors themselves, through regulating the G proteins directly or by altering the signaling efficiency of downstream effectors. One common mechanism for heterologous desensitization is the feedback regulation of receptors by the second-messenger-regulated kinases they activate. For example, β-adrenergic receptors use Gs to activate adenylyl cyclase to synthesize cAMP, which activates protein kinase A (PKA). PKA can (among myriad other things) then phosphorylate the β-adrenergic receptors themselves, even those particular receptor proteins that were not activated by the current stimulation. PKA activated by stimulation of totally distinct receptor types can similarly phosphorylate and alter the responsiveness of β-adrenergic receptors. These PKA-phosphorylated receptors are less able to mount a response to a subsequent application of their own activating agonist. Similar regulation of various types of receptors occurs for GPCR-activated PKA, protein kinase C (PKC), mitogen-activated protein (MAP) kinases, and many other kinases. Along the same lines, second-messenger-activated kinase can also phosphorylate and regulate G protein effectors, such as adenylyl cyclase, phospholipase C, and others, also contributing to the cell's responsiveness to subsequent or concurrent activation (Hamm & Gilchrist 1996, Watson & Arkinstall 1994).

A distinct family of accessory proteins, the RGS proteins, act as GTPase-activating proteins (GAPs) for heterotrimeric G proteins (Berman & Gilman 1998, Dohlman & Thorner 1997, Neubig & Siderovski 2002). Thus, they promote inactivation of GTP-bound G protein α-subunits. The family of these proteins includes at least 25 members, all of which contain a characteristic RGS-homology domain consisting of about 130 amino acid residues. The physiological significance of RGSs in regulating GPCR signaling is still poorly characterized, but recent observations in knockout mice demonstrate the importance of this regulation for at least some brain functions (Rahman et al. 2003, Zachariou et al. 2003). Some G protein effectors also act partly as GAPs to promote G protein inactivation (Hall et al. 1999). Inasmuch as these RGS proteins and other GAPs can also be regulated by receptor activation, they will also contribute to altered signaling in response to prior signaling, and in any case will help to shape the basal responsiveness of the system by their mere presence.

GRKs and Arrestins in Homologous Desensitization

The activated state of GPCRs serves not only as an activator of G proteins, but also as the substrate for protein phosphorylation by a family of protein kinases called GPCR kinases (GRKs). GRKs can discriminate between the inactive and agonist-activated states of the receptor, in part because they are catalytically activated by stimulated receptors. Thus, activated receptor regulation by GRKs results in homologous desensitization (Figure 1) (Claing et al. 2002, Hausdorff et al. 1990, Lefkowitz 1998, Perry & Lefkowitz 2002, Sterne-Marr & Benovic 1995). There are seven known GRK subtypes, which are classified in three subfamilies (GRK1/7, GRK2/3, GRK4/5/6) based on sequence and functional similarity (Benovic et al. 1987, Chen et al. 1999a, Pitcher et al. 1998a, Premont et al. 1995, Willets et al. 2003). One of these families is primarily visual (GRK1/7), whereas one other kinase is expressed primarily in testes (GRK4). Thus four GRK subtypes (GRK2, GRK3, GRK5, GRK6) must account for regulation of most of the GPCRs found throughout the body (Gainetdinov et al. 2000, Pitcher et al. 1998a, Premont et al. 1995). All GRKs share a domain structure of an amino terminal RGS-like domain, a central protein kinase domain, and a variable carboxyl terminal. In the GRK2 subfamily, the RGS-like domain binds to G_q α-subunits but does not facilitate GTP

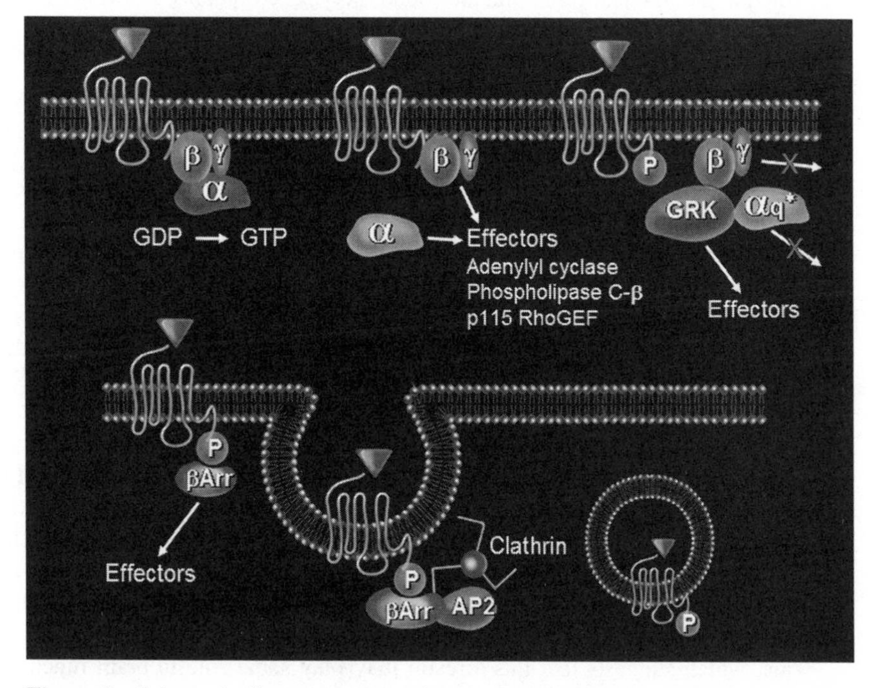

Figure 1 Schematic diagram representing key steps in GPCR signaling and homologous desensitization. See text for details.

hydrolysis (Pitcher et al. 1998a). Thus GRK2 acts not as a GAP but as a signaling dampener by preventing activated G_q α-subunit binding with other effectors, and perhaps by acting as a G_q effector itself. Other GRK subfamilies, though sharing a similar RGS-like domain, do not appear to bind G_q-α-subunits or other G proteins, and the function of this domain remains ill-defined (Pitcher et al. 1998a; Premont et al. 1994, 1996, 1999). In addition, the amino terminal domain has been implicated in recognizing activated receptors, although the recently solved structure of GRK2 tends to discount this (Lodowski et al. 2003). The GRK protein kinase catalytic domains are most similar to the PKA/PKC family (Hanks et al. 1988). The variable carboxyl terminal domains function in proper juxtamembrane localization of the GRKs (Pitcher et al. 1998a; Premont et al. 1995, 1999). In the GRK1 subfamily, the extreme carboxyl terminal is modified by prenylation, proteolysis, and carboxymethylation. GRK1 is farnesylated, whereas GRK7 is geranygeranylated. In GRK1, the farnesyl group is critical for light-regulated membrane association: An unprenylated GRK1 fails to associate with membranes, whereas a geranylgeranylated GRK1 is constitutively membrane-associated. In the GRK2 subfamily, the carboxyl terminal domain is extended and contains a pleckstrin homology (PH) domain. The GRK2 PH domain binds to both PIP_2 and G protein $\beta\gamma$-subunits. The $\beta\gamma$-subunits, released from receptor-activated G proteins, and PIP_2 cooperate to activate the kinase. In the GRK4 subfamily, two types of carboxyl terminal membrane-association motifs are found, polybasic domains allowing PIP_2 binding, and in a subset (GRK4 and the GRK6A splice variant), palmitoylated cysteine residues (Stoffel et al. 1994, 1998). Importantly, GRK activity is a highly regulated process and may be determined not only by expression level and intrinsic activity, but also by subcellular compartmentalization of the kinase (Penn et al. 2000).

Once phosphorylated by a GRK, the activated receptor is bound by a member of another protein family, the arrestins (Figure 1). Arrestins recognize both GRK phosphorylation sites on the receptor and the active conformation of the receptor, so that both together drive robust arrestin association (Luttrell & Lefkowitz 2002, Perry & Lefkowitz 2002). Arrestins interdict further G protein activation despite the continued activation of the receptor by agonist by preventing the receptor from exchanging GTP for GDP on the G protein α-subunit. Arrestins consist of a bi-lobed predominantly β-sheet structure, with a large phosphoprotein binding pocket. The arrestin superfamily in vertebrates includes visual arrestins and βarrestins. Visual arrestins that play an important role in the modulation of phototransduction are expressed almost exclusively in the retina and represented by two members: rod arrestin (S-antigen or arrestin 1) and cone arrestin (CAR, X-arrestin, or arrestin 4) (Chen et al. 1999b). Intriguingly, rod arrestin was found also in pineal gland and in small populations of neurons in the brain, particularly in habenular commissura, amygdala, ventral tegmental area, and superior colliculus, which suggests that this arrestin may play some role in brain functions as well (Sunayashiki-Kusuzaki et al. 1997). However, the two nonvisual arrestins (βarrestins), βarrestin-1 (arrestin 2) and βarrestin-2 (arrestin 3), which are highly expressed all over the body, must account for regulation of the vast majority of GPCRs.

In addition to this role as a receptor desensitization mechanism, the GRK-arrestin system also serves to promote the internalization of inactivated receptors and the subsequent recycling of resensitized receptors back to the cell surface (Ferguson et al. 1996, 1998). GRKs promote receptor internalization primarily by virtue of helping recruit arrestins to the activated receptors. The arrestins themselves bind to the clathrin adaptor protein AP2 and to clathrin itself, which facilitates the entry of desensitized receptors into clathrin-coated pits for subsequent internalization (Goodman et al. 1996, Laporte et al. 1999).

Furthermore, GRKs and arrestins appear to play direct signaling roles (Hall et al. 1999, Luttrell & Lefkowitz 2002). That is, along with the G proteins themselves, the GRKs and arrestins share the ability to recognize and bind to the activated state of the receptor. By virtue of binding to additional cellular proteins, GRKs and arrestins themselves serve as signal transducers by bringing specific signaling molecules into proximity of the activated receptor and the cell membrane and/or by altering their activity. Thus, GRKs have been reported to bind to GIT proteins and PI3-kinases and bring these to the receptor, where they are involved in regulating receptor trafficking and in promulgating further receptor-dependent signals (Hall et al. 1999). Arrestins serve as adaptors to ferry a wide variety of signaling proteins to activated receptors, including c-Src, entire MAP kinase cascades, Mdm2, ARNO, NSF, and others (Gurevich & Gurevich 2003, Luttrell & Lefkowitz 2002, Shenoy & Lefkowitz 2003). The relative contribution of these noncanonical signaling pathways as compared to direct G protein signaling in the CNS is mostly unknown but is an area of active investigation.

Specificity of GPCR Desensitization by GRKs and Arrestins

One major unanswered question regarding the physiological regulation of GPCRs is understanding which GRK(s) and arrestin(s) regulate any given receptor subtype. Studies over the past decade have defined the ability of certain GRKs and arrestins to phosphorylate and desensitize several GPCRs in model systems, but most receptors remain totally uncharacterized.

One approach to assessing the functional specificity of GRKs and arrestins in vivo is to ablate individual GRK or arrestin genes and determine whether this loss of function alters the regulatory properties of GPCRs. Over the past several years, we and others have developed mouse lines bearing deletions of each of the GRK and arrestin genes (Table 1 and references within), which we have used previously to characterize the GPCR regulation machinery in the heart (Rockman et al. 2002). In this way, we now can focus on other GPCR-mediated physiological systems of interest and assess whether any individual GRK or βarrestin (or combination of GRKs and βarrestins) plays an important role in regulating GPCR-mediated functions there.

TABLE 1 Expression pattern of GRKs and arrestins and major phenotypes in mutants lacking these proteins

GRK or arrestin	Expression	Knockout phenotype
GRK1 (Rhodopsin kinase)	Retinal rods and cones	Oguchi Stationary Night Blindness (human) (Yamamoto et al. 1997). Light-dependent retinal degeneration (mice) (Chen et al. 1999a).
GRK2 (β-adrenergic receptor kinase, βARK; βARK1)	Ubiquitous; brain	Embryonic lethal, thin myocardium syndrome in embryos (Jaber et al. 1996), and enhanced basal and adrenergic responses in cardiac function in adult heterozygotes (Rockman et al. 1998).
GRK3 (β-adrenergic receptor kinase 2, βARK2)	Ubiquitous (in the brain lower than GRK2)	Lack olfactory receptor desensitization (Peppel et al. 1997). Altered M2 muscarinic airway regulation (Walker et al. 1999). Reduced tolerance to opioid fentanyl, but not morphine (Terman et al. 2004).
GRK4 (IT11)	Testes; brain (low)	Normal fertility and sperm function (R.T. Premont, unpublished). No obvious phenotype.
GRK5	Ubiquitous; brain	Altered central (Gainetdinov et al. 1999a) and lung (Walker et al. 2004) M2 muscarinic receptor regulation, with normal heart M2 receptor regulation (Walker et al. 2004).
GRK6	Ubiquitous; brain	Altered central dopamine receptor regulation (Gainetdinov et al. 2003a). Deficient lymphocyte chemotaxis (Fong et al. 2002). Increased neutrophil chemotaxis (Kaavelars et al. 2003, Vroon et al. 2004).
GRK7 (Iodopsin kinase)	Retinal cones	⟨Gene is not present in mice, but is present in humans⟩
Rod arrestin (S-antigen or arrestin 1)	Retinal rods	Oguchi Stationary Night Blindness (human) (Yamamoto et al. 1997). Light-dependent retinal degeneration (mice) (Chen et al. 1999b).
βarrestin-1 (arrestin 2)	Ubiquitous; brain	Altered cardiac responses to beta-adrenergic stimulation (Conner et al. 1997).

(Continued)

TABLE 1 *(Continued)*

GRK or arrestin	Expression	Knockout phenotype
βarrestin-2 (arrestin 3)	Ubiquitous (in the brain lower than βarrestin-1)	Enhanced morphine antinociception (Bohn et al. 1999, 2002) and reward (Bohn et al. 2003) and disrupted morphine tolerance (Bohn et al. 2000, 2002); deficient lymphocyte chemotaxis (Fong et al. 2002); impaired asthmatic response (Walker et al. 2003).
Cone arrestin (X-arrestin or arrestin 4)	Retinal cones; pineal gland	Not reported

These studies have demonstrated that loss of individual GRKs or βarrestins under basal conditions mostly produces relatively minor phenotypes, but in the presence of GPCR activators or other forms of stress, the importance of the GRK-arrestin regulation for GPCR is revealed (Table 1). A crucial role of visual rod arrestin and GRK1 in the termination of the light response in photoreceptors has been demonstrated convincingly in mice lacking these regulatory elements (Chen et al. 1999a,b). Among nonvisual GRKs, only GRK2 has proven to be embryonic lethal as a single gene deletion, owing to a developmental defect in the heart (Jaber et al. 1996). Loss of any other GRK or arrestin gene leads to a mouse that appears outwardly normal. Upon addition of exogenous GPCR agonists, however, abnormally supersensitive responses are present in some GRK knockouts, but not others, as measured by accentuated physiological responses. These supersensitive responses indicate that the GRK of interest is important for desensitizing that receptor type, at least in the tissue or system being examined.

Given the large number of GPCR types present throughout the body, only a fraction of GPCRs and receptor-regulated systems have been examined to date for regulation by GRKs and arrestins in any way. The relatively recent derivation of the last of the GRK knockout mouse lines means that the potential involvement of each GRK or arrestin has been examined in detail in vivo in very, very few systems. Our hope is that studies in the near future will begin to map the functional specificity of GRKs and arrestins in regulating many distinct, pharmacologically relevant GPCR types. One clear result of such studies is the realization that there exists a continuum of receptor regulation by GRKs and arrestins, such that some receptors appear to require one particular GRK or arrestin exclusively for their regulation, whereas other receptors are regulated in part by several GRKs or arrestins. Additionally, a given receptor need not be regulated by the same GRK or arrestin in all tissues.

Here, we summarize our experience in investigating the physiological roles and receptor specificity of neuronal GRKs and arrestins using knockout mice as a model. Particularly, results of initial screening for the role of each of these molecules in classic physiological responses mediated by dopamine and μ-opioid GPCRs are presented, since receptors for dopamine (DA) and opiates are among the most clinically important neuronal GPCRs.

Dopamine and μ-Opioid Receptor–Mediated Behaviors as In Vivo Model Systems

Five distinct but related receptor proteins are activated by DA (dopamine D1–D5 receptors) and are classified into two distinct groups based on sequence and functional similarities: D1-like (D1 and D5) and D2-like (D2, D3, D4). Several signaling events can be regulated by DA receptors, including adenylyl cyclase and phospholipase C activity and the opening of various ion channels. The D1-like receptors couple to, or activate, the G_s family of G proteins (primarily G_{olf}) to increase cAMP production by adenylyl cyclase (mainly AC5) and are thought to be found only postsynaptically on dopaminergic target cells (in the striatum, primarily on GABA-ergic medium spiny neurons). The D2-like receptors couple primarily to the G_i family of G proteins to activate K^+ channels and inhibit adenylyl cyclase, and are present both presynaptically on DA-producing cells and postsynaptically on DA target cells. In the brain, the various receptor subtypes display specific distributions with highest density detected in the nigrostriatal and mesolimbic areas, such as caudate-putamen (striatum), nucleus accumbens, amygdala, and frontal cortex (Gardner et al. 2001, Grandy & Civelli 1992, Missale et al. 1998, Picetti et al. 1997, Schwartz et al. 1993, Seeman & Van Tol 1994, Sibley et al. 1999). DA plays a critical role in the control of movement, emotion, affect, and reward and is believed to be involved in brain disorders, such as Parkinson's disease, schizophrenia, addiction, Tourette's syndrome, attention deficit hyperactivity disorder (ADHD), and Huntington's disease (Carlsson 2001, Hornykiewicz 1966).

Cocaine and amphetamine are known to induce psychomotor activation by interfering primarily with the function of the dopamine transporter and thereby leading to elevated levels of DA in the extracellular space (Gainetdinov & Caron 2003b, Jones et al. 1998, Wise & Bozarth 1987). In rodents, the elevated DA levels in the major dopaminergic regions, such as striatum, are manifested behaviorally as locomotor hyperactivity. Sensitivity of DA receptors to endogenous and exogenous ligands is known to be an important modulator of DA-related functions. Supersensitivity of DA signaling has been suggested in human disorders such as schizophrenia (Jenner & Marsden 1987, Pandey et al. 1977), Tourette's syndrome (Singer 1994), and addiction (Hyman & Malenka 2001, Nestler & Aghajanian 1997, Nestler 2001, Robinson & Berridge 1993) and can be easily demonstrated in experimental animals chronically treated with psychostimulants (Laakso et al. 2002, Nestler 2001). This abnormal behavioral sensitization induced by chronic psychostimulants is associated with long-term changes in DA receptor responsiveness as evidenced by exaggerated locomotor responses not only to psychostimulants, but also to direct D1/D2 DA receptor agonists in various tests, such as the characteristic "climbing" response to apomorphine (Wang et al. 1997, Wilcox et al. 1980). It is hypothesized that, among other mechanisms, long-term adaptations in GPCR desensitization can contribute to this phenomenon (Nestler & Aghajanian 1997, Nestler 2001), but the role of specific components of direct DA receptor regulatory mechanisms largely remains unknown.

Similarly, three distinct receptors (called μ, k, and δ opiate receptors) bind opioid peptides and opiate drugs (Kieffer 1999, Snyder & Pasternak 2003). Each couples primarily to the G_i family of G proteins to activate K^+ channels and inhibit adenylyl cyclase and is found in many brain areas such as caudate-putamen, periaqueductal gray, thalamic nuclei, and amygdala. Further, each receptor is distributed differently throughout the CNS. Some areas express all three subtypes (striatum and dorsal horn of the spinal cord), whereas, for example, in the thalamic nuclei only μ-opioid receptor (μOR) is found. Among the opioid receptors, μOR is primarily involved in the antinociceptive activity, but it also has the highest abuse liability (DiChiara & North 1992). Several recent reports in genetically altered animals convincingly demonstrated the predominant role of μOR in the antinociceptive and rewarding properties of morphine. Lack of morphine analgesia, as well as disrupted morphine-induced locomotor activity, hypothermia, respiratory suppression, gastrointestinal disturbances, tolerance to chronic treatment, dependence, and withdrawal, was observed in mice lacking the μOR (Kieffer 1999, Loh et al. 1998, Matthes et al. 1996, Uhl et al. 1999). Therefore, the desensitization of the μOR may present a critical point of regulation of the responsiveness to morphine that could dictate the extent of morphine effects on all of the physiological parameters, including the ones associated with its chronic use such as tolerance and addiction.

To understand the role of GRKs and βarrestins in DA receptor regulation and their contribution to aberrant neuroplasticity induced by chronic drugs of abuse, we have initially examined mice bearing inactivated GRK and arrestin genes for alterations in locomotor responses to cocaine, amphetamine, and/or nonselective dopamine agonist apomorphine. In a similar preliminary screen to assess μOR responsiveness, the effect of morphine on centrally mediated analgesia was assessed in all these mutants using the classic hot-plate antinociception test as described (Bohn et al. 1999).

GRKS AND βARRESTINS IN NEURONAL FUNCTIONS

Neuronal GRKs

GRK2 GRK2 was the first nonvisual GRK to be discovered, and it has been extensively characterized (Benovic et al. 1987, Pitcher et al. 1998a). The widespread expression of this kinase in many tissues in the body (Arriza et al. 1992) suggests that multiple GPCRs are physiological targets of this kinase. It is not surprising therefore that many GPCRs can be phosphorylated by GRK2 in in vitro preparations (Pitcher et al. 1998a, Premont et al. 1995). In the rat brain, GRK2 is expressed in most neuronal populations, both in association with postsynaptic densities and presynaptically within axon terminals. Particularly, GRK2 immunoreactivity is found within cell bodies of neurons, as well as within structures that correspond to dendritic shafts, dendritic spines, and presynaptic axon terminals in most brain regions including those critical for locomotion and antinociception such as striatum, cortex, periaqueductal gray, and thalamus (Arriza et al. 1992). As with other GRKs, the pattern of expression of GRK2 does not correlate with that of any known single neurotransmitter system (Arriza et al. 1992, Erdtmann-Vourliotis et al. 2001). In a recent in situ hybridization study, GRK2 mRNA was found to be distributed in a nearly uniform manner through all cortical layers, the islands of Calleja, the claustrum, the dorsal endopiriform nucleus, the limbic diagonal band, the lateral septal nuclei, the bed nucleus of the stria terminalis, several hypothalamic and thalamic nuclei, hippocampus, the substantia nigra compacta, the ventral tegmental area, the pons, the reticulotegmental nucleus of the pons and the central gray, the cerebellar cortex, the locus coeruleus, and other regions. A significantly lower signal was detected in caudate-putamen (Erdtmann-Vourliotis et al. 2001). The expression levels of GRK2 display a marked increase during the second postpartum week in rat pups, reaching levels comparable to that in adult brain (Penela et al. 2000). GRK2 expression has been found to be altered in some disorders and can be modulated by pharmacological treatments. For example, a recent study reported that major depression may be associated with upregulation of GRK2 in the prefrontal cortex, and antidepressants appear to induce downregulation of the GRK2 protein (Grange-Midroit et al. 2003). Interestingly, acute but not chronic treatment with the norepinephrine transporter selective antidepressant desipramine (but not selective serotonin transporter inhibitor fluoxetine) increased membrane-associated GRK2-like immunoreactivity in the rat frontal cortex, which suggests that the in vivo activation of adrenergic receptors is associated with time-dependent modulation of GRK2 (Miralles et al. 2002). Responsiveness of hippocampal neurons to cannabinoid-mediated presynaptic inhibition of neurotransmission and luteinizing hormone secretion by pituitary gonadotropes is sensitive to GRK2 overexpression (Neill et al. 1999). One interesting aspect of GRK2 physiology that may have potential impact on brain functions is related to its ability to phosphorylate tubulin, thus potentially mediating GPCR effects on the neuronal cytoskeleton (Pitcher et al. 1998b). Furthermore, it has been reported that α and β isoforms of synucleins, proteins highly expressed in the brain and linked to the development of Parkinson's and Alzheimer's diseases, can be potently phosphorylated by GRK2 and GRK5 (Pronin et al. 2000). One intriguing observation indicates that the neuronal calcium sensor-1 (NCS-1) can mediate desensitization of D2 DA receptors via interaction with GRK2 (Kabbani et al. 2002). Furthermore, NCS-1 was found to be elevated in the dorsolateral prefrontal cortex in schizophrenia and bipolar disorder patients, which suggests that abnormalities in NCS-1-dependent desensitization of DA receptor signaling may contribute to these disorders (Koh et al. 2003).

In in vitro cellular systems, overexpressed GRK2 was shown to enhance phosphorylation and regulation of dopamine D1, D2, and D3 receptors (Ito et al. 1999,

Iwata et al. 1999, Kabbani et al. 2002, Kim et al. 2001, Lamey et al. 2002, Tiberi et al. 1996) and opioid receptors (Whistler & von Zastrow 1998, Zhang et al. 1998). Given the fact that this kinase is expressed in virtually all brain regions it is reasonable to expect that this regulation can occur in vivo as well. Deletion of the GRK2 gene in mice results in embryonic lethality due to cardiac hypoplasia (Jaber et al. 1996), so the role of this GRK on adult mouse behaviors cannot be examined fully. At the same time, mice heterozygous for this mutation are viable and do not display any obvious behavioral phenotype. These mice were used to characterize responses to psychostimulants and direct DA agonist apomorphine (Figure 2). As presented in Figure 2, doses of 10, 15, 25, and 30 mg/kg of cocaine induced comparable locomotor activation in wild-type and heterozygous mice, but at 20 mg/kg, cocaine induced significantly enhanced responses. Furthermore, no alterations were found when locomotor-stimulating effects to the indirect DA agonist amphetamine or climbing responses to the nonselective DA agonist apomorphine were analyzed. Thus, the impact of partial deletion of GRK2 on dopamine-mediated responses seems to be minimal. Nevertheless, the inavailability of GRK2 "null" mice does not allow us to exclude the involvement of this kinase in DA receptor regulation. Further studies would be necessary to examine this possibility in more detail.

Interestingly, it has been observed that GRK2 levels were increased in the locus coeruleus of rats chronically treated with morphine, suggesting a role of this kinase in μOR regulation (Terwilliger et al. 1994). Furthermore, both acute and chronic treatment with opioid drugs as well as opioid withdrawal induce an increase in GRK2 levels in the rat cerebral cortex in experimental animals, and membrane-associated GRK2 levels are increased in brains of human opioid addicts (Ozaita et al. 1998). GRK2 immunoreactivity was increased in the cortex of rats treated with opioids and rendered tolerant to the antinociceptive effect of opioids (Hurle 2001). Chronic treatment with the opioid antagonist naltrexone also resulted in significant upregulation of μORs, as well as several GRKs, including GRK2 (Diaz et al. 2002). In addition, it has been shown that GRK2 is highly expressed in nucleus raphe magnus GABAergic neurons projecting to spinal cord, where it appears to mediate desensitization of μORs (Li & Wang 2001). These and other (Fan et al. 2002) findings strongly suggest that GRK2 may contribute to the cellular processes underlying in vivo μOR desensitization and could play an important role in the development of opioid tolerance and withdrawal. In our preliminary investigations, however, we did not see any significant difference in acute morphine-induced analgesia in the hot-plate test between GRK2 heterozygous and control mice (L.M. Bohn, unpublished). Again, further studies are necessary to explore this possibility more fully.

GRK3 The GRK3 protein shares a high structural similarity to GRK2, with over 80% amino acid identity (Arriza et al. 1992). In the periphery, GRK3 is highly expressed in olfactory receptor neurons and dorsal root ganglion (DRG) neurons and has been suggested to mediate homologous desensitization of odorant receptors in olfactory receptor cells and α_2-adrenergic receptors in DRG neurons (Boekhoff et al. 1994, Diverse-Pierluissi et al. 1996). GRK3 mRNA and protein is widely

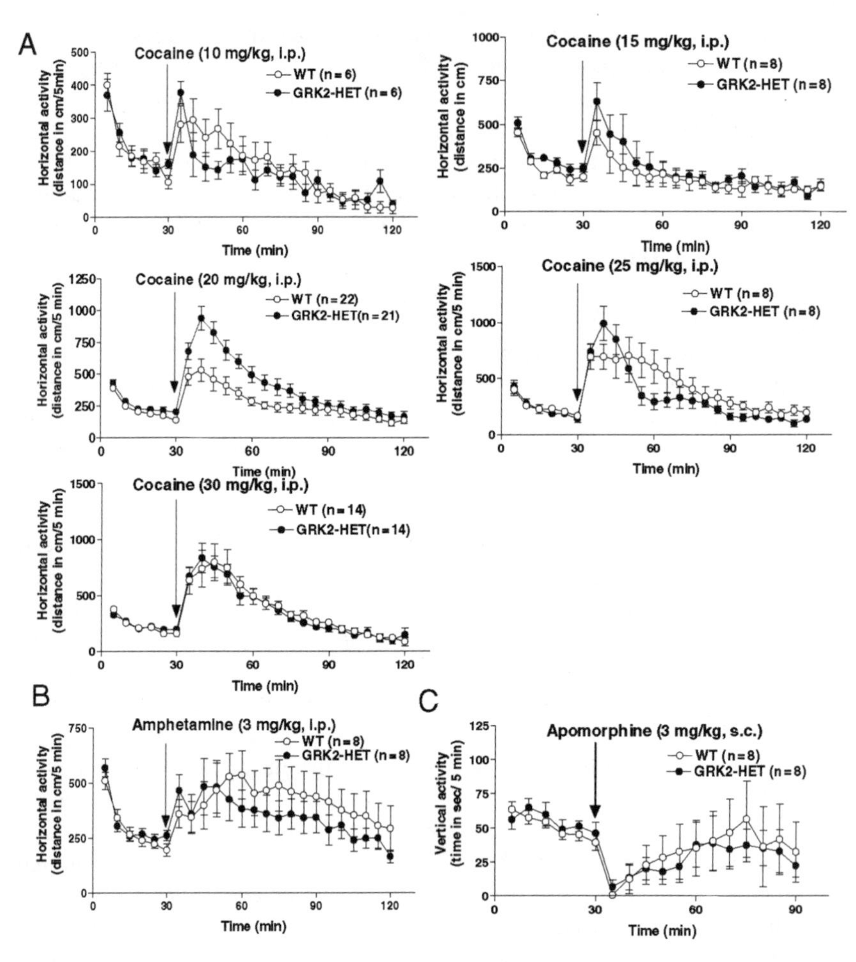

Figure 2 Locomotor responses of GRK2 heterozygous mice to (*A*) cocaine (10–30 mg/kg, i.p.), (*B*) amphetamine (3 mg/kg, i.p.) and (*C*) climbing responses to direct DA agonist apomorphine (3 mg/kg, s.c.). GRK2 heterozygous are significantly different from wild-type (WT) controls only in response to 20 mg/kg cocaine. GRK2 mutant mice (Jaber et al. 1996) were backcrossed 10 generations onto a C57/BL6 background (3–5 months old, both genders).

distributed in the rat brain, with a pattern of expression similar to GRK2, but at lower levels (Arriza et al. 1992, Erdtmann-Vourliotis et al. 2001). Interestingly, GRK3 mRNA was found at relatively low levels in the striatum, the paratenial thalamic nucleus, the bed nucleus of the stria terminalis, and in the periventricular hypothalamic nucleus. In contrast, GRK3 mRNA is expressed at levels higher than that of GRK2 in the islands of Calleja, the substantia nigra compacta, and in the locus coeruleus (Erdtmann-Vourliotis et al. 2001). In vitro studies in cell culture have demonstrated a role of GRK3 in the regulation of numerous neuronal GPCRs, including D1, D2, and D3 dopamine receptors (Kabbani et al. 2002, Kim et al. 2001, Tiberi et al. 1996) and μORs (Celver et al. 2001, Kovoor et al. 1998). A role for GRK3 in the adaptive changes in μOR activity in the brain has been also suggested, based on alteration in expression of this kinase following opiate agonists and antagonists (Diaz et al. 2002, Hurle 2001). However, Terwilliger et al. (1994) reported that chronic morphine, while producing upregulation of GRK2 and βarrestin-1, failed to modify GRK3 levels in the rat locus coeruleus, and in another investigation expression of GRK3 remained unchanged after chronic treatment with opiates (Hurle 2001).

In a genome-wide linkage survey, the region of chromosome 22q12 containing the GRK3 gene was identified as a susceptibility locus for bipolar disorder in humans. Furthermore, GRK3 expression in the frontal cortex was found to be induced by amphetamine in the rats. Finally, transmission disequilibrium analyses indicated that two 5′-UTR/promoter polymorphisms are associated with human bipolar disorder, leading to the hypothesis that a dysregulation in GRK3 expression may alter GPCR desensitization, and thereby predispose affected individuals to the development of this disorder. It has also been suggested that primary candidates for this dysregulation would be DA receptors because DA has long been suspected to play an important role in bipolar disorder (Barrett et al. 2003).

Loss of the GRK3 gene leads to a mouse that appears outwardly normal but has impaired olfactory receptor desensitization and altered M2 muscarinic airway regulation (Peppel et al. 1997, Walker et al. 1999). In a basal locomotor activity test, these mice were not different from wild-type controls (Figure 3). Furthermore, mice lacking GRK3 do not demonstrate enhanced locomotor or climbing responses to either cocaine or apomorphine. In fact, each of these drugs induce somewhat reduced responses in mutant mice. Thus, it seems unlikely that this kinase is directly involved in regulation of dopamine receptors in vivo, at least for their effects on motor behaviors in mice. Rather, this kinase may impact other populations of GPCRs negatively affecting dopamine-related behaviors, such as, for example, receptors for serotonin (Gainetdinov et al. 1999b). Similarly, in a hot-plate analgesia test, acute morphine and fentanyl induced similar analgesia in both wild-type and GRK3-KO mice, but tolerance to the antinociceptive and electrophysiological effects of fentanyl was reduced in mutants. However, analgesic tolerance to morphine was not affected, which suggests that whereas GRK3 may play a role in opioid receptor regulation in response to high efficacy opioids, it is not the case regarding low-efficacy agonist morphine (Terman et al. 2004).

GRK4 The highest level of expression of GRK4 in the body is found in testes (Gainetdinov et al. 2000, Premont et al. 1996). Only a limited expression of GRK4 was detected in the brain, particularly in cerebellar Purkinje cells (Sallese et al. 2000). It has been demonstrated that the metabotropic glutamate 1 (mGlu1) receptor in cerebellar Purkinje cells can be regulated by GRK4, via a mechanism different from that used by GRK2 (Iacovelli et al. 2003). GRK4 may also play

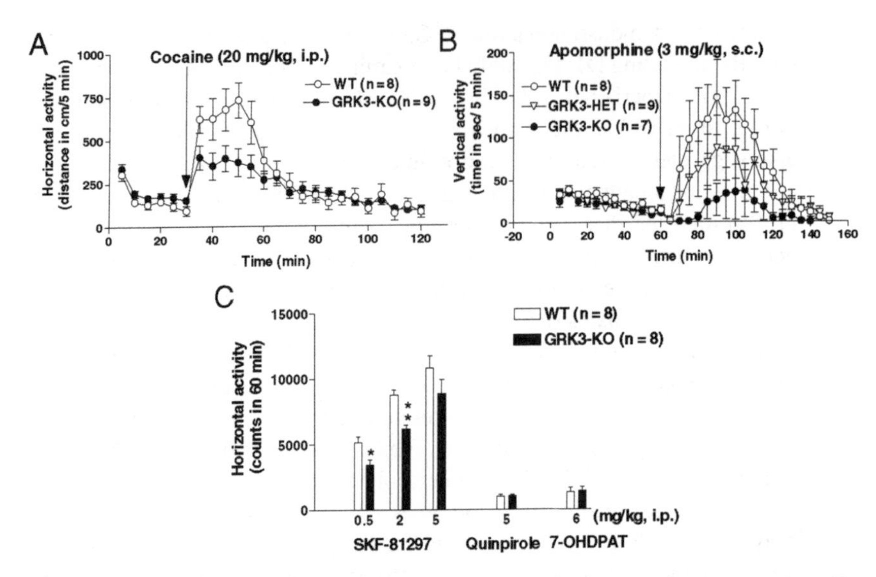

Figure 3 Responsiveness to cocaine (*A*) and direct DA agonists apomorphine (*B*), D1 DA agonist SKF-81,297 and D2/D3 agonists quinpirole and 7-OH-DPAT (*C*) in GRK3 mutant mice. *−p < 0.05 vs. respective WT controls. GRK3-KO mice are significantly less activated after cocaine, apomorphine, and SKF-81,297. GRK3 mutant mice (Peppel et al. 1997) were backcrossed 7 generations onto a C57/BL6 background (3–5 months old, both genders).

an important role in the agonist-promoted desensitization of GABA-B receptor in primary cerebellar granule cells (Perroy et al. 2003). These observations would suggest a role of GRK4 in motor coordination and learning (Sallese et al. 2000).

GRK4-KO mice showed no overt behavioral phenotype. No differences in basal level of locomotor activity or motor coordination in a rotorod test were found in these mutants. As might be expected from the expression pattern, locomotor responses to cocaine were not significantly changed in GRK4-KO mice (Figure 4). Furthermore, no difference in acute morphine-induced analgesia was observed in these mutants (L.M. Bohn, unpublished). Thus, the role of this kinase in the desensitization of brain DA receptors and μORs is unlikely. However, it is possible that GRK4 may be involved in desensitization of renal D1 DA receptors (Watanabe et al. 2002).

GRK5 GRK5 is the best-characterized member of the GRK4 subfamily of GRKs (Premont et al. 1995). The GRK5 mRNA is expressed widely in brain and in peripheral tissues, with highest expression evident in heart, lung, and placenta; however, the complete expression pattern of the GRK5 protein is still lacking. In the brain, GRK5 mRNA was found to be expressed moderately in several limbic regions such as the cingulate cortex, the septohippocampal nucleus, the anterior

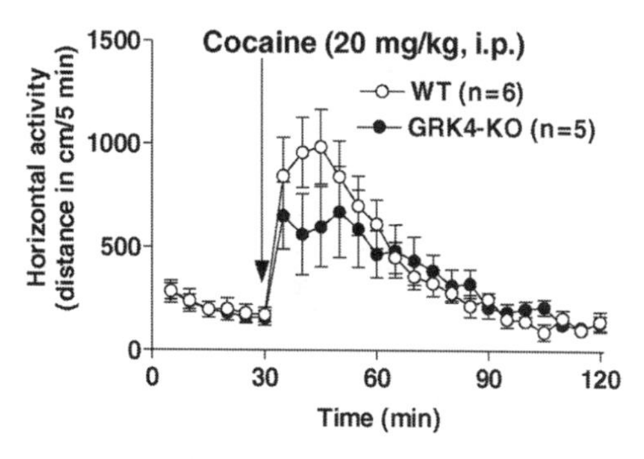

Figure 4 Locomotor responses of GRK4-KO mice to cocaine (20 mg/kg, i.p.) are not significantly changed. GRK4 mutant mice (generation of mice is described elsewhere; R.T. Premont & R.J. Lefkowitz, unpublished) were backcrossed 8 generations onto a C57/BL6 background (3 months old, both genders).

thalamic nuclei, the dentate gyrus of Ammon's horn, the medial habenula, the locus coeruleus, and the cerebellar cortex with the highest level of expression in the lateral septum (Erdtmann-Vourliotis et al. 2001).

In cellular model systems, GRK5 can phosphorylate several neuronal GPCRs, including β_2-adrenergic, M_2-muscarinic, secretin, angiotensin AT_1, and thyroid–stimulating hormone receptors (Kunapuli et al. 1994, Menard et al. 1996, Pitcher et al. 1998a, Premont et al. 1994). D1 and D2 DA receptors (Ito et al. 1999, Tiberi et al. 1996) and μOR (Koovor et al. 1998) can also be regulated by GRK5 in these model systems. Interestingly, investigators report that chronic, but not acute, treatment with cocaine resulted in upregulation of GRK5 mRNA in the septum (Erdtmann-Vourliotis et al. 2001). Furthermore, acute treatment with morphine, as well as spontaneous and naloxone-precipitated morphine withdrawal induced significant changes in GRK5 mRNA levels in several brain regions (Fan et al. 2002). One additional interesting aspect of GRK5 neurobiology arises from studies demonstrating the ability of different calcium sensor proteins and calmodulin to significantly modulate GRK5 activity (Iacovelli et al. 1999, Pronin et al. 1997). The calcium-dependent modulation of GRK5 may represent an important feedback mechanism to modulate homologous desensitization of these receptors in neuronal systems. Indeed, this may account for the curious observation that in transfected cell systems overexpressed GRK5 can phosphorylate and desensitize the angiotensin II AT1a receptor, but transgenic mice with overexpression of GRK5 in myocytes exhibit normal heart contractility to angiotensin II (Rockman et al. 1996).

In mice lacking GRK5, only a very modest phenotype, a slight decrease in basal body temperature, was found (Gainetdinov et al. 1999a). Behavioral analyses were performed after challenging animals with a number of agonists to seek out the specific receptors affected by the loss of GRK5. We found no differences in cocaine-induced locomotor responses and climbing responses following a high

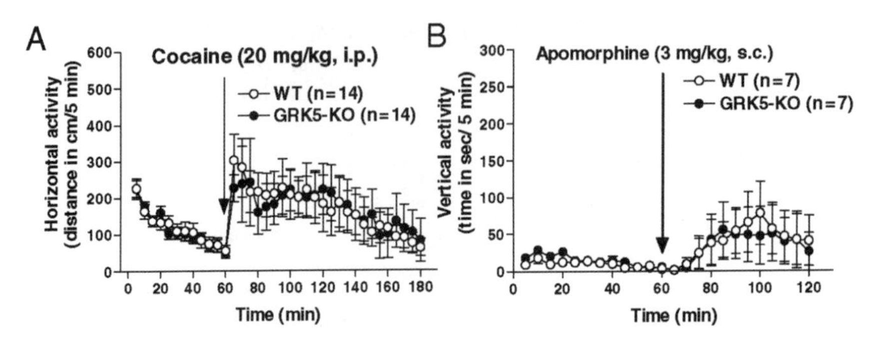

Figure 5 Responsiveness of GRK5-KO mice (Gainetdinov et al. 1999a) to cocaine (20 mg/kg, i.p.) and apomorphine (3 mg/kg, s.c.). Data on apomorphine responses are reproduced with permission from Gainetdinov et al. (1999a).

dose of apomorphine (Figure 5). Furthermore, acute morphine produced comparable analgesia in a hot-plate test in mutant and wild-type mice (L.M. Bohn, unpublished). Similarly, hypothermic responses to stimulation of serotonin 5-HT1A receptor by 8-OH-DPAT did not differ between the genotypes (Gainetdinov et al. 1999a). These data suggest that the relative responsiveness of DA receptors, affected by cocaine and apomorphine, as well as the relevant μORs and 5-HT1A subtype of serotonin receptors, seems to be unchanged by deletion of GRK5. At the same time, these mice demonstrated remarkable exaggeration of central muscarinic cholinergic responses (Gainetdinov et al. 1999a). The classic responses to muscarinic stimulation, such as hypothermia, tremor, salivation, and locomotor suppression were all enhanced and prolonged in mutant mice. Further, the antinociceptive effect of nonselective muscarinic agonist oxotremorine was also significantly potentiated. The lack of oxotremorine-mediated desensitization of muscarinic receptors in membranes from the brain of GRK5-KO mice has been directly demonstrated using the ^{35}[S]GTPγS binding assay. Although lack of selective agents precludes definitive clarification of the muscarinic receptor subtype involved, observations in transgenic animals strongly suggest that the vast majority of these behaviors are mediated by M2 muscarinic receptors (Gomeza et al. 1999). The M2 receptor is widely expressed not only in the brain but also in many peripheral tissues and organs including smooth muscle and the heart. Indeed, recent studies suggest that M2 receptors in airway are supersensitive in the absence of GRK5, whereas M2 receptors in the heart are regulated normally (Walker et al. 2004). This demonstrates that the same receptor, M2 muscarinic in this case, need not be regulated by the same GRK in all tissues. The muscarinic-receptor-mediated hypersalivation response displayed by the GRK5 knockout mice (Gainetdinov et al. 1999a) suggests that M3 receptors are also probable targets for regulation by GRK5 in vivo (Wess 2000). Supersensitivity of muscarinic receptors has been described in several brain disorders, including depression and posttraumatic stress disorder, as well as multiple chemical sensitivities (Janowsky et al. 1994, Overstreet et al. 1996, Sapolsky 1998). Moreover, rodent animal models of behavioral muscarinic supersensitivity have been developed to model these conditions (Overstreet et al. 1996). It would be of interest to explore the role of GRK5-mediated muscarinic receptor desensitization in these manifestations.

GRK6 GRK6 appears to be ubiquitously expressed (Benovic & Gomez 1993, Fehr et al. 1997, Gainetdinov et al. 2000). In the brain, GRK6 is expressed in many areas (Fehr et al. 1997) with the level and pattern of GRK6 mRNA expression somewhat similar to that of GRK2 mRNA. Interestingly, in the caudate putamen, GRK6 mRNA was highest of all GRKs (GRK2, GRK3, and GRK5), which suggests that GRK6 is the predominant receptor kinase in this brain area. Furthermore, GRK6 mRNA is also expressed in primary dopaminergic cell body areas, such as substantia nigra (Erdtmann-Vourliotis et al. 2001). In a recent immunohistochemical investigation (Gainetdinov et al. 2003a) a high expression of GRK6 protein was documented in the GABAergic medium spiny neurons as well as large cholinergic interneurons in the mouse striatum. These neurons represent the major striatal cell groups receiving dopaminergic input and are believed to be critically involved in brain disorders such as addiction, schizophrenia, and Huntington's disease.

GRK6 can regulate several neuronal GPCRs under in vitro conditions including D2 and D3 DA receptors (Gainetdinov et al. 2003a) and δ-opioid receptors (Willets & Kelly 2001). However, little is known about the role of this kinase in native brain tissue. Significant changes in the expression of GRK6 in rat brain were induced by chronic treatment with agonists and antagonists of μORs (Diaz et al. 2002, Hurle 2001), suggesting a role for this kinase in μOR regulation. But, in the hot-plate morphine analgesia test, mice lacking GRK6 were indistinguishable from wild-type littermates (L.M. Bohn, unpublished). Like all the other GRK mutants available, mice lacking GRK6 under basal conditions do not demonstrate any obvious behavioral phenotype. However, GRK6-deficient mice, unlike other GRK mutants, are remarkably supersensitive to the locomotor-stimulating effect of psychostimulants (Gainetdinov et al. 2003a), including cocaine, amphetamine (Figure 6), and endogenous "trace amine" β-phenylethylamine (β-PEA) (Janssen et al. 1999, Premont et al. 2001). In biochemical experiments, these mice demonstrated an enhanced coupling of striatal D2-like DA receptors to G proteins and increased affinity for D2 but not D1 DA receptors. Furthermore, augmented locomotor response to direct dopamine agonists were observed both in intact and in dopamine-depleted animals (Figure 6). These data show that postsynaptic D2-like DA receptors are physiological targets for GRK6. These remarkably altered responses to cocaine and other psychostimulants suggest that a GRK6-dependent regulatory mechanism may contribute to central dopaminergic supersensitivity in various pathological states, such as addiction. One intriguing observation in these mice is that a 50% reduction in GRK6 levels in heterozygote mice produces a phenotype nearly identical to the complete knockout of the gene. This raises the possibility

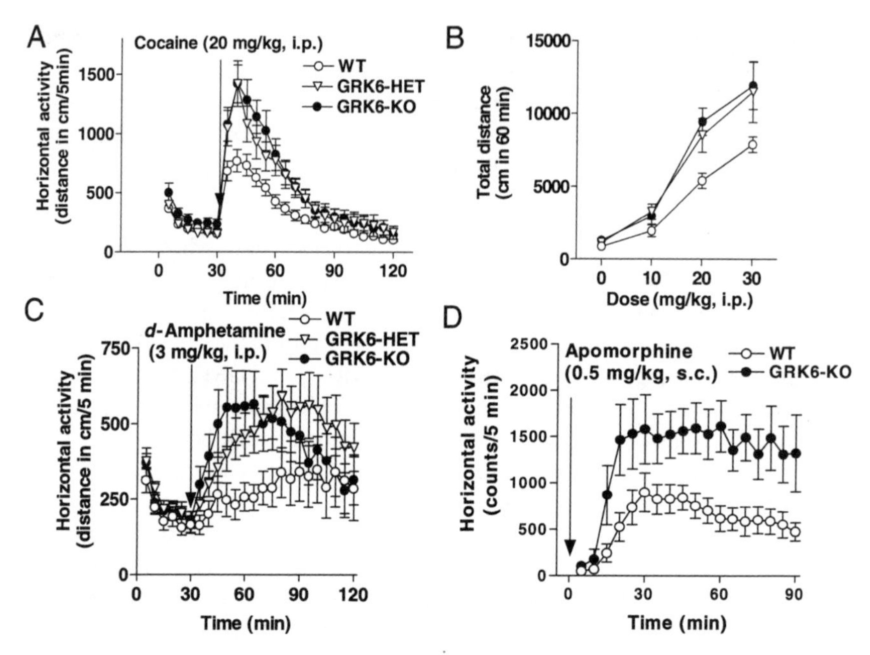

Figure 6 Enhanced locomotor effects of cocaine, amphetamine, and apomorphine in GRK6 heterozygous and knockout mice. (*A*) Dynamics of the effect of cocaine (20 mg/kg, i.p.) on the locomotion of mice. (*B*) Dose-response curve of the effect of cocaine (10–30 mg/kg, i.p.). Both GRK6 heterozygous and GRK6-KO mice are significantly different from WT controls in responses to cocaine. (*C*) Enhanced locomotor responses to amphetamine in GRK6 mutant mice. (*D*) Locomotor responses to direct DA receptor activation are enhanced in DA-depleted GRK6-KO mice. To deplete brain DA, animals were treated with a combination of reserpine (5 mg/kg, i.p.) and α-methyl-*p*-tyrosine (250 mg/kg, i.p). Data are reproduced with permission from Gainetdinov et al. (2003a).

that even subtle allelic variations in the human GRK6 gene or drug-induced alterations in GRK6 expression or activity might contribute to individual sensitivity to drugs of abuse affecting DA function. In addition, supersensitivity to DA agonist stimulation in the absence or reduction of GRK6 suggests that a pharmacological strategy targeted on GRK6 expression or activity may be beneficial in conditions when dopaminergic signaling is limited, such as Parkinson's disorder. Because supersensitivity of DA receptors has been implicated in the pathogenesis or adverse reactions associated with treatment of schizophrenia, Tourette's syndrome, and Parkinson's disease (Jenner & Marsden 1987, Pandey et al. 1977, Singer 1994), a role for GRK6-mediated DA receptor regulation in these conditions should be considered.

β-Arrestins

βARRESTIN-1 βarrestin-1 is expressed from birth in the majority of brain regions (Attramadal et al. 1992, Gurevich et al. 2002, Parruti et al. 1993, Penela et al. 2000). βarrestin-1 mRNA levels change during development, peaking on the fourteenth postnatal day and then somewhat decreasing, whereas protein levels continue to rise until adulthood (Gurevich et al. 2002). In adult rat brain, the highest expression of βarrestin-1 was detected in the cortex, hippocampus, striatum, and in the anterior, intralaminar, and midline nuclei of thalamus. In general, βarrestin-1 seems to be the major arrestin subtype in the rat brain. It has been estimated that in adult rat brain, the concentration of βarrestin-1 mRNA was two- to threefold higher than βarrestin-2 mRNA, whereas the ratio of βarrestin-1 to βarrestin-2 protein was much higher (10–20-fold) (Gurevich et al. 2002). Importantly, expression of βarrestins is particularly strong at postsynaptic densities but is also detectable at spines and nonsynaptic plasma membranes and intracellular organelles (Attramadal et al. 1992). In the spinal cord, immunoreactivity for βarrestin-1 was found in the motorneurons in lamina IX of the ventral horn and elongated cells in the dorsal nucleus of Clarke. Modest expression was detected in the neurons of laminae V and VII/VIII, and somewhat weaker immunoreactivity was observed in laminae III, IV, and X. Interestingly, in the spinal cord, both βarrestin-positive and -negative dendrites were observed, whereas axons and terminal boutons seem to be lacking βarrestins. Like in the brain, strong immunoreactivity for βarrestin-1 was mostly found at postsynaptic densities in the spinal cord (Kittel & Komori 1999).

In transfected cellular systems, βarrestin-1 appears to regulate desensitization of numerous GPCRs (Oakley et al. 2000), including D1 and D2 dopamine receptors (Kim et al. 2001, Oakley et al. 2000) and μORs (Oakley et al. 2000, Schulz et al. 1999, Zhang et al. 1998; but see Cheng et al. 1998). It was found that chronic systemic morphine treatment in rats produced an increase in βarrestin immunoreactivity in the locus coeruleus, measured using an antibody that recognizes both βarrestin-1 and βarrestin-2 (Terwilliger et al. 1994). In addition, chronic morphine treatment of cells expressing μOR receptors resulted in attenuation of βarrestin-1 functions (Eisinger et al. 2002). In another investigation a differential regulation of βarrestin-1 mRNA in the locus coeruleus, periaqueductal gray, and cerebral cortex was observed following acute and systemic administration of morphine, but not during naloxone-precipitated withdrawal (Fan et al. 2003). Altogether, these findings suggest that βarrestin-1 may be important for at least some of the behavioral manifestations of acute and chronic morphine. It is important to note also that a presence of circulating autoantibodies reactive with βarrestin-1 has been described in multiple sclerosis patients (Ohguro et al. 1993).

Mice lacking βarrestin-1 demonstrate altered cardiac responses to β-adrenergic agonists, although these mice were found to be overtly normal (Conner et al. 1997). Analysis of basal locomotor activity also did not reveal any significant alterations in mutant mice (Figure 7). Furthermore, we did not observe the expected

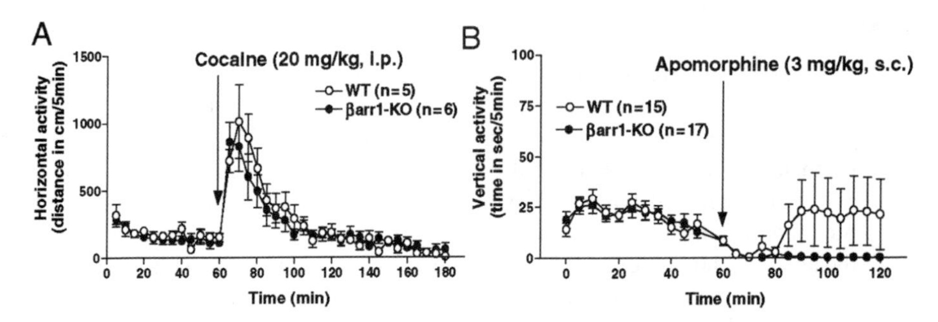

Figure 7 Responsiveness of βarrestin-1-KO mice to cocaine (20 mg/kg, i.p.) and apomorphine (3 mg/kg, s.c.). βarrestin-1-KO mice demonstrate significantly less climbing behavior after apomorphine. βarrestin1 mutant mice (Conner et al. 1997) were backcrossed 10 generations onto a C57/BL6 background (3–5 months old, both genders).

increase in dopamine-mediated locomotor responses. In fact, apomorphine-induced climbing was significantly suppressed (Figure 7). Surprisingly, we have observed no difference in acute morphine-induced antinociception in a hot-plate test in these mutants (L.M. Bohn, unpublished).

βARRESTIN-2 βarrestin-2, like βarrestin-1, is widely expressed in many brain regions from birth (Attramadal et al. 1992, Gurevich et al. 2002). βarrestin-2 mRNA reaches maximal levels in neonatal pups and then somewhat decreases. During postnatal development, βarrestin-2 protein levels change little. In many brain areas there is a significant overlap between the expression of βarrestin-2 and βarrestin-1, suggesting colocalization of these proteins. However, each βarrestin also demonstrated a unique distribution in certain brain areas. Both arrestins were highly expressed in the cortex and hippocampus, but βarrestin-2 was particularly abundant in the medial habenular, in most hypothalamic nuclei, the extended amygdala, and, in the developing brain, in the subventricular zone (Gurevich et al. 2002). In the spinal cord, βarrestin-2 was found through laminae III–X in the order of IX > dorsal nucleus of Clarke > V > VII/VIII > IV > III > X (Kittel & Komori 1999).

Numerous GPCRs are known to interact with and be regulated by βarrestin-2 in transfected cell systems (Ferguson et al. 1998, Luttrell & Lefkowitz 2002, Perry & Lefkowitz 2002). βarrestin-2 in general appears to be more potent than βarrestin-1 in mediating agonist-dependent internalization and promotes faster recycling of many GPCRs, thus making this subtype the most commonly used in in vitro investigations of a GPCR desensitization. In cellular experiments, βarrestin-2 regulates D1, D2, and D3 DA receptors (Kim et al. 2001, Oakley et al. 2000, Zhang et al. 1999) and μORs (Bushell et al. 2002, Celver et al. 2001, Kovoor et al. 1998, Lowe et al. 2002, Oakley et al. 2000, Whistler et al. 1999, Zhang et al. 1998). Interestingly, μOR interaction with βarrestin-2 was found to be dependent on agonist efficacy, and the low efficacy agonist morphine produced receptor desensitization at a significantly slower rate (Kovoor et al. 1998). However, overexpression of GRK2 or βarrestin resulted in robust morphine-triggered μOR internalization and desensitization (Whistler & Von Zastrow 1998, Zhang et al. 1998), suggesting that despite substantial differences in the potency, morphine-activated μORs do not "elude desensitization by βarrestin" as suggested by Whistler & Von Zastrow (1998) but rather are subject to complex yet essential regulation by βarrestins.

A substantial amount of information has been gained on the role of βarrestin-2 in regulating receptor responsiveness to opioids. As discussed earlier, chronic morphine increases βarrestin-1/2 levels in the rat locus coeruleus (Terwilliger et al. 1994). Chronic treatment with μOR agonist that resulted in a development of antinociceptive tolerance, as well as chronic μOR antagonism, caused a significant increase in expression of βarrestin-2 in the cortex and striatum (Diaz et al. 2002, Hurle 2001). Acute and chronic treatment with morphine variably regulates βarrestin-2 mRNA in several brain regions in rats (Fan et al. 2003). Finally, an involvement of βarrestin-2 in modulating spinal antinociception induced by μOR agonists was demonstrated in mice recently by using intrathecally administered βarrestin-2 antibody (Ohsawa et al. 2003) or by using antisense RNA to βarrestin-2 (Przewlocka et al. 2002).

To assess directly if βarrestin-2 can be involved in regulation of μORs in physiologically relevant situations we assessed effects of morphine in mice lacking βarrestin-2. Mice lacking βarrestin-2 do not demonstrate any gross phenotype, although somewhat reduced locomotor activity in a novel environment is noticeable in these mice. However, when these mice were challenged with morphine, a number of characteristic reactions to this drug were remarkably altered. First, the acute antinociceptive effect of morphine in a hot-plate test is enhanced and prolonged in these mutants (Bohn et al. 1999, 2002). Importantly, alterations in the antinociceptive properties of morphine are correlated with enhanced μOR-G protein coupling (Bohn et al. 1999, 2002). Second, tolerance to morphine's antinociceptive effects in this test is attenuated in βarrestin-2-KO mice (Bohn et al. 2000). Taken together, these studies demonstrated that lack of the βarrestin-2 leads to impaired desensitization of the μOR, thus resulting in dramatically attenuated antinociceptive tolerance. Interestingly, manifestations of naltrexone-precipitated physical withdrawal were intact in βarrestin-2-KO mice, which suggests that, although the mutants did not experience antinociceptive tolerance to chronic morphine, they still became physically dependent on this drug (Bohn et al. 2000). It should be emphasized that these dramatic alterations in morphine analgesia in mice lacking βarrestin-2 were detected in the hot-plate test, a paradigm that primarily assesses supraspinal pain responsiveness. Morphine also induces spinal cord–mediated antinociception. In the warm water tail-immersion test, the paradigm primarily assessing spinal reflexes to painful thermal stimuli, the βarrestin-2-KO mice have shown greater basal nociceptive thresholds as well as markedly enhanced sensitivity to morphine. However, although they had a delayed onset of tolerance to chronic morphine in this test, the mutants eventually developed tolerance to this drug. Thus, in the absence of βarrestin-2, the contribution of a previously elucidated PKC-dependent regulatory system to the development of morphine tolerance in this paradigm became apparent (Bohn et al. 2002).

Finally, significant alterations in locomotor and reinforcing properties of morphine were observed in βarrestin-2-KO mice (Bohn et al. 2003). The activation of both μOR and DA receptors is known to play a critical role in the locomotor and reinforcing effects of morphine (Elmer et al. 2002, Kieffer 1999, Koob 1992, Maldonado et al. 1997). Morphine, as well as many other drugs of abuse, increases DA signaling in striatal and mesolimbic brain structures such as the striatum and the nucleus accumbens. In the case of morphine, stimulation of DA systems is indirect, originating from a disinhibition of GABAergic cells in DA cell body regions (substantia nigra and the ventral tegmental area) leading to increased neuronal firing and increased DA release in terminal regions (within striatum and nucleus accumbens) (DiChiara & North 1992). In mice lacking βarrestin-2, morphine produces a greater increase in DA release and induces increased reward as measured in conditioned place preference test. However, acute morphine treatment induced actually less pronounced hyperactivity in βarrestin-2 mutant mice, which potentially indicates an impact of this mutation on neurotransmitter systems other than DA, which are also involved in morphine effect on locomotion. Potential supersensitivity of these systems, such as serotonin (Sills & Fletcher 1997, Tao & Auerbach 1994), which are known to exert a general inhibitory action on DA-dependent hyperactivity (Gainetdinov et al. 1999b), may be involved and therefore needs further exploration. It is also interesting that in both wild-type and βarrestin-2 mutant mice chronic morphine induced comparable behavioral sensitization to this drug in locomotor test, suggesting a minimal role of βarrestin-2-mediated processes in this particular form of neuronal plasticity (Bohn et al. 2003).

In striking contrast to morphine, cocaine did not demonstrate exaggerated responses in these mice in any paradigm tested. In fact, acute cocaine induced somewhat reduced locomotor activation, but mutants demonstrated normal behavioral sensitization to chronic treatment. Similarly, a reduced response to direct DA

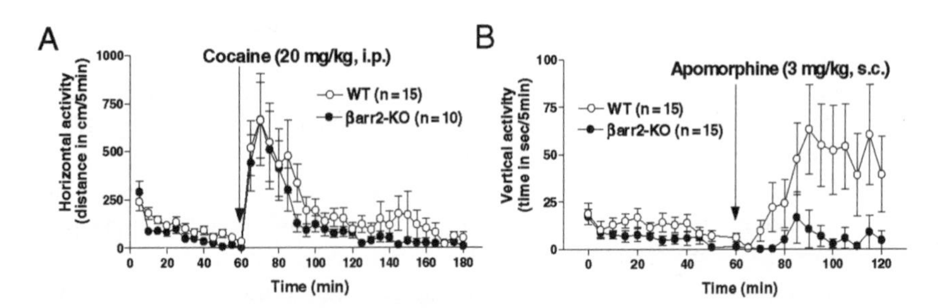

Figure 8 Responsiveness of βarrestin-2-KO mice (Bohn et al. 1999) to cocaine (20 mg/kg, i.p.) and apomorphine (3 mg/kg, s.c.). βarrestin-2-KO mice demonstrate lower basal activities and significantly less climbing after apomorphine. Data on cocaine responses are reproduced with permission from Bohn et al. (2003).

agonist apomorphine was noted (Figure 8). At the same time, cocaine induces normal elevation in extracellular DA and reward in conditioned place preference. Thus it is likely that βarrestin-2 is not critically involved in the regulation of DA receptor desensitization. However, a reduced basal locomotion, as well as responses to cocaine and apomorphine, may suggest that without this regulatory element, DA signaling may be somewhat impaired.

CONCLUSIONS

Seven GRKs and four arrestins have been identified, whereas more than 700 GPCRs have already been cloned (Luttrell & Lefkowitz 2002, Perry & Lefkowitz 2002, Pitcher et al. 1998a, Wise et al. 2004). Thus, redundancy in the mechanisms of GPCR regulation, such as the involvement of each GRK or arrestin in the desensitization of multiple receptors, is to be expected. Most of the initial work attempting to define the specificity of GRKs' and arrestins' action on individual GPCRs has used either overexpression of these proteins in cultured cells or purified GRKs acting on receptor preparations (e.g., in washed rod outer segment membranes or purified receptor proteins reconstituted into lipid vesicles). Other in vitro studies have attempted to assess inhibition of individual GRKs (using heparin or anti-GRK antibodies or expressing GRK fragments or kinase-deficient mutants), but generally they have suffered from incomplete inhibition or uncertainty about the specificity of inhibition. These studies have revealed significant differences in the substrate recognition of GPCRs among the GRK subtypes (Pitcher et al. 1998a). Similarly the specificity of interaction of arrestin family members with GPCRs in cultured cells has been demonstrated using green fluorescent protein-tagged arrestins. In these in vitro studies, analysis of agonist-mediated arrestin translocation to multiple GPCRs identified two major classes of receptors (Class A and Class B), which differ in relative affinity for βarrestin-1 versus βarrestin-2 and in internalization/recycling properties (Oakley et al. 2000). However, whether this specificity could be demonstrated in native tissue in vivo has largely remained unknown. The development of mice lacking specific GRKs or arrestins brings yet a novel tool to address the problem of matching individual GRKs and arrestins with specific GPCRs and the physiology they control.

The observations summarized here demonstrate the functionally important roles of specific GRKs and βarrestins in the regulation of DA and μ-opioid receptors. However, although this initial screening brought to our attention a prominent role of GRK6 in DA-mediated responses and βarrestin-2 in μOR-mediated responses, many questions remain unanswered.

With regard to regulation of DA receptors (Table 2), one important question remains: Which particular receptor subtypes are primarily regulated by which GRK and βarrestin? Functionally, GRK6 clearly appears to be the main GRK regulator of some D2-like receptor, either D2 or D3, or perhaps both. Although assessment of binding characteristics of striatal DA receptors and locomotor responses to

TABLE 2 Summary of locomotor responses to psychostimulants and dopaminergic agonists in mice lacking GRKs and βarrestins

Mutants	Basal level	Cocaine	Apomorphine	Other drugs
GRK2-HET	Normal	Slightly increased	Normal	Normal response to amphetamine
GRK3-KO	Normal	Decreased	Decreased	Decreased response to D1 dopamine agonist
GRK4-KO	Normal	Not different	Not tested	Not tested
GRK5-KO	Normal	Not different	Normal	Not tested
GRK6-HET GRK6-KO	Normal	Markedly increased	Increased	Increased responses to amphetamine; β-PEA, D2 DA agonist; Normal responses to D1 DA agonist
βarrestin-1-KO	Normal	Not different	Decreased	Not tested
βarrestin-2-KO	Decreased	Not different	Decreased	Decreased responses to morphine

selective agonists demonstrated that properties of D2-like DA receptors, but not that of D1-like, are significantly affected in GRK6 mutant mice, the possibility that various other populations of DA receptors might also be regulated by GRK6 cannot be excluded. It would be reasonable to expect also that more than one GRK would be needed for regulation of all five subtypes of DA receptors; however, in our initial screen no obvious alterations in DA responses were found in other GRK mutants. D1-like receptors appear subject to GRK regulation in transfected cell systems, yet no single GRK appears to regulate these receptors in knockout mice. Perhaps D1 can be regulated by several GRKs such that loss of any one does not perturb its regulation significantly. At the same time, the minor alterations found in GRK2 heterozygous mice in locomotor responses to cocaine leaves this GRK as a suspect and suggests that further investigation is needed to test a role for GRK2 in DA signaling. Also important, lack of behavioral alterations in responses to psychostimulants does not necessarily mean that no DA receptors are affected by these mutations. It is also possible that there may be functional substitutes for the missing GRK that allow the animal to maintain a normal phenotype. For example, other downstream signaling elements such as PKA may be involved in regulating the receptors (Hausdorff et al. 1990). Alternatively, RGS proteins, particularly the subtype highly enriched in the striatum, RGS9-2, also could play a role in direct modulation of DA receptor function. RGS proteins are known to accelerate the rate of GTP hydrolysis on activated G proteins, so they can speed up the deactivation of a system after stimulation or dampen signaling during persistent stimulation. In fact, recent observations of a functional interaction between RGS9-2 and D2 DA receptor signaling and the behavioral effects of psychostimulants suggest that psychostimulants may induce RGS9-2 to diminish responsiveness to a drug (Rahman et al. 2003).

Even more intriguing, in this initial screening, no supersensitivity to DA stimulation was found in either the βarrestin-1 or βarrestin-2 mutant mice (Table 2). Furthermore, even mice developed by crossing these two strains, three-allele knockouts, βarrestin-1-KO/βarrestin-2+/− and βarrestin-1+/−/βarrestin-2-KO mice, behave similarly to single knockout lines; that is, they showed somewhat reduced responses to apomorphine (R.R. Gainetdinov, unpublished). An attractive hypothesis is that there is significant functional redundancy between the two βarrestin subtypes so that the complement of one isoform of βarrestin is sufficient to regulate dopamine receptors in the absence of the other isoform. In fact, in transfected cell-culture systems, D1 and D2 DA receptors do interact with both types of βarrestins (Kim et al. 2001, Oakley et al. 2000), and expression of both βarrestins substantially overlaps in primary DA regions. It is also notable that lack of both βarrestin-1 and βarrestin-2 is embryonically lethal in mice (Kohout et al. 2001), suggesting a critical and cooperative role of both these regulatory proteins for normal physiological functions, at least during development. In line with this conclusion, βarrestin-2 appears to be involved in the Wnt5A-stimulated endocytosis of Frizzled 4 (Chen et al. 2003), indicating a direct role of this multifunctional adaptor protein in the regulation of developmental pathways. Further, although we have assessed dopaminergic stimulant function with single drugs, intact animals possess a myriad of other receptors and agonists that may also be subject to regulation by arrestins, and it may be that compensatory alterations in other receptor pathways may mask βarrestin effects on the DA receptors themselves. Alternatively, it is possible that the recently characterized signaling properties of βarrestins (Luttrell & Lefkowitz 2002) may be functionally relevant for DA signaling, and inactivation of βarrestins may contribute directly to a reduction in DA-dependent responses observed in both βarrestin mutant strains.

Similar considerations also should be taken into account when effects of morphine were analyzed in mutants lacking GRKs or βarrestins. βarrestin-2-KO mice have demonstrated remarkable alterations in μOR regulation and morphine responses not seen with βarrestin-1 deletion. Surprisingly, no GRK-KO mice available demonstrated altered antinociceptive responses to morphine, comparable to that found in βarrestin-2 mutants. It is, thus, possible that more than one GRK may be involved in μOR regulation, and other subtypes compensate for lack of any specific GRK. Importantly, the apparent specificity of desensitization of a particular GPCR measured in vitro does not seem to be an accurate predictor of specificity in the intact animal. For example, to demonstrate a role of βarrestin-2 in morphine-induced desensitization of μORs in cellular model systems, overexpression of GRK2 was necessary (Zhang et al. 1998), but in mice lacking βarrestin-2 virtually all physiological responses to morphine are remarkably altered. Among the primary determinants of GPCR-GRK-arrestin specificity, not only intrinsic kinase activity but also the colocalization of the receptor, kinase, and arrestin within individual cells in a tissue or brain region are critical. One expectation is that this specificity will vary considerably, with some receptors being regulated by a single GRK subtype without compensation by other GRKs, whereas other receptors might be regulated more-or-less equally by several GRK/arrestin subtypes. The μOR would appear to be an example of a receptor that can be regulated by several GRKs, such that loss of any one does not appreciably alter receptor function. However, the ability of other GRK subtypes to compensate for the loss of one GRK subtype can be limited, at least for some receptors, as evidenced by the embryonic lethality of the GRK2 knockout mouse. Again, we cannot completely exclude that GRK2 is primarily involved in this regulation and future development of time or region-selective GRK2 mutants would be helpful to address this issue. Interestingly, chronic morphine differentially affects various neuronal populations; that is, some populations display μOR desensitization, whereas other brain regions do not (Noble & Cox 1996, Sim et al. 1996). Furthermore, it has been demonstrated by Haberstock-Debic et al. (2003) that morphine can induce μOR internalization in dendrites but not cell bodies of neurons of the nucleus accumbens, hinting that receptors residing in different compartments within the same neurons may be subject to different means or degrees of regulation. This could, for example, depend on the cellular complement and fine localization of regulatory elements.

Our findings thus far suggest that there is significant degree of specificity in the regulation of the specific GPCRs by GRKs and βarrestins in vivo. Although we have utilized simple screens to assess the most obvious differences between genotypes in response to morphine or cocaine, we do not eliminate the possibility that we may be overlooking changes in responsiveness in other behavioral paradigms. For example, whereas the βarrestin-2-KO mice are supersensitive to morphine in antinociception tests, they are actually less activated in locomotion by morphine (Bohn et al. 2003).

The observations summarized here strongly support the general principle that the inactivation of components of GRK/arrestin-mediated desensitization machinery leads to enhanced GPCR signaling, and thereby, enhanced physiological function mediated by a given receptor (Bohn et al. 1999, 2002, 2003; Gainetdinov et al. 1999a, 2003a). The challenge remaining is to extend this principle to the vast sea of remaining receptors and the physiological responses they control. A further indication of the degree of importance of the desensitization machinery in modulating physiological responsiveness is the fact that both the GRK6 and the βarrestin-2 heterozygous mice demonstrate alterations almost identical to that observed in mice lacking these proteins (Bohn et al. 1999, 2002, 2003; Gainetdinov et al. 2003a). Therefore, deficiency in one allele, or pharmacological inhibition of half of the GRK6 or βarrestin-2 activity, could potentially result in a maximally possible physiological phenotype similar to that observed upon full inactivation of the protein. This could have great therapeutic implications because most pharmacological inhibitors do not result in complete elimination of molecular function.

One brain disorder where GPCR desensitization processes could be particularly important is addiction, where abnormal neuronal plasticity to chronic drug treatment is believed to be a primary cause for compulsive drug abuse. Among several behavioral manifestations of these permanent molecular changes, such phenomena as "sensitization" and "tolerance" directly point to the alterations in sensitivity of a receptor to its agonists. Repeated administration of psychostimulants like cocaine is known to result in a progressive enhancement of psychomotor responses, and this paradigm is classically used in studies of drugs of abuse in experimental animals. This phenomenon, termed behavioral sensitization or reverse tolerance, is believed to be related to permanent neuronal adaptations associated with alterations in responsiveness to DA stimulation (Hyman & Malenka 2001, Laakso et al. 2002, Nestler 2001). The abnormal supersensitivity of DA receptors involves changes in signaling molecules such as transcription factors ΔFosB and the cyclic-AMP response-element-binding protein (CREB), as well as cyclin-dependent kinase 5 (Cdk5) (Bibb et al. 2001, Hyman & Malenka 2001, Nestler 2001). However, the contribution of direct DA receptor regulatory mechanisms in this phenomenon has only begun to be characterized (Gainetdinov et al. 2003a; Rahman et al. 2003). As summarized in the present review, we find that direct regulation of DA receptors by GRK6, but not by other GPCR specific kinases, represents an important determinant by which receptor supersensitivity and responses to drugs of abuse can be controlled. This raises an intriguing possibility that potential alterations in GRK6 levels or activity could be another important determinant of responsiveness to drugs of abuse.

Strikingly, an opposite reaction to chronic drug treatment, termed tolerance and defined as a decrease in the biological effect of the drug despite a constant dose, has also been found affected in mice deficient in homologous desensitization. Particularly, in the best-known paradigm to study this process, morphine-induced antinociceptive tolerance, βarrestin-2-KO mice were deficient while retaining classic withdrawal reactions to chronic morphine. Moreover, it has been observed that morphine is more rewarding to the βarrestin-2-KO mice than to wild-type mice, an observation that correlates with their pain responses (Bohn et al. 2003). It will be very important to further assess the contribution of each of these regulatory elements, specifically, to the development of behaviors associated more directly with drug abuse. Future testing of the GRK and βarrestin KO mice in paradigms that assess drug reinforcement, such as conditioned place preference and self administration, should further our knowledge of the contribution of GPCR regulation to the molecular mechanisms of neuronal plasticity underlying drug addiction.

From this initial characterization of the GRK and βarrestin mutants, it has become evident that under basal, unchallenged conditions, these mice generally demonstrate only a modest phenotype, if any at all. This observation of a weak phenotype suggests that these regulatory elements play a minor role in setting the basal "tone" of the signaling pathway. However, upon challenge with the relevant agonist, a GPCR that cannot be properly regulated results in a mutant animal that is no longer able to compensate to the same degree as wild-type controls. Thus, it is reasonable to expect that a potential pharmacological window of future drugs targeting desensitization machinery would be related to the fine tuning of physiological responses. Having mice available that lack each GRK/arrestin subtype will now permit examination of any individual GPCR and its physiological effects to assess the contribution of each GRK or arrestin to regulation of that receptor. In addition, because altered responsiveness of GPCRs has been found in many diseases, the use of mutant mice will allow evaluation of the potential contribution of these molecules to these conditions. Knowledge of this sort will be critical to promote further specific GRKs or arrestins as potential therapeutic targets. Furthermore, an important and very specific role of GRKs and βarrestins in complex functions, such as locomotion and antinociception, suggests that a precise and safe pharmacology based on targeting these general kinases and arrestins is possible. As one such example, it is conceivable that pharmacological inactivation of GRK6 may provide an effective approach to amplify efficacy of endogenous or exogenous DA stimulation and may be helpful to restore movement in patients with Parkinson's disease. Similarly, suppression or inhibition of βarrestin-2 may become an effective strategy to enhance antinociceptive properties of the "gold standard" analgesic morphine without affecting, or actually reducing, its side effects. Thus, targeting GPCR desensitization machinery may provide a novel principle of pharmacology where fine tuning of GPCR signaling, rather than direct receptor stimulation or blockade, would result in a more precise correction of abnormal physiological processes.

The *Annual Review of Neuroscience* is online at http://neuro.annualreviews.org

LITERATURE CITED

Angers S, Salahpour A, Bouvier M. 2002. Dimerization: an emerging concept for G protein-coupled receptor ontogeny and function. *Annu. Rev. Pharmacol. Toxicol.* 42:409–35

Arriza JL, Dawson TM, Simerly RB, Martin LJ, Caron MG, et al. 1992. The G protein-coupled receptor kinases βARK1 and βARK2 are widely distributed at synapses in rat brain. *J. Neurosci.* 12:4045–55

Attramadal H, Arriza JL, Aoki C, Dawson TM, Codina J, et al. 1992. Beta-arrestin2, a novel member of the arrestin/beta-arrestin gene family. *J. Biol. Chem.* 267:17882–90

Barrett TB, Hauger RL, Kennedy JL, Sadovnick AD, Remick RA, et al. 2003. Evidence that a single nucleotide polymorphism in the promoter of the G protein receptor kinase 3 gene is associated with bipolar disorder. *Mol. Psychiatry* 8:546–57

Benovic JL, Mayor F Jr, Staniszewski C, Lefkowitz RJ, Caron MG. 1987. Purification and characterization of the beta-adrenergic receptor kinase. *J. Biol. Chem.* 262:9026–32

Benovic JL, Gomez J. 1993. Molecular cloning and expression of GRK6. A new member of the G protein-coupled receptor kinase family. *J. Biol. Chem.* 268:19521–27

Berman DM, Gilman AG. 1998. Mammalian RGS proteins: barbarians at the gate. *J. Biol. Chem.* 273:1269–72

Bibb JA, Chen J, Taylor JR, Svenningsson P,

Nishi A, et al. 2001. Effects of chronic exposure to cocaine are regulated by the neuronal protein Cdk5. *Nature* 410:376–80

Boekhoff I, Inglese J, Schleicher S, Koch WJ, Lefkowitz RJ, Breer H. 1994. Olfactory desensitization requires membrane targeting of receptor kinase mediated by beta gamma-subunits of heterotrimeric G proteins. *J Biol. Chem.* 269:37–40

Bohm SK, Grady EF, Bunnett NW. 1997. Regulatory mechanisms that modulate signaling by G-protein-coupled receptors. *Biochem. J.* 322:1–18

Bohn LM, Lefkowitz RJ, Gainetdinov RR, Peppel K, Caron MG, Lin FT. 1999. Enhanced morphine analgesia in mice lacking beta-arrestin 2. *Science* 286:2495–98

Bohn LM, Gainetdinov RR, Lin FT, Lefkowitz RJ, Caron MG. 2000. Mu-opioid receptor desensitization by beta-arrestin-2 determines morphine tolerance but not dependence. *Nature* 408:720–23

Bohn LM, Lefkowitz RJ, Caron MG. 2002. Differential mechanisms of morphine antinociceptive tolerance revealed in (beta)arrestin-2 knock-out mice. *J. Neurosci.* 22: 10494–500

Bohn LM, Gainetdinov RR, Sotnikova TD, Lefkowitz RJ, Dykstra LA, Caron MG. 2003. Enhanced rewarding properties of morphine, but not cocaine, in βarrestin-2 knockout mice. *J. Neurosci.* 23:10265–73

Bushell T, Endoh T, Simen AA, Ren D, Bindokas VP, Miller RJ. 2002. Molecular components of tolerance to opiates in single hippocampal neurons. *Mol. Pharmacol.* 61: 55–64

Carlsson A. 2001. A paradigm shift in brain research. *Science* 294:1021–24

Celver JP, Lowe J, Kovoor A, Gurevich VV, Chavkin C. 2001. Threonine 180 is required for G-protein-coupled receptor kinase 3- and beta-arrestin 2-mediated desensitization of the mu-opioid receptor in Xenopus oocytes. *J. Biol. Chem.* 276:4894–900

Chen CK, Burns ME, Spencer M, Niemi GA, Chen J, et al. 1999a. Abnormal photoresponses and light-induced apoptosis in rods lacking rhodopsin kinase. *Proc. Natl. Acad. Sci. USA* 96:3718–22

Chen J, Simon MI, Matthes MT, Yasumura D, LaVail MM. 1999b. Increased susceptibility to light damage in an arrestin knockout mouse model of Oguchi disease (stationary night blindness). *Invest. Ophthalmol. Vis. Sci.* 40:2978–82

Chen W, ten Berge D, Brown J, Ahn S, Hu LA, et al. 2003. Dishevelled 2 recruits beta-arrestin 2 to mediate Wnt5A-stimulated endocytosis of Frizzled 4. *Science* 301:1391–94

Cheng ZJ, Yu QM, Wu YL, Ma L, Pei G. 1998. Selective interference of beta-arrestin 1 with kappa and delta but not mu opioid receptor/G protein coupling. *J. Biol. Chem.* 273:24328–33

Claing A, Laporte SA, Caron MG, Lefkowitz RJ. 2002. Endocytosis of G protein-coupled receptors: roles of G protein-coupled receptor kinases and beta-arrestin proteins. *Prog. Neurobiol.* 66:61–79

Conner DA, Mathier MA, Mortensen RM, Christe M, Vatner SF, et al. 1997. beta-Arrestin1 knockout mice appear normal but demonstrate altered cardiac responses to beta-adrenergic stimulation. *Circ. Res.* 81:1021–26

Diaz A, Pazos A, Florez J, Ayesta FJ, Santana V, Hurle MA. 2002. Regulation of mu-opioid receptors, G-protein-coupled receptor kinases and beta-arrestin 2 in the rat brain after chronic opioid receptor antagonism. *Neuroscience* 112:345–53

Di Chiara G, North RA. 1992. Neurobiology of opiate abuse. *Trends Pharmacol. Sci.* 13:185–93

Dickey BF, Birnbaumer L, ed. 1993. *GTPases in Biology (Handbook of Experimental Pharmacology).* Berlin: Springer-Verlag. Vol. 108

Diverse-Pierluissi M, Inglese J, Stoffel RH, Lefkowitz RJ, Dunlap K. 1996. G protein-coupled receptor kinase mediates desensitization of norepinephrine-induced Ca^{2+} channel inhibition. *Neuron* 16:579–85

Dohlman HG, Thorner J. 1997. RGS proteins

and signaling by heterotrimeric G proteins. *J. Biol. Chem.* 272:3871–74

Eisinger DA, Ammer H, Schulz R. 2002. Chronic morphine treatment inhibits opioid receptor desensitization and internalization. *J. Neurosci.* 22:10192–200

Elmer GI, Pieper JO, Rubinstein M, Low MJ, Grandy DK, Wise RA. 2002. Failure of intravenous morphine to serve as an effective instrumental reinforcer in dopamine D2 receptor knock-out mice. *J. Neurosci.* 22:RC224

Erdtmann-Vourliotis M, Mayer P, Ammon S, Reichert U, Hollt V. 2001. Distribution of G protein-coupled receptor kinase (GRK) isoforms 2, 3, 5 and 6 mRNA in the rat brain. *Brain Res. Mol. Brain. Res.* 95:129–37

Fan X, Zhang J, Zhang X, Yue W, Ma L. 2002. Acute and chronic morphine treatments and morphine withdrawal differentially regulate GRK2 and GRK5 gene expression in rat brain. *Neuropharmacology* 43:809–16

Fan XL, Zhang JS, Zhang XQ, Yue W, Ma L. 2003. Differential regulation of beta-arrestin 1 and beta-arrestin 2 gene expression in rat brain by morphine. *Neuroscience* 117:383–89

Fehr C, Fickova M, Hiemke C, Reuss S, Dahmen N. 1997. Molecular cloning of rat G-protein-coupled receptor kinase 6 (GRK6) from brain tissue, and its mRNA expression in different brain regions and peripheral tissues. *Brain Res. Mol. Brain Res.* 49:278–82

Ferguson SS, Downey WE 3rd, Colapietro AM, Barak LS, Menard L, Caron MG. 1996. Role of beta-arrestin in mediating agonist-promoted G protein-coupled receptor internalization. *Science* 271:363–66

Ferguson SS, Zhang J, Barak LS, Caron MG. 1998. Molecular mechanisms of G protein-coupled receptor desensitization and resensitization. *Life Sci.* 62:1561–65

Fong AM, Premont RT, Richardson RM, Yu YR, Lefkowitz RJ, Patel DD. 2002. A positive regulatory role for the GRK-βarrestin system in lymphocyte chemotaxis. *Proc. Natl. Acad. Sci. USA* 99:7478–83

Gainetdinov RR, Bohn LM, Walker JKL, Laporte SA, Macrae AD, et al. 1999a. Muscarinic supersensitivity and impaired receptor desensitization in G protein-coupled receptor kinase 5-deficient mice. *Neuron* 24:1029–36

Gainetdinov RR, Wetsel WC, Jones SR, Levin ED, Jaber M, Caron MG. 1999b. Role of serotonin in the paradoxical calming effect of psychostimulants on hyperactivity. *Science* 283:397–401

Gainetdinov RR, Premont RT, Caron MG, Lefkowitz RJ. 2000. Reply: receptor specificity of G protein-coupled receptor kinases. *Trends Pharmacol. Sci.* 21:366–67

Gainetdinov RR, Bohn LM, Sotnikova TD, Cyr M, Laakso A, et al. 2003a. Dopaminergic supersensitivity in G protein-coupled receptor kinase 6-deficient mice. *Neuron* 38:291–303

Gainetdinov RR, Caron MG. 2003b. Monoamine transporters: from genes to behavior. *Annu. Rev. Pharmacol. Toxicol.* 43:261–84

Gardner B, Liu ZF, Jiang D, Sibley DR. 2001. The role of phosphorylation/dephosphorylation in agonist-induced desensitization of D1 dopamine receptor function. *Mol. Pharmacol.* 59:310–21

Gomeza J, Shannon H, Kostenis E, Felder C, Zhang L, et al. 1999. Pronounced pharmacologic deficits in M2 muscarinic acetylcholine receptor knockout mice. *Proc. Natl. Acad. Sci. USA* 96:1692–97

Goodman OB Jr, Krupnick JG, Santini F, Gurevich VV, Penn RB, et al. 1996. Beta-arrestin acts as a clathrin adaptor in endocytosis of the beta2-adrenergic receptor. *Nature* 383:447–50

Grandy DK, Civelli O. 1992. G-protein-coupled receptors: the new dopamine receptor subtypes. *Curr. Opin. Neurobiol.* 2:275–81

Grange-Midroit M, Garcia-Sevilla JA, Ferrer-Alcon M, La Harpe R, Huguelet P, Guimon J. 2003. Regulation of GRK 2 and 6, beta-arrestin-2 and associated proteins in the prefrontal cortex of drug-free and antidepressant drug-treated subjects with major depression. *Brain Res. Mol. Brain Res.* 111:31–41

Gurevich EV, Benovic JL, Gurevich VV. 2002.

Arrestin2 and arrestin3 are differentially expressed in the rat brain during postnatal development. *Neuroscience* 109:421–36

Gurevich VV, Gurevich EV. 2003. The new face of active receptor bound arrestin attracts new partners. *Structure* 11:1037–42

Haberstock H, Wein M, Barrot M, Colago EE, Rahman Z, et al. 2003. Morphine acutely regulates opioid receptor trafficking selectively in dendrites of nucleus accumbens neurons. *J. Neurosci.* 23:4324–32

Hall RA, Premont RT, Lefkowitz RJ. 1999. Heptahelical receptor signaling: beyond the G protein paradigm. *J. Cell Biol.* 145:927–32

Hamm HE, Gilchrist A. 1996. Heterotrimeric G proteins. *Curr. Opin. Cell Biol.* 8:189–96

Hanks SK, Quinn AM, Hunter T. 1988. The protein kinase family: conserved features and deduced phylogeny of the catalytic domains. *Science* 241:42–52

Hausdorff WP, Caron MG, Lefkowitz RJ. 1990. Turning off the signal: desensitization of β-adrenergic receptor function. *FASEB J.* 4:2881–89

Hornykiewicz O. 1966. Dopamine (3-hydroxytyramine) and brain function. *Pharmacol. Rev.* 18:925–64

Hurle MA. 2001. Changes in the expression of G protein-coupled receptor kinases and β-arrestin 2 in rat brain during opioid tolerance and supersensitivity. *J. Neurochem.* 77:486–92

Hyman SE, Malenka RC. 2001. Addiction and the brain: the neurobiology of compulsion and its persistence. *Nature Rev. Neurosci.* 2:695–703

Iacovelli L, Sallese M, Mariggio S, de Blasi A. 1999. Regulation of G-protein-coupled receptor kinase subtypes by calcium sensor proteins. *FASEB J.* 13:1–8

Iacovelli L, Salvatore L, Capobianco L, Picascia A, Barletta E, et al. 2003. Role of G protein-coupled receptor kinase 4 and beta-arrestin 1 in agonist-stimulated metabotropic glutamate receptor 1 internalization and activation of mitogen-activated protein kinases. *J. Biol. Chem.* 278:12433–42

Ito K, Haga T, Lameh J, Sadee W. 1999. Sequestration of dopamine D2 receptors depends on coexpression of G protein-coupled receptor kinases 2 or 5. *Eur. J. Biochem.* 260:112–19

Iwata K, Ito K, Fukuzaki A, Inaki K, Haga T. 1999. Dynamin and rab5 regulate GRK2-dependent internalization of dopamine D2 receptors. *Eur. J. Biochem.* 263:596–602

Jaber M, Koch WJ, Rockman H, Smith B, Bond RA, et al. 1996. Essential role of β-adrenergic receptor kinase 1 in cardiac development and function. *Proc. Natl. Acad. Sci. USA* 93:12974–79

Janowsky DS, Overstreet DH, Nurnberger JI Jr. 1994. Is cholinergic sensitivity a genetic marker for the affective disorders? *Am. J. Med. Genet.* 54:335–44

Janssen PA, Leysen JE, Megens AA, Awouters FH. 1999. Does phenylethylamine act as an endogenous amphetamine in some patients? *Int. J. Neuropsychopharmacol.* 2:229–40

Jenner P, Marsden CD. 1987. Chronic pharmacological manipulation of dopamine receptors in brain. *Neuropharmacology* 26:931–40

Jones SR, Gainetdinov RR, Wightman RM, Caron MG. 1998. Mechanisms of amphetamine action revealed in mice lacking the dopamine transporter. *J. Neurosci.* 18:1979–86

Kabbani N, Negyessy L, Lin R, Goldman-Rakic P, Levenson R. 2002. Interaction with neuronal calcium sensor NCS-1 mediates desensitization of the D2 dopamine receptor. *J. Neurosci.* 22:8476–86

Kavelaars A, Vroon A, Raatgever R, Fong AM, Premont RT, et al. 2003. Increased acute inflammation, leukotriene B4-induced chemotaxis and signaling in mice deficient for GRK6. *J. Immunol.* 171:6128–34

Kieffer BL. 1999. Opioids: first lessons from knockout mice. *Trends Pharmacol. Sci.* 20: 19–26

Kim KM, Valenzano KJ, Robinson SR, Yao WD, Barak LS, Caron MG. 2001. Differential regulation of the dopamine D2 and D3 receptors by G protein-coupled receptor kinases and β-arrestins. *J. Biol. Chem.* 276:37409–14

Kittel A, Komori N. 1999. Ultrastructural localization of beta-arrestin-1 and -2 in rat lumbar spinal cord. *J. Comp. Neurol.* 412:649–55

Koh PO, Undie AS, Kabbani N, Levenson R, Goldman-Rakic PS, Lidow MS. 2003. Up-regulation of neuronal calcium sensor-1 (NCS-1) in the prefrontal cortex of schizophrenic and bipolar patients. *Proc. Natl. Acad. Sci. USA* 100:313–17

Kohout TA, Lin FS, Perry SJ, Conner DA, Lefkowitz RJ. 2001. beta-Arrestin 1 and 2 differentially regulate heptahelical receptor signaling and trafficking. *Proc. Natl. Acad. Sci. USA* 98:1601–6

Koob GF. 1992. Drugs of abuse: anatomy, pharmacology and function of reward pathways. *Trends Pharmacol. Sci.* 13:177–84

Kovoor A, Celver JP, Wu A, Chavkin C. 1998. Agonist induced homologous desensitization of mu-opioid receptors mediated by G protein-coupled receptor kinases is dependent on agonist efficacy. *Mol. Pharmacol.* 54:704–11

Kunapuli P, Onorato JJ, Hosey MM, Benovic JL. 1994. Expression, purification, and characterization of the G protein-coupled receptor kinase GRK5. *J. Biol. Chem.* 269:1099–105

Laakso A, Mohn AR, Gainetdinov RR, Caron MG. 2002. Experimental genetic approaches to addiction. *Neuron* 36:213–28

Lamey M, Thompson M, Varghese G, Chi H, Sawzdargo M, et al. 2002. Distinct residues in the carboxyl tail mediate agonist-induced desensitization and internalization of the human dopamine D1 receptor. *J. Biol. Chem.* 277:9415–21

Laporte SA, Oakley RH, Zhang J, Holt JA, Ferguson SS, et al. 1999. The beta2-adrenergic receptor/betaarrestin complex recruits the clathrin adaptor AP-2 during endocytosis. *Proc. Natl. Acad. Sci. USA* 96:3712–17

Lefkowitz RJ. 1998. G protein-coupled receptors. III. New roles for receptor kinases and beta-arrestins in receptor signaling and desensitization. *J. Biol. Chem.* 273:18677–80

Li AH, Wang HL. 2001. G protein-coupled receptor kinase 2 mediates mu-opioid receptor desensitization in GABAergic neurons of the nucleus raphe magnus. *J. Neurochem.* 77:435–44

Loh HH, Liu HC, Cavalli A, Yang W, Chen YF, Wei LN. 1998. mu Opioid receptor knockout in mice: effects on ligand-induced analgesia and morphine lethality. *Brain Res. Mol. Brain Res.* 54:321–26

Lodowski DT, Pitcher JA, Capel WD, Lefkowitz RJ, Tesmer JJ. 2003. Keeping G proteins at bay: a complex between G protein-coupled receptor kinase 2 and Gbetagamma. *Science* 300:1256–62

Lowe JD, Celver JP, Gurevich VV, Chavkin C. 2002. mu-Opioid receptors desensitize less rapidly than delta-opioid receptors due to less efficient activation of arrestin. *J. Biol. Chem.* 277:15729–35

Luttrell LM, Lefkowitz RJ. 2002. The role of beta-arrestins in the termination and transduction of G-protein-coupled receptor signals. *J. Cell. Sci.* 115:455–65

Maldonado R, Saiardi A, Valverde O, Samad TA, Roques BP, Borrelli E. 1997. Absence of opiate rewarding effects in mice lacking dopamine D2 receptors. *Nature* 388:586–89

Matthes HW, Maldonado R, Simonin F, Valverde O, Slowe S, et al. 1996. Loss of morphine-induced analgesia, reward effect and withdrawal symptoms in mice lacking the mu-opioid-receptor gene. *Nature* 383:819–23

Menard L, Ferguson SS, Barak LS, Bertrand L, Premont RT, et al. 1996. Members of the G protein-coupled receptor kinase family that phosphorylate the beta2-adrenergic receptor facilitate sequestration. *Biochemistry* 35:4155–60

Miralles A, Asensio VJ, Garcia-Sevilla JA. 2002. Acute treatment with the cyclic antidepressant desipramine, but not fluoxetine, increases membrane-associated G protein-coupled receptor kinases 2/3 in rat brain. *Neuropharmacology* 43:1249–57

Missale C, Nash SR, Robinson SW, Jaber M, Caron MG. 1998. Dopamine receptors: from structure to function. *Physiol. Rev.* 78:189–225

Neill JD, Musgrove LC, Duck LW, Sellers JC. 1999. High efficiency method for gene transfer in normal pituitary gonadotropes: adenoviral-mediated expression of G protein-coupled receptor kinase 2 suppresses luteinizing hormone secretion. *Endocrinology* 140:2562–69

Nestler EJ, Aghajanian GK. 1997. Molecular and cellular basis of addiction. *Science* 278:58–63

Nestler EJ. 2001. Molecular basis of long-term plasticity underlying addiction. *Nature Rev. Neurosci.* 2:119–28

Neubig RR, Siderovski DP. 2002. Regulators of G-protein signalling as new central nervous system drug targets. *Nat. Rev. Drug Discov.* 1:187–97

Noble F, Cox BM. 1996. Differential desensitization of mu- and delta-opioid receptors in selected neural pathways following chronic morphine treatment. *Br. J. Pharmacol.* 117:161–69

Oakley RH, Laporte SA, Holt JA, Caron MG, Barak LS. 2000. Differential affinities of visual arrestin, beta arrestin1, and beta arrestin2 for G protein-coupled receptors delineate two major classes of receptors. *J. Biol. Chem.* 275:17201–10

Ohguro H, Chiba S, Igarashi Y, Matsumoto H, Akino T, Palczewski K. 1993. Beta-arrestin and arrestin are recognized by autoantibodies in sera from multiple sclerosis patients. *Proc. Natl. Acad. Sci. USA* 90:3241–45

Ohsawa M, Mizoguchi H, Narita M, Nagase H, Dun NJ, Tseng LF. 2003. Involvement of beta-arrestin-2 in modulation of the spinal antinociception induced by mu-opioid receptor agonists in the mouse. *Neurosci. Lett.* 346:13–16

Overstreet DH, Miller CS, Janowsky DS, Russell RW. 1996. Potential animal model of multiple chemical sensitivity with cholinergic supersensitivity. *Toxicology* 111:119–34

Ozaita A, Escriba PV, Ventayol P, Murga C, Mayor F Jr, Garcia-Sevilla JA. 1998. Regulation of G protein-coupled receptor kinase 2 in brains of opiate-treated rats and human opiate addicts. *J. Neurochem.* 70:1249–57

Pandey GN, Garver DL, Tamminga C, Ericksen S, Ali SI, Davis JM. 1977. Postsynaptic supersensitivity in schizophrenia. *Am. J. Psychiatry* 134:518–22

Parruti G, Peracchia F, Sallese M, Ambrosini G, Masini M, et al. 1993. Molecular analysis of human beta-arrestin-1: cloning, tissue distribution, and regulation of expression. Identification of two isoforms generated by alternative splicing. *J. Biol. Chem.* 268:9753–61

Penela P, Alvarez-Dolado M, Munoz A, Mayor F Jr. 2000. Expression patterns of the regulatory proteins G protein-coupled receptor kinase 2 and beta-arrestin 1 during rat postnatal brain development: effect of hypothyroidism. *Eur. J. Biochem.* 267:4390–96

Penn RB, Pronin AN, Benovic JL. 2000. Regulation of G protein-coupled receptor kinases. *Trends Cardiovasc. Med.* 10:81–89

Peppel K, Boekhoff I, McDonald P, Breer H, Caron MG, Lefkowitz RJ. 1997. G protein-coupled receptor kinase 3 (GRK3) gene disruption leads to loss of odorant receptor desensitization. *J. Biol. Chem.* 272:25425–28

Perroy J, Adam L, Qanbar R, Chenier S, Bouvier M. 2003. Phosphorylation-independent desensitization of GABA(B) receptor by GRK4. *EMBO J.* 22:3816–24

Perry SJ, Lefkowitz RJ. 2002. Arresting developments in heptahelical receptor signaling and regulation. *Trends Cell Biol.* 12:130–38

Picetti R, Saiardi A, Abdel Samad T, Bozzi Y, Baik JH, Borrelli E. 1997. Dopamine D2 receptors in signal transduction and behavior. *Crit. Rev. Neurobiol.* 11:121–42

Pitcher JA, Freedman NJ, Lefkowitz RJ. 1998a. G protein-coupled receptor kinases. *Annu. Rev. Biochem.* 67:653–92

Pitcher JA, Hall RA, Daaka Y, Zhang J, Ferguson SS, et al. 1998b. The G protein-coupled receptor kinase 2 is a microtubule-associated protein kinase that phosphorylates tubulin. *J. Biol. Chem.* 273:12316–24

Premont RT, Koch WJ, Inglese J, Lefkowitz RJ. 1994. Identification, purification, and characterization of GRK5, a member of the family of G protein-coupled receptor kinases. *J. Biol. Chem.* 269:6832–41

Premont RT, Inglese J, Lefkowitz RJ. 1995. Protein kinases that phosphorylate activated G protein-coupled receptors. *FASEB J.* 9:175–82

Premont RT, Macrae AD, Stoffel RH, Chung N, Pitcher JA, et al. 1996. Characterization of the G protein coupled receptor kinase GRK4. Identification of four splice variants. *J. Biol. Chem.* 271:6403–10

Premont RT, Macrae AD, Aparicio SA, Kendall HE, Welch J, Lefkowitz RJ. 1999. The GRK4 subfamily of G protein-coupled receptor kinases: alternative splicing, gene organization and sequence conservation. *J. Biol. Chem.* 274:29381–89

Premont RT, Gainetdinov RR, Caron MG. 2001. Following the trace of elusive amines. *Proc. Natl. Acad. Sci. USA* 98:9474–75

Pronin AN, Satpaev DK, Slepak VZ, Benovic JL. 1997. Regulation of G protein-coupled receptor kinases by calmodulin and localization of the calmodulin binding domain. *J. Biol. Chem.* 272:18273–80

Pronin AN, Morris AJ, Surguchov A, Benovic JL. 2000. Synucleins are a novel class of substrates for G protein-coupled receptor kinases. *J. Biol. Chem.* 275:26515–22

Przewlocka B, Sieja A, Starowicz K, Maj M, Bilecki W, Przewlocki R. 2002. Knockdown of spinal opioid receptors by antisense targeting beta-arrestin reduces morphine tolerance and allodynia in rat. *Neurosci. Lett.* 325:107–10

Rahman Z, Schwarz J, Gold SJ, Zachariou V, Wein MN, et al. 2003. RGS9 modulates dopamine signaling in the basal ganglia. *Neuron* 38:941–52

Robinson TE, Berridge KC. 1993. The neural basis of drug craving: an incentive-sensitization theory of addiction. *Brain Res. Rev.* 18:247–91

Rockman HA, Choi DJ, Rahman NU, Akhter SA, Lefkowitz RJ, Koch WJ. 1996. Receptor-specific in vivo desensitization by the G protein-coupled receptor kinase-5 in transgenic mice. *Proc. Natl. Acad. Sci. USA* 93:9954–59

Rockman HA, Choi DJ, Akhter SA, Jaber M, Giros B, et al. 1998. Control of myocardial contractile function by the level of beta-adrenergic receptor kinase 1 in gene-targeted mice. *J. Biol. Chem.* 273:18180–84

Rockman HA, Koch WJ, Lefkowitz RJ. 2002. Seven-transmembrane-spanning receptors and heart function. *Nature* 415:206–12

Sallese M, Salvatore L, D'Urbano E, Sala G, Storto M, et al. 2000. The G-protein-coupled receptor kinase GRK4 mediates homologous desensitization of metabotropic glutamate receptor 1. *FASEB J.* 14:2569–80

Sapolsky RM. 1998. The stress of Gulf War syndrome. *Nature* 393:308–9

Schulz R, Wehmeyer A, Murphy J, Schulz K. 1999. Phosducin, beta-arrestin and opioid receptor migration. *Eur. J. Pharmacol.* 375: 349–57

Schwartz JC, Levesque D, Martres MP, Sokoloff P. 1993. Dopamine D3 receptor: basic and clinical aspects. *Clin. Neuropharmacol.* 16:295–314

Shenoy SK, Lefkowitz RJ. 2003. Multifaceted roles of beta-arrestins in the regulation of seven-membrane-spanning receptor trafficking and signalling. *Biochem. J.* 375:503–15

Sibley DR. 1999. New insights into dopamine receptor function using antisense and genetically altered animals. *Annu. Rev. Pharmacol. Toxicol.* 39:313–41

Sills TL, Fletcher PJ. 1997. Fluoxetine attenuates morphine-induced locomotion and blocks morphine-sensitization. *Eur. J. Pharmacol.* 337:161–64

Sim LJ, Selley DE, Dworkin SI, Childers SR. 1996. Effects of chronic morphine administration on mu opioid receptor-stimulated [35S]GTPgammaS autoradiography in rat brain. *J. Neurosci.* 16:2684–92

Singer HS. 1994. Neurobiological issues in Tourette syndrome. *Brain Dev.* 16:353–64

Seeman P, Van Tol HH. 1994. Dopamine receptor pharmacology. *Trends Pharmacol. Sci.* 15:264–70

Snyder SH, Pasternak GW. 2003. Historical review: opioid receptors. *Trends Pharmacol. Sci.* 24:198–205

Sterne-Marr R, Benovic JL. 1995. Regulation of G protein-coupled receptors by receptor kinases and arrestins. *Vitam. Horm.* 51:193–234

Stoffel RH, Randall RR, Premont RT, Lefkowitz RJ, Inglese J. 1994. Palmitoylation of G protein-coupled receptor kinase, GRK6: lipid modification diversity in the GRK family. *J. Biol. Chem.* 269:27791–94

Stoffel RH, Inglese J, Macrae AD, Lefkowitz RJ, Premont RT. 1998. Palmitoylation increases the kinase activity of the G protein-coupled receptor kinase, GRK6. *Biochemistry* 37:16053–59

Sunayashiki-Kusuzaki K, Kikuchi T, Wawrousek EF, Shinohara T. 1997. Arrestin and phosducin are expressed in a small number of brain cells. *Brain Res. Mol. Brain Res.* 52:112–20

Tao R, Auerbach SB. 1994. Increased extracellular serotonin in rat brain after systemic or intraraphe administration of morphine. *J. Neurochem.* 63:517–24

Terman GW, Jin W, Cheong YP, Lowe J, Caron MG, et al. 2004. G-protein receptor kinase 3 (GRK3) influences opioid analgesic tolerance but not opioid withdrawal. *Br. J. Pharmacol.* 141:55–64

Terwilliger RZ, Ortiz J, Guitart X, Nestler EJ. 1994. Chronic morphine administration increases β-adrenergic receptor kinase (βARK) levels in the rat locus coeruleus. *J. Neurochem.* 63:1983–86

Tiberi M, Nash SR, Bertrand L, Lefkowitz RJ, Caron MG. 1996. Differential regulation of dopamine D1a responsiveness by various G protein-coupled receptor kinases. *J. Biol. Chem.* 271:3771–78

Uhl GR, Sora I, Wang Z. 1999. The mu opiate receptor as a candidate gene for pain: polymorphisms, variations in expression, nociception, and opiate responses. *Proc. Natl. Acad. Sci. USA* 96:7752–55

Vassilatis DK, Hohmann JG, Zeng H, Li F, Ranchalis JE, et al. 2003. The G protein-coupled receptor repertoires of human and mouse. *Proc. Natl. Acad. Sci. USA* 100:4903–8

Vroon A, Heijnen CJ, Raatgever R, Touw IP, Ploemacher RE, et al. 2004. GRK6 deficiency is associated with enhanced CXCR4-mediated neutrophil chemotaxis in vitro and impaired responsiveness to G-CSF in vivo. *J. Leukoc. Biol.* 75:doi:10.1189/jlb.0703320. In press

Walker JKL, Peppel K, Lefkowitz RJ, Caron MG, Fisher JT. 1999. Altered airway and cardiac responses in mice lacking G protein-coupled receptor kinase 3. *Am. J. Physiol.* 276:R1214–21

Walker JKL, Fong AM, Lawson BL, Savov JD, Patel DD, et al. 2003. Beta-arrestin-2 regulates the development of allergic asthma. *J. Clin. Invest.* 112:566–74

Walker JKL, Gainetdinov RR, Feldman DS, McFawn PK, Caron MG, et al. 2004. G protein-coupled receptor kinase 5 regulates airway responses induced by muscarinic receptor activation. *Am. J. Physiol. Lung Cell. Mol. Physiol.* 286:L312–19

Wang YM, Gainetdinov RR, Fumagalli F, Xu F, Jones SR, et al. 1997. Knockout of the vesicular monoamine transporter 2 gene results in neonatal death and supersensitivity to cocaine and amphetamine. *Neuron* 19:1285–96

Watanabe H, Xu J, Bengra C, Jose PA, Felder RA. 2002. Desensitization of human renal D1 receptors by G protein-coupled receptor kinase 4. *Kidney Int.* 62:790–98

Watson S, Arkinstall S, ed. 1994. *The G-Protein Linked Receptor Facts-Book.* San Diego: Academic

Wess J. 2000. Physiological roles of G-protein-coupled receptor kinases revealed by gene-targeting technology. *Trends Pharmacol. Sci.* 21:364–67

Whistler JL, von Zastrow M. 1998. Morphine-activated opioid receptors elude desensitization by beta-arrestin. *Proc. Natl. Acad. Sci. USA* 95:9914–19

Whistler JL, Chuang HH, Chu P, Jan LY, von Zastrow M. 1999. Functional dissociation of mu opioid receptor signaling and endocytosis: implications for the biology of opiate tolerance and addiction. *Neuron* 23:737–46

Wickman K, Clapham DE. 1995. Ion channel regulation by G proteins. *Physiol. Rev.* 75:865–85

Wilcox RE, Smith RV, Anderson JA, Riffee WH. 1980. Apomorphine-induced stereotypic cage climbing in mice as a model for studying changes in dopamine receptor sensitivity. *Pharmacol. Biochem. Behav.* 12:29–33

Willets J, Kelly E. 2001. Desensitization of endogenously expressed delta-opioid receptors: no evidence for involvement of G protein-coupled receptor kinase 2. *Eur. J. Pharmacol.* 431:133–41

Willets JM, Challiss RA, Nahorski SR. 2003. Non-visual GRKs: Are we seeing the whole picture? *Trends Pharmacol. Sci.* 24:626–33

Wise RA, Bozarth MA. 1987. A psychomotor stimulant theory of addiction. *Psychol. Rev.* 94:469–92

Wise A, Jupe SC, Rees S. 2004. The identification of ligands at orphan G-protein coupled receptors. *Annu. Rev. Pharmacol. Toxicol.* 44:43–66

Yamamoto S, Sippel KC, Berson EL, Dryja TP. 1997. Defects in the rhodopsin kinase gene in the Oguchi form of stationary night blindness. *Nat. Genet.* 15:175–78

Zachariou V, Georgescu D, Sanchez N, Rahman Z, DiLeone R, et al. 2003. Essential role for RGS9 in opiate action. *Proc. Natl. Acad. Sci. USA* 100: 13656–61

Zhang J, Ferguson SS, Barak LS, Bodduluri SR, Laporte SA, et al. 1998. Role for G protein-coupled receptor kinase in agonist-specific regulation of mu-opioid receptor responsiveness. *Proc. Natl. Acad. Sci. USA* 95:7157–62

Zhang J, Barak LS, Anborgh PH, Laporte SA, Caron MG, Ferguson SS. 1999. Cellular trafficking of G protein-coupled receptor/beta-arrestin endocytic complexes. *J. Biol. Chem.* 274:10999–1006

Annu. Rev. Neurosci. 2004. 27:145–67
doi: 10.1146/annurev.neuro.27.070203.144308

First published online as a Review in Advance on February 23, 2004

PLASTICITY OF THE SPINAL NEURAL CIRCUITRY AFTER INJURY*

V. Reggie Edgerton,[1,2,3] Niranjala J.K. Tillakaratne,[1,3] Allison J. Bigbee,[2] Ray D. de Leon,[4] and Roland R. Roy[1,3]
[1]Physiological Science, [2]Neurobiology, and [3]Brain Research Institute, University of California, Los Angeles, California 90095; [4]Kinesiology and Nutritional Science, California State University, Los Angeles, California 90032; email: vre@ucla.edu, nirat@lifesci.ucla.edu, abigbee@ucla.edu, rdeleon@calstatela.edu, rrr@ucla.edu

Key Words automaticity, motor learning, locomotor rehabilitation, therapeutic strategies

■ **Abstract** Motor function is severely disrupted following spinal cord injury (SCI). The spinal circuitry, however, exhibits a great degree of automaticity and plasticity after an injury. Automaticity implies that the spinal circuits have some capacity to perform complex motor tasks following the disruption of supraspinal input, and evidence for plasticity suggests that biochemical changes at the cellular level in the spinal cord can be induced in an activity-dependent manner that correlates with sensorimotor recovery. These characteristics should be strongly considered as advantageous in developing therapeutic strategies to assist in the recovery of locomotor function following SCI. Rehabilitative efforts combining locomotor training pharmacological means and/or spinal cord electrical stimulation paradigms will most likely result in more effective methods of recovery than using only one intervention.

INTRODUCTION

There is a growing awareness that a high level of functional recovery of standing and stepping can be achieved in adult mammals following a complete spinal cord injury (SCI). It is also clear that the level of recovery in motor function is defined,

---

*__Abbreviations__: 5-hydroxytryptamine (5-HT); 2-amino-5-phosphonovaleric acid (AP-5); American Spinal Injury Association (ASIA); brain-derived neurotrophic factor (BDNF); central nervous system (CNS); central pattern generation (CPG); electrical stimulation (ES); electromyography (EMG); extensor digitorum longus (EDL); gamma-aminobutyric acid (GABA); glial fibrillary acidic protein (GFAP); glutamic acid decarboxylase (GAD); long-term potentiation (LTP); long-term depression (LTD); m-chlorophenylpiperazine (m-CPP); medial gastrocnemius (MG); neurotrophin-3 (NT-3); noradrenaline (NA); N-methyl-D-aspartate (NMDA); paw contact (PC); peripheral nervous system (PNS); soleus (SOL); spinal cord injury (SCI); swing-phase force field (SWPFF); tibialis anterior (TA).

to a significant degree, by the level and types of motor training or experience following the injury (Edgerton et al. 2001a, Wernig et al. 1995). These findings suggest strongly that any promising interventional strategies for regeneration and redevelopment of supraspinal-spinal connectivity following SCI are likely to be more successful when the intervention is combined with some type of motor training. Insight into the mechanisms of recovery depends on a basic understanding of the neural control of standing and stepping in the uninjured compared to the injured spinal cord, as well as the properties and adaptations of the force generating tissue, i.e., muscle-tendon units. Much of the insight we have gained in the neural control of locomotion is attributable to the fact that the basic networks that control locomotion in animals that walk, fly, and/or swim are remarkably similar. The physiological properties of the supraspinal and spinal circuits that control posture and locomotion are similar across many vertebrate systems. Glutamate and glycine are essential neurotransmitter systems for central pattern generation (CPG) in virtually all vertebrates. Similarly, noradrenergic and serotonergic neurotransmitter systems are important modulators of the basic rhythmic cycle associated with locomotion among most vertebrates (Grillner 2003). The extensive conservation of the neuromotor control system and its neuromodulators and neurotransmitters enable us to gain a deeper understanding of the functional properties of complex systems by studying simple animal models (Grillner 2003). This in turn has also been an important factor in helping to understand the recovery of locomotion following SCI.

In this review we highlight some features of the normal neuromotor control apparatus and how SCI changes this neural control. An understanding of this control provides a reasonable rationale for the level of plasticity that persists and the highly functional state of posture and locomotion that can be achieved in the neuromusculoskeletal system following SCI. Because the number of papers published on the topic of SCI over the past few years has risen rapidly, we restricted the present review primarily to those papers that include a complete SCI and motor training post-SCI. The two objectives of this review are (*a*) to summarize the level of motor or functional control that can persist in the lumbosacral spinal cord in the absence of supraspinal control, and (*b*) to provide some insight into the physiological and biochemical changes that can occur in the lumbosacral segments, and the sensory and motor adaptations that can account for locomotor recovery after SCI. A number of recent reviews have been written on similar topics (Barbeau et al. 2002; Edgerton et al. 2001a; Edgerton & Roy 2002; Grillner 1981, 2003; Grillner & Wallen 1985; Harkema 2001; Rossignol et al. 2002a,b; Wolpaw & Tennissen 2001).

AUTOMATICITY IN POSTURE AND LOCOMOTION: SOME BASIC NEUROBIOLOGICAL PRINCIPLES OF MOTOR CONTROL BEFORE AND AFTER SCI

In the uninjured state, the spinal cord, along with the brain, plays a considerable role in the automaticity of motor control (Baev 1998, Edgerton et al. 2004, Orlovsky & Feldman 1972). Automaticity describes the concept by which one is able to carry out complex but routine motor tasks without "conscious" thought, such as walking across a room. Descending brain control, central pattern generator activity, and peripheral inputs are components of automaticity. Following SCI, the spinal circuitry below the lesion site does not become silent; rather it continues to maintain active and functional neuronal properties. The potential for the spinal circuitry to generate oscillating coordinated motor patterns remains, and peripheral input continues to flow into the spinal cord and be processed after a complete SCI, although in a modified manner. Spinal automaticity is then comprised of CPG circuitry combined with sensory input from the periphery.

One can ask, how did automaticity with such utility evolve? All animals on Earth with vastly different neuromusculoskeletal structures have evolved within a 1 *g* gravitational environment. Over millions of years the execution of many complex postural and locomotor tasks has evolved to become automatic in many ways within this constant environment (Edgerton et al. 2001b). The automaticity of the neural control of locomotion and posture facilitates the execution of complicated tasks that might otherwise require decision-making processes that would take additional time. Automaticity also frees selected neural networks that can be reserved for less predictable events. From this perspective spinal automaticity seems to have been learned in the evolutionary process, i.e., "evolutionary learning" has played a key role in shaping the mammalian neural systems that control posture and locomotion. A few obvious examples of the automaticity that is built into the neural control of posture and locomotion are shorter steps taken by the inside limb when walking in an arc or by the limb of the side of the body to which the head is turned; increasing the level of excitation of extensor muscles when the body is bearing more load; and enhancing flexion of the ipsilateral leg when stepping over an object versus enhancing extension in the contralateral limb. In the extreme case of SCI where the spinal cord is completely severed in the midthoracic region, there is no supraspinal control. In this case the automaticity within the spinal cord becomes even more critical for the CPG circuitry to successfully process sensory input. The manner in which these features of the control system are affected after a complete SCI is discussed below.

Supraspinal Control of Posture and Locomotion

Shik & Orlovsky (1976) proposed a two-level automatism control system for locomotion. One level provides nonspecific tonic input, which determines the speed of locomotion. The second level is responsible for making fine adjustments in the control of the limbs, including maintaining equilibrium, by interacting with multiple modes of sensory information such as proprioceptive, visual, and auditory inputs (Figure 1, see color insert). Normally there are several descending spinal tracts that project motor commands from the brain to the spinal cord, including the reticulospinal, vestibulospinal, rubrospinal, and corticospinal tracts. Each of these tracts receives multiple modes of input and incorporates peripheral sensory modalities with information from the cerebellum and other supraspinal structures to produce appropriate locomotor responses in many different situations.

The specific control features of each of the descending spinal tracts in controlling locomotion are poorly understood. Also the anatomy of these tracts differs to some degree from species to species. Although the detailed clinical manifestation of a SCI will reflect the amount of tissue damage, it seems that the specific tracts that are damaged are also an important factor. However, there has been relatively little focus on how the loss of specific tracts following SCI relates to the specific kinds and levels of motor function that remain after the injury. For example, the corticospinal tract of quadrupedal animals does not play an essential role in generating the basic locomotor pattern, but it is very important in making fine adjustments (Eidelberg 1981, Grillner 2003). Although damage to this tract will not result in a severe deficit during treadmill locomotion, these fine adjustments in the basic activation patterns during locomotion in a variable environment can be important. In nonhuman primates, however, the basic locomotor patterns can be relatively normal without corticospinal input, with most of the dysfunction occurring primarily in the distal musculature, i.e., in the wrist, ankle, and digits (Eidelberg 1981, Hodgson et al. 2003). After a more severe SCI, however, the integral role of spinally mediated mechanisms in postural and locomotor control becomes more apparent (Figure 1*B*).

Spinal Control of Posture and Locomotion

CENTRAL PATTERN GENERATION CPG is a physiological phenomenon in which oscillatory motor output is generated in the absence of any oscillatory input. In mammalian, nonhuman systems, CPGs within the lumbosacral spinal cord segments represent an important component of the total circuitry that generates and controls posture and locomotion. With an appropriate pharmacological stimulus, a highly coordinated oscillatory pattern, i.e., alternating flexion and extension (fictive locomotion), can be generated in a fictive preparation by the circuitry within the spinal cord in the absence of input from the brain or from the periphery. During this oscillatory behavior, interneurons are active throughout all layers of the gray matter of the spinal cord and over all of the lumbosacral segments (Edgerton et al. 1976, Jordan & Schmidt 2002). Many excellent reviews on CPG and locomotion have been written (Barbeau et al. 2002; Dietz 2003; Edgerton et al. 2001a; Grillner 2002, 2003; Grillner & Wallen 1985; Harkema 2001; Rossignol et al. 2002a,b; Wolpaw & Tennissen 2001) and provide more detailed information. Although there are reports that claim to demonstrate CPG in nonhuman primates (Fedirchuk et al. 1998) and in humans (Calancie et al. 1994, Dimetrievic et al. 1998, Gurfinkel et al. 1997,Gerasimenko et al. 2002), the evidence is not conclusive, largely because of the infeasibility of eliminating all peripheral as well as supraspinal input to the spinal cord, at least in humans.

SENSORY INPUT An intact and functional neuromotor control system cannot rely solely on CPG or on direct supraspinal control to generate effective movements, particularly in a dynamic environment. The profuse sources of peripheral sensory information represent an integral component of the neuromotor control system, allowing the spinal cord to effectively interface with its physical environment. The spinal cord receives sensory input from all of the mechanoreceptors and cutaneous receptors and produces the appropriate temporal response, thus generating many highly adaptable motor tasks. In this context, it is convenient to think of the spinal cord as interpreting the ensemble of sensory information at any given time, as opposed to responding to each receptor in a stereotypically reflexive manner. This feature of the spinal cord makes it "smart" and provides the foundation for a substantial portion of the recovery of postural and locomotor function following SCI.

The intrinsic CPG activity and the sensory input to the spinal circuitry combined allow for rapid adjustment to parameters such as the speed of stepping, the level of load imposed during stepping, and a wide range of unpredictable sensory anomalies (Edgerton et al. 2001a). We propose that the spinal cord processes and interprets proprioception in a manner similar to how our visual system processes information. When we view a painting, the brain interprets the total visual field, as opposed to processing each individual pixel of information independently, and then derives an image. At any instant the spinal cord receives an ensemble of information from all receptors throughout the body that signals a proprioceptive "image" that represents time and space, and it computes "online" which neurons to excite next based on the most recently perceived "images." Those CPG neurons that generate locomotor patterns appear to predict the next sequence of neurons to activate on the basis of the specific groups of neurons that were just activated. The importance of the CPG is not simply its ability to generate repetitive cycles, but also to receive, interpret, and predict the appropriate sequences of actions during any part of the step cycle, i.e., "state dependence." The peripheral input then provides important information from which the probabilities of a given set of neurons being active at any given time can be finely tuned to a given situation during a specific phase of a step cycle. An excellent example of this is when a mechanical stimulus is applied to the dorsum of the paw of a cat. When the stimulus is applied during the swing phase, the flexor muscles of that limb are excited, and the result is enhanced flexion. However, when the same stimulus is applied during stance, the extensors are excited (Forssberg et al. 1975). Thus, the functional connectivity between mechanoreceptors and specific interneuronal populations within the spinal cord varies according to the physiological state. Even the efficacy of the monosynaptic input from muscle spindles to the motoneuron changes readily from one part of the step cycle to another, according to whether a subject is running or walking (Simonsen et al. 1999).

The injured spinal cord interprets sensory changes in load and speed The spinal cord clearly interprets ensembles of information derived from mechanoreceptors in the musculoskeletal system and activates the appropriate motor pools in a precise and highly coordinated manner in response to changes in speed and load. An example of the spinal cord's ability to receive complex proprioceptive information and use it in a functional way is shown in Figure 2.

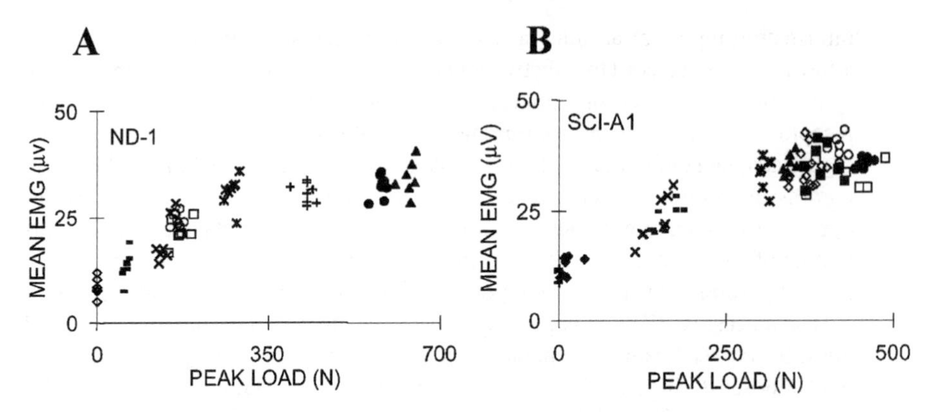

Figure 2 Loading alters the activation levels of motor pools during stepping in human SCI patients. The relationships between soleus EMG mean amplitude (μV) and limb peak load (N) from a nondisabled subject (*A*, ND-1) and a subject with a SCI classified on the ASIA scale as an A, i.e., having a complete injury (*B*, SCI-A1), over a range of loading conditions are shown. Each data point represents one step, and each symbol represents a series of consecutive steps at one level of body-weight support. (Reproduced with permission from Edgerton et al. 2001a.)

The activation level of an extensor muscle, in this case the soleus, is modulated according to the amount of load placed on the lower limbs of a nondisabled human subject (Harkema et al. 1997). The increase in activation (Figure 2*A*), illustrated by the amplitude of the electromyographic (EMG) signal, is directly related to the load imposed on the limb. Data from a similar experiment in a complete SCI patient is also shown (Figure 2*B*). Even though the SCI subject has no voluntary control of any muscles and no sense of sensation from tissues below the lesion, there is a clear similarity in the relationship between the level of loading and the level of motor pool activation as reflected by the EMG amplitude compared to the nondisabled subject.

Similar adaptations have been observed with respect to speed (Forssberg et al. 1980a,b) and force modulation (Figure 3; Lovely et al. 1990) in low thoracic, chronically spinalized cats during assisted treadmill locomotion. Indirect evidence suggests that the spinal cord neural circuits vary the speed of locomotion primarily by modulating the level of excitation of the extensor muscles. For example, there are greater angular excursions, enhanced muscle forces, and shorter EMG durations in the SOL and MG at a faster (Figure 4*B*) than a slower (Figure 4*A*) treadmill speed in complete SCI cats (Figure 4; Lovely et al. 1990). However, experiments in decerebrate cats showed that if treadmill speed is kept constant, increased input into the midbrain has no effect on the duration of the stance phase or step frequency (Shik et al. 1966). This latter observation suggests that the cycle duration is influenced by mechanical factors such as the position or the placement of the hindlimb and, therefore, is mediated by proprioceptive signals.

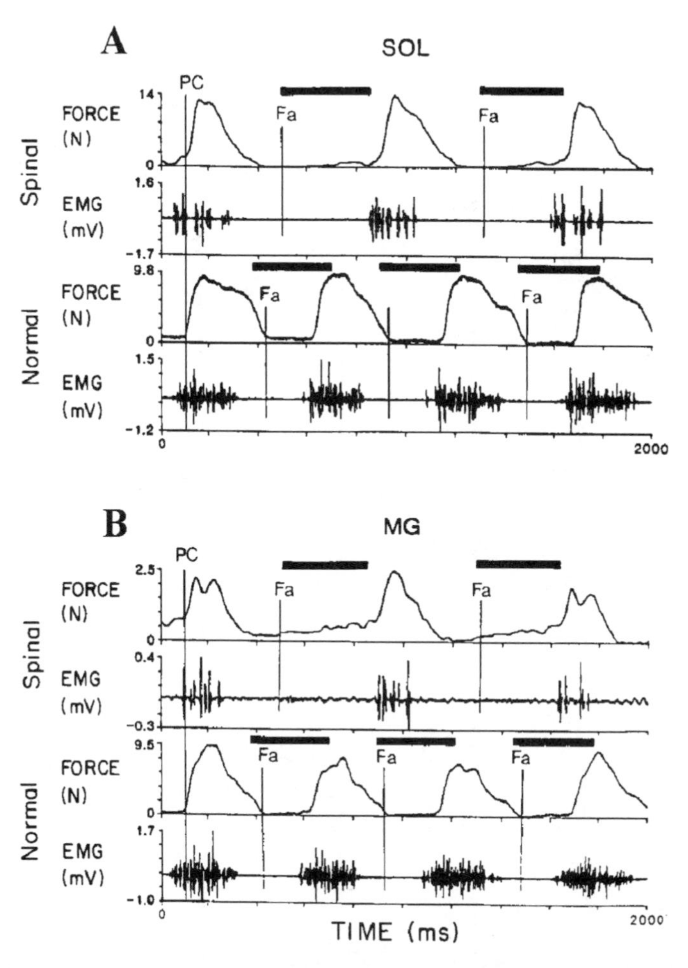

Figure 3 Hindlimb muscle EMG and force production (via tendon force transducers) are prominent during assisted treadmill locomotion in adult spinal cats. Comparisons of the activation patterns of the soleus (SOL) and medial gastrocnemius (MG), and corresponding force production in a normal (*A*) and a complete low thoracic spinal cat that has been trained to step on a treadmill with full weight-bearing (*B*) are shown. The beginning of the stance phase (paw contact; PC) and the beginning of the swing phase (ankle flexion; Fa) are noted. The EMG bursts per step are less robust, and there is a slightly more irregular and more brief force pattern in both the SOL and MG after than before SCI. The peak force levels are also lower after than before SCI, particularly in the MG, which has a higher percentage of larger motoneurons that require more excitatory drive to be activated. However, there are significant forces and activation levels in the muscles of this complete spinal cat. The bold bars indicate the period of contralateral support. (Reproduced with permission from Lovely et al. 1990.)

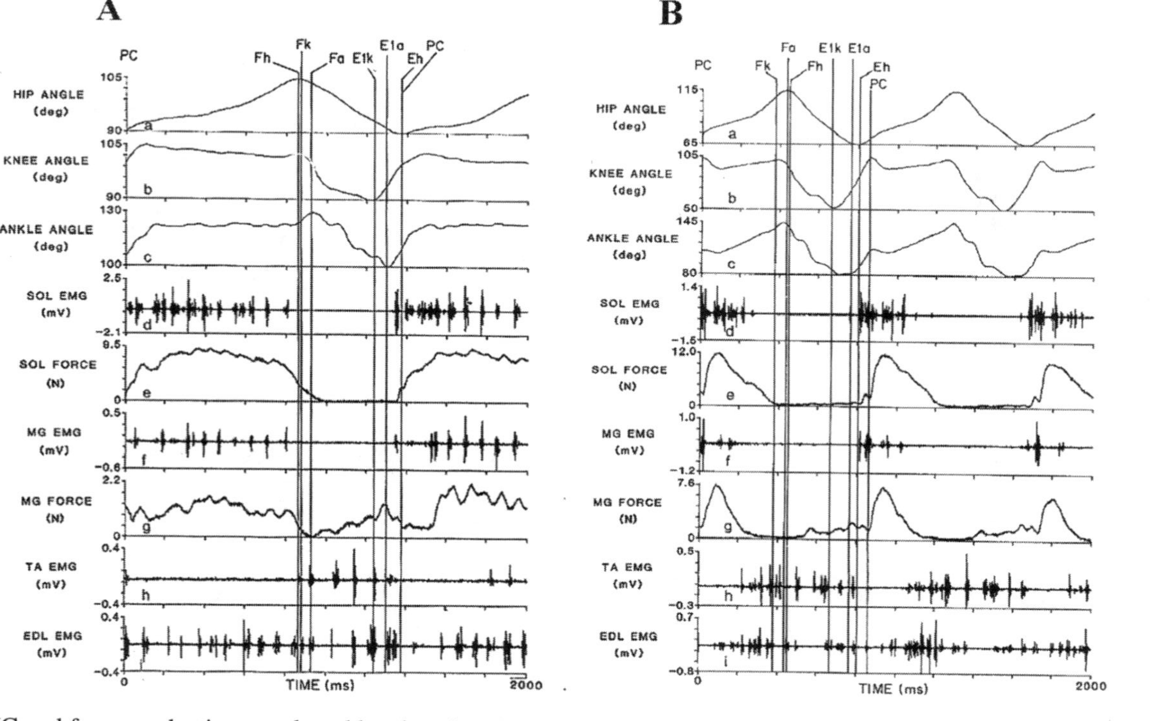

Figure 4 EMG and force production are altered by changing the speed of assisted treadmill locomotion in complete SCI cats. EMG and force patterns of the soleus (SOL), medial gastrocnemius (MG), tibialis anterior (TA), and extensor digitorum longus (EDL) of an adult cat with a complete SCI during stepping at a slow speed (*A*; 0.1 $m \cdot sec^{-1}$) and at a faster, but modest, speed (*B*; 0.6 $m \cdot sec^{-1}$) on a motor-driven treadmill. Clonus is more evident in the EMG and force records of these muscles at the slower than at the faster speed of stepping. Also note the shorter stance period at the faster speed, indicating the modulation of extensor power output (greater recruitment for a shorter period) to walk faster. The vertical lines at the beginning of the SOL force indicate the beginning of stance, i.e., paw contact (PC), as derived from high-speed film (200 fps). Note the different scales for the force and EMG records in *A* and *B*. Hip flexion, Fh; knee flexion, Fk; ankle flexion, Fa; knee extension prior to PC, E1k; ankle extension prior to PC, E1a; hip extension, Eh. (Reproduced with permission from Lovely et al. 1990.)

SCI patients can voluntarily initiate locomotion An understanding of the extensive level of built-in spinal automaticity can be used in developing therapeutic strategies to facilitate functional recovery after SCI. With the increasing success in improving the locomotor potential of laboratory animals, however, the question has been raised as to the functional significance of this motor recovery in humans after a complete SCI, since there is presumably no voluntary initiation of stepping or standing. This presumption, however, is not entirely correct. Both standing and stepping can be initiated voluntarily, although indirectly, in SCI patients. For example, when an ASIA A SCI patient is standing with bilateral weight support, stepping can be initiated (consciously) by shifting the body weight to one leg and moving the head and trunk so that the hip position of the contralateral leg is extended by leaning the body forward (Wernig & Müller 1992). Proprioceptively, this maneuver is similar to unloading a limb at the end of the stance phase of a step, which facilitates the initiation of the swing phase of that leg (Duysens & Pearson 1980, Harkema 2001, Wernig & Mueller 1992). The same events initiate stepping on a moving treadmill belt because the weight-bearing limb moves backward into hip extension. A moving treadmill, however, is not required to successfully initiate stepping because the patient can use the upper body to alternate loading from one side to the other at the appropriate time to facilitate the joint actions bilaterally.

THE INJURED SPINAL CORD IS AN "ALTERED" SPINAL CORD After a SCI, supraspinal and spinal sources of control of movement differ substantially from that which existed prior to the injury, thus resulting in an altered spinal cord. The concept of an altered spinal cord after SCI implies that the spinal cord processes input and generates motor output in a different manner as a result of injury-related adaptations (Edgerton et al. 1997b, 2001b). How, then, does the automaticity of posture and locomotion emerge from the interactions between the sensory inputs and the CPG circuitry in spinalized animals? For these two systems to work in synergy, each system must have intrinsic activation and inhibition patterns that can generate coordinated motor outputs. For example, the sequence of muscle activation patterns associated with stepping can occur only if a critical level of synergy occurs between the peripheral inputs and the cyclic events of the CPG. At any given time "bin" within a step cycle, the sensory information interfaces with the CPG circuitry to (*a*) apprise the CPG circuitry about the present state of the step cycle, (*b*) inform the CPG circuitry about which "bin(s)" preceded that instant, and (*c*) predict which "bin(s)" should occur next. In this scenario a "bin" consists of that sensory input and the stage of activation-inhibition of the neurons that generate the CPG output at that instant. The temporal patterns of peripheral inputs must be matched with the intrinsic CPG activity for locomotion to continue effectively. After a complete SCI, the interaction between the CPG circuitry and the peripheral inputs is critical because a major source of control, the brain, has been eliminated.

Spasticity is a sign of activity in the spinal circuitry Spasticity might be considered as automaticity gone awry. Although extensive muscle contractions occur with spasticity, there is little reciprocal activation of agonistic and antagonistic motor pools, as occurs during locomotion (de Leon et al. 1998a). One possible explanation for a more-or-less random motor pool activation is that the spinal cord has lost the ability to synchronize and interpret the coordinated ensemble of afferent information that produces a predictable motor outcome in nondisabled subjects. This deficit may be due to the absence or rare occurrence of synchronized events normally associated with load-bearing stepping. In the absence of these coordinating events, the spinal cord loses the ability to synchronize input into functional movements of the limbs. The presence of spasticity is, however, a positive sign of the potential for a SCI patient to regain some locomotor ability. A patient who is hyporesponsive to sensory input is less likely to respond to the proprioceptive input associated with load-bearing stepping. Clearly, understanding the physiological mechanisms that underlie spasticity will enhance our efforts to facilitate locomotor recovery in SCI patients.

MOTOR OUTPUT IS ENHANCED BY REPETITIVE TRAINING

Chronic Motor Training Modulates Spinal Plasticity to Enhance Motor Output After SCI

Experiments with spinal cats using chronic locomotor training paradigms suggest that the ability to learn and successfully perform a motor task is dependent on repetitive practice (de Leon et al. 1998a,b; Lovely et al. 1986) and appears to be specific to the task being tested (de Leon et al. 1998a,b). Although training for a specific motor task such as stepping can enhance the capacity to perform that task, it may reduce the capacity to perform a different complex motor task that uses different patterns of information, e.g., standing (de Leon et al. 1998b). Cats that were trained to step perform that motor task well, whereas those trained for weight-bearing standing did not step well. These findings imply that the functional state of the cord is shaped by specific locomotor training regimens. Although the mechanisms underlying locomotor training-enhanced plasticity are not well understood, it is clear that the physiological state of the cord can be affected by activity-dependent processes that can influence its ability to learn and perform a specified motor behavior.

The Spinal Cord Can Respond to Novel, Acute Perturbations

Although spinal cord plasticity can be modulated to alter its motor capacity after a complete SCI, one may argue that this behavior requires long periods of motor training and new neuronal connections to form within the spinal cord. However, the state-dependent property of the spinal circuitry, which enables highly functional

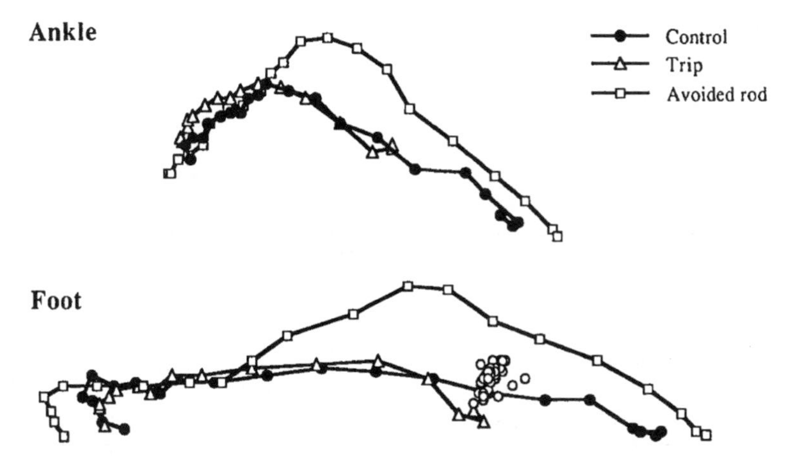

Figure 5 Swing trajectories of the foot (*lower traces*) and ankle (*upper traces*) of a spinal cat responding to the placement of a 1-cm diameter bar in the path of the foot during the swing phase of the step cycle. The control trace shows the trajectory before placement of the bar, the trip trace shows the trajectory when the cat hit the bar, and the remaining trace shows the trajectory when the foot cleared the bar. The bar position for each frame of the digitized video is shown by an open circle. Note the increased elevation of the foot well before it encountered the bar in the step where the rod was avoided. (Reproduced with permission from Hodgson et al. 1994.)

and integrated responses to occur, can also occur in novel, acute situations. For example, when an object is placed in front of a spinal cat stepping on a treadmill during the swing phase of one hind limb, that limb will exhibit a greater degree of flexion during the following steps to avoid the perturbation (Forssberg et al. 1975). This hyperflexion persists for several steps even after the removal of the perturbation (Figure 5; Hodgson et al. 1994, Nakada et al. 1994), which suggests that a learning and memory-type phenomenon may be taking place. This must be considered as more than a momentary adaptation because a memory trace is shown behaviorally in a number of steps immediately following the removal of the perturbation.

More recent evidence of the "smartness" of the spinal cord was demonstrated when a robotic device was used to apply specific forces to rat hindlimbs during certain phases of the step cycle. In one experiment a downward force proportional to the velocity of stepping was applied unilaterally or bilaterally to the ankle(s) of spinal rats. Step timing and kinematics were altered within a few steps, enabling locomotion to continue (Timoszyk et al. 2002). In a separate but similar set of experiments, a robotically induced upward force proportional to the forward velocity of the swing (SWPFF) was exerted at the ankle of one limb during the swing phase, resulting in a visually obvious kinematic disturbance. After as few as 20 steps, the limb adjusted its output to become more kinematically

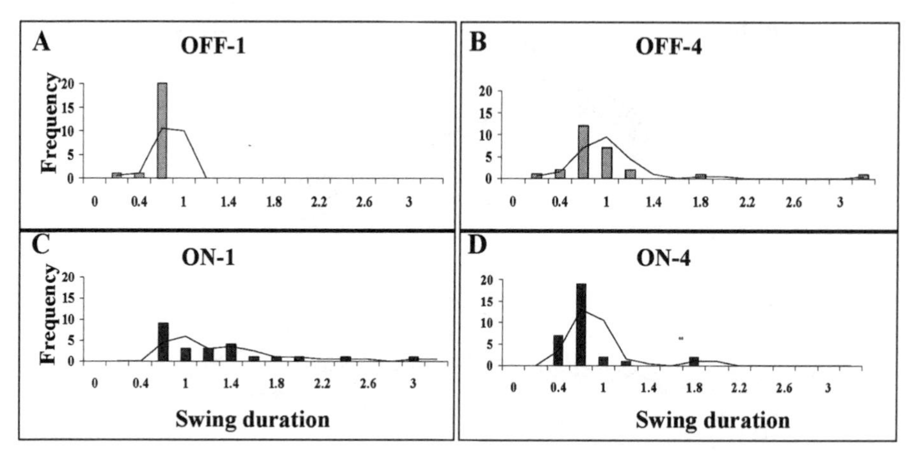

Figure 6 The spinal cord adapts in a functionally appropriate manner to an acute perturbation to normalize step kinematics. A robotic system applied an upward force (swing-phase force field; SWPFF) at one ankle of a complete spinal rat during bipedal, weight-supported stepping on a treadmill. The rat took ~20 steps with the SWPFF on and ~20 with the SWPFF off for 4 consecutive bouts (between 160–200 total steps). The first and fourth SWPFF-on and -off bouts are shown. (*A*) OFF-1 swing phase duration is highly consistent (mean ± SD = 551 ± 127 ms). (*B*) OFF-4 is similar to OFF-1 (683 ± 331 ms). (*C*) ON-1 SWPFF exposure increases the swing-phase duration (1103 ± 610 ms). (*D*) ON-4 swing duration shifted toward OFF-1 and OFF-4 levels (582 ± 326 ms). Thus, step characteristics return toward normal within a short period of time (about 4 min total) upon repeated exposure to the SWPFF perturbation.

similar to that observed prior to the perturbation, although with different rectus femoris EMG patterns (de Leon et al. 2002). Furthermore, with repetitive trials of ~20 steps on-, ~20 steps off × 4 bouts of the SWPFF paradigm, the average swing duration during the "force-field-on" period decreased to "force-field-off" levels by the fourth bout (Figure 6; Liu et al. 2003). These studies show that, in essence, the spinal cord is "solving" problems in real time based on the continually changing state of incoming peripheral information to elicit a nearly constant behavior, even though the means to the endpoint differ.

These studies imply that spinal learning can occur in a very short period of time, and a type of memory trace allows for quicker adaptation upon reexposure to a given perturbation. Although the underlying cellular mechanisms are unknown, studies are in progress to determine if some of the molecules and processes involved in spinal learning are similar to those involved in hippocampal learning, and whether this phenomenon continues for longer periods (hours to days). Indeed, long-term potentiation (LTP) and long-term depression (LTD), which are vital in hippocampal learning, are also exhibited in dorsal horn neurons in the spinal cord in response to nociceptive stimuli (Garraway & Hochman 2001, Ikeda et al. 2003, Rygh et al. 2002). These observations confirm that hippocampal learning–like phenomena can occur in the spinal cord. Whether spinal motor learning occurs via similar processes or affects spinally mediated pain responses remains to be determined, but it will be an interesting topic for future study.

Biochemical and Pharmacological Evidence for Spinal Cord Plasticity After Injury

As shown above, activity-dependent motor training facilitates the recovery of posture and locomotion after a complete SCI in mammals. Because the functionally recovered spinal animals showed no evidence of regeneration of descending pathways (Joynes et al. 1999) or showed minimal changes in hindlimb skeletal muscle properties (Roy et al. 1999, 1998; Roy & Acosta 1986) to account for the recovery characteristics, the functional behavior exhibited by these animals must have been mediated by the plasticity in existing spinal pathways. This plasticity may occur at any of many spinal cord regions or cell types such as motoneurons, premotor pattern-generating neurons, and/or nonneuronal cell types. There could also be anatomically altered synaptic connections, increased active zones of synapses, altered sensitivities of neurotransmitter receptors, or altered production of neurotransmitters.

Pharmacologically induced activation of the spinal cord has been well studied. In the isolated neonatal rodent spinal cord, fictive locomotion can be activated by application of different neurotransmitters such as N-methyl-D-aspartate (NMDA), 5-hydroxytryptamine (5-HT), or dopamine, resulting in a stable rhythmic motor output (Cazalets et al. 1998, Kiehn & Kjaerulff 1998, Kudo & Nishimaru 1998, Schmidt et al. 1998). These patterns closely resemble the activation of hindlimb muscles during locomotion in the intact adult rat (Roy et al. 1991). These data showed a significant role for these neurotransmitter systems in facilitating locomotor activity and thus suggest that these agents might be useful for inducing locomotion in SCI animals. Indeed, the alpha-2-noradrenergic (NA) agonist clonidine induced stepping early (within 1 week postinjury) in adult, nontrained spinal cats that could not otherwise step (Chau et al. 1998). The glutamatergic agonist NMDA had a dramatic positive effect on locomotion at a later stage (months postinjury) in chronic spinal cats. This effect was blocked by the NMDA antagonist Ap5 (Giroux et al. 2003), which demonstrated that the activation of NMDA receptors may play a role in facilitating stepping. Ap5 was, however, ineffective in the induction of stepping in acute spinal cats (Chau et al. 2002), which suggests that the cellular biochemistry of the spinal cord changes over the time course of an injury and is an important consideration when developing therapeutic strategies for SCI. Administration of 5-HT, its precursor 5-hydroxytryptophan (5-HTP), or the 5-HT agonists, quipazine and m-chlorophenylpiperazine (m-CPP), are also effective in improving locomotion in spinal cats (Barbeau & Rossignol 1990, 1991) and rats (Feraboli-Lohnherr et al. 1999, Kim et al. 2001), even though descending 5-HT axons are eliminated by a complete SCI. Consistent with these pharmacological studies, autoradiographic receptor-binding studies showed elevated levels of alpha1- and alpha2-NA and 5-HT1A receptors in selected laminae of the lumbar spinal cord segments of spinal cats at 15–30 days post-SCI, which gradually returned to control values (Giroux et al. 1999, Roudet et al. 1996). It is not known whether the locomotor patterns are stimulated by directly activating the neurons responsible for CPG or by modulating the spinal circuits that process proprioceptive information.

A complete SCI, and step or stand training after SCI, can markedly change the physiological and biochemical state of multiple neurotransmitter-modulator systems, and thereby change the pharmacological properties of the sensorimotor pathways that generate stepping and standing. Thus, the response of spinal animals to externally administered drugs that mimic or antagonize various neurotransmitter molecules provides a clue for the changes in the synaptic milieu (de Leon et al. 1999b). Pharmacological as well as biochemical data show that task-specific adaptive changes occur in spinal neuronal circuits of spinal cats trained to stand or step (de Leon et al. 1999b; Edgerton et al. 1997a; Rossignol et al. 2001; Tillakaratne et al. 2000, 2002). Administration of modest doses of strychnine, a glycine receptor antagonist that did not affect locomotion in step-trained spinal cats, significantly improved the stepping ability of poorly stepping, nontrained or stand-trained spinal cats (de Leon et al. 1999b). Similarly, the stepping ability of poorly stepping spinal cats can be dramatically improved by administration of the $GABA_A$ receptor antagonist bicuculline (Edgerton et al. 1997a). The improvement of stepping upon the administration of strychnine or bicuculline may occur by facilitating neuronal excitation by blocking the abnormally high levels of general inhibition resulting from a complete SCI (de Leon et al. 1999b). Nontrained spinal rats have increased GABA synthetic enzyme GAD_{67} and glycine and $GABA_A$ receptors in the lumbar spinal cord, whereas step-trained spinal rats have near-normal levels (Edgerton et al. 2001a, Tillakaratne et al. 2000). In agreement with this, GAD_{67} protein levels in lumbar spinal cord extracts were higher in spinal cats who were step trained for two months and then not trained for five months (de Leon et al. 1999a, Tillakaratne et al. 2002). After just one week of retraining, however, there was a significant reduction in GAD_{67} protein levels (Tillakaratne et al. 2002). In step-trained cats, the GAD_{67} levels around motoneurons were inversely correlated with stepping ability (Tillakaratne et al. 2002). Training spinal cats to weight-bear also appears to reduce GABA signaling in some spinal interneurons. Cellular mapping of GAD_{67} also suggests that the spinal cord of both stand-trained and nontrained spinal cats have selectively higher inhibitory potential compared to step-trained animals. For example, nontrained and stand-trained cats had more inhibitory GABAergic inputs (increased number of GAD_{67}-positive terminals) around the motoneurons than the step-trained cats. Furthermore, spinal cats trained to stand on one leg had elevated GAD_{67} within the motor pools corresponding to knee flexors in the trained leg but not in the nontrained leg. These findings suggest that the inability of stand-trained spinal animals to step is closely linked to an elevated level of inhibition of flexor motor pools. The specificity of GABAergic changes associated with the specific type of training emphasizes the concept that the activity in the neural networks of

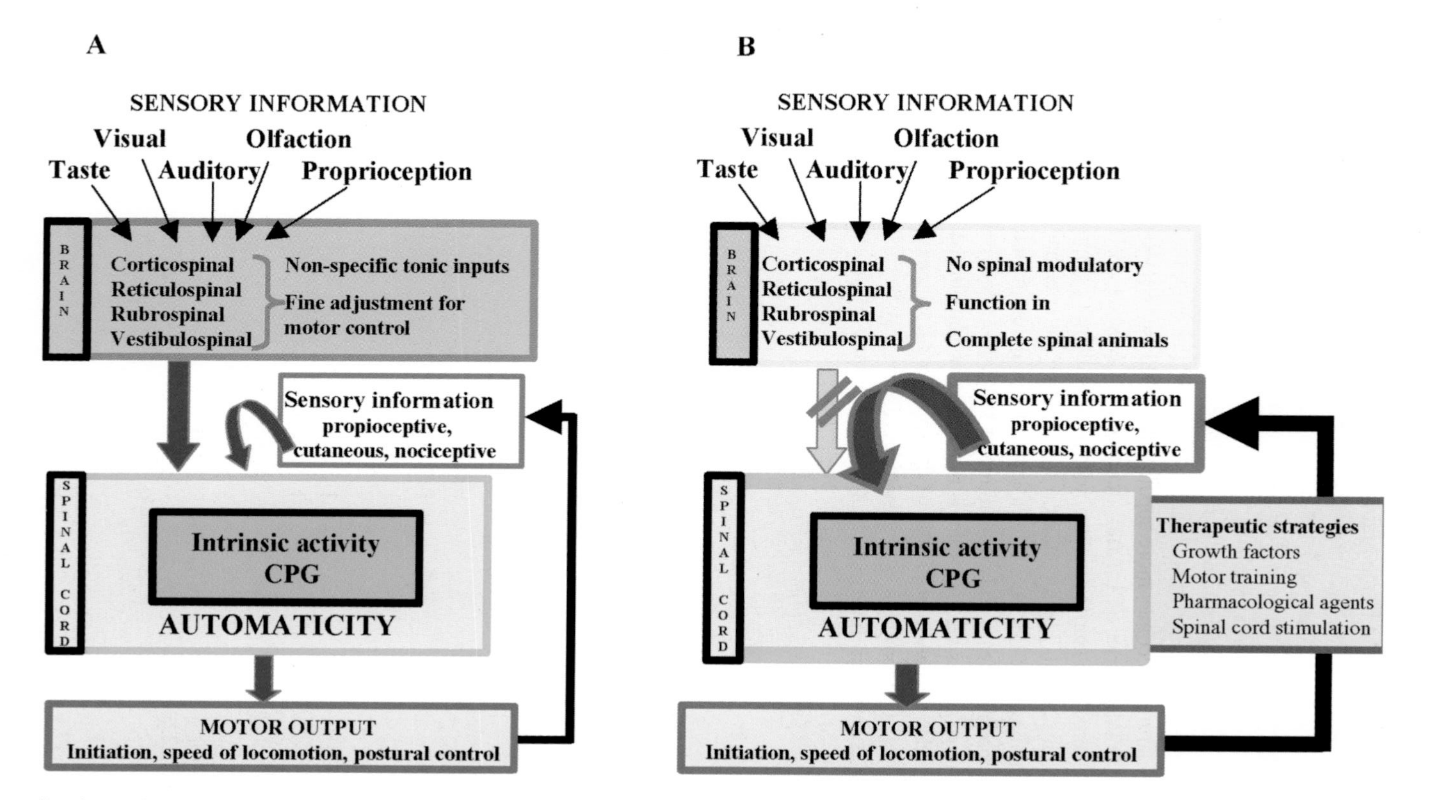

See legend on next page

Figure 1 (*A*) Automaticity is a key feature of neural motor control systems. Many sources of input to the brain (*blue box*) and spinal cord (*yellow box*) are constantly processed as we accommodate to the complexities of an ever-changing physical environment. The multiple levels of sensory processing are eventually manifested in the "decision" as to which and how many motoneurons are activated at any given sensory state. An extensive level of sensory processing and integration by the brain and spinal cord is automatic, resulting in reasonably "smart" and predictable motor output (*pink box*) that is appropriate for the current state. This automaticity and smartness becomes more apparent after SCI (see *B*). (*B*) Spinal automaticity is highly evident after SCI. After a complete SCI, where the supraspinal connections are severed, the spinal circuitry adapts to its altered combination of inputs to facilitate motor output. The locomotion and standing patterns that can be exhibited after training illustrate not only a high level of spinal automaticity but also the spinal cord's capacity to learn and perform motor tasks. The relative importance of the spinal circuitry and its sensory input is much greater than normal (see *A*) as illustrated by the thicker arrows into and out of the spinal cord, whereas the direct supraspinal influence on locomotion is greatly reduced. The gray box lists various therapeutic strategies by which functional motor recovery may be enhanced after SCI.

the lumbar spinal cord is, to a large extent, determined by the pattern of activity that the hindlimbs experience. How these changes are related to increased inhibitory and/or excitatory synaptic activity, and in which neural circuits this is manifested, remains unclear.

SCI also may have an effect on nonneuronal cell types in the spinal cord. For example, compared to control rats, spinal rats showed elevated immunostaining of the $GABA_A$ receptor gamma-2 subunit in astroglial cells, i.e., cells positive for the cellular marker glial fibrillary acidic protein (GFAP) in the ventral horn regions surrounding retrogradely labeled SOL and TA motoneurons (Bravo et al. 2002). This elevated level returns toward nearly undetectable control levels after six weeks of step training. Plasticity of spinal circuits may also be mediated by activity-dependent induction of neurotrophins. For example, voluntary exercise induces an upregulation of BDNF and NT-3 mRNA and protein levels in the spinal cord (Gomez-Pinilla et al. 2002, 2001; Ying et al. 2003). Neurotrophins stimulate axonal growth through the lesion site of an injured spinal cord (Tuszynski et al. 1994), and exogenously applied BDNF can improve stepping in spinal rats (Jakeman et al. 1998). Whether BDNF acts directly on those neurons that induce CPG or whether it plays a facilitating role, such as strychnine and bicuculline, is not known.

Effects of Electrical Stimulation on Locomotor Recovery After SCI

Numerous experiments have demonstrated that nonpatterned electrical stimulation (ES) of the lumbosacral enlargement can induce locomotor patterns and even hindlimb stepping in acute and chronic low-spinal kittens and acute spinalized rats (Iwahara et al. 1992), cats (Edgerton et al. 1976, Grillner & Zangger 1984), and humans (Dimitrijevic et al. 1998; Gerasimenko et al. 2003, 2002). ES differs from direct stimulation of individual muscles via electrodes implanted intramuscularly or around/near peripheral nerves. Only the spinal cord ES approach capitalizes on the coordinating capacity that is built into the spinal circuitry, which remains largely functional after most SCI. Dimitrijevic et al. (1998) showed that bilateral or unilateral epidural spinal cord ES can elicit step-like EMG activity bilaterally or unilaterally, respectively, in chronic, complete paraplegic subjects (ASIA A). Increased stimulation amplitude resulted in increased EMG amplitudes and an increased frequency of rhythmic activity. High frequencies of stimulation (>70 Hz) produced tonic activity in the leg musculature, which suggests that the upper lumbar stimulation may activate neuronal structures that then recruit interneurons involved in CPG.

Gerasimenko et al. (2002) also used spinal cord stimulation in complete SCI patients (thoracic or cervical level) and found that stimulation of the ventral surface of the cord initially produced tonic activity in the motor pools of the leg, eventually giving way to rhythmic activity. The most effective electrode placement for rhythmic activity was at L2, and once begun, it continued throughout the stimulation period and persisted for 1 or 2 cycles after cessation of the stimulus. Kinematically, the movements resembled bicycling, but the frequency and coordination pattern corresponded to stepping. Interestingly, progressive increases in the stimulation intensity produced activation from the proximal to the distal muscles and from tonic to rhythmic activity.

The stimulation site and parameters for generating tonic and locomotor-like activity with spinal cord ES in chronic spinal (T10) cats were generally similar to that in humans (Gerasimenko et al. 2003, 2002). Dorsolateral column lesion experiments and unilateral dorsal rhizotomy experiments resulted in loss of locomotor activity and rhythmic activity of the unlesioned side, respectively, when stimulating, which indicates that afferent inputs, i.e., dorsal root afferents and propriospinal fibers, can play a significant role in initiating and maintaining locomotor activity (Dimitrijevic et al. 1998, Gerasimenko et al. 2002).

Dimitrijevic and colleagues suggest that stimulation of the same lumbar site at different frequencies has distinctly different physiological effects (Pinter et al. 2000). For example, stimulation at 50–100 Hz was effective in controlling spasticity. Stimulating at 25–50 Hz at 7–10 V initiated and maintained rhythmical step-like hip and knee flexion in an ASIA A subject (Dimitrijevic et al. 1998). In addition only extensor movements were produced when stimulating at <15 Hz. Thus, the evidence to date seems to suggest that nonspecific stimulation of the upper lumbar segments with epidural electrodes, combined with the sensory input generated by weight-bearing stepping, has considerable potential as a therapeutic strategy to improve locomotor performance in SCI patients (Barbeau et al. 1999).

Herman et al. (2002) studied the efficacy of spinal cord ES in facilitating functional gait in an ASIA C patient. They found that motor training alone improved the stepping pattern and reduced spasticity, but overground stepping was very slow and tiring. A combination of treadmill training and stimulation, however, markedly improved the quality and quantity of stepping during the training session and resulted in immediate improvement in the quality of overground walking. These encouraging results imply that a generalized, nonspecific electrical stimulation of upper lumbar neurons may be sufficient to generate, or at least facilitate, rhythmic stepping movements of the lower limbs. The results raise the possibility that when spinal cord ES is combined with proprioceptive input associated with weight-bearing stepping, greater levels of activation and coordination can be attained than if either method is applied alone.

Electrical microstimulation of the spinal cord has been used to selectively activate specific hindlimb motor pools to produce predictable multi-joint movements. Multiple fine stimulating electrodes are inserted intraspinally in "activation pools" located in the ventral horn and activate specific muscle groups independently. The feasibility of using this approach to produce functional motor tasks was studied in chronically implanted (lumbar enlargement, 6 months) intact cats (Mushahwar et al. 2000). Stimulation through selective electrodes was used to produce single and multi-joint movements, in some cases producing a standing posture in the intact cat. In spinally intact anesthetized cats, bilateral locomotor-like activity was elicited via stimulation of a few electrodes placed in the ventral horn of the spinal cord, particularly in lamina IX. Whether intraspinal microstimulation of specific "activation pools" involves the CPG circuitry is unknown at this time.

The current strategies to electrically stimulate the spinal cord to initiate, modulate, and control locomotion, show that (*a*) locomotor-like movements of the limbs can be generated after a complete SCI via electrodes placed over the dura of the upper/middle lumbar segments; (*b*) microelectrodes placed in localized regions within the spinal cord gray matter can elicit more defined movements of the entire limb; and (*c*) spinal cord ES may prove to be more efficacious when combined with other interventions such as step training and pharmacological facilitation.

HUMAN SCI: A PERSPECTIVE

Three general principles have emerged largely from the study of spinal cats: (*a*) body weight–supported treadmill training improves the ability of the lumbosacral spinal cord to generate weight-bearing stepping; (*b*) patterns of sensory input provided during locomotor training are critical for driving the plasticity that mediates locomotor recovery; and (*c*) pharmacological treatments can be used to excite the spinal neurons that generate stepping. The question of whether the human spinal cord is capable of the level of adaptability demonstrated in other mammalian systems has not been fully answered. However, there is increasing evidence that the human spinal cord is capable of a significant amount of plasticity and that this plasticity is, to a large extent, driven by activity-dependent processes. A number of clinical studies demonstrate the applications of these lessons learned from the study of spinal cats toward facilitating the recovery of walking in humans with a complete SCI (Curt et al. 2004).

Treadmill Training and the Recovery of Walking Ability After SCI

In humans with incomplete SCI, treadmill training has a profound impact on overground walking ability (Wernig et al. 1995). To date, however, there has not been a similar training effect on overground walking in humans with clinically complete SCI. There is evidence, however, that treadmill training can improve several aspects of walking on a treadmill with some weight-supporting assistance in humans with clinically complete SCI. Dietz and colleagues (1995) reported that after several weeks of treadmill training, the levels of weight bearing that can be imposed on the legs of clinically complete SCI subjects during treadmill walking significantly increases. When stepping on a treadmill with body-weight support, rhythmic leg muscle activation patterns can be elicited in clinically complete subjects who are otherwise unable to voluntarily produce muscle activity in their legs (Maegele et al. 2002). A recent study has demonstrated that the levels of leg extensor muscle activity recorded in clinically complete SCI subjects significantly improved over the course of several weeks of step training (Wirz et al. 2001). Interestingly, the levels of extensor muscle activity decreased over a three-year period of time following the training program. Together, these results indicate that a use-dependent phenomenon may exist for the human spinal cord as has been reported for spinal cats (de Leon et al. 1999a,b). In short, the stepping ability of clinically complete SCI subjects can improve in response to step training, but the level of improvement has not reached a level that allows complete independence from assistance during full weight-bearing.

Use of Pharmacological Therapies in Enhancing Walking after SCI

Some of the pharmacological agents used successfully to induce weight-bearing stepping in spinal cats have also been used in humans with SCI in the hopes of enhancing walking performance, but have produced somewhat disappointing results. The NA agonist clonidine, for example, which is highly effective in inducing full weight-bearing stepping in spinal cats (Barbeau & Rossignol 1987, Chau et al. 1998) has had minimal effects on treadmill walking performance (Barbeau & Norman 2003) and can even have a depressing effect on EMG activity levels during treadmill walking (Dietz et al. 1998a,b) in clinically complete SCI patients. Although any number of factors may account for the lack of an effect of clonidine in humans with complete SCI, clonidine may have a beneficial effect in humans if it is combined with treadmill training, and its effects may depend on the time after the injury. In support of this, Chau and colleagues (1998) reported that a combined treatment of clonidine and treadmill training in spinal cats beginning two days after spinalization had a significant facilitatory effect on treadmill walking ability. Clearly, further studies are necessary to examine the effectiveness of pharmacological agents in combination with motor training. Examining the efficacy of these pharmacological agents over the course of recovery following a SCI will be particularly important. Obviously, there is a great need to know the cellular and synaptic mechanisms through which the pharmacological agonists and antagonists may modulate locomotor function. Presently, however, there is limited availability of experimental preparations in which detailed electrophysiological studies can be performed on the adult mammalian spinal cord.

CONCLUSIONS

There is both supraspinal and spinal automaticity in neuromotor control systems that initiate and generate coordinated motor behavior such as standing and stepping after SCI. This automaticity is clearly reflected in the fact that the spinal cord circuitry, which has access to proprioceptive and cutaneous input, can execute standing and stepping while readily adapting to varying loads, speeds of stepping, turning, and stepping over objects.

The motor patterns that the spinal cord generates after loss of supraspinal input are a function, to a great extent, of the specific sensorimotor events that occur after the injury. Training to stand improves standing ability and training to step improves stepping ability, and these behaviors are associated with motor pool–specific biochemical changes. One of the biochemical consequences of a complete SCI is an upregulation of inhibitory neurotransmitter systems, and step training reverses this effect. Therefore, the physiological and biochemical state of the spinal cord will affect how it responds to any given therapeutic intervention.

Ultimately, the application of any curative intervention for SCI is likely to have a higher level of success if combined with a motor training program that provides proprioceptive and cutaneous information to the spinal cord circuitry associated with the specific motor task being trained.

ACKNOWLEDGMENTS

These studies were supported by awards from the National Institutes of Health (1PO1NS16333; 1RO1NS40917; 1RO1NS42951), Christopher Reeve Paralysis Foundation and Consortium on Spinal Cord Injury (DA1-9902–2), NASA Bioengineering Research Partnership (PAS-99–010), and NASA GSRP (NGT2-52265). Figure 6 is compliments of C. Liu at UCLA.

The *Annual Review of Neuroscience* is online at http://neuro.annualreviews.org

LITERATURE CITED

Baev KV. 1998. *Biological Neural Networks: Hierachical Concept of Brain Function.* Boston: Birkhauser. 273 pp.

Barbeau H, Fung J, Leroux A, Ladouceur M. 2002. A review of the adaptability and recovery of locomotion after spinal cord injury. *Prog. Brain Res.* 137:9–25

Barbeau H, McCrea DA, O'Donovan MJ, Rossignol S, Grill WM, Lemay MA. 1999. Tapping into spinal circuits to restore motor function. *Brain Res. Brain Res. Rev.* 30:27–51

Barbeau H, Norman KE. 2003. The effect of noradrenergic drugs on the recovery of walking after spinal cord injury. *Spinal Cord* 41: 137–43

Barbeau H, Rossignol S. 1987. Recovery of locomotion after chronic spinalization in the adult cat. *Brain Res.* 412:84–95

Barbeau H, Rossignol S. 1990. The effects of serotonergic drugs on the locomotor pattern and on cutaneous reflexes of the adult chronic spinal cat. *Brain Res.* 514:55–67

Barbeau H, Rossignol S. 1991. Initiation and modulation of the locomotor pattern in the adult chronic spinal cat by noradrenergic, serotonergic and dopaminergic drugs. *Brain Res.* 546:250–60

Bravo AB, Bigbee AJ, Roy RR, Edgerton VR, Tobin AJ, Tillakaratne NJK. 2002. Gamma2 subunit of the $GABA_A$ receptor is increased in lumbar astrocytes in neonatally spinal cord transected rats. *Soc. Neurosci. Abstr.* 32: 853.6

Calancie B, Needham-Shropshire B, Jacobs P, Willer K, Zych G, Green BA. 1994. Involuntary stepping after chronic spinal cord injury. Evidence for a central rhythm generator for locomotion in man. *Brain* 117:1143–59

Cazalets JR, Bertrand S, Sqalli-Houssaini Y, Clarac F. 1998. GABAergic control of spinal locomotor networks in the neonatal rat. *Ann. NY Acad. Sci.* 860:168–80

Chau C, Barbeau H, Rossignol S. 1998. Early locomotor training with clonidine in spinal cats. *J. Neurophysiol.* 79:392–409

Chau C, Giroux N, Barbeau H, Jordan L, Rossignol S. 2002. Effects of intrathecal glutamatergic drugs on locomotion I. NMDA in short-term spinal cats. *J. Neurophysiol.* 88:3032–45

Curt A, Schwab ME, Dietz V. 2004. Providing the clinical basis for new interventional therapies: refined diagnosis and assessment of recovery after spinal cord injury. *Spinal Cord* 42:1–6

de Leon RD, Hodgson JA, Roy RR, Edgerton VR. 1998a. Locomotor capacity attributable to step training versus spontaneous recovery after spinalization in adult cats. *J. Neurophysiol.* 79:1329–40

de Leon RD, Hodgson JA, Roy RR, Edgerton VR. 1998b. Full weight-bearing hindlimb standing following stand training in the adult spinal cat. *J. Neurophysiol.* 80:83–91

de Leon RD, Hodgson JA, Roy RR, Edgerton VR. 1999a. Retention of hindlimb stepping ability in adult spinal cats after the cessation of step training. *J. Neurophysiol.* 81:85–94

de Leon RD, Reinkensmeyer DJ, Timoszyk WK, London NJ, Roy RR, Edgerton VR. 2002. Use of robotics in assessing the adaptive capacity of the rat lumbar spinal cord. *Prog. Brain Res.* 137:141–49

de Leon RD, Tamaki H, Hodgson JA, Roy RR, Edgerton VR. 1999b. Hindlimb locomotor and postural training modulates glycinergic inhibition in the spinal cord of the adult spinal cat. *J. Neurophysiol.* 82:359–69

Dietz V. 2003. Spinal cord pattern generators for locomotion. *Clin. Neurophysiol.* 114:1379–89

Dietz V, Colombo G, Jensen L, Baumgartner L. 1995. Locomotor capacity of spinal cord in paraplegic patients. *Ann. Neurol.* 37:574–82

Dietz V, Wirz M, Colombo G, Curt A. 1998a. Locomotor capacity and recovery of spinal cord function in paraplegic patients: a clinical and electrophysiological evaluation. *Electroencephalogr. Clin. Neurophysiol.* 109:140–53

Dietz V, Wirz M, Curt A, Colombo G. 1998b. Locomotor pattern in paraplegic patients: training effects and recovery of spinal cord function. *Spinal Cord* 36:380–90

Dimitrijevic MR, Gerasimenko Y, Pinter MM. 1998. Evidence for a spinal central pattern generator in humans. *Ann. NY Acad. Sci.* 860:360–76

Duysens J, Pearson KG. 1980. Inhibition of flexor burst generation by loading ankle extensor muscles in walking cats. *Brain Res.* 187:321–32

Edgerton VR, de Leon RD, Harkema SJ, Hodgson JA, London N, et al. 2001a. Retraining the injured spinal cord. *J. Physiol.* 533:15–22

Edgerton VR, de Leon RD, Tillakaratne N, Recktenwald MR, Hodgson JA, Roy RR. 1997a. Use-dependent plasticity in spinal stepping and standing. *Adv. Neurol.* 72:233–47

Edgerton VR, Grillner S, Sjostrom A, Zangger P. 1976. Central generation of locomotion in vertebrates. In *Neural Control of Locomotion*, ed. RN Herman, S Grillner, PSG Stein, DG Stuart, pp. 439–64. New York: Plenum

Edgerton VR, Roy RR. 2002. Paralysis recovery in humans and model systems. *Curr. Opin. Neurobiol.* 12:658–67

Edgerton VR, Roy RR, de Leon RD. 2001b. Neural darwinism in the mammalian spinal cord. In *Spinal Cord Plasticity; Alterations in Reflex Function*, ed. MM Patterson, JW Grau, pp. 185–206. Boston: Kluwer

Edgerton VR, Roy RR, de Leon R, Tillakaratne N, Hodgson JA. 1997b. Does motor learning occur in the spinal cord? *Neuroscientist* 3:287–94

Edgerton VR, Tillakaratne NJK, Bigbee AJ, de Leon RD, Roy RR. 2004. Locomotor recovery potential after spinal cord injury. In *Neuro-Behavioral Determinants of Interlimb Coordination*, ed. SP Swinnen, J Duysens. Boston: Kluwer. In press

Eidelberg E. 1981. Consequences of spinal cord

lesions upon motor function, with special reference to locomotor activity. *Prog. Neurobiol.* 17:185–202

Fedirchuk B, Nielsen J, Petersen N, Hultborn H. 1998. Pharmacologically evoked fictive motor patterns in the acutely spinalized marmoset monkey (Callithrix jacchus). *Exp. Brain Res.* 122:351–61

Feraboli-Lohnherr D, Barthe JY, Orsal D. 1999. Serotonin-induced activation of the network for locomotion in adult spinal rats. *J. Neurosci. Res.* 55:87–98

Forssberg H, Grillner S, Halbertsma J. 1980a. The locomotion of the low spinal cat. I. Coordination within a hindlimb. *Acta Physiol. Scand.* 108:269–81

Forssberg H, Grillner S, Halbertsma J, Rossignol S. 1980b. The locomotion of the low spinal cat. II. Interlimb coordination. *Acta Physiol. Scand* 108:283–95

Forssberg H, Grillner S, Rossignol S. 1975. Phase dependent reflex reversal during walking in chronic spinal cats. *Brain Res.* 85:103–7

Garraway SM, Hochman S. 2001. Serotonin increases the incidence of primary afferent-evoked long-term depression in rat deep dorsal horn neurons. *J. Neurophysiol.* 85:1864–72

Gerasimenko YP, Avelev VD, Nikitin OA, Lavrov IA. 2003. Initiation of locomotor activity in spinal cats by epidural stimulation of the spinal cord. *Neurosci. Behav. Physiol.* 33:247–54

Gerasimenko YP, Makarovskii AN, Nikitin OA. 2002. Control of locomotor activity in humans and animals in the absence of supraspinal influences. *Neurosci. Behav. Physiol.* 32:417–23

Giroux N, Chau C, Barbeau H, Reader TA, Rossignol S. 2003. Effects of intrathecal glutamatergic drugs on locomotion. II. NMDA and AP-5 in intact and late spinal cats. *J. Neurophysiol.* 90:1027–45

Giroux N, Rossignol S, Reader TA. 1999. Autoradiographic study of alpha1- and alpha2-noradrenergic and serotonin1A receptors in the spinal cord of normal and chronically transected cats. *J. Comp. Neurol.* 406:402–14

Gomez-Pinilla F, Ying Z, Opazo P, Roy RR, Edgerton VR. 2001. Differential regulation by exercise of BDNF and NT-3 in rat spinal cord and skeletal muscle. *Eur. J. Neurosci.* 13:1078–84

Gomez-Pinilla F, Ying Z, Roy RR, Molteni R, Edgerton VR. 2002. Voluntary exercise induces a BDNF-mediated mechanism that promotes neuroplasticity. *J. Neurophysiol.* 88:2187–95

Grillner S. 1979. Interaction between central and peripheral mechanisms in the control of locomotion. *Prog. Brain Res.* 50:227–35

Grillner S. 1981. The nervous system II. Motor control. In *Handbook of Physiology Section 1*. ed. VB Brooks, pp. 1179–236. Waverly, MD: Am. Physiol. Soc.

Grillner S. 2002. The spinal locomotor CPG: a target after spinal cord injury. *Prog. Brain Res.* 137:97–108

Grillner S. 2003. The motor infrastructure: from ion channels to neuronal networks. *Nat. Rev. Neurosci.* 4:573–86

Grillner S, Wallen P. 1985. Central pattern generators for locomotion, with special reference to vertebrates. *Annu. Rev. Neurosci.* 8:233–61

Grillner S, Zangger P. 1984. The effect of dorsal root transection on the efferent motor pattern in the cat's hindlimb during locomotion. *Acta Physiol. Scand.* 120:393–405

Gurfinkel VS, Levik YuS, Kazennikov OV, Selionov VA. 1997. Involuntary air-stepping induced by the increase of the level of tonic readiness: evidence for CPG in human. In *Brain and Movement*, ed. VS Gurfinkel, YuS Levik, pp. 86–87. Moscow: Proc. Int. Symp.

Harkema SJ. 2001. Neural plasticity after human spinal cord injury: application of locomotor training to the rehabilitation of walking. *Neuroscientist* 7:455–68

Harkema SJ, Hurley SL, Patel UK, Requejo PS, Dobkin BH, Edgerton VR. 1997. Human lumbosacral spinal cord interprets loading during stepping. *J. Neurophysiol.* 77:797–811

Herman R, He J, D'Luzansky S, Willis W, Dilli S. 2002. Spinal cord stimulation facilitates functional walking in a chronic, incomplete spinal cord injured. *Spinal Cord* 40:65–68

Hodgson JA, Ly L, Raven J, Roy RR, Edgerton VR, et al. 2003. The effects of a unilateral corticospinal tract (CST) lesion on motor performance in Rhesus. *Soc. Neurosci. Abstr.* 33:71.5

Hodgson JA, Roy RR, de Leon R, Dobkin B, Edgerton VR. 1994. Can the mammalian lumbar spinal cord learn a motor task? *Med. Sci. Sports Exerc.* 26:1491–97

Ikeda H, Heinke B, Ruscheweyh R, Sandkuhler J. 2003. Synaptic plasticity in spinal lamina I projection neurons that mediate hyperalgesia. *Science* 299:1237–40

Iwahara T, Atsuta Y, Garcia-Rill E, Skinner RD. 1992. Spinal cord stimulation-induced locomotion in the adult cat. *Brain Res. Bull.* 28:99–105

Jakeman LB, Wei P, Guan Z, Stokes BT. 1998. Brain-derived neurotrophic factor stimulates hindlimb stepping and sprouting of cholinergic fibers after spinal cord injury. *Exp. Neurol.* 154:170–84

Jordan LM, Schmidt BJ. 2002. Propriospinal neurons involved in the control of locomotion: potential targets for repair strategies? *Prog. Brain Res.* 137:125–40

Joynes RL, de Leon RD, Tillakaratne NJK, Roy RR, Tobin AJ, Edgerton VR. 1999. Recovery of locomotion in rats after neonatal spinal transection is not attributable to growth across the lesion. *Soc. Nerosci. Abstr.* 29:467.18

Kiehn O, Kjaerulff O. 1998. Distribution of central pattern generators for rhythmic motor outputs in the spinal cord of limbed vertebrates. *Ann. NY Acad. Sci.* 860:110–29

Kim D, Murray M, Simansky KJ. 2001. The serotonergic 5-HT(2C) agonist m-chlorophenylpiperazine increases weight-supported locomotion without development of tolerance in rats with spinal transections. *Exp. Neurol.* 169:496–500

Kudo N, Nishimaru H. 1998. Reorganization of locomotor activity during development in the prenatal rat. *Ann. NY Acad. Sci.* 860:306–17

Liu CC, de Leon RD, Guu JJ, Zdunowski SL, Melikian AH, et al. 2003. Retention of a corrective kinematic response in the rat spinal cord. *Soc. Neurosci. Abstr.* 33:497.13

Lovely RG, Gregor RJ, Roy RR, Edgerton VR. 1986. Effects of training on the recovery of full-weight-bearing stepping in the adult spinal cat. *Exp. Neurol.* 92:421–35

Lovely RG, Gregor RJ, Roy RR, Edgerton VR. 1990. Weight-bearing hindlimb stepping in treadmill-exercised adult spinal cats. *Brain Res.* 514:206–18

Maegele M, Muller S, Wernig A, Edgerton VR, Harkema SJ. 2002. Recruitment of spinal motor pools during voluntary movements versus stepping after human spinal cord injury. *J. Neurotrauma* 19:1217–29

Mushahwar VK, Collins DF, Prochazka A. 2000. Spinal cord microstimulation generates functional limb movements in chronically implanted cats. *Exp. Neurol.* 163:422–29

Nakada K, Hodgson JA, de Leon RD, Roy RR, Edgerton VR. 1994. Prolonged modification of the mechanisms of the step cycle by single and repetitive mechanical stimuli in chronic spinal cats. *Soc. Neurosci. Abstr.* 20:1755

Orlovsky GN, Feldman AG. 1972. Classification of lumbosacral neurons according to their discharge patterns during evoked locomotion neurophysiology. New York: Plenum. (From Russian)

Pinter MM, Gerstenbrand F, Dimitrijevic MR. 2000. Epidural electrical stimulation of posterior structures of the human lumbosacral cord: 3. Control of spasticity. *Spinal Cord* 38:524–31

Roy RR, Acosta L. 1986. Fiber type and fiber size changes in selected thigh muscles six months after low thoracic spinal cord transection in adult cats: exercise effects. *Exp. Neurol.* 92:675–85

Roy RR, Hutchison DL, Pierotti DJ, Hodgson JA, Edgerton VR. 1991. EMG patterns of rat ankle extensors and flexors during treadmill locomotion and swimming. *J. Appl. Physiol.* 70:2522–29

Roy RR, Talmadge RJ, Hodgson JA, Oishi Y, Baldwin KM, Edgerton VR. 1999. Differential response of fast hindlimb extensor and flexor muscles to exercise in adult spinalized cats. *Muscle Nerve* 22:230–41

Roy RR, Talmadge RJ, Hodgson JA, Zhong H, Baldwin KM, Edgerton VR. 1998. Training effects on soleus of cats spinal cord transected (T12-13) as adults. *Muscle Nerve* 21:63–71

Rossignol S, Bouyer L, Barthelemy D, Langlet C, Leblond H. 2002a. Recovery of locomotion in the cat following spinal cord lesions. *Brain Res. Brain Res. Rev.* 40:257–66

Rossignol S, Chau C, Giroux N, Brustein E, Bouyer L, et al. 2002b. The cat model of spinal injury. *Prog. Brain Res.* 137:151–68

Rossignol S, Giroux N, Chau C, Marcoux J, Brustein E, Reader TA. 2001. Pharmacological aids to locomotor training after spinal injury in the cat. *J. Physiol.* 533:65–74

Roudet C, Gimenez Ribotta M, Privat A, Feuerstein C, Savasta M. 1996. Regional study of spinal alpha 2-adrenoceptor densities after intraspinal noradrenergic-rich implants on adult rats bearing complete spinal cord transection or selective chemical noradrenergic denervation. *Neurosci. Lett.* 208:89–92

Rygh LJ, Tjolsen A, Hole K, Svendsen F. 2002. Cellular memory in spinal nociceptive circuitry. *Scand. J. Psychol.* 43:153–59

Schmidt BJ, Hochman S, MacLean JN. 1998. NMDA receptor-mediated oscillatory properties: potential role in rhythm generation in the mammalian spinal cord. *Ann. NY Acad. Sci.* 860:189–202

Shik ML, Orlovsky GN. 1976. Neurophysiology of locomotor automatism. *Physiol. Rev.* 56:465–501

Shik ML, Severin FV, Orlovsky GN. 1966. Control of walking and running by electrical stimulation of the mid brain. *Biophysics* 11:659–66

Simonsen EB, Dyhre-Poulsen P. 1999. Amplitude of the human soleus H reflex during walking and running. *J. Physiol.* 515(Pt. 3):929–39

Tillakaratne NJ, de Leon RD, Hoang TX, Roy RR, Edgerton VR, Tobin AJ. 2002. Use-dependent modulation of inhibitory capacity in the feline lumbar spinal cord. *J. Neurosci.* 22:3130–43

Tillakaratne NJ, Mouria M, Ziv NB, Roy RR, Edgerton VR, Tobin AJ. 2000. Increased expression of glutamate decarboxylase (GAD_{67}) in feline lumbar spinal cord after complete thoracic spinal cord transection. *J. Neurosci. Res.* 60:219–30

Timoszyk WK, de Leon RD, London N, Roy RR, Edgerton VR, Reinkensmeyer DJ. 2002. The rat lumbosacral spinal cord adapts to robotic loading applied during stance. *J. Neurophysiol.* 88:3108–17

Tuszynski MH, Peterson DA, Ray J, Baird A, Nakahara Y, Gage FH. 1994. Fibroblasts genetically modified to produce nerve growth factor induce robust neuritic ingrowth after grafting to the spinal cord. *Exp. Neurol.* 126:1–14

Wernig A, Muller S. 1992. Laufband locomotion with body weight support improved walking in persons with severe spinal cord injuries. *Paraplegia* 30:229–38

Wernig A, Muller S, Nanassy A, Cagol E. 1995. Laufband therapy based on 'rules of spinal locomotion' is effective in spinal cord injured persons. *Eur. J. Neurosci.* 7:823–29

Wirz M, Colombo G, Dietz V. 2001. Long term effects of locomotor training in spinal man. *J. Neurol. Neurosurg. Psychiatry* 71:93–96

Wolpaw JR, Tennissen AM. 2001. Activity-dependent spinal cord plasticity in health and disease. *Annu. Rev. Neurosci.* 24:807–43

Ying Z, Roy RR, Edgerton VR, Gomez-Pinilla F. 2003. Voluntary exercise increases neurotrophin-3 and its receptor TrkC in the spinal cord. *Brain Res.* 987:93–99

Annu. Rev. Neurosci. 2004. 27:169–92
doi: 10.1146/annurev.neuro.27.070203.144230

First published online as a Review in Advance on March 5, 2004

THE MIRROR-NEURON SYSTEM

Giacomo Rizzolatti[1] and Laila Craighero[2]
[1]Dipartimento di Neuroscienze, Sezione di Fisiologia, via Volturno, 3, Università di Parma, 43100, Parma, Italy; email: giacomo.rizzolatti@unipr.it;
[2]Dipartimento SBTA, Sezione di Fisiologia Umana, via Fossato di Mortara, 17/19, Università di Ferrara, 44100 Ferrara, Italy; email: crh@unife.it

Key Words mirror neurons, action understanding, imitation, language, motor cognition

■ **Abstract** A category of stimuli of great importance for primates, humans in particular, is that formed by actions done by other individuals. If we want to survive, we must understand the actions of others. Furthermore, without action understanding, social organization is impossible. In the case of humans, there is another faculty that depends on the observation of others' actions: imitation learning. Unlike most species, we are able to learn by imitation, and this faculty is at the basis of human culture. In this review we present data on a neurophysiological mechanism—the mirror-neuron mechanism—that appears to play a fundamental role in both action understanding and imitation. We describe first the functional properties of mirror neurons in monkeys. We review next the characteristics of the mirror-neuron system in humans. We stress, in particular, those properties specific to the human mirror-neuron system that might explain the human capacity to learn by imitation. We conclude by discussing the relationship between the mirror-neuron system and language.

INTRODUCTION

Mirror neurons are a particular class of visuomotor neurons, originally discovered in area F5 of the monkey premotor cortex, that discharge both when the monkey does a particular action and when it observes another individual (monkey or human) doing a similar action (Di Pellegrino et al. 1992, Gallese et al. 1996, Rizzolatti et al. 1996a). A lateral view of the monkey brain showing the location of area F5 is presented in Figure 1 (see color insert).

The aim of this review is to provide an updated account of the functional properties of the system formed by mirror neurons. The review is divided into four sections. In the first section we present the basic functional properties of mirror neurons in the monkey, and we discuss their functional roles in action understanding. In the second section, we present evidence that a mirror-neuron system similar to that of the monkey exists in humans. The third section shows that in humans, in addition to action understanding, the mirror-neuron system plays a fundamental role in action imitation. The last section is more speculative.

We present there a theory of language evolution, and we discuss a series of data supporting the notion of a strict link between language and the mirror-neuron system (Rizzolatti & Arbib 1998).

THE MIRROR-NEURON SYSTEM IN MONKEYS

F5 Mirror Neurons: Basic Properties

There are two classes of visuomotor neurons in monkey area F5: canonical neurons, which respond to the presentation of an object, and mirror neurons, which respond when the monkey sees object-directed action (Rizzolatti & Luppino 2001). In order to be triggered by visual stimuli, mirror neurons require an interaction between a biological effector (hand or mouth) and an object. The sight of an object alone, of an agent mimicking an action, or of an individual making intransitive (nonobject-directed) gestures are all ineffective. The object significance for the monkey has no obvious influence on the mirror-neuron response. Grasping a piece of food or a geometric solid produces responses of the same intensity.

Mirror neurons show a large degree of generalization. Presenting widely different visual stimuli, but which all represent the same action, is equally effective. For example, the same grasping mirror neuron that responds to a human hand grasping an object responds also when the grasping hand is that of a monkey. Similarly, the response is typically not affected if the action is done near or far from the monkey, in spite of the fact that the size of the observed hand is obviously different in the two conditions.

It is also of little importance for neuron activation if the observed action is eventually rewarded. The discharge is of the same intensity if the experimenter grasps the food and gives it to the recorded monkey or to another monkey introduced in the experimental room.

An important functional aspect of mirror neurons is the relation between their visual and motor properties. Virtually all mirror neurons show congruence between the visual actions they respond to and the motor responses they code. According to the type of congruence they exhibit, mirror neurons have been subdivided into "strictly congruent" and "broadly congruent" neurons (Gallese et al. 1996).

Mirror neurons in which the effective observed and effective executed actions correspond in terms of goal (e.g., grasping) and means for reaching the goal (e.g., precision grip) have been classed as "strictly congruent." They represent about one third of F5 mirror neurons. Mirror neurons that, in order to be triggered, do not require the observation of exactly the same action that they code motorically have been classed as "broadly congruent." They represent about two thirds of F5 mirror neurons.

F5 Mouth Mirror Neurons

The early studies of mirror neurons concerned essentially the upper sector of F5 where hand actions are mostly represented. Recently, a study was carried out on the properties of neurons located in the lateral part of F5 (Ferrari et al. 2003), where, in contrast, most neurons are related to mouth actions.

The results showed that about 25% of studied neurons have mirror properties. According to the visual stimuli effective in triggering the neurons, two classes of mouth mirror neurons were distinguished: ingestive and communicative mirror neurons.

Ingestive mirror neurons respond to the observation of actions related to ingestive functions, such as grasping food with the mouth, breaking it, or sucking. Neurons of this class form about 80% of the total amount of the recorded mouth mirror neurons. Virtually all ingestive mirror neurons show a good correspondence between the effective observed and the effective executed action. In about one third of them, the effective observed and executed actions are virtually identical (strictly congruent neurons); in the remaining, the effective observed and executed actions are similar or functionally related (broadly congruent neurons).

More intriguing are the properties of the communicative mirror neurons. The most effective observed action for them is a communicative gesture such as lip smacking, for example. However, from a motor point of view they behave as the ingestive mirror neurons, strongly discharging when the monkey actively performs an ingestive action.

This discrepancy between the effective visual input (communicative) and the effective active action (ingestive) is rather puzzling. Yet, there is evidence suggesting that communicative gestures, or at least some of them, derived from ingestive actions in evolution (MacNeilage 1998, Van Hoof 1967). From this perspective one may argue that the communicative mouth mirror neurons found in F5 reflect a process of corticalization of communicative functions not yet freed from their original ingestive basis.

The Mirror-Neuron Circuit

Neurons responding to the observation of actions done by others are present not only in area F5. A region in which neurons with these properties have been described is the cortex of the superior temporal sulcus (STS; Figure 1) (Perrett et al. 1989, 1990; Jellema et al. 2000; see Jellema et al. 2002). Movements effective in eliciting neuron responses in this region are walking, turning the head, bending the torso, and moving the arms. A small set of STS neurons discharge also during the observation of goal-directed hand movements (Perrett et al. 1990).

If one compares the functional properties of STS and F5 neurons, two points emerge. First, STS appears to code a much larger number of movements than F5. This may be ascribed, however, to the fact that STS output reaches, albeit indirectly (see below), the whole ventral premotor region and not only F5. Second, STS neurons do not appear to be endowed with motor properties.

Another cortical area where there are neurons that respond to the observation of actions done by other individuals is area 7b or PF of Von Economo (1929) (Fogassi et al. 1998, Gallese et al. 2002). This area (see Figure 1) forms the rostral part of the inferior parietal lobule. It receives input from STS and sends an important output to the ventral premotor cortex including area F5.

PF neurons are functionally heterogeneous. Most of them (about 90%) respond to sensory stimuli, but about 50% of them also have motor properties discharging when the monkey performs specific movements or actions (Fogassi et al. 1998, Gallese et al. 2002, Hyvarinen 1982).

PF neurons responding to sensory stimuli have been subdivided into "somatosensory neurons" (33%), "visual neurons" (11%), and "bimodal (somatosensory and visual) neurons" (56%). About 40% of the visually responsive neurons respond specifically to action observation and of them about two thirds have mirror properties (Gallese et al. 2002).

In conclusion, the cortical mirror neuron circuit is formed by two main regions: the rostral part of the inferior parietal lobule and the ventral premotor cortex. STS is strictly related to it but, lacking motor properties, cannot be considered part of it.

Function of the Mirror Neuron in the Monkey: Action Understanding

Two main hypotheses have been advanced on what might be the functional role of mirror neurons. The first is that mirror-neuron activity mediates imitation (see Jeannerod 1994); the second is that mirror neurons are at the basis of action understanding (see Rizzolatti et al. 2001).

Both these hypotheses are most likely correct. However, two points should be specified. First, although we are fully convinced (for evidence see next section) that the mirror neuron mechanism is a mechanism of great evolutionary importance through which primates understand actions done by their conspecifics, we cannot claim that this is the only mechanism through which actions done by others may be understood (see Rizzolatti et al. 2001). Second, as is shown below, the mirror-neuron system is the system at the basis of imitation in humans. Although laymen are often convinced that imitation is a very primitive cognitive function, they are wrong. There is vast agreement among ethologists that imitation, the capacity to learn to do an action from seeing it done (Thorndyke 1898), is present among primates, only in humans, and (probably) in apes (see Byrne 1995, Galef 1988, Tomasello & Call 1997, Visalberghi & Fragaszy 2001, Whiten & Ham 1992). Therefore, the primary function of mirror neurons cannot be action imitation.

How do mirror neurons mediate understanding of actions done by others? The proposed mechanism is rather simple. Each time an individual sees an action done by another individual, neurons that represent that action are activated in the observer's premotor cortex. This automatically induced, motor representation of the observed action corresponds to that which is spontaneously generated during active action and whose outcome is known to the acting individual. Thus, the mirror system transforms visual information into knowledge (see Rizzolatti et al. 2001).

Evidence in Favor of the Mirror Mechanism in Action Understanding

At first glance, the simplest, and most direct, way to prove that the mirror-neuron system underlies action understanding is to destroy it and examine the lesion effect on the monkey's capacity to recognize actions made by other monkeys. In practice, this is not so. First, the mirror-neuron system is bilateral and includes, as shown above, large portions of the parietal and premotor cortex. Second, there are other mechanisms that may mediate action recognition (see Rizzolatti et al. 2001). Third, vast lesions as those required to destroy the mirror neuron system may produce more general cognitive deficits that would render difficult the interpretation of the results.

An alternative way to test the hypothesis that mirror neurons play a role in action understanding is to assess the activity of mirror neurons in conditions in which the monkey understands the meaning of the occurring action but has no access to the visual features that activate mirror neurons. If mirror neurons mediate action understanding, their activity should reflect the meaning of the observed action, not its visual features.

Prompted by these considerations, two series of experiments were carried out. The first tested whether F5 mirror neurons are able to recognize actions from their sound (Kohler et al. 2002), the second whether the mental representation of an action triggers their activity (Umiltà et al. 2001).

Kohler et al. (2002) recorded F5 mirror neuron activity while the monkey was observing a noisy action (e.g., ripping a piece of paper) or was presented with the same noise without seeing it. The results showed that about 15% of mirror neurons responsive to presentation of actions accompanied by sounds also responded to the presentation of the sound alone. The response to action sounds did not depend on unspecific factors such as arousal or emotional content of the stimuli. Neurons responding specifically to action sounds were dubbed "audio-visual" mirror neurons.

Neurons were also tested in an experimental design in which two noisy actions were randomly presented in vision-and-sound, sound-only, vision-only, and motor conditions. In the motor condition, the monkeys performed the object-directed action that they observed or heard in the sensory conditions. Out of 33 studied neurons, 29 showed auditory selectivity for one of the two hand actions. The selectivity in visual and auditory modality was the same and matched the preferred motor action.

The rationale of the experiment by Umiltà et al. (2001) was the following. If mirror neurons are involved in action understanding, they should discharge also in conditions in which monkey does not see the occurring action but has sufficient clues to create a mental representation of what the experimenter does. The neurons were tested in two basic conditions. In one, the monkey was shown a fully visible action directed toward an object ("full vision" condition). In the other, the monkey saw the same action but with its final, critical part hidden ("hidden" condition). Before each trial, the experimenter placed a piece of food behind the screen so that the monkey knew there was an object there. Only those mirror neurons were studied that discharged to the observation of the final part of a grasping movement and/or to holding.

Figure 2 (see color insert) shows the main result of the experiment. The neuron illustrated in the figure responded to the observation of grasping and holding (*A*, full vision). The neuron discharged also when the stimulus-triggering features (a hand approaching the stimulus and subsequently holding it) were hidden from monkey's vision (*B*, hidden condition). As is the case for most mirror neurons, the observation of a mimed action did not activate the neuron (*C*, full vision, and *D*, hidden condition). Note that from a physical point of view *B* and *D* are identical. It was therefore the understanding of the meaning of the observed actions that determined the discharge in the hidden condition.

More than half of the tested neurons discharged in the hidden condition. Out of them, about half did not show any difference in the response strength between the hidden- and full-vision conditions. The other half responded more strongly in the full-vision condition. One neuron showed a more pronounced response in the hidden condition than in full vision.

In conclusion, both the experiments showed that the activity of mirror neurons correlates with action understanding. The visual features of the observed actions are fundamental to trigger mirror neurons only insomuch as they allow the understanding of the observed actions. If action comprehension is possible on another basis (e.g., action sound), mirror neurons signal the action, even in the absence of visual stimuli.

THE MIRROR-NEURON SYSTEM IN HUMANS

There are no studies in which single neurons were recorded from the putative mirror-neuron areas in humans. Thus, direct evidence for the existence of mirror neurons in humans is lacking. There is, however, a rich amount of data proving, indirectly, that a mirror-neuron system does exist in humans. Evidence of this comes from neurophysiological and brain-imaging experiments.

Neurophysiological Evidence

Neurophysiological experiments demonstrate that when individuals observe an action done by another individual their motor cortex becomes active, in the absence of any overt motor activity. A first evidence in this sense was already provided in the 1950s by Gastaut and his coworkers (Cohen-Seat et al. 1954, Gastaut & Bert 1954). They observed that the desynchronization of an EEG rhythm recorded from central derivations (the so-called mu rhythm) occurs not only during active movements of studied subjects, but also when the subjects observed actions done by others.

This observation was confirmed by Cochin et al. (1998, 1999) and by Altschuler et al. (1997, 2000) using EEG recordings, and by Hari et al. (1998) using magnetoencephalographic (MEG) technique. This last study showed that the desynchronization during action observation includes rhythms originating from the cortex inside the central sulcus (Hari & Salmelin 1997, Salmelin & Hari 1994).

More direct evidence that the motor system in humans has mirror properties was provided by transcranial magnetic stimulation (TMS) studies. TMS is a noninvasive technique for electrical stimulation of the nervous system. When TMS is applied to the motor cortex, at appropriate stimulation intensity, motor-evoked potentials (MEPs) can be recorded from contralateral extremity muscles. The amplitude of these potentials is modulated by the behavioral context. The modulation of MEPs' amplitude can be used to assess the central effects of various experimental conditions. This approach has been used to study the mirror neuron system.

Fadiga et al. (1995) recorded MEPs, elicited by stimulation of the left motor cortex, from the right hand and arm muscles in volunteers required to observe an experimenter grasping objects (transitive hand actions) or performing meaningless arm gestures (intransitive arm movements). Detection of the dimming of a small spot of light and presentation of 3-D objects were used as control conditions. The results showed that the observation of both transitive and intransitive actions determined an increase of the recorded MEPs with respect to the control conditions. The increase concerned selectively those muscles that the participants use for producing the observed movements.

Facilitation of the MEPs during movement observation may result from a facilitation of the primary motor cortex owing to mirror activity of the premotor areas, to a direct facilitatory input to the spinal cord originating from the same areas, or from both. Support for the cortical hypothesis (see also below, Brain Imaging Experiments) came from a study by Strafella & Paus (2000). By using a double-pulse TMS technique, they demonstrated that the duration of intracortical recurrent inhibition, occurring during action observation, closely corresponds to that occurring during action execution.

Does the observation of actions done by others influence the spinal cord excitability? Baldissera et al. (2001) investigated this issue by measuring the size of the H-reflex evoked in the flexor and extensor muscles of normal volunteers during the observation of hand opening and closure done by another individual. The results showed that the size of H-reflex recorded from the flexors increased during the observation of hand opening, while it was depressed during the observation of hand closing. The converse was found in the extensors. Thus, while the cortical excitability varies in accordance with the seen movements, the spinal cord excitability changes in the opposite direction. These findings indicate that, in the spinal cord, there is an inhibitory mechanism that prevents the execution of an observed action, thus leaving the cortical motor system free to "react" to that action without the risk of overt movement generation.

In a study of the effect of hand orientation on cortical excitability, Maeda et al. (2002) confirmed (see Fadiga et al. 1995) the important finding that, in humans, intransitive movements, and not only goal-directed actions, determine motor resonance. Another important property of the human mirror-neuron system, demonstrated with TMS technique, is that the time course of cortical facilitation during action observation follows that of movement execution. Gangitano et al. (2001) recorded MEPs from the hand muscles of normal volunteers while they were observing grasping movements made by another individual. The MEPs were recorded at different intervals following the movement onset. The results showed that the motor cortical excitability faithfully followed the grasping movement phases of the observed action.

In conclusion, TMS studies indicate that a mirror-neuron system (a motor resonance system) exists in humans and that it possesses important properties not observed in monkeys. First, intransitive meaningless movements produce mirror-neuron system activation in humans (Fadiga et al. 1995, Maeda et al. 2002, Patuzzo et al. 2003), whereas they do not activate mirror neurons in monkeys. Second, the temporal characteristics of cortical excitability, during action observation, suggest that human mirror-neuron systems code also for the movements forming an action and not only for action as monkey mirror-neuron systems do. These properties of the human mirror-neuron system should play an important role in determining the humans' capacity to imitate others' action.

Brain Imaging Studies: The Anatomy of the Mirror System

A large number of studies showed that the observation of actions done by others activates in humans a complex network formed by occipital, temporal, and parietal visual areas, and two cortical regions whose function is fundamentally or predominantly motor (e.g., Buccino et al. 2001; Decety et al. 2002; Grafton et al. 1996; Grèzes et al. 1998; Grèzes et al. 2001; Grèzes et al. 2003; Iacoboni et al. 1999, 2001; Koski et al. 2002, 2003; Manthey et al. 2003; Nishitani & Hari 2000, 2002; Perani et al. 2001; Rizzolatti et al. 1996b). These two last regions are the rostral part of the inferior parietal lobule and the lower part of the precentral gyrus plus the posterior part of the inferior frontal gyrus (IFG). These regions form the core of the human mirror-neuron system.

Which are the cytoarchitectonic areas that form these regions? Interpretation of the brain-imaging activations in cytoarchitectonic terms is always risky. Yet, in the case of the inferior parietal region, it is very plausible that the mirror activation corresponds to areas PF and PFG, where neurons with mirror properties are found in the monkeys (see above).

More complex is the situation for the frontal regions. A first issue concerns the location of the border between the two main sectors of the premotor cortex: the ventral premotor cortex (PMv) and the dorsal premotor cortex (PMd). In nonhuman primates the two sectors differ anatomically (Petrides & Pandya 1984, Tanné-Gariepy et al. 2002) and functionally (see Rizzolatti et al. 1998). Of them, PMv only has (direct or indirect) anatomical connections with the areas where there is visual coding of action made by others (PF/PFG and indirectly STS) and, thus, where there is the necessary information for the formation of mirror neurons (Rizzolatti & Matelli 2003).

On the basis of embryological considerations, the border between human PMd and PMv should be located, approximately, at Z level 50 in Talairach coordinates (Rizzolatti & Arbib 1998, Rizzolatti et al. 2002). This location derives from the view that the superior frontal sulcus (plus the superior precentral sulcus) represents the human homologue of the superior branch of the monkey arcuate sulcus. Because the border of monkey PMv and PMd corresponds approximately to the caudal continuation of this branch, the analogous border should, in humans, lie slightly ventral to the superior frontal sulcus.

The location of human frontal eye field (FEF) supports this hypothesis (Corbetta 1998, Kimming et al. 2001, Paus 1996, Petit et al.1996). In monkeys, FEF lies in the anterior bank of the arcuate sulcus, bordering posteriorly the sector of PMv where arm and head movements are represented (area F4). If one accepts the location of the border between PMv and PMd suggested above, FEF is located in a similar position in the two species. In both of them, the location is just anterior to the upper part of PMv and the lowest part of PMd.

The other issue concerns IFG areas. There is a deeply rooted prejudice that these areas are radically different from those of the precentral gyrus and that they are exclusively related to speech (e.g., Grèzes & Decety 2001). This is not so. Already at the beginning of the last century, Campbell (1905) noted clear anatomical similarities between the areas of posterior IFG and those of the precentral gyrus. This author classed both the *pars opercularis* and the *pars triangularis* of IFG together with the precentral areas and referred to them collectively as the "intermediate precentral" cortex. Modern comparative studies indicate that the *pars opercularis* of IFG (basically corresponding to area 44) is the human homologue of area F5 (Von Bonin & Bailey 1947, Petrides & Pandya 1997). Furthermore, from a functional perspective, clear evidence has been accumulating in recent years that human area 44, in addition to speech representation, contains (as does monkey area F5) a motor representation of hand movements (Binkofski et al. 1999, Ehrsson et al. 2000, Gerardin et al. 2000, Iacoboni et al. 1999, Krams et al. 1998). Taken together, these data strongly suggest that human PMv is the homologue of monkey area F4, and human area 44 is the homologue of monkey area F5. The descending branch of the inferior precentral sulcus (homologue to the monkey inferior precentral dimple) should form the approximate border between the two areas (for individual variations of location and extension area 44, see Amunts et al. 1999 and Tomaiuolo et al. 1999).

If the homology just described is correct, one should expect that the observation of neck and proximal arm movements would activate predominantly PMv, whereas hand and mouth movements would activate area 44. Buccino et al. (2001) addressed this issue in an fMRI experiment. Normal volunteers were presented with video clips showing actions performed with the mouth, hand/arm, and foot/leg. Both transitive (actions directed toward an object) and intransitive actions were shown. Action observation was contrasted with the observation of a static face, hand, and foot (frozen pictures of the video clips), respectively.

Observation of object-related mouth movements determined activation of the lower part of the precentral gyrus and of the *pars opercularis* of the inferior frontal gyrus (IFG), bilaterally. In addition, two activation foci were found in the parietal lobe. One was located in the rostral part of the inferior parietal lobule (most likely area PF), whereas the other was located in the posterior part of the same lobule. The observation of intransitive actions activated the same premotor areas, but there was no parietal lobe activation.

Observation of object-related hand/arm movements determined two areas of activation in the frontal lobe, one corresponding to the *pars opercularis* of IFG and the other located in the precentral gyrus. The latter activation was more dorsally located than that found during the observation of mouth movements. As for mouth movements, there were two activation foci in the parietal lobe. The rostral focus was, as in the case of mouth actions, in the rostral part of the inferior parietal lobule, but more posteriorly located, whereas the caudal focus was essentially in the same location as that for mouth actions. During the observation of intransitive movements the premotor activations were present, but the parietal ones were not.

Finally, the observation of object-related foot/leg actions determined an activation of a dorsal sector of the precentral gyrus and an activation of the posterior parietal lobe, in part overlapping with those seen during mouth and hand actions, in part extending more dorsally. Intransitive foot actions produced premotor, but not parietal, activation.

A weakness of the data by Buccino et al. (2001) is that they come from a group study. Data from single individuals are badly needed for a more precise somatotopic map. Yet, they clearly show that both the frontal and the parietal "mirror" regions are somatotopically organized. The somatotopy found in the inferior parietal lobule is the same as that found in the monkey. As far as the frontal lobe is concerned, the data appear to confirm the predictions based on the proposed homology. The activation of the *pars opercularis* of IFG should reflect the observation of distal hand actions and mouth actions, whereas that of the precentral cortex activation should reflect that of proximal arm actions and of neck movements.

It is important to note that the observation of transitive actions activated both the parietal and the frontal node of the mirror-neuron system, whereas the intransitive actions activated the frontal node only. This observation is in accord with the lack of inferior parietal lobule activation found in other studies in which intransitive actions were used (e.g., finger movements; Iacoboni et al. 1999, 2001; Koski et al. 2002, 2003). Considering that the premotor areas receive visual information from the inferior parietal lobule, it is hard to believe that the inferior parietal lobule was not activated during the observation of intransitive actions. It is more likely, therefore, that when an object is present, the inferior parietal activation is stronger than when the object is lacking, and the activation, in the latter case, does not reach statistical significance.

Brain Imaging Studies: Mirror-Neuron System Properties

As discussed above, the mirror-neuron system is involved in action understanding. An interesting issue is whether this is true also for actions done by individuals belonging to other species. Is the understanding by humans of actions done by monkeys based on the mirror-neuron system? And what about more distant species, like dogs?

Recently, an fMRI experiment addressed these questions (Buccino et al. 2004). Video clips showing silent mouth actions performed by humans, monkeys, and dogs were presented to normal volunteers. Two types of actions were shown: biting and oral communicative actions (speech reading, lip smacking, barking). As a control, static images of the same actions were presented.

The results showed that the observation of biting, regardless of whether it was performed by a man, a monkey, or a dog, determined the same two activation foci in the inferior parietal lobule discussed above and activation in the *pars opercularis* of the IFG and the adjacent precentral gyrus (Figure 3, see color insert). The left rostral parietal focus and the left premotor focus were virtually identical for all three species, whereas the right side foci were stronger during the observation of actions made by a human being than by an individual of another species. Different results were obtained with communicative actions. Speech reading activated the left *pars opercularis* of IFG; observation of lip smacking, a monkey communicative gesture, activated a small focus in the right and left *pars opercularis* of IFG; observation of barking did not produce any frontal lobe activation (Figure 4, see color insert).

These results indicated that actions made by other individuals could be recognized through different mechanisms. Actions belonging to the motor repertoire of the observer are mapped on his/her motor system. Actions that do not belong to this repertoire do not excite the motor system of the observer and appear to be recognized essentially on a visual basis without motor involvement. It is likely that these two different ways of recognizing actions have two different psychological counterparts. In the first case the motor "resonance" translates the visual experience into an internal "personal knowledge" (see Merleau-Ponty 1962), whereas this is lacking in the second case.

One may speculate that the absence of the activation of the frontal mirror area reported in some experiments might be due to the fact that the stimuli used (e.g., light point stimuli, Grèzes et al. 2001) were insufficient to elicit this "personal" knowledge of the observed action.

An interesting issue was addressed by Johnson Frey et al. (2003). Using event-related fMRI, they investigated whether the frontal mirror activation requires the observation of a dynamic action or if the understanding of the action goal is sufficient. Volunteers were presented with static pictures of the same objects being grasped or touched. The results showed that the observation of the goals of hand-object interactions was sufficient to activate selectively the frontal mirror region.

In this experiment, *pars triangularis* of IFG has been found active in several subjects (see also Rizzolatti et al. 1996b, Grafton et al. 1996). In speech, this sector appears to be mostly related to syntax (Bookheimer 2002). Although one may be tempted to speculate that this area may code also the syntactic aspect of action (see Greenfield 1991), there is at present no experimental evidence in support of this proposal. Therefore, the presence of activation of *pars triangularis* lacks, at the moment, a clear explanation.

Schubotz & Von Cramon (2001, 2002a,b) tested whether the frontal mirror region is important not only for the understanding of goal-directed actions, but also for recognizing predictable visual patterns of change. They used serial prediction tasks, which tested the participants' performance in a sequential perceptual task without sequential motor responses. Results showed that serial prediction caused activation in premotor and parietal cortices, particularly within the right hemisphere. The authors interpreted these findings as supporting the notion that sequential perceptual events can be represented independent of preparing an intended action toward the stimulus. According to these authors, the frontal mirror-neuron system node plays, in humans, a crucial role also in the representation of sequential information, regardless of whether it is perceptual or action related.

MIRROR-NEURON SYSTEM AND IMITATION

Imitation of Actions Present in the Observer's Repertoire

Psychological experiments strongly suggest that, in the cognitive system, stimuli and responses are represented in a commensurable format (Brass et al. 2000, Craighero et al. 2002, Wohlschlager & Bekkering 2002; see Prinz 2002). When observers see a motor event that shares features with a similar motor event present in their motor repertoire, they are primed to repeat it. The greater the similarity between the observed event and the motor event, the stronger the priming is (Prinz 2002).

These findings, and the discovery of mirror neurons, prompted a series of experiments aimed at finding the neural substrate of this phenomenon (Iacoboni et al. 1999, 2001; Nishitani & Hari 2000, 2002).

Using fMRI, Iacoboni et al. (1999) studied normal human volunteers in two conditions: observation-only and observation-execution. In the "observation-only" condition, subjects were shown a moving finger, a cross on a stationary finger, or a cross on an empty background. The instruction was to observe the stimuli. In the "observation-execution" condition, the same stimuli were presented, but this time the instruction was to lift the right finger, as fast as possible, in response to them.

The most interesting statistical contrast was that between the trials in which the volunteers made the movement in response to an observed action (imitation) and the trials in which the movement was triggered by the cross. The results showed that the activity was stronger during imitation trials than during the other motor trials in four areas: the left *pars opercularis* of the IFG, the right anterior parietal region, the right parietal operculum, and the right STS region (see for this last activation Iacoboni et al. 2001). Further experiments by Koski et al. (2002) confirmed the importance of Broca's area, in particular when the action to be imitated had a specific goal. Grèzes et al. (2003) obtained similar results, but only when participants had to imitate pantomimes. The imitation of object-directed actions surprisingly activated PMd.

Nishitani & Hari (2000, 2002) performed two studies in which they investigated imitation of grasping actions and of facial movements, respectively. The event-related MEG was used. The first study confirmed the importance of the left IFG (Broca's area) in imitation. In the second study (Nishitani & Hari 2002), the authors asked volunteers to observe still pictures of verbal and nonverbal (grimaces) lip forms, to imitate them immediately after having seen them, or to make similar lip forms spontaneously. During lip form observation, cortical activation progressed from the occipital cortex to the superior temporal region, the inferior parietal lobule, IFG (Broca's area), and finally to the primary motor cortex. The activation sequence during imitation of both verbal and nonverbal lip forms was the same as during observation. Instead, when the volunteers executed the lip forms spontaneously, only Broca's area and the motor cortex were activated.

Taken together, these data clearly show that the basic circuit underlying imitation coincides with that which is active during action observation. They also indicate that, in the posterior part of IFG, a direct mapping of the observed action and its motor representation takes place.

The studies of Iacoboni et al. (1999, 2001) showed also activations—superior parietal lobule, parietal operculum, and STS region—that most likely do not reflect a mirror mechanism. The activation of the superior parietal lobule is typically not present when subjects are instructed to observe actions without the instruction to imitate them (e.g., Buccino et al. 2001). Thus, a possible interpretation of this activation is that the request to imitate produces, through backward projections, sensory copies of the intended actions. In the monkey, superior parietal lobule and especially its rostral part (area PE) contains neurons that are active in response to proprioceptive stimuli as well as during active arm movements (Kalaska et al. 1983, Lacquaniti et al. 1995, Mountcastle et al. 1975). It is possible, therefore, that the observed superior parietal activation represents a kinesthetic copy of the intended movements. This interpretation fits well previous findings by Grèzes et al. (1998), who, in agreement with Iacoboni et al. (1999), showed a strong activation of superior parietal lobule when subjects' tasks were to observe actions in order to repeat them later.

An explanation in terms of sensory copies of the intended actions may also account for the activations observed in the parietal operculum and STS. The first corresponds to the location of somatosensory areas hidden in the sylvian sulcus (Disbrow et al. 2000), whereas the other corresponds to higher-order visual areas of the STS region (see above). Thus, these two activations might reflect somatosensory and visual copies of the intended action, respectively.

The importance of the *pars opercularis* of IFG in imitation was further demonstrated using repetitive TMS (rTMS), a technique that transiently disrupts the functions of the stimulated area (Heiser et al. 2003). The task used in the study was, essentially, the same as that of Iacoboni et al. (1999). The results showed that following stimulation of both left and right Broca's area, there was significant impairment in imitation of finger movements. The effect was absent when finger movements were done in response to spatial cues.

Imitation Learning

Broadly speaking, there are two types of newly acquired behaviors based on imitation learning. One is substitution, for the motor pattern spontaneously used by the observer in response to a given stimulus, of another motor pattern that is more adequate to fulfill a given task. The second is the capacity to learn a motor sequence useful to achieve a specific goal (Rizzolatti 2004).

The neural basis of the capacity to form a new motor pattern on the basis of action observation was recently studied by Buccino et al. (G. Buccino, S. Vogt, A. Ritzl, G.R. Fink, K. Zilles, H.J. Freund & G. Rizzolatti, submitted manuscript), using an event-related fMRI paradigm. The basic task was the imitation, by naive participants, of guitar chords played by an expert guitarist. By using an event-related paradigm, cortical activation was mapped during the following events: (*a*) action observation, (*b*) pause (new motor pattern formation and consolidation), (*c*) chord execution, and (*d*) rest. In addition to imitation condition, there were three control conditions: observation without any motor request, observation followed by execution of a nonrelated action (e.g., scratching the guitar neck), and free execution of guitar chords.

The results showed that during the event observation-to-imitate there was activation of a cortical network that coincided with that which is active during observation-without-instruction-to-imitate and during observation in order not to imitate. The strength of the activation was, however, much stronger in the first condition. The areas forming this common network were the inferior parietal lobule, the dorsal part of PMv, and the *pars opercularis* of IFG. Furthermore, during the event observation-to-imitate, but not during observation-without-further-motor-action, there was activation of the superior parietal lobule, anterior mesial areas plus a modest activation of the middle frontal gyrus.

The activation during the pause event in imitation condition involved the same basic circuit as in event observation-of-the-same-condition, but with some important differences: increase of the superior parietal lobule activation, activation of PMd, and, most interestingly, a dramatic increase in extension and strength of the middle frontal cortex activation (area 46) and of the areas of the anterior mesial wall. Finally, during the execution event, not surprisingly, the activation concerned mostly the sensorimotor cortex contralateral to the acting hand.

These data show that the nodal centers for new motor pattern formation coincide with the nodal mirror-neuron regions. Although fMRI experiments cannot give information on the mechanism involved, it is plausible (see the neurophysiological sections) that during learning of new motor patterns by imitation the observed actions are decomposed into elementary motor acts that activate, via mirror mechanism, the corresponding motor representations in PF and in PMv and in the *pars opercularis* of IFG. Once these motor representations are activated,

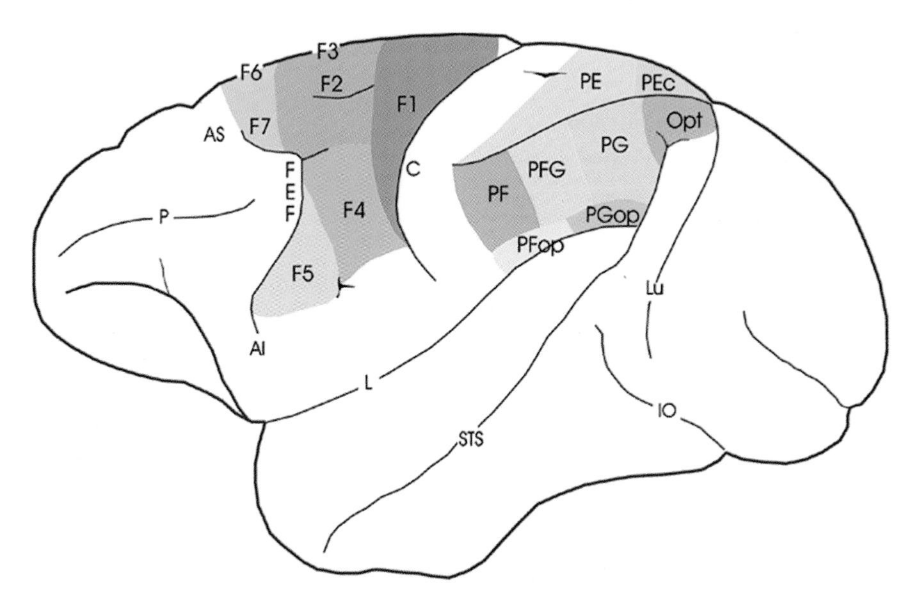

Figure 1 Lateral view of the monkey brain showing, in color, the motor areas of the frontal lobe and the areas of the posterior parietal cortex. For nomenclature and definition of frontal motor areas (F1–F7) and posterior parietal areas (PE, PEc, PF, PFG, PG, PF op, PG op, and Opt) see Rizzolatti et al. (1998). AI, inferior arcuate sulcus; AS, superior arcuate sulcus; C, central sulcus; L, lateral fissure; Lu, lunate sulcus; P, principal sulcus; POs, parieto-occipital sulcus; STS, superior temporal sulcus.

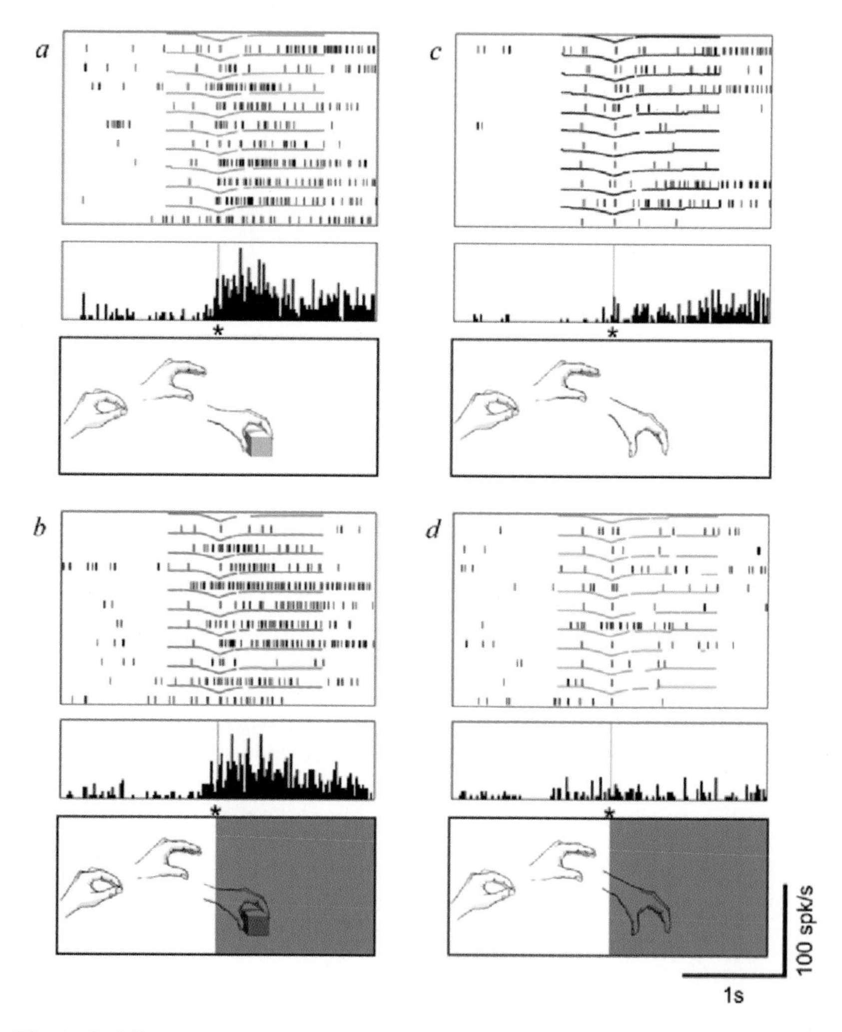

Figure 2 Mirror neuron responses to action observation in full vision (*A* and *C*) and in hidden condition (*B* and *D*). The lower part of each panel illustrates schematically the experimenter's action as observed from the monkey's vantage point. The asterisk indicates the location of a stationary marker attached to the frame. In hidden conditions the experimenter's hand started to disappear from the monkey's vision when crossing this marker. In each panel above the illustration of the experimenter's hand, raster displays and histograms of ten consecutive trials recorded are shown. Above each raster, the colored line represents the kinematics of the experimenter's hand movements expressed as the distance between the hand of the experimenter and the stationary marker over time. Rasters and histograms are aligned with the moment when the experimenter's hand was closest to the marker. *Green vertical line*: movement onset; *red vertical line*: marker crossing; *blue vertical line*: contact with the object. Histograms bin width = 20 ms. The ordinate is in spike/s. (From Umiltà et al. 2001.)

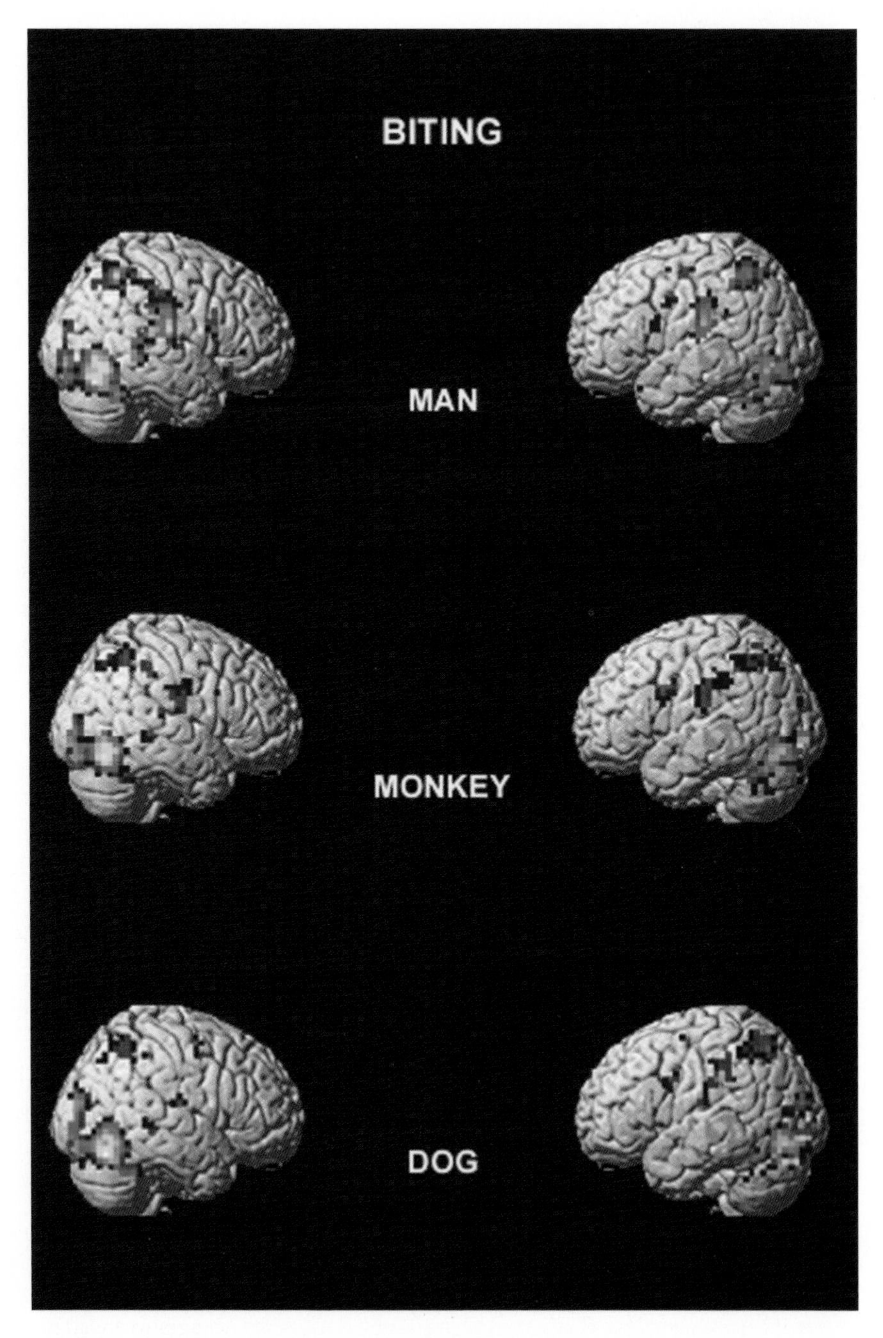

Figure 3 Cortical activations during the observation of biting made by a man, a monkey, and a dog. From Buccino et al. 2004.

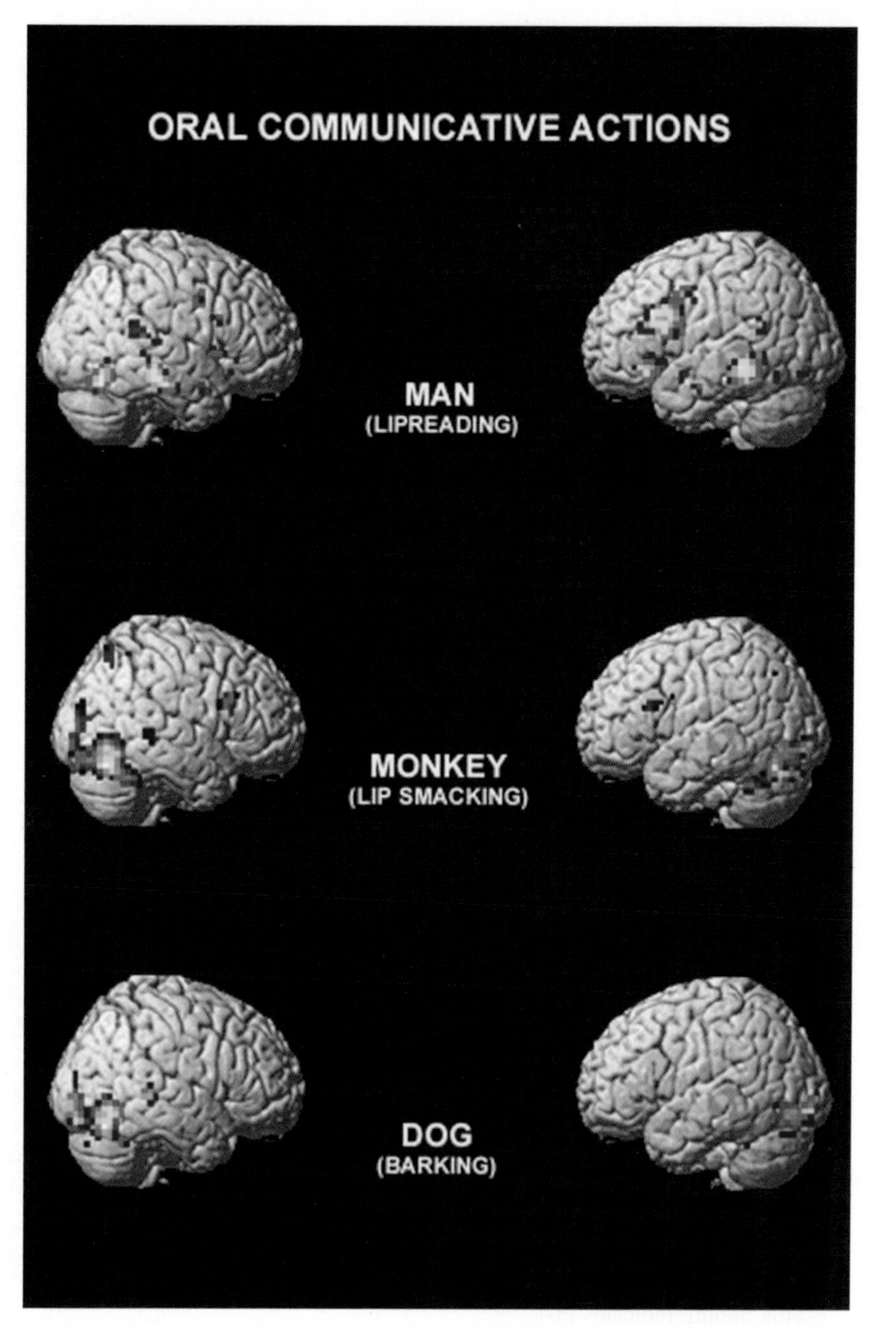

Figure 4 Cortical activations during the observation of communicative actions. For other explanations see text. From Buccino et al. 2004.

they are recombined, according to the observed model by the prefrontal cortex. This recombination occurs inside the mirror-neuron circuit with area 46 playing a fundamental orchestrating role.

To our knowledge, there are no brain-imaging experiments that studied the acquisition of new sequences by imitation from the perspective of mirror neurons. Theoretical aspect of sequential learning by imitation and its possible neural basis have been discussed by Arbib (2002), Byrne (2002), and Rizzolatti (2004). The interested reader can find there an exhaustive discussion of this issue.

MIRROR-NEURON SYSTEM AND COMMUNICATION

Gestural Communication

Mirror neurons represent the neural basis of a mechanism that creates a direct link between the sender of a message and its receiver. Thanks to this mechanism, actions done by other individuals become messages that are understood by an observer without any cognitive mediation.

On the basis of this property, Rizzolatti & Arbib (1998) proposed that the mirror-neuron system represents the neurophysiological mechanism from which language evolved. The theory of Rizzolatti & Arbib belongs to theories that postulate that speech evolved mostly from gestural communication (see Armstrong et al. 1995, Corballis 2002). Its novelty consists of the fact that it indicates a neurophysiological mechanism that creates a common (parity requirement), nonarbitrary, semantic link between communicating individuals.

The mirror-neuron system in monkeys is constituted of neurons coding object-directed actions. A first problem for the mirror-neuron theory of language evolution is to explain how this close, object-related system became an open system able to describe actions and objects without directly referring to them.

It is likely that the great leap from a closed system to a communicative mirror system depended upon the evolution of imitation (see Arbib 2002) and the related changes of the human mirror-neuron system: the capacity of mirror neurons to respond to pantomimes (Buccino et al. 2001, Grèzes et al. 2003) and to intransitive actions (Fadiga et al. 1995, Maeda et al. 2002) that was absent in monkeys.

The notion that communicative actions derived from object-directed actions is not new. Vygotski (1934), for example, explained that the evolution of pointing movements was due to attempts of children to grasp far objects. It is interesting to note that, although monkey mirror neurons do not discharge when the monkey observes an action that is not object directed, they do respond when an object is hidden, but the monkey knows that the action has a purpose (Kohler et al. 2002). This finding indicates that breaking spatial relations between effector and target does not impair the capacity of understanding the action meaning. The precondition for understanding pointing—the capacity to mentally represent the action goal—is already present in monkeys.

A link between object-directed and communicative action was also stressed by other authors (see McNeilage 1998, Van Hoof 1967; for discussion of this link from the mirror neurons perspective, see above).

Mirror Neurons and Speech Evolution

The mirror neuron communication system has a great asset: Its semantics is inherent to the gestures used to communicate. This is lacking in speech. In speech, or at least in modern speech, the meaning of the words and the phono-articulatory actions necessary to pronounce them are unrelated. This fact suggests that a necessary step for speech evolution was the transfer of gestural meaning, intrinsic to gesture itself, to abstract sound meaning. From this follows a clear neurophysiological prediction: Hand/arm and speech gestures must be strictly linked and must, at least in part, share a common neural substrate.

A number of studies prove that this is true. TMS experiments showed that the excitability of the hand motor cortex increases during both reading and spontaneous speech (Meister et al. 2003, Seyal et al. 1999, Tokimura et al. 1996). The effect is limited to the left hemisphere. Furthermore, no language-related effect is found in the leg motor area. Note that the increase of hand motor cortex excitability cannot be attributed to word articulation because, although word articulation recruits motor cortex bilaterally, the observed activation is strictly limited to the left hemisphere. The facilitation appears, therefore, to result from a coactivation of the dominant hand motor cortex with higher levels of language network (Meister et al. 2003).

Gentilucci et al. (2001) reached similar conclusions using a different approach. In a series of behavioral experiments, they presented participants with two 3-D objects, one large and one small. On the visible face of the objects there were either two crosses or a series of dots randomly scattered on the same area occupied by the crosses. Participants were required to grasp the objects and, in the condition in which the crosses appeared on the object, to open their mouth. The kinematics of hand, arm, and mouth movements was recorded. The results showed that lip aperture and the peak velocity of lip aperture increased when the movement was directed to the large object.

In another experiment of the same study Gentilucci et al. (2001) asked participants to pronounce a syllable (e.g., GU, GA) instead of simply opening their mouth. It was found that lip aperture was larger when the participants grasped a larger object. Furthermore, the maximal power of the voice spectrum recorded during syllable emission was also higher when the larger object was grasped.

Most interestingly, grasping movements influence syllable pronunciation not only when they are executed, but also when they are observed. In a recent study (Gentilucci 2003), normal volunteers were asked to pronounce the syllables BA or GA while observing another individual grasping objects of different size. Kinematics of lip aperture and amplitude spectrum of voice was influenced by the grasping movements of the other individual. Specifically, both lip aperture and voice peak amplitude were greater when the observed action was directed to larger objects. Control experiments ruled out that the effect was due to the velocity of the observed arm movement.

Taken together, these experiments show that hand gestures and mouth gestures are strictly linked in humans and that this link includes the oro-laryngeal movements used for speech production.

Auditory Modality and Mirror-Neuron Systems

If the meaning of manual gestures, understood through the mirror-neuron mechanism, indeed transferred, in evolution, from hand gestures to oro-laryngeal gestures, how did that transfer occur?

As described above, in monkeys there is a set of F5 mirror neurons that discharge in response to the sound of those actions that, when observed or executed by the monkey, trigger a given neuron (Kohler et al. 2002). The existence of these audio-visual mirror neurons indicates that auditory access to action representation is present also in monkeys.

However, the audio-visual neurons code only object-related actions. They are similar, in this respect, to the "classical" visual mirror neurons. But, as discussed above, object-related actions are not sufficient to create an efficient intentional communication system. Therefore, words should have derived mostly from association of sound with intransitive actions and pantomimes, rather than from object-directed actions.

An example taken from Paget (1930) may clarify the possible process at work. When we eat, we move our mouth, tongue, and lips in a specific manner. The observation of this combined series of motor actions constitutes the gesture whose meaning is transparent to everybody: "eat." If, while making this action, we blow air through the oro-laryngeal cavities, we produce a sound like "mnyam-mnyam," or "mnya-mnya," words whose meaning is almost universally recognized (Paget 1930). Thus through such an association mechanism, the meaning of an action, naturally understood, is transferred to sound.

It is plausible that, originally, the understanding of the words related to mouth actions occurred through activation of audio-visual mirror neurons related to ingestive behavior (see Ferrari et al. 2003). A fundamental step, however, toward speech acquisition was achieved when individuals, most likely thanks to improved imitation capacities (Donald 1991), became able to generate the sounds originally accompanied by a specific action without doing the action. This new capacity should have led to (and derived from) the acquisition of an auditory mirror system, developed on top of the original audio-visual one, but which progressively became independent of it.

More specifically, this scenario assumes that, in the case discussed above, the premotor cortex became progressively able to generate the sound "mnyam-mnyam" without the complex motor synergies necessary for producing ingestive action, and, in parallel, neurons developed able to both generate the sound and discharge (resonate) in response to that sound (echo-neurons). The incredibly confusing organization of Broca's area in humans, where phonology, semantics, hand actions, ingestive actions, and syntax are all intermixed in a rather restricted neural space (see Bookheimer 2002), is probably a consequence to this evolutive trend.

Is there any evidence that humans possess an echo-neuron system, i.e., a system that motorically "resonates" when the individual listens to verbal material? There is evidence that this is the case.

Fadiga et al. (2002) recorded MEPs from the tongue muscles in normal volunteers instructed to listen carefully to acoustically presented verbal and nonverbal stimuli. The stimuli were words, regular pseudowords, and bitonal sounds. In the middle of words and pseudowords either a double "f" or a double "r" were embedded. "F" is a labio-dental fricative consonant that, when pronounced, requires slight tongue mobilization, whereas "r" is linguo-palatal fricative consonant that, in contrast, requires a tongue movement to be pronounced. During the stimulus presentation the participants' left motor cortices were stimulated.

The results showed that listening to words and pseudowords containing the double "r" determines a significant increase of MEPs recorded from tongue muscles as compared to listening to words and pseudowords containing the double "f" and listening to bitonal sounds. Furthermore, the facilitation due to listening of the "r" consonant was stronger for words than for pseudowords.

Similar results were obtained by Watkins et al. (2003). By using TMS technique they recorded MEPs from a lip (*orbicularis oris*) and a hand muscle (first *interosseus*) in four conditions: listening to continuous prose, listening to nonverbal sounds, viewing speech-related lip movements, and viewing eye and brow movements. Compared to control conditions, listening to speech enhanced the MEPs recorded from the *orbicularis oris* muscle. This increase was seen only in response to stimulation of the left hemisphere. No changes of MEPs in any condition were observed following stimulation of the right hemisphere. Finally, the size of MEPs elicited in the first *interosseus* muscle did not differ in any condition.

Taken together these experiments show that an echo-neuron system exists in humans: when an individual listens to verbal stimuli, there is an activation of the speech-related motor centers.

There are two possible accounts of the functional role of the echo-neuron system. A possibility is that this system mediates only the imitation of verbal sounds. Another possibility is that the echo-neuron system mediates, in addition, speech perception, as proposed by Liberman and his colleagues (Liberman et al. 1967, Liberman & Mattingly 1985, Liberman & Wahlen 2000). There is no experimental evidence at present proving one or another of the two hypotheses. Yet, is hard to believe that the echo-system lost any relation with its original semantic function.

There is no space here to discuss the neural basis of action word semantics. However, if one accepts the evolutionary proposal we sketched above, there should be two roots to semantics. One, more ancient, is closely related to the action mirror-neuron system, and the other, more recent, is based on the echo-mirror-neuron system.

Evidence in favor of the existence of the ancient system in humans has been recently provided by EEG and fMRI studies. Pulvermueller (2001, 2002) compared EEG activations while subjects listened to face- and leg-related action verbs ("walking" versus "talking"). They found that words describing leg actions evoked stronger in-going current at dorsal sites, close to the cortical leg-area, whereas those of the "talking" type elicited the stronger currents at inferior sites, next to the motor representation of the face and mouth.

In an fMRI experiment, Tettamanti et al. (M. Tettamanti, G. Buccino, M.C. Saccuman, V. Gallese, M. Danna, P. Scifo, S.F. Cappa, G. Rizzolatti, D. Perani & F. Fazio, submitted manuscript) tested whether cortical areas active during action observation were also active during listening to action sentences. Sentences that describe actions performed with mouth, hand/arm, and leg were used. The presentation of abstract sentences of comparable syntactic structure was used as a control condition. The results showed activations in the precentral gyrus and in the posterior part of IFG. The activations in the precentral gyrus, and especially that during listening to hand-action sentences, basically corresponded to those found during the observation of the same actions. The activation of IFG was particularly strong during listening of mouth actions, but was also present during listening of actions done with other effectors. It is likely, therefore, that, in addition to mouth actions, in the inferior frontal gyrus there is also a more general representation of action verbs. Regardless of this last interpretation problem, these data provide clear evidence that listening to sentences describing actions engages visuo-motor circuits subserving action representation.

These data, of course, do not prove that the semantics is exclusively, or even mostly, due to the original sensorimotor systems. The devastating effect on speech of lesions destroying the perisylvian region testifies the importance in action understanding of the system based on direct transformation of sounds into speech motor gesture. Thus, the most parsimonious hypothesis appears to be that, during speech acquisition, a process occurs somehow similar to the one that, in evolution, gave meaning to sound. The meaning of words is based first on the old nonverbal semantic system. Subsequently, however, the words are understood even without a massive activation of the old semantic system. Experiments, such as selective inhibition through TMS or electrical stimulation of premotor and parietal areas, are needed to better understand the relative role of the two systems in speech perceptions.

ACKNOWLEDGMENTS

This study was supported by EU Contract QLG3-CT-2002-00746, Mirror, EU Contract IST-2000-28159, by the European Science Foundation, and by the Italian Ministero dell'Università e Ricerca, grants Cofin and Firb RBNEO1SZB4.

The *Annual Review of Neuroscience* is online at http://neuro.annualreviews.org

LITERATURE CITED

Altschuler EL, Vankov A, Hubbard EM, Roberts E, Ramachandran VS, Pineda JA. 2000. Mu wave blocking by observation of movement and its possible use as a tool to study theory of other minds. *Soc. Neurosci.* 68.1 (Abstr.)

Altschuler EL, Vankov A, Wang V, Ramachandran VS, Pineda JA. 1997. Person see, person do: human cortical electrophysiological correlates of monkey see monkey do cell. *Soc. Neurosci.* 719.17 (Abstr.)

Amunts K, Schleicher A, Buergel U, Mohlberg H, Uylings HBM, Zilles K. 1999. Broca's region re-visited: cytoarchitecture and intersubject variability. *J. Comp. Neurol.* 412: 319–41

Arbib MA. 2002. Beyond the mirror system: imitation and evolution of language. In *Imitation in Animals and Artifacts*, ed. C Nehaniv, K Dautenhan, pp. 229–80. Cambridge MA: MIT Press

Armstrong AC, Stokoe WC, Wilcox SE. 1995. *Gesture and the Nature of Language*. Cambridge, UK: Cambridge Univ. Press

Baldissera F, Cavallari P, Craighero L, Fadiga L. 2001. Modulation of spinal excitability during observation of hand actions in humans. *Eur. J. Neurosci.* 13:190–94

Binkofski F, Buccino G, Posse S, Seitz RJ, Rizzolatti G, Freund H. 1999. A fronto-parietal circuit for object manipulation in man: evidence from an fMRI-study. *Eur. J. Neurosci.* 11:3276–86

Bookheimer S. 2002. Functional MRI of language: new approaches to understanding the cortical organization of semantic processing. *Annu. Rev. Neurosci.* 25:151–88

Brass M, Bekkering H, Wohlschlager A, Prinz W. 2000. Compatibility between observed and executed finger movements: comparing symbolic, spatial, and imitative cues. *Brain Cogn.* 44:124–43

Buccino G, Binkofski F, Fink GR, Fadiga L, Fogassi L, et al. 2001. Action observation activates premotor and parietal areas in a somatotopic manner: an fMRI study. *Eur. J. Neurosci.* 13:400–4

Buccino G, Lui F, Canessa N, Patteri I, Lagravinese G, et al. 2004a. Neural circuits involved in the recognition of actions performed by non-conspecifics: an fMRI study. *J. Cogn. Neurosci.* 16:1–14

Byrne RW. 1995. *The Thinking Ape. Evolutionary Origins of Intelligence*. Oxford, UK: Oxford Univ. Press

Byrne RW. 2002. Seeing actions as hierarchically organized structures: great ape manual skills. See Meltzoff & Prinz 2002, pp. 122–40

Campbell AW. 1905. *Histological Studies on the Localization of Cerebral Function*. Cambridge, UK: Cambridge Univ. Press. 360 pp.

Cochin S, Barthelemy C, Lejeune B, Roux S, Martineau J. 1998. Perception of motion and qEEG activity in human adults. *Electroencephalogr. Clin. Neurophysiol.* 107:287–95

Cochin S, Barthelemy C, Roux S, Martineau J. 1999. Observation and execution of movement: similarities demonstrated by quantified electroencephalograpy. *Eur. J. Neurosci.* 11:1839–42

Cohen-Seat G, Gastaut H, Faure J, Heuyer G. 1954. Etudes expérimentales de l'activité nerveuse pendant la projection cinématographique. *Rev. Int. Filmologie* 5:7–64

Corballis MC. 2002. *From Hand to Mouth. The Origins of Language*. Princeton: Princeton Univ. Press. 257 pp.

Corbetta M. 1998. Frontoparietal cortical networks for directing attention and the eye to visual locations: identical, independent, or overlapping neural systems? *Proc. Natl. Acad. Sci. USA* 95:831–38

Craighero L, Bello A, Fadiga L, Rizzolatti G. 2002. Hand action preparation influences the responses to hand pictures. *Neuropsychologia* 40:492–502

Decety J, Chaminade T, Grezes J, Meltzoff AN.

2002. A PET exploration of the neural mechanisms involved in reciprocal imitation. *Neuroimage* 15:265–72

Di Pellegrino G, Fadiga L, Fogassi L, Gallese V, Rizzolatti G. 1992. Understanding motor events: a neurophysiological study. *Exp. Brain Res.* 91:176–80

Disbrow E, Roberts T, Krubitzer L. 2000. Somatotopic organization of cortical fields in the lateral sulcus of homo sapiens: evidence for SII and PV. *J. Comp. Neurol.* 418:1–21

Donald M. 1991. *Origin of the Modern Mind: Three Stages in the Evolution of Culture and Cognition.* Cambridge, MA: Harvard Univ. Press

Ehrsson HH, Fagergren A, Jonsson T, Westling G, Johansson RS, Forssberg H. 2000. Cortical activity in precision- versus power-grip tasks: an fMRI study. *J. Neurophysiol.* 83:528–36

Fadiga L, Craighero L, Buccino G, Rizzolatti G. 2002. Speech listening specifically modulates the excitability of tongue muscles: a TMS study. *Eur. J. Neurosci.* 15:399–402

Fadiga L, Fogassi L, Pavesi G, Rizzolatti G. 1995. Motor facilitation during action observation: a magnetic stimulation study. *J. Neurophysiol.* 73:2608–11

Ferrari PF, Gallese V, Rizzolatti G, Fogassi L. 2003. Mirror neurons responding to the observation of ingestive and communicative mouth actions in the monkey ventral premotor cortex. *Eur. J. Neurosci.* 17:1703–14

Fogassi L, Gallese V, Fadiga L, Rizzolatti G. 1998. Neurons responding to the sight of goal directed hand/arm actions in the parietal area PF (7b) of the macaque monkey. *Soc. Neurosci.* 24:257.5 (Abstr.)

Galef BG. 1988. Imitation in animals: history, definition and interpretation of data from psychological laboratory. In *Comparative Social Learning,* ed. T Zental, BG Galef, pp. 3–28, Hillsdale, NJ: Erlbaum

Gallese V, Fadiga L, Fogassi L, Rizzolatti G. 1996. Action recognition in the premotor cortex. *Brain* 119:593–609

Gallese V, Fogassi L, Fadiga L, Rizzolatti G. 2002. Action representation and the inferior parietal lobule. In *Attention & Performance XIX. Common Mechanisms in Perception and Action*, ed. W Prinz, B Hommel, pp. 247–66. Oxford, UK: Oxford Univ. Press

Gangitano M, Mottaghy FM, Pascual-Leone A. 2001. Phase specific modulation of cortical motor output during movement observation. *NeuroReport* 12:1489–92

Gastaut HJ, Bert J. 1954. EEG changes during cinematographic presentation. *Electroencephalogr. Clin. Neurophysiol.* 6:433–44

Gentilucci M. 2003. Grasp observation influences speech production. *Eur. J. Neurosci.* 17:179–84

Gentilucci M, Benuzzi F, Gangitano M, Grimaldi S. 2001. Grasp with hand and mouth: a kinematic study on healthy subjects. *J. Neurophysiol.* 86:1685–99

Gerardin E, Sirigu A, Lehericy S, Poline JB, Gaymard B, et al. 2000. Partially overlapping neural networks for real and imagined hand movements. *Cereb. Cortex* 10:1093–104

Grafton ST, Arbib MA, Fadiga L, Rizzolatti G. 1996. Localization of grasp representations in humans by PET: 2. Observation compared with imagination. *Exp. Brain Res.* 112:103–11

Grèzes J, Armony JL, Rowe J, Passingham RE. 2003. Activations related to "mirror" and "canonical" neurones in the human brain: an fMRI study. *Neuroimage* 18:928–37

Grèzes J, Costes N, Decety J. 1998. Top-down effect of strategy on the perception of human biological motion: a PET investigation. *Cogn. Neuropsychol.* 15:553–82

Grèzes J, Decety J. 2001. Functional anatomy of execution, mental simulation, observation, and verb generation of actions: a meta-analysis. *Hum. Brain Mapp.* 12:1–19

Grèzes J, Fonlupt P, Bertenthal B, Delon-Martin C, Segebarth C, Decety J. 2001. Does perception of biological motion rely on specific brain regions? *Neuroimage* 13:775–85

Greenfield PM. 1991. Language, tool and brain: the ontogeny and phylogeny of hierarchically organized sequential behavior. *Behav. Brain Sci.* 14:531–95

Hari R, Forss N, Avikainen S, Kirveskari S, Salenius S, Rizzolatti G. 1998. Activation of human primary motor cortex during action observation: a neuromagnetic study. *Proc. Natl. Acad. Sci. USA* 95:15061–65

Hari R, Salmelin R. 1997. Human cortical oscillations: a neuromagnetic view through the skull. *Trends Neurosci.* 20:44–49

Heiser M, Iacoboni M, Maeda F, Marcus J, Mazziotta JC. 2003. The essential role of Broca's area in imitation. *Eur. J. Neurosci.* 17:1123–28

Hyvarinen J. 1982. Posterior parietal lobe of the primate brain. *Physiol. Rev.* 62:1060–129

Iacoboni M, Koski LM, Brass M, Bekkering H, Woods RP, et al. 2001. Reafferent copies of imitated actions in the right superior temporal cortex. *Proc. Natl. Acad. Sci. USA* 98:13995–99

Iacoboni M, Woods RP, Brass M, Bekkering H, Mazziotta JC, Rizzolatti G. 1999. Cortical mechanisms of human imitation. *Science* 286:2526–28

Jeannerod M. 1994. The representing brain. Neural correlates of motor intention and imagery. *Behav. Brain Sci.* 17:187–245

Jellema T, Baker CI, Wicker B, Perrett DI. 2000. Neural representation for the perception of the intentionality of actions. *Brain Cogn.* 442:280–302

Jellema T, Baker CI, Oram MW, Perrett DI. 2002. Cell populations in the banks of the superior temporal sulcus of the macaque monkey and imitation. See Meltzoff & Prinz 2002, pp. 267–90

Johnson Frey SH, Maloof FR, Newman-Norlund R, Farrer C, Inati S, Grafton ST. 2003. Actions or hand-objects interactions? Human inferior frontal cortex and action observation. *Neuron* 39:1053–58

Kalaska JF, Caminiti R, Georgopoulos AP. 1983. Cortical mechanisms related to the direction of two-dimensional arm movements: relations in parietal area 5 and comparison with motor cortex. *Exp. Brain Res.* 51:247–60

Kimmig H, Greenlee MW, Gondan M, Schira M, Kassubek J, Mergner T. 2001. Relationship between saccadic eye movements and cortical activity as measured by fMRI: quantitative and qualitative aspects. *Exp. Brain Res.* 141:184–94

Kohler E, Keysers C, Umiltà MA, Fogassi L, Gallese V, Rizzolatti G. 2002. Hearing sounds, understanding actions: action representation in mirror neurons. *Science* 297:846–48

Koski L, Iacoboni M, Dubeau MC, Woods RP, Mazziotta JC. 2003. Modulation of cortical activity during different imitative behaviors. *J. Neurophysiol.* 89:460–71

Koski L, Wohlschlager A, Bekkering H, Woods RP, Dubeau MC. 2002. Modulation of motor and premotor activity during imitation of target-directed actions. *Cereb. Cortex* 12: 847–55

Krams M, Rushworth MF, Deiber MP, Frackowiak RS, Passingham RE. 1998. The preparation, execution and suppression of copied movements in the human brain. *Exp. Brain Res.* 120:386–98

Lacquaniti F, Guigon E, Bianchi L, Ferraina S, Caminiti R. 1995. Representing spatial information for limb movement: role of area 5 in the monkey. *Cereb. Cortex* 5:391–409

Liberman AM, Cooper FS, Shankweiler DP, Studdert-Kennedy M. 1967. Perception of the speech code. *Psychol. Rev.* 74:431–61

Liberman AM, Mattingly IG. 1985. The motor theory of speech perception revised. *Cognition* 21:1–36

Liberman AM, Whalen DH. 2000. On the relation of speech to language. *Trends Cogn. Neurosci.* 4:187–96

MacNeilage PF. 1998. The frame/content theory of evolution of speech production. *Behav. Brain Sci.* 21:499–511

Maeda F, Kleiner-Fisman G, Pascual-Leone A. 2002. Motor facilitation while observing hand actions: specificity of the effect and role of observer's orientation. *J. Neurophysiol.* 87:1329–35

Manthey S, Schubotz RI, von Cramon DY. 2003. Premotor cortex in observing erroneous action: an fMRI study. *Brain Res. Cogn. Brain Res.* 15:296–307

Meister IG, Boroojerdi B, Foltys H, Sparing R, Huber W, Topper R. 2003. Motor cortex hand area and speech: implications for the development of language. *Neuropsychologia* 41:401–6

Meltzoff AN, Prinz W. 2002. *The Imitative Mind. Development, Evolution and Brain Bases*. Cambridge, UK: Cambridge Univ. Press

Merleau-Ponty M. 1962. *Phenomenology of Perception*. Transl. C Smith. London: Routledge (From French)

Mountcastle VB, Lynch JC, Georgopoulos A, Sakata H, Acuna C. 1975. Posterior parietal association cortex of the monkey: command functions for operations within extrapersonal space. *J. Neurophysiol.* 38:871–908

Nishitani N, Hari R. 2000. Temporal dynamics of cortical representation for action. *Proc. Natl. Acad. Sci. USA* 97:913–18

Nishitani N, Hari R. 2002. Viewing lip forms: cortical dynamics. *Neuron* 36:1211–20

Paget R. 1930. *Human Speech*. London: Kegan Paul, Trench

Patuzzo S, Fiaschi A, Manganotti P. 2003. Modulation of motor cortex excitability in the left hemisphere during action observation: a single and paired-pulse transcranial magnetic stimulation study of self- and non-self action obervation. *Neuropsychologia* 41:1272–78

Paus T. 1996. Location and function of the human frontal eye-field: a selective review. *Neuropsychologia* 34:475–83

Perani D, Fazio F, Borghese NA, Tettamanti M, Ferrari S, et al. 2001. Different brain correlates for watching real and virtual hand actions. *Neuroimage* 14:749–58

Perrett DI, Harries MH, Bevan R, Thomas S, Benson PJ, et al. 1989. Frameworks of analysis for the neural representation of animate objects and actions. *J. Exp. Biol.* 146:87–113

Perrett DI, Mistlin AJ, Harries MH, Chitty AJ. 1990. Understanding the visual appearance and consequence of hand actions. In *Vision and Action: The Control of Grasping*, ed. MA Goodale, pp. 163–342. Norwood, NJ: Ablex

Petit L, Orssaud C, Tzourio N, Crivello F, Berthoz A, Mazoyer B. 1996. Functional anatomy of a prelearned sequence of horizontal saccades in humans. *J. Neurosci.* 16:3714–26

Petrides M, Pandya DN. 1984. Projections to the frontal cortex from the posterior parietal region in the rhesus monkey. *J. Comp. Neurol.* 228:105–16

Petrides M, Pandya DN. 1997. Comparative architectonic analysis of the human and the macaque frontal cortex. In *Handbook of Neuropsychology*, ed. F Boller, J Grafman, pp. 17–58. New York: Elsevier. Vol. IX

Prinz W. 2002. Experimental approaches to imitation. See Meltzoff & Prinz 2002, pp. 143–62

Pulvermueller F. 2001. Brain reflections of words and their meaning. *Trends Cogn. Sci.* 5:517–24

Pulvermueller F. 2002. *The Neuroscience of Language*. Cambridge, UK: Cambridge Univ. Press. 315 pp.

Rizzolatti G. 2004. The mirror-neuron system and imitation. In *Perspectives on Imitation: From Mirror Neurons to Memes*, ed. S Hurley, N Chater. Cambridge, MA: MIT Press. In press

Rizzolatti G, Arbib MA. 1998. Language within our grasp. *Trends Neurosci.* 21:188–94

Rizzolatti G, Fadiga L, Fogassi L, Gallese V. 1996a. Premotor cortex and the recognition of motor actions. *Cogn. Brain Res.* 3:131–41

Rizzolatti G, Fadiga L, Matelli M, Bettinardi V, Paulesu E, et al. 1996b. Localization of grasp representation in humans by PET: 1. Observation versus execution. *Exp. Brain Res.* 111:246–52

Rizzolatti G, Fogassi L, Gallese V. 2001. Neurophysiological mechanisms underlying the understanding and imitation of action. *Nat. Rev. Neurosci.* 2:661–70

Rizzolatti G, Fogassi L, Gallese V. 2002. Motor and cognitive functions of the ventral premotor cortex. *Curr. Opin. Neurobiol.* 12:149–54

Rizzolatti G, Luppino G. 2001. The cortical motor system. *Neuron* 31:889–901

Rizzolatti G, Luppino G, Matelli M. 1998.

The organization of the cortical motor system: new concepts. *Electroencephalogr. Clin. Neurophysiol.* 106:283–96

Rizzolatti G, Matelli M. 2003. Two different streams form the dorsal visual system. *Exp. Brain Res.* 153:146–57

Salmelin R, Hari R. 1994. Spatiotemporal characteristics of sensorimotor neuromagnetic rhythms related to thumb movement. *Neuroscience* 60:537–50

Schubotz RI, von Cramon DY. 2001. Functional organization of the lateral premotor cortex: fMRI reveals different regions activated by anticipation of object properties, location and speed. *Brain Res. Cogn. Brain Res.* 11:97–112

Schubotz RI, von Cramon DY. 2002a. A blueprint for target motion: fMRI reveals perceived sequential complexity to modulate premotor cortex. *Neuroimage* 16:920–35

Schubotz RI, von Cramon DY. 2002b. Predicting perceptual events activates corresponding motor schemes in lateral premotor cortex: an fMRI study. *Neuroimage* 15:787–96

Seyal M, Mull B, Bhullar N, Ahmad T, Gage B. 1999. Anticipation and execution of a simple reading task enhance corticospinal excitability. *Clin. Neurophysiol.* 110:424–29

Strafella AP, Paus T. 2000. Modulation of cortical excitability during action observation: a transcranial magnetic stimulation study. *NeuroReport* 11:2289–92

Tanné-Gariepy J, Rouiller EM, Boussaoud D. 2002. Parietal inputs to dorsal versus ventral premotor areas in the monkey: evidence for largely segregated visuomotor pathways. *Exp. Brain. Res.* 145:91–103

Thorndyke EL. 1898. Animal intelligence: an experimental study of the associative process in animals. *Psychol. Rev. Monogr.* 2:551–53

Tokimura H, Tokimura Y, Oliviero A, Asakura T, Rothwell JC. 1996. Speech-induced changes in corticospinal excitability. *Ann. Neurol.* 40:628–34

Tomaiuolo F, MacDonald JD, Caramanos Z, Posner G, Chiavaras M, et al. 1999. Morphology, morphometry and probability mapping of the pars opercularis of the inferior frontal gyrus: an in vivo MRI analysis. *Eur. J. Neurosci.* 11:3033–46

Tomasello M, Call J. 1997. *Primate Cognition.* Oxford, UK: Oxford Univ. Press

Umiltà MA, Kohler E, Gallese V, Fogassi L, Fadiga L, et al. 2001. "I know what you are doing": a neurophysiological study. *Neuron* 32:91–101

Van Hoof JARAM. 1967. The facial displays of the catarrhine monkeys and apes. In *Primate Ethology*, ed. D Morris, pp. 7–68. London: Weidenfield & Nicolson

Visalberghi E, Fragaszy D. 2001. Do monkeys ape? Ten years after. In *Imitation in Animals and Artifacts*, ed. K Dautenhahn, C Nehaniv. Boston, MA: MIT Press

Von Bonin G, Bailey P. 1947. *The Neocortex of Macaca Mulatta.* Urbana: Univ. Ill. Press. 136 pp.

Von Economo C. 1929. *The Cytoarchitectonics of the Human Cerebral Cortex.* London: Oxford Univ. Press. 186 pp.

Vygotsky LS. 1934. *Thought and Language.* Cambridge, MA: MIT Press

Watkins KE, Strafella AP, Paus T. 2003. Seeing and hearing speech excites the motor system involved in speech production. *Neuropsychologia* 41:989–94

Whiten A, Ham R. 1992. On the nature and evolution of imitation in the animal kingdom: reappraisal of a century of research. In *Advances in the Study of Behavior*, ed. PBJ Slater, JS Rosenblatt, C Beer, M Milinski, pp. 239–83. San Diego: Academic

Wohlschlager A, Bekkering H. 2002. Is human imitation based on a mirror-neurone system? Some behavioural evidence. *Exp. Brain Res.* 143:335–41

Annu. Rev. Neurosci. 2004. 27:193–222
doi: 10.1146/annurev.neuro.27.070203.144212

First published online as a Review in Advance on March 2, 2004

GENETIC APPROACHES TO THE STUDY OF ANXIETY

Joshua A. Gordon[1] and Rene Hen[2]
Center for Neurobiology and Behavior, Departments of Psychiatry[1] and Pharmacology[2], Columbia University and the New York State Psychiatric Institute, New York, NY 10032; email: rh95@columbia.edu, jg343@columbia.edu

Key Words knockout mouse, fear, serotonin, GABA, Benzodiazepines

■ **Abstract** Anxiety and its disorders have long been known to be familial. Recently, genetic approaches have been used to clarify the role of heredity in the development of anxiety and to probe its neurobiological underpinnings. Twin studies have shown that a significant proportion of the liability to develop any given anxiety disorder is due to genetic factors. Ongoing efforts to map anxiety-related loci in both animals and humans are underway with limited success to date. Animal models have played a large role in furthering our understanding of the genetic basis of anxiety, demonstrating that the genetic factors underlying anxiety are complex and varied. Recent advances in molecular genetic techniques have allowed increasing specificity in the manipulation of gene expression within the central nervous system of the mouse. With this increasing specificity has come the ability to ask and answer precise questions about the mechanisms of anxiety and its treatment.

ANXIETY AND ANXIETY DISORDERS

Wariness of threatening aspects of the environment is a protective, if at times uncomfortable, trait. In humans the experience of such wariness is called anxiety, and it is typically accompanied by characteristic autonomic responses and defensive behaviors. Whereas anxiety in many settings is protective, excessive anxiety can prove disabling. Individuals who suffer from anxiety disorders are often crippled by overwhelming emotional and physical symptoms. As many as 25% of adults will, at one point in their lives, suffer from one of the six described forms of anxiety disorder (Kessler & Greenberg 2002): panic disorder, characterized by unpredictable, rapid-fire attacks of intense anxiety; generalized anxiety disorder, marked by excessive worry in multiple areas; social anxiety disorder, characterized by fear and avoidance of social situations; specific phobia, notable for intense fear of a specific trigger; posttraumatic stress disorder, characterized by intrusive, anxiety-provoking memories of trauma; and obsessive-compulsive disorder, marked by anxious obsessions and anxiety-reducing compulsive behaviors. In aggregate these disorders account for tremendous morbidity, and the collective economic cost of these disorders is estimated to be over $40 billion per year

0147-006X/04/0721-0193$14.00

(Kessler & Greenberg 2002). Although effective treatments for anxiety disorders are available, many patients are left with residual symptoms or experience side effects that limit their adherence to prescribed regimens (Gorman et al. 2002). Improved understanding of the neural mechanisms of anxiety and its treatments, therefore, could reduce the individual and societal costs of anxiety disorders by pointing the way toward improved treatments.

In order to attempt an understanding of neural mechanisms, numerous investigators have turned to animal models of anxiety. Like humans, animals respond to the potential presence of a threat with characteristic autonomic responses and defensive behaviors. These physiological and behavioral parameters can be measured with relative ease and have been validated with drugs known to be anxiolytic or anxiogenic in humans. This approach has been used to characterize numerous models of anxiety-like behavior in laboratory animals (Borsini et al. 2002). Paradigms such as fear-conditioned freezing and fear-potentiated startle rely on learned associations between innocuous and painful stimuli; pairing of, say, a particular sound with a mild shock causes the animal to display anxiety-like defensive behaviors such as freezing (periods of attentive stillness) or enhanced startle (heightened reactivity to a loud, sudden noise) in response to later presentations of the sound (Davis et al. 2003, LeDoux 2000). Other animal models of anxiety depend upon innate, species-specific danger signals. The elevated plus maze and novelty-suppressed feeding tasks, for example, measure a rodent's avoidance of exposed or novel areas, in which they might be more vulnerable to predation (Rodgers & Dalvi 1997, Santarelli et al. 2003). Anxiolytic drugs, such as benzodiazepines, reduce the expression of anxiety-like behaviors in each of these paradigms (Conti et al. 1990, Pellow & File 1986, Shephard & Broadhurst 1982).

Pharmacological and neurobiological analysis of anxiety disorders and animal models of anxiety have helped to identify neurotransmitters and neural systems involved in the expression of anxiety. The most widely prescribed classes of anxiolytic drugs, the benzodiazepines and the selective serotonin reuptake inhibitors (SSRIs), modulate γ-aminobutyric acid (GABA)- and serotonin-mediated neurotransmission, respectively (Gorman et al. 2002). Likewise, injection of drugs that alter GABAergic or serotonergic neurotransmission alter anxiety-like behavior in animals (Borsini et al. 2002, Griebel 1995). Pharmacological challenge studies in anxiety disorder patients have also implicated the noradrenergic system, especially in panic disorder (Sullivan et al. 1999). Studies of stress physiology in both animals and humans have suggested a powerful role for the adrenocortical system and its neuropeptide regulator, corticotrophin-releasing hormone (CRH), in anxiety disorders and anxiety-like behavior (Clark & Kaiyala 2003). Imaging studies in humans, as well as lesion and microinjection studies in animals, have in turn implicated various brain structures in anxiety, including the amygdala, the hippocampus, the cingulate cortex, the hypothalamus, and various brainstem areas (Charney & Drevets 2002). Hypothetical anxiety circuits have been proposed to explain particular forms of anxiety-like behavior, such as those suggested by Davis et al. (2003) and LeDoux (2000), to underlie the learning and expression of conditioned-fear behaviors (see Figure 1).

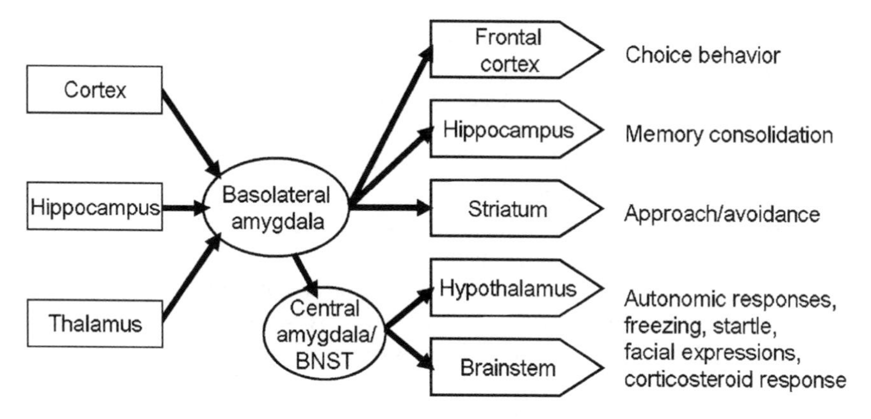

Figure 1 Schematic diagram of a proposed fear circuit. Adapted from Davis et al. (2003).

It has long been known that anxiety disorders, like most psychiatric illnesses, have important heritable components. Hopes that linkage studies might reveal single genes predisposing for specific disorders have gone unfulfilled thus far (Finn et al. 2003). Single-gene knockout studies, on the other hand, have led to the identification of numerous genes seemingly required for normal anxiety-like behavior in mouse models (Finn et al. 2003, Lesch 2001b). Indeed, the diversity of single-gene deletions that result in anxiety phenotypes in animals is so great that it calls into question the specificity of these manipulations. Here we examine the impact of recent forays into genetic approaches to anxiety. We highlight particular studies that best illustrate the capabilities and, in some cases, limitations of particular genetic approaches. These results suggest areas of focus for future use of genetic technology and point out that harnessing the full power of genetics will likely require applying more traditional neuroscience techniques in order to fully characterize the impact of genetic manipulations.

GENETIC APPROACHES TO HUMAN ANXIETY DISORDERS

Numerous studies have demonstrated that anxiety disorders run in families (Fyer et al. 1995, Merikangas & Pine 2002, Noyes et al. 1987). Individuals with first-degree relatives who suffer from a given anxiety disorder are 4–6 times more likely to develop that disorder than individuals without such relatives (Hettema et al. 2001a). Twin studies have since confirmed that the great majority of this familial risk is due to genetic heritability. Comparisons of concordance rates in dizygotic and monozygotic twins have established that 30%–50% of the individual variability in risk to develop a given anxiety disorder is due to genetic factors (Hettema et al. 2001b, Kendler 2001, Roy-Byrne et al. 2002, Skre et al. 2000). The remainder of this variability is accounted for by environmental factors specific to the individual twin, as opposed to environmental experiences shared within the family (Kendler 2001). These data demonstrate that anxiety disorders have a moderate genetic predisposition strongly influenced by environmental interactions.

The nature of this moderate predisposition is not entirely clear, however. Twin studies have consistently demonstrated a genetic predisposition that accounts for 30%–35% of the liability to develop posttraumatic stress disorder (PTSD) following exposure to traumatic events (Goldberg et al. 1990, True et al. 1993). Intriguingly, there is also a genetic predisposition for exposure to certain types of trauma; thus, genetic factors account for 20% and 35% of the liability to exposure to assaultative and combat-related trauma, respectively (Lyons et al. 1993, Stein et al. 2002). With regard to civilian assaults, a high correlation between genetic effects on exposure to trauma and development of PTSD symptoms suggest that a single genetic factor increases the likelihood of both exposure and pathological reaction to extreme trauma (Stein et al. 2002). These data point out that the straightforward interpretation of genetic factors as grounded in strictly neurobiological mechanisms ignores the fact that genetic factors influence how individuals interact with their environment. Indeed, Kendler (2001) has suggested that up to 20% of the genetic influences in psychiatric disorders could be mediated by such indirect mechanisms.

Attempts to map the loci underlying these genetic factors using genome-wide linkage scans have shown scattered promise. Panic disorder has been the focus of most of the genetic studies because of its relatively distinctive symptomatology. Findings to date have been mostly weak and inconclusive (Crowe et al. 2001, Knowles et al. 1998, Weissman 1993), although recently panic disorder has been tentatively linked with sites on 7p (Logue et al. 2003) and 9q (Thorgeirsson et al. 2003). Attempts to take advantage of possible genetic syndromes involving co-transmission of panic disorder and joint laxity or kidney/bladder disease have suggested that specific subtypes of panic might be more amenable to traditional mapping techniques (Gratacos et al. 2001, Hamilton et al. 2003, Weissman et al. 2000), but these studies have yet to be replicated. The data for other anxiety disorders are sparse. One recent study revealed a model-dependent link to a single marker on chromosome 14 in simple phobia (Gelernter et al. 2003). An initial genome-wide scan for linkage to obsessive compulsive disorder failed to reveal any significant associations (Hanna et al. 2002).

Candidate Genes and Gene-Environment Interactions

Numerous attempts have been made to link the genetic predisposition toward anxiety disorders with specific candidate genes (for a recent review, see Finn et al. 2003). Taking their cues from effective treatments for anxiety, investigators have targeted genes encoding components of the monoaminergic and GABAergic systems. Whereas a genetic association between the gene encoding for the serotonin transporter (SERT) and anxiety traits has been found (see below), no such association was found with panic disorder or obsessive-compulsive disorder (Frisch et al. 2000, Hamilton et al. 1999). A longer, more active form of the monoamine oxidase A promoter has been found with higher frequency in female panic disorder patients than in normal controls in one study (Deckert et al. 1999), but this result did not survive replication (Hamilton et al. 2000). Similarly, contradictory data have been reported for catecholamine-O-methyltransferase in obsessive-compulsive disorder and panic disorder (Alsobrook et al. 2002, Hamilton et al. 2002, Karayiorgou et al. 1999, Ohara et al. 1998) and the serotonin 1B (5-HT1B) receptor in obsessive-compulsive disorder (Di Bella et al. 2002, Mundo et al. 2000). The gene encoding the GABA-synthetic enzyme GAD65 has been associated with the panic disorder–related trait of behavioral inhibition in children (Smoller et al. 2001b), but none of the eight genes encoding GABA receptors tested thus far has been linked to panic disorder (Finn & Smoller 2001, van den Heuvel et al. 2000). Several other attempts to link various other specific genes with specific anxiety disorders have been mostly negative (Finn & Smoller 2001, Finn et al. 2003).

Several studies, but not all, have found a link between anxiety traits and a variation in the gene encoding the SERT (Greenberg et al. 1999, Gustavsson et al. 1999, Jorm et al. 1998, Lesch 2001a, Lesch et al. 1996). A simple repeat sequence lies within the SERT gene promoter. There are two different repeat lengths in the caucasian population: a 14 repeat short (*s*) allele and a 16 repeat long (*l*) allele (Lesch et al. 1996). Adults who are homozygous for the *s* allele have lower levels of SERT activity and score higher for "neuroticism" on an anxiety test battery than heterozygous *s*/*l* or homozygous *l*/*l* adults (Greenberg et al. 1999, Lesch et al. 1996). Infants homozygous for the *s* allele have similar increases in anxiety-related measures (Auerbach et al. 1999). Intriguingly, recent functional imaging studies have found that adults homozygous for the *s* allele have greater amygdala activation during the observation of fearful faces (Hariri et al. 2002). Nonetheless, the effect of the *s* allele on trait anxiety is modest at best, accounting for approximately 4% of the phenotypic variance (Lesch et al. 1996).

The low predictive value of the SERT polymorphism genotype allows for multiple interacting mechanisms, including, of course, additional genes. Recent research supports the possibility that gene/environment interactions account for a portion of the additional variance. In humans, the *s* allele increases the risk of major depressive disorder but only in those individuals with a previous history of major life stressors and/or childhood trauma (Caspi et al. 2003). In monkeys, as in humans, short and long versions of the SERT promoter repeat exist (Lesch et al. 1997). Also in monkeys, the *s* allele is associated with an increase in the serotonin metabolite, 5-HIAA (consistent with decreased SERT activity), and an increase in emotional reactivity (Champoux et al. 2002). These effects of the SERT polymorphism in monkeys are dependent on early rearing environment. Monkeys reared by their mothers show normal levels of 5-HIAA regardless of SERT genotype, whereas nursery-reared monkeys have increased 5-HIAA if they carry the *l*/*s* genotype, but not if they carry the *l*/*l* allele (Bennett et al. 2002). An identical rearing by genotype interaction was seen in an anxiety-related behavioral variable (Champoux et al. 2002). These data demonstrate that physiological and behavioral effects of the

SERT polymorphism are dependent on a permissive rearing environment in both humans and nonhuman primates. They provide a clear example of an interaction between genes and environment in the regulation of anxiety-related behaviors.

ANIMAL MODELS OF ANXIETY

Whereas human studies have established the genetic basis of anxiety, animal studies have been used to attempt to further clarify its genetic determinants. There are numerous animal models of anxiety-like behavior. The relevance of these models to human anxiety disorders is derived first and foremost from pharmacological validity. Benzodiazepines, the archetypal anxiolytic drugs, reduce the level of anxiety-like behavior in virtually every animal model of anxiety (Borsini et al. 2002). The development of the SSRIs as the preferred pharmacological treatment for anxiety has called into question the direct relevance of many of these models to specific anxiety disorders, as the effects of SSRIs and other nonbenzodiazepine anxiolytics on these models are at best inconsistent (Borsini et al. 2002). Nonetheless many of these models have been useful when combined with genetic approaches to the study of anxiety.

Fear Versus Anxiety: Learned Fear Paradigms

As models for anxiety disorders are developed and refined, a key issue that arises is the difference between fear and anxiety. Davis (1998) has suggested that fear is the response to a specific, stimulus-linked threat, whereas anxiety is a nonstimulus-linked defensive state. Gray & McNaughton (2000) similarly proposed that fear is the defensive response to the actual presence of a threat, whereas anxiety is the defensive response to the potential presence of a threat. These two definitions both characterize fear as tied to specific stimuli and anxiety as more generally linked to situations or environments. Both may have their parallels in human anxiety disorders. Specific phobia and PTSD, for example, are disorders in which emotional and behavioral responses are often tied to specific stimuli, such as snake phobias, or flashbacks triggered by loud noises. Patients with generalized anxiety disorder and panic disorder, on the other hand, experience symptoms without regard to the presence of particular stimuli.

Animal models of stimulus-linked defensive behaviors have been extensively studied, especially with regard to the mechanisms by which stimuli and responses are associated. Fear-conditioned freezing and fear-potentiated startle involve pairing of a stimulus (typically a tone or a specific context) with shock. The learned association between the stimulus and the shock then results in freezing behavior or an increased startle response. Benzodiazepines decrease the time spent freezing after presentation of a footshock (Conti et al. 1990) or after exposure to a conditioned stimulus (Beck & Fibiger 1995, Pletnikov et al. 1996, Quartermain et al. 1993). The conditioned stimulus-induced potentiation of the startle response is also decreased by benzodiazepines (Davis 1979). Acute treatment with SSRIs has been reported to inhibit both the learning and expression of conditioned fear responses (Inoue et al. 1996). Although chronic treatment with the SSRI fluovoxamine has been reported to have no effect on conditioned freezing, the duration of treatment (15 d) may have been inadequate (Li et al. 2001).

Genetic approaches to learned fear have received considerable attention, most prominently with regard to genetically manipulated mice. Numerous knockouts have been shown to have altered fear conditioning, although most studies focus on understanding the learning component itself (Ammassari-Teule et al. 2001, Chen et al. 1994, Crestani et al. 2002, Frankland et al. 2001, Howe et al. 2002, Pape & Stork 2003, Stork et al. 2003, Wei et al. 2002). Differences in inbred and selectively bred strains have been characterized (Balogh & Wehner 2003, Bolivar et al. 2001, Falls et al. 1997, Nie & Abel 2001, Paterson et al. 2001), and attempts have been made to map genetic loci that modify fear-conditioning behaviors (Fernandez-Teruel et al. 2002, Radcliffe et al. 2000, Wehner et al. 1997). Fear conditioning can be readily performed in humans, and a recent twin study suggested that genetic factors account for approximately 35%–45% of the variance in a nonclinical population (Hettema et al. 2003). This proportion is within the range of those described for anxiety disorders (see above). Attempts to characterize fear-potentiated startle in clinical populations have thus far yielded only subtle differences between patients and controls (Grillon et al. 1994, Morgan et al. 1995).

Models of Innate Anxiety

Much of the use of genetic approaches to study anxiety have focused on tests of innate anxiety, rather than fear learning. Perhaps the best characterized model of innate anxiety is the elevated plus maze. The maze consists of an elevated, plus-shaped platform, two arms of which are enclosed by high walls (Figure 2). The other two arms are unenclosed or surrounded by a small lip. Consistent with the idea

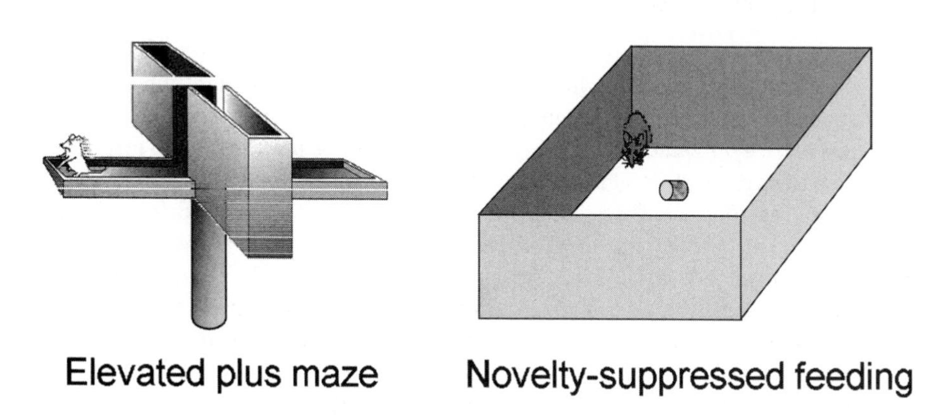

Figure 2 Animal models of innate anxiety.

that the open arms provoke an anxiety-like behavioral state, rats and mice exposed to the maze will make fewer entries into and spend less time in the open arms than the closed arms (Lister 1987, Pellow et al. 1985). The validity of these measures as a model of anxiety has been confirmed pharmacologically, ethologically, and physiologically. Benzodiazepines increase the number of open-arm entries and time spent in the open arms, whereas yohimbine and other anxiogenic agents decrease these parameters (Lister 1987, Pellow et al. 1985, Pellow & File 1986). Ethological analysis has demonstrated that animals perform more defensive and anxiety-related behaviors in the open arms than in the closed arms (Cruz et al. 1994, Pellow et al. 1985, Rodgers & Johnson 1995). Animals confined to the open arms have higher plasma corticosterone concentrations and produce more fecal boli than those confined to the closed arms (Pellow et al. 1985). A common variant of the elevated plus maze is the elevated zero maze, in which a circular platform is divided into four quadrants, two of which are enclosed and two of which are open, eliminating the potentially problematic center of the elevated plus (Shepherd et al. 1994). Several other tests rely on relatively "safe" and "unsafe" aspects of the testing environment, including the open field (in which the rodents prefer the periphery to the center) and light/dark test (in which rodents prefer the dark area). These tests are also generally responsive to benzodiazepine anxiolytics (Borsini et al. 2002).

The novelty-suppressed feeding task is a newer test that depends on the suppression of feeding induced by a novel environment (Figure 2). Animals are food-deprived and placed in a novel chamber; the latency to eat from a pellet of food placed in the center is among the parameters measured. Benzodiazepines reduce this latency (Bodnoff et al. 1989, Shephard & Broadhurst 1982), as does chronic (but not acute) treatment with SSRIs (Bodnoff et al. 1988, 1989; Merali et al. 2003; Santarelli et al. 2003). Novelty-suppressed feeding is, therefore, one of the few tests of anxiety that mimics the time course of anxiolytic efficacy seen in humans.

Interstrain Differences in Anxiety-Like Behaviors

Initial attempts to explore the genetic basis of anxiety with animal models focused on examining differences in anxiety behaviors between different inbred rodent strains. It has been clearly established that genetically different strains of mice and rats behave differently in a variety of tests for anxiety, including those mentioned above (Avgustinovich et al. 2000, Cook et al. 2001, Griebel et al. 2000, Rex et al. 1996, Tarantino et al. 2000, Trullas & Skolnick 1993, van Gaalen & Steckler 2000); there are also strain-dependent differences in the responses to anxiolytics in these tests (Crawley & Davis 1982, Griebel et al. 2000). The relative order of mouse strains, in terms of level of anxiety-like behaviors, differs depending on the task. For example, Griebel et al. (2000) found that mice of the NZB strain were among the most anxious in the elevated plus maze and the least anxious in the light/dark test. Although some studies suggest anxiety-like behavior in certain tests is correlated [for example, novelty-suppressed feeding and elevated plus behavior in eight strains of mice (Trullas & Skolnick 1993)], most suggest that different behavioral tests measure different aspects of anxiety (Avgustinovich et al. 2000, Griebel et al. 2000, van Gaalen & Steckler 2000). In rats, a factor analysis of elevated plus and open field behavior in an F1 intercross between two rat strains suggested that anxiety-related behaviors in these two tests loaded onto separate factors (Ramos et al. 1998). A six-strain, four test-factor analysis in rats places anxiety-related behaviors on two factors and locomotion on a third (Ramos et al. 1997). More extensive multitest and multistrain analyses have yet to be implemented.

A recent study of the behavior of several strains in the elevated zero test suggest one potential source of interstrain differences. Cook et al. (2001) found that three inbred mouse strains (C3H, CBA, and FVB) that carry a mutation (*rd1*) leading to retinal degeneration demonstrate uncharacteristically low levels of anxiety-like behavior in this test. For one of these strains (C3H), a congenic strain with the wild-type allele at this locus had elevated levels of anxiety-like behavior compared to the original strain, arguing strongly that the *rd1*-induced visual deficits account for the elevated anxiety levels in the original strain. Because this study used only this one test of anxiety, it is not clear, nor is it likely, that visual deficits can explain the inconsistent strain-dependent results across multiple anxiety models. Nonetheless, this study makes explicit the notion that animal models involve many behaviors and systems other than those directly involved in the expression of anxiety.

As an alternative to relying on undirected inbreeding to establish genetically based variability in anxiety measures, several investigators have used directed breeding to purposefully derive lines of mice or rats expressing high or low levels of anxiety-related behaviors (Broadhurst 1975, Goto et al. 1993, Landgraf & Wigger 2002, Liebsch et al. 1998a, Liebsch et al. 1998b, Suaudeau et al. 2000). Some of these models suffer from the same inconsistencies across anxiety-related tasks as do inbred lines (Broadhurst 1975, Chaouloff et al. 1994). Interestingly, one group has reported selectively breeding rat lines with high (HA) and low (LA) levels of anxiety-like behavior in the elevated plus maze (Liebsch et al. 1998a,b). Further characterization of these lines revealed that HA rats exhibit higher levels of anxiety across several different tasks (Landgraf & Wigger 2002, Liebsch et al. 1998a). Yet the inbred strain and factor analysis studies listed above suggest that different factors govern performance in the different anxiety models. How does one explain that a line of rats bred for behavior on a single anxiety-related task simultaneously inherits multiple seemingly independent anxiety-predisposing factors? Comparing factor analyses of HA, LA, and control strains across different tests, Ohl et al. (2001) found that in the HA and LA rats, anxiety-related factors overwhelmed other factors, explaining a greater proportion of the variance as compared to control rats. These data argue that the genetic predisposition toward increased anxiety in the elevated plus maze inherited by the HA rats is actually a generalized predisposition, trumping other factors and leading to increased anxiety-like behavior across several models.

Attempts to map the genetic factors involved in the regulation of anxiety-like behaviors have primarily utilized the quantitative trait loci (QTL) mapping technique, in which the degree of association between genetic loci and quantitative measures are estimated (Belknap et al. 1997). Numerous studies that intercross selectively bred lines (Aguilar et al. 2002; Fernandez-Teruel et al. 2002; Flint et al. 1995; Turri et al. 2001a,b), inbred lines (Clement et al. 1997, Cohen et al. 2001, Gershenfeld & Paul 1998, Plomin et al. 1991, Yoshikawa et al. 2002), or outbred lines (Talbot et al. 1999) have implicated various loci in various anxiety-related tasks (for review, see Flint 2003). In mice, a locus on chromosome 1 has been the most consistently reported as being associated with anxiety-related behaviors (Flint et al. 1995, Gershenfeld & Paul 1998, Talbot et al. 1999, Turri et al. 2001a). A careful analysis of QTLs relevant to several different anxiety-related tasks, however, suggests that this locus may link more closely to variables related to exploration, rather than to anxiety per se (Turri et al. 2001a). For example, the chromosome 1 site was associated with both open-arm entries and closed-arm entries, whereas another site in that study (on chromosome 15) was more specifically associated with open-arm entries. QTLs have also been identified in studies of the rat, but these do not share homology with any of the identified regions in the mouse (Fernandez-Teruel et al. 2002). As of yet, linkage studies have not led to the identification of any specific genes involved in the regulation of anxiety-like behaviors.

The Role of Environment

Recent, compelling studies provide concrete evidence that the environment plays a crucial role in the establishment of anxiety-like behaviors. Francis et al. (2003) set out to determine whether and when maternal behaviors might influence the development of strain-specific differences in anxiety-like behavior. In their hands, BALB/cJ (BALB) mice behave as if they are more anxious than C57BL/6J (C57) mice, making fewer entries into and spending less time in the open arms of the elevated plus maze. To test the effects of prenatal and postnatal experience, C57 embryos were transferred into BALB or C57 dams. After birth, the pups were then cross fostered to BALB or C57 dams and tested in the elevated plus at three months of age. Only those C57 pups raised by BALB dams both in utero and postnatally took on the more anxious phenotype of their BALB foster mothers. Neither pre- nor postnatal maternal influences alone were sufficient to cause the C57 pups to develop BALB-like levels of anxiety-related behaviors (Figure 3).

A number of studies have demonstrated that maternal behavior has an important influence on adult anxiety-like behavior. Maternal separation for several hours a day during the early postnatal period results in adults that display increased anxiety-like behaviors as well as increased hormonal reactivity to stress (Huot et al. 2001, Kalinichev et al. 2002). Briefer separation actually leads to decreased anxiety-like behaviors and hormonal reactivity (Meaney et al. 1996, Vallee et al. 1997). This latter effect is primarily due to the maternal behavioral response to such separation. Dams respond to brief separation with higher levels of licking, grooming, and arched-back nursing (LG-ABN) (Liu et al. 1997). Indeed, pups

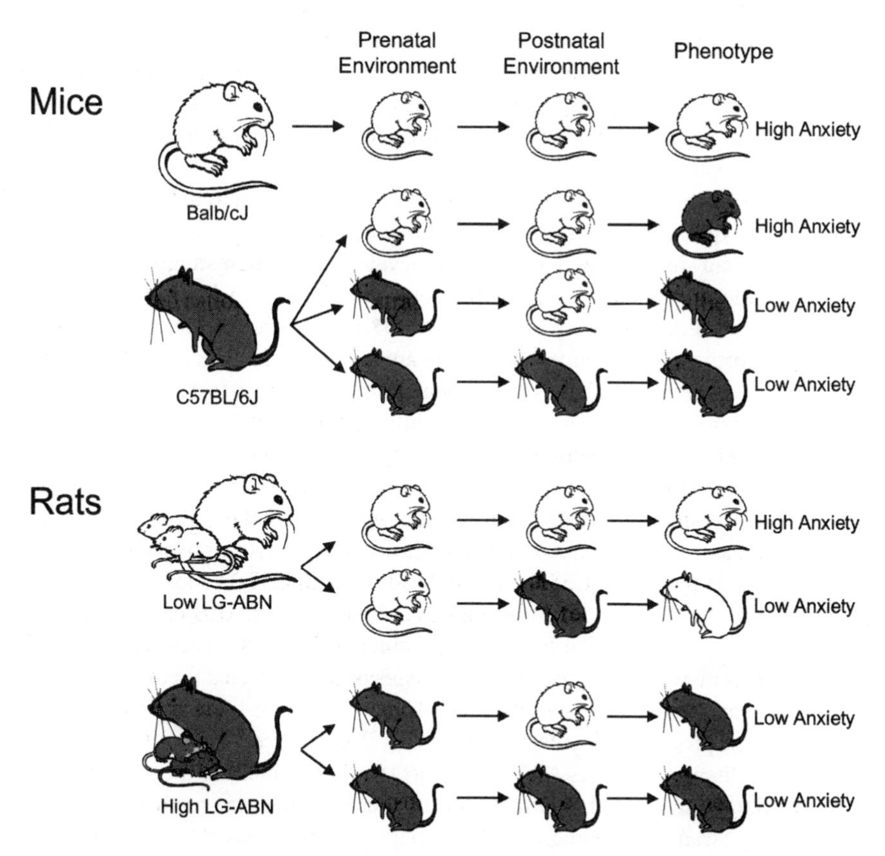

Figure 3 Epigenetic modes of inheritance. In mice and rats, anxiety-like behavior is influenced by pre- and postnatal maternal rearing. In mice (*upper pane*), both prenatal and postnatal cross fostering are required to convert a low anxiety C57BL/6J pup into a high-anxiety adult (Francis et al. 2003). In rats (*lower pane*), postnatal cross fostering is sufficient to convert a high-anxiety, low licking, grooming, and arched-back nursing (LG-ABN) pup into a low-anxiety, high LG-ABN adult. As in mice, postnatal cross-fostering alone does not convert a low-anxiety pup into a high-anxiety adult (Liu et al. 2000). *Shading* indicates genotype, *posture* indicates phenotype.

that have not been separated, but are simply raised by mothers that display high LG-ABN behaviors, have lower levels of anxiety as adults than pups raised by low LG-ABN mothers (Caldji et al. 1998). Cross fostering offspring of low LG-ABN mothers to high LG-ABN mothers causes the pups to develop low-anxiety-like behavior as adults (Liu et al. 2000). However, the converse is not true. Offspring of high LG-ABN mothers raised by low LG-ABN mothers do not show high-anxiety-like behavior, which suggests that prenatal experience or genetic endowment in the high LG-ABN offspring protect them from the effects of low LG-ABN mothering

(Figure 3). Of note, Francis et al. (1999) have shown that the effect of high licking-and-grooming can be inherited. Females cross-fostered to high LG-ABN mothers themselves become high LG-ABN mothers and go on to produce low-anxiety offspring regardless of whether their biological mother showed low or high LG-ABN behavior.

These studies on pre- and postnatal inheritance of maternal behavior and anxiety phenotype suggest that adult anxiety-like behavior is heavily influenced by events that occur during pre- and postnatal development. Interestingly, the data suggest that the dependency on pre- and postnatal experience is not symmetric. From the mouse data, making low-anxiety C57 pups more anxious requires both pre- and postnatal experience with a more anxious dam (Francis et al. 2003). This is in rough agreement with the rat data, where postnatal cross fostering does not convert low-anxiety pups into high-anxiety adults (Liu et al. 2000). High-anxiety rat pups, however, can be converted to low-anxiety adults through postnatal fostering alone (Liu et al. 2000). It would be interesting to know if the same is true in mice; the effects on anxiety of cross fostering high-anxiety BALB pups with low-anxiety C57 dams have not yet been reported. It is known, however, that postnatal cross fostering of BALB pups with C57 dams yields adults with C57-like learning abilities. The reverse, postnatal cross fostering of C57 pups with BALB dams, does not yield adults with BALB-like learning (Anisman et al. 1998). Thus the data so far are consistent with the hypothesis that aspects of the prenatal environment in C57 mice, and high LG-ABN rats, not only establish long-lasting behavioral patterns, but also protect against later exposure to BALB mice, or low LG-ABN rats. C57 mice display high LG-ABN behaviors, further tightening the analogy to the rat work (Anisman et al. 1998). This hypothesis leads to the prediction that prenatal embryo transfer of high LG-ABN rat pups into the uterus of low LG-ABN mothers would yield adults with increased anxiety-like behavior; and that postnatal cross fostering of BALB pups with C57 dams would yield low levels of anxiety-like behavior.

MOLECULAR GENETIC MANIPULATIONS

Maternal behavior tells only part of the story. Cross-fostering experiments with the HA- and LA-selected rat lines discussed above (Landgraf & Wigger 2002) clearly demonstrate a genetic mode of transmission, as pups from either line retain their biological parents' anxiety phenotype (Wigger et al. 2001). The QTL mapping data also strongly argue for genetic control of anxiety in both rats and mice. Finally, numerous genetic manipulations have direct and profound effects on anxiety-like behaviors (for a recent comprehensive review see Finn et al. 2003).

Indeed, the dizzying array of knockouts and transgenics that demonstrate abnormal levels of anxiety-like behaviors is rather daunting. Many fit neatly into preexisting theoretical constructs of anxiety. For example, mice deficient in the GABA-synthetic enzyme GAD-65 have reduced levels of the neurotransmitter and show increased levels of anxiety-like behavior in the elevated zero maze and open field (Kash et al. 1999). GAD-65 knockouts are also less sensitive to the behavioral effects of benzodiazepines (Kash et al. 1999). These data are consistent with the known role of GABA in the anxiolytic effect of benzodiazepines and the hypothetical role (based on this pharmacological evidence) for endogenous GABA in the regulation of anxiety. Likewise, mice with alterations in various components of the GABAergic, serotonergic, noradrenergic, and adrenocortical systems have been found to have abnormal levels of anxiety-like behavior. Several reviewers have capably organized and discussed these and other knockouts in a comprehensive fashion (Clement et al. 2002, Finn et al. 2003, Flint 2003, Holmes 2001, Toth 2003).

Rather than confirming or extending existing hypotheses, data from other mutant mouse lines are more surprising. Anxiety-related phenotypes have been found in mice lacking such diverse genes as those encoding substance P receptor (Santarelli et al. 2001), preproenkephalin (Konig et al. 1996), neuropeptide Y (Inui et al. 1998), interferon (Kustova et al. 1998), dopamine D4 receptor (Dulawa et al. 1999), glucocorticoid receptor (Tronche et al. 1999), the α-isoform of calcium/calmodulin-dependent protein kinase II (αCaMKII) (Chen et al. 1994), protein kinase Cγ (Bowers et al. 2000), apolipoprotein E (Raber et al. 2000), neuronal cell-adhesion molecule (NCAM) (Stork et al. 2000, Stork et al. 1999), and dystrophin (Vaillend & Ungerer 1999). This group of genes is comprised of elements of virtually all the diverse functions of neurons and neuronal assemblies, from signaling molecules, through receptors and intracellular effectors, to cytoskeletal and extracellular matrix molecules (Wood & Toth 2001). The overall message one takes away from these studies is that normal anxiety requires normal neuronal functioning. Disrupt such functioning in any of a number of different ways and anxiety-like behavior is likely to be disrupted—not a very specific or informative conclusion.

Part of the specificity problem stems from the nature of knockout technology, and part from the limited degree to which most of these knockouts have been studied. Using the straightforward, first-generation knockout technology, genes are deleted wholesale from the mouse genome, eliminating expression throughout the life cycle and throughout the entire brain (not to mention the rest of the mouse). In most of these studies, therefore, it is unclear whether lack of the molecule during development or in the adult causes the phenotype. It is also unclear where in the brain (or elsewhere) expression of the gene product is required for normal anxiety-like behavior. It is important to know what neural systems the knockout affects because compensatory changes in any number of systems might result from the wholesale deletion of a gene during development or adulthood. Reviewed here are studies of four different lines of genetically modified mice, which, through use of innovative technology and/or careful study, have avoided or even taken advantage of the limitations of traditional knockout studies. The results from each enrich our understanding of anxiety and its treatment and illustrate the kinds of questions that genetic manipulation is perhaps best suited to answer.

The α-Subunit of the $GABA_A$ Receptor: Using Knock-in Technology to Demonstrate Isoform Specificity

Members of the benzodiazepine class of anxiolytic drugs derive their efficacy from binding and modulating the activity of the ionotropic $GABA_A$ receptor. In addition to reducing anxiety, these drugs also cause sedation, motor impairment, muscle relaxation, and amnesia, side effects that limit the use of these powerful anxiolytics (Buffett-Jerrott & Stewart 2002). Benzodiazepines bind the α-subunit of the pentameric receptor, enhancing the efficacy of GABA in activating the receptor (Mohler et al. 2001). Of the six isoforms of the α-subunit, two are insensitive to benzodiazepines; these two lack a conserved histidine residue found in each of the four benzodiazepine-sensitive isoforms (Mohler et al. 2002). Benzodiazepine-sensitive isoforms can be rendered insensitive by replacing this histidine with arginine (Wieland et al. 1992). In order to test the subunit specificity of the various behavioral effects of benzodiazepines, the histidine to arginine mutation was introduced by homologous recombination into the genes encoding each of the $\alpha 1$, $\alpha 2$, and $\alpha 3$ subunits (Low et al. 2000, Rudolph et al. 1999). In each case, the knock-in mutation rendered the mutated subunit unable to bind the benzodiazepine diazepam without affecting the expression of the mutated or other α isoforms.

Behavioral testing revealed that the $\alpha 2$ isoform is responsible for the anxiolytic effects of benzodiazepines in mice. In the elevated plus maze and light/dark choice tests, rendering the $\alpha 1$ or $\alpha 3$ isoform insensitive to diazepam via the histidine to arginine substitution did not prevent the drug from exerting its anxiety-reducing effect (Low et al. 2000, Rudolph et al. 1999). However, the same substitution in the $\alpha 2$ subunit eliminated the anxiolytic-like effect of diazepam in these tests (Low et al. 2000). Moreover, mice carrying the mutated $\alpha 2$ subunit were still sensitive to the sedating and motor-impairing effects of the drug (Low et al. 2000). Mice carrying the mutated $\alpha 1$ subunit were insensitive to the sedating and amnestic influences of the drug (Rudolph et al. 1999). These data demonstrate that the $\alpha 2$-containing $GABA_A$ receptor mediates the anxiolytic, but not the sedative or amnestic, effects of diazepam. These results have direct clinical relevance, in that they encourage the search for isoform-specific medications; a drug specific for the $\alpha 2$ subunit would be expected to reduce anxiety with little or no sedation (Mohler et al. 2002).

Importantly, the results also have implications for the basic neurobiology of anxiety, as the distribution of the receptor indicates where in the brain benzodiazepines might exert their anxiolytic effect. The $\alpha 2$ isoform is expressed mostly in the limbic system and cortex (Fritschy & Mohler 1995, Pirker et al. 2000). The subregional and subcellular localization of the isoform may be of even greater significance. Within the amygdala, the $\alpha 2$ isoform is expressed primarily in the central nucleus (Kaufmann et al. 2003). Analysis of subregional expression in other areas would help refine hypotheses regarding the circuits capable of modulating anxiety. Within pyramidal neurons, the $\alpha 2$ isoform is preferentially localized within the axon initial segment, where it can play a crucial role in the output from these neurons (Fritschy & Mohler 1995, Nusser et al. 1996). A simple hypothesis consistent with these data is that output from the central nucleus promotes anxiety-like behavior; benzodiazepines, by enhancing inhibition onto the axonal initial segment, reduce that output. As tools develop for manipulating genes with increasing tissue specificity, one could imagine testing this hypothesis, perhaps by directing expression of a wildtype $\alpha2$ isoform to the amygdala of the knock-in mouse carrying the $\alpha2$ arginine substitution. Demonstrating anxiolytic efficacy in such a mouse would directly prove a role for the amygdala in the action of benzodiazepines.

Protein Kinase Cε: Studying the Downstream Effects of a Knockout Reveals Influences on Other Systems

Combining multiple neurobiological approaches to the study of a knockout of the ε isoform of protein kinase C (PKCε) has helped identify a novel mechanism in the regulation of anxiety. PKCε is a serine/threonine kinase expressed widely in the nervous system and other tissues (Saito et al. 1993). Because of data suggesting a possible role of PKCε in mediating the effects of alcohol, the sensitivity of PKCε knockout mice to alcohol was measured and found to be increased (Hodge et al. 1999). Further study revealed that these mice were supersensitive to various allosteric modulators of the $GABA_A$ receptor, including benzodiazepines, barbiturates, and neurosteroids (Hodge et al. 1999, 2002). These data suggest that PKCε acts to desensitize the $GABA_A$ receptor to its modulators. To determine if this putative effect of the kinase was dependent on adult or developmental activity, the authors used a peptide that inhibits translocation of the kinase to the periphery, blocking its activity (Yedovitzky et al. 1997). Inhibiting PKCε with this peptide significantly increased the benzodiazepine-induced enhancement of $GABA_A$ function in cortical microsacs derived from wildtype, but not knockout, mice, arguing strongly that the kinase acts in the adult to diminish sensitivity to $GABA_A$ receptor modulators.

Intriguingly, when the PKCε knockouts were placed in the elevated plus maze and open field, they made more entries to the open arms and center of the open field, respectively, as if they were less anxious (Hodge et al. 2002). Low doses of bicuculline normalized the behavior of the knockouts in these tests but had no effect on wildtype control animals (Hodge et al. 2002). The anxiety phenotype of these mice, therefore, is likely related to the enhancing effect of the gene deletion on $GABA_A$ receptor function. The simplest interpretation of these data is that the knockout enhances $GABA_A$ receptor sensitivity to an endogenous modulator (Gordon 2002, Hodge et al. 2002). The most likely candidates for such an endogenous modulator are the neurosteroids (Baulieu et al. 2001, Gordon 2002). Those neurosteroids that enhance $GABA_A$ receptor function are also anxiolytic in the elevated plus maze (Melchior & Ritzmann 1994); the anxiety-related effects of neurosteroids can be prevented by the $GABA_A$ antagonist, picrotoxin (Baulieu et al. 2001). Abnormal levels of neurosteroids have even been detected in patients suffering from anxiety and depression (Bicikova et al. 2000, Spivak et al. 2000). The combination of behavioral, pharmacological, and neurophysiological approaches to the study of this genetically manipulated organism allowed the authors to go beyond the simple statement that the knockout was less anxious. This combination of approaches led to the recognition that PKCε plays a role in regulating the sensitivity to endogenous modulators of anxiety and suggests the kinase as a target for anxiolytic therapy.

CRH Receptor 1: Using Tissue Specificity to Localize Function

Stress and anxiety have long been thought of as tightly linked, although numerous exceptions can be found to this general rule (File 1996). In many cases, anxiety-provoking situations increase, and benzodiazepines decrease, the activation of the hypothalamus-pituitary-adrenal (HPA) axis, the primary mediator of stress hormone release (Le Fur et al. 1979, Pellow et al. 1985). Corticotrophin-releasing hormone (CRH) is the primary regulator of HPA activity. Stress induces the release of CRH from the hypothalamus, which causes the release of adrenocorticotropin (ACTH) from the pituitary into the general circulation. ACTH then induces the release of glucocorticoid stress hormones from the adrenal glands (Miller & O'Callaghan 2002). CRH also modulates anxiety when injected directly into the brain (Dunn & Swiergiel 1999, Eckart et al. 1999, Martins et al. 1997). CRH activity is mediated by at least two receptors, CRH receptors 1 and 2 (CRH-R1 and CRH-R2) (Eckart et al. 2002). Both receptors also bind urocortin, a peptide related to CRH (Vaughan et al. 1995). CRH, urocortin, and its receptors are expressed in numerous parts of the central nervous system, and there is evidence to suggest they have roles in processes other than the stress response (Chang & Opp 2001, Drolet & Rivest 2001, Eckart et al. 1999, Vaughan et al. 1995, Wood & Toth 2001).

The relationship between CRH and anxiety has been tested in several different lines of genetically modified mice (Finn et al. 2003). Mice lacking CRH behave similarly to wildtype controls in the elevated plus maze and open field (Dunn & Swiergiel 1999). Mice overexpressing CRH, however, have increased anxiety-like behaviors in the elevated plus maze and light/dark test (Heinrichs et al. 1997, Stenzel-Poore et al. 1994, van Gaalen et al. 2002). Two different lines of mice lacking CRH-R1 have been produced, and both have decreased anxiety-like behavior in the elevated plus and light/dark tests (Smith et al. 1998, Timpl et al. 1998). Data on anxiety measures in CRH-R2 knockout mice are inconsistent across three different laboratories and several different tests (Bale et al. 2000, 2002; Coste et al. 2000; Finn et al. 2003; Kishimoto et al. 2000). The effect on anxiety of deleting the gene for urocortin has also been inconsistent. One group reported an anxiogenic effect in the elevated plus maze and open field, but not the light/dark test (Vetter et al. 2002), whereas a second group reported no effect on these three tests (Wang et al. 2002). Although the reasons are unclear why different groups report different results, the data do suggest that CRH or a related ligand acts on CRH-R1 to increase anxiety-related behaviors.

These studies left open the long-standing question of the localization of the role for the CRH system in anxiety: Are the CRH effects on anxiety due to activation of the HPA axis or to CRH neurotransmission in the forebrain? To address this question, Müller et al. (2003) used a tissue-specific knockout strategy to delete the CRH-R1 gene from the forebrain, including limbic areas, but not from the HPA axis. Using the *cre/loxP* system, the investigators used the αCaMKII promoter to drive expression of the Cre recombinase to delete exons 9–13 of the CRH-R1 gene from the forebrain, sparing the hypothalamus and pituitary (although the authors did not specifically show high-magnification views of the HPA axis in the paper). Unlike the original CRH-R1 knockout, which has a dramatically reduced HPA axis response to stress, the tissue-specific knockout has slightly increased ACTH and corticosterone responses to stress (Müller et al. 2003, Timpl et al. 1998). Nonetheless, the tissue-specific knockout has the same anxiety phenotype as the complete knockout, with substantial increases in entries into the open arms of the elevated plus maze and decreased latency to enter the dark area in the light/dark test (Müller et al. 2003). These data differentiate the role of the CRH system in regulating the HPA axis response to stress, which involves primarily pituitary CRH-R1 and its role in regulating anxiety-like behavior, involving primarily forebrain CRH-R1. Confirmation of this dichotomy with a tissue-specific knockout of pituitary CRH-R1 would be a logical and satisfying next step.

5-HT1A Receptor: Temporal Specificity Defines a Developmental Critical Period

Of the several knockouts of individual serotonin receptors, the 5-HT1A receptor knockout has been the most informative with respect to anxiety. Three independent laboratories have produced 5-HT1A receptor knockout mice independently, each on a different background strain (Heisler et al. 1998, Parks et al. 1998, Ramboz et al. 1998). In each case, the knockout mice behave as if they are more anxious in the elevated plus (or zero) maze and the open field test. These data are consistent with the generally anxiolytic effects of 5-HT1A agonists (Griebel 1995). Antagonists of the 5-HT1A receptor, however, are not generally anxiogenic (Canto-de-Souza et al. 2002), which raises the question of whether the knockout effect is indeed due simply to the absence of the adult receptor rather than to some developmental or compensatory effect.

In order to clarify the dependence of the anxiety phenotype on the presence of the receptor in the adult, and to begin to define in which tissues expression of the receptor is required, Gross (2002) and colleagues developed a tissue-specific, inducible rescue of the 5-HT1A knockout. They first directed the insertion of a cassette containing a promoter responsive to the tetracycline-regulated transcriptional activator protein (tTA) into the 5′ leader sequence of the 5-HT1A receptor gene. This insertion ablated expression of the receptor by its native promoter. The line carrying this insertion was then crossed with a line carrying a transgene with the tTA protein expressed under the control of the αCaMKII promoter. The tTA protein successfully induced expression of the 5-HT1A receptor in postsynaptic target tissues, such as the hippocampus and cortex, but not in the serotonergic neurons of the dorsal raphe nucleus. Rescue in this manner restored normal anxiety-like behavior, demonstrating that the lack of postsynaptic receptors, rather than presynaptic, causes the anxiety phenotype in the original knockout line.

To directly test when expression of the postsynaptic 5-HT1A receptor is required, Gross (2002) treated animals at various ages with tetracycline, which prevents activation via the tTA protein. When tetracycline was given only during adulthood, allowing 5-HT1A receptor expression only during development, the mice behaved like wildtype animals. When tetracycline was given throughout gestation and weaning, adult expression of the receptor was not able to rescue the anxiety phenotype, and the mice behaved like knockout animals (Figure 4). These surprising findings demonstrate that stimulation of postsynaptic 5-HT1A receptors during a developmental critical period is required to establish normal patterns of anxiety-like behavior that then persist into adulthood. They help clarify why 5-HT1A antagonists might not cause anxiety when given to adult animals. Moreover, they leave open the question of the relevance of adult 5-HT1A expression to anxiety.

Then why are 5-HT1A agonists anxiolytic in the adult? Intriguing new data combining further analysis of the 5-HT1A knockouts with more old-fashioned

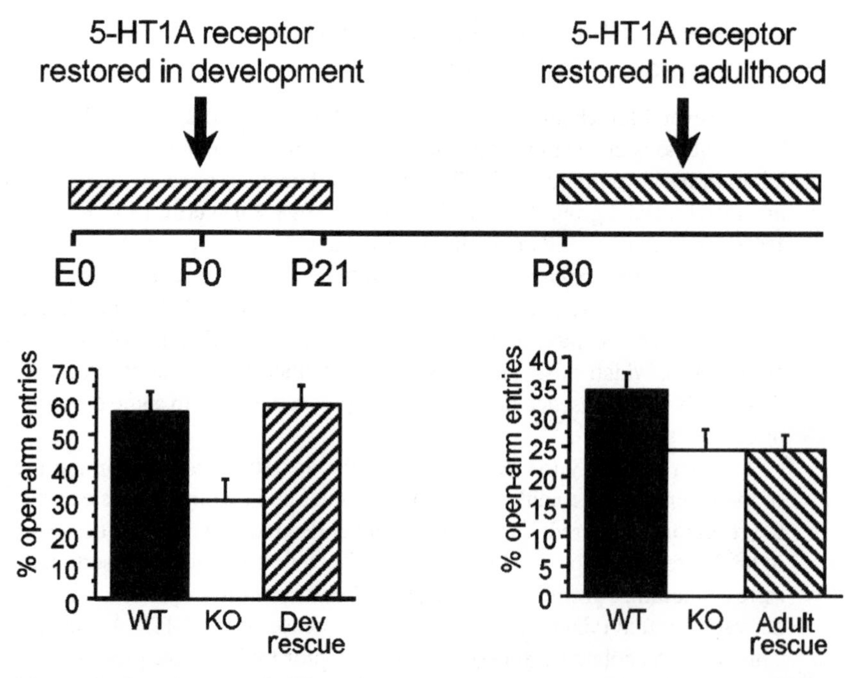

Figure 4 Developmental rescue of a knockout anxiety phenotype. Timeline of rescue of 5-HT1A receptor expression (*upper panel*). In the elevated plus maze, developmental rescue (*lower left*) but not adult rescue (*lower right*) restores wildtype levels of anxiety-related behavior (Gross et al. 2002).

manipulations bears directly on this question (Santarelli et al. 2003). SSRIs induce neurogenesis, an effect that has been hypothesized to rely on activation of the 1A receptor (Gould 1999, Radley & Jacobs 2002). In wildtype mice, chronic, but not acute, fluoxetine or 8-OH-DPAT increases neurogenesis in the dentate gyrus and exerts anxiolytic-like effects in the novelty suppressed feeding paradigm (Bodnoff et al. 1988, 1989; Malberg et al. 2000; Santarelli et al. 2003). The neurogenic and behavioral effects of both agents are blocked in the 5-HT1A receptor knockouts (Figure 3) (Santarelli et al. 2003). Santarelli et al. (2003) also used hippocampal-specific x-irradiation to ablate neurogenesis without otherwise altering serotonergic neurotransmission. They found that ablation of hippocampal neurogenesis prevented the behavioral effects of fluoxetine. These data suggest that chronic SSRI treatments exert their anxiolytic effect through the activation of 5-HT1A receptors, which in turn increase neurogenesis in the dentate gyrus of the hippocampus. It is unclear as of yet whether the plasticity-inducing effects of stimulating 5-HT1A receptors in the adult are related to those of stimulating the receptors during development.

Lessons from Genetically Modified Mice: Power Depends on Specificity and Depth

These selected studies demonstrate the power of using genetically modified mice to probe the neurobiological mechanisms underlying anxiety. Increasing tissue and temporal specificity allows one to test specific hypotheses—the relevance of the HPA axis to the effects of CRH on anxiety, or the isoform-dependence of the benzodiazepine effect—that would be difficult if not impossible to study with traditional pharmacological or neurobiological tools. At the same time, applying these traditional tools to the study of genetically modified animals allows one to begin understanding why a given genetic lesion causes a given behavioral phenotype. The explanatory power of a given genetic model, therefore, depends on the specificity of that model, the depth of knowledge of any secondary effects, and, of course, the specificity of the question being asked.

FROM MOUSE, BACK TO HUMAN

One advantage of the genetic approach to anxiety is that animal model data can be readily used to probe human anxiety disorders. One recent attempt tested the hypothesis that QTLs regulating anxiety-related behavior in the mouse would also be associated with anxiety disorders in humans (Smoller et al. 2001a). Mouse QTLs from a variety of studies of anxiety models were used to determine the homologous regions on human chromosomes. Polymorphisms from these regions were then genotyped in members of a large, multigenerational pedigree with a high incidence of panic disorder. Two regions of possible association were found: one on chromosome 12, corresponding to the region of mouse chromosome 15, linked to open-arm entries in the elevated plus maze (Turri et al. 2001a); and a weaker, broader region on chromosome 1, corresponding to the same chromosome in the mouse, the region most consistently reported in mouse studies (Flint et al. 1995; Gershenfeld & Paul 1998; Talbot et al. 1999; Turri et al. 2001a,b). This same group has utilized a candidate gene approach in a similar fashion. Five candidate genes were chosen from the broad array of knockouts that result in abnormal anxiety and were used to search for linkage to the anxiety-related trait of behavioral inhibition in children (Smoller et al. 2001a, Smoller et al. 2004). Associations were found between behavioral inhibition and the genes for GAD-65 and CRH. While these initial studies await confirmation, they demonstrate the potential power of genetics to cross species barriers.

Of course the principle goal of bridging animal and human studies involves the development of new treatments. Some of the studies mentioned above have the potential to result in new treatments for anxiety. Agents specific to α2-subunits of the $GABA_A$ receptor are already under study, in an effort to improve upon benzodiazepines (Atack 2003; Collins et al. 2002; Griebel et al. 2001, 2003). CRH antagonists are under study as well, as are specific CRH-R1 antagonists (Li et al. 2003, Oshima et al. 2003, Seymour et al. 2003). Looking into the future, specific PKCε inhibitors might treat anxiety by making $GABA_A$ receptors more sensitive to endogenous anxiolytics, and novel agents, which promote hippocampal neurogenesis, might be used to treat anxiety. Further into the future, one can imagine developing a sophisticated understanding of the circuit- and systems-level effects of the absence of the 5-HT1A receptor during development and harnessing that understanding by developing agents that could counteract those effects. These possibilities are a sampling of the potential contributions of genetic approaches to the study of anxiety.

The *Annual Review of Neuroscience* is online at http://neuro.annualreviews.org

LITERATURE CITED

Aguilar R, Gil L, Flint J, Gray JA, Dawson GR, et al. 2002. Learned fear, emotional reactivity and fear of heights: a factor analytic map from a large F_2 intercross of roman rat strains. *Brain Res. Bull.* 57:17–26

Alsobrook JP 2nd, Zohar AH, Leboyer M, Chabane N, Ebstein RP, Pauls DL. 2002. Association between the COMT locus and obsessive-compulsive disorder in females but not males. *Am. J. Med. Genet.* 114:116–20

Ammassari-Teule M, Restivo L, Pietteur V, Passino E. 2001. Learning about the context in genetically-defined mice. *Behav. Brain Res.* 125:195–204

Anisman H, Zaharia MD, Meaney MJ, Merali Z. 1998. Do early-life events permanently alter behavioral and hormonal responses to stressors? *Int. J. Dev. Neurosci.* 16:149–64

Atack JR. 2003. Anxioselective compounds acting at the GABA(A) receptor benzodiazepine binding site. *Curr. Drug Target. CNS Neurol. Disord.* 2:213–32

Auerbach J, Geller V, Lezer S, Shinwell E, Belmaker RH, et al. 1999. Dopamine D4 receptor (D4DR) and serotonin transporter promoter (5-HTTLPR) polymorphisms in the determination of temperament in 2-month-old infants. *Mol. Psychiatry* 4:369–73

Avgustinovich DF, Lipina TV, Bondar NP,

Alekseyenko OV, Kudryavtseva NN. 2000. Features of the genetically defined anxiety in mice. *Behav. Genet.* 30:101–9

Bale TL, Contarino A, Smith GW, Chan R, Gold LH, et al. 2000. Mice deficient for corticotropin-releasing hormone receptor-2 display anxiety-like behaviour and are hypersensitive to stress. *Nat. Genet.* 24:410–14

Bale TL, Picetti R, Contarino A, Koob GF, Vale WW, Lee KF. 2002. Mice deficient for both corticotropin-releasing factor receptor 1 (CRFR1) and CRFR2 have an impaired stress response and display sexually dichotomous anxiety-like behavior. *J. Neurosci.* 22:193–99

Balogh SA, Wehner JM. 2003. Inbred mouse strain differences in the establishment of long-term fear memory. *Behav. Brain Res.* 140:97–106

Baulieu EE, Robel P, Schumacher M. 2001. Neurosteroids: beginning of the story. *Int. Rev. Neurobiol.* 46:1–32

Beck CH, Fibiger HC. 1995. Conditioned fear-induced changes in behavior and in the expression of the immediate early gene c-fos: with and without diazepam pretreatment. *J. Neurosci.* 15:709–20

Belknap JK, Dubay C, Crabbe JC, Buck KJ. 1997. Mapping quantitative trait loci for behavioral traits in the mouse. In *Handbook of Psychiatric Genetics*, ed. K Blum, EP Noble, pp. 435–53. Boca Raton, FL: CRC Press

Bennett AJ, Lesch KP, Heils A, Long JC, Lorenz JG, et al. 2002. Early experience and serotonin transporter gene variation interact to influence primate CNS function. *Mol. Psychiatry* 7:118–22

Bicikova M, Tallova J, Hill M, Krausova Z, Hampl R. 2000. Serum concentrations of some neuroactive steroids in women suffering from mixed anxiety-depressive disorder. *Neurochem. Res.* 25:1623–27

Bodnoff SR, Suranyi-Cadotte B, Aitken DH, Quirion R, Meaney MJ. 1988. The effects of chronic antidepressant treatment in an animal model of anxiety. *Psychopharmacology (Berl.)* 95:298–302

Bodnoff SR, Suranyi-Cadotte B, Quirion R, Meaney MJ. 1989. A comparison of the effects of diazepam versus several typical and atypical anti-depressant drugs in an animal model of anxiety. *Psychopharmacology (Berl.)* 97:277–79

Bolivar VJ, Pooler O, Flaherty L. 2001. Inbred strain variation in contextual and cued fear conditioning behavior. *Mamm. Genome* 12:651–56

Borsini F, Podhorna J, Marazziti D. 2002. Do animal models of anxiety predict anxiolytic-like effects of antidepressants? *Psychopharmacology (Berl.)* 163:121–41

Bowers BJ, Collins AC, Tritto T, Wehner JM. 2000. Mice lacking PKC gamma exhibit decreased anxiety. *Behav. Genet.* 30:111–21

Broadhurst PL. 1975. The Maudsley reactive and nonreactive strains of rats: a survey. *Behav. Genet.* 5:299–319

Buffett-Jerrott SE, Stewart SH. 2002. Cognitive and sedative effects of benzodiazepine use. *Curr. Pharm. Des.* 8:45–58

Caldji C, Tannenbaum B, Sharma S, Francis D, Plotsky PM, Meaney MJ. 1998. Maternal care during infancy regulates the development of neural systems mediating the expression of fearfulness in the rat. *Proc. Natl. Acad. Sci. USA* 95:5335–40

Canto-de-Souza A, Luiz Nunes-de-Souza R, Rodgers RJ. 2002. Anxiolytic-like effect of WAY-100635 microinfusions into the median (but not dorsal) raphe nucleus in mice exposed to the plus-maze: influence of prior test experience. *Brain Res.* 928:50–59

Caspi A, Sugden K, Moffitt TE, Taylor A, Craig IW, et al. 2003. Influence of life stress on depression: moderation by a polymorphism in the 5-HTT gene. *Science* 301:386–89

Champoux M, Bennett A, Shannon C, Higley JD, Lesch KP, Suomi SJ. 2002. Serotonin transporter gene polymorphism, differential early rearing, and behavior in rhesus monkey neonates. *Mol. Psychiatry* 7:1058–63

Chang FC, Opp MR. 2001. Corticotropin-releasing hormone (CRH) as a regulator of waking. *Neurosci. Biobehav. Rev.* 25:445–53

Chaouloff F, Castanon N, Mormede P. 1994. Paradoxical differences in animal models of anxiety among the roman rat lines. *Neurosci. Lett.* 182:217–21

Charney DS, Drevets WC. 2002. The neurobiological basis of anxiety disorders. See Davis et al. 2002, pp. 901–30

Chen C, Rainnie DG, Greene RW, Tonegawa S. 1994. Abnormal fear response and aggressive behavior in mutant mice deficient for alpha-calcium-calmodulin kinase II. *Science* 266:291–94

Clark MS, Kaiyala KJ. 2003. Role of corticotropin-releasing factor family peptides and receptors in stress-related psychiatric disorders. *Semin. Clin. Neuropsychiatry* 8:119–36

Clement Y, Calatayud F, Belzung C. 2002. Genetic basis of anxiety-like behaviour: a critical review. *Brain Res. Bull.* 57:57–71

Clement Y, Proeschel MF, Bondoux D, Girard F, Launay JM, Chapouthier G. 1997. Genetic factors regulate processes related to anxiety in mice. *Brain Res.* 752:127–35

Cohen RM, Kang A, Gulick C. 2001. Quantitative trait loci affecting the behavior of A/J and CBA/J intercross mice in the elevated plus maze. *Mamm. Genome* 12:501–7

Collins I, Moyes C, Davey WB, Rowley M, Bromidge FA, et al. 2002. 3-heteroaryl-2-pyridones: benzodiazepine site ligands with functional selectivity for alpha 2/alpha 3-subtypes of human GABA(A) receptor-ion channels. *J. Med. Chem.* 45:1887–900

Conti LH, Maciver CR, Ferkany JW, Abreu ME. 1990. Footshock-induced freezing behavior in rats as a model for assessing anxiolytics. *Psychopharmacology (Berl.)* 102:492–97

Cook MN, Williams RW, Flaherty L. 2001. Anxiety-related behaviors in the elevated zero-maze are affected by genetic factors and retinal degeneration. *Behav. Neurosci.* 115:468–76

Coste SC, Kesterson RA, Heldwein KA, Stevens SL, Heard AD, et al. 2000. Abnormal adaptations to stress and impaired cardiovascular function in mice lacking corticotropin-releasing hormone receptor-2. *Nat. Genet.* 24:403–9

Crawley JN, Davis LG. 1982. Baseline exploratory activity predicts anxiolytic responsiveness to diazepam in five mouse strains. *Brain Res. Bull.* 8:609–12

Crestani F, Keist R, Fritschy JM, Benke D, Vogt K, et al. 2002. Trace fear conditioning involves hippocampal alpha5 GABA(A) receptors. *Proc. Natl. Acad. Sci. USA* 99:8980–85

Crowe RR, Goedken R, Samuelson S, Wilson R, Nelson J, Noyes R Jr. 2001. Genomewide survey of panic disorder. *Am. J. Med. Genet.* 105:105–9

Cruz AP, Frei F, Graeff FG. 1994. Ethopharmacological analysis of rat behavior on the elevated plus-maze. *Pharmacol. Biochem. Behav.* 49:171–76

Davis KL, Charney DS, Coyle JT, Nemeroff C, eds. 2002. *Neuropsychopharmacology: The Fifth Generation of Progress.* Philadelphia: Lippincott Williams and Wilkins

Davis M. 1979. Diazepam and flurazepam: effects on conditioned fear as measured with the potentiated startle paradigm. *Psychopharmacology (Berl.)* 62:1–7

Davis M. 1998. Are different parts of the extended amygdala involved in fear versus anxiety? *Biol. Psychiatry* 44:1239–47

Davis M, Walker DL, Myers KM. 2003. Role of the amygdala in fear extinction measured with potentiated startle. *Ann. NY Acad. Sci.* 985:218–32

Deckert J, Catalano M, Syagailo YV, Bosi M, Okladnova O, et al. 1999. Excess of high activity monoamine oxidase A gene promoter alleles in female patients with panic disorder. *Hum. Mol. Genet.* 8:621–24

Di Bella D, Cavallini MC, Bellodi L. 2002. No association between obsessive-compulsive disorder and the 5-HT_{1DB} receptor gene. *Am. J. Psychiatry* 159:1783–85

Drolet G, Rivest S. 2001. Corticotropin-releasing hormone and its receptors; an evaluation at the transcription level in vivo. *Peptides* 22:761–67

Dulawa SC, Grandy DK, Low MJ, Paulus MP, Geyer MA. 1999. Dopamine D4 receptor-knock-out mice exhibit reduced exploration of novel stimuli. *J. Neurosci.* 19:9550–56

Dunn AJ, Swiergiel AH. 1999. Behavioral

responses to stress are intact in CRF-deficient mice. *Brain Res.* 845:14–20

Eckart K, Jahn O, Radulovic J, Radulovic M, Blank T, et al. 2002. Pharmacology and biology of corticotropin-releasing factor (CRF) receptors. *Recept. Channels* 8:163–77

Eckart K, Radulovic J, Radulovic M, Jahn O, Blank T, et al. 1999. Actions of CRF and its analogs. *Curr. Med. Chem.* 6:1035–53

Falls WA, Carlson S, Turner JG, Willott JF. 1997. Fear-potentiated startle in two strains of inbred mice. *Behav. Neurosci.* 111:855–61

Fernandez-Teruel A, Escorihuela RM, Gray JA, Aguilar R, Gil L, et al. 2002. A quantitative trait locus influencing anxiety in the laboratory rat. *Genome Res.* 12:618–26

File SE. 1996. Recent developments in anxiety, stress, and depression. *Pharmacol. Biochem. Behav.* 54:3–12

Finn CT, Smoller JW. 2001. The genetics of panic disorder. *Curr. Psychiatry Rep.* 3:131–37

Finn DA, Rutledge-Gorman MT, Crabbe JC. 2003. Genetic animal models of anxiety. *Neurogenetics* 4:109–35

Flint J. 2003. Animal models of anxiety and their molecular dissection. *Semin. Cell Dev. Biol.* 14:37–42

Flint J, Corley R, DeFries JC, Fulker DW, Gray JA, et al. 1995. A simple genetic basis for a complex psychological trait in laboratory mice. *Science* 269:1432–35

Francis D, Diorio J, Liu D, Meaney MJ. 1999. Nongenomic transmission across generations of maternal behavior and stress responses in the rat. *Science* 286:1155–58

Francis DD, Szegda K, Campbell G, Martin WD, Insel TR. 2003. Epigenetic sources of behavioral differences in mice. *Nat. Neurosci.* 6:445–46

Frankland PW, O'Brien C, Ohno M, Kirkwood A, Silva AJ. 2001. Alpha-CaMKII-dependent plasticity in the cortex is required for permanent memory. *Nature* 411:309–13

Frisch A, Michaelovsky E, Rockah R, Amir I, Hermesh H, et al. 2000. Association between obsessive-compulsive disorder and polymorphisms of genes encoding components of the serotonergic and dopaminergic pathways. *Eur. Neuropsychopharmacol.* 10:205–9

Fritschy JM, Mohler H. 1995. $GABA_A$-receptor heterogeneity in the adult rat brain: differential regional and cellular distribution of seven major subunits. *J. Comp. Neurol.* 359:154–94

Fyer AJ, Mannuzza S, Chapman TF, Martin LY, Klein DF. 1995. Specificity in familial aggregation of phobic disorders. *Arch. Gen. Psychiatry* 52:564–73

Gelernter J, Page GP, Bonvicini K, Woods SW, Pauls DL, Kruger S. 2003. A chromosome 14 risk locus for simple phobia: results from a genomewide linkage scan. *Mol. Psychiatry* 8:71–82

Gershenfeld HK, Paul SM. 1998. Towards a genetics of anxious temperament: from mice to men. *Acta Psychiatr. Scand. Suppl.* 393:56–65

Goldberg J, True WR, Eisen SA, Henderson WG. 1990. A twin study of the effects of the Vietnam War on posttraumatic stress disorder. *JAMA* 263:1227–32

Gordon JA. 2002. Anxiolytic drug targets: beyond the usual suspects. *J. Clin. Invest.* 110:915–17

Gorman JM, Kent JM, Coplan JD. 2002. Current and emerging therapeutics of anxiety and stress disorders. See Davis et al. 2002, pp. 967–80

Goto SH, Conceicao IM, Ribeiro RA, Frussa-Filho R. 1993. Comparison of anxiety measured in the elevated plus-maze, open-field and social interaction tests between spontaneously hypertensive rats and wistar EPM-1 rats. *Br. J. Med. Biol. Res.* 26:965–69

Gould E. 1999. Serotonin and hippocampal neurogenesis. *Neuropsychopharmacology* 21:46S–51

Gratacos M, Nadal M, Martin-Santos R, Pujana MA, Gago J, et al. 2001. A polymorphic genomic duplication on human chromosome 15 is a susceptibility factor for panic and phobic disorders. *Cell* 106:367–79

Gray JA, McNaughton N. 2000. *The Neuropsychology of Anxiety*. New York: Oxford Univ. Press. 424 pp.

Greenberg BD, Tolliver TJ, Huang SJ, Li Q,

Bengel D, Murphy DL. 1999. Genetic variation in the serotonin transporter promoter region affects serotonin uptake in human blood platelets. *Am. J. Med. Genet.* 88:83–87

Griebel G. 1995. 5-hydroxytryptamine-interacting drugs in animal models of anxiety disorders: more than 30 years of research. *Pharmacol. Ther.* 65:319–95

Griebel G, Belzung C, Perrault G, Sanger DJ. 2000. Differences in anxiety-related behaviours and in sensitivity to diazepam in inbred and outbred strains of mice. *Psychopharmacology (Berl.)* 148:164–70

Griebel G, Perrault G, Simiand J, Cohen C, Granger P, et al. 2001. Sl651498: an anxioselective compound with functional selectivity for alpha2- and alpha3-containing gamma-aminobutyric acid(A) (GABA(A)) receptors. *J. Pharmacol. Exp. Ther.* 298:753–68

Griebel G, Perrault G, Simiand J, Cohen C, Granger P, et al. 2003. Sl651498, a $GABA_A$ receptor agonist with subtype-selective efficacy, as a potential treatment for generalized anxiety disorder and muscle spasms. *CNS Drug Rev.* 9:3–20

Grillon C, Ameli R, Goddard A, Woods SW, Davis M. 1994. Baseline and fear-potentiated startle in panic disorder patients. *Biol. Psychiatry* 35:431–39

Gross C, Zhuang X, Stark K, Ramboz S, Oosting R, et al. 2002. Serotonin 1A receptor acts during development to establish normal anxiety-like behavior in the adult. *Nature* 416:396–400

Gustavsson JP, Nothen MM, Jonsson EG, Neidt H, Forslund K, et al. 1999. No association between serotonin transporter gene polymorphisms and personality traits. *Am. J. Med. Genet.* 88:430–36

Hamilton SP, Fyer AJ, Durner M, Heiman GA, Baisre de Leon A, et al. 2003. Further genetic evidence for a panic disorder syndrome mapping to chromosome 13q. *Proc. Natl. Acad. Sci. USA* 100:2550–55

Hamilton SP, Heiman GA, Haghighi F, Mick S, Klein DF, et al. 1999. Lack of genetic linkage or association between a functional serotonin transporter polymorphism and panic disorder. *Psychiatr. Genet.* 9:1–6

Hamilton SP, Slager SL, Heiman GA, Deng Z, Haghighi F, et al. 2002. Evidence for a susceptibility locus for panic disorder near the catechol-O-methyltransferase gene on chromosome 22. *Biol. Psychiatry* 51:591–601

Hamilton SP, Slager SL, Heiman GA, Haghighi F, Klein DF, et al. 2000. No genetic linkage or association between a functional promoter polymorphism in the monoamine oxidase-A gene and panic disorder. *Mol. Psychiatry* 5:465–66

Hanna GL, Veenstra-VanderWeele J, Cox NJ, Boehnke M, Himle JA, et al. 2002. Genome-wide linkage analysis of families with obsessive-compulsive disorder ascertained through pediatric probands. *Am. J. Med. Genet.* 114:541–52

Hariri AR, Mattay VS, Tessitore A, Kolachana B, Fera F, et al. 2002. Serotonin transporter genetic variation and the response of the human amygdala. *Science* 297:400–3

Heinrichs SC, Min H, Tamraz S, Carmouche M, Boehme SA, Vale WW. 1997. Anti-sexual and anxiogenic behavioral consequences of corticotropin-releasing factor overexpression are centrally mediated. *Psychoneuroendocrinology* 22:215–24

Heisler LK, Chu HM, Brennan TJ, Danao JA, Bajwa P, et al. 1998. Elevated anxiety and antidepressant-like responses in serotonin 5-HT1A receptor mutant mice. *Proc. Natl. Acad. Sci. USA* 95:15049–54

Hettema JM, Annas P, Neale MC, Kendler KS, Fredrikson M. 2003. A twin study of the genetics of fear conditioning. *Arch. Gen. Psychiatry* 60:702–8

Hettema JM, Neale MC, Kendler KS. 2001a. A review and meta-analysis of the genetic epidemiology of anxiety disorders. *Am. J. Psychiatry* 158:1568–78

Hettema JM, Prescott CA, Kendler KS. 2001b. A population-based twin study of generalized anxiety disorder in men and women. *J. Nerv. Ment. Dis.* 189:413–20

Hodge CW, Mehmert KK, Kelley SP, McMahon T, Haywood A, et al. 1999.

Supersensitivity to allosteric GABA(A) receptor modulators and alcohol in mice lacking PKCε. *Nat. Neurosci.* 2:997–1002

Hodge CW, Raber J, McMahon T, Walter H, Sanchez-Perez AM, et al. 2002. Decreased anxiety-like behavior, reduced stress hormones, and neurosteroid supersensitivity in mice lacking protein kinase Cepsilon. *J. Clin. Invest.* 110:1003–10

Holmes A. 2001. Targeted gene mutation approaches to the study of anxiety-like behavior in mice. *Neurosci. Biobehav. Rev.* 25:261–73

Howe DG, Wiley JC, McKnight GS. 2002. Molecular and behavioral effects of a null mutation in all PKA C beta isoforms. *Mol. Cell. Neurosci.* 20:515–24

Huot RL, Thrivikraman KV, Meaney MJ, Plotsky PM. 2001. Development of adult ethanol preference and anxiety as a consequence of neonatal maternal separation in long evans rats and reversal with antidepressant treatment. *Psychopharmacology (Berl.)* 158:366–73

Inoue T, Hashimoto S, Tsuchiya K, Izumi T, Ohmori T, Koyama T. 1996. Effect of citalopram, a selective serotonin reuptake inhibitor, on the acquisition of conditioned freezing. *Eur. J. Pharmacol.* 311:1–6

Inui A, Okita M, Nakajima M, Momose K, Ueno N, et al. 1998. Anxiety-like behavior in transgenic mice with brain expression of neuropeptide Y. *Proc. Assoc. Am. Physicians* 110:171–82

Jorm AF, Henderson AS, Jacomb PA, Christensen H, Korten AE, et al. 1998. An association study of a functional polymorphism of the serotonin transporter gene with personality and psychiatric symptoms. *Mol. Psychiatry* 3:449–51

Kalinichev M, Easterling KW, Plotsky PM, Holtzman SG. 2002. Long-lasting changes in stress-induced corticosterone response and anxiety-like behaviors as a consequence of neonatal maternal separation in Long-Evans rats. *Pharmacol. Biochem. Behav.* 73:131–40

Karayiorgou M, Sobin C, Blundell ML, Galke BL, Malinova L, et al. 1999. Family-based association studies support a sexually dimorphic effect of COMT and MAOA on genetic susceptibility to obsessive-compulsive disorder. *Biol. Psychiatry* 45:1178–89

Kash SF, Tecott LH, Hodge C, Baekkeskov S. 1999. Increased anxiety and altered responses to anxiolytics in mice deficient in the 65-kDa isoform of glutamic acid decarboxylase. *Proc. Natl. Acad. Sci. USA* 96:1698–703

Kaufmann WA, Humpel C, Alheid GF, Marksteiner J. 2003. Compartmentation of alpha 1 and alpha 2 GABA(A) receptor subunits within rat extended amygdala: implications for benzodiazepine action. *Brain Res.* 964:91–99

Kendler KS. 2001. Twin studies of psychiatric illness: an update. *Arch. Gen. Psychiatry* 58:1005–14

Kessler RC, Greenberg PE. 2002. The economic burden of anxiety and stress disorders. See Davis et al. 2002, pp. 981–92

Kishimoto T, Radulovic J, Radulovic M, Lin CR, Schrick C, et al. 2000. Deletion of Crhr2 reveals an anxiolytic role for corticotropin-releasing hormone receptor-2. *Nat. Genet.* 24:415–19

Knowles JA, Fyer AJ, Vieland VJ, Weissman MM, Hodge SE, et al. 1998. Results of a genome-wide genetic screen for panic disorder. *Am. J. Med. Genet.* 81:139–47

Konig M, Zimmer AM, Steiner H, Holmes PV, Crawley JN, et al. 1996. Pain responses, anxiety and aggression in mice deficient in preproenkephalin. *Nature* 383:535–38

Kustova Y, Sei Y, Morse HC Jr, Basile AS. 1998. The influence of a targeted deletion of the ifngamma gene on emotional behaviors. *Brain. Behav. Immun.* 12:308–24

Landgraf R, Wigger A. 2002. High vs low anxiety-related behavior rats: an animal model of extremes in trait anxiety. *Behav. Genet.* 32:301–14

LeDoux JE. 2000. Emotion circuits in the brain. *Annu. Rev. Neurosci.* 23:155–84

Le Fur G, Guilloux F, Mitrani N, Mizoule J, Uzan A. 1979. Relationships between plasma corticosteroids and benzodiazepines in stress. *J. Pharmacol. Exp. Ther.* 211:305–8

Lesch KP. 2001a. Molecular foundation of

anxiety disorders. *J. Neural Transm.* 108: 717–46

Lesch KP. 2001b. Mouse anxiety: the power of knockout. *Pharmacogenomics J.* 1:187–92

Lesch KP, Bengel D, Heils A, Sabol SZ, Greenberg BD, et al. 1996. Association of anxiety-related traits with a polymorphism in the serotonin transporter gene regulatory region. *Science* 274:1527–31

Lesch KP, Meyer J, Glatz K, Flugge G, Hinney A, et al. 1997. The 5-HT transporter gene-linked polymorphic region (5-HTTLPR) in evolutionary perspective: alternative biallelic variation in rhesus monkeys. Rapid communication. *J. Neural Transm.* 104:1259–66

Li XB, Inoue T, Hashimoto S, Koyama T. 2001. Effect of chronic administration of flesinoxan and fluvoxamine on freezing behavior induced by conditioned fear. *Eur. J. Pharmacol.* 425:43–50

Li YW, Hill G, Wong H, Kelly N, Ward K, et al. 2003. Receptor occupancy of nonpeptide corticotropin-releasing factor 1 antagonist DMP696: correlation with drug exposure and anxiolytic efficacy. *J. Pharmacol. Exp. Ther.* 305:86–96

Liebsch G, Linthorst AC, Neumann ID, Reul JM, Holsboer F, Landgraf R. 1998a. Behavioral, physiological, and neuroendocrine stress responses and differential sensitivity to diazepam in two wistar rat lines selectively bred for high- and low-anxiety-related behavior. *Neuropsychopharmacology* 19:381–96

Liebsch G, Montkowski A, Holsboer F, Landgraf R. 1998b. Behavioural profiles of two wistar rat lines selectively bred for high or low anxiety-related behaviour. *Behav. Brain Res.* 94:301–10

Lister RG. 1987. The use of a plus-maze to measure anxiety in the mouse. *Psychopharmacology (Berl.)* 92:180–85

Liu D, Diorio J, Day JC, Francis DD, Meaney MJ. 2000. Maternal care, hippocampal synaptogenesis and cognitive development in rats. *Nat. Neurosci.* 3:799–806

Liu D, Diorio J, Tannenbaum B, Caldji C, Francis D, et al. 1997. Maternal care, hippocampal glucocorticoid receptors, and hypothalamic-pituitary-adrenal responses to stress. *Science* 277:1659–62

Logue MW, Vieland VJ, Goedken RJ, Crowe RR. 2003. Bayesian analysis of a previously published genome screen for panic disorder reveals new and compelling evidence for linkage to chromosome 7. *Am. J. Med. Genet.* 121B:95–99

Low K, Crestani F, Keist R, Benke D, Brunig I, et al. 2000. Molecular and neuronal substrate for the selective attenuation of anxiety. *Science* 290:131–34

Lyons MJ, Goldberg J, Eisen SA, True W, Tsuang MT, et al. 1993. Do genes influence exposure to trauma? A twin study of combat. *Am. J. Med. Genet.* 48:22–27

Malberg JE, Eisch AJ, Nestler EJ, Duman RS. 2000. Chronic antidepressant treatment increases neurogenesis in adult rat hippocampus. *J. Neurosci.* 20:9104–10

Martins AP, Marras RA, Guimaraes FS. 1997. Anxiogenic effect of corticotropin-releasing hormone in the dorsal periaqueductal grey. *Neuroreport* 8:3601–4

Meaney MJ, Diorio J, Francis D, Widdowson J, LaPlante P, et al. 1996. Early environmental regulation of forebrain glucocorticoid receptor gene expression: implications for adrenocortical responses to stress. *Dev. Neurosci.* 18:49–72

Melchior CL, Ritzmann RF. 1994. Dehydroepiandrosterone is an anxiolytic in mice on the plus maze. *Pharmacol. Biochem. Behav.* 47:437–41

Merali Z, Levac C, Anisman H. 2003. Validation of a simple, ethologically relevant paradigm for assessing anxiety in mice. *Biol. Psychiatry* 54:552–65

Merikangas KR, Pine D. 2002. Genetic and other vulnerability factors for anxiety and stress disorders. See Davis et al. 2002, pp. 867–82

Miller DB, O'Callaghan JP. 2002. Neuroendocrine aspects of the response to stress. *Metabolism* 51:5–10

Mohler H, Crestani F, Rudolph U. 2001.

GABA(A)-receptor subtypes: a new pharmacology. *Curr. Opin. Pharmacol.* 1:22–25

Mohler H, Fritschy JM, Rudolph U. 2002. A new benzodiazepine pharmacology. *J. Pharmacol. Exp. Ther.* 300:2–8

Morgan CA 3rd, Grillon C, Southwick SM, Davis M, Charney DS. 1995. Fear-potentiated startle in posttraumatic stress disorder. *Biol. Psychiatry* 38:378–85

Müller MB, Zimmermann S, Sillaber I, Hagemeyer TP, Deussing JM, et al. 2003. Limbic corticotropin-releasing hormone receptor 1 mediates anxiety-related behavior and hormonal adaptation to stress. *Nat. Neurosci.* 6:1100–7

Mundo E, Richter MA, Sam F, Macciardi F, Kennedy JL. 2000. Is the 5-ht(1dbeta) receptor gene implicated in the pathogenesis of obsessive-compulsive disorder? *Am. J. Psychiatry* 157:1160–61

Nie T, Abel T. 2001. Fear conditioning in inbred mouse strains: an analysis of the time course of memory. *Behav. Neurosci.* 115:951–56

Noyes R Jr, Clarkson C, Crowe RR, Yates WR, McChesney CM. 1987. A family study of generalized anxiety disorder. *Am. J. Psychiatry* 144:1019–24

Nusser Z, Sieghart W, Benke D, Fritschy JM, Somogyi P. 1996. Differential synaptic localization of two major gamma-aminobutyric acid type A receptor alpha subunits on hippocampal pyramidal cells. *Proc. Natl. Acad. Sci. USA* 93:11939–44

Ohara K, Nagai M, Suzuki Y, Ochiai M. 1998. No association between anxiety disorders and catechol-*O*-methyltransferase polymorphism. *Psychiatry Res.* 80:145–48

Ohl F, Toschi N, Wigger A, Henniger MS, Landgraf R. 2001. Dimensions of emotionality in a rat model of innate anxiety. *Behav. Neurosci.* 115:429–36

Oshima A, Flachskamm C, Reul JM, Holsboer F, Linthorst AC. 2003. Altered serotonergic neurotransmission but normal hypothalamic-pituitary-adrenocortical axis activity in mice chronically treated with the corticotropin-releasing hormone receptor type 1 antagonist NBI 30775. *Neuropsychopharmacology* 28:2148–59

Pape HC, Stork O. 2003. Genes and mechanisms in the amygdala involved in the formation of fear memory. *Ann. NY Acad. Sci.* 985:92–105

Parks CL, Robinson PS, Sibille E, Shenk T, Toth M. 1998. Increased anxiety of mice lacking the serotonin$_{1A}$ receptor. *Proc. Natl. Acad. Sci. USA* 95:10734–39

Paterson A, Whiting PJ, Gray JA, Flint J, Dawson GR. 2001. Lack of consistent behavioural effects of Maudsley reactive and non-reactive rats in a number of animal tests of anxiety and activity. *Psychopharmacology (Berl.)* 154:336–42

Pellow S, Chopin P, File SE, Briley M. 1985. Validation of open:closed arm entries in an elevated plus-maze as a measure of anxiety in the rat. *J. Neurosci. Methods* 14:149–67

Pellow S, File SE. 1986. Anxiolytic and anxiogenic drug effects on exploratory activity in an elevated plus-maze: a novel test of anxiety in the rat. *Pharmacol. Biochem. Behav.* 24:525–29

Pirker S, Schwarzer C, Wieselthaler A, Sieghart W, Sperk G. 2000. GABA(A) receptors: immunocytochemical distribution of 13 subunits in the adult rat brain. *Neuroscience* 101:815–50

Pletnikov MV, Storozheva ZI, Sherstnev VV. 1996. Relationship between memory and fear: developmental and pharmacological studies. *Pharmacol. Biochem. Behav.* 54:93–98

Plomin R, McClearn GE, Gora-Maslak G, Neiderhiser JM. 1991. Use of recombinant inbred strains to detect quantitative trait loci associated with behavior. *Behav. Genet.* 21:99–116

Quartermain D, Clemente J, Shemer A. 1993. 5-HT1A agonists disrupt memory of fear conditioning in mice. *Biol. Psychiatry* 33:247–54

Raber J, Akana SF, Bhatnagar S, Dallman MF, Wong D, Mucke L. 2000. Hypothalamic-pituitary-adrenal dysfunction in *APOE* (–/–) mice: possible role in behavioral and metabolic alterations. *J. Neurosci.* 20:2064–71

Radcliffe RA, Lowe MV, Wehner JM. 2000. Confirmation of contextual fear conditioning qtls by short-term selection. *Behav. Genet.* 30:183–91

Radley JJ, Jacobs BL. 2002. 5-HT1A receptor antagonist administration decreases cell proliferation in the dentate gyrus. *Brain Res.* 955:264–67

Ramboz S, Oosting R, Amara DA, Kung HF, Blier P, et al. 1998. Serotonin receptor 1A knockout: an animal model of anxiety-related disorder. *Proc. Natl. Acad. Sci. USA* 95:14476–81

Ramos A, Berton O, Mormede P, Chaouloff F. 1997. A multiple-test study of anxiety-related behaviours in six inbred rat strains. *Behav. Brain Res.* 85:57–69

Ramos A, Mellerin Y, Mormede P, Chaouloff F. 1998. A genetic and multifactorial analysis of anxiety-related behaviours in Lewis and SHR intercrosses. *Behav. Brain Res.* 96:195–205

Rex A, Sondern U, Voigt JP, Franck S, Fink H. 1996. Strain differences in fear-motivated behavior of rats. *Pharmacol. Biochem. Behav.* 54:107–11

Rodgers RJ, Dalvi A. 1997. Anxiety, defence and the elevated plus-maze. *Neurosci. Biobehav. Rev.* 21:801–10

Rodgers RJ, Johnson NJ. 1995. Factor analysis of spatiotemporal and ethological measures in the murine elevated plus-maze test of anxiety. *Pharmacol. Biochem. Behav.* 52:297–303

Roy-Byrne P, Afari N, Ashton S, Fischer M, Goldberg J, Buchwald D. 2002. Chronic fatigue and anxiety/depression: a twin study. *Br. J. Psychiatry* 180:29–34

Rudolph U, Crestani F, Benke D, Brunig I, Benson JA, et al. 1999. Benzodiazepine actions mediated by specific gamma-aminobutyric acid(A) receptor subtypes. *Nature* 401:796–800

Saito N, Itouji A, Totani Y, Osawa I, Koide H, et al. 1993. Cellular and intracellular localization of epsilon-subspecies of protein kinase C in the rat brain; presynaptic localization of the epsilon-subspecies. *Brain Res.* 607:241–48

Santarelli L, Gobbi G, Debs PC, Sibille ET, Blier P, et al. 2001. Genetic and pharmacological disruption of neurokinin 1 receptor function decreases anxiety-related behaviors and increases serotonergic function. *Proc. Natl. Acad. Sci. USA* 98:1912–17

Santarelli L, Saxe M, Gross C, Surget A, Battaglia F, et al. 2003. Requirement of hippocampal neurogenesis for the behavioral effects of antidepressants. *Science* 301:805–9

Seymour PA, Schmidt AW, Schulz DW. 2003. The pharmacology of CP-154,526, a non-peptide antagonist of the CRH1 receptor: a review. *CNS Drug Rev.* 9:57–96

Shephard RA, Broadhurst PL. 1982. Hyponeophagia and arousal in rats: effects of diazepam, 5-methoxy-*N*,*N*-dimethyltryptamine, *d*-amphetamine and food deprivation. *Psychopharmacology (Berl.)* 78:368–72

Shepherd JK, Grewal SS, Fletcher A, Bill DJ, Dourish CT. 1994. Behavioural and pharmacological characterisation of the elevated "zero-maze" as an animal model of anxiety. *Psychopharmacology (Berl.)* 116:56–64

Skre I, Onstad S, Torgersen S, Philos DR, Lygren S, Kringlen E. 2000. The heritability of common phobic fear: a twin study of a clinical sample. *J. Anxiety Disord.* 14:549–62

Smith GW, Aubry JM, Dellu F, Contarino A, Bilezikjian LM, et al. 1998. Corticotropin releasing factor receptor 1-deficient mice display decreased anxiety, impaired stress response, and aberrant neuroendocrine development. *Neuron* 20:1093–102

Smoller JW, Acierno JS Jr, Rosenbaum JF, Biederman J, Pollack MH, et al. 2001a. Targeted genome screen of panic disorder and anxiety disorder proneness using homology to murine QTL regions. *Am. J. Med. Genet.* 105:195–206

Smoller JW, Rosenbaum JF, Biederman J, Kennedy J, Dai D, et al. 2004. Association of a genetic marker at the corticotropin releasing hormone locus with behavioral inhibition. *Biol. Psychiatry* 54:1376–81

Smoller JW, Rosenbaum JF, Biederman J, Susswein LS, Kennedy J, et al. 2001b. Genetic association analysis of behavioral inhibition using candidate loci from mouse models. *Am. J. Med. Genet.* 105:226–35

Spivak B, Maayan R, Kotler M, Mester R, Gil-Ad I, et al. 2000. Elevated circulatory level of GABA(A)–antagonistic neurosteroids in patients with combat-related post-traumatic stress disorder. *Psychol. Med.* 30:1227–31

Stein MB, Jang KL, Taylor S, Vernon PA, Livesley WJ. 2002. Genetic and environmental influences on trauma exposure and posttraumatic stress disorder symptoms: a twin study. *Am. J. Psychiatry* 159:1675–81

Stenzel-Poore MP, Heinrichs SC, Rivest S, Koob GF, Vale WW. 1994. Overproduction of corticotropin-releasing factor in transgenic mice: a genetic model of anxiogenic behavior. *J. Neurosci.* 14:2579–84

Stork O, Welzl H, Wolfer D, Schuster T, Mantei N, et al. 2000. Recovery of emotional behaviour in neural cell adhesion molecule (NCAM) null mutant mice through transgenic expression of NCAM180. *Eur. J. Neurosci.* 12:3291–306

Stork O, Welzl H, Wotjak CT, Hoyer D, Delling M, et al. 1999. Anxiety and increased 5-HT1A receptor response in NCAM null mutant mice. *J. Neurobiol.* 40:343–55

Stork O, Yamanaka H, Stork S, Kume N, Obata K. 2003. Altered conditioned fear behavior in glutamate decarboxylase 65 null mutant mice. *Genes Brain Behav.* 2:65–70

Suaudeau C, Rinaldi D, Lepicard E, Venault P, Crusio WE, et al. 2000. Divergent levels of anxiety in mice selected for differences in sensitivity to a convulsant agent. *Physiol. Behav.* 71:517–23

Sullivan GM, Coplan JD, Kent JM, Gorman JM. 1999. The noradrenergic system in pathological anxiety: a focus on panic with relevance to generalized anxiety and phobias. *Biol. Psychiatry* 46:1205–18

Talbot CJ, Nicod A, Cherny SS, Fulker DW, Collins AC, Flint J. 1999. High-resolution mapping of quantitative trait loci in outbred mice. *Nat. Genet.* 21:305–8

Tarantino LM, Gould TJ, Druhan JP, Bucan M. 2000. Behavior and mutagenesis screens: the importance of baseline analysis of inbred strains. *Mamm. Genome* 11:555–64

Thorgeirsson TE, Oskarsson H, Desnica N, Kostic JP, Stefansson JG, et al. 2003. Anxiety with panic disorder linked to chromosome 9q in Iceland. *Am. J. Hum. Genet.* 72:1221–30

Timpl P, Spanagel R, Sillaber I, Kresse A, Reul JM, et al. 1998. Impaired stress response and reduced anxiety in mice lacking a functional corticotropin-releasing hormone receptor 1. *Nat. Genet.* 19:162–66

Toth M. 2003. 5-HT(1A) receptor knockout mouse as a genetic model of anxiety. *Eur. J. Pharmacol.* 463:177–84

Tronche F, Kellendonk C, Kretz O, Gass P, Anlag K, et al. 1999. Disruption of the glucocorticoid receptor gene in the nervous system results in reduced anxiety. *Nat. Genet.* 23:99–103

True WR, Rice J, Eisen SA, Heath AC, Goldberg J, et al. 1993. A twin study of genetic and environmental contributions to liability for posttraumatic stress symptoms. *Arch. Gen. Psychiatry* 50:257–64

Trullas R, Skolnick P. 1993. Differences in fear motivated behaviors among inbred mouse strains. *Psychopharmacology (Berl.)* 111:323–31

Turri MG, Datta SR, DeFries J, Henderson ND, Flint J. 2001a. QTL analysis identifies multiple behavioral dimensions in ethological tests of anxiety in laboratory mice. *Curr. Biol.* 11:725–34

Turri MG, Henderson ND, DeFries JC, Flint J. 2001b. Quantitative trait locus mapping in laboratory mice derived from a replicated selection experiment for open-field activity. *Genetics* 158:1217–26

Vaillend C, Ungerer A. 1999. Behavioral characterization of mdx3cv mice deficient in C-terminal dystrophins. *Neuromuscul. Disord.* 9:296–304

Vallee M, Mayo W, Dellu F, Le Moal M, Simon H, Maccari S. 1997. Prenatal stress induces high anxiety and postnatal handling induces low anxiety in adult offspring: correlation with stress-induced corticosterone secretion. *J. Neurosci.* 17:2626–36

van den Heuvel OA, van de Wetering BJ, Veltman DJ, Pauls DL. 2000. Genetic studies of panic disorder: a review. *J. Clin. Psychiatry* 61:756–66

van Gaalen MM, Steckler T. 2000. Behavioural analysis of four mouse strains in an anxiety test battery. *Behav. Brain Res.* 115:95–106

van Gaalen MM, Stenzel-Poore MP, Holsboer F, Steckler T. 2002. Effects of transgenic overproduction of CRH on anxiety-like behaviour. *Eur. J. Neurosci.* 15:2007–15

Vaughan J, Donaldson C, Bittencourt J, Perrin MH, Lewis K, et al. 1995. Urocortin, a mammalian neuropeptide related to fish urotensin I and to corticotropin-releasing factor. *Nature* 378:287–92

Vetter DE, Li C, Zhao L, Contarino A, Liberman MC, et al. 2002. Urocortin-deficient mice show hearing impairment and increased anxiety-like behavior. *Nat. Genet.* 31:363–69

Wang X, Su H, Copenhagen LD, Vaishnav S, Pieri F, et al. 2002. Urocortin-deficient mice display normal stress-induced anxiety behavior and autonomic control but an impaired acoustic startle response. *Mol. Cell. Biol.* 22:6605–10

Wehner JM, Radcliffe RA, Rosmann ST, Christensen SC, Rasmussen DL, et al. 1997. Quantitative trait locus analysis of contextual fear conditioning in mice. *Nat. Genet.* 17:331–34

Wei F, Qiu CS, Liauw J, Robinson DA, Ho N, et al. 2002. Calcium calmodulin-dependent protein kinase IV is required for fear memory. *Nat. Neurosci.* 5:573–79

Weissman MM. 1993. Family genetic studies of panic disorder. *J. Psychiatr. Res.* 27(Suppl. 1):69–78

Weissman MM, Fyer AJ, Haghighi F, Heiman G, Deng Z, et al. 2000. Potential panic disorder syndrome: clinical and genetic linkage evidence. *Am. J. Med. Genet.* 96:24–35

Wieland HA, Luddens H, Seeburg PH. 1992. A single histidine in $GABA_A$ receptors is essential for benzodiazepine agonist binding. *J. Biol. Chem.* 267:1426–29

Wigger A, Loerscher P, Weissenbacher P, Holsboer F, Landgraf R. 2001. Cross-fostering and cross-breeding of HAB and LAB rats: a genetic rat model of anxiety. *Behav. Genet.* 31:371–82

Wood SJ, Toth M. 2001. Molecular pathways of anxiety revealed by knockout mice. *Mol. Neurobiol.* 23:101–19

Yedovitzky M, Mochly-Rosen D, Johnson JA, Gray MO, Ron D, et al. 1997. Translocation inhibitors define specificity of protein kinase C isoenzymes in pancreatic beta-cells. *J. Biol. Chem.* 272:1417–20

Yoshikawa T, Watanabe A, Ishitsuka Y, Nakaya A, Nakatani N. 2002. Identification of multiple genetic loci linked to the propensity for "behavioral despair" in mice. *Genome Res.* 12:357–66

Annu. Rev. Neurosci. 2004. 27:223–46
doi: 10.1146/annurev.neuro.27.070203.144317

UBIQUITIN-DEPENDENT REGULATION OF THE SYNAPSE

Aaron DiAntonio[1] and Linda Hicke[2]

[1]*Department of Molecular Biology and Pharmacology, Washington University School of Medicine, St. Louis, Missouri 63110 email: diantonio@wustl.edu*
[2]*Department of Biochemistry, Molecular Biology, and Cell Biology, Northwestern University, Evanston, Illinois 60208; email: l-hicke@northwestern.edu*

Key Words protein degradation, proteasome, synaptic transmission, synaptic plasticity, axon guidance

■ **Abstract** Posttranslational modification of cellular proteins by the covalent attachment of ubiquitin regulates protein stability, activity, and localization. Ubiquitination is rapid and reversible and is a potent mechanism for the spatial and temporal control of protein activity. By sculpting the molecular composition of the synapse, this versatile posttranslational modification shapes the pattern, activity, and plasticity of synaptic connections. Synaptic processes regulated by ubiquitination, as well as ubiquitination enzymes and their targets at the synapse, are being identified by genetic, biochemical, and electrophysiological analyses. This work provides tantalizing hints that neuronal activity collaborates with ubiquitination pathways to regulate the structure and function of synapses.

INTRODUCTION

Synapses are the primary site of communication in the nervous system. Understanding how synapses form, how they grow throughout development, and how their structure and function are molded by experience are the fundamental issues facing the synaptic biologist. A satisfying explanation for how we learn, remember, and think—and insights into how neurological and psychiatric diseases disrupt these processes—requires a deep understanding of synaptic development, function, and plasticity.

Many molecules and mechanisms participate in the life history of a synapse. Key regulatory systems include, but are not limited to, the level of intracellular calcium, the activity of protein kinase and phosphatase cascades, and the activity-dependent control of gene transcription and protein translation. Recent studies demonstrate that ubiquitination is another potent regulatory element that shapes the structure and function of the synapse. The covalent attachment of ubiquitin to target proteins is a

0147-006X/04/0721-0223$14.00

rapid and reversible modification that regulates not only protein stability, but also protein activity and localization. As such, it is a potent mechanism for sculpting the synapse. In this article we review ubiquitin and ubiquitin-binding proteins, focusing on the machinery of ubiquitination and the function of the ubiquitin modification. We then discuss the experimental evidence that ubiquitination is a potent regulator of the synapse.

INTRODUCTION TO UBIQUITIN

Ubiquitin is a member of a family of small proteins that are structurally similar and have the unusual property of forming a stable chemical bond with other proteins. Ubiquitin carries a C-terminal glycine carboxy group that forms an isopeptide bond with the ε-amino group of lysine residues, or less commonly, with the amino group at the N-terminus of a substrate protein. The attachment (conjugation) of ubiquitin to a protein can regulate a protein in different ways. Known or suspected mechanisms of ubiquitin action include altering protein stability, location, binding partners, or physical conformation. Ubiquitin modifies proteins involved in many, if not all, cellular functions (Glickman & Ciechanover 2002, Hicke & Dunn 2003, Muratani & Tansey 2003, Peng et al. 2003).

Proteins can be modified by a single ubiquitin moiety or by polymeric ubiquitin chains, referred to as mono- and polyubiquitination, respectively. Multiple lysines within ubiquitin are used to form polyubiquitin chains that carry out distinct molecular and cellular functions. Recent evidence indicates that ubiquitin chains in the cell can be formed through all seven lysines within ubiquitin [K6, K11, K27, K29, K33, K48, and K63; (Peng et al. 2003)], and many of these chains have been synthesized in vitro, suggesting that the forms of ubiquitin that could regulate ubiquitin function are numerous.

Lys48-linked polyubiquitin chains have been studied extensively, and they regulate nuclear, cytosolic, and endoplasmic reticulum membrane proteins by targeting these proteins for degradation by the 26S proteasome (Glickman & Ciechanover 2002). Lys63-linked chains are known to regulate DNA repair, signal transduction, and endocytosis and are likely to control other basic cellular processes as well (Fisk & Yaffe 1999, Galan & Haguenauer-Tsapis 1997, Hoege et al. 2002 and references therein). Monoubiquitination also serves as an important regulatory signal for many cellular functions (Schnell & Hicke 2003). Ubiquitin-like proteins, such as SUMO and Nedd8, that share ubiquitin's three-dimensional structure and conjugation properties seem to function primarily as mono-modifiers and, like ubiquitin, regulate a wide variety of proteins and processes (Schwartz & Hochstrasser 2003). In addition, there are proteins that carry ubiquitin-like domains called UBL (UBQ/UBD) and UBX domains (Buchberger 2002). These domains are not conjugatable and are found in larger, multidomain proteins. UBX domains are usually found at the C-terminus and are structurally similar to ubiquitin, although UBX domains and ubiquitin do not share significant sequence similarity. UBL domains are usually located at the N-terminus of a protein and are related in both sequence and structure to ubiquitin. UBL domains bind to the proteasome, and UBL-containing proteins appear to be intimately connected to the ubiquitin-proteasome pathway.

One way that cells interpret and transmit the information conferred by ubiquitin signals is through proteins that bind to ubiquitin and ubiquitinated proteins noncovalently. Ubiquitin-binding proteins identified to date carry small, modular domains that recognize ubiquitin primarily through hydrophobic interactions. Several distinct ubiquitin-binding domains have been characterized, and these are discussed below.

UBIQUITIN-BINDING PROTEINS

The first insight into how ubiquitin signals are interpreted came from the identification of a hydrophobic patch on the ubiquitin surface (Leu8, Ile44, Val70), which is required for binding to the S5a subunit of the proteasome (Beal et al. 1996). This same hydrophobic patch and a second patch on a different molecular surface, near Phe4, are critical for endocytic transport and vegetative growth in *Saccharomyces cerevisiae* (Sloper-Mold et al. 2001). Thus, the surface regions of ubiquitin appear to be three-dimensional signals that are interpreted by the cell through interaction with distinct protein domains. Domains that directly bind to monoubiquitin and/or polyubiquitin chains are being identified at a rapid rate, and at present at least seven have been described (reviewed in Di Fiore et al. 2003, Schnell & Hicke 2003). Ubiquitin-interacting motifs (UIMs), UBA (ubiquitin-associated) domains, and CUE (similar to Cue1) domains are the best-characterized ubiquitin-binding motifs (Schnell & Hicke 2003). All three domains bind to mono- and polyubiquitin and interact with the Ile44 hydrophobic surface of ubiquitin.

In some cases, distinct members of a domain family bind ubiquitin with different affinities (e.g., Mizuno et al. 2003, Shih et al. 2003) and, in the cell, may bind to different types of ubiquitin modifications. These differences may be due to specific sequence variations within the degenerate domain families or because context within a protein affects a domain's ubiquitin-binding properties. Most of the domains bind monoubiquitin with low affinity and can bind to polyubiquitin as well. For several domains the affinity for polyubiquitin is higher, even when genetic or physiological data suggest that monoubiquitin is the binding partner in the cell (e.g., Polo et al. 2002, Shih et al. 2002, Terrell et al. 1998). In vitro, a polyubiquitin chain presents the binding domain with hydrophobic surfaces from each ubiquitin moiety, allowing for the possibility of a multivalent interaction. However, in vivo the only available binding partner may be monoubiquitin, and low-affinity interactions with monoubiquitin would ensure that the interaction is readily reversible.

In addition to binding monoubiquitin, UIM and CUE domains promote the monoubiquitination of proteins within which they are carried (Davies et al. 2003, Klapisz et al. 2002, Oldham et al. 2002, Polo et al. 2002, Shih et al. 2003). These domains do not posses ubiquitination activity per se, but they do possibly recruit enzymes required for ubiquitination (Polo et al. 2002). The observation that multiple ubiquitin-binding proteins are themselves monoubiquitinated hints that the pairing of ubiquitin binding and ubiquitination may be an important regulatory mechanism. Ubiquitin binding by and ubiquitination of the same protein could provide one type of signal for the sequential assembly of a protein network (Di Fiore et al. 2003). Alternatively, an appended monoubiquitin may bind to a ubiquitin-binding domain within the same protein. Sites of ubiquitination lie outside the UIM and CUE domains (Klapisz et al. 2002, Polo et al. 2002; S. Francis & L. Hicke, unpublished data), consistent with the possibility of an intramolecular interaction between the conjugated ubiquitin and the ubiquitin-binding domain. In any event, the consequences of ubiquitination coupled to ubiquitin binding are likely to be intimately connected with the functions of ubiquitin and ubiquitin-binding domains.

FUNCTIONS OF UBIQUITIN

Ubiquitin Regulates Protein Stability

The first molecular function described for ubiquitin was as a signal to target substrates for degradation by the multisubunit proteolytic particle known as the proteasome. Polyubiquitin chains at least four units long that are linked through Lys48 commonly target substrates to the proteasome. In contrast, Lys63-linked chains appear to be used primarily for nonproteasome-dependent regulation of processes. The ability of other types of polyubiquitin chains to signal proteasomal degradation has not been well characterized.

The proteasome is a large 26 S particle composed of a 20 S barrel, or cylindrical chamber, capped on either end by a 19 S regulatory complex (Glickman & Ciechanover 2002). The 20 S barrel is a complex of 14 different proteins, three of which have active proteolytic sites, exposed only on the interior of the barrel. Unfolded substrates are transferred through a gated channel at either end of the barrel into the proteasome interior, where they are degraded. The ubiquitin tag is removed from the substrates prior to entering the barrel and recycled by the action of deubiquitinating enzymes associated with the proteasome. In rare cases the proteasome can degrade proteins without a ubiquitin tag.

Some ubiquitinated proteins undergo limited proteolysis by the proteasome as a processing step; however, generally a polyubiquitinated proteasome substrate is degraded to small peptides. In addition to polyubiquitin chains, *trans*-acting proteins, including chaperones, facilitate the recognition of substrates by the proteasome. Several specific inhibitors, such as lactacystin, can block the proteolytic activities of the proteasome. These inhibitors can also deplete the level of free ubiquitin in the cell; thus inhibitor experiments done in the presence of overexpressed ubiquitin ensure that an observed effect is actually due to proteasome inhibition.

Ubiquitin-dependent proteasome degradation regulates many crucial cell biological processes, including progression through the cell cycle, gene regulation, and signal transduction. As discussed below, in neurons proteasomes play a role in neuronal connectivity, synaptic transmission, synaptic plasticity, and protection against neurodegenerative diseases.

Nontraditional Functions of Ubiquitin

Although proteasome-mediated degradation is the most thoroughly characterized ubiquitin function, recent observations indicate that ubiquitin plays a number of key regulatory roles in the cell that are proteasome-independent.

UBIQUITIN REGULATES PROTEIN LOCATION One of the best-characterized nontraditional functions for ubiquitin is as a regulated localization signal for the transport of integral membrane proteins (reviewed in Hicke & Dunn 2003). Monoubiquitination is a necessary and sufficient signal for internalization of many cell-surface proteins into the endocytic pathway. Later in the pathway as proteins are delivered from the plasma membrane to the lysosome for degradation, monoubiquitin is also used as a signal for the sorting of transmembrane proteins into vesicles that bud into the lumen of a late endosomal compartment known as the multivesicular body or endosome (MVB/MVE). In a similar way, proteins from the biosynthetic pathway are sorted into MVE vesicles by a ubiquitin signal (reviewed in Katzmann et al. 2002, Raiborg et al. 2003).

Ubiquitin-dependent endosomal sorting is crucial for the downregulation of internalized plasma membrane proteins. For instance, activated signaling receptors that cannot enter the MVE pathway are recycled back to the cell surface rather than being degraded in the lysosome. Some membrane proteins that are internalized or sorted in a ubiquitin-dependent manner are not themselves ubiquitinated, but they interact with a ubiquitinated transport modifier that dictates the trafficking of the protein. Monoubiquitin transport signals most likely act by recruiting ubiquitin-binding components of the endocytic machinery, e.g., epsin, that behave as adaptor proteins to link ubiquitinated cargo to membrane vesicle budding machinery (Schnell & Hicke 2003, Wendland 2002).

Ubiquitin can also regulate protein localization within the nucleus. For example, monoubiquitination of the Fanconi anemia (FA) protein FANCD2 during the S phase of the cell cycle stimulates FANCD2 translocation into discrete nuclear foci (Gregory et al. 2003). It is not yet known whether monoubiquitin is a common signal for intranuclear localization, or how this ubiquitin-dependent localization occurs.

UBIQUITIN REGULATES PROTEIN ACTIVITY In addition to serving as a cargo sorting signal, ubiquitination regulates the activity of components of the endocytic machinery (Dunn & Hicke 2001, van Kerkhof et al. 2000 and references therein). Several members of the sorting and vesicle budding machinery are monoubiquitinated, and genetic evidence from *Drosophila* suggests that deubiquitination is also important for controlling at least one of these proteins, epsin (Chen et al. 2002 and references therein). Thus both ubiquitination and deubiquitination are emerging as mechanisms for regulating the temporal and spatial activity of endocytic proteins.

Monoubiquitin regulates the activity of transcription factors in the nucleus (Muratani & Tansey 2003). Genetic evidence from yeast suggests that monoubiquitination is required for normal transactivation activity of many transcription factors, and in these proteins the transcriptional activation domain overlaps with sequences that promote rapid proteasomal degradation (degrons). A polyubiquitin chain may be extended on the active, monoubiquitinated transcription factor to rapidly turn off transcriptional activity when it is no longer necessary.

Ubiquitination is proposed not only to activate proteins but also to auto-inhibit substrates. Phosphorylation of substrates by a kinase in an immune system signal transduction pathway (MEKK1; Ben-Neriah 2002) is inhibited by kinase auto-ubiquitination. Because this kinase also carries two ubiquitin-binding domains, kinase activity may be inhibited by an intramolecular interaction between the conjugated ubiquitin and the ubiquitin-binding domains. The regulation of proteins such as epsin and MEKK1 by ubiquitin needs to be defined in mechanistic detail to provide paradigms for how ubiquitin signals control protein activity. In addition, there are likely to be many other ubiquitin-dependent regulatory mechanisms waiting to be discovered.

UBIQUITINATION AND DEUBIQUITINATION MACHINERY

Because ubiquitin signals are often sufficient to change the location or activity of a modified protein, the timing and location of ubiquitin conjugation is crucial for appropriate control of biological events. Ubiquitin conjugation occurs by the sequential action of three enzymes: a ubiquitin-activating enzyme (E1), a ubiquitin-conjugating enzyme (E2), and a ubiquitin ligase (E3) (Figure 1, see color insert; reviewed in Weissman 2001). First, the E1 activates and forms a high-energy thiolester bond with ubiquitin. Ubiquitin is then passed to the E2, and this enzyme, usually in cooperation with an E3, transfers ubiquitin to a lysine side chain of the substrate. E3s specify the timing and substrate selection of ubiquitination reactions. Therefore, ubiquitin ligases are the key regulatory determinants in the ubiquitination reaction, analogous to kinases in phosphorylation reactions.

Ubiquitin ligases comprise two major families (reviewed in Pickart 2001). The first family, characterized by the zinc-binding RING (Really Interesting New Gene) finger domain and related domains, promotes ubiquitination by simultaneously binding the substrate and an E2 (Joazeiro & Weissman 2000). The second family, defined by the HECT (homologous to E6-AP carboxy-terminus) domain, participates directly in catalysis by forming an obligate thiolester bond with ubiquitin during the ubiquitination reaction (Huibregtse et al. 1995). Because E3s carry the specificity information of the ubiquitination machinery, understanding the function of this class of enzymes is critical for understanding events regulated by ubiquitin in vivo.

HECT Domain Ubiquitin Ligases

HECT (homologous to E6-AP C-terminus) domain ligases share a conserved carboxy terminal ~350 amino acid catalytic domain. An essential cysteine within this domain forms a thiolester bond with ubiquitin during transfer of ubiquitin from an E2 to a substrate. The N-terminal regions of HECT ligases are diverse and probably carry domains involved in substrate recognition. E6-AP is the founding member of the HECT domain ligase family that is linked to the severe neurological disorder Angelman syndrome. Although several cellular substrates for E6-AP are known, the substrate(s) responsible for Angelman syndrome has not been identified.

One subclass of HECT domain ligases implicated in axonal guidance is the Nedd4 (neuronally expressed developmentally downregulated 4) family of E3s, which share a highly similar structural organization and are required for or implicated in regulating membrane protein transport (reviewed in Rotin et al. 2000). At the N-terminus of these proteins is a C2 domain, a protein module that can bind calcium, proteins, and lipids, followed by two to four copies of the WW protein-protein interaction domain that binds to proline-rich target sequences or to phosphoserine/threonine (Sudol & Hunter 2000). At the carboxyl terminus is the HECT catalytic domain that binds ubiquitin-conjugating enzymes and contains the catalytic cysteine residue.

RING Finger Ubiquitin Ligases

Ubiquitin ligases that carry a RING finger domain exist as multisubunit complexes, or as monomers with substrate binding information and E3 activity built into the same molecule (Joazeiro & Weissman 2000). Genome sequencing suggests that RING finger E3s may far outnumber their HECT domain counterparts, and the identification of E3 activity in domains related to the RING finger, the PHD (plant homeodomains), and U box (UFD2-homology domain) lengthens the list (e.g., Coscoy et al. 2001, Hatakeyama et al. 2001). The RING finger directly interacts with an E2 and the substrate. RING finger ligases do not function as enzymes per se but instead activate E2s to modify specific substrates. RING finger ligases have emerged as key regulators of neuronal function in many places, including Highwire in synaptic development and Parkin in protein aggregation in Parkinson's disease (DiAntonio et al. 2001, Mizuno et al. 2001).

For regulated ubiquitination by both HECT and RING finger ligases, substrate phosphorylation is a common, though not the only, trigger for substrate binding. In addition, ligases themselves can be regulated by phosphorylation. One significant question about which little is known is how the type of ubiquitin modification that occurs on a particular substrate is specified. This question is important because changes in the extent of ubiquitination (mono versus poly) on both soluble and membrane proteins can alter their fate. HECT and RING family E3s can catalyze both monoubiquitination and polyubiquitination. Therefore, this choice must be regulated by factors other than the identity of the E3. The type of modification may depend on posttranslational modification of either the ubiquitin ligase or regulatory E3-binding proteins. Mono or polyubiquitination may also depend on whether the E3 is stably or transiently associated with a substrate because stable association may allow sufficient time for extension of a polyubiquitin chain (Di Fiore et al. 2003). Regulation of monoubiquitination might occur by deubiquitination enzymes that trim polyubiquitin chains or by capping proteins that bind to a monoubiquitin-conjugated substrate to prevent chain extension.

Deubiquitinating Enzymes

Ubiquitination is a reversible process, and eukaryotic genomes encode a large number of putative and confirmed deubiquitinating enzymes (DUBs) that hydrolyze ubiquitin-protein isopeptide bonds. Because DUBs remove ubiquitin signals, they are analogous to phosphatases and are likely to be crucial regulatory proteins. For instance, a *Drosophila* DUB known as Fat facets has been shown to regulate synaptic development at the neuromuscular junction and acts as an antagonist to the putative ubiquitin ligase Highwire (DiAntonio et al. 2001). At least one DUB, UCH-L1, appears to have the ability not only to remove ubiquitin from a protein, but also to transfer the ubiquitin moiety to another protein. UCH-L1 can modify α-synuclein, a protein that accumulates in protein aggregates in brain cells of Parkinson's disease patients, and polymorphisms in UCH-L1 are correlated with differential risk of Parkinson's disease (Liu et al. 2002) and age of onset of Huntington's disease (Naze et al. 2002).

UBIQUITIN AND THE NERVOUS SYSTEM

As described above, ubiquitin and the ubiquitination machinery are potent regulators of protein stability, localization, and activity. Their function in the nervous system has been investigated for many years, although the primary focus has been on the role of ubiquitination in neurodegenerative diseases (reviewed in Giasson & Lee 2003, Mayer et al. 1992). Recently, a number of studies have linked ubiquitin modification to the development and function of synapses in the healthy nervous system. In fact, ubiquitin's role in neurodegeneration may result, in part, from the impairment of ubiquitin-dependent regulation of normal neuronal physiology. Below we focus on the role of ubiquitin in the healthy brain, reviewing the emerging evidence that ubiquitin and ubiquitination enzymes are key regulators of the development, function, and plasticity of synaptic connections.

UBIQUITIN AND NEURONAL CONNECTIVITY

Wiring the developing brain is an essential part of building a functional nervous system. Establishing the appropriate connections is a complex, multistep task (reviewed in Huber et al. 2003, Yu & Bargmann 2001). The axons of developing

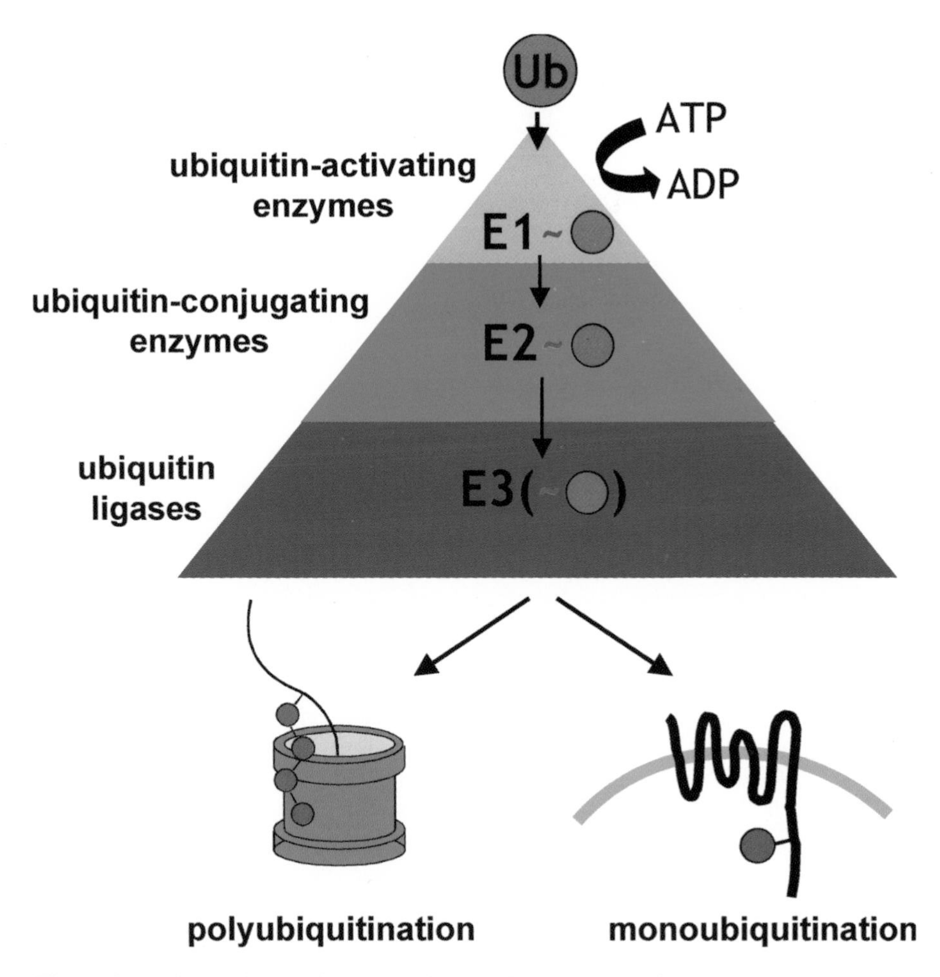

Figure 1 Ubiquitin is transferred to substrates by the action of three enzymes, E1s, E2s, and E3s. In a cell there are typically one to several E1s, multiple E2s, and many E3s. Ubiquitin modifications include monoubiquitin, the addition of a single ubiquitin molecule to a substrate, and chains of ubiquitin molecules that are linked through ubiquitin lysine residues.

neurons must navigate through their environment, often guided to numerous intermediate targets, before reaching their final destination. The growth cone, a sensorimotor structure at the tip of the outgrowing axon, is the central player mediating these axon guidance decisions. The growth cone senses guidance cues in its environment and translates this information into changes in cytoskeleton that determine the direction of outgrowth. Upon reaching the appropriate target, the growth cone differentiates into a presynaptic terminal as synapse formation begins. Once formed, synaptic connections are themselves dynamically regulated by neuronal activity, competition with other synapses, and intrinsic factors that remain to be defined (reviewed in Sanes & Lichtman 1999, Zhang & Poo 2001). If inappropriate connections have been made, or if connections are no longer needed, then an orchestrated program exists to retract axon arbors (reviewed in Kantor & Kolodkin 2003). Alternatively, appropriate connections are strengthened and new synapses grow. Although these events occur during development, this plasticity continues throughout the life of the organism and may underlie the mechanism of learning and memory.

Great insights into the molecular mechanisms that regulate neuronal connectivity have been made in the past decade. Numerous molecules have been identified that guide axons to their targets and promote synaptogenesis. Much work in the field is now focused on how these signaling molecules affect the behavior of growth cones and synapses. Key players identified to date include kinases and phosphatases, small G proteins, cyclic nucleotides, and intracellular calcium (Doherty et al. 2000, Korey & Van Vactor 2000). In addition, local protein translation has been implicated in the regulation of growth cones and synapses—the machinery for translating proteins is present in axons and dendrites and is regulated by guidance cues and neuronal activity (Brittis et al. 2002, Campbell & Holt 2001, Jiang & Schuman 2002). All of these molecules can act locally in the growth cone or synapse and can be regulated rapidly and reversibly. Such local mechanisms are essential because growth cones and synapses are often located far from their neuronal cell bodies.

Ubiquitination is another rapid, reversible, and local mechanism that can change the levels, localization, or activity of target proteins. A flurry of recent studies provides evidence that the ubiquitin and the proteasome are important regulators of neuronal connectivity. Below we describe the key experiments that link ubiquitin-dependent events to axon guidance, axonal pruning, and synapse formation and growth.

Axon Guidance

The first clue that ubiquitin might regulate neuronal connectivity came from the analysis of the *Drosophila* mutant *bendless*. *bendless* was identified more than 20 years ago in an elegant screen for flies with an altered escape response, and it sparked great interest because it disrupted only a single neuronal connection (Thomas & Wyman 1984). The giant fiber axon, which is essential for the escape response, failed to turn and make a synapse with its motoneuron target. Cloning of the responsible gene identified bendless as an E2 ubiquitin conjugating enzyme (Muralidhar & Thomas 1993, Oh et al. 1994). At this time in the early 1990s, it was not possible to distinguish a housekeeping role for bendless-dependent ubiquitination from a more active, dynamic regulation of proteins necessary for neural development. More recent analysis of semaphorin signaling in the giant fiber demonstrates that changing levels of semaphorin give a similar phenotype to that of the *bendless* mutant (Murphey et al. 2003), consistent with the hypothesis that *bendless* may be regulating the levels of key signaling molecules on the growth cone (Muralidhar & Thomas 1993, Oh et al. 1994). This idea would be a satisfying conclusion to a story that has been underappreciated for many years.

In the decade since *bendless* introduced the field of neurodevelopment to ubiquitination, great progress has been made in identifying signals, such as the netrins, ephrins, semaphorins, and slits, that guide axons (Huber et al. 2003). With the identification of these ligands and their cognate receptors, interest in the field is now focused on how these guidance cues are translated into changes in cytoskeletal dynamics and growth cone behavior. A study by Campbell & Holt using *Xenopus* retinal growth cones demonstrates that ubiquitin-dependent proteasomal degradation is part of the answer (Campbell & Holt 2001). The growth cones of *Xenopus* retinal neurons in culture contain ubiquitin, the E1 ubiquitin-activating enzyme, and proteasomes. Netrin-1 and L-a-lysophosphatidic acid (LPA) are guidance molecules that use the ubiquitination machinery to regulate the behavior of these growth cones. Within five minutes of encountering netrin or LPA, the levels of ubiquitin-protein conjugates in the growth cone double. Pharmacological inhibition of the proteasome blocks both netrin-dependent growth cone turning and LPA-induced growth cone collapse. In this system, local protein translation is also required for appropriate growth cone dynamics. Hence guidance cues likely control the concentration of proteins within the growth cone by regulating the balance of protein synthesis and degradation. Such rapid and local changes in protein composition are a powerful mechanism for altering the behavior of growth cones.

Guidance receptors not only regulate ubiquitination, but also they are themselves targets of ubiquitin. At the *Drosophila* midline, the repellant slit and its receptor Robo regulate an axon's decision to cross the midline. Axons that do not cross the midline, and those that have crossed once, express high levels of Robo on their surface. To cross the midline, an axon must downregulate Robo. Commissureless (Comm), a transmembrane protein, is required for this downregulation (Kidd et al. 1998, 1999). Comm regulates the surface levels of Robo by recruiting the ubiquitin ligase Nedd4, which binds to and ubiquitinates Comm (Myat et al. 2002). This interaction with Nedd4 is necessary for Comm to regulate Robo surface expression. In this system Comm acts as a transport modifier that binds Robo and is ubiquitinated by Nedd4. The ubiquitinated Comm presumably serves as an endocytosis signal, triggering the removal of Robo from the cell surface and changing the axon's sensitivity to slit during pathfinding. A complementary study from the Dickson lab proposed a somewhat different model in which Comm diverts Robo from the golgi to the lysosome, thereby keeping Robo from reaching the cell surface (Keleman et al. 2002). Although this study did not implicate ubiquitin in this trafficking event, work in yeast has shown that ubiquitin can regulate the choice for sorting to either the plasma membrane or lysosome (Helliwell et al. 2001). These results raise many questions. Are other guidance receptors regulated in a similar fashion? Axons must change their response to guidance cues as they confront intermediate targets along their path. The ubiquitin-dependent downregulation of guidance receptors would allow for rapid and local changes in the growth cone's response to signaling molecules. If this mechanism is widely used, then what proteins link ubiquitin ligases with guidance receptors? Homologs of Comm have not been identified in vertebrates. Perhaps functionally similar but structurally distinct proteins regulate vertebrate guidance decisions. Finally, what does this insight into Comm function teach us about synapse formation? At the *Drosophila* neuromuscular junction, the downregulation of Comm from the muscle surface is necessary for synapse formation (Wolf et al. 1998). Perhaps ubiquitinated Comm regulates the surface expression of a protein that inhibits this event.

Axonal Pruning

Wiring the nervous system requires more than guiding axons to their targets. Often, the initial connections are exuberant and must be pruned (reviewed in O'Leary & Koester 1993). Later in development, some connections may lose their usefulness and are eliminated. Examples include the large-scale rearrangement of the insect nervous system during metamorphosis (Truman 1990) or the competitive elimination of motoneuron branches at the vertebrate neuromuscular junction (Sanes & Lichtman 1999). During these rearrangements single neurons lose certain axonal branches and retain others. The high degree of specificity observed in many examples of axon pruning suggests that the underlying mechanism is carefully regulated both temporally and spatially. Genetic analysis of axon pruning during *Drosophila* metamorphosis demonstrates a requirement for ubiquitin and proteasomes.

The *Drosophila* mushroom body (MB) is a key associative area for olfactory learning. Using single-cell mosaic analysis of MB projections, Luo and colleagues have defined the stereotyped rearrangements that occur in identified classes of MB neurons during metamorphosis (Jefferis et al. 2002). MB γ neurons extend a single axonal peduncle that bifurcates into a dorsal and medial branch to form larval projections. During metamorphosis, the dorsal and medial branches are pruned. In the pupae, the medial branch reextends to form an adult-specific projection. In this system it is possible to generate mutations of candidate genes in the small subsets of neurons under study. This genetic analysis demonstrates that the pruning process requires both ubiquitination and proteasomal machinery (Watts et al. 2003). Pruning is blocked by loss-of-function mutations in the E1 ubiquitin-activating enzyme, demonstrating a requirement for ubiquitination. This ubiquitination is likely used as a signal for proteasome-dependent proteolysis because mutations in two different subunits of the proteasome also block axonal pruning. Furthermore, mutations in clathrin heavy chain, which should block monoubiquitination-dependent endocytosis, do not inhibit pruning. Taken together, these data suggest that degradation by the proteasome is essential for axonal pruning.

What is the role of the ubiquitin-dependent proteolysis in axonal pruning? Two models have been suggested. It may participate in the initiation of the pruning mechanism, degrading negative regulators of the pruning machinery. Alternatively, it may act downstream of these regulators and execute the pruning process by degrading the structural components of the axon. Luo and colleagues (Watts et al. 2003) favor the initiation model because the very earliest morphological change seen during pruning, the disruption of the microtubule cytoskeleton, requires a functional E1 and proteasome. Interestingly, another form of axonal loss, Wallerian degeneration, shows similar characteristics. Following axon injury, the distal axon fragments and degenerates. As with axon pruning in *Drosophila*, disruption of the microtubule cytoskeleton is the first morphological change observed, and proteasome inhibitors block the degeneration (Zhai et al. 2003). These similarities suggest that the rearrangement of neuronal connectivity that occurs during development and with neurodegenerative disease may share similar molecular mechanisms (Mack et al. 2001, Saigoh et al. 1999).

Synapse Development

Once an axon has found its target, the growth cone differentiates into a presynaptic terminal, and synapse formation begins. If the connection is stabilized, the synapse may grow during development. Work from the *Drosophila* neuromuscular junction (NMJ) implicates ubiquitination as a key regulator of synapse formation and growth (DiAntonio et al. 2001). The deubiquitinating enzyme *fat facets* (*faf*) was identified in a genetic screen for molecules controlling the size and strength of synapses. *faf* is a negative regulator of ubiquitin-dependent events, removing ubiquitin from target proteins (Huang et al. 1995). Overexpression of either *faf* or a yeast deubiquitinating enzyme in the developing nervous system leads to dramatic synaptic overgrowth and impaired synaptic transmission. Hence, genetically antagonizing ubiquitination disrupts the normal synaptic growth control mechanisms, implying that ubiquitin-dependent events normally regulate synaptic growth.

In order to identify other molecules in this pathway, DiAntonio et al. screened for enhancers of the *faf* overexpression phenotype. They identified multiple alleles of *highwire* (*hiw*), a negative regulator of synaptic growth and a putative ubiquitin ligase (Wan et al. 2000). The loss-of-function phenotype of *hiw* is nearly identical to the gain-of-function phenotype of *faf*; there is massive synaptic overgrowth and impaired synaptic transmission. In addition, loss-of-function *faf* mutants partially suppress the physiological deficits in *hiw* mutants, demonstrating a role for endogenous *faf* at the synapse (DiAntonio et al. 2001). Highwire is a huge synaptic protein (>500 kd) that contains the RING-H2 domain characteristic of a class of ubiquitin ligases (Joazeiro & Weissman 2000). The genetic interactions between a putative ubiquitin ligase and a deubiquitinating enzyme suggest that a balance between the ubiquitination and deubiquitination of a key target molecule regulates synaptic growth. This mechanism may be evolutionarily conserved; mutations in *rpm-1*, the *C. elegans* homology of *highwire*, disrupt synaptic morphology in the worm (Schaefer et al. 2000, Zhen et al. 2000). In mammals the single *highwire* homolog, PAM, is strongly expressed in the brain and the RING domain is the most highly conserved region of the protein (Guo et al. 1998).

The finding that ubiquitination can regulate synapse development raises many questions. First, does *highwire* function as a ubiquitin ligase, and is this activity necessary for its in vivo function? Second, what are the substrates of ubiquitination that mediate the regulation of synaptic growth? In the *Drosophila* eye, the ubiquitin-binding protein epsin is ubiquitinated and is a substrate for *faf* (Chen et al. 2002), and therefore is a candidate for regulating the synapse. Identification of functionally relevant substrates may come from genetic experiments in *Drosophila*; for example, mutations in targets of *highwire* may suppress the synaptic overgrowth seen in the *highwire* mutant. Finally, what is the role of the mammalian homolog of *highwire* in vertebrate synaptic development? If its function is conserved, it may act to maintain stable synapses and inhibit synaptic plasticity. Genetic analysis in the mouse, fly, and worm should define the *highwire* pathway and the role of ubiquitination in the regulation of synaptic growth and development.

UBIQUITIN AND SYNAPTIC TRANSMISSION

At the majority of synapses in the brain, information flow occurs via the regulated secretion of chemical neurotransmitters from the presynaptic cell and the reception of the signal by transmitter receptors on the postsynaptic cell. Altering neuronal excitability, presynaptic release machinery and postsynaptic sensitivity to transmitter can all change the efficacy of synaptic transmission. Numerous cellular mechanisms regulate the machinery of synaptic transmission including the intracellular levels of calcium and cyclic nucleotides, protein phosphorylation and dephosphorylation, and the localization and translation of synaptic proteins. Hence, the activity, location, and levels of key synaptic proteins are central determinants of synaptic efficacy. Because protein ubiquitination regulates the amount and subcellular location of target proteins, it is a candidate mechanism to regulate synaptic transmission. Recent work validates this prediction, demonstrating that ubiquitination regulates presynaptic neurotransmitter release and postsynaptic neurotransmitter receptors.

Presynaptic Function

Some of the best evidence for the local and rapid regulation of neurotransmitter release by ubiquitin comes from the glutamatergic *Drosophila* neuromuscular junction (Speese et al. 2003). Broadie and colleagues find that the ubiquitination machinery and proteasomes are present in the presynaptic terminals of the NMJ.

Furthermore, these proteasomes are functional. A conditional proteasome substrate expressed at the synapse is degraded with a half-life of ~30 min. Genetic and pharmacological inhibitors of the proteasome can be used at the *Drosophila* NMJ. Genetic inhibitors, such as dominant, temperature-sensitive mutations in proteasome subunits, allow for tissue-specific reductions in proteasome function. Pharmacological inhibitors provide for rapid control of proteasome activity. Application of the proteasome inhibitors lactacystin or epoxomicin induces a rapid strengthening of synaptic transmission. In less than an hour, the amplitude of evoked neurotransmission increases by more than 50% with either inhibitor. This response is likely due to a presynaptic increase in the number of vesicles released because there is no change in quantal size, the postsynaptic response to single synaptic vesicles. Analysis of the levels of proteins mediating presynaptic transmitter release reveals a potential mechanism. Inhibition of the proteasome with genetic or pharmacological inhibitors leads to a doubling of levels of synaptic DUNC-13, a molecule known to control synaptic strength by regulating synaptic vesicle priming. Furthermore, DUNC-13 is ubiquitinated in vivo (Aravamudan & Broadie 2003). Other presynaptic proteins, such as syntaxin and synaptotagmin, are unaffected by these manipulations. This study is significant because it provides good evidence that the ubiquitin-dependent proteolysis acts locally and rapidly in the presynaptic terminal to regulate neurotransmitter release, and it identifies a candidate substrate, DUNC-13. Future work will focus on the identification of the E3 ubiquitin ligase that acts on DUNC-13 and on the regulation of the synaptic ubiquitination. For example, is the proteasome-dependent regulation of DUNC-13 constitutive, or might it be altered by changes in neuronal activity?

Genetic analysis in the mouse provides additional evidence that ubiquitin-dependent events regulate synaptic transmission (Wilson et al. 2002). Mice homozygous for the ataxia^J (ax^J) mutation develop severe tremors and early death. Electrophysiological analysis of the ax^J mutant demonstrates that neurotransmitter release at the neuromuscular junction is impaired because the mutant synapse releases only half as many synaptic vesicles following an action potential as does the wild-type synapse. Synaptic transmission in the CNS is also affected. In the hippocampus, baseline transmission is normal, but multiple forms of short-term synaptic plasticity such as paired-pulse and post-tetanic potentiation are impaired. Positional cloning of the responsible gene demonstrates that ubiquitin-specific protease 14 (Usp14) encodes a protein that cleaves ubiquitin from target proteins. Because Usp14 can process short ubiquitin side chains but not polyubiquitin, it may regulate the monoubiquitination state, and hence the localization or activity, of target proteins that regulate synaptic transmission. To date, the subcellular localization of Usp14 is unknown. It will be of great interest to determine if Usp14 or other deubiquitinating enzymes act locally within the synapse to regulate dynamically the ubiquitination state of synaptic proteins.

Electrophysiological data demonstrating elegantly that ubiquitination regulates synaptic transmission come from the analysis of DUNC-13 in *Drosophila* and Usp14 in the mouse. However, there are also biochemical data that are consistent with a role for ubiquitin in regulating neurotransmitter release. For example, syntaxin is a SNARE protein that plays an essential role in synaptic vesicle fusion with the plasma membrane and thus in neurotransmitter release. A RING-finger E3 ubiquitin-ligase, Staring, recruits the E2 ubiquitin-conjugating enzyme UbcH8 to syntaxin 1, promoting its ubiquitination and subsequent degradation by the proteasome (Chin et al. 2002). Similarly, the RING-domain proteins Siah-1A and Siah-2 bind the synaptic vesicle protein synaptophysin and facilitate its ubiquitination and degradation (Wheeler et al. 2002). The Siah proteins colocalize with synaptophysin to synaptic vesicles, putting them in position to regulate transmitter release. Staring and Siah proteins are excellent candidates to mediate ubiquitin-dependent regulation of synaptic transmission; however, confirmation of their role awaits an in vivo analysis of their function.

Postsynaptic Function

Changing the number of neurotransmitter receptors in the postsynaptic membrane is one means by which synaptic strength is regulated. Neuronal activity controls the abundance of these receptors at the synapse, likely by regulating their insertion and removal from the membrane (Malinow & Malenka 2002). Ubiquitin-dependent endocytosis can trigger the removal of membrane proteins from the cell surface (reviewed in Hicke & Dunn 2003) and is a candidate mechanism for regulating receptor levels at the synapse. Analysis of the inhibitory glycine and $GABA_A$ receptors supports this possibility. In *Xenopus* oocytes, glycine receptors can be ubiquitinated, internalized, and cleaved in the lysosome, demonstrating that the ubiquitination machinery could function to limit the abundance of synaptic receptors (Buttner et al. 2001). The $GABA_A$ receptor cycles between surface and internal pools, and its levels are regulated by proteasomal degradation (Bedford et al. 2001). $GABA_A$ levels are also regulated by Plic-1, a protein that carries a UBL domain and binds to and stabilizes the $GABA_A$ receptor. It is attractive to speculate that, with its ubiquitin-like domain, Plic-1 may inhibit the polyubiquitination and subsequent degradation of the $GABA_A$ receptor (Bedford et al. 2001). In this model, proteins such as Plic-1 act as antagonists of ubiquitination, joining deubiquitinating enzymes as negative regulators of the ubiquitination machinery.

Work with inhibitory receptors suggests that ubiquitin regulates levels of synaptic receptors, and analysis of excitatory glutamate receptors provides an in vivo demonstration that this is the case. Work with a *C. elegans* glutamate receptor shows that it is ubiquitinated in vivo (Burbea et al. 2002). Mutations that inhibit ubiquitination of the receptor increase its levels at the synapse. Behavioral analysis demonstrates that this increase in synaptic receptor leads to an increase in synaptic strength. Reciprocal findings were obtained by genetic manipulations that promote ubiquitination: There is a decrease in the synaptic abundance of glutamate receptors via a mechanism that requires endocytosis (Burbea et al. 2002). At vertebrate hippocampal synapses, homologous AMPA receptors are also regulated by ubiquitination. A brief (<5 min) application of proteasome inhibitors blocks the internalization of AMPA receptors induced by either agonist or increased synaptic activity (Patrick et al. 2003). How might neuronal activity regulate ubiquitination to control receptor abundance? Possibilities include an increased access to substrates (e.g., phosphorylation of substrates to provide E3 binding sites) or an upregulation of ligase activity.

UBIQUITIN AND SYNAPTIC PLASTICITY

Once functional synaptic connections are formed, the strength of synapses can be modified by activity within the nervous system. Such experience-dependent synaptic plasticity likely underlies the refinement of neural circuits during development and the capacity to learn and remember in the adult. The strength of synapses changes with alterations in their morphology and physiology. Morphological changes occur with the growth or retraction of synapses, whereas physiological changes may include alterations in presynaptic transmitter release characteristics or the postsynaptic sensitivity to neurotransmitter. What are the molecular mechanisms that bring about long-term synaptic plasticity? The prevailing model postulates that neuronal activity, likely acting through intracellular calcium and kinases, regulates gene expression and protein translation (Kandel 2001, Jiang & Schuman 2002, West et al. 2001). Much evidence supports this biosynthetic model of plasticity; however, the model is incomplete. If addition of proteins can reshape synapses, then it is likely that removal of proteins will also sculpt the synapse. Recent data support this view. Studies of long-term facilitation in Aplysia, a classic model of long-term synaptic change, indicate multiple functions for the ubiquitin proteasome system. Similar findings are emerging from vertebrate systems, most notably in the experience-dependent remodeling of postsynaptic proteins.

Long-Term Facilitation in Aplysia

The sensorimotor synapses that underlie a simple reflex in Aplysia, gill withdrawal following a gentle touch, can be reconstituted in vitro. These synapses display short- and long-term plasticity that is similar to the behavioral plasticity in this reflex exhibited by the intact animal. A single application of serotonin (5-HT) sensitizes the reflex and produces a short-term increase in synaptic strength. Repeated applications of serotonin result in a long-lasting increase in synaptic strength [long-term facilitation (LTF)] and the growth of new connections in a process requiring both mRNA and protein synthesis (Kandel 2001). Ubiquitin modification modulates the response to serotonin and plays a key role in converting this plasticity from short to long term (Hegde & DiAntonio 2002).

The cAMP-dependent protein kinase (PKA) is activated by serotonin signaling and promotes synaptic strengthening. PKA is a heterotrimer with two catalytic subunits (PKA-C) and two regulatory subunits (PKA-R). Binding of the regulatory subunits to cAMP dissociates the C and R subunits activating the enzyme. However with LTF, PKA remains active well beyond the period of cAMP elevation. This is achieved through the ubiquitin-dependent degradation of the PKA-R subunits (Hegde et al. 1993). This degradation is regulated in two ways. First, the R subunits are only substrates for degradation after binding cAMP. This is an example of how signaling pathways can control the timing and specificity of ubiquitin-dependent proteolysis. Second, the ubiquitin hydrolase Ap-uch is an immediate early gene activated during, and important for, LTF (Hegde et al. 1997). This enzyme promotes proteasomal degradation by recycling ubiquitin from the polyubiquitin chains of degraded proteins. Together, these mechanisms persistently activate PKA by down-regulating the R subunits and help convert a short-term signal into a long-lasting synaptic change.

The findings described above lead to the model that ubiquitin-dependent proteolysis removes inhibitory constraints on synaptic strengthening. However, recent studies with pharmacological inhibitors of the proteasome in Aplysia culture demonstrate that the proteasome itself functions as an inhibitor of synaptic strengthening (Zhao et al. 2003). Inhibiting the proteasome produces a long-lasting (>24 h) increase in synaptic strength at the sensory-to-motoneuron synapse and potentiates serotonin-dependent LTF. Inhibiting the proteasome also increases the growth of synaptic connections between these two neurons. The effects are both pre- and postsynaptic. Blocking the proteasome in isolated presynaptic sensory neurons increases neurite outgrowth, whereas blockade of the proteasome in isolated postsynaptic motoneurons increases the sensitivity to glutamate. Taken together, these data demonstrate that ubiquitin-dependent proteolysis inhibits synaptic strengthening by modifying multiple substrates.

The ubiquitin-dependent regulation of long-term synaptic plasticity also occurs in the vertebrate CNS. The E6-AP ubiquitin ligase is mutated in the human disease Angelman syndrome, which is characterized by mental retardation (Kishino et al. 1997, Matsuura et al. 1997). In mice mutant for E6-AP there is a deficit in hippocampal long-term potentiation and context-dependent memory (Jiang et al. 1998). Pharmacological inhibition of the proteasome in the rat blocks two behaviors that likely depend on long-term synaptic plasticity, the development of long-term memories (Lopez-Salon et al. 2001), and neuropathic pain (Moss et al. 2002).

Remodeling the Postsynaptic Density

At the postsynaptic membrane of vertebrate excitatory synapses, glutamate receptors are linked via scaffolding molecules to a host of signaling and regulatory proteins (reviewed in Sheng 2001). This specialization is termed the postsynaptic density (PSD), and it is a key substrate for synaptic plasticity. The activity-dependent insertion and removal of glutamate receptors from the PSD are likely fundamental mechanisms for regulating synaptic strength (Malinow & Malenka 2002). Whereas many of the molecular constituents of the PSD have been identified, much less is known about how these molecules are regulated by activity. A recent study by

Ehlers demonstrates that neuronal activity shapes the composition and activity of the PSD via ubiquitination and proteasomal degradation (Ehlers 2003).

When vertebrate neurons are confronted with long-term alterations in neuronal activity, they can respond with compensatory, homeostatic changes that reestablish the initial level of excitability (Turrigiano 1999). The PSD is a likely locus for this change because postsynaptic responsiveness to transmitter is a key physiological compensation for changes in activity. Whereas previous studies have investigated the physiological response to altered activity, Ehlers performed a detailed, biochemical investigation of the molecular response to activity change (Ehlers 2003). In cultured hippocampal neurons, the molecular constituents of the PSD are regulated by synaptic activity. Ensembles of glutamate receptors, scaffolding proteins, and signaling molecules are coregulated bidirectionally and reversibly. Activity likely changes the composition of proteins in the PSD via ubiquitin-dependent proteolysis. Neuronal activity controls the half-life of synaptic proteins and regulates the ubiquitination of PSD proteins. Synaptic activity increases and inactivity decreases ubiquitin conjugation in the PSD. Moreover, pharmacological inhibitors of the proteasome block the molecular rearrangements induced by increased synaptic activity and mimic the effects of decreased activity. These results link synaptic activity and protein turnover to structural plasticity of synapses.

A striking finding in the Ehlers study is that groups of synaptic proteins are coregulated. The change in levels for each protein in an ensemble is very similar in both magnitude and time course. Ehlers suggests that these protein modules are regulated by the ubiquitination state of "master organizing molecules" (2003).

TABLE 1 Ubiquitination machinery and substrates regulating the synapse

Protein	Function	Synaptic action	Reference
Ubiquitination machinery			
bendless	E2 ubiquitin conjugase	Axon guidance	33, 44
Nedd4	E3 ubiquitin ligase	Axon guidance	33, 44
fat facets	Deubiquitinating enzyme	Synapse development	33
Usp14	Deubiquitinating enzyme	Synaptic transmission	33
Staring	E3 ubiquitin ligase	Synaptic transmission	33
Siah	E3 ubiquitin ligase	Synaptic transmission	33
Ap-uch	Deubiquitinating enzyme	Synaptic plasticity	33
E6-AP	E3 ubiquitin ligase	Synaptic plasticity	33, 44
Ubiquitinated substrates			
Commissureless	Regulates Robo levels	Axon guidance	33, 44
Unc-13	Synaptic vesicle release	Synaptic transmission	33, 44
Syntaxin	Synaptic vesicle release	Synaptic transmission	33
Synaptophysin	Synaptic vesicle protein	Synaptic transmission	33
Glycine, GABA, & glutamate receptors	Neurotransmitter receptors	Synaptic transmission	33, 44, 55, 66
PKA-R	Regulatory subunit of PKA	Synaptic plasticity	33
Shank, GKAP, & AKAP79	Postsynaptic scaffolding proteins	Synaptic plasticity	33

While many PSD proteins are regulated by the proteasome, only a few are ubiquitinated. In particular, the scaffolding proteins Shank, GKAP, and AKAP79 are substrates for ubiquitination. These scaffolding proteins are excellent candidates to be these master regulators (Ehlers 2003). The regulated removal of such a scaffolding protein could destabilize all of its associated proteins, resulting in the coregulation of a protein cohort.

CONCLUDING REMARKS

The addition of ubiquitin to target proteins is a potent mechanism for regulating protein stability, activity, and localization. The nervous system has harnessed the power of this versatile regulatory system to shape the development, function, and plasticity of neural circuits. Much like phosphorylation, ubiquitination is a rapid, local, and reversible protein modification. Unlike phosphorylation, the importance of the ubiquitin pathway in regulating the synapse has only recently been appreciated. To date, a few components of the ubiquitination machinery at the synapse have been described and a handful of substrates identified (see Table 1). It is likely that many more synaptic E3 ubiquitin ligases, deubiquitinating enzymes, and their substrates remain to be identified. The repertoire of regulatory possibilities will expand once there is a deeper understanding of the role of ubiquitin-like proteins and proteins carrying ubiquitin-like domains, such as SUMO and Plic-1. Finally, the key challenge is to identify a clear mechanism linking neuronal activity to changes in the activity of the ubiquitination machinery. Countless experiments demonstrate how neuronal activity regulates calcium, kinases, transcription, and translation. For the ubiquitin pathway, this work is just beginning (Ehlers 2003). When this connection is firmly established, ubiquitination will take its place in the pantheon of synaptic regulatory systems.

ACKNOWLEDGMENTS

We wish to thank Rebecca Dunn for help with Figure 1. Our work was supported by funding from the NIH (DK53257 and 61299) to L.H., and from the Keck, McKnight, and Burroughs Wellcome Foundations and the NIH (NS043071) to A.D.

The *Annual Review of Neuroscience* is online at http://neuro.annualreviews.org

LITERATURE CITED

Aravamudan B, Broadie K. 2003. Synaptic Drosophila UNC-13 is regulated by antagonistic G-protein pathways via a proteasome-dependent degradation mechanism. *J. Neurobiol.* 54:417–38

Beal R, Deveraux Q, Xia G, Rechsteiner M, Pickart C. 1996. Surface hydrophobic residues of multiubiquitin chains essential for proteolytic targeting. *Proc. Natl. Acad. Sci. USA* 93:861–66

Bedford FK, Kittler JT, Muller E, Thomas P, Uren JM, et al. 2001. GABA(A) receptor cell surface number and subunit stability are regulated by the ubiquitin-like protein Plic-1. *Nat. Neurosci.* 4:908–16

Ben-Neriah Y. 2002. Regulatory functions of ubiquitination in the immune system. *Nat. Immunol.* 3:20–26

Brittis PA, Lu Q, Flanagan JG. 2002. Axonal protein synthesis provides a mechanism for localized regulation at an intermediate target. *Cell* 110:223–35

Buchberger A. 2002. From UBA to UBX: new words in the ubiquitin vocabulary. *Trends Cell Biol.* 12:216–21

Burbea M, Dreier L, Dittman JS, Grunwald ME, Kaplan JM. 2002. Ubiquitin and AP180 regulate the abundance of GLR-1 glutamate receptors at postsynaptic elements in C. elegans. *Neuron* 35:107–20

Buttner C, Sadtler S, Leyendecker A, Laube B, Griffon N, et al. 2001. Ubiquitination precedes internalization and proteolytic cleavage of plasma membrane-bound glycine receptors. *J. Biol. Chem.* 276:42978–85

Campbell DS, Holt CE. 2001. Chemotropic responses of retinal growth cones mediated by rapid local protein synthesis and degradation. *Neuron* 32:1013–26

Chen X, Zhang B, Fischer JA. 2002. A specific protein substrate for a deubiquitinating enzyme: Liquid facets is the substrate of fat facets. *Genes Dev.* 16:289–94

Chin LS, Vavalle JP, Li L. 2002. Staring, a novel E3 ubiquitin-protein ligase that targets syntaxin 1 for degradation. *J. Biol. Chem.* 277:35071–79

Coscoy L, Sanchez DJ, Ganem D. 2001. A novel class of herpesvirus-encoded membrane-bound E3 ubiquitin ligases regulates endocytosis of proteins involved in immune recognition. *J. Cell Biol.* 155:1265–74

Davies BA, Topp JD, Sfeir AJ, Katzmann DJ, Carney DS, et al. 2003. Vps9p CUE domain ubiquitin binding is required for efficient endocytic protein traffic. *J. Biol. Chem.* 278:19826–33

DiAntonio A, Haghighi AP, Portman SL, Lee JD, Amaranto AM, Goodman CS. 2001. Ubiquitination-dependent mechanisms regulate synaptic growth and function. *Nature* 412:449–52

Di Fiore PP, Polo S, Hofmann K. 2003. When ubiquitin meets ubiquitin receptors: a signalling connection. *Nat. Rev. Mol. Cell Biol.* 4:491–97

Doherty P, Williams G, Williams EJ. 2000. CAMs and axonal growth: a critical evaluation of the role of calcium and the MAPK cascade. *Mol. Cell Neurosci.* 16:283–95

Dunn R, Hicke L. 2001. Multiple roles for Rsp5p-dependent ubiquitination at the internalization step of endocytosis. *J. Biol. Chem.* 276:25974–81

Ehlers MD. 2003. Activity level controls postsynaptic composition and signaling via the ubiquitin-proteasome system. *Nat. Neurosci.* 6:231–42

Fisk HA, Yaffe MP. 1999. A role for ubiquitination in mitochondrial inheritance in *Saccharomyces cerevisiae*. *J. Cell Biol.* 145:1199–208

Galan JM, Haguenauer-Tsapis R. 1997. Ubiquitin Lys63 is involved in ubiquitination and endocytosis of a yeast plasma membrane protein. *EMBO J.* 16:5847–54

Giasson BI, Lee VM. 2003. Are ubiquitination pathways central to Parkinson's disease? *Cell* 114:1–8

Glickman MH, Ciechanover A. 2002. The ubiquitin-proteasome proteolytic pathway: destruction for the sake of construction. *Physiol. Rev.* 82:373–428

Gregory RC, Taniguchi T, D'Andrea AD. 2003. Regulation of the Fanconi anemia pathway by monoubiquitination. *Semin. Cancer Biol.* 13:77–82

Guo Q, Xie J, Dang CV, Liu ET, Bishop JM. 1998. Identification of a large Myc-binding protein that contains RCC1-like repeats. *Proc. Natl. Acad. Sci. USA* 95:9172–77

Hatakeyama S, Yada M, Matsumoto M, Ishida N, Nakayama KI. 2001. U box proteins as a new family of ubiquitin-protein ligases. *J. Biol. Chem.* 276:33111–20

Hegde AN, DiAntonio A. 2002. Ubiquitin and the synapse. *Nat. Rev. Neurosci.* 3:854–61

Hegde AN, Goldberg AL, Schwartz JH. 1993. Regulatory subunits of cAMP-dependent protein kinases are degraded after conjugation to ubiquitin: a molecular mechanism underlying long-term synaptic plasticity. *Proc. Natl. Acad. Sci. USA* 90:7436–40

Hegde AN, Inokuchi K, Pei W, Casadio A, Ghirardi M, et al. 1997. Ubiquitin C-terminal hydrolase is an immediate-early gene essential for long-term facilitation in Aplysia. *Cell* 89:115–26

Helliwell SB, Losko S, Kaiser CA. 2001. Components of a ubiquitin ligase complex specify polyubiquitination and intracellular trafficking of the general amino acid permease. *J. Cell Biol.* 153:649–62

Hicke L, Dunn R. 2003. Regulation of membrane protein transport by ubiquitin and ubiquitin-binding proteins. *Annu. Rev. Cell Dev. Biol.* 19:141–72

Hoege C, Pfander B, Moldovan GL, Pyrowolakis G, Jentsch S. 2002. RAD6-dependent DNA repair is linked to modification of PCNA by ubiquitin and SUMO. *Nature* 419:135–41

Huang Y, Baker RT, Fischer-Vize JA. 1995. Control of cell fate by a deubiquitinating enzyme encoded by the fat facets gene. *Science* 270:1828–31

Huber AB, Kolodkin AL, Ginty DD, Cloutier JF. 2003. Signaling at the growth cone: ligand-receptor complexes and the control of axon growth and guidance. *Annu. Rev. Neurosci.* 26:509–63

Huibregtse JM, Scheffner M, Beaudenon S, Howley PM. 1995. A family of proteins structurally and functionally related to the E6-AP ubiquitin-protein ligase. *Proc. Natl. Acad. Sci. USA* 92:2563–67

Jefferis GS, Marin EC, Watts RJ, Luo L. 2002. Development of neuronal connectivity in Drosophila antennal lobes and mushroom bodies. *Curr. Opin. Neurobiol.* 12:80–86

Jiang C, Schuman EM. 2002. Regulation and function of local protein synthesis in neuronal dendrites. *Trends Biochem. Sci.* 27:506–13

Jiang YH, Armstrong D, Albrecht U, Atkins CM, Noebels JL, et al. 1998. Mutation of the Angelman ubiquitin ligase in mice causes increased cytoplasmic p53 and deficits of contextual learning and long-term potentiation [see comments]. *Neuron* 21:799–811

Joazeiro CA, Weissman AM. 2000. RING finger proteins: mediators of ubiquitin ligase activity. *Cell* 102:549–52

Kandel ER. 2001. The molecular biology of memory storage: a dialogue between genes and synapses. *Science* 294:1030–38

Kantor DB, Kolodkin AL. 2003. Curbing the excesses of youth: molecular insights into axonal pruning. *Neuron* 38:849–52

Katzmann DJ, Odorizzi G, Emr SD. 2002. Receptor downregulation and multivesicular-body sorting. *Nat. Rev. Mol. Cell Biol.* 3:893–905

Keleman K, Rajagopalan S, Cleppien D, Teis D, Paiha K, et al. 2002. Comm sorts Robo to control axon guidance at the Drosophila midline. *Cell* 110:415–27

Kidd T, Bland KS, Goodman CS. 1999. Slit is the midline repellent for the Robo receptor in Drosophila. *Cell* 96:785–94

Kidd T, Russell C, Goodman CS, Tear G. 1998. Dosage-sensitive and complementary functions of roundabout and commissureless control axon crossing of the CNS midline. *Neuron* 20:25–33

Kishino T, Lalande M, Wagstaff J. 1997. UBE3A/E6-AP mutations cause Angelman syndrome. *Nat. Genet.* 15:70–73

Klapisz E, Sorokina I, Lemeer S, Pijnenburg M, Verkleij AJ, van Bergen en Henegouwen PM. 2002. A ubiquitin-interacting motif (UIM) is essential for Eps15 and Eps15R ubiquitination. *J. Biol. Chem.* 277:30746–53

Korey CA, Van Vactor D. 2000. From the growth cone surface to the cytoskeleton: one journey, many paths. *J. Neurobiol.* 44:184–93

Liu Y, Fallon L, Lashuel HA, Liu Z, Lansbury PT Jr. 2002. The UCH-L1 gene encodes two opposing enzymatic activities that affect alpha-synuclein degradation and Parkinson's disease susceptibility. *Cell* 111:209–18

Lopez-Salon M, Alonso M, Vianna MR, Viola H, Souza TME, et al. 2001. The ubiquitin-proteasome cascade is required for mammalian long-term memory formation. *Eur. J. Neurosci.* 14:1820–26

Mack TG, Reiner M, Beirowski B, Mi W, Emanuelli M, et al. 2001. Wallerian degeneration of injured axons and synapses is delayed by a Ube4b/Nmnat chimeric gene. *Nat. Neurosci.* 4:1199–206

Malinow R, Malenka RC. 2002. AMPA receptor trafficking and synaptic plasticity. *Annu. Rev. Neurosci.* 25:103–26

Matsuura T, Sutcliffe JS, Fang P, Galjaard RJ, Jiang YH, et al. 1997. De novo truncating mutations in E6-AP ubiquitin-protein ligase gene (UBE3A) in Angelman syndrome. *Nat. Genet.* 15:74–77

Mayer RJ, Laszlo L, Middleton A, Landon M, Hope J, Lowe J. 1992. Ubiquitin, lysosomes and neurodegenerative diseases. *Biochem. Soc. Trans.* 20:645–48

Mizuno E, Kawahata K, Kato M, Kitamura N, Komada M. 2003. STAM proteins bind ubiquitinated proteins on the early endosome via the VHS domain and ubiquitin-interacting motif. *Mol. Biol. Cell* 14:3675–89

Mizuno Y, Hattori N, Mori H, Suzuki T, Tanaka K. 2001. Parkin and Parkinson's disease. *Curr. Opin. Neurol.* 14:477–82

Moss A, Blackburn-Munro G, Garry EM, Blakemore JA, Dickinson T, et al. 2002. A role of the ubiquitin-proteasome system in neuropathic pain. *J. Neurosci.* 22:1363–72

Muralidhar MG, Thomas JB. 1993. The Drosophila bendless gene encodes a neural protein related to ubiquitin-conjugating enzymes. *Neuron* 11:253–66

Muratani M, Tansey WP. 2003. How the ubiquitin-proteasome system controls transcription. *Nat. Rev. Mol. Cell Biol.* 4:192–201

Murphey RK, Froggett SJ, Caruccio P, Shan-Crofts X, Kitamoto T, Godenschwege TA. 2003. Targeted expression of shibire ts and semaphorin 1a reveals critical periods for synapse formation in the giant fiber of Drosophila. *Development* 130:3671–82

Myat A, Henry P, McCabe V, Flintoft L, Rotin D, Tear G. 2002. Drosophila Nedd4, a ubiquitin ligase, is recruited by commissureless to control cell surface levels of the roundabout receptor. *Neuron* 35:447–59

Naze P, Vuillaume I, Destee A, Pasquier F, Sablonniere B. 2002. Mutation analysis and association studies of the ubiquitin carboxy-terminal hydrolase L1 gene in Huntington's disease. *Neurosci. Lett.* 328:1–4

Oh CE, McMahon R, Benzer S, Tanouye MA. 1994. bendless, a Drosophila gene affecting neuronal connectivity, encodes a ubiquitin-conjugating enzyme homolog. *J. Neurosci.* 14:3166–79

Oldham CE, Mohney RP, Miller SL, Hanes RN, O'Bryan JP. 2002. The ubiquitin-interacting motifs target the endocytic adaptor protein epsin for ubiquitination. *Curr. Biol.* 12:1112–16

O'Leary DD, Koester SE. 1993. Development of projection neuron types, axon pathways, and patterned connections of the mammalian cortex. *Neuron* 10:991–1006

Patrick GN, Bingol B, Weld HA, Schuman EM. 2003. Ubiquitin-mediated proteasome activity is required for agonist-induced endocytosis of GluRs. *Curr. Biol.* 13:2073–81

Peng J, Schwartz D, Elias JE, Thoreen CC, Cheng D, et al. 2003. A proteomics approach to understanding protein ubiquitination. *Nat. Biotechnol.* 21:921–26

Pickart CM. 2001. Mechanisms underlying ubiquitination. *Annu. Rev. Biochem.* 70:503–33

Polo S, Sigismund S, Faretta M, Guidi M, Capua MR, et al. 2002. A single motif responsible for ubiquitin recognition and monoubiquitination in endocytic proteins. *Nature* 416:451–55

Raiborg C, Rusten TE, Stenmark H. 2003. Protein sorting into multivesicular endosomes. *Curr. Opin. Cell Biol.* 15:446–55

Rotin D, Staub O, Haguenauer-Tsapis R. 2000. Ubiquitination and endocytosis of plasma membrane proteins: role of Nedd4/Rsp5p family of ubiquitin-protein ligases. *J. Membr. Biol.* 176:1–17

Saigoh K, Wang YL, Suh JG, Yamanishi T, Sakai Y, et al. 1999. Intragenic deletion in the gene encoding ubiquitin carboxy-terminal hydrolase in gad mice. *Nat. Genet.* 23:47–51

Sanes JR, Lichtman JW. 1999. Development of the vertebrate neuromuscular junction. *Annu. Rev. Neurosci.* 22:389–442

Schaefer AM, Hadwiger GD, Nonet ML. 2000. rpm-1, a conserved neuronal gene that regulates targeting and synaptogenesis in C. elegans. *Neuron* 26:345–56

Schnell JD, Hicke L. 2003. Non-traditional functions of ubiquitin and ubiquitin-binding proteins. *J. Biol. Chem.* 278:35857–60

Schwartz DC, Hochstrasser M. 2003. A superfamily of protein tags: ubiquitin, SUMO and related modifiers. *Trends Biochem. Sci.* 28:321–28

Sheng M. 2001. Molecular organization of the postsynaptic specialization. *Proc. Natl. Acad. Sci. USA* 98:7058–61

Shih SC, Katzmann KJ, Schnell JD, Sutanto M, Emr SC, Hicke LH. 2002. Epsins and Vps27/Hrs contain ubiquitin-binding domains that function in receptor endocytosis. *Nat. Cell Biol.* 4:389–93

Shih SC, Prag G, Francis SA, Sutanto MA, Hurley JH, Hicke L. 2003. A ubiquitin-binding motif required for intramolecular ubiquitylation, the CUE domain. *EMBO J.* 22:1273–81

Sloper-Mould KE, Pickart CM, Hicke L. 2001. Distinct functional surface regions on ubiquitin. *J. Biol. Chem.* 276:30483–89

Speese SD, Trotta N, Rodesch CK, Aravamudan B, Broadie K. 2003. The ubiquitin proteasome system acutely regulates presynaptic protein turnover and synaptic efficacy. *Curr. Biol.* 13:899–910

Sudol M, Hunter T. 2000. NeW wrinkles for an old domain. *Cell* 103:1001–4

Terrell J, Shih S, Dunn R, Hicke L. 1998. A function for monoubiquitination in the internalization of a G protein-coupled receptor. *Mol. Cell* 1:193–202

Thomas JB, Wyman RJ. 1984. Mutations altering synaptic connectivity between identified neurons in Drosophila. *J. Neurosci.* 4:530–38

Truman JW. 1990. Metamorphosis of the central nervous system of Drosophila. *J. Neurobiol.* 21:1072–84

Turrigiano GG. 1999. Homeostatic plasticity in neuronal networks: the more things change, the more they stay the same. *Trends Neurosci.* 22:221–27

van Kerkhof P, Govers R, Alves Dos Santos CM, Strous GJ. 2000. Endocytosis and degradation of the growth hormone receptor are proteasome-dependent. *J. Biol. Chem.* 275:1575–80

Wan HI, DiAntonio A, Fetter RD, Bergstrom K, Strauss R, Goodman CS. 2000. Highwire regulates synaptic growth in Drosophila. *Neuron* 26:313–29

Watts RJ, Hoopfer ED, Luo L. 2003. Axon pruning during Drosophila metamorphosis: evidence for local degeneration and requirement of the ubiquitin-proteasome system. *Neuron* 38:871–85

Weissman A. 2001. Themes and variations on ubiquitylation. *Nat. Rev. Mol. Cell Biol.* 2:169–78

Wendland B. 2002. Epsins: adaptors in endocytosis? *Nat. Rev. Mol. Cell Biol.* 3:971–77

West AE, Chen WG, Dalva MB, Dolmetsch RE, Kornhauser JM, et al. 2001. Calcium regulation of neuronal gene expression. *Proc. Natl. Acad. Sci. USA* 98:11024–31

Wheeler TC, Chin LS, Li Y, Roudabush FL, Li L. 2002. Regulation of synaptophysin degradation by mammalian homologues of seven in absentia. *J. Biol. Chem.* 277:10273–82

Wilson SM, Bhattacharyya B, Rachel RA, Coppola V, Tessarollo L, et al. 2002. Synaptic defects in ataxia mice result from a mutation in Usp14, encoding a ubiquitin-specific protease. *Nat. Genet.* 32:420–25

Wolf B, Seeger MA, Chiba A. 1998. Commissureless endocytosis is correlated with initiation of neuromuscular synaptogenesis. *Development* 125:3853–63

Yu TW, Bargmann CI. 2001. Dynamic regulation of axon guidance. *Nat. Neurosci.* 4(Suppl.):1169–76

Zhai Q, Wang J, Kim A, Liu Q, Watts R, et al. 2003. Involvement of the ubiquitin-proteasome system in the early stages of Wallerian degeneration. *Neuron* 39:217–25

Zhang LI, Poo MM. 2001. Electrical activity and development of neural circuits. *Nat. Neurosci.* 4(Suppl.):1207–14

Zhao Y, Hegde AN, Martin KC. 2003. The ubiquitin proteasome system functions as an inhibitory constraint on synaptic strengthening. *Curr. Biol.* 13:887–98

Zhen M, Huang X, Bamber B, Jin Y. 2000. Regulation of presynaptic terminal organization by C. elegans RPM-1, a putative guanine nucleotide exchanger with a RING-H2 finger domain. *Neuron* 26:331–43

Annu. Rev. Neurosci. 2004. 27:247–78
doi: 10.1146/annurev.neuro.27.070203.144303

First published online as a Review in Advance on February 23, 2004

CELLULAR MECHANISMS OF NEURONAL POPULATION OSCILLATIONS IN THE HIPPOCAMPUS IN VITRO

Roger D. Traub,[1] Andrea Bibbig,[1] Fiona E.N. LeBeau,[2] Eberhard H. Buhl,[2*] and Miles A. Whittington[2]

[1]*Department of Physiology and Pharmacology, State University of New York Downstate Medical Center, Brooklyn, New York 11203; email: roger.traub@downstate.edu, andrea.bibbig@downstate.edu*

[2]*School of Biomedical Sciences, The Worsley Building, University of Leeds, Leeds LS2 9NQ, United Kingdom; email: f.e.n.lebeau@leeds.ac.uk, m.a.whittington@leeds.ac.uk*

Key Words gamma oscillation, 40 Hz, theta oscillation, ripple, gap junctions

■ **Abstract** A variety of population oscillations, at frequencies ~5 Hz up to 200 Hz and above, can be induced in hippocampal slices either by (*a*) manipulation of the ionic environment, or (*b*) by stimulation of metabotropic receptors; brief oscillations can even occur spontaneously. In this review, we consider in vitro theta (4–12 Hz), gamma/beta (15–70 Hz), and very fast oscillations (VFO) (>70 Hz). Many in vitro oscillations are gated by synaptic inhibition but are influenced by electrical coupling as well; one type depends solely on electrical coupling. For some oscillations dependent upon inhibition, the detailed firing patterns of interneurons can influence long-range synchronization. Two sorts of electrical coupling are important in modulating or generating various in vitro oscillations: (*a*) between interneurons, primarily between dendrites; and (*b*) between axons of pyramidal neurons. VFO can exist in isolation or can act as generators of gamma frequency oscillations. Oscillations at gamma frequencies and below probably create conditions under which synaptic plasticity can occur, between selected neurons—even those separated by significant axonal conduction delays.

IMPORTANCE OF STUDYING OSCILLATIONS IN VITRO

Network oscillations in vivo have been proposed to be important in sensory processing, for example in the olfactory system (Laurent & Davidowitz 1994, Freeman 1972), in sensory and perceptual binding (Gray et al. 1989, Miltner et al. 1999, Rodriguez et al. 1999, Roelfsema et al. 1994, Singer & Gray 1995), in motor programming (Murthy & Fetz 1996), in associative learning (Buzsáki 2002, Larson &

*Deceased.

0147-006X/04/0721-0247$14.00

Lynch 1988), and in epileptogenesis (Grenier et al. 2001, 2003; Traub et al. 2001b). One key proposal (Singer & Gray 1995) has been that oscillations provide a temporal framework, with regard to which neurons may either fire in synchrony or not, and that the synchrony of firing carries meaning, in that such synchrony (or its absence) defines whether particular groups of neurons are devoted to a common task. Furthermore, if this idea is correct, both local and long-range synchronization are important to study, as both could have functional consequences. In vitro oscillations may provide clues to the cellular mechanisms of in vivo oscillations. In addition, study of in vitro oscillations can lead to discoveries that are possibly significant in functions other than oscillations; one example is electrical coupling between axons, which was hypothesized to occur after study of in vitro 200-Hz spontaneous oscillations (Draguhn et al. 1998, Schmitz et al. 2001, Traub et al. 1999c).

SYNCHRONIZATION: LOCAL AND LONG-RANGE

Oscillations provide one means by which specific phase relations can be established between the firing times of nearby neurons. In particular, assemblies of nearby neurons may be prone to fire in synchrony during an oscillation. (This is convenient experimentally because it allows the oscillation to be detected with a field potential electrode.) In addition, synchronization can sometimes develop in a spatially distributed region, or between two relatively discrete but interconnected sites. As noted above, both local synchrony and two-site (or multisite and distributed) synchrony are likely important: The collective synaptic output of a synchronized population could be especially salient for downstream and within-assembly neurons ["coincidence detection" (Abeles 1982, König et al. 1996)]; and, as argued below, synchronization between regions creates conditions favorable to synaptic plasticity.

From a mechanistic point of view, two problems exist, however. The first problem is understanding synchrony within a local population of neurons; here, "local" means that axonal conduction delays can be ignored. Second is the problem of understanding synchrony between sites separated by significant conduction delays. The second problem depends on the solution to the first but is inherently more difficult. As we shall see, in tetanically induced oscillations, long-range synchrony depends critically on the phasic properties of interneuron EPSCs (excitatory postsynaptic conductances) and on details in the firing patterns of interneurons. The hippocampal slice is a useful preparation for investigating this problem, in part because conduction delays are surprisingly long. Thus, Schaffer collateral conduction velocities are estimated to be about 0.25 to 0.5 mm/ms (Andersen et al. 1978, Knowles et al. 1987); furthermore, an obliquely cut rat hippocampal slice might have 3.5 mm of CA1 tissue. Maximal conduction delays as long as ~8.75 ms are therefore possible, at least in principle, and still longer delays might be obtained in longitudinally cut slices.

METHODS OF INDUCING AND EXPERIMENTALLY CHARACTERIZING OSCILLATIONS

There are four main experimental paradigms in which hippocampal in vitro network oscillations have been studied. We list them here briefly; a full discussion of specific types of oscillation follows later. The reason for an initial consideration at this point is this: Knowing the experimental paradigm helps one to focus on which system parameters might be relevant for generating and structuring the oscillation.

First Paradigm

Oscillations may occur spontaneously, in normal artificial cerebrospinal fluid (ACSF).

Second Paradigm

Drugs may be puffed (pressure ejected) onto the tissue or introduced into the bath. Drugs used for this purpose include activators of metabotropic glutamate and kainate receptors, as well as muscarinic receptors. In this paradigm, other subtypes of receptors (including AMPA/kainate, NMDA, and $GABA_B$) might be pharmacologically blocked.

Third Paradigm

Tetanic (repetitive electrical) stimulation can be briefly applied. Such stimulation activates neurons (including axons and presynaptic terminals) directly and, additionally, leads to release of neuroactive substances (glutamate, GABA, acetylcholine, etc.). It is the neuroactive substances that induce the oscillation after a pause (about 100 ms), and the oscillation is not locked to the electrical stimulation. Tetanic stimulation can be combined with drug application to reduce neuronal excitability or to open gap junctions. It can also be combined with surgical disruption of intrahippocampal pathways. The tetanic paradigm can be applied to two tissue sites at once and therefore is valuable for the study of oscillatory synchronization between two sites.

Fourth Paradigm

The ionic milieu can be altered, either via changing bath ionic composition (for instance, removing calcium ions), or by pressure ejection of hypertonic solutions (for instance, hypertonic K^+ solution). To some extent, the ejection of hypertonic K^+ solution overlaps the tetanic paradigm, in that such a solution is expected to depolarize at least some neurons (including axons) and to release neuroactive substances such as GABA (Kaila et al. 1997).

NETWORK MODELS OF OSCILLATIONS

Network models have played a pivotal role in understanding the physical principles involved in in vitro oscillations and in generating experimentally testable predictions (Traub et al. 1999a). One depends on an accurate model for grasping how an oscillation occurs at all, and how particular structural features of the network determine the frequency and number of cells participating on each wave; one can also see, in simulations, how phase relations between different neuronal types arise. Our approach has been to use a model with the number of cells roughly equal to the number of cells in the experimental system and to employ models of individual neurons that contain branching dendrites and a segment of axon, along with a repertoire of Na^+, Ca^{2+}, and K^+ currents (Traub et al. 1999a). Models of this sort, once they are known to be consistent with experiment, can act as a springboard for the construction of far simpler reduced models that are amenable to mathematical analysis (Ermentrout & Kopell 1998, Kopell et al. 2000). Network models are most important in complex experimental situations where the biological network is spatially distributed (giving rise to a continuum of conduction delays) and where there are multiple types of between-neuron interactions: for example, where there are both excitatory and inhibitory synaptic interactions, with different time courses in different neurons; and/or where both chemical synapses and electrical coupling are important (Hormuzdi et al. 2001; Traub et al. 2001, 2003a).

Relevant Neuronal Intrinsic Properties

Amid the vast complexity of firing repertoires (and subthreshold repertoires) displayed by neurons, and amid the large ensemble of membrane currents responsible for this complexity, certain issues in neuronal intrinsic properties stand out for their clear relevance to oscillations. Included in these features, we list the following:

1. How, and in what part of the neuronal membrane, do spontaneous action potentials arise? For example, kainate is known to increase interneuron axonal excitability and could give rise to ectopic axonal spikes (Semyanov & Kullmann 2001); principal cell ectopic axonal spikes are, in turn, important in the generation of very fast oscillations and of certain gamma oscillations. It is interesting that GABA can, at least in certain conditions, excite pyramidal cell axons (Avoli et al. 1998, Traub et al. 2003b).
2. Which processes allow metabotropic receptors to excite cells, depolarizing them and blocking selected K^+ currents? The loss of K^+ currents can reduce accommodation, allowing cells to fire on each wave of an oscillation, whereas the depolarization can promote action potentials during the relaxation phase of IPSPs (inhibitory postsynaptic potentials) (Whittington et al. 1997a). These considerations are relevant in both tetanic and persistent gamma oscillations.

3. To what extent is dendritic electrogenesis present in the oscillation condition? Thus, pyramidal cell dendritic Ca^{2+} spikes contribute to one type of in vitro theta (Gillies et al. 2002) and may also play a role in vivo (Kamondi et al. 1998, Penttonen et al. 1998); Na^+ dendritic electrogenesis may influence the gain of interneuron EPSPs and their tendency to fire doublets (Martina et al. 2000, Traub & Miles 1995).
4. How is intrinsic rhythmicity generated? This process appears important in in vitro theta, and its dependence on one or more h currents (hyperpolarization-activated cation currents) in stratum oriens interneurons (Gillies et al. 2002).

Chemical Synaptic Interactions

In the oscillations discussed below, when chemical synapses play a role, that role falls (roughly) into one of two categories. First are slow (hundreds of ms to seconds) or persistent (hours) effects mediated by metabotropic glutamate, muscarinic, and kainate receptors. These effects are probably mediated through actions on multiple sites in neurons, leading to inward currents, to blocking of various ionic currents, and to effects on neurotransmitter release. Second are relatively fast phasic synaptic actions, mediated primarily by $GABA_A$ and AMPA receptors. (NMDA receptors, important in CA1 in vitro theta, lie in an intermediate category.)

Certain particular aspects of chemical synaptic interactions require special note, as these aspects are of direct relevance to population oscillations:

1. After tetanic stimulation, metabotropic glutamate receptors contribute to large-amplitude slow EPSPs in both pyramidal cells and interneurons, EPSPs that promote action potentials in both cell types (Whittington et al. 1995, 1997a). MGluRs additionally act to suppress frequency accommodation in pyramidal neurons (Whittington et al. 1997a) by virtue of their block of selected K^+ currents (Charpak et al. 1990); suppression of accommodation, in turn, allows action potentials to follow gamma waves 1:1. Presynaptic actions of mGluRs tend to reduce fluctuations in synaptic conductances that might otherwise occur, wave to wave, during a gamma oscillation (discussed in Traub et al. 1999a).
2. Again, after tetanic stimulation, pyramidal cell spike afterhyperpolarizations (AHPs) are reduced for some hundreds of ms, during the evoked gamma oscillation. This effect may also be secondary to mGluR activation. Recovery of spike AHPs facilitates the transition from gamma to beta frequencies (Whittington et al. 1997b, Traub et al. 1999b; see also below).
3. Kainate receptor activation has been reported to increase the excitability of interneuron axons (Semyanov & Kullmann 2001); kainate also elevates the excitability of pyramidal cell axons, although this effect may be indirectly mediated by another neuroactive substance, such as GABA, whose release is triggered by kainate acting elsewhere (for example, on interneurons) (Traub et al. 2003b) (see also Avoli et al. 1998, Pinault & Pumain 1989, Sakatani

et al. 1992, Stasheff et al. 1993). Axonal excitability—in particular, the ability to generate spontaneous action potentials—appears to play a key role in the generation of persistent gamma oscillations (Traub et al. 2000).

4. The time course of $GABA_A$-receptor-mediated IPSCs is an important factor in determining the frequency of gamma oscillations. The way in which this occurs is dependent on the experimental model, however. Consider first gamma oscillations in pharmacologically isolated interneuron networks, so-called ING (interneuron network gamma). Population interneuronal IPSCs in this model have a decay time constant of about 10 ms, which, in simulations, readily leads to the observed network frequencies of around 40 Hz (Traub et al. 1996a; see also Wang & Buzsáki 1996). On the other hand, Bartos et al. (2001, 2002) have shown that unitary basket cell/basket cell IPSCs actually have two time constants; the larger, faster component has a decay time constant of 1–2 ms, the smaller, slower component about 8 ms; the kinetics of population IPSCs during ING were not measured in this study. In their simulations of a ring of interneurons containing inhibitory and also gap-junctional coupling (Bartos et al. 2002), a wide range of frequencies could be observed, wider than was possible to obtain experimentally. It is possible that experimental ING is controlled by a class of interneurons that includes, but is not restricted to, basket cells; it is also possible that the large, fast IPSC depresses during the course of the oscillation so that the frequency is largely determined by the slower time constant. Second, consider gamma oscillations that depend on both pyramidal cells and interneurons (including tetanic and persistent gamma, see below). In this case, it is believed that the time constant of IPSCs in pyramidal cells is the parameter of primary importance in determining frequency (Fisahn et al. 1998, Traub et al. 2000).
5. The respective amplitudes and time courses of EPSCs in pyramidal cells, as compared with interneurons, are also functionally relevant for in vitro oscillations. Pyramidal cell EPSCs increase during the course of tetanically induced gamma oscillations and appear to contribute to the structure and synchrony of beta oscillations (Whittington et al. 1997b, Traub et al. 1999b). These EPSCs are probably also important in the generation of synchronized epileptiform bursts that sometimes follow tetanic beta (Traub et al. 1999a) or that occur interspersed with kainate-induced gamma oscillations in connexin36 knockout mice (Pais et al. 2003). Large-amplitude rapid EPSCs in interneurons (Geiger et al. 1997, Miles 1990) are important for the generation of spike doublets in interneurons, which occur during gamma oscillations elicited by simultaneously tetanizing two interconnected sites. Interneuron doublets contribute to the ability of the two sites to oscillate synchronously, despite axonal conduction delays (Ermentrout & Kopell 1998, Fuchs et al. 2001, Traub et al. 1996b); synaptic plasticity of interneuron EPSCs during the course of a gamma oscillation is also believed to contribute to producing synchrony (Bibbig et al. 2001).

6. Finally, synaptic inhibition can, in principle, regulate the functional extent of electrical coupling between pyramidal cell axons via a shunting effect (Traub et al. 2000; Traub et al. 2003b).

Electrical Coupling

Two types of electrical coupling in the hippocampus influence population oscillations: (*a*) between interneurons, with gap junctions located primarily between dendrites (Fukuda & Kosaka 2000, Kosaka 1983); and (*b*) between pyramidal cells (and also probably dentate granule cells), with the site of coupling in proximal axons (Schmitz et al. 2001), but with gap junctions themselves not yet identified ultrastructurally. The effects on oscillations that these two types of coupling exert appear to be quite different (Traub et al. 2003a), as shall be discussed further below. Here we consider, from a basic physical point of view, why the effects might be different.

At the level of the putative gap junctions themselves, coupling between interneurons and between pyramidal neurons appears to be physically similar: The coupling appears to behave as a resistance between the interiors of nearby neurons. Because, however, coupling between interneurons is electrically close to the soma—a fraction of a space constant removed from it—and because this coupling is relatively strong [between cortical interneurons, values of about 0.7–1.6 nS have been estimated (Galarreta & Hestrin 1999, Gibson et al. 1999)], the coupling can be detected with DC measurements during an experiment with dual simultaneous impalements of two coupled neurons. That is, a slow hyperpolarization or depolarization in one cell produces a detectable slow hyperpolarization or depolarization, respectively, in the connected cell. What this implies is potential fluctuations in one cell can induce corresponding, albeit smaller, potential fluctuations in the second cell that are independent of the active membrane properties of the second cell.

In contrast, in the case of axonal electrical coupling, at least so far, dual impalements of nearby principal neurons have not provided evidence of DC coupling. Although this negative evidence might suggest a rarity, or even absence, of the coupling in question, other evidence does indeed indicate the existence of coupling; we would suggest that the lack of DC coupling seen between principal neuron pairs results instead from the electrical distance of the putative gap junction from the recording sites.

How, then, can the existence of electrical coupling be inferred? MacVicar & Dudek (1982) used antidromic stimulation of dentate granule cell axons to suggest the existence of electrical coupling between these neurons, based on the appearance of spikelets in neurons at slightly longer latency than that of antidromic spikes. Knowles & Schwartzkroin (1981) used oriens/alveus (i.e., antidromic) stimulation to demonstrate the existence in CA1 pyramidal neurons of spikelets, or fast prepotentials.

Schmitz et al. (2001) used antidromic stimulation to induce spikelets in hippocampal principal neurons (CA1 and CA3 pyramidal cells and dentate granule cells), mostly but not entirely in low [Ca^{2+}] media, combined with single and dual patch recordings and with evidence of dye coupling between axons. Their data provide evidence that spikelets arise from electrical coupling between axons. The data, and some theoretical implications, have been reviewed by Traub et al. (2002).

An earlier dye-coupling study supports the notion of sparse electrical coupling between pyramidal neurons. Church & Baimbridge (1991) estimated that CA1 pyramidal neurons were coupled to 1.6 others, on average, at physiological pH, a low density. This is the density of coupling that we use in most of our network simulations. The average number of coupled neurons increases to 3.2 at alkaline pH (7.9), as would be expected if dye coupling is mediated by gap junctions (Spray et al. 1981). Note that the density of pyramidal cell electrical coupling is low, in comparison with synaptic coupling, and is especially low in comparison to interneuron electrical coupling (at least in neocortex). To see this, observe first that between CA1 pyramidal cells the estimated probability of a synaptic connection is about 1% (Deuchars & Thomson 1996), so that in a population of thousands of neurons (as would exist in a hippocampal slice) each neuron should receive dozens of synaptic inputs. Observe second that cortical interneurons are electrically coupled to more than 50% of nearby interneurons (Gibson et al. 1999, Venance et al. 2000).

In addition, connexin36 is a major neuronal gap-junction protein, which is located primarily on interneurons in mature animals (Condorelli et al. 1998, Venance et al. 2000). On the other hand, dye coupling between interneurons is difficult to demonstrate (Gibson et al. 1999). Given that dye coupling does exist between principal hippocampal neurons (Church & Baimbridge 1991, MacVicar & Dudek 1980, Perez Velazquez et al. 1997, Schmitz et al. 2001), one suspects that electrical coupling between principal neurons is mediated by some protein other than connexin36. Furthermore, very fast oscillations involving hippocampal pyramidal neurons, which are known to depend on electrical coupling (Draguhn et al. 1998), have a structure in connexin36 knockout mice that is indistinguishable from wild-type mice (Hormuzdi et al. 2001), which further suggests that pyramidal neuron electrical coupling does not depend on connexin36. Finally, spikelets occur in pyramidal neurons from connexin36 knockout mice (Pais et al. 2003).

In summary, hippocampal pyramidal cells appear to be functionally coupled, electrically, at axonal sites. This coupling involves active membrane properties, although not necessarily voltage-dependent properties of the junctional site itself. The low density of coupling, along with the involvement of active membrane processes, distinguishes axonal coupling from interneuron/dendritic coupling. For oscillations to be generated in axonal networks, it is necessary that spikes be able to cross from axon to axon (Traub et al. 1999c).

SELECTED EXAMPLES OF IN VITRO HIPPOCAMPAL NETWORK OSCILLATIONS

Figure 1 summarizes the interactions between neurons that are important, in various combinations, for hippocampal oscillations in vitro: recurrent synaptic excitation of pyramidal cells and of interneurons; synaptic inhibition of pyramidal

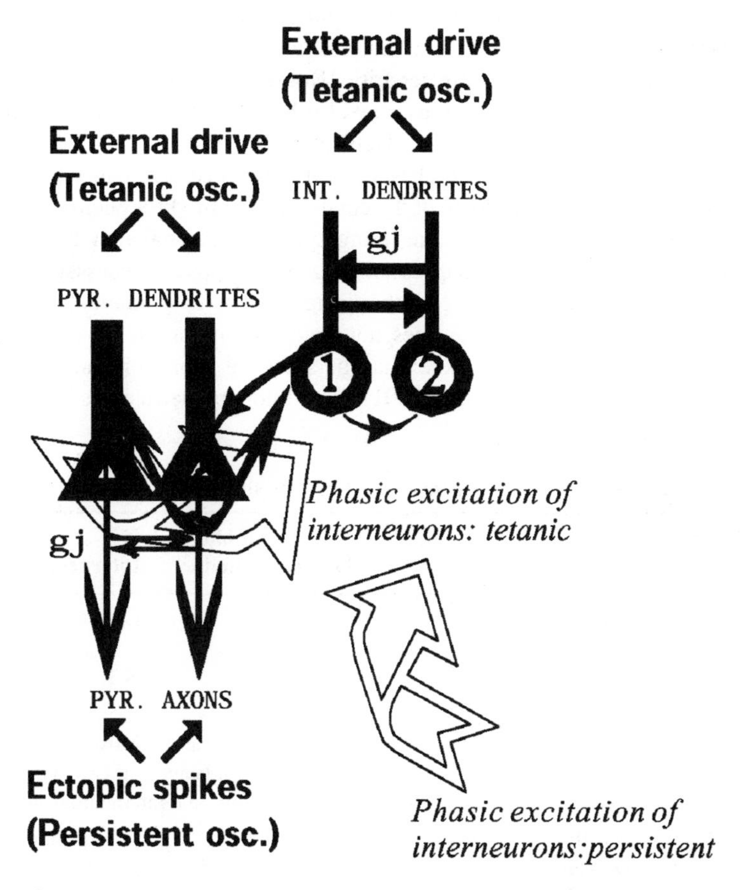

Figure 1 Structural features of oscillating hippocampal networks. Not all structural features are pertinent, however, to each given oscillation. There are pyramidal neurons and interneurons; the interneurons are of different types, but only one type is illustrated. The pyramidal cells synaptically excite each other and the interneurons (*curved arrows*). The interneurons synaptically inhibit the pyramidal neurons and each other (*curved arrows*). Gap junctions (gj) occur between dendrites of selected interneurons. Pyramidal cell axons are electrically coupled as well, presumably by morphological gap junctions (although this has not been proven). During, for example, tetanically driven oscillations, pyramidal cells and interneurons are depolarized by neuroactive substances, such as glutamate acting on metabotropic glutamate receptors. Phasic excitation of interneurons, in this case, comes about via orthodromic spikes in the pyramidal cells (*left open arrow*). During persistent oscillations, ectopic spikes are believed to occur in the axons of pyramidal cells; such spikes can apparently spread from axon to axon. The phasic excitation of interneurons can thus occur directly from the pyramidal cell axonal plexus without the necessity of orthodromic pyramidal cell spikes (*right open arrow*).

cells and interneurons; electrical coupling (presumably gap junction–mediated) between pyramidal cell axons; and electrical coupling between interneurons that is apparently predominantly dendritic (Kosaka 1983, Tamas et al. 2000). [Axonal coupling between interneurons may also be present in certain circumstances, as suggested by indirect evidence (Traub et al. 2001a).]

Table 1 lists some of the basic features for a series of in vitro hippocampal oscillations, including some of the conditions required for eliciting the oscillation (and for suppressing other, interfering, types of neuronal activity). The table also lists whether chemical synapses and/or gap junctions are necessary for generation of the oscillation and presents the firing patterns of pyramidal cell somata and of interneurons. The table does not state one important set of conditions, but we wish to emphasize it here: that the slices be kept moist and not excessively stimulated (Whittington et al. 2001). In addition, preperfusion of the animal (prior

TABLE 1 Properties of some in vitro hippocampal oscillations

Oscillation type	Conditions	Chemical synapses	GJ	Pyramidal cell somata	Interneurons
ING	mGluR + iGluR block	Yes	Yes	—	Each wave
	$[K^+]$ puff + iGluR block	Yes	Yes	—	N.T.
tetanic γ	Elect. stim.	Yes	No?	Variable: can be most beats	Most/all beats; doubles if two-site
				[pyr. cells and interneurons in phase]	
tetanic β	Elect stim. (two-site, strong)	Yes	No?	Skip beats	γ frequency
"pure" VFO	spontaneous or low $[Ca^{2+}]$	No	Yes	Some waves; spikelets	—
persistent γ	kainate, mAChR, mGluR	Yes	Yes	Rare firing	Fire 1/2 to most waves (axo-axonic even more)
				[pyr. cells slightly lead interneurons]	
puff γ	Local hypertonic $[K^+]$	Yes	Yes	Rare firing	N.T.
theta (CA1)	mGluR + AMPA block	Yes	Yes	Rare firing	S. oriens interneurons (O/LM) fire each wave
theta (CA3)	mAChR	Yes (not $GABA_A$)	N.T.	May fire little, or may burst	Probably not necessary

Abbreviations: N.T., not tested; puff, pressure ejection; VFO, very fast oscillation (>70 Hz).

to removal of the brain) and preparation of slices in sucrose solutions may help to preserve interneuron function (Kuenzi et al. 2000), a factor of likely importance in oscillations that depend critically on synaptic inhibition.

INTERNEURON NETWORK GAMMA Interneuron network gamma (ING) is the prototypical in vitro gamma ($\sim$30–$\sim$70 Hz) oscillation (Whittington et al. 1995), although the in vivo relevance of ING remains unclear. By definition, ING is the type of gamma that occurs when an interneuron population is pharmacologically disconnected from pyramidal neurons by blockade of ionotropic glutamate receptors, but in which the interneurons are still excited enough to fire. Between-interneuron interactions—mediated by $GABA_A$ receptors (possibly along with gap junctions)—also remain intact. The physical principles underlying ING are similar to those described by (among others) Rinzel and colleagues (Wang & Rinzel 1993) in a model of spindle oscillations in the surgically isolated nucleus reticularis thalami, a structure consisting entirely of GABAergic neurons, in which spindle oscillations have been postulated to arise in vivo (Steriade et al. 1987). [See also Chow et al. 1998, Lytton & Sejnowski 1991, van Vreeswijk et al. 1994, and White et al. 1998.]

Several experimental models exist, differing primarily in the means by which the interneurons are excited enough that they can fire repeatedly. Figure 2 shows oscillatory field potentials from three such models, all during blockade of AMPA receptors, and in some cases with additional blockade of NMDA and $GABA_B$ receptors. The three models are pressure ejection of hypertonic K^+ solution, pressure ejection of glutamate (which, during blockade of ionotropic glutamate receptors, acts as a metabotropic glutamate agonist), and bath application of the Group I metabotropic glutamate receptor agonist DHPG.

Both in model and in experiment (Traub et al. 1996a) it can be shown that ING frequency depends on particular parameters in predictable ways: the conductance of $GABA_A$ receptor-gated channels, their decay time constant for the conductance, and the driving currents (or tonic excitatory conductance) delivered to the interneurons. Two technical issues in this earlier analysis are of possible relevance to later studies of ING: (*a*) Gap junctions were not included in the model at that time; and (*b*) the decay time constant for $GABA_A$ receptor IPSCs was determined from population IPSCs as they occurred during the course of the oscillation—the time constants so measured are about 10 ms. Such IPSCs are not necessarily identical to those determined from interneuron pair recordings (see also above), in that metabotropic receptors are activated during the oscillation, repetitive activation of the receptors is occurring, and the population IPSC probably reflects contributions from different types of presynaptic interneurons.

Another factor of importance in the structure of ING is so-called heterogeneity, including dispersion in the excitabilities of the interneurons. Indeed, ING appears to be extremely sensitive to heterogeneity of tonic excitation to different cells (Wang & Buzsáki 1996, White et al. 1998). Other types of heterogeneity also interfere with ING structure, including nonuniformities introduced by axonal conduction delays

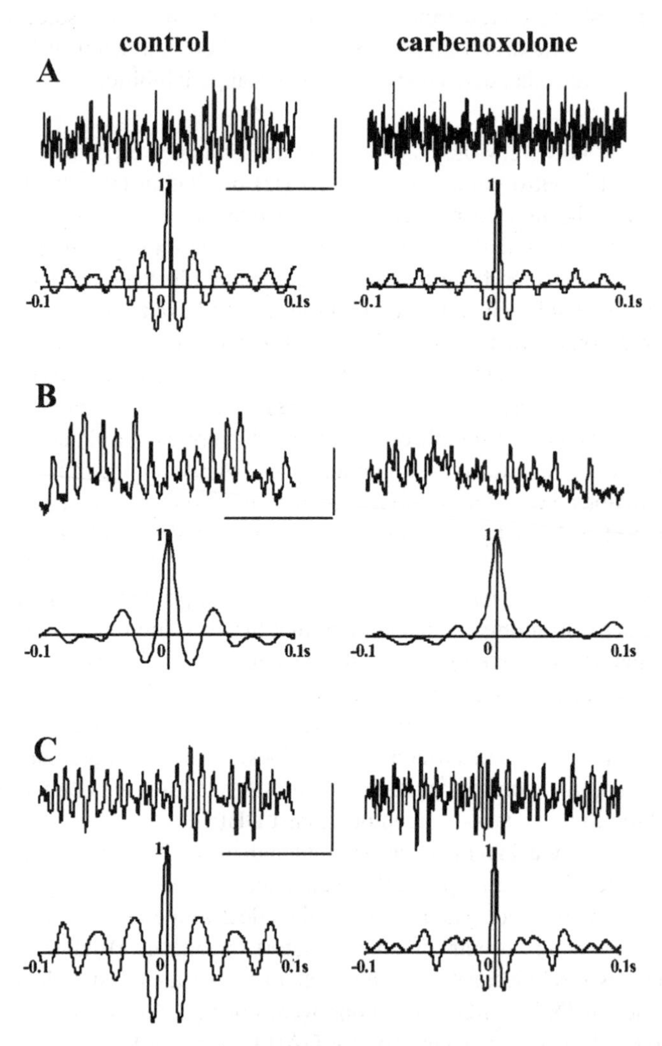

Figure 2 Three examples of interneuron network gamma (ING). Each example shows the oscillating field potential or synaptic current (and its autocorrelation, 0.5–1 s of data), with and without the gap junction–blocking compound carbenoxolone (0.1–0.2 mM), in order to demonstrate loss of coherence of the oscillation when gap junctions are blocked. (*A*) Pressure ejection of 1.5 M potassium solution to CA3 s. radiatum, showing s. radiatum field potential; 0.2 mV, 200 ms. Ionotropic glutamate receptors blocked. (*B*) Pressure ejection of glutamate to CA1 s. pyramidale, showing IPSCs in a CA1 pyramidal cell; 0.3 nA, 200 ms. Ionotropic glutamate receptors and $GABA_B$ receptors blocked. (*C*) Bath application of the metabotropic agonist DHPG, along with NBQX. CA1 s. pyramidale field potential (filtered at 30–90 Hz); 0.1 mV, 200 ms. From Traub et al. (2001c), with permission.

and spatial restrictions on axonal spread (Traub et al. 2001c). Again, however, the earlier analyses were performed without taking into account dendritic gap junctions between interneurons. It appears that even a low density of such gap junctions, with conductance a fraction of a nS, can, at least to some extent, compensate for heterogeneity (Traub et al. 2001c). An experimental demonstration of this can be seen in Figure 2, where, in each of the three experimental ING models, gap-junction blockade with carbenoxolone significantly degrades the rhythmicity of the oscillation. Nevertheless, even without gap-junction blockade, ING does not appear to be able to synchronize in hippocampal CA1 in vitro over distances of 1.5 mm or more (see Figure 7 in Traub et al. 1999b).

TETANIC GAMMA/BETA One of the motivations for studying in vitro gamma oscillations was to understand better in vivo gamma oscillations—not just ongoing hippocampal gamma (Bragin et al. 1995, Csicsvari et al. 2003), but also transient gamma oscillations evoked in neocortex by, for example, sensory stimulation (Eckhorn et al. 1988, Gray & Singer 1989, Gray et al. 1989). During a neocortical oscillation, both pyramidal neurons and interneurons participate, as evidenced by the field potentials and also by the occurrence of mixed trains of EPSPs and IPSPs during intracellular recordings (Jagadeesh et al. 1992). Network simulations furthermore showed that long-distance synchronization of gamma oscillations—something that occurs in vivo and is believed to be of functional significance (Gray et al. 1989)—could occur when both pyramidal neurons and interneurons participated, provided certain conditions are met (for example on the time course of EPSCs in interneurons—see below). It therefore became essential to develop experimental in vitro models of gamma oscillations in which pyramidal neurons participated, and in which it would be possible to examine long-range synchrony. Two-site tetanic stimulation in CA1 was the first experimental preparation that allowed such study in the hippocampal slice; the sites are typically 1.5–3.5 mm apart. Tetanically evoked oscillations are, however, more complex than ING for several reasons: (*a*) both pyramidal cells and interneurons are involved, hence, both synaptic excitation and inhibition contribute [and it should be noted that the time course of EPSCs is slower in pyramidal cells than in interneurons (Deuchars & Thomson 1996, Geiger et al. 1997)]; (*b*) the tetanic stimulation releases neuroactive substances as well as exciting cells directly; (*c*) axon conduction delays matter, in that conduction delays of 4 ms are easily possible, with stimulation sites 2 mm apart and axon conduction velocity 0.5 mm/ms or less (Andersen et al. 1978, Knowles et al. 1987); (*d*) many technical details in performance of the experiments also matter, such as fluid level in the chamber and slice preparation in sucrose; (*e*) strong two-site stimulation evokes not only gamma oscillations, but also a so-called gamma/beta shift (or just beta shift); and (*f*) the neuronal network itself is altered by the stimulation in such a way as to influence the patterning of future oscillations. Some of these complexities are briefly considered below.

Figure 3*A* illustrates the appearance of gamma-frequency field potentials during two-site stimulation of CA1 in vitro, using stimuli that are close to threshold for evoking a response: The oscillation under these conditions is transient, lasting hundreds of ms, and consists of population spikes. Provided the stimulus intensity is at least approximately balanced between sites, the oscillations at the two sites are synchronized to within ~1 ms, as can occur in vivo (Gray et al. 1989; Traub et al. 1996b; Whittington et al. 1997a,b). The oscillations occur at long (50–150 ms) and variable latency after the end of the stimulus, a figure similar to in vivo pontine-tegmentum-evoked gamma oscillations in cats (Steriade & Amzica 1996). Whereas multiunit activity occurs in vivo during a visually evoked gamma oscillation (Gray & Singer 1989), indicating the participation of multiple principal neurons, the participation of principal neurons in the in vitro oscillation is more extreme (but see below).

Intracellular recordings (Whittington et al. 1997a) provide further information on the tetanic oscillation. First, the tetanus evokes a slow (hundreds of ms) depolarization, larger than 10 mV, in both pyramidal cells and interneurons. Most, but not all, of this depolarization is blocked in pyramidal cells when blockers of metabotropic glutamate receptors are applied (see also Congar et al. 1997, Pozzo Miller et al. 1995). The remainder of the depolarization in pyramidal cells appears to involve depolarizing GABA responses, perhaps induced by $[K^+]_o$ rises (Kaila et al. 1997), as well as by cholinergic depolarization. Nearby pyramidal neurons and interneurons fire closely in phase with each other.

Figure 3*B* shows that stronger, but still balanced, two-site stimulation leads also to a transient gamma oscillation, but the gamma is followed by a period of slower (beta) oscillation, during which low-amplitude gamma continues. How this frequency transition is believed to occur is discussed below. Figure 3*C* illustrates a "memory" feature of the system: After gamma/beta has been evoked once with twice-threshold stimulation, a later threshold stimulus can evoke gamma/beta as well, even though a threshold stimulus given to the naive system evokes only gamma. Possible loci of the requisite plasticity are analyzed in Bibbig et al. (2001).

Critical for understanding tetanic gamma is the observation that the tetanic stimulus produces ING in the interneuron population, under conditions when pyramidal cell somatic firing is greatly reduced, as with BP554, a $5HT1_A$ agonist, for example (Whittington et al. 2001). The resulting IPSPs are prominent in pyramidal cells and can, as shown previously (Whittington et al. 1995), gate the timing of pyramidal cell action potentials.

In view of this observation, is tetanic gamma simply an inconsequential modification of ING? This is not the case: Tetanic gamma can synchronize over distances as long as 3.5 mm (Traub et al. 1996b), whereas ING does not; and, furthermore, the ability of tetanic gamma to synchronize over distance depends on AMPA receptors (Traub et al. 1999b, Whittington et al. 1997a). A combination of modeling, mathematical analysis (Ermentrout & Kopell 1998), and experiment indicate why this may be so. The mechanism depends on the intrinsic properties of many interneurons, specifically the ability to generate narrow action potentials that follow

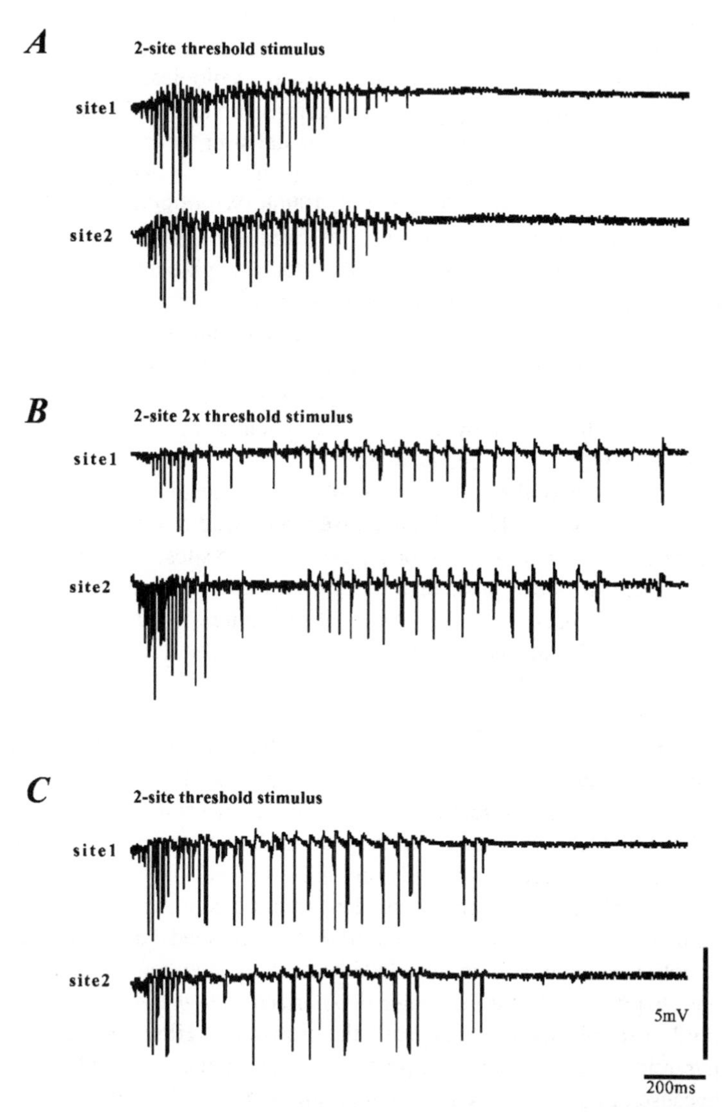

Figure 3 Tetanic gamma/beta oscillation, showing memory. Tetanic stimulation artifacts removed. Recordings are field potentials in the rat CA1 region, s. pyramidale, in vitro, at 2 sites 1.5 mm apart. (*A*) Threshold stimuli given simultaneously at the 2 sites elicit gamma frequency population spikes that last several hundred ms. The oscillations are synchronized (*not shown*). (*B*) When the stimulus strength is doubled, a brief period of gamma is followed by beta-frequency (about 15 Hz in this case) oscillations, also synchronized between sites. (*C*) Following the large stimulus of B, if a threshold stimulus is again given, gamma followed by beta is elicited. From Traub et al. (1999b), with permission.

each other at short intervals; and it depends on the extremely rapid time course and relatively large amplitude of EPSCs in interneurons (Geiger et al. 1997, Miles 1990). A nonintuitive consequence of these physiological properties is this: AMPA receptor–mediated excitation of interneurons at one site, deriving from action potentials in pyramidal neurons at the other site after the intersite conduction delay, can drive action potentials that are precisely timed, and that give rise to interneuron doublets (double spikes) (Traub et al. 1996b, Whittington et al. 1997a); the first spike in the interneuron doublet arises from intrinsic excitability and/or locally generated excitatory inputs. The within-doublet spike interval in turn provides to pyramidal cells at one site information about the firing time of pyramidal cells at the other site. At least under certain circumstances, this information acts as a feedback signal that tends to stabilize the firing pattern in which the two sites fire synchronously (Ermentrout & Kopell 1998). [Interestingly, with long enough conduction, firing patterns exist that are synchronized without a necessity for interneuron doublets (Bibbig et al. 2001).]

A prediction of the interneuron doublet mechanism is this: that small prolongations of interneuron EPSC time course would lead to characteristic instability in the relative timing of population spikes at the two sites, and that the relative phase would shift between positive and negative values. This prediction appears to be borne out (Figure 4) in studies of a genetically altered mouse line, in which GluR-B (GluR2) AMPA receptor subunits are overexpressed in interneurons (Fuchs et al. 2001)—normally, the level of expression of GluR-B in interneurons is low (Racca et al. 1996).

Let us return now to the mechanism of the gamma/beta shift (Figure 3*B*,*C*). It has been shown experimentally (Whittington et al. 1997b) that during tetanically induced gamma, recurrent EPSPs in pyramidal neurons are small, and pyramidal cell spike AHPs are attenuated. Prior to and during the transition from gamma to beta, EPSPs enlarge dramatically, and spike AHPs also increase. Simulation studies indicate that these two time-dependent alterations in parameters are, in principle, sufficient to account for the gamma/beta shift, provided that (*a*) interneurons remain depolarized enough to produce ING beats even when few pyramidal neurons fire; and (*b*) pyramidal neurons remain depolarized enough to fire on alternate (or every third or fourth) ING beats, despite the growth in AHPs. The physical idea is that pyramidal cell beat-skipping (relative to the interneuron rhythm) starts to occur because of time-dependent growth of one or more K^+ conductances; but because of the enhanced recurrent excitation between pyramidal cells, the pyramidal cells tend to fire on the same ING beats. The correlated firing of pyramidal cells on only some ING beats is what gives rise to the beta rhythm superimposed on the underlying gamma rhythm (Traub et al. 1999b). Simulations also suggest that the growth of EPSPs during gamma could be Hebbian (Bibbig et al. 2001); that beta should be capable of synchronizing with longer conduction delays than gamma (Bibbig et al. 2002, Kopell et al. 2000); and that Hebbian synaptic plasticity in inhibitory circuits could allow switching from an asynchronous gamma rhythm to a synchronous beta rhythm (Bibbig et al. 2002).

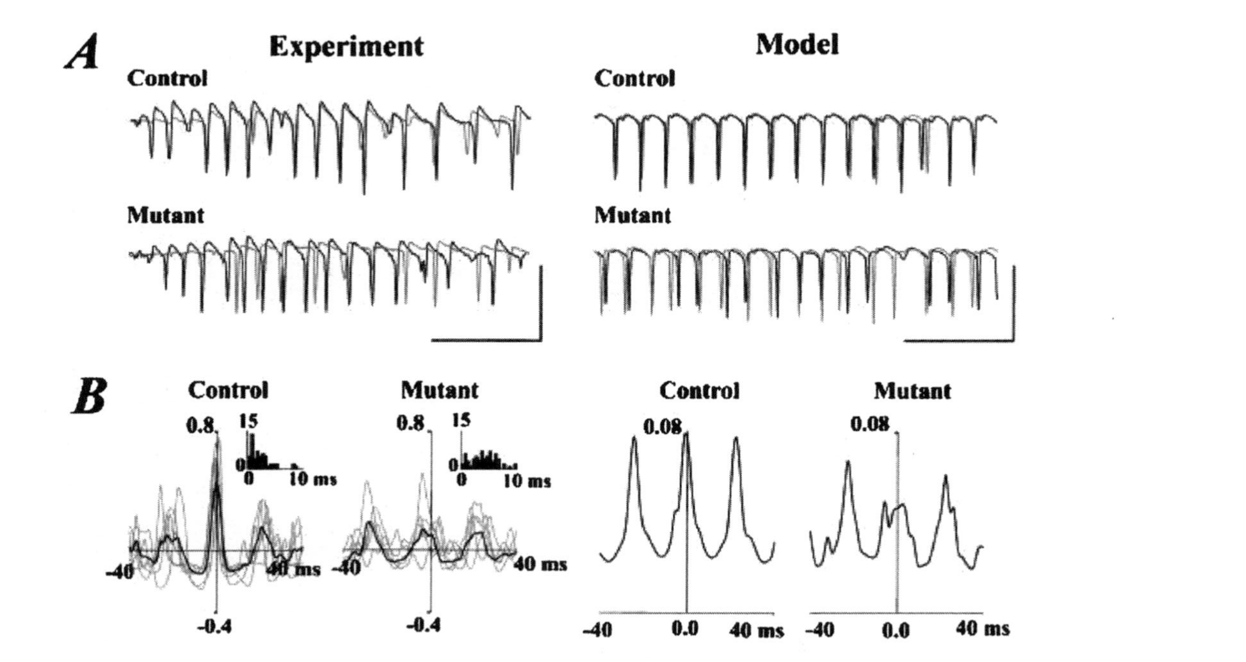

Figure 4 Small alterations in EPSC kinetics in interneurons disrupt two-site synchrony of tetanic gamma oscillations in mouse hippocampal slices (CA1), as predicted by the network model. For the experiments, a transgenic mouse was developed (mutant) with enhanced expression of GluR-B in interneurons; this leads to a small increase in the number of interneuron EPSCs with slowed kinetics and allows single shocks to elicit spike doublets in interneurons, instead of the usual single spikes. For the model, a distributed network was used to simulate gamma oscillations, in which interneuron EPSCs were proportional to an alpha-function of time; the time constant was either 0.95 ms (wild-type) or 1.05 ms (mutant). (*A*) Superimposed field potentials from the wild-type overlap in experiment and model; but they jitter back and forth in the mutant. (*B*) Cross-correlations of the field potential signals, pooled data plus average cross-correlation (*thick line*) in the experimental case. The mechanism of synchronization-disruption involves the tendency of interneurons to generate doublets and triplets, rather than singlets and doublets (Traub et al. 1996b). From Fuchs et al. (2001), with permission.

VERY FAST (>70 Hz) OSCILLATIONS GENERATED IN THE AXONAL PLEXUS VIA ELECTRICAL COUPLING BETWEEN AXONS VFO are of interest in themselves because they exemplify a novel means of generation of a population rhythm, one whose frequency is not gated by a synaptic time constant (like ING), nor by a membrane kinetic time constant (such as the relaxation of the slow AHP), but rather by the topological properties of a sparsely connected network (Traub et al. 1999c, 2001b, 2002). VFO are of further interest because, remarkably, they appear to drive persistent gamma oscillations (see below).

Spontaneous VFO in hippocampal slices were discovered by Draguhn et al. (1998). Such VFO were enhanced by low $[Ca^{2+}]_o$ media (that blocked synaptic transmission) and were also enhanced by alkalinization (which opens gap junctions). The VFO were suppressed by a variety of blockers of gap-junction conductances (carbenoxolone, halothane, octanol). Finally, the VFO were not simply the result of single-unit activity because of spread over $>100\ \mu m$, and because small population spikes could be correlated with spikelets in an individual pyramidal cell.

Because (*a*) spikelets were coincident with population spikes, and (*b*) population spikes are caused by synchronous firing of sets of pyramidal neurons, and (*c*) the whole phenomenon appears to depend on gap junctions, it was hypothesized that spikelets represented coupling potentials, produced in one neuron via electrical coupling to one or more other coupled firing neurons. The problem was to determine the site of coupling. Simulations (Draguhn et al. 1998) indicated that the rapid upstroke and downstroke of the spikelet would not be replicated by electrical coupling between somata or dendrites (see discussion of dendritic coupling above). On the other hand, when the coupling site was between axons it was found that if the between-axonal conductance was large enough, then the active properties of the axon would "sharpen" the spikelet: Na^+ conductance would accelerate the upstroke, and K^+ conductance(s) the downstroke. The evidence that such axonal coupling actually exists (Schmitz et al. 2001) is reviewed above.

A critical conceptual issue in VFO is this: Even if it is true that axonal electrical coupling exists, how does an oscillation come about, and what physical factors determine the frequency? This seems especially puzzling if the axonal coupling density is about the same as dye-coupling density, that is, each cell couples to less than two others, on average. One clue derives from earlier studies on epileptogenesis mediated primarily by recurrent synaptic excitation, which exists at relatively low densities in hippocampus (although still much higher than dye-coupling density). Thus, it was found that highly localized stimuli could reliably lead to long-latency population bursts in the disinhibited CA3 region in vitro; such an observation could be explained if it was assumed that a burst in one CA3 neuron could induce a burst in synaptically coupled neurons (Traub & Wong 1982), a prediction later verified experimentally (Miles & Wong 1986). In an analogous way, simulations have shown that a background of spontaneous ectopic axonal action potentials can lead to VFO, in a very sparse network, if one assumes that a single axonal action potential can, under appropriate conditions, induce an action potential in a coupled axon, with latency a fraction of a ms (Traub et al. 1999c,

2001b, 2002). [It turns out that such a latency is physiologically reasonable, in light of the data of MacVicar & Dudek (1982), who demonstrated a brief (<1 ms) increase of latency to an antidromic response, as a decreased strength of axonal stimulation was used. Axonal coupling, however, was not directly demonstrated in that study.]

Even if action potentials cross from axon to axon, why would an oscillation take place? To gain an intuitive feel for this, it helps to consider a so-called sparse random graph (Erdös & Rényi 1960; see also Traub et al. 2002). A random graph is a mathematical structure that consists of a set of nodes and a randomly chosen set of links between the nodes. (Think of the nodes as representing axons and the links as representing electrical coupling sites, putatively gap junctions.) It turns out that if the average number of links coming out of each node is >1, then in a sufficiently big random graph, there is guaranteed to be exactly one large cluster, a substructure of the original graph whose size is of the same order as the size of the original graph. If one of the nodes were to fire, and if this firing could spread across the links from node to node, then the total amount of firing would first grow and then recede. The time it takes to grow to its maximum depends on (*a*) the time it takes for firing to spread from node to node, and (*b*) the topological structure of the network (Traub et al. 1999c; see also Lewis & Rinzel 2000). This gives an initial intuition as to how an oscillation is possible. In addition, if the spike-crossing time is, say, 0.25 ms, then one predicts that in a graph with several thousand nodes, and 1.6 links emanating from each node on average, oscillation periods of 4 or 5 ms are expected. Thus, the period of VFO is at least approximately predicted from these relatively simple considerations.

Yet another question concerns the origin of spontaneous action potentials in the axonal plexus. Recent data (Traub et al. 2003b) suggest that GABA, acting via $GABA_A$ receptors, may be the neuroactive substance that is directly responsible for enhancing axonal excitability. This is consistent with earlier results on CA3 pyramidal cell axons (Avoli et al. 1998, Stasheff et al. 1993), as well as optic nerve axons (Sakatani et al. 1992) and thalamocortical projection neurons (Pinault & Pumain 1989).

Persistent gamma oscillations are so named because bath application of appropriate pharmacological agents can induce oscillations lasting for hours. Example agents that induce this oscillatory effect include kainate (<1 μM) and carbachol (tens of μM) (Figure 5), as well as the Group I metabotropic agonist DHPG (Fisahn et al. 1998, 2002; Hormuzdi et al. 2001). The power spectrum of field potentials associated with these oscillations often shows, in addition to the gamma peak, a second smaller peak at 70–80 Hz or more that is not a harmonic of the gamma peak (Figure 5*B*). Persistent gamma oscillations are of interest for two reasons: (*a*) They may be a reasonable experimental model of hippocampal gamma that persists during the theta state (Sik et al. 1995, Ylinen et al. 1995b); and (*b*) the generation and structure of persistent gamma depends on chemical synapses and also on electrical coupling, both axonal (principal cell) and dendritic (interneuron). In order to understand the contributions of chemical synapses and gap junctions (see

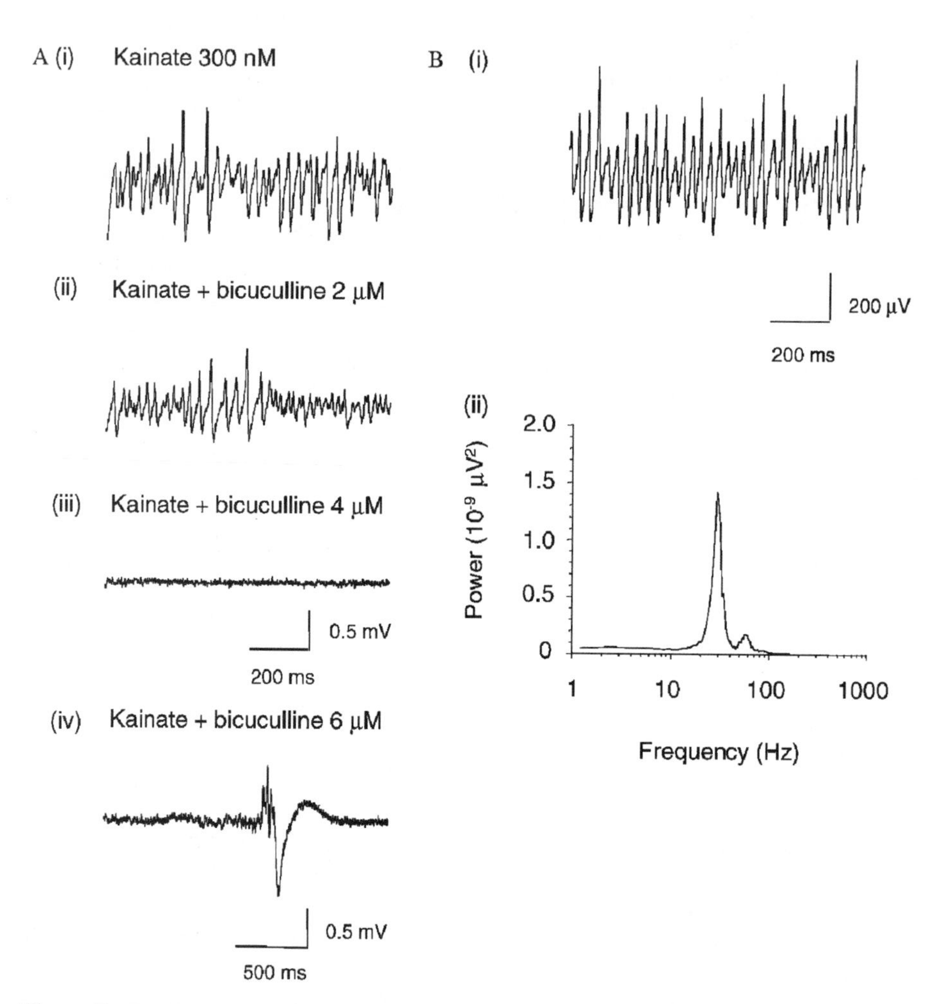

Figure 5 Persistent gamma oscillations depend on $GABA_A$ receptors and exhibit activity at high frequency, in addition to gamma frequency. (*Left*) (*A*) S. radiatum field potential recordings from rat CA3 region in vitro. *Ai* shows gamma frequency field potential oscillations induced by 300 nM kainate. The oscillations are reduced and then abolished by bicuculline, 2 μM and 4 μM, respectively (*Aii and Aiii*). A higher concentration of bicuculline (6 μM) induces epileptiform discharges, again without evidence of gamma oscillation (*Aiv*). (*B*) (*i*) Extracellular field recording from s. radiatum in the CA3 subfield of the rat hippocampus shows a persistent oscillation evoked following bath application of carbachol (20 μM). (*ii*) The power spectra shows a main peak in the gamma frequency range at ~30 Hz, but a second higher frequency component (~60 Hz) is also present. (*Left*) From Pais et al. (2003), with permission. (*Right*) F.E.N. LeBeau, E.H. Buhl, & M.A. Whittington, unpublished data.

Figure 1), it is first necessary to summarize the basic phenomenology of persistent oscillations, including what does happen as well as what does not happen:

1. There is a rhythmic field potential, <1 mV in amplitude (Figure 5).
2. Intracellular recordings show in pyramidal cells that there is no large depolarization (unlike tetanic gamma); although there are rhythmic gamma-frequency synaptic potentials, the pyramidal cell somata fire only rarely, on ~5% of the waves, typically (Fisahn et al. 1998).
3. Intracellular recordings in unidentified s. pyramidale and s. oriens interneurons also show an absence of strong underlying depolarization, along with an action potential on approximately half of the gamma waves [sometimes interneuron doublets, in kainate (E.H. Buhl, unpublished data)]. On average, pyramidal cell action potentials lead interneuron action potentials by a few ms (Fisahn et al. 1998), as if interneuron action potentials are driven by the pyramidal cells, although pyramidal cell somata can fire at only ~2 Hz.
4. Bicuculline abolishes persistent gamma before sharp waves are induced (Figure 5*A*); evidently, the firing of interneurons is actually necessary for the oscillation.
5. Blockade of AMPA receptors also abolishes the oscillation (Fisahn et al. 1998). Thus, persistent gamma is not simply a form of ING. Instead, the firing of pyramidal cells, or at least their axons, if not their somata, is important. NMDA receptors and $GABA_B$ receptors are not necessary for persistent gamma (Fisahn et al. 1998).
6. Blockade of gap junctions, pharmacologically, abolishes persistent gamma (Pais et al. 2003, Traub et al. 2000).
7. Removal of interneuron dendritic gap junctions, through a genetic knockout of connexin36, does not abolish persistent gamma, although it does reduce the power. In this same knockout, VFO that occur in the presence of low $[Ca^{2+}]_o$ media (Draguhn et al. 1998) are indistinguishable from VFO in wild-type mice (Hormuzdi et al. 2001). In the knockout mouse, electrical coupling between pairs of interneurons is not detectable.
8. Pharmacological blockade of gap junctions in the connexin36 knockout mouse eliminates persistent gamma (Pais et al. 2003); presumably this occurs because of a blockade of axonal electrical coupling, superimposed on the genetic loss of interneuron electrical coupling (Traub et al. 2003a).

How is one to make sense of these observations, particularly the puzzling one that pyramidal cells fire rarely, while nevertheless AMPA receptor blockade eliminates the oscillation? We have developed a network model (Traub et al. 2000, 2003a) that accounts for all of the above experimental data. The basic structure is as shown in Figure 1, and the key idea is that the oscillation is driven by the plexus of electrically coupled pyramidal cell axons: Spontaneous action potentials in these axons pass from axon to axon, generating waves of firing, most of which do not propagate antidromically to somata because of synaptic inhibition and shunting. Hence the firing of pyramidal cell somata is rare. (The need for spikes to spread throughout the axonal plexus accounts for the ability of pharmacological blockade of gap junctions to wipe out the oscillation.) But the firing does propagate orthodromically, exciting interneurons, hence the requirement for AMPA receptors. Because interneuron firing gates the gamma oscillation, $GABA_A$ receptors are required. Finally, interneuron gap junctions act to tighten the gamma oscillation, in part by leading to higher interneuron firing rates, hence the reduction in gamma power when interneuron gap junctions are removed transgenically. This network model is consistent with data showing that kainate excites interneuron axons (Semyanov & Kullmann 2001), with GABA release perhaps exciting pyramidal cell axons as well (see above). It also exhibits the experimentally observed lead in pyramidal cell spikes over interneuron spikes. Finally, the model explains the high-frequency component in the power spectrum: It results from short segments of axonally generated VFO, segments that are "chopped up" by the gamma-frequency IPSPs.

GAMMA OSCILLATIONS EVOKED BY PRESSURE EJECTION OF HYPERTONIC POTASSIUM SOLUTION, "PUFF GAMMA" Pressure ejection of molar concentrations of K^+ solution into any hippocampal stratum evokes gamma oscillations that last for some seconds (LeBeau et al. 2002, Towers et al. 2002). Although investigations have not extended in as much detail for puff gamma as for persistent gamma, the basic phenomenology—other than oscillation duration—appears to be quite similar: Pyramidal cells fire rarely, $GABA_A$ and AMPA receptors are both necessary, and gap junctions are also necessary. It is possible that the pressure ejection of K^+ solution excites pyramidal cell axons briefly, either directly or indirectly, and evokes a population oscillation quite like persistent gamma in its cellular mechanisms. One indirect mechanism by which this could happen is that the hypertonic solution excites interneurons, which in turn release GABA, and thereby excite axons.

Metabotropically activated theta oscillations can be evoked in area CA1 under somewhat unexpected conditions (Figure 6; Gillies et al. 2002). The Group I metabotropic glutamate receptor agonist DHPG (20–200 μM), applied in the bath, elicits gamma oscillations in CA1 (Figure 6*A*) and a mixture of gamma and theta oscillations in CA3 (Figure 6*B*). Superimposed pharmacological blockade of AMPA/kainate receptors with NBQX (20 μM) blocks the gamma in both CA1 and CA3; and, remarkably, in CA1—but not in CA3—this superimposed blockade also leads to a prominent theta oscillation. [In CA3, NBQX diminishes theta power (Figure 6*B*). See also MacVicar & Tse 1989, and Williams & Kauer 1997. In vitro CA3 theta-frequency oscillations, induced by carbachol, continue during block of $GABA_A$ receptors (MacVicar & Tse 1989); it is not clear, therefore, how CA3 in vitro theta oscillations relate to in vivo theta.]

The phenomenology of CA1 in vitro theta is complicated. At least two populations of interneurons, with different postsynaptic actions, are involved, and the participation of pyramidal cell dendrite intrinsic membrane properties is an

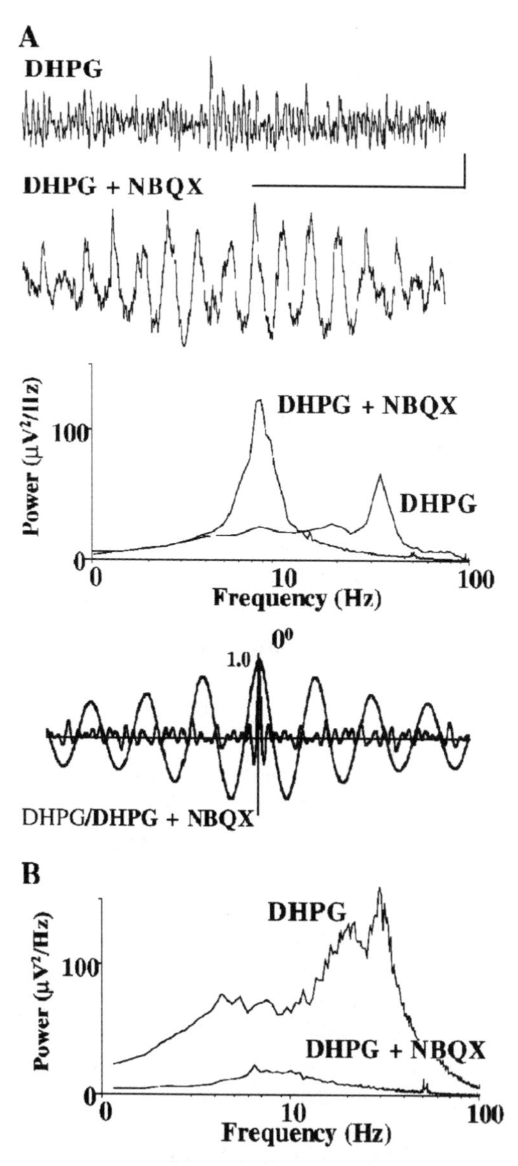

Figure 6 Metabotropic glutamate receptor activation elicits theta oscillations, when AMPA receptors are blocked, in the CA1 region in vitro, but not in CA3. (*A*) CA1 s. pyramidale field potential recordings. The Group I metabotropic GluR agonist DHPG (*upper trace*) elicits gamma oscillations. Further addition of the AMPA/kainate receptor blocker NBQX transforms the oscillation to theta frequency. Power spectra are shown below. Calibration: 1 s, 50 uV. (*B*) Power spectra of analogous field potential recordings from CA3 in vitro. DHPG induces a mixture of theta and gamma activity, but further addition of NBQX suppresses both. From Gillies et al. (2002), with permission.

important aspect. A preliminary model of the oscillation exists (Rotstein et al. 2002). Here, we summarize some of the phenomenology:

1. The oscillation requires $GABA_A$ receptors; nevertheless, thiopental has little effect on it.
2. Two classes of population IPSPs occur during CA1 in vitro theta, in pyramidal cells: relatively fast ones (decay time constant about 15 ms), at gamma frequency; and relatively slow ones (decay time constant about 25–30 ms). The slower IPSPs are prominent in apical dendrites. Remarkably, even in the gamma state (DHPG alone, without NBQX), the slower dendritic IPSPs have similar appearance to those that occur during theta. As oriens/lacunosum-moleculare interneurons (O/LM cells) fire at theta frequency (Maccaferri & McBain 1996), and themselves exhibit slower IPSPs, O/LM cells are the ones generating slower IPSPs in pyramidal cell dendrites, in this model. This notion is supported by the observation that O/LM cells possess an h-current; and specific blockade of h-current, with bath-applied ZD7288, abolishes in vitro CA1 theta. O/LM interneurons are also involved in theta rhythm in vivo (Klausberger et al. 2003).
3. Pyramidal cell somata fire rarely, if at all (unless artificially depolarized with injected current). Nevertheless, pyramidal cell apical dendrites generate broad, putative Ca^{2+} spikes at theta frequencies. Application of the calcium channel blocker Ni^{2+} (0.1 mM) not only suppresses the dendritic spikes, but also abolishes the theta oscillation, as it is measured with a field potential electrode. Despite this abolition, theta frequency slower IPSPs persist in pyramidal cell apical dendrites. The AHP following dendritic putative Ca^{2+} spikes appears to contribute to setting the oscillation period, in addition to contributions from IPSPs.
4. The firing of different sorts of interneurons—s. pyramidale and s. oriens—occurs out of phase with each other. There are, however, characteristic, generally fixed, relative phase relations. As "seen" from a pyramidal cell apical dendrite, the phase relations would be these: S. oriens cells fire and produce slower IPSPs in the dendrites. As these IPSPs decay, a broad dendritic spike occurs, followed by an intrinsic AHP. Then, s. pyramidale interneurons fire, producing a train of fast IPSPs that prolong the AHP. And so the cycle repeats.

The above-described CA1 in vitro theta appears after a manipulation that seems nonphysiological, the application of NBQX. In order to make a case that in vitro CA1 theta could be relevant to in vivo theta, it is necessary to understand why NBQX allows the oscillation to occur. [Note that NBQX is not simply "unmasking" theta, as theta power in CA1 grows significantly in the presence of the drug, as compared with its absence.] What follows is one hypothesis. We presume that axonal activity in the pyramidal cell axonal plexus is essential for the oscillation because (*a*) pyramidal cell somata fire rarely, if at all; (*b*) the theta oscillation is nevertheless blocked by AP5, an NMDA receptor antagonist; and (*c*) theta is suppressed by gap-junction blockade (M. Gillies, M.A. Whittington, F.E.N. LeBeau, unpublished data). It seems unlikely that AMPA receptor–mediated input to pyramidal cells would prevent CA1 theta (this input is not very large), nor input to s. oriens interneurons [slow IPSPs in pyramidal cell dendrites look similar during gamma (without NBQX) as compared to theta (with NBQX)]. Therefore, we would guess that it is AMPA receptor input to s. pyramidale interneurons, including basket cells, that keeps gamma oscillations going (consistent with our discussion of persistent gamma above), and that the existence of excessive gamma actively interferes with the generation of theta oscillations. If this is true, then some limitation of gamma activity in s. pyramidale interneurons could be achieved by a variety of mechanisms, including modulation of their intrinsic properties. Note, however, that in vivo gamma presumably must be limited, rather than suppressed: Such gamma does indeed exist during the theta state (Bragin et al. 1995, Soltesz & Deschênes 1993), and basket cell firing is tightly locked to this gamma (Sik et al. 1995).

OSCILLATIONS AND SYNAPTIC PLASTICITY: A TWO-WAY STREET

Experiments have shown that tetanic (transient) in vitro oscillations in the hippocampus are associated with synaptic potentiation, occurring on the timescale of the oscillation itself, hundreds of ms. Stimuli intense enough to elicit gamma → beta oscillations cause increases in pyramidal cell EPSPs (Bibbig et al. 2001, Traub et al. 1999b, Whittington et al. 1997b), which grow over the course of the gamma oscillation and appear necessary for the stability of the beta oscillation that follows (Faulkner et al. 1999). The increase in pyramidal cell EPSPs is evident in intracellular recordings from hyperpolarized neurons, as well as in field EPSPs recorded in s. oriens. Indirect evidence indicates that during a gamma/beta oscillation, recurrent excitation becomes effective enough (in at least some instances) so that a single action potential in one pyramidal neuron can elicit an action potential in a connected neuron (Whittington et al. 1997b). [In addition, pyramidal cell/interneuron EPSPs also become potentiated during tetanic gamma → beta oscillations (Bibbig et al. 2001).]

Simulations indicate that synaptic plasticity in turn influences the form of tetanic oscillations. As noted above, potentiation of recurrent synaptic excitation is necessary for beta oscillations to occur at all. Models can replicate the tetanic gamma → beta transition if one assumes Hebbian plasticity at pyramidal cell synapses onto other pyramidal cells and onto interneurons, along with time-dependent increases in pyramidal cell AHPs (Bibbig et al. 2001). Finally, Hebbian plasticity in interneuron circuits (specifically including pyramidal cell → interneuron and interneuron → pyramidal cell connections) is predicted to allow a switch from two-site antiphase firing to two-site in-phase firing, in the presence of long conduction delays (Bibbig et al. 2002).

TABLE 2 Expected patterns of synaptic plasticity in two types of in vitro oscillation

Tetanic γ	
	Many spikes in postsynaptic e-cells + temporally correlated EPSPs $\rightarrow$ e$\rightarrow$e LTP
	Many spikes in postsynaptic i-cells + temporally correlated EPSPs $\rightarrow$ e$\rightarrow$i LTP?
Persistent γ	
	Rare spikes in postsynaptic e-cells + temporally correlated EPSPs $\rightarrow$ e$\rightarrow$e LTD?
	Many spikes in postsynaptic i-cells + temporally correlated EPSPs $\rightarrow$ e$\rightarrow$i LTP?

Whether synaptic plasticity takes place during persistent gamma oscillations, in vitro, is not yet known. If such plasticity does indeed occur, and is Hebbian, then one expects LTD (long-term depression) at pyramidal/pyramidal connections and LTP (long-term potentiation) at pyramidal/interneuron connections (Table 2). Such an arrangement would be ideal for maintaining synchronized gamma without generating beta activity.

IN VIVO CORRELATIONS

It is probably not possible to draw a precise correspondence between any of the in vitro oscillations and in vivo oscillations at present. We shall offer, however, our present views on what correlations seem most likely; here, we follow Table 1.

1. The in vivo correlate of interneuron network gamma remains unclear.
2. Tetanic gamma, and gamma/beta, appear to resemble most closely transient gamma and gamma/beta EEG sequences evoked in neocortex by sensory stimulation (Gray & Singer 1989, Haenschel et al. 2000, Tallon-Baudry et al. 1999). When tetanic oscillations are evoked in vitro, in hippocampus, without drugs, the degree of pyramidal cell firing is much greater than is the case in vivo; the degree of pyramidal cell firing can, however, be reduced pharmacologically without altering the basic structure of the oscillation, at least of the gamma portion (Whittington et al. 2001). To investigate this issue further, it would help to determine if metabotropic glutamate receptors are required for sensory-activated gamma to occur.
3. VFO occur in the hippocampus in vivo, superimposed on physiological sharp waves, and at highest amplitude in s. pyramidale, as is true in vitro. In vivo VFO, or "ripples," are suppressed by the gap-junction blocker halothane, although sharp waves themselves are not (Buzsaki et al. 1992, Ylinen et al. 1995a). In connexin36 knockout mice, studied in vivo, high-frequency ripples are intact, as one might expect from in vitro data (Buhl et al. 2003, Hormuzdi et al. 2001). We suspect, therefore, that the basic mechanisms

(including the requirement for axonal electrical coupling) of in vitro and in vivo VFO are similar or identical.

4. Persistent in vitro gamma resembles hippocampal in vivo gamma, in that pyramidal cells fire at much lower rates than interneurons in both circumstances, and pyramidal cells slightly lead interneurons in both (Csicsvari et al. 2003). Knockout of connexin36 reduces hippocampal gamma power in vivo, as it does in vitro (Buhl et al. 2003, Hormuzdi et al. 2001). Whether hippocampal gamma requires electrical coupling between pyramidal cell axons (as is the case in vitro for persistent gamma), and whether many of the somatic action potentials during in vivo hippocampal gamma are antidromic remains to be determined.
5. In vitro CA1 theta may resemble in vivo atropine-resistant theta, which is hypothesized to depend on NMDA receptors (Buzsáki 2002), perhaps in a manner similar to in vitro CA1 theta (Gillies et al. 2002). What would correspond in vivo to the in vitro requirement for AMPA receptor block needs to be investigated.

ACKNOWLEDGMENTS

Supported by NINDS, NIH; the Volkswagen Stiftung; the MRC (U.K.); and the Wellcome Trust. We wish to thank our long-standing colleagues Nancy Kopell and Hannah Monyer, and we are grateful to Robert Walkup of IBM Corp. for help with computing systems.

The *Annual Review of Neuroscience* is online at http://neuro.annualreviews.org

LITERATURE CITED

Abeles M. 1982. Role of the cortical neuron. Integrator or coincidence detector. *Israel J. Med. Sci.* 18:83–92

Andersen P, Silfvenius H, Sundberg S, Sveen O, Wigström H. 1978. Functional characteristics of unmyelinated fibres in the hippocampal cortex . *Brain Res.* 144:11–18

Avoli M, Methot M, Kawasaki H. 1998. GABA-dependent generation of ectopic action potentials in the rat hippocampus. *Eur. J. Neurosci.* 10:2714–22

Bartos M, Vida I, Frotscher F, Geiger JRP, Jonas P. 2001. Rapid signaling at inhibitory synapses in a dentate gyrus interneuron network. *J. Neurosci.* 21:2687–98

Bartos M, Vida I, Frotscher M, Meyer A, Monyer A, et al. 2002. Fast synaptic inhibition promotes synchronized gamma oscillations in hippocampal interneuron networks. *Proc. Natl. Acad. Sci. USA* 99:13222–27

Bibbig A, Faulkner HJ, Whittington MA, Traub RD. 2001. Self-organized synaptic plasticity contributes to the shaping of γ and β oscillations in vitro. *J. Neurosci.* 21:9053–67

Bibbig A, Traub RD, Whittington MA. 2002. Characteristics of long-range synchronization of γ and β oscillations and the plasticity of excitatory and inhibitory synapses: a network model. *J. Neurophysiol.* 88:1634–54

Bragin A, Jandó G, Nádasdy Z, Hetke J, Wise K, Buzsáki G. 1995. Gamma (40–100 Hz) oscillation in the hippocampus of the behaving rat. *J. Neurosci.* 15:47–60

Buhl DL, Harris KD, Hormuzdi SG, Monyer

H, Buzsáki G. 2003. Selective impairment of hippocampal gamma oscillations in connexin-36 knock-out mouse *in vivo*. *J. Neurosci.* 23:1013–18

Buzsáki G. 2002. Theta oscillations in the hippocampus. *Neuron* 33:325–40

Buzsáki G, Horváth Z, Urioste R, Hetke J, Wise K. 1992. High-frequency network oscillation in the hippocampus. *Science* 256:1025–27

Charpak S, Gähwiler B, Do KQ, Knöpfel T. 1990. Potassium conductances in hippocampal neurons blocked by excitatory amino-acid transmitters. *Nature* 347:765–67

Chow CC, White JA, Ritt J, Kopell N. 1998. Frequency control in synchronized networks of inhibitory neurons. *J. Comput. Neurosci.* 5:407–20

Church J, Baimbridge KG. 1991. Exposure to high-pH medium increases the incidence and extent of dye coupling between rat hippocampal CA1 pyramidal neurons in vitro. *J. Neurosci.* 11:3289–95

Condorelli DF, Parenti R, Spinella F, Salinaro AT, Belluardo N, et al. 1998. Cloning of a new gap junction gene (Cx36) highly expressed in mammalian brain neurons. *Eur. J. Neurosci.* 10:1202–8

Congar P, Leinekugel X, Ben-Ari Y, Crépel V. 1997. A long-lasting calcium-activated non-selective cationic current is generated by synaptic stimulation or exogenous activation of group I metabotropic glutamate receptors in CA1 pyramidal neurons. *J. Neurosci.* 17:5366–79

Csicsvari J, Jamieson B, Wise KD, Buzsáki G. 2003. Mechanisms of gamma oscillations in the hippocampus of the behaving rat. *Neuron* 37:311–22

Deuchars J, Thomson AM. 1996. CA1 pyramid-pyramid connections in rat hippocampus *in vitro*: dual intracellular recordings with biocytin filling. *Neuroscience* 74: 1009–18

Draguhn A, Traub RD, Schmitz D, Jefferys JGR. 1998. Electrical coupling underlies high-frequency oscillations in the hippocampus *in vitro*. *Nature* 394:189–92

Eckhorn R, Bauer R, Jordan W, Brosch M, Kruse W, et al. 1988. Coherent oscillations: a mechanism of feature linking in the visual cortex? *Biol. Cybern.* 60:121–30

Ermentrout GB, Kopell N. 1998. Fine structure of neural spiking and synchronization in the presence of conduction delays. *Proc. Natl. Acad. Sci. USA* 95:1259–64

Erdös P, Rényi A. 1960. On the evolution of random graphs. *Publ. Math. Instit. Hungar. Acad. Sci.* 5:17–61

Faulkner HJ, Traub RD, Whittington MA. 1999. Anaesthetic/amnesic agents disrupt beta frequency oscillations associated with potentiation of excitatory synaptic potentials in the rat hippocampal slice. *Br. J. Pharmacol.* 128:1813–25

Fisahn A, Pike FG, Buhl EH, Paulsen O. 1998. Cholinergic induction of network oscillations at 40 Hz in the hippocampus *in vitro*. *Nature* 394:186–89

Fisahn A, Yamada M, Duttaroy A, Gan J-W, Deng C-X, et al. 2002. Muscarinic induction of hippocampal gamma oscillations requires coupling of the M1 receptor to two mixed cation currents. *Neuron* 33:615–24

Freeman WJ. 1972. Measurement of oscillatory responses to electrical stimulation in olfactory bulb of cat. *J. Neurophysiol.* 35:762–79

Fuchs E, Doheny HC, Faulkner HJ, Caputi A, Traub RD, et al. 2001. Genetically altered AMPA-type glutamate receptor kinetics in interneurons disrupt long-range synchrony of gamma oscillation. *Proc. Natl. Acad. Sci. USA* 98:3571–76

Fukuda T, Kosaka T. 2000. Gap junctions linking the dendritic network of GABAergic interneurons in the hippocampus. *J. Neurosci.* 20:1519–28

Galarreta M, Hestrin S. 1999. A network of fast-spiking cells in the neocortex connected by electrical synapses. *Nature* 402:72–75

Geiger JRP, Lübke J, Roth A, Frotscher M, Jonas P. 1997. Submillisecond AMPA receptor-mediated signaling at a principal neuron-interneuron synapse. *Neuron* 18:1009–23

Gibson JR, Beierlein M, Connors BW. 1999. Two networks of electrically coupled inhibitory neurons in neocortex. *Nature* 402:75–79

Gillies M, Traub RD, LeBeau FEN, Davies CH, Gloveli T, et al. 2002. A model of atropine-resistant theta oscillations in hippocampal area CA1. *J. Physiol.* 543:779–93

Gray CM, König P, Engel AK, Singer W. 1989. Oscillatory responses in cat visual cortex exhibit inter-columnar synchronization which reflects global stimulus properties. *Nature* 338:334–37

Gray CM, Singer W. 1989. Stimulus-specific neuronal oscillations in orientation columns of cat visual cortex. *Proc. Natl. Acad. Sci. USA* 86:1698–702

Grenier F, Timofeev I, Steriade M. 2001. Focal synchronization of ripples (80–200 Hz) in neocortex and their neuronal correlates. *J. Neurophysiol.* 86:1884–98

Grenier F, Timofeev I, Steriade M. 2003. Neocortical very fast oscillations (ripples, 80–200 Hz) during seizures: intracellular correlates. *J. Neurophysiol.* 89:841–52

Haenschel C, Baldeweg T, Croft RJ, Whittington M, Gruzelier J. 2000. Gamma and beta frequency oscillations in response to novel auditory stimuli: a comparison of human electroencephalogram (EEG) data with *in vitro* models. *Proc. Natl. Acad. Sci. USA* 97:7645–50

Hormuzdi SG, Pais I, LeBeau FEN, Towers SK, Rozov A, et al. 2001. Impaired electrical signaling disrupts gamma frequency oscillations in connexin 36-deficient mice. *Neuron* 31:487–95

Jagadeesh B, Gray CM, Ferster D. 1992. Visually evoked oscillations of membrane potential in cells of cat visual cortex. *Science* 257:552–54

Kaila K, Lamsa K, Smirnov S, Taira T, Voipio J. 1997. Long-lasting GABA-mediated depolarization evoked by high-frequency stimulation in pyramidal neurons of rat hippocampal slice is attributable to a network-driven, bicarbonate-dependent K^+ transient. *J. Neurosci.* 17:7662–72

Kamondi A, Acsády L, Wang X-J, Buzsáki G. 1998. Theta oscillations in somata and dendrites of hippocampal pyramidal cells in vivo: activity-dependent phase-precession of action potentials. *Hippocampus* 8:244–61

Klausberger T, Magill PJ, Márton LF, Roberts JDB, Cobden PM, et al. 2003. Brain-state- and cell-type-specific firing of hippocampal interneurons *in vivo*. *Nature* 421:844–48

Knowles WD, Schwartzkroin PA. 1981. Axonal ramifications of hippocampal CA1 pyramidal cells. *J. Neurosci.* 1:1236–41

Knowles WD, Traub RD, Strowbridge BW. 1987. The initiation and spread of epileptiform bursts in the *in vitro* hippocampal slice. *Neuroscience* 21:441–55

König P, Engel AK, Singer W. 1996. Integrator or coincidence detector? The role of the cortical neuron revisited. *Trends Neurosci.* 19:130–37

Kopell N, Ermentrout GB, Whittington MA, Traub RD. 2000. Gamma rhythms and beta rhythms have different synchronization properties. *Proc. Natl. Acad. Sci. USA* 97:1867–72

Kosaka T. 1983. Gap junctions between non-pyramidal cell dendrites in the rat hippocampus (CA1 and CA3 regions). *Brain Res.* 271:157–61

Kuenzi FM, Fitzjohn SM, Morton RA, Collingridge GL, Seabrook GR. 2000. Reduced long-term potentiation in hippocampal slices prepared using sucrose-based artificial cerebrospinal fluid. *J. Neurosci. Methods* 100:117–22

Larson J, Lynch G. 1988. Role of N-methyl-D-aspartate receptors in the induction of synaptic potentiation by burst stimulation patterned after hippocampal θ-rhythm. *Brain Res.* 441:111–18

Laurent G, Davidowitz H. 1994. Encoding of olfactory information with oscillating neural assemblies. *Science* 265:1872–75

LeBeau FEN, Towers SK, Traub RD, Whittington MA, Buhl EH. 2002. Fast network oscillations induced by potassium transients in the rat hippocampus *in vitro*. *J. Physiol.* 542:167–79

Lewis TJ, Rinzel J. 2000. Self-organized synchronous oscillations in a network of excitable cells coupled by gap junctions. *Network: Comput. Neural Syst.* 11:299–320

Lytton WW, Sejnowski TJ. 1991. Simulations of cortical pyramidal neurons synchronized by inhibitory interneurons. *J. Neurophysiol.* 66:1059–79

Maccaferri G, McBain CJ. 1996. The hyperpolarization-activated current (I_h) and its contribution to pacemaker activity in rat CA1 hippocampal stratum oriens-alveus interneurones. *J. Physiol.* 497:119–30

MacVicar BA, Dudek FE. 1980. Dye-coupling between CA3 pyramidal cells in slices of rat hippocampus. *Brain Res.* 196:494–97

MacVicar BA, Dudek FE. 1982. Electrotonic coupling between granule cells of the rat dentate gyrus: physiological and anatomical evidence. *J. Neurophysiol.* 47:579–92

MacVicar BA, Tse FWY. 1989. Local neuronal circuitry underlying cholinergic rhythmical slow activity in CA3 area of rat hippocampal slices. *J. Physiol.* 417:197–212

Martina M, Vida I, Jonas P. 2000. Distal initiation and active propagation of action potentials in interneuron dendrites. *Science* 287:295–300

Miles R. 1990. Synaptic excitation of inhibitory cells by single CA3 hippocampal pyramidal cells of the guinea-pig *in vitro*. *J. Physiol.* 428:61–77

Miles R, Wong RKS. 1986. Excitatory synaptic interactions between CA3 neurones in the guinea-pig hippocampus. *J. Physiol.* 373: 397–418

Miltner WHR, Braun C, Arnold M, Witte H, Taub E. 1999. Coherence of gamma-band EEG activity as a basis for associative learning. *Nature* 397:434–36

Murthy VN, Fetz EE. 1996. Oscillatory activity in sensorimotor cortex of awake monkeys: synchronization of local field potentials and relation to behavior. *J. Neurophysiol.* 76:3949–67

Pais I, Hormuzdi SG, Monyer H, Traub RD, Wood IC, et al. 2003. Sharp wave-like activity in the hippocampus *in vitro* in mice lacking the gap junction protein connexin 36. *J. Neurophysiol.* 89:2046–54

Penttonen M, Kamondi A, Acsády L, Buzsáki G. 1998. Gamma frequency oscillation in the hippocampus: intracellular analysis *in vivo*. *Eur. J. Neurosci.* 10:718–28

Perez Velazquez JL, Han D, Carlen PL. 1997. Neurotransmitter modulation of gap junctional communication in the rat hippocampus. *Eur. J. Neurosci.* 9:2522–31

Pinault D, Pumain R. 1989. Antidromic firing occurs spontaneously on thalamic relay neurons: triggering of somatic intrinsic burst discharges by ectopic action potentials. *Neuroscience* 31:625–37

Pozzo Miller LD, Petrozzino JJ, Connor JA. 1995. G protein-coupled receptors mediate a fast excitatory postsynaptic current in CA3 pyramidal neurons in hippocampal slices. *J. Neurosci.* 15:8320–30

Racca C, Catania MV, Monyer H, Sakmann B. 1996. Expression of AMPA-glutamate receptor B subunit in rat hippocampal GABAergic neurons. *Eur. J. Neurosci.* 8:1580–90

Rodriguez E, George N, Lachaux J-P, Martinerie J, Renault B, Varela FJ. 1999. Perception's shadow: long-distance synchronization of human brain activity. *Nature* 397: 430–33

Roelfsema PR, König P, Engel AK, Sireteanu R, Singer W. 1994. Reduced synchronization in the visual cortex of cats with strabismic amblyopia. *Eur. J. Neurosci.* 6:1645–55

Rotstein HG, Gillies MJ, Whittington MA, Buhl EH, Kopell N. 2002. A model of an inhibition-based atropine-resistant theta frequency oscillation in CA1 *in vitro*. *Abstr. Soc. Neurosci.* 753.10

Sakatani K, Black JA, Kocsis JD. 1992. Transient presence and functional interaction of endogenous GABA and $GABA_A$ receptors in developing rat optic nerve. *Proc. R. Soc. Lond. B* 247:155–61

Schmitz D, Schuchmann S, Fisahn A, Draguhn A, Buhl EH, et al. 2001. Axo-axonal coupling: a novel mechanism for ultrafast neuronal communication. *Neuron* 31:831–40

Semyanov A, Kullmann DM. 2001. Kainate receptor-dependent axonal depolarization and action potential initiation in interneurons. *Nat. Neurosci.* 4:718–23

Sik A, Penttonen M, Ylinen A, Buzsáki G. 1995. Hippocampal CA1 interneurons: an *in vivo* intracellular labeling study. *J. Neurosci.* 15:6651–65

Singer W, Gray CM. 1995. Visual feature integration and the temporal correlation hypothesis. *Annu. Rev. Neurosci.* 18:555–86

Soltesz I, Deschênes M. 1993. Low- and high-frequency membrane potential oscillations during theta activity in CA1 and CA3 pyramidal neurons of the rat hippocampus under ketamine-xylazine anesthesia. *J. Neurophysiol.* 70:97–116

Spray DC, Harris AL, Bennett MVL. 1981. Gap junctional conductance is a simple and sensitive function of intracellular pH. *Science* 211:712–15

Stasheff SF, Mott DD, Wilson WA. 1993. Axon terminal hyperexcitability associated with epileptogenesis in vitro. II. Pharmacological regulation by NMDA and $GABA_A$ receptors. *J. Neurophysiol.* 70:976–84

Steriade M, Amzica F. 1996. Intracortical and corticothalamic coherency of fast spontaneous oscillations. *Proc. Natl. Acad. Sci. USA* 93:2533–38

Tallon-Baudry C, Kreiter A, Bertrand O. 1999. Sustained and transient oscillatory responses in the gamma and beta bands in a visual short-term memory task in humans. *Vis. Neurosci.* 16:449–59

Tamás G, Buhl EH, Lörincz A, Somogyi P. 2000. Proximally targeted GABAergic synapses and gap junctions precisely synchronize cortical interneurons. *Nat. Neurosci.* 3:366–71

Towers SK, LeBeau FEN, Gloveli T, Traub RD, Whittington MA, Buhl EH. 2002. Fast network oscillations in the rat dentate gyrus in vitro. *J. Neurophysiol.* 87:1165–68

Traub RD, Bibbig A, Fisahn A, LeBeau FEN, Whittington MA, Buhl EH. 2000. A model of gamma-frequency network oscillations induced in the rat CA3 region by carbachol *in vitro*. *Eur. J. Neurosci.* 12:4093–106

Traub RD, Bibbig A, Piechotta A, Draguhn A, Schmitz D. 2001a. Synaptic and nonsynaptic contributions to giant IPSPs and ectopic spikes induced by 4-aminopyridine in the hippocampus in vitro. *J. Neurophysiol.* 85:1246–56

Traub RD, Cunningham MO, Gloveli T, LeBeau FEN, Bibbig A, et al. 2003b. GABA-enhanced collective behavior in neuronal axons underlies persistent gamma frequency oscillations. *Proc. Natl. Acad. Sci. USA* 100:11047–52

Traub RD, Draguhn A, Whittington MA, Baldeweg T, Bibbig A, et al. 2002. Axonal gap junctions between principal neurons: a novel source of network oscillations, and perhaps epileptogenesis. *Rev. Neurosci.* 13:1–30

Traub RD, Jefferys JGR, Whittington MA. 1999a. *Fast Oscillations in Cortical Circuits.* Cambridge, MA: MIT Press

Traub RD, Kopell N, Bibbig A, Buhl EH, LeBeau FEN, Whittington MA. 2001c. Gap junctions between interneuron dendrites can enhance long-range synchrony of gamma oscillations. *J. Neurosci.* 21:9478–86

Traub RD, Miles R. 1995. Pyramidal cell-to-inhibitory cell spike transduction explicable by active dendritic conductances in inhibitory cell. *J. Comput. Neurosci.* 2:291–98

Traub RD, Pais I, Bibbig A, LeBeau FEN, Buhl EH, et al. 2003a. Contrasting roles of axonal (pyramidal cell) and dendritic (interneuron) electrical coupling in the generation of gamma oscillations in the hippocampus in vitro. *Proc. Natl. Acad. Sci. USA* 100:1370–74

Traub RD, Schmitz D, Jefferys JGR, Draguhn A. 1999c. High-frequency population oscillations are predicted to occur in hippocampal pyramidal neuronal networks interconnected by axoaxonal gap junctions. *Neuroscience* 92:407–26

Traub RD, Whittington MA, Buhl EH, Jefferys JGR, Faulkner HJ. 1999b. On the mechanism of the $\gamma \rightarrow \beta$ frequency shift in neuronal oscillations induced in rat hippocampal slices by tetanic stimulation. *J. Neurosci.* 19:1088–105

Traub RD, Whittington MA, Buhl EH, LeBeau FEN, Bibbig A, et al. 2001b. A possible role for gap junctions in generation of very fast EEG oscillations preceding the onset of, and perhaps initiating, seizures. *Epilepsia* 42:153–70

Traub RD, Whittington MA, Colling SB, Buzsáki G, Jefferys JGR. 1996a. Analysis of gamma rhythms in the rat hippocampus *in vitro* and *in vivo*. *J. Physiol.* 493:471–84

Traub RD, Whittington MA, Stanford IM, Jefferys JGR. 1996b. A mechanism for generation of long-range synchronous fast oscillations in the cortex. *Nature* 383:621–24

Traub RD, Wong RKS. 1982. Cellular mechanism of neuronal synchronization in epilepsy. *Science* 216:745–47

van Vreeswijk C, Abbott LF, Ermentrout GB. 1994. When inhibition not excitation synchronizes neural firing. *J. Comput. Neurosci.* 1:313–21

Venance L, Rozov A, Blatow M, Burnashev N, Feldmeyer D, Monyer H. 2000. Connexin expression in electrically coupled postnatal rat brain neurons. *Proc. Natl. Acad. Sci. USA* 97:10260–65

Wang X-J, Buzsáki G. 1996. Gamma oscillation by synaptic inhibition in a hippocampal interneuronal network model. *J. Neurosci.* 16:6402–13

Wang X-J, Rinzel J. 1993. Spindle rhythmicity in the reticularis thalami nucleus: synchronization among mutually inhibitory neurons. *Neuroscience* 53:899–904

White JA, Chow CC, Ritt J, Soto-Treviño C, Kopell N. 1998. Synchronization and oscillatory dynamics in heterogeneous, mutually inhibited neurons. *J. Comput. Neurosci.* 5:5–16

Whittington MA, Doheny HC, Traub RD, LeBeau FEN, Buhl EH. 2001. Differential expression of synaptic and non-synaptic mechanisms during stimulus-induced gamma oscillations *in vitro*. *J. Neurosci.* 21:1727–38

Whittington MA, Stanford IM, Colling SB, Jefferys JGR, Traub RD. 1997a. Spatiotemporal patterns of γ frequency oscillations tetanically induced in the rat hippocampal slice. *J. Physiol.* 502:591–607

Whittington MA, Traub RD, Faulkner HJ, Stanford IM, Jefferys JGR. 1997b. Recurrent excitatory postsynaptic potentials induced by synchronized fast cortical oscillations. *Proc. Natl. Acad. Sci. USA* 94:12198–203

Whittington MA, Traub RD, Jefferys JGR. 1995. Synchronized oscillations in interneuron networks driven by metabotropic glutamate receptor activation. *Nature* 373:612–15

Williams JH, Kauer JA. 1997. Properties of carbachol-induced oscillatory activity in rat hippocampus. *J. Neurophysiol.* 78:2631–40

Ylinen A, Bragin A, Nádasdy Z, Jandó G, Szabó I, et al. 1995a. Sharp wave-associated high frequency oscillation (200 Hz) in the intact hippocampus: network and intracellular mechanisms. *J. Neurosci.* 15:30–46

Ylinen A, Soltész I, Bragin A, Penttonen M, Sik A, Buzsáki G. 1995b. Intracellular correlates of hippocampal theta rhythm in identified pyramidal cells, granule cells and basket cells. *Hippocampus* 5:78–90

Annu. Rev. Neurosci. 2004. 27:279–306
doi: 10.1146/annurev.neuro.27.070203.144130
First published online as a Review in Advance on March 5, 2004

THE MEDIAL TEMPORAL LOBE*

Larry R. Squire,[1,2,3,4] Craig E.L. Stark,[5] and Robert E. Clark[1,2]

[1]Veterans Affairs Healthcare System, San Diego, California 92161; email: lsquire@ucsd.edu, reclark@ucsd.edu;
[2]Department of Psychiatry, [3]Department of Neurosciences, [4]Department of Psychology, University of California, San Diego, La Jolla, California 92093;
[5]Department of Psychological and Brain Sciences, Johns Hopkins University, Baltimore, Maryland 21205; email: cstark@jhu.edu

Key Words memory, declarative, hippocampus, perirhinal cortex, amnesia, fMRI

■ **Abstract** The medial temporal lobe includes a system of anatomically related structures that are essential for declarative memory (conscious memory for facts and events). The system consists of the hippocampal region (CA fields, dentate gyrus, and subicular complex) and the adjacent perirhinal, entorhinal, and parahippocampal cortices. Here, we review findings from humans, monkeys, and rodents that illuminate the function of these structures. Our analysis draws on studies of human memory impairment and animal models of memory impairment, as well as neurophysiological and neuroimaging data, to show that this system (*a*) is principally concerned with memory, (*b*) operates with neocortex to establish and maintain long-term memory, and (*c*) ultimately, through a process of consolidation, becomes independent of long-term memory, though questions remain about the role of perirhinal and parahippocampal cortices in this process and about spatial memory in rodents. Data from neurophysiology, neuroimaging, and neuroanatomy point to a division of labor within the medial temporal lobe. However, the available data do not support simple dichotomies between the functions of the hippocampus and the adjacent medial temporal cortex, such as associative versus nonassociative memory, episodic versus semantic memory, and recollection versus familiarity.

INTRODUCTION

A link between the medial temporal lobe and memory function can be found in clinical case material more than a century old (von Bechterew 1900), but the point became firmly established only when the profound effects of medial temporal lobe resection on memory were documented systematically by Brenda Milner in a patient who became known as H.M. (Milner 1972, Scoville & Milner 1957). At the time that H.M. was first described, the anatomy of the medial temporal lobe was poorly understood, and it was not known what specific damage within this large region was responsible for H.M.'s memory impairment.

Ultimately, an animal model of human memory impairment was developed in the nonhuman primate (Mishkin et al. 1982, Squire & Zola-Morgan 1983). Cumulative behavioral work with the animal model over a 10-year period, together with neuroanatomical studies, succeeded in identifying the anatomical components of the medial temporal lobe memory system (Squire & Zola-Morgan 1991): the hippocampal region (the CA fields, the dentate gyrus, and the subicular complex) and the adjacent entorhinal, perirhinal, and parahippocampal cortices that make up much of the parahippocampal gyrus. The anatomical studies described the boundaries and connectivity of these areas, initially in the monkey and subsequently in the rat (Burwell et al. 1995, Insausti et al. 1987, Lavenex & Amaral 2000, Suzuki & Amaral 1994). In outline, the hippocampus lies at the end of a cortical processing hierarchy, and the entorhinal cortex is the major source of its cortical projections. In the monkey, nearly two thirds of the cortical input to the entorhinal cortex originates in the adjacent perirhinal and parahippocampal cortices, which in turn receive widespread projections from unimodal and polymodal areas in the frontal, temporal, and parietal lobes as well as from retrosplenial cortex (Figure 1).

During the period in which this memory system was identified, it became understood that the system is specifically important for declarative memory (Schacter & Tulving 1994, Squire 1992). Declarative memory supports the capacity to recollect facts and events and can be contrasted with a collection of nondeclarative memory abilities including skills and habits, simple forms of conditioning, and the phenomenon of priming. What is acquired by nondeclarative memory is expressed through performance rather than recollection. Different forms of nondeclarative

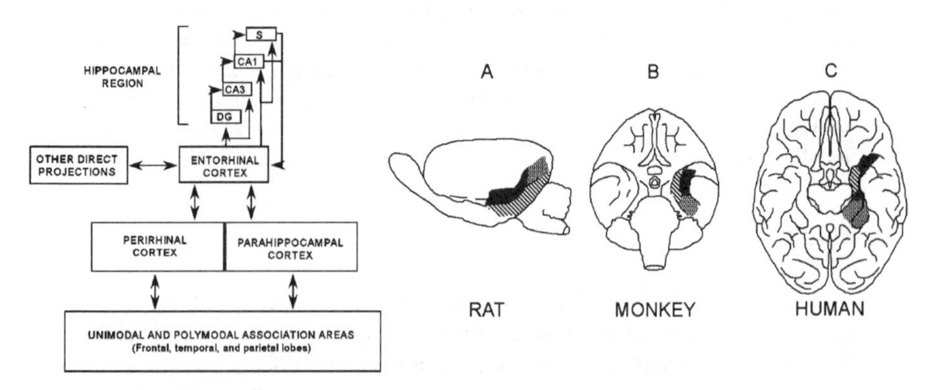

Figure 1 (*Left*) A schematic view of the medial temporal lobe structures important for declarative memory. S, subicular complex; DG, dentate gyrus; CA1, CA3, the CA fields of the hippocampus. (*Right*) Lateral view of the rat brain (*A*) and ventral views of the monkey (*B*) and human (*C*) brain showing the borders of perirhinal cortex (*gray*), entorhinal cortex (*diagonal stripes*), and parahippocampal cortex (*mottled shading*). In the rat, the parahippocampal cortex is termed postrhinal cortex. (Adapted from Burwell et al. 1996.)

memory depend on the integrity of specific brain systems, including for example the neostriatum, the amygdala, and the cerebellum (Eichenbaum & Cohen 2001, Squire & Knowlton 1999).

As a first approximation, one can say that all the structures of the medial temporal lobe contribute in some way to declarative memory. The evidence for this is twofold. First, patients with histological evidence of damage limited primarily to the hippocampal region (patients R.B., G.D., L.M., and W.H.; Rempel-Clower et al. 1996, Zola-Morgan et al. 1986) have a moderately severe memory impairment but one that is less severe than is found in patients with larger lesions that include the medial temporal cortex adjacent to the hippocampus (patients H.M. and E.P.; Corkin et al. 1997, Stefanacci et al. 2000). Second, in systematic comparisons of monkeys with medial temporal lobe lesions, monkeys with limited hippocampal lesions were moderately impaired, whereas monkeys with lesions that included the hippocampal region as well as adjacent cortex were severely impaired (Zola-Morgan et al. 1994). These simple facts are not an argument for the idea that the structures of the medial temporal lobe must work together as a single and uniform functional unit (see Zola-Morgan et al. 1994), though this view has sometimes been attributed to us (e.g., Murray & Bussey 2001).

THE NATURE OF THE DISORDER

The hallmark of the impairment following medial temporal lobe lesions is profound forgetfulness. There are three important aspects of this condition. First, the impairment is multimodal. Memory is affected regardless of the sensory modality in which information is presented (Levy et al. 2003, Milner 1972, Murray & Mishkin 1984, Squire et al. 2001). This finding accords with the fact that the structures of the medial temporal lobe constitute one of the convergent zones of cortical processing and receive input from all the sensory modalities (Lavenex & Amaral 2000).

Second, immediate memory is intact. Even patients with large medial temporal lobe lesions have a normal digit span (Drachman & Arbit 1966). Further, when the material to be learned can be easily rehearsed (as in the case of 2 or 3 digits), retention may succeed even after several minutes (Milner et al. 1998). In contrast, when the material to be learned is difficult to rehearse (as in the case of complex designs), an impairment may be evident within 10 s (Buffalo et al. 1998).

Although intact immediate memory and impaired delayed memory are easy to demonstrate in humans, during the period that animal models were being developed it was a matter of some debate whether this same finding could be obtained in experimental animals (Horel 1994, Ringo 1991). Experimental work eventually established this point unambiguously (Alvarez et al. 1994, Clark et al. 2001, Overman et al. 1991). For example, rats with hippocampal lesions learned the delayed nonmatching-to-sample task at a normal rate using a short delay (4 s) between sample and choice (Clark et al. 2001). Yet performance was impaired when the delay was increased to 1 or 2 min. Further, during delayed testing, performance remained fully intact when 4-s delay trials were introduced intermittently, thereby indicating both retention of the nonmatching rule and intact short-term memory. Finally, even when extended training was given at a 1-min delay, exceeding the training given at the 4-s delay, performance remained intact at the short delay and impaired at the long delay.

The third notable aspect of the disorder is that memory impairment can occur against a background of intact perceptual abilities and intellectual functions. This finding is most unambiguous in patients with damage limited to the hippocampal formation (the hippocampal region plus entorhinal cortex) (Schmolck et al. 2000, 2002). Findings for patient H.M. are also relevant. Patient H.M. has bilateral damage to the hippocampal formation as well as perirhinal cortex, and he scores normally on intelligence tests as well as on many tests of perceptual function and lexical knowledge (Kensinger et al. 2001, Milner et al. 1968). It is true that H.M. exhibits mild impairment on a few tests of semantic knowledge (Schmolck et al. 2002), but H.M. has some bilateral damage to anterolateral temporal cortex, lateral to the medial temporal lobe (Corkin et al. 1997), and it is possible that the lateral damage is responsible for this impairment. Indeed, in patients with variable damage to lateral temporal cortex, the severity of impaired semantic knowledge was related to the extent of lateral temporal damage (Schmolck et al. 2002).

During the 1960s and 1970s, when human amnesia began to be widely and systematically studied, there was considerable interest in whether the disorder was principally one of storage or retrieval (cf. Squire 1982). The weight of the evidence since then has favored storage and the idea that the hippocampus and related structures are needed to establish representations in long-term memory. In the case of short-lasting episodes of severe memory impairment, such as transient global amnesia (Kritchevsky et al. 1988), the events that occur during the period of amnesia are not subsequently remembered after recovery from amnesia. New learning again becomes possible, but events from the amnesic episode itself do not return to memory. Accordingly, when the medial temporal lobe is damaged, representations in neocortex that are initially established in immediate memory must reach some abnormal fate. If the medial temporal lobe is not functional at the time of learning, memory is not established in a usable way and is not available later. The most direct evidence for this idea comes from combined single-cell recording and lesion work in monkeys (Higuchi & Miyashita 1996; see below).

To take a storage view of memory impairment is not to suppose that the medial temporal lobe is the permanent repository of memory. Because remote memory is spared even in patients with large medial temporal lobe lesions (see Retrograde Memory), permanent memory must be stored elsewhere. For example, higher visual area TE, lateral to the medial temporal region, is thought to be a site for long-term, visual memory storage (Mishkin 1982, Miyashita 1993). The medial temporal lobe works in conjunction with area TE to establish long-term visual memory. In monkeys, changes in single-cell activity have been observed within the medial temporal lobe during and after the formation of long-term, associative memory (Messinger et al. 2001, Wirth et al. 2003, Naya et al. 2003), but it is not known how long such changes persist or for how long they may be needed for the expression of long-term memory.

It was also found that a signal appears in the medial temporal lobe during the recall of visual associative information. When monkeys performed an associative memory task (when A is presented as a sample, choose A′, not B; conversely, when A′ is presented as a sample, choose A, not C), perceptual signals (e.g., in response to stimulus A) were evident in area TE before they were evident in perirhinal cortex (Naya et al. 2001). In contrast, memory-retrieval signals, initiated during the short delay between the sample and the choice stimuli, appeared first in perirhinal cortex and then in TE. Further, lesions of perirhinal and entorhinal cortex abolished the ability of neurons in area TE to represent associations between stimulus pairs (Higuchi & Miyashita 1996). The implication is that neurons in area TE are part of long-term memory representations and that the connections from the medial temporal lobe to temporal neocortex are needed to maintain previously acquired representations and to establish new ones.

ALTERNATIVE VIEWS OF MEDIAL TEMPORAL LOBE CORTEX

Studies in the monkey have led to the proposal that severe memory impairment results not from damage to medial temporal cortex but from conjoint interruption of axons in the fornix, in fibers of passage through the amygdala, and in the temporal stem (Gaffan 2002). Yet, in patient H.M., the fornix and the temporal stem are almost entirely intact (Corkin et al. 1997). Further, histologically confirmed damage limited to the hippocampal formation can cause clinically significant, disabling memory impairment despite sparing of amygdala, temporal stem, and fornix (patients G.D. and L.M.; Rempel-Clower et al. 1996). Also, temporal stem white matter lesions in monkeys severely impair pattern discrimination learning (Zola-Morgan et al. 1982). Yet, the large medial temporal lobe lesions that established a model of amnesia in the monkey spare the learning and retention of pattern discriminations (Zola-Morgan et al. 1982). Humans with medial temporal lobe lesions can learn a two-choice discrimination in one or two trials, like normal individuals, but they then forget what they learned (Squire et al. 1988). In contrast, normal monkeys learn the pattern discrimination task gradually during several hundred trials in a manner reminiscent of skill learning (Iversen 1976). Accordingly, pattern discrimination learning in the monkey is better viewed as a task of nondeclarative memory (visual habit formation) (Mishkin et al. 1984). Pattern discrimination learning is dependent on an inferior temporal lobe–neostriatal pathway, and impaired pattern discrimination learning in the monkey does not model the impairment in human amnesia (Fernandez-Ruiz et al. 2001, Phillips et al. 1988, Teng et al. 2000).

Other work has asked whether perirhinal cortex supports primarily memory functions or whether it might also have a role in perceptual processing of visual information, for example, in the ability to identify complex objects (Buckley & Gaffan 1998, Buckley et al. 2001, Eacott et al. 1994). The proposal that perirhinal cortex is important for visual identification of objects is founded primarily on lesion studies in monkeys. Yet, it is difficult to test experimental animals for the ability to identify objects independent of their ability to learn about the objects (Bussey et al. 2003, Buffalo et al. 1999). Accordingly, some impairments attributed to a perceptual deficit could have resulted from impaired learning.

The perirhinal cortex lies medial and adjacent to unimodal visual area TE. The evidence that perirhinal cortex supports primarily memory functions comes, in part, from the sharp distinction that is found between the effects of perirhinal and TE lesions in the monkey (Buffalo et al. 1999). Perirhinal lesions cause multimodal (visual and tactual) memory impairment at long retention delays but not at short delays. In contrast, TE lesions cause visual (not tactual) memory impairment that appears even at very short retention delays, as would be predicted for a structure that supports visual information processing. For example, in the visual paired-comparison task, monkeys see two identical displays and then, after a variable delay, see the old display and a new one. Normal monkeys prefer to look at the new display, thereby indicating that they remember the familiar and presumably less interesting old display. At the shortest (1-s) delay, perirhinal lesions spared performance (66.4% preference for the new display; controls, 66.4%) but impaired performance at delays from 10 s to 10 min (56.6% preference; controls, 64.6%). In contrast, monkeys with TE lesions exhibited no preference at any delay (1 s to 10 min; 50.2% preference; chance = 50%).

Hampton & Murray (2002) systematically evaluated the perceptual abilities of monkeys with perirhinal lesions. Monkeys first acquired a number of visual discriminations and then were given probe trials to assess their ability to perform when the stimuli were manipulated. Monkeys with perirhinal lesions performed as well as controls when the stimuli were rotated (30°–120°), enlarged, shrunk, presented with color removed, or degraded with masks. The various manipulations did reduce performance, but performance was reduced to the same extent in operated and control animals.

When perceptual tasks were given to three patients with complete damage to perirhinal cortex bilaterally, performance was good on seven different discrimination tasks, including four that had revealed impairment in monkeys (Stark & Squire 2000). It is also notable that all three patients have some damage lateral to the perirhinal cortex, for example, in the fusiform gyrus and (for two of the patients) in inferolateral temporal cortex. Accordingly, some impairments found in these patients, for example, impairments in long-established semantic knowledge, may be due to lateral temporal damage rather than perirhinal damage (Schmolck et al. 2002). Indeed, this possibility seems likely, inasmuch as all three patients have complete damage to perirhinal cortex, but they differ in the severity of their semantic knowledge deficits and, correspondingly, in the extent of their lateral temporal damage.

The most recent proposal attributing perceptual functions to perirhinal cortex is more narrow than the idea that the perirhinal cortex is important for object identification or all perceptually difficult discriminations. Rather, the perirhinal cortex is proposed to be important for visual discriminations that have a high degree of feature ambiguity, that is, discriminations where a feature is rewarded when it appears as part of one object but not when part of another (Bussey et al. 2003). Yet, it is noteworthy that patient H.M. performs normally on a number of perceptual tasks, despite his extensive damage to perirhinal cortex (Milner et al. 1968). It would be useful to compare monkeys and patients with perirhinal damage on comparable tests of perceptual ability.

ACQUISITION OF KNOWLEDGE DESPITE SEVERE AMNESIA

Memory-impaired patients with medial temporal lobe lesions often have some capacity for learning about facts and events (Tulving et al. 1991, Westmacott & Moscovitch 2001). In these cases, the question of interest concerns the kind of learning that occurs. Is learning supported by whatever declarative memory remains and by residual medial temporal lobe tissue, or is some other kind of (nondeclarative) memory, and some other brain system, able to support performance?

In one study, the severely amnesic patient E.P. successfully acquired fact-like information in the form of novel three-word sentences, e.g., venom caused fever (Bayley & Squire 2002). Across 12 weeks and 24 study sessions, E.P. gradually improved his performance on both cued-recall tests (e.g., venom caused ???) and forced-choice tests. A number of observations indicated that what E.P. had learned was not factual knowledge in the ordinary sense. First, he never indicated that he believed he was producing correct answers, and he never made reference to the learning sessions. Second, his confidence ratings were the same for his correct answers as for his incorrect answers. Third, he failed altogether (1 of 48 correct) when the second word in each sentence was replaced by a synonym (venom induced ???). Thus, what E.P. learned was rigidly organized, unavailable as conscious knowledge, and in every respect exhibited the characteristics of nondeclarative memory, perhaps something akin to perceptual learning.

When medial temporal lobe damage is as extensive as it is in E.P. (Stefanacci et al. 2000), whatever fact-like learning can be acquired probably occurs directly in neocortex. Yet, there are many reported cases of less severe memory impairment where patients successfully acquire factual information as declarative knowledge (i.e., patients are aware of what they have learned, and their confidence ratings are commensurate with their successful performance) (Hamann & Squire 1995, Shimamura & Squire 1988, Westmacott & Moscovitch 2001). In these cases, structures remaining intact within the medial temporal lobe are likely responsible for the successful learning.

DIVISION OF LABOR WITHIN THE MEDIAL TEMPORAL LOBE

There has been extended exploration of the possibility that different structures within the medial temporal lobe may contribute to declarative memory in different ways. Discussion of this issue properly begins with a neuroanatomical perspective.

Information from neocortex enters the medial temporal lobe at different points. Thus, perirhinal cortex receives stronger projections from unimodal visual areas than does parahippocampal cortex, and parahippocampal cortex receives prominent projections from dorsal stream areas, including retrosplenial cortex, area 7a of posterior parietal cortex, and area 46 (Suzuki & Amaral 1994). Correspondingly, in monkeys, visual memory is more dependent on perirhinal cortex than on parahippocampal cortex (Squire & Zola-Morgan 1996), whereas spatial memory is more dependent on parahippocampal cortex (Malkova & Mishkin 2003, Parkinson et al. 1988). Similar results have been obtained in human neuroimaging studies, with the additional finding that the hippocampus is active in relation to both visual and spatial memory (Bellgowan et al. 2003).

The hippocampus is ultimately a recipient of convergent projections from these adjacent cortical structures, which are located earlier in the hierarchy of information processing. Accordingly, the hippocampus may have a special role in tasks that depend on relating or combining information from multiple sources, such as tasks that ask about specific events (episodic memory) or associative memory tasks that require different elements to be remembered as a pair (e.g., a name and a face). A related idea is that tasks that do not have these requirements, such as tasks that ask about general facts (semantic memory) or tasks that ask for judgments of familiarity about recently presented single items, may be supported by the cortex adjacent to the hippocampus (Brown & Aggleton 2001, Tulving & Markowitsch 1998). These ideas lead to two kinds of predictions. First, lesions limited to the hippocampus should disproportionately impair tasks of episodic memory and tasks of associative memory, relative to tasks of semantic memory or single-item memory. Second, limited hippocampal lesions should largely spare memory tasks that do not have these characteristics because such tasks can be supported by the perirhinal and parahippocampal cortices. Although these ideas have been prominent in discussions of medial temporal lobe function, the experimental work reviewed in the following sections provides little support for such sharp distinctions.

Episodic and Semantic Memory

The ability to acquire semantic memory has often been observed to be quite good, and better than the ability to acquire episodic memory, in single-case studies of memory-impaired patients (e.g., Hayman et al. 1993, Verfaellie et al. 2000). Because the general knowledge that makes up semantic memory can be based on multiple learning events, and because episodic memory is, by definition, unique to a single event, it is not surprising that semantic memory should usually be better than episodic memory. So long as memory impairment is not absolute, patients will always do better after many repetitions of material than after a single encounter, just as healthy individuals do. In the present context, one can begin with the observation that patients with limited hippocampal lesions have difficulty learning about single events. The question of interest is whether the acquisition of semantic information is also impaired when damage is limited to the hippocampal region.

A recent study of five patients with limited hippocampal damage found marked deficits in knowledge about events in the news that occurred after the onset of amnesia (Manns et al. 2003b). Memory for remote events (11 to 30 years before amnesia) was intact, and time-limited retrograde amnesia was apparent during the several years before amnesia (see Retrograde Memory). A second study of the same patient group showed that impaired semantic memory was not an indirect result of impaired episodic memory. Healthy individuals were not aided in their recall of facts by being able to recollect episodic details about the circumstances in which they acquired their knowledge (Manns et al. 2003b). Other studies of smaller numbers of patients also found memory for facts to be impaired (Holdstock et al. 2002, Kapur & Brooks 1999, Reed & Squire 1998). Thus, semantic memory and episodic memory are both dependent on the hippocampal region.

Patients with developmental amnesia, who sustained limited hippocampal damage early in life, may provide an exception to this generalization (Baddeley et al. 2001, Vargha-Khadem et al. 1997). The best studied of these cases (Jon) has above-average intelligence and performs normally in language and other scholastic tests, despite having marked day-to-day memory problems since early childhood. It is possible that Jon's capacity for semantic learning is disproportionate to his ability to remember events from day to day, perhaps because early hippocampal damage (but not adult-onset amnesia) affords an opportunity for functional reorganization or compensation through learned strategies. However, an alternative possibility is that patients with developmental amnesia, given sufficient effort and repetition, may be able to acquire considerable semantic knowledge but only in proportion to what would be expected from their day-to-day episodic memory ability. Direct comparisons between adult-onset and developmental amnesia should clarify these issues.

Learning Associations and Learning Single Items

The idea that the ability to combine two unrelated items into memory depends more on the hippocampal region than the ability to learn single items has seemed plausible, and there is some support for this view (Kroll et al. 1996, Giovanello et al. 2003). Two other studies found that performance on these two kinds of tasks was similarly impaired. In the first study, patients saw a continuous stream of two-component stimuli (e.g., word pairs). On each trial, the item could be entirely novel (two new words), novel but with one repeated component (an old word and a new word), familiar but in a novel pairing (two old words recombined to form a new pair), or a true repetition (Stark & Squire 2003). The task was to endorse the true repetitions as having been encountered before and to reject the other items. In five separate experiments, patients with limited hippocampal lesions were impaired similarly across all item types. There was no suggestion that hippocampal damage selectively (or disproportionately) impaired the ability to reject recombined stimulus elements.

A second study evaluated the ability of the same patient group to remember faces, houses, or face-house pairs (Stark et al. 2002). The patients were impaired in all three conditions. To evaluate the severity of the impairment across conditions, the patients were allowed eight presentations of the study list instead of one. Their performance now matched control performance on both the single-item and the associative tests. Thus, the findings from these patient studies suggest that the hippocampus is similarly important for single-item and associative memory.

Recognition Memory: Recollection and Familiarity

One focus of discussion about possible division of labor within the medial temporal lobe has concerned recognition memory (the capacity to identify an item as one that was recently encountered). Recognition memory is widely viewed as consisting of two components, a recollective (episodic) component and a familiarity component. Recollection provides information about the episode in which an item was encountered, and familiarity provides information that an item was encountered but does not provide any knowledge about the learning context. It has been proposed that recollection depends especially on the hippocampus and that familiarity depends more on the adjacent cortex (Brown & Aggleton 2001, Rugg & Yonelinas 2003, Yonelinas 1998).

Studies of both humans and experimental animals are relevant to this proposal. In the case of patients with lesions limited to the hippocampal region, the remember/know technique has been used to assess recollection and familiarity, respectively. When a recently presented item evokes a recollection of the learning episode itself, one is said to remember. When a recently presented item is experienced simply as familiar, one is said to know (Tulving 1985). In one study of seven patients with limited hippocampal damage, the capacity for knowing was unmistakably impaired, and this impairment was as severe as the impairment in remembering (Manns et al. 2003a).

A disadvantage of the remember/know technique is that it is ultimately subjective, and there is disagreement about how reliably it can index recollection and familiarity (Donaldson 1996). Participants must judge the quality of what they retrieve, and how that judgment is made is open to interpretation by each participant. Perhaps the subjectivity of the method can explain why one patient with apparently limited hippocampal lesions (Holdstock et al. 2002), and three other memory-impaired patients for whom anatomical information was not available (Yonelinas et al. 2002; but see Wixted & Squire 2004), were reported to have a relatively preserved capacity for knowing (familiarity). Further work will be needed to decide this issue.

The distinction between recollection and familiarity is difficult, if not impossible, to apply to experimental animals because methods are not available in animals to reveal a capacity for the mental time travel that is central to episodic recollection (Tulving 2002; for a demonstration of episodic-like memory in scrub jays, see Clayton 1998). Nevertheless, it is of interest that studies of both monkeys and rodents have typically found recognition memory impairment following restricted hippocampal lesions (for monkeys, Beason-Held et al. 1999, Zola et al. 2000; for an exception in a study involving two-stage lesions and a different testing method, see Murray & Mishkin 1998). For rodents, the findings are rather clear when the lesions are sufficiently large and the retention delay is sufficiently long. Large hippocampal lesions impaired performance in the delayed nonmatching to sample task (Clark et al. 2001; Mumby et al. 1992, 1995, but see Mumby et al. 1996) and in the novel object recognition task (Clark et al. 2000, Gould et al. 2002). Similar findings were obtained after intrahippocampal injections of APV (Baker & Kim 2002) and in mice lacking the NMDAR-1 subunit in the hippocampal region (Rampon et al. 2000).

Impaired recognition performance has sometimes not been observed following subtotal hippocampal lesions (Gaskin et al. 2003), even in the same rats that exhibited impaired spatial learning in the standard water maze task (Duva et al. 1997). This finding can be potentially understood as an effect of lesion size. The water maze task is quite sensitive to hippocampal lesions. Damage to the dorsal hippocampus involving either 30% to 50% or 40% to 60% of total hippocampal volume severely impaired water maze performance (Broadbent et al. 2003, Moser et al. 1995). In contrast, neither a dorsal nor a ventral lesion that damaged 50% of total hippocampal volume affected performance on a recognition memory task (novel object recognition). Yet, as the hippocampal lesion increased in size from 50% to 100% of total hippocampal volume, a deficit gradually appeared, and the deficit was severe after a complete lesion (Broadbent et al. 2003).

These findings show that recognition memory in the rat depends on the integrity of the hippocampus, but less hippocampal tissue is needed to support object recognition than is needed to support spatial learning. Spatial memory tasks have much in common with tasks of free recall, and establishing a representation that can support unaided recall may require more hippocampal circuitry than establishing a representation sufficient to support recognition.

The novel object recognition task, as given to rodents, may be useful for considering the distinction between recollection and familiarity. In this task, animals initially explore two identical objects, and later they explore a new object and a copy of the old object. Recognition memory is demonstrated when animals explore the novel object more than the old object. This task depends on a spontaneous tendency to seek novelty and would seem to depend less on recollection of a previous event and more on the simple detection of familiarity. To the extent that rats and mice do base their performance in this task on the ability to discriminate between familiarity and novelty, the impairment in this task after hippocampal lesions provides direct evidence for the importance of the hippocampus in detecting familiarity.

Neurophysiology of Recognition

Recordings from single cells during recognition performance in rodents, monkeys, and humans are broadly consistent with the lesion data and also suggest ways in which the contribution of the hippocampus can be different than the contribution of adjacent cortex (Suzuki & Eichenbaum 2000). Neurons in perirhinal and entorhinal cortex typically respond in a stimulus-specific manner during recognition testing, modifying their firing rate if the stimulus is familiar rather than relatively novel (Suzuki et al. 1997, Young et al. 1997). In contrast, in the hippocampus, neurons tend to signal familiarity versus novelty irrespective of stimulus identity (that is, they provide an abstract recognition signal). These abstract recognition signals can report the familiar/novel status of single stimuli, or they can report the status of single stimuli in conjunction with other task features, such as the spatial position of the stimulus (conjunctive coding) (Wood et al. 1999).

These features of recognition signals in hippocampus and adjacent cortex are not absolute. Stimulus-specific responses can be found in the hippocampus, at least when relatively complex visual stimuli are used (Fried et al. 1997, Wirth et al. 2003). Thus, in humans, neurons in both hippocampus and entorhinal cortex responded selectively to faces and objects and also responded differently according to whether the stimulus was old or new (Fried et al. 1997). Most of the cells responded to familiarity in conjunction with other task features, such as the gender of a face or the facial expression.

The neurophysiological data suggest that all the anatomical components of the medial temporal lobe signal information relevant to recognition memory performance. The conjunctive recognition signals prominent in hippocampus and the stimulus-specific signals commonly found in adjacent cortex may contribute differently to recognition. It is also possible that both kinds of signals are needed to achieve intact recognition performance on most tasks. This possibility may explain why it has been difficult to demonstrate qualitatively distinct effects on recognition following damage to hippocampus or adjacent cortex.

NEUROIMAGING STUDIES OF THE MEDIAL TEMPORAL LOBE

During the past few years, there has been an explosion of interest in using neuroimaging techniques to study medial temporal lobe function and the possible division of labor within this region. While useful information has been obtained from neuroimaging studies, these techniques present certain challenges, which are useful to consider before reviewing recent findings. Principal among these are (*a*) the correlational nature of the data, (*b*) the lack of a true baseline, and (*c*) the technical problem of localizing results across participants to particular brain structures.

First, because neuroimaging techniques provide correlational data, neuroimaging cannot provide evidence about the necessity (or the importance) of a particular structure for a particular function. Early studies of eyeblink conditioning in the rabbit make this point nicely. Multiple-unit activity in the hippocampus develops robustly during delay conditioning in response to the conditioned stimulus, and this activity precedes and predicts the behavioral response (Berger et al. 1980). Yet, hippocampal lesions have no effect on the acquisition or retention of the conditioned response (Mauk & Thompson 1987, Schmaltz & Theios 1972). These findings show that the hippocampus is not important for this conditioned behavior, even though the hippocampus exhibits activity correlated with the behavior. Hippocampal activity would presumably prove itself important if measures of declarative memory were taken, for example, if the animal needed to report when or where the conditioning occurred. The work on eyeblink conditioning makes the simple point that the activity observed in a particular brain region in a neuroimaging study may be incidental to the task that individuals are performing. Further, greater activity in a particular region during task A than during task B does not necessarily mean that this region is more important for task A than for task B.

The second challenge is that neuroimaging techniques are contrastive, and the choice of control task (that is, the baseline task) can determine whether activity increases, decreases, or does not change in association with the task of interest (Gusnard & Raichle 2001). For example, when interspersed 3-s periods of rest were used as a baseline task, neither novel nor familiar scenes elicited activity in the hippocampal region (Stark & Squire 2001a). Yet, rest periods are times of active mental activity. If the rest condition were the only baseline available, one might conclude that familiar and novel pictures do not activate the hippocampus. However, when a tedious, repetitive task served as baseline (judging digits as odd or even), robust hippocampal activity was observed bilaterally in response to both familiar and novel scenes.

The third challenge is the need to accurately align activity in medial temporal lobe structures across participants. Traditional techniques that optimize whole-brain alignment [e.g., aligning to the atlas of Talairach & Tournoux (1988)] do not adequately account for variations in location and shape of medial temporal lobe structures. For example, if one overlays the medial temporal lobes from two brains, each of which has been segmented manually to identify subregions of the medial temporal lobe and then aligned to the Talairach atlas, about half of the voxels are segmented differently in the two brains. For example, a voxel may be segmented as entorhinal cortex in one brain and in perirhinal cortex or outside the medial temporal lobe in the other brain (Stark & Okado 2003).

Investigators have taken three approaches to this issue, all of which have advantages over simply aligning all brains to a common brain atlas (Talairach & Tournoux 1988). One method is to collapse activity for each participant across all the voxels that fall within anatomically defined regions of interest. A second method is to adapt cortical unfolding techniques to map data for each participant onto a two-dimensional map and then to align the individual maps (Zeineh et al. 2000). A third method is to align participants by maximizing the overlap of anatomically defined regions of interest (e.g. hippocampus, perirhinal, entorhinal, and parahippocampal cortex) at the expense of whole-brain alignment [the ROI-AL (region of interest-based alignment) method; Stark & Okado 2003]. All three methods include the essential ingredient of identifying anatomical boundaries in individual brains rather than basing localization on where voxels end up after transforming each brain into a standard atlas.

Specialization of Function Within the Medial Temporal Lobe

A theme of many early neuroimaging studies concerned whether encoding and retrieval are associated with distinct loci of activity in the medial temporal lobe. One complication in looking for such a contrast is that encoding occurs not only when items are first presented for study but also when the same study items and new (foil) items are presented together in a retrieval test. Indeed, activity in the hippocampus and adjacent cortex during retrieval predicts how well the new items will be remembered in a postscan memory test (Stark & Okado 2003). A second complication is that many of the relevant studies make only a coarse division between anterior and posterior regions of the medial temporal lobe so that it is difficult to relate findings to anatomical structures and connectivity.

Although the literature is somewhat mixed, the available work does not suggest any simple, large-scale division of labor for encoding and retrieval (Schacter & Wagner 1999). For example, in one study, encoding and retrieval of face-name pairs activated the full longitudinal axis of the hippocampus (aligned using anatomical ROIs), and the pattern of activity was similar during encoding and retrieval (Small et al. 2001). Further, in a study of picture recognition, both encoding and retrieval were associated with activity in the hippocampal region, the perirhinal cortex, and the parahippocampal cortex (aligned using the ROI-AL technique) (Stark & Okado 2003).

Distinct patterns of activity have sometimes been observed within the medial temporal lobe. In one study, activity was observed in perirhinal cortex during encoding of picture pairs but not during retrieval, whereas activity in the hippocampus and parahippocampal cortex was observed during both encoding and retrieval (Pihlajamäki et al. 2003). In another study, the above-mentioned unfolding technique was used to map activity related to encoding and retrieval of face-name pairs (Zeineh et al. 2003). Encoding but not retrieval was associated with above-baseline (fixation) activity in hippocampal fields CA2 and CA3 and in the dentate gyrus. It is unclear why activity was not observed in field CA1, which is an anatomical bottleneck essential to hippocampal function and to human memory (Zola-Morgan et al. 1986). In contrast, retrieval (and, to a lesser extent, encoding) was associated with above-baseline activity in the posterior subiculum. The right parahippocampal cortex also appeared to exhibit activity during encoding (Figure 3*A* in Zeineh et al. 2003). (For a finding of parahippocampal cortex and subiculum activity in encoding and retrieval, respectively, see Gabrieli et al. 1997). Additional studies using techniques that permit fine-scale anatomical distinctions will be useful. A continuing challenge for all such studies is the need to standardize test protocols and to use carefully selected baseline tasks so that specific aspects of memory and cognition can be isolated and findings can be compared across laboratories.

Another theme of recent neuroimaging studies has concerned the possibility that the hippocampal region (the CA fields, dentate gyrus, and subicular complex) may have identifiably distinct functions relative to the adjacent medial temporal cortex.

For example, as discussed earlier, there has been interest in the idea that the hippocampal region might be especially active during the recollective or associative aspects of declarative memory. Accordingly, some studies have contrasted activity in the medial temporal lobe in association with "remember" responses and "know" responses, which are meant to index recollection and familiarity. Greater activity associated with "remember" than "know" responses may index the recollective aspects of declarative memory (but may also reflect simple differences in the amount of information retrieved or in one's confidence that what is retrieved is correct). Other studies contrasted activity associated with forming or retrieving associations (e.g., face-name pairs) with the activity associated with forming or retrieving single items (e.g., faces or names alone). Greater activity during the successful encoding of a face-name pair than during the successful encoding of a face and a name (but not their association) may index the formation of associations per se.

Although dissociations between recollective or associative memory and familiarity or single-item memory have been reported, the findings do not reveal a sharp distinction between the hippocampal region and adjacent cortex. Recent studies have implicated both the hippocampal region (Davachi et al. 2003; Dobbins et al. 2002; Düzel et al. 2003; Eldridge et al. 2000; Henke et al. 1997, 1999; Kirwan & Stark 2004; Ranganath et al. 2003; Small et al. 2001; Sperling et al. 2001a, 2003; Stark & Squire 2001b; Yonelinas et al. 2001) and the parahippocampal cortex (Davachi et al. 2003; Dobbins et al. 2002; Düzel et al. 2003; Eldridge et al. 2000; Henke et al. 1997, 1999; Ranganath et al. 2003; Kirwan & Stark 2004; Yonelinas et al. 2001) in recollective memory and in the encoding and retrieval of associations. Additionally, several studies have implicated the perirhinal cortex or the entorhinal cortex in these same processes (Dobbins et al. 2002, Düzel et al. 2003, Kirwan & Stark 2004, Pihlajämaki et al. 2003, Sperling et al. 2003). Accordingly, it would be an oversimplification to conclude that the hippocampal region has a specific or unique role in associative or recollective aspects of declarative memory. The same patterns of activity observed in the hippocampal region have been observed in adjacent cortex (most often in parahippocampal cortex). Likewise, it would be an oversimplification to conclude that the cortex adjacent to the hippocampus has a specific or unique role in nonassociative forms of declarative memory. Although there is evidence for nonassociative or familiarity-based activity in the entorhinal and perirhinal cortices (Dobbins et al. 2002, Davachi et al. 2003, Henson et al. 2003, Kirwan & Stark 2004, Ranganath et al. 2003), nonassociative or familiarity-based activity can also be observed in the parahippocampal cortex (Kirwan & Stark 2004) as well as in the hippocampal region (Henson et al. 2003, Small et al. 2001, Stark & Squire 2000, 2001b).

Thus, the considerable data available from recent neuroimaging studies do not lead to any simple conclusions about division of labor within the medial temporal lobe. Although activity in the hippocampal region has been correlated with the associative, recollective, and contextual aspects of declarative memory, activity in the posterior parahippocampal gyrus (parahippocampal cortex) has been correlated with these same aspects of memory. Further, although the perirhinal cortex has been linked to nonassociative (single-item) memory, this region is sometimes active as well in relation to associative memory. Lastly, activity in the hippocampal region has been correlated with nonassociative memory. Neuroimaging techniques present a number of challenges for the objective of illuminating the functional anatomy of memory. Future work will benefit from carrying out more rigorous anatomic localization. In addition, it is striking to what extent results can differ across studies that ostensibly attack the same problem, and it appears that almost any methodological variation will affect what is found. Gains can be obtained by reducing the differences between studies in design, test materials, and data analysis.

SPATIAL MEMORY

Since the discovery of hippocampal place cells in the rodent (O'Keefe & Dostrovsky 1971), an influential idea has been that the hippocampus creates and uses spatial maps and that its predominant function is to support spatial memory (O'Keefe & Nadel 1978). Place cells are best observed in empty environments. When a task is introduced, the same cells come to be activated in relation to the significant features of the task (Eichenbaum et al. 1999). In one study, more than half of the neurons that exhibited task-related activity fired in relation to nonspatial variables (Wood et al. 1999). It is also true that selective hippocampal lesions impair nonspatial memory in rodents (Bunsey & Eichenbaum 1996), monkeys (Doré et al. 1998), and humans (Squire et al. 2001, Levy et al. 2003). Accordingly, spatial memory can be viewed as a subset, a good example, of a broader category (declarative memory), with the idea that this broader category is the province of the hippocampus and related structures (Eichenbaum & Cohen 2001, Squire 1992).

The development of neuroimaging techniques and virtual reality environments has afforded the opportunity to study spatial learning and memory in humans in some detail. Learning one's way through a virtual environment (Hartley et al. 2003, Maguire et al. 1998, Shelton & Gabrieli 2002) or recalling complex routes through a city (Maguire et al. 1997) activated the posterior parahippocampal gyrus (parahippocampal cortex) and sometimes the hippocampus, as well. Activity was often bilateral, but sometimes right unilateral, presumably depending on the strategy that participants used during learning. These activations often appeared to be specifically spatial. For example, activation was greater during wayfinding than when following a well-learned path (Hartley et al. 2003), greater during route learning than when learning via an aerial view (Shelton & Gabrieli 2002), and greater when recalling spatial layouts and landmarks than when recalling nontopographical information (Maguire et al. 1997). These findings imply an important role of the parahippocampal cortex, and possibly hippocampus, in spatial memory. Alternatively, a more abstract formulation is also possible, namely, that the hippocampus is important in both spatial and nonspatial tasks where new information must be acquired and associated in ways that allow it to be used flexibly to guide behavior (McNamara & Shelton 2003).

RETROGRADE MEMORY

Damage to the medial temporal lobe almost always results in some loss of memory for information acquired before the damage occurred. When damage is limited to the hippocampus, entorhinal cortex, or fornix, the retrograde memory impairment is temporally graded, impairing recent memory and sparing more remote memory. In the case of experimental animals, more than a dozen studies have demonstrated this phenomenon, typically across a time course of ~30 days (see Squire et al. 2004). For example, in a study of trace classical eyeblink conditioning in the rat (Takehara et al. 2003), hippocampal lesions 1 day after learning nearly abolished the conditioned response, but the lesion had no effect after 4 weeks. In contrast, lesions of medial prefrontal cortex affected the conditioned response only marginally 1 day after learning but severely affected performance after 4 weeks and had an intermediate effect after 2 weeks. These findings are consistent with the results from 2-deoxyglucose studies following spatial discrimination learning in mice. Metabolic activity decreased in the hippocampus from 5 to 25 days after learning but increased in frontal, anterior cingulate, and temporal neocortex (Bontempi et al. 1999).

Temporal gradients of retrograde amnesia have also been well described in patients with damage limited to the hippocampal region (Kapur & Brooks 1999, Manns et al. 2003b). Here, the amnesia extends across a period of a few years rather than a few weeks and spares more remote memory. Interestingly, there is sparing of remote memory for facts (semantic memory) as well as sparing of remote episodic memory for autobiographical events (Bayley et al. 2003). In one study, 8 patients, including 2 with large medial temporal lobe lesions (E.P. and G.P.), and 25 age-matched controls attempted to recollect early memories, specific to time and place, in response to each of 24 different cue words (e.g., river, bottle). Overall, the recollections of the patients and the controls contained a similar number of details (±5%) and were comparable by several other measures as well. A few memory-impaired patients have been found to have difficulty recalling autobiographical episodes, even from their early life (Bayley & Squire 2003, Cipolotti et al. 2001, Moscovitch et al. 2000). In one study (Bayley & Squire 2003), such patients had significant reductions in brain volume in the frontal and/or temporal lobes. Therefore, it seems likely that, as the anatomy of memory-impaired patients comes to be described more completely, those who fail at remote autobiographical recollection will prove to have damage outside the medial temporal lobe.

Neuroimaging studies of retrograde memory have also been reported, but the results are mixed and difficult to interpret (Haist et al. 2001, Maguire et al. 2001, Maguire & Frith 2003, Niki & Luo 2002, Ryan et al. 2001). One difficulty is that, when individuals in the scanner are asked about an old memory that they have not thought about for some time, they can almost always, after the scanning session, remember whatever they were able to retrieve during the scanning session. Accordingly, activity observed during memory retrieval may be related not to retrieval but to the encoding of new information into long-term memory.

The consolidation view of temporally graded retrograde amnesia begins with the principle that long-term memory is stored as outcomes of processing and in the same regions of neocortex that are specialized for what is to be remembered (Mishkin 1982, Squire 1987). By this view, the hippocampus initially works together with the neocortex to allow memory to be encoded and then to be accessible. Through a gradual process of reorganization, connections among the cortical regions are progressively strengthened until the cortical memory can be accessed independently of the hippocampus (McClelland et al. 1995, Squire & Alvarez 1995).

The available data appear to discount an alternative proposal, which states that the hippocampus and related structures are always necessary for recalling the richness of detail available in autobiographical recollections (Nadel & Moscovitch 1997). Furthermore, the alternative view proposes that temporal gradients are a byproduct of incomplete lesions. Yet, in experimental animals temporal gradients frequently have been reported after complete hippocampal lesions (Clark et al. 2002, Kim et al. 1995, Winocur et al. 2001).

One feature of human retrograde amnesia, scarcely explored in animal studies, is that retrograde memory loss is extensive when the damage includes not only hippocampus but also the adjacent cortex (Reed & Squire 1998). Thus, patient E.P. has intact recollections of his early life but nevertheless has retrograde amnesia covering several decades. One possibility is that the perirhinal and parahippocampal cortices do not have the temporary role in memory storage that has been attributed to the hippocampus itself. Physiological changes and changes in the distribution of divergent projections from TE have been described in perirhinal cortex after the learning of visual-paired associates (Naya et al. 2003, Yoshida et al. 2003), but it is not yet known whether these changes are needed to guide memory storage in the adjacent area TE (Higuchi & Miyashiata 1996) or whether these changes are themselves part of an essential long-term memory store. Interestingly, temporal gradients of retrograde amnesia have been reported in rats following selective lesions of perirhinal cortex (Kornecook et al. 1999, Wiig et al. 1996), and studies of larger lesions would be informative, for example lesions that include both hippocampus and postrhinal cortex.

One area of continuing uncertainty concerns the status of spatial memory following medial temporal lobe lesions. In humans, remote spatial memory is spared even following large medial temporal lesions. Thus, patient E.P. could recall the spatial layout of the region where he grew up and from which he moved away as a young adult more than 50 years earlier (Teng & Squire 1999; for a description of patient K.C., who could also navigate in his home environment, see Rosenbaum et al. 2000). E.P. performed as well as age-matched controls who had grown up in the same region and also moved away. He could mentally navigate, construct novel routes, and point correctly to landmarks while imagining himself at various locations. Yet E.P. has no knowledge of the neighborhood where he moved in 1993, after he became amnesic; and although he lives within two miles of the Pacific Ocean, he cannot when asked point in the direction of the ocean. These findings show that the human medial temporal lobe is needed to acquire new spatial knowledge but is not the repository of remotely acquired spatial maps.

The few studies available involving remote spatial memory in rodents have led to mixed results. Although two studies of rats with hippocampal lesions found evidence for sparing of remote spatial memory (64-day-old memory, Ramos 1998; 98-day-old memory, Kubie et al. 1999), in studies involving the most widely used test of spatial memory (the Morris water maze) remote spatial memory was impaired (Bolhuis et al. 1994, Mumby et al. 1999, Sutherland et al. 2001). One study (Clark et al. 2003) used three different tests of spatial memory: the standard water maze; the annular water maze, which removes the need for spatial navigation (Hollup et al. 2001); and a dry-land version of the water maze. In all three tasks, animals with large hippocampal lesions exhibited impaired spatial memory, even when the learning-surgery interval reached 98 days.

It is unclear why the findings in rodents differ from the findings in humans. One difference between the two kinds of studies is that spatial learning in rodents occurred during a limited period of time when the animals were adults, whereas the patient studies involved information acquired over many years while the patients were growing up. Studies of spatial learning in very young rodents would be informative. A second possibility is that typical tests of remote spatial memory in the rodent may require some new learning ability because the animal must continually update its location in space to succeed at the retention test (Knowlton & Fanselow 1998). In contrast, patients do not need to acquire new information in order to answer questions about their remote spatial memory. Additional work with reversible lesions in rodents, introduced early or late after learning, should be useful in resolving this issue (Riedel et al. 1999).

CONCLUSION

Study of the medial temporal lobe and memory is benefited by the possibility of addressing similar questions in humans, monkeys, and rodents using a variety of techniques: lesions, neuroanatomical tract-tracing, neuroimaging, single-cell recording, and manipulation of gene expression. Many questions are currently under debate, but the clearest path to settling these questions, and answering the next ones, lies in approaches that begin with thorough anatomical information about lesions, neuroimages, and electrode sites. Promising directions include parallel lesion studies in patients and neuroimaging studies in healthy volunteers, parallel approaches to neuroimaging of humans and single-cell recording in monkeys, and genetic studies of mice that build on what is learned from humans and monkeys. In all these endeavors, there is a need for standardized behavioral test protocols as well as programs of work that build cumulatively from study to study. Two additional topics have not been considered here because the work is too new to fully appreciate its significance. The first topic is the phenomenon of neurogenesis in the dentate gyrus and its possible relevance to behavioral plasticity (Kemperman

2002). The second topic is the concept of what has been newly termed reconsolidation (Nader et al. 2000), which revives older claims (Misanin et al. 1968) that reactivation of a consolidated memory can sometimes make information vulnerable to interference or disruption. These and many other topics will occupy students of memory in the years to come.

ACKNOWLEDGMENTS

We thank P. Bayley, N. Broadbent, M. Conroy, J. Gold, A. Shelton, C. Smith, W. Suzuki, and J. Wixted for comments and discussion. The authors have been supported by the Medical Research Service of the Department of Veterans Affairs, the Metropolitan Life Foundation, the James S. McDonnell Foundation, and grants from NIMH and NSF.

The *Annual Review of Neuroscience* is online at http://neuro.annualreviews.org

LITERATURE CITED

Alvarez P, Zola-Morgan S, Squire LR. 1994. The animal model of human amnesia: long-term memory impaired and short-term memory intact. *Proc. Natl. Acad. Sci. USA* 91: 5637–41

Baddeley A, Vargha-Khadem F, Mishkin M. 2001. Preserved recognition in a case of developmental amnesia: implications for the acquisition of semantic memory? *J. Cogn. Neurosci.* 13:357–69

Baker KB, Kim JJ. 2002. Effects of stress and hippocampal NMDA receptor antagonism on recognition memory in rats. *Learn. Mem.* 9:58–65

Bayley PJ, Hopkins RO, Squire LR. 2003. Successful recollection of remote autobiographical memories by amnesic patients with medial temporal lobe lesions. *Neuron* 37: 135–44

Bayley PJ, Squire LR. 2002. Medial temporal lobe amnesia: gradual acquisition of factual information by nondeclarative memory. *J. Neurosci.* 22:5741–48

Bayley PJ, Squire LR. 2003. The neuroanatomy of remote autobiographical memory. *Soc. Neurosci.* 514.15 (Abstr.)

Beason-Held LL, Rosene DL, Killiany RJ, Moss MB. 1999. Hippocampal formation lesions produce memory impairment in the rhesus monkey. *Hippocampus* 9:562–74

Bellgowan PSF, Buffalo EA, Bodurka J, Martin A. 2003. High resolution imaging of the anterior medial temporal lobe during object and spatial memory. *Soc. Neurosci.* 556.1 (Abstr.)

Berger TW, Laham RI, Thompson RF. 1980. Hippocampal unit-behavior correlations during classical conditioning. *Brain Res.* 193: 229–48

Bolhuis J, Stewart CA, Forrest EM. 1994. Retrograde amnesia and memory reactivation in rats with ibotenate lesions to the hippocampus or subiculum. *Q. J. Exp. Psychol.* 47:129–50

Bontempi B, Laurent-Demir C, Destrade C, Jaffard R. 1999. Time-dependent reorganization of brain circuitry underlying long-term memory storage. *Nature* 400:671–75

Broadbent NJ, Clark RE, Squire LR. 2003. Small dorsal hippocampal lesions impair spatial memory whereas large hippocampal lesions are required to impair novel object recognition. *Soc. Neurosci.* 938.1 (Abstr.)

Brown MW, Aggleton JP. 2001. Recognition memory: What are the roles of the perirhinal cortex and hippocampus? *Nat. Rev. Neurosci.* 2:51–61

Buckley MJ, Booth MC, Rolls ET, Gaffan D. 2001. Selective perceptual impairments after perirhinal cortex ablation. *J. Neurosci.* 21:9824–36

Buckley MJ, Gaffan D. 1998. Perirhinal cortex ablation impairs visual object identification. *J. Neurosci.* 18:2268–75

Buffalo EA, Ramus SJ, Clark RE, Teng E, Squire LR, Zola SM. 1999. Dissociation between the effects of damage to perirhinal cortex and area TE. *Learn. Mem.* 6:572–99

Buffalo EA, Reber PJ, Squire LR. 1998. The human perirhinal cortex and recognition memory. *Hippocampus* 8:330–39

Bunsey M, Eichenbaum H. 1996. Conservation of hippocampal memory function in rats and humans. *Nature* 379:255–57

Burwell RD, Suzuki W, Amaral DG. 1996. Some observations on the perirhinal and parahippocampal cortices in the rat, monkey and human brains. In *Perception, Memory and Emotion: Frontiers in Neuroscience*, ed. T Ono, BL McNaughton, S Molotchnikoff, ET Rolls, H Nishijo, pp. 95–110. New York: Elsevier. 620 pp.

Burwell RD, Witter MP, Amaral DG. 1995. Perirhinal and postrhinal cortices of the rat: a review of the neuroanatomical literature and comparison with findings from the monkey brain. *Hippocampus* 5:390–408

Bussey TJ, Saksida LM, Murray EA. 2003. Impairments in visual discrimination after perirhinal cortex lesions: testing 'declarative' vs. 'perceptual-mnemonic' view of perirhinal cortex function. *Eur. Neurosci.* 17:649–60

Cipolotti L, Shallice T, Chan D, Fox N, Scahill R, et al. 2001. Long-term retrograde amnesia...the crucial role of the hippocampus. *Neuropsychologia* 39:151–72

Clark RE, Broadbent NJ, Squire LR. 2003. An examination of remote spatial memory with three spatial tasks following complete hippocampal lesions in rats. *Soc. Neurosci.* 938.2 (Abstr.)

Clark RE, Broadbent NJ, Zola SM, Squire LR. 2002. Anterograde amnesia and temporally-graded retrograde amnesia for a nonspatial memory task following lesions of hippocampus and subiculum. *J. Neurosci.* 22:4663–69

Clark RE, West AN, Zola SM, Squire LR. 2001. Rats with lesions of the hippocampus are impaired on the delayed nonmatching-to-sample task. *Hippocampus* 11:176–86

Clark RE, Zola SM, Squire LR. 2000. Impaired recognition memory in rats after damage to the hippocampus. *J. Neurosci.* 20:8853–60

Clayton RS, Dickinson A. 1998. Episodic-like memory during cache recovery by scrub jays. *Nature* 395:272–74

Corkin S, Amaral DG, Gonzalez RG, Johnson KA, Hyman BT. 1997. H.M.'s medial temporal lobe lesion: findings from magnetic resonance imaging. *J. Neurosci.* 17:3964–80

Davachi L, Mitchell JP, Wagner AD. 2003. Multiple routes to memory: distinct medial temporal lobe processes build item and source memories. *Proc. Natl. Acad. Sci. USA* 100:2157–62

Dobbins IG, Rice HJ, Wagner AD, Schacter D. 2002. Memory orientation and success: separable neurocognitive components underlying episodic recognition. *Neuropsychologia* 41:318–33

Donaldson W. 1996. The role of decision processes in remembering and knowing. *Mem. Cognit.* 18:23–30

Doré FY, Thornton JA, White NM, Murray EA. 1998. Selective hippocampal lesions yield nonspatial memory impairment in rhesus monkeys. *Hippocampus* 4:323–29

Drachman DA, Arbit J. 1966. Memory and the hippocampal complex. II. Is memory a multiple process? *Arch. Neurol.* 15:52–61

Duva CA, Floresco SB, Wunderlich GR, Lao TL, Pinel JPJ, Phillips AG. 1997. Disruption of spatial but not object-recognition memory by neurotoxic lesions of the dorsal hippocampus in rats. *Behav. Neurosci.* 111:1184–96

Düzel E, Habib R, Rotte M, Guderian S, Tulving E, Heinze HJ. 2003. Human hippocampal and parahippocampal activity during visual associative recognition memory for spatial and nonspatial stimulus configurations. *J. Neurosci.* 23:9439–44

Eacott MJ, Gaffan D, Murray EA. 1994. Preserved recognition memory for small sets, and impaired stimulus identification for large sets, following rhinal cortex ablations in monkeys. *Eur. J. Neurosi.* 6:1466–78

Eichenbaum H, Cohen NJ. 2001. *From Conditioning to Conscious Recollection: Memory Systems of the Brain.* New York: Oxford Univ. Press

Eichenbaum H, Dudchenko P, Wood E, Shapiro M, Tanila H. 1999. The hippocampus, memory, and place cells: Is it spatial memory or a memory space? *Neuron* 23:209–26

Eldridge LL, Knowlton BJ, Furmanski CS, Bookheimer SY, Engel SA. 2000. Remembering episodes: a selective role for the hippocampus during retrieval. *Nat. Neurosci.* 3:1149–52

Fernandez-Ruiz J, Wang J, Aigner TG, Mishkin M. 2001. Visual habit formation in monkeys with neurotoxic lesions of the ventrocaudal neostriatum. *Proc. Natl. Acad. Sci. USA* 98:4196–201

Fried I, MacDonald KA, Wilson CL. 1997. Single neuron activity in human hippocampus and amygdala during recognition of faces and objects. *Neuron* 18:753–65

Gabrieli JDE, Brewer JB, Desmond JE, Glover GH. 1997. Separate neural bases of two fundamental memory processes in the human medial temporal lobe. *Science* 11:264–66

Gaffan D. 2002. Against memory systems. *Philos. Trans. R. Soc. London* 357:1111–21

Gaskin S, Tremblay A, Mumby DG. 2003. Retrograde and anterograde object recognition in rats with hippocampal lesions. *Hippocampus* 13:962–69

Giovanello KS, Verfaellie M, Keane MM. 2003. Disproportionate deficit in associative recognition relative to item recognition in global amnesia. *Cogn. Affect. Behav. Neurosci.* 3:186–94

Gould TJ, Rowe WB, Heman KL, Mesches MH, Young DA, et al. 2002. Effects of hippocampal lesions on patterned motor learning in the rat. *Brain Res. Bull.* 58:581–86

Gusnard D, Raichle ME. 2001. Searching for a baseline: functional imaging and the resting human brain. *Nat. Rev. Neurosci.* 2:685–94

Haist F, Gore J, Mao H. 2001. Consolidation of human memory over decades revealed by functional magnetic resonance imaging. *Nat. Neurosci.* 4:1139–45

Hamann SB, Squire LR. 1995. On the acquisition of new declarative knowledge in amnesia. *Behav. Neurosci.* 109:1027–44

Hampton RR, Murray EA. 2002. Learning of discriminations is impaired, but generalization to altered views is intact, in monkeys (Macaca mulatta) with perirhinal cortex removal. *Behav. Neurosci.* 116:363–77

Hartley T, Maguire EA, Spiers HJ, Burgess N. 2003. The well-worn route and the path less traveled: distinct neural bases of route following and wayfinding in humans. *Neuron* 37:877–88

Hayman CA, MacDonald CA, Tulving E. 1993. The role of repetition and associative interference in new semantic learning in amnesia: a case experiment. *J. Cogn. Neurosci.* 5:375–89

Henke K, Buck A, Weber B, Wieser HG. 1997. Human hippocampus establishes associations in memory. *Hippocampus* 7:249–56

Henke K, Weber DA, Kneifel L, Wieser HG, Buck A. 1999. Human hippocampus associates information in memory. *Proc. Natl. Acad. Sci. USA* 96:5884–89

Henson RN, Cansino S, Herron J, Robb W, Rugg MD. 2003. A familiarity signal in human anterior medial temporal cortex? *Hippocampus* 13:301–4

Higuchi S, Miyashita Y. 1996. Formation of mnemonic neural responses to visual paired associates in inferotemporal cortex is impaired by perirhinal and entorhinal lesions. *Proc. Natl. Acad. Sci.USA* 93:739–43

Holdstock JS, Mayes AR, Roberts N, Cezayirli E, Isaac CL, et al. 2002. Under what conditions is recognition spared relative to recall after selective hippocampal damage in humans? *Hippocampus* 12:341–51

Hollup SA, Kjelstrup KG, Hoff J, Moser MB, Moser EI. 2001. Impaired recognition of the goal location during spatial navigation in rats with hippocampal lesions. *J. Neurosci.* 21:4505–13

Horel J. 1994. Some comments on the special cognitive functions claimed for the hippocampus. *Cortex* 30:269–80

Insausti R, Amaral DG, Cowan WM. 1987. The entorhinal cortex of the monkey: II. Cortical afferents. *J. Comp. Neurol.* 264:356–95

Iversen SD. 1976. Do hippocampal lesions produce amnesia in animals? *Int. Rev. Neurobiol.* 19:1–49

Kapur N, Brooks DJ. 1999. Temporally-specific retrograde amnesia in two cases of discrete bilateral hippocampal pathology. *Hippocampus* 9:247–54

Kemperman G. 2002. Why new neurons? Possible functions for adult hippocampal neurogenesis. *J. Neurosci.* 22:635–38

Kensinger E, Ullman MT, Corkin S. 2001. Bilateral medial temporal lobe damage does not affect lexical or grammatical processing: evidence from amnesic patient H.M. *Hippocampus* 11:347–60

Kim JJ, Clark RE, Thompson RF. 1995. Hippocampectomy impairs the memory of recently, but not remotely, acquired trace eyeblink conditioned responses. *Behav. Neurosci.* 109:195–203

Kirwan C, Stark CEL. 2004. Medial temporal lobe activation during encoding and retrieval of novel face-name pairs. *Hippocampus.* In press

Knowlton BJ, Fanselow MS. 1998. The hippocampus, consolidation and on-line memory. *Curr. Opin. Neurobiol.* 8:293–96

Kornecook TJ, Anzarut A, Pinel JPJ. 1999. Rhinal cortex, but not medial thalamic, lesions cause retrograde amnesia for objects in rats. *NeuroReport* 10:2853–58

Kritchevsky M, Squire LR, Zouzounis JA. 1988. Transient global amnesia: characterization of anterograde and retrograde amnesia. *Neurology* 38:213–19

Kroll NE, Knight RT, Metcalfe J, Wolf ES, Tulving E. 1996. Cohesion failure as a source of memory illusions. *J. Mem. Lang.* 35:176–96

Kubie JL, Sutherland RJ, Miller RU. 1999. Hippocampal lesions produce a temporally graded retrograde amnesia on a dry version of the Morris swimming task. *Psychobiology* 27:313–30

Lavenex P, Amaral DG. 2000. Hippocampal-neocortical interaction: a hierarchy of associativity. *Hippocampus* 10:420–30

Levy DA, Manns JR, Hopkins RO, Gold JJ, Broadbent NJ, Squire LR. 2003. Impaired visual and odor recognition memory span in patients with hippocampal lesions. *Learn. Mem.* 10:531–36

McNamara TP, Shelton AL. 2003. Cognitive maps and the hippocampus. *Trends Cogn. Sci.* 7:333–35

Maguire EA, Frackowiak RS, Frith CD. 1997. Recalling routes around London: activation of the right hippocampus in taxi drivers. *J. Neurosci.* 17:7103–10

Maguire EA, Frith CD. 2003. Lateral asymmetry in the hippocampal response to the remoteness of autobiographical memories. *J. Neurosci.* 23:5302–7

Maguire EA, Frith CD, Burgess N, Donnett J, O'Keefe J. 1998. Knowing where things are: parahippocampal involvement in encoding objection locations in a virtual large-scale space. *J. Cogn. Neurosci.* 10:61–76

Maguire EA, Henson RN, Mummery C, Frith CD. 2001. Activity in prefrontal cortex, not hippocampus, varies parametrically with the increasing remoteness of memories. *NeuroReport* 12:441–44

Malkova L, Mishkin M. 2003. One-trial memory for object-place associations after separate lesions of hippocampus and posterior parahippocampal region in the monkey. *J. Neurosci.* 23:1956–65

Manns JR, Hopkins RO, Reed JM, Kitchener EG, Squire LR. 2003a. Recognition memory and the human hippocampus. *Neuron* 37:1–20

Manns JR, Hopkins RO, Squire LR. 2003b. Semantic memory and the human hippocampus. *Neuron* 37:127–33

Mauk MD, Thompson RF. 1987. Retention of classically conditioned eyelid responses following acute decerebration. *Brain Res.* 403:89–95

McClelland JL, McNaughton BL, O'Reilly RC. 1995. Why there are complementary learning systems in the hippocampus and neocortex: insights from the successes and failures of connectionist models of learning and memory. *Psychol. Rev.* 102:419–57

McNamara TP, Shelton AL. 2003. Cognitive maps and the hippocampus. *Trends Cogn. Sci.* 7:333–35

Messinger A, Squire LR, Zola SM, Albright TD. 2001. Neuronal representations of stimulus associations develop in the temporal lobe during learning. *Proc. Natl. Acad. Sci. USA* 98:12239–44

Milner B. 1972. Disorders of learning and memory after temporal lobe lesions in man. *Clin. Neurosurg.* 19:421–66

Milner B, Corkin S, Teuber HL. 1968. Further analysis of the hippocampal amnesic syndrome: 14 year follow-up study of H.M. *Neuropsychologia* 6:215–34

Milner B, Squire LR, Kandel ER. 1998. Cognitive neuroscience and the study of memory. *Neuron* 20:445–68

Misanin JR, Miller RR, Lewis DJ. 1968. Retrograde amnesia produced by electroconvulsive shock after reactivation of a consolidated memory trace. *Science* 160:554–58

Mishkin M. 1982. A memory system in the monkey. *Philos. R. Soc. London [Biol.]* 298: 85–92

Mishkin M, Malamut B, Bachevalier J. 1984. Memories and habits: two neural systems. In *Neurobiology of Learning and Memory*, ed. G Lynch, JL McGaugh, NM Weinberger, pp. 65–77. New York: Guilford. 528 pp.

Mishkin M, Spiegler BJ, Saunders RC, Malamut BJ. 1982. An animal model of global amnesia. In *Alzheimer's Disease: A Report of Progress Research*, ed. S Corkin, KL Davis, JH Growdon, EJ Usdin, RJ Wurtman, pp. 235–47. New York: Raven. 525 pp.

Miyashita Y. 1993. Inferior temporal cortex: where visual perception meets memory. *Annu. Rev. Neurosci.* 16:245–63

Moscovitch M, Yaschyshyn T, Ziegler M, Nadel L. 2000. Remote episodic memory and retrograde amnesia: Was Endel Tulving right all along? In *Memory, Consciousness and the Brain: The Tallinn Conference*, ed. E Tulving, pp. 331–45. Philadelphia: Psychol. Press/Taylor & Francis. 397 pp.

Moser M-B, Moser EI, Forrest E, Andersen P, Morris RGM. 1995. Spatial learning with a minislab in the dorsal hippocampus. *Proc. Natl. Acad. Sci. USA* 92:9697–701

Mumby DG, Asturm RS, Weisand MP, Sutherland RJ. 1999. Retrograde amnesia and selective damage to the hippocampal formation: memory for places and object discriminations. *Behav. Brain Res.* 106:97–107

Mumby DG, Pinel JPJ, Kornecook TJ, Shen MJ, Redila VA. 1995. Memory deficits following lesions of hippocampus or amygdala in rat: assessment by an object-memory test battery. *Psychobiology* 23:26–36

Mumby DG, Wood ER, Pinel JPJ. 1992. Object-recognition memory is only mildly impaired in rats with lesions of the hippocampus and amygdala. *Psychobiology* 20:18–27

Mumby DG, Wood ER, Duva CA, Kornecook TJ, Pinel JPJ, Phillips AG. 1996. Ischemia-induced object-recognition deficits in rats are attenuated by hippocampal ablation before or soon after ischemia. *Behav. Neurosci.* 110:266–81

Murray EA, Bussey TJ. 2001. Consolidation and the medial temporal lobe revisited: methodological considerations. *Hippocampus* 11:1–7

Murray EA, Mishkin M. 1984. Severe tactual as well as visual memory deficits following combined removal of the amygdala and hippocampus in monkeys. *J. Neurosci.* 4:2565–80

Murray EA, Mishkin M. 1998. Object recognition and location memory in monkeys with excitotoxic lesions of the amygdala and hippocampus. *J. Neurosci.* 18:6568–82

Nadel L, Moscovitch M. 1997. Memory consolidation, retrograde amnesia and the hippocampal complex. *Curr. Opin. Neurobiol.* 7:217–27

Nader I, Schafe GE, LeDoux JE. 2000. Fear memories require protein synthesis in the

amygdala for reconsolidation after retrieval. *Nature* 406:722–26
Naya Y, Yoshida M, Miyashita Y. 2001. Backward spreading of memory-retrieval signal in the primate temporal cortex. *Science* 291: 661–64
Naya Y, Yoshida M, Miyashita Y. 2003. Forward processing of long-term associative memory in monkey inferotemporal cortex. *J. Neurosci.* 23:2861–71
Niki K, Luo J. 2002. An fMRI study on the time-limited role of the medial temporal lobe in long-term topographical autobiographic memory. *J. Cogn. Neurosci.* 14:500–7
O'Keefe J, Dostrovsky J. 1971. *The hippocampus as a spatial map: preliminary evidence from unit activity in the freely-moving rat. Brain Res.* 34:171–75
O'Keefe J, Nadel L. 1978. *The Hippocampus as a Cognitive Map*. Oxford, UK: Oxford Univ. Press
Overman WH, Ormsby G, Mishkin M. 1991. Picture recognition vs. picture discrimination learning in monkeys with medial temporal removals. *Exp. Brain Res.* 79:18–24
Parkinson JK, Murray E, Mishkin M. 1988. A selective mnemonic role for the hippocampus in monkeys: memory for the location of objects. *J. Neurosci.* 8:4159–67
Phillips RR, Malmut B, Bachevalier J, Mishkin M. 1988. Dissociation of the effects of inferior temporal and limbic lesions on object discrimination learning with 24-h intertrial intervals. *Behav. Brain Res.* 27:99–107
Pihlajamäki M, Hanninen T, Kononen M, Mikkonen M, Jalkanen V, et al. 2003. Encoding of novel picture pairs activates the perirhinal cortex: an fMRI study. *Hippocampus* 13:67–80
Ramos J. 1998. Retrograde amnesia for spatial information: dissociation between intra and extramaze cues following hippocampus lesions in rats. *Eur. J. Neurosci.* 10:3295–301
Rampon C, Ya-Ping T, Goodhouse J, Shimizu E, Kyin M, Tsien JZ. 2000. Enrichment induces structural changes and recovery from nonspatial memory deficits in CA1 NMDAR1-knockout mice. *Nat. Neurosci.* 3: 238–44
Ranganath C, Yonelinas AP, Cohen MX, Dy CJ, Tom SM, D'Esposito MD. 2003. Dissociable correlates of recollection and familiarity within the medial temporal lobes. *Neuropsychologia* 42:2–13
Reed JM, Squire LR. 1998. Retrograde amnesia for facts and events: findings from four new cases. *J. Neurosci.* 18:3943–54
Rempel-Clower N, Zola SM, Squire LR, Amaral DG. 1996. Three cases of enduring memory impairment following bilateral damage limited to the hippocampal formation. *J. Neurosci.* 16:5233–55
Riedel G, Micheau J, Lam AGM, Roloff EvL, Martin SJ, et al. 1999. Reversible neural inactivation reveals hippocampal participation in several memory processes. *Nat. Neurosci.* 2:898–905
Ringo JL. 1991. Memory decays at the same rate in macaques with and without brain lesions when expressed in d' or arcsine terms. *Behav. Brain Res.* 34:123–34
Rosenbaum DL, Priselac S, Kohler S, Black S, Gao F, et al. 2000. Remote spatial memory in an amnesic person with extensive bilateral hippocampal lesions. *Nat. Neurosci.* 3:1044–48
Rugg MD, Yonelinas AP. 2003. Human recognition memory: a cognitive neuroscience perspective. *Trends Cogn. Sci.* 7:313–19
Ryan L, Nadel L, Keil D, Putnam K, Schuyner D, et al. 2001. Hippocampal complex and retrieval of recent and very remote autobiographical memories: evidence from functional magnetic resonance imaging in neurologically intact people. *Hippocampus* 11:707–14
Schacter DL, Tulving E. 1994. What are the memory systems of 1994? In *Memory Systems 1994*, ed. DL Schacter, E Tulving, pp. 1–38. Cambridge, MA: MIT Press
Schacter DL, Wagner AD. 1999. Medial temporal lobe activations in fMRI and PET studies of episodic encoding and retrieval. *Hippocampus* 9:7–24
Schmaltz LW, Theios J. 1972. Acquisition

and extinction of a classically conditioned response in hippocampectomized rabbits (Oryctologaus cuniculus). *J. Comp. Physiol. Psychol.* 79:328–33

Schmolck H, Kensinger E, Corkin S, Squire LR. 2002. Semantic knowledge in patient H.M. and other patients with bilateral medial and lateral temporal lobe lesions. *Hippocampus* 12:520–33

Schmolck H, Stefanacci L, Squire LR. 2000. Detection and explanation of sentence ambiguity are unaffected by hippocampal lesions but are impaired by larger temporal lobe lesions. *Hippocampus* 10:759–70

Scoville WB, Milner B. 1957. Loss of recent memory after bilateral hippocampal lesions. *J. Neurol. Neurosurg. Psychiatry* 20:11–21

Shelton A, Gabrieli JDE. 2002. Neural correlates of encoding space from route and survey perspectives. *J. Neurosci.* 22:2711–17

Shimamura AP, Squire LR. 1988. Long-term memory in amnesia: cued recall, recognition memory, and confidence ratings. *J. Exp. Psychol. Learn. Mem. Cogn.* 14:763–70

Small S, Arun A, Perera G, Delapaz R, Mayeaux R, Stern Y. 2001. Circuit mechanisms underlying memory encoding and retrieval in the long axis of the hippocampal formation. *Nat. Neurosci.* 4:442–49

Sperling R, Bates J, Cocchiarella A, Schacter DL, Rosen BR, Albert M. 2001. Encoding novel face-name associations: a functional MRI study. *Hum. Brain Mapp.* 14:129–39

Sperling R, Chua E, Cocchiarella A, Rand-Giovannetti E, Poldrack RA, et al. 2003. Putting names to faces: Successful encoding of associative memories activates the anterior hippocampal formation. *Neuroimage* 20:1400–10

Squire LR. 1982. The neuropsychology of human memory. *Annu. Rev. Neurosci.* 5:241–73

Squire LR. 1987. *Memory and Brain.* New York: Oxford Univ. Press

Squire LR. 1992. Memory and the hippocampus: a synthesis from findings with rats, monkeys, and humans. *Psychol. Rev.* 99: 195–231

Squire LR, Alvarez P. 1995. Retrograde amnesia and memory consolidation: a neurobiological perspective. *Curr. Opin. Neurobiol.* 5:169–77

Squire LR, Clark RE, Bayley PJ. 2004. Medial temporal lobe function and memory. In *The Cognitive Neurosciences*, ed. M Gazzaniga. Cambridge, MA: MIT Press. 3rd ed. In press

Squire LR, Knowlton B. 1999. The medial temporal lobe, the hippocampus, and the memory systems of the brain. In *The Cognitive Neurosciences*, ed. M Gazzaniga, pp. 765–799. Cambridge, MA: MIT Press. 1276 pp. 2nd ed.

Squire LR, Schmolck H, Stark S. 2001. Impaired auditory recognition memory in amnesic patients with medial temporal lobe lesions. *Learn. Mem.* 8:252–56

Squire LR, Zola SM. 1996. Structure and function of declarative and nondeclarative memory systems. *Proc. Natl. Acad. Sci. USA* 93:13515–22

Squire LR, Zola-Morgan S. 1983. The neurology of memory: the case for correspondence between the findings for human and nonhuman primates. In *The Physiological Basis of Memory*, ed. JA Deutsch, pp. 199–268. New York: Academic. 431 pp.

Squire LR, Zola-Morgan S. 1991. The medial temporal lobe memory system. *Science* 253:1380–86

Squire LR, Zola-Morgan S, Chen K. 1988. Human amnesia and animal models of amnesia: performance of amnesic patients on tests designed for the monkey. *Behav. Neurosci.* 11:210–21

Stark CEL, Bayley PJ, Squire LR. 2002. Recognition memory for single items and for associations is similarly impaired following damage limited to the hippocampal region. *Learn. Mem.* 9:238–42

Stark CEL, Okado H. 2003. Making memories without trying: medial temporal lobe activity associated with incidental memory formation during recognition. *J. Neurosci.* 23:6748–53

Stark CEL, Squire LR. 2000. fMRI activity in the hippocampal region during recognition memory. *J. Neurosci* 20:7776–81

Stark CEL, Squire LR. 2001a. When zero is not zero: the problem of ambiguous baseline conditions in fMRI. *Proc. Natl. Acad. Sci. USA* 98:12760–66

Stark CEL, Squire LR. 2001b. Simple and associative recognition in the hippocampal region. *Learn. Mem.* 8:190–97

Stark CEL, Squire LR. 2003. Hippocampal damage equally impairs memory for single items and memory for conjunctions. *Hippocampus* 13:281–92

Stefanacci L, Buffalo EA, Schmolck H, Squire LR. 2000. Profound amnesia following damage to the medial temporal lobe: a neuroanatomical and neuropsychological profile of patient E.P. *J. Neurosci.* 20:7024–36

Sutherland RJ, Weisend MP, Mumby D, Astur RS, Hanlon FM, et al. 2001. Retrograde amnesia after hippocampal damage: recent vs. remote memories in two tasks. *Hippocampus* 11:27–42

Suzuki WA, Amaral DG. 1994. Topographic organization of the reciprocal connections between the monkey entorhinal cortex and the perirhinal and parahippocampal cortices. *J. Neurosci.* 14:1856–77

Suzuki WA, Eichenbaum H. 2000. The neurophysiology of memory. *Ann. NY. Acad. Sci.* 911:175–91

Suzuki WA, Miller EK, Desimone R. 1997. Object and place memory in macaque entorhinal cortex. *J. Neurophysiol.* 78:1062–81

Takehara K, Kawahara S, Kirino Y. 2003. Time-dependent reorganization of the brain components underlying memory retention in trace eyeblink conditioning. *J. Neurosci.* 23:9896–905

Talairach J, Tournoux P. 1988. *A Co-planar Stereotaxic Atlas of the Human Brain.* Stuttgard, Germany: Thieme

Teng E, Squire LR. 1999. Memory for places learned long ago is intact after hippocampal damage. *Nature* 400:675–77

Teng E, Stefanacci L, Squire LR, Zola SM. 2000. Contrasting effects on discrimination learning following hippocampal lesions or conjoint hippocampal-caudate lesions in monkeys. *J. Neurosci.* 20:3853–63

Tulving E. 1985. Memory and consciousness. *Canad. Psychologist* 26:1–12

Tulving E. 2002. Episodic memory: from mind to brain. *Annu. Rev. Psychol.* 53:1–25

Tulving E, Hayman CAG, MacDonald CA. 1991. Long-lasting perceptual priming and semantic learning in amnesia: a case experiment. *J. Exp. Psychol. Learn. Mem. Cogn.* 17:595–617

Tulving E, Markowitsch HJ. 1998. Episodic and declarative memory: role of the hippocampus. *Hippocampus* 8:198–204

Vargha-Khadem F, Gaffan D, Watkins KE, Connelly A, Van Paesschen W, Mishkin M. 1997. Differential effects of early hippocampal pathology on episodic and semantic memory. *Science* 277:376–80

Verfaellie M, Koseff P, Alexander MP. 2000. Acquisition of novel semantic information in amnesia: effects of lesion location. *Neuropsychologia* 38:484–92

von Bechterew W. 1900. Demonstration eines gehirns mit zerstorung der vorderen und inneren theile der hirnrinde beider schlafenlappen. *Neurologisch. Zentralbl.* 19:990–91

Westmacott R, Moscovitch M. 2001. Names and words without meaning: incidental postmorbid semantic learning in a person with extensive bilateral medial temporal lobe damage. *Neuropsychologia* 15:586–96

Wiig KA, Cooper LN, Bear MF. 1996. Temporally graded retrograde amnesia following separate and combined lesions of the perirhinal cortex and fornix in the rat. *Learn. Mem.* 3:313–25

Winocur G, McDonald RM, Moscovitch M. 2001. Anterograde and retrograde amnesia in rats with large hippocampal lesions. *Hippocampus* 11:18–26

Wirth S, Yanike M, Frank LM, Smith AC, Brown EN, Suzuki WA. 2003. Single neurons in the monkey hippocampus and learning of new associations. *Science* 300:1578–81

Wixted JT, Squire LR. 2004. Recall and recognition are equally impaired in patients with selective hippocampal damage. *Cogn. Affect. Behav. Neurosci.* In press

Yonelinas AP, Hopfinger JB, Buonocore MH, Kroll NE, Baynes K. 2001. Hippocampal, parahippocampal and occipital-temporal contributions to associative and item recognition memory: an fMRI study. *NeuroReport* 12:359–63

Yonelinas AP, Kroll NE, Dobbins I, Lazzara M, Knight RT. 1998. Recollection and familiarity deficits in amnesia: Convergence of remember-know, process dissociation, and receiver operating characteristic data. *Neuropsychology* 12:323–39

Yonelinas AP, Kroll NE, Quamme JR, Lazzara MM, Sauve MJ, et al. 2002. Effects of extensive temporal lobe damage or mild hypoxia on recollection and familiarity. *Nat. Neurosci.* 5:1236–41

Yoshida M, Naya Y, Miyashita Y. 2003. Anatomical organization of forward fiber projections from area TE to perirhinal neurons representing visual long-term memory in monkeys. *Proc. Natl. Acad. Sci. USA* 100: 4257–62

Young BJ, Otto T, Fox GD, Eichenbaum H. 1997. Memory representation within the parahippocampal region. *J. Neurosci.* 17: 5183–95

Zeineh MM, Engel SA, Bookheimer SY. 2000. Application of cortical unfolding techniques to functional MRI of the human hippocampal region. *NeuroImage* 11:668–83

Zeineh MM, Engel SA, Thompson PM, Bookheimer SY. 2003. Dynamics of the hippocampus during encoding and retrieval of face-name pairs. *Science* 299:577–80

Zola-Morgan S, Squire LR, Amaral DG. 1986. Human amnesia and the medial temporal region: enduring memory impairment following a bilateral lesion limited to field CA1 of the hippocampus. *J. Neurosci.* 6:2950–67

Zola-Morgan S, Squire LR, Mishkin M. 1982. The neuroanatomy of amnesia: amygdala-hippocampus versus temporal stem. *Science* 218:1337–39

Zola-Morgan S, Squire LR, Ramus SJ. 1994. Severity of memory impairment in monkeys as a function of locus and extent of damage within the medial temporal lobe memory system. *Hippocampus* 4:483–95

Zola SM, Squire LR, Teng E, Stefanacci L, Buffalo EA, Clark RE. 2000. Impaired recognition memory in monkeys after damage limited to the hippocampal region. *J. Neurosci.* 20:451–63

Annu. Rev. Neurosci. 2004. 27:307–40
doi: 10.1146/annurev.neuro.27.070203.144247

First published online as a Review in Advance on February 26, 2004

THE NEURAL BASIS OF TEMPORAL PROCESSING

Michael D. Mauk[1] and Dean V. Buonomano[2]
[1]Department of Neurobiology and Anatomy, University of Texas, Houston Medical School, Houston, Texas 77030; email: m.mauk@uth.tmc.edu;
[2]Departments of Neurobiology and Psychology, University of California, Los Angeles, California 90095-1761; email: dbuono@ucla.edu

Key Words timing, cerebellum, cortex, dynamics, interval, conditioning

■ **Abstract** A complete understanding of sensory and motor processing requires characterization of how the nervous system processes time in the range of tens to hundreds of milliseconds (ms). Temporal processing on this scale is required for simple sensory problems, such as interval, duration, and motion discrimination, as well as complex forms of sensory processing, such as speech recognition. Timing is also required for a wide range of motor tasks from eyelid conditioning to playing the piano. Here we review the behavioral, electrophysiological, and theoretical literature on the neural basis of temporal processing. These data suggest that temporal processing is likely to be distributed among different structures, rather than relying on a centralized timing area, as has been suggested in internal clock models. We also discuss whether temporal processing relies on specialized neural mechanisms, which perform temporal computations independent of spatial ones. We suggest that, given the intricate link between temporal and spatial information in most sensory and motor tasks, timing and spatial processing are intrinsic properties of neural function, and specialized timing mechanisms such as delay lines, oscillators, or a spectrum of different time constants are not required. Rather temporal processing may rely on state-dependent changes in network dynamics.

INTRODUCTION

In his chapter "The Problem of Serial Order in Behavior," Karl Lashley (1951) was among the first neurophysiologists to broach the issue of temporal processing.

> Temporal integration is not found exclusively in language; the coordination of leg movements in insects, the song of birds, the control of trotting and pacing in a gaited horse, the rat running the maze, the architect designing a house, and the carpenter sawing a board present a problem of sequences of action which cannot be explained in terms of succession of external stimuli.

Lashley emphasized the inherently temporal nature of our environment. He explains that without an understanding of the neural mechanisms underlying our ability to process the order, interval, and duration of sensory and motor events, it is not possible to gain insight into how the brain processes complex real-world stimuli.

All sensory and motor processing ultimately relies on spatial-temporal patterns of action potentials. For the purpose of this review it is useful to draw clear distinctions between spatial and temporal processing. We use the former term to refer to the processing of stimuli defined by which sensory neurons are activated. For example, in the visual domain the orientation of a bar of light can be determined based on a static snapshot of active retinal ganglion neurons. Similarly, the discrimination of the pitch of two high-frequency tones (that activate different populations of hair cells in the cochlea), or the color of a bar of light, or the position of a needle prick to the skin, can be discriminated solely upon the spatial patterns of activation, that is, by which afferent fibers are active. In contrast, other stimuli, such as the duration of a flashed bar of light or the interval between two tones, cannot be characterized by a snapshot of neural activity. These stimuli require the nervous system to process the temporal pattern of incoming action potentials. We refer to the analysis of these stimuli as temporal processing. In contrast to these simple examples, most sensory stimuli are not purely spatial or temporal but, like speech and motion processing, require analysis of the spatial-temporal patterns of activity produced at the sensory layers.

In the 50 years since Lashley's chapter, much progress has been made on understanding the neural basis of sensory and motor processing; however, much of this progress has been made regarding the spatial components of processing. Hebb's postulate, published two years before Lashley's chapter on temporal integration, plays a fundamental role in our understanding of spatial processing. Hebbian or associative synaptic plasticity presents a means by which neurons can develop selectivity to spatial input patterns, and it provides the underlying basis for the emergence of self-organizing maps (e.g., von der Malsburg 1973, Bienenstock et al. 1982, Miller et al. 1989, Buonomano & Merzenich 1998a). In contrast, associative plasticity alone cannot underlie the discrimination of a 100- or 125-ms presentation of a vertical bar or a 2-kHz tone.

Here we review the behavioral, electrophysiological, and theoretical data on temporal processing. We first define the different timescales over which the brain processes information and then focus on temporal processing in the range of a few milliseconds (ms) up to a second.

SCALES AND TYPES OF TEMPORAL PROCESSING

The terms temporal processing, temporal integration, and timing are used to describe a number of different phenomena. One source of ambiguity is that these terms are used to refer to a wide range of timescales over which animals process time or generate timed responses. This range spans at least 12 orders of magnitude—from microseconds to circadian rhythms. Based on the relevant timescales and the presumed underlying neural mechanisms, we categorize

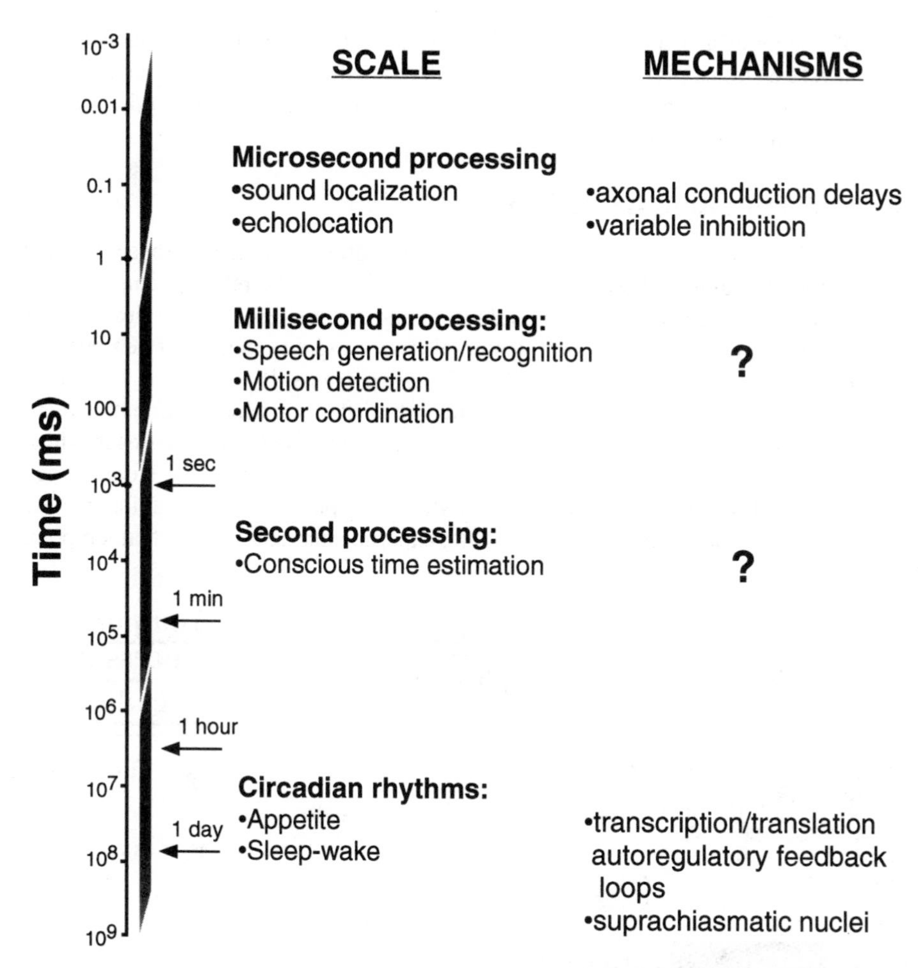

Figure 1 Timescales of temporal processing. Humans process temporal information over a scale of at least 12 orders of magnitude. On one extreme we detect the delay required for sound to travel from one ear to the other. These delays are on the order of tens to hundreds of microseconds. On the other extreme, we exhibit daily physiological oscillations, such as our sleep-wake cycle. These circadian rhythms are controlled by molecular/biochemical oscillators. Temporal processing on the scale of tens and hundreds of ms is probably the most sophisticated and complex form of temporal processing and is fundamental for speech processing and fine motor coordination. Time estimation refers to processing in the range of seconds and minutes and is generally seen as the conscious perception of time.

temporal processing into four different time scales (Figure 1): microseconds (Carr 1993, Covey & Casseday 1999), milliseconds (Buonomano & Karmarkar 2002), seconds (Gibbon et al. 1997), and circadian rhythms (King & Takahashi 2000). These general classes are not meant to represent purely nonoverlapping types of processing or indivisible categories. Rather, they probably reflect the minimal set of categories that serve different functions and rely on different mechanisms yet, nevertheless, exhibit significant overlap. Although there are numerous issues of interest at all these scales, here we focus on temporal processing on the scale of tens to hundreds of ms.

Temporal Processing Versus Temporal Coding

Another important distinction and source of confusion is the difference between temporal coding and temporal processing (Figure 2). We refer to temporal processing as the decoding of temporal information or the generation of timed motor responses. In its simplest form, temporal processing may consist of neurons that respond selectively to the interval between two events. By definition, to process temporal information, one must start with spike patterns in which information is encoded in the temporal domain. In the sensory domain we focus on cases in which the temporally encoded information arises directly from external stimuli (e.g., duration discrimination, Morse code, rhythm perception, etc.). In addition to these external temporal codes, theoretical and experimental data suggest that temporal

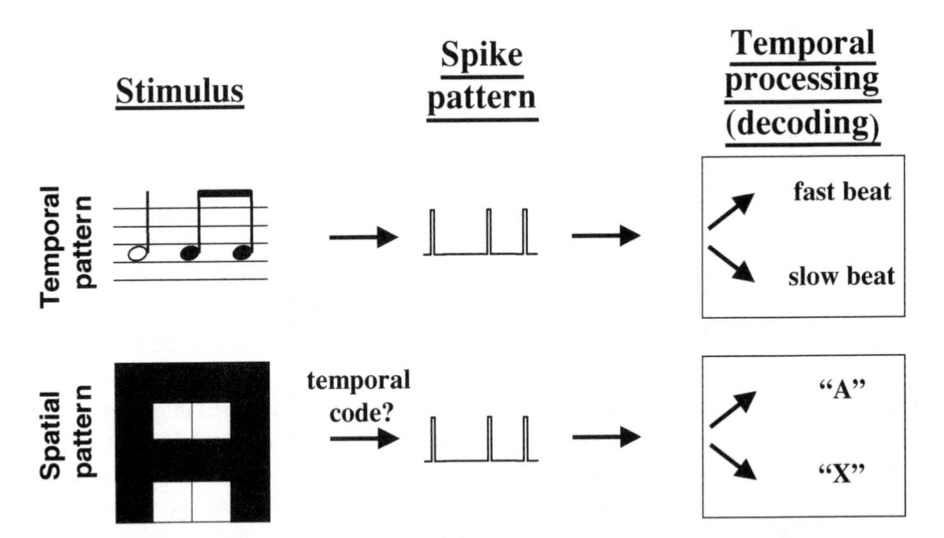

Figure 2 Temporal processing and temporal coding. (*Upper panel*) Temporal processing refers to decoding of temporal information arriving from environmental stimuli such as music (*left*). A stimulus such as a piece of music will generate temporal patterns of action potentials that follow the beat of the music (*middle*). These action potential patterns must be decoded in order to decide whether the stimulus was played at a fast or slow tempo (*right*). (*Lower panel*) Spatial stimuli such as a statically flashed image of a letter (*left*) generate spatial patterns of action potentials. Even in response to a rapid spatial stimulus, all neurons will not fire in synchrony, and it is possible that temporal codes for spatial stimuli may be generated at early states of sensory processing (*middle*). In principle, this temporal encoding of spatial stimuli might be used by the brain for stimulus processing. However, the temporal code would also have to be decoded (*right*) as with stimuli that are inherently temporal in nature.

codes may also be internally generated. That is, static or steady-state stimuli may be partially encoded in the temporal patterns of spikes (e.g., Richmond et al. 1990, McClurkin et al. 1991, Middlebrooks et al. 1994, Laurent et al. 1996, Rieke et al. 1996, Mechler et al. 1998, Prut et al. 1998). For example, by taking into account the temporal structure of neuronal responses to static Walsh patterns there is more information about the stimuli than there is in the firing rate alone (McClurkin et al. 1991). Mechler et al. (1998) have shown that there is significant information about the contrast of transient stimuli in the temporal pattern of V1 neuron firing. Internally generated temporal codes may provide a means to increase the bandwidth (Rieke et al. 1996) or to perform computations such as invariant pattern recognition (Buonomano & Merzenich 1999, Wyss et al. 2003).

Although the studies above suggest that in some cases there is information in the temporal pattern of action potentials generated internally, there are few data showing that the brain uses this information (see, however, Stopfer et al. 1997). If internal temporal codes are generated by the brain, they must be decoded or processed, like the external temporal patterns discussed here.

SENSORY TIMING

Temporal information in the range of tens to hundreds of ms is fundamental to many forms of sensory processing. Motion processing is a ubiquitous example in the auditory, somatosensory, and visual domains of a task that requires temporal information. However, it is arguably in the auditory domain that timing is most prominent, owing to its importance in vocalization and speech recognition.

A good example of the ability of the auditory system to process temporal signals is Morse code, in which language is reduced to temporal code. First, Morse code requires discriminating the duration of single tones (short versus long) and the interval between them (element, letter, and word pauses). Second, it requires perception of a sequence of tones, which represent auditory objects (letters and words). Third, the timing of the stimuli is not absolute but rather a function of the speed of transmission. At 15 words per minute (wpm), each dot and dash and interelement and intercharacter pause are 80, 240, 80, and 240 ms, respectively. Experts can understand Morse code at rates of 40–80 wpm; at 40 wpm the above elements' values are 30, 90, 30, and 90 ms, respectively. Thus, Morse code requires discrimination of continuous streams of sounds and discrimination of the duration, interval, number, and sequence of elements, as well as temporal invariance. The complexity of this analysis provides an example of the sophistication of temporal processing on the timescale of tens to hundreds of ms.

Speech Recognition

To nonexperts, Morse code at high speed sounds much like noise, and considerable training is required to understand it. However, in many ways it is a simpler task than speech recognition, which shares much of the temporal richness of Morse code but exhibits additional features such as prosody, spectral information, and speaker-specific recognition. During continuous speech, syllables are generated every 200–400 ms. The sequential arrangement of syllables is important in speech recognition (e.g., "la-dy" × "de-lay"). The pauses between syllables or words are also critical for parsing, as in "black bird" × "blackbird," or for example, the ambiguity in the mondegreen "kiss the sky" × "kiss this guy" can be decreased by longer interword intervals. The temporal structure within each syllable and phoneme also contributes to speech recognition. Specifically, temporal features are fundamental for phoneme discrimination. These features include voice-onset time (the time between air release and vocal cord vibration), which contributes to the "ba" × "pa" discrimination (Lisker & Abramson 1964), the duration of frequency transitions (e.g., "ba" × "wa"; Liberman et al. 1956), and the silent time between consonants and vowels (e.g., "sa" × "sta"; Dorman et al. 1979). Additionally, prosodic cues such as pauses and duration of speech segments are used to determine semantic content (Lehiste et al. 1976).

Owing to the multiple levels and scales of temporal information in addition to spatial information, speech is one of the most complex forms of pattern recognition and requires both spatial and temporal processing (Shannon et al. 1995, Tallal 1994, Doupe & Kuhl 1999). Various lines of evidence have revealed the degree to which speech recognition relies on temporal information. Indeed, in some cases it can rely primarily on the temporal structure. For example, experiments with cochlear implants show it is possible to achieve good levels of speech comprehension with 2–4 electrodes (Dorman et al. 1989, Dorman et al. 1997). Additionally, Shannon et al. (1995) showed that speech recognition could be achieved with relatively little spectral information. Near-perfect recognition of vowels, consonants, and sentences was observed with four broad spectral bands, and significant recognition of consonants and vowels was seen with a single band, in which only temporal and amplitude information was available.

Given the importance of temporal information in speech and language it would be expected that deficits in temporal processing would produce language deficits. Indeed, it has been suggested that certain forms of language-based learning disabilities may be caused by generalized sensory deficits in temporal processing (Livingstone et al. 1991, Eden et al. 1996, Tallal & Piercy 1973; for a review see Farmer & Klein 1995). However, even if some forms of language-based learning disabilities result from generalized sensory deficits, it is not yet clear whether those deficits are specific to timing or to more general features such as complex stimuli or rapidly changing stimuli.

MOTOR TIMING

Because movements involve changes in muscle length over time, motor control and timing are inextricably related. Most movements involve the coordinated activation of agonist muscles to initiate motion and antagonist muscles as a brake. These activations require accurate timing on the order of tens of ms. Indeed, pathologies that disrupt the timing between agonist and antagonist actions lead to dysmetric or inaccurate movements. Lesions of the cerebellum, for example, tend to delay the activation of antagonist muscles, which causes movements to be hypermetric or to overshoot (e.g., Hore et al. 1991). Cerebellar patients often display oscillating-like tremors during movements as they make a series of overshoots and corrections. A recent study shows that for saccade eye movements, which also involve agonist muscles to initiate and antagonist muscles to brake, the activity of populations of cerebellar Purkinje cells precisely encodes the onset and offset of a saccade (Thier et al. 2000). Motor control represents a clear example of an inherently timing-intensive computation in the range of tens to hundreds of ms.

Numerous studies focusing on timing have made use of repetitive movements as their readout. In particular, Keele, Ivry, and others have used such movements as rhythmic tapping of the finger to pursue the hypothesis that the cerebellum is a general-purpose timer in the tens-to-hundreds-of-ms range (e.g., Ivry & Keele 1989). In the prototypical experiment, subjects are first asked to tap their finger in time with a metronome (say at 400-ms intervals). After a brief training period, the subject continues tapping without the metronome. The main dependent measure is variability in the intertap intervals. This and similar paradigms have been used as screens to find brain regions for which damage disturbs the timing of the taps. These and related findings are discussed in more detail below in the section on the cerebellum.

Timed Conditioned Responses

One of the more experimentally tractable forms of motor timing is seen in the precise learned timing of classically conditioned eyelid responses. In a typical eyelid-conditioning experiment, training consists of repeated presentation of a tone paired with a reinforcing stimulus such as an air puff directed at the eye. Over the course of 100–200 of such trials the animals acquire conditioned eyelid responses: The eyelids close in response to the tone (Figure 3*a*, see color insert). The time interval between the onsets of the tone and the puff influences the nature of this learning (Figure 3*b*). Conditioned responses are acquired only when the tone onset precedes the puff by at least 100 ms and by less than ~3 s. Within this range, the timing of the conditioned responses is also affected by the tone-puff time interval. Short intervals promote the learning of responses with short latencies to onset and fast rise times. As the interval increases, the learned responses have longer latencies to onset and slower rise times. The result is that, in general, the responses peak near the time at which the puff is presented.

Several studies have demonstrated that these responses are a genuine example of timing and exclude the previously generally accepted alternative that response timing derives from response strength. For example, Millenson et al. (1977) and Mauk & Ruiz (1992) trained animals by presenting the puff on alternate trials at two different times during the tone. The responses the animals learn have two peaks, each corresponding to one of the times at which the puff was presented.

PSYCHOPHYSICAL STUDIES

The predominate working hypothesis in the psychophysical literature has been a centralized internal clock model (Creelman 1962, Treisman 1963; for a review see Allan 1979), in which an oscillator beating at a fixed frequency generates tics that are detected by a counter. These models often assume that timing is centralized, that is, the brain uses the same circuitry to determine the duration of an auditory tone and for the duration of a visual flash. The alternate view is that timing is distributed, meaning that many brain areas are capable of temporal processing and that the area or areas involved depend on the task and modality being used. In addition to the question of centralized versus distributed mechanisms, there is the issue of timescale specificity. A universal clock (of which there could be a single instantiation or multiple instantiations) could be the sole timing mechanism for all intervals/durations, or there could be a set of dedicated circuits, each specific to given lengths of time (referred to as interval-based mechanisms; Ivry 1996).

Interval and Duration Discrimination

The best-studied temporal tasks in humans are interval and duration discrimination (Divenyi & Danner 1977, Getty 1975, Wright et al. 1997). In a typical interval discrimination task two brief tones separated by a standard interval (T, e.g., 100 ms) or longer interval ($T + \Delta T$) are presented to the subject. The presentation order of the short and long intervals is randomized. The subject may be asked to make a judgment as to whether the longer interval was the first or second. ΔT can be varied adaptively to estimate the interval discrimination threshold. Duration discrimination tasks are similar, except each stimulus is a continuous tone (filled interval).

The relationship between the threshold and the standard interval constrains the underlying mechanisms. Figure 4 shows the relationship between threshold and the standard interval for a compilation of interval and duration discrimination studies in the range of tens of ms to one second. In untrained subjects the threshold for a 100-ms standard interval is ~20 ms (Weber fraction of 20%). Note that although in absolute terms the threshold increases with increasing intervals, the Weber fraction (threshold/standard interval) decreases for short intervals (50 to 200 ms). For intervals from 200 to 1000 ms, the Weber fraction is fairly constant, perhaps suggesting that different neural mechanisms are responsible for interval discrimination at these intervals.

INTERMODAL TIMING Psychophysical studies have attempted to address the issue of centralized versus distributed timing by comparing performance on intra- versus intermodal tasks. In the intermodal tasks a standard interval may be demarcated by a tone at 0 ms and a flash of light at 100 ms. Performance on the intermodal condition is then compared to pure auditory and visual discrimination. The first observation that comes from these studies is that interval discrimination

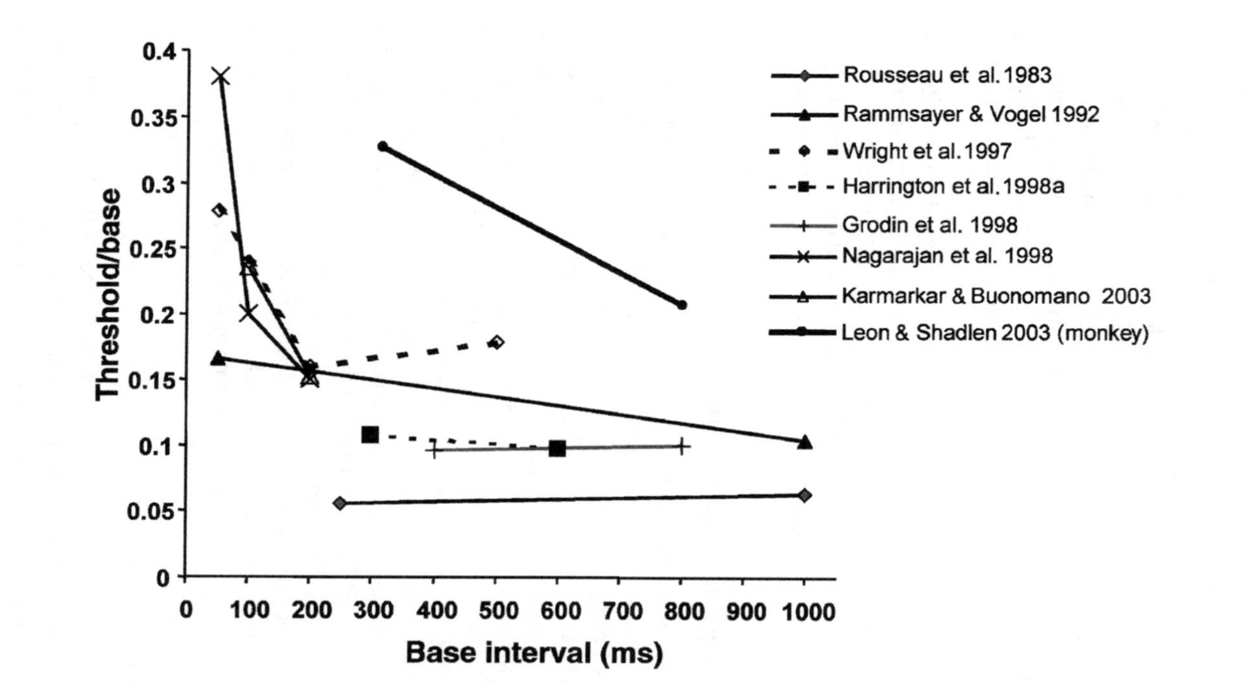

Figure 4 Cross-study interval discrimination thresholds. The standard or base interval is represented on the X axis, and thresholds are plotted in the Y axis as Weber fractions (threshold/standard interval). Thresholds are calculated differently in different studies, thus comparing absolute thresholds across studies is not appropriate. Lines join the thresholds from different intervals within studies. These data indicate that, at short intervals, temporal discrimination thresholds do not follow a Weber fraction. However, at longer intervals, above 200 ms, thresholds are fairly constant in relation to the standard interval.

in the auditory modality is better then that in the visual modality (Rousseau et al. 1983, Grondin & Rosseau 1991). Additionally, these studies show that interval discrimination between modalities is significantly worse than that within modalities (Rousseau et al. 1983, Grondin & Rousseau 1991, Westheimer 1999). Specifically for standard intervals in the range of 100–250 ms, the threshold for tone-light discrimination can be 50%–300% worse than for light-light discriminations. Interestingly, Rousseau et al. (1983) showed that intermodal discrimination was significantly more effected for a 250-ms interval as compared to a 1-s interval. Within a modality, changing stimulus features also decreases performance. If the first tone is played at 1 kHz and the second tone is played at 4 kHz, interval discrimination is significantly worse than if both tones were played at the same pitch (Divenyi & Danner 1977).

These data are consistent with the notion of distributed timers. Specifically, because the stimulus features that delimit the interval in a cross-modality task are arriving at different timers, performance is decreased. However, an alternative explanation is that timing is still centralized, but intermodal timing is simply a more difficult task because it requires a shift of attention from one modality to the other.

Psychopharmacology of Temporal Processing

On the timescale of seconds, dopamine antagonists produce temporal overshoot, and stimulants such as methamphetamine produce temporal undershoot (for a review see Meck 1996). On the timescale of a second and below, Rammsayer (1999) has shown in human psychophysical experiments that the dopaminergic antagonist, haloperidol, significantly impaired discrimination thresholds for 100-ms and 1-s intervals. Remoxipride, a dopamine antagonist more selective for D2 receptors, impaired processing on the scale of a second but not for a 50-ms interval (Rammsayer 1997). Experiments with benzodiazepines also support the dissociation between millisecond and second processing by showing that performance in a 50- or 100-ms task is unaffected, whereas performance in a 1-s task is significantly worse (Rammsayer 1992, 1999). Together these results show that two distinct drug classes (dopaminergic antagonists and benzodiazepines) can selectively interfere with second but not with millisecond processing. Future experiments will be necessary to determine whether the above results are due to direct action on a timing mechanism or to more nonspecific actions on arousal and/or cognition.

Interval Discrimination Learning

Can temporal resolution improve with practice? One of the first studies on this issue reported no perceptual learning (Rammsayer 1994). In this study, subjects were trained on 50-ms intervals for 10 min a day for 4 weeks. Subsequent studies revealed robust learning with training (Wright et al. 1997, Nagarajan et al. 1998, Karmarkar & Buonomano 2003). In these studies subjects were generally trained for an hour a day (400–800 trials) for 10 days.

GENERALIZATION OF INTERVAL DISCRIMINATION The perceptual learning studies, in addition to suggesting that the neural mechanisms underlying timing can be fine-tuned with experience, provide a means to examine the issue of central versus distributed timing. We can ask, after training on 100-ms intervals using 1-kHz tones, if performance improves for different intervals and frequencies.

Generalization studies reveal that interval discrimination learning is specific to the temporal domain, and generalization occurs in the spatial domain (Wright et al. 1997, Nagarajan et al. 1998, Westheimer 1999, Karmarkar & Buonomano 2003). Figure 5 shows the results from a study in which subjects were trained on a 100-ms–1-kHz interval discrimination task. Subjects were pre- and posttested on conditions that varied across the temporal and spatial domain: 100-ms–4-kHz, 200-ms–1-kHz, and a 100-ms–1-kHz continuous tone condition. Generalization to the 100-ms–4-kHz tone was virtually complete, and there was no generalization to the 200-ms interval. This eliminates the possibility that learning was due to a nonspecific improvement such as task familiarization.

Interval learning has also been reported to generalize across modalities. Nagarajan et al. (1998) show that training on a somatosensory task can produce

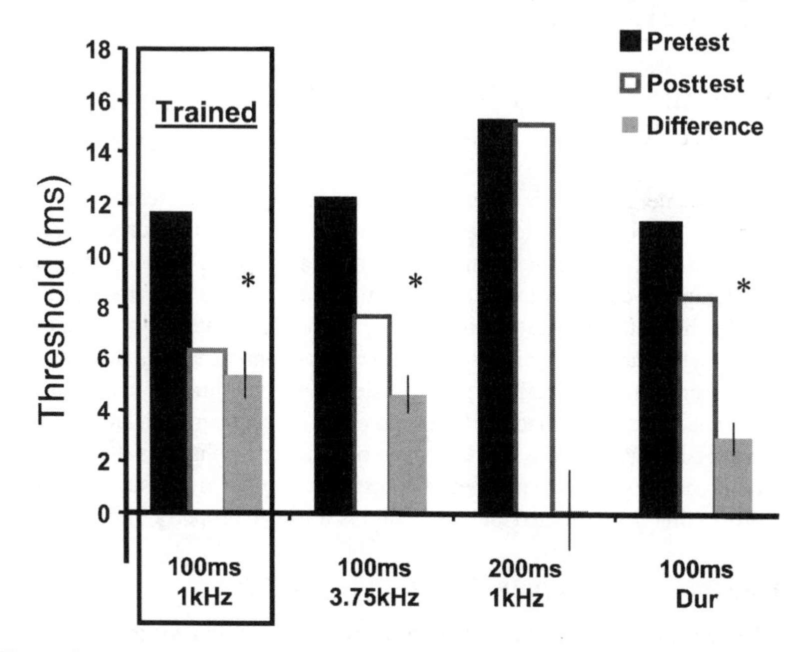

Figure 5 Generalization of interval discrimination learning. A group of 10 subjects underwent training on a 100-ms–1-kHz interval discrimination task. After 10 days of training (an hour a day), they exhibited significant learning (*left bars*). Pre- and posttests on 3 different conditions revealed generalization to the same interval played at a different frequency, as well as to the duration discrimination task (continuous tone) at the same absolute time (100 ms). However, no generalization to novel intervals was observed. Modified from Karmarkar & Buonomano 2003.

improvement on an auditory interval discrimination task similar to the interval used for somatosensory training. Even more surprising, training on an auditory task appears to result in an interval-specific improvement in a motor task requiring that the subjects tap their fingers to mark specific intervals (Meegan et al. 2000).

The simplest interpretation of these data is that centralized circuits exist for each interval, and with training, either the temporal accuracy or the downstream processing of these circuits undergoes plasticity. In this interpretation, timing is centralized but interval based. However, it is possible that in these tasks learning occurs as a result of interval-specific cognitive processes other than temporal processing per se. For example, because interval discrimination tasks require comparing the test interval and a standard interval, improvement could rely on better representation of the standard interval or improved storage or retrieval from working or short-term memory. Such alternative explanations would be consistent with the generalization across different stimulus markers and modalities, as well as the lack of generalization to novel intervals. Alternatively, it could be argued that, although many circuits are capable of temporal processing, the relatively simple nature of these temporal tasks allows the brain to use multimodal pathways and a single timing circuit.

TEMPORAL SELECTIVITY AND ANATOMICAL LOCALIZATION

A fundamental step in understanding the neural basis of temporal processing is finding neurons that are selective to the temporal features of sensory stimuli or responsible for the generation of timed motor responses. To date, interval, duration, or temporal-combination sensitive neurons have been described in a variety of different systems. These findings range from simple interval or duration-sensitive cells in bats and amphibians to more complex temporal-combination sensitive cells involved in song-selectivity in birds. Below we examine the electrophysiological and anatomical data that address the potential mechanisms and location of temporal processing. We believe that the range of tasks and behaviors that rely on temporal processing, and the number of areas putatively involved, suggest that temporal processing is distributed and a ubiquitous intrinsic property of neural circuits.

Brainstem: Frogs and Bats

To communicate, some anuran amphibians (frogs and toads) use vocalizations rich in temporal information. The temporal structure of some frog calls is used to discriminate between vocalizations (Klump & Gerhardt 1987, Rose & Brenowitz 2002). Specifically, calls can be distinguished based on the number and frequency of pulses. Alder & Rose (1998, 2000) show that neurons in the auditory midbrain can be tuned to both the frequency and the number of auditory pulses. Selectivity was not sensitive to intensity. Neurons exhibited a preferred pulse frequency (e.g., 80 Hz) at which they would produce their maximal number of spikes. Lower or higher frequencies elicited fewer or no spikes. These studies provide an elegant example of temporal tuning curves, a temporal analog to orientation tuning curves in V1 neurons. It is not yet known whether the temporal tuning arises primarily from synaptic/cellular or network properties.

Neurons in the bat auditory brainstem also respond selectively to specific temporal features such as the pulse-echo delay and sound duration (Covey & Cassidy 1999). Neurons in the inferior colliculus can be tuned to pulse-echo delays or to sounds of specific durations. Temporal tuning in these cells is known to rely on inhibition (Casseday et al. 1994, Saitoh & Suga 1995). One hypothesis is that stimulus onset produces inhibition, and the offset of inhibition causes rebound depolarization. If this rebound coincides with the second excitatory input (produced by sound offset), a duration-specific response can be generated. However, this mechanism may be a specialized brainstem process, and it is not clear if it will generalize to more complex patterns (see below).

Temporal Selectivity in Songbirds

One of the best-studied systems regarding temporal processing is in songbirds. Similar to human language the songs of birds are rich in temporal structure and composed of complex sequences of individual syllables. Each individual syllable and the interval between syllables is on the order of tens of ms to 200 ms. The areas involved in the generation and learning of song have been identified (Bottjer & Arnold 1997, Doupe & Kuhl 1999). Song selectivity is often established by comparing the response to the normal song against the same song in reverse or reversing the syllable order. Recordings in the HVc (Margoliash 1983, Margoliash & Fortune 1992, Mooney 2000) and in the anterior forebrain nuclei (Doupe & Konishi 1991, Doupe 1997) reveal neurons that are selective to playback of the birds own song, specifically syllable sequences played in the correct order. Additionally, song selectivity of neurons in cmHV can be modified by a behavioral task requiring song discrimination (Gentner & Margoliash 2003). Thus, experience can lead to selectivity of complex temporal-spatial stimuli in adult birds.

Figure 6 shows an example of an order-sensitive cell in the HVc (Lewicki & Arthur 1996). Two syllables (A and B) are presented in all combinations with a fixed interval between them. The cell is selective to the AB sequence, and it does not respond well to either syllable individually or to BA. The order selectivity in neurons from HVc has been well established. Interval and duration selectivity have been less studied. Although, in some cases the neurons are also sensitive to the interval between sounds (Margoliash 1983, Margoliash & Fortune 1992). The mechanisms underlying this selectivity are not understood. Unlike simple detection of the interval between two tones, these cells are selective to both the spatial-temporal structure within each syllable, as well as to the sequence in which these elements are put together. This selectivity emerges in stages because neurons in earlier auditory areas of the songbird respond selectively to syllables but not to the sequence (Lewicki & Arthur 1996).

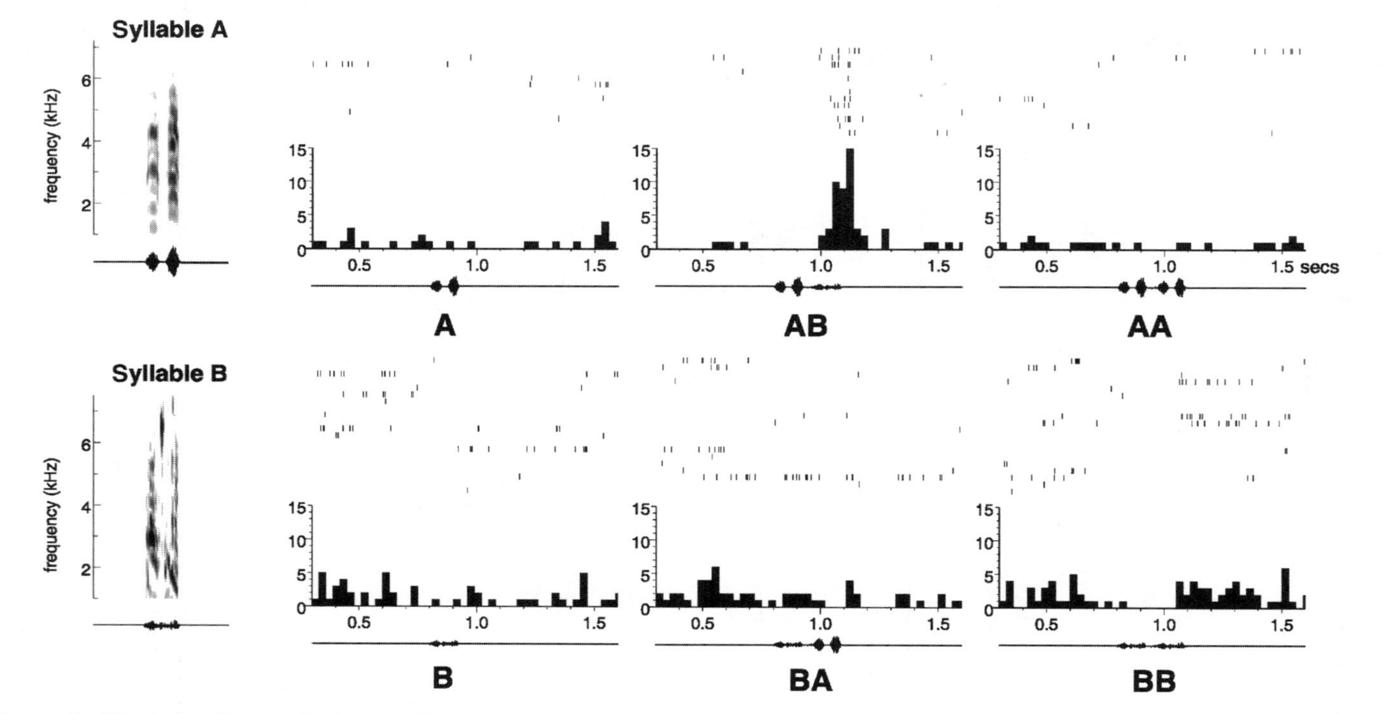

Figure 6 Example of a temporally combination-sensitive neuron of a songbird. Recordings from a neuron in HVc in response to two syllables from the bird's own song. Spectrograms of each syllable are shown to the left. Above each histogram is the response of the neuron to each presentation of the stimuli. This cell is selective to syllable B only if preceded by syllable A. Modified from Lewicki & Arthur 1996.

Because HVc neurons can respond selectively to the auditory presentation of songs (these studies are generally done under anesthesia), these neurons are clearly sensitive to temporal information in the sensory domain. However, these same cells are also active during singing and can be activated at precise times during song production. A subset of HVc neurons may be responsible for generating the timed responses that drive the sequence of syllable production (Hahnloser et al. 2002). Whether or not this is true, it is clear that the song circuity is capable of temporal processing because cross correlations with peaks in the tens-to-hundreds-of-ms range have been reported (Hahnloser et al. 2002, Kimpo et al. 2003).

Basal Ganglia

There are numerous studies suggesting the basal ganglia is involved in timing; however, most of the data focus on the timescale of seconds rather than in the range of tens to hundreds of ms. Much of these data relies on pharmacology studies. Specifically, drugs that act on the dopaminergic system interfere with timing. Because the basal ganglia is important in the dopaminergic system, the basal ganglia is likely involved in temporal processing (for a review, see Meck 1996). Studies of Parkinson patients, who in some cases have shown specific deficits in temporal tasks, support this claim (Artieda et al. 1992, Harrington et al. 1998a, Riesen & Schnider 2003).

Imaging studies have reported changes in BOLD signals in the basal ganglia. Rao et al. (2001) showed an increase in the BOLD signal in the basal ganglia during a duration discrimination task of 1.2 s. No significant basal ganglia activation was observed during a control frequency discrimination task using a similar stimulus protocol. Similarly, an fMRI study by Nenadic et al. (2003) revealed activation of the basal ganglia (putamen) during a 1-s duration discrimination task compared to a frequency discrimination task. This study also revealed activation of the ventrolateral prefrontal and insular cortex, but not the cerebellum, in the temporal condition.

Thus the basal ganglia likely plays a role in timing of sensory and motor events on the timescale of seconds. However, to date, there are few data that suggests involvement of the basal ganglia in temporal processing in the range of tens to hundreds of ms.

Cerebellum

Although the cerebellum is generally viewed as primarily a motor structure, it has also been proposed to be a general-purpose interval timer in the range of tens to hundreds of ms. "General purpose" in this sense encompasses both sensory and motor timing. One advantage of such a theory is that the synaptic organization and physiology of the cerebellum are known. Much is known about the relationships between the cerebellum and forms of motor learning such as eyelid conditioning and adaptation of the vestibulo-ocular reflex (Raymond et al. 1996; Boyden et al. 2004, in this volume).

Support for the role of the cerebellum in timing is based on both motor and sensory timing experiments. Ivry and others presented a variety of evidence demonstrating cerebellar involvement in timing tasks. The fundamental observation was made in experiments in which the task required human subjects to make rhythmic taps with their finger. Analysis was based on a hypothetical construct that divides errors (tapping at the wrong time) into those attributable to motor execution versus those attributable to a timer (Wing & Kristofferson 1973). Ivry et al. (1988) showed that patients with lesions of the medial cerebellum have increased motor errors, whereas lesions that were more lateral increased timer errors. Cerebellar patients also display deficits in interval discrimination (Ivry & Keele 1989) and are impaired at judging the speed of moving visual targets (Ivry & Diener 1991, Nawrot & Rizzo 1995). Ackermann and colleagues (1997) observed that patients with lateral cerebellar lesions are impaired in their ability to discriminate phonemes that differ only in the timing of consonants. Imaging studies also suggest a potential connection between timing and the lateral neo-cerebellum in humans. PET imaging was used to detect activation in lateral portions of the cerebellum during an interval discrimination (Jueptner et al. 1995).

The timing hypotheses of cerebellar function attempt to explain the various tasks for which the cerebellum is engaged or is necessary in terms of the need to gauge the explicit timing between events in the hundreds-of-ms range. Despite the intent that these theories build on a computational base, supporting data remain mostly task-based. Most data involve demonstrations that the cerebellum is activated during, or is required for, tasks that we view as examples of timing.

CEREBELLUM IN TIMING OF CONDITIONED RESPONSES Lesions and reversible inactivation studies have shown that learned response timing of conditioned eyelid responses is mediated by the cerebellar cortex. Perrett et al. (1993) used a within-subject design to demonstrate the effect of cerebellar cortex lesions on eyelid response timing. Animals were trained to make a fast response to one tone and a slower response to a second tone. Using this two-interval procedure, it was demonstrated that lesions of the cerebellar cortex in already trained animals spare conditioned responses but abolish response timing (Figure 3*c*). The results demonstrated that the lesions do not produce a fixed shift in timing. Rather, the postlesion timing defaults to a short, fixed latency independent of the prelesion timing. Subsequent studies have replicated this effect on response timing using reversible inactivation techniques. Garcia & Mauk (1998) showed that disconnection of the cerebellar cortex with infusion of a GABA antagonist into the cerebellar interpositus nucleus (the downstream target of the relevant region of cerebellar cortex) also cause response timing to default to very short latency (Figure 3*d*). Recent studies have demonstrated similar results with infusions of lidocaine in the cerebellar cortex (W.L. Nores, T. Ohyama & M.D. Mauk, manuscript in preparation).

The implications of conditioned eyelid response timing involve much more than the finding that the cortex of the cerebellum is necessary. Eyelid conditioning is an especially useful tool for studying the input/output computations of the cerebellum, owing to the relatively direct ways in which eyelid conditioning engages the cerebellum. Several decades of research, beginning with the studies of Thompson and his colleagues (e.g., Thompson 1986) have solidified three important findings in this regard (see Figure 7, see color insert):

1. During eyelid conditioning the conditioned stimulus, often a tone, is conveyed to the cerebellum via activation of mossy fiber afferents from the pons.
2. Similarly, the reinforcing or unconditioned stimulus, usually a mild shock around the eye from a puff of air directed at the eye, is conveyed to the cerebellum via climbing fiber afferents from the inferior olive.
3. Output from the cerebellum, in the form of increased activity of particular neurons in the cerebellar interpositus nucleus, drives the efferent pathways responsible for the expression of the learned responses.

Because of these three findings, the extensively characterized behavioral properties of eyelid conditioning can be applied as a first approximation of what the cerebellum computes (Mauk & Donegan 1997, Medina et al. 2000, Medina & Mauk 2000, Ohyama et al. 2003).

The involvement of the cerebellum in both interval timing tasks and in the timing of learned responses raises the question: Is the computation performed by the cerebellum best understood as an interval timer or clock, or does cerebellar involvement in eyelid conditioning reveal a more learning-related computation? Based on recent evidence we support the latter. Specifically, cerebellar involvement in both tasks can be explained by the hypothesis that the computation performed by the cerebellum is a learned, feed-forward prediction. Additionally, the temporal portion of the computation would not rely on fixed timers or clocks but instead on network mechanisms that can perform both temporal and spatial computations. Several authors have argued that the cerebellum makes a feed-forward prediction, or generates forward models (e.g., Ito 1970, Kawato & Gomi 1992). Here we focus on the feed-forward computation itself and implications of its temporal specificity. Although it is easier to introduce the feed-forward prediction idea in the context of motor control, the computation is presumably applicable to nonmotor tasks influenced by the cerebellum as well (see Schmahmann 1997).

FEED-FORWARD PREDICTION AND THE CEREBELLUM To help make movements accurate, sensory input can be used in two general ways: feedback and feed-forward. Feedback is like a thermostat; outputs are produced by comparing sensory input with a target. When input from its thermometer indicates the room is too cold, a thermostat engages the heater. Although accuracy is easily achieved with feedback, it has the inherent disadvantage of being slow. Adjustments are only possible once errors have already occurred.

In contrast, feed-forward use of sensory input can operate quickly but at the cost of requiring experience through learning. To react to a command to change room temperature quickly, a hypothetical feed-forward thermostat would predict the heater blast required from current sensory input. This prediction would draw upon previous experience and require associative learning in which error signals were used to adjust decision parameters for errant outputs. If our hypothetical feed-forward thermostat undershoots the target temperature, then learning from the error signal should adjust the connections of recently activated inputs so that in subsequent similar situations the heater is activated a little longer. Thus, through associative, error-driven learning it is possible to acquire the experience necessary to make accurate feed-forward predictions.

Eyelid conditioning reveals that cerebellar learning displays precisely these properties (see Mauk & Donegan 1997, Ohyama et al. 2003). Learning associated with feed-forward prediction should be associative, and there should be a precise timing to the association. An error signal indicates that the prediction just made was incorrect. For example, an error signal activated by stubbing one's toe when walking indicates that in similar circumstances the leg should be lifted higher. Thus, error signals should modify feed-forward predictions for the inputs that occurred approximately 100 ms prior (Figure 8*a*). This means the results of the learning will be timed to occur just prior to the time error signals arrive. Eyelid conditioning displays these properties. The conditioned responses are timed to occur just before the time at which the error signal (puff to the eye) normally occurs (Figure 8*b*).

The timing displayed by conditioned eyelid responses reveals both temporal specificity and flexibility to this associative learning, both in ways that are useful for feed-forward prediction. Timing specificity is revealed in the way conditioned eyelid responses are time locked to occur just before the arrival of the puff. This is consistent with what feed-forward associative learning must accomplish. When a climbing fiber error signal arrives, learning should selectively alter the cerebellar output that contributed to the faulty movement. Thus, learning should produce changes in output that are time locked to occur around 100 ms prior to the climbing fiber input, as is seen in the timing of eyelid responses. The flexibility of the timing is revealed by the way in which eyelid conditioning can occur with a range of time intervals between the onsets of the tone and puff. Even though learning can occur for mossy fiber inputs that begin 100 to ~2500 ms prior to the climbing fiber input, the changes in output remain time locked to occur just before the climbing fiber input (Figure 8*b*). To accomplish this, the learning must have the capacity to delay the responses with respect to the onset of the mossy fiber input—again, as eyelid conditioning reveals. These examples show the utility for feed-forward control of learning that is time locked to occur just before error signals (when the decisions actually have to be made) but that can vary with respect to the timing of predictive sensory signals (see Ohyama et al. 2003).

TEMPORALLY SPECIFIC FEED-FORWARD PREDICTION AND TIMING Considering cerebellar function in terms of its feed-forward computation provides an example of the cerebellum's role in timing. Feed-forward prediction helps determine the force required for agonist muscles and the force and timing of activating

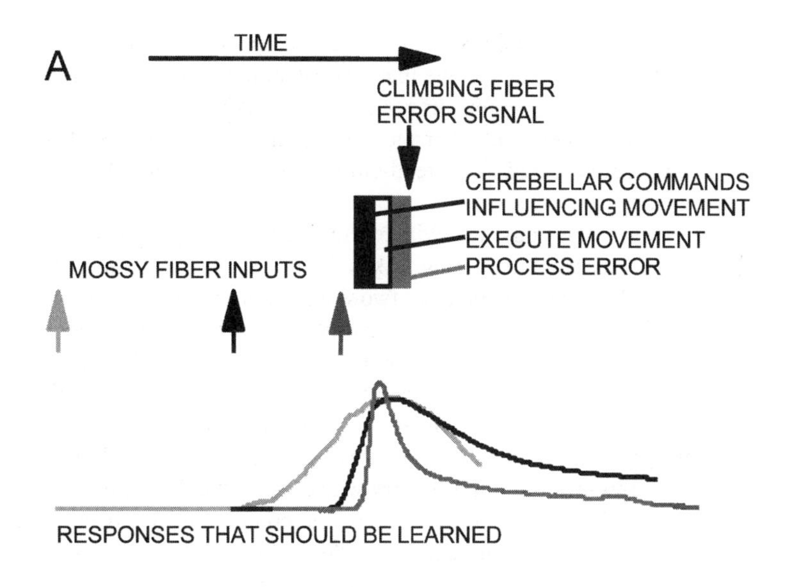

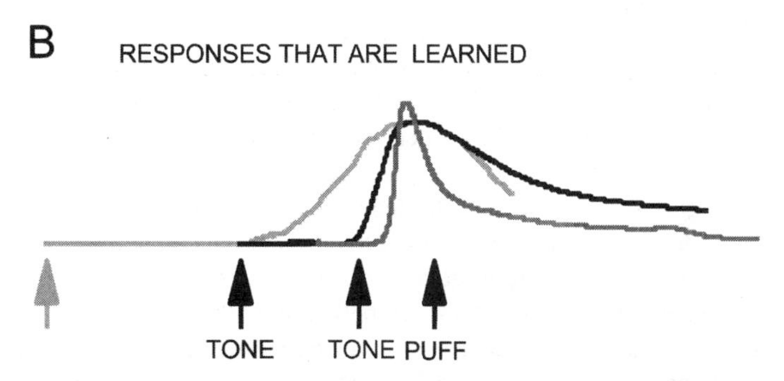

Figure 8 Feed-forward learning is enhanced by temporal specificity. (*A*) A schematic representation of the timing required for error-driven associative learning supporting feed-forward predictions. A climbing fiber input to the cerebellum (*gray*) signals movement error as detected by an inappropriate consequence (e.g., stubbing the toe while walking). The cerebellar output that contributed to this errant movement (*black*) occurred approximately 100 ms prior, owing to the time required to execute the movement (*white*) and the time required to detect the error and convey the signal to the cerebellum. To improve subsequent performance, learning must alter cerebellar output for the time indicated by the black region. Because mossy fiber inputs that predict this error may occur at varying intervals prior to the output commands (*light gray, black, and dark gray*), the cerebellar learning mechanism must be able to delay learned responses elicited by the mossy fiber input so that they can be time locked to occur just before arrival of the error signal (*corresponding light gray, black, and dark gray traces*). (*B*) The learned timing of eyelid responses indicates that cerebellar learning displays temporal specificity in its learning. Response timing is delayed with respect to the tone (mossy fiber) onset so that it can be time locked to peak when the puff (climbing fiber) occurs.

antagonist muscles. Deficits from the absence of this contribution would be especially notable for movements that involve stopping and starting, as in the timing experiments that require finger tapping. This is consistent with the deficits seen from medial cerebellar damage (vermal and intermediate cerebellum), whose outputs contribute relatively directly to movement execution through descending pathways.

This view is also consistent with recent findings that apparent timing deficits are specific to discontinuous timing tasks relative to continuous ones. Spencer et al. (2003) tested cerebellar patients on two similar timing tasks. Two groups of subjects were required to draw circles at regular intervals. The "discontinuous" group was required to keep a beat by pausing at the top of each circle. The "continuous" group was instructed to keep a beat by drawing circles using a steady continuous motion. Cerebellar damage affected discontinuous drawing and not continuous. The authors interpret these findings as evidence that the cerebellum is required for tasks where timing is explicitly represented, as in the discontinuous task. In this view, the cerebellum is not required by the continuous task because timing can be implicit—that is, timing can be produced by maintaining a constant angular velocity. Alternatively, such findings can be seen as examples of the contributions of feed-forward prediction in the starting and stopping of movements. Holmes (1939) made a similar observation (see also Dow & Moruzzi 1958). He asked a patient to first draw squares with the hand affected by the cerebellar lesion and then by the unaffected hand. Holmes found that the motor deficit of the affected hand was most notable at the corners of the square, where stopping and starting movements are required.

Although more speculative, the feed-forward computation of the cerebellum may provide a way to understand the activation of the cerebellum in many timing tasks and explain the timing deficits observed with lateral cerebellar damage. Feed-forward prediction in lateral cerebellum may be a mechanism for predicting when the next tap should occur in a timing experiment. The cerebellum therefore underlies some forms of motor timing. This timing relies on distributed network mechanisms as opposed to a dedicated clock or timer (see below).

CORTEX

The cortex has also been proposed to be the the primary site for temporal processing. If the cortex is involved in timing, whether virtually all cortical areas can processes time, or if specialized cortical areas devoted to temporal processing exist, is a fundamental issue.

Anatomy

Based on data from stroke patients Harrington et al. (1998b) suggested the right parietal cortex may be involved in temporal processing. Specifically, right hemisphere, but not left hemisphere, lesions produced a deficit for 300- and 600-ms

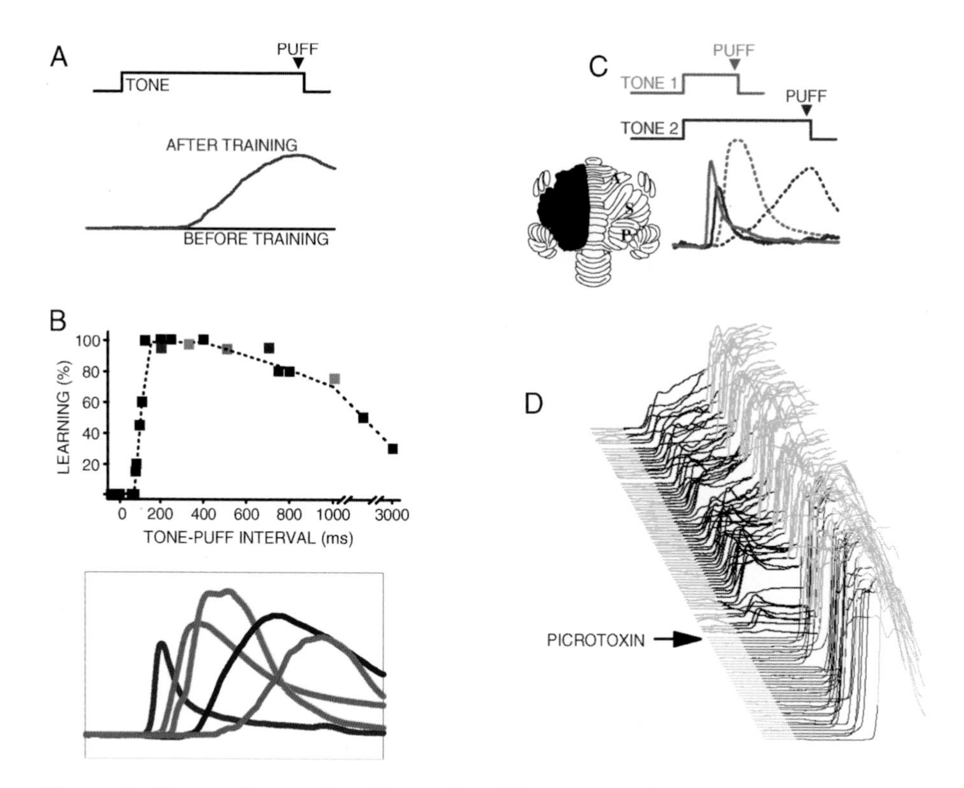

Figure 3 Temporal properties of learned eyelid responses. Classical or Pavlovian eyelid conditioning displays learned timing. (*A*) In a typical experiment, training involves presentation of a neutral stimulus, such as a tone, paired with a reinforcing stimulus, such as a puff of air directed at the eye. (*Lower traces*) Repeated presentation of such trials leads to the acquisition of learned eyelid responses. Before training the tone does not elicit an eyelid response, whereas after training the upward deflection of the trace indicates that the tone elicits learned eyelid closure. In this case the tone-puff interval is 500 ms. (*B*) The time delay between the onsets of the tone and puff influences learning in two ways. First, learning only occurs for delays between approximately 100 and 3000 ms. Best learning is produced by delays ranging from 200 to 1000 ms. The tone-puff delay also determines the timing of the learned responses. These are sample learned responses for animals trained with the delays coded by the color of the points in the graph. (*C*) Lesions of the cerebellar cortex disrupt learned response timing. Animals trained using two tones and two tone-puff delays were then subjected to lesions of the cerebellar cortex (*example shown in inset*). The lesions produced a short and relatively fixed latency-to-onset interval independent of prelesion timing. Modified from Perret et al. 1993. (*D*) Reversible lesions or disconnection of the cerebellar cortex produce the same effect on timing. These are example responses from a training session in which the cerebellar cortex was functionally disconnected via infusion of the GABA antagonist picrotoxin into the cerebellar interpositus nucleus. The darker portion of each trace indicates the tone; responses are chronologically organized front to back. Modified from Medina et al. 2000.

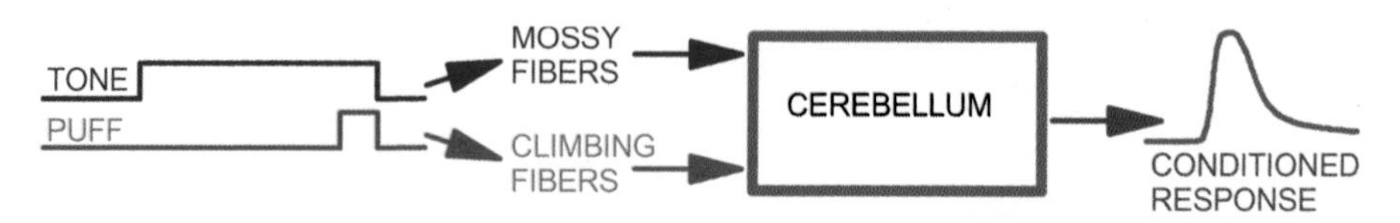

Figure 7 Eyelid conditioning engages the cerebellum relatively directly. This is a schematic representation of the relationship between eyelid conditioning and the cerebellum. Output of the cerebellum via its anterior interpositus nucleus drives the expression of conditioned responses. Stimuli such as tones are conveyed to the cerebellum via activation of mossy fiber inputs. Reinforcing stimuli such as the puff of air directed at the eye are conveyed to the cerebellum via activation of climbing fibers.

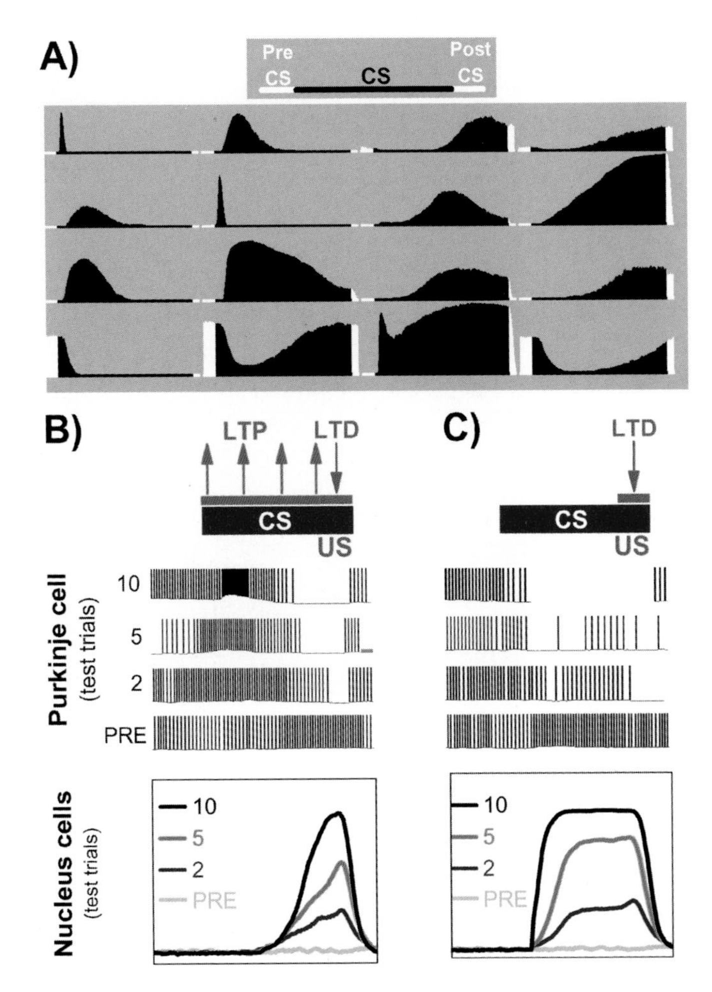

Figure 9 Mechanisms of timing-specific learning in the cerebellum. Computer simulations of the cerebellum in the context of eyelid conditioning suggest mechanisms for learned response timing. (*A*) Peri-stimulus histograms of simulated granule cells for the presentation of a tone-like mossy fiber input to the cerebellum. This sample shows how different granule cells respond at different times during this stimulus. These simulated granule cells have identical temporal properties; these differently timed responses arise from network interactions with mossy fiber inputs and with cerebellar Golgi cells. (*B*, *C*) The simulations suggest that learned timing is enhanced by competitive learning within each trial. Proper timing requires mechanisms both for learning (LTD) responses, when a climbing fiber is present, and unlearning (LTP) responses, when it is absent. (*B*) Through these two mechanisms, the simulated cerebellar Purkinje cells can learn well-timed modulation of their activity during learning. (*C*) In simulations with unlearning disabled, timing of Purkinje cell response and of the learned responses of the simulation is impaired. Modified from Medina & Mauk 2001.

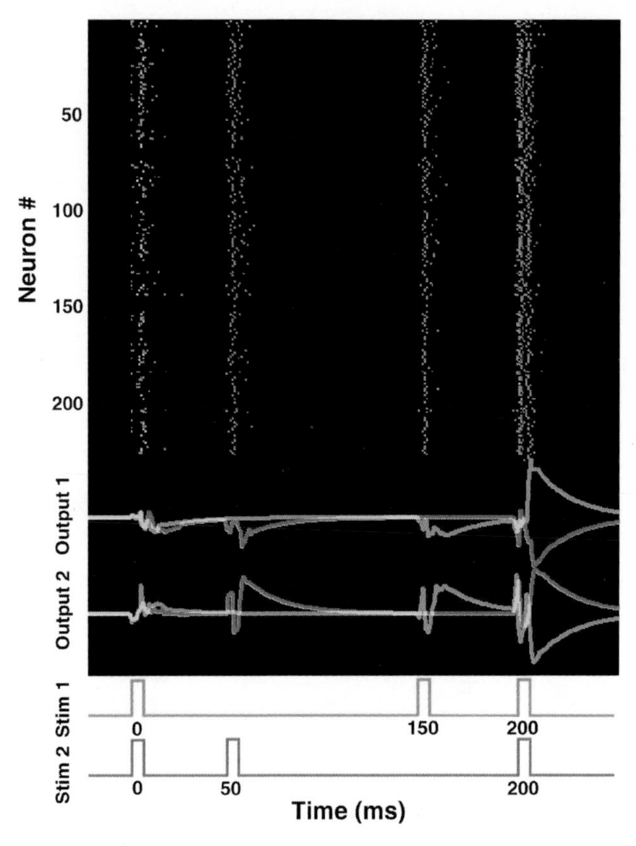

Figure 10 State-dependent model of sequence recognition. The model is composed of excitatory and inhibitory neurons. The connectivity and synaptic weights are randomly assigned, the synapses exhibit short-term synaptic plasticity, and a slow-IPSP is present. The time constant of the short-term plasticity and slow IPSP is the same for all synapses in the network. The raster plot shows which excitatory neurons fired to the long-short stimulus (*green*) and to the short-long stimulus (*red*). If the neuron responded at the same time to both stimuli the spike is plotted in yellow. Note that there is more yellow in response to the first pulse than to the last (all points in response to the first pulse are not yellow because of intrinsic noise). Each pulse of a stimulus will activate a population of neurons and trigger short-term plasticity; thus at the arrival of the second pulse the network will be in a different state, depending on whether the second pulse arrived at 150 (*green*) or 50 ms (*red*). For both stimuli (long-short or short-long) the third pulse arrives at 200 ms; however, the network will be in a different state depending on the stimulus, allowing the network to respond differently to the same pulse. The two lower traces represent the voltage of two output neurons that receive input from all the excitatory neurons above. The weights on the output neurons were set by training (using a nontemporal learning rule) on different stimulus set presentations. Outputs 1 and 2 respond selectively to the long-short and short-long stimuli, respectively.

interval discrimination. Imaging studies also reported changes in blood flow during temporal tasks in various cortical areas. In a PET study Belin et al. (2002) report activity in the right fronto-parietal network and prefrontal cortex during a 300-ms duration discrimination task. However, this study did not include a control task, and thus activation could be related to any form of processing. A second PET study in the visual modality reported activation in a number of cortical areas during a 700-ms duration discrimination task but no significant difference regarding an intensity discrimination task (Maquet et al. 1996). Onoe et al. (2001) showed activation of the dorsolateral prefrontal cortex in a monkey PET study. This study used a visual duration discrimination task in the range of 400 to 1500 ms. They report activation of the dorsolateral prefrontal cortex. Although there was no control task, they did report that bicuculline administration to the dorsolateral prefrontal cortex impaired duration discrimination more so than position discrimination.

Two fMRI studies revealed specific increases in BOLD signal, and both reported activation of the right parietal and dorsolateral prefrontal cortex (Rao et al. 2001, Nenadic et al. 2003). In both these studies the increases were in comparison to a pitch discrimination task using stimuli in the 1-s range. As mentioned above, both these studies also revealed increased signal attributed to temporal processing in the basal ganglia but not in the cerebellum.

Electrophysiology

In addition to imaging data a few studies attempted to find, in the mammalian cortex, neurons that respond selectivity to temporal features. Vocalization-sensitive neurons were reported in primary auditory cortex of marmoset monkeys (Wang et al. 1995). Neurons responded more robustly to conspecific vocalizations compared to the same vocalization played in reverse. Additionally, vocalization-sensitive neurons were also reported in early auditory areas of Rhesus monkeys (Rauschecker et al. 1995). Creutzfeldt et al. (1989) described speech-specific neural responses in the human lateral temporal lobe. However, to date, no areas have been described in which the neurons exhibit the same degree of selectivity to vocalizations as that observed in songbirds. Other investigators have looked for combination or interval-sensitive neurons using tone pairs or sequences. Selectivity has been observed in primary auditory areas in cat (McKenna et al. 1989, Brosch & Schreiner 1997) and monkey (Riquimaroux 1994). Kilgard & Merzenich (1998, 2002) characterized the temporal selectivity of auditory cortical neurons to sequences of tones. In one study three element sequences such as high tone (H), low tone (L), noise burst (N) were paired with basal forebrain stimulation in awake rats (Kilgard & Merzenich 2002). A significant increase was reported in the number of sites that exhibited facilitated responses to the target sequence, indicating experience-dependent plasticity. For example, after training in H-L-N sequence, an enhanced response to N preceded by H–L was reported, as compared to N alone. The enhanced responses often generalized to degraded stimuli such as L-H-N. The temporal feature selectivity of cortical neurons undergoes experience-dependent plasticity. However, future research is necessary to determine the degree of selectivity and whether these areas represent the primary locus for features such as interval, duration, and order.

To date, one study has looked for neurons that may code for time in awake-behaving monkeys. Leon & Shadlen (2003) recorded in the lateral intraparietal cortex in two monkeys trained on a duration discrimination task in the visual modality. Two standard durations were examined: 316 and 800 ms. The individual neurons contained information about time from stimulus onset. Time from stimulus onset was encoded in the instantaneous firing rate, which changed predictably with time. The encoding was very dynamic; specifically, the same neuron would show an upward or downward ramping of its firing rate depending on the location of the short or long target used for the response. Additionally the rate of change was slower for long durations than for short durations. Thus timing might be achieved by complex network mechanisms capable of dynamically changing firing rates in a context-specific manner. Whether the same neurons would contain temporal information if the task was auditory, or whether neurons in other areas contained the same information, has not been determined.

In Vitro Studies

It has been proposed that cortical neural networks are intrinsically capable of processing temporal information (Buonomano & Merzenich 1995). If this is the case it may be possible to observe timed responses in vitro. In vitro studies cannot address whether the observed timing is behaviorally relevant. They can, however, establish whether neurons and neural circuits are capable of processing temporal information or whether specialized mechanisms are present. Long-latency timed action potentials in response to continuous synaptic stimulation (Beggs et al. 2000), or in response to single stimuli (Buonomano 2003), have been observed. In organotypic cortical slices, neurons can respond reliably at latencies of up to 300 ms after a single stimulus (Buonomano 2003). Thus cortical circuits are intrinsically capable of generating timed responses on timescales well above monosynaptic transmission delays. Mechanistically, timing relied on network dynamics, specifically, activity propagated throughout functionally defined polysynaptic pathways. The propagation path was a complex function of the functional connectivity within the network and was not simply a result of spatial wave-like propagation.

To date, relatively few studies have revealed cortical neurons strongly tuned to the interval or duration of tones or to complex sounds on the scale of hundreds of ms. These data contrast sharply with the tuning of cortical neurons to spatial stimuli such as orientation, ocular dominance, tonotopy, and somatotopy. It is more difficult to study temporal selectivity because temporally tuned neurons may not be topographically organized. In the visual cortex, if we record from a cell selective to vertical bars, the neighboring cells may also be tuned to vertical bars. Given the vast number of possible spatio-temporal stimuli, and the potential absence of chronotopy, it may prove difficult to localize temporal selective neurons with conventional extracellular techniques.

NEURAL MECHANISMS AND MODELS OF TIMING

Analyses of the neural basis of timing have generally focused on three general computational strategies: mechanisms based on neural clocks, mechanisms based on arrays of elements that differ in terms of some temporal parameter, or mechanisms that emerge from the dynamics of neural networks. In general, these models must accomplish some variant of the same computational task. They must recode the temporal information present in an input into a spatial code. That is, in some way different cells must respond selectively to temporal features of the stimulus. For example, to discriminate differences in the duration of two stimuli, there must be differential neuronal responses to each duration.

Clock Models

When considering the mechanisms of timing it is perhaps most intuitive to think in terms of clocks or interval timers. The basic computational unit of clock theories involves an oscillator and a counter (Creelman 1962, Treisman 1962). Conceptually, the oscillator beats at some constant frequency, and each beat would then be counted by some sort of neural integrator. These ideas have not yet been expressed concretely in terms of the synaptic organization of a specific brain region. Indeed, in its simplest form, if such a clock were used for the discrimination of 100-ms intervals (and allowed the discrimination of a 100- and 105-ms interval) the period of the oscillator would have to be at least 200 Hz. At the neurophysiological level, oscillating at this frequency, as well as accurately counting each beat, seems unlikely. However, as proposed by Meck and colleagues, clock-like mechanisms could be involved in timing on the scale of seconds and minutes (Meck 1996, Matell & Meck 2000).

OSCILLATOR-PHASE MODELS In addition to the oscillator/counter models mentioned above, more sophisticated models based on oscillators have been proposed (Ahissar et al. 1997, Ahissar 1998, Hooper 1998). These include the use of oscillators placed in phase-locked loop circuits. Specifically, Ahissar and colleagues have proposed (Ahissar et al. 1997, Ahissar 1998) that the thalamo → cortical → thalamo loop may use dynamic oscillators (oscillators that can change their period in an adaptive manner) to decode temporal information from the vibrissa during whisking in rodents.

Spectral Models

Many of the proposed models share the characteristic of decoding time using arrays of neural elements that differ in terms of some temporal property. The most generic of these is the spectral timing model of Grossberg and colleagues (Grossberg & Schmajuk 1989), which has been expressed in varying forms. The original model assumed a population of cells that react to a stimulus with an array of differently timed responses. Two variants of this motif have also appeared. One is a variant of clock models: Stimuli activate arrays of cells that oscillate at different frequencies and phases. By doing so, points in time following the onset of a stimulus can be encoded by activity in a subset of neurons that differs, at least somewhat, from the subsets of cells active at other times (Miall 1989, Gluck et al. 1990). In another model generally referred to as tapped delay lines, simple assumptions about connectivity lead to a sequential activation of different neurons at different times following a stimulus (Desmond & Moore 1988, Moore 1992, Moore & Choi 1997).

A number of studies propose biologically plausible implementations of spectral models. In these models all elements share a common implementation, but at least one of the variables is set to a different value, which allows each unit to respond selectively to different intervals. A wide range of biological variables have been proposed, including the kinetic constants of the metabotropic receptor pathway (Fiala et al. 1996), the time constant of slow membrane conductances (Hooper et al. 2002; see also Beggs et al. 2000), the decay time of inhibitory postsynaptic potentials (IPSPs) (Sullivan 1982, Saitoh & Suga 1995), short-term synaptic plasticity (Buonomano 2000, Fortune & Rose 2001), or even cell thresholds (Antón et al. 1991).

Spectral models have the advantage of encoding the time since the arrival of a stimulus by having different subsets of cells active at different times. Combined with simple learning rules where a teaching or error signal modifies connections for only active cells, spectral models can learn outputs that are properly timed and can even show the Weber effect of increased variance with increased delay. However, to date, neither arrays of elements with different time constants, arrays of elements that oscillate at different phases and frequencies, nor connectivity that supports tapped delay lines are supported by identified properties of neurons or networks. Additionally, these models focused on simple forms of temporal discrimination and may not generalize well to more complex forms of temporal processing without additional network layers (see below).

Network or State-Dependent Models

The above models represent top-down approaches where timing is addressed by inferring a computation and then implementing the computation with neurons. An alternative bottom-up approach is to start with biologically realistic assumptions and then to ask the extent to which temporal processing can be found as an emergent property. These models have no built-in temporal processing or selectivity with ad hoc assumptions. That is, they do not rely on explicitly setting oscillators, synaptic or current-time constants, or some other variable that, in effect, functions as a delay line.

CORTICAL MODEL It has been proposed that cortical networks are inherently able to process temporal information because information about the recent input history is inherently captured by time-dependent changes in the state of the network (Buonomano & Merzenich 1995, Buonomano 2000, Maass et al. 2002). One set of studies has examined how interval selectivity can be encoded in a population of cortical neurons (Buonomano & Merzenich 1995, Buonomano 2000). In an interval discrimination task, when the first of a pair of tones arrives in a cortical network, it will stimulate hundreds of excitatory and inhibitory neurons, a subset of which will fire. In addition to producing action potentials in some neurons, a series of time-dependent processes will also be engaged. In this model the time-dependent properties were short-term synaptic plasticity (Deisz & Prince 1989, Stratford et al. 1996, Reyes et al. 1998, Zucker 1989) and slow IPSPs (Newberry & Nicoll 1984, Buonomano & Merzenich 1998b), but it could include many other time-dependent properties. In this model all synapses exhibit the same short-term plasticity temporal profile, as opposed to spectral models. Because of these time-dependent properties, the network will be in different states at 50, 100, and 200 ms. Thus, at the arrival of a second event at 100 ms, the same stimulus that arrived at 0 ms will arrive in a different network state. That is, some synapses will be facilitated/depressed, and some neurons may be hyperpolarized by slow IPSPs. As a result, the same input can activate different subpopulations of neurons dependent on the recent stimulus history of the network. The differences in the population activity produced by the second and first pulse can be used to code for the 100-ms interval. Given the high dimensionality and abundance of time-dependent properties of cortical networks, this type of model could provide a realistic means to decode complex temporal and spatial-temporal patterns of sensory information (see below).

CEREBELLAR MODEL The evidence from the cerebellum illustrates how timing and performance on experimental tasks designed to study timing are mediated by computations that include temporal processing. For example, cerebellar-mediated, feed-forward prediction may be the computational basis for the temporal processing responsible for timing tasks in the millisecond range.

Buonomano & Mauk (1994) used the correspondence between eyelid conditioning and the cerebellum to test the timing capabilities of a network model of the cerebellar cortex. Although this model failed in many of its key properties, it showed how the connectivity of the cerebellar cortex could represent the time since the onset of a stimulus with subsets of different granule cells that become active at different times (Figure 9*A*, see color insert). This time-varying stimulus representation was similar in many respects to the activity assumed in certain of the spectral timing models described above. The key mechanistic difference was that this activity was the natural consequence of the sparse, distributed, and recurrent connectivity of the cerebellar cortex.

By incorporating a more complete representation of the connectivity of the olivo-cerebellar circuitry, and by including recent findings regarding the specific synaptic conductances found in cerebellar neurons, a second-generation model now accounts for all key temporal properties of eyelid conditioning (Medina & Mauk 2000). As shown in Figure 9*B*, the timing of conditioned eyelid responses was partly derived from a competitive learning mechanism that increases the temporal specificity of the cerebellar learning was one of the key findings from these simulations (Medina et al. 2000). The key process involves the bidirectional learning in the cerebellum that eyelid conditioning and other forms of learning reveal (Raymond et al. 1996).

Thus, computer simulations and related eyelid conditioning experiments suggest that timing mechanisms in the cerebellar cortex involve three interacting processes (Figure 9). First, sparse recurrent interactions between cerebellar Golgi and granule cells lead to the activation of different granule cells at different times during a stimulus. The activity in granule cells therefore not only codes stimuli, as suggested in seminal theories of cerebellum (Marr 1969), but also codes time elapsed during stimuli. With this temporal code it is then possible for a coincidence-based form of plasticity, such as cerebellar LTD (see Hansel et al. 2001), to mediate learned responses that can be specific for certain times during a stimulus. Finally, competition between excitatory and inhibitory learning sharpens the temporal resolution of the timed responses.

In these network or state-dependent models, timing does not arise from clocks or even from brain systems specifically dedicated to temporal processing. Rather, the evidence from the cerebellum, for example, illustrates how timing and performance on experimental tasks designed to study timing may be mediated by computations that include temporal processing but that are not accurately characterized as interval timers or clocks.

FUTURE CHALLENGES: COMPLEX STIMULI

Most of the experimental and theoretical studies discussed above have focused on relatively simple stimuli. In particular, much of the work has been on the discrimination of the interval or duration of stimuli or on the generation of a single, timed motor response. The mechanisms underlying speech and music recognition, as well as the ability to process Morse code, require sophisticated mechanisms that can process multiple temporal cues in parallel and sequences composed of a continuous stream of elements with no a priori first and last element. Thus, a fundamental issue, particularly in relation to the computational models, is whether these models are sufficiently robust to account for more complex data. Indeed, if a model is limited to the discrimination of simple first-order stimuli (interval and duration), then this model is unlikely to represent the biological mechanisms underlying temporal processing in the range of tens to hundreds of ms.

Higher-Order Stimuli

Consider the stimuli shown in Figure 10 (see color insert), in which a subject must discriminate between 2 sequences composed of 2 intervals (3 tones): 50–150 and 150–50. In reality, in this task one would include 50–50- and 150–150-ms stimulus conditions to prevent the use of simple strategies. In clock or spectral models, neurons would have to respond selectively to the 50- and 150-ms intervals. Additionally, because both stimuli would activate the 50- and 150-ms interval detectors, another circuit would have to keep track of the order of activation, to discriminate between (50–150 and 150–50). Thus as sequences become more complicated, additional circuitry is generally required to keep track of the higher-order features.

Reset Problem

The processing of sequences, as opposed to a single interval or duration, also imposes another constraint on the potential mechanisms underlying temporal processing. Let us consider how a spectral model will perform in response to the sequences shown in Figure 10. In a model based on a slow conductance such as an IPSP, the first tone will activate an IPSP of a different duration in each cell. If the second pulse arrives at 50 ms, the 50-ms detector will fire (owing to the interaction between IPSP offset and arrival of the second stimulus). However, the second pulse is also the first pulse of the second interval, and thus to detect the subsequent 150-ms interval, the second pulse would essentially have to reset the inhibitory conductance. We refer to this as the reset problem. When stimulus elements arrive on the same timescale as the intervals being processed, discrimination requires that the event that marks the end of one interval engage the initiation of the timing of the next interval. Resetting of synaptic conductances, in particular, is unlikely. In spectral models, a potential solution for this problem is to look at the above task as detecting two intervals 50–200 (50 + 150) versus 150–200 (150 + 50). In this manner the second pulse would not have to reset the timer because all timing would be relative to the first pulse. Nevertheless, the second pulse could not interfere with the ongoing computation of the 200-ms interval. This could perhaps be achieved by assuming that the first pulse saturated or depleted the mechanisms responsible for inhibition. However, we believe it is unlikely that spectral models are robust enough to generalize to complex temporal processing involved in speech and music recognition and complex motor patterns.

In contrast, models based on network dynamics may better generalize to the processing of more complex temporal patterns. In state-dependent network models (see above; Buonomano & Merzenich 1995, Buonomano 2000, Maass et al. 2002), the current state of the network is always dependent on the recent history of activity. Thus, in the above example, if the third input arrives at 200 ms, the network will be in a different state depending on whether the second pulse arrived at 50 or 150 ms. In these models, time-dependent properties, such as short-term synaptic plasticity, slow PSPs (e.g., $GABA_B$ or NMDA-dependent currents), or, potentially, slow conductance, function as state-dependent memory traces of the recent stimulus history. In contrast to single-cell models, these time-dependent properties are not tuned for any particular interval; rather these states are expressed as changes in the probability of different neurons becoming activated.

Figure 10 shows results from a state-dependent network model capable of discriminating intervals as well as simple sequences (Buonomano 2000). The network was composed of 400 excitatory and 100 inhibitory units; all synapses exhibited short-term synaptic plasticity, and a slow IPSP was also present. As a result of the time-dependent properties, the network is in a different state at 50 and 150 ms; thus different neurons will respond to the second pulse depending on its arrival time. Because different neurons responded to the second pulse, state-dependent change will be cumulative and alter the response to the third pulse in a different manner depending on the stimulus history. There are two potential shortcomings of state-dependent networks. First, the network must be in a specific regime that allows that expression of the state-dependent changes, which can be nontrivial because a balance between excitation and inhibition is required. Specifically, inhibition must enable excitatory neurons to fire while preventing run-away excitation. Second, because these networks encode time as relative to previous stimuli, they would be least effective at identifying specific intervals embedded in sequences, for example, comparing a 100-ms interval defined by two tones with a 100-ms stimulus embedded within a sequence of tones.

CONCLUSIONS

The study of the neural basis of temporal processing is in its infancy. Few agree on whether temporal processing is centralized or distributed and which structures are involved. Indeed, if all neural circuits can intrinsically process temporal information, then virtually any circuit could be involved, and the location of temporal processing would depend on the nature and modality of the task at hand. Despite the fact that these important questions remain unanswered, the studies, to date, allow several insights into the nature of timing. First, although researchers do not agree on which areas are involved in sensory timing, it seems clear that the cerebellum is responsible for some forms of motor timing. Whether it is the sole source of motor timing and whether it is involved in sensory processing remain open to debate. Second, much evidence indicates that distinct neural mechanisms underlie millisecond and second timing.

Many models of timing have focused on specialized synaptic and cellular mechanisms aimed specifically at processing temporal information, and investigators assumed that spatial and temporal information are essentially processed separately. Given the inherent temporal nature of our sensory environment, and the continuous, real-time motor interaction with our environment, we favor the view that temporal and spatial information are generally processed together by the same circuits, and that there is no centralized clock for temporal processing on the scale of tens to hundreds of ms. Additionally, we propose that temporal processing does not rely on specialized mechanisms, such as oscillators or arrays of elements, as with a spectrum of different time constants. Rather, we believe that neural circuits are inherently capable of processing temporal information as a result of state-dependent changes in network dynamics.

ACKNOWLEGMENTS

We thank Uma Karmarkar and Tatsuya Ohyama for helpful comments on this manuscript, and Michael Lewicki for Figure 6. This work was supported by NIH grant MH60163 to D.V.B., and MH57051 and MH46904 to M.D.M.

The *Annual Review of Neuroscience* is online at http://neuro.annualreviews.org

LITERATURE CITED

Alder TB, Rose GJ. 1998. Long-term temporal integration in the anuran auditory system. *Nat. Neurosci.* 1:519–23

Alder TB, Rose GJ. 2000. Integration and recovery processes contribute to the temporal selectivity of neurons in the midbrain of the northern leopard frog, *Rana pipiens*. *J. Comp. Physiol. A* 186:923–37

Ahissar E. 1998. Temporal-code to rate-code conversion by neuronal phase-locked loops. *Neural Comput.* 10:597–650

Ahissar E, Haidarliu S, Zacksenhouse M. 1997. Decoding temporally encoded sensory input by cortical oscillations and thalamic phase comparators. *Proc. Natl. Acad. Sci. USA* 94: 11633–38

Allan LG. 1979. The perception of time. *Percept. Psychophys.* 26:340–54

Antón PS, Lynch G, Granger R. 1991. Computation of frequency-to-spatial transform by olfactory bulb glomeruli. *Biol. Cybern.* 65: 407–14

Artieda J, Pastor MA, Lacuz F, Obeso JA. 1992. Temporal discrimination is abnormal in Parkinson's disease. *Brain* 115:199–210

Beggs JM, Moyer JR, McGann JP, Brown TH. 2000. Prolonged synaptic integration in perihinal cortical neurons. *J. Neurophys.* 83:3294–98

Belin P, McAdams S, Thivard L, Smith B, Savel S, et al. 2002. The neuroanatomical substrate of sound duration discrimination. *Neuropsychologia* 40:1956–64

Bienenstock EL, Cooper LN, Munro PW. 1982. Theory for the development of neuron selectivity: orientation specificity and binocular interaction in visual cortex. *J. Neurosci.* 2: 32–48

Bottjer SW, Arnold AP. 1997. Developmental plasticity in neural circuits of a learned behavior. *Annu. Rev. Neurosci.* 20:459–81

Boyden E, Katoh A, Raymond J. 2004. Multiple plasticity mechanisms and information coding strategies contribute to the flexibility of cerebellum-dependent learning. *Annu. Rev. Neurosci.* 27:581–609

Brosch M, Schreiner CE. 1997. Time course of forward masking tuning curves in cat primary auditory cortex. *J. Neurophysiol.* 77: 923–43

Buonomano DV. 2000. Decoding temporal information: a model based on short-term synaptic plasticity. *J. Neurosci.* 20: 1129–41

Buonomano DV. 2003. Timing of neural responses in cortical organotypic slices. *Proc. Natl. Acad. Sci. USA* 100:4897–902

Buonomano DV, Karmarkar UR. 2002. How do we tell time? *Neuroscientist* 8:42–51

Buonomano DV, Mauk MD. 1994. Neural network model of the cerebellum: temporal discrimination and the timing of motor responses. *Neural Comput.* 6:38–55

Buonomano DV, Merzenich MM. 1995. Temporal information transformed into a spatial code by a neural network with realistic properties. *Science* 267:1028–30

Buonomano DV, Merzenich MM. 1998a. Cortical plasticity: from synapses to maps. *Annu. Rev. Neurosci.* 21:149–86

Buonomano DV, Merzenich MM. 1998b. Net interaction between different forms of short-term synaptic plasticity and slow-IPSPs in the hippocampus and auditory cortex. *J. Neurophysiol.* 80:1765–74

Buonomano DV, Merzenich MM. 1999. A

neural network model of temporal code generation of position invariant pattern recognition. *Neural Comput.* 11:103–16

Carr CE. 1993. Processing of temporal information in the brain. *Annu. Rev. Neurosci.* 16:223–43

Covey E, Casseday JH. 1999. Timing in the auditory system of the bat. *Annu. Rev. Physiol.* 61:457–76

Casseday JH, Ehrlich D, Covey E. 1994. Neural tuning for sound duration: role of inhibitory mechanisms in the inferior colliculus. *Science* 264:847–50

Creelman CD. 1962. Human discrimination of auditory duration. *J. Acoust. Soc. Am.* 34: 582–93

Creutzfeldt O, Ojemann G, Lettich E. 1989. Neuronal activity in the human lateral temporal lobe. I. Responses to speech. *Exp. Brain Res.* 77:451–75

Deisz RA, Prince DA. 1989. Frequency-dependent depression of inhibition in guinea-pig neocortex in vitro by $GABA_B$ receptor feed-back on GABA release. *J. Physiol.* 412:513–41

Desmond JE, Moore JW. 1988. Adaptive timing in neural networks: the conditioned response. *Biol. Cybern.* 58:405–15

Divenyi P, Danner WF. 1977. Discrimination of time intervals marked by brief acoustic pulses of various intensities and spectra. *Percept. Psychophys.* 21:125–42

Dorman J, Dankowski K, McCandless G, Smith L. 1989. Consonant recognition as a function of the number of channels of stimulation by patients who use the Symbion cochlear implant. *Ear Hear.* 10:288–91

Dorman JF, Loizou PC, Rainey D. 1997. Speech intelligibility as a function of the number of channels of stimulation of signal processors using sine-wave and noise band outputs. *J. Acoust. Soc. Am.* 102:2403–11

Dorman JF, Raphael LJ, Liberman AM. 1979. Some experiments on the sound of silence in phonetic perception. *J. Acoust. Soc. Am.* 65:1518–32

Doupe A. 1997. Song- and order-selective neurons in the songbird anterior forebrain and their emergence during vocal development. *J. Neurosci.* 17:1147–67

Doupe AJ, Konishi M. 1991. Song-selective auditory circuits in the vocal control system of the zebra finch. *Proc. Natl. Acad. Sci. USA* 88:11339–43

Doupe AJ, Kuhl PK. 1999. Birdsong and human speech: common themes and mechanisms. *Annu. Rev. Neurosci.* 22:567–631

Dow RS, Moruzzi G. 1958. *The Physiology and Pathology of the Cerebellum.* Minneapolis: Univ. Minn. Press

Eden G, VanMeter JW, Rumsey JM, Maisog JM, Woods RP, Zeffiro TA. 1996. Abnormal processing of visual motion in dyslexia revealed by functional brain imaging. *Nature* 382:66–69

Farmer ME, Klein RM. 1995. The evidence for a temporal processing deficit linked to dyslexia: a review. *Psychon. Bull. Rev.* 2:460–93

Fiala JC, Grossberg S, Bullock D. 1996. Metabotropic glutamate receptor activation in cerebellar Purkinje Cells as substrate for adaptive timing of the classically conditioned eye-blink response. *J. Neurosci.* 16:3760–74

Fortune ES, Rose GJ. 2001. Short-term synaptic plasticity as a temporal filter. *Trends Neurosci.* 24:381–85

Garcia KS, Mauk MD. 1998. Pharmacological analysis of cerebellar contributions to the timing and expression of conditioned eyelid responses. *Neuropharmacology* 37(4–5):471–80

Gentner TQ, Margoliash D. 2003. Neuronal populations and singles representing learned auditory objects. *Nature* 424:669–74

Getty DJ. 1975. Discrimination of short temporal intervals: a comparison of two models. *Percept. Psychophys.* 18:1–8

Gibbon J, Malapani C, Dale CL, Gallistel CR. 1997. Toward a neurobiology of temporal cognition: advances and challenges. *Curr. Opin. Neurobiol.* 7:170–84

Gluck MA, Reifsnider ES, Thompson RF. 1990. Adaptive signal processing in the cerebellum: models of classical conditioning and VOR adaptation. In *Neuroscience and*

Connectionist Theory, ed. MA Gluck, DE Rumelhart, pp. 131–86. Hillsdale, NJ: Erlbaum

Grondin S, Meilleur-Wells G, Ouellette C, Macar F. 1998. Sensory effects on judgements of short time-intervals. *Psychol. Res.* 61:261–68

Grondin S, Rousseau R. 1991. Judging the duration of multimodal short empty time intervals. *Percept. Psychophys.* 49:245–56

Grossberg S, Schmajik NA. 1989. Neural dynamics of adaptive timing and temporal discrimination during associative learning. *Neural Networks* 2:79–102

Hahnloser RHR, Kozhevnikov AA, Fee MS. 2002. An ultra-sparse code underlies the generation of neural sequence in a songbird. *Nature* 419:65–70

Hansel C, Linden DJ, D'Angelo E. 2001. Beyond parallel fiber LTD: the diversity of synaptic and nonsynaptic plasticity in the cerebellum. *Nat. Neurosci.* 4:467–75

Harrington DL, Haaland KY, Hermanowicz N. 1998a. Temporal processing in the basal ganglia. *Neuropsychology* 12:3–12

Harrington DL, Haaland KY, Knight RT. 1998b. Cortical networks underlying mechanisms of time perception. *J. Neurosci.* 18:1085–95

Holmes G. 1939. The cerebellum of man. *Brain* 50:385–88

Hooper SL. 1998. Transduction of temporal patterns by single neurons. *Nat. Neurosci.* 1:720–26

Hooper SL, Buchman E, Hobbs KH. 2002. A computational role for slow conductances: single-neuron models that measure duration. *Nat. Neurosci.* 5:551–56

Hore J, Wild B, Diener HC. 1991. Cerebellar dysmetria at the elbow, wrist, and fingers. *J. Neurophysiol.* 65(3):563–71

Ito M. 1970. Neurophysiological aspects of the cerebellar motor control system. *Int. J. Neurol.* 2:162–76

Ivry R. 1996. The representation of temporal information in perception and motor control. *Curr. Opin. Neurobiol.* 6:851–57

Ivry RB, Diener HC. 1991. Impaired velocity perception in patients with lesions of the cerebellum. *J. Cogn. Neurosci.* 3:355–66

Ivry RB, Keele SW. 1989. Timing functions of the cerebellum. *J. Cogn. Neurosci.* 1: 136–52

Ivry RB, Keele SW, Diener HC. 1988. Dissociation of the lateral and medial cerebellum in movement timing and movement execution. *Exp. Brain Res.* 73:167–80

Jueptner M, Rijntjes C, Weiller C, Faiss JH, Timmann D, et al. 1995. Localization of a cerebellar timing process using PET. *Neurology* 45:1540–45

Karmarkar U, Buonomano DV. 2003. Temporal specificity of perceptual learning in an auditory discrimination task. *Learn. Mem.* 10:141–47

Kawato M, Gomi H. 1992. The cerebellum and VOR/OKR learning models. *Trends Neurosci.* 15:445–53

Kilgard MP, Merzenich MM. 1998. Plasticity of temporal information processing in the primary auditory cortex. *Nat. Neurosci.* 1:727–31

Kilgard MP, Merzenich MM. 2002. Order-sensitive plasticity in adult primary auditory cortex. *Proc. Natl. Acad. Sci. USA* 99:3205–9

Kimpo RR, Theunissen FE, Doupe AJ. 2003. Propagation of correlated activity through multiple stages of a neural circuit. *J. Neurosci.* 23:5750–61

King DP, Takahashi JS. 2000. Molecular genetics of circadian rhythms in mammals. *Annu. Rev. Neurosci.* 23:713–42

Klump GM, Gerhardt HC. 1987. Use of non-arbitrary acoustic criteria in mate choice by female gray tree frogs. *Nature* 326:286–88

Lashley K. 1960 [1951]. The problem of serial order in behavior. In *The Neuropsychology of Lashley*, ed. FA Beach, DO Hebb, CT Morgan, HW Nissen, pp. 506–21. New York: McGraw-Hill

Laurent G, Wehr M, Davidowitz H. 1996. Temporal representation of odors in an olfactory network. *J. Neurosci.* 16:3837–47

Lehiste I, Olive JP, Streeter LA. 1976. Role of duration in disambiguating syntactically ambiguous sentences. *J. Acoust. Soc. Am.* 60:1199–202

Leon MI, Shadlen MN. 2003. Representation of time by neurons in the posterior parietal cortex of the macaque. *Neuron* 38:317–27

Lewicki MS, Arthur BJ. 1996. Hierarchical organization of auditory temporal context sensitivity. *J. Neurosci.* 16:6987–98

Liberman AM, Delattre PC, Gerstman LJ, Cooper FS. 1956. Tempo of frequency change as a cue for distinguishing classes of speech sounds. *J. Exp. Psychol.* 52:127–37

Lisker L, Abramson AS. 1964. A cross language study of voicing in initial stops: acoustical measurements. *Word* 20:384–422

Livingstone MS. 1998. Mechanisms of direction selectivity in macaque V1. *Neuron* 20:509–26

Livingstone MS, Rosen GD, Drislane FW, Galaburda AM. 1991. Physiological and anatomical evidence for a magnocellular defect in developmental dyslexia. *Proc. Natl. Acad. Sci. USA* 88:7943–47

Maquet P, Lejeune H, Pouthas V, Bonnet M, Casini L, et al. 1996. Brain activation induced by estimation of duration: a PET study. *Neuroimage* 3:119–26

Margoliash D. 1983. Acoustic parameters underlying the responses of song-specific neurons in the white-crowned sparrow. *J. Neurosci.* 3:133–43

Margoliash D, Fortune ES. 1992. Temporal and harmonic combination-sensitive neurons in the Zebra Finch's HVc. *J. Neursoci.* 12: 4309–26

Maass W, Natschläger T, Markram H. 2002. Real-time computing without stable states: a new framework for neural computation based on perturbations. *Neural Comput.* 14:2531–60

Marr D. 1969. A theory of cerebellar cortex. *J. Physiol.* 202:437–70

Matell MS, Meck WH. 2000. Neuropsychological mechanisms of interval timing behavior. *BioEssays* 22:94–103

Mauk MD, Donegan NH. 1997. A model of Pavlovian eyelid conditioning based on the synaptic organization of the cerebellum. *Learn. Mem.* 3:130–58

Mauk MD, Ruiz BP. 1992. Learning-dependent timing of Pavlovian eyelid responses: differential conditioning using multiple interstimulus intervals. *Behav. Neurosci.* 106(4):666–81

McClurkin JW, Optican LM, Richmond BJ, Gawne TJ. 1991. Concurrent processing and complexity of temporally encoded neuronal messages in visual perception. *Science* 253:675–77

McKenna TM, Weinberger NW, Diamond DM. 1989. Responses of single auditory cortical neurons to tone sequences. *Brain Res.* 481:142–53

Mechler R, Victor JD, Purpura KP, Shapley R. 1998. Robust temporal coding of contrast by V1 neurons for transient but not for steady-state stimuli. *J. Neurosci.* 18:6583–98

Meck WH. 1996. Neuropharmacology of timing and time perception. *Cogn. Brain Res.* 3:227–42

Medina JF, Mauk MD. 2000. Computer simulation of cerebellar information processing. *Nat. Neurosci.* 3:1205–11

Medina JF, Garcia KS, Nores WL, Taylor NM, Mauk MD. 2000. Timing mechanisms in the cerebellum: testing predictions of a large-scale computer simulation. *J. Neurosci.* 20:5516–25

Meegan DV, Aslin RN, Jacobs RA. 2000. Motor timing learned without motor training. *Nat. Neurosci.* 3:860–62

Miall C. 1989. The storage of time intervals using oscillating neurons. *Neural Comput.* 1:359–71

Middlebrooks JC, Clock AE, Xu L, Green DM. 1994. A panoramic code for sound location by cortical neurons. *Science* 264: 842–44

Millenson JR, Kehoe EJ, Gormezano I. 1977. Classical conditioning of the rabbit's nictitating membrane response under fixed and mixed CS-US intervals. *Learn. Motiv.* 8:351–66

Miller KD, Keller JB, Stryker MP. 1989.

Ocular dominance column development: analysis and simulation. *Science* 245:605–15

Mooney R. 2000. Different subthreshold mechanisms underlie song selectivity in identified HVc neurons of the zebra finch. *J. Neurosci.* 20:5420–36

Moore JW. 1992. A mechanism for timing conditioned responses. In *Time, Action, and Cognition*, ed. E Macar, pp 229–38. Dordrecht, The Neth.: Kluwer

Moore JW, Choi JS. 1997. The TD model of classical conditioning: response topography and brain implementation. In *Neural-Network Models of Cognition, Biobehavioral Foundations, Advances in Psychology*, ed. JW Donahoe, VP Dorsel, pp. 387–405. Amsterdam, The Neth.: North-Holland/Elsevier. Vol. 121

Nagarajan SS, Blake DT, Wright BA, Byl N, Merzenich MM. 1998. Practice-related improvements in somatosensory interval discrimination are temporally specific but generalize across skin location, hemisphere, and modality. *J. Neurosci.* 18:1559–70

Nawrot M, Rizzo M. 1995. Motion perception deficits from midline cerebellar lesions in human. *Vision Res.* 35:723–31

Nenadic I, Gaser C, Volz H-P, Rammsayer T, Häger F, Sauer H. 2003. Processing of temporal information and the basal ganglia: new evidence from fMRI. *Exp. Brain Res.* 148:238–46

Newberry NR, Nicoll NA. 1984. A bicuculline-resistant inhibitory post-synaptic potential in rat hippocampal pyramidal cells in vitro. *J. Physiol.* 348:239–54

Ohyama T, Nores WL, Murphy M, Mauk MD. 2003. What the cerebellum computes. *Trends Neurosci.* 26(4):222–27

Onoe H, Komori M, Onoe K, Takechi H, Tsukada H, Wtanabe Y. 2001. Cortical networks recruited for time perception: a monkey positron emission tomography (PET) study. *NeuroImage* 12:37–45

Perrett SP, Ruiz BP, Mauk MD. 1993. Cerebellar cortex lesions disrupt learning-dependent timing of conditioned eyelid responses. *J. Neurosci.* 13:1708–18

Prut Y, Vaadia E, Berman H, Haalman I, Solvin H, Abeles H. 1998. Spatiotemporal structure of cortical activity: properties and behavioral relevance. *J. Neurophysiol.* 2857–74

Rammsayer T. 1992. Effects of benzodiazepine-induced sedation on temporal processing. *Hum. Psychopharmacol.* 7:311–18

Rammsayer TH. 1994. Effects of practice and signal energy on duration discrimination of brief auditory intervals. *Percept. Psychophys.* 55:454–64

Rammsayer TH. 1997. Are there dissociable roles of the mesostriatal and mesolimbocortical dopamine systems on temporal information processing in humans? *Biol. Psychol./Pharmacopsychol.* 35:36–46

Rammsayer TH. 1999. Neuropharmacological evidence for different timing mechanisms in humans. *Q. J. Exp. Psychol.* 52B:273–86

Rammsayer TH, Vogel WH. 1992. Pharmacological properties of the internal clock underlying time perception in humans. *Neuropsychobiology* 26:71–80

Rao SM, Mayer AR, Harrington DL. 2001. The evolution of brain activation during temporal processing. *Nat. Neurosci.* 4:317–23

Rauschecker JP, Tian B, Hauser M. 1995. Processing of complex sounds in the macaque nonprimary auditory cortex. *Science* 268:111–14

Raymond J, Lisberger SG, Mauk MD. 1996. The cerebellum: a neuronal learning machine? *Science* 272:1126–32

Reyes A, Lujan R, Burnashev N, Somogyi P, Sakmann B. 1998. Target-cell-specific facilitation and depression in neocortical circuits. *Nat. Neurosci.* 1:279–85

Richmond BJ, Optican LM, Sptizer H. 1990. Temporal encoding of two-dimensional patterns by single units in primate visual cortex. I. Stimulus-response relations. *J. Neurophysiol.* 64:351–68

Rieke FD, Warland R, de Ruyter van Steveninck WB. 1996. *Spikes: Exploring the Neural Code*. Cambridge, MA: MIT Press

Riquimaroux H. 1994. Neuronal auditory science analysis? *Trans. Tech. Commun. Psychol. Physiol. Acoust. Soc. Jpn.* H-94-28:1–8

Riesen JM, Schnider A. 2001. Time estimation in Parkinson's disease: normal long duration estimation despite impaired sort duration discrimination. *J. Neurol.* 248:27–35

Rose GJ, Brenowitz EA. 2002. Pacific treefrogs use temporal integration to differentiate advertisement from encounter calls. *Anim. Behav.* 63:1183–90

Rousseau R, Poirier J, Lemyre L. 1983. Duration discrimination of empty time intervals marked by intermodal pulses. *Percept. Psychophys.* 34:541–48

Saitoh I, Suga N. 1995. Long delay lines for ranging are created by inhibition in the inferior colliculus of the mustached bat. *J. Neurophysiol.* 74:1–11

Schmahmann JD. 1997. *The Cerebellum and Cognition.* New York: Academic

Shannon RV, Zeng FG, Kamath V, Wygonski J, Ekelid M. 1995. Speech recognition with primarily temporal cues. *Science* 270: 303–4

Spencer R, Zelaznick H, Diedrichsen J, Ivry RB. 2003. Disrupted timing of discontinuous but not continuous movements by cerebellar lesions. *Science* 300:1437–39

Stopfer M, Bhagavan S, Smith BH, Laurent G. 1997. Impaired odour discrimination on desynchronization of odour-encoding neural assemblies. *Nature* 390:70–74

Stratford KJ, Tarczy-Hornoch K, Martin KAC, Bannister NJ, Jack JJB. 1996. Excitatory synaptic inputs to spiky stellate cells in cat visual cortex. *Nature* 382:258–61

Sullivan WE. 1982. Possible neural mechanisms of target distance coding in the auditory system of the echolocating bat Myotis lucifugus. *J. Neurophysiol.* 48:1033–47

Tallal P. 1994. In the perception of speech time is of the essence. In *Temporal Coding in the Brain*, ed. G Buzsaki, R Llinas, W Singer, A Berthoz, Y Christen, pp. 291–99. Berlin: Springer-Verlag

Tallal P, Piercy M. 1973. Defects of non-verbal auditory perception in children with developmental aphasia. *Nature* 241:468–69

Thier P, Dicke PW, Haas R, Barash S. 2000. Encoding of movement time by populations of cerebellar Purkinje cells. *Nature* 405(6782):72–76

Thompson RF. 1986. The neurobiology of learning and memory. *Science* 233:941–47

Treisman M. 1963. Temporal discrimination and the indifference interval: implications for a model of the 'internal clock'. *Psychol. Monogr.* 77:1–31

von der Malsburg C. 1973. Self-organization of orientation sensitive cells in the striata cortex. *Kybernetik* 14:84–100

Wang X, Merzenich MM, Beitel R, Schreiner CE. 1995. Representation of a species-specific vocalization in the primary auditory cortex of the common marmoset: temporal and spectral characteristics. *J. Neurophysiol.* 74:2685–706

Westheimer G. 1999. Discrimination of short time intervals by the human observer. *Exp. Brain Res.* 129:121–26

Wing AM, Kristofferson AB. 1973. Response delays and the timing of discrete motor responses. *Percept. Psychophys.* 14:5–12

Wright BA, Buonomano DV, Mahncke HW, Merzenich MM. 1997. Learning and generalization of auditory temporal-interval discrimination in humans. *J. Neurosci.* 17: 3956–63

Wyss R, König P, Verschure PFMJ. 2003. Invariant representations of visual patterns in a temporal population code. *Proc. Natl. Acad. Sci. USA* 100:324–29

Zucker RS. 1989. Short-term synaptic plasticity. *Annu. Rev. Neurosci.* 12:13–31

Annu. Rev. Neurosci. 2004. 27:341–68
doi: 10.1146/annurev.neuro.27.070203.144340

First published online as a Review in Advance on February 23, 2004

THE NOGO SIGNALING PATHWAY FOR REGENERATION BLOCK

Zhigang He and Vuk Koprivica
Division of Neuroscience, Children's Hospital, Harvard Medical School, Boston, Massachusetts 02115; Program in Neuroscience, Harvard Medical School, Boston, Massachusetts 02115; email: zhigang.he@tch.harvard.edu, koprivic@fas.harvard.edu

Key Words axon regeneration, Nogo-A, myelin-associated glycoprotein, oligodendrocyte myelin glycoprotein, chondroitin sulfate proteoglycan, p75, RhoA

■ **Abstract** A hostile environment and decreased regenerative capacity may contribute to the failure of axon regeneration in the adult central nervous system. Recent studies leading to the identification of several myelin-associated inhibitors and their signaling molecules provide opportunitities to assess the contribution of these inhibitory molecules in restricting axon regeneration. These findings may ultimately allow for the development of strategies to alleviate the inhibitory effects of such molecules in an effort to encourage axon regeneration after spinal cord and brain injury.

INTRODUCTION

As neuronal networks form in the developing nervous system of mammals, axons progressively cease growing. In the central nervous system (CNS), lesions that occur at or around the perinatal period can trigger some degree of regeneration, whereas the majority of lesioned axons in a postnatal organism are not repaired, resulting in devastating and permanent functional deficits. This is in stark contrast to axons of the peripheral nervous system (PNS), where regeneration occurs even in the adult. To date, the underlying mechanisms that account for this regeneration failure in the adult CNS are poorly understood. However, because the proximal ends of lesioned adult axons can reform growth cones that appear fundamentally equivalent to those of developing fibers, it is conceivable that axon regeneration in adults may reinitiate mechanisms that normally orchestrate the intricate wiring processes that occur during nervous system development. During development, the precise trajectory of an individual axon is determined by the balanced effects of various guidance cues in the vicinity of the projecting nerve terminal as well as the intracellular signaling state within the responding axon (Dickson 2002, Huber et al. 2003, Song & Poo 2001, Tessier-Lavigne & Goodman 1996, Yu & Bargmann 2001). Extracellular cues exert their actions by engaging specific cell-surface receptors and activating downstream signaling pathways that ultimately lead to the cytoskeletal rearrangements that underlie various axonal behaviors. However, the final response of an axon to certain cues can be modulated, and it is at least partially dictated by the axon's intracellular state. Identifying the environmental cues and signaling mechanisms involved in regeneration may allow us to assess whether the principles that govern axon pathfinding during development also apply to the regeneration of axons in adults. An important recent breakthrough in the field of axon regeneration is the identification of several inhibitory molecules in the adult CNS and some of the signaling components mediating the inhibitory activity in mature neurons. Many reviews have covered several aspects of axon regeneration (David & Lacroix 2003, Filbin 2003, Goldberg & Barres 2000, Horner & Gage 2000, McGee & Strittmatter 2003, Schwab & Bartholdi 1996). Thus we focus our discussion on the functional properties and signaling mechanisms of myelin-associated inhibitors of axon regeneration, with an emphasis on a comparison of the mechanisms involved in developmental axon guidance.

INHIBITORY INFLUENCES ON AXON REGENERATION

Although the failure of mature axons to regenerate was initially thought to result from a loss of the intrinsic regenerative capacity of adult neurons, some early studies suggested a possible involvement of the adult CNS environment. In 1911, Tello (Ramon & Cajal 1928) showed that lesioned adult CNS neurons could extend axonal processes within a permissive environment such as that of a peripheral nerve. Pieces of previously denervated peripheral nerve were implanted into the cortex of rabbits, and two to four weeks after transplantation, silver staining revealed that bundles of axons had entered the peripheral nerve grafts and grown along Schwann cell bands. This initial observation was strengthened by the elegant demonstration decades later that adult CNS neurons could indeed form long projections through peripheral nerve grafts (Benfey & Aguayo 1982). Using retrograde tracing techniques, investigators showed that a substantial number of fibers from CNS neurons had extended for several millimeters into the graft tissue (David & Aguayo 1981, Richardson et al. 1980). However, only a few fibers grew out of the graft, and their growth stopped soon after re-entering the spinal cord (David & Aguayo 1981). These results suggest that at least some CNS neurons can regenerate their axons under favorable conditions and that the adult CNS environment is less permissive for axon growth than is the Schwann cell–dominated microenvironment.

To further dissect whether the difference between the CNS and PNS environments is due to a lack of trophic support or the presence of inhibitory molecules in the CNS, Schwab & Thoenen (1985) compared neurite growth from dissociated perinatal sympathetic neurons that were in contact with adult PNS (sciatic nerve) or CNS (optic nerve) tissue explants in the presence of nerve growth factor (NGF). After two weeks in culture, massive fiber ingrowth into the sciatic nerve was observed, although few or no axons had grown in the optic nerve. These findings further suggest the presence of inhibitory influences in the adult CNS environment.

PUTATIVE MYELIN INHIBITORS

The most apparent difference between the CNS and the PNS is the presence of specific glial cell types, namely astrocytes, oligodendrocytes, and microglial cells, in the CNS. Some of the essential functions of CNS-specific glial cells are known. For example, astrocytes participate in the formation and maintenance of the blood-brain barrier (Risau & Wolburg 1990), ion homeostasis (Bekar & Walz 1999), and neurotransmitter transport (Schousboe & Westergaard 1995), as well as in the production of the extracellular matrix (Liesi & Silver 1988). Oligodendrocytes are responsible for the myelination of axons within the CNS (Bunge et al. 1968). Microglial cells are the resident phagocytic cells that maintain a resting phenotype during nonpathological conditions in the CNS (Banati & Graeber 1994, Gehrmann 1996).

In most instances where CNS regeneration has been described following injury, regenerating nerve fibers either bypass or stop at the neuroglial scar that arises from the lesioned tissue. This scar is composed chiefly of the components from glial cells (primarily astrocytes) and connective tissue elements (reviewed by Reier et al. 1983). Thus, it acts as a mechanical barrier that is impenetrable by regenerating axons. However, many lines of study have since demonstrated that the glial scar contains components that can inhibit axon growth, including the glycoprotein tenascin-C (TEN-C) (Probstmeier et al. 2000) and chondroitin sulfate proteoglycans (CSPG) (reviewed by Fawcett & Asher 1999, Fitch & Silver 1998, Morgenstern et al. 2002). The expression of several different CSPGs in astrocytes is increased following CNS injury (Davies et al. 1997, 1999; McKeon et al. 1991; Rudge & Silver 1990), and enzymatic removal of these molecules from the lesion sites using chondroitinase ABC has been reported to improve axon regeneration and functional recovery in various lesion models (Bradbury et al. 2002, Moon et al. 2001).

In addition to these glial scar–associated inhibitors, CNS myelin may represent another major source of inhibitors of axon regeneration. Berry (1982) first proposed the concept of CNS myelin inhibiting axon regeneration. He noted that nonmyelinated axons in the CNS regenerate after chemical axotomy, which does not damage nearby myelinated fibers, but not after mechanical axotomy, which always damages some myelinated axons. Because damage to myelinated tracts results in the degeneration of myelin structures, Berry hypothesized that products of injured CNS myelin are inhibitory to axonal regeneration. This premise is supported by several studies of axon regeneration that used animals at different developmental stages. For example, in chicken, opossum, and rat, the end of the regeneration-permissive period correlates temporally with the appearance of the first differentiated oligodendrocytes and the onset of myelination (Bregman et al. 1993, Hasan et al. 1993, Kalil & Reh 1982, Saunders et al. 1998, Tolbert & Der

1987, Treherne et al. 1992). Either by preventing the development of oligodendrocytes via the use of X-ray irradiation in rats or by delaying the start of myelination via the application of antibodies against GalC, together with complement in chickens, the growth-permissive period could be extended (Keirstead et al. 1992, Savio & Schwab 1990). Direct experimental evidence of CNS myelin's ability to inhibit neurite outgrowth was later provided by Schwab and other researchers who performed elegant cell and tissue culture experiments to show that CNS myelin, but not PNS myelin, possesses potent neurite growth inhibitory activity (Carbonetto et al. 1987, Crutcher 1989, Khan et al. 1990, Sagot et al. 1991, Savio & Schwab 1989). These studies showed that neurite outgrowth on frozen sections from developing and adult PNS and CNS tissue was clearly worse on white matter than it was on gray matter. Likewise, in cocultures of dissociated glial cells and neurons, mature oligodendrocytes were strictly avoided by neurons (Caroni & Schwab 1988a, Bandtlow et al. 1990, Moorman 1996). Subsequent studies from several laboratories have resulted in the identification of a number of putative inhibitors associated with CNS myelin.

Nogo-A

Experiments involving size fractionation of rat brain myelin proteins identified two membrane protein fractions containing peptides of molecular weight 35 kDa and 250 kDa that had potent inhibitory properties when used as tissue-culture substrates or reconstituted in liposomes and added to growing neurites (Caroni & Schwab 1988a). These proteins were named NI-35 and NI-250, respectively (Caroni & Schwab 1988b). A monoclonal antibody, IN-1, raised against NI-250, reacted with both NI-250 and NI-35 and blocked the inhibitory activity of these protein fractions and of myelin (Caroni & Schwab 1988a). Strikingly, this IN-1 antibody was also able to induce regeneration of some lesioned CNS axons in vivo (Bregman et al. 1995; Schnell & Schwab 1990, 1993; Schnell et al. 1994). Because the IN-1 antibody belongs to the IgM isotype family, it is not suitable for conventional antibody-based affinity chromatography. Thus, through conventional biochemical purification, Schwab's group was able to purify the bovine NI-250 and obtain the amino acid sequences of six peptides from this protein (Spillmann et al. 1998). The peptide sequence information allowed Schwab's group (Chen et al. 2000) and two other labs (GrandPré et al. 2000, Prinjha et al. 2000) to clone the full-length mammalian gene, termed Nogo. At least three major transcripts (Nogo-A, -B, and -C) originate from this gene by alternative splicing and promoter use. All three isoforms share a common C-terminal domain of 188 amino acids. The longest isoform is termed Nogo-A and contains a unique amino-terminal region. Nogo-A is highly expressed in oligodendrocytes but not in Schwann cells, consistent with the notion that Nogo-A is a CNS myelin–associated inhibitor. Surprisingly, Nogo-A is also expressed in many types of central and peripheral neurons (Huber et al. 2002, Josephson et al. 2001, Wang et al. 2002c), but the role of neuronal Nogo-A remains unclear. In contrast, no detectable Nogo-B and Nogo-C expression was found in myelin-forming oligodendrocytes. Thus, most of the studies to date have focused on Nogo-A.

Amino acid sequence analysis suggests that none of the Nogo isoforms possess an amino-terminal signal sequence that would allow the protein to be presented on the cell surface. Instead, they exhibit a dilysine endoplasmic reticulum retention signal (-KRKAE) within the conserved carboxyl terminus. Consistent with this finding, overexpression studies suggest that most of the protein is localized within transfected non-neuronal cells (Chen et al. 2000, Oertle et al. 2003, Prinjha et al. 2000). The question is how much, if any, of the Nogo protein is present at the cell surface. The immunocytochemical identification of Nogo-A in plasma membranes of oligodendrocytes in white matter (Huber et al. 2002, Wang et al. 2002a) and in culture (Oertle et al. 2003) strongly suggests that at least some of the molecules can be targeted to the cell surface. In white matter, Nogo-A can be seen on the innermost loop of the myelin membrane that is adjacent to the axon (Huber et al. 2002, Wang et al. 2002c). This is not unusual because at least two other myelin proteins, OMP-22 and MAL, also harbor endoplasmic reticulum retention signals and yet are able to reach the cell surface. Thus, it is of future interest to investigate how these myelin proteins might be differentially targeted in myelin-forming cells as well as other cells.

If Nogo-A is involved in oligodendrocyte-mediated growth cone collapse or neurite-outgrowth inhibition, it will be important to clarify which region of the Nogo-A protein is exposed to the cell surface. The initial structure/function studies by Strittmatter and colleagues suggest that within the conserved C-terminal region of the full-length Nogo-A protein there are two transmembrane domains separated by a region of 66 amino acids, termed Nogo-66 (Prinjha et al. 2000). Thus both the N and C termini of Nogo-A are cytosolic, whereas the short 66–amino acid loop (Nogo-66) may be extracellular (Figure 1, see color insert). In vitro studies, however, show that both recombinant N-terminal Nogo-A (amino-Nogo) and Nogo-66 proteins inhibit neurite outgrowth (Chen et al. 2000, Fournier et al. 2001, Prinjha et al. 2000). A recent study further suggests the presence of two different active sites in the N-terminal Nogo-A: a region involved in the inhibition of fibroblast spreading localized at the far N terminus, and a stretch encoded by the Nogo-A-specific exon that restricts neurite outgrowth and cell spreading and induces growth cone collapse (Oertle et al. 2003). Moreover, at least some of the protein's Nogo-A-specific active domain can be detected on the cell surface. Therefore, it remains somewhat unclear how and why Nogo-A can take such different membrane topologies.

Although the inhibitory activity of Nogo has been clearly demonstrated in a variety of in vitro studies, the contribution of Nogo in restricting axon regeneration in vivo has not been fully delineated. Axon regeneration is inhibited in the peripheral nerves of transgenic mice expressing Nogo-A under the control of a P0 PNS cell–specific promoter (Pot et al. 2002). Similarly, using an OCT-6 promoter to drive transgenic Nogo-C expression in Schwann cells also results in delayed PNS axon regeneration (Kim et al. 2003a). Because the shared region of Nogo-A and Nogo-C is the extracellular Nogo-66 region, these results are consistent with an inhibitory role of Nogo-66 in vivo. It will be interesting to compare the extent of the delay in axon regeneration in the PNS of mice expressing similar levels of Nogo-A and Nogo-C. These studies may then provide insights into the relative contribution of the amino-Nogo and Nogo-66 regions in vivo.

More recently, three groups analyzed the axon regeneration phenotypes of Nogo knockout mice and obtained surprisingly different results (Kim et al. 2003, Simonen et al. 2003, Zheng et al. 2003). Significant improvement in regeneration was found only in the young mice lacking Nogo-A/B (Kim et al. 2003). In contrast, Zheng et al. (2003) found no evidence of axon regeneration in the two lines of mutant mice studied, one lacking all three Nogo proteins and the other lacking Nogo-A/B. In the line lacking Nogo-A but with increased expression of Nogo-B (Simonen et al. 2003), there was a statistically significant improvement in axonal growth, but the number of axons that grew in these mice was small. All these studies involve a lesion model in which the regeneration of descending corticospinal-tract fibers is monitored, so it remains unclear whether the regeneration of other axonal tracts might be improved in these Nogo mutants. Nevertheless, these results emphasize the fact that additional inhibitory molecules may be important in restricting axon regeneration in vivo.

Myelin-Associated Glycoprotein (MAG)

MAG is a transmembrane protein with five immunoglobulin-like domains in its extracellular domain that is expressed in both PNS and CNS (Figure 1) (Lai et al. 1987, Salzer et al. 1987). MAG was initially implicated in the formation and maintenance of myelin sheaths (reviewed by Schachner & Bartsch 2000). In attempts to identify myelin-associated inhibitors, McKerracher et al. (1994) applied detergent-solubilized myelin proteins onto an anion exchange column and monitored via a neurite-outgrowth assay the inhibitory activity of each fraction. They found that MAG was enriched in a fraction with high inhibitory activity and that removal of MAG reduced this activity. Independently, Filbin and colleagues (DeBellard et al. 1996, Mukhopadhy et al. 1994) found that MAG is an inhibitory molecule for many types of mature neurons in vitro. MAG appears to be bi-functional. Depending on the age of the neuron, it can either promote or inhibit neurite outgrowth (DeBellard et al. 1996). For retinal ganglion neurons and spinal neurons, the switch occurs by the time of birth: Outgrowth of embryonic retinal ganglion neurons and spinal neuron axons is promoted by MAG, whereas postnatal neurite growth is inhibited (DeBellard et al. 1996, Salzer et al. 1990, Turnley & Bartlett 1998). For dorsal root ganglion neurons, the switch occurs postnatally with a sharp transition from promotion to inhibition by MAG at postnatal day 3–4 (Debellard et al. 1996, Johnson et al. 1989, Mukhopadhyay et al. 1994). Although the MAG knockout mouse does not exhibit any axon regeneration in the spinal cord (Bartsch et al. 1995), transgenic mice expressing MAG under the control of the p75 promoter show retarded axon regeneration in the PNS (Shen et al. 1998). Along the same line, axon regeneration in the PNS was improved in the MAG knockout mice (Schafer et al. 1996). Why does axon regeneration occur in the PNS even though MAG is expressed in PNS myelin? A possible explanation is that macrophages in the PNS can rapidly get rid of myelin debris after an injury, which is not the case in the CNS (David et al. 1990).

Oligodendrocyte Myelin Glycoprotein (OMgp)

Many membrane-associated axon guidance cues can be in the form of transmembrane proteins or glycosylphosphatidylinositol (GPI)-linked proteins (Huber et al. 2003, Tessier-Lavigne & Goodman 1996). During a search for GPI-linked inhibitors in CNS myelin, Wang et al. (2002a) identified OMgp as a putative inhibitor of neurite outgrowth. Recombinant OMgp can induce growth cone collapse and inhibit neurite outgrowth (Barton et al. 2003, Wang et al. 2002a). OMgp contains a leucine-rich repeat (LRR) domain followed by a C-terminal domain with serine/threonine repeats (Mikol et al. 1990). Its expression in oligodendrocytes correlates with the onset of myelination, and the protein can be detected on the surface of oligodendrocytes and on axon-adjacent myelin layers (Habib et al. 1998a, Mikol & Stefansson 1988, Mikol et al. 1990). Early experiments using chromatographic separation of neurite-outgrowth inhibitors in CNS myelin suggest the existence of other inhibitory molecules in addition to MAG (McKerracher et al. 1994). In a follow-up study, Kottis et al. (2002) identified OMgp as an inhibitor of neurite outgrowth responsible for this inhibitory activity. However, the contribution of OMgp in inhibiting axon regeneration in vivo has not yet been determined. Because the mouse OMgp gene is embedded within an intron of the large tumor suppressor gene NF1 (Habib et al. 1988b) and overexpression of OMgp in NIH3T3 cells inhibits cell proliferation (Habib et al. 1998b), OMgp may affect the development of neurofibromatosis type I (McCormick 1995). It will be interesting to examine whether OMgp acts as a ligand or receptor in this case.

Other Possible Inhibitors

Although Nogo-A, MAG, and OMgp may account for a significant fraction of the in vitro inhibitory activity in CNS myelin, many lines of evidence suggest that other myelin components may also play a role in regeneration inhibition. For example, outgrowth-inhibiting CSPGs are a minor component of CNS myelin (Niederost et al. 1999). A recent study suggests that the transmembrane semaphorin Sema4D/CD100 is expressed on oligodendrocytes and can be detected in CNS myelin (Moreau-Fauvarque et al. 2003). Recombinant proteins containing the extracellular domain of Sema4D/CD100 are able to induce growth cone collapse and repel growing axons in a stripe assay. Following spinal cord injury, Sema4D/CD100 is strongly upregulated in oligodendrocytes at the periphery of the lesion. Thus, even though Sema4D/CD100 may not be an abundant CNS myelin component, its injury-elicited upregulation may make it more relevant to axon regeneration. Examining whether other axonal-repulsive molecules expressed in oligodendrocytes may also be injury inducible and whether they participate in restricting axon regeneration will be interesting. As has been suggested, the relative contributions of different types of myelin-associated inhibitors have not been fully characterized. Overall, these studies suggest that the effects of multiple inhibitors may be necessary to elicit complete blockage of axon regeneration upon injury in the adult CNS.

Other Functions of Myelin-Associated Inhibitors

INITIATION AND MAINTENANCE OF MYELINATION During myelination, MAG is expressed in Schwann cells and both MAG and OMgp are expressed in oligodendrocytes, where they are enriched in the periaxonal myelin layers. Because of this expression pattern, investigators thought these molecules had a role in myelination (reviewed by Mikol et al. 1990, Schachner & Bartsch 2000). Although it remains to be determined whether a myelination defect occurs in the absence of OMgp, the onset of myelination in mice lacking MAG (Li et al. 1994, Montag et al. 1994) or Nogo (Simonen et al. 2003) appears normal. However, a more detailed analysis reveals an interesting difference in the PNS and the CNS of aging mice lacking MAG. Although the PNS myelin is largely normal, morphological and biochemical signs of axon degeneration are found in MAG-deficient animals (Fruttiger et al. 1995, Yin et al. 1998). In contrast, the most obvious phenotype in the CNS is the degeneration of periaxonal oligodendroglial myelin, with no detectable alteration of the axons of these mice (Lassmann et al. 1997). Although the precise mechanisms are still elusive, these observations suggest a critical role for MAG in maintaining functional and structural relationships between glial processes and axons.

STRUCTURAL PLASTICITY Because many of the above-mentioned myelin proteins such as Nogo-A and OMgp are expressed in projection neurons in addition to oligodendrocytes, these inhibitory molecules may participate in other aspects of neuronal function. Some evidence suggests a possible role for these myelin proteins in regulating the structural plasticity of the nervous system. For example, myelin-associated inhibitors have been implicated in the structural remodeling of the Purkinje intracortical plexus (Gianola et al. 2003). During the first postnatal week, the Purkinje axons in rat cerebellum form additional collaterals that will undergo in the following week a reorganization process with extensive pruning of collateral branches and remodeling of terminal arbors. Interfering with normal myelinogenesis by killing oligodendrocyte precursors with 5′-azacytidine or applying neutralizing IN-1 antibody causes Purkinje axons to retain exuberant branches. As a result, the terminal plexus spans the entire extent of the granule layer. Thus, although intrinsic determinants affect the formation of Purkinje axon collaterals, myelin-associated inhibitors/components may contribute to the growth and distribution of the axons (Gianola et al. 2003). Along the same line, Thallmair and colleagues (1998) found that IN-1-antibody treatment dramatically improved the recovery of forelimb function in the rat following a unilateral transection of the corticospinal tract at the brain stem level. Such an effect of IN-1 does not seem to be caused by the promotion of axon regeneration but is more likely a result of significant collateral sprouting of corticospinal and corticobulbar nerve fibers across the midline at several levels (Thallmair et al. 1998). Hence, these inhibitors may serve to restrict structural plasticity by suppressing sprouting and spurious synapse formation in the CNS. Furthermore, CSPGs, another major class of axon regeneration inhibitors associated with the glial scar at a CNS lesion site, have been recently implicated in the regulation of ocular dominance column plasticity (Pizzorusso et al. 2002). Therefore, examining the role such inhibitory molecules as Nogo and OMgp play in terms of structural plasticity in the adult CNS may also be of interest.

RECEPTORS OF MYELIN INHIBITORS

NgR Is a Common Receptor for Three Myelin Inhibitors

Using an expression-cloning strategy, Strittmatter and colleagues (Fournier et al. 2001) identified a GPI-linked axonal surface protein, termed Nogo-66 receptor (NgR), that can bind Nogo-66 with high affinity (Figure 1). The protein is predicted to contain 473 amino acids, with a conventional amino-terminal membrane translocation signal sequence. This is followed by eight LRR motifs, and an LRR carboxy-terminal motif (LRRCT). In the carboxyl terminus, there is a unique region located prior to the GPI anchor. Enzymatic cleavage of NgR renders neurons unresponsive to inhibition by Nogo-66. In addition, forced expression of NgR is sufficient to confer sensitivity of normally unresponsive neurons to each of these inhibitors. Surprisingly, by the same expression-cloning strategy, the same NgR molecule has been identified as a high-affinity OMgp-binding protein (Wang et al. 2002a). In both loss-of-function and gain-of-function experiments similar to those for NgR/Nogo-66 interaction (Fournier et al. 2001), NgR has also been implicated as a required receptor component for OMgp in neurite-outgrowth inhibition (Wang et al. 2002a). Although the C-terminal serine/threonine repeat-containing region can also bind weakly to NgR, the LRR-containing domain of OMgp appears to be sufficient for binding to NgR and inhibiting neurite outgrowth. Moreover, two independent studies later suggested that NgR is also involved in the inhibitory activity of MAG (Domeniconi et al. 2002, Liu et al. 2002). Liu et al. (2002) identified MAG as an NgR-binding protein, using soluble NgR to screen for binding partners via expression cloning. Following up their previous observation that a number of cellular proteins bound to MAG-Fc, Domeniconi et al. (2002) found NgR to be a MAG-binding protein. Each of these studies also provided functional evidence for the requirement of NgR in mediating the activity of MAG.

Such a convergence of the three inhibitors on the same receptor molecule is unprecedented. Because an NgR knockout mouse has not been available, none of the studies mentioned above could offer unambiguous loss-of-function data.

However, it is possible that some of the unusual biophysical properties of NgR [for example, a surface enriched in charged residues (He et al. 2003, Barton et al. 2003)] may allow NgR to bind other proteins easily. Although the final answer will rely on functional analyses of NgR knockout mice, currently available structural and functional data argue against a nonspecific interaction for NgR with these ligands. The results show that the N-terminal region of NgR, which harbors eight LRR motifs and an LRRCT, is sufficient to bind each of these ligands (Barton et al. 2003, He et al. 2003, Liu et al. 2002, Wang et al. 2002b). Although the precise binding sites for these three distinct ligands have not yet been defined, it is likely that these ligands bind to overlapping but distinct regions of NgR. For example, the LRRCT region appears to be sufficient for binding weakly to Nogo-66, but not to OMgp (Wang et al. 2002a). More recently, a 44–amino acid synthetic peptide based on Nogo-66 (NEP1-40) has been shown to selectively block the inhibitory activity of Nogo-66, but not of MAG (GrandPré et al. 2002).

The Structure of NgR Ligand-Binding Domain

An approach to understanding the molecular interaction of NgR with its ligands and coreceptors is to determine the fine structure of NgR. He et al. (2003) reported the 1.5 Å crystal structure of a soluble extracellular domain of the human NgR. They found that NgR adopts an LRR module structure whose concave exterior surface contains a broad region of evolutionarily conserved patches of aromatic residues, offering a possible structural explanation for such extraordinary promiscuity for ligand binding. A deep cleft at the C-terminal base of the LRR appears to play a role in NgR's association with its coreceptor(s) (He et al. 2003). Barton et al. (2003) also reported similar results in which two proteins, termed NgR2 and NgR3, were found to share a high sequence similarity with NgR but were unable to bind all the known NgR ligands, rendering the possibility of redundant binding proteins for myelin inhibitors less probable.

Do Other Receptors of Myelin Inhibitors Exist?

Recent in situ hybridization studies suggest that NgR may not be expressed in all the neurons in the adult CNS (Hunt et al. 2002, Josephson et al. 2001). For example, high levels of NgR expression can be detected in only 20–25% of dorsal root ganglion neurons (Hunt et al. 2002). Why do those neurons without or with low expression of NgR still fail to regenerate their lesioned axons? One possibility is that neurons lacking NgR may be insensitive to identified NgR ligands but are inhibited by other inhibitory molecules in the CNS. For example, the N-terminal region of Nogo-A (amino-Nogo) acts in an NgR-independent manner (Figure 1, Liu et al. 2002, Niederost et al. 2002). Alternatively, adult neurons lacking NgR expression may still be sensitive to these myelin inhibitors, but they could be signaling via another receptor. The ultimate proof of the role of NgR signaling in myelin inhibition will come from a clear loss-of-function study, which, it should be noted, has not been done yet. Although at least two proteins with similarity to

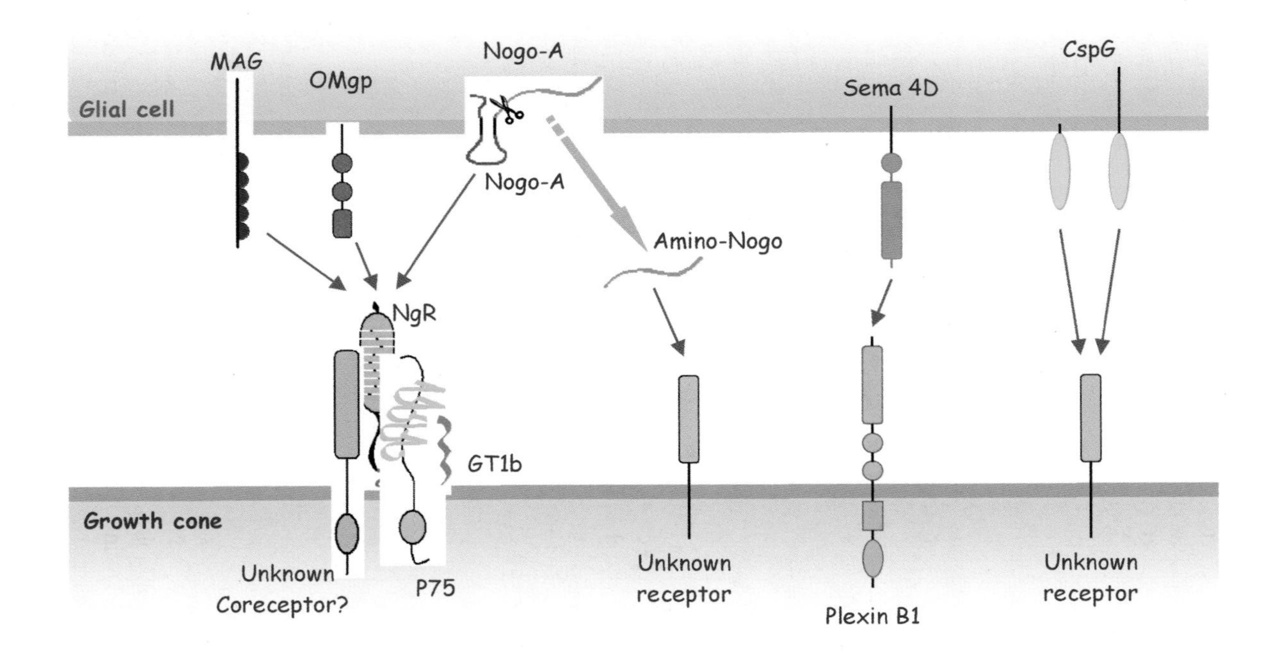

Figure 1 Schematic representation of known myelin-associated inhibitors and their receptor complexes. Three major myelin inhibitors, including the extracellular domain of Nogo-A (Nogo-66), MAG, and OMgp, all bind to NgR with high affinity. This complex in turn associates with p75 and other possible coreceptors to transduce the signal across the plasma membrane. In addition, the N-terminal region of Nogo-A (amino-Nogo) and CSPG can inhibit neurite outgrowth independent of NgR. Sema4D can be upregulated by injury and may act through its receptor plexinB1.

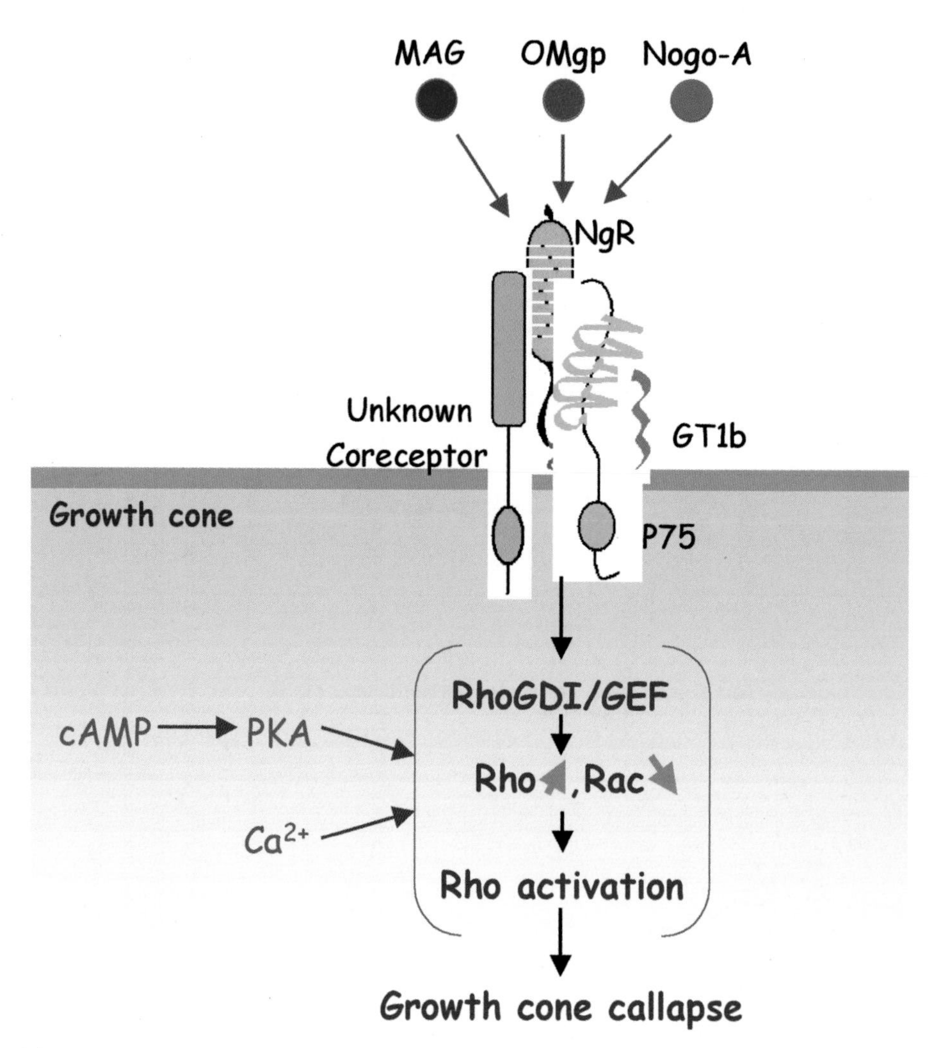

Figure 2 Schematic diagram showing two types of signaling molecules functioning in the signaling pathways that mediate the growth cone collapse activity of three major myelin-associated inhibitors. Whereas the small GTPases act as direct signaling mediators, cAMP/PKA and calcium may indirectly modulate the final responses of responding axons to these inhibitors.

NgR have been cloned, these proteins do not seem to bind any of the known NgR ligands, making it unlikely that NgR2 and NgR3 act as receptors (Barton et al. 2003, He et al. 2003). On the other hand, MAG is a sialic acid–binding SIGLEC (sialic acid–dependent immunoglobulin-like family member lectin) protein with an affinity for such gangliosides as GT1b and GD1a (Collins et al. 1997, Tang et al. 1997, Vinson et al. 2001). These gangliosides might also serve as axonal MAG receptors (Vinson et al. 2001, Vyas et al. 2002). In support of this, function-blocking antibodies against GT1b have been found to neutralize the inhibitory activity of MAG (Vinson et al. 2001, Vyas et al. 2002). Moreover, neurons from a mutant mouse that lacks complex gangliosides, including GT1b, are insensitive to inhibition by MAG (Vyas et al. 2002). Although it seems that the binding of MAG with NgR is sialic acid independent (Domeniconi et al. 2002, Liu et al. 2002), the exact role of GT1b and other gangliosides in the signaling pathway of MAG and other inhibitors remains unclear.

Finally, the bi-functional property of MAG and perhaps other myelin components may also suggest the presence of different receptor molecules for these proteins (Cai et al. 2001, Johnson et al. 1989). As discussed above, these molecules can inhibit the regeneration of mature neurons but promote neurite outgrowth from immature cells. Although NgR is widely expressed in the nervous system during development (Hunt et al. 2002), it remains unclear whether NgR or other NgR-independent interactions are needed to account for the neurite outgrowth–promoting activity of these myelin components.

CORECEPTOR(S) OF NgR

P75 Is an NgR Coreceptor for Myelin Inhibitors

Because NgR is GPI linked and lacks an intracellular signaling domain, it must rely on additional transmembrane coreceptor(s) to transduce the inhibitory signal. Overexpression of a truncated form of NgR lacking its unique C-terminal region (CT) allows extensive neurite outgrowth on myelin substrates, suggesting the CT may be involved in engaging the coreceptor(s) (Domeniconi et al. 2002, Wang et al. 2002b). Different lines of evidence suggest that p75, initially identified as a low-affinity receptor for neurotrophins, could function as a coreceptor for NgR (Wang et al. 2002b, Wong et al. 2002). The initial hint came from a study that revealed that neurons isolated from mutant mice lacking p75 are not inhibited by MAG (Yamashita et al. 2002). This study also showed that p75 colocalizes with sites of MAG binding, even though direct binding between MAG and p75 could not be demonstrated. Following these observations, it was further demonstrated that p75 is required for mediating the inhibitory activity of all three myelin-associated inhibitors (Wang et al. 2002b). By using both cell-surface binding assays and coimmunoprecipitation, investigators have shown that p75 forms a physical receptor complex with NgR and that such an interaction is potentiated by pretreating the cells with such ligands as MAG-Fc (Wang et al. 2002b, Wong et al. 2002). Binding studies using purified recombinant proteins also demonstrated that the full-length, but not the truncated, NgR protein is required to bind p75 (He et al. 2003). This is consistent with the prediction that the CT of NgR participates in the coreceptor interaction (Wang et al. 2002b, Wong et al. 2002). In support of p75's function as a coreceptor, researchers found that introducing a truncated form of p75 lacking the intracellular domain into postnatal cerebellar granule neurons allows these cells to extend neurites when grown on myelin or its individual inhibitory components (Wang et al. 2002b). Moreover, both p75 and NgR appear to be required for the activation of the small GTPase RhoA, as elicited by MAG-Fc and other ligands (Yamashita et al. 2002, Wang et al. 2002b). In spite of this functional and biochemical evidence, it is unknown whether p75 is sufficient to mediate the inhibitory activity of myelin-associated inhibitors. How do these observations reconcile with the involvement of p75 in neurotrophin signaling? Although a certain degree of cross talk probably exists between signaling pathways for myelin inhibitors and neurotrophins, evidence from both in vitro and in vivo experiments suggests that p75 is a growth-inhibiting molecule (Hanilla & Kawaja 1999, Kohn et al. 1999, Walsh et al. 1999). In cultured sympathetic neurons that normally express TrkA and p75 (but not TrkB), NGF and brain-derived neurotrophic factor have functionally antagonistic actions on neurite outgrowth and target innervation: NGF acts via TrkA to promote growth, and brain-derived neurotrophic factor acts via p75 to inhibit growth. Consistent with the notion that p75 functions as an inhibitory molecule, sympathetic neurons from p75 knockout mice grow more robustly in response to NGF (Kohn et al. 1999). Although the axon regeneration phenotype of p75 knockout mice has not been determined, previous studies show that axons of adult sympathetic neurons from p75 mutant mice overexpressing NGF can grow extensively in myelinated portions of the cerebellum (Walsh et al. 1999) and myelinated optic-nerve tracts (Hanilla & Kawaja 1999), areas that typically do not support axon growth in wild-type mice. These studies provide further evidence that p75 is involved in myelin-mediated inhibition.

Other NgR Coreceptors

In general, p75 is highly expressed in the nervous system in both glial cells and neurons during development and early postnatal stages, but the expression level declines to background levels in the adult (reviewed by Chao 2003, Hempstead 2002, Ernfors et al. 1988). There are many examples of elevated p75 expression in the adult brain and spinal cord after injury, inflammation, and stress (reviewed by Dechant & Barde 2002, Dobrowsky & Carter 2000, Roux et al. 1999). For example, in adult Purkinje cells, p75 upregulation can be induced by traumatic injury (Martinez-Murillo et al. 1998). However, it remains to be determined whether p75 is expressed in all adult CNS neurons. For example, using in situ hybridization Suzuki et al. (1998) clearly demonstrated that p75 is expressed in subpopulations of retinal ganglion cells (RGCs) in adult rats. However, immunostaining failed to detect p75 expression in adult RGCs (Ding et al. 2001), suggesting that p75 may be either absent or expressed at low levels in some adult CNS neurons. The extent of p75 and NgR coexpression also remains to be determined.

How is the NgR signal transduced across the axon membrane in the absence of p75? An answer to this may lie in the existence of other p75 homologues, which has indeed been demonstrated at least in *Xenopus* (Hutson et al. 2001). Alternatively, other transmembrane molecules may interact with NgR and act in transducing the signals. As anti-GT1b antibodies neutralize the inhibitory activity of MAG (Vinson et al. 2001, Vyas et al. 2002) and p75 associates with GT1b (Yamashita et al. 2002), other ganglioside-binding proteins may function in lieu of p75 in at least some types of neurons in the adult CNS.

Receptor Complexes Associated with Lipid Rafts

Although additional components of the NgR receptor complex probably exist and remain to be identified, a common feature of these known molecules, including NgR, p75, and gangliosides, is their physical association with lipid rafts (Vinson et al. 2003), specialized signaling microdomains that are enriched in cholesterol and glycosphingolipids (Simons & Ikonen 1997). Rafts are believed to function in cellular signaling by concentrating or separating specific molecules in a unique lipid environment (Galbiati et al. 2001). For example, glial cell–derived neurotrophic factor (GDNF) transduces its signal through a receptor complex that consists of a transmembrane receptor tyrosine kinase and a GPI-anchored coreceptor, GFRα1, which localizes to lipid rafts (Tansey et al. 2000). Upon GDNF stimulation, GFRα1 recruits receptor tyrosine kinase to the lipid rafts, resulting in strong and continuous signaling (Tansey et al. 2000). Thus, myelin inhibitors may trigger a similar relocalization of the receptor complex components, resulting in the activation of inhibitory pathways. A recent report suggests that phosphorylation of p75 by protein kinase A (Higuchi et al. 2003), a well-characterized regulator of axon guidance molecules including MAG and other myelin inhibitors, regulates its localization to lipid rafts. Because ligands like MAG-Fc could potentiate the association of NgR with p75 (Wang et al. 2002b, Wong et al. 2002), such an interaction may reflect a recruitment of p75 into the lipid rafts where most GPI-linked surface molecules, including NgR, are localized, a scenario reminiscent of the GDNF receptor complex (Tansey et al. 2000). This model predicts that interfering with the function of other components in the same lipid raft complex may change the integrity of the signaling complex, thereby affecting indirectly the signaling of NgR and its ligands. An anti-GT1b antibody, for example, may disrupt the ligand-induced interaction of NgR with other receptor complex components. It will be interesting to examine whether the inhibitory activity of other NgR ligands in addition to MAG are also affected by anti-GT1b antibody treatment. A future challenge is to isolate and analyze such a receptor complex in the lipid rafts to understand how the signals are transduced and how specificity is conferred.

INTRACELLULAR SIGNALING EVENTS OF MYELIN INHIBITORS

Our knowledge of the signaling mechanisms that underlie myelin inhibition has largely been derived from two different approaches. First, the identification of the individual inhibitors and their receptor components has provided the molecular tools for identifying the intracellular signaling components. Second, because myelin-associated inhibitors and other types of axonal repellents like Semaphorins and Slits can elicit such similar axonal responses as growth cone collapse, the information from axon guidance studies can provide valuable insights into the mechanisms of axon regeneration. Although we are still at the early stages of determining the detailed signaling mechanisms of myelin inhibition, several important intracellular molecules appear to be directly involved in myelin-elicited pathways. Among these molecules are small GTPases, cyclic nucleotides, and intracellular calcium. Whereas the small GTPases are likely to be direct signal mediators, the cyclic nucleotides may act more indirectly by modulating the inhibitory signaling pathways (Figure 2, see color insert).

Small GTPases as Critical Signaling Mediators

Small GTPases of the Rho family, of which the most widely expressed members include RhoA, Rac1, and Cdc42, are known regulators of the actin cytoskeleton in all eukaryotic cells (Dickson 2001, Ettienne-Manneville & Hall 2002, Luo 2000). In general, these molecules cycle between active (GTP-bound) and inactive (GDP-bound) states through the binding of guanine nucleotides. In resting conditions, these proteins are kept in an inactive state by guanine dissociation inhibitors (GDI). In addition, the cycling of Rho-GTPases is also regulated by either enzymes that enhance GTP-binding activity [guanine nucleotide exchange factors (GEFs)] or proteins that increase GTP hydrolysis (GTPase-activating proteins). When GTP bound, these proteins can interact with a series of intracellular effectors to reorganize the cytoskeleton in two ways: first, by regulating actin filament assembly and disassembly by controlling actin polymerization, branching, and depolymerization; and second, by directing actin-myosin-dependent contractility to control the retrograde transport of F actin within the growth cone (reviewed by Dickson et al. 2001; Huber et al. 2003; Luo 2000, 2002).

In neuronal cell lines, activation of RhoA stimulates actinomyosin contractility and stress fiber formation, resulting in growth cone collapse, whereas the induction of Cdc42 and Rac1 leads to the extension of filopodia and lamellipodia, respectively (Dickson et al. 2001, Hall et al. 1998, Luo et al. 2002). The first direct biochemical demonstration of RhoA activation by axonal repellents came from the finding that Ephrin signaling results in robust RhoA activation (Wahl et al. 2000). This activation appears to involve a Dbl family Rho-GEF, Ephexin (Eph-interacting exchange factor) (Shamah et al. 2001). Although the resting Eph receptor binds to Ephexin, resulting in activation of RhoA, Rac1, and Cdc42 in vitro, Ephrin A binding stimulates Ephexin's activity toward RhoA but inhibits its activity toward Rac1 and Cdc42 (Shamah et al. 2001). In a similar way, both biochemical and genetic studies have provided evidence that the Semaphorins, members of another axonal repellent family, also function through modulating the activity of small GTPases. Semaphorins repel axons by using their receptor molecules, plexins (He et al. 2002). The cytoplasmic domains of plexins bind directly to Rac-GTP in both vertebrates (Vikis et al. 2000) and *Drosophila* (Driessens et al. 2001), and to RhoA in *Drosophila* (Hu et al. 2001). Genetically, Semaphorin-induced axon repulsion can be suppressed by reducing the Rac protein levels and enhanced by reducing RhoA levels, consistent with the idea that plexin activates RhoA but inhibits Rac (Hu et al. 2001). Recent studies have shown that at least Plexin B1 regulates RhoA activity through the GEFs: leukemia-associated Rho-GEF and PDZ Rho-GEF (Aurandt et al. 2002, Perrot et al. 2002, Swiercz et al. 2002). On the other hand, RhoA activation by Slits, another family of axon-repulsive cues, involves specific GTPase-activating-protein molecules (Wong et al. 2001). This evidence suggests that axonal repellents/inhibitors act through RhoA activation as well as Rac1 and Cdc42 inhibition.

Recent evidence suggests a similar involvement by small GTPases in the signaling pathways downstream of myelin-associated inhibitors. Researchers first suggested that RhoA activation was required to produce the inhibitory activity of MAG and other myelin components upon observing that the inhibition of RhoA activity by either C3 transferase or dominant-negative Rho in neurons allows neurites to grow on inhibitory substrates (Jin & Strittmatter 1997, Lehman et al. 1999, Kuhn et al. 1999). Subsequent studies provide direct biochemical evidence for RhoA activation by MAG and other myelin inhibitors (Dergham et al. 2002, Vinson et al. 2001, Winton et al. 2002). RhoA-associated kinase (ROCK), a major RhoA effector, has also been implicated in the myelin-mediated reorganization of cytoskeleton structures. Similar to C3, a synthetic inhibitor of ROCK called Y27632 can also block the inhibitory activity of myelin inhibitors (Fournier et al. 2003, Vinson et al. 2001). This finding adds ROCK to the list of major downstream components of the NgR-mediated inhibitory signal. In vivo experiments in rats and mice have even indicated that inactivation of RhoA or ROCK promotes axon regeneration and functional recovery after spinal cord injury (Dergham et al. 2002, Fournier et al. 2003, Lehmann et al. 1999).

Like the function of Ephrins (Shamah et al. 2001, Wahl et al. 2000) and perhaps other guidance molecules (Yuan et al. 2003), myelin inhibitors induce the activation of RhoA and the concomitant inhibition of Rac1 (Niederost et al. 2002). Current evidence suggests at least two possible avenues for the functional interaction between these two classes of functionally distinct small GTPases. First, ROCK can increase the phosphorylation of the myosin light chain (MLC) by phosphorylating MLC phosphatase (Kimura et al. 1996) as well as MLC (Amano et al. 1996). In contrast, active Rac1 can signal to its effector protein PAK1, which inhibits MLC kinase, resulting in decreased phosphorylation of MLC and then decreased actinomyosin contractility (Sanders et al. 1999). Thus, Rac1 and RhoA may have antagonistic effects on myosin contractility. Second, Rac1 and RhoA may also affect actin polymerization in an opposing manner. Rac1 can activate its effector PI4P5K (phosphatidylinositol-4-phosphate 5-kinase), which generates phosphatidylinositol-4,5-biphosphate (PIP2) (Tolias et al. 1995, 1998). PIP2 stimulates actin assembly by binding to the neural Wiskott-Aldrich syndrome protein and promoting the extension of existing filaments by inhibiting barbed-end capping proteins (Janmey & Stossel 1987).

How Are the Activities of These GTPases Regulated by Myelin Components?

Although it is unclear whether p75 is sufficient for the transduction of inhibitory myelin signals across the axonal membrane, an early study demonstrates that overexpression of p75 is sufficient for the activation of endogenous RhoA activity in both neuronal and non-neuronal cells (Yamashita et al. 1999), suggesting a possible link between p75 and RhoA. Recent in vivo studies also point to a critical role for p75 in RhoA activation at the lesion site after spinal cord injury (Dubreuil et al. 2003). By using a method of in situ Rho-GTP detection, researchers showed that both glial cells and neurons exhibit RhoA activation after spinal cord injury. Experiments with p75 knockout mice revealed that at least an early phase of RhoA activation after spinal cord injury is p75 dependent (Dubreuil et al. 2003). This observation suggests that the components of myelin-breakdown products after a lesion trigger a signaling cascade that leads to the inhibition of axon regeneration.

What is the molecular mechanism underlying p75-mediated RhoA activation? Some insights come from studies using cells expressing p75 (Yamashita & Tohyama 2003; Yamashita et al. 1999, 2002). Yamashita et al. (1999) showed that an interaction between p75 and RhoA can be detected via a yeast two-hybrid screen, but they failed to detect a direct interaction between p75 and RhoA in neurons. A recent paper by Yamashita & Tohyama (2003) proposes Rho-GDI as the missing link between p75 and RhoA. Unlike Rho-GEFs or GTPase-activating proteins, Rho-GDIs function to keep RhoA in an inactive state by sequestering it in the cytoplasm and inhibiting the formation of active Rho-GTP. The binding of Rho-GDI by p75 prevents Rho-GDI from inhibiting RhoA, thereby enabling GEFs to activate RhoA. Such an interaction between p75 and Rho-GDI appears to be particularly relevant to myelin inhibition, as myelin-associated inhibitors, but not NGF, strengthen the interaction between Rho-GDI and p75 (Yamashita & Tohyama 2003). On the basis of this, a peptide that can inhibit the interaction between Rho-GDI and p75 was found to neutralize myelin-elicited inhibition of neurite outgrowth (Yamashita & Tohyama 2003). Because these studies predict that an unidentified GEF is needed to fully activate RhoA after its dissociation from Rho-GDI, identifying such a GEF molecule and determining whether it is in fact recruited by the NgR-p75 receptor complex to activate RhoA is important. Moreover, whether the same GEF or other molecules mediate the inactivation of Rac and/or Cdc42 in response to myelin inhibitors remains unknown.

Are Other Pathways Involved?

The fact that so many repulsive and inhibitory factors seem to signal through RhoA necessitates the question of how specificity is conferred. One possibility involves the presence of other signaling pathways, in addition to RhoA activation, that are simultaneously triggered, the combined action of which results in the inhibitory action of myelin-associated factors on neurite outgrowth and axon regeneration. This appears to be the case for other guidance molecules, as the repulsive effects of Slit/Robo and Semaphorin/plexin signaling have been demonstrated to involve the enabled (Bashaw et al. 2000) and LIM kinase (Aizawa et al. 2001) pathways, respectively, in addition to RhoA activation. Indeed, inhibiting the phosphatidylinositol 3-kinase (PI3K) enzyme can block MAG-elicited growth cone turning responses at least in *Xenopus* spinal neurons (Ming et al. 1999). How PI3 K and other possible components are involved in the signaling pathways of myelin inhibitors, however, remains to be resolved.

Modulation by cAMP

In understanding the signaling mechanisms of guidance effects, it is important to consider other modulatory pathways that may participate indirectly in determining the specificity of response as suggested for axon guidance responses (Song & Poo 2001). A well-documented example of this is the effect of intracellular cAMP levels on growth cones (Cai et al. 1999, Neumann et al. 2002, Qiu et al. 2002, Song et al. 1998). By using a *Xenopus* spinal neuron–based growth cone turning assay, researchers (Ming et al. 1997; Song et al. 1997, 1998) found that, depending on the level of cyclic nucleotides within a neuron, the response of the growth cone to many guidance cues can be either attractive or repulsive, with high levels favoring attraction and low levels favoring repulsion. Independently, Cai et al. (1999) found that pretreatment (priming) of responding neurons with neurotrophins can block the inhibitory effects of MAG and perhaps other components of CNS myelin. In addition, they showed that this priming procedure elevates cAMP levels and activates PKA, thus providing a mechanism for overcoming neurite-outgrowth inhibition. How cAMP/PKA impinges on the inhibitory pathway of myelin components, however, remains unclear. Although studies from non-neuronal cells suggest that PKA may phosphorylate RhoA and regulate its activity (Ellerbroek et al. 2003), there is no definitive evidence for such in the myelin inhibitory pathways. As mentioned above, cAMP-dependent PKA can phosphorylate p75 and affect its translocation to lipid rafts (Higuchi et al. 2003). A more recent study showed that elevated cAMP could also upregulate Arginase I and thus enhance the synthesis of polyamines (Cai et al. 2002). Overexpression of either Arginase I or exogenous polyamines allows neurons to overcome the inhibition by MAG and myelin, thus offering a mechanism by which intracellular cAMP levels may mediate myelin inhibition. It is intriguing to note that endogenous levels of Arginase I are high in young neurons, but they spontaneously drop at an age when the effects of MAG and myelin are switched from promotion to inhibition (Cai et al. 2002). It will be interesting to examine how these direct and indirect pathways are integrated to exert their inhibitory effects on neurite outgrowth.

Calcium

The critical involvement of calcium (Ca^{2+}) in mediating the axonal responses to environmental cues has been well documented in the past decade (reviewed by Gomez & Spitzer 2000). In cultured *Xenopus* spinal neurons, a variety of guidance cues, such as netrin-1 and MAG, can induce a rise of intracellular Ca^{2+} level in the growth cone. Also, preventing an elevation in cytoplasmic Ca^{2+} levels abolishes the growth cone turning response in gradients of individual cues (Hong et al. 2000, Ming et al. 1997, Wong et al. 2002, Zheng 2000, Zheng et al. 1994). Moreover, photolytic release of caged Ca^{2+} or induction of Ca^{2+} release from internal stores with an extracellular gradient of ryanodine (in the absence of guidance cues) is sufficient to induce growth cone turning (Hong et al. 2000). Purified myelin fractions enriched in NI-35 also trigger a large and rapid increase of Ca^{2+} levels in the responding growth cone of rat dorsal root ganglion neurons (Bandtlow et al. 1993). Depletion of the caffeine-sensitive intracellular Ca^{2+} stores prevents the increase in $[Ca^{2+}]i$ evoked by NI-35, thus demonstrating the involvement of Ca^{2+} release from intracellular stores in the signaling pathways that mediate NI-35-induced growth cone collapse (Bandtlow et al. 1993). Recent studies also demonstrate that recombinant MAG also triggers a Ca^{2+} influx (Wong et al. 2002). Even in non-neuronal cells expressing both NgR and p75, MAG-Fc can trigger a robust Ca^{2+} influx (Wong et al. 2002), suggesting that p75 may be sufficient to trigger the signal for Ca^{2+}influx as elicited by myelin inhibitors.

Despite ample evidence of calcium's role in the signaling of myelin inhibition, it remains to be determined how the signal is transmitted from p75 to Ca^{2+} influx and how Ca^{2+} influx affects growth cone behavior. Recent studies implicate two known Ca^{2+} effectors in influencing axonal growth behaviors (Graef et al. 2003, Robles et al. 2003). By analyzing the phenotypes of mice lacking the expression of three isoforms of the Ca^{2+}-dependent transcription factor NFAT, Graef et al. (2003) demonstrated that calcineurin/NFAT signaling pathways are required for the neurite outgrowth–promoting activity of neurotrophins and the chemoattractant netrin-1 to occur. Investigators (Robles et al. 2003) also believe that a class of Ca^{2+}-activated proteases, termed calpains, respond to filopodial Ca^{2+}transients to regulate growth cone motility (Robles et al. 2003). It will be interesting to examine whether these or other Ca^{2+} downstream effectors act in the signaling pathways of myelin inhibition. In addition to these guidance molecules, many pathways, particularly those involved in neuronal activity, can affect the intracellular Ca^{2+} level (Spitzer et al. 2002). Thus Ca^{2+} may function, at least in some circumstances, as a coincidence detector for different signaling pathways that affect the final axonal responses to myelin inhibitors and other cues.

POSSIBLE AVENUES TO STIMULATE AXON REGENERATION

With advances in our understanding of the signaling mechanisms that mediate the inhibitory activity of myelin and other potential inhibitors, we may conceive of a number of strategies that block the activity of myelin-associated inhibitors and thereby stimulate axon regeneration in vivo. At the level of the axonal surface, blocking the activity of individual inhibitors by using such neutralizing antibodies as the IN-1 antibody (Schnell & Schwab 1993, Schnell et al. 1994, Bregmann et al. 1995) and anti-myelin antibodies (Huang et al. 1999) is a strategy that has been examined for many years. The observation that these structurally distinct molecules converge on a single receptor suggests a common set of downstream signaling mechanisms mediating myelin inhibition and offers the promising prospect that affecting a single target may be all that is required to block this inhibition. For example, NEP1-40, a small peptide that consists of the first 40 residues of the Nogo-66 sequence and that can purportedly compete with Nogo-66 for binding without inducing inhibition, enhances axonal regeneration of descending corticospinal tract fibers following spinal cord hemisection (GrandPré et al. 2002). Other potentially useful reagents include recombinant soluble NgR proteins that block the inhibitory activity of CNS myelin at least in in vitro assays (Domeniconi et al. 2002, Fournier et al. 2002, Liu et al. 2002), recombinant soluble p75 protein that can block the interaction of NgR and p75 and reduce myelin inhibition (Wang et al. 2002b, Wong et al. 2002), and truncated NgR that when overexpressed can bind all ligands but not coreceptor(s) (Domeniconi et al. 2002, Wang et al. 2002b).

In addition to these ligand/receptor interactions, intracellular signaling components are also potential targets for alleviating the inhibitory influences associated with CNS myelin. For example, C3 transferase (a RhoA inhibitor) and Y27632 (a ROCK inhibitor) permit a certain degree of axon regeneration when applied to both optic nerve crush and spinal cord injury models (Dergham et al. 2002, Fournier et al. 2003, Lehmann et al. 1999). Both RhoA and ROCK have also been implicated in the inhibitory signaling pathways of CSPGs (Monnier et al. 2003). Nevertheless, as these signaling molecules are also components of other cellular pathways, the application of these agents in different CNS lesion models may reflect more than just their effects on myelin-mediated inhibition. For example, a recent study suggests an additional role for RhoA activation in cell death after CNS injury (Dubreuil et al. 2003). With the identification of other signaling components involved in myelin-inhibitor signaling pathways, it is certain that additional relevant reagents will be considered for similar use.

Is eliminating the inhibitory activity of these myelin components sufficient to allow axon regeneration? Because varying degrees of axon regeneration are seen after different treatments (described previously), assorted mature neurons may express distinct receptors and thus differ in their sensitivity to individual inhibitors. An alternative but not mutually exclusive possibility is that, even though these neurons express the same set of receptors, they may possess varying regenerative capacities and thus behave differently in response to individual treatments. Indeed, recent in vitro studies suggest that in the retina only a subpopulation of mature retinal ganglion neurons can regrow axons (Goldberg et al. 2002). An informative step in clarifying these issues will be the phenotypic analysis of NgR knockout mice. However, it is unlikely that blocking NgR-mediated pathways will be enough for all axon regeneration in vivo, when one considers the other inhibitory contributors in CNS myelin, such as amino-Nogo and the Semaphorins, as well as the inhibitory molecules in the glial scar. Perhaps neutralization of these inhibitory influences with concomitant stimulation of the intrinsic regenerative capacity can create conditions that allow axon regeneration to occur. The identification of both myelin inhibitors and their signaling molecules will allow us to take a step-by-step analytical approach to studying the relative contribution of these factors in restricting axon regeneration in vivo.

ACKNOWLEDGMENTS

We are grateful to J. Wang for helping with the preparation of the diagrams and to Glenn Yiu for critically reading the manuscript. Our work is supported by grants from the National Institutes of Health, International Spinal Research Trust, and The John Merck Fund.

The *Annual Review of Neuroscience* is online at http://neuro.annualreviews.org

LITERATURE CITED

Aizawa H, Wakatsuki S, Ishii A, Moriyama K, Sasaki Y, et al. 2001. Phosphorylation of cofilin by LIM-kinase is necessary for semaphorin 3A-induced growth cone collapse. *Nat. Neurosci.* 4:367–73

Amano M, Ito M, Kimura K, Fukata Y, Chihara K, et al. 1996. Phosphorylation and activation of myosin by Rho-associated kinase (Rho-kinase). *J. Biol. Chem.* 271:20246–49

Aurandt J, Vikis HG, Gutkind JS, Ahn N, Guan KL. 2002. The semaphorin receptor plexin-B1 signals through a direct interaction with the Rho-specific nucleotide exchange factor, LARG. *Proc. Natl. Acad. Sci. USA* 99:12085–90

Banati RB, Graeber MB. 1994. Surveillance, intervention and cytotoxicity: is there a protective role of microglia? *Dev. Neurosci.* 16:114–27

Bandtlow C, Zachleder T, Schwab ME. 1990. Oligodendrocytes arrest neurite growth by contact inhibition. *J. Neurosci.* 10:3837–48

Bandtlow CE, Schmidt MF, Hassinger TD, Schwab ME, Kater SB. 1993. Role of intracellular calcium in NI-35-evoked collapse of neuronal growth cones. *Science* 259:80–83

Barton WA, Liu BP, Tzvetkova D, Jeffrey PD, Fournier AE, et al. 2003. Structure and axon outgrowth inhibitor binding of the Nogo-66 receptor and related proteins. *EMBO J.* 22:3291–302

Bartsch U, Bandtlow CE, Schnell L, Bartsch S, Spillmann AA, et al. 1995. Lack of evidence that myelin-associated glycoprotein is a major inhibitor of axonal regeneration in the CNS. *Neuron* 15:1375–81

Bashaw GJ, Kidd T, Murray D, Pawson T, Goodman CS. 2000. Repulsive axon guidance: Abelson and Enabled play opposing roles downstream of the roundabout receptor. *Cell* 101:703–15

Bekar LK, Walz W. 1999. Evidence for chloride ions as intracellular messenger substances in astrocytes. *J. Neurophysiol.* 82:248–54

Benfey M, Aguayo AJ. 1982. Extensive elongation of axons from rat brain into peripheral nerve grafts. *Nature* 296:150–52

Berry M. 1982. Post-injury myelin-breakdown products inhibit axonal growth: an hypothesis to explain the failure of axonal regeneration in the mammalian central nervous system. *Bibl. Anat.* 23:1–11

Bradbury EJ, Moon LD, Popat RJ, King VR, Bennett GS, et al. 2002. Chondroitinase ABC promotes functional recovery after spinal cord injury. *Nature* 416:636–40

Bregman BS, Kunkel-Bagden E, Reier PJ, Dai HN, McAtee M, Gao D. 1993. Recovery of function after spinal cord injury: mechanisms underlying transplant-mediated recovery of function differ after spinal cord injury in newborn and adult rats. *Exp. Neurol.* 123:3–16

Bregman BS, Kunkel-Bagden E, Schnell L, Dai HN, Gao D, Schwab ME. 1995. Recovery from spinal cord injury mediated by antibodies to neurite growth inhibitors. *Nature* 378:498–501

Bunge RP. 1968. Glial cells and the central myelin sheath. *Physiol. Rev* . 48:197–251

Cai D, Deng K, Mellado W, Lee J, Ratan RR, Filbin MT. 2002. Arginase I and polyamines act downstream from cyclic AMP in overcoming inhibition of axonal growth MAG and myelin in vitro. *Neuron* 35:11–19

Cai D, Qiu J, Cao Z, McAtee M, Bregman BS, Filbin MT. 2001. Neuronal cyclic AMP controls the developmental loss in ability of axons to regenerate. *J. Neurosci.* 21:4731–39

Cai D, Shen Y, De Bellard M, Tang S, Filbin MT. 1999. Prior exposure to neurotrophins blocks inhibition of axonal regeneration by MAG and myelin via a cAMP-dependent mechanism. *Neuron* 22:89–101

Carbonetto S, Evans D, Cochard P. 1987. Nerve fiber growth in culture on tissue substrata from central and peripheral nervous systems. *J. Neurosci.* 7:610–20

Caroni P, Schwab ME. 1988a. Antibody against myelin-associated inhibitor of neurite growth neutralizes nonpermissive substrate properties of CNS white matter. *Neuron* 1:85–96

Caroni P, Schwab ME. 1988b. Two membrane protein fractions from rat central myelin with inhibitory properties for neurite growth and fibroblast spreading. *J. Cell Biol.* 106:1281–88

Chao MV. 2003. Neurotrophins and their receptors: a convergence point for many signalling pathways. *Nat. Rev. Neurosci* . 4:299–309

Chen MS, Huber AB, van der Haar ME, Frank M, Schnell L, et al. 2000. Nogo-A is a myelin-associated neurite outgrowth inhibitor and an antigen for monoclonal antibody IN-1. *Nature* 403:434–39

Collins BE, Yang LJ, Mukhopadhyay G, Filbin MT, Kiso M, et al. 1997. Sialic acid specificity of myelin-associated glycoprotein binding. *J. Biol. Chem.* 272:1248–55

Crutcher KA. 1989. Tissue sections from the mature rat brain and spinal cord as substrates for neurite outgrowth in vitro: extensive growth on gray matter but little growth on white matter. *Exp. Neurol.* 104:39–54

David S, Aguayo AJ. 1981. Axonal elongation into peripheral nervous system "bridges" after central nervous system injury in adult rats. *Science* 214:931–33

David S, Bouchard C, Tsatas O, Giftochristos N. 1990. Macrophages can modify the nonpermissive nature of the adult mammalian central nervous system. *Neuron* 5:463–69

David S, Lacroix S. 2003. Molecular approaches to spinal cord repair. *Annu. Rev. Neurosci.* 26:411–40

Davies SJ, Fitch MT, Memberg SP, Hall AK, Raisman G, Silver J. 1997. Regeneration of adult axons in white matter tracts of the central nervous system. *Nature* 390:680–83

Davies SJ, Goucher DR, Doller C, Silver J.

1999. Robust regeneration of adult sensory axons in degenerating white matter of the adult rat spinal cord. *J. Neurosci.* 19:5810–22

Dechant G, Barde YA. 2002. The neurotrophin receptor p75(NTR): novel functions and implications for diseases of the nervous system. *Nat. Neurosci.* 5:1131–36

DeBellard ME, Tang S, Mukhopadhyay S, Shen Y, Filbin MT. 1996. Myelin-associated glycoprotein inhibits axon regeneration from a variety of neurons via interactions with a sialoglycoprotein. *Mol. Cell. Neurosci.* 7:89–101

Dergham P, Ellezam B, Essagian C, Avedissian H, Lubell WD, McKerracher L. 2002. Rho signaling pathway targeted to promote spinal cord repair. *J. Neurosci.* 22:6570–77

Dickson BJ. 2001. Rho GTPases in growth cone guidance. *Curr. Opin. Neurobiol.* 11:103–10

Dickson BJ. 2002. Molecular mechanisms of axon guidance. *Science* 298:1959–64

Ding J, Hu B, Tang LS, Yip HK. 2001. Study of the role of the low-affinity neurotrophin receptor p75 in naturally occurring cell death during development of the rat retina. *Dev. Neurosci.* 23:390–98

Domeniconi M, Cao Z, Spencer T, Sivasankaran R, Wang K, et al. 2002. Myelin-associated glycoprotein interacts with the nogo66 receptor to inhibit neurite outgrowth. *Neuron* 35:283–90

Dobrowsky RT, Carter BD. 2000. p75 neurotrophin receptor signaling: mechanisms for neurotrophic modulation of cell stress? *J. Neurosci. Res.* 61:237–43

Driessens MH, Hu H, Nobes CD, Self A, Jordens I, et al. 2001. Plexin-B semaphorin receptors interact directly with active Rac and regulate the actin cytoskeleton by activating Rho. *Curr. Biol.* 11:339–44

Dubreuil CI, Winton MJ, McKerracher L. 2003. Rho activation patterns after spinal cord injury and the role of activated Rho in apoptosis in the central nervous system. *J. Cell Biol.* 162:233–43

Ellerbroek SM, Wennerberg K, Burridge K. 2003. Serine phosphorylation negatively regulates RhoA in vivo. *J. Biol. Chem.* 278:19023–31

Ernfors P, Hallbook F, Ebendal T, Shooter EM, Radeke MJ, et al. 1988. Developmental and regional expression of beta-nerve growth factor receptor mRNA in the chick and rat. *Neuron* 1:983–96

Ettienne-Manneville S, Hall A. 2002. Rho GTPases in cell biology. *Nature* 420:629–35

Fawcett JW, Asher RA. 1999. The glial scar and central nervous system repair. *Brain Res. Bull.* 49:377–91

Filbin MT. 2003. Myelin-associated inhibitors of axonal regeneration in the adult mammalian CNS. *Nat. Rev. Neurosci.* 4:703–13

Fitch MT, Silver J. 1999. Beyond the glial scar. In *CNS Regeneration-Basic Science and Clinical Advances*. ed. MH Tuszynski & J Kordower, pp. 55–87. New York: Academic

Fournier AE, GrandPré T, Strittmatter SM. 2001. Identification of a receptor mediating Nogo-66 inhibition of axonal regeneration. *Nature* 409:341–46

Fournier AE, Gould GC, Liu BP, Strittmatter SM, 2002. Truncated soluble Nogo receptor binds Nogo-66 and blocks inhibition of axon growth by myelin. *J. Neurosci.* 22:8876–83

Fournier AE, Takizawa BT, Strittmatter SM, 2003. Rho kinase inhibition enhances axonal regeneration in the injured CNS. *J. Neurosci.* 23:1416–23

Galbiati F, Razani B, Lisanti MP. 2001. Emerging themes in lipid rafts and caveolae. *Cell* 106:403–11

Gehrmann J. 1996. Microglia: a sensor to threats in the nervous system? *Res. Virol.* 147:79–88

Gianola S, Savio T, Schwab ME, Rossi F. 2003. Cell-autonomous mechanisms and myelin-associated factors contribute to the development of Purkinje axon intracortical plexus in the rat cerebellum. *J. Neurosci.* 23:4613–24

Goldberg JL, Barres BA. 2000. The relationship between neuronal survival and regeneration. *Annu. Rev. Neurosci.* 23:579–612

Goldberg JL, Klassen MP, Hua Y, Barres BA. 2002. Amacrine-signaled loss of intrinsic axon growth ability by retinal ganglion cells. *Science* 296:1860–64

Gomez TM, Spitzer NC. 2000. Regulation of growth cone behavior by calcium: new dynamics to earlier perspectives. *J. Neurobiol.* 44:174–83

GrandPré T, Li S, Strittmatter SM. 2002. Nogo-66 receptor antagonist peptide promotes axonal regeneration. *Nature* 417:547–51

GrandPré T, Nakamura F, Vartanian T, Strittmatter SM. 2000. Identification of the Nogo inhibitor of axon regeneration as a Reticulon protein. *Nature* 403:439–44

Graef IA, Wang F, Charron F, Chen L, Neilson J, et al. 2003. Neurotrophins and netrins require calcineurin/NFAT signaling to stimulate outgrowth of embryonic axons. *Cell* 113:657–70

Habib AA, Marton LS, Allwardt B, Gulcher JR, Mikol DD, et al. 1998a. Expression of the oligodendrocyte-myelin glycoprotein by neurons in the mouse central nervous system. *J. Neurochem.* 70:1704–11

Habib AA, Gulcher JR, Hognason T, Zheng L, Stefansson K. 1988b. The OMgp gene, a second growth suppressor within the NF1 gene. *Oncogene* 16:1525–31

Hall A. 1998. Rho GTPases and the actin cytoskeleton. *Science* 279:509–14

Hannila SS, Kawaja MD. 1999. Nerve growth factor-induced growth of sympathetic axons into the optic tract of mature mice is enhanced by an absence of p75NTR expression. *J. Neurobiol.* 39:51–66

Hasan SJ, Keirstead HS, Muir GD, Steeves JD. 1993. Axonal regeneration contributes to repair of injured brainstem-spinal neurons in embryonic chick. *J. Neurosci.* 13:492–507

He XL, Bazan JF, McDermott G, Park JB, Wang K, et al. 2003. Structure of the nogo receptor ectodomain. A recognition module implicated in myelin inhibition. *Neuron* 38:177–85

He Z, Wang KC, Koprivica V, Ming G, Song HJ. 2002. Knowing how to navigate: mechanisms of semaphorin signaling in the nervous system. *Sci. STKE* 119:RE1

Hempstead BL. 2002. The many faces of p75NTR. *Curr. Opin. Neurobiol.* 12:260–67

Higuchi H, Yamashita T, Yoshikawa H, Tohyama M. 2003. PKA phosphorylates the p75 receptor and regulates its localization to lipid rafts. *EMBO J.* 22:1790–800

Hong K, Nishiyama M, Henley J, Tessier-Lavigne M, Poo M. 2000. Calcium signalling in the guidance of nerve growth by netrin-1. *Nature* 403:93–98

Horner PJ, Gage FH. 2000. Regenerating the damaged central nervous system. *Nature* 407:963–70

Hu H, Marton TF, Goodman CS. 2001. Plexin B mediates axon guidance in Drosophila by simultaneously inhibiting active Rac and enhancing RhoA signaling. *Neuron* 32:39–51

Huang DW, McKerracher L, Braun PE, David S. 1999. A therapeutic vaccine approach to stimulate axon regeneration in the adult mammalian spinal cord. *Neuron* 24:639–47

Huber AB, Weinmann O, Brösamle C, Oertle T, Schwab ME. 2002. Patterns of Nogo mRNA and protein expression in the developing and adult rat and after CNS lesions. *J. Neurosci.* 22:3553–67

Huber AB, Kolodkin AL, Ginty DD, Cloutier JF. 2003. Signaling at the growth cone: ligand-receptor complexes and the control of axon growth and guidance. *Annu. Rev. Neurosci.* 26:509–63

Hunt D, Mason MR, Campbell G, Coffin R, Anderson PN. 2002. Nogo receptor mRNA expression in intact and regenerating CNS neurons. *Mol. Cell Neurosci.* 20:537–52

Hutson LD, Bothwell M. 2001. Expression and function of Xenopus laevis p75(NTR) suggest evolution of developmental regulatory mechanisms. *J. Neurobiol.* 49:79–98

Janmey PA, Stossel TP. 1987. Modulation of gelsolin function by phosphatidylinositol 4,5-bisphosphate. *Nature* 325:362–64

Jin Z, Strittmatter SM. 1997. Rac1 mediates collapsin-1-induced growth cone collapse. *J. Neurosci.* 17:6256–63

Johnson PW, Abramow-Newerly W, Seilheimer B, Sadoul R, Tropak MB, et al. 1989. Recombinant myelin-associated glycoprotein confers neural adhesion and neurite outgrowth function. *Neuron* 3:377–85

Josephson A, Widenfalk J, Widmer HW, Olson L, Spenger C. 2001. NOGO mRNA expression in adult and fetal human and rat nervous tissue and in weight drop injury. *Exp. Neurol.* 169:319–28

Kalil K, Reh T. 1982. A light and electron microscopic study of regrowing pyramidal tract fibers. *J. Comp. Neurol.* 211:265–75

Keirstead HS, Hasan SJ, Muir GD, Steeves JD. 1992. Suppression of the onset of myelination extends the permissive period for the functional repair of embryonic spinal cord. *Proc. Natl. Acad. Sci. USA.* 89:11664–68

Khan U, Starega U, Seeley PJ. 1990. Selective growth of hippocampal neurites on cryostat sections of rat brain. *Brain Res. Dev. Brain Res.* 54:87–92

Kim JE, Li S, GrandPréT, Qiu D, Strittmatter SM. 2003. Axon regeneration in young adult mice lacking nogo-a/b. *Neuron* 38:187–99

Kim JE, Bonilla IE, Qiu D, Strittmatter SM. 2003a. Nogo-C is sufficient to delay nerve regeneration. *Mol. Cell Neurosci.* 23:451–59

Kimura K, Ito M, Amano M, Chihara K, Fukata Y, et al. 1996. Regulation of myosin phosphatase by Rho and Rho-associated kinase (Rho-kinase). *Science* 273:245–48

Kohn J, Aloyz RS, Toma JG, Haak-Frendscho M, Miller FD. 1999. Functionally antagonistic interactions between the TrkA and p75 neurotrophin receptors regulate sympathetic neuron growth and target innervation. *J. Neurosci.* 19:5393–408

Kottis V, Thibault P, Mikol D, Xiao ZC, Zhang R, et al. 2002. Oligodendrocyte-myelin glycoprotein (OMgp) is an inhibitor of neurite outgrowth. *J. Neurochem.* 82:1566–69

Fruttiger M, Montag D, Schachner M, Martini R. 1995. Crucial role for the myelin-associated glycoprotein in the maintenance of axon-myelin integrity. *Eur. J. Neurosci.* 7:511–15

Kuhn TB, Brown MD, Wilcox CL, Raper JA, Bamburg JR. 1999. Myelin and collapsin-1 induce motor neuron growth cone collapse through different pathways: inhibition of collapse by opposing mutants of rac1. *J Neurosci.* 19:1965–75

Lai C, Watson JB, Bloom FE, Sutcliffe JG, Milner RJ. 1987. Neural protein 1B236/myelin-associated glycoprotein (MAG) defines a subgroup of the immunoglobulin superfamily. *Immunol. Rev.* 100:129–51

Lassmann H, Bartsch U, Montag D, Schachner M. 1997. Dying-back oligodendrogliopathy: a late sequel of myelin-associated glycoprotein deficiency. *Glia* 19:104–10

Lehmann M, Fournier A, Selles-Navarro I, Dergham P, Sebok A, et al. 1999. Inactivation of Rho signaling pathway promotes CNS axon regeneration. *J. Neurosci.* 19:7537–47

Li C, Tropak MB, Gerlai R, Clapoff S, Abramow-Newerly W, et al. 1994. Myelination in the absence of myelin-associated glycoprotein. *Nature* 369:747–50

Liesi P, Silver J. 1988. Is astrocyte laminin involved in axon guidance in the mammalian CNS? *Dev. Biol.* 130:774–85

Liu BP, Fournier A, GrandPré T, Strittmatter SM. 2002. Myelin-associated glycoprotein as a functional ligand for the Nogo-66 receptor. *Science* 297:1190–93

Luo L. 2000. Rho GTPases in neuronal morphogenesis. *Nature Rev. Neurosci.* 1:173–80

Luo L. 2002. Actin cytoskeleton regulation in neuronal morphogenesis and structural plasticity. *Annu. Rev. Cell Dev. Biol.* 18:601–35

Martinez-Murillo R, Fernandez AP, Bentura ML, Rodrigo J. 1998. Subcellular localization of low-affinity nerve growth factor receptor-immunoreactive protein in adult rat purkinje cells following traumatic injury. *Exp. Brain Res.* 119:47–57

McCormick F. 1995. Ras signaling and NF1. *Curr. Opin. Genet. Dev.* 5:51–55

McGee AW, Strittmatter SM. 2003. The Nogo-66 receptor: focusing myelin inhibition of axon regeneration. *Trends Neurosci.* 26:193–98

McKeon RJ, Schreiber RC, Rudge JS, Silver J. 1991. Reduction of neurite outgrowth in a model of glial scarring following CNS injury is correlated with the expression of inhibitory molecules on reactive astrocytes. *J. Neurosci.* 11:3398–411

McKerracher L, David S, Jackson DL, Kottis

V, Dunn RJ, Braun PE. 1994. Identification of myelin-associated glycoprotein as a major myelin-derived inhibitor of neurite growth. *Neuron* 13:805–11

Mikol DD, Stefansson K. 1988. A phosphatidylinositol-linked peanut agglutinin-binding glycoprotein in central nervous system myelin and on oligodendrocytes. *J. Cell Biol.* 106:1273–79

Mikol DD, Gulcher JR, Stefansson K. 1990. The oligodendrocyte-myelin glycoprotein belongs to a distinct family of proteins and contains the HNK-1 carbohydrate. *J. Cell Biol.* 110:471–79

Ming GL, Song HJ, Berninger B, Holt CE, Tessier-Lavigne M, Poo MM. 1997. cAMP-dependent growth cone guidance by netrin-1. *Neuron* 19:1225–35

Ming G, Song H, Berninger B, Inagaki N, Tessier-Lavigne M, Poo M. 1999. Phospholipase C-gamma and phosphoinositide 3-kinase mediate cytoplasmic signaling in nerve growth cone guidance. *Neuron* 23:139–148

Monnier PP, Sierra A, Schwab JM, Henke-Fahle S, Mueller BK. 2003. The Rho/ROCK pathway mediates neurite growth-inhibitory activity associated with the chondroitin sulfate proteoglycans of the CNS glial scar. *Mol. Cell Neurosci.* 22:319–30

Montag D, Giese KP, Bartsch U, Martini R, Lang Y, et al. 1994. Mice deficient for the myelin-associated glycoprotein show subtle abnormalities in myelin. *Neuron* 13:229–46

Moreau-Fauvarque C, Kumanogoh A, Camand E, Jaillard C, Barbin G, et al. 2003. The transmembrane semaphorin Sema4D/CD100, an inhibitor of axonal growth, is expressed on oligodendrocytes and upregulated after CNS lesion. *J. Neurosci.* 23:9229–39

Moon LD, Asher RA, Rhodes KE, Fawcett JW. 2001. Regeneration of CNS axons back to their target following treatment of adult rat brain with chondroitinase ABC. *Nat. Neurosci.* 4:465–66

Moorman SJ. 1996. The inhibition of motility that results from contact between two oligodendrocytes in vitro can be blocked by pertussis toxin. *Glia* 16:257–65

Morgenstern DA, Asher RA, Fawcett JW. 2002. Chondroitin sulphate proteoglycans in the CNS injury response. *Prog. Brain Res.* 137:313–32

Mukhopadhyay G, Doherty P, Walsh FS, Crocker PR, Filbin MT. 1994. A novel role for myelin-associated glycoprotein as an inhibitor of axonal regeneration. *Neuron* 13:757–67

Niederost BP, Zimmermann DR, Schwab ME, Bandtlow CE. 1999. Bovine CNS myelin contains neurite growth-inhibitory activity associated with chondroitin sulfate proteoglycans. *J. Neurosci.* 19:8979–89

Niederost B, Oertle T, Fritsche J, McKinney RA, Bandtlow CE. 2002. Nogo-A and myelin-associated glycoprotein mediate neurite growth inhibition by antagonistic regulation of RhoA and Rac1. *J. Neurosci.* 22:10368–76

Neumann S, Bradke F, Tessier-Lavigne M, Basbaum AI. 2002. Regeneration of sensory axons within the injured spinal cord induced by intraganglionic cAMP elevation. *Neuron* 34:885–93

Oertle T, van der Haar ME, Bandtlow CE, Robeva A, Burfeind P, et al. 2003. Nogo-A inhibits neurite outgrowth and cell spreading with three discrete regions. *J Neurosci.* 23:5393–406

Perrot V, Vazquez-Prado J, Gutkind JS. 2002. Plexin B regulates Rho through the guanine nucleotide exchange factors leukemia-associated Rho GEF (LARG) and PDZ-Rho GEF. *J. Biol. Chem.* 277:43115–20

Pizzorusso T, Medini P, Berardi N, Chierzi S, Fawcett JW, Maffei L. 2002. Reactivation of ocular dominance plasticity in the adult visual cortex. *Science* 298:1248–51

Pot C, Simonen M, Weinmann O, Schnell L, Christ F, et al. 2002. Nogo-A expressed in Schwann cells impairs axonal regeneration after peripheral nerve injury. *J. Cell Biol.* 159:29–35

Prinjha R, Moore SE, Vinson M, Blake S,

Morrow R, et al. 2000. Inhibitor of neurite outgrowth in humans. *Nature* 403:383–84
Probstmeier R, Stichel C, Muller HW, Asou H, Pesheva P. 2000. Chondroitin sulfates expressed on oligodendrocyte-derived tenascin-R are involved in neural cell recognition. Functional implications during CNS development and regeneration. *J. Neurosci. Res.* 60:21–36
Qiu J, Cai D, Dai H, McAtee M, Hoffman PN, et al. 2002. Spinal axon regeneration induced by elevation of cyclic AMP. *Neuron* 34:895–903
Ramon Y, Cajal S. 1991 [1928]. *Degeneration and Regeneration of the Nervous System.* Transl. RM May. New York: Oxford Univ. Press. 976 pp. From Spanish
Reier PJ, Stensaas LJ, Guth L. 1983. The astrocytic scar as an impediment to regeneration in the central nervous system. In *Spinal Cord Reconstruction*, ed. CC Kao, RP Bunge, PJ Reier, pp. 163–95. New York: Raven
Richardson PM, McGuinness UM, Aguayo AJ. 1980. Axons from CNS neurons regenerate into PNS grafts. *Nature* 284:264–65
Risau W, Wolburg H. 1990. Development of the blood-brain barrier. *Trends Neurosci.* 13:174–78
Robles E, Huttenlocher A, Gomez TM. 2003. Filopodial calcium transients regulate growth cone motility and guidance through local activation of calpain. *Neuron* 38:597–609
Roux PP, Colicos MA, Barker PA, Kennedy TE. 1999. p75 neurotrophin receptor expression is induced in apoptotic neurons after seizure. *J. Neurosci.* 19:6887–96
Rudge JS, Silver J. 1990. Inhibition of neurite outgrowth on astroglial scars in vitro. *J. Neurosci.* 10:3594–603
Sagot Y, Swerts JP, Cochard P. 1991. Changes in permissivity for neuronal attachment and neurite outgrowth of spinal cord grey and white matters during development: a study with the 'cryoculture' bioassay. *Brain Res.* 543:25–35
Salzer JL, Holmes WP, Colman DR. 1987. The amino acid sequences of the myelin-associated glycoproteins: homology to the immunoglobulin gene superfamily. *J. Cell Biol.* 104:957–65
Salzer JL, Pedraza L, Brown M, Struyk A, Afar D, Bell J. 1990. Structure and function of the myelin-associated glycoproteins. *Ann. NY Acad. Sci.* 605:302–12
Sanders LC, Matsumura F, Bokoch GM, de Lanerolle P. 1999. Inhibition of myosin light chain kinase by p21-activated kinase. *Science* 283:2083–85
Saunders NR, Kitchener P, Knott GW, Nicholls JG, Potter A, Smith TJ. 1998. Development of walking, swimming and neuronal connections after complete spinal cord transection in the neonatal opossum, Monodelphis domestica. *J. Neurosci.* 18:339–55
Savio T, Schwab ME. 1989. Rat CNS white matter, but not gray matter, is nonpermissive for neuronal cell adhesion and fiber outgrowth. *J. Neurosci.* 9:1126–33
Savio T, Schwab ME. 1990. Lesioned corticospinal tract axons regenerate in myelin-free rat spinal cord. *Proc. Natl. Acad. Sci. USA.* 87:4130–33
Schachner M, Bartsch U. 2000. Multiple functions of the myelin-associated glycoprotein MAG (siglec-4a) in formation and maintenance of myelin. *Glia* 29:154–65
Schafer M, Fruttiger M, Montag D, Schachner M, Martini R. 1996. Disruption of the gene for the myelin-associated glycoprotein improves axonal regrowth along myelin in C57BL/Wlds mice. *Neuron* 16:1107–13
Schnell L, Schneider R, Kolbeck R, Barde YA, Schwab ME. 1994. Neurotrophin-3 enhances sprouting of corticospinal tract during development and after adult spinal cord lesion. *Nature* 367:170–73
Schnell L, Schwab ME. 1990. Axonal regeneration in the rat spinal cord produced by an antibody against myelin-associated neurite growth inhibitors. *Nature* 343:269–72
Schnell L, Schwab ME. 1993. Sprouting and regeneration of lesioned corticospinal tract fibers in the adult rat spinal cord. *Eur. J. Neurosci* 5:1156–72
Schousboe A, Westergaard N. 1995. Transport

of neuroactive amino acids in astrocytes. In *Neuroglia,* ed. H Kettenmann, BR Ransom, pp. 246–58. New York: Oxford Univ. Press

Schwab ME, Thoenen H. 1985. Dissociated neurons regenerate into sciatic but not optic nerve explants in culture irrespective of neurotrophic factors. *J. Neurosci.* 5:2415–23

Schwab ME, Bartholdi D. 1996. Degeneration and regeneration of axons in the lesioned spinal cord. *Physiol. Rev.* 76:319–70

Shamah SM, Lin MZ, Goldberg JL, Estrach S, Sahin M, et al. 2001. EphA receptors regulate growth cone dynamics through the novel guanine nucleotide exchange factor ephexin. *Cell* 105:233–44

Shen YJ, DeBellard ME, Salzer JL, Roder J, Filbin MT. 1998. Myelin-associated glycoprotein in myelin and expressed by Schwann cells inhibits axonal regeneration and branching. *Mol. Cell Neurosci.* 12:79–91

Simonen M, Pedersen V, Weinmann O, Schnell L, Buss A, et al. 2003. Systemic deletion of the myelin-associated outgrowth inhibitor nogo-a improves regenerative and plastic responses after spinal cord injury. *Neuron* 38:201–11

Simons K, Ikonen E. 1997. Functional rafts in cell membranes. *Nature* 387:569–72

Song HJ, Ming GL, Poo MM. 1997. cAMP-induced switching in turning direction of nerve growth cones. *Nature* 388:275–79

Song HJ, Ming GL, He Z, Lehmann M, McKerracher L, et al. 1998. Conversion of neuronal growth cone responses from repulsion to attraction by cyclic nucleotides. *Science* 281:1515–18

Song H, Poo M. 2001. The cell biology of neuronal navigation. *Nat. Cell Biol.* 3:E81–88

Spillmann AA, Bandtlow CE, Lottspeich F, Keller F, Schwab ME. 1998. Identification and characterization of a bovine neurite growth inhibitor (bNI-220). *J. Biol Chem.* 273:19283–93

Spitzer NC, Kingston PA, Manning TJ, Conklin MW. 2002. Outside and in: development of neuronal excitability. *Curr. Opin. Neurobiol.* 12:315–23

Suzuki A, Nomura S, Morii E, Fukuda Y, Kosaka J. 1998. Localization of mRNAs for trkB isoforms and p75 in rat retinal ganglion cells. *J. Neurosci. Res.* 54:27–37

Swiercz JM, Kuner R, Behrens J, Offermanns S. 2002. Plexin-B1 directly interacts with PDZ-RhoGEF/LARG to regulate RhoA and growth cone morphology. *Neuron* 35:51–63

Tang S, Shen YJ, DeBellard ME, Mukhopadhyay G, Salzer JL, et al. 1997. Myelin-associated glycoprotein interacts with neurons via a sialic acid binding site at ARG118 and a distinct neurite inhibition site. *J. Cell Biol.* 138:1355–66

Tansey MG, Baloh RH, Milbrandt J, Johnson EM Jr, 2000. GFRalpha-mediated localization of RET to lipid rafts is required for effective downstream signaling, differentiation, and neuronal survival. *Neuron* 25:611–23

Tessier-Lavigne M, Goodman CS. 1996. The molecular biology of axon guidance. *Science* 274:1123–33

Thallmair M, Metz GAS, Z'Graggen WJ, Raineteau O, Kartje GL, Schwab ME. 1998. Neurite growth inhibitors restrict structural plasticity and functional recovery following corticospinal tract lesions. *Nat. Neurosci.* 1:124–31

Tolbert DL, Der T. 1987. Redirected growth of pyramidal tract axons following neonatal pyramidotomy in cats. *J. Comp. Neurol.* 260:299–311

Tolias KF, Cantley LC, Carpenter CL. 1995. Rho family GTPases bind to phosphoinositide kinases. *J. Biol. Chem.* 270:17656–59

Tolias KF, Couvillon AD, Cantley LC, Carpenter CL. 1998. Characterization of a Rac1- and RhoGDI-associated lipid kinase signaling complex. *Mol. Cell Biol.* 18:762–70

Treherne JM, Woodward SK, Varga ZM, Ritchie JM, Nicholls JG. 1992. Restoration of conduction and growth of axons through injured spinal cord of neonatal opossum in culture. *Proc. Natl. Acad. Sci. USA.* 89:431–34

Turnley AM, Bartlett PF. 1998. MAG and

MOG enhance neurite outgrowth of embryonic mouse spinal cord neurons. *Neuroreport.* 9:1987–90

Yin X, Crawford TO, Griffin JW, Tu P, Lee VM, et al. 1998. Myelin-associated glycoprotein is a myelin signal that modulates the caliber of myelinated axons. *J. Neurosci.* 18:1953–62

Yu TW, Bargmann CI. 2001. Dynamic regulation of axon guidance. *Nat. Neurosci.* 4 (Suppl.):1169–76

Vikis HG, Li W, He Z, Guan KL. 2000. The semaphorin receptor plexin-B1 specifically interacts with active Rac in a ligand-dependent manner. *Proc. Natl. Acad. Sci. USA* 97:12457–62

Vinson M, Strijbos PJLM, Rowles A, Facci L, Moore SE, et al. 2001. Myelin-associated glycoprotein interacts with ganglioside GT1b. *J. Biol. Chem.* 276:20280–85

Vinson M, Rausch O, Maycox PR, Prinjha RK, Chapman D, et al. 2003. Lipid rafts mediate the interaction between myelin-associated glycoprotein (MAG) on myelin and MAG-receptors on neurons. *Mol. Cell Neurosci.* 22:344–52

Vyas AA, Patel HV, Fromholt SE, Heffer-Lauc M, Vyas KA, et al. 2002. Gangliosides are functional nerve cell ligands for myelin-associated glycoprotein (MAG), an inhibitor of nerve regeneration. *Proc. Natl. Acad. Sci. USA.* 99:8412–17

Wahl S, Barth H, Ciossek T, Aktories K, Mueller BK. 2000. Ephrin-A5 induces collapse of growth cones by activating Rho and Rho kinase. *J. Cell Biol* 149:2630–70

Walsh GS, Krol KM, Crutcher KA, Kawaja MD. 1999. Enhanced neurotrophin-induced axon growth in myelinated portions of the CNS in mice lacking the p75 neurotrophin receptor. *J. Neurosci.* 19:4155–68

Wang KC, Koprivica V, Kim JA, Srivasankaran R, Guo Y, et al. 2002a. Oligodendrocyte-myelin glycoprotein is a Nogo receptor ligand that inhibits neurite outgrowth. *Nature* 417:941–44

Wang KC, Kim J, Srivasankaran R, Segal R, He Z. 2002b. p75 interacts with the Nogo receptor as a co-receptor for Nogo, MAG and OMgp. *Nature* 420:74–78

Wang X, Chun SJ, Treloar H, Vartanian T, Greer CA, Strittmatter SM. 2002c. Localization of Nogo-A and Nogo-66 receptor proteins at sites of axon-myelin and synaptic contact. *J. Neurosci.* 22:5505–15

Winton MJ, Dubreuil CI, Lasko D, Leclerc N, McKerracher L. 2002. Characterization of new cell permeable C3-like proteins that inactivate Rho and stimulate neurite outgrowth on inhibitory substrates. *J. Biol. Chem.* 277:32820–29

Wong K, Ren XR, Huang YZ, Xie Y, Liu G, et al. 2001. Signal transduction in neuronal migration: roles of GTPase activating proteins and the small GTPase Cdc42 in the Slit-Robo pathway. *Cell* 107:209–21

Wong ST, Henley JR, Kanning KC, Huang K, Bothwell M, Poo M. 2002. A p75NTR and Nogo receptor complex mediates repulsive signaling by myelin-associated glycoprotein. *Nat. Neurosci.* 5:1305–8

Yamashita T, Tucker KL, Barde Y. 1999. Neurotrophin binding to the p75 receptor modulates Rho activity and axonal outgrowth. *Neuron* 24:585–93

Yamashita T, Higuchi H, Tohyama M. 2002. The p75 receptor transduces the signal from myelin-associated glycoprotein to Rho. *J. Cell Biol.* 157:565–70

Yamashita T, Tohyama M. 2003. The p75 receptor acts as a displacement factor that releases Rho from Rho-GDI. *Nat. Neurosci.* 6:461–67

Yuan XB, Jin M, Xu X, Song YQ, Wu CP, et al. 2003. Signalling and crosstalk of Rho GTPases in mediating axon guidance. *Nat Cell Biol.* 5:38–45

Zheng B, Ho C, Li S, Keirstead H, Steward O, Tessier-Lavigne M. 2003. Lack of enhanced spinal regeneration in nogo-deficient mice. *Neuron* 38:213–24

Zheng JQ, Felder M, Connor JA, Poo MM. 1994. Turning of nerve growth cones induced by neurotransmitters. *Nature* 368:140–44

Zheng JQ. 2000. Turning of nerve growth cones induced by localized increases in intracellular calcium ions. *Nature* 403:89–93

Annu. Rev. Neurosci. 2004. 27:369–92
doi: 10.1146/annurev.neuro.27.070203.144226

MAPS IN THE BRAIN: What Can We Learn from Them?

Dmitri B. Chklovskii and Alexei A. Koulakov
Cold Spring Harbor Laboratory, Cold Spring Harbor, New York 11724; email: mitya@cshl.edu, akula@cshl.edu

Key Words cerebral cortex, visual processing, wiring economy, optimization, cortical map

■ **Abstract** In mammalian visual cortex, neurons are organized according to their functional properties into multiple maps such as retinotopic, ocular dominance, orientation preference, direction of motion, and others. What determines the organization of cortical maps? We argue that cortical maps reflect neuronal connectivity in intracortical circuits. Because connecting distant neurons requires costly wiring (i.e., axons and dendrites), there is an evolutionary pressure to place connected neurons as close to each other as possible. Then, cortical maps may be viewed as solutions that minimize wiring cost for given intracortical connectivity. These solutions can help us in inferring intracortical connectivity and, ultimately, in understanding the function of the visual system.

INTRODUCTION

Wiring distant neurons in the brain is costly to an organism (Ramón y Cajal 1999). The cost of wiring arises from its volume (Cherniak 1992, Mitchison 1991), metabolic requirements (Attwell & Laughlin 2001), signal delay and attenuation (Rall et al. 1992, Rushton 1951), or possible guidance defects in development (Tessier-Lavigne & Goodman 1996). Whatever the origin of the wiring cost, it must increase with the distance between connected neurons. Then, among various functionally equivalent arrangements of neurons, the one having connected neurons as close as possible is most evolutionarily fit and, therefore, likely to be selected. This argument is known as the wiring optimization principle, or the wiring economy principle, and is rooted in the laws of economy of space, time, and conductive matter postulated by Ramón y Cajal in the nineteenth century (Ramón y Cajal 1999). Since then, the wiring optimization principle has been used to answer many questions about brain organization: why there are separate visual cortical areas (Mitchison 1991), why neocortex folds in a characteristic species-specific pattern (Van Essen 1997), why ocular dominance patterns exist (Chklovskii & Koulakov 2000; Koulakov & Chklovskii 2003; Mitchison 1991, 1992), why orientation preference patterns are present in the visual cortex (Durbin

& Mitchison 1990, Koulakov & Chklovskii 2001, Mitchison 1991), why axonal and dendritic arbors have particular dimensions (Cherniak et al. 1999, Chklovskii 2000b, Chklovskii & Stepanyants 2003) and branching angles (Cherniak 1992, Cherniak et al. 1999), why gray and white matter segregate in the cerebral cortex (Murre & Sturdy 1995, Ruppin et al. 1993), why axons and dendrites occupy a certain fraction of the gray matter (Chklovskii et al. 2002, Stepanyants et al. 2002), and why cortical areas in mammals and ganglia in *C. elegans* are arranged as they are (Cherniak 1994, Cherniak 1995; Klyachko & Stevens 2003).

The circumstantial evidence in favor of wiring optimization is complemented by the smoking gun, which shows the principle in action. When neurons are grown in low-density culture, their shapes are often strikingly regular (Figure 1). Neurite branches are almost straight, similar to stretched rubber bands (Bray 1979). Because a straight line is the shortest trajectory connecting points on a plane, this is exactly what wiring optimization would predict. Straightness is not always a result of linear growth (Katz 1985) because, under some conditions, neurites straighten out only after they reach their targets (Shefi et al. 2002). A likely biophysical mechanism for straight segments and, hence, for wiring optimization is tension along neurites (Bray 1979). In addition to straightening neurites, tension pulls on cell bodies, resulting in effective attraction between synaptically connected neurons. Such attraction is difficult to counteract in cultures (Zeck & Fromherz 2001) and may be responsible for the formation of gyri and sulci in the cortex (Van Essen 1997). Although tension along neurites exists also in vivo (Condron & Zinn 1997), axons and dendrites are not always straight in dense neuropil. The reason for

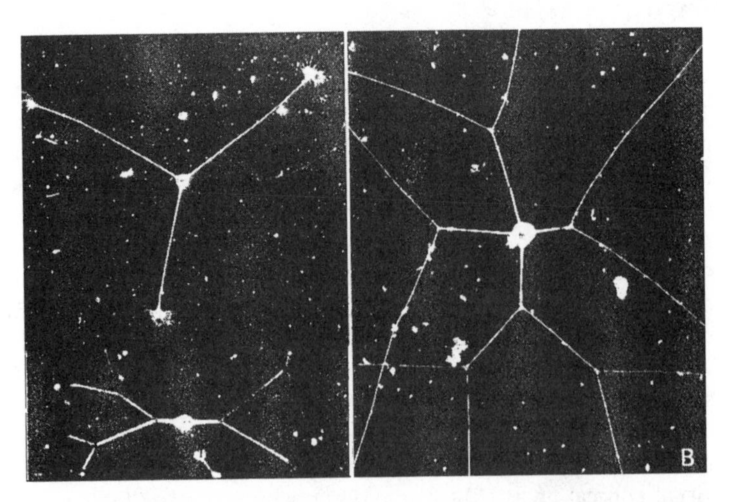

Figure 1 The smoking gun illustrating wiring optimization in action. Neurons grown at low density have regular shape with straight neurite branches consistent with wiring optimization. Reproduced neurons were taken from chick dorsal root ganglia and grown on a glass surface in culture medium with nerve growth factor (Bray 1979).

tortuous shapes is that optimization of different axons and dendrites comes into conflict owing to volume exclusion. Therefore, finding examples of neurons whose axons or dendrites are not straight (Young & Scannell 1996) does not disprove wiring optimization but rather emphasizes the importance of choosing an appropriate subset of the brain for applying wiring optimization.

In this review, we discuss the application of the wiring optimization principle to cortical maps. We stress the role of wiring optimization in establishing a link between the appearance of cortical maps and the connectivity of intracortical circuits. The review is organized as follows. First, we recapitulate classical explanations for the existence of retinotopic maps and multiple cortical areas, which rely on the wiring optimization principle. Second, we argue that wiring optimization may be a general principle of the cortical map organization and apply this principle to explain the appearance of ocular dominance patterns, orientation preference maps, and direction preference maps. Third, we address frequently asked questions about wiring optimization in the context of cortical maps.

RETINOTOPY INDICATES THAT CORTICAL PROCESSING IS LOCAL IN VISUAL SPACE

A visual cortical area is called retinotopic (or said to contain a retinotopic map) if nearby cortical neurons receive inputs from nearby retinal neurons. Retinotopy is common among visual cortical areas and is even integral to their definition, with mirror reflection of a retinotopic map often being used to establish area boundaries. Why does retinotopic organization of cortical areas occur? One might think that retinotopy is inherited from the neuronal order in the retina and preserved by the axons in the optic nerve. However, experiments in adult cats (Horton et al. 1979) and in juvenile and adult rats (Simon & O'Leary 1991) show that axons from adjacent retinal ganglion cells are scattered in the optic nerve, yet they reassemble and project retinotopically in the lateral geniculate nucleus (LGN) and V1. Therefore, retinotopy requires another explanation.

The reason retinotopic maps exist is because "... visuotopic organization would permit neurons representing adjacent parts of the visual field to interact over short axonal and dendritic pathways" (Allman & Kaas 1974). The explanation goes as follows (see also Cowey 1979 and Nelson & Bower 1990). Because of spatial correlations in the external world, the early stages of the visual system combine information coming from the adjacent points in the visual field. For example, recognition of a face on a portrait (Figure 2, see color insert) relies on exchanging information about facial features represented by neurons with adjacent receptive fields. Such processing requires connecting such neurons by costly wiring, i.e., axons and dendrites. To minimize the wiring length, neurons with adjacent receptive fields should be placed as close to each other as possible. This is exactly what an ordered retinotopic map (as defined above) accomplishes (Figure 2). Thus, the reason retinotopic maps exist is to minimize the total length of intracortical connections that are required for processing local features of the visual space.

In addition to providing an explanation for the existence of retinotopic maps, the wiring optimization principle possesses predictive power. For example, one may wonder whether processing within a given visual cortical area is local in visual space. It is difficult to answer such questions without direct knowledge of the intracortical neuronal connectivity. Yet, the wiring optimization principle suggests that the neuronal connectivity may be inferred from the spatial map of that cortical area. If the representation of the visual field in that cortical area is retinotopic, then by using the above wiring optimization explanation in reverse one can infer that the processing is likely local in the visual space. If the representation is nonretinotopic, then the processing is not likely local in the visual space. The utility of wiring optimization is not limited to visual cortical areas and applies to other topographic maps in the cortex: auditory, somatosensory, motor, and others. Just as in the visual cortex, wiring optimization provides a link between the topography of a cortical map and the locality of processing in a sensory space.

EXISTENCE OF MULTIPLE VISUAL AREAS REFLECTS MODULARITY OF VISUAL PROCESSING

Visual cortex in primates contains almost 30 different visual areas, see e.g., Felleman & Van Essen 1991. Why do they exist in such numbers? Indeed, one could imagine a visual superarea (Barlow 1986) that combines the function of, say, both V1 and V2. In this hypothetical superarea, neurons from V1 and V2 would be finely intermingled but would preserve retinotopy. All the connections between neurons, and, hence, the function of the superarea, would be exactly the same as separate V1 and V2. Why do primates lack such a superarea? Mitchison (1991) argues that such a superarea would be detrimental, provided neurons have more connections with neurons of the same area (e.g., V1 to V1) than with neurons of the other (e.g., V1 to V2). His argument can be appreciated best if we first assume that neurons connect only with neurons of the same area. Then, inserting V2 neurons in between V1 neurons will push them farther apart from each other. In turn, longer distances between neurons of the same area lead to longer connections between them, hence increasing the wiring cost. Now, let us include the connections between V1 and V2 neurons. These connections would get shorter in the merger process because retinotopy is preserved. So, the result depends on the relative balance between intra- and inter-area connections. Mitchison (1991) predicted that for the typical numbers of connections in the mammalian brain, separate visual areas are advantageous from the wiring point of view. This explanation highlights a general principle of brain organization: If two sets of neurons connect mostly within their own set, they are better kept separate.

There is an alternative explanation for the existence of multiple cortical areas (Barlow 1986), which also relies on wiring minimization. This explanation assumes that, because of different functional requirements, the connectivity in intracortical circuits is different from area to area. Then the optimal layouts are also different from area to area, leading to conflicting demands on spatial organization if the areas were combined into a superarea. Although this explanation seems plausible in principle, it is difficult to see how it applies to early levels of visual processing. Indeed, the most salient feature of spatial organization, retinotopy, is shared by all early visual areas.

In addition to explaining the existence of multiple visual areas, the wiring optimization principle can help us understand visual processing as illustrated by the following example. Theories based on the efficient coding hypothesis, together with statistics of natural images, predict orientationally selective receptive fields, similar to those observed in V1 (Bell & Sejnowski 1997, Olshausen & Field 1996). Unfortunately, the predictive power of these theories is limited because they do not specify why the orientationally selective receptive fields should exist in V1 but not in the LGN or V4. This shortcoming can be corrected by introducing an additional parameter, distinguishing different areas. Such a parameter should appear naturally in the wiring optimization approach because it explains the existence of multiple visual areas. This example illustrates how wiring optimization can help the construction of visual processing theories.

WIRING OPTIMIZATION ESTABLISHES A LINK BETWEEN CORTICAL MAPS AND INTRACORTICAL CIRCUITS

Retinotopy and multiple visual areas are examples of how the spatial organization of cortical areas and the connectivity of neuronal circuits may be linked by the wiring optimization principle. In fact, the wiring optimization principle plays a unique role in establishing such links (Swindale 2001). Imagine taking a cortical area containing a map and scrambling neurons in that area, while preserving all the connections between neurons. Because the circuit is unchanged, the functional properties of the neurons remain intact. Then, the scrambled region without a map is functionally identical to the original one with the map. Thus, if we neglect possible interactions through nonsynaptic diffusion of messengers, the only remaining parameter that can differentiate candidate layouts for fixed neuronal connectivity is the length of wiring. If cortical maps are selected in the course of evolution to improve the fitness of the organism, they can only be chosen on the basis of the length of connections. Therefore, it is hard (if not impossible) to justify the existence of systematic cortical maps without invoking the cost of making long neuronal connections. This argument is not limited to maps and leads to the conclusion that "the principle of minimizing wire length appears to be a general factor governing the connections of nervous systems" (Allman 1999).

The link between cortical maps and circuits is significant for understanding the brain because data on spatial organization are often available, whereas data on neuronal connectivity are usually scarce and hard to obtain. In turn, knowledge

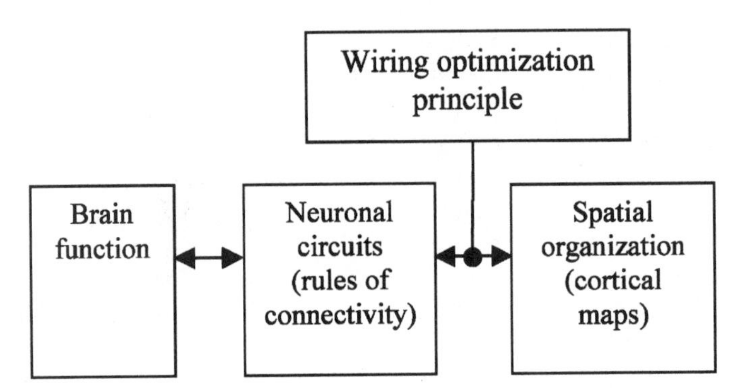

Figure 3 The wiring optimization principle helps establish a link between neuronal circuitry and spatial organization of neurons. This link allows one to infer neuronal circuitry, which is usually poorly understood, from the spatial organization of neurons, which is often better understood. Knowledge of neuronal circuitry is, in turn, essential for understanding brain function.

of connectivity in neuronal circuits is essential for understanding brain function (Figure 3). To provide the link between intracortical connectivity and maps, the wiring optimization principle must be formulated in a rigorous quantitative manner. Next, we illustrate such formulation on several examples of cortical maps, starting with the simplest: the ocular dominance pattern (ODP).

OCULAR DOMINANCE PATTERNS ARISE FROM SAME-EYE BIAS IN CONNECTIVITY

What is the functional significance of the ocular dominance pattern (ODP), i.e., clustering of neurons based on their ocular dominance? Mitchison (1991) proposed that ODP is an adaptation that minimizes the length of intracortical connections involved in visual processing. The main idea is similar to the explanation for the existence of multiple visual areas. If there is bias in connectivity toward neurons with the same OD (i.e., the number of connections between neurons of the same OD exceeds the number of connections between neurons of different OD), the formation of OD clusters reduces the total length of connections. The original (Mitchison 1991) and more recent theories (Chklovskii & Koulakov 2000, Koulakov & Chklovskii 2003) explain why some mammals have ODPs and others do not, and why monocular regions have different appearances (stripes as opposed to patches) between different parts of V1 in macaque (Horton & Hocking 1996) and *Cebus* monkeys (Rosa et al. 1992) (Figure 4). In particular, wiring optimization predicts a transition from stripes to patches when the fraction of neurons dominated by one eye drops below 40%, in agreement with the ODP observed in macaque (Figure 4).

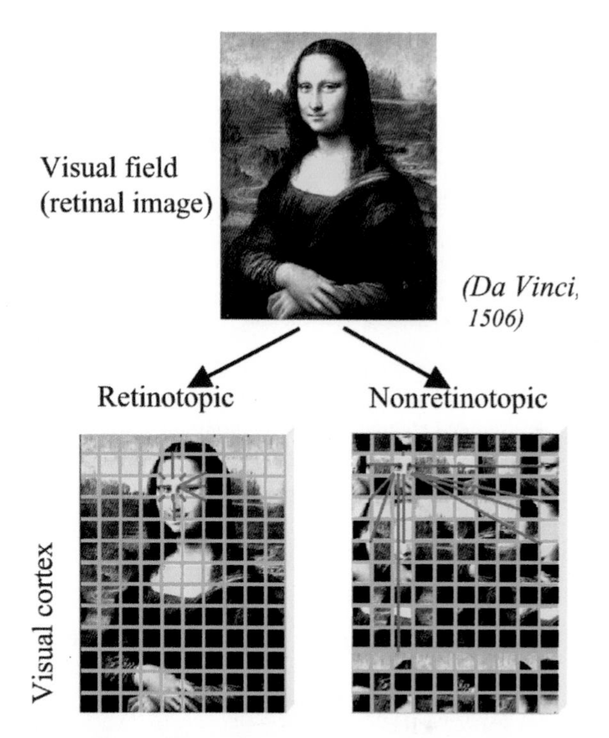

Figure 2 A retinotopic map minimizes the cost of connecting neurons that have adjacent receptive fields. (*Top*) Retinal image. (*Bottom*) Two candidate layouts of cortical neurons: retinotopic (*left*), nonretinotopic (*right*). *Squares* represent receptive fields of cortical neurons at corresponding cortical locations. Retinotopic layout minimizes the length of intracortical connections (*red*) that are required for processing local features of the image.

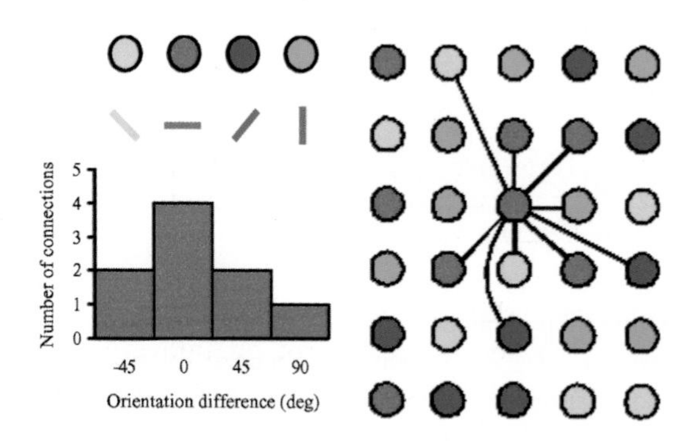

Figure 9 The wiring optimization model of orientation preference maps in a simplified case of only four different orientation values.

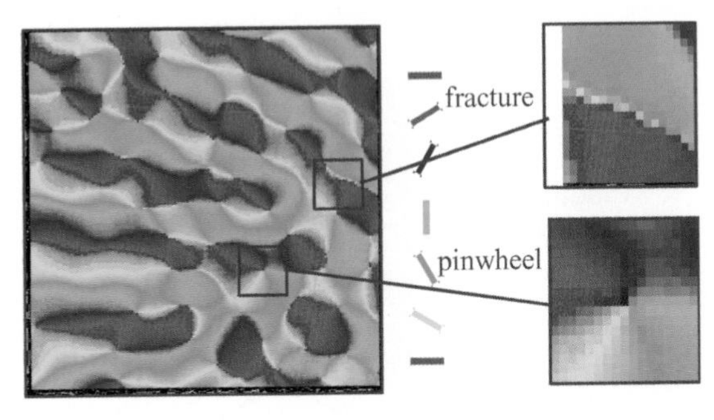

Figure 8 An orientation preference map in tree shrew primary visual cortex (Bosking et al. 1997). Colors code for orientation preference according to the legend. Orientation preference changes smoothly along the cortex with the exception of singularities: pinwheels and fractures (*right*). Notice a pinwheel-free region in the left part of the map. Based on our wiring optimization theory, we predict that the connectivity rules in intracortical neuronal circuits are different between that region and the rest of the map.

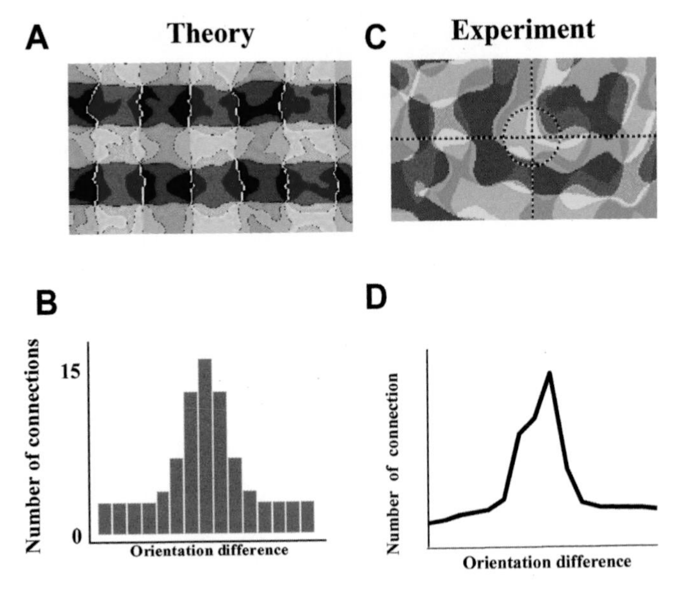

Figure 11 Results of wiring optimization are consistent with experimental data. (*A*) Orientation map obtained by wiring minimization from the connection function shown in (*B*). (*C*, *D*) Experimentally obtained orientation map and anatomically measured connection function (Yousef et al. 1999). Although the theoretical map is more orderly, it contains all the main features of the experimental data: linear regions and pinwheels. Similarity between the theoretical and the anatomically measured connection functions supports our approach and suggests using wiring optimization to infer connectivity from map appearance.

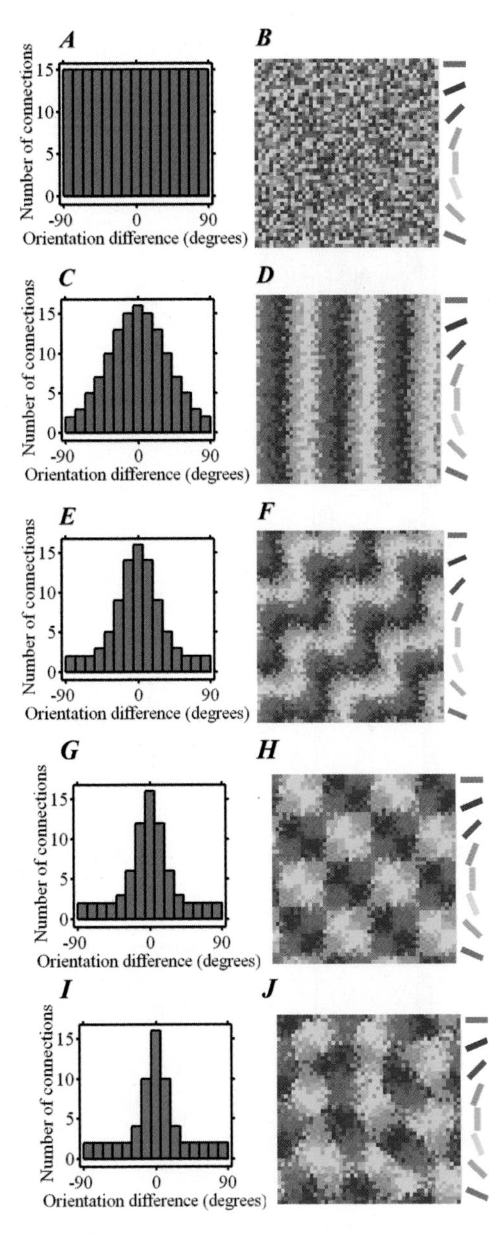

Figure 10 Numbers of intracortical neuronal connections in the visual cortex (*left column*) and corresponding orientation preference patterns (*right column*) obtained by minimizing the length of these connections (Koulakov & Chklovskii 2001). Based on these results, we propose that the differences in orientation preference patterns [within one animal (Figure 10) and between species] reflect the differences in the connectivity rules of intracortical circuits. In addition, we suggest that the functional significance of pinwheels and fractures may be in minimizing wiring length for certain intracortical connection rules.

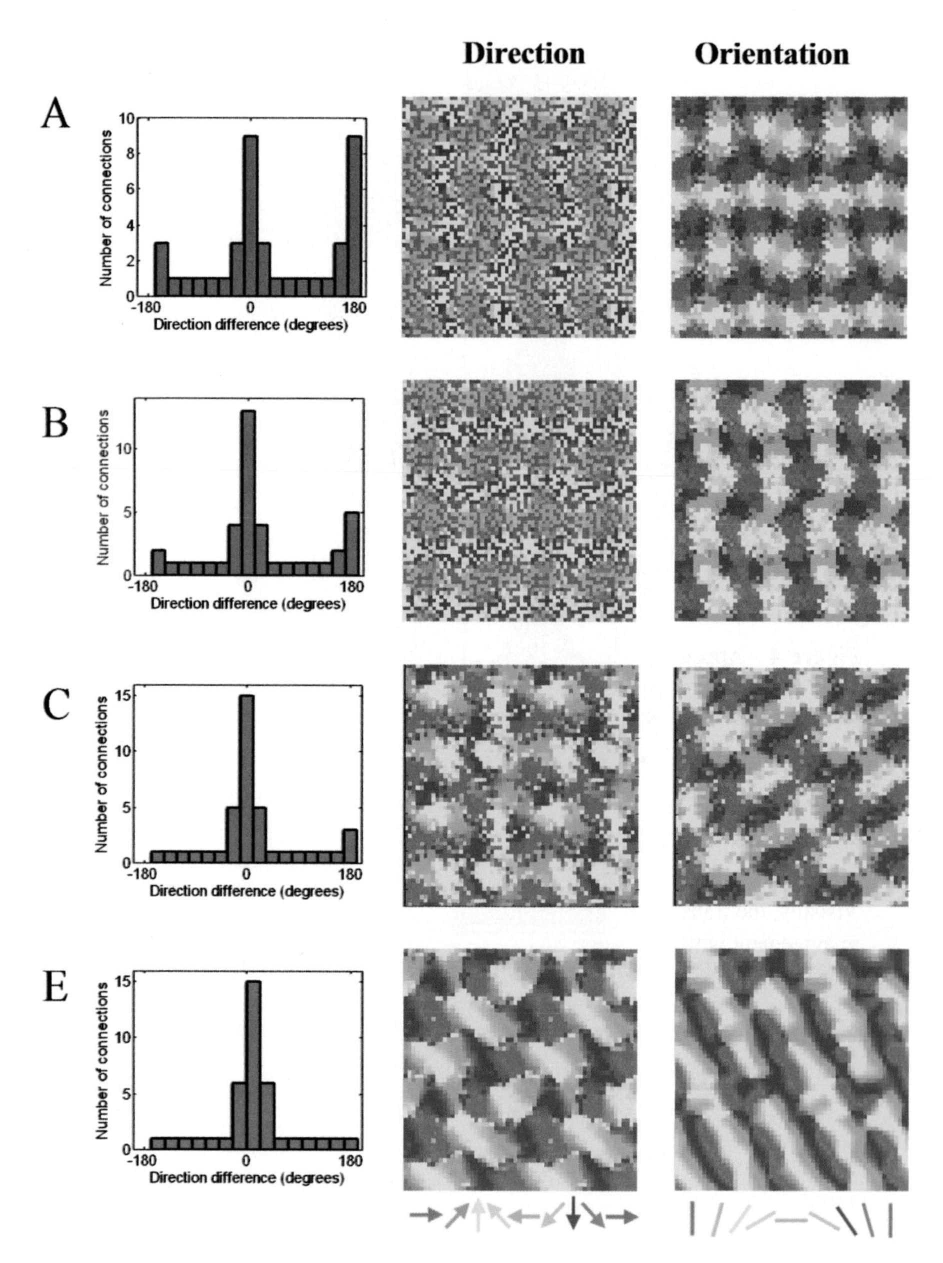

Figure 12 Wiring optimization relates intracortical connectivity (*left* column) to direction and orientation preference maps (*middle* and *right* columns, respectively). Different connection functions correspond to different direction and orientation preference maps (*A–D*).

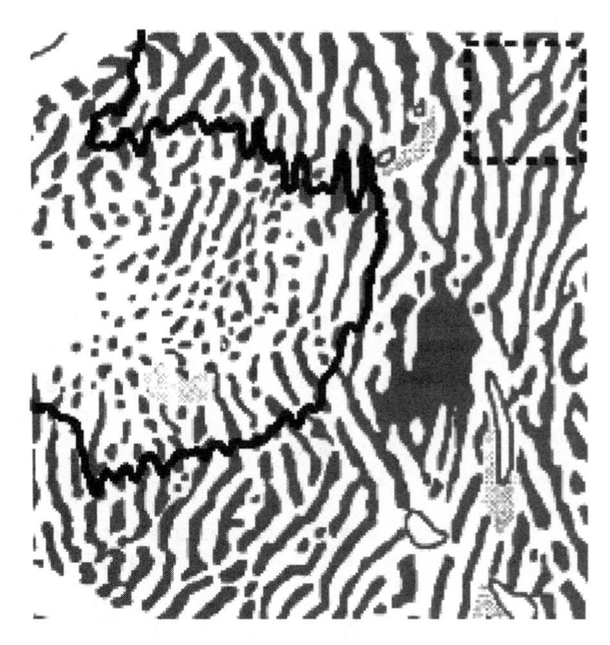

Figure 4 Appearance of the ODP in macaque visual cortex (Horton & Hocking 1996) is consistent with the wiring optimization theory (Chklovskii & Koulakov 2000, Koulakov & Chklovskii 2003). Gray regions show neurons dominated by the left eye, and white regions show neurons dominated by the right eye. This fragment shows extrafoveal representation in V1 (the large gray spot is the representation of the optic disc). The black line is the theoretical prediction for the location of the transition between *patchy* and *stripy* patterns, based on the fraction of left-eye neurons being about 40% (as averaged over an area of the size shown in the upper right corner). Visually, the transition from *stripy* to *patchy* patterns takes place near the black line, in agreement with the theory.

To illustrate how wiring optimization leads to these findings, we consider a simplified model. If the number of connections per neuron is small, solving the optimal layout problem does not require a computer. The problem is to find the layout of neurons, which minimizes the length of connections for given connection rules (Figure 5). Solutions shown in Figure 5 have been confirmed in computer simulations involving large numbers of connections per neuron (Chklovskii & Koulakov 2000, Koulakov & Chklovskii 2003). Such simulations are essential to establish a link with actual cortical circuits where each neuron receives thousands of inputs. In addition, for the large numbers of connections, the choice of the lattice does not affect the results.

As argued in the previous section, the only difference between various layouts with the same connectivity is the cost of wiring. Therefore, any theory of the map appearance must invoke wiring optimization. Hence, theories of the map

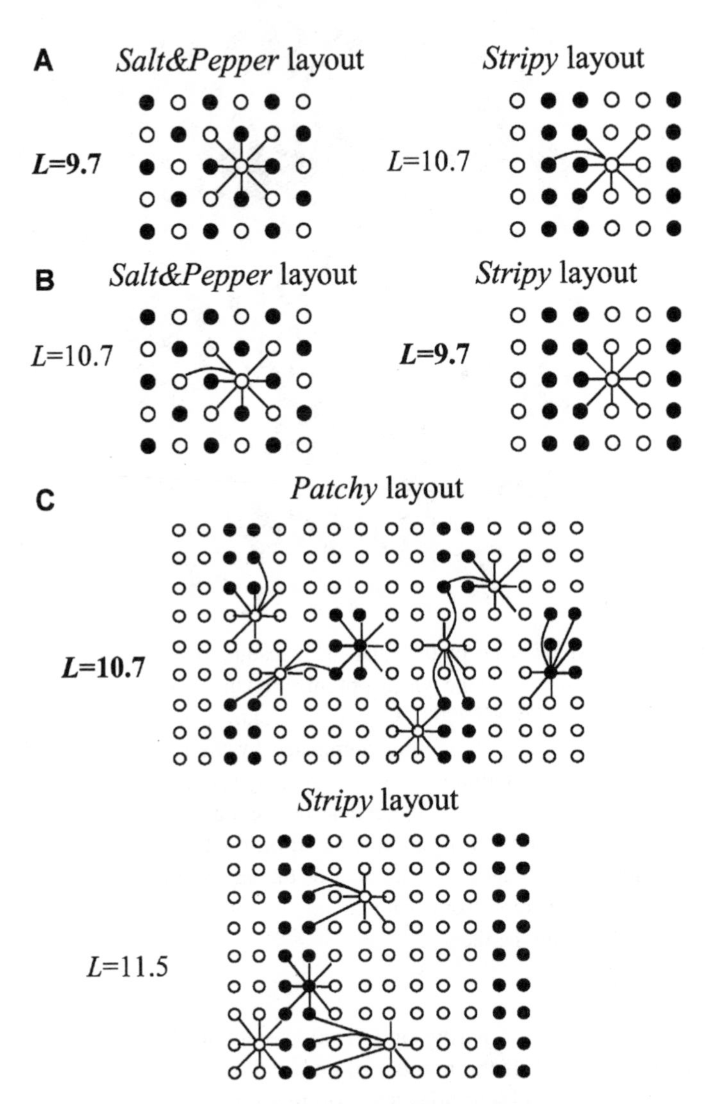

Figure 5 Wiring model of ODP. White circles are neurons dominated by the right eye, and black circles are neurons dominated by the left eye. (*A*) For equal numbers of same-eye and opposite-eye connections (four versus four) *Salt&Pepper* layout (*right*) gives a lower wire length per neuron, *L* (minimal lengths are in bold). (*B*) If there is a connectivity bias, i.e., same-eye connections are more abundant than the opposite-eye ones (five versus three), the ODP pattern, such as *Stripes*, provides a smaller wire length. (*C*) If left-eye neurons are less abundant, the same connection rules as in (*B*) result in a smaller wire length for *Patches*. For each layout, connections for all types of neurons are shown. All neurons in (*C*) satisfy the same (five versus three) connection rule as in (*B*) (Chklovskii & Koulakov 2000).

formation that do not explicitly rely on wiring optimization invoke it implicitly, usually requiring the locality of intracortical connections. Here, we discuss the models that invoke explicitly the wiring cost.

Mitchison (1991) compared wiring cost between ODP and *Salt&Pepper* layout, assuming that interneuronal connections are determined both by ocular dominance and retinotopy. He found that the answer depends on the rules of axonal branching. If at branch points all axonal segments have the same caliber, ODPs are advantageous for some connectivity. However, if at branch points the total axonal cross-sectional area is conserved, the ODP do not save wiring volume relative to *Salt&Pepper*. The actual situation is probably somewhere in between the two cases (Chklovskii & Stepanyants 2003). Because the wiring optimization model presented here assumes separate connections for each neuron, it corresponds to the case of the cross-sectional area conservation. However, unlike Mitchison, we find that ODPs are more efficient than *Salt&Pepper*. This happens because, in accordance with experimental data (Katz et al. 1989), the intracortical connectivity rules do not rigidly enforce the retinotopy of connections. This assumption simplifies the theory, allowing us to map out the complete phase diagram and make experimentally testable predictions (Chklovskii & Koulakov 2000, Koulakov & Chklovskii 2003). The full theory of the ODP will require a more detailed analysis of axonal branching (Chklovskii & Stepanyants 2003).

Jones et al. (1991) proposed an explanation for why ODP have either *Stripy* or *Patchy* appearance. They assumed that neurons are already segregated into the ODP (by considering units whose size equals the width of monocular regions) and found that the difference between *Stripy* and *Patchy* appearances of the ODP could be due to the different shape of V1 in different species. Although the correlation between the shape of V1 and the ODP layout is observed, this model (Jones et al. 1991) does not explain why peripheral representation of macaque V1 has patchy ODP or why ocular dominance stripes run perpendicular to the long axis of V1 in some parts of V1 but not in others. Most important, unlike Jones et al. (1991), the wiring optimization theory presents a unified theory of ODP, including *Salt&Pepper*, *Stripy*, and *Patchy* layouts, and relates ODP appearance to the connectivity rules.

Another model related to wiring length minimization is the elastic net (Durbin & Mitchison 1990, Durbin & Willshaw 1987, Goodhill & Sejnowski 1997, Goodhill & Willshaw 1990). The original formulation of the model minimized the cost function, which penalized nearby placement of neurons whose activity was not correlated. This choice was justified by computational convenience. Unlike the wiring optimization theory, this penalty does not increase beyond a distance called cortical interaction. Because of this distance, the elastic net often has solutions, in which the cortical area is partitioned into two large monocular domains, although annealing procedure does not yield them. Later the elastic net model was generalized by the introduction of a *C*-measure (Goodhill & Sejnowski 1997). The wiring optimization theory presented here can be viewed as a subset of models described by *C*-measure. The virtue of wiring optimization is that it assigns a clear biological cost for placing neurons far from each other (the cost of wiring). Moreover, wiring optimization establishes a link between cortical maps and connectivity. Because both maps and connectivity can be obtained experimentally, the wiring optimization theory makes clear experimentally testable predictions.

OCULAR DOMINANCE STRIPES ALIGN WITH THE DISPARITY DIRECTION

ODP in primate V1 is not random: Orientation of the ocular dominance stripes on the cortical surface follows systematic trends found in macaque (Horton & Hocking 1996, LeVay et al. 1985) and *Cebus* monkeys (Rosa et al. 1992). These trends are easiest to describe when the ODP is transformed back into the visual field coordinates (Hubel & Freeman 1977, LeVay et al. 1985, Rosa et al. 1992, von Berg 1997) by dividing all cortical distances by the local magnification factor. [The magnification factor is defined as distance along the cortex (in millimeters), which corresponds to a 1° separation on the retina (Daniel & Whitteridge 1961).] The transformed ODP shows two major trends: In the parafoveal region stripes tend to run horizontally, whereas farther from the fovea, stripes follow concentric circles (Figure 6).

What is the functional significance of these trends in the orientation of ocular dominance stripes on the cortical surface? Chklovskii (2000a) proposed that the trends in the ODP reflect the properties of the binocular stereopsis circuitry as a result of wiring optimization. Visual information arriving through the two eyes is initially recombined in V1. Because V1 contains retinotopic maps, binocular disparity of a visual object leads to a separation between cortical representations of the same object (Figure 7). Therefore, recombining information coming from the two eyes requires horizontal intracortical connections (Gilbert et al. 1996). The length of such connections is minimized if the ocular dominance stripes are aligned with the direction of disparity (Figure 7*B*). The direction of disparity was determined for every point in the visual field by calculating the distribution of disparity and taking the dominant direction. Disparity being mainly horizontal in the parafoveal region and mainly cyclotorsional in the peripheral region (Figure 6) explains the main trends in stripe orientation (Chklovskii 2000a).

ORIENTATION PREFERENCE MAPS REFLECT CONNECTIVITY BIAS

Cortical maps of orientation preference, as obtained by optical imaging, exhibit linear zones, whereas orientation preference changes smoothly, with occasional singularities such as pinwheels and fractures (Figure 8, see color insert) (Blasdel 1992, Bonhoeffer & Grinvald 1991). Why these singularities exist has remained a mystery from the time they were discovered, which led some authors to suggest that pinwheels and fractures are developmental defects (Wolf & Geisel 1998). We

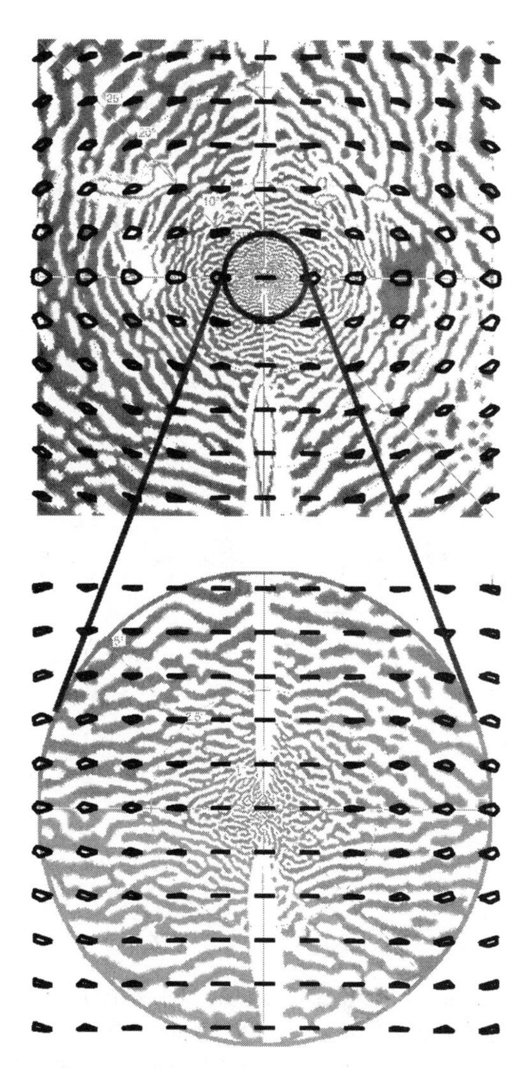

Figure 6 Global structure of the ODP is consistent with theoretical results (Chklovskii 2000a). Gray & white stripes show ODP (Horton & Hocking 1996) warped back into the visual field (von Berg 1997). *(Top)* Visual field up to 25° eccentricity. *(Bottom)* Magnification of the foveal region up to 5° eccentricity. Black polar plots show theoretically obtained frequency distributions of binocular disparity direction for corresponding locations in the visual field. The dominant direction of binocular disparity at each point of the visual field is given by the major axis (longest dimension) of the corresponding plot. Theory based on wiring optimization suggests that ocular dominance stripes should follow the dominant disparity direction. Experimental data are consistent with this prediction with the exception of the regions in the upper left and upper right corners of the top figure and the very center of the bottom figure (for possible reasons behind the discrepancy see Chklovskii 2000a).

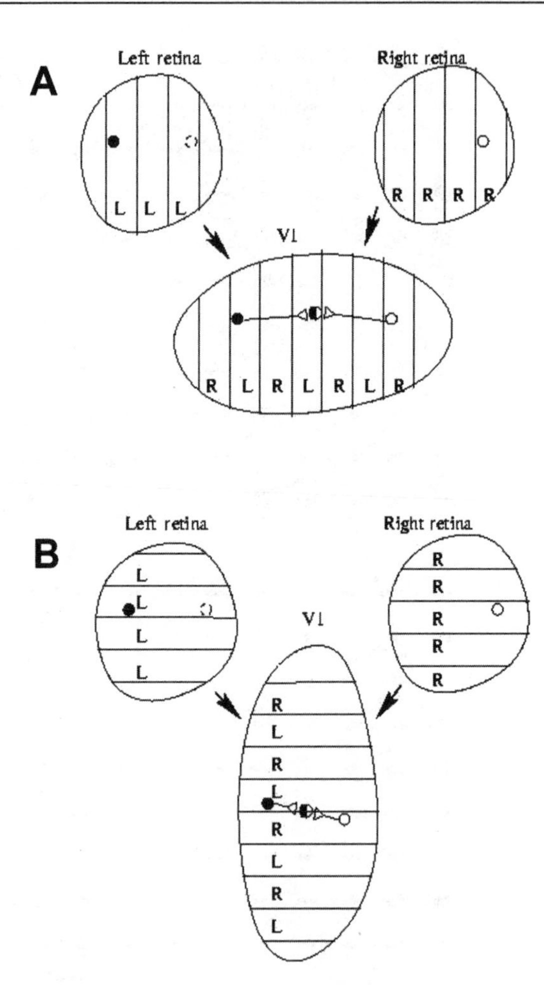

Figure 7 Schematic illustration showing a binocular neuron in V1 (*semifilled circle*) recombining corresponding information from the two eyes via long-range horizontal connections. Connections are longer if the ocular dominance stripes run orthogonally to the disparity direction (*A*) than if they run parallel to that direction (*B*). This finding suggests that the ocular dominance stripes should align with the dominant disparity direction (Chklovskii 2000a).

show that, for some intracortical circuits, singularities are needed to minimize the intracortical wire length, as previously surmised by Swindale (1996, p. 235). Therefore, we suggest that pinwheels and fractures are evolutionary adaptations that keep the cortical volume to a minimum. In addition, we propose that differences in map structure (both between species and within one animal) reflect differences in underlying intracortical neuronal circuits. The latter may be related to differences in natural scene statistics or in behavioral tasks solved by a particular animal.

The wiring model of the orientation preference patterns is an extension of the model for ODPs described in the previous section. Instead of having just two categories of neurons (left-eye dominated and right-eye dominated) we introduce several categories of neurons differing by their preferred orientation. This idea is illustrated on a toy model that includes four different orientation preference categories (Figure 9, see color insert). As in the OD model, functionality of the network requires each neuron to make certain numbers of connections with neurons of other categories. But instead of there being only two numbers in the OD model, now there are as many numbers as there are categories. We refer to these numbers collectively as the connection function. The results of layout optimization with fifteen categories of orientation preference are shown in Figure 10 (see color insert).

Maps presented in Figure 10 reflect bias in connectivity toward neurons of similar orientations (Koulakov & Chklovskii 2001). This bias favors layouts in which neurons of close orientation reside next to each other, thus resulting in smooth maps. If the connectivity bias is strong, as in Figures 10*C–F*, the corresponding maps are completely smooth and free of singularities (pinwheels or fractures). As the bias decreases in strength, as in Figures 10*G–J*, singularities appear in the orientation maps, thus leading to *Pinwheels* or *Fractures* layouts (Figures 10*H* and *J*, respectively). The emergence of singularities is related to the unbiased, or uniform, component in the connection function, which favors close placement of neurons with dissimilar orientation. If the connection function contains only the unbiased, or uniform, component (Figure 10*A* and *B*), singularities proliferate, and the *Salt&Pepper* layout becomes optimal.

These results imply that for each shape of the connection function (*left column*) there is a layout that minimizes wiring length (*right column*). Therefore, interspecies differences in orientation maps or variations in the layout within a single map could be traced back to the intracortical connectivity. This hypothesis can be tested experimentally. For example, the connection function for the pinwheel layout (Figure 11*A*, see color insert) should be of the form shown in Figure 11*B*. This is consistent with experimental data (Yousef et al. 1999) (Figure 11*C,D*). Another prediction is based on the observation that the rodent V1 has *Salt&Pepper* layout (Girman et al. 1999). Theories based on wiring optimization suggest that the connection function in rodents is different from the one in cats and monkeys, despite the similarity in cell tuning properties (for details see Koulakov & Chklovskii 2001).

Another approach to inferring intracortical connectivity from orientation maps is based on the locality of connections (Das & Gilbert 1999, Schummers et al. 2002). To determine the origin of intracortical inputs to a neuron, one draws a circle around that neuron and counts the numbers of neurons in that circle with various orientation preferences. This procedure implicitly relies on wiring optimization because it assumes that connections are made with neurons within a certain distance only. In fact, the set of connected neurons in the wiring optimization theory presented here is usually close to a circle (Koulakov & Chklovskii 2001). One virtue of the circle procedure is its prediction of the difference in connectivity for neurons located in pinwheels and orientation domains (Schummers et al. 2002). This result does not contradict the wiring optimization theory because the uniformity of connection function among neurons is a simplifying assumption, not a conclusion, of the theory. It is relatively straightforward to generalize the wiring optimization theory by including an additional category of neurons with different connection function. The circle procedure does not render the wiring optimization theory useless. First, the result of this procedure depends strongly on the radius of the circle. Whereas using a circle with a short radius produces a strongly biased connection function, long radius produces a uniform connection function. Because the radius of the circle must be guessed, arbitrariness is introduced into the procedure. Second, this procedure does not explain the existence of the maps. The connectivity is determined from maps and not the other way around.

There is an important difference between wiring optimization and related theories such as the locality principle (Das & Gilbert 1999, Schummers et al. 2002) and the elastic net (Durbin & Mitchison 1990, Goodhill & Cimponeriu 2000). These theories make rigid assumptions about intracortical connectivity, such as requiring connections within a circle or with the nearest neighbors. Conversely, in the wiring optimization approach, the intracortical connectivity can vary from case to case. In an attempt to establish a link between maps and connectivity, we consider as many connection functions as needed to explain the variability of observed maps (the most up-to-date database of connectivity and corresponding layouts is maintained at http://koulakovlab.cshl.edu/anneal). The only assumption made in these models is that the link between connectivity and maps is established through wiring optimization.

DIRECTION OF MOTION PREFERENCE MAPS

In addition to being selective to stimulus orientation, many neurons in visual cortex show preference for a particular direction of motion. Spatial organization of neurons according to preferred direction is reflected in direction preference maps (DPMs), which have been studied electrophysiologically (Albright et al. 1984, Diogo et al. 2003, Swindale et al. 1987) and optically (Kim et al. 1999, Malonek et al. 1994, Shmuel & Grinvald 1996, Weliky et al. 1996). The organization of the DPMs differs between visual areas and species. In some cases, such as ferret area 17 (Weliky et al. 1996) and cat area 18 (Shmuel & Grinvald 1996, Swindale et al. 1987), DPMs contain clusters of neurons with the same direction preference. These clusters are correlated with the orientation maps present in these areas: The preferred direction of motion is orthogonal to the preferred orientation. In other cases, such as cat area 17, no robust directional clustering is observed (Bonhoeffer et al. 1995; but see Swindale et al. 1990), despite the presence of directionally selective cells (Hubel & Wiesel 1962).

As with the orientation preference maps, the differences in the DPMs could be due to the differences in the intracortical circuitry. The link between maps and connectivity is established through wiring optimization. Some preliminary considerations (Koulakov & Chklovskii 2001, 2002) show promising results in explaining map diversity. Although solving layout optimization problems for direction maps is very similar to the case of orientation maps discussed above, the parameter space of direction-sensitive connection functions is larger. Thus, we limit ourselves to a few examples, which show reasonable similarity to DPMs observed in visual cortices. The connectivity is characterized by the connection function similar to that used for orientation maps, but it depends on the difference in direction preference. According to experimental data (Roerig & Kao 1999), the connection function has two peaks: One is around 0° direction difference (connections with neurons of similar preferred direction), and another is around 180° direction difference (connections with neurons of opposite preferred direction). We vary the relative amplitude of the two peaks and search for the optimal layout, which minimizes the total length of connections.

First, we use the connection function with two peaks of the same magnitude (Figure 12*A*, see color insert). In this case the direction map (Figure 12*A*, center) is disordered, i.e., a *Salt&Pepper* layout is formed. At the same time, the orientation map, obtained by assuming orthogonality between direction and orientation preference (Swindale et al. 2003), shows a regular lattice of pinwheels, surrounded by areas with smooth orientation. This finding implies that the direction map is not entirely random: The neurons with similar orientation form clusters where neurons of opposite direction preference are intermixed. Intrinsic optical imaging of such a map should show a weak direction signal because it averages responses over a small cortical area. Because in the *Salt&Pepper* layout such an area is expected to contain an equal mixture of units with opposite preferred directions (same orientation), no robust direction map results. At the same time the orientation map is well defined. These features are consistent with observations in cat area 17 (Bonhoeffer et al. 1995, Hubel & Wiesel 1962). Although this explanation seems plausible, we cannot rule out the possibility that the direction preference signal could be weaker than orientation preference because direction tuning is weaker than orientation tuning.

Next, we use the connection function with reduced second peak, where the ratio between the magnitudes of the same direction and opposite direction peaks is about 5 to 1 (Figure 12*C*). We find that in the optimized layout, neurons form an axis of motion clusters (Figure 12*C*). These clusters segregate into several regions of opposite direction preference. Boundaries between these regions zigzag through clusters. In the cortical plane, direction preference changes frequently by 180°, whereas the axis of motion remains more continuous. This feature is similar to electrophysiological measurements done in macaque area MT (Albright et al. 1984). In optical imaging experiments, such maps should result in extended areas, where direction signal is weak, intermixed with the areas where direction is well defined. Whether intrinsic optical imaging supports these results in animals, such as owl monkey (Malonek et al. 1994), where MT is readily accessible on the brain surface, would be interesting.

Finally, if the second peak in the connection function is very weak or nonexistent (Figure 12*D*), the direction preference map is continuous with occasional fractures. These fractures terminate at pinwheels. When going around a pinwheel, direction preference changes gradually by 180° and jumps by 180° at the fracture (Swindale et al. 1987). Because preferred orientation is orthogonal to preferred direction, the orientation map is continuous across the fracture. The orientation map contains only one type of singularity: 180° pinwheels. These results are consistent with observations in ferret area 17 (Weliky et al. 1996) and cat area 18 (Shmuel & Grinvald 1996).

FREQUENTLY ASKED QUESTIONS ABOUT WIRING OPTIMIZATION AND MAPS

Q: If wiring optimization is so important, why does not brain size shrink to zero, presumably, the most economical solution?

A: Although decreasing brain volume to zero would reduce the wiring cost, it would also reduce brain's functionality. Because functionality and cost contribute to fitness with different signs, such a radical solution is not favored by animal evolution. In the wiring optimization approach, functionality is fixed by considering various layouts with the same connectivity.

Q: The minimization of wiring cost in the course of evolution proceeded in parallel with functionality maximization. That said, why is it correct to minimize wiring cost for fixed functionality?

A: From the mathematical point of view, fitness is a function of many variables including wiring cost and circuit functionality. Evolution tends to maximize this multivariable function. Yet, to explore that maximum varying only one variable, the wiring cost, while fixing functionality by specifying neuronal connectivity is a valid approach. This approach permits one to answer many questions of brain design. When a quantitative description of brain functionality is available, it can be included in the optimization approach, along with the wiring cost.

Q: If wiring volume is so costly, why not make all the wires infinitesimally thin?

A: Thinner axons and dendrites are detrimental for brain functionality because they conduct signals slower and with more attenuation. If brain functionality is fixed, then wires can be made thinner only if they get shorter as well. This solution would be possible if the brain contained only wiring components. However, when the brain contains nonwire components (i.e., those that do not shrink with brain size), such as synapses, cell bodies, and blood vessels, its cost is minimized when nonwire components compose 40% of the volume (Chklovskii et al. 2002).

Q: If two candidate layouts have a small difference, say 10%, in wiring cost, should one layout be favored?

A: In principle, an animal with a suboptimal layout with 10% more wiring than the existing one could exist if brain's functionality is reduced. However, if brain's functionality is exactly the same the existence of such an animal is close to impossible. Indeed, imagine that an external object, such as a blood vessel, is introduced in a certain area of the gray matter. In this case some of the neuronal connections would have to go around the vessel and, therefore, become longer. If the nerve pulses are to be delivered at the original speed and intensity, the elongated axons and dendrites have to be made thicker, to increase the pulse propagation speed and decrease dendritic attenuation. This procedure leads to more obstacles for other neuronal connections and so on. Thus, introduction of a new blood vessel leads to an infinite series of axonal and dendritic reconstructions. Such an infinite series can diverge, which implies that the connection volume increases indefinitely. In this case the new blood vessel can never be inserted without sacrificing significantly the brain function. Mammalian cortex is on the verge of this so-called wiring catastrophe (Chklovskii et al. 2002) so that it becomes increasingly more difficult to accommodate excess volume in the nerve tissues. Thus, even 10% increase, resulting from wasteful neuronal positioning, may be important.

Q: Many theories produce realistic ocular dominance and orientation preference maps. What is special about the wiring optimization theory?

A: Most theories of map formation generate maps by repetitive application of postulated learning rules that emulate development (Swindale 1996). Although understanding development is clearly important, our current level of knowledge is, in most cases, insufficient to formulate mechanistic theories of map formation, on the level required for testing these theories experimentally. Conversely, the wiring optimization theory bypasses development and provides a link between map structure and intracortical connectivity, both experimentally measurable quantities.

Q: If the wiring optimization theory does not predict the outcome of developmental manipulations, how can it be tested?

A: The role of theory is to make predictions and ask questions; these predictions and questions make further experimental work more effective. By using an assumption of wiring optimization, we predict that the difference in cortical map appearance reflects the difference in intracortical connectivity. These predictions can be tested by comparing maps and connectivity in different animals or in different parts of the same animal. For example, despite a similarity in the orientation tuning, the intracortical connectivity in cats and rats should be different. Testing these predictions will provide necessary feedback for further development of the theory, which will, in turn, generate new predictions.

Q: What can wiring optimization say about developmental mechanisms?

A: The wiring optimization principle aids studies of development by imposing a constraint on the possible forms of the mechanisms. Indeed, to produce an evolutionarily fit organism, developmental rules should, if possible, respect the wiring constraint. A link between wiring optimization and developmental rules can be illustrated with the following example. In the theory of ocular dominance patterns (Chklovskii & Koulakov 2000), formation of ocular dominance columns minimizes wiring length for certain neuronal circuits. In addition, by performing a gradient descent on the wiring length cost function we derive a learning rule that has a "Mexican hat" appearance. Such a learning rule has been used to model development of ocular dominance patterns (Swindale 1980) and also appears in a more biologically realistic model (Miller et al. 1989). Therefore, "Mexican hat" developmental rules can minimize wiring length. A nice bonus of the wiring minimization approach is that the exact shape of the Mexican hat can be derived without any ad hoc parameters.

Q: Why do wiring optimization theories of maps in V1 involve intracortical connections only? Is not ODP set up by thalamic afferents, and is not feedback from V2 important?

A: The current wiring optimization theory considers intracortical connections only because they constitute the majority of wiring in the cortex. Inclusion of the extracortical connections should yield only small corrections. Also, the wiring optimization theory bypasses development: Whatever the developmental mechanism, it should strive to maximize the fitness of an organism. Therefore, wiring optimization results are immune to the exact developmental mechanism responsible for ODP, whatever it turns out to be.

Q: Does not existence of singularities in the orientation map follow from dimension reduction? Specifically, neurons in V1 represent retinotopic coordinates and the orientation variable. But the mapping from a three-dimensional (3D) feature space onto a two-dimensional (2D) cortical sheet cannot be accomplished continuously. Therefore, singularities in orientation maps, such as pinwheels and fractures, are inevitable.

A: Although a mapping from 3D to 2D cannot be accomplished continuously, this is not sufficient to explain singularities in orientation maps because any two coordinates of the 3D space can be mapped continuously in 2D. For example, imagine an ensemble of neurons, each characterized by a 2D position of the receptive field (X and Y coordinates) and the preferred orientation (θ). These neurons can be arranged in 2D so that the X coordinate increases continuously along one axis, while orientation increases continuously in the other direction. Of course, the Y coordinate cannot be mapped continuously in this case. Therefore, dimension reduction is not sufficient to explain singularities in orientation maps. Moreover, experimental observations of continuous orientation maps (see Figure 10 or Shmuel & Grinvald 2000) prove that the presence of singularities is not necessary. That the wiring optimization principle yields orientation maps with and without singularities depending on the connection function is one of the theory's virtues (Figure 10).

Q: Many theories of orientation maps assume the locality of connections based on the distance along the cortical sheet. What is the advantage of wiring optimization compared to such theories?

A: Although some theories of map formation may not explicitly mention the wiring optimization principle, they use it implicitly, usually in requiring the locality of intracortical connections. The locality of connections is likely to be a consequence of wiring minimization. In most cases the locality of connections and the wiring optimization principle yield the same predictions. A notable exception is the existence of long-range horizontal connections (Gilbert et al. 1996, Gilbert & Wiesel 1989), which cannot be accounted for by the locality. Yet, the wiring optimization approach leads naturally to the appearance of long-range horizontal connections. Indeed, imagine that the composition of intracortical circuits is dominated by the connections between neurons with adjacent receptive fields, with a few connections between neurons with remote receptive fields. Then, the optimal layout includes a retinotopic map with local connections and a few long-range horizontal connections. A hypothetical rearrangement of neurons that minimizes the long-range projections would not save wiring because it would elongate the local connections. Therefore, wiring optimization seems more fundamental than locality of connections.

Q: How can one prove that the wiring is optimal in the brain?

A: Although evolution is known to optimize many aspects of animal design (Alexander 1996, Weibel 2000), the claim of design optimality is fraught with caveats. For example, animals may have been optimized for a lifestyle different from the one they lead today, or many design features are inherited from the ancestors; also, the role of chance in evolution cannot be excluded. However, the goal of the wiring optimization approach is not to prove the wiring is optimal, but rather to use the idea of optimality to understand as much brain architecture as possible (Parker & Maynard Smith 1990) without resorting to "historical accident" explanations, which often have an agnostic flavor. Understanding of brain design based on optimization satisfies our scientific curiosity and helps build a harmonic and self-consistent view of neurobiology.

Q: Wiring cost may play some role in brain design, but it cannot be the only factor. In view of this, how can one make any predictions based on wiring optimization alone?

A: Wiring optimization does not insist that wiring cost is the only constraint. It suggests that, other things being equal, evolution prefers the layout with minimum wiring cost. That this cost can be expressed quantitatively allowing one to generate experimentally testable predictions is a virtue of this theory. In those cases, where these predictions are confirmed, wiring optimization is likely to be the crucial factor. In other cases, where wiring optimization predictions disagree with the experimental results, some other factors may be more important.

To differentiate these cases, it is crucial to solve the optimal layout problems and compare their solutions with the experiment results. Therefore, wiring optimization is only the first step in the program of unraveling principles of brain design. Structural plasticity is another factor that has emerged already in this program (Stepanyants et al. 2002).

Q: If wiring optimization is so important, why is V1 in the back of the brain, as far as possible from the eyes?

A: Sensory input to the cortex (with the exception of olfaction) is required to pass through the thalamus, which is located roughly in the center of the head. Because all the locations on the cortex are roughly equidistant from the thalamus, placing V1 in the occipital lobe is not particularly detrimental. Moreover, the extreme posterior location of V1 may make sense because it connects with fewer visual areas (Felleman & Van Essen 1991).

Q: Could the length of connections be maximized (rather than minimized) by evolution?

A: One could imagine a task, such as binaural sound localization, that requires detecting short time intervals by introducing conduction delays. In such a system, the dynamic range would be broader if wires were longer. Although this may suggest wiring maximization, there is no contradiction with the wiring optimization theory. This theory optimizes wiring length among various functionally identical layouts, meaning the dynamic range should be fixed from the outset.

CONCLUSION

In this review, we argue that cortical maps reflect the connectivity properties of intracortical circuits as a consequence of wiring optimization. By formulating and solving optimal layout problems, we show that wiring optimization may account for the existence of cortical maps. The variety of map appearances follows naturally from the differences in intracortical connectivity. Therefore, the wiring optimization approach may provide a general unifying framework, which will help organize the multitude of experimental facts about cortical architecture. Although predictions of the wiring optimization theory are mostly consistent with experimental data, further testing is desirable. Detection of mismatches between the wiring optimization theory and experiment results will help investigators discover other principles of brain design.

ACKNOWLEDGMENTS

The authors thank Charles Stevens for his encouragement and support, Marcello Rosa, Nicholas Swindale and Geoffrey Goodhill for helpful comments on the manuscript, and numerous anonymous reviewers for forcing us to sharpen our arguments and for providing material for the FAQ. This research was supported by the Lita Annenberg Hazen Foundation, the Sloan Foundation, and the David and Lucile Packard Foundation.

The *Annual Review of Neuroscience* is online at http://neuro.annualreviews.org

LITERATURE CITED

Albright TD, Desimone R, Gross CG. 1984. Columnar organization of directionally selective cells in visual area MT of the macaque. *J. Neurophysiol.* 51:16–31

Alexander R. 1996. *Optima for Animals.* Princeton, NJ: Princeton Univ. Press

Allman JM. 1999. *Evolving Brains.* New York: Freeman

Allman JM, Kaas JH. 1974. The organization of the second visual area (V II) in the owl monkey: a second order transformation of the visual hemifield. *Brain Res.* 76:247–65

Attwell D, Laughlin SB. 2001. An energy budget for signaling in the grey matter of the brain. *J. Cereb. Blood Flow Metab.* 21:1133–45

Barlow HB. 1986. Why have multiple cortical areas? *Vision Res.* 26:81–90

Bell AJ, Sejnowski TJ. 1997. The "independent components" of natural scenes are edge filters. *Vision Res.* 37:3327–38

Blasdel GG. 1992. Orientation selectivity, preference, and continuity in monkey striate cortex. *J. Neurosci.* 12:3139–61

Bonhoeffer T, Grinvald A. 1991. Iso-orientation domains in cat visual cortex are arranged in pinwheel-like patterns. *Nature* 353:429–31

Bonhoeffer T, Kim DS, Malonek D, Shoham D, Grinvald A. 1995. Optical imaging of the layout of functional domains in area 17 and across the area 17/18 border in cat visual cortex. *Eur. J. Neurosci.* 7:1973–88

Bosking WH, Zhang Y, Schofield B, Fitzpatrick D. 1997. Orientation selectivity and the arrangement of horizontal connections in tree shrew striate cortex. *J. Neurosci.* 17:2112–27

Bray D. 1979. Mechanical tension produced by nerve cells in tissue culture. *J. Cell Sci.* 37:391–410

Cherniak C. 1992. Local optimization of neuron arbors. *Biol. Cybern.* 66:503–10

Cherniak C. 1994. Component placement optimization in the brain. *J. Neurosci.* 14:2418–27

Cherniak C. 1995. Neural component placement. *Trends Neurosci.* 18:522–27

Cherniak C, Changizi M, Won Kang D. 1999. Large-scale optimization of neuron arbors. *Phys. Rev. E Stat. Phys. Plasmas Fluids Relat. Interdiscip. Topics* 59:6001–9

Chklovskii DB. 2000a. Binocular disparity can explain the orientation of ocular dominance stripes in primate primary visual area (V1). *Vision Res.* 40:1765–73

Chklovskii DB. 2000b. Optimal sizes of dendritic and axonal arbors in a topographic projection. *J. Neurophysiol.* 83:2113–19

Chklovskii DB, Koulakov AA. 2000. A wire length minimization approach to ocular dominance patterns in mammalian visual cortex. *Physica A* 284:318–34

Chklovskii DB, Schikorski T, Stevens CF. 2002. Wiring optimization in cortical circuits. *Neuron* 34:341–47

Chklovskii DB, Stepanyants A. 2003. Power-law for axon diameters at branch point. *BMC Neurosci.* 4:18

Condron BG, Zinn K. 1997. Regulated neurite tension as a mechanism for determination of neuronal arbor geometries in vivo. *Curr. Biol.* 7:813–16

Cowey A. 1979. Cortical maps and visual perception: the Grindley Memorial Lecture. *Q. J. Exp. Psychol.* 31:1–17

Daniel PM, Whitteridge D. 1961. *J. Physiol. (Lond.)* 159:203–21

Das A, Gilbert CD. 1999. Topography of contextual modulations mediated by short-range interactions in primary visual cortex. *Nature* 399:655–61

Diogo AC, Soares JG, Koulakov A, Albright TD, Gattass R. 2003. Electrophysiological imaging of functional architecture in the cortical middle temporal visual area of Cebus apella monkey. *J. Neurosci.* 23:3881–98

Durbin R, Mitchison G. 1990. A dimension reduction framework for understanding cortical maps. *Nature* 343:644–47

Durbin R, Willshaw D. 1987. An analogue approach to the travelling salesman problem using an elastic net method. *Nature* 326:689–91

Felleman DJ, Van Essen DC. 1991. Distributed hierarchical processing in the primate cerebral cortex. *Cereb. Cortex* 1:1–47

Gilbert CD, Das A, Ito M, Kapadia M, Westheimer G. 1996. Spatial integration and cortical dynamics. *Proc. Natl. Acad. Sci. USA* 93:615–22

Gilbert CD, Wiesel TN. 1989. Columnar specificity of intrinsic horizontal and corticocortical connections in cat visual cortex. *J. Neurosci.* 9:2432–42

Girman SV, Sauve Y, Lund RD. 1999. Receptive field properties of single neurons in rat primary visual cortex. *J. Neurophysiol.* 82:301–11

Goodhill GJ, Cimponeriu A. 2000. Analysis of the elastic net model applied to the formation of ocular dominance and orientation columns. *Network* 11:153–68

Goodhill GJ, Lowel S. 1995. Theory meets experiment: correlated neural activity helps determine ocular dominance column periodicity. *Trends Neurosci.* 18:437–39

Goodhill GJ, Sejnowski TJ. 1997. A unifying objective function for topographic mappings. *Neural Comput.* 9:1291–303

Goodhill GJ, Willshaw DJ. 1990. Application of the elastic net algorithm to the formation of ocular dominance stripes. *Network* 1:41–59

Horton JC, Greenwood MM, Hubel DH. 1979. Non-retinotopic arrangement of fibres in cat optic nerve. *Nature* 282:720–22

Horton JC, Hocking DR. 1996. Intrinsic variability of ocular dominance column periodicity in normal macaque monkeys. *J. Neurosci.* 16:7228–39

Hubel DH, Freeman DC. 1977. Projection into the visual field of ocular dominance columns in macaque monkey. *Brain Res.* 122:336–43

Hubel DH, Wiesel TN. 1962. Receptive fields, binocular interaction and functional architecture in the cat's visual cortex. *J. Physiol.* 160:106–54

Jones DG, Van Sluyters RC, Murphy KM. 1991. A computational model for the overall pattern of ocular dominance. *J. Neurosci.* 11:3794–808

Katz LC, Gilbert CD, Wiesel TN. 1989. Local circuits and ocular dominance columns in monkey striate cortex. *J. Neurosci.* 9:1389–99

Katz MJ. 1985. How straight do axons grow? *J. Neurosci.* 5:589–95

Kim DS, Matsuda Y, Ohki K, Ajima A, Tanaka S. 1999. Geometrical and topological relationships between multiple functional maps in cat primary visual cortex. *Neuroreport* 10:2515–22

Klyachko VA, Stevens CF. 2003. Connectivity optimization and the positioning of cortical areas. *Proc. Natl. Acad. Sci. USA* 100:7937–41

Koulakov A, Chklovskii D. 2003. Ocular dominance patterns and the wire length minimization: a numerical study. http://arxiv.org/abs/q-bio.NC/0311027

Koulakov AA, Chklovskii DB. 2001. Orientation preference patterns in mammalian visual cortex: a wire length minimization approach. *Neuron* 29:519–27

Koulakov AA, Chklovskii DB. 2002. Direction of motion maps in the visual cortex: a wire length minimization approach. *Neurocomputing* 44–46:489–94

LeVay S, Connolly M, Houde J, Van Essen DC. 1985. The complete pattern of ocular dominance stripes in the striate cortex and visual field of the macaque monkey. *J. Neurosci.* 5:486–501

Malonek D, Tootell RB, Grinvald A. 1994. Optical imaging reveals the functional architecture of neurons processing shape and motion in owl monkey area MT. *Proc. R. Soc. Lond. B Biol. Sci.* 258:109–19

Miller KD, Keller JB, Stryker MP. 1989. Ocular dominance column development: analysis and simulation. *Science* 245:605–15

Mitchison G. 1991. Neuronal branching patterns and the economy of cortical wiring. *Proc. R. Soc. Lond. B Biol. Sci.* 245:151–58

Mitchison G. 1992. Axonal trees and cortical architecture. *Trends Neurosci.* 15:122–26

Murre JM, Sturdy DP. 1995. The connectivity of the brain: multi-level quantitative analysis. *Biol. Cybern.* 73:529–45

Nelson ME, Bower JM. 1990. Brain maps and parallel computers. *Trends Neurosci.* 13:403–8

Olshausen BA, Field DJ. 1996. Emergence of simple-cell receptive field properties by learning a sparse code for natural images. *Nature* 381:607–9

Parker G, Maynard Smith J. 1990. Optimality theory in evolutionary biology. *Nature* 348:27–33

Rall W, Burke RE, Holmes WR, Jack JJ, Redman SJ, Segev I. 1992. Matching dendritic neuron models to experimental data. *Physiol. Rev.* 72:S159–86

Ramón y Cajal S. 1999. *Texture of the Nervous System of Man and the Vertebrates.* New York: Springer. 631 pp.

Roerig B, Kao JP. 1999. Organization of intracortical circuits in relation to direction preference maps in ferret visual cortex. *J. Neurosci.* 19:RC44

Rosa MG, Gattass R, Fiorani M, Soares JG. 1992. Laminar, columnar and topographic aspects of ocular dominance in the primary visual cortex of Cebus monkeys. *Exp. Brain Res.* 88:249–64

Ruppin E, Schwartz EL, Yeshurun Y. 1993. Examining the volume efficiency of the cortical architecture in a multi-processor network model. *Biol. Cybern* 70:89–94

Rushton WA. 1951. Theory of the effects of fibre size in medullated nerve. *J. Physiol.* 115:101–22

Schummers J, Marino J, Sur M. 2002. Synaptic integration by V1 neurons depends on location within the orientation map. *Neuron* 36:969–78

Shefi O, Harel A, Chklovskii DB, Ben Jacob E, Ayali A. 2002. Growth morphology of two-dimensional insect neural networks. *Neurocomputing* 44–46:635–43

Shmuel A, Grinvald A. 1996. Functional organization for direction of motion and its relationship to orientation maps in cat area 18. *J. Neurosci.* 16:6945–64

Shmuel A, Grinvald A. 2000. Coexistence of linear zones and pinwheels within orientation maps in cat visual cortex. *Proc. Natl. Acad. Sci. USA* 97:5568–73

Simon DK, O'Leary DD. 1991. Relationship of retinotopic ordering of axons in the optic pathway to the formation of visual maps in central targets. *J. Comp. Neurol.* 307:393–404

Stepanyants A, Hof PR, Chklovskii DB. 2002. Geometry and structural plasticity of synaptic connectivity. *Neuron* 34:275–88

Swindale NV. 1980. A model for the formation of ocular dominance stripes. *Proc. R. Soc. Lond. B Biol. Sci.* 208:243–64

Swindale NV, Cynader MS, Matsubara J. 1990. Cortical cartography: a two-dimensional view. In *Computational Neuroscience*, ed. E Schwartz, pp. 232–341. Cambridge, MA: MIT Press

Swindale NV. 1996. The development of topography in the visual cortex: a review of models. *Netw.: Comput. Neural Syst.* 7:161–247

Swindale NV. 2001. Keeping the wires short: a singularly difficult problem. *Neuron* 29:316–17

Swindale NV, Grinvald A, Shmuel A. 2003. The spatial pattern of response magnitude and selectivity for orientation and direction in cat visual cortex. *Cereb. Cortex* 13:225–38

Swindale NV, Matsubara JA, Cynader MS. 1987. Surface organization of orientation and direction selectivity in cat area 18. *J. Neurosci.* 7:1414–27

Tessier-Lavigne M, Goodman CS. 1996. The molecular biology of axon guidance. *Science* 274:1123–33

Van Essen DC. 1997. A tension-based theory

of morphogenesis and compact wiring in the central nervous system. *Nature* 385:313–18

von Berg J. 1997. Mapping of the cortical ocular dominance pattern onto the visual field http://www.informatik.uni-hamburg.de/GRK/Personen/Jens/online-poster-ecvp97/poster.html

Weibel E. 2000. *Symmorphosis: On Form and Function in Shaping Life*. Cambridge, MA: Harvard Univ. Press

Weliky M, Bosking WH, Fitzpatrick D. 1996. A systematic map of direction preference in primary visual cortex. *Nature* 379:725–28

Wolf F, Geisel T. 1998. Spontaneous pinwheel annihilation during visual development. *Nature* 395:73–78

Young MP, Scannell JW. 1996. Component-placement optimization in the brain. *Trends Neurosci.* 19:413–15

Yousef T, Bonhoeffer T, Kim DS, Eysel UT, Toth E, Kisvarday ZF. 1999. Orientation topography of layer 4 lateral networks revealed by optical imaging in cat visual cortex (area 18). *Eur. J. Neurosci.* 11:4291–308

Zeck G, Fromherz P. 2001. Noninvasive neuroelectronic interfacing with synaptically connected snail neurons immobilized on a semiconductor chip. *Proc. Natl. Acad. Sci. USA* 98:10457–62

Annu. Rev. Neurosci. 2004. 27:393–418
doi: 10.1146/annurev.neuro.26.041002.131128

First published online as a Review in Advance on February 26, 2004

ELECTRICAL SYNAPSES IN THE MAMMALIAN BRAIN

Barry W. Connors[1] and Michael A. Long[2]
[1]Department of Neuroscience, Brown University, Providence, Rhode Island 02912; email: BWC@Brown.edu
[2]Department of Brain and Cognitive Sciences, Massachusetts Institute of Technology, Cambridge, Massachusetts 02139; email: miclong@mit.edu

Key Words gap junction, electrotonic synapse, electrical coupling, connexin, connexin36

■ **Abstract** Many neurons in the mammalian central nervous system communicate through electrical synapses, defined here as gap junction–mediated connections. Electrical synapses are reciprocal pathways for ionic current and small organic molecules. They are often strong enough to mediate close synchronization of subthreshold and spiking activity among clusters of neurons. The most thoroughly studied electrical synapses occur between excitatory projection neurons of the inferior olivary nucleus and between inhibitory interneurons of the neocortex, hippocampus, and thalamus. All these synapses require the gap junction protein connexin36 (Cx36) for robust electrical coupling. Cx36 appears to interconnect neurons exclusively, and it is expressed widely along the mammalian neuraxis, implying that there are undiscovered electrical synapses throughout the central nervous system. Some central neurons may be electrically coupled by other connexin types or by pannexins, a newly described family of gap junction proteins. Electrical synapses are a ubiquitous yet underappreciated feature of neural circuits in the mammalian brain.

INTRODUCTION

The mammalian brain excels at rapid information processing, thanks to its massively parallel architecture and the high operational speed (by biological standards) of its neuronal elements. Neurons generate rapid signals by controlling ionic current flow across their membranes. The most common mechanism for signaling between neurons is the neurotransmitter-releasing chemical synapse. Faster and simpler signaling can be achieved with electrical synapses, specialized junctions that allow ionic current to flow directly between neurons. The vast majority of electrical synapses are membrane-to-membrane appositions called gap junctions, which are clusters of transcellular channels composed of protein subunits termed connexins. This type of electrical synapse, also known as the electrotonic synapse, can mediate electrical coupling between cells, and it has functional properties strikingly different from those of chemical synapses (Bennett 1977). Most

0147-006X/04/0721-0393$14.00

notably, vertebrate electrical synapses are bidirectional. With their speed, simplicity, and reciprocity, electrical synapses are a unique feature of neuronal circuits in the mammalian brain. Nevertheless, the prevalence of electrical synapses has been recognized only in the past few years.

The notion that neurons communicate electrically is almost as old as the idea of bioelectricity itself (Eccles 1982, Cowan & Kandel 2001). Studies of crayfish and shrimp neurons offered the first compelling evidence for electrical synapses (Furshpan & Potter 1957, Watanabe 1958), and electrical synapses were revealed in the vertebrate central nervous system of teleost fish soon after (Bennett et al. 1959). Electrical synapses in mammalian brains proved much harder to find. Single-cell recordings provided the first strong evidence for mammalian electrical synapses in the mesencephalic nucleus of cranial nerve V (Hinrichsen 1970, Baker & Llinás 1971), the vestibular nucleus (Wylie 1973, Korn et al. 1973), and the inferior olivary nucleus (Llinás et al. 1974). The most convincing way to demonstrate electrotonic coupling is to record intracellularly from two neighboring cells simultaneously, a procedure that is exceptionally difficult to perform in the intact brain. Studies of electrical synapses in the mammalian brain languished for years, even as breakthroughs were made in the general physiology of gap junctions (Spray et al. 1979, 1981) and the molecular biology of connexins (Willecke et al. 2002).

Electrical synapses are now being intensively examined in mammals, thanks to recent technical developments in electrophysiology, isolated brain slice preparations, cell labeling and imaging, molecular cloning, and transgenics. In this review, we examine the burgeoning evidence that electrical synapses function throughout the mammalian brain and describe pertinent studies of nonmammalian and nonneuronal gap junctions. We do not review the roles of gap junctions in early neural development and the functions of glia, and we do not discuss gap junction–free forms of electrical communication (Jefferys 1995).

STRUCTURE AND MOLECULAR COMPOSITION OF ELECTRICAL SYNAPSES

Gap Junctions and Connexins

The most visible and common structural correlate of electrical synapses is the gap junction, which is seen most readily with electron microscopy using the freeze-fracture technique. Gap junctions are clusters of connexin-containing channels that are coextensive across regions of apposing membranes of coupled cells (Evans & Martin 2002). A gap of extracellular space separates the two membrane leaflets, usually by ~2–3 nm. Gap junctions between neurons have been observed in the majority of cases where electrophysiological evidence for electrical synapses is well established.

Neuronal gap junctions are often synonymous with electrical synapses, but there are apparent limitations to this assumption. Tissue preparation can both destroy bona fide gap junctions and create spurious gap junction–like structures (Brightman & Reese 1969). Gap junctions are notoriously difficult to observe, and the absence of evidence for gap junctions cannot be construed as evidence for the absence of electrical coupling. Gap junction channels may be functional even when they are too widely dispersed to form conventional gap junctions (Williams & DeHaan 1981). Conversely, electrotonic coupling may be absent despite the presence of virtually normal-looking gap junctions (Raviola & Gilula 1973, De Zeeuw et al. 2003).

Connexins are a family of proteins with ~20 isoforms in humans and mice (Willecke et al. 2002). Across species there are ~40 connexin orthologues. In the most commonly used nomenclature, connexins are named for their predicted molecular weights (e.g., Cx36 has a mass of ~36 kDa). Each gap junction hemichannel, also known as a connexon, is a connexin hexamer. Most cells can express multiple connexins, and connexons can be homomeric or heteromeric. Only some combinations of homomeric connexons can form functional heterotypic channels; Cx36 may function only homotypically (Al-Ubaidi et al. 2000, Teubner et al. 2000).

Approximately half of the mammalian connexins are abundant in the central nervous system. Several are strongly expressed in astrocytes (Cx26, Cx30, Cx43) and oligodendrocytes (Cx29, Cx32, Cx47) (Nagy et al. 2001, Menichella et al. 2003, Odermatt et al. 2003), but most are not expressed in neurons (Rash et al. 2001a,b; Odermatt et al. 2003). Cx36, which is the mammalian orthologue of fish Cx35 (O'Brien et al., 1998), is expressed widely in the brain (Condorelli et al. 2000). Data from freeze-fracture replica immunolabeling indicate that Cx36 protein appears in neuron-neuron gap junctions but not in gap junctions between astrocytes and/or oligodendrocytes (Rash et al. 2001a,b). Single-cell reverse transcription (RT)-PCR shows that the Cx36 message is consistently present in neurons of the hippocampus and neocortex (Venance et al. 2000). When the gene for Cx36 is knocked out in mice, electrical coupling that normally occurs between certain neurons in the retina (Deans et al. 2002) and in the neocortex (Deans et al. 2001, Blatow et al. 2003), hippocampus (Hormuzdi et al. 2001), thalamic reticular nucleus (TRN) (Landisman et al. 2002), and inferior olivary nucleus (Long et al. 2002, DeZeeuw et al. 2003) is eliminated or profoundly reduced. If Cx36 is not the exclusive connexin involved in coupling these neurons, it is at least a necessary constituent.

As prevalent as Cx36 seems to be, it is probably not the only connexin involved in mammalian electrical synapses. Neurons known to be electrically coupled, but which may not express Cx36, include those of the locus coeruleus (Christie et al. 1989, Alvarez et al. 2002), the horizontal cells of the retina (Deans & Paul 2001), and perhaps the pyramidal cells in the hippocampus (MacVicar & Dudek 1981, 1982; Draguhn et al. 1998; Schmitz et al. 2001). Cx45 is expressed in neurons of olfactory epithelium and bulb (Zhang & Restrepo 2002), in horizontal cells (D.L. Paul, personal communication), and in other brain regions (Maxeiner et al. 2003), and it may be a neuronal gap junction protein. Despite an initial claim that Cx47 is expressed by central neurons (Teubner et al. 2001), it is now clear that Cx47 is a product of oligodendrocytes and not of neurons (Menichella et al. 2003, Odermatt et al. 2003).

Innexins and Pannexins

Connexins may not be the only channel-forming proteins in the electrical synapses of the mammalian brain. Clues for this possibility originated with studies of invertebrates, where electrical synapses are pervasive and indeed where they were discovered. The nervous system of the nematode *Caenorhabditis elegans*, for example, has 302 neurons with ~600 gap junctions interconnecting them (White et al. 1986). Invertebrate gap junctions are similar to vertebrate gap junctions, yet connexins have been found only in vertebrate species. The genomes of *Drosophila melangaster* and *C. elegans* have now been entirely sequenced, and they do not have connexin-like sequences.

A family of genes unrelated to connexins, called the innexins (invertebrate connexins), codes for the proteins in the gap junctions of *Drosophila*; *C. elegans*; and species of Mollusca, Annelida, and Platyhelminthes (Phelan & Starich 2001). Innexin proteins form functional gap junction channels (Landesman et al. 1999). More recently, innexin-like genes were discovered in mammals (Panchin et al. 2000). The mammalian genes are called pannexins (Px) (pan meaning everything). Apart from two conserved cysteine residues in their extracellular loops, connexins have little sequence similarity to the innexins and pannexins, yet the overall topologies of connexin and pannexin subunits are remarkably alike (Hua et al. 2003). The function of pannexins in mammals is currently unknown. There is, however, clear expression of Px1 and Px2 mRNA in certain neurons, including pyramidal cells and interneurons of the hippocampus, and expression of Px1 in pairs of *Xenopus* oocytes forms robust and nearly voltage-independent intercellular channels (Bruzzone et al. 2003). Whether pannexin-dependent electrical synapses exist among vertebrate neurons remains to be determined.

DISTRIBUTION OF ELECTRICAL SYNAPSES IN THE BRAIN

On the basis of the distribution of Cx36 expression alone, it seems likely that electrical synapses occur in every major region of the central nervous system (Condorelli et al. 2000), although compelling functional and morphological data have been collected for only a few areas. Here we highlight examples for which the evidence is strongest and the prospective functions are most interesting (Figure 1).

Inferior Olivary Nucleus

Neurons in the inferior olivary nucleus are the source of climbing fiber input to the cerebellar cortex. Thirty years ago, using both ultrastructural (Sotelo et al. 1974) and indirect electrophysiological (Llinás et al. 1974) evidence, researchers made a strong case for electrotonic interconnections between neurons of the cat inferior olive. Subsequent work used paired intracellular recordings to demonstrate

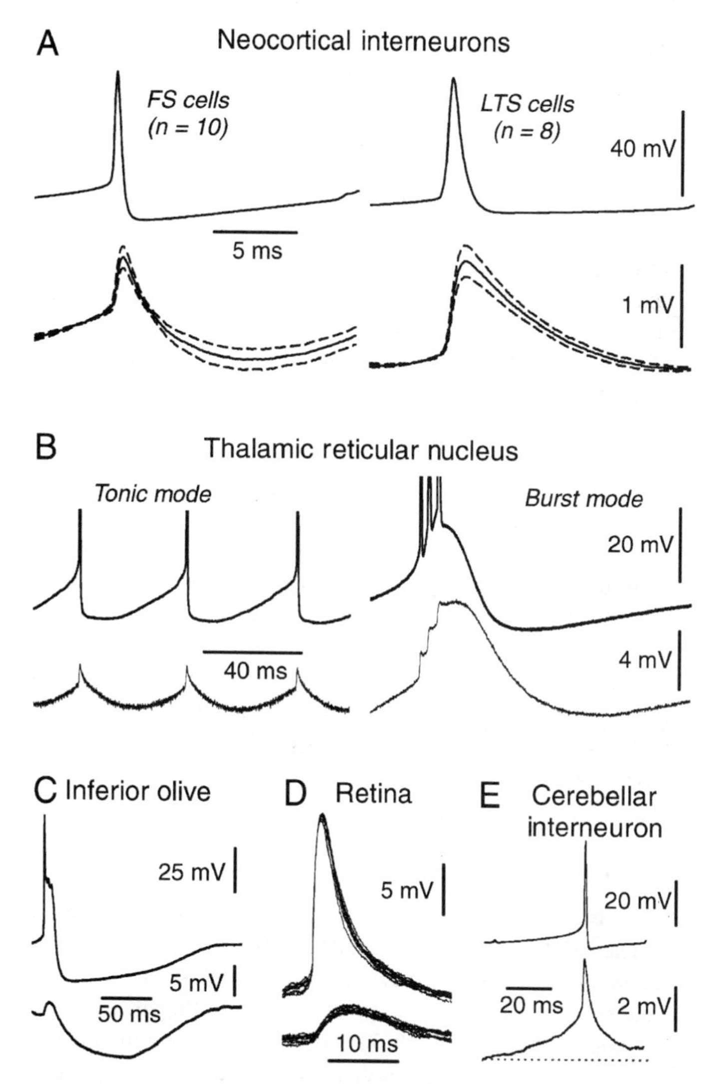

Figure 1 Electrical postsynaptic potentials in neurons from the mammalian brain. Each case illustrates simultaneous intracellular recordings from a pair of similar neurons, with presynaptic action potentials above and electrical postsynaptic potentials below. (*A*) Two types of inhibitory interneurons from the neocortex: fast-spiking (FS) cells and low threshold–spiking (LTS) cells. Traces are averaged from ten and eight neuron pairs, respectively, and the dashed lines are ±SD (J.R. Gibson, M. Beierlein, B.W. Connors, unpublished report). (*B*) Recordings from a pair of thalamic reticular neurons in tonically spiking mode (*left*) and bursting mode (*right*). Action potentials are truncated (from Long et al. 2004). (*C*) Inferior olivary neurons (MA Long, unpublished report). (*D*) AII amacrine neurons from the retina (Veruki & Hartveit 2002a). (*E*) Cerebellar interneurons (Mann-Metzer & Yarom 1999).

electrical synapses directly in the inferior olive of guinea pig (Llinás & Yarom 1981), rat (Devor & Yarom 2002), and mouse (Long et al. 2002) in vitro. Injections of Lucifer yellow (Benardo & Foster 1986) or neurobiotin (Devor & Yarom 2002) yielded dye coupling. The gap junctions between olivary neurons appear to contain Cx36, as demonstrated by both freeze-fracture immunolabeling (Rash et al. 2000) and thin-section electron microscopic immunocytochemistry (Teubner et al. 2000).

Many inferior olivary neurons have an unusual propensity to generate large, spontaneous, synchronous, subthreshold fluctuations of membrane voltage at 2–8 Hz (Figure 2*A*) (Benardo & Foster 1986, Devor & Yarom 2002). Indirect evidence implies that these rhythms emerged from an electrically coupled network of neurons that are, when uncoupled, inherently quiescent (Manor et al. 2000). This was recently tested in mice with a null mutation for Cx36. Both the prevalence and the strength of electrical synapses were dramatically reduced in Cx36 knockout mice, as expected, but spontaneous subthreshold rhythms were similar in size, shape, and frequency to those of wild-type animals (see Figure 2*A*,*B*,*C*) (Long et al. 2002, De Zeeuw et al. 2003). Subthreshold rhythms and the spikes they evoked were strongly synchronized among neighboring cells in the wild-type olive; however, synchrony was abolished in the Cx36 knockout mouse (Figure 2*D*). These results imply that electrical synapses are necessary for the synchronization, but not the generation, of olivary rhythms. An alternate interpretation is that the uncoupled olivary neurons of the developing Cx36 knockout mouse express an abnormal complement of ion channels that yields intrinsic rhythmicity in individual neurons (De Zeeuw et al. 2003).

Thalamic Reticular Nucleus

The TRN is a thin layer of GABAergic neurons that surround and inhibit the relay nuclei of the dorsal thalamus. The TRN receives excitatory synapses from both thalamocortical and corticothalamic collaterals. It can influence the activity of the entire thalamocortical system, and it participates in rhythmic forebrain activity (McCormick & Bal 1997). TRN neurons have mutually inhibitory connections, but evidence suggests that this is not the whole story. TRN neurons have dendritic bundles and specialized dendrodendritic contacts (Ohara 1988). In situ hybridization showed heavy labeling for Cx36 mRNA in the TRN but not in relay nuclei (Condorelli et al. 2000). Cx36-like immunoreactivity was seen in all neurons of the TRN, but gap junctions remain elusive (Liu & Jones 2003). The most definitive evidence for electrical synapses was obtained from simultaneous recordings of neighboring pairs of rat and mice TRN neurons (Figure 1*B*) (Landisman et al. 2002). The prevalence, strength, biophysical properties, and Cx36-dependence of electrical connections between TRN neurons (Landisman et al. 2002, Long et al. 2004) are almost indistinguishable from those between inhibitory interneurons of the neocortex (Gibson et al. 1999, Deans et al. 2001).

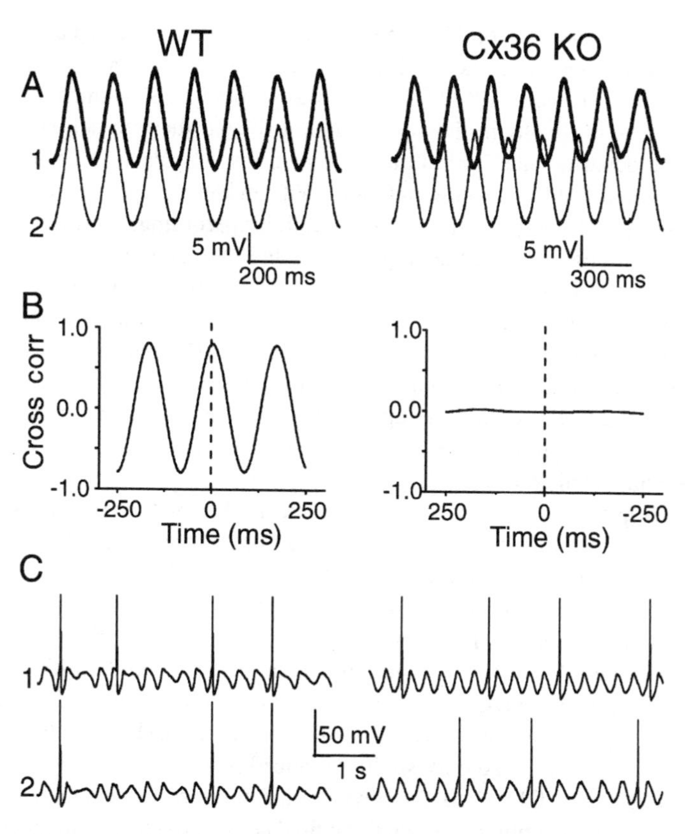

Figure 2 Subthreshold and suprathreshold synchrony in electrically coupled inferior olivary neurons. Left panels are from wild-type (WT) mice. Right panels are from Cx36 knockout (KO) mice. (*A*) Paired whole-cell recordings showing spontaneous subthreshold rhythmic activity in WT and KO cells. (*B*) Cross-correlograms demonstrate very high synchrony among the WT cells and no synchrony among the KO cells. (*C*) Subthreshold rhythms evoke occasional, but highly correlated, spikes in the WT and entirely uncorrelated spikes in the KO cells (from Long et al. 2002).

Electrical coupling in the TRN seems to be restricted to cells no more than 40 μm apart (Long et al. 2003). In this sense, coupling in the TRN resembles the spatially localized coupling in the inferior olivary nucleus (Devor & Yarom 2002). Furthermore, when subthreshold rhythms of ~10 Hz were induced in TRN neurons with an agonist of metabotropic glutamate receptors, rhythms were synchronized only among small, closely adjacent clusters of coupled neurons (Long et al. 2003). Electrical synapses may coordinate local groups of TRN neurons, whereas more distant interactions within the TRN may occur via inhibitory connections.

Hippocampus

The excitatory cells of the hippocampus have been implicated in a variety of electronic interactions, some of which may involve electrical synapses (Jefferys 1995). Spencer & Kandel (1961) observed "fast prepotentials"—small, spike-like events of a few millivolts in the soma—and attributed them to action potentials from distal dendritic sites within the same neurons. MacVicar & Dudek (1981, 1982) suggested a different interpretation. With evidence from dye coupling, indirect electrophysiological tests, dual-intracellular recordings, and freeze-fracture gap junction observations, they argued that there are dendrodendritic and dendrosomatic electrical synapses between pyramidal cells in the CA3 region as well as between granule cells of the dentate gyrus. More recently, computer simulations of high-frequency oscillations led to suggestions that pyramidal cells may be electrotonically coupled through axoaxonal gap junctions (Draguhn et al. 1998, Traub et al. 2002). Physiological support for this idea now includes dye coupling and the successive propagation of antidromic spikelets from axon to soma to dendrite (Schmitz et al. 2001).

Despite its long history, the hypothesis that mature pyramidal and granule cells interact via electrical synapses is not quite compelling. Recent studies using paired whole-cell recordings have repeatedly revealed electrical synapses among inhibitory interneurons (see below) but not among excitatory cells. Although Cx36 mRNA has been observed in CA3 pyramidal cells (Condorelli et al. 1998, Venance et al. 2000), Cx36 protein has not been found. Fast hippocampal oscillations thought to require axoaxonal coupling were unaffected by the absence of Cx36 in two different studies of one knockout mouse (Hormuzdi et al. 2001, Buhl et al. 2003) and were only slightly reduced in a different Cx36 knockout mouse (Maier et al. 2002). It is conceivable that some other connexin, or a pannexin, accounts for coupling between pyramidal cells. Axoaxonal gap junctions have not been observed in the hippocampus (or any other cerebral cortical region). Although ultrastructural evidence for gap junctions between interneurons is abundant, reports of gap junctions between excitatory cells of the hippocampus are scant and suffer from uncertainties about the cell types involved. In short, the evidence that hippocampal pyramidal cells are electrically coupled is intriguing but limited.

Evidence that inhibitory interneurons of the hippocampus are interconnected by electrical synapses is more persuasive. Dendrodendritic gap junctions between interneurons are frequently seen in areas CA1 and CA3 (Kosaka & Hama 1985) and in the dentate gyrus (Kosaka 1983). Several types of gap junction–coupled interneurons have been identified (Katsumaru et al. 1988, Fukuda & Kosaka 2000), and dye coupling has been observed between inhibitory cells of the hilus and the CA1 (Michelson & Wong 1994). Paired-interneuron recordings have shown electrical coupling directly, and single-cell RT-PCR revealed Cx36 mRNA (Venance et al. 2000). Furthermore, although electrical coupling between pairs of interneurons was abundant in the CA3 and dentate regions of wild-type mice, it was absent in cells of Cx36 knockout mice (Hormuzdi et al. 2001).

The functions of electrical synapses between hippocampal interneurons are unknown. Most studies have focused on the possibility that they play a role in generating or modulating synchronous oscillations or seizure-like activity. Some researchers used gap junction–blocking drugs as a diagnostic test for gap junctions (e.g., Draguhn et al. 1998, Skinner et al. 1999, Jahromi et al. 2002), but the interpretation of these experiments must be tempered because of the notoriously nonspecific effects of the drugs (see below). Measurements in Cx36 knockout mice have implicated electrical synapses in the generation of gamma-frequency (30–70 Hz) rhythms, but not of fast ripples (140–200 Hz) or slower theta (5–10 Hz) rhythms (Hormuzdi et al. 2001, Traub et al. 2003, Buhl et al. 2003).

Neocortex

In the mature neocortex, as in the hippocampus, the case for electrical synapses between interneurons is currently much stronger than the case for electrical synapses between excitatory principal neurons (Galarreta & Hestrin 2001a). The first evidence for neocortical coupling came from studies of the mature primate sensorimotor cortex. Beautiful electron micrographs by Sloper (1972) and Sloper & Powell (1978) show numerous dendrodendritic and dendrosomatic gap junctions with features that strongly suggest inhibitory interneurons. In an elegant corroboration of this work, Fukuda & Kosaka (2003) recently described dendrodendritic gap junctions between the parvalbumin-immunolabeled interneurons of primary somatosensory, auditory, and visual areas in mature rats. Despite decades of ultrastructural investigations of the neocortex (Peters 2002), convincing gap junctions involving excitatory neurons in the neocortex have been described very rarely (see Peters 1980).

Recent electrophysiological studies strongly reinforce the ultrastructural evidence: Inhibitory cells are often coupled, whereas mature excitatory cells are rarely (if ever) coupled. Paired whole-cell recordings in rat neocortex showed that parvalbumin-expressing, fast-spiking (FS) interneurons are coupled to one another frequently ($>60\%$ of tested pairs) and strongly (mean coupling coefficients ~ 0.07, ranging as high as 0.4; mean cell-cell coupling conductances ~ 1.6 nS, ranging up to 5.5 nS) (Figure 1*A*) (Galarreta & Hestrin 1999, Gibson et al. 1999). Electrical coupling between FS cells is not limited to the adolescent ages commonly studied in vitro; it persists in fully mature mice (Galarretta & Hestrin 2002). Using intracellular recording and staining followed by electron microscopy, Tamás and colleagues (Tamás et al. 2000) confirmed that electrically coupled FS interneurons, as well as electrically coupled non-FS interneurons (called regular-spiking non-pyramidal cells by Szabadics et al. 2001), formed dendrodendritic gap junctions.

A striking feature of neocortical circuitry is the variety of distinct types of GABAergic interneurons in the neocortex. In both rats (Gibson et al. 1999) and mice (Deans et al. 2001), somatostatin-expressing interneurons of a type called low threshold–spiking (LTS) cells were often electrically coupled to each other (Figure 1*A*), but they were coupled only occasionally and weakly to the FS interneurons. Another type of interneuron, termed the multipolar bursting (MB) cell, was also coupled to cells of the same type but not to FS cells (Blatow et al. 2003). In layer I of the neocortex, the late-spiking (LS) inhibitory interneuron made electrical synapses to other LS cells 83% of the time, but it coupled to non-LS interneurons only 2% of the time (Chu et al. 2003). Thus, the surprising evidence to date suggests that the large majority of neuronal gap junctions in the neocortex interconnect inhibitory interneurons of the same type. In effect, each type of interneuron forms an extended, gap junction–coupled network that is nearly independent, electrotonically, of each other type of coupled interneuron network.

Electrically coupled networks of interneurons are large. The probability and strength of coupling falls with intersomatic distance, and beyond 200 μm no coupling has been detected between pairs of either FS or LTS cells (Amitai et al. 2002). Considering this spatial profile of coupling along with measures of the density of interneurons, one infers that each interneuron is coupled to between 20 and 40 neighboring interneurons. It is not known whether such interneuron syncytia extend indefinitely across the cortical mantle, or whether they have distinct boundaries.

Direct tests of electrical coupling between pairs of excitatory cells (i.e., pyramidal cells or spiny stellate cells) or between pairs of excitatory and inhibitory cells in the neocortex have yielded generally negative results (Galarretta & Hestrin 1999, Gibson et al. 1999), and other studies using closely spaced paired-cell recordings from excitatory neurons have not reported coupling (e.g., Thomson & Deuchars 1997). This is consistent with the meager ultrastructural evidence for pyramid-pyramid gap junctions. The presence of Cx36 mRNA in some spiny neocortical neurons (Venance et al. 2000) suggests that Cx36 may serve as the substrate for the rare electrical coupling seen between interneuron-pyramidal cell pairs in the immature cortex (Venance et al. 2000, Meyer et al. 2002; cf. Galarretta & Hestrin 1999, Gibson et al. 1999). An early report showed dye coupling between mature pyramidal neurons (Gutnick & Prince 1981), but studies during development emphasized its postnatal transience (Connors et al. 1983, Peinado et al. 1993, Rorig et al. 1995, Bittman et al. 1997). Furthermore, dye coupling may not be a reliable measure of gap junction coupling among postnatal cortical neurons (Knowles et al. 1982, Connors et al. 1984, Gutnick et al. 1985, Rorig et al. 1996).

Cx36 appears to be necessary for the large majority of electrical coupling among the three types of neocortical interneurons tested to date. Single-cell RT-PCR showed consistent expression of Cx36 in several types of interneurons (Venance et al. 2000), and a histochemical reporter enzyme driven by the Cx36 promoter identified a variety of interneurons that included both parvalbumin- and somatostatin-expressing cells (Deans et al. 2001). When electrical coupling was tested in Cx36 knockout mice, it was nearly absent among FS interneurons (Deans et al. 2001, Hormuzdi et al. 2001), LTS interneurons (Deans et al. 2001), and MB interneurons (Blatow et al. 2003). Occasional weak coupling detected in some interneuron pairs of Cx36 knockout mice implies that another connexin or a pannexin may account for a small fraction of the electrical synapses in the neocortex.

FS interneurons in vivo often display remarkably tight synchrony of spiking under physiological conditions (Swadlow 2003), and electrical coupling may account for some of this synchrony. When coupled pairs of FS neurons are stimulated in vitro, in the absence of functional chemical synapses, their electrical synapses alone can mediate robust, temporally close spike synchrony (Galarreta & Hestrin 1999, Gibson et al. 1999, Mancilla et al. 2003). Many FS pairs are connected by both electrical and GABAergic synapses, and this combination may actually facilitate synchronous states under some conditions (Támas et al. 2000, Galarreta & Hestrin 2001b, Lewis & Rinzel 2003). Synchronous activity can also occur among much larger cohorts of spatially extended interneurons. For example, agonists of mGluRs or muscarinic cholinergic receptors excite LTS cells and induce close spike synchrony and more widely correlated subthreshold rhythmic fluctuations (Beierlein et al. 2000). This type of synchronous activity is greatly reduced among the LTS cells of Cx36 knockout mice, implying that it depends on electrical synapses (Deans et al. 2001). Similarly, muscarinic receptor–induced rhythms in MB interneurons also require electrical coupling and Cx36, although synchronous activity in this system also depends on intact GABAergic synapses (Blatow et al. 2003).

Other Regions of the Central Nervous System

The retina is a famously coupled place, where nearly all types of neurons participate in electrical networks (Vaney 2002). Both homologous and heterologous electrical synapses occur in the retina. Paired-cell recordings of AII amacrine cells showed that their electrical synapses are functionally similar to those between neocortical interneurons (Figure 1*D*) (Veruki & Hartveit 2002a). Heterologously coupled pairs of AII amacrine cells and ON bipolar cells have a functional asymmetry in electrical coupling strength that is most likely due to the different input impedances of the two cell types (Veruki & Hartveit 2002b). When Cx36 is knocked out in mice, scotopic vision is strongly impaired (Guldenagel et al. 2001) because obligatory Cx36-dependent electrical synapses are lost between rods and cones and between AII amacrine cells and bipolar cells (Deans et al. 2002).

The olfactory bulb is, like the retina, a wonderland of extraordinary synaptic relationships that include electrical synapses. Neurons in both the bulb and the olfactory epithelium express Cx36 (Zhang & Restrepo 2003), and gap junctions appear between granule cells (Reyher et al. 1991). Mitral cells with dendrites in the same glomerulus of the bulb are probably coupled, whereas mitral cells projecting to different glomeruli are never coupled (Schoppa & Westbrook 2002). Electrical coupling in the olfactory bulb may play a role in coordinating oscillations (Friedman & Strowbridge 2003).

The locus coeruleus is a small cluster of widely projecting noradrenergic brainstem neurons implicated in the modulation of arousal and attention (Usher et al.

1999). There is strong evidence that electrical synapses synchronize subthreshold rhythms generated in the locus coeruleus of rodents (Christie et al. 1989, Alverez et al. 2002).

Electrical synapses occur at all levels of the mammalian motor system. Cx36 immunoreactivity is present within the neuronal gap junctions of the immature rat spinal cord (Rash et al. 2001b). Young motor neurons are electrically coupled (Fulton et al. 1980) in functionally relevant clusters (Walton & Navarrete 1991). Although motor neuron coupling declines steeply during the first postnatal week, Cx36 is still expressed in the adult (Chang et al. 1999). The progressive loss of gap junctions between developing motor neurons may reduce their correlated activity, which in turn may trigger synaptic competition between neuromuscular inputs (Personius & Balice-Gordon 2001). Electrical synapses may also help to synchronize spinal locomotor rhythms (Tresch & Kiehn 2000) and respiratory rhythm-generating neurons in the brainstem (Rekling et al. 2000). Higher in the motor system, within the neostriatum, electrical synapses occur between GABAergic local interneurons (Koos & Tepper 1999) and between the output cells (the medium spiny neurons) (Venance et al. 2003). The inhibitory interneurons of the cerebellar cortex are also coupled (Sotelo & Llinás 1972, Mann-Metzer & Yarom 1999). In each of these circuits, electrical synapses may enhance synchronous neuronal activity.

The literature abounds with additional claims for electrical synapses in the brain. Most of the evidence for these synapses is less than airtight. On the basis of the trends described above, however, persuasive evidence for electrical synapses may soon accrue for all brain regions.

FUNCTIONAL PROPERTIES OF GAP JUNCTIONS

Connexin Channels

The gating and permeation characteristics of gap junction channels have been extensively studied and reviewed (e.g., Harris 2001, Sáez et al. 2003). The salient properties of these channels are that the pores are wide (12–14 Å), single-channel conductances are variable and often large (10–300 pS), ion selectivity is relatively poor, and moderately large organic molecules (including tracers such as Lucifer yellow, neurobiotin, and fluorescien derivatives, and endogenous substances such as cAMP, cGMP, IP_3, glucose, and Ca^{2+}) can often permeate. Each of these traits varies broadly across connexin subtypes.

The central neuron-specific connexin, Cx36, has the smallest single-channel conductance of any connexin described, ~10–15 pS (Srinivas et al. 1999). Cx36 channels may be particularly impermeable to positively charged dyes such as neurobiotin (287 Da) (Teubner et al. 2000). Most attempts to produce dye coupling with Lucifer yellow or neurobiotin in demonstrably electrically coupled central neurons have failed (e.g., Gibson et al. 1999, Landisman et al. 2002), although some dye coupling between neocortical interneurons has been reported (Connors et al. 1983, Benardo 1997). Dye coupling across Cx36-dependent gap junctions may be limited to systems that are strongly coupled, such as the AII amacrine cells of the retina (Deans et al. 2002).

Estimates based on biophysical data and ultrastructure imply that the number of connexin channels open at any given time is low (Lin & Faber 1988, Pereda et al. 2003). For example, on the basis of their size, the gap junctions interconnecting mature interneurons of the neocortex have ~150–380 connexin channels (Fukuda & Kosaka 2003). If we assume these are Cx36 channels (Deans et al. 2001) with a unitary channel conductance of 14 pS (Teubner et al. 2000) and a mean junctional conductance of ~0.2 nS in the mature cortex (Galarretta & Hestrin 2002), only ~4–9% of junctional channels are generally open. Immature (2–3 weeks of age) rat neocortical interneurons are more strongly coupled, with a mean estimated junctional conductances of 0.7–1.6 nS (Galarreta & Hestrin 1999, Gibson et al. 1999), but the size of their gap junctions is unknown.

Gap junction channels are gated by transjunctional voltage (V_j), the difference between the internal voltages of the interconnected cells (Furshpan & Potter 1957, Spray et al. 1979, Harris 2001). For most connexin channels, conductance is maximal when $V_j = 0$ and it declines symmetrically with deviations in either direction. The gating process is quite slow. Among connexins, Cx36 channels are the least voltage-dependent. Even with very large deviations of V_j ($\pm$100 mV), junctional conductance declines by less than half (Srinivas et al. 1999, Al-Ubaidi et al. 2000). This is consistent with measurements from pairs of neocortical interneurons, where no voltage-dependence was apparent when V_j was varied by $\pm$40 mV (Gibson et al. 1999). It seems unlikely that the slow and very weak voltage-dependence of Cx36 channel gating has biological significance.

Some connexins may function outside gap junctions. Hexameric hemichannels, or connexons, serve as nonjunctional, plasma membrane–spanning pores (Goodenough & Paul 2003, Sáez et al. 2003). The first reliable evidence for this came from neurons, the horizontal cells of catfish retina (DeVries & Schwartz 1992). In catfish horizontal cells, Cx26 connexons may mediate an unusual mechanism of feedback regulation of cone synapses (Kamermans et al. 2001). Cx36, however, appears unable to form functional hemichannels (Al Ubaidi et al. 2000), and the horizontal cells of mammalian retina may contain neither Cx36 nor Cx26 (Deans & Paul 2001). Although Cx43 hemichannels have interesting functions in astrocyte membranes (Ye et al. 2003), to date there is no evidence for operational hemichannels in native neurons of the mammalian brain.

Modulation and Regulation

One of the most important properties of chemical synapses is their ability to change strength as a function of prior activity and chemical regulation. Gap junctions can also be modified in diverse ways, although little is known about the specific mechanisms that regulate the electrical synapses of the brain.

ACTIVITY-DEPENDENT PLASTICITY Evidence of activity-dependent plasticity of mammalian electrical synapses is essentially nonexistent. Notable experiments in fish, however, show that the electrical synapses formed between auditory nerve endings (club endings) and Mauthner cells (large reticulospinal neurons) can either increase or decrease their junctional conductance as a function of prior neural activity (Yang et al. 1990, Pereda & Faber 1996, Pereda et al. 1998). The electrical synapses at club endings operate in parallel with excitatory, glutamatergic chemical synapses located within the same terminals. High-frequency stimulation of the auditory axons tends to enhance the strength of both the electrical and chemical components of these mixed synapses. Enhancement may last for hours, and it seems to depend on a close interaction between Cx35 channels and neighboring glutamatergic receptor–channels within the same synaptic endings (Pereda et al. 2003, Smith & Pereda 2003). Potentiation of the gap junctions depends on postsynaptic NMDA (N-methyl-D-aspartate) receptors and an increase of postsynaptic $[Ca^{2+}]_i$ (Yang et al. 1990, Pereda & Faber 1996), which leads to the activation of Ca^{2+}/calmodulin-dependent protein kinase II (CaM-KII) (Pereda et al. 1998). Thus, there are remarkable mechanistic parallels between the long-term potentiation of mammalian glutamatergic synapses (Nicoll 2003) and that of fish electrical synapses.

The prospects for finding plasticity in mammalian electrical synapses are very good. Although the molecular basis for gap junction plasticity is unknown, the sequences of fish Cx35 and mammalian Cx36 are similar (O'Brien et al. 1998) and include several shared consensus phophorylation sites (Mitropoulou & Bruzzone 2003). Mixed electrical-chemical synapses have been observed in the mammalian spinal cord (Rash et al. 1996), neocortex (Sloper & Powell 1978, Fukuda & Kosaka 2003), brainstem (Sotelo et al. 1974, Rash et al. 2000), and elsewhere.

PH AND [Ca] SENSITIVITY The conductance of gap junction channels is reduced when intracellular $[H^+]$ or $[Ca^{2+}]$ increases. Whether either or both of these constitutes a physiological mechanism of channel regulation has long been debated (Rose & Rick 1978, Rozental et al. 2001). In most coupled systems of cells, Ca^{2+} is a much less potent regulator of gap junction conductance than $[H^+]$ is, and in general, $[Ca^{2+}]_i$ must rise to pathologically high concentrations for gap junctions to close.

Neural activity can either acidify or alkalinize the intracellular pH of central mammalian neurons by several tenths of a pH unit (Chesler 2003). The conductance of many gap junctions is exquisitely sensitive to the pH of the cytoplasm and nearly insensitive to extracellular pH (Spray et al. 1981). Cytoplasmic acidification tends to close channels, whereas alkalinization tends to open them. The effect of pH_i on junctional conductance varies widely across connexin types. In some cases the relationship is very steep and centered on the normal resting pH_i, suggesting physiological relevance. The regulation of central neuronal gap junctions by pH has not been studied in detail. Electrical coupling between HeLa cells transfected with Cx36 is readily abolished by acidification with 100% CO_2 (Teubner et al. 2000), but this was not further quantified. Acidification of central neurons reduces the incidence of dye coupling in some cases (Church & Baimbridge 1991, Rorig et al. 1996) but not in others (Connors et al. 1984), whereas alkalinization may increase dye coupling (Church & Baimbridge 1991). Dye coupling is an imperfect assay of gap junction function, however, and direct tests of the pH sensitivity of central electrical synapses are needed.

NEUROTRANSMITTER MODULATION AND CONNEXIN PHOSPHORYLATION A variety of endogenous substances can modulate gap junctions. A few of these are well known as neurotransmitters. Impermeable extracellular agents almost always influence gap junctions via intracellular second messengers rather than by affecting channel properties directly. In the best-studied cases, gap junction channels are modified by kinases that phosphorylate one or more sites on the cytoplasmic domains of the connexins. All connexin subtypes have multiple phosphorylation sites for several types of kinases, and these may influence channel gating directly or regulate the assembly, trafficking, and turnover of gap junction channels (Lampe & Lau 2000).

Studies of retinal neurons have provided the most comprehensive evidence that the modulation of electrical synapses plays a physiologically important role (Piccolino et al. 1984, DeVries & Schwartz 1989). The action of dopamine in the retina is particularly well understood, although details differ depending on the species studied (Weiler et al. 2000). In general, an increase in ambient light triggers release of dopamine from amacrine or interplexiform cells. Dopamine then binds to D_1 receptors on horizontal cells and AII amacrine cells and activates their adenylyl cyclase. cAMP concentration then increases and activates cyclic AMP-dependent protein kinase (PKA), and PKA-mediated phosphorylation of connexins reduces the probability of channel opening, thus lowering gap junction conductance (McMahon et al. 1989). Because of extensive coupling with adjacent horizontal cells, receptive fields of horizontal cells in the absence of dopamine are larger than the spread of their dendrites. When their electrical synapses are suppressed by dopamine, receptive fields narrow considerably. Although most of this work was performed on fish and reptilian retinas, similar effects of dopamine have also been observed in mammalian retinas (e.g., Hampson et al. 1994). Dopamine is certainly not the only endogenous modulator of electrical synapses in retinal neurons. Nitric oxide, arachidonic acid, retinoic acid, and low intracellular pH all reduce gap junction coupling between horizontal cells (DeVries & Schwartz 1989, Miyachi et al. 1994, Weiler et al. 2000).

In an interesting twist, the electrical synapses between club endings and Mauthner cells in goldfish are actually enhanced by activation of dopamine receptors and PKA (Pereda et al. 1992). Fish Cx35 and mammalian Cx36 share a similar PKA consensus site (Mitropoulou & Bruzzone 2003), so there is a good chance that Cx36—and the mammalian electrical synapses constructed of it—are also modulated by PKA activation.

Dopamine and other neurotransmitters may also regulate neuronal coupling in the mammalian brain (Roerig & Feller 2000), although most evidence is indirect.

In the nucleus accumbens and striatum, activation of D_1 receptors tends to decrease dye coupling, whereas D_2 receptors often enhance coupling (O'Donnell & Grace 1993). In the supraoptic nucleus of the hypothalamus, a variety of manipulations including hormones, physiological state (i.e., lactation or dehydration), local synaptic activation, elevation of nitric oxide or cGMP, and histamine receptors increase dye coupling, whereas cAMP reduces it (Cobbett & Hatton 1984, Yang & Hatton 1987, Hatton 1998).

Pharmacology

Drugs are indispensable tools in experimental neurobiology. The substances that influence the function of gap junctions and hemichannels are chemically diverse (reviewed by Harris 2001, Rozental et al. 2001). Lipophilic compounds such as long-chain alcohols (heptanol and octanol) and the gaseous anesthetic halothane have long been known to reduce gap junction function (Johnston et al. 1980, Rozental et al. 2001). More recently, fatty acid amides such as oleamide, anandamide, and arachidonic acid were found to have a similar effect (Venance et al. 1995, Boger et al. 1998). All these drugs tend to have only partial efficacy, poor selectivity for different connexins, and significant effects on other cellular processes. Some of the derivatives of glycyrrhetinic acid, including carbenoxolone (originally isolated from licorice root), reversibly reduce connexin channel conductances (Davidson & Baumgarten 1988), albeit with partial efficacy and variable selectivity. Carbenoxolone has a reputation for being reasonably specific, but this is undeserved. The generally toxic effects of the drug are illustrated by the symptoms of licorice overindulgence, which include hypertension and hypokalemia due to interference with cortisol degradation (Walker & Edwards 1994). Several studies have reported that carbenoxolone does not influence neuronal membrane or chemical synaptic functions (e.g., Yang & Michelson 2001, Kohling et al. 2001, Schmitz et al. 2001). However, others have described significant effects of carbenoxolone on processes other than gap junctions, including increased action potential threshold and reduced repetitive firing rates (Rekling et al. 2000, Rouach et al. 2003).

A few other gap junction blockers have recently shown promise. All-*trans* retinoic acid potently reduced electrical coupling between horizontal cells of fish retina (Zhang & McMahon 2000) and other gap junction–coupled systems. Quinine, the antimalarial drug, selectively blocks Cx36 and Cx50 and moderately reduces Cx45. It has little effect on other connexins (Srinivas et al. 2001), but it does have a variety of nonjunctional effects. Mefloquine, a quinine derivative, is 100 times more potent than quinine in blocking Cx36 and seems much more specific (Srinivas & Spray 2003).

To summarize, most of the currently available blockers of connexins tend to be low in potency, only partially effective, and poorly selective. Caution is required when using them as experimental tools. Octanol, halothane, and carbenoxolone in particular have been widely used in neurophysiology, and their positive effects were often the primary evidence for implicating electrical synapses in the phenomena under study. On the basis of the well-documented effects of these drugs on a wide range of nonconnexin ion channels, receptors, and enzymes, however, this is a dubious practice.

FUNCTIONS OF ELECTRICAL SYNAPSES

The generic capabilities of electrical synapses have been extensively reviewed (e.g., Bennett 1977, 1997; Jaslove & Brink 1987; Galarreta & Hestrin 2001a). Electrical synapses are faster than chemical synapses, but this advantage is minimized at mammalian body temperatures, where chemical synaptic delays are only 150 μsec (Sabatini & Regehr 1996). Perhaps the most singular attribute of electrical synapses is bidirectionality. Electrical synapses in the brain often interconnect neurons of similar type, size, and input resistance, using Cx36-dependent gap junctions (e.g., Galarreta & Hestrin 1999, Gibson et al. 1999, Deans et al. 2001, Hormuzdi et al. 2001, Landisman et al. 2002, Long et al. 2002); these features lead to coupling strengths that are closely symmetrical. Thus, defining which cell is pre- or postsynaptic often depends on circumstances or can even be ambiguous. This is fundamentally different from the unidirectional (or in rare cases highly asymmetrical) operations of nearly all chemical synapses. The dynamics of electrical synapses are also unique. They tend to be more reliable than the generally stochastic chemical synapses, but because of membrane capacitance and dendritic cable properties, they also closely resemble first-order low-pass electrical filters (Galarreta & Hestrin 1999, Landisman et al. 2002). Relatively small signals that are also slow, such as afterhyperpolarizations (Figure 1*C*), burst envelopes (Figure 1*B*), or subthreshold oscillations (Figure 2*A*), are communicated more effectively than are action potentials, which are much larger but briefer (Figure 1). Chemical synapses have interesting short-term dynamics as well, but they are widely variable and distinctly different from the dynamics of electrical synapses (Zucker & Regehr 2002).

When two or more electrically coupled neurons are active under laboratory conditions, the most consistent and robust outcome is synchronization. The speed and reciprocity of gap junctions allow each coupled cell to influence, by the rapid transfer of small currents, voltage deflections in its coupled neighbors. In the mammalian brain, both action potentials (Galarreta & Hestrin 1999, Gibson et al. 1999, Mann-Metzer & Yarom 1999, Landisman et al. 2002) and subthreshold fluctuations (Benardo & Foster 1986; Christie et al. 1989; Beierlein et al. 2000; Long et al. 2002, 2004) readily synchronize in many neuronal types, even with moderate electrical coupling strengths. Synchronization is decidedly not the whole story (Marder 1998). Computational models of coupled neurons predict that weak coupling can sometimes lead to antiphasic or asynchronous spike firing, and the stability of the synchronous and antisynchronous states may depend strongly on firing frequency and the detailed properties of the neurons (e.g., Sherman & Rinzel 1992, Chow & Kopell 2000, Lewis & Rinzel 2003, Pfeuty et al. 2003). The fact that many neural networks have chemical synapses (inhibitory, excitatory, or both) operating in parallel with electrical synapses immensely complicates analysis of their dynamics.

The function of electrical synapses may not be entirely electrical. Neuronal gap junctions could well be more important for the specific neuron-to-neuron passage of small organic signaling molecules than for conducting ionic current. Dye coupling supports the feasibility of this idea (Hatton 1998, Roerig & Feller 2000), but there is almost no direct evidence for physiologically relevant chemical coupling in a mammalian system (cf. Dunlap et al. 1987).

One way to test the function of a neural element is to eliminate it, and Cx36 knockout mice have provided the best opportunity to date for understanding the functions of electrical synapses (Deans et al. 2001, Guldenagel et al. 2001, Hormuzdi et al. 2001). The cellular phenotype of the mutant animal is exquisite: Electrical synapses were all but eliminated in the normally well-coupled neurons of the neocortex, hippocampus, TRN, inferior olive, and retina (described above). Neuronal and chemical synaptic properties, to the extent they have been measured, are minimally altered in knockout mice. In the intact knockout mouse, forebrain rhythms are subtly affected (Buhl et al. 2003), but there is no obvious seizure disorder. If the mouse has a behavioral phenotype (apart from that due to retinal deficits), it is not an immediately obvious one (Kistler et al. 2002). No studies of cognitive or affective effects have yet been published. As with most mutant models, the lack of palpable deficits in the Cx36 knockout may be due to compensatory changes during development. If that is the case, the compensations are quite interesting because they do not involve the simple replacement of Cx36 with another connexin type. Compensation instead must involve rewiring the brain with chemical synapses or altering its excitability by changing intrinsic membrane properties (De Zeeuw et al. 2003). Definitive behavioral tests of electrical synaptic functions await the application of more discriminating genetic manipulations or more selective gap junction blockers.

ACKNOWLEDGMENTS

We thank our colleagues for their many contributions, and Jessica Kostarides for editorial assistance. Our research was supported by the National Institutes of Health, the American Epilepsy Foundation, the Defense Advanced Research Projects Agency, and a Fox Postdoctoral Fellowship.

The *Annual Review of Neuroscience* is online at http://neuro.annualreviews.org

LITERATURE CITED

Al-Ubaidi MR, White TW, Ripps H, Poras I, Avner P, et al. 2000. Functional properties, developmental regulation, and chromosomal localization of murine connexin36, a gap-junctional protein expressed preferentially in retina and brain. *J. Neurosci. Res.* 59:813–26

Alvarez VA, Chow CC, Van Bockstaele EJ, Williams JT. 2002. Frequency-dependent synchrony in locus ceruleus: role of electrotonic coupling. *Proc. Natl. Acad. Sci. USA* 99:4032–36

Amitai Y, Gibson JR, Beierlein M, Patrick SL, Ho AM, et al. 2002. The spatial dimensions of electrically coupled networks of interneurons in the neocortex. *J. Neurosci.* 22:4142–52

Baker R, Llinás R. 1971. Electrotonic coupling between neurons in the rat mesencephalic nucleus. *J. Physiol.* 212:45–63

Beierlein M, Gibson JR, Connors BW. 2000. A network of electrically coupled interneurons drives synchronized inhibition in neocortex. *Nat. Neurosci.* 3:904–10

Benardo LS. 1997. Recruitment of GABAergic inhibition and synchronization of inhibitory interneurons in rat neocortex. *J. Neurophysiol.* 77:3134–44

Benardo LS, Foster RE. 1986. Oscillatory behavior in inferior olive neurons: mechanism, modulation, cell aggregates. *Brain Res. Bull.* 17:773–84

Bennett MVL. 1977. Electrical transmission: a functional analysis and comparison with chemical transmission. In *Cellular Biology of Neurons, Handbook of Physiology, The Nervous System*, ed. ER Kandel 1(1):357–416. Baltimore: Williams & Wilkins. 717 pp.

Bennett MV. 1997. Gap junctions as electrical synapses. *J. Neurocytol.* 26:349–66

Bennett MVL, Crain SM, Grundfest H. 1959. Electrophysiology of supramedullary neurons in *Spheroides maculates*: I. Orthodromic and antidromic responses. *J. Gen. Physiol.* 43:159–88

Bittman K, Owens DF, Kriegstein AR, LoTurco JJ. 1997. Cell coupling and uncoupling in the ventricular zone of developing neocortex. *J. Neurosci.* 17:7037–44

Blatow M, Rozov A, Katona I, Hormuzdi SG, Meyer AH, et al. 2003. A novel network of multipolar bursting interneurons generates theta frequency oscillations in neocortex. *Neuron* 38:805–17

Boger DL, Patterson JE, Guan X, Cravatt BF, Lerner RA, Gilula NB. 1998. Chemical requirements for inhibition of gap junction communication by the biologically active lipid oleamide. *Proc. Natl. Acad. Sci. USA* 95:4810–15

Brightman MW, Reese TS. 1969. Junctions between intimately apposed cell membranes in the vertebrate brain. *J. Cell Biol.* 40:648–77

Bruzzone R, Hormuzdi SG, Barbe M, Herb A, Monyer H. 2003. Pannexins, a novel family of gap junction proteins expressed in the brain. *Proc. Natl. Acad. Sci. USA* 100:13644–49

Buhl DL, Harris KD, Hormuzdi SG, Monyer H, Buzsaki G. 2003. Selective impairment of hippocampal gamma oscillations in connexin-36 knock-out mouse in vivo. *J. Neurosci.* 23:1013–18

Chang Q, Gonzalez M, Pinter MJ, Balice-Gordon RJ. 1999. Gap junctional coupling and patterns of connexin expression among neonatal rat lumbar spinal motor neurons. *J. Neurosci.* 19:10813–28

Chesler M. 2003. Regulation and modulation of pH in the brain. *Physiol. Rev.* 83:1183–221

Chow CC, Kopell N. 2000. Dynamics of spiking neurons with electrical coupling. *Neural Comput.* 12:1643–78

Christie MJ, Williams JT, North RA. 1989. Electrical coupling synchronizes subthreshold activity in locus coeruleus neurons in vitro from neonatal rats. *J. Neurosci.* 9:3584–89

Chu Z, Galarreta M, Hestrin S. 2003. Synaptic interactions of late-spiking neocortical neurons in layer 1. *J. Neurosci.* 23:96–102

Church J, Baimbridge KG. 1991. Exposure to high-pH medium increases the incidence and extent of dye coupling between rat hippocampal CA1 pyramidal neurons in vitro. *J. Neurosci.* 11:3289–95

Cobbett P, Hatton GI. 1984. Dye coupling in hypothalamic slices: dependence on in vivo hydration state and osmolality of incubation medium. *J. Neurosci.* 4:3034–38

Condorelli DF, Belluardo N, Trovato-Salinaro A, Mudo G. 2000. Expression of Cx36 in mammalian neurons. *Brain Res. Brain Res. Rev.* 32:72–85

Condorelli DF, Parenti R, Spinella F, Trovato

Salinaro A, Belluardo N, et al. 1998. Cloning of a new gap junction gene (Cx36) highly expressed in mammalian brain neurons. *Eur. J. Neurosci.* 10:1202–8

Connors BW, Benardo LS, Prince DA. 1983. Coupling between neurons of the developing rat neocortex. *J. Neurosci.* 3:773–82

Connors BW, Benardo LS, Prince DA. 1984. Carbon dioxide sensitivity of dye-coupling among glia and neurons of the neocortex. *J. Neurosci.* 4:1324–30

Cowan WM, Kandel ER. 2001. A brief history of synapses and synaptic transmission. In *Synapses*, ed. WM Cowan, TC Südhof, CF Stevens, pp. 1–88. Baltimore: Johns Hopkins Univ. Press. 767 pp.

Davidson JS, Baumgarten IM. 1988. Glycyrrhetinic acid derivatives: a novel class of inhibitors of gap-junctional intercellular communication. Structure-activity relationships. *J. Pharmacol. Exp. Ther.* 246:1104–07

Deans MR, Gibson JR, Sellitto C, Connors BW, Paul DL. 2001. Synchronous activity of inhibitory networks in neocortex requires electrical synapses containing connexin 36. *Neuron* 31:477–85

Deans MR, Paul DL. 2001. Mouse horizontal cells do not express connexin26 or connexin36. *Cell. Commun. Adhes.* 8:361–66

Deans MR, Volgyi B, Goodenough DA, Bloomfield SA, Paul DL. 2002. Connexin36 is essential for transmission of rod-mediated visual signals in the mammalian retina. *Neuron* 36:703–12

De Zeeuw CI, Chorev E, Devor A, Manor Y, Van Der Giessen RS, et al. 2003. Deformation of network connectivity in the inferior olive of connexin 36-deficient mice is compensated by morphological and electrophysiological changes at the single neuron level. *J. Neurosci.* 23:4700–11

Devor A, Yarom Y. 2002. Electrotonic coupling in the inferior olivary nucleus revealed by simultaneous double patch recordings. *J. Neurophysiol.* 87:3048–58

DeVries SH, Schwartz EA. 1989. Modulation of an electrical synapse between solitary pairs of catfish horizontal cells by dopamine and second messengers. *J. Physiol.* 414:351–75

DeVries SH, Schwartz EA. 1992. Hemi-gap-junction channels in solitary horizontal cells of the catfish retina. *J. Physiol.* 445:201–30

Draguhn A, Traub RD, Schmitz D, Jefferys JG. 1998. Electrical coupling underlies high-frequency oscillations in the hippocampus in vitro. *Nature* 394:189–92

Dunlap K, Takeda K, Brehm P. 1987. Activation of a calcium-dependent photoprotein by chemical signalling through gap junctions. *Nature* 325:60–62

Eccles JC. 1982. The synapse: from electrical to chemical transmission. *Annu. Rev. Neurosci.* 5:325–39

Evans WH, Martin PE. 2002. Gap junctions: structure and function (review). *Mol. Membr. Biol.* 19:121–36

Friedman D, Strowbridge BW. 2003. Both electrical and chemical synapses mediate fast network oscillations in the olfactory bulb. *J. Neurophysiol.* 89:2601–10

Fukuda T, Kosaka T. 2000. Gap junctions linking the dendritic network of GABAergic interneurons in the hippocampus. *J. Neurosci.* 20:1519–28

Fukuda T, Kosaka T. 2003. Ultrastructural study of gap junctions between dendrites of parvalbumin-containing GABAergic neurons in various neocortical areas of the adult rat. *Neuroscience* 120:5–20

Fulton BP, Miledi R, Takahashi T. 1980. Electrical synapses between motoneurons in the spinal cord of the newborn rat. *Proc. R. Soc. London Ser. B* 208:115–10

Furshpan EJ, Potter DD. 1957. Mechanism of nerve-impulse transmission at a crayfish synapse. *Nature* 180:342–43

Galarreta M, Hestrin S. 1999. A network of fast-spiking cells in the neocortex connected by electrical synapses. *Nature* 402:72–75

Galarreta M, Hestrin S. 2001a. Electrical synapses between GABA-releasing interneurons. *Nat. Rev. Neurosci.* 2:425–33

Galarreta M, Hestrin S. 2001b. Spike transmission and synchrony detection in networks

of GABAergic interneurons. *Science* 292: 2295–99
Galarreta M, Hestrin S. 2002. Electrical and chemical synapses among parvalbumin fast-spiking GABAergic interneurons in adult mouse neocortex. *Proc. Natl. Acad. Sci. USA* 99:12438–43
Gibson JR, Beierlein M, Connors BW. 1999. Two networks of electrically coupled inhibitory neurons in neocortex. *Nature* 402:75–79
Goodenough DA, Paul DL. 2003. Beyond the gap: functions of unpaired connexon channels. *Nat. Rev. Mol. Cell Biol.* 4:285–94
Guldenagel M, Ammermuller J, Feigenspan A, Teubner B, Degen J, et al. 2001. Visual transmission deficits in mice with targeted disruption of the gap junction gene connexin36. *J. Neurosci.* 21:6036–44
Gutnick MJ, Lobel-Yaakov R, Rimon G. 1985. Incidence of neuronal dye-coupling in neocortical slices depends on the plane of section. *Neuroscience* 15:659–66
Gutnick MJ, Prince DA. 1981. Dye coupling and possible electrotonic coupling in the guinea pig neocortical slice. *Science* 211:67–70
Hampson EC, Weiler R, Vaney DI. 1994. pH-gated dopaminergic modulation of horizontal cell gap junctions in mammalian retina. *Proc. R. Soc. London Ser. B* 255:67–72
Harris AL. 2001. Emerging issues of connexin channels: biophysics fills the gap. *Q. Rev. Biophys.* 34:325–472
Hatton GI. 1998. Synaptic modulation of neuronal coupling. *Cell Biol. Int.* 22:765–80
Hinrichsen CF. 1970. Coupling between cells of the trigeminal mesencephalic nucleus. *J. Dent. Res.* 49(Suppl.):1369–73
Hormuzdi SG, Pais I, LeBeau FE, Towers SK, Rozov A, et al. 2001. Impaired electrical signaling disrupts gamma frequency oscillations in connexin 36-deficient mice. *Neuron* 31:487–95
Hua VB, Chang AB, Tchieu JH, Kumar NM, Nielsen PA, Saier MH Jr. 2003. Sequence and phylogenetic analyses of four TMS junctional proteins of animals: connexins, innexins, claudins and occludins. *J. Membr. Biol.* 194:59–76
Jahromi SS, Wentlandt K, Piran S, Carlen PL. 2002. Anticonvulsant actions of gap junctional blockers in an in vitro seizure model. *J. Neurophysiol.* 88:1893–902
Jaslove SW, Brink PR. 1987. Electrotonic coupling in the nervous system. In *Cell-to-Cell Communication*, ed. WC De Mello, pp. 103–47. New York: Plenum
Jefferys JG. 1995. Nonsynaptic modulation of neuronal activity in the brain: electric currents and extracellular ions. *Physiol. Rev.* 75:689–723
Johnston MF, Simon SA, Ramon F. 1980. Interaction of anaesthetics with electrical synapses. *Nature* 286:498–500
Kamermans M, Fahrenfort I, Schultz K, Janssen-Bienhold U, Sjoerdsma T, Weiler R. 2001. Hemichannel-mediated inhibition in the outer retina. *Science* 292:1178–80
Katsumaru H, Kosaka T, Heizmann CW, Hama K. 1988. Gap junctions on GABAergic neurons containing the calcium-binding protein parvalbumin in the rat hippocampus (CA1 region). *Exp. Brain Res.* 72:363–70
Kistler WM, De Jeu MT, Elgersma, Y, Van Der Giessen RS, Hensbroek R, et al. 2002. Analysis of Cx36 knockout does not support tenet that olivary gap junctions are required for complex spike synchronization and normal motor performance. *Ann. NY Acad. Sci.* 978:391–404
Knowles WD, Funch PG, Schwartzkroin PA. 1982. Electrotonic and dye coupling in hippocampal CA1 pyramidal cells in vitro. *Neuroscience* 7:1713–22
Kohling R, Gladwell SJ, Bracci E, Vreugdenhil M, Jefferys JG. 2001. Prolonged epileptiform bursting induced by 0-Mg^{2+} in rat hippocampal slices depends on gap junctional coupling. *Neuroscience* 105:579–87
Koos T, Tepper JM. 1999. Inhibitory control of neostriatal projection neurons by GABAergic interneurons. *Nat. Neurosci.* 2:467–72
Korn H, Sotelo C, Crepel F. 1973. Electronic coupling between neurons in the rat lateral

vestibular nucleus. *Exp. Brain Res.* 16:255–75

Kosaka T. 1983. Neuronal gap junctions in the polymorph layer of the rat dentate gyrus. *Brain Res.* 277:347–51

Kosaka T, Hama K. 1985. Gap junctions between non-pyramidal cell dendrites in the rat hippocampus (CA1 and CA3 regions): a combined Golgi-electron microscopy study. *J. Comput. Neurol.* 231:150–61

Lampe PD, Lau AF. 2000. Regulation of gap junctions by phosphorylation of connexins. *Arch. Biochem. Biophys.* 384:205–15

Landesman Y, White TW, Starich TA, Shaw JE, Goodenough DA, Paul DL. 1999. Innexin-3 forms connexin-like intercellular channels. *J. Cell Sci.* 112:2391–96

Landisman CE, Long MA, Beierlein M, Deans MR, Paul DL, Connors BW. 2002. Electrical synapses in the thalamic reticular nucleus. *J. Neurosci.* 22:1002–9

Lewis TJ, Rinzel J. 2003. Dynamics of spiking neurons connected by both inhibitory and electrical coupling. *J. Comput. Neurosci.* 14:283–309

Lin JW, Faber DS. 1988. Synaptic transmission mediated by single club endings on the goldfish Mauthner cell. I. Characteristics of electrotonic and chemical postsynaptic potentials. *J. Neurosci.* 8:1302–12

Liu XB, Jones EG. 2003. Fine structural localization of connexin-36 immunoreactivity in mouse cerebral cortex and thalamus. *J. Comp. Neurol.* 466:457–67

Llinás RR, Baker Sotelo C. 1974. Electrotonic coupling between neurons in cat inferior olive. *J. Neurophysiol.* 37:560–71

Llinás R, Yarom Y. 1981. Electrophysiology of mammalian inferior olivary neurones in vitro. Different types of voltage-dependent ionic conductances. *J. Physiol.* 315:549–67

Long MA, Deans MR, Paul DL, Connors BW. 2002. Rhythmicity without synchrony in the electrically uncoupled inferior olive. *J. Neurosci.* 22:10898–905

Long MA, Landisman CE, Connors BW. 2004. Small clusters of electrically coupled neurons generate synchronous rhythms in the thalamic reticular nucleus. *J. Neurosci.* 24:341–49

MacVicar BA, Dudek FE. 1981. Electrotonic coupling between pyramidal cells: a direct demonstration in rat hippocampal slices. *Science* 213:782–85

MacVicar BA, Dudek FE. 1982. Electrotonic coupling between granule cells of rat dentate gyrus: physiological and anatomical evidence. *J. Neurophysiol.* 47:579–92

Maier N, Guldenagel M, Sohl G, Siegmund H, Willecke K, Draguhn A. 2002. Reduction of high-frequency network oscillations (ripples) and pathological network discharges in hippocampal slices from connexin 36-deficient mice. *J. Physiol.* 541:521–28

Mancilla JG, Lewis TJ, Pinto DJ, Rinzel J, Connors BW. 2003. Synchrony of firing in coupled pairs of inhibitory interneurons in neocortex. *Soc. Neurosci. Abstr.* 173.4

Mann-Metzer P, Yarom Y. 1999. Electrotonic coupling interacts with intrinsic properties to generate synchronized activity in cerebellar networks of inhibitory interneurons. *J. Neurosci.* 19:3298–306

Manor Y, Yarom Y, Chorev E, Devor A. 2000. To beat or not to beat: a decision taken at the network level. *J. Physiol.* 94:375–90

Marder E. 1998. Electrical synapses: beyond speed and synchrony to computation. *Curr. Biol.* 8:R795–97

Maxeiner S, Kruger O, Schilling K, Traub O, Urschel S, Willecke K. 2003. Spatiotemporal transcription of connexin45 during brain development results in neuronal expression in adult mice. *Neuroscience* 119:689–700

McCormick DA, Bal T. 1997. Sleep and arousal: thalamocortical mechanisms. *Annu. Rev. Neurosci.* 20:185–215

McMahon DG, Knapp AG, Dowling JE. 1989. Horizontal cell gap junctions: single-channel conductance and modulation by dopamine. *Proc. Natl. Acad. Sci. USA* 86:7639–43

Menichella DM, Goodenough DA, Sirkowski E, Scherer SS, Paul DL. 2003. Connexins are critical for normal myelination in the CNS. *J. Neurosci.* 23:5963–73

Meyer AH, Katona I, Blatow M, Rozov

A, Monyer H. 2002. In vivo labeling of parvalbumin-positive interneurons and analysis of electrical coupling in identified neurons. *J. Neurosci.* 22:7055–64

Michelson HB, Wong RK. 1994. Synchronization of inhibitory neurones in the guinea-pig hippocampus in vitro. *J. Physiol.* 477(Pt. 1): 35–45

Miyachi E, Kato C, Nakaki T. 1994. Arachidonic acid blocks gap junctions between retinal horizontal cells. *NeuroReport* 5:485–88

Mitropoulou G, Bruzzone R. 2003. Modulation of perch connexin35 hemi-channels by cyclic AMP requires a protein kinase A phosphorylation site. *J. Neurosci. Res.* 72:147–57

Nagy JI, Li X, Rempel J, Stelmack G, Patel D, et al. 2001. Connexin26 in adult rodent central nervous system: demonstration at astrocytic gap junctions and colocalization with connexin30 and connexin43. *J. Comp. Neurol.* 441:302–23

Nicoll RA. 2003. Expression mechanisms underlying long-term potentiation: a postsynaptic view. *Philos. Trans. R. Soc. London Ser. B* 358:721–26

Odermatt B, Wellershaus K, Wallraff A, Seifert G, Degen J, et al. 2003. Connexin 47 (Cx47)-deficient mice with enhanced green fluorescent protein reporter gene reveal predominant oligodendrocytic expression of Cx47 and display vacuolized myelin in the CNS *J. Neurosci.* 23:4549–59

O'Brien J, Bruzzone R, White TW, Al-Ubaidi MR, Ripps H. 1998. Cloning and expression of two related connexins from the perch retina define a new subgroup of the connexin family. *J. Neurosci.* 18:7625–37

O'Donnell P, Grace AA. 1993. Dopaminergic modulation of dye coupling between neurons in the core and shell regions of the nucleus accumbens. *J. Neurosci.* 13:3456–71

Ohara PT. 1988. Synaptic organization of the thalamic reticular nucleus. *J. Electron Microsc. Tech.* 10:283–92

Panchin Y, Kelmanson I, Matz M, Lukyanov K, Usman N, Lukyanov S. 2000. A ubiquitous family of putative gap junction molecules. *Curr. Biol.* 10:R473–74

Peinado A, Yuste R, Katz LC. 1993. Extensive dye coupling between rat neocortical neurons during the period of circuit formation. *Neuron* 10:103–14

Pereda AE, Bell TD, Chang BH, Czernik AJ, Nairn AC, et al. 1998. Ca^{2+}/calmodulin-dependent kinase II mediates simultaneous enhancement of gap-junctional conductance and glutamatergic transmission. *Proc. Natl. Acad. Sci. USA* 95:13272–77

Pereda AE, Faber DS. 1996. Activity-dependent short-term enhancement of intercellular coupling. *J. Neurosci.* 16:983–92

Pereda A, O'Brien J, Nagy JI, Bukauskas F, Davidson KG, et al. 2003. Connexin35 mediates electrical transmission at mixed synapses on Mauthner cells. *J. Neurosci.* 23:7489–503

Pereda A, Triller A, Korn H, Faber DS. 1992. Dopamine enhances both electrotonic coupling and chemical excitatory postsynaptic potentials at mixed synapses. *Proc. Natl. Acad. Sci. USA* 89:12088–92

Personius KE, Balice-Gordon RJ. 2001. Loss of correlated motor neuron activity during synaptic competition at developing neuromuscular synapses. *Neuron* 31:395–408

Peters A. 1980. Morphological correlates of epilepsy: cells in the cerebral cortex. In *Antiepileptic Drug—Mechanism of Action, Advances in Neurology*, ed. OH Glaser, JK Penry, DM Woodbury, 27:21–48. New York: Raven

Peters A. 2002. Examining neocortical circuits: some background and facts. *J. Neurocytol.* 31:183–93

Pfeuty B, Mato G, Golomb D, Hansel D. 2003. Electrical synapses and synchrony: the role of intrinsic currents. *J. Neurosci.* 23:6280–94

Phelan P, Starich TA. 2001. Innexins get into the gap. *BioEssays* 23:388–96

Piccolino M, Neyton J, Gerschenfeld HM. 1984. Decrease of gap junction permeability induced by dopamine and cyclic adenosine 3′:5′-monophosphate in horizontal cells of turtle retina. *J. Neurosci.* 4:2477–88

Rash JE, Dillman RK, Bilhartz BL, Duffy

HS, Whalen LR, Yasumura T. 1996. Mixed synapses discovered and mapped throughout mammalian spinal cord. *Proc. Natl. Acad. Sci. USA* 93:4235–39

Rash JE, Staines WA, Yasumura T, Patel D, Furman CS, et al. 2000. Immunogold evidence that neuronal gap junctions in adult rat brain and spinal cord contain connexin-36 but not connexin-32 or connexin-43. *Proc. Natl. Acad. Sci. USA* 97:7573–78

Rash JE, Yasumura T, Davidson KG, Furman CS, Dudek FE, Nagy JI. 2001a. Identification of cells expressing Cx43, Cx30, Cx26, Cx32 and Cx36 in gap junctions of rat brain and spinal cord. *Cell Commun. Adhes.* 8:315–20

Rash JE, Yasumura T, Dudek FE, Nagy JI. 2001b. Cell-specific expression of connexins and evidence of restricted gap junctional coupling between glial cells and between neurons. *J. Neurosci.* 21:1983–2000

Raviola E, Gilula NB. 1973. Gap junctions between photoreceptor cells in the vertebrate retina. *Proc. Natl. Acad. Sci. USA* 70:1677–81

Rekling JC, Shao XM, Feldman JL. 2000. Electrical coupling and excitatory synaptic transmission between rhythmogenic respiratory neurons in the preBotzinger complex. *J. Neurosci.* 20:RC113

Reyher CK, Lubke J, Larsen WJ, Hendrix GM, Shipley MT, Baumgarten HG. 1991. Olfactory bulb granule cell aggregates: morphological evidence for interperikaryal electrotonic coupling via gap junctions. *J. Neurosci.* 11:1485–95

Roerig B, Feller MB. 2000. Neurotransmitters and gap junctions in developing neural circuits. *Brain Res. Rev.* 32:86–114

Rorig B, Klausa G, Sutor B. 1995. Beta-adrenoreceptor activation reduces dye-coupling between immature rat neocortical neurones. *NeuroReport* 6:1811–15

Rorig B, Klausa G, Sutor B. 1996. Intracellular acidification reduced gap junction coupling between immature rat neocortical pyramidal neurones. *J. Physiol.* 490:31–49

Rose B, Rick R. 1978. Intracellular pH, intracellular free Ca, and junctional cell-cell coupling. *J. Membr. Biol.* 44:377–415

Rouach N, Segal M, Koulakoff A, Giaume C, Avignone E. 2003. Carbenoxolone blockade of neuronal network activity in culture is not mediated by an action on gap junctions. *J. Physiol.* 553:729–45

Rozental R, Srinivas M, Spray DC. 2001. How to close a gap junction channel: efficacies and potencies of uncoupling agents. In *Methods in Molecular Biology, Connexin Methods and Protocols*, ed. R Bruzzone, C Giaume, 154:447–76. Totowa, New Jersey: Humana

Sabatini BL, Regehr WG. 1996. Timing of neurotransmission at fast synapses in the mammalian brain. *Nature* 384:170–72

Sáez JC, Berthoud VM, Brañes MC, Martínez AD, Beyer EC. 2003. Plasma membrane channels formed by connexins: their regulation and functions. *Physiol. Rev.* 83:1359–400

Schmitz D, Schuchmann S, Fisahn A, Draguhn A, Buhl EH, et al. 2001. Axo-axonal coupling: a novel mechanism for ultrafast neuronal communication. *Neuron* 31:831–40

Schoppa NE, Westbrook GL. 2002. AMPA autoreceptors drive correlated spiking in olfactory bulb glomeruli. *Nat. Neurosci.* 5:1194–202

Sherman A, Rinzel J. 1992. Rhythmogenic effects of weak electrotonic coupling in neuronal models. *Proc. Natl. Acad. Sci. USA* 89:2471–74

Skinner FK, Zhang L, Velazquez JL, Carlen PL. 1999. Bursting in inhibitory interneuronal networks: a role for gap-junctional coupling. *J. Neurophysiol.* 81:1274–83

Sloper JJ. 1972. Gap junctions between dendrites in the primate neocortex. *Brain Res.* 44:641–46

Sloper JJ, Powell TP. 1978. Gap junctions between dendrites and somata of neurons in the primate sensori-motor cortex. *Proc. R. Soc. London Ser. B* 203:39–47

Smith M, Pereda AE. 2003. Chemical synaptic activity modulates nearby electrical synapses. *Proc. Natl. Acad. Sci. USA* 100: 4849–54

Sotelo C, Llinás R. 1972. Specialized membrane junctions between neurons in the vertebrate cerebellar cortex. *J. Cell Biol.* 53:271–89

Sotelo C, Llinás R, Baker R. 1974. Structural study of inferior olivary nucleus of the cat: morphological correlates of electrotonic coupling. *J. Neurophysiol.* 37:541–59

Spencer WA, Kandel ER. 1961. Electrophysiology of hippocampal neurons. IV. Fast prepotentials. *J. Neurophysiol.* 24:272–85

Spray DC, Harris AL, Bennett MV. 1979. Voltage dependence of junctional conductance in early amphibian embryos. *Science* 204:432–34

Spray DC, Harris AL, Bennett MV. 1981. Gap junctional conductance is a simple and sensitive function of intracellular pH. *Science* 211:712–15

Srinivas M, Hopperstad MG, Spray DC. 2001. Quinine blocks specific gap junction channel subtypes. *Proc. Natl. Acad. Sci. USA* 98:10942–47

Srinivas M, Rozental R, Kojima T, Dermietzel R, Mehler M, et al. 1999. Functional properties of channels formed by the neuronal gap junction protein connexin36. *J. Neurosci.* 19:9848–55

Srinivas M, Spray DC. 2003. Specific block of connexin36 gap junction channels. *Soc. Neurosci. Abstr.* 370.10

Swadlow HA. 2003. Fast-spike interneurons and feedforward inhibition in awake sensory neocortex. *Cereb. Cortex* 13:25–32

Szabadics J, Lorincz A, Tamás G. 2001. Beta and gamma frequency synchronization by dendritic gabaergic synapses and gap junctions in a network of cortical interneurons. *J. Neurosci.* 21:5824–31

Tamás G, Buhl EH, Lorincz A, Somogyi P. 2000. Proximally targeted GABAergic synapses and gap junctions synchronize cortical interneurons. *Nat. Neurosci.* 3:366–71

Teubner B, Degen J, Sohl G, Guldenagel M, Bukauskas FF, et al. 2000. Functional expression of the murine connexin 36 gene coding for a neuron-specific gap junctional protein. *J. Membr. Biol.* 176:249–62

Teubner B, Odermatt B, Guldenagel M, Sohl G, Degen J, et al. 2001. Functional expression of the new gap junction gene connexin47 transcribed in mouse brain and spinal cord neurons. *J. Neurosci.* 21:1117–26

Thomson AM, Deuchars J. 1997. Synaptic interactions in neocortical local circuits: dual intracellular recordings in vitro. *Cereb. Cortex* 7:510–22

Traub RD, Draguhn A, Whittington MA, Baldeweg T, Bibbig A, et al. 2002. Axonal gap junctions between principal neurons: a novel source of network oscillations, and perhaps epileptogenesis. *Rev. Neurosci.* 13:1–30

Tresch MC, Kiehn O. 2000. Motor coordination without action potentials in the mammalian spinal cord. *Nat. Neurosci.* 3:593–99

Usher M, Cohen JD, Servan-Schreiber D, Rajkowski J, Aston-Jones G. 1999. The role of locus coeruleus in the regulation of cognitive performance. *Science* 283:549–54

Vaney DI. 2002. Retinal neurons: cell types and coupled networks. *Prog. Brain Res.* 136: 239–54

Venance L, Piomelli D, Glowinski J, Giaume C. 1995. Inhibition by anandamide of gap junctions and intercellular calcium signaling in striatal astrocytes. *Nature* 376:590–94

Venance L, Rozov A, Blatow M, Burnashev N, Feldmeyer D, Monyer H. 2000. Connexin expression in electrically coupled postnatal rat brain neurons. *Proc. Natl. Acad. Sci. USA* 97:10260–65

Venance L, Vandecasteele M, Glowinski J, Giaume C. 2003. Striatal output neurons are connected through electrical and unidirectional chemical synapses in rat brain slices. *Soc. Neurosci. Abstr.* 370.9

Veruki ML, Hartveit E. 2002a. AII (rod) amacrine cells form a network of electrically coupled interneurons in the mammalian retina. *Neuron* 33:935–46

Veruki ML, Hartveit E. 2002b. Electrical synapses mediate signal transmission in the rod pathway of the mammalian retina. *J. Neurosci.* 22:10558–66

Walker BR, Edwards CR. 1994. Licorice-induced hypertension and syndromes of

apparent mineralocorticoid excess. *Endocrinol. Metab. Clin. N. Am.* 23:359–77
Walton KD, Navarrete R. 1991. Postnatal changes in motorneuron electrotonic coupling studied in the in vitro rat lumbar spinal cord. *J. Physiol.* 433:283–305
Watanabe A. 1958. The interaction of electrical activity among neurons of lobster cardiac ganglion. *Jpn. J. Physiol.* 8:305–18
Weiler R, Pottek M, He S, Vaney DI. 2000. Modulation of coupling between retinal horizontal cells by retinoic acid and endogenous dopamine. *Brain Res. Rev.* 32:121–29
White JG, Southgate E, Thomson JN, Brenner S. 1986. The structure of the nervous system of the nematode *Caenorhabditis elegans*. *Philos. Trans. R. Soc. London Ser. B* 314:1–340
Willecke K, Eiberger J, Degen J, Eckardt D, Romualdi A, et al. 2002. Structural and functional diversity of connexin genes in the mouse and human genome. *Biol. Chem.* 383:725–37
Williams EH, DeHaan RL. 1981. Electrical coupling among heart cells in the absence of ultrastructurally defined gap junctions. *J. Membr. Biol.* 60:237–48
Wylie RM. 1973. Evidence of electrotonic transmission in the vestibular nuclei of the rat. *Brain Res.* 50:179–83
Yang QZ, Hatton GI. 1987. Dye coupling among supraoptic nucleus neurons without dendritic damage: differential incidence in nursing mother and virgin rats. *Brain Res. Bull.* 19:559–65
Yang XD, Korn H, Faber DS. 1990. Long-term potentiation of electrotonic coupling at mixed synapses. *Nature* 48:542–45
Yang QZ, Michelson HB. 2001. Gap junctions synchronize the firing of inhibitory interneurons in guinea pig hippocampus. *Brain Res.* 907:139–43
Ye ZC, Wyeth MS, Baltan-Tekkok S, Ransom BR. 2003. Functional hemichannels in astrocytes: a novel mechanism of glutamate release. *J. Neurosci.* 23:3588–96
Zhang DQ, McMahon DG. 2000. Direct gating by retinoic acid of retinal electrical synapses. *Proc. Natl. Acad. Sci. USA* 97:14754–59
Zhang C, Restrepo D. 2002. Expression of connexin 45 in the olfactory system. *Brain Res.* 929:37–47
Zhang C, Restrepo D. 2003. Heterogeneous expression of connexin 36 in the olfactory epithelium and glomerular layer of the olfactory bulb. *J. Comput. Neurol.* 459:426–439
Zucker RS, Regehr WG. 2002. Short-term synaptic plasticity. *Annu. Rev. Physiol.* 64:355–405

Annu. Rev. Neurosci. 2004. 27:419–51
doi: 10.1146/annurev.neuro.27.070203.144152

NEURONAL CIRCUITS OF THE NEOCORTEX

Rodney J. Douglas and Kevan A.C. Martin
Institute of Neuroinformatics, University/ETH Zurich, Zurich 8057, Switzerland; email: kevan@ini.phys.ethz.ch, rjd@ini.phys.ethz.ch

Key Words network, computation, model, excitation, inhibition

■ **Abstract** We explore the extent to which neocortical circuits generalize, i.e., to what extent can neocortical neurons and the circuits they form be considered as canonical? We find that, as has long been suspected by cortical neuroanatomists, the same basic laminar and tangential organization of the excitatory neurons of the neocortex is evident wherever it has been sought. Similarly, the inhibitory neurons show characteristic morphology and patterns of connections throughout the neocortex. We offer a simple model of cortical processing that is consistent with the major features of cortical circuits: The superficial layer neurons within local patches of cortex, and within areas, cooperate to explore all possible interpretations of different cortical input and cooperatively select an interpretation consistent with their various cortical and subcortical inputs.

INTRODUCTION

The enormous strides made in understanding the formation and operation of the neocortical circuits have been matched by the detailed analyses of the cellular and synaptic physiology of the elements that make up the neocortical circuits. Rapid advances in theory have also begun to clarify the nature of the computations carried out by the neocortical microcircuits. There are now many different models of cortical circuits, based on experimental data or theoretical considerations. Perhaps unsurprisingly, given their different explanatory and descriptive purposes, these model circuits differ greatly in form and content. For historical reasons, most biologically defensible circuits rely heavily on data from cat and primate visual cortex, but cell types and patterns of connections have also been described in many other cortical areas, with the contribution from rodent somatosensory cortex being perhaps the most prominent in recent years.

Here we explore to what extent the neocortical circuits discovered in primary sensory areas generalize, i.e., to what extent can neocortical neurons and the circuits they form be considered as canonical? A similar question has been applied to every aspect of the vertebrate brain and spinal cord, and it is especially relevant to questions of evolution, development, and homology of form and function. Most reasonable people would agree that evolution has been conservative, so one is not surprised to find close similarities in basic organization across vertebrate brains. At another level, one is not surprised to find that the neocortices of different mammals contain recognizably similar cell types (Ramon y Cajal 1911), or that many measures of neocortical anatomy scale with brain size. Even within single brains, there are systematic changes at the microstructural level, for example, from posterior to anterior cortical areas in primate the pyramidal cells increase in size, complexity of dendritic branching, and number of spines (Lund et al. 1993, Elston & Rosa 1998, Elston et al. 1999, Elston 2003). However, it might reasonably be argued that the similarities and regularities are incidental to the many different mappings of input to output that are evident in the different neocortical areas and across species, and that the specific functional requirements of a given area generate an experience-dependent circuit adapted for that requirement. Can general lessons really be learned from the neocortical circuits? The answer we give is an unequivocal yes. It does not require the eye of faith to be struck, as the early anatomists were, by features of the three-dimensional circuits that are common across neocortex.

LAMINATION

Despite its evident idealization, the notion of mammalian neocortex as a six-layered structure is widely accepted and widely used as the reference for describing a wide range of anatomical and physiological data. This is significant because the six layers provide perhaps the only commonly agreed upon framework with which to explore models of cortical circuits. From early on, the lamination has prompted thoughts as to its function. Application of the Golgi-staining technique had shown laminar-specific projections of cortical neurons and presumed afferents (Ramon y Cajal 1911). Based on studies of brains of patients and experimental material from animals, Campbell & Bolton (Bolton 1910, Campbell 1905) believed that the superficial cortical layers were principally concerned with "receptive and associative" functions, whereas the deep layers had "corticofugal and commissural" functions. [It was Campbell who studied the detailed histology of the brains of two chimpanzees and an orangutan, whose motor cortices had been mapped electrophysiologically by Grünbaum & Sherrington (1901)]. Although many electrophysiological and degeneration methods had shown that cortical afferents terminated in particular layers and projection pathways had their origins in particular layers, the fine degree of laminar organization of the projection pathways was only fully appreciated with the introduction of retrograde tracers, such as horseradish peroxidase or fast blue. Applied to many cortical areas in different species, these techniques show clearly the detailed laminar specificity of the cells of origin of the efferent fiber systems (e.g., Gilbert & Kelly 1975, Jones & Wise 1977, Lund et al. 1975, Wise 1975). Thus, all the players in the cortical circuitry—afferents, intrinsic neurons, and projection neurons—organize themselves with respect to laminae. Not only that, but rules by which the afferents and efferents organize seem to be universal for all neocortical areas (Creutzfeldt 1993, Jones 1999, Powell 1973, Powell 1981).

Unlike cortical structures, such as hippocampus or cerebellum, neocortex has its "crowning mystery," layer 1 (Hubel 1982), which consists mostly of the distal tufts of pyramidal cell apical dendrites, a sprinkling of GABAergic neurons, and many axon terminations. This layer is one major target of "feedback" connections between cortical areas and also receives input from subcortical nuclei. Why this layer has specialized for the connection between distal dendrites and cortical and subcortical inputs is one of its mysteries, but the existence of layer 1 points to a general purpose of cortical lamination, which is to generate a scaffold that constrains the way in which neurons can connect. This principle alone may be the reason why the brain makes such extensive use of cortical structures. A corollary of this is that cortical structures may allow neurons to connect with each other with the minimum use of wire (Chklovskii et al. 2002, Mead 1990, Mitchison 1991). Mitchison has shown theoretically that if there were but one cortical area instead of hundreds, the volume of cortex required to form the same circuits would be an order of magnitude larger (Mitchison 1991, Mitchison 1992).

If lamination is to be the means of defining the cortical circuits, then the axons of cortical afferents and neurons must not distribute randomly through the layers, but must show biases, preferably strong biases, for particular laminae. Vertical asymmetries in the dendrites can add a further degree of specificity. The laminar preferences of axons and dendrites were exploited in most early models of cortical circuits, which were based on the premise that where dendrites and axons overlap, there must be synaptic connections between them. Ramon y Cajal's solutions of the circuits of laminated structures, such as the retina, cerebellar cortex, hippocampus, and olfactory bulb, recommends this assumption. However, applied to the cortex, this method was evidently hindered by the plethora of different cell types found in cortex, which formed "impenetrable thickets" (Ramon y Cajal 1937) of connections. Thus, even in the neoclassical era of Golgi studies of neocortex, few attempts were made to use the data to develop circuits. One exception was Szentágothai (Szentágothai 1978) who produced several axonometric drawings of the cortical column showing the connections between different cell types inferred from data from electron microscopy and Golgi-stained material.

EXCITATORY CIRCUITS

Fundamental Intrinsic Circuits

Gilbert & Wiesel provided one of the first functional interpretations of a defined anatomical circuit (Gilbert 1983, Gilbert & Wiesel 1983) based on their intracellular recordings and reconstructions of individual cells filled with horseradish peroxidase (HRP) in cat visual cortex. The completeness of the axons revealed with intracellular injections of HRP was a revelation for eyes used to the immature or incomplete adult structures offered by the Golgi-stains. Here the laminar preferences of the axons of different types of neurons were revealed unambiguously. By using the simple rule that axons connected to neurons whose somata were located in the layer to which the axons project, Gilbert & Wiesel developed a simple circuit for cat area 17 (V1) that was consistent with the hypothetical circuits developed by Hubel & Wiesel (Hubel & Wiesel 1962) two decades earlier on the basis of receptive field structures. Obviously, Gilbert & Wiesel's simplification is not strictly correct: If instead of taking the target neurons to be a point, the full dendritic tree of the target neuron is considered, the spatially separated apical and basal dendrites come into play as extended connecting elements and more elaborate circuits are generated. However, because the basal dendrites radiating from the cell body form about 90% of the dendritic length of any cortical pyramidal neuron (Larkman 1991), statistically at least, their circuit shows the majority view.

In Gilbert & Wiesel's circuit (see Figures 1 and 2), the thalamic input arrives in layer 4. The excitatory cells in layer 4 project to the superficial layers. The superficial pyramidal neurons project to layer 5, which in turn projects to layer 6, and the loop is closed by a projection from layer 6 to the input layer 4. This was a landmark achievement that has yet to be matched for any other cortical area.

The other great simplification offered by the Gilbert & Wiesel circuit is that it inferred only the connections of the spiny, excitatory neurons, thus eliminating at a stroke the complications offered by the different types of smooth neurons. The spiny neurons as a class provide most of the inter-laminar connections within a cortical area, whereas the axons of smooth neurons principally arborize locally within their layer of origin. Overall, the spiny cells provide the basic framework of long-distance excitation in both the vertical and lateral dimensions, which is then moulded by local inhibitory neurons.

Their excitatory circuit for the cat now has to be modified by the addition of several pyramidal cell types in the deep layers (see Figure 1). A class of layer 5A pyramidal cells project to the superficial cortical layers, as shown by Lund et al. in their Golgi study (Lund et al. 1979) and confirmed by intracellular injections of HRP (Martin & Whitteridge 1984). Other modifications include a class of layer 6 pyramidal cells that project principally to layer 3 (Hirsch et al. 1998), as in the tree shrew striate cortex (Usrey & Fitzpatrick 1996), and a class of layer 6 pyramidal cells that project within layer 6 (Katz 1987). Many elements of this same basic cat pattern of excitatory circuits have been identified in area 17 of macaque monkey (Anderson et al. 1993, Blasdel et al. 1985, Callaway 1998, Fisken et al. 1975, Fitzpatrick et al. 1985, Lund et al. 1979). This same pattern is repeated in other primate cortical areas, for example, auditory cortex (Ojima et al. 1991, Ojima et al. 1992) and motor cortex (Ghosh et al. 1988, Ghosh & Porter 1988, Huntley & Jones 1991).

Data regarding the pattern of inter-laminar projections of the rodent barrel cortex is more limited than in the cat, tree shrew, or monkey visual cortex. Nonetheless, the basic pattern of projections of the spiny neurons in the barrel cortex follows that of the cat and monkey visual cortex (Bernardo et al. 1990a,b; Chapin et al. 1987; Hoeflinger et al. 1995; Gottlieb & Keller 1997; Schubert et al. 2003; Zhang

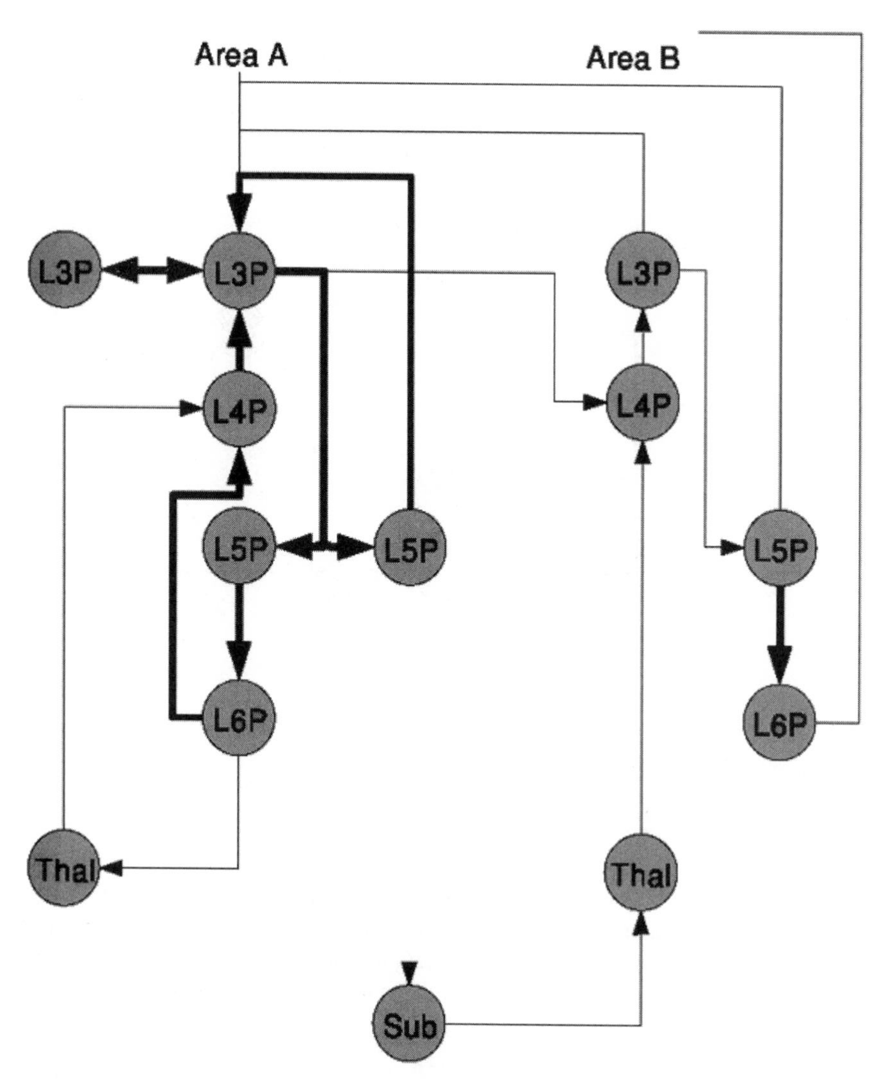

Figure 1 Graph of the dominant interactions between significant excitatory cell types in neocortex and their subcortical relations. The nodes of the graph are organized spatially; vertical corresponds to the layers of cortex and horizontal to its lateral extent. Directed edges (*arrows*) indicate the direction of excitatory action. Thick edges indicate the relations between excitatory neurons in a local patch of neocortex, which are essentially those described originally by Gilbert & Wiesel (Gilbert & Wiesel 1983, Gilbert 1983) for visual cortex. Thin edges indicate excitatory connections to and from subcortical structures and inter-areal connections. Each node is labeled for its cell type. For cortical cells, *Lx* refers to the layer in which its soma is located. *P* indicates that it is an excitatory neuron (generally of pyramidal morphology). *Thal* denotes the thalamus and *Sub* denotes other subcortical structures, such as the basal ganglia.

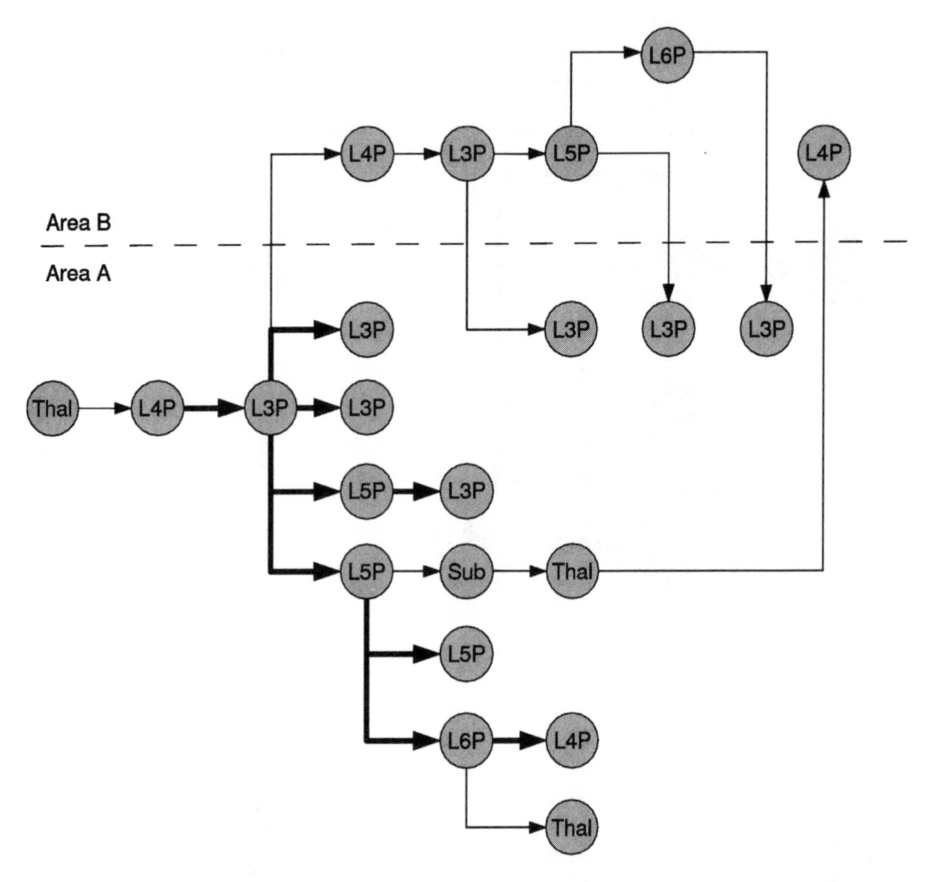

Figure 2 Graph of the temporal interactions between the cell types shown in Figure 1. Time unfolds toward the right. Each edge represents one synaptic delay. A temporal path ends when it is no longer unique; that is, further possible paths from that end node can be traced by selecting other nodes in the graph of the same cell type. For additional description, see Figure 1.

& Deschenes 1997). The one interesting variant on the vertical circuit seen in cat and monkey visual cortex is a projection from layer 6 to the upper tier of layer 5 of barrel cortex (Zhang & Deschenes 1997).

Matching of Thalamic and Layer 6 Pyramidal Arborizations

The tree shrew has been an invaluable model system for studying the intrinsic excitatory connections in neocortex. Fitzpatrick (1996) has noted that the stratification of intrinsic axonal arbors in the tree shrew visual cortex reflects the organization of parallel functional streams and that in the tree shrew it seems to do so more emphatically than in most species. Thus, the intrinsic interlaminar connections match with an exquisite sublamination the different afferent and efferent projections and their functional properties (e.g., magno-, parvo-, and koniocellular thalamic inputs; On and Off pathways; and binocular and monocular). Matching these stratified patterns of thalamic afferent input to the middle layers are the stratified arbors of the axons of subtypes of the layer 6 pyramidal cells, most of which form axonal arborizations into the middle layers. Outside layer 4 and 6, the tree shrew also shows an interesting sublamination. The interlaminar projection from upper layer 3 pyramidal cells is to lower layer 5, whereas the pyramidal cells in the lower part of layer 3 arborize in the upper tier of layer 5 (Fitzpatrick 1996, Fitzpatrick & Raczkowski 1990, Muly & Fitzpatrick 1992, Usrey et al. 1992, Usrey & Fitzpatrick 1996).

The variants in the pattern of thalamic afferent input and the matching variants in the axonal arborizations of the layer 6 pyramidal cells seen in the tree shrew are also present in macaque monkey area 17. There are stratified patterns of arborizations formed by the pyramidal cells of layer 6, which, as in the tree shrew, follow the pattern of afferent input stratification. In all, eight different subtypes of layer 6 pyramidal cells have been distinguished in the macaque monkey area 17 (Briggs & Callaway 2001). Two of the subtypes, called class I, have axons that arborize mainly in layer 6. The other six subtypes, called class II, have axons arborizing in various subdivisions of layer 4 (Wiser & Callaway 1996). In both classes, the apical dendrite formed more side branches in the same layer as the axonal arborization. Despite their apparent diversity, these eight subtypes of layer 6 pyramidal cells are only variations on the same two morphological themes seen in other species.

In the cat visual cortex (Katz 1987) and rat barrel cortex (Zhang & Deschenes 1997), layer 6 pyramidal cells form two basic types: One forms an axonal arbor in the middle layers, principally layer 4. In the cat, many of this type project back to the lateral geniculate nucleus. The other has a laterally extending axon that aborizes in layers 5 and 6. In the cat, these cells also project to the claustrum. The output of layer 6 is also expressed in the position of the cell somata. In monkey area 17, the pyramidal cells that project back to the magnocellular layers of the lateral geniculate nucleus (LGN) are located in the lower part of layer 6, whereas the pyramidal cells that project to the parvocellular layers of the LGN tend to be concentrated in the upper parts of layer 6, with some also in the lower tier. Pyramidal cells whose somata form the middle of layer 6 do not project to the LGN (Fitzpatrick et al. 1994). In the rat somatosensory cortex, the pyramidal cells that project to the ventral posteromedial nucleus (VPm) alone are concentrated in the upper part of the lamina, whereas those that project to VPm and the posterior group (Po) are concentrated lower in the lamina. Layer 6 pyramidal neurons that project to other cortical regions are found in far larger numbers in the rat than in cat or monkey area 17, and they are distributed through the lamina (Zhang & Deschenes 1997).

The examples of the macaque monkey and the tree shrew indicate an elaboration of the somewhat simpler patterns seen in ferret, rat, or cat. The elaboration seems to correlate best with an increased segregation in some input stream, although this may not be the sole reason for the differences between primate and tree shrew on one hand, and ferret, cat, and rat on the other. One desirable consequence of the increased stratification or sublamination is that it allows a higher selectivity in both the inputs that a particular neuron receives and ipso facto, a greater diversity of what is passed on to other brain regions (see the related argument of Malach below).

However, we should emphasize again that the differences seen between cat and monkey, for example, are like the differences between a grandfather clock and a Swiss chronometer. If more precision is required for functional streaming, or if the outputs need to be diversified, then Nature's solution has not been to build entirely different circuits, but rather to focus axonal and dendritic arbors into a sublamina and duplicate or triplicate the same basic pattern of interlaminar excitatory connections.

Static Connectivity

While it is a modest achievement to construct a simple diagram of the basic interlaminar connections, constructing a circuit based on a quantitative assessment of the numbers of neuronal types and their synaptic connections is much more difficult. Missing from the literature are quantitative estimates of the proportion of synapses any given class of spiny neurons contributes to a particular lamina. This missing factor makes the interpretation of connections on the basis of functional assays especially difficult. For example, in the cat, layer 5 pyramidal cells and layer 4 spiny stellate cells can both be activated monosynaptically by electrically stimulating the Y-type thalamic afferents, which form arbors principally in layer 4 with an additional collateral projection to layer 6 (Bullier & Henry 1979, Martin & Whitteridge 1984). It seems likely from simple geometric considerations that many more Y-type synapses are formed with the spiny stellate dendrites than on the layer 5 pyramidal neurons (Freund et al. 1985), which have only a short segment of their apical dendrite in the zone of thalamic termination. Yet the relative difference in connection strength is not differentiated by a simple electrical stimulus.

Callaway's study with caged glutamate in the rat (Callaway 2002) shows that glutamate stimulation apparently does reveal stronger or weaker projections. For example, glutamate stimulation of layer 4 provides more activation in layer 3 than it does in layer 5. This may indicate that there are numerically many more layer 4 neurons than layer 5 neurons projecting to layer 3. Alternatively, the results may indicate that uncaging glutamate in layer 4 activates both layer 5 neurons (e.g., via their apical dendrites) and layer 4 neurons, so that the net excitatory effect on layer 3 pyramidal cells is the sum of layer 4 and 5 effects.

Another interpretation for differences in strengths of the activation is that different types of excitatory neurons engage in two distinctly different operations, one being a driving function, the other modulating. The proposal for two basic excitatory connections, termed drivers and modulators, has come from a consideration of the long-distance synaptic projections from cortex to thalamus (Crick & Koch 1998, Sherman & Guillery 1996). The driving inputs are defined by their ability to affect the qualitative aspects of the receptive field. For example the center-surround organization of dLGN neurons is given by driving retino-thalamic inputs. The modulating inputs, however, alter the quantitative aspects of the response of their target neurons, not the qualitative structure of the receptive field. They (Crick & Koch 1998, Sherman & Guillery 1996) also propose that there are direct morphological correlates of the two functional types. The axons of the drivers form terminals with thick branches and grape-like clusters of large boutons and usually originate from layer 5 pyramidal cells, whereas the modulators form thin axons with tiny terminal boutons protruding from the main branch and usually originate from layer 6 pyramidal cells. According to Sherman & Guillery these two types of corticothalamic boutons are likely to occur in all thalamic nuclei, indicating that they form part of a stereotyped output from neocortex (Sherman & Guillery 2001). The driving projections to thalamus would thus provide a significant alternative path for inter-areal communication. It should also be noted, however, that the same two morphological types are also seen in the direct inter-areal projections in the primate visual cortex (Rockland 1996), but their possible functional roles remain unexplored.

The pattern of activation following the uncaging of glutamate indicates that the target layers that are maximally activated are largely those expected from the laminar pattern of connections. The one major anomaly is the lack of strong activation of layer 3 pyramidal cells from their neighbors, but this is simply explained by the difficulty of preventing the artifact of direct activation of the recorded neuron by nearby photo-stimulation. Thus the functional maps derived from the glutamate uncaging do not accurately reflect the extent of the contribution of neighboring neurons.

What has yet to be established is whether the varying strengths of the activation patterns revealed by methods such as photo-stimulation quantitatively reflect the anatomy or whether there are other factors, such as those suggested by Sherman & Guillery, that are crucial to understanding the basic functional interactions between the components of the circuits. If the numerically superior projections predominate in the responses evoked by electrical stimulation or uncaging glutamate, then it is clear that our understanding of the physical connections needs to be leavened with our knowledge of the functional consequences of the projection. It seems clear already that numerically small projections need not necessarily reflect functional impotence. On the contrary, it is the numerically small projections, such as the thalamic afferents or afferents from other cortical areas, that are thought to dominate or strongly modulate the response properties of their target areas in many instances. For example, synapses from the lateral geniculate nucleus form less than 10% of the excitatory synapses in layer 4 of area 17 in cats and monkeys (Ahmed et al. 1994, Garey & Powell 1971, Latawiec et al. 2000, Winfield & Powell 1983), yet they clearly provide sufficient excitation to drive the cortex. Similarly, the inter-areal projections also form a few percent of the synapses in their target layers, yet both the feedforward and feedback inter-areal circuits are thought to be functionally powerful.

There are various explanations for this apparent disparity between functional efficacy and actual numbers of synapses. One possibility is that there are large variations in synaptic strength, so that the thalamic synapses formed with layer 4 neurons, for example, have very strong synapses relative to the 10- or 20-fold more numerous excitatory synapses deriving from cortical neurons. This seems not to be the case: While thalamocortical synapses do appear exceptional in having a very low quantal variance (Bannister et al. 2002, Gil et al. 1999, Stratford et al. 1996, Tarczy-Hornoch et al. 1999), the peak amplitudes of the excitatory postsynaptic potentials (epsp's) are at most a factor of 2 greater than say, spiny stellate excitatory synapses. In fact, across all species, the peak amplitudes of epsp's or inhibitory postsynaptic potentials (ipsp's) evoked by single presynaptic neurons are small, despite a wide variety of target neurons and sites of synapse. One solution to this apparent discrepancy between structure and function is that the relatively small inputs from the thalamus or from inter-areal connections are amplified by recurrent circuits (Douglas et al. 1989). Thus, numerically small inputs with moderate synaptic strengths could, through the actual configuration of the recurrent cortical microcircuits, play a key role. The interpretation of the physical and functional anatomy thus depends crucially on an understanding of configuration of the circuits themselves.

Lateral Connections

It is clear that the interlaminar connections not only have characteristic patterns but that for the most part, the contribution of a particular spiny neuron to the interlaminar connections exceeds that of its intralaminar connections. For example, most layer 5 and layer 6 pyramidal cells connect outside their layer of origin. While some layer 4 spiny neurons do arborize extensively within layer 4, the major projection of layer 4 spiny neurons is to layer 3. It is only in the superficial layers that the pyramidal cells make extensive arborizations within the same layers, so monosynaptic recurrent connections between layer 2 and 3 pyramidal cells are likely to predominate more than in any other layer. It is these intralaminar connections that are of particular interest for our consideration of lateral excitatory connections.

We know more of the pattern of lateral connections of the superficial layer pyramidal cells than for any other layer. Early evidence from retrograde labeling indicated that discrete patches of neurons projected to a single point (e.g., Jones & Wise 1977). Evidence of "patchy" local axonal connections was most clearly seen in reconstructions of individual pyramidal cells filled intracellularly with a label in cat and monkey, or after bulk injections of tracers into the superficial layers in cat (Gilbert & Wiesel 1989, Kisvarday & Eysel 1992, Lowel & Singer 1992); tree shrew (Chisum et al. 2003, Rockland et al. 1982); or monkey somatosensory, motor, and visual areas (Huntley & Jones 1991, Juliano et al. 1990, Levitt et al.

1993, Lund et al. 1993, Malach 1992, Rockland & Lund 1983, Yoshioka et al. 1992), and prefrontal cortex (Kritzer & Goldman-Rakic 1995, Melchitzky et al. 1998, Pucak et al. 1996, Selemon & Goldman-Rakic 1988). Interestingly, similar injections into the rat visual and somatosensory cortex did not generate the same patchy connections (Lund et al. 1993), although Burkhalter & Charles (Burkhalter 1989, Burkhalter & Charles 1990) have described periodicities in the density in layers 2–6 after bulk injections of tracers in rat V1.

The interpretation of the patchy labeling is not straightforward. The tracers commonly used (HRP, biocytin, phaseolus vulgaris lectin, biotinylated dextrose amine) are taken up by cells that project from the injection site and by distant cells that send their axons to the injection site. The result is that the labeled patches are a mix of cells and axons, with some axons belonging to local cells and others coming from neurons at distant sites, including the injection site. One implication of the observation that cells and axons are colocalized is that patches or stripes of the superficial layer neurons, 200–500 microns in diameter or width, make reciprocal connections with each other.

In a comparative study, Lund et al. made similar-sized injections in visual area V1, V2, and V4; somatosensory areas 3b, 1, and 2; motor area 4, and prefrontal cortical areas 9 and 46 (Lund et al. 1993). In all the areas studied, there was a dense central core of labeled axons and cells at the injection site surrounded by fingers of label that separate into isolated patches at the furthest distances from the injection site. Comparing the patterns in the different areas of macaque monkey indicated that layer 3 was the major source of the label, with a lesser contribution from layer 2 neurons. The average size of a patch or stripe and the spacing between them varied from area to area, but a close correlation was found between the patch size and the spacing between them (Figure 3); this was true also for cat and tree shrew visual cortex. For all three species, the inter-patch spacing was roughly double that of the patch diameter in a given area. Injections into area TE in the macaque inferotemporal cortex also showed patches that were larger and less orderly than those seen in area 17 (Fujita 2002, Fujita & Fujita 1996).

Obviously, these lateral connections are not generated without regard to the functional architecture of the particular area in which they are found. Quite what the clusters relate to and how they develop has been a question avidly pursued by experimentalists and theorists alike since the layer 3 patches were first analyzed in tangential sections (Callaway & Katz 1990, Koulakov & Chklovskii 2001, Lowel & Singer 1992, Mitchison & Crick 1982, Rockland et al. 1982, Rockland & Lund 1983, Swindale 1992). In macaque area 17, Livingstone & Hubel (1984) showed that neurons located in the cytochrome oxidase–rich blobs formed reciprocal connections. Similarly, the cytochrome oxidase–poor compartments also formed reciprocal clustered connections. In the macaque monkey, Malach et al. (1993) found that domains with like eye dominance (monocular versus binocular) tend to be linked, as were domains with like orientation preference. The same trend was noted by Yoshioka et al. (1996). In both studies, the intensity of the label tailed off with distance from the injection, suggesting that neurons in neighboring

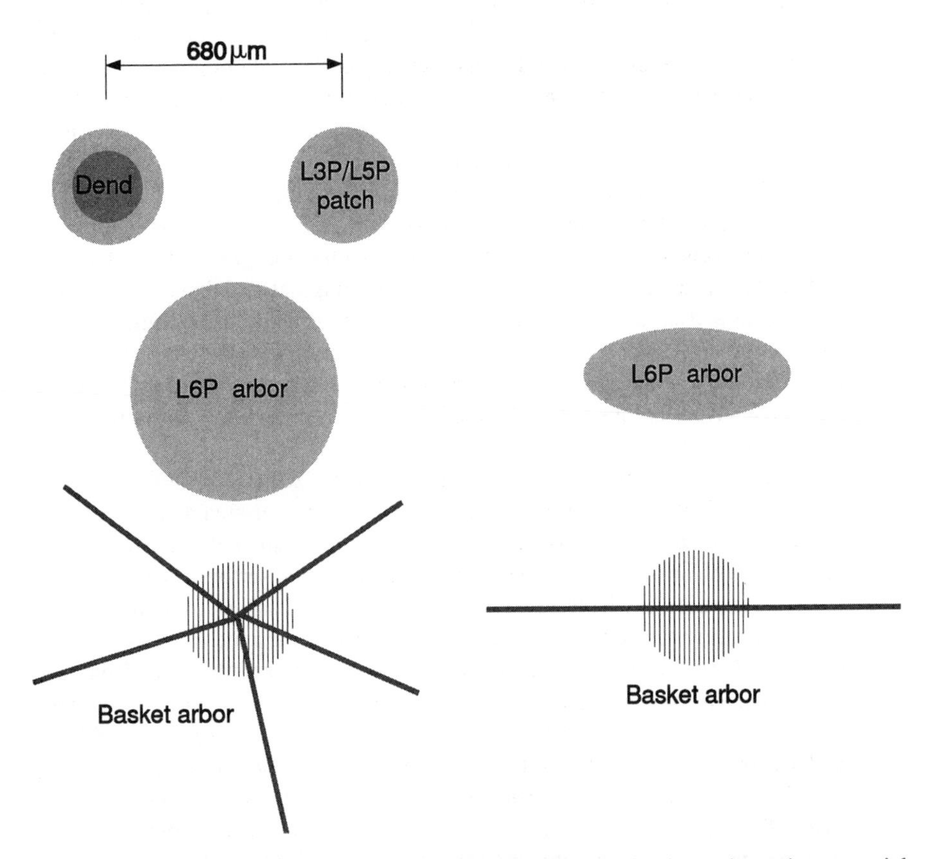

Figure 3 Approximate sizes of some important axonal arborizations, shown in tangential (*left*) and vertical (*right*) sections. Top: Diameter of layer 3 (or layer 5) patches (*light gray disk*, 320 μm diameter), compared with that of the basal dendrites of a layer 3 pyramidal cell (*dark gray disk*, 200 μm diameter). The inter-patch distance is 680 μm. Patch data were averaged over multiple animals and cortical areas (derived from Amir et al. 1993, Blasdel 1992, Burkhalter 1989, Fitzpatrick et al. 1998, Fujita & Fujita 1996, Kaas et al. 1989, Kisvárday et al. 1997, Kisvarday & Eysel 1992, Levitt et al. 1994, Luhmann et al. 1986, Lund et al. 1993, Malach et al. 1997, Rockland et al. 1982, Yoshioka et al. 1992). Middle: Diameter of layer 6 pyramid axonal arborization (*light gray*, 590 μm diameter) in layer 4. (Data derived from Hirsch et al. 1998, Martin & Whitteridge 1984, Usrey & Fitzpatrick 1996, Wiser & Callaway 1996, Zhang & Deschenes 1997.) Bottom: Diameter of perisomatic arborization (*vertical hatched disk*, 280 μm diameter) and approximately 5 long radial branches (*dark gray lines*, 650 μm radius) of large L3 basket cell. (Data estimated from Kawaguchi & Kondo 2002; Kisvárday et al. 1987, 1997, 2002; Somogyi et al. 1983.)

patches are more strongly connected than distant ones. Eye dominance specific connections of layer 3 pyramidal cells are not seen in normal cats, but are seen in strabismic cats (Schmidt et al. 1997, Trachtenberg & Stryker 2001). Like-to-like connections are only seen at some distance from the injections site (400–1000 microns) (Kisvárday et al. 1997, Malach et al. 1993, Stettler et al. 2002). At short range, the correlation breaks down and the axons freely innervate regions with quite diverse functional properties.

The view that the patches link neurons having similar functional characteristics invites a tempting simplification: Like connects to like. This pattern is explained by the equally beguiling slogan: Cells that fire together, wire together. But this interpretation is not entirely convincing, because it neglects the spatial organization of the dendritic trees of the target neurons. It turns out that the patch diameter correlates closely with the spread of the basal dendrites of the layer 3 pyramidal cells (Luhmann et al. 1986, Lund et al. 1993, Rockland et al. 1982). Thus, the largest patches were found in the macaque motor cortex (481 microns diameter), which also has the largest pyramidal cells. Malach (Malach 1992) has explored the significance of this scaling of patch and the size of individual cell dendrites. His idea is that the matching of size maximizes the diversity of a neuron's connections. Thus, assuming there is no selection of inputs, only neurons located in the middle of patches will have a pure sample of the patch inputs (see Figure 3), whereas those lying in the middle of a nonpatch receive pure nonpatch inputs. Neurons in-between have mixed inputs. This model generates a simple continuum from pure patch properties to pure nonpatch properties, so that more often, like does not connect to like. His model provides a convenient hypothesis to explain the size matching of patch and dendritic arbor, assuming equal sampling of inputs and unbiased dendritic arbors. His assumption of no dendritic bias seems to be valid for most neurons in cat area 17 (Anderson et al. 1999) but may not hold for some neurons in layer 4 of macaque area 17 (Katz et al. 1989) and the layer 4 of rat barrel cortex (Petersen & Sakmann 2000).

Questions of the meaning of the lateral connections extend to the overall distribution of the patches, which usually shows some degree of bias. In area 17 of New World monkeys, for example, the distribution of patches is roughly ellipsoid, with the long axis extending 1.7 times further from the injection site than the minor axis (Sincich & Blasdel 2001). The interpretation of the bias depends obviously on how much is known of the functional architecture of the cortical area concerned. For example, the layer 3 projections in area MT of the owl monkey are ellipsoid and asymmetrical relative to the injection site, but apart from a tendency for domains of similar orientation preference to be connected, the overall pattern of patches in area MT do not correlate with any obvious feature of the functional architecture (Malach et al. 1997). In inferotemporal cortex, adjacent injections in area TE gave rise to irregular patches that either were not adjacent to each other or overlapped extensively, which suggested that the underlying functional map is itself patchy (Fujita 2002, Fujita & Fujita 1996). A similar pattern is seen in prefrontal cortex (Lewis et al. 2002).

Even in a functionally well-defined area like macaque area 17, where the maps of functional properties are continuous, attempts to map the overall biases in the distribution of the lateral patches onto some functional attribute such as orientation have had mixed fortunes. The early papers on macaque referred to above did not detect any correlations with receptive field properties, such as orientation selectivity, and suggested that the bias reflected the anisotropy in the visual field map (Yoshioka et al. 1996). Indeed, the anisotropy of the retinotopic map in macaque area 17 is considerable. Most studies find a ratio of 1.6:1 for the magnification factor measured parallel and perpendicular to the V1/V2 borders, which represents the vertical meridian (Tootell et al. 1988, Van Essen et al. 1984). Blasdel & Campbell (2001) have shown by functional imaging that this anistropy is most likely generated by the ocular dominance columns, which would mask any correlations of the bias in lateral connections with the functional architecture. In the tree shrew, ocular dominance columns are absent, the visual field map is quite isotropic, the receptive field sizes are very large, and here the lateral projections of the layer 3 pyramidal cells do show a striking extension along the axis of the map of visual space that corresponds to preferred orientation of the pyramidal cells (Bosking et al. 1997, Chisum et al. 2003). Subsequently, a similar but much weaker bias was found for the pattern of lateral connections in area 17 of New World monkey (Sincich & Blasdel 2001). In general, the lateral projections of layer 3 pyramidal cells in Old and New World monkey area 17 cover only a few degrees of the visuotopic representation compared to the 5–20 degrees covered in cat and tree shrews.

Biases accepted, the question remains, what are lateral connections there for, and why do they seem so similar across cortex? In the visual cortex, one answer is that they are responsible for the physiological effects expressed in the "nonclassical" receptive field (see Fitzpatrick 2000). For example, a bar presented outside the classical receptive field can facilitate the response of a bar presented to the classical receptive field. This notion that the lateral connections mediate contextual interactions is an attractive hypothesis. But even here it is still not clear to what extent the lateral projection in visual cortex provides a catchment area that is larger than anticipated from the dimensions of the classical receptive field. On one hand, there is the claim that the lateral connections in macaque area 17 extend eight times the size of a classical receptive field (Gilbert 1992, Stettler et al. 2002); however, Yoshioka et al. (1996) and Sincich & Blasdel (2001) claim that the dimensions of the labeled patches match the extent expected from the classical receptive field size. All these estimates consider only the extent of the monosynaptic connections within layer 3 and, of course, neglect the possible role of intra- and inter-areal connections in providing additional contextual information. The question of what they are for remains largely unanswered. It is clear that yards of ignorance remain at even the most basic level; e.g., it remains to be discovered what determines the number of patches, the extent of their distribution, how individual neurons contribute to the input to these patches, and where the neurons that constitute a patch send all their outputs.

INTER-AREAL The layer 3 pyramidal cells are also the main source of the feedforward projections to other cortical areas, where they terminate in the middle layers, much as the thalamic afferents do. The feedback projections, by contrast, originate mainly in layers 5 and 6 and terminate outside layer 4 (Rockland & Pandya 1979). This has proved an invaluable simplification in generating an anatomical hierarchy of visual areas (Felleman & Van Essen 1991). However, the hierarchies generated by this method are very under-constrained, and enormous numbers of equally plausible hierarchies can be generated (Hilgetag et al. 1996). One invaluable means of constraining the solution has proved to be the quantitative measure of the proportion of layer 2 and 3 neurons that contribute to the feedforward and feedback pathways. Surprisingly, these proportions change in a regular manner from area to successive area and so suggest the existence of a "distance rule" (Kennedy & Bullier 1985, Rockland 1997, Barone et al. 2000). In this rule, the higher the proportion of superficial layer neurons (SLN%) of the total neurons that contribute to a projection to another area, the closer are the areas in the hierarchy. For example, after injections into area V4, 100% of the labeled neurons in V1 (area 17) are in layers 2 and 3, compared to 93% in V2 and 60% in V3A. By ranking the areas according to the SLN%, a single hierarchy emerges. Interestingly, it shows some striking differences from previous hierarchies. For example, the area called frontal eye field (FEF) lies at level 8 of the Felleman–Van Essen hierarchy. With the SLN% ranking it lies only at level 4, together with areas V3 and V3A, which again lie on different levels in the Felleman–Van Essen hierarchy.

The functional significance of these revisions of the cortical hierarchy have yet to be explored. Nevertheless, the evidence from the study of areal connections in the macaque visual cortex indicates that the superficial and deep layers of the cortex vary inversely in their projections to any other cortical area and thus in their respective influences on the local circuits in their target areas. The distance rule governing the hierarchical relationships of cortical areas may well be an organizing principle of monkey neocortex. Evidence supporting this view comes from the laminar organization of afferents in the frontal lobe (Barbas 1986) and the somatosensory cortex (Batardiere et al. 1998). These results have potentially important consequences for how we view the influence and role of the inter-areal connections to the local cortical circuits. These issues are considered in the final section.

INHIBITORY CIRCUITS

Smooth neurons may not be well-endowed with dendritic spines, but they are richly endowed with names. Almost all types individually bear more than one name, and as a group they are referred to exchangeably (albeit not always completely accurately) as aspiny, nonpyramidal, inhibitory, GABAergic, or inter-neurons. Following Ramon y Cajal, the neoclassical school of anatomists have generally preferred the evocative descriptive images of "double-bouquet," "basket," or "chandelier," based chiefly on the gestalt of the axonal arbor, but the modern trend is toward multivariate classification schemes in which factors such as morphometrics, biophysics, synaptic dynamics, synaptic targets, and neurochemical markers are employed to subdivide ever more finely, if arbitrarily, the population of smooth neurons.

The time constant of cortical evolution being somewhat longer than that of scientific nomenclature, the morphology of the smooth neurons has been remarkably conserved. For example, the smooth neurons in the primary visual cortex of marsupials and macaques are recognizably similar (Tyler et al. 1998) even though the two principal marsupial lines diverged from the eutherian line over 135 million years ago. Although from Ramon y Cajal onward there have been claims for neuronal types that are unique to the neocortex of humans or great apes [most recently from Nimchimsky et al. (1999)] there is nevertheless a great similarity in the proportion of the GABAergic neurons and their patterns of connection across cortical areas that have widely different functions. Remarkable too is that while the proportion of GABAergic neurons in a given area varies, between 10%–20% of the synapses found in any neocortical area in all species examined are the symmetric variety formed by GABAergic boutons. In one recent count of neocortical areas as different as human anterolateal temporal cortex and the hindlimb area of rat somatosensory cortex, symmetric synapses formed 11.5% and 10.7% of the population, respectively (DeFelipe et al. 2002). Generally, the ratio of symmetric to asymmetric synapses is remarkably constant, despite large regional variations in the average number of spines (the main site of asymmetric synapses) born by a pyramidal cell (Elston et al. 2001; Elston & Rosa 1998, 2000). While some morphological features, such as spine numbers on pyramidal cells, can vary widely from area to area or species to species (see Elston 2002), other parameters, such as the overall density of synapses, shows remarkably little variance from area to area in different species (Cragg 1967, O'Kusky & Colonnier 1982, Rakic et al. 1986, Schuz & Palm 1989).

Morphological Types of Smooth Neurons

There seems to be broad agreement over time and place that about ten morphologically distinct varieties of smooth neurons can be distinguished and that examples of the basic forms are found in all species. Although there are claims that the proportion of double bouquet cells greatly increases in primates, double bouquet cells are certainly seen in other species, including cat (Peters & Regidor 1981, Somogyi & Cowey 1981, Szentágothai 1973) and rodent (Connor & Peters 1984, Kawaguchi & Kubota 1997, Peters & Harriman 1988). In the monkey, double bouquet and bipolar cells have been lumped together by Lund & Wu on the grounds that their axon collaterals form narrow columns regardless of whether the dendritic morphology is bipolar or multipolar (Lund & Wu 1997).

DeFelipe (2002) has conveniently divided the smooth neurons into just three basic groups on the basis of the clustering of the axon. Those neurons that have local arbors include the neurogliaform (also called spider web); small basket (also called clewed, or clutch); chandelier cell (also called axo-axonic); and common type, i.e., a type with no particularly distinguishing features. Those forming vertically oriented axons include the neurons with axonal arcades, Martinotti cells, bipolar cells, and double bouquet cells. Only the large, or wide arbor basket cells, and the medium arbor cells have axons that extend horizontally. Large basket cells have the most extensive horizontal axons, but the extensions typically consist of 4 or 5 long branches that extend a few 100 microns laterally from the cell body but, unlike the horizontal projections of pyramidal cells, do not form dense bouton clusters (Figure 3). Thus, the diversity of the GABAergic neurons, currently much emphasized, is perhaps no greater than that of pyramidal cells, e.g., there are eight morphological variants alone of layer 6 pyramidal cells in the macaque area 17 (Wiser & Callaway 1996).

Immunoreactivity of Smooth Neurons

All GABAergic cells show immunoreactivity to calcium-binding proteins, as well as to neuropeptides, such as cholecystokinin, somatostatin, vasoactive intestinal polypeptide, neuropeptide Y, and corticotropin-releasing factor (Demeulemeester et al. 1988, 1991; Hendry et al. 1984; Schmechel et al. 1984; Somogyi et al. 1984). The profile of immunoreactivity expressed by a neuron depends on its laminar location and morphological type. There is considerable overlap in expression of a particular peptide or calcium-binding protein between cell types, and the profile depends also on the cortex's state of embryological development. Nevertheless, the immunoreactivity of cells to these markers provides another basis for the classification of smooth cell subtypes (e.g., Wang et al. 2002). VIP and substance P are also associated with cholinergic axons (Eckenstein & Baughman 1984, Vincent et al. 1983). Some spiny neurons also express immunoreactivity for calbindin, cholecystokinin, and somatostatin, but their immunoreactivity for these molecules is weaker, particularly in more mature animals, and the expression of different peptides may vary over development.

Although the proportion of neurons expressing a particular calcium-binding protein differs widely between areas, even in the same species, the basic laminar pattern of the neurons that express the calcium-binding proteins is conserved. For example, the pattern of calbindin immunoreactivity is similar for all cortical areas in species as diverse as rat, cat, and monkey. In the macaque somatic sensory, auditory, and extrastriate visual cortex, most immunoreactive cell bodies are in layer 2 and the upper part of 3, with a lower density in layer 5. One variant occurs in area 17, where there is an additional tier of neurons in layer 4. Curiously, the macaque motor cortex has few calbindin-immunoreactive cells (DeFelipe et al. 1990), whereas in the rat (Sun et al. 2002) and cat motor cortex (Porter et al. 2000), the calbindin immunoreactive cells have a similar distribution to macaque somatosensory, auditory, and extrastriate areas. The majority of immunoreactive synaptic boutons are distributed in the superficial layers with only a sparse distribution in the deep layers. In macaque the synapses formed in the somatic sensory areas (1, 2, 3a, 3b) are 40% on spines and 60% on small-caliber dendritic shafts, thought to be the distal portions of the basal and apical dendritic branches of pyramidal cells (DeFelipe et al. 1989a). Similar proportions and targets were found for the output of tachykinin-positive double bouquet cells in the macaque auditory cortex (DeFelipe et al. 1990).

Proportions of Morphological Types

One difficult question, only partially answered, is how many basket cells, chandelier cells, double bouquets, etc. are there in a given area? One attempt at an answer has come by correlating morphological features of the smooth neurons with the expression of various peptides and calcium-binding proteins. Unfortunately, because there is no one-to-one correlation of morphology with immunochemical identity, quantitative estimates of the proportions of different morphological types of smooth neurons are difficult to obtain. For example, the most widely used immunochemical markers for subdividing the smooth neurons are the three calcium-binding proteins (parvalbumin, calbindin, calretinin), which between them label virtually all the variants of GABAergic neurons. This means that no single morphological type can be identified and counted exclusively on the basis of its expression of a particular calcium-binding protein. As a means of classifying different types of smooth neurons, calcium-binding protein and peptide immunochemistry needs to be interpreted judiciously because the expression of these markers in a particular morphological type varies across areas and species (DeFelipe 1993).

Parvalbumin labels small and large basket cells (Blumcke et al. 1990, Hendry et al. 1989, Van Brederode et al. 1990) and chandelier cells (DeFelipe et al. 1989b, Lewis & Lund 1990). Calretinin labels a heterogenous group of neurons (Meskenaite 1997, Rogers 1992, Rogers & Resibois 1992), which include the Cajal-Retzius cells of layer one and a morphologically heterogenous group of neurons with vertically oriented axons, among them a small group of double bouquet cells (Conde et al. 1994, Gabbott & Bacon 1996a, Meskenaite 1997). Interestingly, calretinin-positive neurons form synapses mainly with other smooth neurons (possibly calbindin-positive neurons) in layer 3, but with pyramidal cells in layer 5 (Meskenaite 1997). In an interesting parallel, the hippocampus calretinin-positive neurons also form synapses mainly with other calretinin- and calbindin-positive neurons (Gulyas et al. 1996). In the macaque visual cortex, calbindin labels mainly, but not exclusively, the double bouquet cells (Hendry et al. 1989) and in the prefrontal cortex it labels double bouquet, Martinotti, and neurogliaform cells. Similarly, in rat frontal cortex, some Martinotti cells, which also express somatostatin, are immunoreactive for calbindin, as are the double bouquet cells (Kawaguchi & Kubota 1997).

Most estimates of the proportions of the different neurons are necessarily indirect, but one useful number has come from the study of calbindin-immunoreactive neurons in macaque cortex. Calbindin is expressed in the axons as well as the cell body and dendrites of double bouquet cells. In tangential sections of area 17 (Peters & Sethares 1997) and auditory cortex (DeFelipe et al. 1990), the tight columnar bundles of calbindin-immunoreactive axon collaterals form an apparently regular array spaced 25 microns apart. Thus under each square millimeter of surface there are 2500 vertical bundles. On average, each double bouquet cell makes more than one bundle to give a ratio of 0.7 double bouquet cells per bundle (Peters & Sethares 1997). This gives 1750 calbindin neurons in the superficial layers under each square millimeter of cortical surface. Because the number of superficial layer neurons under each square millimeter of macaque area 17 is known to be 52,000, 17% of which are GABAergic (Beaulieu et al. 1992), it follows that only 20% of the GABAergic neurons in the superficial layers are calbindin-positive. This is comparable with actual counts made in the macaque frontal cortex (Gabbott & Bacon 1996a) and cat visual cortex, where 20%–30% of the GABAergic neurons are calbindin-positive (Demeulemeester et al. 1989, Hogan et al. 1992, Huxlin & Pasternak 2001).

In macaque area 17, the majority of the GABAergic neurons must be parvalbumin-positive because the calretinin population forms only 14% of GABAergic neurons (Meskenaite 1997). The parvalbumin population consists of the chandelier cells, the large basket cells, and five or six other types, all of which have axonal arborizations that surround the dendritic tree (Jones 1975, Lund & Wu 1997). In the macaque prefrontal cortex, calretinin neurons are in the majority, forming 45% of the GABAergic neurons, compared to 24% for parvalbumin and 20% for calbindin (Conde et al. 1994, Gabbott & Bacon 1996b). The remaining 11% of GABAergic neurons did not stain strongly for any of the three calcium-binding proteins, but technical issues cannot be ruled out in accounting for these negative results.

Many of these difficulties in determining the precise proportions of the different morphological types will disappear as new molecular markers are developed to differentiate the types of cortical neurons, but it seems unlikely that there will be any great surprises in the future as to the known morphological variants of the smooth neurons. Probably, the major classes will still be those revealed to us over many years through application of the traditional Golgi stain. What this method has shown us, and what subsequent techniques such as intracellular labeling, immunostaining, and electron microscopy have confirmed, is that there are two basic modes of connection of the smooth neurons (Figure 4). The first, horizontal class, exemplified by the basket and chandelier cells, are neurons with local axonal arbors that target the proximal portions of the dendritic tree of spiny cells. This targeting is very obvious in the parvalbumin immunostained material. The second, vertical class, most elegantly exemplified by the double bouquet cells, have vertically oriented axonal arbors and target the more distal portions of the dendritic trees of spiny cells, principally pyramidal cells. Because the smooth neurons are inhibitory, it is evident that these two broad divisions give rise to different functional possibilities. In the first class, the principal targets cluster around the integration and the output region of the neuron: the axon initial segment, soma, and proximal

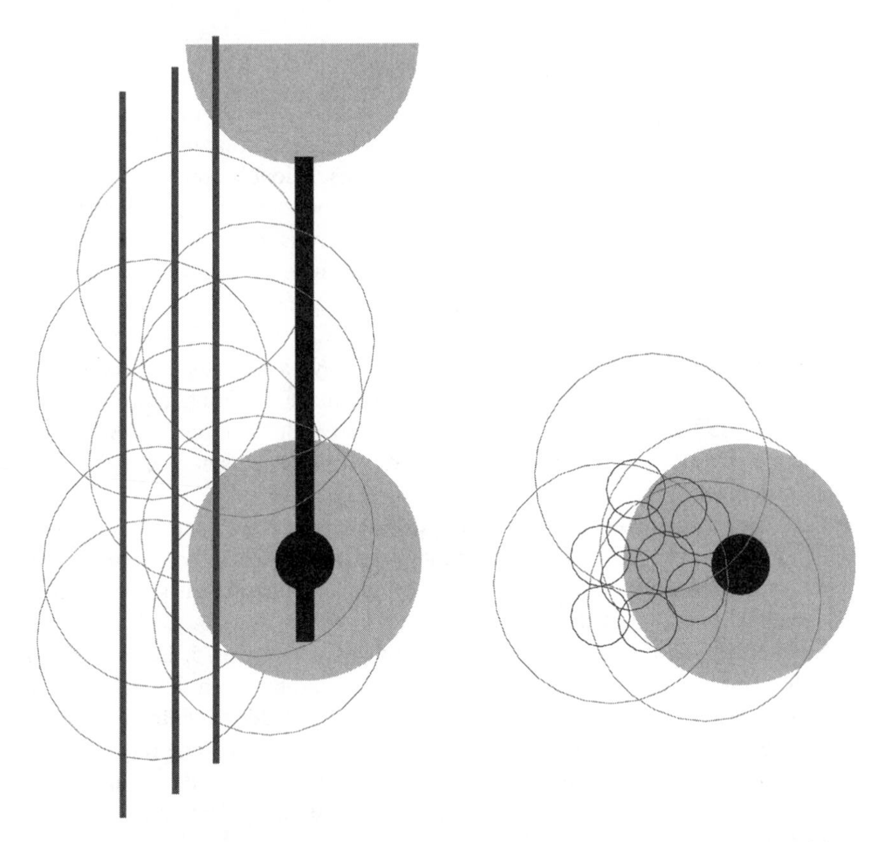

Figure 4 Schematic showing the proposed distinction between the effects of horizontal and vertical smooth cells. Vertical section through superficial layers on left; tangential section on right. The parvalbumin positive horizontal smooth cells make multiple synaptic contacts on the crucial dendritic output path (apical dendrite, soma, and initial segment; *black-gray*) of a representative superficial pyramidal neuron, whose apical and basal dendritic fields are shown as light gray regions. The dendritic fields of some overlapping neighboring pyramids are indicated as light gray circles. The trajectories of three double bouquet axons (*left*, *dark gray lines*) pass vertically through the dendritic fields, making contact with some of them at various locations ranging from proximal to distal. Typical zones of influence of some vertical double bouquet axons are shown in the tangential view (*right*, *small dark gray circles*). These zones are spaced at 25 μm (see text).

dendrites. In the second class, the targets are the main regions of excitatory input: the distal basal dendrites and branches of pyramidal cell apical dendrites.

The differential laminar distribution of the various smooth cell types has an important consequence: The relative proportions of chandelier, basket cell, double bouquet cell, etc. synapses that a spiny cell receives will depend critically on its laminar location. Thus, for example, layer 3 pyramidal cells receive far more input from chandelier cells than those of layer 6 simply because layer 3 is more richly endowed with chandelier cells (Farinas & DeFelipe 1991, Sloper & Powell 1979).

FUNCTION FROM STRUCTURE

As has long been suspected by many cortical neuroanatomists, the same basic laminar and tangential organization of the excitatory neurons of the neocortex, the spiny neurons, is evident wherever it has been sought (Figures 1, 2, and 5). The inhibitory neurons similarly show characteristic morphology and patterns of connections throughout the cortex. Here, we have simply identified the constant elements of this circuit and pointed to the existence of some of the variants in both the spiny and smooth populations. These variants, occurring across areas in the same species and between species, are unsurprising given the widely different uses of the different cortical areas and the different activities of their owners. In terms of their amino acid neurotransmitters, the spiny and the smooth neurons are monogamous. The former use glutamate, the latter GABA. Variations in the immunochemistry arise with secondary markers, such as the calcium-binding proteins and peptides. In the case of the smooth neurons, this is perhaps due to their long migration from the median eminence, as in their travels they are likely to meet with remarkably different microenvironments that could induce the expression of different genes. Nevertheless, all things considered, many crucial aspects of morphology, laminar distribution, and synaptic targets are very well conserved between areas and between species.

One of the most intriguing consistencies of neocortical structure is the presence of patchy connections made by the pyramidal cells, particularly of the superficial cortical layers. Localized injections of tracers produce labeling of 10–30 patches, usually regularly spaced and appearing as petals of a daisy (Figure 5). The axons of individual pyramidal cells form far fewer patches than the collective, indicating that within any patch there is a heterogenous cluster of pyramidal cells, which distribute their output to different subsets of the total complement of patches. The patches also receive additional excitatory input from pyramidal cells projecting from other cortical areas (Gilbert & Wiesel 1989). Double-labeling experiments (Bullier et al. 1984, Kennedy & Bullier 1985, Perkel et al. 1986) show that the axons of individual projection neurons rarely innervate more than one other cortical area, despite the fact that each cortical area connects to many other cortical areas (Zeki 1978a,b). This means that the fan-out from one patch of projection neurons to other cortical areas is organized on similar principles to the fan out to other clusters within a cortical area. Thus, pyramidal cells lying within a single patch may receive inputs from other patches in its own area, or from other areas, that are most likely quite heterogenous in functional properties. Their individual responses, even within a single patch, might be much more varied than suggested by the like-to-like simplification. What could the possible role of such an arrangement be?

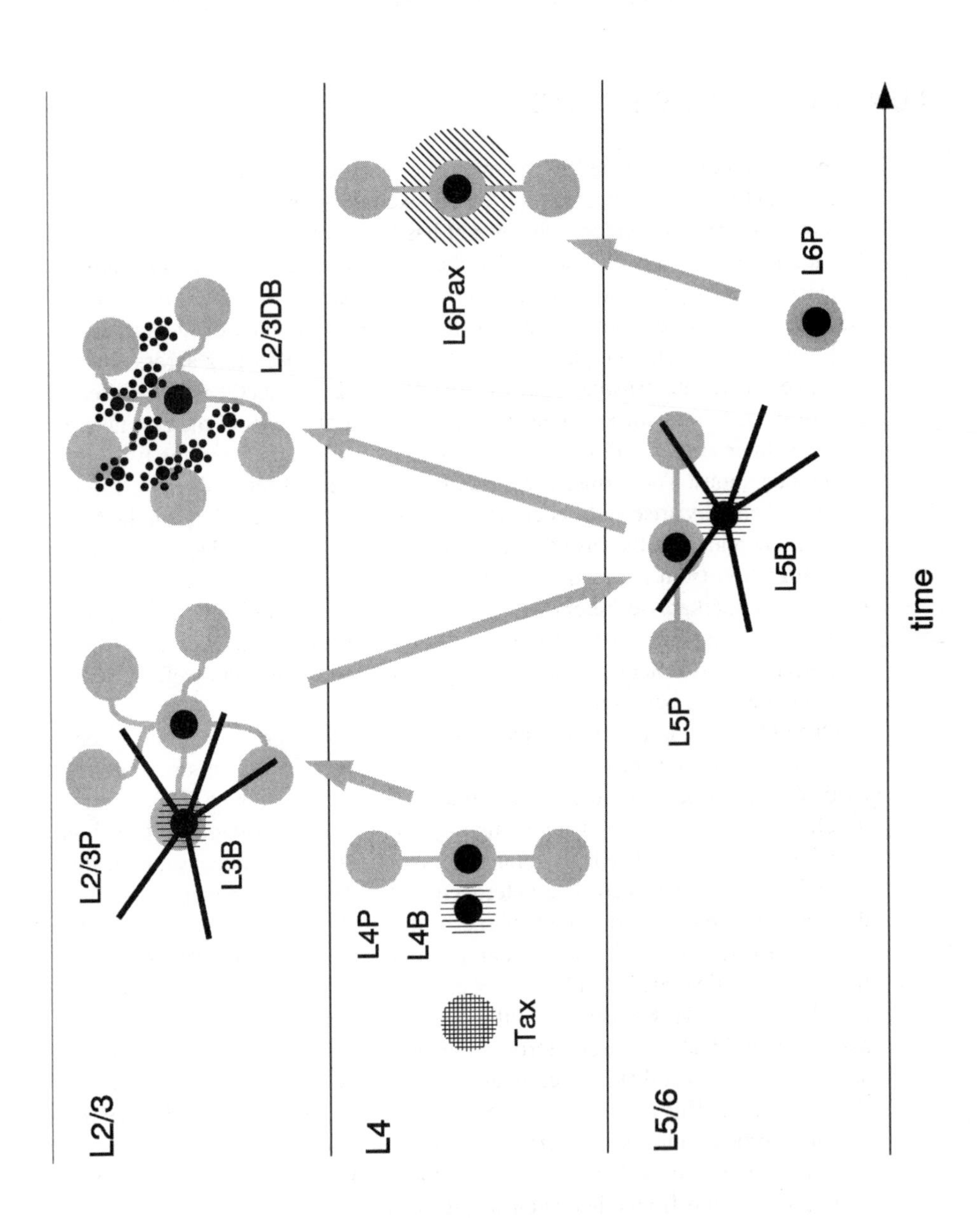
L2/3
L4
L5/6
Tax
L4P
L4B
L2/3P
L3B
L2/3DB
L6Pax
L6P
L5P
L5B
time

An Anatomical Model of Cortical Function

A simple model of cortical processing, consistent with the major features of cortical circuits discussed in this review, is as follows (see Figure 6): A patch of superficial pyramidal neurons receive feedforward excitatory input from subcortical, inter-areal, and intra-areal sources. In addition to their interactions with their close neighbors within their patch, the members of this patch also receive feedback from a number of sources: from deep pyramidal cells immediately beneath their patch, from other close patches within the superficial layers, and from subcortical inter-areal connections. Thus, the neurons of a superficial patch, taken as a group, receive a sample of thalamic input (some preprocessed by layer 4), a sample of surrounding and remote superficial patches, and a sample of the output from their corresponding deep pyramidal neurons.

All of these inputs are processed by the dendrites of the superficial pyramids whose signal transfer properties can be adjusted dynamically by the pattern of the vertical inputs from smooth cells (e.g., double bouquet cells). The superficial

Figure 5 Schematic showing the laminar-temporal evolution of interactions between some important neuronal types, following an input from the thalamus. Time unfolds toward the right and has a duration of 5 synaptic crossings. Neuronal elements are shown in plan (tangential) view, but located in their laminae (*L2/3*, *L4*, and *L5/6*). The relative sizes of axonal arbors, patches, inter-patch distances, etc. conform as reasonably as possible to Figure 3. The patchy axonal arborizations of excitory neurons are shown as connected gray disks. Their dendritic arborizations are denoted as smaller black discs superimposed on the central axonal clusters. The dendritic arbors of inhibitory neurons are also shown in black, but superimposed on vertical hatching, which denotes their dominate axonal arborization. Basket cells have in addition a few thin radial axons (*radiating black lines*). Double bouquet cells have a small black dendritic arborization surrounded by black dots that denote their vertically oriented axonal arbors (see also Figures 1, 3, and 4). Excitatory interlaminar effects are indicated by gray arrows. Intralaminar effects are unmarked. Thalamic afferents (*Tax*) activate spiny stellate (*L4P*) and small basket (*L4B*) neurons in layer 4. The stellates activate pyramidal cells (*L2/3P*) and basket cells (*L3B*) in layer 3. Note the region of influence of the large basket cell (*L3B*) relative to the patches of the pyramidal cells. The primary arbor of the basket cell matches approximately the pyramidal patch size, but unlike the pyramidal patches, the basket cell's thin radial arborizations focus their longer range inhibitory effect along restricted, nearly radial paths. The superficial pyramidal cells activate the pyramidal cells of layer 5 (*L5P*), which in turn activate those of layer 6 (*L6P*). Both the superficial and deep pyramidal cells activate superficial neurons. Importantly, they also activate the vertically disposed double bouquet cells. The pattern of activation of double bouquet cells could dynamically determine the input-output relations computed by the dendrites of the various L2/3Ps that they contact (see Figures 4 and 6). The layer 6 pyramidal cells project to layer 4, where their wide arbors (*L6Pax*) combine with thalamic afferents to shape activation of layer 4 spiny stellates.

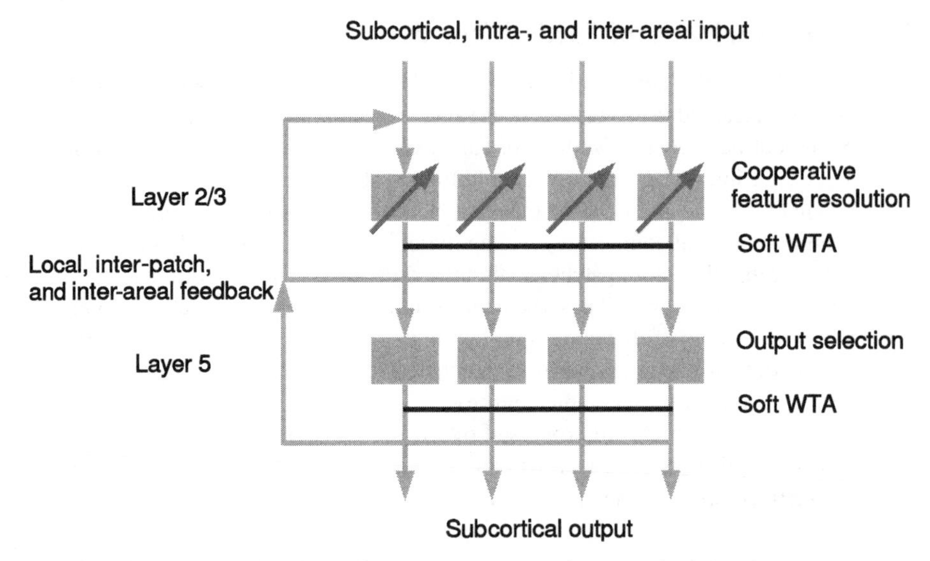

Figure 6 Simple model of cortical processing incorporating the principal features of cortical circuits. A patch of superficial pyramidal neurons receive feedforward input from subcortical, inter-areal, and intra-areal excitatory sources. They also receive recurrent input from other local superficial and deep pyramidal cells. These inputs are processed by dendrites of the superficial pyramidal neurons (*upper gray rectangles*, layer 2/3) whose signal transfer properties are adjusted dynamically by the pattern of vertical smooth cell inputs (*oblique dark gray arrows*). The outputs of the superficial pyramids participate in a selection network (e.g., soft winner-take-all mechanism) mediated by the horizontal smooth cells (*upper horizontal dark gray line*). These outputs of the superficial pyramids adjust the pattern of vertical smooth cell activation. In this way, the superficial layer neurons within and between patches, and within and between areas, cooperate to resolve a consistent interpretation. The layer 5 pyramids (*lower gray rectangles*) have a similar soft selection configuration (*lower dark gray line*) to process local superficial signals and decide on the output to motor structures.

pyramids collectively participate in a selection network, mediated by the horizontal inputs from the smooth cells that control their outputs (e.g., basket and chandelier cells). The selection mechanism is a soft winner-take-all or soft MAX mechanism, which are important elements of many neuronal network models (Maass 2000, Riesenhuber & Poggio 1999, Yuille & Geiger 2003). The outputs of the selected superficial pyramids feed back to adapt the pattern of vertical smooth cell activation. In this way, the superficial layer neurons within and between patches, and within and between areas, cooperate to explore all possible interpretations of input, and so select an interpretation consistent with their various subcortical inputs.

The superficial layers are organized to distribute and explore possible interpretations, whereas the deeper layers are organized to exploit the evolving interpretations. The pyramidal cells of layer 5 that drive subcortical structures involved in action (e.g., basal ganglia, colliculus, ventral spinal cord) decide the output of the cortical circuits. The same layer 5 pyramidal cells influence the ongoing input by their connection to layer 6 pyramidal cells that connect to the thalamic input layers. The explorative processing in the superficial layers is constrained via the recurrent projection from other layer 5 pyramidal cells to conform to the output that has already been decided. These layer 5 pyramidal cells are also the origin of the feedback projections to the superficial layers of other cortical areas. In this way, they also provide additional contextual information to the evolving interpretations occurring in the superficial layers of other cortical areas.

Clearly, this model is a tentative hypothesis of how the generic circuits might express themselves functionally. However, its strength is that it casts antomical data in a way that is accessible to theoreticians and systems physiologists. The investigation of neocortical structure and its development has entered an exciting phase in which the detailed organization is accessible to experiment and essential to the theoretical understanding of cortical computation. It is thus a curious paradox that while molecular biology has long recognized the central importance of detailed structural studies for understanding function, the same cannot be said for contemporary neuroscience.

ACKNOWLEDGMENTS

We thank our colleagues John Anderson and Tom Binzegger for their collaboration; Klaus Hepp, Marie-Claude Hepp-Reymond, and Christof Koch for reading drafts; and the Human Frontiers Science Program, the European Union, and the Körber Foundation for financial support.

The *Annual Review of Neuroscience* is online at http://neuro.annualreviews.org

LITERATURE CITED

Ahmed B, Anderson JC, Douglas RJ, Martin KAC, Nelson JC. 1994. Polyneuronal innervation of spiny stellate neurons in cat visual cortex. *J. Comp. Neurol.* 341:39–49

Amir Y, Harel M, Malach R. 1993. Cortical hierarchy reflected in the organization of intrinsic connections in macaque monkey visual cortex. *J. Comp. Neurol.* 334:19–46

Anderson JC, Binzegger T, Kahana O, Martin KA, Segev I. 1999. Dendritic asymmetry cannot account for directional responses of neurons in visual cortex. *Nat. Neurosci.* 2:820–24

Anderson JC, Martin KA, Whitteridge D. 1993. Form, function, and intracortical projections of neurons in the striate cortex of the monkey macacus nemestrinus. *Cereb. Cortex* 3:412–20

Bannister NJ, Nelson JC, Jack JJ. 2002. Excitatory inputs to spiny cells in layers 4 and 6 of cat striate cortex. *Philos. Trans. R. Soc. London B* 357:1793–808

Barbas H. 1986. Pattern in the laminar origin of corticocortical connections. *J. Comp. Neurol.* 252:415–22

Barone P, Batardiere A, Knoblauch K, Kennedy H. 2000. Laminar distribution of neurons in extrastriate areas projecting to visual areas

V1 and V4 correlates with the hierarchical rank and indicates the operation of a distance rule. *J. Neurosci.* 20:3263–81

Batardiere A, Barone P, Dehay C, Kennedy H. 1998. Area-specific laminar distribution of cortical feedback neurons projecting to cat area 17: quantitative analysis in the adult and during ontogeny. *J. Comp. Neurol.* 396:493–510

Beaulieu C, Kisvárday Z, Somogyi P, Cynader M, Cowey A. 1992. Quantitative distribution of GABA-immunopositive and -immunonegative neurons and synapses in the monkey striate cortex (area 17). *Cereb. Cortex* 2:295–309

Bernardo KL, McCasland JS, Woolsey TA. 1990a. Local axonal trajectories in mouse barrel cortex. *Exp. Brain Res.* 82:247–53

Bernardo KL, McCasland JS, Woolsey TA, Strominger RN. 1990b. Local intra- and interlaminar connections in mouse barrel cortex. *J. Comp. Neurol.* 291:231–55

Blasdel G, Campbell D. 2001. Functional retinotopy of monkey visual cortex. *J. Neurosci.* 21:8286–301

Blasdel GG. 1992. Orientation selectivity, preference, and continuity in monkey striate cortex. *J. Neurosci.* 12:3139–61

Blasdel GG, Lund JS, Fitzpatrick D. 1985. Intrinsic connections of macaque striate cortex: axonal projections of cells outside lamina 4C. *J. Neurosci.* 5:3350–69

Blumcke I, Hof PR, Morrison JH, Celio MR. 1990. Distribution of parvalbumin immunoreactivity in the visual cortex of Old World monkeys and humans. *J. Comp. Neurol.* 301:417–32

Bolton J. 1910. A contribution to the localization of cerebral function, based on the clinico-pathological study of mental disease. *Brain* 22:26–147

Bosking WH, Zhang Y, Schofield B, Fitzpatrick D. 1997. Orientation selectivity and the arrangement of horizontal connections in tree shrew striate cortex. *J. Neurosci.* 17:2112–27

Briggs F, Callaway EM. 2001. Layer-specific input to distinct cell types in layer 6 of monkey primary visual cortex. *J. Neurosci.* 21:3600–8

Bullier J, Henry GH. 1979. Laminar distribution of first-order neurons and afferent terminals in cat striate cortex. *J. Neurophysiol.* 42:1271–81

Bullier J, Kennedy H, Salinger W. 1984. Branching and laminar origin of projections between visual cortical areas in the cat. *J. Comp. Neurol.* 228:329–41

Burkhalter A. 1989. Intrinsic connections of rat primary visual cortex: laminar organization of axonal projections. *J. Comp. Neurol.* 279:171–86

Burkhalter A, Charles V. 1990. Organization of local axon collaterals of efferent projection neurons in rat visual cortex. *J. Comp. Neurol.* 302:920–34

Callaway EM. 1998. Local circuits in primary visual cortex of the macaque monkey. *Annu. Rev. Neurosci.* 21:47–74

Callaway EM. 2002. Cell type specificity of local cortical connections. *J. Neurocytol.* 31:231–37

Callaway EM, Katz LC. 1990. Emergence and refinement of clustered horizontal connections in cat striate cortex. *J. Neurosci.* 10:1134–53

Campbell A. 1905. *Histological Studies on the Localization of Cerebral Function.* Cambridge, UK: Cambridge Univ. Press

Chapin JK, Sadeq M, Guise JL. 1987. Corticocortical connections within the primary somatosensory cortex of the rat. *J. Comp. Neurol.* 263:326–46

Chisum HJ, Mooser F, Fitzpatrick D. 2003. Emergent properties of layer 2/3 neurons reflect the collinear arrangement of horizontal connections in tree shrew visual cortex. *J. Neurosci.* 23:2947–60

Chklovskii DB, Schikorski T, Stevens CF. 2002. Wiring optimization in cortical circuits. *Neuron* 34:341–47

Conde F, Lund JS, Jacobowitz DM, Baimbridge KG, Lewis DA. 1994. Local circuit neurons immunoreactive for calretinin, calbindin D-28k or parvalbumin in monkey prefrontal cortex: distribution and morphology. *J. Comp. Neurol.* 341:95–116

Connor JR, Peters A. 1984. Vasoactive intestinal polypeptide-immunoreactive neurons in rat visual cortex. *Neuroscience* 12:1027–44

Cragg BG. 1967. The density of synapses and neurones in the motor and visual areas of the cerebral cortex. *J. Anat.* 101:639–54

Creutzfeldt O. 1993. *Cortex Cerebri: Performance, Structural and Functional Organization of the Cortex.* Berlin: Springer Verlag

Crick F, Koch C. 1998. Constraints on cortical and thalamic projections: the no-strong-loops hypothesis. *Nature* 391:245–50

DeFelipe J. 1993. Neocortical neuronal diversity: chemical heterogeneity revealed by colocalization studies of classic neurotransmitters, neuropeptides, calcium-binding proteins, and cell surface molecules. *Cereb. Cortex* 3:273–89

DeFelipe J. 2002. Cortical interneurons: from Cajal to 2001. *Prog. Brain Res.* 136:215–38

DeFelipe J, Alonso-Nanclares L, Arellano JI. 2002. Microstructure of the neocortex: comparative aspects. *J. Neurocytol.* 31:299–316

DeFelipe J, Hendry SH, Hashikawa T, Molinari M, Jones EG. 1990. A microcolumnar structure of monkey cerebral cortex revealed by immunocytochemical studies of double bouquet cell axons. *Neuroscience* 37:655–73

DeFelipe J, Hendry SH, Jones EG. 1989a. Synapses of double bouquet cells in monkey cerebral cortex visualized by calbindin immunoreactivity. *Brain Res.* 503:49–54

DeFelipe J, Hendry SH, Jones EG. 1989b. Visualization of chandelier cell axons by parvalbumin immunoreactivity in monkey cerebral cortex. *Proc. Natl. Acad. Sci. USA* 86:2093–97

Demeulemeester H, Arckens L, Vandesande F, Orban GA, Heizmann CW, Pochet R. 1991. Calcium binding proteins and neuropeptides as molecular markers of GABAergic interneurons in the cat visual cortex. *Exp. Brain Res.* 84:538–44

Demeulemeester H, Vandesande F, Orban GA, Brandon C, Vanderhaeghen JJ. 1988. Heterogeneity of GABAergic cells in cat visual cortex. *J. Neurosci.* 8:988–1000

Demeulemeester H, Vandesande F, Orban GA, Heizmann CW, Pochet R. 1989. Calbindin D-28K and parvalbumin immunoreactivity is confined to two separate neuronal subpopulations in the cat visual cortex, whereas partial coexistence is shown in the dorsal lateral geniculate nucleus. *Neurosci. Lett.* 99:6–11

Douglas R, Martin K, Witteridge D. 1989. A canonical microcircuit for neocortex. *Neural Comput.* 1:480–88

Eckenstein F, Baughman RW. 1984. Two types of cholinergic innervation in cortex, one colocalized with vasoactive intestinal polypeptide. *Nature* 309:153–55

Elston GN. 2002. Cortical heterogeneity: implications for visual processing and polysensory integration. *J. Neurocytol.* 31:317–35

Elston GN. 2003. The pyramidal neuron in occipital, temporal and prefrontal cortex of the owl monkey (*Aotus trivirgatus*): regional specialization in cell structure. *Eur. J. Neurosci.* 17:1313–18

Elston GN, Benavides-Piccione R, DeFelipe J. 2001. The pyramidal cell in cognition: a comparative study in human and monkey. *J. Neurosci.* 21:RC163

Elston GN, Rosa MG. 1998. Morphological variation of layer III pyramidal neurones in the occipitotemporal pathway of the macaque monkey visual cortex. *Cereb. Cortex* 8:278–94

Elston GN, Rosa MG. 2000. Pyramidal cells, patches, and cortical columns: a comparative study of infragranular neurons in TEO, TE, and the superior temporal polysensory area of the macaque monkey. *J. Neurosci.* 20(RC117):1–5

Elston GN, Tweedale R, Rosa MG. 1999. Cellular heterogeneity in cerebral cortex: a study of the morphology of pyramidal neurones in visual areas of the marmoset monkey. *J. Comp. Neurol.* 415:33–51

Farinas I, DeFelipe J. 1991. Patterns of synaptic input on corticocortical and corticothalamic cells in the cat visual cortex. II. the axon initial segment. *J. Comp. Neurol.* 304:70–77

Felleman DJ, Van Essen DC. 1991. Distributed hierarchical processing in the primate cerebral cortex. *Cereb. Cortex* 1:1–47

Fisken RA, Garey LJ, Powell TP. 1975. The intrinsic, association and commissural connections of area 17 on the visual cortex. *Philos. Trans. R. Soc. London B* 272:487–536

Fitzpatrick D. 1996. The functional organization of local circuits in visual cortex: insights from the study of tree shrew striate cortex. *Cereb. Cortex* 6:329–41

Fitzpatrick D. 2000. Seeing beyond the receptive field in primary visual cortex. *Curr. Opin. Neurobiol.* 10:438–43

Fitzpatrick D, Lund JS, Blasdel GG. 1985. Intrinsic connections of macaque striate cortex: afferent and efferent connections of lamina 4C. *J. Neurosci.* 5:3329–49

Fitzpatrick D, Raczkowski D. 1990. Innervation patterns of single physiologically identified geniculocortical axons in the striate cortex of the tree shrew. *Proc. Natl. Acad. Sci. USA* 87:449–53

Fitzpatrick D, Usrey WM, Schofield BR, Einstein G. 1994. The sublaminar organization of corticogeniculate neurons in layer 6 of macaque striate cortex. *Vis. Neurosci.* 11:307–15

Fitzpatrick DC, Olsen JF, Suga N. 1998. Connections among functional areas in the mustached bat auditory cortex. *J. Comp. Neurol.* 391:366–96

Freund TF, Martin KA, Somogyi P, Whitteridge D. 1985. Innervation of cat visual areas 17 and 18 by physiologically identified X- and Y-type thalamic afferents. II. Identification of postsynaptic targets by gaba immunocytochemistry and golgi impregnation. *J. Comp. Neurol.* 242:275–91

Fujita I. 2002. The inferior temporal cortex: architecture, computation, and representation. *J. Neurocytol.* 31:359–71

Fujita I, Fujita T. 1996. Intrinsic connections in the macaque inferior temporal cortex. *J. Comp. Neurol.* 368:467–86

Gabbott PLA, Bacon SJ. 1996a. Local circuit neurons in the medial prefrontal cortex (areas 24a,b,c, 25 and 32) in the monkey: I. Cell morphology and morphometrics. *J. Comp. Neurol.* 364:567–608

Gabbott PLA, Bacon SJ. 1996b. Local circuit neurons in the medial prefrontal cortex (areas 24a,b,c, 25 and 32) in the monkey: II. Quantitative areal and laminar distributions. *J. Comp. Neurol.* 364:609–36

Garey L, Powell T. 1971. An experimental study of the termination of the lateral geniculo-cortical pathway in the cat and monkey. *Proc. R. Soc. London B* 179:21–40

Ghosh S, Fyffe RE, Porter R. 1988. Morphology of neurons in area 4 gamma of the cat's cortex studied with intracellular injection of HRP. *J. Comp. Neurol.* 277:290–312

Ghosh S, Porter R. 1988. Morphology of pyramidal neurones in monkey motor cortex and the synaptic actions of their intracortical axon collaterals. *J. Physiol.* 400:593–615

Gil Z, Connors BW, Amitai Y. 1999. Efficacy of thalamocortical and intracortical synaptic connections: quanta, innervation, and reliability. *Neuron* 23:385–97

Gilbert CD. 1983. Microcircuitry of the visual cortex. *Annu. Rev. Neurosci.* 6:217–47

Gilbert CD. 1992. Horizontal integration and cortical dynamics. *Neuron* 9:1–13

Gilbert CD, Kelly JP. 1975. The projections of cells in different layers of the cat's visual cortex. *J. Comp. Neurol.* 163:81–105

Gilbert CD, Wiesel TN. 1983. Functional organization of the visual cortex. *Prog. Brain Res.* 58:209–18

Gilbert CD, Wiesel TN. 1989. Columnar specificity of intrinsic horizontal and corticocortical connections in cat visual cortex. *J. Neurosci.* 9:2432–42

Gottlieb JP, Keller A. 1997. Intrinsic circuitry and physiological properties of pyramidal neurons in rat barrel cortex. *Exp. Brain Res.* 115:47–60

Gulyás AI, Hájos N, Freund TF. 1996. Interneurons containing calretinin are specialized to control other interneurons in the rat hippocampus. *J. Neurosci.* 16:3397–411

Hendry SH, Jones EG, Emson PC. 1984. Morphology, distribution, and synaptic relations of somatostatin- and neuropeptide

Y-immunoreactive neurons in rat and monkey neocortex. *J. Neurosci.* 4:2497–517

Hendry SH, Jones EG, Emson PC, Lawson DE, Heizmann CW, Streit P. 1989. Two classes of cortical GABA neurons defined by differential calcium binding protein immunoreactivities. *Exp. Brain Res.* 76:467–72

Hilgetag CC, O'Neill MA, Young MP. 1996. Indeterminate organization of the visual system. *Science* 271:776–77

Hirsch JA, Gallagher CA, Alonso JM, Martinez LM. 1998. Ascending projections of simple and complex cells in layer 6 of the cat striate cortex. *J. Neurosci.* 18:8086–94

Hoeflinger BF, Bennett-Clarke CA, Chiaia NL, Killackey HP, Rhoades RW. 1995. Patterning of local intracortical projections within the vibrissae representation of rat primary somatosensory cortex. *J. Comp. Neurol.* 354:551–63

Hogan D, Terwilleger ER, Berman NE. 1992. Development of subpopulations of GABAergic neurons in cat visual cortical areas. *NeuroReport* 3:1069–72

Hubel D, Wiesel T. 1962. Receptive fields, binocular interaction and functional architecture in the cat's visual cortex. *J. Physiol.* 160:106–54

Hubel DH. 1982. Cortical neurobiology: a slanted historical perspective. *Annu. Rev. Neurosci.* 5:363–70

Huntley GW, Jones EG. 1991. Relationship of intrinsic connections to forelimb movement representations in monkey motor cortex: a correlative anatomic and physiological study. *J. Neurophysiol.* 66:390–413

Huxlin KR, Pasternak T. 2001. Long-term neurochemical changes after visual cortical lesions in the adult cat. *J. Comp. Neurol.* 429: 221–41

Jones E. 1975. Varieties and distribution of nonpyramidal cells in the somatic sensory cortex of the squirrel monkey. *J. Comp. Neurol.* 160:205–68

Jones EG. 1999. Making brain connections: neuroanatomy and the work of TPS Powell, 1923–1996. *Annu. Rev. Neurosci.* 22:49–103

Jones EG, Wise SP. 1977. Size, laminar and columnar distribution of efferent cells in the sensory-motor cortex of monkeys. *J. Comp. Neurol.* 175:391–438

Juliano SL, Friedman DP, Eslin DE. 1990. Corticocortical connections predict patches of stimulus-evoked metabolic activity in monkey somatosensory cortex. *J. Comp. Neurol.* 298:23–39

Kaas JH, Krubitzer LA, Johanson KL. 1989. Cortical connections of areas 17 (V-I) and 18 (V-II) of squirrels. *J. Comp. Neurol.* 281: 426–46

Katz LC. 1987. Local circuitry of identified projection neurons in cat visual cortex brain slices. *J. Neurosci.* 7:1223–49

Katz LC, Gilbert CD, Wiesel TN. 1989. Local circuits and ocular dominance columns in monkey striate cortex. *J. Neurosci.* 9:1389–99

Kawaguchi Y, Kondo S. 2002. Parvalbumin, somatostatin and cholecystokinin as chemical markers for specific GABAergic interneuron types in the rat frontal cortex. *J. Neurocytol.* 31:277–87

Kawaguchi Y, Kubota Y. 1997. GABAergic cell subtypes and their synaptic connections in rat frontal cortex. *Cereb. Cortex* 7:476–86

Kennedy H, Bullier J. 1985. A double-labeling investigation of the afferent connectivity to cortical areas V1 and V2 of the macaque monkey. *J. Neurosci.* 5:2815–30

Kisvárday ZF, Eysel UT. 1992. Cellular organization of reciprocal patchy networks in layer III of cat visual cortex (area 17). *Neuroscience* 46(2):275–86

Kisvárday ZF, Ferecskó AS, Kovács K, Buzás P, Budd JM, Eysel UT. 2002. One axon—multiple functions: specificity of lateral inhibitory connections by large basket cells. *J. Neurocytol.* 31:255–64

Kisvárday ZF, Martin K, Friedlander M, Somogyi P. 1987. Evidence for interlaminar inhibitory circuits in the striate cortex of the cat. *J. Comp. Neurol.* 260:1–19

Kisvárday ZF, Toth E, Rausch M, Eysel UT. 1997. Orientation-specific relationship between populations of excitatory and inhibitory lateral connections in the visual cortex of the cat. *Cereb. Cortex* 7:605–18

Koulakov AA, Chklovskii DB. 2001. Orientation preference patterns in mammalian visual cortex: a wire length minimization approach. *Neuron* 29:519–27

Kritzer MF, Goldman-Rakic PS. 1995. Intrinsic circuit organization of the major layers and sublayers of the dorsolateral prefrontal cortex in the rhesus monkey. *J. Comp. Neurol.* 359:131–43

Larkman AU. 1991. Dendritic morphology of pyramidal neurones of the visual cortex of the rat: I. Branching patterns. *J. Comp. Neurol.* 306:307–19

Latawiec D, Martin KA, Meskenaite V. 2000. Termination of the geniculocortical projection in the striate cortex of macaque monkey: a quantitative immunoelectron microscopic study. *J. Comp. Neurol.* 419:306–19

Levitt JB, Lewis DA, Yoshioka T, Lund JS. 1993. Topography of pyramidal neuron intrinsic connections in macaque monkey prefrontal cortex (areas 9 and 46). *J. Comp. Neurol.* 338:360–76

Levitt JB, Yoshioka T, Lund JS. 1994. Intrinsic cortical connections in macaque visual area V2: evidence for interaction between different functional streams. *J. Comp. Neurol.* 342:551–70

Lewis DA, Lund JS. 1990. Heterogeneity of chandelier neurons in monkey neocortex: corticotropin-releasing factor- and parvalbumin-immunoreactive populations. *J. Comp. Neurol.* 293:599–615

Lewis DA, Melchitzky DS, Burgos GG. 2002. Specificity in the functional architecture of primate prefrontal cortex. *J. Neurocytol.* 31:265–76

Livingstone MS, Hubel DH. 1984. Specificity of intrinsic connections in primate primary visual cortex. *J. Neurosci.* 4:2830–35

Lowel S, Singer W. 1992. Selection of intrinsic horizontal connections in the visual cortex by correlated neuronal activity. *Science* 255:209–12

Luhmann HJ, Martinez-Millan L, Singer W. 1986. Development of horizontal intrinsic connections in cat striate cortex. *Exp. Brain Res.* 63:443–48

Lund JS, Henry GH, MacQueen CL, Harvey AR. 1979. Anatomical organization of the primary visual cortex (area 17) of the cat. A comparison with area 17 of the macaque monkey. *J. Comp. Neurol.* 184:599–618

Lund JS, Lund RD, Hendrickson AE, Bunt AH, Fuchs AF. 1975. The origin of efferent pathways from the primary visual cortex, area 17, of the macaque monkey as shown by retrograde transport of horseradish peroxidase. *J. Comp. Neurol.* 164:287–303

Lund JS, Wu CQ. 1997. Local circuit neurons of macaque monkey striate cortex: IV. Neurons of laminae 1-3A. *J. Comp. Neurol.* 384:109–26

Lund JS, Yoshioka T, Levitt JB. 1993. Comparison of intrinsic connectivity in different areas of macaque monkey cerebral cortex. *Cereb. Cortex* 3:148–62

Maass W. 2000. On the computational power of winner-take-all. *Neural Comput.* 12:2519–35

Malach R. 1992. Dendritic sampling across processing streams in monkey striate cortex. *J. Comp. Neurol.* 315:303–12

Malach R, Amir Y, Harel M, Grinvald A. 1993. Relationship between intrinsic connections and functional architecture revealed by optical imaging and in vivo targeted biocytin injections in primate striate cortex. *Proc. Natl. Acad. Sci. USA* 90:10469–73

Malach R, Schirman TD, Harel M, Tootell RB, Malonek D. 1997. Organization of intrinsic connections in owl monkey area MT. *Cereb. Cortex* 7(4):386–93

Martin K, Whitteridge D. 1984. Form, function and intracortical projection of spiny neurones in the striate visual cortex of the cat. *J. Physiol.* 353:463–504

Mead C. 1990. Neuromorphic electronic systems. *Proc. IEEE* 78:1629–36

Melchitzky DS, Sesack SR, Pucak ML, Lewis DA. 1998. Synaptic targets of pyramidal neurons providing intrinsic horizontal connections in monkey prefrontal cortex. *J. Comp. Neurol.* 390:211–24

Meskenaite V. 1997. Calretinin-immunoreactive local circuit neurons in the area 17 of the cynomolgus monkey, *Macaca fascicularis*. *J. Comp. Neurol.* 379:113–32

Mitchison G. 1991. Neuronal branching patterns and the economy of cortical wiring. *Proc. R. Soc. London B* 245:151–58

Mitchison G. 1992. Axonal trees and cortical architecture. *TINS* 15:122–26

Mitchison G, Crick F. 1982. Long axons within the striate cortex: their distribution, orientation, and patterns of connection. *Proc. Natl. Acad. Sci. USA* 79:3661–65

Muly EC, Fitzpatrick D. 1992. The morphological basis for binocular and on/off convergence in tree shrew striate cortex. *J. Neurosci.* 12:1319–34

Nimchinsky EA, Gilissen E, Allman JM, Perl DP, Erwin JM, Hof PR. 1999. A neuronal morphologic type unique to humans and great apes. *Proc. Natl. Acad. Sci. USA* 96:5268–73

Ojima H, Honda CN, Jones EG. 1991. Patterns of axon collateralization of identified supragranular pyramidal neurons in the cat auditory cortex. *Cereb. Cortex* 1:80–94

Ojima H, Honda CN, Jones EG. 1992. Characteristics of intracellularly injected infragranular pyramidal neurons in cat primary auditory cortex. *Cereb. Cortex* 2:197–216

O'Kusky J, Colonnier M. 1982. A laminar analysis of the number of neurons, glia, and synapses in the adult cortex (area 17) of adult macaque monkeys. *J. Comp. Neurol.* 210:278–90

Perkel DJ, Bullier J, Kennedy H. 1986. Topography of the afferent connectivity of area 17 in the macaque monkey: a double-labelling study. *J. Comp. Neurol.* 253:374–402

Peters A, Harriman KM. 1988. Enigmatic bipolar cell of rat visual cortex. *J. Comp. Neurol.* 267:409–32

Peters A, Regidor J. 1981. A reassessment of the forms of nonpyramidal neurons in area 17 of cat visual cortex. *J. Comp. Neurol.* 203:685–716

Peters A, Sethares C. 1997. The organization of double bouquet cells in monkey striate cortex. *J. Neurocytol.* 26:779–97

Petersen CC, Sakmann B. 2000. The excitatory neuronal network of rat layer 4 barrel cortex. *J. Neurosci.* 20:7579–86

Porter LL, Matin D, Keller A. 2000. Characteristics of GABAergic neurons and their synaptic relationships with intrinsic axons in the cat motor cortex. *Somatosens. Mot. Res.* 17:67–80

Powell T. 1973. The organization of the major functional areas of the cerebral cortex. *Symp. Zool. Soc. London* 33:235–52

Powell T. 1981. Certain aspects of the intrinsic organisation of the cerebral cortex. In *Brain Mechanisms and Perceptual Awareness*, ed. O Pompeiano, CA Marsan, pp. 1–19. New York: Raven

Pucak ML, Levitt JB, Lund JS, Lewis DA. 1996. Patterns of intrinsic and associational circuitry in monkey prefrontal cortex. *J. Comp. Neurol.* 376:614–30

Rakic P, Bourgeois JP, Eckenhoff MF, Zecevic N, Goldman-Rakic PS. 1986. Concurrent overproduction of synapses in diverse regions of the primate cerebral cortex. *Science* 232:232–35

Ramon y Cajal S. 1911. *Histologie du Systeme Nerveux de l'Homme et des Vertebres*. Vol. 2. Paris: Maloine

Ramon y Cajal S. 1937. *Recollections of My Life*. Transl. EH Craigie, J Cano, 1989. Philadelphia, PA: Am. Philos. Soc.

Riesenhuber M, Poggio T. 1999. Hierarchical models of object recognition in cortex. *Nat. Neurosci.* 2:1019–25

Rockland K. 1997. Elements of cortical architecture hierarchy revisited. In *Cerebral Cortex*, ed. K Rockland, J Kaas, A Peters, Vol. 12, pp. 243–93. New York/London: Plenum Press

Rockland KS. 1996. Two types of corticopulvinar terminations: round (type 2) and elongate (type 1). *J. Comp. Neurol.* 368:57–87

Rockland KS, Lund JS. 1983. Intrinsic laminar lattice connections in primate visual cortex. *J. Comp. Neurol.* 216:303–18

Rockland KS, Lund JS, Humphrey AL. 1982.

Anatomical binding of intrinsic connections in striate cortex of tree shrews (*Tupaia glis*). *J. Comp. Neurol.* 209:41–58

Rockland KS, Pandya DN. 1979. Laminar origins and terminations of cortical connections of the occipital lobe in the rhesus monkey. *Brain Res.* 179:3–20

Rogers JH. 1992. Immunohistochemical markers in rat cortex: co-localization of calretinin and calbindin-D28k with neuropeptides and GABA. *Brain Res.* 587:147–57

Rogers JH, Resibois A. 1992. calretinin and calbindin-D28k in rat brain: patterns of partial co-localization. *Neuroscience* 51:843–65

Schmechel DE, Vickrey BG, Fitzpatrick D, Elde RP. 1984. GABAergic neurons of mammalian cerebral cortex: widespread subclass defined by somatostatin content. *Neurosci. Lett.* 47:227–32

Schmidt KE, Kim DS, Singer W, Bonhoeffer T, Lowel S. 1997. Functional specificity of long-range intrinsic and interhemispheric connections in the visual cortex of strabismic cats. *J. Neurosci.* 17:5480–92

Schubert D, Kotter R, Zilles K, Luhmann HJ, Staiger JF. 2003. Cell type-specific circuits of cortical layer IV spiny neurons. *J. Neurosci.* 23:2961–70

Schuz A, Palm G. 1989. Density of neurons and synapses in the cerebral cortex of the mouse. *J. Comp. Neurol.* 286:442–55

Selemon LD, Goldman-Rakic PS. 1988. Common cortical and subcortical targets of the dorsolateral prefrontal and posterior parietal cortices in the rhesus monkey: evidence for a distributed neural network subserving spatially guided behavior. *J. Neurosci.* 8:4049–68

Sherman SM, Guillery R. 2001. *Exploring the Thalamus*. San Diego: Academic

Sherman SM, Guillery RW. 1996. Functional organization of thalamocortical relays. *J. Neurophysiol.* 76:1367–95

Sincich LC, Blasdel GG. 2001. Oriented axon projections in primary visual cortex of the monkey. *J. Neurosci.* 21:4416–26

Sloper JJ, Powell TP. 1979. A study of the axon initial segment and proximal axon of neurons in the primate motor and somatic sensory cortices. *Philos. Trans. R. Soc. London B* 285:173–97

Somogyi P, Cowey A. 1981. Combined golgi and electron microscopic study on the synapses formed by double bouquet cells in the visual cortex of the cat and monkey. *J. Comp. Neurol.* 195:547–66

Somogyi P, Hodgson AJ, Smith AD, Nunzi MG, Gorio A, Wu JY. 1984. Different populations of GABAergic neurons in the visual cortex and hippocampus of cat contain somatostatin- or cholecystokinin-immunoreactive material. *J. Neurosci.* 4:2590–603

Somogyi P, Kisvárday Z, Martin K, Whitteridge D. 1983. Synaptic connections of morphologically identified and physiologically characterized large basket cells in the striate cortex of the cat. *Neuroscience* 10:261–94

Stettler DD, Das A, Bennett J, Gilbert CD. 2002. Lateral connectivity and contextual interactions in macaque primary visual cortex. *Neuron* 36:739–50

Stratford KJ, Tarczy-Hornoch K, Martin KAC, Bannister NJ, Jack JJ. 1996. Excitatory synaptic inputs to spiny stellate cells in cat visual cortex. *Nature* 382:258–61

Sun XZ, Takahashi S, Cui C, Inoue M, Fukui Y. 2002. Distribution of calbindin-D28K immunoreactive neurons in rat primary motor cortex. *J. Med. Invest.* 49:35–39

Swindale NV. 1992. A model for the coordinated development of columnar systems in primate striate cortex. *Biol. Cybern.* 66:217–30

Szentágothai J. 1973. Synaptology of the visual cortex. In *Handbook of Sensory Physiology: Central Processing of Visual Information. Part B.*, ed. R Jung, vol. II, pp. 269–324. Berlin-Heidelberg, New York: Springer Verlag

Szentágothai J. 1978. The neuron network of the cerebral cortex: a functional interpretation. *Proc. R. Soc. London B* 201:219–48

Tarczy-Hornoch K, Martin KAC, Stratford KJ, Jack JJ. 1999. Intracortical excitation of spiny neurons in layer 4 of cat striate cortex in vitro. *Cereb. Cortex* 9:833–43

Tootell RB, Switkes E, Silverman MS, Hamilton SL. 1988. Functional anatomy of macaque striate cortex. II. Retinotopic organization. *J. Neurosci.* 8:1531–68

Trachtenberg JT, Stryker MP. 2001. Rapid anatomical plasticity of horizontal connections in the developing visual cortex. *J. Neurosci.* 21:3476–82

Tyler CJ, Dunlop SA, Lund RD, Harman AM, Dann JF, et al. 1998. Anatomical comparison of the macaque and marsupial visual cortex: common features that may reflect retention of essential cortical elements. *J. Comp. Neurol.* 400:449–68

Usrey WM, Fitzpatrick D. 1996. Specificity in the axonal connections of layer VI neurons in tree shrew striate cortex: evidence for distinct granular and supragranular systems. *J. Neurosci.* 16:1203–18

Usrey WM, Muly EC, Fitzpatrick D. 1992. Lateral geniculate projections to the superficial layers of visual cortex in the tree shrew. *J. Comp. Neurol.* 319:159–71

Van Brederode JF, Mulligan KA, Hendrickson AE. 1990. Calcium-binding proteins as markers for subpopulations of GABAergic neurons in monkey striate cortex. *J. Comp. Neurol.* 298:1–22

Van Essen DC, Newsome WT, Maunsell JH. 1984. The visual field representation in striate cortex of the macaque monkey: asymmetries, anisotropies, and individual variability. *Vis. Res.* 24:429–48

Vincent SR, Satoh K, Armstrong DM, Fibiger HC. 1983. Substance P in the ascending cholinergic reticular system. *Nature* 306:688–91

Wang Y, Gupta A, Toledo-Rodriguez M, Wu CZ, Markram H. 2002. Anatomical, physiological, molecular and circuit properties of nest basket cells in the developing somatosensory cortex. *Cereb. Cortex* 12:395–410

Winfield DA, Powell TP. 1983. Laminar cell counts and geniculo-cortical boutons in area 17 of cat and monkey. *Brain Res.* 277:223–29

Wise SP. 1975. The laminar organization of certain afferent and efferent fiber systems in the rat somatosensory cortex. *Brain Res.* 90:139–42

Wiser AK, Callaway EM. 1996. Contributions of individual layer 6 pyramidal neurons to local circuitry in macaque primary visual cortex. *J. Neurosci.* 16:2724–39

Yoshioka T, Blasdel GG, Levitt JB, Lund JS. 1996. Relation between patterns of intrinsic lateral connectivity, ocular dominance, and cytochrome oxidase-reactive regions in macaque monkey striate cortex. *Cereb. Cortex* 6:297–310

Yoshioka T, Levitt JB, Lund JS. 1992. Intrinsic lattice connections of macaque monkey visual cortical area V4. *J. Neurosci.* 12:2785–802

Yuille AL, Geiger D. 2003. Winner-take-all networks. In *The Handbook of Brain Theory and Neural Networks*, ed. M Arbib, pp. 1228–31. Cambridge, MA: MIT Press

Zeki SM. 1978a. The cortical projections of foveal striate cortex in the rhesus monkey. *J. Physiol.* 277:227–44

Zeki SM. 1978b. Functional specialisation in the visual cortex of the rhesus monkey. *Nature* 274:423–28

Zhang ZW, Deschenes M. 1997. Intracortical axonal projections of lamina VI cells of the primary somatosensory cortex in the rat: a single-cell labeling study. *J. Neurosci.* 17:6365–79

Annu. Rev. Neurosci. 2004. 27:453–85
doi: 10.1146/annurev.neuro.27.070203.144255

THE NEUROBIOLOGY OF THE ASCIDIAN TADPOLE LARVA: Recent Developments in an Ancient Chordate

Ian A. Meinertzhagen,[1,2] Patrick Lemaire,[3] and Yasushi Okamura[4,5]

[1]*Marine Biological Laboratory, Woods Hole, Massachusetts 02543*
[2]*Life Sciences Centre, Dalhousie University, Halifax, Canada B3H 4J1; email: iam@dal.ca*
[3]*Laboratoire de Génétique et Physiologie du Développement, IBDM, CNRS/INSERM/Université de la Méditerranée/AP de Marseille, Case 907, Campus de Luminy, F-13288 Marseille Cedex 09, France; email: lemaire@ibdm.univ-mrs.fr*
[4]*Section of Developmental Neurophysiology, Okazaki Institute for Integrative Bioscience, Nishigonaka 38, Myodaiji, Okazaki 444-8585, Aichi, Japan*
[5]*Molecular Neurobiology Group, Neuroscience Research Institute, National Institute of Advanced Industrial Science, Higashi 1–1, Tsukuba, Ibaraki 305, Japan; email: maboya@dc4.so-net.ne.jp*

Key Words chordate evolution, neural plate, neurulation, cell lineage, ion channels

■ **Abstract** With little more than 330 cells, two thirds within the sensory vesicle, the CNS of the tadpole larva of the ascidian *Ciona intestinalis* provides us with a chordate nervous system in miniature. Neurulation, neurogenesis and its genetic bases, as well as the gene expression territories of this tiny constituency of cells all follow a chordate plan, giving rise in some cases to frank structural homologies with the vertebrate brain. Recent advances are fueled by the release of the genome and EST expression databases and by the development of methods to transfect embryos by electroporation. Immediate prospects to test the function of neural genes are based on the isolation of mutants by classical genetics and insertional mutagenesis, as well as by the disruption of gene function by morpholino antisense oligo-nucleotides. Coupled with high-speed video analysis of larval swimming, optophysiological methods offer the prospect to analyze at single-cell level the function of a CNS built on a vertebrate plan.

INTRODUCTION

> "Nothing in biology makes sense except in the light of evolution."
>
> (Dobzhansky 1973)

Animals in which a back and belly can be recognized, so-called bilaterians, fall into two major groups. Protostomes include model systems such as *Drosophila*

0147-006X/04/0721-0453$14.00

and *Caenorhabditis elegans*, and deuterostomes include the vertebrates such as ourselves. Three main groups of deuterostomes, echinoderms, hemichordates, and chordates, are widely held to have diverged from a common ancestor during the Cambrian explosion around 550 mya (Valentine et al. 1996). Although the debate about vertebrate ancestry, whether a hypothetical ancestor arose from a free-swimming ancestor or by a process of pedomorphosis from sessile forms like adult sea squirts, is still ongoing, many recent accounts (e.g., Wada 1998, Swalla et al. 2000) favor descent from a basal, free-swimming chordate.

The chordate lineage, which comprises two sibling groups, cephalochordates such as amphioxus and true chordates such as the more familiar vertebrates, is distinguished by possessing a larval notochord and dorsal tubular central nervous system (CNS). Urochordates, the sibling ancestral group, survive as a number of forms. Larvaceans, such as *Oikopleura*, develop directly from an embryo that possibly arose from a urochordate ancestor by neoteny (e.g., Garstang 1928). In contrast, sea squirts, or ascidians, have a metamorphic development, the adult stage of the life cycle arising from a transient free-swimming tadpole larva. The larval body comprises a trunk or head region and a motile tail and has a miniature CNS with many chordate credentials (Katz 1983, Meinertzhagen & Okamura 2001). Finally, the salps and doliolids, closely related to ascidians, are metamorphic but live a pelagic life (Bone 1998) and are not accessible to most laboratories. Analysis of the development and physiology of ascidian tadpoles can thus shed light on the origins of the chordate phylum.

Given its phylogenetic significance, the ascidian tadpole larva has long attracted the attention of biologists. Nevertheless, many studies on ascidian larvae have remained somewhat specialized, at least until recently. A number of new developments are rapidly reversing this longstanding neglect (Corbo et al. 2001, Holland 2002), however, especially for the neurobiology of this simple animal. Ascidian larvae indeed have features favorable to neurobiological study (Bone & Mackie 1982, Burighel & Cloney 1997), chief among which is that cell number is very small and largely, if not wholly, fixed. As a result, their simple nervous system should enable us to study the neuroethology of a chordate larva, cell by cell, by the identified-neuron strategy widely adopted in invertebrate systems (e.g., Hoyle 1983). Three recent topics, in particular, underpin significant research themes in ascidian larval neurobiology.

First is the recent release of the genome sequences for the common sea squirt *Ciona intestinalis* (Dehal et al. 2002) and of the related *Ciona savignyi* (http://www-genome.wi.mit.edu/annotation/ciona/). These simple genomes are 20 times smaller than the mouse genome, yet they harbor most gene families found in vertebrates. They are thought to be related to the ancestral condition of the chordate genome before it underwent the two complete duplications that are proposed to have occurred in the evolution of craniates (Holland et al. 1994). The simplicity of the *Ciona* genomes thus parallels the anatomical simplicity of their larva. These genomic milestones consummate earlier studies to provide access to the genes that underlie the organization, function, and development of the larval nervous system and to new molecular tools to analyze these.

Second is the long history of experimental embryology of the ascidian embryo, initiated by Chabry's (1887) cell ablation experiments and by Conklin's pioneering work on cell lineage (Conklin 1905) and now revisited with molecular approaches (Satoh 2001).

Third is the appearance of analytical studies on larval behavior, especially on larval swimming and phototaxis, that have started to provide insights into the outcome of neural development and function, as well as methods to assay perturbations induced in these. Analysis of the underlying circuits and their neurophysiology has, by contrast, been traditionally frustrated by the problems of the small sizes of larval nerve cells and the tough tunic that surrounds the larva, from which urochordates derive their alternative name, tunicates. Some recent progress has, however, occurred in this field as well.

Despite numerous important reports on various other species, there is a progressive canonization of *Ciona* (*intestinalis* or *savigny*) and *Halocynthia roretzi* in most recent work. Each has particular advantages, and the question of species is still vexed. Adult *Halocynthia roretzi* are more robust, and their embryos are larger and more suited to experimental embryology. Both blastomere injection and recombination experiments are more routine and allow cell lineage to be defined from blastomere injections (Nishida & Satoh 1985, Nishida 1987). The larva is also better suited to electrophysiology (Ohmori & Sasaki 1977). A database (MAGEST: http://www.genome.ad.jp/magest) provides data on DNA sequences and expression patterns of expressed sequence tags (ESTs) from maternal mRNAs (Kawashima et al. 2000, 2002). The commercial fishery of this species provides an additional incentive to its study, even if its Pacific distribution effectively restricts experimental work to Japan. *Ciona*, by contrast, is pandemic in temperate waters. Both the anatomy (Nicol & Meinertzhagen 1991) and behavior (e.g., Kawakami et al. 2002) of its larva are reported in great detail. The recent draft release of the *Ciona* genome has been complemented by large-scale cDNA and EST projects (Satou et al. 2002a,b, 2003b; Satoh et al. 2003), as well as by expression profiles of genes at different developmental stages (e.g., tailbud embryos, Satou et al. 2001; swimming larvae, Kusakabe et al. 2002). Transient, *en masse* transfection by electroporation (Corbo et al. 1997b) can be used to label individual cells, including neurons (Okada et al. 2001), to examine gene function (Hotta et al. 2000), and to execute large- or small-scale promoter analyses (Corbo et al. 1997b, Harafuji et al. 2002). Investigators have recently reported transposon-mediated germline transgenesis using the Minos transposon (Sasakura et al. 2003a). Thanks to the relatively short life cycle (three months), hermaphroditism, and success of sperm cryopreservation techniques, mutagenesis screens (Moody et al. 1999; Nakatani et al. 1999; Sordino et al. 2000a,b, 2001) carry the prospect of isolating developmental mutants of the nervous system. Finally, injection of morpholino antisense oligonucleotides (Satou et al. 2001) provides an alternative way to analyze loss of gene function.

Collectively, these technical developments in *Ciona* provide powerful tools for future studies on the development and function of the larval nervous system. The current mood is very confident, which is why much of the emphasis presented in this review is on *Ciona*. Ascidian larvae are highly diverse, however, comprising ~3000 species (Jefferey 1997), of which *Ciona* has a relatively simple and unspecialized larval form. Therefore, one should perhaps remember that the prime mover in much of the recent attention paid to *Ciona*, which for most workers is to understand evolution in chordates, ultimately may not be well served by undue focus on this one form.

As *Ciona*'s larva becomes a model in neurobiology, it seems inevitable that strange hands will use her to their own ends, different investigators pursuing different goals. Two important goals emerging are (*a*) to provide a basis to analyze the action of vertebrate neural specification and neurogenesis genes in a more accessible embryo and (*b*) to examine the genetic determination/regulation of neural circuits underlying chordate behavior. The latter can be facilitated by the analysis of behavioral and connectivity mutants.

STRUCTURAL ORGANIZATION

The larva of *Halocynthia* contains an estimated 3000 cells (Yamada & Nishida 1999); that of *Ciona* contains even fewer, with ~2600 cells (Satoh 1999). In the *Ciona* larva, these include 40 notochord and 36 muscle cells, which exhibit clear constancy between individuals. The CNS has, by contrast, over 330 cells (Nicol & Meinertzhagen 1991), and the constancy of this number is less clear. These are the progeny of no more than 13 divisions (compared with, for example, the *Drosophila* embryo, which undergoes 13 nuclear cleavages prior to blastoderm formation).

The CNS: Structural Divisions and Their Genetic Bases

Many investigations have sought to define the structural divisions of the larval brain and, latterly, the gene expression territories within these. The motivation is obvious: Like searching the family photo album for likenesses among remote relatives, such studies aim to establish the identity of ancestral chordate features.

ANTERO-POSTERIOR DIVISIONS OF THE NEURAXIS Unlike other invertebrates, but much like vertebrates, the CNS of ascidian larvae develops from a neural plate. The caudo-rostral pattern by which it rolls up to form the neural tube resembles that seen during amphibian neurulation. The ascidian neural plate differs from its amphibian homologue by having fewer blastomeres and by undergoing relatively more of its cleavages and neurulation before sinking beneath the ectoderm than in vertebrate embryos (Nicol & Meinertzhagen 1988). In contrast to vertebrates, however, the anteriormost region of the neural plate does not roll up and internalize but contributes to the so-called dorso-anterior epidermis, which includes the adhesive organs, or palps, head sensory neurons, and pharynx (Nishida 1987). The posterior part of the neural plate gives rise to three structural divisions of the larval ascidian CNS that have long been recognized (reviewed in Bone & Mackie 1982). These are (*a*) a rostral sensory vesicle containing the sensory receptor systems, with a pigmented ocellus (Dilly 1964) and, anterior to it, a pigmented otolith (Dilly 1962) (this region constitutes most of the neural cells, totaling around 215 in *Ciona*); (*b*) a caudal nerve cord (65 cells, mostly ependymal); and, between these, (*c*) a visceral ganglion containing the motoneurons (about 45 cells). In addition to these three, a slender neck region connecting the sensory vesicle and visceral ganglion has since been distinguished in *Ciona* and comprises six cells (Nicol & Meinertzhagen 1991). The caudal nerve cord is the simplest geometric configuration: a row of ventral keel cells, left and right lateral rows, and a dorsal row of capstone cells, thus four cells in cross section (Nicol & Meinertzhagen 1991). Additional cells derived from the lateral cell rows augment the cell population of the visceral ganglion and its cross section; in the sensory vesicle the cells are far more numerous, especially in the posterior wall of the vesicle, and are not easily related to the caudal four-cell cross section (Cole & Meinertzhagen 2004). Each of these antero-posterior subdivisions of the ascidian CNS is itself patterned along the dorso-ventral axis. This patterning, which is not easily seen at the morphologic level, is revealed by the expression of specific marker genes (see below).

UROCHORDATE BRAINS COMPARED Although they generally retain the above tripartite organization, ascidian brains are as diverse as ascidian larvae themselves, with considerable variation in the ocellus and otolith of the sensory vesicle, for example (Berrill 1950, Vorontsova 1988), and the smallest numbers in the larvae of solitary forms such as *Ciona*. Cell number in urochordate brains is generally small, but numbers in other urochordate groups differ somewhat from those in ascidian larvae. In the third urochordate group, the salps, the numbers of cells are not as reduced in the symmetrical dorsal ganglion, which is thus more easily compared with the more complex brains of chordates (Lacalli & Holland 1998). By contrast, in larvaceans such as *Oikopleura* and *Fritillaria*, cell number is even smaller than in ascidians such as *Ciona*. For example, with a body length at least one order of magnitude greater than in a *Ciona* larva, and with far greater behavioral complexity (Bone 1985), the CNS in an adult *Oikopleura* nevertheless has only about a quarter the number of cells found in *Ciona*.

The following sections focus on the nervous systems of larval *Ciona intestinalis* and *Halocynthia roretzi*. The reader should remember, however, that a closer study of other tunicate brains will be required to gain a global view of the organization and evolution of these early chordate nervous systems.

COMPARISONS BETWEEN UROCHORDATE AND CRANIATE BRAINS Ascidian orthologues of developmental genes are expressed in individual territories along the neuraxis, in patterns that resemble closely those that regionalize the vertebrate brain. Their expression patterns are similar in both *Ciona* and *Halocynthia*, providing the most authoritative basis to compare brain regions in different groups. Even though structure alone is insufficient to arbitrate such homologies with vertebrate brain divisions, these have in fact been assumed in some early reports (e.g.,

Katz 1983). Recent reviews (Meinertzhagen & Okamura 2001, Lemaire et al. 2002) summarize details for many sections that follow.

Initially the patterns of *Otx*, *Pax-2/5/8*, and *Hox* genes indicated the tripartite organization of the neuraxis (Wada et al. 1998) and thus intimated the homology of these three regions to the vertebrate fore-, mid-, and hindbrain, respectively. Proceeding rostrally along the neuraxis, patterns of correspondence support homology between the anterior region of the caudal nerve cord and the anterior vertebrate spinal cord (*Hox5*), between the posterior visceral ganglion and rhombomeres 5–8 of the vertebrate brain (*Hox3*), and between the anterior visceral ganglion and rhombomere 4 (*Hox 1*). Thus although ascidian motoneurons originate in the visceral ganglion, the gene-expression code of this region is more similar to the vertebrate rhombencephalon than to that of the vertebrate spinal cord, the motoneurons occupying instead the same genetic territory as vertebrate-descending brainstem reticulospinal neurons.

The neck region of the ascidian CNS is of particular interest. Sandwiched between the *Otx* territory of the sensory vesicle and the *Hox3* territory of the anterior visceral ganglion, it expresses *Pax2/5/8*, *engrailed*, and *FGF8/17/18* (Imai et al. 2002a, Jiang & Smith 2002) in a manner similar to the vertebrate isthmus or mid-hindbrain boundary (MHB) (reviewed in Wurst & Bally-Cuif 2001), although in a somewhat different rostro-caudal sequence. Note, the neck region is reported to comprise only six cells (Nicol & Meinertzhagen 1991). Regardless of where one draws the borders, the correct attribution of *Pax2/5/8*, *Fgf9/16/20*, and *Fgf8/17/18* coexpression sites among these cells must therefore have a very sharp focus. This is particularly striking because the current version of the sequenced genome lacks a clear orthologue for *Gbx*, one of the genes that helps position the MHB in vertebrates. Although the genetic program of the neck appears very much similar to that of the MHB organizer, so far nothing indicates that the neck transmits signals that organize its neighboring regions.

Finally, the posterior sensory vesicle is similar to the metencephalon (*Fgf9/16/20*; *engrailed*), whereas the anterior sensory vesicle corresponds to the di-/mesencephalon brain (*Otx*). On the other hand, if the neck is the MHB, i.e., is the mes-/met-encephalon boundary, then the posterior sensory vesicle is part of the mesencephalon, which fits with the fact that it expresses *Otx* (see Hudson et al. 2003). Only one telencephalon marker, *emx*, so far has been examined and is expressed in the dorso-anterior epidermal territory originating from the anteriormost aspect of the neural plate. This finding suggests that ascidians, like amphioxus (Holland & Holland 2001), lack a true homologue of the telencephalon. Analysis of the patterning of the diffuse, superficial nervous system of a more basal deuterostome, the hemichordate *Saccoglossus kowalevskii* (Lowe et al. 2003), suggests that the chordate CNS was generated by the internalization of a well-patterned superficial neural network. The epidermal expression domain of *emx* may thus correspond to the forerunner of the telencephelon, the formation of a CNS from an ancestral, diffuse network occurring in several steps. Thus, although the enormous overgrowth of the forebrain is a vertebrate invention, these territories may have arisen from an ancestral territory that preexisted in urochordates.

DORSO-VENTRAL DIVISIONS OF THE NEURAXIS Similarities and/or homologies between the ascidian larval CNS and the vertebrate brain also extend to dorso-ventral patterning. The ventralmost row of cells in the caudal neural tube expresses *HNF3β* (Corbo et al. 1997a) and *sonic hedgehog* (Takatori et al. 2002) and is thus comparable to the vertebrate floor plate. In more lateral territories, expression of *Ciona gsx*, like vertebrate *Gsh*, is detected in intermediate cells of the neural tube, though expression of the ascidian orthologue is restricted to the anterior territories. Finally, two ascidian bone morphogenetic protein (*Bmp*) genes, orthologues of *Bmp5/7/8* and *Bmp2/4*, are expressed in the midline epidermal territories flanking or abutting the neural region. This finding parallels the situation in vertebrate embryos, suggesting the existence of a neural crest–like identity (see below). Expression of *Pax3/7*, *Snail*, and *distal-less* in the dorsal neural tube or flanking epidermis reinforces both that dorso-ventral regionalization of the vertebrate and ascidian neural tissues is very similar and that the dorsal territories display a genetic program similar to that in vertebrate neural crest.

The Ascidian Peripheral Nervous System

Like the CNS, the peripheral nervous system (PNS) of ascidians exhibits similarities with that of vertebrates. Ascidian larvae have a system of 30–40 epithelial neurons in both the tail (dorso- and ventro-caudal epidermal neurons) and head, or trunk (apical- and rostral-trunk epidermal neurons), which form a simple PNS (Takamura 1998). At least some of these peripheral neurons likely are mechanoreceptors. They are embedded in the epithelium (Jia 1987), extend long cilia into the tunic of the tail (Crowther & Whittaker 1994), and contribute or are connected to nerves that run back to the sensory vesicle. In neither *Ciona* nor *Halocynthia* do the tail epidermal neurons derive from the neural plate but rather from the midline epidermal territories (Nishida 1987). The dorsal epidermal tail sensory neurons originate from a territory that expresses a neural crest–like genetic program; this is not the case for the ventral sensory neurons.

HOMOLOGIES OF ASCIDIAN BRAIN STRUCTURES

These similarities between urochordate and vertebrate brains are at the level of the patterning of the nervous system and do not necessarily reflect the orthology of their differentiated structures. For the neuromuscular apparatus, in particular, Bone (1992) has spoken against considering the ascidian tadpole larva as a chordate prototype. For the larval CNS, too, there are major differences. The larval brain lacks clear signs of segmentation and laminae and retains the character of a hollow epithelial tube, whereas chordate, especially craniate, brains undergo lamination by radial migration. Given this difference, the fact that *Ciona* has the gene *reelin*, which in mammals plays a role in organizing the brain's layers (Rice & Curran 2001) is interesting, and it remains to be seen what ancestral functions of lamination reelin may mediate in urochordates, possibly in the more complex brain of adults. Associated with the lack of lamination is a lack of stem cells in the developing brain of ascidian embryos, in which cells arise through equal cleavages, at least as so far identified. The relative absence of radial cell migration and of massive cell death also distinguish this simple brain from its more complex vertebrate counterpart. Likewise, the absence of myelinated nerves and axons in both fiber tracts of the CNS (Katz 1983) and peripheral nerves (Torrence 1983), confirmed by the absence of myelin-related genes in the *Ciona* genome (Dehal et al. 2002), bespeaks a lack of rapid conduction pathways and a simpler, possibly ancestral chordate organization. The latter possibly correlates with, in the case of myelination genes, (*a*) the lack of neuregulins, which are involved in axon-oligodendrocyte signaling (Canoll et al. 1996), and of orthologues for the oligodendrocyte determinants *Olig1* and *2* (Zhou et al. 2000); and (*b*) the lack of neurotrophins and their receptors (Dehal et al. 2002), the presence of which promotes survival and neurite extension. Despite these differences, there are many obvious homologies, which we consider next.

Structural Homologies Between Ascidian and Vertebrate Brains

The line of chordate descent has bequeathed to vertebrates a number of structural homologues that must already have been present in the larval brains of the common ancestor with ascidian (Meinertzhagen & Okamura 2001). Some structures reflect the corresponding patterns of homeobox gene expression along the neuraxis, reported above.

SOME FRANK STRUCTURAL HOMOLOGUES In addition to the most obvious embryonic homologue, the neural plate itself, many structures of the differentiated CNS show clear structural similarities to those in vertebrates.

1. An epithelium of ciliated ependymal cells lines the neural canal (Mackie & Bone 1976, Katz 1983), as in the vertebrate CNS.
2. Secreted by these ependymal cells, a thread running within the neural canal is the claimed homologue of Reissner's fiber in vertebrates (Olsson 1972). This complex of glycoproteins from the vertebrate subcommissural organ binds monoamines present in the ventricular cerebrospinal fluid and transports them along the central canal (Rodriguez & Caprile 2001); such a function is untested for the ascidian larval CNS, in which the homology is solely based upon ultrastructural criteria.
3. Coronet cells form a group of ~18 cells on the left side of the sensory vesicle (Dilly 1969, Eakin & Kuda 1971, Nicol & Meinertzhagen 1991). These cells structurally resemble rather closely cells of the saccus vasculosus in the vertebrate hypothalamus (Svane 1982). From an old theory of Dammerman, the latter may function as hydrostatic pressure detectors (Kühlenbeck 1977), a function also proposed for their ascidian counterparts (Eakin & Kuda 1971). Originally designated as an alternate photoreceptor system (Dilly 1969), in fact no evidence exists that coronet cells have a sensory function at all

(Torrence 1983), nor do hydrostatic pressure increases influence larval swimming (Tsuda et al. 2003).

4. Pigment spots of the ocellus and otolith both contain melanin and its key synthetic enzyme tyrosinase (Sato & Yamamoto 2001), expression of which is an early marker for these cells (Whittaker 1973, 1979). Melanin is typical of vertebrate sensory organs, where in visual organs it screens light, and may play a more general role in sensory transduction (Dräger & Balkema 1987). The melanin is distributed as granules in the ocellus and as a single large spherical granule in the ocellus. Sato & Yamamoto (2001) present a comprehensive summary of ascidian pigment cells.
5. The 18 or so photoreceptors, 3 lens cells, and pigment cell of the right-sided ocellus form an obvious candidate homologue of the vertebrate eye. As in vertebrates, ascidian photoreceptors are ciliary in origin (Eakin & Kuda 1971) and hyperpolarize to light (Gorman et al. 1971). The single ocellus expresses the eye determination gene *Pax-6* (Glardon et al. 1997), but differing interpretations have been placed upon which eye(s) is/are its vertebrate homologue. Early interpretations homologized the ascidian ocellus and the vertebrate epiphysis (pineal), in keeping with the lateral position of the pigment cell precursor in the neural plate (Nishida & Satoh 1989). Photoreceptors in *Ciona* express *Ci-Opsin1*, which shows highest homology to both vertebrate retinal and pineal opsins (Kusakabe et al. 2001). On the basis of similarities in sequence data and development, Kusakabe et al. (2001) propose that the median eye of basal vertebrates and the ascidian ocellus may represent the ancestral state of chordate photoreceptors. Comparative studies on larval development have suggested an alternative view: that the ascidian ocellus is a homologue of what in vertebrates gave rise to the right lateral eye (Sorrentino et al. 2000).

OTHER CASES ARE STRUCTURALLY MORE CRYPTIC In addition to such obvious cases of structural homology, expression patterns of certain genes in ascidian embryos are likely harbingers of those seen during early developmental stages of a number of vertebrate neural structures. Such patterns may have therefore already existed in the last common ancestors of all chordates, even if their representation in ascidians is structurally cryptic. Vertebrate neural features with such representation in the ascidian larval brain include the following: placodes, neural crest, and the neurohypophysis.

Placodes and neural crest are two defining structures of the vertebrate nervous system (Northcutt & Gans 1983), and their derivatives ramify throughout the vertebrate body. Even though structural counterparts have long been thought lacking in ascidians, there is evidence for the existence of gene expression patterns ancestral to those structures, which suggests that placodes and neural crest arose by stages and that these stages first arose before the divergence of urochordates and chordates. For example, epidermal sensory neurons in the trunk, which apparently derive from progeny of the anterior animal a8.26 cell pair, line up along the edge of the presumptive neural tube (Ohtsuka et al. 2001), in a region corresponding topologically to the territory of the neural crest in vertebrate embryos, and where the ascidian homologue of *Pax3/7* and *snail* also act (Wada et al. 1997, Wada & Saiga 1999). Thus the development of ascidian epidermal sensory neurons shares these features with that of vertebrate placodes and neural crest.

Wada (2001) suggests that neural crest may have arisen from ancestral ectoderm represented by ascidian dorsal midline ectoderm, to which were added properties of pluripotency, delamination, and migration, and the possession of antero-posterior positional information. Gostling & Shimeld (2003) propose further that the evolution of a dorsal neural expression domain for genes of the vertebrate *Zic* family was an important step in the evolution of the neural crest. Indeed, an ascidian *Zic* orthologue is expressed in the neural plate, commencing during neurulation (see below).

Different placodes are thought to have had different evolutionary origins (Graham 2000, Shimeld & Holland 2000), with at least two of the vertebrate's sensory placodes present already in the last ancestors of ascidians and chordates. They are

1. the acousticolateralis system. The atrial siphon primordium of adult ascidians is a proposed homologue of the otic placode (Shimeld & Holland 2000), insofar as it gives rise to sensory neurons resembling hair cells of the acousticolateralis system (Bone & Ryan 1978) and expresses the ascidian orthologue of the *Pax-2/5/8* gene family (Wada et al. 1998). On the other hand, the mechanoreceptors of the cupular organ are primary sensory neurons and thus structurally less well qualified as candidate homologues for vertebrate hair cells than the anaxonal secondary neurons of the coronal organ in *Botryllus*, which have a peripheral synapse and an afferent sensory axon arising from a central neuron (Burighel et al. 2003).
2. the pituitary. Confirming a historical suggestion (Willey 1894), the neurohypophysial duct in larval ascidians is a proposed homologue of the vertebrate olfactory/adenohypophyseal/hypothalamic placode (Ruppert 1990, Manni et al. 2001), which gives rise to the neurohypophysis. The duct forms from an anterior prolongation of the neural tube (Willey 1894, Manni et al. 1999, Cole & Meinertzhagen 2001), which, in a duality reminiscent of the pituitary, becomes abutted by an ectodermal invagination from the pharyngeal primordium, the cells of which express the ascidian *pituitary homeobox* gene, *Pitx* (Shimeld & Holland 2000, Christiaen et al. 2002). Progeny of the duct, moreover, express immunoreactivity to GnRH (Mackie 1995), as do progeny of the olfactory placode.

NEURAL DEVELOPMENT

Cleavage and Cell Lineage

Ascidian embryos undergo early determinate cleavages that are radial, rapid (initially every ~30 min in *Ciona*), and symmetric about the midline. Cleavage 1 divides the embryo into left and right, cleavage 2 into anterior and posterior, and cleavage 3 into animal hemispheres (lower-case letters a and b) and vegetal hemispheres (upper-case letters A and B). Each blastomere has a two-part number: the generation number (1–14), followed by the individual blastomere number within this generation, which, in the early generations at least, decreases with proximity to the vegetal pole (Conklin 1905). At each division, the progeny of a given blastomere are assigned individual blastomere numbers that double those of its progenitor: For example, the daughters of a4.2 are a5.3 and a5.4; those of B7.4 are B8.8 and B8.7. The invariant pattern of cleavage allows the fate of each cell to be followed to a point in the 110-cell stage in *Halocynthia* when most contribute to a single tissue type (Nishida 1987) and to be extended to 226 of the ~330 CNS cells in the larva of *Ciona* (Cole & Meinertzhagen 2004). It is remarkable that the lineage is so very well conserved among ascidians, even between phylogenetically distant forms such as *Halocynthia* and *Ciona* (Swalla et al. 2000). This conserved pattern of cleavage contrasts with the situation in nematodes, in which, for example, *C. elegans* develops with a fixed cleavage pattern, whereas marine nematodes such as *Enoplus brevis* show no signs of such invariance (Lemaire & Marcellini 2003).

Investigators first took the existence of an invariant cell lineage to indicate autonomous differentiation of blastomeres, and consequently they saw the ascidian embryo as a typical example of mosaic development (Lemaire & Marcellini 2003). This interpretation is, however, wrong for many lineages including some that give rise to neural tissue. Cell lineage can play both a topographical and a typological role in the assignment of cell fate (Meinertzhagen 2002). For example, the a-lineage from a4.2, which gives rise to anterior neural tube, requires inductive mechanisms like those in vertebrates (see below). In this case, the fixed cleavage pattern provides the precise positioning of the induced blastomere with respect to its inducing neighbors. In contrast, in the A-lineage from A4.1, which gives rise to posterior neural tube, a generic neural identity is achieved cell autonomously, and the role of the fixed cleavage is to partition maternal determinants precisely into different blastomeres.

The overall descriptive sequence of cleavage has been summarized elsewhere (Nicol & Meinertzhagen 1988, Satoh 1994, Lemaire et al. 2002), and its details are not fundamental for this review. A recent reexamination using confocal imaging of wholemount embryos (Cole & Meinertzhagen 2004) indicates the following: The caudal nerve cord derives from b9.37 (dorsal), A9.32 and A9.29 (lateral), and A10.29 and A10.25 (ventral); the visceral ganglion derives from b9.38 (dorsal), A9.29, A9.30, and A10.31 (lateral), and A10.30 (ventral); and the posterior sensory vesicle derives from b9.38 (dorsal), A10.32 (lateral), and A10.26 (ventral); except for four ventral cells from A9.14, the remaining sensory vesicle and neurohypophysis derives from a-line blastomeres.

Initial Specification of a Neural Cell Fate

AUTONOMOUS SPECIFICATION OF NEURAL IDENTITY IN THE A-LINE The posterior A-line neural tissue originates from two pairs of blastomeres at the 32-cell stage: the A6.2 and 6.4 pairs. At the next cleavage (64-cell stage), each of these blastomeres contributes one daughter restricted to the notochord and one restricted to the posterior nerve cord. The fate distinction between notochord and posterior nerve cord precursors results from the reception by the notochord precursor of a combination of basic fibroblast growth factor (FGF)-like and BMP signals between the 32- and 64-cell stages (Nishida 2002).

Although a generic neural identity is acquired cell-autonomously in the A-line, not all posterior neural fates are acquired in the absence of cell communication. As in vertebrates (Holowacz & Sokol 1999), formation of the posteriormost neural cells requires externally regulated kinase (ERK) signaling. ERK is activated in the presumptive posterior neural cells by the 32-cell stage, and the inhibition of MAP-ERK kinase (MEK) signaling at this stage respecifies the fate of visceral ganglion and tail nerve cord precursors into posterior sensory vesicle (Hudson et al. 2003).

INDUCTION OF ANTERIOR NEURAL PLATE FATES IN a-LINE ECTODERM The anterior neural plate is derived from the a4.2 blastomeres. Unlike A4.1, a4.2 requires instruction from its vegetal neighbors to adopt a neural fate (Reverberi et al. 1960; reviewed in Lemaire et al. 2002). The neural-inducing signal originates principally from the A4.1 progeny, and the competence to form anterior neural tissue is restricted to the a4.2 progeny (Lemaire et al. 2002, Hudson et al. 2003). Little is known at present about the molecular mechanisms that spatially restrict animal competence (see Lemaire et al. 2002), although more is known about the inducing signals and their temporal response. These signals show both parallels to and differences from the situation in vertebrate embryos, in which the FGF and BMP signaling pathways both play a crucial role in the binary decision between neural and epidermal fates (Wilson & Edlund 2001). In ascidian embryos, antagonizing the BMP pathway appears neither necessary nor sufficient to specify a CNS fate (Darras & Nishida 2001). Although these experiments fail to exclude a role for the inhibition of BMP signaling in the anteriormost neural plate fates (the palps), they do at least suggest that the role of the BMP pathway is much less important in ascidian neural induction than it is in vertebrates.

Ascidian embryos more closely resemble vertebrates when considering the role of the FGF signaling pathway. According to gain- and loss-of-function analyses of the FGF pathway, this pathway is crucial to initiate a stable program of anterior neural differentiation in a-line blastomeres (Lemaire et al. 2002). Recent investigations confirm the molecular identification of FGF9/16/17 as the vegetally secreted ascidian early neural inducer (Bertrand et al. 2003). Furthermore, analysis of the *Cis*-regulatory regions of *Ci-otx*, a direct FGF target, reveals that the action of the neural inducer is mediated by two maternal transcription factors, ETS1/2 and GATAa. The former mediates FGF responsiveness in both animal and vegetal lineages, whereas the activity of the latter is restricted to the animal hemisphere, thus preventing the formation of neural tissue in vegetal territories.

Finally, trypsin-like serine proteases of the subtillisin family are necessary and sufficient to trigger neural induction in isolated ascidian animal explants (Ortolani et al. 1979, Okado & Takahashi 1993). Whether, or how, this finding relates to the role of FGF 9/16/20, and whether it reflects an ascidian peculiarity or a more general phenomenon, is unclear at present.

TAIL EPIDERMAL SENSORY NEURONS AND DORSAL NEURAL TUBE ARISE FROM b-LINE ECTODERM The b-line ectoderm gives rise to two types of neural tissue: the dorsalmost row of cells in the neural tube (Nishida 1987, Nicol & Meinertzhagen 1988), and the midline tail epidermis, which includes the tail epidermal sensory neurons (Y. Ohtsuka & Y. Okamura, unpublished observations). Very little is known about the mechanisms presiding over the adoption of the dorsal neural tube fate, in part owing to the lack of specific markers; this problem should find a solution in several current large-scale *in situ* hybridization screens (e.g., Ogasawara et al. 2001). Formation of the tail epidermal sensory neurons is better understood and requires an induction between the b-line ectoderm and B-line vegetal blastomeres (Hudson & Lemaire 2001, Ohtsuka et al. 2001). The b-line cells lose competence between the 110-cell and neurula stages, but induction *in vivo* may take place earlier. Blastomere isolation experiments, and identification of the natural inducer, will help solve this issue. Although the action of the natural inducer can be mimicked by FGF (Hudson & Lemaire 2001, Ohtsuka et al. 2001), inhibition of ERK activation with pharmacological MEK inhibitors fails to prevent the formation of tail epidermal sensory neurons (Hudson et al. 2003). This finding suggests that an additional class of (posterior) neural inducers exists. Recombination experiments indicate that these factors are secreted from B-line, but not A-line, blastomeres (Ohtsuka et al. 2001).

Neural Specification in Ascidians and Vertebrates

Summarizing the previous section, comparing the specification of neural fate in ascidians and vertebrates allows us to pinpoint some likely ancestral strategies that led to a tripartite chordate neural tube. In both systems, rostral neural tissue is induced, an induction that involves FGF signaling and an initial binary decision between a fate as either epidermis or anterior neural plate. More posteriorly, a binary decision takes place in ascidians between trunk/tail mesoderm and the posterior neural plate, which forms cell-autonomously. Although this seems at odds with current vertebrate models, mutation of the *Tbx6* gene in the mouse leads to the conversion of somitic tissue into spinal cord (Chapman & Papaioannou 1998). Furthermore, dorsal mesoderm (prechordal plate and notochord) in zebrafish embryos that are mutant for Nodal signaling changes its identity to become neural (reviewed in Harland 2000 and Schier & Talbot 2001). Finally, in mice, localized inhibition of broadly dispersed trunk mesoderm-inducing signals is vital to allow the formation of a neural plate (Perea-Gomez et al. 2002). A neural-to-mesoderm binary decision may thus be a shared feature of all chordates. Finally, in both systems, formation of the caudal neural tube requires ERK signaling.

One obvious difference between ascidians and vertebrates involves the role of BMP signaling. This difference may concern epidermis specification more than neural induction, however. In contrast to vertebrates in which it is induced by BMPs, epidermis differentiates autonomously in ascidians (Nishida 2002). Assuming that the role of BMP inhibition in vertebrates is to prevent epidermis from forming, this function is likely to be accomplished by a different process in ascidians.

Genes Acting Downstream of the Acquisition of a Generic Neural Identity

FROM NEURAL SPECIFICATION TO NEURONAL DIFFERENTIATION Even though ascidian embryos are renowned for specifying embryonic fates earlier than in vertebrates, the a-lineage may not acquire a stable neural fate until the end of gastrulation. *Zic* and *Sox* genes participate in the early gene network that links neural induction and neurogenesis in vertebrates (Sasai 1998). In ascidians, inhibiting the function of *Halocynthia HrZicN* prevents neural differentiation in both A- and a-line blastomeres. In the A-line, in which *HrZicN* is first expressed during cleavage stages, early neural markers are never activated. In the a-line, in which *HrZicN* is first expressed during gastrulation, inhibiting *HrZicN* leads to the progressive loss of a-line neural markers by the mid-gastrula (Wada & Saiga 2002). Conversely, the PNS is not as affected by *ZicN* loss of function as is the CNS. Other ascidian species are likely similar. For example, *Cs-ZicL* is also expressed pan-neurally during gastrula stages in *Ciona savignyi* and is likewise required in that species to express at the tailbud stage the pan-neural marker *ETR* (Imai et al. 2002b).

Sox 2 and *3* genes are expressed throughout the neural plate in vertebrate embryos. In contrast, the single *Halocynthia* orthologue of the SoxB class, *HrSoxB1*, is restricted to the posterior neural plate from late gastrula stages. As for its vertebrate counterpart, overexpressing *HrSoxB1* is not sufficient to drive production of ectopic neural tissue in *Halocynthia*. In vertebrates, *Sox2* possibly acts to modify the competence to form neural tissue, and to test this idea in ascidians would be interesting. The effects of loss of function for this gene have yet to be reported in tunicates.

Finally, the Notch cascade plays a crucial role in the lateral inhibition between neighboring cells that controls neurogenesis in both flies and vertebrates. Single copies of the *Notch*, *delta*, and *serrate/jagged* genes exist in the *Ciona* genome (Satou et al. 2003a). Furthermore, overexpression of a constitutively active form of Notch in *Halocynthia* inhibits the formation of PNS neurons, which suggests that lateral inhibition is used by ascidians to specify at least some of their larval neurons (Akanuma et al. 2002).

FINE-SCALE REGIONALIZATION OF THE NEURAL TUBE In parallel with the determination of neuronal identity, the neural plate is progressively regionalized during gastrulation. This regionalization is exemplified by the discovery that a number of genes are expressed by the late gastrula in either a single or a few bilateral pairs of cells in the neural plate (e.g., *gsx*; Hudson & Lemaire 2001). In parallel, individual classes of neurons and glia begin to differentiate from regionalized neural plate cells. Compared with early embryonic development, the differentiation of frank phenotypes among cells in the developing larval brain has been studied little, and the recent identification of a panel of neuronal markers (Mochizuki et al. 2003) will help to classify and identify different cell types. Although our understanding of the mechanisms that lead to the regionalization of the neural plate is still very poor, in general the different steps of pigment cell formation form a paradigm for further studies.

Following neural induction and the activation of general markers such as otx, the tyrosinase-related protein gene (*TRP*) is first broadly expressed in the neural plate and becomes progressively restricted to the pigment-cell lineage during gastrulation. This gene is directly controlled by *Otx* (Wada et al. 2002). As *Otx* is expressed throughout the anterior neural plate, however, restriction of *TRP* expression to the dorsal pigment cell lineage must involve additional regulators.

The pigment cells originate from the progeny of a8.25, the lateralmost cells of neural plate row III in the 110-cell stage, and their precursors are in contact with cells expressing *bmpb*, orthologous to vertebrate *Bmp2/4*. As in vertebrates, during gastrulation *Bmp2/4* acts as a dorsalizing signal in some ascidian neural plate cells, including the pigment-cell lineage (Darras & Nishida 2001). Whereas a combination of *Ci-otx* and BMP signaling could account for the activation and progressive restriction of *TRP* expression, embryos treated during gastrulation with a pharmacological MEK inhibitor also fail to develop pigment cells (Hudson et al. 2003), which suggests the involvement of multiple pathways in this lineage. By the neurula stage, the left and right a8.25 cells express *TRP* and can both form either ocellus or otolith. The decision of which sensory organule to make, ocellus or otolith, is made by left-right interactions between the cell pair progeny (Nishida 1987) after the neural tube closes and the precursors align along the anteroposterior axis (Nishida & Satoh 1989). This decision again involves an antagonism between BMP2/4 and its antagonist CHORDIN (Darras & Nishida 2001). The former specifically induces the differentiation of the anterior pigment cell into the otolith, whereas the latter, which suppresses BMP activity in the posterior cell, allows an ocellus to differentiate.

STRUCTURAL DIFFERENTIATION AND THE FORMATION OF CONNECTIONS Few studies address this important aspect of larval brain development (Meinertzhagen & Okamura 2001). The greatest need is for markers to distinguish between, and determine the onset of expression in, each of the cell types in the CNS because existing markers, such as immunoreactivity to UA301 (Takamura 1998) and β-tubulin (Miya & Satoh 1997), are expressed pan-neuronally. The form of larval neurons in *Ciona* was revealed using electroporation (Corbo et al. 1997b) to transfect neural-plate progeny transiently with a green fluorescent protein (*GFP*) gene (Okada et al. 2001). The neurons have a simple form with either no or few dendrites, an axon, and a simple terminal, features confirmed by serial-EM reconstructions (Stanley MacIsaac 1999). Unlike GFP transfection methods, the latter sample all neurons, leaving us with a familiar hard choice: one larva in its entirety at resolution sufficient to analyze synaptic connectivity, or single neurons at lower resolution from many larvae. Consistent with the simple morphology of ascidian neurons, genomic evidence suggests that axon pathfinding may be much simpler than in vertebrate embryos (Dehal et al. 2002). Even so, sensory neurons or motoneurons of the tail must navigate the two main tracts, left and right, some time during the second half of tail elongation because such axons are lacking at the mid-tailbud stage (Burighel & Cloney 1997). Twitching movements of the tail prior to larval hatching indicate, however, that axon growth must be complete and at least some motor circuits formed by that time.

Previous studies identify synaptic contacts in hatched larvae, in the sensory vesicle (Barnes 1971), and in the visceral ganglion (Stanley MacIsaac 1999). They form, in vertebrate pattern, on the cell body of the postsynaptic neuron and are revealed by simple active zones at which a small cumulus of presynaptic vesicles clusters opposite membrane densities at postsynaptic sites. Little structural differentiation exists among the synapses of the visceral ganglion (Stanley MacIsaac 1999), so as to reveal different types. The late development of neuromuscular transmission in *Halocynthia* includes a rapid increase in sensitivity to acetylcholine (ACh) at ~63% of embryonic development. The appearance at ~80% of embryonic development of giant excitatory junctional potentials (ejps) is interpreted as random synchronized presynaptic activity, currently providing the only evidence for the timing of synaptogenesis. These giant ejps are followed in a few hours by the miniature ejps typical of the free-swimming larva (Ohmori & Sasaki 1977).

Attainment of Cell Number and the Role of Cell Death

The final differentiation of neuronal cell types involves regulation of the pool of each type of cell. Cell death plays a crucial role in shaping the development of vertebrate brains, which incorporates a regressive step, using cell death to sculpt cell number from a larger number of blastomere progeny (Kuan et al. 2000). Core components of the cell death machinery exist in the *Ciona* genome (Terajima et al. 2003), and apoptoses, dependent on the actions of caspases and signaled by cell death markers such as TUNEL, are conspicuous during tail resorption (Chambon et al. 2002, Cole & Meinertzhagen 2004), which involves cell death in the nerve cord. In the embryo, by contrast, TUNEL labeling has not been seen among neural plate progeny (Cole & Meinertzhagen 2004). In molgulids, one claim suggests that programmed cell death is initiated in the CNS and epidermis but that affected cells do not die until metamorphosis, their apoptosis requiring the action of *FoxA5* and *Manx* genes (Jeffery 2002a). In *Styela*, embryos treated with antisense oligonucleotides for proliferating cell nuclear antigen exhibit nuclear DNA fragmentation typical of programmed cell death and become boomerang-shaped larvae that swim in circles (Jeffery 2002b). From these reports, however, it is not clear whether cell death actually shapes the number of neural cells in hatchling larvae. In fact, embryonic neural cell death was thought unlikely in *Ciona* from the outset (Nicol

& Meinertzhagen 1988), and BrdU incorporations provide no support for an initial production of excess cells (Bollner & Meinertzhagen 1993). Thus, at least for *Ciona*, ascidian embryos develop by constructive mitotic progression, which implies that all cells born in the embryonic CNS are found in the hatchling larva.

Metamorphosis

The ascidian larva is an incubator for the tissue primordia of the next generation of adults, transporting these away from the parent, and is thereby the means by which one sea squirt gives rise to the next. Metamorphosis from the larva to the juvenile stage of the life cycle begins when a polarized wave of regression originates at the tip of the tail. The wave is caspase-dependent and associated with activation of the ERK protein (Chambon et al. 2002), and it coincides with, or is followed by, the cessation of swimming, retraction of the papillae, and eventual resorption of the entire tail (Cloney 1978). These changes all have profound repercussions for the larval nervous system, including massive cell death (Jeffery 2002a, Chambon et al. 2002). The neurohypophysis of the larval brain undergoes extensive mitoses to generate the cells of the adult brain, the neural ganglion, but the extent to which other cells of the larval CNS survive, to transform and incorporate into the neural ganglion, is unclear.

This review restricts itself to the period when the larva settles and commences metamorphosis. We choose to view the larva as the gifted offspring of a dull parent that serves only to bring embryos into the world. To be fair, however, the adult exhibits some remarkable phenomena also, not the least of which is the ability to survive the extirpation of its neural ganglion and to regenerate an entire new ganglion in its place (Bollner et al. 1995). The neurobiology of this, of ascidian neural stem cells, and other details of the adult nervous system are considered elsewhere (e.g., Thorndyke et al. 2001).

GENOMICS AND NEURAL GENES

The genome of *Ciona intestinalis* is compact and simple, and the recently released entire genome database indicates that it contains an estimated 15,852 protein-coding genes (Dehal et al. 2002). Genome searches, for developmentally relevant (Satou & Satoh 2003) and other genes, indicate the genes that may have existed in ancestral forms prior to the divergence of urochordates from their chordate siblings. For the nervous system, 56 CNS-specific genes have been reported recently (Mochizuki et al. 2003).

Genes for Neuronal Channel Proteins

The following discussion summarizes a recent survey of the *Ciona* genome database (Y. Okamura, Y. Satou, Y. Mori, N. Satoh, et al., unpublished manuscript). To support neuronal excitability, *Ciona* has a minimum set of prototypical genes, such as for Nav-, Cav-, and Kv-channels, and transmitter-gated ion channels. This is hardly surprising and implies that diversity in these channel species in mammals depended on replication from a restricted set of ancestral genes. Similarly, the *Ciona* genome contains sets of prototype orthologues for most subfamilies of vertebrate transmitter-gated channel genes, as for glutamate receptor genes, which suggests that the main framework of molecular diversity of transmitter-gated channels was already established prior to the origin of ascidians. *Ciona* does have varied acetylcholine subunit genes, but they arose independently of their diversification in vertebrates, whereas *Ciona* has multiple subfamilies of ionotropic GABA receptors, which indicates that the diversification of these receptors predated the split between deuterostomes and protostomes. *Ciona* resembles vertebrates in lacking both histamine- and glutamate-gated chloride channels, but it lacks ATP-gated channels and orthologues of genes that function in vertebrate pain pathways; these genes, therefore, must have arisen after the urochordate-chordate split. The number of transmitter-gated channel genes is fewer than that of *C. elegans* and *Drosophila*, which probably reflects the greater simplicity of the ascidian nervous system.

Genes for Gap Junction Proteins

Gap junctions are also present in the CNS (Barnes 1971, Stanley MacIsaac 1999). Close membrane appositions, tentative sites of coupling, exist between neurons in the visceral ganglion (Stanley MacIsaac 1999). The *Ciona* genome contains genes encoding proteins associated with tight junctions, adherens junctions, focal adhesions, and hemidesmosomes, as well as vertebrate gap junctions (Sasakura et al. 2003b). There are 17 putative connexin genes in *Ciona* (Sasakura et al. 2003b). In a clear dichotomy, however, there are no clear orthologues of innexins (Sasakura et al. 2003b), the genes that code for gap junction channel proteins in protostomes (Phelan et al. 1998). Thus the last common ancestor of ascidians and vertebrates essentially had the same systems of cell junctions for intercellular communication as modern-day vertebrates.

Neurophysiology

Neurophysiological studies are currently a weak link between the anatomy and behavior of the larva.

Sensory Physiology

The receptor systems of the larva are the subject of recent laser microablation experiments (Tsuda et al. 2003). Ablation of the anterior pigment cell of the otolith, while sparing the posterior pigment cell of the ocellus, abolishes the negative geotaxis response of the early larva, which suggests that upward swimming toward the light is driven by the otocyte, not by the ocellus. Conversely, ablation of the posterior pigment cell, while sparing the anterior, abolishes negative phototaxis in the late larva, which suggests that downward movement away from the light prior to larval settlement is driven by photoreceptor input (Tsuda et al. 2003). This finding confirms the opsin-like action spectrum (Nakagawa et al. 1999). Gene knockdown experiments using antisense morpholino-oligonucleotides show that phototaxis depends on Ci-OPSIN-1, the product of one of three opsin-like genes (Inada et al. 2003). Changes in hydrostatic pressure do not affect the pattern of locomotion (Tsuda et al. 2003), which suggests that the role of hydrostatic pressure receptors (Eakin & Kuda 1971) has been incorrectly assigned to coronet cells. Among other sensory modalities, the finding that no pain sensory system exists was based on the absence of anatomically discrete sensory neurons and associated channel proteins (Y. Okamura, Y. Satou, Y. Mori, N. Satoh, unpublished manuscript; Dib-Hajj 2002). There also may be no olfactory system because the *Ciona* genome lacks recognizable genes for olfactory receptors (Dib-Hajj 2002). Whether mechanosensory feedback from the body surface, similar to tactile sensation in vertebrates (Di Prisco et al. 1990), modulates the patterns of symmetrical swimming in ascidian larvae remains unclear. In *Halocynthia* (Y. Ohtsuka & Y. Okamura, unpublished observations) and maybe *Ciona* (Stanley MacIsaac 1999), morphological evidence proves that epidermal sensory neurons innervate the putative area of motoneurons. However, isolated, truncated tails with the visceral ganglion region intact can still beat symmetrically (Q. Bone, Y. Okamura & E. Brown, unpublished observations), which suggests that the activity of epidermal sensory neurons in the trunk does not influence larval locomotion. Future knockdown experiments for genes specifically expressed in epidermal sensory neurons, or laser ablation of individual sensory neurons, will test this possibility.

Electrophysiology

Sharp-electrode recordings of ascidian larval neurons, from presumed photoreceptors in the ocellus (Gorman et al. 1971) and motor units at the base of the tail (Bone 1992), provide rare electrophysiological glimpses of neuronal activity in the larval brain so difficult to obtain that they are unlikely to be precedents for many others. These recordings indicate that traditional approaches to electrophysiology are technically possible, but to hope for rapid progress in classical neuroethological approaches to the ascidian larva would be unduly optimistic.

OPTOPHYSIOLOGICAL APPROACHES The dimensions of the ascidian embryo and its larva, and the relative transparency of both, would seem better suited to more recent optophysiological methods. Two types of approach merit particular attention. Calcium imaging of the CNS in ascidian larvae, and the correlation between activity in identified neurons and high-speed videos of tail movements, on the basis of similar successful approaches in zebrafish larvae (e.g., Ritter et al. 2001), provide one promising opportunity. In a complementary approach, microablation especially of cells expressing GFP under neuron-specific promoters, can help test directly the function of neurons having identified patterns of activity during swimming. Although there are still few champions for these methods, and no published reports, the rapid growth of the field now makes this approach not only a pressing imperative but also a growing reality.

Evidence for Neurotransmitter and Neuropeptide Function

Genes for the synthesis and metabolism of all common amine neurotransmitters are present in the ascidian genome, with the exception of those for histamine (Dehal et al. 2002), but have yet to be surveyed in detail and are in any case no guarantee that the larva actually uses the neurotransmitter. Immunoreactivity to GABA is localized to about six cells in the visceral ganglion of *Ciona* (Johnson et al. 1987; T. Koropatnick & I.A. Meinertzhagen, unpublished observations), which also expresses a corresponding metabolic enzyme, GABA aminotransferase (Mochizuki et al. 2003). It is therefore interesting that larval locomotion is sensitive to GABA and picrotoxin (Brown et al. 2004). Larval expression of immunoreactivity to other neurotransmitters can be problematic (Vorontsova et al. 1997). ACh-associated genes and their expression are considered above, in connection with the larval motoneurons.

At the time of writing, detailed surveys for the genes encoding neuropeptides and their receptors still must be undertaken, but they are anticipated in many cases, on the basis of immunoreactivity to peptide antisera in the adult CNS (reviewed in Pestarino 1991). Conversely, in the larval visceral ganglion, dense-core synaptic vesicles either are absent or few, which suggests that few neuropeptides may be expressed in the CNS at this stage of the life cycle. One exception is the expression of Met-enkephalin-like immunoreactivity in a single tail neuron connecting between dorsal and ventral fiber bundles (Jia 1987; cf. Figure 2*F* in Takamura 1998).

An Orexin-R-like receptor gene is present in *Ciona*. Insofar as the orexin-orexinR system is involved in regulating mammalian feeding and sleeping behaviors (Willie et al. 2001), the phylogenetic origin of sleep mechanisms will be an interesting topic for future study in ascidians.

GnRH is a key regulator of the reproductive axis in vertebrates. GnRH hormone proteins occur in the gonads of adult *Ciona* (Di Fiore et al. 2000), and GnRH induces spawning of eggs when injected into the gonads (Terakado 2001). Searches identify six putative genes in the *Ciona* genome (Adams et al. 2003), and GnRH is expressed in subsets of neurons in the neural complex (Tsutsui et al. 1998), which could be regulated by light exposure (Ohkuma et al. 2000). Although ascidians exhibit circadian rhythms (Ryland 1990), clear orthologues of the *period* and *clock* genes are lacking in *Ciona*, as are genes that code for the enzymes that synthesize melatonin (Dehal et al. 2002).

Conventional neurotransmitters such as epinephrine and serotonin may also play a role in regulating later steps of neural development and metamorphosis in ascidians. Even though *Ciona* lacks the machinery to synthesize epinephrine (Dehal et al. 2002), this catecholamine induces tail resorption during metamorphosis of *Ciona* larvae (Kimura et al. 2003), possibly by acting at adrenergic receptors.

Immunoreactivity to antidopamine-hydroxylase, the enzyme regulating norepinephrine synthesis, is localized around the sensory vesicle of larvae during metamorphosis, indicating their likely capacity to synthesize norepinephrine. Serotonin-like immunoreactivity is expressed in larval *Phallusia mammillata*, and treatment of embryos with a 5HTA antagonist results in morphogenetic abnormalities, changing the morphology of anterior neural structures in the resultant larvae (Pennati et al. 2001).

Behavior

A first glimpse of a tadpole larva swimming provides compelling evidence for its chordate character. Considerable variations exist between species in the overall pattern and strength of swimming, however. *Ciona*, for example, swims intermittently, whereas larger larvae of other species swim more strongly. Svane & Young (1989) report other details.

Tail Beat and Swimming Behavior

Alternate contractions of three longitudinal muscle bands along each flank, each comprising a fixed number of muscle cells, generate the tail movements in *Ciona*. There are 36 of these in *Ciona* and 42 in *Halocynthia*; the arrangements in other species can differ (Jeffery & Swalla 1992, Turon 1992). Larval striated muscle consists of uninucleate cells that communicate electrically via gap junctions, as in amphioxus skeletal muscle. Although such electrical coupling probably enables synchronous triggering in the contraction of the entire muscle band, the bands are innervated differentially (Bone 1992) with both polyneuronal and multiterminal innervation (Stanley McIsaac 1999).

Myofibrils are confined to the superficial layer of cytoplasm between the plasma membrane and mitochondria in the deeper region of the muscle cells (Burighel et al. 1977), although the detailed arrangement and orientation of myofibrils along the muscle band are unknown. Whether myofibril orientation is related to swimming pattern, such as for helical motion (McHenry 2001), awaits future investigation.

Cultured, cleavage-arrested blastomeres of the muscle lineage exhibit caffeine-sensitive Ca^{2+} release dependent on Ca^{2+} influx (Nakajo et al. 1999). Because this Ca^{2+} release is totally dependent on external Ca^{2+}, excitation-contraction coupling in ascidian larval muscle is of an invertebrate type, rather than the more-sophisticated, voltage-dependent mechanism found in vertebrate skeletal muscle (Okamura et al. 2003). *Halocynthia* muscle blastomeres also exhibit a significant shift in the threshold of the active membrane potential. Muscle in larvae, prior to hatching, fires action potentials after small depolarizing current injections, whereas muscle cells at later larval stages require larger depolarizing currents (Nakajo & Okamura 2002). This may reflect the fact that, prior to hatching, young muscle fires spontaneous action potentials, which causes the late embryo to twitch, but that activity in mature muscle needs to be controlled more tightly by synaptic input. Single muscle cells exhibit membrane potential oscillations, resulting from coupling between Ca^{2+} release from ryanodine receptors and plasma membrane Ca-activated K^+ channels. The frequency of these oscillations is 15–20 Hz, close to the tail beat frequency, which suggests that developing muscle cells acquire intrinsic membrane properties suitable for regular and rhythmic larval locomotion.

Muscle action during the larval tail beat is rapid, and the tail undergoes wide angular excursions, first analyzed cinematographically by Bone (1992). Ascidian larvae are small and subject to appreciable viscous drag, for which their tail movements are adapted to intermediate Reynolds' numbers of sea water, as modeled recently by McHenry et al. (2003). Flexion occurs at the point of tail insertion at the larval trunk, and the tail's excursion is passive, like a whip. This pattern is unlike swimming movements of the craniate tail, as in teleost fry, which are actively propagated down the length of the tail by successive myotomes. Forward progression occurs in a helical trajectory in forms such as *Distaplia* through an asymmetry in the tail beat, possibly arising from the tail's bent shape (McHenry 2001), but helical movement is not obvious in *Ciona*.

TWO TYPES OF TAIL MOVEMENT The larval tail exhibits two forms of flexion: It either oscillates symmetrically or is flicked asymmetrically (Bone 1992). Spontaneous swimming by symmetrical tail movements is mediated by a glutamate-sensitive pathway with a firing frequency of about 20 Hz (Brown et al. 2004). Symmetrical flexions result in forward propulsion of the larva, whereas the asymmetrical tail flicks alter the angular orientation of the trunk, typically by 45°, thereby determining the direction of forward propulsion. Tail flicks also appear early, in the embryo, and aid the larva in hatching from the chorion.

RESPONSES TO SENSORY STIMULI In the general swimming pattern, larvae exhibit first a negative geotaxis, taking the larva up to the surface and away from the parent(s), followed then by a negative phototaxis, taking the larva back down to the substratum for settlement (Svane & Young 1989). In *Aplidium* this difference in larval ontogeny is achieved by altering the patterns of rotation about anteroposterior and dorsoventral axes so as to change the helical trajectory of the larva (McHenry & Strother 2003). Analysis in other species is lacking and may be more demanding in small larvae, such as *Ciona*, with a more fusiform shape.

Recently larval visual behavior has been studied intensively in *Ciona*. Larvae exhibit various swimming responses to light, and these responses change during the life of the larva. After hatching, the average swimming speed is initially rapid (>1 mm/sec), and the larva shows no response to light increases (Tsuda et al. 2001). Later its speed slows, and the larva progressively exhibits a dimming response: an increase in swimming velocity at lights 'off', reminiscent of that seen in young *Xenopus* tadpoles (Jamieson & Roberts 2000). Lights "off" activates a pathway that is insensitive to ionotropic glutamate receptor blockers and extended by GABA receptor antagonists, and which cross-inhibits the spontaneous swimming pathway (Brown et al. 2004). These observations are generalized behavioral responses averaged from data on larvae, in groups of 50 (Tsuda et al. 2001); the behavior of individual larvae is more variable and is less well characterized. In the future, to assay the responses of individual larvae to isolate behavioral mutants during screens for neural genes may become critical.

Defining the Neural Mechanisms of Behavior

The ascidian larval brain seems ideally suited to analysis of the neural bases of chordate swimming and its regulation, similar to the way in which lamprey swimming (Grillner et al. 1995) has served as an acraniate model or the functional organization of the spinal cord in young *Xenopus* tadpoles (Roberts 2000) has served as a vertebrate model. In the lamprey, a central pattern generator (CPG) contains mutually inhibiting contralaterally projecting interneurons that constitute pathways of mutual reciprocal inhibition with a delay, thus establishing alternating patterns of excitation among interneurons innervating the motoneurons of the spinal cord. This form of CPG circuit reasonably could be anticipated in the CNS of larval *Ciona* but in fact has not been found among the circuits of the visceral ganglion, in which the numbers of descending inputs from the sensory vesicle predominate (Stanley MacIsaac 1999). Possibly CPG circuits reside wholly within the sensory vesicle or may coextend between the neurons of the visceral ganglion and sensory vesicle. This finding will make their identification and analysis more difficult than had they been located entirely within the visceral ganglion.

NEUROMUSCULAR MECHANISMS Ultimately, patterned motor output during swimming, either alternating or ipsilateral, converges upon motoneurons in the visceral ganglion, which number five pairs in *Ciona* (Stanley MacIsaac 1999) but only three in *Halocynthia* (Okada et al. 2002). Pharmacological evidence indicates that cholinergic neuromuscular transmission acts by a nicotinic postsynaptic mechanism (Ohmori & Sasaki 1977), which is familiar enough at vertebrate neuromuscular junctions but is also found in echinoderm muscles (Devlin 2001) and therefore is possibly typical of deuterostomes. Correspondingly, probes against two phenotypic markers for ACh, cholineacetyltransferase, and a specific vesicular transporter, VAChTr, label cells in the visceral ganglion (Takamura et al. 2002), but the numbers of these so labeled are not consistent. We do not know whether any of the visceral ganglion's GABA-immunoreactive neurons (see above) directly innervate the muscle bands, too, in the manner of peripheral neuromuscular innervation in arthropods (Otsuka et al. 1966), larvaceans (Bollner et al. 1991), and echinoderms (Devlin 2001). If such direct innervation were in fact to exist in larval ascidians, it could possibly constitute an ancestral character. Such peripheral inhibition would be unlike muscle innervation in chordates proper, however; moreover neuromuscular transmission in ascidian larvae is completely blocked by D-tubocurarine, extended by the cholinesterase inhibitor eserine, and has a reversal potential at −10 mV (Ohmori & Sasaki 1977), compatible with cholinergic mechanisms providing the sole mode of innervation.

NEUROID CONDUCTION Given the numerical simplicity of the CNS in the ascidian larva, it is doubtful whether we can account fully for the complexity of urochordate behavior by means of a neural substrate. In adult larvaceans, neuroid conduction in the external epithelia (Mackie 1970) clearly is a major contributor to behavioral complexity (Bone 1985). The extent to which other urochordates have access to neuroid conduction pathways in ectodermal tissues, which might augment more focused activation in the CNS, is unclear, at least in both *Ciona* and *Halocynthia*. In favor of the possibility of neuroid conduction, cleavage-arrested epidermally differentiated blastomeres from *Halocynthia* fire long-lasting Ca-action potentials (Hirano & Takahashi 1987). Moreover, epidermal cells in *Ciona* and *Halocynthia* express a set of genes related to evoked exocytosis for synaptic transmission, such as the P/Q type Cav channel, syntaptobrevin, and syntaxin. For the latter, however, one possible interpretation is that active membrane excitability of epidermal cells functions for both tunic secretion and neuroid conduction.

PROSPECTS FOR FUTURE STUDIES

Future experiments, in which direct electrical recording from identified neurons marked by cell-specific promoters, will help link neuroanatomy with neurophysiology in the ascidian larval brain. *Ciona*'s genes lack excessive redundancy, and its larvae hatch quickly. Using either gene knockdown methods with morpholino oligo-nucleotides (e.g., Inada et al. 2003) or overexpression of dominant-negative channels (e.g., Ono et al. 1999), it should therefore be possible also to define in vivo the physiological roles in larval behavior of, for example, ion channels and receptors. The redundancy of vertebrate genes, by contrast, makes such approaches more problematic in mice, for which the kinship of *Ciona* offers an alternative that is superior to studies on more distant invertebrates such as *C. elegans* or *Drosophila*.

Studies on ascidian brains tend toward one of two extremes: those for which the ascidian larva is an unpopulated chordate brain, a simple version—albeit at some remove—of the real thing; and those for whom the ascidian larval brain is itself the real thing. If the latter suffer from being too specialized in focus, the former can merely repeat at higher resolution in ascidians what is already well established in vertebrates. The increased integration now provided by molecular and genomic approaches and the recruitment to ascidian studies of workers from other fields, especially from *Drosophila* and vertebrate development, have served to unify the field considerably in the last decade and to smooth these two extremes. These developments are now beginning to accelerate the rate of postgenomic analysis by providing the tools to interrupt both cellular function and gene action in the nervous system. A bright future is yet to come.

ACKNOWLEDGMENTS

The authors' original research reported here was supported by NSERC grant OPG 0000065, the Killam Trust of Dalhousie University, and the John Simon Guggenheim Memorial Foundation (I.A.M.); by the Naito Foundation and a Grant-in-Aid for Scientific Research from the Ministry of Education, Culture, Sports, Science, and Technology, in Japan (Y.O.); and by grants in aid from the CNRS, French Ministry of Research (ACI Program), the HFSPO, the European Community, the AFM, and the ARC (P.L.).

The *Annual Review of Neuroscience* is online at http://neuro.annualreviews.org

LITERATURE CITED

Adams BA, Tello JA, Erchegyi J, Warby C, Hong DJ, et al. 2003. Six novel gonadotropin-releasing hormones are encoded as triplets on each of two genes in the protochordate, *Ciona intestinalis*. *Endocrinology* 144:1907–19

Akanuma T, Hori S, Darras S, Nishida H. 2002. Notch signaling is involved in nervous system formation in ascidian embryos. *Dev. Genes Evol.* 212:459–72

Barnes SN. 1971. Fine structure of the photoreceptor and cerebral ganglion of the tadpole larva of *Amaroucium constellatum* (Verrill) (Subphylum: Urochordata; Class: Ascidiacea). *Z. Zellforsch.* 117:1–16

Berrill NJ. 1950. *The Tunicata with an Account of the British Species*. London: Ray Soc.

Bertrand V, Hudson C, Caillol D, Popovici C, Lemaire P. 2003. Neural tissue in ascidian embryos is induced by FGF9/16/20 acting via maternal GATA and ETS factors. *Cell* 115:615–67

Bollner T, Howalt S, Thorndyke MC, Beesley PW. 1995. Regeneration and post-metamorphic development of the central nervous system in the protochordate *Ciona intestinalis*: a study with monoclonal antibodies. *Cell Tissue Res.* 279:421–32

Bollner T, Meinertzhagen IA. 1993. The patterns of bromodeoxyuridine incorporation in the nervous system of a larval ascidian, *Ciona intestinalis*. *Biol. Bull.* 184:277–85

Bollner T, Storm-Mathisen J, Ottersen OP. 1991. GABA-like immunoreactivity in the nervous system of *Oikopleura dioica* (Appendicularia). *Biol. Bull.* 180:119–24

Bone Q. 1985. Locomotor adaptations of some gelatinous zooplankton. *Symp. Soc. Exp. Biol.* 39:487–520

Bone Q. 1992. On the locomotion of ascidian tadpole larvae. *J. Mar. Biol. Assoc. U.K.* 72:161–86

Bone Q, ed. 1998. *The Biology of Pelagic Tunicates*. New York: Oxford Univ. Press

Bone Q, Mackie GO. 1982. Urochordata. In *Electrical Conduction and Behaviour in 'Simple' Invertebrates*, ed. GAB Shelton, pp. 473–535. Oxford: Clarendon

Bone Q, Ryan KP. 1978. Cupular sense organs in *Ciona* (Tunicata: Ascidiacea). *J. Zool. (London)* 186:417–29

Brown ER, Bone Q, Okamura Y. 2004. Properties of the neural network that controls locomotion in the larvae of the ascidian (*C. intestinalis*); combining physiological and genomic approaches. *Comp. Biochem. Physiol. Part A* 137:S3

Burighel P, Cloney RA. 1997. Urochordata: Ascidiacea. In *Microscopic Anatomy of Invertebrates (Vol. 15), Hemichordata, Chaetognatha, and the Invertebrate Chordates*, ed. FW Harrison, EE Ruppert, pp. 221–347. New York: Wiley-Liss

Burighel P, Lane NJ, Gasparini F, Tiozzo S, Zaniolo G, et al. 2003. Novel, secondary sensory cell organ in ascidians: in search of the ancestor of the vertebrate lateral line. *J. Comp. Neurol.* 461:236–49

Burighel P, Nunzi MG, Schiaffino S. 1977. A comparative study of the organization of the sarcotubular system in ascidian muscle. *J. Morphol.* 153:205–24

Canoll PD, Musacchio JM, Hardy R, Reynolds R, Marchionni MA, et al. 1996. GGF/neuregulin is a neuronal signal that promotes the proliferation and survival and inhibits the differentiation of oligodendrocyte progenitors. *Neuron* 17:229–43

Chabry L. 1887. Embryologie normale et tératologique des Ascidies simples. *J. Anat. Physiol.* (*Paris*) 23:167–319

Chambon JP, Soule J, Pomies P, Fort P, Sahuquet A, et al. 2002. Tail regression in *Ciona intestinalis* (Prochordate) involves a caspase-dependent apoptosis event associated with ERK activation. *Development* 129:3105–14

Chapman DL, Papaioannou VE. 1998. Three neural tubes in mouse embryos with mutations in the T-box gene Tbx6. *Nature* 391:695–97

Christiaen L, Burighel P, Smith WC, Vernier P, Bourrat F, et al. 2002. Pitx genes in Tunicates provide new molecular insight into the evolutionary origin of pituitary. *Gene* 287:107–13

Cloney RA. 1978. Ascidian metamorphosis: review and analysis. In *Settlement and Metamorphosis of Marine Invertebrate Larvae*, ed. F-S Chia, ME Rice, pp. 255–82. New York: Elsevier

Cole AG, Meinertzhagen IA. 2001. Tail-bud embryogenesis and the development of the neurohypophysis in the ascidian *Ciona intestinalis*. In *Biology of Ascidians*, ed. H Sawada, H Yokosawa, CC Lambert, pp. 137–41. Tokyo: Springer-Verlag

Cole AG, Meinertzhagen IA. 2004. The central nervous system of the ascidian larva: mitotic history of cells forming the larval neural tube in *Ciona intestinalis*. *Dev. Biol.* In press

Conklin EG. 1905. Organization and cell-lineage of the ascidian egg. *J. Acad. Natl. Sci. Phila.* 13:1–119

Corbo JC, Di Gregorio A, Levine M. 2001. The ascidian as a model organism in developmental and evolutionary biology. *Cell* 106:535–38

Corbo JC, Erives A, Di Gregorio A, Chang A, Levine M. 1997a. Dorsoventral patterning of the vertebrate neural tube is conserved in a protochordate. *Development* 124:2335–44

Corbo JC, Levine M, Zeller RW. 1997b. Characterization of a notochord-specific enhancer from the Brachyury promoter region of the ascidian, *Ciona intestinalis*. *Development* 124:589–602

Crowther RJ, Whittaker JR. 1994. Serial repetition of cilia pairs along the tail surface of an ascidian larva. *J. Exp. Zool.* 268:9–16

Darras S, Nishida H. 2001. The BMP/CHORDIN antagonism controls sensory pigment cell specification and differentiation in the ascidian embryo. *Dev. Biol.* 236:271–88

Dehal P, Satou Y, Campbell RK, Chapman J, Degnan B, et al. 2002. The draft genome of *Ciona intestinalis*: insights into chordate and vertebrate origins. *Science* 298:2157–67

Devlin CL. 2001. The pharmacology of gamma-aminobutyric acid and acetylcholine receptors at the echinoderm neuromuscular junction. *J. Exp. Biol.* 204:887–96

Dib-Hajj S, Black JA, Cummins TR, Waxman SG. 2002. NaN/Nav1.9: a sodium channel with unique properties. *Trends Neurosci.* 25:253–59

Di Fiore MM, Rastogi RK, Ceciliani F, Messi E, Botte V, et al. 2000. Mammalian and chicken I forms of gonadotropin-releasing hormone in the gonads of a protochordate, *Ciona intestinalis*. *Proc. Natl. Acad. Sci. USA* 97:2343–48

Di Prisco GV, Wallen P, Grillner S. 1990. Synaptic effects of intraspinal stretch receptor neurons mediating movement-related feedback during locomotion. *Brain Res.* 530: 161–66

Dilly PN. 1962. Studies on the receptors in the cerebral vesicle of the ascidian tadpole. 1. The otolith. *Q. J. Microsc. Sci.* 103:393–98

Dilly PN. 1964. Studies on the receptors in the cerebral vesicle of the ascidian tadpole. 2. The ocellus. *Q. J. Microsc. Sci.* 105:13–20

Dilly PN. 1969. Studies on the receptors in the cerebral vesicle of the ascidian tadpole. 3. A second type of photoreceptor in the tadpole larva of *Ciona intestinalis*. *Z. Zellforsch.* 96:63–65

Dobzhansky T. 1973. Nothing in biology makes sense except in the light of evolution. *Am. Biol. Teacher* 35:125–29

Dräger UC, Balkema GW. 1987. Does melanin do more than protect from light? *Neurosci. Res. Suppl.* 6:S75–86

Eakin RM, Kuda A. 1971. Ultrastructure of sensory receptors in ascidian tadpoles. *Z. Zellforsch.* 112:287–312

Garstang W. 1928. The morphology of the Tunicata, and its bearings on the phylogeny of the Chordata. *Quart. J. Microsc. Sci.* 72:51–187

Glardon S, Callaerts P, Halder G, Gehring WJ. 1997. Conservation of Pax-6 in a lower chordate, the ascidian *Phallusia mammillata. Development* 124:817–25

Gorman ALF, McReynolds JS, Barnes SN. 1971. Photoreceptors in primitive chordates: fine structure, hyperpolarizing receptor potentials, and evolution. *Science* 172:1052–54

Gostling NJ, Shimeld SM. 2003. Protochordate Zic genes define primitive somite compartments and highlight molecular changes underlying neural crest evolution. *Evol. Dev.* 5: 136–44

Graham A. 2000. The evolution of the vertebrates—genes and development. *Curr. Opin. Genet. Dev.* 10:624–28

Grillner S, Deliagina T, Ekebuerg Ö, El Manira A, Hill RH, et al. 1995. Neural networks that co-ordinate locomotion and body orientation in lamprey. *Trends Neurosci.* 18:270–80

Harafuji N, Keys DN, Levine M. 2002. Genome-wide identification of tissue-specific enhancers in the *Ciona* tadpole. *Proc. Natl. Acad. Sci. USA* 99:6802–5

Harland R. 2000. Neural induction. *Curr. Opin. Genet. Dev.* 10:357–62

Hirano T, Takahashi K. 1987. Development of ionic channels and cell-surface antigens in the cleavage-arrested one-cell embryo of an ascidian. *J. Physiol.* 386:113–33

Holowacz T, Sokol S. 1999. FGF is required for posterior neural patterning but not for neural induction. *Dev. Biol.* 205:296–308

Holland LZ, Holland ND. 2001. Evolution of neural crest and placodes: amphioxus as a model for the ancestral vertebrate? *J. Anat.* 199:85–98

Holland PWH. 2002. *Ciona. Curr. Biol.* 12: R609

Holland PWH, Garcia-Fernàndez J, Williams NA, Sidow A. 1994. Gene duplications and the origins of vertebrate development. *Development* 120S:125–33

Hotta K, Takahashi H, Asakura T, Saitoh B, Takatori N, et al. 2000. Characterization of Brachyury-downstream notochord genes in the *Ciona intestinalis* embryo. *Dev. Biol.* 224:69–80

Hoyle G. 1983. On the way to neuroethology: the identified neuron approach. In *Neuroethology and Behavioral Physiology*, ed. F Huber, H Markl, pp. 9–25. Berlin/Heidelberg: Springer-Verlag

Hudson C, Darras S, Caillol D, Yasuo H, Lemaire P. 2003. A conserved role for the MEK signalling pathway in neural tissue specification and posteriorisation in the invertebrate chordate, the ascidian *Ciona intestinalis. Development* 130:147–59

Hudson C, Lemaire P. 2001. Induction of anterior neural fates in the ascidian *Ciona intestinalis. Mech. Develop.* 100:189–203

Imai KS, Satoh N, Satou Y. 2002a. Region specific gene expressions in the central nervous system of the ascidian embryo. *Mech. Dev.* 119 (Suppl.) 1:S275–77

Imai KS, Satou Y, Satoh N. 2002b. Multiple functions of a Zic-like gene in the differentiation of notochord, central nervous system and muscle in *Ciona savignyi* embryos. *Development* 129:2723–32

Inada K, Horie T, Kusakabe T, Tsuda M. 2003. Targeted knockdown of an opsin gene inhibits the swimming behaviour photoresponse of ascidian larvae. *Neurosci. Lett.* 347:167–70

Jamieson D, Roberts A. 2000. Responses of young *Xenopus laevis* tadpoles to light dimming: possible roles for the pineal eye. *J. Exp. Biol.* 203:1857–67

Jeffery WR. 1997. Evolution of ascidian development. *Bioscience* 47:417–25

Jeffery WR. 2002a. Programmed cell death in the ascidian embryo: modulation by FoxA5 and Manx and roles in the evolution of larval development. *Mech. Develop.* 118:111–24

Jeffery WR. 2002b. Role of PCNA and ependymal cells in ascidian neural development. *Gene* 287:97–105

Jeffery WR, Swalla BJ. 1992. Evolution of alternate modes of development in ascidians. *Bioessays* 14:219–26

Jia W-G. 1987. *Investigations on the larval nervous system and its development in the ascidian,* Ciona intestinalis. MSc thesis, Dalhousie Univ.

Jiang D, Smith WC. 2002. An ascidian engrailed gene. *Dev. Genes Evol.* 212:399–402

Johnson CD, Crowther RJ, Whittaker JR. 1987. Immunocytochemical studies of an ascidian larvae. *Soc. Neurosci. Abstr.* 13:1142

Katz MJ. 1983. Comparative anatomy of the tunicate tadpole *Ciona intestinalis*. *Biol. Bull.* 164:1–27

Kawakami I, Shiraishi S, Tsuda M. 2002. Photoresponse and learning behavior of ascidian larvae, a primitive chordate, to repeated stimuli of step-up and step-down of light. *J. Biol. Phys.* 28:549–59

Kawashima T, Kawashima S, Kanehisa M, Nishida H, Makabe KW. 2000. MAGEST: maboya gene expression patterns and expression sequence tags. *Nucl. Acid Res.* 28:133–35

Kawashima T, Kawashima S, Kohara Y, Kanehisa M, Makabe KW. 2002. Update of MAGEST: maboya gene expression patterns and sequence tags. *Nucl. Acids Res.* 30:119–20

Kimura Y, Yoshida M, Morisawa M. 2003. Interaction between noradrenaline or adrenaline and the beta 1-adrenergic receptor in the nervous system triggers early metamorphosis of larvae in the ascidian, *Ciona savignyi*. *Dev. Biol.* 258:129–40

Kuan CY, Roth KA, Flavell RA, Rakic P. 2000. Mechanisms of programmed cell death in the developing brain. *Trends Neurosci.* 23:291–97

Kühlenbeck H. 1977. *The Central Nervous System of Vertebrates.* Vol. 5, Part I: *Derivatives of the Prosencephalon*: *Diencephalon and Telencephalon.* Basel, Switz.: Karger

Kusakabe T, Kusakabe R, Kawakami I, Satou Y, Satoh N, et al. 2001. *Ci-opsin1*, a vertebrate-type opsin gene, expressed in the larval ocellus of the ascidian *Ciona intestinalis*. *FEBS Lett.* 506:69–72

Kusakabe T, Yoshida R, Kawakami I, Kusakabe R, Mochizuki Y, et al. 2002. Gene expression profiles in tadpole larvae of *Ciona intestinalis*. *Dev. Biol.* 242:188–203

Lacalli TC, Holland LZ. 1998. The developing dorsal ganglion of the salp *Thalia democratica*, and the nature of the ancestral chordate brain. *Phil. Trans. R. Soc. London B* 353:1–25

Lemaire P, Bertrand V, Hudson C. 2002. Early steps in the formation of neural tissue in ascidian embryos. *Dev. Biol.* 252:151–69

Lemaire P, Marcellini. 2003. Early animal embryogenesis: Why so much variability? *Biologist* 50:136–40

Lowe CJ, Wu M, Salic A, Evans L, Lander E, et al. 2003. Anteroposterior patterning in hemichordates and the origins of the chordate nervous system. *Cell* 113:853–65

Mackie GO. 1970. Neuroid conduction and the evolution of conducting tissues. *Q. Rev. Biol.* 45:319–32

Mackie GO. 1995. On the "visceral nervous system" of *Ciona*. *J. Mar. Biol. Assoc. UK* 75:141–51

Mackie GO, Bone Q. 1976. Skin impulses and locomotion in an ascidian tadpole. *J. Mar. Biol. Assoc. UK* 56:751–68

Manni L, Lane NJ, Burighel P, Zaniolo G. 2001. Are neural crest and placodes exclusive to vertebrates? *Evol. Dev.* 3:297–98

Manni L, Lane NJ, Sorrentino M, Zaniolo G, Burighel P. 1999. Mechanism of neurogenesis during the embryonic development of a tunicate. *J. Comp. Neurol.* 412:527–41

McHenry MJ. 2001. Mechanisms of helical swimming: asymmetries in the morphology, movement and mechanics of larvae of the ascidian *Distaplia occidentalis*. *J. Exp. Biol.* 204:2959–73

McHenry MJ, Azizi E, Strother JA. 2003. The hydrodynamics of locomotion at intermediate Reynolds numbers: undulatory swimming in ascidian larvae (*Botrylloides* sp.). *J. Exp. Biol.* 206:327–43

McHenry MJ, Strother JA. 2003. The kinematics of phototaxis in larvae of the ascidian *Aplidium constellatum*. *Mar. Biol.* 142:173–84

Meinertzhagen IA. 2002. Cell lineage. In *Encyclopedia of Evolution*, ed. M Pagel, pp. 142–44. New York: Oxford Univ. Press. Vol. 1

Meinertzhagen LA, Okamura Y. 2001. The larval ascidian nervous system: the chordate brain from its small beginnings. *Trends Neurosci.* 24:401–10

Miya T, Satoh N. 1997. Isolation and characterization of cDNA clones for beta-tubulin genes as a molecular marker for neural cell differentiation in the ascidian embryo. *Int. J. Dev. Biol.* 41:551–57

Mochizuki Y, Satou Y, Satoh N. 2003. Large-scale characterization of genes specific to the larval nervous system in the ascidian *Ciona intestinalis*. *Genesis* 36:62–71

Moody R, Davis SW, Cubas F, Smith WC. 1999. Isolation of developmental mutants of the ascidian *Ciona savignyi*. *Mol. Gen. Genet.* 262:199–206

Nakagawa M, Miyamoto T, Ohkuma M, Tsuda M. 1999. Action spectrum for the photophobic response of *Ciona intestinalis* (Ascidieacea, Urochordata) larvae implicates retinal protein. *Photochem. Photobiol.* 70:359–62

Nakajo K, Chen L, Okamura Y. 1999. Cross-coupling between voltage-dependent Ca^{2+} channels and ryanodine receptors in developing ascidian muscle blastomeres. *J. Physiol.* 515(3):695–710

Nakajo K, Okamura Y. 2002. Caffeine sensitive transient outward current confers oscillatory nature of muscle cell in developing ascidians. *Neurosci. Res.* 26(Suppl.):S16

Nakatani Y, Moody R, Smith WC. 1999. Mutations affecting tail and notochord development in the ascidian *Ciona savignyi*. *Development* 126:3293–301

Nicol D, Meinertzhagen IA. 1988. Development of the central nervous system of the larva of the ascidian, *Ciona intestinalis* L. II. Neural plate morphogenesis and cell lineages during neurulation. *Dev. Biol.* 130:737–66

Nicol D, Meinertzhagen IA. 1991. Cell counts and maps in the larval central nervous system of the ascidian *Ciona intestinalis* (L.). *J. Comp. Neurol.* 309:415–29

Nishida H. 1987. Cell lineage analysis in ascidian embryos by intracellular injection of a tracer enzyme. III. Up to the tissue restricted stage. *Dev. Biol.* 121:526–41

Nishida H. 2002. Specification of developmental fates in ascidian embryos: molecular approach to maternal determinants and signaling molecules. *Int. Rev. Cytol.* 217:227–76

Nishida H, Satoh N. 1985. Cell lineage analysis in ascidian embryos by intracellular injection of a tracer enzyme. II. The 16- and 32-cell stages. *Dev. Biol.* 110:440–54

Nishida H, Satoh N. 1989. Determination and regulation in the pigment cell lineage of the ascidian embryo. *Dev. Biol.* 132:355–67

Northcutt RG, Gans C. 1983. The genesis of neural crest and epidermal placodes: a reinterpretation of vertebrate origins. *Q. Rev. Biol.* 58:1–28

Ogasawara M, Minokawa T, Sasakura Y, Nishida H, Makabe KW. 2001. A large-scale whole-mount *in situ* hybridization system: rapid one-tube preparation of DIG-labeled RNA probes and high throughput hybridization using 96-well silent screen plates. *Zool. Sci.* 18:187–93

Ohkuma M, Katagiri Y, Nakagawa M, Tsuda M. 2000. Possible involvement of light regulated gonadotropin-releasing hormone neurons in biological clock for reproduction in the cerebral ganglion of the ascidian, *Halocynthia roretzi*. *Neurosci Lett.* 293:5–8

Ohmori H, Sasaki S. 1977. Development of neuromuscular transmission in a larval tunicate. *J. Physiol.* 269:221–54

Ohtsuka Y, Obinata T, Okamura Y. 2001. Induction of ascidian peripheral neuron by vegetal blastomeres. *Dev. Biol.* 239:107–17

Okado H, Takahashi K. 1993. Neural differentiation in cleavage-arrested ascidian blastomeres induced by a proteolytic enzyme. *J. Physiol.* 463:269–90

Okada T, Katsuyama Y, Ono F, Okamura Y. 2002. The development of three identified motor neurons in the larva of an ascidian, *Halocynthia roretzi*. *Dev. Biol.* 244:278–92

Okada T, Stanley MacIsaac S, Katsuyama Y, Okamura Y, Meinertzhagen IA. 2001. Neuronal form in the central nervous system of the tadpole larva of the ascidian *Ciona intestinalis*. *Biol. Bull.* 200:252–56

Okamura Y, Izumi-Nakaseko H, Nakajo K, Ohtsuka Y, Ebihara T. 2003. The ascidian dihydropyridine-resistant calcium channel as the prototype of chordate L-type calcium channel. *Neurosignals* 12:142–58

Olsson R. 1972. Reissner's fiber in ascidian tadpole larvae. *Acta Zool. (Stockh.)* 53:17–21

Ono F, Katsuyama Y, Nakajo K, Okamura Y. 1999. Subfamily-specific posttranscriptional mechanism underlies K(+) channel expression in a developing neuronal blastomere. *J. Neurosci.* 19:6874–86

Ortolani G, Patricolo E, Mansueto C. 1979. Trypsin-induced cell surface changes in ascidian embryonic cells: regulation of differentiation of a tissue-specific protein. *Exp. Cell Res.* 122:137–47

Otsuka M, Iversen LL, Hall ZW, Kravitz EA. 1966. Release of gamma-aminobutyric acid from inhibitory nerves of lobster. *Proc. Natl. Acad. Sci. USA* 56:1110–15

Pennati R, Groppelli S, Sotgia C, Zega G, Pestarino M, et al. 2001. Serotonin localization in *Phallusia mammillata* larvae and effects of 5-HT antagonists during larval development. *Dev. Growth Differ.* 43:647–56

Perea-Gomez A, Vella FD, Shawlot W, Oulad-Abdelghani M, Chazaud C, et al. 2002. Nodal antagonists in the anterior visceral endoderm prevent the formation of multiple primitive streaks. *Dev. Cell* 3:745–56

Pestarino M. 1991. The neuroendocrine and immune systems in Protochordates. *Adv. Neuroimmunol.* 1:114–23

Phelan P, Bacon JP, Davies JA, Stebbings LA, Todman MG, et al. 1998. Innexins: a family of invertebrate gap-junction proteins. *Trends Genet.* 14:348–49

Reverberi G, Ortolani G, Farinella-Ferruzza N. 1960. The causal formation of the brain in the ascidian larva. *Acta Embryol. Morphol. Exp.* 3:296–336

Rice DS, Curran T. 2001. Role of the reelin signaling pathway in central nervous system development. *Annu. Rev. Neurosci.* 24:1005–39

Ritter DA, Bhatt DH, Fetcho JR. 2001. *In vivo* imaging of zebrafish reveals differences in the spinal networks for escape and swimming movements. *J. Neurosci.* 21:8956–65

Roberts A. 2000. Early functional organisation of spinal neurons in developing lower vertebrates. *Brain Res. Bull.* 53:585–93

Rodriguez S, Caprile T. 2001. Functional aspects of the subcommissural organ–Reissner's fiber complex with emphasis in the clearance of brain monoamines. *Microsc. Res. Technol.* 52:564–72

Ruppert EE. 1990. Structure, ultrastructure and function of the neural gland complex of *Ascidia interrupta* (Chordata, Ascidiacea)—clarification of hypotheses regarding the evolution of the vertebrate anterior pituitary. *Acta Zool. (Stockh.)* 71:135–49

Ryland JS. 1990. A circadian rhythm in the tropical ascidian *Diplosoma virens* (Ascidiacea, Didemnidae). *J. Exp. Mar. Biol. Ecol.* 138:217–25

Sasai Y. 1998. Identifying the missing links: genes that connect neural induction and primary neurogenesis in vertebrate embryos. *Neuron* 21:455–58

Sasakura Y, Awazu S, Chiba S, Satoh N. 2003a. Germ-line transgenesis of the Tc1/mariner superfamily transposon Minos in *Ciona intestinalis*. *Proc. Natl. Acad. Sci. USA* 100: 7726–30

Sasakura Y, Shoguchi E, Takatori N, Wada S, Meinertzhagen IA, et al. 2003b. A genome-wide survey of developmentally relevant genes in *Ciona intestinalis*: X. Genes for cell junctions and extracellular matrix. *Dev. Genes Evol.* 213:303–13

Sato S, Yamamoto H. 2001. Development of pigment cells in the brain of ascidian tadpole larvae: insights into the origins of vertebrate pigment cells. *Pigment Cell Res.* 14:428–36

Satoh N. 1994. *Developmental Biology of Ascidians*. Cambridge, UK: Cambridge Univ. Press

Satoh N. 1999. Cell fate determination in the ascidian embryo. In *Cell-Lineage and Fate Determination*, ed. SA Moody, pp. 59–74. London: Academic

Satoh N. 2001. Ascidian embryos as a model system to analyze expression and function of developmental genes. *Differentiation* 68:1–12

Satoh N, Satou Y, Davidson B, Levine M. 2003. *Ciona intestinalis*: an emerging model for whole-genome analyses. *Trends Genet.* 19:376–81

Satou Y, Imai KS, Satoh N. 2001. Action of morpholinos in *Ciona* embryos. *Genesis* 30:103–6

Satou Y, Kawashima T, Kohara Y, Satoh N. 2003b. Large scale EST analyses in *Ciona intestinalis*: its application as Northern blot analyses. *Dev. Genes Evol.* 213:314–18

Satou Y, Sasakura Y, Yamada L, Imai KS, Satoh N, et al. 2003a. A genomewide survey of developmentally relevant genes in *Ciona intestinalis*: V. Genes for receptor tyrosine kinase pathway and Notch signaling pathway. *Dev. Genes Evol.* 213:254–63

Satou Y, Satoh N. 2003. Genomewide surveys of developmentally relevant genes in *Ciona intestinalis*. *Dev. Genes Evol.* 213:211–12

Satou Y, Takatori N, Fujiwara S, Nishikata T, Saiga H, et al. 2002b. *Ciona intestinalis* cDNA projects: expressed sequence tag analyses and gene expression profiles during embryogenesis. *Gene* 287:83–96

Satou Y, Takatori N, Yamada L, Mochizuki Y, Hamaguchi M, et al. 2001. Gene expression profiles in *Ciona intestinalis* tailbud embryos. *Development* 128:2893–904

Satou Y, Yamada L, Mochizuki Y, Takatori N, Kawashima T, et al. 2002a. A cDNA resource from the basal chordate *Ciona intestinalis*. *Genesis* 33:153–54

Schier AF, Talbot WS. 2001. Nodal signaling and the zebrafish organizer. *Int. J. Dev. Biol.* 45:289–97

Shimeld SM, Holland PWH. 2000. Vertebrate innovations. *Proc. Natl. Acad. Sci. USA* 97:4449–52

Sordino P, Belluzzi L, De Santis R, Smith WC. 2001. Developmental genetics in primitive chordates. *Phil. Trans. R. Soc. B* 356:1573–82

Sordino P, Heisenberg CP, Cirino P, Toscano A, Giuliano P, et al. 2000a. A mutational approach to the study of development of the protochordate *Ciona intestinalis* (Tunicata, Chordata). *Sarsia* 85:173–76

Sordino P, Heisenberg CP, Cirino P, Toscano A, Giuliano P, et al. 2000b. A mutational approach to the study of development of the protochordate *Ciona intestinalis* (Tunicata, Chordata). *Sarsia* 85:353

Sorrentino M, Manni L, Lane NJ, Burighel P. 2000. Evolution of cerebral vesicles and their sensory organs in an ascidian larva. *Acta Zool. (Stockh.)* 81:243–58

Stanley MacIsaac S. 1999. *Ultrastructure of the visceral ganglion in the ascidian larva* Ciona intestinalis: *cell circuitry and synaptic distribution*. MSc thesis, Dalhousie Univ.

Svane I. 1982. Possible ascidian counterpart to the vertebrate saccus vasculosus with reference to *Pyura tessellata* (Forbes) and *Boltenia echinata* (L.). *Acta Zool. (Stockh.)* 63:85–89

Svane I, Young CM. 1989. The ecology and behavior of ascidian larvae. *Oceanogr. ar. Biol. Annu. Rev.* 27:45–90

Swalla BK, Cameron CB, Corley LS, Garey JR. 2000. Urochordates are monophyletic within the deuterostomes. *Syst. Biol.* 49:52–64

Takamura K. 1998. Nervous network in larvae of the ascidian *Ciona intestinalis*. *Dev. Genes Evol.* 208:1–8

Takamura K, Egawa T, Ohnishi S, Okada T, Fukuoka T. 2002. Developmental expression of ascidian neurotransmitter synthesis genes. I. Choline acetyltransferase and acetylcholine transporter genes. *Dev. Genes Evol.* 212:50–53

Takatori N, Satou Y, Satoh N. 2002. Expression of hedgehog genes in *Ciona intestinalis* embryos. *Mech. Develop.* 116:235–38

Terajima D, Shida K, Takada N, Kasuya A, Rokhsar D, et al. 2003. Identification of candidate genes encoding the core components of the cell death machinery in the *Ciona intestinalis* genome. *Cell Death Differ.* 10:749–53

Terakado K. 2001. Induction of gamete release by gonadotropin-releasing hormone in a protochordate, *Ciona intestinalis*. *Gen. Comp. Endocrinol.* 124:277–84

Thorndyke MC, Patruno M, Chen WC, Beesley PW. 2001. Stem cells and regeneration in invertebrate Deuterostomes. *Symp. Soc. Exp. Biol.* 53:107–20

Torrence SA. 1983. *Ascidian larval nervous system: anatomy, ultrastructure and metamorphosis*. PhD thesis, Univ. Washington, Seattle

Tsuda M, Kawakami I, Miyamoto T, Nakagawa M, Shiraishi S, et al. 2001. Photoresponse and habituation of swimming behavior of ascidian larvae, *Ciona intestinalis*. In *Biology of Ascidians*, ed. H Sawada, H Yokosawa, CC Lambert, pp. 153–57. Tokyo: Springer-Verlag

Tsuda M, Sakurai D, Goda M. 2003. Direct evidence for the role of pigment cells in the brain of ascidian larvae by laser ablation. *J. Exp. Biol.* 206:1409–17

Tsutsui H, Yamamoto N, Ito H, Oka Y. 1998. GnRH-immunoreactive neuronal system in the presumptive ancestral chordate, *Ciona intestinalis* (Ascidian). *Gen. Comp. Endocrinol.* 112:426–32

Turon X. 1992. Morfología del sistema caudal en las larvas de tres especies de ascidias (Tunicata). *Hist. Animal.* 1:71–86

Valentine JW, Erwin DH, Jablonski D. 1996. Developmental evolution of metazoan bodyplans: the fossil evidence. *Dev. Biol.* 173: 373–81

Vorontsova MN. 1988. *Morfologia i ultrastructura lichinok ascidii (P/T Tunicata, T Chordata) i ih organov chuvstv* (*Morphology and fine structure of larvae of several ascidian species, emphasizing the structure and evolution of larval sense organs*). PhD thesis, Inst. Evol. Morphol., Russ. Acad. Sci., Moscow

Vorontsova MN, Nezlin LP, Meinertzhagen IA. 1997. Nervous system of the larva of the ascidian *Molgula citrina* (Alder and Hancock, 1848). *Acta Zool. (Stockh.)* 78:177–85

Wada H. 1998. Evolutionary history of free-swimming and sessile lifestyles in urochordates as deduced from 18S rDNA molecular phylogeny. *Mol. Biol. Evol.* 15:1189–94

Wada H. 2001. Origin and evolution of the neural crest: a hypothetical reconstruction of its evolutionary history. *Dev. Growth Differ.* 43:509–20

Wada H, Holland PWH, Sato S, Yamamoto H, Satoh N. 1997. Neural tube is partially dorsalized by overexpression of HrPax-37: the ascidian homologue of Pax-3 and Pax-7. *Dev. Biol.* 187:240–52

Wada H, Saiga H, Satoh N, Holland PW. 1998. Tripartite organization of the ancestral chordate brain and the antiquity of placodes: insights from ascidian *Pax-2/5/8*, *Hox* and *Otx* genes. *Development* 125:1113–22

Wada S, Saiga H. 1999. Cloning and embryonic expression of Hrsna, a snail family gene of the ascidian *Halocynthia roretzi*: implication in the origins of mechanisms for mesoderm specification and body axis formation in chordates. *Dev. Growth Differ.* 41:9–18

Wada S, Saiga H. 2002. HrzicN, a new Zic family gene of ascidians, plays essential roles in the neural tube and notochord development. *Development* 129:5597–608

Wada S, Toyoda R, Yamamoto H, Kobayashi K, Di Gregorio A, et al. 2002. Ascidian otx gene Hroth activates transcription of the brain-specific gene HrTRP. *Dev. Dynam.* 225:46–53

Whittaker JR. 1973. Tyrosinase in the presumptive pigment cells of ascidian embryos: tyrosine accessibility may initiate melanin synthesis. *Dev. Biol.* 30:441–54

Whittaker JR. 1979. Quantitative control of end products in the melanocyte lineage of the ascidian embryo. *Dev. Biol.* 73:76–83

Willey A. 1894. Studies on the protochordata. II. The development of the neurohypophysial system in *Ciona intestinalis* and

Clavelina lepadiformis, with an account of the origin of the sense-organs in *Ascidia mentula*. *Q. J. Microsc. Sci.* 35:295–316

Willie JT, Chemelli RM, Sinton CM, Yanagisawa M. 2001. To eat or to sleep? Orexin in the regulation of feeding and wakefulness. *Annu. Rev. Neurosci.* 24:429–58

Wilson SI, Edlund T. 2001. Neural induction: toward a unifying mechanism. *Nat. Neurosci.* 4(Suppl.):1161–68

Wurst W, Bally-Cuif L. 2001. Neural plate patterning: upstream and downstream of the isthmic organizer. *Nat. Rev. Neurosci.* 2:99–108

Yamada A, Nishida H. 1999. Distinct parameters are involved in controlling the number of rounds of cell division in each tissue during ascidian embryogenesis. *J. Exp. Zool.* 284:379–91

Zhou Q, Wang S, Anderson DJ. 2000. Identification of a novel family of oligodendrocyte lineage-specific basic helix-loop-helix transcription factors. *Neuron* 25:331–43

Annu. Rev. Neurosci. 2004. 27:487–507
doi: 10.1146/annurev.neuro.27.070203.144233

First published online as a Review in Advance on March 25, 2004

CORTICAL NEURAL PROSTHETICS

Andrew B. Schwartz
Departments of Neurobiology and Bioengineering, University of Pittsburgh, Pittsburgh, Pennsylvania 15203; email: abs21@pitt.edu

Key Words neuronal populations, chronic microelectrodes, prostheses, arm control, extraction algorithms

■ **Abstract** Control of prostheses using cortical signals is based on three elements: chronic microelectrode arrays, extraction algorithms, and prosthetic effectors. Arrays of microelectrodes are permanently implanted in cerebral cortex. These arrays must record populations of single- and multiunit activity indefinitely. Information containing position and velocity correlates of animate movement needs to be extracted continuously in real time from the recorded activity. Prosthetic arms, the current effectors used in this work, need to have the agility and configuration of natural arms. Demonstrations using closed-loop control show that subjects change their neural activity to improve performance with these devices. Adaptive-learning algorithms that capitalize on these improvements show that this technology has the capability of restoring much of the arm movement lost with immobilizing deficits.

INTRODUCTION

Microelectrodes embedded chronically in the cerebral cortex hold promise for using neural activity to control devices with enough speed and agility to replace natural, animate movements in paralyzed individuals. Known as cortical neural prostheses (CNPs), devices based on this technology are a subset of neural prosthetics, a larger category that includes stimulating, as well as recording, electrodes. For many years, patients have been implanted with stimulation-based devices designed to activate neurons in different parts of the CNS. This class of neural prostheses is now used extensively in applications to restore hearing and alleviate the symptoms of Parkinson's disease. However, the mirror technology used to record signals from neurons has been applied rarely to human patients and to this point is found only in research settings. These devices, called brain-computer interfaces (BCIs), link the brain to the external world by computer processing the recorded neural signal to extract the subject's command to control an external device. For those who are movement impaired, recording-based neural prostheses may enable communication or movement. Likely beneficiaries of this evolving technology include people paralyzed by head or spinal-cord trauma or those with deficits caused by stroke, amyotrophic lateral sclerosis (ALS), cerebral palsy, and multiple sclerosis. Their

0147-006X/04/0721-0487$14.00

paralyses may range from complete—with no respiratory or eye movements—to quadri- and paraplegia. Whereas some BCI devices are designed for communication only (i.e., electroencephalography (EEG)-based, word-spelling programs), CNPs, by using single-cell activity, aim to restore movement as well. The scope of this review is limited to CNPs, specifically those designed to record signals in a form that can be used by the subject to control arm movement.

BACKGROUND

The basic theorem in this field states that within the discharge pattern of cortical neurons, there exists a rather direct representation of the desired movement. This movement image has been documented only in the last 20 years. Before this, investigators assumed that the primary motor cortex, the cortical region most closely related to movement, drove muscle activation, directly. Even though this region seemed to be anatomically and topographically organized into a body map on the cortical surface, recorded signals from a purely muscle-based coordinate system would be computationally difficult to transform into natural movements of the limb. Certainly the transformation of muscle activation to muscle force alone is a difficult nonlinear problem. Given a force and a muscle length, the way in which a limb is displaced following contraction of a specific muscle is dependent on limb geometry, the orientation of the limb relative to external forces (loads and gravity), and the history (inertia) of the moving segments at the time of contraction. Factor in the complexity of a redundant muscle system with many effectors operating simultaneously, and the problem of calculating the hand's trajectory from a sample of muscle activations becomes a very difficult engineering problem. Yet these are precisely the problems confronting designers of functional electrical stimulation (FES) systems. Electrical stimulation of paralyzed muscles must produce coordinated shortening of the muscles around the limb's joints that move the attached segments to achieve the proper end point displacement. Current CNPs provide a control signal in end point coordinates (see below). This signal, combined with the problems just enumerated and the general difficulty inherent in long-term electrical activation of limb muscles, is why initial implementations have been based on the control of artificial rather than real arms.

End point (the end of the arm or the hand) movement is represented simply in the activity of motor cortical cells. Georgopoulos and his colleagues showed that motor cortical discharge rate was directionally tuned as monkeys made reaching movements (Georgopoulos et al. 1982, 1986). Movement direction was measured at the hand when all seven degrees of freedom in the shoulder, arm, and wrist were free to move. The tuning function (cosine-shaped) relating discharge rate to direction is broad, covering all movement directions, and shows that each cell changes its discharge rate for all directions, or conversely, that all cells actively code each direction (simultaneous activity). By itself, the tuning function of a single cell is not very useful for decoding direction because a single direction will correspond to more than one discharge rate, and the broadness of the function means that small fluctuations of discharge rate will correspond to large changes in direction. However, specific directions were well predicted if weighted responses from many cells were added together vectorially (Georgopoulos et al. 1984) using a linear method termed the population vector algorithm (PVA). This direction was instantaneously and continuously represented in the cortical activity throughout movement (Schwartz 1992). The magnitude as well as the direction of this neural vector representation was highly correlated with movement velocity (Georgopoulos et al. 1988, Moran & Schwartz 1999). With these properties, the movement trajectory (time pattern of hand positions) could be extracted from the population activity for reaching movements and for a variety of drawing tasks (Schwartz 1993, Schwartz 1994, Schwartz & Moran 1999).

Why is it so desirable to extract a trajectory signal from the brain? The trajectory of the end point contains natural characteristics of animate motion. Examples of these invariant features are the bell-shaped velocity profile (Morasso 1981) of reaching movement and the two-thirds power law pertaining to drawing and handwriting (Viviani & Terzuolo 1982). Although prosthetic devices can be effective without operating like natural limbs, the embodiment of these characteristics is desirable in terms of biomechanical compatibility with other body parts, ease of control, and aesthetics.

APPLICATION

These experimental results show that accurate predictions of arm movement can be generated by recoding activity from a population of cortical neurons. For this to be a real-time control signal, parallel recordings must be made from multiple electrodes. In addition, this signal should be dependable—units (preferably the same ones) should be able to be discriminated for years from each electrode. Recording single-unit activity this way requires chronic intracortical implantation of microelectrodes. Alternatively, noninvasive EEG scalp electrodes can record electric fields useful for prosthetic control. However, aside from a brief comparison to EEG prostheses, this review concentrates on intracortical, single-unit activity used for CNPs.

Most of the work in this field presumes that movement-related information in the recorded unit activity is encoded as firing rate. For single units, this rate is the inverse of the time interval between action potentials of a single neuron, and for multiunit clusters the analogous interval is measured between amplitude crossings of the group's summed electrical signal. Alternatively, power within a frequency ban of multiunit activity may be taken as an activity measure. Although information may be contained in the synchronous activity of neurons (Butler et al. 1992, Hatsopoulos et al. 1998, Mazurek & Shadlen 2002), so far the amount of information in such a code seems relatively small. Extraction algorithms convert firing rate to movement displacement. Typically, firing rates from many different units recorded simultaneously are used as input to the algorithm, and hand position is the output. Real-time decoding is needed for prosthetic control, so instantaneous firing rate calculated in a bin (e.g., 10–100 ms wide) is fed continuously to the algorithm, producing a continuous stream of hand positions. CNPs operate by recording multiple channels of single-unit activity simultaneously, conditioning these signals, usually by discriminating spike activity from the recorded signal, processing the spike trains with an extraction algorithm to generate a movement trajectory, and finally, feeding the extracted movement trajectory to a computer graphics display or a robot arm controller. Note, the control scheme does not end with movement generation. Rather, this is an example of closed-loop control where subjects observe the generated movement and modify their underlying neural activity to change the movement in a continuous fashion.

Every CNP is composed of three building blocks. The first step is to record the type of neural activity from which a consistent control signal can be extracted. Extraction is based on the concept of a neural code—recorded signals need to be deciphered and related to a desired movement. Finally, the control made possible by the extracted signal is implemented either in a computer display, an active prosthetic arm or other mechanical device, or by using electrical activation of the subject's muscles. In summary, the three CNP components are

1. microelectrodes and recording electronics. Chronic electrodes provide many individual recording sites implanted permanently in the cerebral cortex. The recording electronics condition and discriminate the recorded signal. An excellent review on this technology is available (Schmidt 1999).
2. extraction algorithms. These are computer programs running in real time that take the conditioned data (e.g., action potential events or spike times) and convert them to end point positions.
3. actuators. These can be animated computer displays, movement of a robot arm, or activation of muscles in a subject's own arm.

This review addresses the first two topics. The control of virtual-reality, computer displays and tele-robotic actuators are beyond the scope of this review.

ELECTRODES

Microwires

The first chronic recording electrodes were microwires. Developed over the last 40 years, these electrodes consist of fine wires 20 to 50 microns in diameter. They are generally composed of stainless steel or tungsten and insulated with teflon or polyimide. The tips may be etched or ground, but more often they are simply cut with a pair of scissors, leaving a planar recording surface. The wires can be arranged as arrays (for instance, in two rows of eight wires) by soldering them to a small connector (Williams et al. 1999, Nicolelis et al. 1999). Spacing between wires (100–300 microns) is maintained either with polyethylene glycol or methyl methacrylate. Wire arrays are surgically implanted in the anesthetized animal during a procedure that takes 8 to 10 h (rhesus monkey, four arrays, 64 electrodes). The cortex is exposed through a hole in the skull. After investigators remove the dura, the array is advanced slowly with a micromanipulator (100 microns/min) to minimize dimpling of the cortical surface. Although some anesthetics (i.e., ketamine) permit enough spontaneous or sensory-invoked activity in motor cortex to serve as an indicator of how far to advance the arrays, with gas inhalants such as fluorothane, spontaneous neural activity is not detectable with microwires in the motor cortex, making the optimal insertion depth uncertain. This uncertainty may leave the tips in a layer of cortex where it is difficult to record unitary activity and may be a key reason for subsequent recording failures. For exposed macaque cortex, the best depths are $\leq$2 mm below the surface. The shafts of the arrays are glued to the bone, so the depth of the electrode tip relative to the skull is permanently fixed. However, the cortical surface may move after the surgery, perhaps rebounding, if it had been dimpled, or shrinking, if swelling had taken place. The causes and dynamics of this phenomenon are not understood, but shrinking or expanding the brain, however slight, is a key issue because it will change the relative position of the electrode tip, perhaps moving it to different cortical layers or into the underlying white matter.

Silicon Micromachined Microprobes

A number of silicon substrate microprobes have been developed (Hetke & Anderson 2002, Jones et al. 1992). Two types are reviewed here. The first are planar devices from the University of Michigan, and the second is the array developed at the University of Utah.

The Michigan probe is somewhat unique, in that boron diffusion is used as an initial processing step of the silicon wafer to delineate the shape of the probe. A number of steps are used to deposit silicon dioxide and silicon nitride for insulation, and this is followed by photolithography to pattern the interconnects and recording sites. Iridium is deposited on the exposed recording sites as the electrode surface. This fabrication allows for a wide variety of probe shapes and configurations. A standard probe for chronic implants consists of four parallel, dagger-like shanks connected to a microsilicon ribbon cable. The shanks are 15 microns thick and 50 to 100 microns wide, with shank tips spaced 150 microns apart. The probe designed for monkey recording is 3.8 mm long and has 4 recording shafts placed along the shaft. The ribbon cable is flexible and has a connector at the end. Probes are implanted with a pair of forceps through the open dura. The connector is glued to the skull, but in contrast to the microwires, the semiflexible ribbon cable allows the probes to "float" in the brain (e.g., move up and down with the cortex as it pulses). Because the multiple recording sites are placed along the shaft, at least some of the sites will be situated at cortical depths desirable for good extracellular recordings.

Investigators used a completely different approach for the probe designed at the University of Utah, which is now commercially available through Cyberkinetics, Inc. The fabrication of this device begins with a solid block of silicon. Checkerboard slices with a microsaw are cut most of the way through the block. Etching of the block then results in a three-dimensional, 10 × 10 array of needles on a 4 × 4–mm square. Additional processing applies metal and insulation layers. The final array has a recording site at the tip of each shank with its interconnect running down the shank and through the back of the block, where gold pads are located for wire bonding to leads projecting to a skull connector. The 35- to 75-micron-long recording tips are platinum, with an impedance of 100 to 500 kOhms. The shank lengths range from 1.0 to 1.5 mm. Implantation of the array is achieved by injecting the array through the reflected dura with a special high-speed device that overcomes the inertia of the cortex. Leads are flexible enough to allow the array to "float" on the cortical surface. This design has the advantage of placing a relatively large number of recording sites in a compact volume of cortex. Furthermore, conventional wisdom suggests that a recording site at the tip is ideal in terms of sampling the potential field of an action potential and is the place least likely to experience tissue damage from electrode insertion (see below). However, with a single recording site at a fixed cortical depth, the Utah array suffers from the same placement problem as microwires. The length of the shanks is limited to 1.5 mm because of the way the device is fabricated from a single block of silicon.

Tissue Reactions

One common problem of all chronically implanted electrodes is that of the tissue-electrode interface. Any object inserted into the brain damages the parenchyma. During insertion, blood vessels are disrupted and microhemorrhage is common. Neurons are either ripped or sliced as the electrode is inserted. Microglia derived from monocytes are activated, and astrocytes begin to proliferate, which forms a loose encapsulation around the electrode for a considerable distance (100–200 microns). A poorly understood cascade of signaling events stemming from disruption of vessels and the blood-brain barrier takes place, which leads to an infiltration of nonlocal cellular elements, immune components, and epithelial cell proliferation. Local changes in the extracellular concentration of potassium and calcium may silence the activity of nearby neurons through a local mechanism similar to spreading depression (Somjen 2001). Most histological studies report that neuron density is near normal at distances within 100 microns of the electrode after several weeks. However, mechanical considerations would suggest that the kill zone around the electrode could be larger if the electrode is not inserted exactly at 90° perpendicular to the cortical surface (Edell et al. 1992). Initially, few neurons could be found within 50 to 100 microns of the implant site, but the cells looked normal outside this zone. After four to six weeks, this sparse zone decreased, with healthy-appearing neurons apparent closer to the electrode. Early on, astrocytes, identified by glial fibrillary acetic protein (GFAP) could be seen in the margins of the sheath, with processes infiltrating and surrounding the implant site. As neurons filled in around the implant, the perielectrode sheath became more compact with heavy GFAP staining, but cell bodies and processes could not be readily distinguished in the highly compressed tissue.

Astrocytes, the most prevalent type of glial cell, play a role in supporting brain tissue and nourishing neurons (for instance, they supply neurons with lactate derived from glucose). They also play a role in scar formation. Another type of glia, the microglia, are CNS analogs of macrophages. These are mobile cells that engulf fragments of damaged cells. Microglia are the other major component of the electrode encapsulation (Szarowski et al. 2003). Another component of reactive astrocyte response, identified by the marker vimentin, was present in the encapsulation (Szarowski et al. 2003). Vimentin is thought to be specific to immature astrocytes, and in this study, cells staining positive for this marker had long thin processes extending more than 100 microns into the sheath. These astrocytes were found much less frequently than those positive for GFAP, and they formed a thinner layer than their counterparts. Organization of the vimentin layer followed a slightly different time course than the GFAP astrocytes and microglia, both of which had similar morphology within the sheath over time.

This multicomponent process following electrode insertion is still not well understood. The shape and size of the electrode, and the way it is inserted, are probably critical factors in the type of damage imparted (Edell et al. 1992). A cylindrical shape may be ideal for pushing blood vessels and cells away without damage. However, because of the filamentous nature of neural tissue, forces from this displacement may propagate through the tissue, resulting in tearing, stretching, and compression. Alternatively, a conical electrode shape with a sharp tip will cut a hole through the tissue, leaving adjacent tissue intact as the electrode slips by. But a pressure band will eventually build in front of the electrode as some of the tissue is compressed, and this pressure will be transmitted to nearby tissue. A flat, sword-like tip may be preferable because its slicing action would cause minimal tissue compression. Recently, investigators compared different electrode shapes and insertion techniques (Szarowski et al. 2003). Although the size and shape of the electrode made a difference in the reaction around the electrode in the first one to two weeks after implantation, histology taken at longer intervals showed little difference except for the volume of tissue affected. Insertion technique appeared to make little difference.

The second component of the implant reaction is a chronic process, taking place on a slower timescale after implantation, and results in the formation of a tight cellular sheath around the electrode (Turner et al. 1999). Whereas the sheath was completely formed around the insertion site, composed of loosely packed cells and 2 layers thick in the first 2 weeks postinsertion, by 6 to 12 weeks, the sheath had become thinner and tighter with 4 to 6 layers of compact, dense cells with small nuclei, effectively isolating the probe from the brain tissue (Szarowski et al. 2003). A study using impedance spectroscopy (Williams 2001) showed that electrode impedance was well correlated with sheath density. This increase in tip impedance following implantation is one of the reasons that microwire recordings have been successful. The diameter of the wire's exposed surface is 50 microns, which is generally too large for good isolation of single units. As the sheath forms, the amount of exposed surface is reduced, which raises the electrode impedance. These higher impedance recordings are more in range with conventional extracellular recordings and make it possible to isolate single action-potential waveforms. This principle may apply to the Utah arrays, which have relatively large recording sites at the shank tips. In contrast, this type of encapsulation is an apparent problem for the Michigan probes, which, in the past, tended to record good action potentials for the first one to three weeks after implantation, after which time the signal degraded.

Some of the these issues are being addressed with new technology. Ideally, one would like to regulate the extraelectrode environment so that sprouting neurites are attracted to the vicinity of the recording surface before encapsulation takes place (Kennedy et al. 2000). Presently, investigators are studying modifications to the electrode surface using hydrogels, silk-like polymers, and nanotubes. These structures can be bound to bioactive compounds, such as neurotrophins, that attract growing neurites (Cui et al. 2001). Conversely, the inflammatory response can be reduced. Lipid microtubes (Meilander et al. 2001, Zhong et al. 2001) can deliver molecules to block transcription factors and downregulate genes controlling release of proinflammatory cytokines (Manna & Aggarwal 1998). Systemic and local dexamethasone administration was effective at reducing the density of the peri-electrode sheath, whereas cyclosporin A was found ineffective (Shain et al. 2003). Investigators made no attempt to correlate this with improved recording conditions. Dexamethasone inhibits astrocyte hyperplasia and can also be incorporated into coatings applied to the electrode. The diffusion of these molecules into the tissue can be regulated so that a therapeutic dose is maintained over time. This technology may achieve the desired electrode-neuron-glial sheath sandwich that could lead to permanent long-term recording of neural activity.

Experience with these probes suggests that encapsulation is a major factor in the deterioration of recording conditions. In the past, this deterioration may have been due to the configuration of the probe, the planar recording sites on the side of the shaft, and/or the size of the recording site (Hetke & Anderson 2002, Schmidt et al. 1997). That the recording deterioration was due to encapsulation is supported by the observation that the recording sites could be reactivated by passing current through the electrode, which reduces the impedance by removing tissue or opening tunnels through the encapsulation (Schmidt et al. 1993a,b). Better results have been reported, recently. Investigators have maintained good recordings in the guinea pig for more than one year (Kipke et al. 2003). In our laboratory we have been recording good units for more than ten months from implanted probes in monkey cortex. One consistent finding in comparing the success of chronic recordings in rhesus monkeys, as compared to either guinea pigs or rats, is that it is easier to obtain consistent high-yield recordings in the rodent. This finding may be related to differences in the reaction to implantation or to the differences in cortical folding between the rodent (lissencephalic) and primate (gyrencephalic).

EXTRACTION ALGORITHMS

To generate a prosthetic movement signal, information contained in the parallel recordings is transformed from the domain of spikes/sec to extrinsic (i.e., Cartesian) coordinates. Approaches for extracting information divide into two broad categories: inferential methods and classifiers. Inferential methods are model based—they depend on some understanding of underlying mechanisms. To illustrate this, we use a somewhat unrelated example from the vestibular system. The response generated by receptors in the semicircular canals to angular acceleration was predicted by modeling the endolymph, cupula, and canal as a torsional pendulum. On the basis of this mechanical model, the canal input, angular acceleration, is integrated to become angular velocity at the ouput. This model was confirmed by recordings of axonal activity in the eighth nerve, which were correlated with angular velocity. In contrast, classifier methods need not consider any mechanism. Rather they rely on a consistent representation of the parameter to be extracted. Imagine a set of three neurons recorded as a monkey moves to targets in eight directions. Considering each neuron as a binary element (on or off), these three neurons could encode each direction unambiguously. Each combination of the neuronal pattern would be assigned to one target direction. Then, observation of the neurons as a pattern obtained by simultaneously recording could be used to predict movement direction. In this ideal example, no assumptions of mechanism, or even of continuity between patterns, are necessary. However, if the task was changed, for instance by adding or removing targets, there is no guarantee that the eight-target code would still apply. In reality, neurons are not bistable but change their firing rates continuously. Their firing rate–movement tuning functions are unimodal, broad, and noisy, with no clear transition between on and off states. Furthermore, in practice, to find a perfect set of neurons that encode a single parameter in an orthogonal manner (i.e., 3 neurons for direction in 3D space) is difficult. These realistic conditions have led to more sophisticated classifiers than the ideal example given here.

Inferential Methods

A population of neural activity measured as spike occurrences can be represented as a vector. In this example, each dimension of the vector corresponds to a particular neuron, and the value or magnitude of each dimension is proportional to that cell's firing rate. This vector then would have a direction and magnitude in a neural space. The purpose of the extraction algorithm is to transform this vector into a corresponding vector in movement space. The direction of each dimension of the neural vector has no physical meaning other than acting as a neuron label. Models (for instance, based on directional tuning functions) are used to transform this neural vector into a movement vector.

POPULATION VECTOR ALGORITHM As introduced earlier, the relation between movement direction and firing rate in the motor cortex can be described with the cosine function

$$D - b_o = A \cdot \cos\theta \qquad 1.$$

This function is equivalent to the expression

$$D - b_o = b_x m_x + b_y m_y + b_z m_z \qquad 2.$$

or to the dot product of the two vectors **B** and **M**, where D is the discharge rate of the studied unit, A is the amplitude of the tuning function, θ is the angle between the cell's preferred direction, the movement direction **B** is a vector in the unit's preferred direction with a magnitude equal to the amplitude of the tuning function, and **M** is a unit vector in the movement direction. The cell's maximum firing rate is the tuning function amplitude plus b_o. The coefficients of the **B** vector (the unit's preferred direction) are typically found by regressing movement direction to discharge rate. Most commonly, this relation comes from data gathered during a task where the subject reaches from a center-start position to a set of targets arranged circumferentially around the start position so that the reaches occur uniformly in different directions.

The population vector is formed by combining weighted contributions from each unit along its preferred direction. For a particular movement, the **B** vector of each unit is normalized to the firing rate of the cell during that movement. This contributory vector, $\mathbf{C}_i$, of the ith unit, is added with those of the other *N* units in the population to form the population vector.

The population vector can be generated in small bins (i.e., 20 ms) and correlates very well with the hand's velocity throughout an arm movement. These vectors can be added together, tip to tail, to form a neural trajectory that predicts and matches the arm's trajectory (Georgopoulos et al. 1988). In the past, these experiments were executed by recording units one at a time, summing their responses together to form the population vector. For prosthetic control, the vector output must be generated in real time by recording from electrode arrays using electronics capable of processing many units simultaneously.

The population vector algorithm can be an efficient decoder if two ideal conditions are met: The preferred directions of the recorded units should be distributed uniformly in space with radially symmetric tuning functions (Georgopoulos et al. 1988). Considering the small number of units recorded simultaneously with chronic electrodes, these conditions usually are not met. Often units are not well tuned, and those that are do not form a symmetric distribution of preferred directions. These problems are accentuated in prosthetic control, where trials cannot be averaged and movement commands are calculated in small bins. These issues were addressed in the modified population vector algorithm used by

Taylor et al. (2002). Instead of weighting each unit's contribution to the population vector only by its discharge rate, a number of additional factors were used. As subjects used their neural activity to direct a cursor to center-out targets, *X*, *Y*, and *Z* weighting coefficients were calculated iteratively. This procedure was considered coadaptive because the weighting coefficients were adjusted as the subject was learning to make brain-controlled movements. By accounting for changes in the tuning characteristics of the units as the subject learned, this algorithm was very effective with relatively few units. Unlike the original population vector algorithm, this algorithm reduced the contribution of poorly tuned units by scaling the coefficients according to the unit's cosine fit. One way that units commonly deviated from the cosine function was that firing rates in the preferred direction were not always 180° from the antipreferred direction. Using separate coefficients when rates were above or below the mean compensated for this noncosine behavior. Nonuniformity of the directional distribution was addressed directly by normalizing the *X*, *Y*, and *Z* contributions and was addressed indirectly by emphasizing coefficients that produced good movements in all parts of the workspace.

Other extraction techniques are based on more formalized optimization algorithms. The optimized linear estimator (Salinas & Abbott 1994) is similar to the population vector algorithm in that individual cell responses are added vectorially to give a single population vector. The difference is that each unit's preferred direction is calculated using a multiple regression across all the units simultaneously to give optimal fits to the sampled movement directions. A correlation matrix between the firing rates of all cells is inverted to give the global least-squares fit of the population. Compared to the population vector, this method has the advantage of correcting for nonuniform direction distributions. Other investigators used a similar linear approach with time-shifted firing rates regressed to hand position (Wessberg et al. 2000).

KALMAN FILTER The methods described to this point are considered static models because they consider the velocity at each time step to be independent. Dynamic state models have been developed to account for correlations between nearby velocities (or movement increments).

These methods have great potential because they use features of the movement or behavior that are predictable in their own right, in addition to predictions based on neural activity. Because many features of movement are regular (speed changes smoothly, direction tends to be constant at high speed, etc.), these state conditions can lend great power to the overall prediction. The prior hand velocities combined with a history of firing rates are incorporated into the Kalman filter (Paninski et al. 2004) to predict future hand velocity. The state model for velocity is often very simple, i.e., random walk. Used this way, velocities change smoothly—the velocity of the hand does not change much from one instant to the next and is characteristic of trajectories during reaching movements. Formally, x_k is the velocity at time step *k*, and x_{k+1} is the velocity at the next time step. The discharge rate at *k* is z_k. The state and observation equations are

$$x_k = Ax_{k-1} + w_k \tag{3}$$

and

$$z_k = Hx_k + v_k. \tag{4}$$

Velocity can be expressed by a three-dimensional vector (*X*, *Y*, and *Z*); $\mathbf{z_k}$ is a vector of the current firing rates (at time step *k*) for the *N* simultaneously recorded units. *A* is a matrix, for example, and is an identity matrix if the state model is a random walk. *H* is a matrix of coefficients to convert the velocity into discharge rates and consists of an *X*, *Y*, and *Z* coefficient for each recorded unit (i.e., preferred direction). The error in velocity between the estimated (Equation 3) and actual value is w_k with a distribution specified with a covariance matrix Θ. For the discharge rates, the error is v_k with a covariance of Λ.

Of course, the state model can be more complex, for instance by taking into account the bell-shaped velocity profile and the tendency for straight arm movements during reaching or by including terms for position and acceleration in addition to velocity.

The Kalman filter works iteratively in steps. It begins by guessing the velocity from an initial distribution specified by a mean $\hat{x}_0$ and covariance Σ_0. This choice is then used to get $\hat{x}_k^-$, the state estimate (Equation 5).

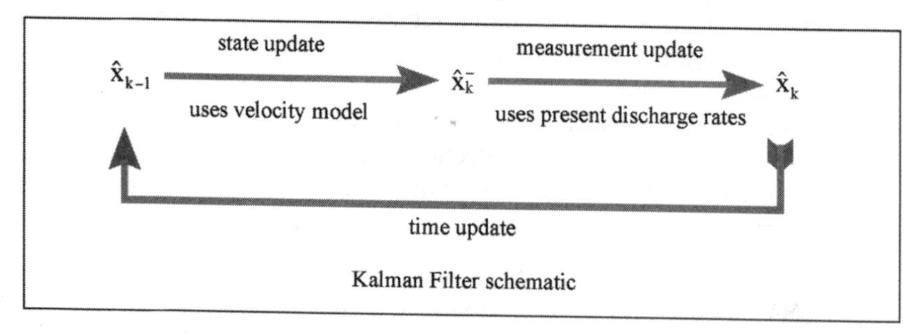

Kalman Filter schematic

$\hat{x}_{k-1}$ is the estimate of velocity in the previous time step $k - 1$ and has a covariance of Σ_{k-1}.

The current velocity estimate at step *k* uses $\hat{x}_{k-1}$ and the velocity model, *A*:

$$\hat{x}_k^- = A\hat{x}_{k-1}. \tag{5}$$

The covariance of estimate errors in the state estimate of velocity is

$$P_k^- = A\Sigma_{k-1}A^T + \Theta. \tag{6}$$

The prediction of velocity using $\hat{x}_k^-$ and present discharge rates is

$$\hat{x}_k = \hat{x}_k^- + K_k(z_k - H\hat{x}_k^-), \tag{7}$$

where

$$K_k = P_K^- H^T(HP_k^- H^T + \Lambda)^{-1}. \tag{8}$$

The state estimate is then updated with the spike activity to give the new estimate of velocity, $\hat{x}_k$, using Equation 7. This measurement update uses the difference between the actual, z_k, and estimated, $H\hat{x}_k^-$, discharge rates multiplied by the gain factor, *K*, defined in Equation 8. The $(\ )^{-1}$ in this equation denotes pseudoinverse, and $(\ )^T$ signifies transpose. The current velocity estimate, $\hat{x}_k$, is then used as the time update and becomes $\hat{x}_{k-1}$ in the next iteration.

Investigators have elaborated this general procedure using statistical methods. The estimates of the Kalman filter can be considered the peak value of a probability density function. For example, the density functions for each step of the recursion steps can be defined in the following way: $\hat{x}_{k-1}$ is the mean of $p(x_{k-1}|z_{k-1}, z_{k-2}, \ldots z_1)$, $\hat{x}_k^-$ is the mean of $p(x_k|z_{k-1}, z_{k-2}, \ldots z_1)$, and $\hat{x}_k$ is the mean of $p(x_k|z_k, z_{k-1}, \ldots z_1)$.

The generation of these density functions for nonlinear, non-Gaussian dynamic state models is a current area of interest in statistical research (Brockwell et al. 2004).

Classifiers

The methods described above generate predictions based on prior knowledge of how neural activity is related to movement and, in some cases, the general structure of arm movements. Algorithms based on pattern recognition do not require such prior knowledge. The self-organizing feature map (SOFM) is an example of a classfier. These artificial networks depend on a consistent relation between neural firing rates and movement. SOFMs can be visualized as a single layer of elements or nodes (artificial neurons), each of which is connected to an input vector with a set of connection weights. In one example of this scheme (Lin et al. 1997), each of the *n* weights corresponded to a recorded unit's discharge rate (*n* recorded units in the sampled population). Initially, each of the network element's *n*-dimensional weight vectors was set randomly. In an initial step, the input vector consisting of the recorded neurons' firing rates was compared to each of the artificial element's weight vectors. The element with the weight vector closest to the input vector was declared as the winner and modified to resemble the input vector. Its neighbors' weights were also moved closer to the input vector. This process was repeated for successive time points until clusters of similar elements were created on the surface. These clusters had similar weight vectors and were distinct from other clusters. In an identification step, each cluster was assigned to a direction. At this point the network is considered trained. New input vectors were fed in, and the predicted movement direction was chosen as the labeled direction of the cluster closest to the input vector.

Other networks using back-propagation (Wessberg et al. 2000) and nonlinear maximum-likelihood estimation (Pouget et al. 1998) have been used successfully to convert populations of neural activity to predicted movement.

Another type of decoder was developed using snippets of arm trajectory and time windows of recorded discharge rates (Isaacs et al. 2000). Each 200-ms window contained 10 sequential discharge rates from a single neuron and corresponded to a trajectory snippet. Adjacent windows from the same neuron and windows between different neurons recorded simultaneously were compared in a covariance matrix and categorized with principal components. Used as a decoder, an eigenvector calculated from the training data was multiplied by a novel window of discharge rates to give a principal component. This component was compared with a dictionary of principal components from the training set, and the closest match was declared the winner. The corresponding trajectory snippet was then taken as the current trajectory prediction.

Practical Considerations

Extraction algorithm development can be summarized with a few statements. First, linear methods are very effective in extracting movement information from the recorded population. Second, methods that account for asymmetrical samples can effectively compensate for small sample size. Third, methods relying on training sets must produce robust outputs. The training set should be generated from a range of movements that represent the same range of movements for which the prosthetic device will be used. This problem can be addressed with algorithms that are adaptive, for example by changing weighting coefficients to maximize success rates. Finally, algorithms that capture the features of natural movement such as smoothness, segmentation, and curvature-speed tradeoffs are desirable because these more natural movements may be inherently more controllable.

The issue of controllability is important. Prosthetic devices are closed-loop devices. Subjects generate an output (i.e., neural signal) and then watch the device move. Observation of the device's performance closes the control loop, and subjects have the opportunity to change their output to advance the device to the goal. Learning, in the form of modified neural output, can dramatically improve prosthetic performance (Carmena et al. 2003, Serruya et al. 2002, Taylor et al. 2002) compared to open-loop decoding (in which the subject does not observe movement of the device). The output of an extraction algorithm needs to be understood by the subject, who should be able to alter neural activity in a consistent way to achieve a predictable change in the way the prosthesis moves. This consideration is likely to be more important than the details and performance of a given extraction algorithm developed and evaluated with open-loop data.

OTHER BCIs

Recording individual action potentials requires an invasive surgical procedure to place the electrodes. EEG, which uses scalp electrodes to record signs of electrical brain activity noninvasively, is one type of BCI technology being implemented presently in human subjects. Although many electrodes are placed around the head, often only the electrodes over the sensorimotor cortex record signals useful for brain control. EEG activity is the complex sum of many neuronal currents with complex geometries filtered through the brain, skull, and scalp. Several methods are used to extract movement-related intentions from these signals, including frequency decomposition, recognition of movement-related evoked potentials, and slow cortical potentials. These methods allow subjects to control computer cursors in one dimension in word-spelling and sentence-construction tasks and may be useful in developing a communication interface. Although it is still an open question whether this technology can generate signals for more complex movements, a recent report showed that a well-trained subject could use the EEG approach to perform a two-dimensional, center-out task (Wolpaw & McFarland 2003). A detailed review of this subject has been written recently (Wolpaw et al. 2002).

More localized electrical activity can be recorded when electrodes are placed between the dura and pia, as practiced in cortical mapping prior to epilepsy surgery. Information derived from these signals holds the promise of controlling more complex prosthetic movements (Leuthardt et al. 2003, Rohde et al. 2002).

This work implies that the more closely the recorded signal represents activity of single units, the more useful it is for movement control.

PAST AND CURRENT CNP STATUS

In a lecture at Oxford in 1963 W. Gray Walter (1963), a pioneer in the use of EEG and mobile automata, reported having used brain signals recorded from human motor cortex to operate a slide projector. This was the first example of CNP feasibility. Much later, in 1996, investigators demonstrated continuous movement control with initial, open-loop monkey experiments (Perepelkin & Schwartz 1996, Schwartz et al. 1996), which showed that populations of units recorded simultaneously with microwires in both right and left hemispheres could be used to generate population vectors for prosthetic arm control. In the ensuing four years, investigators continued most of this work in open-loop conditions, calculating population vectors posthoc or using real-time population vectors to drive a robot arm without feedback to the monkey (Isaacs et al. 2000, Wessberg et al. 2000).

More recently, a closed-loop CNP was first used in a rat to move a simple lever for reward (Chapin et al. 1999). Since then, CNPs using small populations of cell activity were shown effective for controlling closed-loop, two- (Serruya et al. 2002) and three-dimensional (Taylor et al. 2002) movements. As described in the previous section, subjects saw the result of the extraction process in real time and learned to modify their neural activity to improve their performance on the task (moving a computer cursor to specified targets). This, compared to open-loop control, dramatically improved their overall ability to reach the target quickly and accurately. In the Taylor et al. (2002) study, monkeys moved a cursor in three-dimensional, virtual-reality targets, performing the task either by moving their hands (hand control) or with their arms restrained (brain control). When switching between these tasks, the preferred directions of the chronically recorded units changed. Although there was no global pattern to these shifts, they were consistent from day to day, and the size of the shifts increased over days as the animals' performance improved. The better linear fit of the activity patterns to the cosine-tuning function was also coincident with this improvement. A coadaptive algorithm designed to track these learning-induced changes in neural activity was very effective. One animal had a success rate of more than 80% in daily training sessions (including periods of inattention) with a population of 64 units. This animal, using brain control, performed consistently for many minutes, reached novel targets on the first attempt, and moved with a speed and accuracy approaching normal arm movements.

This closed-loop task was modified to include an electro-mechanical robot arm in the control loop (Taylor et al. 2003). Instead of signaling cursor movement, the extracted cortical velocity predictions were streamed to a robot controller. Position of the robot arm was tracked three-dimensionally, and these movements were fed back to the graphics routine so that the animal saw the movement of the robot arm as cursor movement in the virtual reality display. Because the robot arm did not move exactly as commanded, the cursor movement was perturbed. However, the monkey learned to correct these perturbations using visual feedback during the movement to achieve a high level of performance. Similar results were found in a later study of two-dimensional movements combined with isometric grip regulation (Camena et al. 2003).

Currently, work is underway to demonstrate brain control of reaching and grasping using direct vision of an anthropomorphic robot arm (Helms Tillery et al. 2003, Schwartz 2003). A child-sized motorized prosthetic arm with a 3-degrees-of-freedom (DOF) shoulder and a 1-DOF elbow is mounted near the monkey's shoulder. The monkey will reach out to a piece of food at different locations in the three-dimensional workspace and grasp it with a simple gripper before bringing the food back to its mouth. For training purposes, portions of the task can be automated. For instance, the arm can be computer-guided to reach for and grasp the food, followed by a brain-controlled retrieval by the monkey.

Recording technology continues to improve. Currently, the Michigan silicon probes continue to record good unit activity following their implantation more than six months ago in macaque cortex. The Utah probe has been modified with new insulating materials and connector technology (Cyberkinetics Inc.), leading to an increase in the number of units recorded per array. Finally, the microwire approach has been extended with new fabrication techniques (Nicolelis et al. 2003). This field is now at the point where more reliable implants can be combined with the control of elaborate prosthetic effectors capable of producing near-natural arm movement.

FUTURE WORK AND PROSPECTS

On the basis of a number of studies, cortical neural prostheses will be feasible in generating natural movements either with artificial effectors or intrinsic muscle activation. The information contained in the recorded signal, consisting of individual action potentials recorded simultaneously from the cerebral cortex, can be extracted in real time and be used to make purposeful movements. Several groups are now preparing studies to apply this technology to humans.

The largest remaining obstacle to the successful implementation of cortical neural prostheses is the chronic recording electrode. The electrode should be able to record single-unit activity reliably for many years, be relatively easy to implant, and be capable of recording many action potentials from different cells in a small volume of cortex. Designs such as those produced by the University of Michigan, where multiple recording sites are placed on each shank, have the advantage of such dense sampling. On the other hand, the Utah probe can achieve high-density sampling by spacing many shanks close together. These probes have an ideal placement of the recording surface at the shank tip but do not have multiple sites along each shaft. A potential compromise would be to arrange planar silicon probes in a high-density array, a project that has produced several prototypes to date (Hetke & Anderson 2002).

As described above, technology is being developed to regulate the tissue/electrode interface associated with arrays implanted in the brain. Investigators have demonstrated the potential of this technology in human patients with the cone electrode (Kennedy et al. 2000). This electrode is a capillary tube filled with growth factor or peripheral nerve extract. Also in the tube are the exposed ends of two microwires, which act as differential electrodes. Neurites that sprout in response to the electrode penetration are attracted to the interior of the tube, through which they grow and form synaptic connections to other neurons. The axon is permanently trapped next to the recording electrode. Although only a few channels of multiunit data were recorded, this activity was used by locked-in ALS patients for communicating with a spelling/letter-board program. One patient used this method for more than a year.

An ethical issue arises as we work toward implementing these devices in disabled patients. Paralyzed patients are motivated to volunteer as experimental subjects. Surgeons and researchers are eager to implant chronic electrodes. The question persists, at what point should this imperfect technology be applied? This common issue in bioengineering has been addressed with other neural prostheses (for example, cochlear implants, deep brain stimulators, and visual prostheses). Presently, an informal survey (A. Schwartz, personal communication) of the laboratories using CNP suggests that, on average, a chronic electrode implanted in monkey cortex has only a 40% to 60% chance of recording unit activity. Although each lab has an example of an all-star animal with good recordings for multiple years, electrode recordings usually deteriorate after several months. How much improvement in this technology is needed before human experimentation is warranted? Considering the rapid improvements in electrode technology, a better understanding of the biology associated with electrode interaction within the brain, the financial race of investment-driven development, and the desires of patients and researchers to try this technology, humans likely will be implanted with these chronic devices in the next few years. The technology used in these devices, the care taken to develop the correct surgical procedures, and the peripheral technologies associated with the electrodes—connectors, telemetry, spike conditioning, and real-time computing—should be well considered, not only by regulatory bodies but also by the scientists, engineers, and patients who will be using them. The marketing aspect of this technology is rarely considered. There is a trade-off between the severity of a patient's deficit and the relative efficacy of a CNP. Locked-in patients will benefit from any device that allows them to communicate better, whereas patients with C5 spinal cord lesions may still have arm mobility and would only benefit from a device that restored natural arm movements combined with a degree of grasping. How many patients within these categories would be willing to undergo surgical implantation before this technology is commercially viable? This answer depends, to a large degree, on the viability of the electrodes, the information content of the recorded signals, and the engineering needed to make natural, agile effectors. Furthermore, it will be important to develop effective training procedures for patients to acquire the skills to use these devices. Certainly, companies have already been formed with the expectation that these conditions will be met. In the mean time, those of us working in the laboratory have the exciting prospect of direct access to the previously inaccessible neuronal substrate of human skill, with the potential generation of scientific discovery of fundamental aspects of learning and cognition.

ACKNOWLEDGMENTS

The author thanks Dawn Taylor, Stephen Helms Tillery, and Rob Isaacs for their contributions to the experimental results cited here. Futhermore, the Harrington Department of Bioengineering at Arizona State University, the Neural Prosthesis Program at NIH, and the Whitaker Foundation supported the author's research discussed in this review.

The *Annual Review of Neuroscience* is online at http://neuro.annualreviews.org

LITERATURE CITED

Brockwell AE, Rojas AL, Kass RE. 2004. Recursive Bayesian decoding of motor cortical signals by particle filtering. *J. Neurophysiol.* In press

Butler EG, Horne MK, Churchward PR. 1992. A frequency analysis of neuronal activity in monkey thalamus, motor cortex and electromyograms in wrist oscillations. *J. Physiol. (London)* 445:49–68

Carmena JM, Lebedev MA, Crist RE, O'Doherty, Santucci DM, et al. 2003. Learning to control a brain-machine interface for reaching and grasping by primates. *PLoS* 1:193–208

Chapin JK, Moxon KA, Markowitz RS, Nicolelis MA. 1999. Real-time control of a robot arm using simultaneously recorded neurons in the motor cortex. *Nat. Neurosci.* 2:583–84

Cui X, Lee VA, Raphael Y, Wiler JA, Hetke JF, et al. 2001. Surface modification of neural recording electrodes with conducting polymer/biomolecule blends. *J. Biomed. Mater. Res.* 56(2):261–72

Edell DJ, Toi VV, McNeil VM, Clark LD. 1992. Factors influencing the biocompatibility of insertable silicon microshafts in cerebral cortex. *IEEE Trans. Biomed. Eng.* 39(6):635–43

Georgopoulos AP, Kalaska JF, Caminiti R, Massey JT. 1982. On the relations between the direction of two-dimensional arm movements and cell discharge in primate motor cortex. *J. Neurosci.* 2(11):1527–37

Georgopoulos AP, Kalaska JF, Crutcher MD, Caminiti R, Massey JT. 1984. The representation of movement direction in the motor cortex: single cell and population studies. In *Dynamic Aspects of Neocortical Function*, ed. GM Edelman, WE Goll, WM Cowan, 16:501–24. New York: Neurosci. Res. Found.

Georgopoulos AP, Kettner RE, Schwartz AB. 1988. Primate motor cortex and free arm movements to visual targets in three-dimensional space. II. Coding of the direction of movement by a neuronal population. *J. Neurosci.* 8:2928–37

Georgopoulos AP, Schwartz AB, Kettner RE. 1986. Neuronal population coding of movement direction. *Science* 233:1357–440

Hatsopoulos NG, Ojakangas CL, Paninski L, Donoghue JP. 1998. Information about movement direction obtained from synchronous activity of motor cortical neurons. *Proc. Natl. Acad. Sci. USA* 95(26):15706–11

Helms Tillery SI, Taylor DM, Schwartz AB. 2003. The general utility of a neuroprosthetic device under direct cortical control. *Proc. Eng. Med. Biol. Soc. 25th Int. Conf.* (Abstr.) 2043–46

Hetke JF, Anderson DJ. 2002. Silicon microelectrodes for extracellular recording. In *Handbook of Neuroprosthetic Methods*, ed. WE Finn, PG LoPresti, 7:163–91. Boca Raton, FL: CRC

Isaacs RE, Weber DJ, Schwartz AB. 2000. Work toward real-time control of a cortical neural prosthesis. *IEEE Trans. Rehabil. Eng.* 8:196–98

Jones KE, Campbell PK, Normann RA. 1992. A glass/silicon composite intracortical electrode array. *Ann. Biomed. Eng.* 20:423–37

Kennedy PR, Bakay RAE, Moore MM, Adams K, Goldwaithe J. 2000. Direct control of a computer from the human central nervous system. *IEEE Trans. Rehabil. Eng.* 8:198–202

Kipke DR, Vetter RJ, Williams JC, Hetke JF. 2003. Silicon-substrate intracortical microelectrode arrays for long-term recording of neuronal spike activity in cerebral cortex. *IEEE Trans. Neural Syst. Rehabil. Eng.* 11:151–55

Leuthardt EC, Schalk G, Chicoine M, Wolpaw J, Ojemann J, Moran D. 2003. Developing a brain computer interface using electrocorticographic signals from subdural arrays in humans. *Soc. Neurosci.* 607.6 (Abstr.)

Lin S, Si J, Schwartz AB. 1997. Self-organization of firing activities in monkey's motor cortex: trajectory computation from spike signals. *Neural Comput.* 9(3):607–21

Manna SK, Aggarwal BB. 1998. Alpha-melanocyte-stimulating hormone inhibits the nuclear transcription factor NF-kappa beta activation induced by various inflammatory agents. *J. Immunol.* 161:2873–80

Mazurek ME, Shadlen MN. 2002. Limits to the temporal fidelity of cortical spike rate signals. *Nat. Neurosci.* 5(5):463–71

Meilander NJ, Yu X, Ziats NP, Bellamkonda RV. 2001. Lipid-based microtubular drug delivery vehicles. *J. Control. Release* 12:141–52

Moran DW, Schwartz AB. 1999. Motor cortical representation of speed and direction during reaching. *J. Neurophysiol.* 82:2676–92

Morasso P. 1981. Spatial control of arm movements. *Exp. Brain Res.* 42:223–27

Nicolelis MA, Dimitrov D, Camena JM, Crist R, Lehew G, et al. 2003. Chronic, multisite, multielectrode recordings in macaque monkeys. *Proc. Natl. Acad. Sci. USA* 100:11041–46

Nicolelis MAL, Stambaugh CR, Brisben A, Laubach M. 1999. Methods for simultaneous multisite neural ensemble recordings in behaving primates. In *Methods for Neural Ensemble Recordings*, ed. MAL Nicolelis, 7:121–56. Boca Raton, FL: CRC

Paninski L, Fellows MR, Hatsopoulos NG, Donoghue JP. 2004. Spatiotemporal tuning of motor cortical neurons for hand position and velocity. *J. Neurophysiol.* 91:515–32

Perepelkin PD, Schwartz AB. 1996. Simultaneous populations of single-cell activity recorded bilaterally in primate motor cortex. *Soc. Neurosci. Abstr.* 23:1139

Pouget A, Zhang K, Deneve S, Latham PE. 1998. Statistically efficient estimation using population coding. *Neural. Comput.* 10:373–401

Rohde MM, BeMent SL, Huggins JE, Levine SP, Kushwaha RK, Schuh LA. 2002. Quality estimation of subdurally recorded, event-related potentials based on signal-to-noise ratio. *IEEE Trans. Biomed. Eng.* 49:31–40

Salinas E, Abbott LF. 1994. Vector reconstruction from firing rates. *J. Comput. Neurosci.* 1:89–107

Schmidt EM. 1999. Electrodes for many single neuron recordings. In *Methods for Neural Ensemble Recordings*, ed. MA Nicolelis, pp. 1–23. Boca Raton, FL: CRC

Schmidt EM, Heetderks WJ, Camesi-Cole DM. 1993a. Chronic neural recording with multicontact silicon microprobes: effects of electrode bias. *Soc. Neurosci. Abstr.* 20:982

Schmidt EM, Heetderks WJ, Camesi DM. 1993b. Chronic recording from cortical areas with multicontact silicon microprobes. *Soc. Neurosci. Abstr.* 19:781

Schmidt EM, Heetderks WJ, Hambrecht FT. 1997. The relationship between recorded spike amplitude and microelectrode surface area. *Soc. Neurosci. Abstr.* 23:1552

Schwartz AB. 1992. Motor cortical activity during drawing movements. Single-unit activity during sinusoid tracing. *J. Neurophysiol.* 68:528–41

Schwartz AB. 1993. Motor cortical activity during drawing movements: population response during sinusoid tracing. *J. Neurophysiol.* 70:28–36

Schwartz AB. 1994. Direct cortical representation of drawing. *Science* 265:540–42

Schwartz AB. 2003. Natural arm movements with a brain-controlled interface. *Soc. Neurosci.* No. 650.2 (Abstr.)

Schwartz AB, Kipke DR, Perepelkin PD. 1996. Cortical control for prosthetic devices. *Proc. SPIE Int. Soc. Opt. Eng.* 2718:530–39

Schwartz AB, Moran DW. 1999. Motor cortical activity during drawing movements: population representation during lemniscate tracing. *J. Neurophysiol.* 82:2705–18

Serruya MD, Hatsopoulos NG, Paninski L, Fellows MR, Donoghue JP. 2002. Instant neural control of a movement signal. *Nature* 416(6877):141–42

Shain W, Spataro L, Dilgen J, Haverstick K, Retterer S, et al. 2003. Controlling cellular reactive responses around neural prosthetic devices using peripheral and local intervention strategies. *IEEE Trans. Neural Syst. Rehabil. Eng.* 11(2):186–88

Somjen GC. 2001. Mechanisms of spreading depression and hypoxic spreading depression-like depolarization. *Physiol. Rev.* 81:1065–96 (Abstr.)

Szarowski DH, Andersen MD, Retterer S, Spence AJ, Issacson M, et al. 2003. Brain responses to micro-machined silicon devices. *Brain Res.* 983:23–35

Taylor DM, Helms Tillery SI, Schwartz AB. 2002. Direct cortical control of 3D neuroprosthetic devices. *Science* 296:1829–32

Taylor DM, Helms Tillery SI, Schwartz AB. 2003. Information conveyed through brain-control: cursor versus robot. *IEEE Trans. Neural Syst. Rehabil. Eng.* 11:195–99

Turner JA, Shain W, Szarowski DH, Andersen M, Martins S, et al. 1999. Cerebral astrocyte response to micromachined silicon implants. *Exp. Neurol.* 156:33–49

Viviani P, Terzuolo C. 1982. Trajectory determines movement dynamics. *Neuroscience* 7(2):431–37

Walter WG. 1991 [1963]. Presentation to the Osler Society, Oxford University. In *Consciousness Explained*, ed. DC Dennett, p. 167. New York, NY: Penguin

Wessberg J, Stambaugh CR, Kralik JD, Beck PD, Laubach M, et al. 2000. Real-time prediction of hand trajectory by ensembles of cortical neurons in primates. *Nature* 408:361–65

Williams JC. 2001. *Performance of chronic neural implants: measurement, modeling and intervention strategies*. PhD thesis. Ariz. State Univ.

Williams JC, Rennaker RL, Kipke DR. 1999. Long-term neural recording characteristics of wire microelectrode arrays implanted in cerebral cortex. *Brain Res. Brain Res. Protoc.* 4(3):303–13

Wolpaw JR, Birbaumer N, McFarland DJ, Pfurtscheller G, Vaughan TM. 2002. Brain-computer interfaces for communication and control. *Clin. Neurophysiol.* 113:767–91

Wolpaw JR, McFarland DJ. 2003. Two-dimensional movement control by scalp-recorded sensorimotor rhythms in humans. *Soc. Neurosci.* 607.2 (Abstr.)

Zhong Y, Yu X, Gilbert R, Bellamkonda RV. 2001. Stabilizing electrode-host interfaces: a tissue engineering approach. *J. Rehabil. Res. Dev.* 38(6):627–32

Annu. Rev. Neurosci. 2004. 27:509–47
doi: 10.1146/annurev.neuro.26.041002.131412

First published online as a Review in Advance on March 12, 2004

THE SYNAPTIC VESICLE CYCLE

Thomas C. Südhof
The Center for Basic Neuroscience, Department of Molecular Genetics, Howard Hughes Medical Institute, The University of Texas Southwestern Medical Center, Dallas, Texas 75390-9111; email: Thomas.Sudhof@UTSouthwestern.edu

Key Words neurotransmitter release, synaptotagmin, active zone, synaptic plasticity, RIM, calcium channel

■ **Abstract** Neurotransmitter release is mediated by exocytosis of synaptic vesicles at the presynaptic active zone of nerve terminals. To support rapid and repeated rounds of release, synaptic vesicles undergo a trafficking cycle. The focal point of the vesicle cycle is Ca^{2+}-triggered exocytosis that is followed by different routes of endocytosis and recycling. Recycling then leads to the docking and priming of the vesicles for another round of exo- and endocytosis. Recent studies have led to a better definition than previously available of how Ca^{2+} triggers exocytosis and how vesicles recycle. In particular, insight into how Munc18-1 collaborates with SNARE proteins in fusion, how the vesicular Ca^{2+} sensor synaptotagmin 1 triggers fast release, and how the vesicular Rab3 protein regulates release by binding to the active zone proteins RIM1α and RIM2α has advanced our understanding of neurotransmitter release. The present review attempts to relate these molecular data with physiological results in an emerging view of nerve terminals as macromolecular machines.

NEUROTRANSMITTER RELEASE AND THE SYNAPTIC VESICLE CYCLE

Synaptic transmission is initiated when an action potential triggers neurotransmitter release from a presynaptic nerve terminal (Katz 1969). An action potential induces the opening of Ca^{2+} channels, and the resulting Ca^{2+} transient stimulates synaptic vesicle exocytosis (Figure 1, see color insert). After exocytosis, synaptic vesicles undergo endocytosis, recycle, and refill with neurotransmitters for a new round of exocytosis. Nerve terminals are secretory machines dedicated to repeated rounds of release. Most neurons form >500 presynaptic nerve terminals that are often widely separated from the neuronal cell bodies. Action potentials, initiated in the neuronal cell body, travel to all of the cell body's nerve terminals to be transformed into synaptic secretory signals. Nerve terminals do not convert reliably every action potential into a secretory signal but are "reliably unreliable" (Goda & Südhof 1997). In most terminals, only 10%–20% of action potentials trigger release. The relationship between action potentials and release in a nerve terminal is regulated by intracellular messengers and extracellular modulators and is dramatically altered by repeated use of a synapse. Thus in addition to secretory machines, nerve terminals are computational units where the relation of input (action potential) to output (neurotransmitter release) continuously changes in response to extra- and intracellular signals.

All presynaptic functions, directly or indirectly, involve synaptic vesicles. Synaptic vesicles undergo a trafficking cycle in the nerve terminal (Figure 1) that can be divided into sequential steps: First, neurotransmitters are actively transported into synaptic vesicles (step 1), and synaptic vesicles cluster in front of the active zone (step 2). Then synaptic vesicles dock at the active zone (step 3), where the vesicles are primed (step 4) to convert them into a state of competence for Ca^{2+}-triggered fusion-pore opening (step 5). After fusion-pore opening, synaptic vesicles endocytose and recycle probably by three alternative pathways: (*a*) Vesicles are reacidified and refilled with neurotransmitters without undocking, thus remaining in the readily releasable pool (step 6, called "kiss-and-stay"); (*b*) vesicles undock and recycle locally (step 7, called "kiss-and-run") to reacidify and refill with neurotransmitters (back to steps 1 and 2); or (*c*) vesicles endocytose via clathrin-coated pits (step 8) and reacidify and refill with neurotransmitters either directly or after passing through an endosomal intermediate (step 9). In the two-dimensional representation depicted here (Figure 1), each step in the vesicle cycle is illustrated by a shift in the position of the vesicle. In reality, however, most successive steps occur without much vesicle movement except for docking (step 3) and recycling (steps 7–9). Investigators sometimes propose that different types of release reactions exist that differ in fusion-pore dynamics (e.g., "kiss-and-run" is used as a description of exocytosis instead of recycling). However, synaptic vesicles are so small (radius 17–22 nm) that even an unstable fusion pore is likely to empty the vesicle rapidly, as reflected in the fast rise times of spontaneous release events ($<$100 μs). The dynamics of fusion pores are thus unlikely to influence release but are probably of vital importance for endocytosis: Fast recycling may preferentially utilize transient fusion pores, whereas slow recycling likely involves a full collapse of the vesicle into the plasma membrane.

A major goal in neurobiology in recent years has been to gain insight into the molecular machinery that mediates neurotransmitter release. More than 1000 proteins function in the presynaptic nerve terminal, and hundreds are thought to participate in exocytosis. In this protein zoo, which proteins are actually important, and which are only bystanders? How do proteins collaborate in shaping the vesicle cycle, and how can we understand the functions of so many proteins? To approach this fundamental problem, I posit that all of the presynaptic functions ultimately converge on the vesicle cycle and that all steps in the vesicle cycle collaborate, directly or indirectly, to make possible rapid, regulated, and repeated rounds of release (Südhof 1995). Because release is mediated by the interaction of synaptic vesicles with the active zone during exocytosis, this interaction is the common final pathway of all nerve terminal functions. In the discussion below, I therefore pursue a "vesicocentric" perspective that focuses on synaptic vesicles as the central organelle of neurotransmitter release. This perspective is guided by the promise that synaptic vesicles as small, relatively simple organelles are in principle amenable to a complete molecular analysis.

Instead of providing a comprehensive overview of the entire literature (which would be impossible within the constraints of this review), I concentrate on a limited number of proteins for which key functions have been proposed. For many topics, other recent reviews provide a good summary [e.g., see Jahn et al. (2003) for a review of membrane fusion, Slepnev & De Camilli (2000) or Galli & Haucke (2001) for reviews on endocytosis, and Zucker & Regehr (2002) or von Gersdorff & Borst (2002) for reviews on short-term synaptic plasticity]. Furthermore, I primarily focus on work done in mammals because space constraints do not allow a discussion of the extensive and outstanding work in flies and worms (reviewed in Richmond & Broadie 2002). Before discussing the molecular machines that are anchored on synaptic vesicles and drive the vesicle cycle, one must review the salient properties of the two principal limbs of the cycle—exocytosis versus endocytosis and recycling—and the methodology used to elucidate how a given molecule could participate in the cycle.

Ca^{2+} TRIGGERING OF NEUROTRANSMITTER RELEASE

In preparation for neurotransmitter release, synaptic vesicles dock at the active zone and are primed to become Ca^{2+} responsive (steps 3 and 4; Figure 1). When an action potential invades a nerve terminal, voltage-gated Ca^{2+} channels open, and the resulting pulse of intracellular Ca^{2+} triggers fusion-pore opening of release-ready vesicles (step 5). In most synapses, release is stimulated by Ca^{2+} influx through P/Q- ($Ca_V2.1$) or N-type Ca^{2+} channels ($Ca_V2.2$), whereas the related R- ($Ca_V2.3$) or the more distant l-type Ca^{2+} channels (Ca_V1 series) are involved only rarely (e.g., see Dietrich et al. 2003). Even at rest, synapses have a finite but low probability of release, causing spontaneous events of exocytosis that are reflected in electrophysiological recordings as miniature postsynaptic currents (Katz 1969). Ca^{2+} influx triggers at least two components of release that are probably mechanistically distinct: A fast, synchronous phasic component is induced rapidly, in as little as 50 μs after a Ca^{2+} transient develops (Sabatini & Regehr 1996), and a slower asynchronous component continues for >1 s as an increase in the rate of spontaneous release after the action potential (Barrett & Stevens 1972, Geppert et al. 1994a, Goda & Stevens 1994, Atluri & Regehr 1998). Both components of release are strictly Ca^{2+} dependent but change differentially upon repetitive stimulation (Hagler & Goda 2001).

The best physiological description of how an action potential induces phasic, synchronous neurotransmitter release was obtained for the synapse formed by the calyx of Held, the only synapse for which models are available that accurately account for all properties of release (reviewed in Meinrenken et al. 2003). The calyx of Held forms a large nerve terminal ($\sim$15 μm diameter) that envelops the soma of the postsynaptic neuron like a cup, hence its name. The calyx terminal makes

~500–600 synaptic contacts with the postsynaptic cell (in young rats at ~P10, where most studies were performed). The synaptic contacts are distributed over the entire inner surface of the calyx and account for 2% of the plasma membrane area (Sätzler et al. 2002). Each synaptic contact acts like an independent synapse. A single action potential triggers release at ~200 synaptic contacts, which suggests that approximately one third of the synapses fire per Ca^{2+} signal (Bollmann et al. 2000). As a result, the calyx is effectively coupled to the postsynaptic cell by hundreds of parallel synapses that guarantee a reliable postsynaptic response to a presynaptic action potential. This is important physiologically because the calyx synapse is part of the auditory pathway where reliable fast transmission is essential for sound localization.

Serial electron micrographs revealed that, similar to other synapses, active zones in the calyx of Held synapse measure ~0.05–0.10 μm^2 (Sätzler et al. 2002, Taschenberger et al. 2002). Each active zone is associated with a cluster of vesicles (125 $\pm$ 82 per active zone). On average, only 2–3 vesicles are docked, and 5 vesicles are within 20 nm of an active zone (Sätzler et al. 2002, Taschenberger et al. 2002). Compared to cerebellar, cortical, or hippocampal synapses where vesicle clusters usually include >200 vesicles, and where more than 8–10 docked vesicles per active zone are generally reported (Harris et al. 1992, Schikorski & Stevens 2001, Xu-Friedman et al. 2001), the calyx thus contains fewer docked and free vesicles.

The size of the calyx allows direct electrophysiological recordings from a terminal, making it possible to monitor pre- and postsynaptic events simultaneously (Figure 2). Major advances in our understanding of neurotransmitter release were made in the calyx of Held by relating the precise concentration and dynamics of Ca^{2+} to synaptic vesicle exocytosis (reviewed in Schneggenburger et al. 2002, Meinrenken et al. 2003). When an action potential invades the calyx terminal, the Ca^{2+} current begins at the peak of the action potential and ends before the calyx terminal is fully repolarized (Figure 2; Helmchen et al. 1997). During the repolarization phase of the action potential, the Ca^{2+} current is much smaller than the K^+ current, which suggests that the Ca^{2+} current does not contribute markedly to the waveform of the action potential. The size of the Ca^{2+} current indicates that if all Ca^{2+} channels were located in active zones (probably an overestimate), each action potential would open ~20 Ca^{2+} channels per active zone (Sätzler et al. 2002), a value similar to estimates for cortical synapses (Koester & Sakmann 2000). Ca^{2+} influx results in a Ca^{2+} transient lasting ~400–500 μsec whose time course closely follows that of the Ca^{2+} current because Ca^{2+} is buffered effectively and rapidly (Meinrenken et al. 2002). Ca^{2+} thus invades the nerve terminal as a brief Ca^{2+} pulse that triggers fast release with a short delay (50–500 μsec depending on the synapse and temperature). The immediate dissipation of the Ca^{2+} transient rapidly terminates release, while the bolus of neurotransmitters released into the synaptic cleft activates postsynaptic receptors to elicit a postsynaptic reponse (Figure 2).

To determine how much Ca^{2+} is necessary to trigger neurotransmitter release, photolysis of caged Ca^{2+} was used (Bollmann et al. 2000; Schneggenburger &

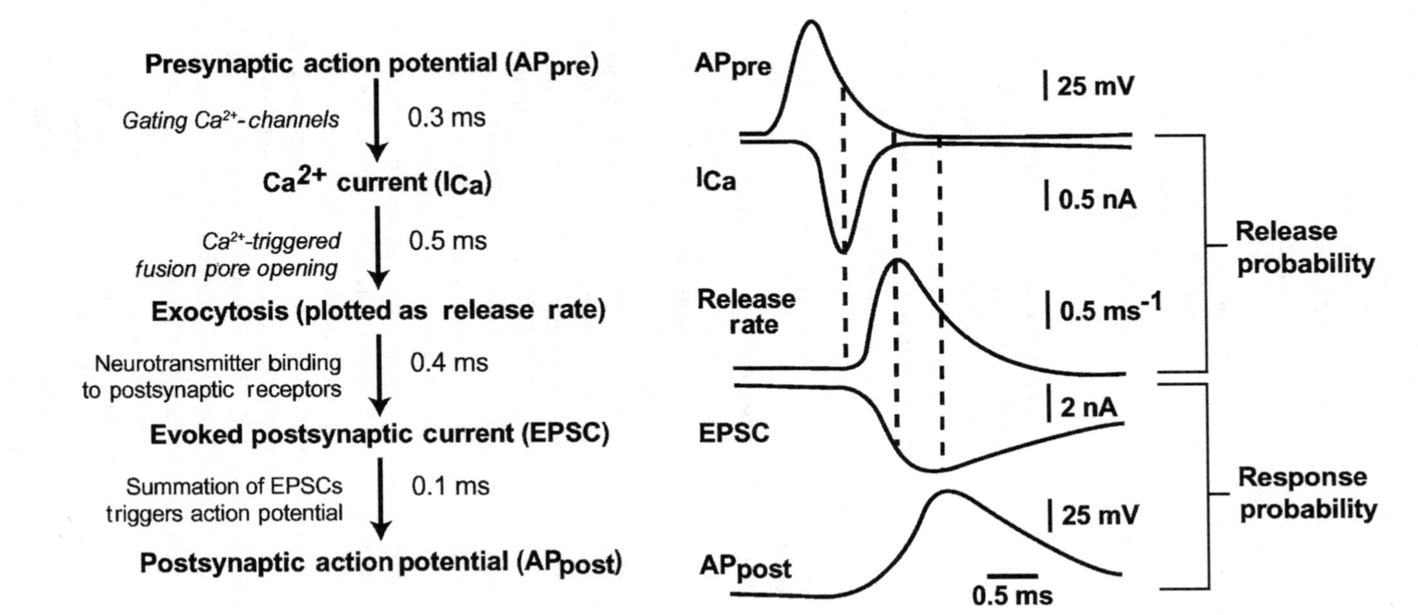

Figure 2 Reaction sequence and timing of synaptic transmission. The principal reactions with the associated time constants are shown on the left, and traces from the corresponding reactions in the calyx of Held synapses are illustrated on the right (modified from Meinrenken et al. 2003). The time calibration bar at the bottom applies to all traces.

Neher 2000). Ca^{2+} uncaging creates a uniform cytosolic Ca^{2+} signal that can be directly measured, which thereby circumvents the problem that local Ca^{2+} concentrations usually cannot be determined and that vesicles may see different Ca^{2+} concentrations during an action potential because they are separated from the mouth of the Ca^{2+} channels by a variable distance. Upon photolysis of caged Ca^{2+}, Ca^{2+} triggers release at the calyx with a high degree of cooperativity, similar to other synapses. Release was undetectable at Ca^{2+} concentrations of $<1\ \mu M$, became measurable at 1–2 μM Ca^{2+}, resembled the release observed during a normal action potential at $>5\ \mu M$ Ca^{2+}, and saturated at $>20\ \mu M$ Ca^{2+} (Bollmann et al. 2000). Thus fast release, at least at this synapse, is triggered by Ca^{2+} binding to a highly cooperative Ca^{2+} sensor with relatively high apparent Ca^{2+} affinity (K_d 5–25 μM).

The precise Ca^{2+} dependence of release was fitted by different models that led to qualitatively similar but numerically distinct conclusions. In both models, the Ca^{2+} sensor contains 5 Ca^{2+}-binding sites (Bollmann et al. 2000, Schneggenburger & Neher 2000). No intrinsic heterogeneity in the Ca^{2+} responsiveness of readily releasable vesicle was observed when release was stimulated by Ca^{2+} uncaging, as opposed to action potential–induced release (Sakaba & Neher 2001). The model by Bollmann et al. (2000) predicts that an action potential induces a roughly Gaussian Ca^{2+} transient with a median duration of 400 μs and that an average readily releasable vesicle sees 9 μM Ca^{2+}. In this model, an action potential results in a 25% release probability per readily releasable vesicle during an action potential with a pool size of 800 vesicles. Schneggenburger & Neher's model (2000) predicts a broader Ca^{2+} transient (median duration of 500 μs) with a peak amplitude of 28 μM Ca^{2+} that induces a release probability of 10% per readily releasable vesicle (pool size = 2000 vesicles). More recent studies in which release induced by uncaged Ca^{2+} was monitored by capacitance recordings also suggested that half-maximal release (measured 50 ms after Ca^{2+} uncaging) requires 5 μM free Ca^{2+}, with saturation at 10 μM Ca^{2+} (Wölfel & Schneggenburger 2003). Although this release probably includes not just the readily releasable pool, it gives a good estimate of the aggregate apparent affinity of all Ca^{2+} sensors. Under these conditions, 3–5 vesicles were estimated to fuse at maximal Ca^{2+} in <1 ms, which corresponds to the number of vesicles docked at the active zone by electron microscopy (Sätzler et al. 2002). As a result of the high cooperativity of the Ca^{2+} sensor, the relationship of the Ca^{2+} concentration to release is supralinear. This supralinearity restricts release to the brief time period during which the Ca^{2+} transient is above threshold and terminates release rapidly even though residual Ca^{2+} remains (Meinrenken et al. 2003). These studies quantitatively described the Ca^{2+}-binding properties of a Ca^{2+} sensor only for fast release but provided little information about the Ca^{2+} sensor for asynchronous release or the adaptations of the release machinery during repetitive stimulation.

One of the key questions about how an action potential induces fast release is how Ca^{2+} channels are organized with respect to the vesicles. At the calyx, multiple Ca^{2+} channels contribute to the Ca^{2+} transient that triggers exocytosis of a readily releasable vesicle (Borst & Sakmann 1996). The relatively slow Ca^{2+}

buffer EGTA, at 10 mM concentration, inhibits release 50% (Borst et al. 1995), which suggests that the majority of the vesicles are not linked to Ca^{2+} channels. Modeling of release observed in the presence of various Ca^{2+} buffers (Meinrenken et al. 2002) indicated that vesicles may be randomly distributed in an active zone but Ca^{2+} channels clustered. According to this model, the distance of a vesicle to the Ca^{2+} channel cluster ranges from 30 to 300 nm (average ~100 nm), and vesicles in different locations are exposed to different Ca^{2+} transients and exhibit different release probabilities (from <0.01 to 1). Consistent with this hypothesis, vesicles exhibit a heterogeneous release probability during an action potential (Sakaba & Neher 2001). The clustering of Ca^{2+} channels accounts for the reliability of the Ca^{2+} signal: Even though Ca^{2+} channels open stochastically during an action potential, individual action potentials seem to create a similar Ca^{2+} signal. The model proposed by Meinrenken et al. (2002) predicts that the only step in the signaling cascade whose speed depends on the Ca^{2+} concentration is Ca^{2+} binding to the Ca^{2+} sensor, which explains why the synaptic delay (i.e., the time between the action potential and release) is relatively independent of the Ca^{2+} concentration, whereas the magnitude of release is supralinearly dependent on the Ca^{2+} concentration. However, an alternative model for the Ca^{2+} dynamics during an action potential that postulates additional intrinsic heterogeneity of vesicles also explains the characteristics of release (Trommershäuser et al. 2003).

The speed with which Ca^{2+} triggers release (<400 μsec) suggests that Ca^{2+} binding to the Ca^{2+} sensor only induces fusion-pore opening and does not initiate a complex reaction cascade. Fusion pores form an aqueous connection across fusing bilayers and are likely to be at least partly lipidic (reviewed in Jahn et al. 2003). Because of the small size of synaptic vesicles, fusion pores cannot be readily measured but have been studied extensively in cells with large secretory vesicles (e.g., see Breckenridge & Almers 1987, Zimmerberg et al. 1987). Not surprisingly, the probability of fusion-pore opening depends on the tension and composition of the participating membranes. Thus any change, even indirect, that alters the tension of the plasma membrane influences neurotransmitter release. This is most prominently observed when neurotransmitter release is triggered by hyperosmotic solutions or by simply stretching the membrane (see Katz 1969). Changes in fusion-pore dynamics are also observed upon overexpression of a large number of proteins in nonneuronal cells [e.g., transfection of CSP (Graham & Burgoyne 2000), Munc18 (Fisher et al. 2001), Complexins (Archer et al. 2002), and synaptotagmin 1 (Wang et al. 2001a)], possibly because the plasma membrane tension is altered in the transfected cells.

SYNAPTIC VESICLE ENDOCYTOSIS AND RECYCLING

More than 30 years ago, biochemical experiments with synaptosomes (Barker et al. 1972) suggested that after exocytosis, vesicles undergo endocytosis and refill rapidly, and that a subpopulation of vesicles associated with the active zone recycles locally (the pathway referred to as kiss-and-stay; Figure 3). In parallel studies, Ceccarelli et al. (1973) demonstrated at the neuromuscular junction that

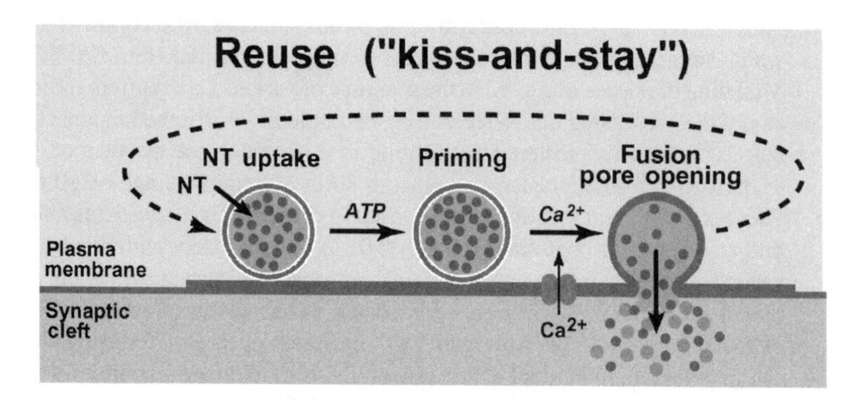
Reuse ("kiss-and-stay")
NT uptake
NT
Priming
Fusion pore opening
ATP
Ca2+
Plasma membrane
Synaptic cleft
Ca2+

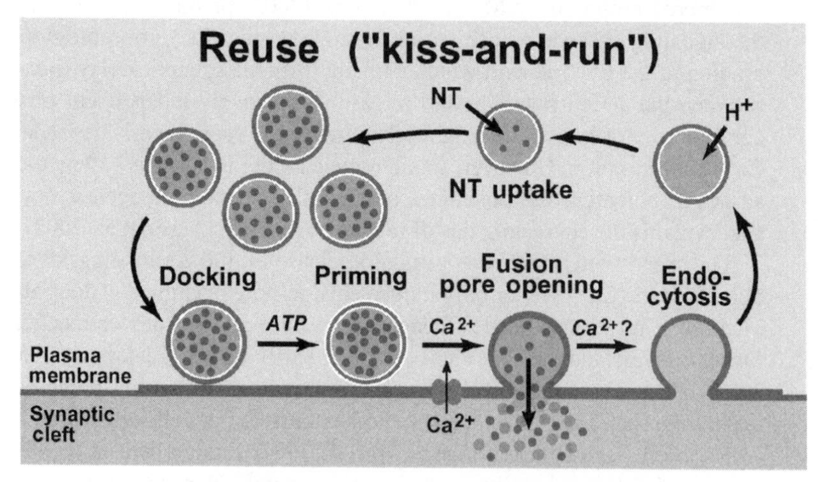
Reuse ("kiss-and-run")
NT
H+
NT uptake
Docking
Priming
Fusion pore opening
Endo-cytosis
ATP
Ca2+
Ca2+?
Plasma membrane
Synaptic cleft
Ca2+

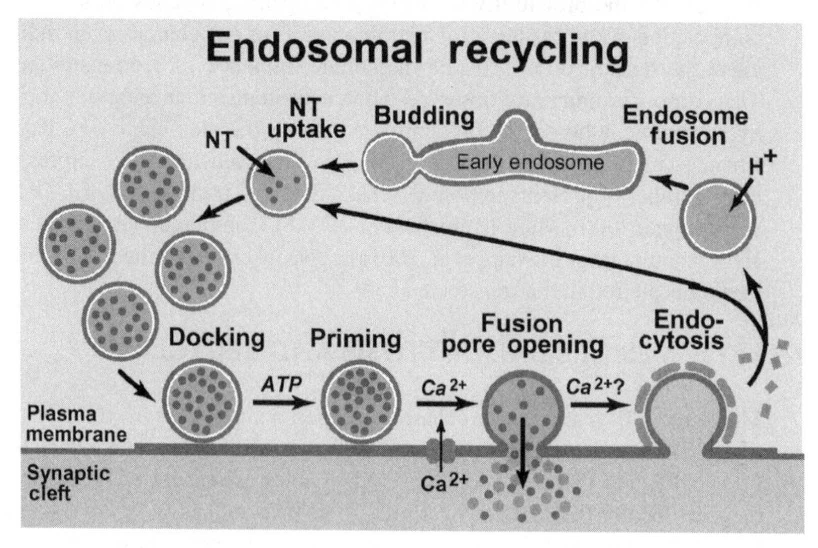
Endosomal recycling
NT uptake
NT
Budding
Early endosome
Endosome fusion
H+
Docking
Priming
Fusion pore opening
Endo-cytosis
ATP
Ca2+
Ca2+?
Plasma membrane
Synaptic cleft
Ca2+

vesicles endocytose and recycle rapidly without a clathrin-coated intermediate (referred to as kiss-and-run because the vesicles did not remain attached to the active zone), whereas Heuser & Reese (1973) described that extensive stimulation of the neuromuscular junctions causes vesicles to endocytose via parasynaptic cisternae and coated pits (endosomal recycling; Figure 3). On the basis of these three observations and subsequent studies (e.g., see Koenig & Ikeda 1996, Pyle et al. 2000, Richards et al. 2000, Wucherpfennig et al. 2003), three vesicle recycling pathways are proposed here (Figure 3): two fast pathways in which the vesicles either remain at the active zone for refilling (kiss-and-stay) or are recycled locally without clathrin-mediated endocytosis (kiss-and-run), and a slower pathway that involves clathrin-mediated endocytosis. The fast pathway is used preferentially to recycle vesicles rapidly into the readily releasable pool at low stimulation frequencies, whereas the slow clathrin-dependent pathway kicks in at higher stimulation frequencies (Koenig & Ikeda 1996, Pyle et al. 2000, Richards et al. 2000).

How fast synaptic vesicle endocytosis happens after exocytosis is a critical question. Recently, this question was addressed directly in the calyx of Held using capacitance measurements (Sun et al. 2002). After single-vesicle exocytosis during spontaneous release, endocytosis was extremely fast (56 ms time constant). When ~200 vesicles were stimulated by a single action potential, or thousands of vesicles by multiple action potentials at low frequency (<2 Hz), the time constant of endocytosis was only two-fold slower (~115 ms), which is not a significant increase considering that many more vesicles were endocytosed after action potential–induced release than were after spontaneous release events. However, when the stimulation frequency was increased, endocytosis slowed down dramatically. After 10 stimuli at 20 or 333 Hz, the time constants of endocytosis increased to 2.3 and 8.3 s, respectively (Sun et al. 2002). The decrease in endocytosis rate did not depend on the cytosolic Ca^{2+} concentration or number of stimuli but on the net increase in membrane area produced by the sum of exo- and endocytosis (Sun et al. 2002). Thus the number of unretrieved vesicles at a given time determined the speed of endocytosis, which suggests that local membrane tension may drive not only exocytosis (see discussion above) but also endocytosis. The speed of endocytosis measured by Sun et al. (2002) implies that vesicle recycling is very rapid, even for the slow component of endocytosis. Nevertheless, this speed does not mean that endocytosis cannot be carried out by a clathrin-independent mechanism because the molecular machinery for clathrin-mediated endocytosis in nerve terminals is highly developed, and could potentially achieve this speed (Brodin

←

Figure 3 Synaptic vesicle recycling pathways. Three pathways are proposed: a pathway in which vesicles endocytose by closure of the fusion pore and are refilled with neurotransmitters while remaining docked to the active zone (kiss-and-stay); a local recycling pathway that is clathrin independent but results in mixing vesicles with the reserve pool after endocytosis (kiss-and-run); and a pathway whereby vesicles undergo clathrin-mediated endocytosis and recycle either directly or via endosomes.

et al. 2000, Slepnev & De Camilli 2000). Gandhi & Stevens (2003) confirmed in principle the presence of activity-dependent modes of endocytosis using optical recordings, although the time constants measured were much slower, possibly because cultured neurons were investigated and because the signal-to-noise ratio in optical recordings makes resolution of fast responses more difficult.

Whereas endocytosis can be directly measured by capacitance recordings, recycling can be assessed only indirectly by the uptake of tracers such as fluorescent dyes. Clearly, most vesicles recycle directly without passing through an endosomal intermediate. In fact, endosomes are almost never observed by electron microscopy in normal nerve terminals, raising doubts about the endosomal recycling pathway. However, multiple lines of evidence unequivocally establish that the endosomal pathway is physiologically relevant for synaptic vesicle recycling under some conditions (Figure 3).

1. Purified synaptic vesicles contain as a stoichiometric component (i.e., a component present on all vesicles and not only a small subset) the SNARE protein Vti1aβ (Antonin et al. 2000). Vti1aβ is a neuronal splice variant of the ubiquitous SNARE protein Vti1aβ, which, like other SNARE proteins, functions in membrane fusion (reviewed in Jahn et al. 2003; see discussion below). Vti1a participates in fusion reactions involving endosomes and the trans-Golgi network but not the plasma membrane. Its presence on all synaptic vesicles implies that all vesicles at one point fuse with intracellular membranes and not only the plasma membrane.
2. Rab5 is also an obligatory synaptic vesicle component (Fischer von Mollard et al. 1994). Rab5 is a ubiquitous protein involved in endosomal fusion and recruits effector proteins such as phosphatidylinositol-3-kinases and rabenosyn to transport vesicles destined to fuse into endosomes (Nielsen et al. 2000). The high concentration of Rab5 on synaptic vesicles suggests a function in endosome fusion during the vesicle cycle. This suggestion is supported by the finding that Rab5 mutations interfere with efficient release during repetitive stimulation (Wucherpfennig et al. 2003).
3. The absence of endosomes from most nerve terminals at steady state is not surprising considering the transient nature of endosomal organelles. For example, in *Drosophila* neuromuscular junctions, endosomes were revealed as an obligatory component of the vesicle pathway in all synapses using endosome-specific green fluorescent protein (GFP)-labeled markers, even though regular electron microscopy did not detect them easily (Wucherpfennig et al. 2003).
4. Pharmacologic inhibition of endosome fusion in frog neuromuscular junctions using phosphatidylinositol 3-kinase blockers potently impaired neurotransmitter release and depleted synaptic vesicles (Rizzoli & Betz 2002).
5. The nerve terminal is endowed richly with proteins that specifically function in clathrin-mediated endocytosis (reviewed by Brodin et al. 2000, Slepnev

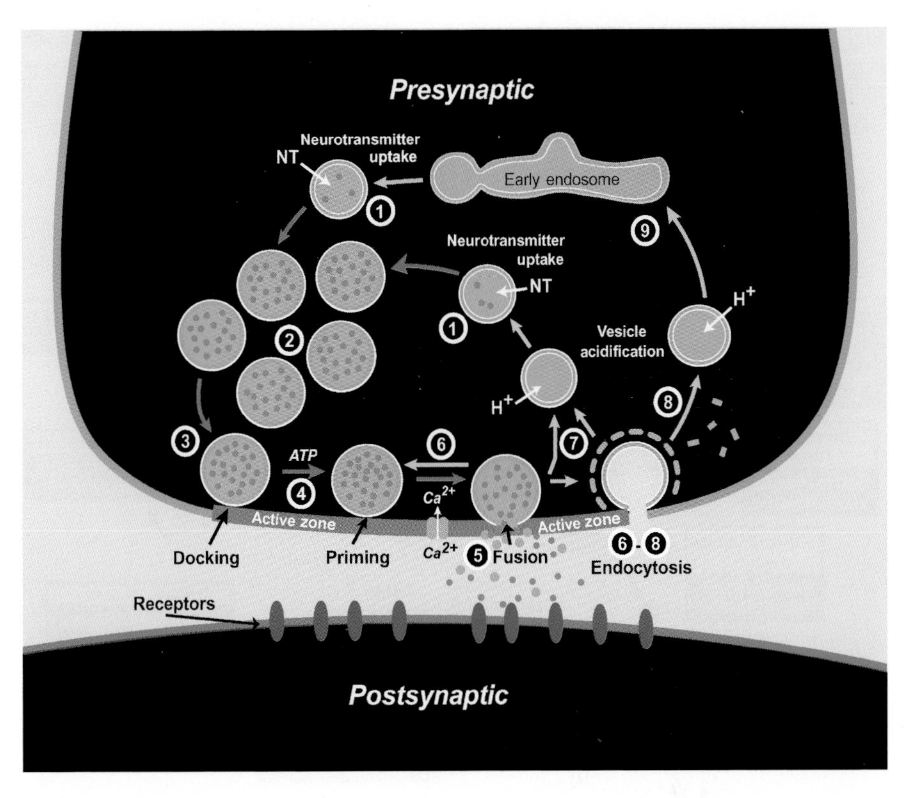

Figure 1 The synaptic vesicle cycle. Synaptic vesicles are filled with neurotransmitters by active transport (step 1) and form the vesicle cluster that may represent the reserve pool (step 2). Filled vesicles dock at the active zone (step 3), where they undergo a priming reaction (step 4) that makes them competent for Ca^{2+} triggered fusion-pore opening (step 5). After fusion-pore opening, synaptic vesicles undergo endocytosis and recycle via several routes: local reuse (step 6), fast recycling without an endosomal intermediate (step 7), or clathrin-mediated endocytosis (step 8) with recycling via endosomes (step 9). Steps in exocytosis are indicated by red arrows and steps in endocytosis and recycling by yellow arrows.

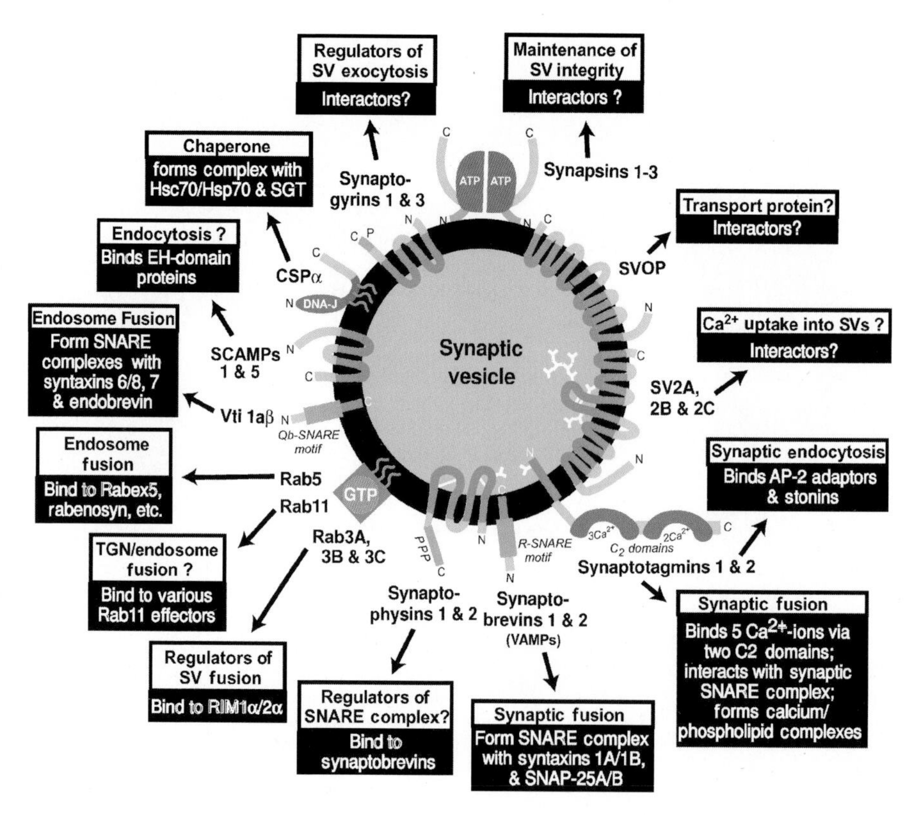

Figure 4 Structures, proposed interactions, and putative functions of synaptic vesicle trafficking proteins. Proteins are shown schematically (*green*, intravesicular sequences; *orange*, transmembrane regions; *blue*, phosphorylation domains; *pink*, SNARE motifs; *red* and *gray*, folded domains; *yellow*, other sequences). The white connecting lines in the intravesicular space identify disulfide bonds, and the branched white lines indicate sugar residues. In the boxes corresponding to the individual proteins, proposed functions are shown on a white background and purported interactions on a black background.

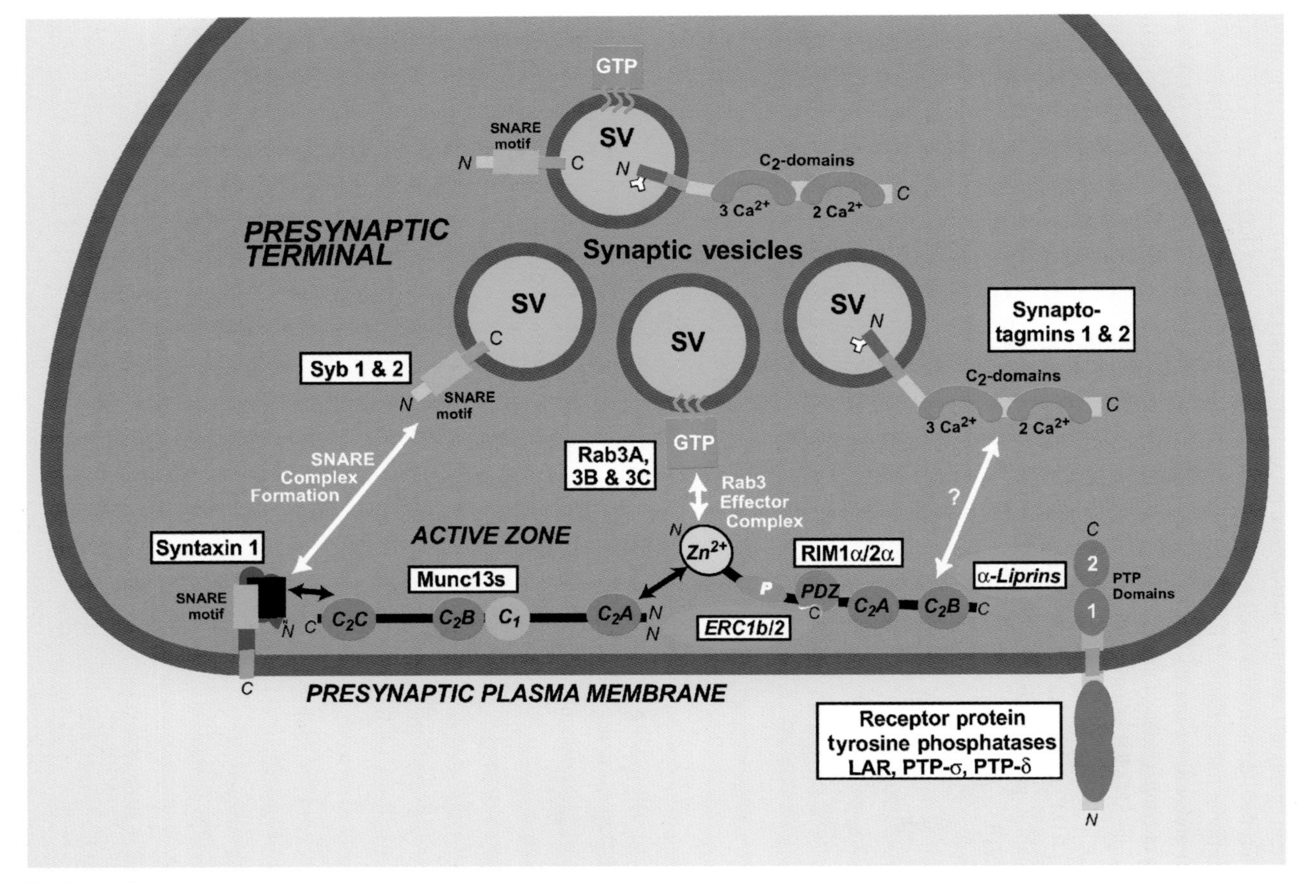

See legend on next page

Figure 5 Protein-protein interactions at the active zone: putative mechanisms of synaptic vesicle attachment. The active zone includes, among others, an interaction of the SNARE protein syntaxin 1 with the active zone component Munc13-1 (*left*), a direct binding of the N-terminal domain of Munc13-1 to the N-terminal zinc-finger of RIM1α/2α (*center*), and additional interactions of RIM1α/2α with the coiled-coil proteins ERC1a and ERC2 (*center*) and with α-liprins that in turn bind to receptor-tyrosine phosphatases (*right*). Two direct connections of synaptic vesicles with the active zone are established: binding of the vesicle SNARE synaptobrevin to syntaxin and SNAP-25 (*not shown*), and binding of the vesicle Rab proteins Rab3A, 3B, 3C, and 3D to the same N-terminal domain of RIM1α and RIM2α that also binds to Munc13-1. In addition, an interaction of the C_2 domains of synaptotagmins with the C_2B domain of RIMs has been observed but is not yet validated (*question mark*).

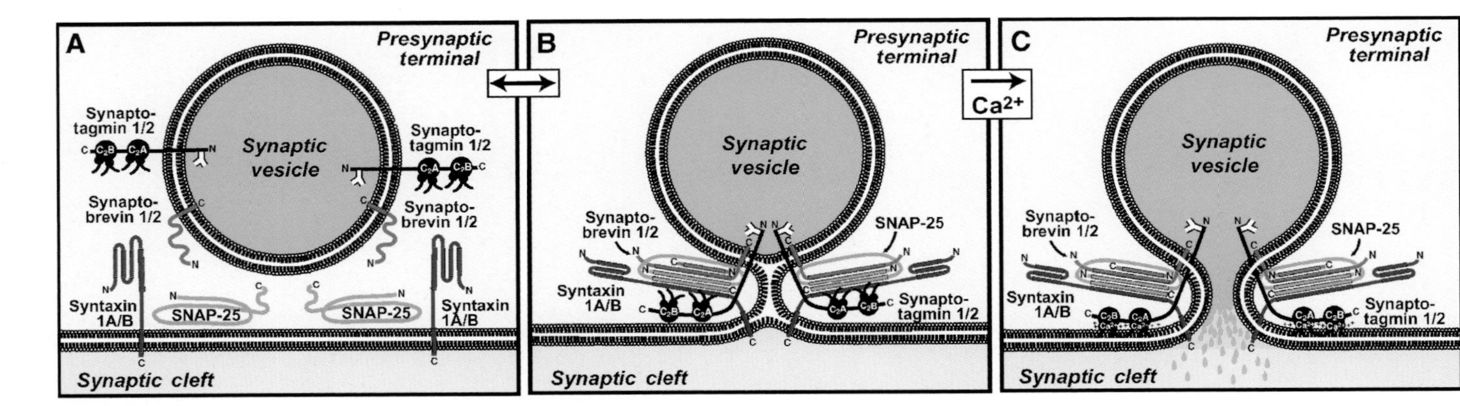

Figure 6 Model for the functions of SNARE proteins, complexins, and synaptotagmins 1 and 2 in synaptic vesicle exocytosis. In docked vesicles (*panel A*), SNAREs and synaptotagmins are not engaged in direct interactions. During priming (*panel B*), SNARE complexes form, complexins (*green*) are bound to fully assembled complexes, and synaptotagmins constitutively associate with the assembled SNARE complexes. The synaptic vesicle membrane and plasma membranes are forced into close proximity by SNARE complex assembly, which results in an unstable intermediate that is shown as a speculative fusion stalk. Ca^{2+} influx (*panel C*) further destabilizes the fusion intermediate by triggering the C_2 domains of synaptotagmin to partially insert into the phospholipids. This action is proposed to cause a mechanical perturbation that opens the fusion pore. Note that the nature and stability of the putative fusion intermediate is unclear and that SNARE complex assembly in panel B is suggested to be reversible, whereas Ca^{2+} triggering is not.

& De Camilli 2000). Particularly striking is the nerve terminal abundance of proteins that are dedicated to accelerating this clathrin-mediated endocytosis, consistent with the relatively fast time constants observed for "slow" endocytosis in capacitance measurements (Sun et al. 2002).

Why clathrin-dependent endosomal recycling is important, and why it is activated selectively upon high-frequency stimulation, are obvious questions. Although no conclusive answers are available, at least two hypotheses are consistent with the data. One hypothesis is that the capacity for fast local recycling is too low to maintain continuous steady-state release at high frequency. According to this hypothesis, the endosomal recycling pathway, although slower, makes up for its lack of speed because it has a much larger capacity. A second hypothesis is that endosomes serve as a sorting station to separate "defective" and "healthy" vesicles. According to this hypothesis, such sorting is particularly important when a synapse is stressed. These two hypotheses are not mutually exclusive and may only partially explain the phenomena.

POOLS AND PLASTICITY

Synaptic Vesicle Pools

When a nerve terminal is stimulated repeatedly at a high rate, release drops dramatically and eventually reaches a lower steady-state level. The use-dependent initial synaptic depression during high-frequency stimulation reflects the depletion of vesicles in the readily releasable pool. The steady-state level of release corresponds to the rate with which vesicles are replenished into the readily releasable pool by recycling or by recruitment from a reserve pool. The concept of equating release rates with vesicle pools has been useful, and different pools of synaptic vesicles were defined on the basis of the rates of release under various stimulation conditions. The size of the readily releasable pool that can be exocytosed by high-frequency stimulation generally agrees well with the amount of release obtained upon application of hypertonic sucrose as a mechanical stimulus (Rosenmund & Stevens 1996) or with the number of vesicles that can be measured as "docked" by electron microscopy (Schikorski & Stevens 2001, Sätzler et al. 2002). The total number of vesicles that participate in exo- and endocytosis during prolonged stimulation are referred to as the recycling pool. This pool is composed of the readily releasable pool and the reserve pool, which serves to replenish the readily releasable pool upon its depletion. In cultured hippocampal neurons, extensive stimulation in the presence of the fluorescent dye FM1-43 (which is taken up with endocytosed vesicles) allows labeling of the total recycling pool of vesicles. This pool has been estimated at only 21–25 vesicles per synapse, with ~4–8 vesicles in the readily releasable and ~17–20 vesicles in the reserve pool (Murthy & Stevens 1999). Thus in this synapse, the size of the recycling pool is surprisingly small compared to the total number of vesicles in the terminal (>200 vesicles), which suggests that a third, very large "resting pool" of vesicles exists (reviewed in Südhof 2000).

However, in other synapses the number and size of vesicle pools are different. For example, at the neuromuscular junction, 20% of all vesicles are in the readily releasable pool and 80% in the reserve pool, without any resting vesicles (Richards et al. 2003). In contrast, at the calyx, two different types of readily releasable pools—immediately and reluctantly releasable vesicles—were identified (Sakaba & Neher 2001), in addition to a large reserve pool (J. Sun, personal communication). Although the concept of vesicle pools thus far has been useful to describe the amount of release observed under different conditions, the differences between synapses in the size and nature of these pools suggest that their definition is operational and may not always reflect separate types of vesicles. This potential problem is shown clearly for the calyx of Held where estimates of the readily releasable pool, determined by summing up the synaptic responses during trains of action potentials and corrected for postsynaptic receptor desensitization, led to pool sizes of 900 vesicles (Bollmann et al. 2000). In contrast, continuous stimulation of the terminals, either by extended depolarizations or by photolysis of caged Ca^{2+}, reveals a readily releasable pool of 3000–5000 vesicles (Schneggenburger & Neher 2000, Sun & Wu 2001, Wölfel & Schneggenburger 2003).

The difference between the two estimates may be partly due to the fact that the second stimulation method also includes asynchronous release and that the capacitance measurements used for some of these studies may be more accurate. However, the difference is too large to be explained by technical factors and suggests an intrinsic problem of trying to define pools on the basis of the amount of release obtained under defined stimulation conditions. The concept of pools depends on the idea that defined numbers of vesicles are in different states of releasability and that, once released, the respective vesicle pools are refilled at a specific rate. If vesicles were present in distinct states that are in an equilibrium that can be quickly shifted in both directions, the pool concept would be difficult to apply. For example, equating the synchronous and asynchronous components of release with distinct vesicle pools would imply that they are physically different populations of vesicles. Equally possible, however, is that all vesicles are in principle components of both pools but are stimulated for release by distinct Ca^{2+} sensors: a low-affinity sensor for fast release, and a high-affinity sensor for slow release. Then only the vicinity of the vesicles to Ca^{2+} channels and the ambient Ca^{2+} concentration would determine its pool identity, and each vesicle would have a finite probability for both pools. Similar arguments apply for other pools of vesicles.

Presynaptic Plasticity

Release is dramatically modulated by signals impinging on presynaptic terminals and changes induced by repetitive stimulation. Two principal points of regulating release exist: (*a*) the peak Ca^{2+} concentration produced by an action potential, i.e., the conversion of an action potential to a Ca^{2+} current (Figure 2); and (*b*) the release probability per given Ca^{2+} concentration, i.e., the conversion of a Ca^{2+} signal to exocytosis. Both types of regulation are important. The peak Ca^{2+} concentration depends in principle on the shape of the action potential, the open probability of the Ca^{2+} channels, and the Ca^{2+} concentration present at the time when the Ca^{2+} channels open. All of these three parameters can be regulated. For example, action-potential duration modulates neurotransmitter release (Qian & Saggau 1999), presynaptic receptors for endocannabinoids inhibit N-type Ca^{2+} channels (Wilson et al. 2001), and residual Ca^{2+} during repetitive action potentials causes short-term plasticity of release (reviewed in von Gersdorff & Borst 2002, Zucker & Regehr 2002). The release probability per peak Ca^{2+} concentration is also highly variable and depends on two principal parameters: the number of release-ready vesicles, and the Ca^{2+} responsiveness of these vesicles. The release probability per Ca^{2+} concentration varies characteristically between different types of synapses (e.g., Xu-Friedmann et al. 2001) and changes during cAMP-dependent presynaptic long-term potentiation (LTP) (Regehr & Tank 1991, Kamiya et al. 2002). Furthermore, the number of release-ready vesicles decreases during repetitive stimulation, causing short-term depression with a magnitude that depends on the pool size and kinetics (see discussion above). Pinpointing the mechanisms involved in regulating release beyond Ca^{2+} triggering has been difficult, and whether a difference in Ca^{2+} responsiveness at a synapse depends on the size of the readily releasable pool or on the Ca^{2+} responsiveness of synaptic vesicles is largely uncertain.

APPROACHES TO ANALYZING SYNAPTIC FUNCTION

Understanding synaptic transmission proceeds via a dialogue between physiological description and molecular analysis. The physiology describes the phenomena that the molecular biology tries to explain, and thus the physiology is always ahead. At this point, rapid physiological advances have produced a rich phenomenology of presynaptic events whose molecular basis is often obscure. That the criteria for what constitutes an explanation, as opposed to an epiphenomenal effect, differ dramatically from lab to lab is a major problem of the molecular analysis of presynaptic events.

A convergence of two approaches is required for a molecular explanation of presynaptic physiology. First, important molecules need to be characterized biochemically and structurally, i.e., their protein-protein interactions, cellular localizations, and atomic structures have to be determined. Second, the functions of such molecules need to be perturbed in vivo, and the consequences of such perturbations must be measured. In the vast literature on presynaptic molecules, fundamental limitations restrict the interpretation of many experiments. For example, many studies examine synaptic protein-protein interactions using a combination of GST (glutathione S-transferase)-fusion protein pulldowns, yeast two-hybrid assays, and immunoprecipitations. Such interactions have been most extensively reported for the SNARE protein syntaxin 1 with >40 purported binding partners. However, for a "sticky" protein such as syntaxin 1, this data may be insufficient. As recently shown for the apparent interaction of syntaxin 1 with K^+ channels, all of these methods can be positive and still give a wrong answer (Fletcher et al. 2003). A similar limitation applies to the location and activities of proteins deduced by analogy. For example, on the basis of the paradigm of synaptotagmin 1, results of synaptotagmin 4 transfection studies were interpreted to reflect a Ca^{2+} sensor function of synaptotagmin 4 on secretory vesicles, even though the localization of synaptotagmin 4 or its Ca^{2+}-binding properties was not established (Wang et al. 2003). A molecular analysis first requires a description of the molecules that is often cumbersome and not publishable but without which the results of "functional" studies are not interpretable.

Three types of techniques are used to perturb a molecule to analyze its function: overexpression of a protein (full-length, truncated, or mutant), introduction of a chemical agent (drug) or peptide, and genetic manipulations with or without expression of wild-type or mutant molecules. Each technique has its own limitations:

1. Overexpression, the easiest approach, often gives the most dramatic results but is the most difficult to interpret, especially when performed on a wild-type background. For example, overexpression of wild-type Rab3A appears to block Ca^{2+}-induced exocytosis (Holz et al. 1994, Johannes et al. 1994) but in fact acts indirectly by activating constitutive exocytosis of secretory vesicles (Schlüter et al. 2002). This action depletes the vesicles and thereby abolishes Ca^{2+}-induced exocytosis indirectly.
2. Introduction of a chemical agent/peptide is potentially the best method because it allows direct and fast interference with a protein's function. However, few chemicals are known to interfere specifically with a function; even peptides are drugs that have no inherent specificity. For example, peptides from Rab3A activate exocytosis not by interfering with release, as proposed by Oberhauser et al. (1992), but by influencing Ca^{2+} signaling (Piiper et al. 1995). Overexpression and chemical/peptide interference experiments are only interpretable when the chemical or peptide involved has a validated mechanism of action.
3. Genetic approaches are the most definitive but often suffer from the failure to delineate the specificity of the changes caused by a genetic manipulation. Analysis of genetic changes is lengthy and laborious. For example, knockout (KO) of CaM kinase IIα creates an array of phenotypes that can be sorted out only when the pre- and postsynaptic contributions of CaM kinase IIα are separated (see Hinds et al. 2003). Thus a genetic manipulation is only interpretable when the effects are selective and when the particular genetic manipulation examined does not broadly alter downstream processes.

In view of these limitations, it is clear that the completeness of the analysis—and not the techniques used—determines the interpretability of the results obtained.

MOLECULAR ARCHITECTURE OF SYNAPTIC VESICLES AND INTERACTING PARTNERS

All activities in the presynaptic nerve terminal are directed toward neurotransmitter release, from single release events to repeated rounds of regulated release. As a result, all processes in a nerve terminal influence, directly or indirectly, the interaction of synaptic vesicles with the presynaptic active zone. Understanding the composition of synaptic vesicles and of the active zone is a first step toward insight into the molecular mechanisms of release.

Synaptic Vesicles

Synaptic vesicles are uniformly small (~20-nm radius), abundant organelles whose only known function is to take up and release neurotransmitters. They are relatively simple because only a limited number of proteins fit into a sphere of 40-nm diameter. Purified vesicles have a protein:phospholipid ratio of 1:3 with an unremarkable lipid composition (40% phosphatidylcholine, 32% phosphatidylethanolamine, 12% phosphatidylserine, 5% phosphatidylinositol, 10% cholesterol, wt/wt; Benfenati et al. 1989). Many proteins that have been associated with synaptic vesicles are probably present only on a subset of vesicles or bind transiently to the vesicles, and the number of proteins that are constitutive parts of all synaptic vesicles may be comparatively small (i.e., less than 50).

Synaptic vesicles contain two classes of obligatory components: transport proteins involved in neurotransmitter uptake (step 1 of the vesicle cycle; Figure 1), and trafficking proteins that participate in synaptic vesicle exo- and endocytosis and recycling (steps 2–9). Transport proteins are composed of a vacuolar-type proton pump that generates the electrochemical gradient, which fuels neurotransmitter uptake and neurotransmitter transporters that mediate the actual uptake. The trafficking proteome of synaptic vesicles is complex. It includes intrinsic membrane proteins, proteins associated via posttranslational lipid modifications, and peripherally bound proteins (Figure 4, see color insert). These proteins do not share a characteristic that would make them identifiable as synaptic vesicle proteins, and little is known about how these proteins are specifically deposited into synaptic vesicles. As summarized in Figure 4, many but not all of the known synaptic vesicle proteins interact with nonvesicular proteins and are linked to specific functions.

Active Zones and Presynaptic Plasma Membranes

The active zone is composed of an electron-dense, biochemically insoluble material located at the presynaptic plasma membrane precisely opposite the synaptic cleft. Electron microscopy revealed that active zones of central synapses are disk-like structures containing a hexagonal grid; synaptic vesicles are embedded in the depressions of the grid (Akert et al. 1971). In contrast, active zones of neuromuscular synapses are composed of an elongated ridge containing vesicles lined up like beads on a string (Harlow et al. 2001).

Investigators identified six large nonmembrane proteins that are each encoded by multiple genes, bind to each other, and probably form a humongous single complex at the active zone (reviewed in Dresbach et al. 2001, Lonart, 2002; see Figure 5, see color insert): Munc13s (Munc13-1, -2, and -3; Brose et al. 1995) and RIMs (Rab3-interacting molecules) (RIM1α, 2$\alpha/\beta/\gamma$, 3γ, and 4γ; Wang et al. 1997a, 2000; Wang & Südhof 2003) are multidomain proteins that interact with each other and many other synaptic components. Piccolo and Bassoon, homologous and very large proteins (Cases-Langhoff et al. 1996, tom Dieck et al. 1998), are part of the cytomatrix. ERCs (ELKS/Rab3-interacting molecule/CAST) (ERC1b and ERC2; Wang et al. 2002, Ohtsuka et al. 2002) are coiled-coil proteins whose C-terminus binds to RIMs, and RIM-BPs (RIM-BP1, 2, and 3; Wang et al. 2000) are SH3-domain proteins that also bind to RIMs. Finally, α-liprins (Liprin $\alpha 1$–$\alpha 4$) bind to RIMs (Schoch et al. 2003), ERCs (Ko et al. 2003), and receptor protein tyrosine phosphatases (Serra-Pages et al. 1998).

RIMs are central elements of active zones because they bind directly or indirectly to many other synaptic proteins. The largest RIM variants (RIM1α and RIM2α) include three types of domains: (*a*) an N-terminal zinc-finger domain that interacts with the synaptic vesicle protein Rab3 and the active zone protein Munc13-1 (Wang et al. 1997a, 2000; Betz et al. 2001; Wang et al. 2001b; Schoch et al. 2002); (*b*) a central PDZ domain that binds the C-terminus of ERCs (Ohtsuka et al. 2002, Wang et al. 2002); and (*c*) two C-terminal C_2 domains, of which the second, so-called C_2B domain binds to α-liprins (Schoch et al. 2002). The *RIM2* gene (but not the *RIM1* gene) produces two additional transcripts from internal promoters besides the RIM2α transcript: the RIM2β transcript, which lacks the zinc-finger domain but contains all other domains of RIM2α, and the RIM2γ transcript, which is composed only of the C_2B domain preceded by a short conserved sequence (Wang & Südhof 2003). The *RIM3* and *RIM4* genes produce only γ transcripts (RIM3γ and 4γ). RIMs directly bind to Munc13, ERCs, RIM-BPs, and α-liprins; in addition, all ERCs bind to α-liprins, Piccolo and Bassoon (see references cited above, and Y. Wang and T.C. Südhof, unpublished observations). The resulting macromolecular complex may be linked to synaptic vesicles via two interactions (Figure 5): the GTP-dependent binding of Rab3 to RIM1α/2α (Wang et al. 1997a, 2000), and the Ca^{2+}-dependent association of synaptotagmin 1 with RIM1α/2α/2β (Coppola et al. 2001, Schoch et al. 2002). Although conceptually attractive, the RIM/synaptotagmin interaction has not been validated, and whether this interaction is physiologically relevant is not clear. In addition, the complex may be linked to SNARE proteins via binding of Munc13 to syntaxin 1 and of RIMs to SNAP-25 (Betz et al. 1997, Coppola et al. 2001), but again these interactions have not been validated physiologically.

NEUROTRANSMITTER UPTAKE INTO SYNAPTIC VESICLES

Synaptic vesicles accumulate and store neurotransmitters at high concentrations by active transport, driven by a vacuolar proton pump whose activity establishes an electrochemical gradient across the vesicle membrane (Maycox et al. 1988). The vesicle proton pump is a large multiprotein complex (~1 million Dalton) that contains at least 13 subunits and measures $14 \times 14 \times 24$ nm (Arata et al. 2002), ~10% of the vesicle volume. The proton pump is the single largest vesicle component that extends from the vesicle by >15 nm, more than half of the vesicle radius; most vesicles contain only a single proton pump molecule (Stadler & Tsukita 1984). The proton pump is composed of a larger peripheral complex called V_1, which includes the ATPase activity, and an integral membrane complex called V_0, which mediates proton translocation. The two parts may be connected by the 116-kDa subunit of the pump, its largest component (Perin et al. 1991a). Although the generic composition of vacuolar proton pumps is well established, whether the subunits or properties of the synaptic vesicle proton pump differ from those of other vacuolar proton pumps (e.g., the endosomal or the Golgi proton pumps) is unclear.

The proton pump establishes an electrochemical gradient that drives all neurotransmitter uptake (reviewed by Fykse & Fonnum 1996). Vesicular uptake is mediated by only seven different transporters representing four distinct uptake systems. Glutamate is taken up into synaptic vesicles by three differentially expressed transporters (Fremeau et al. 2002, Gras et al. 2002, Schafer et al. 2002, Takamori et al. 2002, and references cited therein) and all monoamines (catecholamines, histamine, and serotonin) are taken up by two differentially expressed transporters (Erickson et al. 1992, Liu et al. 1992). A single transporter was identified for GABA and glycine (McIntyre et al. 1997, Sagne et al. 1997) and for acetylcholine (Alfonso et al. 1993, Roghani et al. 1994, Varoqui et al. 1994). The four families of transporters are distantly related to each other, but differ mechanistically. For some transporters (e.g., VGlut1–VGlut3), the main driving force for vesicular uptake is the membrane potential, whereas for other transporters (e.g., VGat), the membrane potential and the proton gradient both contribute to uptake (reviewed by Fykse & Fonnum 1996).

Expression of a particular transporter type is probably a major determinant of the type of neurotransmitter used by a neuron. When VGlut1 was transfected into GABAergic neurons, their synapses became glutamatergic, in addition to remaining GABAergic (Takamori et al. 2000). Thus simple expression of the glutamatergic transporter was enough to specify neurotransmitter type. VGlut3 is present in many neurons not previously considered glutamatergic (e.g., the cholinergic interneurons of the striatum; Fremeau et al. 2002, Schafer et al. 2002), which suggests that coexpression of glutamate transporters with other neurotransmitter transporters may confer onto a small set of neurons the ability to use multiple classical transmitters in violation of Dale's principle.

Whether and how the proton pump and/or transmitter transporters are regulated is a major question. The amount of transmitter released per synaptic vesicle exocytosis varies (reviewed in van der Kloot 1991). One explanation of this variation is that vesicles contain the same concentration of transmitters but have different sizes because the size variation of vesicles matches the variation in the amount of release mediated by a single vesicle (Bekkers et al. 1990, Bruns et al. 2000). Another explanation, though not mutually exclusive, is that neurotransmitter uptake is regulated via the activity of either the proton pump or the transporters. Overexpression of the vesicular monoamine transporter increases the amount of release mediated by exocytosis of a single vesicle (Pothos et al. 2000), whereas hemizygous deletion of this transporter dramatically alters monoaminergic signaling in brain, which suggests that the levels of the transporter are physiologically important (Wang et al. 1997b). At least in the calyx, glutamate transport into vesicles is not saturated but can be enhanced simply by increasing the cytosolic glutamate concentration (Yamashita et al. 2003). Regulation of the proton pump is also important for understanding the vesicle cycle. For example, in the time between synaptic vesicle exo- and endocytosis—which can last seconds—the pump is presumably turned off to avoid pumping protons into the synaptic cleft, but this regulation has not actually been demonstrated. How fast neurotransmitter uptake operates is also unknown. Given the small size of the vesicles, only few protons and neurotransmitter molecules need to be pumped to fill a vesicle, which suggests that vesicle refilling after endocytosis could occur in milliseconds.

MEMBRANE FUSION DURING EXOCYTOSIS: SNARES AND COMPANY

Intracellular membrane fusion generally involves SNARE proteins that are present on both fusing membranes before fusion and that associate into tight core complexes during fusion (reviewed in Chen & Scheller 2001, Jahn et al. 2003). SNARE proteins are characterized by a homologous 70-residue sequence called the SNARE motif. The core complex is formed when four SNARE motifs (present in three or four separate SNARE proteins because some SNAREs contain two SNARE motifs) assemble into a parallel four-helical bundle, with the transmembrane regions of the SNAREs emerging on the C-terminus. Core-complex formation may force the membranes on which the SNAREs reside into close proximity, thereby initiating membrane fusion. Four different classes of SNARE motifs exist (R, Qa, Qb, and Qc SNARE motifs), and stable SNARE complexes only form when the four-helical bundle contains one SNARE motif from each class (reviewed in Jahn et al. 2003). The interaction of SNAREs is otherwise promiscuous in vitro, although in vivo they form highly specific complexes. The specificity is likely achieved, at least in part, via the sequences of SNARE proteins outside of their SNARE motifs.

Synaptic exocytosis is mediated by three SNARE proteins: synaptobrevin (also called vesicle-associated membrane protein) on synaptic vesicles, and syntaxin 1 and SNAP-25 on the presynaptic plasma membrane (Söllner et al. 1993). The synaptic core complex is formed by the R-SNARE motif from synaptobrevin, the Qa-SNARE motif from syntaxin 1, and the Qb- and Qc-SNARE motifs from SNAP-25 (which contains two SNARE motifs). The model shown in Figure 6 (see color insert) proposes that by pulling the synaptic vesicle and plasma membranes close together, the SNARE complex creates an unstable intermediate but does not open the fusion pore. The unstable intermediate can progress to a full-blown fusion pore or regress to the docked state of synaptic vesicles that do not contain engaged SNAREs. Complexins, small neuronal proteins, bind to assembled synaptic core complexes (McMahon et al. 1995). Complexins insert into a groove in the C-terminal half of the SNARE complex that is absent from partially assembled SNARE complexes (Chen et al. 2002a). KO mice revealed that complexin is not essential for SNARE function or synaptic vesicle fusion (Reim et al. 2001) but promotes the action of synaptotagmin 1 (see discussion below).

SNARE complex formation at the synapse and in other intracellular fusion reactions is probably controlled by a class of essential fusion proteins called SM proteins for Sec1/Munc18-like proteins (reviewed in Jahn et al. 2003). SM proteins often interact with syntaxin-like SNAREs. Munc18-1, the SM protein that controls synaptic fusion, binds to a conformation of syntaxin that is closed (Dulubova et al. 1999) and blocks its SNARE motif from participating in SNARE complexes. Thus Munc18-1 must dissociate from syntaxin for SNARE complexes to form. However, regulation of SNARE complex formation by a steric mechanism is not the general function of SM proteins because several SM proteins bind to their cognate syntaxin SNARE proteins in the open conformation (Yamaguchi et al. 2002, Dulubova et al. 2002). It is striking that although SM proteins are generally required for membrane fusion and usually interact with syntaxin-like SNAREs, the interactions themselves vary between fusion reactions.

Several synaptic proteins may regulate the assembly of SNARE complexes. Two soluble proteins called tomosyn (Fujita et al. 1998) and amisyn (Scales et al. 2002) have a C-terminal R-SNARE motif that can substitute for synaptobrevin in the synaptic SNARE complex. Tomosyn and amysin exhibit no other similarity—tomosyn is a relatively large protein of >1100 residues that belongs to the family of lethal-giant-larvae proteins, whereas amisyn is a small protein of 222 residues—but both proteins, when introduced exogenously into a secretory system, inhibit exocytosis by interfering with the action of synaptobrevins (Scales et al. 2002, Hatsuzawa et al. 2003).

Another class of proteins that may regulate SNARE function at the synapse are synaptophysins, abundant synaptic vesicle proteins that bind directly to synaptobrevin (Johnston & Südhof 1990, Calakos & Scheller 1994, Edelmann et al. 1995, Washbourne et al. 1995). Synaptobrevin cannot simultaneously bind to synaptophysins and participate in the SNARE complex, which suggests that binding of synaptobrevin to synaptophysin restricts the availability of synaptobrevin for fusion (Edelmann et al. 1995, Becher et al. 1999). Consistent with this hypothesis, the synaptobrevin/synaptophysin complex is upregulated during development (Becher et al. 1999). Chronic blockade of glutamate receptors caused an increase in neurotransmitter release but a decrease in the synaptobrevin/synaptophysin complex (Bacci et al. 2001). In addition to heteromultimers, synaptophysin and synaptobrevin each assemble into homooligomers via their transmembrane regions (Thomas et al. 1988, Laage & Langosch 1997). Only dimers of synaptobrevin were observed, whereas synaptophysin associates into hexamers and larger oligomers (Johnston & Südhof 1990). The synaptic vesicle proteins synaptobrevin and synaptophysin thus engage in at least three complexes on the surface of synaptic vesicles: separate homooligomers and common heterooligomers.

There is general agreement that SNARE proteins function in fusion, but the precise nature of their activity is uncertain. Because most SNARE proteins contain other essential, conserved sequences besides the SNARE motif, it is important to differentiate between SNARE functions mediated by their SNARE motifs (e.g., core complexes) and SNARE functions mediated by their other sequences (e.g., interactions with Sec1/Munc18-like proteins). Two major hypotheses for SNARE function were advanced: (*a*) SNAREs are minimal fusion machines that single-handedly drive fusion via their SNARE motifs, and all other interactions are secondary; (*b*) SNAREs are components of fusion machines that include other proteins. At least at the synapse, two key observations support the second hypothesis. First, deletion of synaptobrevin caused a major impairment of synaptic vesicle exocytosis but left approximately 10% of exocytosis intact (Schoch et al. 2001). Ca^{2+}-triggered exocytosis was much more severely impaired than was spontaneous exocytosis or exocytosis triggered by hypertonic sucrose. No closely related R-SNARE was present in the affected synapses, which suggests that there is no simple redundancy of SNAREs. Similar results were reported for the SNAP-25 KO (Washbourne et al. 2002). Thus at least at the synapse, the proper SNARE complex was only essential for efficient, physiologically regulated fusion, but not for fusion as such. Second, although the synaptobrevin deletion did not abolish exocytosis, deletion of Munc18-1 eliminated release completely (Verhage et al. 2000). Thus at the mammalian synapse, Munc18-1 is more fundamental for fusion than is synaptobrevin. Munc18-1 probably acts by binding to the plasma membrane SNARE protein syntaxin 1 (Hata et al. 1993) and coupling it to as yet unidentified cytosolic factors to organize SNARE complex assembly (reviewed in Jahn et al. 2003).

Viewed together, the current data suggest that SNAREs still are the best candidates for initiating fusion by inducing transition states that ultimately lead to fusion-pore opening (Figure 6*B*). However, the critical step of bringing membranes close together may, at least in part, be substituted for by other cellular mechanisms. Besides forcing membranes close together, SNAREs likely perform additional functions that are as important, for example ensuring the temporal and spatial specificity of fusion reactions by embedding the approximation of membranes into an ordered sequence of reactions. This embedding is presumably guided by sequences of SNAREs outside of the SNARE motif and may, among others, require SM proteins. The best example for this embedding is the role of Munc18-1, which acts by binding to syntaxin 1 in fusion as discussed above.

In addition to synaptobrevin, synaptic vesicles contain the Qb-SNARE protein Vti1aβ (Antonin et al. 2000; Figure 4). Although as Qb- and R-SNAREs, Vti1aβ and synaptobrevin could, in principle, participate in the same SNARE complex, Vti1aβ functions in endosome and Golgi fusion reactions without interacting with synaptobrevin (Jahn et al. 2003). The selective interactions of SNARE proteins in vivo reinforces the notion that SNARE complex formation must be tightly regulated by mechanisms that are independent of the SNARE motifs of the participating SNARE proteins.

TRIGGERING FUSION VIA Ca^{2+} BINDING TO SYNAPTOTAGMINS

Studies of synaptic vesicle exocytosis in the calyx terminal showed that fast exocytosis is triggered by Ca^{2+} binding to a Ca^{2+} sensor with at least five Ca^{2+}-binding sites of micromolar affinity (reviewed in Meinrenken et al. 2003). Evidence accumulated over the past 13 years demonstrates that two synaptic vesicle proteins called synaptotagmins 1 and 2 (Figure 4) are such Ca^{2+} sensors.

Synaptotagmins 1 and 2 as Ca^{2+} Sensors for Fast Exocytosis

Synaptotagmins are composed of a short N-terminal intravesicular sequence, a single transmembrane region, a short linker sequence, and two cytoplasmic C_2 domains (the C_2A and C_2B domains; Perin et al. 1990, Geppert et al. 1991). Synaptotagmins 1 and 2 are abundant synaptic vesicle proteins that are differentially expressed (Geppert et al. 1991, Ullrich et al. 1994). Most experiments were performed on synaptotagmin 1, but the limited studies on synaptotagmin 2, its high degree of sequence similarity with synaptotagmin 1, and the reciprocal distributions of synaptotagmins 1 and 2 suggest that synaptotagmins 1 and 2 are functionally similar. Besides their differential distributions, the only major difference between synaptotagmins 1 and 2 is an approximately twofold difference in apparent Ca^{2+} affinity (Sugita et al. 2002; for synaptotagmins 3–15, see discussion below, which describes their substantially different properties).

The C_2A domain of synaptotagmin 1 binds three Ca^{2+} ions (Ubach et al. 1998), and the C_2B domain two Ca^{2+} ions (Fernandez et al. 2001). The intrinsic Ca^{2+} affinities of the C_2 domains are very low (0.5–5 mM) because the coordination spheres for the Ca^{2+} ions are incomplete. The apparent Ca^{2+} affinity of the C_2 domains increases dramatically (up to 1000-fold) when the C_2 domains bind to phospholipid membranes whose negatively charged headgroups provide additional coordination sites for the bound Ca^{2+} ions (Fernandez-Chacon et al. 2001). The number of Ca^{2+} ions bound by synaptotagmin 1 and its apparent Ca^{2+} affinity thus correspond to the Ca^{2+} sensor for exocytosis at the calyx (Bollmann et al. 2000, Schneggenburger & Neher 2000). Two experiments demonstrated that synaptotagmin 1 is an essential Ca^{2+} sensor for fast exocytosis but not for the slow component of exocytosis or for membrane fusion in general. First, a KO of synaptotagmin 1 in mice produced a lethal phenotype that consisted of a selective loss of fast Ca^{2+}-triggered exocytosis both in hippocampal synapses (Geppert et al. 1994a) and in chromaffin cells (Voets et al. 2001). No other parameter of synaptic membrane traffic examined was impaired. Second, a point mutation in the synaptotagmin 1 C_2A domain that alters the overall apparent Ca^{2+} affinity of synaptotagmin 1 induced an identical shift in the apparent Ca^{2+} affinity of exocytosis (Fernandez-Chacon et al. 2001). Experiments using flash photolysis of caged Ca^{2+} in chromaffin cells showed that this mutation shifted the apparent Ca^{2+} affinity of fast exocytosis (but not of the slow phase) in precisely the same manner as the apparent Ca^{2+}-binding affinity of the cytoplasmic double C_2 domain fragment (Sorensen et al. 2003). The Ca^{2+}-binding properties and mouse mutants of synaptotagmin 1 established its function as a Ca^{2+} sensor for fast exocytosis but raised two new questions: How does it work, and what Ca^{2+} sensors are responsible for the slow phase of exocytosis?

Mechanism of Action of Synaptotagmin 1

In addition to forming Ca^{2+}-dependent complexes with phospholipids (which are essential for achieving a physiologically apparent Ca^{2+} affinity), synaptotagmins 1 and 2 also bind to SNARE complexes. Binding to SNARE complexes is partly Ca^{2+} independent (i.e., observed in the presence of an excess of EGTA) and partly Ca^{2+} dependent (Bennett et al. 1992, Chapman et al. 1995, Li et al. 1995a, Shin et al. 2003). In addition, synaptotagmins 1 and 2 form homomultimers both Ca^{2+}-independently (via its N-terminal transmembrane region; Perin et al. 1991b) and Ca^{2+}-dependently (via its C_2B domain; Chapman et al. 1996, Sugita et al. 1996). The Ca^{2+}-dependent interaction of synaptotagmin 1 with the SNARE complex is not essential for Ca^{2+} triggering of release because Sr^{2+} can substitute for Ca^{2+} in triggering release, but it is unable to stimulate SNARE binding (Shin et al. 2003). This finding established that triggering of fast exocytosis by Ca^{2+} binding to synaptotagmin 1 is not executed via Ca^{2+}-induced binding of synaptotagmin 1 to SNAREs but does not mean that SNARE binding is not important. In fact, Ca^{2+}-independent binding of synaptotagmin 1 could serve to position synaptotagmin 1 next to SNARE complexes after these have been assembled but before the fusion pore has opened (see model in Figure 6*B*). Even Ca^{2+}-dependent binding of synaptotagmin 1 to SNAREs could be functionally important, for example in recruiting vesicles into the readily releasable pool.

A possible model derived from these observations is based on the notion that SNARE complex formation forces synaptic vesicles into an unstable fusion intermediate that is stabilized by complexin binding to the SNARE complexes (Figure 6*B*). Synaptotagmin 1 binds to the SNARE complex in the absence of Ca^{2+} but switches to the phospholipid membrane as soon as Ca^{2+} enters via Ca^{2+} channels. Binding of synaptotagmin 1 C_2 domains to the phospholipids, with partial insertion of hydrophobic amino acids from synaptotagmin 1 and mechanical stress induced by the binding, may then destabilize the fusion intermediate and open the fusion pore (Figure 6*C*). This model has the advantage of explaining several key features of release. First is speed. Because all Ca^{2+} does in binding to synaptotagmin 1 is cause a switch in binding partners, which results in a mechanical stress, Ca^{2+} binding to synaptotagmin 1 is a fast reaction. Second is cooperativity. Synaptotagmin 1 binds five Ca^{2+} ions as predicted in the calyx model and thus has the requisite number of Ca^{2+}-binding sites. Third is the synaptobrevin KO phenotype (Schoch et al. 2001). Ca^{2+}-triggered release was much more severely depressed in synaptobrevin KO mice than in Ca^{2+}-independent release, which indicates that the proper SNARE complex may not be absolutely essential for fusion but is essential for Ca^{2+} triggering. Fourth is the complexin KO phenotype

(Reim et al. 2001). The complexin KO mouse exhibits a phenotype that has the properties of a milder version of the synaptotagmin 1 KO, consistent with the notion that complexin stabilizes the SNARE complex that is essential for the proper positioning of synaptotagmin 1 (see model in Figure 6*B*). In spite of this supporting evidence, however, the model is far from proven, and alternative interpretations are possible. For example, no mutant of synaptotagmin 1 that still binds Ca^{2+} but is unable to interact with SNARE complexes was reported, and the question of whether synaptotagmin 1 can bind to the SNARE complex simultaneously with complexin has not been investigated. Thus, although it seems assured that Ca^{2+} binding to synaptotagmin 1 (and synaptotagmin 2 in more caudal synapses) triggers fast release, the mechanism of triggering is probably intimately connected to SNARE proteins and likely requires clarification of SNARE function.

Ca^{2+} Sensors for Asynchronous or Delayed Exocytosis

A second major question considers how Ca^{2+} triggers slow, asynchronous release. Synaptotagmin 1 is part of a gene family containing 15 members in vertebrates (reviewed in Südhof 2002). Of these, only synaptotagmins 1–7 and 9–11 have sequences consistent with Ca^{2+} binding to at least one of the two C_2 domains. However, direct measurements revealed that not all C_2 domains of synaptotagmins with predicted Ca^{2+}-binding sites in fact bind Ca^{2+}. For example, the C_2B domains of synaptotagmin 3 and synaptotagmin 9, although including all predicted Ca^{2+}-binding residues, are inactive (Sugita et al. 2002; Shin et al. 2004). Despite the incomplete information, it is likely that other synaptotagmins function as Ca^{2+} sensors for the slow component of exocytosis, either alone or in collaboration with synaptotagmin 1. Support for this hypothesis comes from the observation that the "other" synaptotagmins 3, 6, and 7 exhibit higher apparent Ca^{2+} affinities (Sugita et al. 2001, 2002). Asynchronous release, although intrinsically slower than the fast component, is probably induced by lower Ca^{2+} concentrations. The Ca^{2+} transient induced by an action potential has only a very short (~400 μs) lifetime sufficient to trigger fast release, but the residual Ca^{2+} decays more slowly and still activates asynchronous release at lower Ca^{2+} concentrations. Consistent with this hypothesis, the C_2 domains from high-affinity synaptotagmins 3 and 7 are potent inhibitors of slow release in permeabilized neuroendocrine cells, whereas the C_2 domains of synaptotagmin 1 and 2 are inactive (Sugita et al. 2001, 2002).

REGULATING EXOCYTOSIS VIA Rab3

Rab proteins form a large set of GTP-binding proteins that regulate intracellular transport. Synaptic vesicles contain members of at least three families of rab proteins: Rab3 (Rab3A, 3B, 3C, and 3D; Schlüter et al. 2002), Rab5 (Fischer von Mollard et al. 1994), and Rab11 (Khvotchev et al. 2003). Of these, Rab3 proteins are the most abundant; Rab3A alone accounts for ~25% of the total Rab GTP binding in brain (Geppert et al. 1994b). In addition to synaptic vesicles, Rab3 proteins are also found on other secretory vesicles in a variety of cell types and are the only mammalian Rab proteins that are specific for such vesicles (reviewed in Darchen & Goud 2000).

The Rab3 Cycle

Rab3 undergoes a cycle of synaptic vesicle association and dissociation in parallel with synaptic vesicle exo- and endocytosis (Fischer von Mollard et al. 1991). Rab3 is attached to synaptic vesicles in the GTP-bound state via covalently linked geranylgeranyl moieties (Johnston et al. 1991). During or after synaptic vesicle fusion, GTP on Rab3 is hydrolyzed to GDP, and the resulting GDP-bound Rab3 is dissociated from synaptic vesicles by GDI (named GDP dissociation inhibitor, although its general function is to dissociate rab proteins from membranes; Araki et al. 1990). The soluble GDI/GDP-Rab3 complex is then reattached to synaptic vesicles by a poorly understood process that involves GDP to GTP exchange. Rab3 dissociation from vesicles depends on Ca^{2+}-triggered exocytosis of synaptic vesicles (Fischer von Mollard et al. 1991), which suggests that the Rab3 cycle ensures directional interactions of Rab3 with effector proteins during exocytosis.

Analysis of KO mice revealed a discrete but important function of Rab3A in a late step of exocytosis. Two major phenotypes were observed:

1. In synapses of the hippocampal CA1 region, short-term plasticity was altered without a change in the readily releasable vesicle pool (Geppert et al. 1994b, 1997). Different from other Rab proteins that normally function in vesicle transport or docking (Jahn et al. 2003), Rab3A was found to be selectively essential for a late step in exocytosis that follows transport and docking. However, other Rab3 isoforms are still expressed in the Rab3A KO mice, and the phenotype may reflect only the essential part of Rab3A function, whereas other parts of Rab3A function may be redundant among isoforms.
2. In the hippocampal CA3 region, mossy-fiber LTP was abolished, but no change in short-term plasticity was detected (Castillo et al. 1997). Mossy fiber–type LTP is expressed as a presynaptic increase in release, triggered by stimulation of cAMP-dependent protein kinase A (PKA), which causes a direct modification of the secretory machinery (reviewed in Nicoll & Malenka 1995, Hansel et al. 2001). Mossy fiber–type LTP is expressed not only in hippocampal mossy fiber synapses in the CA3 region (Huang et al. 1994, Weisskopf et al. 1994), but also in corticothalamic and corticostriatal (Castro-Alamancos & Calcagnotto 1999, Spencer & Murphy 2002) and cerebellar parallel fiber synapses (Salin et al. 1996, Linden 1997). Rab3A is not a substrate for PKA, which suggests that although Rab3A is essential for mossy fiber–type LTP, LTP is not simply induced by Rab3A phosphorylation.

The presence of distinct phenotypes in synapses lacking Rab3A suggests that excitatory synapses fall into two principal classes: synapses capable of PKA-dependent presynaptic LTP where Rab3A is only required for LTP, and synapses that lack this form of LTP and exhibit changes only in short-term plasticity upon deletion of Rab3A.

Rab3 Effectors

Two classes of Rab3 effectors that bind only to GTP-Rab3 but not to GDP-Rab3 have been identified, rabphilin (Shirataki et al. 1993, Li et al. 1994) and RIM1α/2α (Wang et al. 1997a, 2000; Wang & Südhof 2003). Both effectors have a similar N-terminal zinc-finger domain that interacts with all Rab3 isoforms, include central phosphorylation sites for PKA, and contain two C-terminal C_2 domains. Otherwise, however, rabphilin and RIM1α/2α are very different (Wang et al. 1997a). Rabphilin is a soluble protein that requires Rab3 for binding to synaptic vesicles (Geppert et al. 1994b, Li et al. 1994) and binds Ca^{2+} via its C_2 domains (Ubach et al. 1998). RIM1α/2α, in contrast, are larger, biochemically insoluble active-zone proteins whose C-terminal C_2 domains lack predicted Ca^{2+}-binding sites.

Rabphilin exhibits biologically interesting properties (Ca^{2+} binding, cycling on- and off-synaptic vesicles in a manner dependent on Rab3, stimulation-dependent phosphorylation by multiple kinases; see Shirataki et al. 1993, Li et al. 1994, Ubach et al. 1999). It was thus disappointing that deletion of rabphilin in mice failed to cause a major impairment (Schlüter et al. 1999). The rabphilin KO mice exhibited none of the phenotypic traits of Rab3A KO mice. The absence of a rabphilin KO phenotype may be due to a hidden redundancy, but no rabphilin isoform that binds to Rab3 and is expressed in neurons has been identified.

A very different picture emerged for RIM1α: KO of RIM1α caused a major phenotype in synaptic transmission that went beyond the features of the Rab3A KO phenotype (Castillo et al. 2002, Schoch et al. 2002). RIM1α KO mice displayed a large decrease in synaptic strength, exhibited major changes in short-term synaptic plasticity, and lacked mossy fiber–type LTP both in CA3 synapses in the hippocampus and in parallel fiber synapses in the cerebellum. In RIM1α KO mice but not Rab3A KO mice, the levels of the active-zone protein Munc13-1 (which binds to the N-terminal zinc-finger of RIMs; see discussion above in Molecular Architecture of Synaptic Vesicles and Interacting Partners) were substantially depressed (Schoch et al. 2002). The binding of Rab3A on synaptic vesicles to RIM1α in the active zone evokes a docking function (Figure 5), but RIM1α KO mice did not exhibit a change in the number of docked vesicles (Schoch et al. 2002), consistent with a lack of change in docking in the Rab3A KO mice (Geppert et al. 1997). Viewed together, these data suggest that RIM1α (and probably RIM2α) regulates neurotransmitter release via interactions of its N-terminal domain with Rab3 and Munc13-1, and possibly via interactions of its PDZ domains with ERCs and its C-terminal C_2 domain with α-liprins and synaptotagmin 1 (Betz et al. 2001, Ohtsuka et al. 2002, Schoch et al. 2002, Wang et al. 2002).

Possibly the most interesting question is how Rab3A and RIM1α participate in mossy fiber–type LTP. Recent studies showed that RIM1α is phosphorylated by PKA in vivo and that phosphorylation of RIM1α at a central PKA site is essential for mossy fiber–type LTP (Lonart et al. 2003). Thus the long-lasting increase in release upon LTP induction appears to require the coincidence of two signals: GTP-dependent binding of Rab3A to the N-terminal zinc-finger domain of RIM1α, and phosphorylation of RIM1α by PKA at a site immediately C-terminal to the zinc-finger domain (Figure 5). Although the mechanism by which this convergence of signals induces LTP is unknown, this mechanism may include a restructuring of the active zone that leads to an increase either in the number of release-ready vesicles or in the efficiency of Ca^{2+} triggering of release. Interestingly, binding of Munc13-1 and Rab3 compete for binding to the N-terminal zinc-finger of RIM1α (Betz et al. 2001). A possible pathway is that Rab3 binding to RIMs inactivates Munc13 by displacing it from RIM and that this action potentiates release, possibly by allowing Munc13 to recruit additional RIM molecules to the active zone and thereby to increase the size of the active zone.

SV2s MAY REGULATE Ca^{2+} LEVELS

SV2, identified in a monoclonal antibody screen for synaptic vesicle proteins (Buckley & Kelly 1985), is a component of synaptic vesicles and neuroendocrine secretory granules in all vertebrates but is not conserved in invertebrates. Three SV2 genes in vertebrates encode highly homologous proteins referred to as SV2A, SV2B, and SV2C (see Janz & Südhof 1999, and references cited therein). SV2 proteins contain 12 potential transmembrane regions with cytoplasmic N- and C-termini and exhibit significant homology with carbohydrate transport proteins in eukaryotes and bacteria. Most loops connecting the transmembrane regions are short except for one large intravesicular loop that is highly glycosylated and may supply an ionic matrix to the synaptic vesicle interior (Buckley & Kelly 1985, Scranton et al. 1993). Comparisons between SV2 isoforms show that the transmembrane regions and the cytoplasmic loops are highly conserved, whereas the N-terminal cytoplasmic sequence and the intravesicular loops exhibit little homology (Janz & Südhof 1999).

The three SV2 proteins are differentially expressed in brain. SV2A is present in almost all neurons, SV2B exhibits a more restricted distribution (Bajjalieh et al. 1994), and SV2C is present only in a small subset of neurons in the basal forebrain and caudal brain regions (Janz & Südhof 1999). In addition to SV2 proteins, synaptic vesicles contain a distantly related protein called SVOP (Janz et al. 1998). SVOP has a similar transmembrane structure as SV2 but lacks the long glycosylated intravesicular loop. Furthermore, SVOP is highly conserved in invertebrates, whereas SV2 is not (Janz et al. 1998). These differences suggest that SV2 proteins may be evolutionarily late, vertebrate-specific descendants of SVOP.

Initially, investigators thought that SV2A may be a neurotransmitter transporter, but its presence in nerve terminals with distinct neurotransmitters dispelled this idea (Bajjalieh et al. 1994). Analysis of KO mice revealed that SV2B knockout mice are phenotypically unremarkable, whereas SV2A and SV2A/SV2B double knockout mice exhibit severe seizures and die postnatally (Janz et al. 1999a). A similar lethal phenotype was observed in independently generated SV2A KO mice (Crowder et al. 1999). In electrophysiological recordings from cultured hippocampal neurons, SV2A- or SV2B-deficient cells exhibited no decrease in evoked release or in the size of the readily releasable pool (Janz et al. 1999a). Neurons lacking both SV2 isoforms, however, experienced sustained increases in Ca^{2+}-dependent synaptic transmission when two or more action potentials were triggered in succession. These increases could be reversed by a membrane-permeable EGTA-ester that is hydrolyzed to EGTA in cells where the EGTA then chelates free Ca^{2+} ions. This observation suggests that the deletion of SV2 caused presynaptic Ca^{2+} accumulation during consecutive action potentials (Janz et al. 1999a).

The phenotype of the SV2-deficient synapses indicates that SV2 may be a Ca^{2+} transporter in synaptic vesicles, which is corroborated by the presence of a pair of conserved negatively charged residues in the first transmembrane region (Janz et al. 1998, Janz & Südhof 1999). However, a separate analysis of SV2A-deficient mice came to a very different conclusion (Xu & Bajjalieh 2002). In chromaffin cells, the Ca^{2+}-induced exocytotic burst of chromaffin granule exocytosis, thought to correspond to the readily releasable pool of vesicles, was significantly decreased in mice lacking SV2A, and in brain, detergent-resistant SNARE complexes were reduced. One possible hypothesis to explain the difference in electrophysiological phenotype—a reduction in the readily releasable pool in chromaffin cells but not in synapses—is based on the role of intravesicular Ca^{2+} in stabilizing the chromaffin granule structure (Südhof 1983). A reduction in the Ca^{2+} content of chromaffin granules in the SV2 KO mice could lead to a destabilization of vesicles, which in turn would manifest as a loss of the readily releasable pool of vesicles without actually exhibiting a docking and priming phenotype. The reduction in SNARE complexes in the brain from SV2A KO mice (that still contain SV2B and SV2C) is more difficult to reconcile with the synaptic phenotype and is hard to explain mechanistically. The structure of SV2 is that of a transporter, with few sequences exposed to the cytoplasm. The reported interaction of the N-terminal region of SV2A with synaptotagmin 1 (Schivell et al. 1996) does not help to explain the purported SNARE-related phenotype because the deletion of synaptotagmin 1 does not change SNARE complexes or reduce the readily releasable pool of synaptic vesicles that is dependent on SNARE complexes (Geppert et al. 1994a). Thus it seems likely that SV2 is indeed a transporter, although the transport of Ca^{2+} remains to be demonstrated directly and other cations could potentially be substrates for the transport activity of SV2.

CONTROLLING POOLS VIA SYNAPSINS

Synapsins are abundant synaptic vesicle proteins that coat the vesicle surface as peripheral membrane proteins. Although synapsins were the first synaptic vesicle proteins identified when they were discovered as neuronal substrates for cAMP- and Ca^{2+}/calmodulin-activated kinases (reviewed in Greengard et al. 1994), their precise function remains unclear.

Three synapsin genes in mammals express alternatively spliced protein variants that have similar N-terminal and central domains but distinct C-terminal sequences (Südhof et al. 1989, Hosaka & Südhof 1998a, Kao et al. 1998, Porton et al. 1999). Most synapses express synapsin 1 and 2, whereas synapsin 3 variants are of lesser abundance. All synapsins contain (*a*) a short N-terminal domain (~20 residues) that features a conserved phosphorylation site for PKA and Ca^{2+}/calmodulin-dependent protein kinase I, (*b*) a linker sequence, and (*c*) a large central C domain (~300 residues) that accounts for more than half of the sequence of most synapsins. C-terminally, synapsins are composed of variable, alternatively spliced sequences that in each synapsin gene include one variant with a single conserved domain (the E domain).

The central C domain of synapsins forms a constitutive dimer and binds ATP with a high affinity (Hosaka & Südhof 1998a,b, 1999; Esser et al. 1998). The crystal structure of the C domain of synapsin 1 revealed a striking similarity to a family of ATPases (Esser et al. 1998). This family includes glutathione synthetase and D-alanine:D-alanine ligase, which suggests that synapsins may be ATP-utilizing enzymes. The ATP-binding domains of different synapsins exhibit distinct properties: Synapsin 1 binds ATP only in the presence of Ca^{2+}, synapsin 2 binds ATP irrespective of Ca^{2+}, and synapsin 3 binds ATP only in the absence of Ca^{2+} (Hosaka & Südhof 1998a,b).

Several mechanisms of binding synapsins to synaptic vesicles were proposed. Originally, investigators thought that synapsin 1 binds to vesicles via Ca^{2+}/calmodulin-dependent protein kinase II (Greengard et al. 1994). However, synapsin 2 does not bind to this kinase but is found on vesicles, and Ca^{2+}/calmodulin-dependent kinase II is not a regular component of synaptic vesicles. Transgenic experiments demonstrated that the N-terminal domains of synapsin 1 are sufficient for synaptic vesicle targeting (Geppert et al. 1994c). Synapsins avidly bind to lipid surfaces (Benfenati et al. 1989) via their short N-terminal phosphorylated domain (Hosaka et al. 1999). The N-terminal domain is phosphorylated by PKA and Ca^{2+}/calmodulin-dependent protein kinase I, and phosphorylation abolishes binding of synapsins to synaptic vesicles. As a result, synapsins cycle on and off vesicles in a stimulation-dependent manner (Hosaka et al. 1999). However, whether synapsins cycle on and off vesicles normally during exo- and endocytosis, or only upon extensive stimulation, is unclear. Some studies observed a depletion of synapsins from synaptic vesicles as the vesicles became docked (Pieribone et al. 1995), whereas others demonstrated that synapsins remain associated with the vesicles during exo- and endocytosis (Torri-Tarelli et al. 1990).

Analysis of KO mice showed that synapsins 1 and 2 are not essential for mouse survival or for synaptic vesicle exocytosis but are required to maintain normal numbers of synaptic vesicles and to regulate short-term synaptic plasticity (Rosahl et al. 1993, 1995; Li et al. 1995b). Deletion of synapsins decreased the number of vesicles without altering vesicle clustering or docking (Rosahl et al. 1995). The synapsin 2 KO phenotype was more severe than was the synapsin 1 KO phenotype, and the double KO exhibited the biggest changes, including a 50%

decrease in vesicle numbers (Rosahl et al. 1995). The loss of synaptic vesicles in the synapsins KO mice indicates a maintenance function of synapsins, possibly via an ATP-dependent activity. Deletion of synapsins also caused dramatic changes in short-term synaptic plasticity (Rosahl et al. 1993, 1995). Short-term plasticity involves changes in release when two or several action potentials are triggered within milliseconds. The short time frame of short-term synaptic plasticity suggests that this type of plasticity regulates the interaction of synaptic vesicles with the active zone. By inference, synapsins thus may regulate this interaction, but the mechanism of such regulation and its relation to the maintenance function of synapsins remains unclear. In addition, deletion of synapsins was found to delay axonal outgrowth and/or synaptogenesis of cultured neurons (e.g., see Ferreira et al. 1998). However, double KO mice lacking both synapsins exhibited no changes in the density of synapses and the appropriateness of synaptic connections (Rosahl et al. 1995). Synapsin 3 KO mice have also been analyzed, but only in isolation without crossing to synapsin 1 and 2 KO mice (Feng et al. 2002). These mice exhibited discrete changes, for example a 5% increase in synaptic vesicle density and a change in evoked GABAergic synaptic responses. The observation of distinct effects by deletion of various synapsins is corroborated by the differential regulation of ATP binding to synapsins by Ca^{2+}, which either stimulates ATP binding (synapsin 1), has no effect on ATP binding (synapsin 2), or inhibits ATP binding (synapsin 3).

What do synapsins actually do? Investigators have reported innumerable in vitro binding activities. Most prominently, synapsins bind to various elements of the cytoskeleton, especially actin, leading to the hypothesis that synapsins may anchor synaptic vesicles in the presynaptic vesicle cluster to the cytoskeleton (reviewed in Greengard 1994). However, with the realization that the mature vesicle cluster contains virtually no cytoskeleton (Dunaevsky & Connor 2000, Morales et al. 2000, Zhang & Benson 2001), this hypothesis is unlikely to explain synapsin function. At present there is a gap between the structural definition of synapsins as ATP-binding molecules and their functional role in maintaining vesicle integrity and regulating release. Key questions are whether ATP binding to synapsins is functionally important, whether synapsin phosphorylation plays a physiological role, and at what point exactly synapsins work in the synaptic vesicle cycle. Addressing these questions will provide crucial advances in understanding these enigmatic proteins.

PERSPECTIVE

In this review, I attempt to evaluate how well the molecular characterization of synaptic vesicle proteins and their interaction partners can account for the physiological properties of the vesicle cycle. Although significant progress has been made (e.g., in the definition of synaptotagmin 1 as the Ca^{2+} sensor for fast release), more progress will be required until investigators achieve a reasonable mechanistic understanding of the cycle. The gap between the physiological description and the molecular understanding of the vesicle cycle is immense. Two principal problems make it difficult to close this gap. First, much of the physiological description is protocol dependent, and does not measure parameters that correspond to a unitary molecular event. One reason why our understanding of Ca^{2+} triggering of release is so advanced is that the physiological description of Ca^{2+} triggering is very precise. If one compares Ca^{2+} triggering, for example, with endocytosis and recycling where different protocols measure distinct time constants, it is clear that defining the molecular basis for endocytosis and recycling will be more difficult. Second, most proteins function in interaction networks that are difficult to define and validate, and many proteins participate in more than one function. Unraveling these protein networks and overlapping functions will be a major challenge, but it will be necessary to achieve a complete molecular understanding. In the molecular mosaic that composes the nerve terminal, many pieces do not yet have an appropriate fit—e.g., the vesicle proteins SVOP or synaptogyrin—and many more pieces are still missing. Nevertheless, the presynaptic nerve terminal promises to be one of the best systems to explore the relationship between the structure of a protein, its biochemical and cell-biological properties, and its physiological role.

ACKNOWLEDGMENTS

I thank Drs. J. Sun, E. Neher, R. Schneggenburger, and E. Kavalali for helpful comments. Work in my laboratory is supported by grants from the NIH and by an investigatorship from the Howard Hughes Medical Institute.

The *Annual Review of Neuroscience* is online at http://neuro.annualreviews.org

LITERATURE CITED

Akert K, Moor H, Pfenninger K. 1971. Synaptic fine structure. *Adv. Cytopharmacol.* 1:273–90

Alfonso A, Grundahl K, Duerr JS, Han HP, Rand JB. 1993. The Caenorhabditis elegans unc-17 gene: a putative vesicular acetylcholine transporter. *Science* 261:617–19

Antonin W, Riedel D, von Mollard GF. 2000. The SNARE Vti1a-beta is localized to small synaptic vesicles and participates in a novel SNARE complex. *J. Neurosci.* 20:5724–32

Araki S, Kikuchi A, Hata Y, Isomura M, Takai Y. 1990. Regulation of reversible binding of smg p25A to synaptic plasma membranes and vesicles by GDP dissociation inhibitor. *J. Biol. Chem.* 265:13007–15

Arata Y, Nishi T, Kawasaki-Nishi S, Shao E, Wilkens S, Forgac M. 2002. Structure, subunit function and regulation of the coated vesicle and yeast vacuolar (H^+)-ATPases. *Biochim. Biophys. Acta* 1555:71–74

Archer DA, Graham ME, Burgoyne RD. 2002. Complexin regulates the closure of the fusion pore during regulated vesicle exocytosis. *J. Biol. Chem.* 277:18249–52

Atluri PP, Regehr WG. 1998. Delayed release of neurotransmitter from cerebellar granule cells. *J. Neurosci.* 18:8214–27

Barker LA, Dowdall MJ, Whittaker VP. 1972. Choline metabolism in the cerebral cortex of guinea pigs. Stable-bound acetylcholine. *Biochem. J.* 130:1063–75

Augustin I, Rosenmund C, Südhof TC, Brose N. 1999. Munc13-1 is essential for fusion competence of glutamatergic synaptic vesicles. *Nature* 400:457–61

Bacci A, Coco S, Pravettoni E, Schenk U, Armano S, et al. 2001. Chronic blockade of glutamate receptors enhances presynaptic release and downregulates the interaction between synaptophysin-synaptobrevin 2. *J. Neurosci.* 21:6588–96

Bajjalieh SM, Frantz GD, Weimann JM, McConnell SK, Scheller RH. 1994. Differential expression of synaptic vesicle protein 2 (SV2) isoforms. *J. Neurosci.* 14:5223–35

Barrett EF, Stevens CF. 1972. The kinetics of transmitter release at the frog neuromuscular junction. *J. Physiol.* 227:691–708

Becher A, Drenckhahn A, Pahner I, Margittai M, Jahn R, Ahnert-Hilger G. 1999. The synaptophysin-synaptobrevin complex: a hallmark of synaptic vesicle maturation. *J. Neurosci.* 19:1922–31

Bekkers JM, Richerson GB, Stevens CF. 1990. Origin of variability in quantal size in cultured hippocampal neurons and hippocampal slices. *Proc. Natl. Acad. Sci. USA* 87:5359–62

Bennett MK, Calakos N, Scheller RH. 1992. Syntaxin: a synaptic protein implicated in docking of synaptic vesicles at presynaptic active zones. *Science* 257:255–59

Benfenati F, Greengard P, Brunner J, Bahler M. 1989. Electrostatic and hydrophobic interactions of synapsin I and synapsin I fragments with phospholipid bilayers. *J. Cell Biol.* 108:1851–62

Betz A, Okamoto M, Benseler F, Brose N. 1997. Direct interaction of the rat unc-13 homologue Munc13-1 with the N terminus of syntaxin. *J. Biol. Chem.* 272:2520–26

Betz A, Thakur P, Junge HJ, Ashery U, Rhee JS, et al. 2001. Functional interaction of the active zone proteins Munc13-1 and RIM1 in synaptic vesicle priming. *Neuron* 30:183–96

Bollmann JH, Sakmann B, Gerard J, Borst G. 2000. Calcium sensitivity of glutamate release in a calyx-type terminal. *Science* 289:953–57

Borst JG, Sakmann B. 1996. Calcium influx and transmitter release in a fast CNS synapse. *Nature* 383:431–34

Borst JGG, Helmchen F, Sakmann B. 1995. Pre- and postsynaptic whole-cell recordings in the medial nucleus of the trapezoid body of the rat. *J. Physiol.* 489:825–40

Breckenridge LJ, Almers W. 1987. Currents through the fusion pore that forms during exocytosis of a secretory vesicle. *Nature* 328:814–17

Brodin L, Low P, Shupliakov O. 2000. Sequential steps in clathrin-mediated synaptic vesicle endocytosis. *Curr. Opin. Neurobiol.* 10:312–20

Brose N, Hofmann K, Hata Y, Südhof TC. 1995 Mammalian homologues of C. elegans unc-13 gene define novel family of C_2-domain proteins. *J. Biol. Chem.* 270:25273–80

Bruns D, Riedel D, Klingauf J, Jahn R. 2000. Quantal release of serotonin. *Neuron* 28:205–20

Buckley K, Kelly RB. 1985. Identification of a transmembrane glycoprotein specific for secretory vesicles of neural and endocrine cells. *J. Cell Biol.* 100:1284–94

Calakos N, Scheller R. 1994. Vesicle-associated membrane protein and synaptophysin are associated on the synaptic vesicle. *J. Biol. Chem.* 269:24534–37

Cases-Langhoff C, Voss B, Garner AM, Appeltauer U, Takei K, et al. 1996. Piccolo, a novel 420 kDa protein associated with the presynaptic cytomatrix. *Eur. J. Cell Biol.* 69:214–23

Castillo PE, Janz R, Südhof TC, Tzounopoulos T, Malenka RC, Nicoll RA. 1997. Rab3A is essential for mossy fibre long-term potentiation in the hippocampus. *Nature* 388:590–93

Castillo PE, Schoch S, Schmitz F, Südhof TC, Malenka RC. 2002. RIM1α is required for presynaptic long-term potentiation. *Nature* 415:327–30

Castro-Alamancos MA, Calcagnotto ME. 1999. Presynaptic long-term potentiation in corticothalamic synapses. *J. Neurosci.* 19: 9090–97

Ceccarelli B, Hurlbut WP, Mauro A. 1973. Turnover of transmitter and synaptic vesicles at the frog neuromuscular junction. *J. Cell Biol.* 57:499–524

Chapman ER, Hanson PI, An S, Jahn R. 1995. Ca^{2+} regulates the interaction between synaptotagmin and syntaxin 1. *J. Biol. Chem.* 270:23667–71

Chapman ER, An S, Edwardson JM, Jahn R. 1996. A novel function for the second C_2 domain of synaptotagmin Ca^{2+} triggered dimerization. *J. Biol. Chem.* 271:5844–49

Chen X, Tomchick DR, Kovrigin E, Arac D, Machius M, et al. 2002a. Three-dimensional structure of the complexin/SNARE complex. *Neuron* 33:397–409

Chen YA, Scheller RH. 2001. SNARE-mediated membrane fusion. *Nat. Rev. Mol. Cell Biol.* 2:98–106

Coppola T, Magnin-Luthi S, Perret-Menoud V, Gattesco S, Schiavo G, Regazzi R. 2001. Direct interaction of the Rab3 effector RIM with Ca^{2+} channels, SNAP-25, and synaptotagmin. *J. Biol. Chem.* 276:32756–62

Crowder KM, Gunther JM, Jones TA, Hale BD, Zhang HZ, et al. 1999. Abnormal neurotransmission in mice lacking synaptic vesicle protein 2A (SV2A). *Proc Natl. Acad. Sci. USA* 96:15268–73

Darchen F, Goud B. 2000. Multiple aspects of Rab protein action in the secretory pathway: focus on Rab3 and Rab6. *Biochimie* 82:375–84

Dietrich D, Kirschstein T, Kukley M, Pereverzev A, von der Brelie C, et al. 2003. Functional specialization of presynaptic Cav2.3 Ca^{2+} channels. *Neuron* 39:483–96

Dresbach T, Qualmann B, Kessels MM, Garner CC, Gundelfinger ED. 2001. The presynaptic cytomatrix of brain synapses. *Cell. Mol. Life Sci.* 58:94–116

Dulubova I, Sugita S, Hill S, Hosaka M, Fernandez I, et al. 1999. A conformational switch in syntaxin during exocytosis. *EMBO J.* 18:4372–82

Dulubova I, Yamaguchi T, Min SW, Gao Y, Südhof TC, Rizo J. 2002. How Tlg2p/syntaxin16 "snares" Vps45. *EMBO J.* 21: 3620–31

Dunaevsky A, Connor EA. 2000. F-actin is concentrated in nonrelease domains at frog neuromuscular junctions. *J. Neurosci.* 20:6007–12

Edelmann L, Hanson PI, Chapman ER, Jahn R. 1995. Synaptobrevin binding to synaptophysin: a potential mechanism for controlling the exocytotic fusion machine. *EMBO J.* 14:224–31

Erickson JD, Eiden LE, Hoffman BJ. 1992. Expression cloning of a reserpine-sensitive vesicular monoamine transporter. *Proc. Natl. Acad. Sci. USA* 89:10993–97

Esser L, Wang CR, Hosaka M, Smagula CS, Südhof TC, Deisenhofer J. 1998. Synapsin I is structurally homologous to ATP-utilizing enzymes. *EMBO J.* 17:977–84

Feng J, Chi P, Blanpied TA, Xu Y, Magarinos AM, et al. 2002. Regulation of neurotransmitter release by synapsin III. *J. Neurosci.* 22:4372–80

Fernandez-Chacon R, Konigstorfer A, Gerber SH, Garcia J, Matos MF, et al. 2001. Synaptotagmin I functions as a calcium regulator of release probability. *Nature* 410:41–49

Fernandez-Chacon R, Shin O, Königstorfer A, Matos MF, Meyer AC, et al. 2002. Structure/function analysis of Ca^{2+} binding to the C_2A domain of synaptotagmin I. *J. Neurosci.* 22:8438–46

Fernandez I, Arac D, Ubach J, Gerber SH, Shin O, et al. 2001. Three-dimensional structure of the synaptotagmin 1 C2B-domain: synaptotagmin 1 as a phospholipid binding machine. *Neuron* 32:1057–69

Ferreira A, Chin LS, Li L, Lanier LM, Kosik KS, Greengard P. 1998. Distinct roles of synapsin I and synapsin II during neuronal development. *Mol. Med.* 4:22–28

Fischer von Mollard G, Südhof TC, Jahn R. 1991. A small GTP-binding protein (rab3A) dissociates from synaptic vesicles during exocytosis. *Nature* 349:79–81

Fischer von Mollard G, Stahl B, Walch-Solimena C, Takei K, Daniels L, et al. 1994. Localization of rab5 to synaptic vesicles identifies endosomal intermediate in synaptic vesicle recycling pathway. *Eur. J. Cell Biol.* 65:319–26

Fisher RJ, Pevsner J, Burgoyne RD. 2001. Control of fusion pore dynamics during exocytosis by Munc18. *Science* 291:875–78

Fletcher S, Bowden SE, Marrion NV. 2003. False interaction of syntaxin 1A with a Ca^{2+} activated K^+ channel revealed by co-immunoprecipitation and pull-down assays: implications for identification of protein-protein interactions. *Neuropharmacology* 44:817–27

Fremeau RT Jr, Burman J, Qureshi T, Tran CH, Proctor J, et al. 2002. The identification of vesicular glutamate transporter 3 suggests novel modes of signaling by glutamate. *Proc. Natl. Acad. Sci. USA* 99:14488–93

Fujita Y, Shirataki H, Sakisaka T, Asakura T, Ohya T, et al. 1998. Tomosyn: a syntaxin-1-binding protein that forms a novel complex in the neurotransmitter release process. *Neuron* 20:905–15

Fykse EM, Fonnum F. 1996. Amino acid neurotransmission: dynamics of vesicular uptake. *Neurochem. Res.* 21:1053–60

Galli T, Haucke V. 2001. Cycling of synaptic vesicles: How far? How fast! *Sci. STKE* 88:RE1

Gandhi SP, Stevens CF. 2003. Three modes of synaptic vesicular recycling revealed by single-vesicle imaging. *Nature* 423:607–13

Geppert M, Archer BT III, Südhof TC. 1991. Synaptotagmin II: a novel differentially distributed form of synaptotagmin. *J. Biol. Chem.* 266:13548–52

Geppert M, Goda Y, Hammer RE, Li C, Rosahl TW, et al. 1994a. Synaptotagmin I: a major Ca^{2+} sensor for transmitter release at a central synapse. *Cell* 79:717–27

Geppert M, Bolshakov VY, Siegelbaum SA, Takei K, De Camilli P, et al. 1994b. The role of Rab3A in neurotransmitter release. *Nature* 369:493–97

Geppert M, Ullrich B, Green DG, Takei K, Daniels L, et al. 1994c. Synaptic targeting domains of synapsin I revealed by transgenic expression in photoreceptor cells. *EMBO J.* 13:3720–27

Geppert M, Goda Y, Stevens CF, Südhof TC. 1997. The small GTP-binding protein Rab3A regulates a late step in synaptic vesicle fusion. *Nature* 387:810–14

Goda Y, Südhof TC. 1997. Calcium regulation of neurotransmitter release: reliably unreliable. *Curr. Opin. Cell Biol.* 9:513–18

Goda Y, Stevens CF. 1994. Two components of transmitter release at a central synapse. *Proc. Natl. Acad. Sci. USA* 91:12942–46

Graham ME, Burgoyne RD. 2000. Comparison of cysteine string protein (Csp) and mutant alpha-SNAP overexpression reveals a role for Csp in late steps of membrane fusion in dense-core granule exocytosis in adrenal chromaffin cells. *J. Neurosci.* 20:1281–89

Gras C, Herzog E, Bellenchi GC, Bernard V, Ravassard P, et al. 2002. A third vesicular glutamate transporter expressed by cholinergic and serotoninergic neurons. *J. Neurosci.* 22:5442–51

Greengard P, Benfenati F, Valtorta F. 1994. Synapsin I, an actin-binding protein regulating synaptic vesicle traffic in the nerve terminal. Adv second messenger. *Phosphoprotein Res.* 29:31–45

Hagler DJ, Goda Y. 2001. Properties of synchronous and asynchronous release during pulse train depression in cultured hippocampal neurons. *J. Neurophysiol.* 85:2324–34

Hansel C, Linden DJ, D'Angelo E. 2001. Beyond parallel fiber LTD: the diversity of synaptic and non-synaptic plasticity in the cerebellum. *Nat. Neurosci.* 4:467–75

Harlow LH, Ress D, Stoschek A, Marshall RM, McMahan UJ. 2001. The architecture of active zone material at the frog's neuromuscular junction. *Nature* 409:479–84

Harris KM, Jensen FE, Tsao B. 1992. Three-dimensional structure of dendritic spines and synapses in rat hippocampus (CA1) at postnatal day 15 and adult ages: implications for the maturation of synaptic physiology and long-term potentiation. *J. Neurosci.* 12:2685–705

Helmchen F, Borst JGG, Sakmann B. 1997. Calcium dynamics associated with a single action potential in a CNS presynaptic terminal. *Biophys. J.* 72:1458–71

Hata Y, Slaughter CA, Südhof TC. 1993. Synaptic vesicle fusion complex contains unc-18 homologue bound to syntaxin. *Nature* 366:347–51

Hatsuzawa K, Lang T, Fasshauer D, Bruns D, Jahn R. 2003. The R-SNARE motif of tomosyn forms SNARE core complexes with syntaxin 1 and SNAP-25 and down-regulates exocytosis. *J. Biol. Chem.* 278:31159–66

Heuser JE, Reese TS. 1973. Evidence for recycling of synaptic vesicle membrane during transmitter release at the frog neuromuscular junction. *J. Cell Biol.* 57:315–44

Hinds HL, Goussakov I, Nakazawa K, Tonegawa S, Bolshakov VY. 2003. Essential function of α-Ca^{2+}/calmodulin-dependent protein kinase II in neurotransmitter release at a glutamatergic central synapse. *Proc. Natl. Acad. Sci. USA* 100:4275–80

Holz RW, Brondyk WH, Senter RA, Kuizon L, Macara IG. 1994. Evidence for the involvement of Rab3A in Ca^{2+} dependent exocytosis from adrenal chromaffin cells. *J. Biol. Chem.* 269:10229–34

Hosaka M, Südhof TC. 1998a. Synapsin III, a novel synapsin with an unusual regulation by Ca^{2+}. *J. Biol. Chem.* 273:13371–74

Hosaka M, Südhof TC. 1998b. Synapsins I and II are ATP-binding proteins with differential Ca^{2+} regulation. *J. Biol. Chem.* 273:1425–29

Hosaka M, Südhof TC. 1999. Homo- and heterodimerization of synapsins. *J. Biol. Chem.* 274:16747–53

Hosaka M, Hammer RE, Südhof TC. 1999. A phospho-switch controls the dynamic association of synapsins with synaptic vesicles. *Neuron* 24:377–87

Huang YY, Li XC, Kandel ER. 1994. cAMP contributes to mossy fiber LTP by initiating both a covalently mediated early phase and macromolecular synthesis-dependent late phase. *Cell* 79:69–79

Jahn R, Lang T, Südhof TC. 2003. Membrane fusion. *Cell* 112:519–33

Janz R, Goda Y, Geppert M, Missler M, Südhof TC. 1999a. SV2A and SV2B function as redundant Ca^{2+} regulators in neurotransmitter release. *Neuron* 24:1003–16

Janz R, Hofmann K, Südhof TC. 1998. SVOP, an evolutionarily conserved synaptic vesicle protein, suggests novel transport functions of synaptic vesicles. *J. Neurosci.* 18:9269–81

Janz R, Südhof TC. 1999. SV2C is a synaptic vesicle protein with an unusually restricted localization: anatomy of a synaptic vesicle protein family. *Neuroscience* 94:1279–90

Janz R, Goda Y, Geppert M, Missler M, Südhof TC. 1999a. SV2A and SV2B function as redundant Ca^{2+} regulators in neurotransmitter release. *Neuron* 24:1003–16

Johannes L, Lledo PM, Roa M, Vincent JD, Henry JP, Darchen F. 1994. The GTPase Rab3a negatively controls calcium-dependent exocytosis in neuroendocrine cells. *EMBO J.* 13:2029–37

Johnston PA, Archer BT III, Robinson K, Mignery GA, Jahn R, Südhof TC. 1991. Rab3A attachment to the synaptic vesicle membrane mediated by a conserved polyisoprenylated carboxy-terminal sequence. *Neuron* 7:101–9

Johnston PA, Südhof TC. 1990. The multisubunit structure of synaptophysin. Relationship between disulfide bonding and homo-oligomerization. *J. Biol. Chem.* 265:8869–73

Kamiya H, Umeda K, Ozawa S, Manabe T. 2002. Presynaptic Ca^{2+} entry is unchanged during hippocampal mossy fiber long-term potentiation. *J. Neurosci.* 22:10524–28

Kao HT, Porton B, Czernik AJ, Feng J, Yiu G, et al. 1998. A third member of the synapsin gene family. *Proc. Natl. Acad. Sci. USA* 95:4667–72

Katz B. 1969. *The Release of Neural Transmitter Substances*. Liverpool: Liverpool Univ. Press

Khvotchev MV, Ren M, Takamori S, Jahn R, Südhof TC. 2003. Divergent functions of neuronal Rab11b in Ca^{2+} regulated vs. constitutive exocytosis. *J. Neurosci.* 23:10531–39

Ko J, Na M, Kim S, Lee JR, Kim E. 2003. Interaction of the ERC family of RIM-binding proteins with the liprin-α family of multidomain proteins. *J. Biol. Chem.* 278:42377–85

Koenig JH, Ikeda K. 1996. Synaptic vesicles have two distinct recycling pathways. *J. Cell Biol.* 135:797–808

Koester H, Sakmann B. 2000. Calcium dynamics associated with action potentials in single nerve terminals of pyramidal cells in layer 2/3 of the young rat neocortex. *J. Physiol.* 529:625–46

Laage R, Langosch D. 1997. Dimerization of the synaptic vesicle protein synaptobrevin (vesicle-associated membrane protein) II depends on specific residues within the transmembrane domain. *Eur. J. Biochem.* 249:540–46

Li C, Takei K, Geppert M, Daniell L, Stenius K, et al. 1994. Synaptic targeting of rabphilin-3A, a synaptic vesicle Ca^{2+}/phospholipid-binding protein, depends on rab3A/3C. *Neuron* 13:85–98

Li C, Ullrich B, Zhang JZ, Anderson RGW, Brose N, Südhof TC. 1995a. Ca^{2+} dependent and Ca^{2+} independent activities of neural and nonneural synaptotagmins. *Nature* 375:594–99

Li L, Chin LS, Shupliakov O, Brodin L, Sihra TS, et al. 1995b. Impairment of synaptic vesicle clustering and of synaptic transmission, and increased seizure propensity, in synapsin I-deficient mice. *Proc. Natl. Acad. Sci. USA* 92:9235–39

Linden DJ. 1997. Long-term potentiation of glial synaptic currents in cerebellar culture. *Neuron* 18:983–94

Liu Y, Peter D, Roghani A, Schuldiner S, Prive GG, et al. 1992. A cDNA that suppresses MPP^{+} toxicity encodes a vesicular amine transporter. *Cell* 70:539–51

Lonart G, Schoch S, Käser PS, Larkin CJ, Südhof TC, Linden DJ. 2003. Phosphorylation of RIM1α by PKA triggers presynaptic long-term potentiation at cerebellar parallel fiber synapses. *Cell* 115:49–60

Lonart G. 2002. RIM1: an edge for presynaptic plasticity. *Trends Neurosci.* 25:329–32

Maycox PR, Deckwerth T, Hell JW, Jahn R. 1988. Glutamate uptake by brain synaptic vesicles. Energy dependence of transport and functional reconstitution in proteoliposomes. *J. Biol. Chem.* 263:15423–28

McIntire SL, Reimer RJ, Schuske K, Edwards RH, Jorgensen EM. 1997. Identification and characterization of the vesicular GABA transporter. *Nature* 389:870–76

McMahon HT, Missler M, Li C, Südhof TC. 1995. Complexins: cytosolic proteins that regulate SNAP receptor function. *Cell* 83:111–19

Meinrenken CJ, Borst JGG, Sakmann B. 2002. Calcium secretion coupling at calyx of Held governed by non-uniform channel-vesicle topography. *J. Neurosci.* 22:1648–67

Meinrenken CJ, Borst JG, Sakmann B. 2003. Local routes revisited: the space and time dependence of the Ca^{2+} signal for phasic transmitter release at the rat calyx of Held. *J. Physiol.* 547:665–89

Morales M, Colicos MA, Goda Y. 2000. Actin-dependent regulation of neurotransmitter release at central synapses. *Neuron* 27:539–50

Murthy VN, Stevens CF. 1999. Reversal of synaptic vesicle docking at central synapses. *Nat. Neurosci.* 2:503–7

Nicoll RA, Malenka RC. 1995. Contrasting properties of two forms of long-term potentiation in the hippocampus. *Nature* 377:115–18

Nielsen E, Christoforidis S, Uttenweiler-Joseph S, Miaczynska M, Dewitte F, et al. 2000. Rabenosyn-5, a novel Rab5 effector, is complexed with hVPS45 and recruited to endosomes through a FYVE finger domain. *J. Cell Biol.* 151:601–12

Oberhauser AF, Monck JR, Balch WE, Fernandez JM. 1992. Exocytotic fusion is activated by Rab3a peptides. *Nature* 360:270–73

Ohtsuka T, Takao-Rikitsu E, Inoue E, Inoue M, Takeuchi M, et al. 2002. Cast, a novel protein of the cytomatrix at the active zone of synapses that forms a ternary complex with RIM1. *J. Cell Biol.* 158:577–90

Perin MS, Fried VA, Mignery GA, Jahn R, Südhof TC. 1990. Phospholipid binding by a synaptic vesicle protein homologous to the regulatory region of protein kinase C. *Nature* 345:260–63

Perin MS, Fried VA, Stone DK, Xie X-S, Südhof TC. 1991a. Structure of the 116 kDa polypeptide of the clathrin-coated vesicle/synaptic vesicle proton pump. *J. Biol. Chem.* 266:3877–81

Perin MS, Brose N, Jahn R, Südhof TC. 1991b. Domain structure of synaptotagmin (p65). *J. Biol. Chem.* 266:623–29

Pieribone VA, Shupliakov O, Bordin L, Hilfiker-Rothenfluh S, Czernik AJ, Greengard P. 1995. Distinct pools of synaptic vesicles in neurotransmitter release. *Nature* 375:493–97

Piiper A, Stryjek-Kaminska D, Jahn R, Zeuzem S. 1995. Stimulation of inositol 1,4,5-trisphosphate production by peptides corresponding to the effector domain of different Rab3 isoforms and cross-linking of an effector domain peptide target. *Biochem. J.* 309:621–27

Pothos EN, Larsen KE, Krantz D, Liu Y-J, Haycock JW, et al. 2000. *J. Neurosci.* 20:7297–306

Pyle JL, Kavalali E, Piedras-Renteria ES, Tsien RW. 2000. Rapid reuse of readily releasable pool vesicles at hippocampal synapses. *Neuron* 28:221–31

Qian J, Saggau P. 1999. Modulation of transmitter release by action potential duration at the hippocampal CA3-CA1 synapse. *J. Neurophysiol.* 81:288–98

Regehr WG, Tank DW. 1991. The maintenance of LTP at hippocampal mossy fiber synapses is independent of sustained presynaptic calcium. *Neuron* 7:451–59

Reim K, Mansour M, Varoqueaux F, McMahon HT, Südhof TC, et al. 2001. Complexins regulate a late step in Ca^{2+} dependent neurotransmitter release. *Cell* 104:71–81

Richards DA, Guatimosim C, Betz WJ. 2000. Two endocytic recycling routes selectively fill two vesicle pools in frog motor nerve terminals. *Neuron* 27:551–59

Richards DA, Guatimosim C, Rizzoli SO, Betz WJ. 2003. Synaptic vesicle pools at the frog neuromuscular junction. *Neuron* 39:529–41

Richmond JE, Broadie KS. 2002. The synaptic vesicle cycle: exocytosis and endocytosis in Drosophila and C. elegans. *Curr. Opin. Neurobiol.* 12:499–507

Rizzoli SO, Betz WJ. 2002. Effects of 2-(4-morpholinyl)-8-phenyl-4H-1-benzopyran-4-one on synaptic vesicle cycling at the frog neuromuscular junction. *J. Neurosci.* 22:10680–89

Roghani A, Feldman J, Kohan SA, Shirzadi A, Gundersen CB, et al. 1994. Molecular cloning of a putative vesicular transporter for acetylcholine. *Proc. Natl. Acad. Sci. USA* 91:10620–24

Rosahl TW, Spillane D, Missler M, Herz J, Selig DK, et al. 1995. Essential functions of synapsins I and II in synaptic vesicle regulation. *Nature* 375:488–93

Rosahl TW, Geppert M, Spillane D, Herz J, Hammer RE, et al. 1993. Short term synaptic plasticity is altered in mice lacking synapsin I. *Cell* 75:661–70

Rosenmund C, Stevens CF. 1996. Definition of the readily releasable pool of vesicles at hippocampal synapses. *Neuron* 16:1197–201

Sabatini BL, Regehr WG. 1996. Timing of neurotransmission at fast synapses in the mammalian brain. *Nature* 384:170–72

Sagne C, El Mestikawy S, Isambert MF, Hamon M, Henry JP, et al. 1997. Cloning of a functional vesicular GABA and glycine transporter by screening of genome databases. *FEBS Lett.* 417:177–83

Sakaba T, Neher E. 2001. Quantitative relationship between transmitter release and calcium current at the calyx of held synapse. *J. Neurosci.* 21:462–76

Salin PA, Malenka RC, Nicoll RA. 1996. Cyclic AMP mediates a presynaptic form of LTP at cerebellar parallel fiber synapses. *Neuron* 16:797–803

Sätzler K, Söhl LF, Bollmann JH, Borst JGG, Frotscher M, et al. 2002. Three-dimensional reconstruction of a calyx of Held and its postsynaptic principal neuron in the medial nucleus of the trapezoid body. *J. Neurosci.* 22:10567–79

Scales SJ, Hesser BA, Masuda ES, Scheller RH. 2002. Amisyn, a novel syntaxin-binding protein that may regulate SNARE complex assembly. *J. Biol. Chem.* 277:28271–79

Schafer MK, Varoqui H, Defamie N, Weihe E, Erickson JD. 2002. Molecular cloning and functional identification of mouse vesicular glutamate transporter 3 and its expression in subsets of novel excitatory neurons. *J. Biol. Chem.* 277:50734–48

Schikorski T, Stevens CF. 2001. Morphological correlates of functionally defined synaptic vesicle populations. *Nat. Neurosci.* 4:391–95

Schivell AE, Batchelor RH, Bajjalieh SM. 1996. Isoform-specific, calcium-regulated interaction of the synaptic vesicle proteins SV2 and synaptotagmin. *J. Biol. Chem.* 271:27770–75

Schlüter OM, Schnell E, Verhage M, Tzonopoulos T, Nicoll RA, et al. 1999. Rabphilin knock-out mice reveal rat rabphilin is not required for rab3 function in regulating neurotransmitter release. *J. Neurosci.* 19:5834–46

Schlüter OM, Khvotchev M, Jahn R, Südhof TC. 2002. Localization versus function of Rab3 proteins: evidence for a common regulatory role in controlling fusion. *J. Biol. Chem.* 277:40919–29

Schneggenburger R, Neher E. 2000. Intracellular calcium dependence of transmitter release rates at a fast central synapse. *Nature* 406:889–93

Schneggenburger R, Sakaba T, Neher E. 2002. Vesicle pools and short-term synaptic depression: lessons from a large synapse. *Trends Neurosci.* 25:206–12

Schoch S, Deak F, Konigstorfer A, Mozhayeva M, Sara Y, et al. 2001. SNARE function analyzed in synaptobrevin/VAMP knockout mice. *Science* 294:1117–22

Schoch S, Castillo PE, Jo T, Mukherjee K, Geppert M, et al. 2002. RIM1α forms a protein scaffold for regulating neurotransmitter release at the active zone. *Nature* 415:321–26

Scranton TW, Iwata M, Carlson SS. 1993. The SV2 protein of synaptic vesicles is a keratan sulfate proteoglycan. *J. Neurochem.* 61:29–44

Serra-Pages C, Medley QG, Tang M, Hart A, Streuli M. 1998. Liprins, a family of LAR transmembrane protein-tyrosine phosphatase-interacting proteins. *J. Biol. Chem.* 273:15611–20

Shin OH, Rhee JS, Tang J, Sugita S, Rosenmund C, Südhof TC. 2003. Binding to the Ca^{2+} binding site of the synaptotagmin 1 C2B domain triggers fast exocytosis without stimulating SNARE interactions. *Neuron* 37:99–108

Shin O-H, Maximov A, Lim BK, Rizo J, Südhof TC. 2004. Unexpected Ca^{2+}-binding properties of synaptotagmin 9. *Proc. Natl. Acad. Sci. USA.* In press

Shirataki H, Kaibuchi K, Sakoda T, Kishida S, Yamaguchi T, et al. 1993. Rabphilin-3A, a putative target protein for smg p25A/rab3A small GTP-binding protein related to synaptotagmin. *Mol. Cell Biol.* 13:2061–68

Slepnev VI, De Camilli P. 2000. Accessory factors in clathrin-dependent synaptic vesicle endocytosis. *Nat. Rev. Neurosci.* 1:161–72

Söllner T, Whiteheart SW, Brunner M, Erdjument-Bromage H, Geromanos S, et al. 1993. SNAP receptors implicated in vesicle targeting and fusion. *Nature* 362:318–24

Sorensen JB, Fernandez-Chacon R, Südhof TC, Neher E. 2003. Examining Synaptotagmin 1 function in dense core vesicle exocytosis under direct control of Ca^{2+}. *J. Gen. Physiol.* 122:265–76

Spencer JP, Murphy KP. 2002. Activation of cyclic AMP-dependent protein kinase is required for long-term enhancement at corticostriatal synapses in rats. *Neurosci. Lett.* 329:217–21

Stadler H, Tsukita S. 1984. Synaptic vesicles contain an ATP-dependent proton pump and show 'knob-like' protrusions on their surface. *EMBO J.* 3:3333–37

Südhof TC. 1983. Evidence for a divalent cation dependent catecholamine storage complex in chromaffin granules. *Biochem. Biophys. Res. Comm.* 116:663–68

Südhof TC, Czernik AJ, Kao H, Takei K, Johnston PA, et al. 1989. Synapsins: mosaics of shared and individual domains in a family of synaptic vesicle phosphoproteins. *Science* 245:1474–80

Südhof TC. 2000. The synaptic vesicle cycle revisited. *Neuron* 28:317–20

Südhof TC. 1995. The synaptic vesicle cycle: a cascade of protein-protein interactions. *Nature* 375:645–53

Südhof TC. 2002. Synaptotagmins: Why so many? *J. Biol. Chem.* 277:7629–32

Sugita S, Hata Y, Südhof TC. 1996. Distinct Ca^{2+} dependent properties of the first and second C_2-domains of synaptotagmin I. *J. Biol. Chem.* 271:1262–65

Sugita S, Han W, Butz S, Liu X, Fernandez-Chacon R, et al. 2001. Synaptotagmin VII as a plasma membrane Ca^{2+} sensor in exocytosis. *Neuron* 30:459–73

Sugita S, Shin OH, Han W, Lao Y, Südhof TC. 2002. Synaptotagmins form a hierarchy of exocytotic Ca^{2+} sensors with distinct Ca^{2+} affinities. *EMBO J.* 21:270–80

Sun JY, Wu LG. 2001. Fast kinetics of exocytosis revealed by simultaneous measurements of presynaptic capacitance and postsynaptic currents at a central synapse. *Neuron* 30:171–82

Sun JY, Wu XS, Wu LG. 2002. Single and multiple vesicle fusion induce different rates of endocytosis at a central synapse. *Nature* 417:555–59

Takamori S, Rhee JS, Rosenmund C, Jahn R. 2000. Identification of a vesicular glutamate transporter that defines a glutamatergic phenotype in neurons. *Nature* 407:189–94

Takamori S, Malherbe P, Broger C, Jahn R. 2002. Molecular cloning and functional characterization of human vesicular glutamate transporter 3. *EMBO Rep.* 3:798–803

Taschenberger H, Leao RM, Rowland KC, Spirou GA, von Gersdorff H. 2002. Optimizing synaptic architecture and efficiency for high-frequency transmission. *Neuron* 36:1127–43

Thomas L, Hartung K, Langosch D, Rehm H, Bamberg E, et al. 1988. Identification of synaptophysin as a hexameric channel protein of the synaptic vesicle membrane. *Science* 242:1050–53

tom Dieck S, Sanmarti-Vila L, Langnaese K, Richter K, Kindler S, et al. 1998. Bassoon, a novel zinc-finger CAG/glutamine-repeat protein selectively localized at the active zone of presynaptic nerve terminals. *J. Cell Biol.* 142:499–509

Torri-Tarelli F, Villa A, Valtorta F, De Camilli P, Greengard P, Ceccarelli B. 1990. Redistribution of synaptophysin and synapsin I during α-latrotoxin-induced release of neurotransmitter at the neuromuscular junction. *J. Cell Biol.* 110:449–59

Trommershäuser J, Schneggenberger R, Zippelius A, Neher E. 2003. Heterogeneous presynaptic release probabilities: functional relevance for short-term plasticity. *Biophys. J.* 84:1563–79

Ullrich B, Li C, Zhang JZ, McMahon H, Anderson RGW, et al. 1994. Functional properties of multiple synaptotagmins in brain. *Neuron* 13:1281–91

Ubach J, Zhang X, Shao X, Südhof TC, Rizo J. 1998. Ca^{2+} binding to synaptotagmin: How many Ca^{2+} ions bind to the tip of a C_2-domain? *EMBO J.* 17:3921–30

Ubach J, Garcia J, Nittler MP, Südhof TC, Rizo J. 1999. Structure of the janus-faced C_2B domain of rabphilin. *Nature Cell Biol.* 1:106–12

van der Kloot W. 1991. The regulation of quantal size. *Prog. Neurobiol.* 36:93–130

Varoqui H, Diebler MF, Meunier FM, Rand JB, Usdin TB, et al. 1994. Cloning and expression of the vesamicol binding protein from the marine ray Torpedo. Homology with the putative vesicular acetylcholine transporter UNC-17 from Caenorhabditis elegans. *FEBS Lett.* 342:97–102

Verhage M, Maia AS, Plomp JJ, Brussaard AB, Heeroma JH, et al. 2000. Synaptic assembly of the brain in the absence of neurotransmitter secretion. *Science* 287:864–69

Voets T, Toonen RF, Brian EC, de Wit H, Moser T, et al. 2001. Munc18-1 promotes large dense-core vesicle docking. *Neuron* 31:581–91

von Gersdorff H, Borst JG. 2002. Short-term plasticity at the calyx of Held. *Nat. Rev. Neurosci.* 3:53–64

Wang Y, Sugita S, Südhof TC. 2000. The RIM/NIM family of neuronal C_2 domain proteins. Interactions with Rab3 and a new class of Src homology 3 domain proteins. *J. Biol. Chem.* 275:20033–44

Wang CT, Grishanin R, Earles CA, Chang PY, Martin TF, et al. 2001a. Synaptotagmin modulation of fusion pore kinetics in regulated exocytosis of dense-core vesicles. *Science* 294:1111–15

Wang X, Hu B, Zimmermann B, Kilimann MW. 2001b. Rim1 and rabphilin-3 bind Rab3-GTP by composite determinants partially related through N-terminal α-helix motifs. *J. Biol. Chem.* 276:32480–88

Wang CT, Lu JC, Bai J, Chang PY, Martin TF, et al. 2003. Different domains of synaptotagmin control the choice between kiss-and-run and full fusion. *Nature* 424:943–47

Wang Y, Liu X, Biederer T, Südhof TC. 2002. A family of RIM-binding proteins regulated by alternative splicing: implications for the genesis of synaptic active zones. *Proc. Natl. Acad. Sci. USA* 99:14464–69

Wang Y, Okamoto M, Schmitz F, Hofman K, Südhof TC. 1997a. RIM: a putative Rab3-effector in regulating synaptic vesicle fusion. *Nature* 388:593–98

Wang YM, Gainetdinov RR, Fumagalli F, Xu F, Jones SR, et al. 1997b. Knockout of the vesicular monoamine transporter 2 gene results in neonatal death and supersensitivity to cocaine and amphetamine. *Neuron* 19:1285–96

Wang Y, Südhof TC. 2003. Genomic definition of RIM proteins: evolutionary amplification of a family of synaptic regulatory proteins. *Genomics* 81:126–37

Washbourne P, Schiavo G, Montecucco C. 1995. Vesicle-associated membrane protein-2 (synaptobrevin-2) forms a complex with synaptophysin. *Biochem. J.* 305:721–24

Washbourne P, Thompson PM, Carta M, Costa ET, Mathews JR, et al. 2002. Genetic ablation of the t-SNARE SNAP-25 distinguishes mechanisms of neuroexocytosis. *Nat. Neurosci.* 5:19–26

Weisskopf MG, Castillo PE, Zalutsky RA, Nicoll RA. 1994. Mediation of hippocampal mossy fiber long-term potentiation by cyclic AMP. *Science* 265:1878–82

Wilson RI, Kunos G, Nicoll RA. 2001. Presynaptic specificity of endocannabinoid signaling in the hippocampus. *Neuron* 31:453–62

Wölfel M, Schneggenburger R. 2003. Presynaptic capacitance measurements and Ca^{2+} uncaging reveal submillisecond exocytosis kinetics and characterize the Ca^{2+} sensitivity of vesicle pool depletion at a fast CNS synapse. *J. Neurosci.* 23:7059–68

Wucherpfennig T, Wilsch-Brauninger M, Gonzalez-Gaitan M. 2003. Role of Drosophila Rab5 during endosomal trafficking at the synapse and evoked neurotransmitter release. *J. Cell Biol.* 161:609–24

Xu-Friedman MA, Harris KM, Regehr WG. 2001. Three-dimensional comparison of ultrastructural characteristics at depressing and facilitating synapses onto cerebellar Purkinje cells. *J. Neurosci.* 21:6666–72

Xu T, Bajjalieh SM. 2001. SV2 modulates the size of the readily releasable pool of secretory vesicles. *Nat. Cell Biol.* 3:691–98

Yamaguchi T, Dulubova I, Min SW, Chen X, Rizo J, Südhof TC. 2002. Sly1 binds to Golgi and ER syntaxins via a conserved N-terminal peptide motif. *Develop. Cell* 2:295–305

Yamashita T, Ishikawa T, Takahashi T. 2003. Developmental increase in vesicular glutamate content does not cause saturation of AMPA receptors at the calyx of held synapse. *J. Neurosci.* 23:3633–38

Zhang W, Benson DL. 2001. Stages of synapse development defined by dependence on F-actin. *J. Neurosci.* 21:5169–81

Zimmerberg J, Curran M, Cohen FS, Brodwick M. 1987. Simultaneous electrical and optical measurements show that membrane fusion precedes secretory granule swelling during exocytosis of beige mouse mast cells. *Proc. Natl. Acad. Sci. USA* 84:1585–89

Zucker RS, Regehr WG. 2002. Short-term synaptic plasticity. *Annu. Rev. Physiol.* 64: 355–405

Annu. Rev. Neurosci. 2004. 27:549–79
doi: 10.1146/annurev.neuro.27.070203.144327

CRITICAL PERIOD REGULATION

Takao K. Hensch
Laboratory for Neuronal Circuit Development, Critical Period Mechanisms Research Group, RIKEN Brain Science Institute, 2-1 Hirosawa, Wako-shi, Saitama 351-0198, Japan; email: hensch@riken.jp

Key Words NMDA, GABA, experience, environment, attention

■ **Abstract** Neuronal circuits are shaped by experience during critical periods of early postnatal life. The ability to control the timing, duration, and closure of these heightened levels of brain plasticity has recently become experimentally accessible, especially in the developing visual system. This review summarizes our current understanding of known critical periods across several systems and species. It delineates a number of emerging principles: functional competition between inputs, role for electrical activity, structural consolidation, regulation by experience (not simply age), special role for inhibition in the CNS, potent influence of attention and motivation, unique timing and duration, as well as use of distinct molecular mechanisms across brain regions and the potential for reactivation in adulthood. A deeper understanding of critical periods will open new avenues to "nurture the brain"—from international efforts to link brain science and education to improving recovery from injury and devising new strategies for therapy and lifelong learning.

INTRODUCTION

Critical periods in development have been recognized by biologists for nearly a century. Chemicals applied at particular times to a developing embryo produce specific malformations, with the most rapidly growing tissues being most sensitive to the change in conditions (Stockard 1921). Stimulated by the external world, the postnatal nervous system responds further to natural sensory experience. Time windows exist when brain circuits that subserve a given function are particularly receptive to acquiring certain kinds of information or even need that instructive signal for their continued normal development. Beginning with the behavioral observations of Konrad Lorenz (1958), this concept has profoundly influenced not only biologists but also psychologists such as Freud, philosophers, physicians, policy makers, parents, and educators. The mechanisms, power, and inherent hazards presented by these special phases in brain development carry a social impact far beyond basic neuroscience.

Although critical periods have been well documented for sensory systems, one concern is that the concept is being used too broadly for too many types of learning, without rigorous demonstration that such windows exist. The purpose of this review is to identify key principles emerging from the study and regulation of known critical periods (Table 1). It should be stated at the outset that transiently heightened levels of brain plasticity after birth do not preclude the possibility for lifelong learning. A critical period is an extreme form of a more general sensitivity, when neuronal properties are particularly susceptible to modification by experience. Through a deeper understanding of these exceptional developmental stages, we hope to gain insight into the process, as well as to design better strategies by which the older brain can learn.

This review presents evidence from each sensory modality, as well as higher order, multimodal brain systems (Table 1). Although, to date, the depth of study in each area varies widely, the seeds of a general theory are evident. Nine facets of critical periods can now be distinguished and guide the discussions to follow:

First is the functional competition between inputs. Genetic specification admirably determines much of the basic structure and function of the nervous system. But, the environment and physical characteristics of the individual into which the brain is born cannot be encoded in the genome. A process by which neurons select their permanent repertoire of inputs (or maps) from a wider array of possibilities is required for proper brain function. Indeed, the tailoring of neuronal circuits custom fitted to each individual is the main purpose of critical periods.

Second is the particular role for electrical activity. Neurons communicate by the transmission of nerve impulses as a reflection of external stimuli or spontaneous, internal states. The various inputs from which the nervous system can choose during the critical period are ultimately encoded in the discharge of action potentials. Most cellular models of plasticity are now based on the ability to potentiate or depress transmission at individual synapses through their pattern of activation. Whether competing spike trains actually instruct who the "winner" shall be, or merely "permit" other processes to adjudicate, is a fundamental question that must be addressed for each system.

Third is a structural consolidation of selected pathways. Early experience specifies a neural commitment to one of a number of possible patterns of connectivity. The magnitude and permanence of anatomical changes—from dendritic spine motility to large-scale rewiring—distinguish developmental plasticity from adult learning. A critical period may be defined in systems where structural modification becomes essentially irreversible beyond a certain age. Continued growth maintains sensitivity to environmental influence throughout life.

Fourth is the regulation of critical period onset and duration not simply by age, but rather by experience. If appropriate neural activation is not provided at all, then developing circuits remain in a waiting state until such input is available. Alternatively, enriched environments may prolong plasticity. In other words, the critical period is itself use-dependent. Understanding the cellular mechanism of this effect will greatly influence strategies for lifelong learning.

Fifth is the unique timing and duration of critical periods across systems. Not all brain regions develop with the same time course. There are both rostro-caudal

TABLE 1 Known critical periods and molecular mechanisms across systems (see text for references)

System	Age	Confirmed regulators	Delay[a]	Species[b]
Neuromuscular junction	<P12	ACh	+	mouse
Climbing fibers (CBL)	P15–16	NMDA, mGluR1, G_q, PLCβ, PKCγ	nd	mouse
LGN layers	<P10	Retinal ACh, cAMP; MAO-A, NO, MHC-I, CREB	nd	mouse, ferret, cat
Ocular dominance	P3 weeks-months	GABA, NMDA, PKA, ERK, CaMKII, CREB, BDNF, tPA, protein synthesis, NE, ACh	+	cat, rat, mouse, ferret
Orientation bias	<P28	NR1, NR2A, PSD95	+	cat, mouse
Whisker-barrel map formation	<P7	NR1, MAOA, $5HT_{1B}$, cAMP mGluR5, PLCβ, FGF8	nd	mouse
Whisker RF tuning	P14–16		nd	rat
Tonotopic map (cortex)	P16–50	ACh	+	rat
Absolute pitch	<7 years		nd	human
Taste/olfaction	none	GABA, mGluR2, NO, neurogenesis	+	mouse
Imprinting	14–42 hrs	Catecholamines	+	chick
Stress/anxiety	<P21	Hormones, $5HT_{1A}$	nd	rat, mouse
Slow-wave sleep	P40–60	NMDA	+	cat, mouse
Sound localization	<P200	GABA, NMDA	+	barn owl
Birdsong	<P100	GABA, hormones, neurogenesis	+	zebrafinch
Human language	0–12 years		nd	human

[a]Potential for critical period delay by altered experience. +, yes; nd, not determined.

[b]Primary species for elucidation of molecular mechanism.

gradients of maturation across modalities and hierarchical levels of processing within a given pathway. Intuitively, the critical period for one stage cannot begin unless its input from a preceding stage is ready. Cascades of critical periods and their cumulative sequence at different ages and levels of processing shape each brain function as the relevant neural pathways develop to a point where they can support plasticity.

Sixth is a diversity of molecular mechanisms across systems or even at various stages along the same pathway. Simply being regulated by neuronal activity in the neonatal brain, or contributing to popular plasticity models does not automatically establish a molecule's role in the critical period. The detailed mediators of plasticity will vary for individual connections and impede the search for canonical plasticity factors (Table 1). Moreover, the initial formation of maps during ontogeny may use molecules differently from the neuronal remodeling of established circuits.

Seventh is a particular role for inhibition in the central nervous system (CNS). Apart from certain rare instances where competing inputs directly impinge upon the same identifiable target cell, most neuronal circuits in the brain are intricately interconnected. In these tangled forests, inhibitory interneurons are rapidly emerging as a vital arbiter of neuronal plasticity (Hensch & Fagiolini 2003). Indeed, they may generally contribute to the onset, offset, or expression of critical periods throughout the brain.

Eighth is the potent influence of attention and motivation. As any teacher can attest, attention in a classroom setting is perhaps the most critical determinant of whether learning succeeds; so too at the neuronal level, where arousal state is translated into the level of aminergic and cholinergic transmission. These modulatory systems may be more actively engaged in the infant brain, allowing the seemingly effortless plasticity not seen in adults. Rekindling attentional mechanisms offers a key to critical period regulation.

Finally, it is the potential for reactivation in adulthood that confirms the very existence of critical periods. Depending on whether neuronal growth is rigidly limited to a critical period, different strategies for therapy, recovery from injury, and continuing education will apply to distinct brain regions.

PRIMARY MODALITIES

Motor Systems

Both in the periphery and in the brain, appropriate critical period development is prerequisite to proper motor control and coordinated movement later in life. Competition among multiple motor axons for a single target muscle fiber eliminates synapses at the neonatal neuromuscular junction (Sanes & Lichtman 1999a). The accessibility of this classic preparation makes it the prototypical synaptic model of critical period plasticity in the nervous system. Direct visualization of the interaction and removal of supernumerary motor axons during the first two postnatal weeks in rodents outlines a progression of synaptic events from reinforcing functional efficacy to eventually consolidating its structure (Lichtman & Colman 2000, Walsh & Lichtman 2003). The period of refinement is slowed or accelerated by the chronic blockade or enhancement of neuromuscular activity, respectively (see Thompson 1985).

In contrast, junctions initially form in the total absence of neurotransmission, as seen in mice lacking the acetylcholine-synthesizing enzyme, choline acetyltransferase (ChAT) (Brandon et al. 2003). Conditional deletion of ChAT in a small subset of axons elegantly demonstrates that better excitation of the target muscle fiber biases the competition in favor of the genetically enhanced inputs (Buffelli et al. 2003). All synapses at which branches of the same two axons compete proceed to the same outcome (Kasthuri & Lichtman 2003). More extensively branched motor units are at a competitive disadvantage, as their larger size dilutes the limited resources at individual terminals. Moreover, a similar stage of synapse elimination is reached concurrently (with the same "winner" at each site), indicating that the pace of rearrangement is highly stereotyped once initiated.

Competition thus occurs globally rather than locally, driven by presynaptic activity that is directly adjudicated by the postsynaptic muscle fiber. In the CNS, it is difficult to similarly isolate inputs onto individual target cells. Climbing fiber axons from the brainstem inferior olivary nucleus terminating onto cerebellar Purkinje cell bodies are one rare example. Multiple climbing fibers present at birth are pruned to a powerful one-to-one relationship over the first few weeks of life (Crepel 1982). As for the neuromuscular junction (Lichtman & Colman 2000), a growing disparity in potency of synaptic excitation precedes eventual elimination (Hashimoto & Kano 2003), with the postsynaptic cell deciding which input to retain.

Blockade of the *N*-methyl-D-aspartate (NMDA)-type glutamate receptor defines a remarkably sharp two-day (P15–16 in mice) critical period for climbing fiber refinement (Kakizawa et al. 2001). This is curious because the Purkinje cell itself is devoid of functional NMDA receptors (Farrant & Cull-Candy 1991). It is likely that NMDA receptor–mediated mossy fiber excitation of underlying granule cells and their parallel fibers provides a permissive level of tonic activation to Purkinje cells that is required for the process. Subsequently, a molecular cascade including type 1 metabotropic glutamate receptor (mGluR) activation coupled to $G_{q\alpha}$-type G proteins, phospholipase Cβ (PLCβ), and the γ-isoform of protein kinase C (PKCγ) is recruited, as revealed by the systematic analysis of poly-innervation by climbing fibers in gene-targeted mice (Hashimoto et al. 2000, Ichise et al. 2000, Kano et al. 1995).

Finally at the neocortical level, refinement is influenced by appropriate sensory feedback during early life. Trimming the whiskers on a rat's snout from birth (but not as an adult) produces a significantly smaller, contralateral motor area that evokes abnormal patterns of movement (Huntley 1997). The basis for such crossmodal effects is likely to be complex.

Visual System I: Retino-Geniculate Connections

The visual system has long been favored for developmental study because there are only two discrete inputs into the system. Proper binocular fusion and stereoscopic vision depend on the correct processing of information from the two eyes (Daw 1995, Wiesel 1982). Multiple critical periods have been defined based on the timing when activity deprivation is effective in disrupting binocular representations along the pathway. There is a logical sequence of critical periods, ending earlier for functions dealt with at lower levels of the system.

The convergence of right- and left-eye input begins in the dorsal thalamus of mammals. Rather than crossing the optic chiasm, ventro-temporal ganglion cell axons are directed ipsilaterally by early expression of the Zic2 transcription factor in the retina (Herrera et al. 2003). Initially, overlapping nasal axons from the opposite eye gradually segregate within the lateral geniculate nucleus (LGN), resulting in the formation of layers or eye-specific domains (Wong 1999). Spontaneous neuronal discharge underlies this process, which occurs well before eye-opening and visual experience. In particular, rhythmic bursts of synchronized activity propagate across the neonatal retina of ferrets and rodents. Blockade of these waves driven initially by cholinergic amacrine cells—either by antagonist injection or targeted disruption of nicotinic receptor subunits—prevents lamina formation and defines the first critical period in the visual stream (Penn et al. 1998, Rossi et al. 2001). Despite the absence of layers, patchy segregation still occurs (Huberman et al. 2002, Muir-Robinson et al. 2002), demonstrating a distinct process that may reflect additional mechanisms such as later glutamatergic waves in the retina (Wong et al. 2000).

The segregation of input is clearly competitive and instructed by retinal activity, as seen when binocular innervation is forced in the tectum (Constantine-Paton & Law 1978, Ruthazer et al. 2003). Increasing the frequency of retinal waves in one eye by elevating cyclic AMP expands that eye's representation in the LGN at the expense of the normally active input (Stellwagen & Shatz 2002). At these monosynaptic connections from ganglion cells, it is reasonable to expect mechanisms of synaptic long-term potentiation (LTP) to be directly engaged (Sanes & Lichtman 1999b). Signaling by major histocompatibility complex (MHC)–related molecules is important (Huh et al. 2000). These may interact with LTP mechanisms in more complex ways, such as retrograde signaling or stimulus frequency dependence (Boulanger et al. 2001), or through unrelated processes, such as neurite outgrowth and adhesion, that remain to be clarified.

The cyclic AMP response element binding protein (CREB) is briefly upregulated in the LGN during the critical period for layer formation in mice. Gene-targeted disruption of CREB impairs segregation (Pham et al. 2001). In this context, monoamines play an interesting transient role in the visual thalamus unrelated to their later function in arousal. Excess serotonin (5-HT) in monoamine oxidase (MAO-A) knockout mice prevents segregation (Upton et al. 1999). Postysnaptic CREB levels could then be adjusted indirectly through 5-HT receptors.

After the eyes open, another round of synapse elimination reduces down to a powerful few multiple ganglion cell contacts onto each thalamic relay cell (Chen & Regehr 2000), similar to the neuromuscular junction or climbing fiber system. This pruning can contribute to the establishment of physiological properties such as ON- and OFF-center responses (Wong 1999). The LGN sublaminae into which these cells are typically sorted also fail to form when retinal activity is blocked earlier. Here, NMDA receptors play a role (Hahm et al. 1991), as does nitric oxide (Cramer et al. 1996).

Visual System II: Cortex

Over forty years ago, Wiesel & Hubel (1963) first described the loss of responsiveness to an eye deprived of vision in the primary visual cortex of kittens, providing the premier physiological model of critical period plasticity. Rapid functional effects of monocular deprivation (MD) are soon accompanied by anatomical rewiring of horizontal connections and thalamic afferents (Antonini & Stryker 1993, Trachtenberg & Stryker 2001). Altogether, these processes follow competitive interactions between the two eyes for the control of cortical territory (Daw 1995, Wiesel 1982). Although the present review focuses on plasticity due to abnormal vision during the critical period, note that the synaptic rearrangement underlying the initial formation of ocular dominance columns has recently been proposed to be genetically predetermined (Crowley & Katz 2002). To date, no such eye-specific molecules have been found. Instead shadows cast by individual retinal blood vessels or early manipulation of intracortical spread of activity are reflected in columnar architecture consistent with a competitive segregation process (Adams & Horton 2002, Hensch & Stryker 2004).

If both eyes are sutured during the critical period, no imbalance of input occurs, and the relative ability to drive visual responses is unchanged. Conversely, strabismus causes the two eyes never to see the same visual field, leading to an instructive decorrelation of retinal activity and loss of binocular responses. When the cortical target is silenced by inhibitory $GABA_A$ receptor agonists or blockade of excitatory NMDA receptors (Bear et al. 1990, Hata & Stryker 1994, Hata et al. 1999), the more active afferents serving the open eye are paradoxically instructed to retract, allowing the better-matched, deprived-eye connections to remain.

As a direct consequence of shifts in cortical ocular dominance, the weakened input becomes amblyopic: Visual acuity is strongly reduced and contrast sensitivity blunted even when no physical damage to the retina exists (Daw 1995, 1998; Dews & Wiesel 1970; Maurer et al. 1999). Importantly, the loss of behavioral visual acuity occurs only during a transient developmental critical period reflecting that measured by single-unit electrophysiology (Hubel & Wiesel 1970, Prusky & Douglas 2003). The rules of activity-dependent competition and timing have been confirmed across a variety of species (Berardi et al. 2000, Gordon & Stryker 1996). Interestingly, critical period duration is tightly correlated with average life expectancy. In all cases, plasticity gradually peters out rather than ceasing abruptly.

In rodents and cats, plasticity is low at eye opening, peaks around four weeks of age, and declines over several weeks to months (Daw 1995, Fagiolini et al. 1994, Gordon & Stryker 1996, Hubel & Wiesel 1970). In humans, amblyopia is set by the age of eight (see Daw 1995). Notably, the critical period is not a simple, age-dependent maturational process but is rather a series of events itself controlled in a use-dependent manner. Animals reared in complete darkness from birth express a delayed onset profile with plasticity persisting into adulthood (Fagiolini et al. 1994, Iwai et al. 2003, Mower 1991).

It is attractive to think of the loss of deprived-eye input as a long-term depression (LTD) or gain of open-eye input as LTP (see Heynen et al. 2003). But, manipulations based on advancing knowledge of their molecular mechanism have frustratingly failed to influence plasticity in vivo (Daw 2003, Bartoletti et al. 2002, Hensch 2003, Renger et al. 2002, Shimegi et al. 2003). For instance, endogenous brain-derived neurotrophic factor (BDNF) prevents LTD in the visual cortex (Jiang et al. 2003) but does not block the loss of deprived-eye input in transgenic mice overexpressing it (Huang et al. 1999). Conversely, early LTP and LTD that remain in the presence of protein synthesis inhibitors are inadequate to sustain shifts of ocular dominance in vivo (Frey et al. 1993, Taha & Stryker 2002). Such mechanistic dissociations between plasticity in vitro and in vivo have also been reported for hippocampal learning (Martin et al. 2000, Sanes & Lichtman 1999b). A role for LTP/LTD models is obviously not ruled out but rather placed at a secondary stage in the critical period process.

Excessive emphasis on homosynaptic mechanism provides incomplete insight into the competitive nature of ocular dominance plasticity (Miller 1996) and can be misleading for two reasons. First, it neglects the anatomical consequences of MD. Second, unlike motor axons competing for a single target muscle fiber, sensory input to the neocortex must be integrated by complex local circuit interactions in vivo. By treating the visual cortex as a monosynaptic connection from the eyes, one loses sight of its physiological function, namely vision. Although measures of subthreshold synaptic activity are residually sensitive to sensory manipulation in older animals (Sawtell et al. 2003), it is the ability to fire cortical action potentials through either eye that accurately reflects the visual capabilities of the system and defines the critical period (Daw 1995, Dews & Wiesel 1970, Prusky & Douglas 2003).

An unbiased perspective on intrinsic local circuit behavior has proven more fruitful (Hensch & Fagiolini 2003). Molecular cascades set in motion by a unique excitatory-inhibitory balance may lead to a structural consolidation that eventually terminates the critical period. Pharmacological attempts to disrupt the balance grossly hyperexcite or shut down the cortex (Hata & Stryker 1994, Ramoa et al. 1988, Shaw & Cynader 1984, Videen et al. 1986), yielding little insight into the normal function of local circuits during plasticity. Instead, by taking advantage of gene-targeting technology, gentle titration of endogenous GABA release or subtle prolongation of glutamatergic currents yields a similar shift of balance in favor of excitation that disrupts ocular dominance plasticity in the same way.

Mice carrying a deletion of the 65-kDa isoform of glutamic acid decarboxylase (GAD65), found primarily in inhibitory terminals (Soghomonian & Martin 1998), exhibit a significant reduction of stimulated GABA release and show no shift in responsiveness toward the open eye following brief MD (Hensch et al. 1998). Accentuated excitation, by preventing the natural developmental switch of NMDA receptor subunit composition, also weakens the response to MD (Fagiolini et al. 2003). Composed of a principal NR1 subunit and distinct modulatory NR2 partners, NMDA current decay is truncated by the insertion of NR2A subunits after eye opening (Nase et al. 1999). Synaptic NMDA responses remain prolonged in the absence of NR2A, yielding increased charge transfer (Fagiolini et al. 2003). Nevertheless, the critical period ends normally, contrary to expectation from an LTP view that predicts greater plasticity when NMDA receptor function is enhanced (Fox 1995, Tang et al. 1999).

Restoration of plasticity to both GAD65 and NR2A knockout mice by acute infusion of benzodiazepine agonists demonstrates a decisive role for excitatory-inhibitory balance (Fagiolini et al. 2003, Hensch et al. 1998). Drugs like diazepam selectively increase the open probability and channel conductance of a limited subset of $GABA_A$ receptors in a use-dependent manner, since they are inert in the absence of synaptic GABA release (Cherubini & Conti 2001, Sieghart 1995). Benzodiazepine binding sites are not associated with thalamocortical axons or other subcortical inputs (Shaw et al. 1987), making detailed local circuit analysis possible (Fagiolini et al. 2004).

A competitive outcome of MD can readily be understood by strategically placed inhibition. Specific spike timing–dependent windows for synaptic plasticity rely upon physiologically realistic, millisecond-scale changes in the temporal order of pre- and postsynaptic action potentials (Bi & Poo 2001, Froemke & Dan 2002). Inhibitory regulation of spike-timing could then instruct the direction of plasticity (Song et al. 2000). In contrast, classical models of LTP induced by changes in mean firing rate are indiscriminately blocked by benzodiazepines (Trepel & Racine 2000). Among the vast diversity of GABAergic interneurons in neocortex, parvalbumin-containing cells target the axon initial segment and soma (DeFelipe 1997, Somogyi et al. 1998), where they can control spike initiation (Chandelier cells) or back-propagation (basket cells), respectively, required for synaptic plasticity in the dendritic arbor.

Maturation of parvalbumin-positive interneurons parallels critical period onset (Del Rio et al. 1994, Gao et al. 2000), and when accelerated by transgenic overexpression, BDNF shifts the critical period earlier in time (Huang et al. 1999). Large basket cells, in particular, extend a wide, horizontal axonal arbor that can span ocular dominance columns in cat visual cortex (Buzas et al. 2001). Moreover, these electrically coupled networks of fast-spiking cells offer a system exquisitely sensitive to timing (Connors 2004, in this volume; Galaretta & Hestrin 2001). Only $GABA_A$ receptors containing the $\alpha 1$ subunit drive visual cortical plasticity and are preferentially enriched at somatic synapses opposite parvalbumin-positive large basket cell terminals (Fagiolini et al. 2004, Klausberger et al. 2002). Taken together, specific inhibitory circuits may be ideally suited to detect and discriminate synchronized signals coming from the eyes.

Excitatory-inhibitory balance determines the neural coding of sensory input and tightly regulates prolonged spike discharge in both GAD65 and NR2A mutants. In the former, this is observed throughout life, and the critical period awaits diazepam treatment even in adulthood (Fagiolini & Hensch 2000). Dark-rearing from birth also impedes the normal maturation of inhibition and naturally delays critical period onset (Morales et al. 2002, Mower 1991), which can be prevented by brief diazepam infusion in the dark (Iwai et al. 2003). Conversely, premature ocular dominance shifts can be triggered in wild-type animals by diazepam as early as eye opening (Fagiolini & Hensch 2000, Fagiolini et al. 2004). Thus, critical period machinery lies dormant until set in motion by proper excitatory-inhibitory levels. Homeostatic scaling of synaptic input to maintain this balance observes a critical period by cortical layer in visual cortex, potentially contributing to ocular dominance plasticity in supragranular layers (Desai et al. 2002).

To fully saturate plasticity requires several days of experience (Gordon & Stryker 1996), whereas it is triggered by less than 48 h of diazepam treatment (Iwai et al. 2003). Accordingly, a cascade downstream of excitatory-inhibitory balance leading toward protracted structural consolidation is gradually being elucidated (Figure 1). Both protein synthesis and extracellular proteolysis via the tissue-type plasminogen activator (tPA)-plasmin axis are required for even brief MD to be effective (Mataga et al. 2002, Taha & Stryker 2002). These are well-positioned downstream of NMDA and $GABA_A$ receptors, calcium-calmodulin-dependent protein kinase II (CaMKII), protein kinase A (PKA), extracellular-regulated protein kinase (ERK), and CREB, which have all been found to affect ocular dominance shifts measured electrophysiologically (reviewed in Berardi et al. 2003). Ultimately, deprivation produces an age-limited increase in dendritic spine motility then rearrangement of thalamo-cortical axons (Antonini & Stryker 1993, Majewska & Sur 2003).

A role for neurotrophic factors may be found at multiple stages (Figure 1). Whereas gene-regulated overexpression promotes critical period onset (Huang et al. 1999), exogenous infusion of BDNF blocks both anatomical and physiological plasticity of excitatory connections (Hata et al. 2000, Jiang et al. 2003, Riddle et al. 1995). BDNF may first establish the proper milieu for plasticity by promoting the maturation of GABA circuits, then later participate directly in the plasticity process of neurite outgrowth or survival (Huang & Reichardt 2001, Berardi et al. 2003). For instance, an increase of proteolytic activity by tPA in visual cortex observed after a few days of MD may be triggered by BDNF in cortical neurons (Fiumelli et al. 1999, Mataga et al. 2002).

Similarly, the multifaceted actions of PKA upon both excitatory and inhibitory transmission must be considered carefully before interpreting its role in ocular dominance plasticity (Beaver et al. 2001, Heynen et al. 2003, McDonald et al. 1998). Moreover, behavioral state potently influences plasticity and can tap into the cAMP cascade (Figure 1). Brief MD produces ocular dominance shifts within

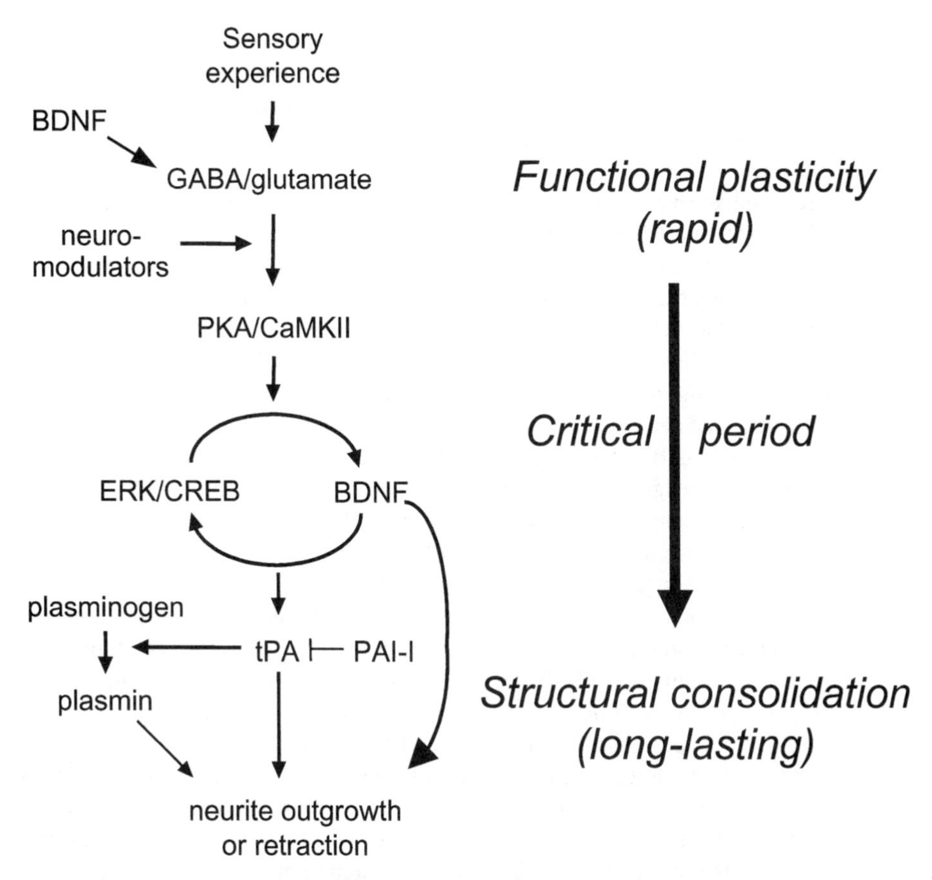

Figure 1 Identified molecular regulators underlying critical period plasticity in the primary visual cortex. A cascade of events, leading from functional excitatory-inhibitory balance to structural change, is delineated. See text for references.

two days in kittens (Mower 1991) but fails to do so under anesthesia (Imamura & Kasamatsu 1991). Conversely, interposed slow-wave sleep enhances the pace of plasticity (Frank et al. 2001). Neuromodulation lies in diffuse projections of norepinephrine and acetylcholine, whose pharmacological depletion prevents plasticity during the critical period (Bear & Singer 1986) and activation partially restores it to adults (Imamura et al. 1999, but see Beaver et al. 2001).

Sensitivity to sensory experience may ultimately disappear as the ability of neurites to navigate the extracellular matrix (ECM) is reduced. A decline in dendritic spine motility in vivo mirrors the end of the critical period in visual cortex (Grutzendler et al. 2002). Active growth inhibition by CNS myelination (Schwab & Caroni 1988) is also age-dependent in cortex (Daw 1995, Schoop et al. 1997). Conversely, injection of immature astrocytes rejuvenates adult cat visual cortex (Muller & Best 1989), perhaps by providing a permissive substrate or otherwise creating a favorable environment for plasticity through releasable nerve growth factors (Gu et al. 1994). Taken together, such evidence indicates that the refractoriness of mature neuronal substrates to change may be "consolidated" in the extracellular milieu.

To reactivate plasticity in the visual cortex would then require drastic disruption of ECM structure. Chondroitin sulfate proteoglycans are key components that condense around neuronal somata and dendrites in the form of perineuronal nets near the end of the critical period (Celio et al. 1998). Degradation of these structures by repeated local chondroitinase injection reveals shifts of responsiveness toward the open eye following MD in adult rats (Pizzorusso et al. 2002). Intriguingly, mature, fast-spiking, parvalbumin-positive neurons are predominantly encapsulated by perineuronal nets (Hartig et al. 1999). To reopen the critical period, a resetting of its original GABAergic trigger may be needed. Cortical or bilateral retinal lesions by photocoagulation also produce enlargement and shifts of receptive field location in the adult primary visual cortex (Gilbert 1998). Rapid disruption of local excitatory-inhibitory balance may be followed by protracted outgrowth of horizontal intracortical connections over months.

Various response properties apart from ocular dominance emerge along the visual stream, and each has its own critical period. For instance, orientation and direction selectivity precede that of ocular dominance (see Daw 1995). In general, a property processed at higher levels of the system has a critical period that lasts longer than one processed at a lower level. Perceptual learning in the adult visual system can be highly specific for certain stimulus configurations (Karni & Sagi 1991), predicting changes in the responses of neurons in primary visual cortex (Gilbert 1998). However, neurons early in the processing stream fail to exhibit strong effects (Ghose et al. 2002). Extensive training on an orientation discrimination task does not alter selectivity of neurons in V1 or V2, but rather is mirrored by responses at later stages (V4) of the macaque visual cortex (Yang & Maunsell 2004). Consistent with this plasticity hierarchy, neurons in inferotemporal cortex at the highest level of the ventral stream may be substantially influenced into adult life by visual experience (see Kobatake et al. 1998). Although V1 comprises a stable representation in adults, plasticity may be progressively less constrained at subsequent cortical levels.

Finally, molecular mechanisms underlying experience-dependent plasticity of individual response properties need not overlap. Whereas the NR2A subunit and its anchoring postsynaptic density protein (PSD95) are dispensable for ocular dominance, they are essential for orientation preference to develop (Fagiolini et al. 2003). Targeted disruption of tPA or GAD65 has no influence upon the competitive segregation of retino-geniculate axons at earlier stages (Y. Tsuchimoto, A. Rebsam, S. Fujishima, M. Fagiolini, T.K. Hensch, submitted), and nitric oxide is unimportant for later ocular dominance plasticity in neocortex (Ruthazer et al. 1996). Moreover, simply because a molecule is regulated by activity does not mean it is required for plasticity, as shown for the immediate early gene *zif268* in primary visual cortex (Mataga et al. 2001). Molecules first identified by subtractive screens in one area will need to be verified individually in other regions by restricted conditional manipulation.

Primary Auditory System

Lasting effects of early auditory experience have long been suspected (see Human Language, below), but their critical periods have only recently been defined in the primary auditory pathway (see Sound Localization). Synapse elimination during the first week of rodent life in the brainstem lateral superior olive results in a two-fold sharpening of functional topography when glycine/GABA synapses are still excitatory (Kim & Kandler 2003). A massive tonotopic map refinement is manifest later in auditory cortex between P16 and P50 in two ways (Chang & Merzenich 2003). Preference for high-frequency tones is converted into a more balanced representation from low to high values, while total cortical surface area contracts. Selective exposure to a particular frequency during this time yields a competitive overrepresentation in the final map. When no particular tone stands out in a uniformly noisy environment, then the critical period is delayed and rapidly appears even in adulthood once the noise is removed.

The functional consequence of intense auditory training is seen in the brains of musicians who are exposed to music before the age of seven. A significantly greater left hemisphere asymmetry of planum temporale activation arises in those who develop absolute pitch (Schlaug et al. 1995); an increased cortical representation of piano tones in highly skilled musicians also exists that is not seen for pure tones (Pantev et al. 1998). Similar enlargement of the cortical representation of the fingers of the left hand (except thumb) in string players, anterior corpus callosum, and cerebellar activation further reflects their unique dexterity (Elbert et al. 1995, Schlaug et al. 1995). Age of commencement of musical training is important to develop these changes.

As in visual cortex, an unexpected late developmental peak of GABA neurons (P60 in ferrets) is observed that is higher overall in auditory cortex. In particular, the proportion of parvalbumin-containing neurons remains immature until late (Gao et al. 2000), which may provide an important substrate for critical period plasticity. Large-scale changes in primary auditory cortex can be rekindled in the adult rat when tones are paired with electrical stimulation of the basal forebrain (Kilgard & Merzenich 1998). Acetylcholine release within cortex increases the salience of concomitantly presented stimuli, yielding map refinements that rival those occurring passively during the critical period (Chang & Merzenich 2003). Tapping these attentional mechanisms may improve function more effectively when treating learning disorders or injury with therapy or cochlear implants (Merzenich et al. 1996).

Somatosensory System

Major topographical reorganization persists in the adult somatosensory system. Consistent with the behavioral improvement following perceptual learning in primates, a trained skin region can activate a cortical area up to three times larger than untrained fingers (Recanzone et al. 1992). Both the thalamus and cortex are extensively reshaped by somatosensory deafferentation (Jones 2000), in contrast to retinal lesions that produce few changes in the LGN (Gilbert 1998). Sensory stages that first embody a response property are more likely to observe a critical period. Tactile stimuli are represented selectively at peripheral sensory receptors, whereas orientation or binocular tuning is first found in cortex. Hence, even across modalities in blind subjects, the tactile response to Braille reading extends into primary visual cortex only if vision is lost within a critical period of 14–16 years (Sadato et al. 2002).

Evidence for critical periods in the somatosensory system is found in the enticing point-to-point anatomy representing rodent whisker mosaics at all levels of the neuraxis. The formation of cortical "barrels," long immune to pharmacological disruption of neuronal activity (O'Leary et al. 1994), is regulated by a variety of molecules revealed by gene targeting (Erzurumlu & Kind 2001). Global removal of the NMDA receptor is lethal at birth yet sufficient to disrupt barrellette structures in the brainstem (Iwasato et al. 1997). Conditional NR1 subunit deletion restricted to the neocortex yields viable animals with a similar degradation of most of the S1 barrel field (Iwasato et al. 2000). Targeted disruption of a group of molecules related to cAMP signaling also produces 'barrelless' mice, including MAO-A, 5-HT_{1B} receptor, adenylate cyclase, mGluR5, and PLCβ (reviewed in Erzurumlu & Kind 2001, Lu et al. 2003). Instead, ectopic overexpression of fibroblast growth factor 8 (FGF8) can produce additional, ectopic barrel domains (Fukuchi-Shimogori & Grove 2001).

Critical period plasticity in response to sensory manipulation appears distinct from the initial formation of barrels. Direct damage to the whisker follicle is required to produce gross changes in the nascent barrel field. The anatomical shrinkage of deprived cortical barrels is strictly limited to whisker cauterization before the end of the first postnatal week (Van der Loos & Woolsey 1973). Unlike barrel formation, however, this plasticity does not require NMDA receptor function because whisker cautery causes barrel shrinkage even in cortex-specific NR1 knockout mice (Datwani et al. 2002). As in the visual cortex, the critical period ends regardless of NMDA receptor kinetics owing to NR2A subunit expression (Lu et al. 2001). When vibrissae are carefully removed, rather than ablated, a similar critical period is detectable by electrophysiological techniques (Fox 1992). The area of cortex driven by stimulation of the spared whisker enlarges, while the anatomical map retains its layout. Functional plasticity decreases rapidly in layer 4 between P0 and P4 and is NMDA-sensitive (Fox et al. 1996), mirrored by a declining ability to induce NMDA-dependent forms of LTP or LTD at thalamo-cortical inputs (Crair & Malenka 1995, Feldman et al. 1998).

Total whisker deprivation during a sharp critical period in the second postnatal week (P12–14) produces profoundly abnormal layer 2/3 receptive fields (Stern et al. 2001). Both the onset of active whisking (Welker 1964) and emergence of mature inhibition (Kiser et al. 1998) anticipate this plastic period, which further corresponds to a peak in experience-dependent motility of dendritic spines on pyramidal cells (Lendvai et al. 2000). However, direct analogy to ocular dominance cannot be made because the competitive situation of partial whisker deprivation has not been tested.

In contrast, expansion of spared whisker responses persists in upper cortical layers throughout life (Fox 1992), involving αCaMKII and CREB (Glazewski et al. 1999, 2000) and an LTD-like depression within the principal barrel of deprived whiskers (Allen et al. 2003). The direction of map plasticity is surprisingly modulated by the whisking experience. Active exploration of novel environments during learning paradoxically contracts the spared input (Polley et al. 1999). Interestingly, whisker trimming can regulate cortical GABA receptor binding at any age (Fuchs & Salazar 1998), which may permit the strong competitive plasticity throughout life. Abnormal lateral inhibitory interactions between trimmed and intact inputs also arise within thalamus (Simons & Land 1994). Early disuse of the whiskers can nevertheless have latent effects, degrading receptive field plasticity later in adulthood (Rema et al. 2003).

Taste/Olfactory System

Olfactory preferences learned early in life will affect the sexual behavior of adult rodents and are mediated by type 2 MHC in mice (Fillion & Blass 1986, Yamazaki et al. 1988). As opposed to these pheromonal actions, unilateral naris closure for one to six months can cause the deprived olfactory bulbs to atrophy even in adult mice (Maruniak et al. 1989). Conversely, taste deprivation or selective taste exposure during the suckling period produces no differences on adult preference for flavored solutions (Bernstein et al. 1986). As in the whisker-barrel system, direct damage to the receptors on the tongue at P2 is required to alter the formation of the gustatory pathways that process particular tastes (Lasiter & Kachele 1990). Such evidence indicates no clear developmental critical period for taste and olfaction.

The potential for strong olfactory plasticity throughout life is behaviorally adaptive. In the well-known "Bruce effect," a pheromonal memory is formed in female mice at the time of mating that programs spontaneous abortion when a novel male intruder appears (Brennan et al. 1990). This process requires nitric oxide and mGluR2 activation that disinhibits mitral cells (Kaba et al. 1994, Kendrick et al. 1997). Within 4 h after parturition, ewes learn to recognize the odor of their lamb. Activation of muscarinic acetylcholine receptors is important for memory formation during a critical period of less than 16 h postpartum (Ferreira et al. 1999). A mother rodent's ability to recognize and nurture her young reflects a boost in new neurons that are integrated into the olfactory bulb (Shingo et al. 2003). Neurogenesis rates jump during pregnancy by 65%, peaking near the seventh day of gestation and again after delivery.

Constant neurogenesis may hold a key to olfactory plasticity throughout life and offers an exception to prove the rule about inhibition in other systems. Neural stem cells migrate from the subventricular zone and connect to the established circuitry in the olfactory bulb (Carleton et al. 2003), where they are activated in response to smells. The new neurons are GABAergic granule cells, which sharpen the evoked pattern of neural activity through lateral inhibition of mitral cells (Yokoi et al. 1995). Mutant mice with deficient olfactory neurogenesis, and therefore relatively few granule cells, have difficulty with odor discrimination but not memory (Gheusi et al. 2000)—reminiscent of the inability to discriminate unbalanced visual input prerequisite to ocular dominance plasticity (Hensch et al. 1998). Both odor enrichment and the hormone prolactin during pregnancy trigger neurogenesis (Rochefort et al. 2002, Shingo et al. 2003).

MULTIMODAL FUNCTIONS

Imprinting

The classic example of a sharp critical period, famously described by Lorenz (1958), is the imprinting of certain precocial birds on a parental figure within hours after hatching. This is a complex behavior that integrates many sensory modalities including vision, taste, olfaction, and audition. It is now known to be a two-step process consisting of a predisposition to approach stimuli with characteristics of the natural mother followed by actual learning, or filial imprinting (Bolhuis & Honey 1998). Neonatal human infants too are predisposed to find face-like stimuli more attractive than nonface-like stimuli, but gradually learn through experience in the first months of life to finely discriminate human faces while grouping others (monkeys) into one general category (Pascalis et al. 2002).

Many sorts of nonspecific stimuli are capable of inducing the predisposition. Chicks reared in darkness without such experience (e.g., handling, running on a wheel, exposure to taped maternal calls) fail to develop a predisposition to approach a bird-like figure (see Bolhuis & Honey 1998). Critical period onset can be delayed by a catecholaminergic neurotoxin, DSP4. Whereas lesions to the intermediate and medial hyperstriatum ventrale (IMHV) have no effect on predisposition, the IMHV is crucial for learning the ensemble of characteristics belonging to one object, which is imprinting. Spine density of IMHV neurons following imprinting is reduced by half, which suggests that synapse elimination may end plasticity (Scheich 1987).

Stress and Anxiety

Harlow (1959) identified the importance of maternal care on developing social behavior in infants. Rhesus monkeys isolated at birth and supplied with dummy mothers invariably choose cloth-covered figures over uncomfortable wire mesh constructions equipped with a functional nursing bottle (Harlow & Zimmermann 1959). Monkeys raised in this way eventually grow up to become uniformly poor mothers, neglecting and punishing their offspring. More than food reward, the quality of maternal care is a crucial nutrient for the development of complex social behavior (Scott 1962). Recent research with rodents has provided insight into the neurobiological underpinnings of such effects.

Rat mothers engaged in high licking and grooming behaviors raise pups that have higher synaptic density, BDNF content, mature NMDA receptors, and spatial learning scores in the hippocampus (Liu et al. 2000), as well as greater tolerance to stress as adults (see Meaney 2001). Reduced hippocampal spatial learning and elevated anxiety in the mature offspring of poor mothers can be rescued by fostering to good mothers or enriched environments (Francis et al. 2002). Moreover, female pups raised by good natural or foster mothers grow up to become good mothers. These effects also become hardwired into better-functioning glucocorticoid systems in these rats (Meaney 2001). Early postnatal experience may thus overcome genetic predispositions.

Anxiety-like behavior results from appropriate serotonin receptor activation during a critical period of early postnatal life (Gross et al. 2002). Whereas global deletion of the 5-HT_{1A} receptor subtype yields increased anxiety in the adult mouse (Sibille et al. 2000), its conditional restoration in the hippocampus and cortex (but not brainstem raphe nucleus) rescues the phenotype. However, the receptor must be activated between P5 and P21 in order to be successful, demarcating a critical period for the development of lifelong anxiety behavior in mouse. Appropriate 5-HT_{1A} receptor activation may ultimately regulate $GABA_A$ signaling (Sibille et al. 2000). Therapeutic intervention for anxiety by serotonin agonists is most effective in chronic paradigms that recognize this critical period (Hen & Gordon 2004, in this volume).

Sleep

Both the amount and quality of sleep changes with age. Initially, rapid-eye movement (REM)-like sleep states predominate, followed by the emergence of slow-wave sleep (SWS), which eventually occupies most of adult sleep (Jouvet-Mounier et al. 1970, Roffwarg et al. 1966). Increasing evidence points toward a replay of daily events during these sleep states for the consolidation of memories (Maquet et al. 2003). Indeed, SWS enhances critical period plasticity in the visual system (see above). Conversely, sleep itself is shaped by experience (Miyamoto et al. 2003). During a critical period from P40 to P60 in cats and mice, total visual deprivation reduces the slow-wave EEG activity selectively in visual cortex, yielding regional inhomogeneities of sleep quality over the brain surface. Sleep plasticity reflects the degree of NMDA receptor activation, and its recovery after dark-rearing reveals a delay of the critical period by the absence of experience. If sleep function includes neuronal replay for brain plasticity, then its own developmental plasticity reveals an intricate interplay between systems.

Sound Localization

Sound localization is a complex task that beautifully exhibits several aspects of critical period development (Knudsen et al. 2000). A map of auditory and visual space must be perfectly aligned within the barn owl tectum for appropriately targeted flight. Shifting the visual scene with displacing prisms causes a mismatch between receptive field position and interaural time differences (ITD) from a sound source. If prisms are mounted during a critical period in early life, birds can learn to remap (over 6 to 8 weeks) their ITD representation to best fit the new visual environment. Large-scale adaptation is possible only in juvenile birds, gradually declining by 150 days after birth. Interestingly, in this system critical period closure can be delayed by sensory and social enrichment (Brainard & Knudsen 1998).

A predisposition for the normal alignment of sensory maps is present throughout life, upon which newly acquired ITD maps are superimposed. The new connections are mediated by NMDA receptors (Feldman et al. 1996) and accompanied by two structural changes: sprouting of excitatory inputs in the adaptive direction for the learned ITD and formation of novel GABAergic circuits that actively suppress the competing, unused map (DeBello et al. 2001, Zheng & Knudsen 1999). In this way, critical period plasticity can etch multiple maps in the developing tectum, which coexist into adulthood even after the prisms have been removed. These early learning experiences are thus not forgotten, as restoring the same goggles to previously trained birds smoothly switches ITD maps at an age when plasticity typically does not occur (Knudsen 1998). Early experience can thus expand the repertoire of neural substrates available throughout life and offers an explanation why, for instance, early language learning in humans is so effective (see Human Language).

A potential for reactivating plasticity after the critical period has recently been tested using incremental training in the adult (Linkenhoker & Knudsen 2002). Rather than a single, large prismatic displacement (>20°), which is normally ineffective in mature barn owls, several smaller steps (~5°) produce gradual ITD adjustments that accumulate over time. Eventually, large shifts are attained on the order of juvenile levels, which subsequently become accessible to single, large displacing prisms. Whether structural changes occur in the adult remains to be seen, but the findings indicate that lifelong learning is possible through the right training regimen.

Birdsong

The three-step process by which certain birds acquire a single, stereotyped song highlights the serial nature of multiple critical periods underlying a complex brain function (Doupe & Kuhl 1999). Young birds first memorize the song of a tutor during a sensory acquisition phase, followed by sensorimotor practice when the bird actively matches its own vocal output to the memorized template. These two phases may overlap somewhat (zebra finch) or be separated by several hundred days (swamp sparrows), indicating the length of memory (Brainard & Doupe 2002). Learning ends in song crystallization when note structure and sequence become highly stereotyped. Seasonal singers repeat this process annually, whereas other species such as the zebra finch learn one song for life during a cumulative critical period over the first 100 days posthatching.

The neurobiological substrate has focused on hypertrophied brain nuclei present in the male zebra finch, as females do not learn to sing (see Mooney 1999). Two pathways are delineated from the higher vocal center (HVc), which receives the auditory information: An anterior forebrain pathway (AFP) subserves song learning, and a ventral motor pathway through the nucleus robustus archistrialis (RA) produces vocalization. Normal song development depends upon an intact AFP because lesions to the lateral portion of the magnocellular nucleus of the anterior neostriatum (LMAN) result in abnormal song in normally tutored pupils. Cells in LMAN acquire precise response tuning to the bird's own song (Brainard & Doupe 2002) and exhibit a decline in dendritic spine density with development (Wallhausser-Franke et al. 1995). Surprisingly though, no AFP nucleus tested so far has emerged as an obvious repository of the tutor's memory trace.

Young birds raised in isolation (with no tutor) or deafened (with no auditory feedback) produce abnormal vocalizations as adults (Brainard & Doupe 2002). Isolation also delays the critical period for learning beyond the normal chronological age, which too is disrupted by LMAN lesions. As in other sensory systems, attention has fallen upon maturational changes of NMDA receptors in LMAN with similar conclusions (see White 2001). Receptor incorporation of NR2A subunits and shortening current decay times are delayed by isolation rearing, but the typical decline in NMDA receptor number is not. Moreover, current kinetics eventually accelerate even in isolation, failing to explain the delayed learning still possible thereafter (Livingston et al. 2000).

A potential site of sensorimotor integration lies in RA (see Mooney 1999), where the two pathways meet upon individual cells. Interestingly, LMAN axons innervate RA first, utilizing mainly NMDA receptors, perhaps to guide the later-arriving HVc axons at song onset, after which LMAN inputs are pruned by half (Doupe & Kuhl 1999, Mooney & Rao 1994). During the ensuing sensorimotor phase, GABA cell number peaks in the RA of males but not females (Sakaguchi 1996). Here again, an optimal excitatory-inhibitory balance may be required to correctly compare distinct aspects of the bird's own song with a memorized internal template. Indeed, the plasticity of individual song details (syllable phonology versus sequences) may be differentially delayed by masking noise throughout this period (Funabiki & Konishi 2003).

Termination of the critical period is strictly linked to hormonal control of sexual maturation (Brainard & Doupe 2002). Testosterone reduces density of dendritic spines in LMAN and prematurely crystallizes song, whereas castration yields inconsistent song throughout life (see White 2001). Seasonal learners have waxing and waning hormone levels. In addition, neurogenesis differs across species. Male canaries learn new song elements every year and their HVc shrinks (during nonbreeding periods) and enlarges accordingly (Alvarez-Buylla et al. 1988). Zebra finches add large numbers of neurons to their HVcs only when they are young and learning their one courtship song for life. Adult song can still degrade over months after deafening (see Brainard & Doupe 2002). Interestingly, the double insult of concurrent LMAN lesion prevents this slow decay, revealing a lifelong role for the AFP in auditory feedback.

Singing is a well-characterized behavior that varies between species. Song length, spectral complexity, and periodic structure reflect geographical area, much like human language (Brainard & Doupe 2002). When given a choice within a range of tutors, young birds show an innate predisposition for conspecific song—learning is more rapid and accurate (Doupe & Kuhl 1999, White 2001). Attentional and motivational factors further influence sensory acquisition, as demonstrated by a higher degree of copying by birds trained to actively peck a key in order to hear a tape of a tutor's song. Intriguingly, birds learn best when playback is limited to 30 seconds daily, and more exposure leads to less copying (Tchernichovski et al. 1999). Expression of the immediate early gene *zif268* is also highest for conspecific song and reflects the amount of copying but declines when the same song is repeatedly presented (Bolhuis et al. 2000, Mello et al. 1995).

The expression of *zif268*, though not necessarily involved in plasticity itself (Mataga et al. 2001), vividly reveals the behavioral state of the animal (Tononi & Cirelli 2001). Norepinephrine gates the auditory response from HVc to RA naturally as birds fall asleep to allow neuronal replay (Dave & Margoliash 2000). Directed singing in waking toward a companion activates the AFP differently from when the bird is singing alone (Hessler & Doupe 1999, Jarvis et al. 1998), and learning from a live tutor is more potent than tape learning (Doupe & Kuhl 1999, White 2001). Neuromodulation by social context reflects the ultimate aim of singing: communication.

Human Language

Synaptogenesis (Huttenlocher & Dabholkar 1997) and metabolic changes (Chugani 1998) differ with age across brain regions in the human infant. Each of these predict separate critical periods for various brain functions, but evidence is largely limited to anecdote except for language acquisition. Lenneberg (1967) originally proposed that a critical period for language ends around puberty. Intense debate has since centered on whether native languages are truly crystallized or merely interfere with second-language learning later in life. Although detailed cellular and structural substrates are difficult to identify in humans, developmental psychology and advanced brain-imaging techniques are revealing developmental milestones that may underlie a critical period.

Full-term neonates already exhibit left-hemisphere dominance (by optical topography) for normal speech (Pena et al. 2003) and segregate concurrent streams of sound (as detected by electrical mismatch negativity, MMN) like adults (Winkler et al. 2003). Functional MRI confirms the precursors of adult cortical language areas at three months of age (Dehaene-Lambertz et al. 2002). A baby's speech then emerges through a series of stages much like the different types of vocalization (subsong, plastic song, full song) observed in birds (Doupe & Kuhl 1999).

Exploiting statistical properties of language input (see Kuhl 2000), the auditory perceptual map is refined by six months of age to eliminate nonnative phoneme distinctions (like "r" from "l" in Japanese). Interestingly, vowel sound discrimination can be taught even during sleep (Cheour et al. 2002). Once vision matures sufficiently, by ten months, information about the speaker's face is combined with the concurrent acoustic signal, leading to perceptual illusions such as the McGurk effect (dubbing mismatch). Ultimately, the cumulative critical period for language ends with the ability to properly discriminate subtle grammatical errors by the age of 12 years (Newport et al. 2001). Other linguistic features, such as semantics, can be learned throughout life.

Bilingual subjects reveal a sequential neural commitment to competing stimuli. Originally sensitive to all speech sounds, the MMN is observed only for native language contrasts after 12 months of age (Cheour et al. 1998), reflecting the great difficulty to hear, as well as to produce, nonnative phonetic distinctions later in life. Interference effects are minimal before adolescence and several different languages can be acquired, perhaps by switching inhibition of unused maps (see barn owl discussion above). Functional MRI reveals, when both languages are learned before age 11, overlapping regions of Broca's area are activated, whereas second languages acquired later must employ two distinct areas (Kim et al. 2002). Age of acquisition also affects the cortical representation of grammatical (but not semantic) processes (Wartenburger et al. 2003).

Social context is instrumental. In the rare cases when children have been raised in isolation, or in children with autism, language skills and social deficits are tightly coupled (Leonard 1998). Conversely, in normal nine-month-old American infants, limited (5 h) exposure to Chinese speakers spaced over one month prevents the loss of Mandarin speech sound distinctions (Kuhl et al. 2003). The reward of a live tutor is essential, since similar exposure to taped instructors has no rescuing effect. From personal anecdote, learning is further facilitated if each tutor speaks exclusively one of the two languages rather than both.

Training paradigms based on exaggerated acoustic cues characteristic of motherese, multiple instances by many talkers, and mass listening experience may succeed to incrementally rewire the brain of adults as it already has for learning-disabled children (Kuhl 2000, Temple et al. 2003). By paradoxically overcoming our "mature" cognition, we may one day learn new skills more naturally and efficiently, as our children do so effortlessly during their critical periods.

The *Annual Review of Neuroscience* is online at http://neuro.annualreviews.org

LITERATURE CITED

Adams DL, Horton JC. 2002. Shadows cast by retinal blood vessels mapped in primary visual cortex. *Science* 298:572–76

Allen CB, Celikel T, Feldman DE. 2003. Long-term depression induced by sensory deprivation during cortical map plasticity in vivo. *Nat. Neurosci.* 6:291–99

Alvarez-Buylla A, Theelen M, Nottebohm F.

1988. Birth of projection neurons in the higher vocal center of the canary forebrain before, during, and after song learning. *Proc. Natl. Acad. Sci. USA* 85:8722–26

Antonini A, Stryker MP. 1993. Rapid remodeling of axonal arbors in the visual cortex. *Science* 260:1819–21

Bartoletti A, Cancedda L, Reid SW, Tessarollo L, Porciatti V, et al. 2002. Heterozygous knock-out mice for brain-derived neurotrophic factor show a pathway-specific impairment of long-term potentiation but normal critical period for monocular deprivation. *J. Neurosci.* 22:10072–77

Bear MF, Kleinschmidt A, Gu QA, Singer W. 1990. Disruption of experience-dependent synaptic modifications in striate cortex by infusion of an NMDA receptor antagonist. *J. Neurosci.* 10:909–25

Bear MF, Singer W. 1986. Modulation of visual cortical plasticity by acetylcholine and noradrenaline. *Nature* 320:172–76

Beaver CJ, Ji Q, Fischer QS, Daw NW. 2001. Cyclic AMP-dependent protein kinase mediates ocular dominance shifts in cat visual cortex. *Nat. Neurosci.* 4:159–63

Berardi N, Pizzorusso T, Maffei L. 2000. Critical periods during sensory development. *Curr. Opin. Neurobiol.* 10:138–45

Berardi N, Pizzorusso T, Ratto GM, Maffei L. 2003. Molecular basis of plasticity in the visual cortex. *Trends Neurosci.* 26:369–78

Bernstein IL, Fenner DP, Diaz J. 1986. Influence of taste stimulation during the suckling period on adult taste preference in rats. *Physiol. Behav.* 36:913–19

Bi G, Poo M. 2001. Synaptic modification by correlated activity: Hebb's postulate revisited. *Annu. Rev. Neurosci.* 24:139–66

Bolhuis JJ, Honey RC. 1998. Imprinting, learning and development: from behavior to brain and back. *Trends Neurosci.* 21:306–11

Bolhuis JJ, Zijlstra GG, den Boer-Visser AM, Van Der Zee EA. 2000. Localized neuronal activation in the zebra finch brain is related to the strength of song learning. *Proc. Natl. Acad. Sci. USA* 97:2282–85

Boulanger LM, Huh GS, Shatz CJ. 2001. Neuronal plasticity and cellular immunity: shared molecular mechanisms. *Curr. Opin. Neurobiol.* 11:568–78

Brainard MS, Doupe AJ. 2002. What songbirds teach us about learning. *Nature* 417:351–58

Brainard MS, Knudsen EI. 1998. Sensitive periods for visual calibration of the auditory space map in the barn owl optic tectum. *J. Neurosci.* 18:3929–42

Brandon EP, Lin W, D'Amour KA, Pizzo DP, Dominguez B, et al. 2003. Aberrant patterning of neuromuscular synapses in choline acetyltransferase-deficient mice. *J. Neurosci.* 23:539–49

Brennan P, Kaba H, Keverne EB. 1990. Olfactory recognition: a simple memory system. *Science* 250:1223–26

Buffelli M, Burgess RW, Feng G, Lobe CG, Lichtman JW, Sanes JR. 2003. Genetic evidence that relative synaptic efficacy biases the outcome of synaptic competition. *Nature* 424:430–34

Buzas P, Eysel UT, Adorjan P, Kisvarday ZF. 2001. Axonal topography of cortical basket cells in relation to orientation, direction, and ocular dominance maps. *J. Comp. Neurol.* 437:259–85

Carleton A, Petreanu LT, Lansford R, Alvarez-Buylla A, Lledo PM. 2003. Becoming a new neuron in the adult olfactory bulb. *Nat. Neurosci.* 6:507–18

Celio MR, Spreafico R, De Biasi S, Vitellaro-Zuccarello L. 1998. Perineuronal nets: past and present. *Trends Neurosci.* 21:510–15

Chang EF, Merzenich MM. 2003. Environmental noise retards auditory cortical development. *Science* 300:498–502

Chen C, Regehr WG. 2000. Developmental remodeling of the retinogeniculate synapse. *Neuron* 28:955–66

Cheour M, Ceponiene R, Lehtokoski A, Luuk A, Allik J, et al. 1998. Development of language-specific phoneme representations in the infant brain. *Nat. Neurosci.* 1:351–53

Cheour M, Martynova O, Naatanen R, Erkkola R, Sillanpaa M, et al. 2002. Speech sounds learned by sleeping newborns. *Nature* 415:599–600

Cherubini E, Conti F. 2001. Generating diversity at GABAergic synapses. *Trends Neurosci.* 24:155–62

Chugani HT. 1998. A critical period of brain development: studies of cerebral glucose utilization with PET. *Prev. Med.* 27:184–88

Connors B. 2004. Electrical synapses in the mammalian brain. *Annu. Rev. Neurosci.* 27: 393–418

Constantine-Paton M, Law MI. 1978. Eye-specific termination bands in tecta of three-eyed frogs. *Science* 202:639–41

Crair MC, Malenka RC. 1995. A critical period for long-term potentiation at thalamocortical synapses. *Nature* 375:325–28

Cramer KS, Angelucci A, Hahm JO, Bogdanov MB, Sur M. 1996. A role for nitric oxide in the development of the ferret retinogeniculate projection. *J. Neurosci.* 16:7995–8004

Crepel F. 1982. Regression of functional synapses in the immature mammalian cerebellum. *Trends Neurosci.* 5:266–69

Crowley JC, Katz LC. 2002. Ocular dominance development revisited. *Curr. Opin. Neurobiol.* 12:104–9

Datwani A, Iwasato T, Itohara S, Erzurumlu RS. 2002. Lesion-induced thalamocortical axonal plasticity in the S1 cortex is independent of NMDA receptor function in excitatory cortical neurons. *J. Neurosci.* 22:9171–75

Dave AS, Margoliash D. 2000. Song replay during sleep and computational rules for sensory motor vocal learning. *Science* 290:812–16

Daw NW. 1995. *Visual Development.* New York: Plenum

Daw NW. 1998. Critical periods and amblyopia. *Arch. Ophthalmol.* 116:502–5

Daw NW. 2003. Mechanisms of plasticity in the visual cortex. In *The Visual Neurosciences*, ed. LM Chalupa, JS Werner, pp. 126–45. Cambridge: MIT Press

DeBello WM, Feldman DE, Knudsen EI. 2001. Adaptive axonal remodeling in the midbrain auditory space map. *J. Neurosci.* 21:3161–74

DeFelipe J. 1997. Types of neurons, synaptic connections and chemical characteristics of cells immunoreactive for calbindin-D28K, parvalbumin and calretinin in the neocortex. *J. Chem. Neuroanat.* 14:1–19

Dehaene-Lambertz G, Dehaene S, Hertz-Pannier L. 2002. Functional neuroimaging of speech perception in infants. *Science* 298:2013–15

Del Rio JA, De Lecea L, Ferrer I, Soriano E. 1994. The development of parvalbumin-immunoreactivity in the neocortex of the mouse. *Dev. Brain Res.* 81:247–59

Desai NS, Cudmore RH, Nelson SB, Turrigiano GG. 2002. Critical periods for experience-dependent synaptic scaling in visual cortex. *Nat. Neurosci.* 5:783–89

Dews PB, Wiesel TN. 1970. Consequences of monocular deprivation on visual behavior in kittens. *J. Physiol.* 206:437–55

Doupe AJ, Kuhl PK. 1999. Birdsong and human speech: common themes and mechanisms. *Annu. Rev. Neurosci.* 22:567–631

Elbert T, Panter C, Wienbruch C, Rockstroh B, Taub E. 1995. Increased cortical representation of the fingers of the left hand in string players. *Science* 270:305–7

Erzurumlu RS, Kind PC. 2001. Neural activity: sculptor of 'barrels' in the neocortex. *Trends Neurosci.* 24:589–95

Fagiolini M, Fritschy J-M, Löw K, Möhler H, Rudolph U, et al. 2004. Specific $GABA_A$ circuits for visual cortical plasticity. *Science* 303:1681–83

Fagiolini M, Hensch TK. 2000. Inhibitory threshold for critical-period activation in primary visual cortex. *Nature* 404:183–86

Fagiolini M, Katagiri H, Miyamoto H, Mori H, Grant SGN, et al. 2003. Separable features of visual cortical plasticity revealed through *N*-Methyl-D-aspartate receptor 2A signaling. *Proc. Natl. Acad. Sci. USA* 100:2854–59

Fagiolini M, Pizzorusso T, Berardi N, Domenici L, Maffei L. 1994. Functional postnatal development of the rat primary visual cortex and the role of visual experience: dark rearing and monocular deprivation. *Vision Res.* 34:709–20

Farrant M, Cull-Candy SG. 1991. Excitatory amino acid receptor-channels in Purkinje cells in thin cerebellar slices. *Proc. R. Soc. London Ser. B* 244:179–84

Feldman DE, Brainard MS, Knudsen EI. 1996. Newly learned auditory responses mediated by NMDA receptors in the owl inferior colliculus. *Science* 271:525–28

Feldman DE, Nicoll RA, Malenka RC, Isaac JT. 1998. Long-term depression at thalamocortical synapses in developing rat somatosensory cortex. *Neuron* 21:347–57

Ferreira G, Gervais R, Durkin TP, Levy F. 1999. Postacquisition scopolamine treatments reveal the time course for the formation of lamb odor recognition memory in parturient ewes. *Behav. Neurosci.* 113:136–42

Fillion TJ, Blass EM. 1986. Infantile experience with suckling odors determines adult sexual behavior male rats. *Science* 231:729–31

Fiumelli H, Jabaudon D, Magistretti PJ, Martin JL. 1999. BDNF stimulates expression, activity and release of tissue-type plasminogen activator in mouse cortical neurons. *Eur. J. Neurosci.* 11:1639–46

Fox K. 1992. A critical period for experience-dependent synaptic plasticity in rat barrel cortex. *J. Neurosci.* 12:1826–38

Fox K. 1995. The critical period for long-term potentiation in primary sensory cortex. *Neuron* 15:485–88

Fox K, Schlaggar BL, Glazewski S, O'Leary DDM. 1996. Glutamate receptor blockade at cortical synapses disrupts development of thalamocortical and columnar organization in somatosensory cortex. *Proc. Natl. Acad. Sci. USA* 93:5584–89

Francis DD, Diorio J, Plotsky PM, Meaney MJ. 2002. Environmental enrichment reverses the effects of maternal separation on stress reactivity. *J. Neurosci.* 22:7840–43

Frank MG, Issa NP, Stryker MP. 2001. Sleep enhances plasticity in the developing visual cortex. *Neuron* 30:275–87

Frey U, Huang YY, Kandel ER. 1993. Effects of cAMP simulate a late-stage of LTP in hippocampal CA1 neurons. *Science* 260:1661–64

Froemke RC, Dan Y. 2002. Spike-timing-dependent synaptic modification induced by natural spike trains. *Nature* 416:433–38

Fuchs JL, Salazar E. 1998. Effects of whisker trimming on GABA(A) receptor binding in the barrel cortex of developing and adult rats. *J. Comp. Neurol.* 395:209–16

Fukuchi-Shimogori T, Grove EA. 2001. Neocortex patterning by the secreted signaling molecule FGF8. *Science* 294:1071–74

Funabiki Y, Konishi M. 2003. Long memory in song learning by zebra finches. *J. Neurosci.* 23:6928–35

Galaretta M, Hestrin S. 2001. Spike transmission and synchrony detection in networks of GABAergic interneurons. *Science* 292:2295–99

Gao WJ, Wormington AB, Newman DE, Pallas SL. 2000. Development of inhibitory circuitry in visual and auditory cortex of postnatal ferrets: immunocytochemical localization of calbindin- and parvalbumin-containing neurons. *J. Comp. Neurol.* 422:140–57

Gheusi G, Cremer H, McLean H, Chazal G, Vincent JD, Lledo PM. 2000. Importance of newly generated neurons in the adult olfactory bulb for odor discrimination. *Proc. Natl. Acad. Sci. USA* 97:1823–28

Ghose GM, Yang T, Maunsell JH. 2002. Physiological correlates of perceptual learning in monkey V1 and V2. *J. Neurophysiol.* 87:1867–88

Gilbert CD. 1998. Adult cortical dynamics. *Physiol. Rev.* 78:467–85

Glazewski S, Barth AL, Wallace H, McKenna M, Silva A, Fox K. 1999. Impaired experience-dependent plasticity in barrel cortex of mice lacking the alpha and delta isoforms of CREB. *Cereb. Cortex* 9:249–56

Glazewski S, Giese KP, Silva A, Fox K. 2000. The role of alpha-CaMKII autophosphorylation in neocortical experience-dependent plasticity. *Nat. Neurosci.* 3:911–18

Gordon JA, Stryker MP. 1996. Experience-dependent plasticity of binocular responses in the primary visual cortex of the mouse. *J. Neurosci.* 16:3274–86

Gross C, Zhuang X, Stark K, Ramboz S, Oosting R, et al. 2002 Serotonin$_{1A}$ receptor acts during development to establish normal anxiety-like behaviour in the adult. *Nature* 416:396–400

Grutzendler J, Kasthuri N, Gan WB. 2002. Long-term dendritic spine stability in the adult cortex. *Nature* 420:812–16

Gu Q, Liu Y, Cynader MS. 1994. Nerve growth factor-induced ocular dominance plasticity in adult cat visual cortex. *Proc. Natl. Acad. Sci. USA* 91:8408–12

Hahm JO, Langdon RB, Sur M. 1991. Disruption of retinogeniculate afferent segregation by antagonists to NMDA receptors. *Nature* 351:568–70

Harlow HF. 1959. Love in infant monkeys. *Sci. Am.* 200:68–74

Harlow HF, Zimmermann RR. 1959. Affectional responses in the infant monkey; orphaned baby monkeys develop a strong and persistent attachment to inanimate surrogate mothers. *Science* 130:421–32

Hartig W, Derouiche A, Welt K, Brauer K, Grosche J, et al. 1999. Cortical neurons immunoreactive for the potassium channel Kv3.1b subunit are predominantly surrounded by perineuronal nets presumed as a buffering system for cations. *Brain Res.* 842: 15–29

Hashimoto K, Kano M. 2003. Functional differentiation of multiple climbing fiber inputs during synapse elimination in the developing cerebellum. *Neuron* 38:785–96

Hashimoto K, Watanabe M, Kurihara H, Offermanns S, Jiang H, et al. 2000. Climbing fiber synapse elimination during postnatal cerebellar development requires signal transduction involving G alpha q and phospholipase C beta 4. *Prog. Brain Res.* 124:31–48

Hata Y, Ohshima M, Ichisaka S, Wakita M, Fukuda M, Tsumoto T. 2000. Brain-derived neurotrophic factor expands ocular dominance columns in visual cortex in monocularly-deprived and non-deprived kittens but does not in adult cats. *J. Neurosci.* 20:RC57–61

Hata Y, Stryker MP. 1994. Control of thalamocortical afferent rearrangement by postsynaptic activity in developing visual cortex. *Science* 265:1732–35

Hata Y, Tsumoto T, Stryker MP. 1999. Selective pruning of more active afferents when cat visual cortex is pharmacologically inhibited. *Neuron* 22:375–81

Hen R, Gordon J. 2004. Genetic approaches to the study of anxiety. *Annu. Rev. Neurosci.* 27:In press

Hensch TK, Stryker MP. 2004. Columnar architecture sculpted by GABA circuits in developing cat visual cortex. *Science* 303:1678–81

Hensch TK. 2003. Controlling the critical period. *Neurosci. Res.* 47:17–22

Hensch TK, Fagiolini M, eds. 2003. *Excitatory-Inibitory Balance: Synapses, Circuits, Systems*. New York: Kluwer/Plenum

Hensch TK, Fagiolini M, Mataga N, Stryker MP, Baekkeskov, Kash SF. 1998. Local GABA circuit control of experience-dependent plasticity in developing visual cortex. *Science* 282:1504–08

Herrera E, Brown L, Aruga J, Rachel RA, Dolen G, et al. 2003. Zic2 patterns binocular vision by specifying the uncrossed retinal projection. *Cell* 114:545–57

Hessler NA, Doupe AJ. 1999. Social context modulates singing in the songbird forebrain. *Nat. Neurosci.* 2:209–11

Heynen AJ, Yoon BJ, Liu CH, Chung HJ, Huganir RL, et al. 2003. Molecular mechanism for loss of visual cortical responsiveness following brief monocular deprivation. *Nat. Neurosci.* 6:854–62

Huang EJ, Reichardt LF. 2001. Neurotrophins: roles in neuronal development and function. *Annu. Rev. Neurosci.* 24:677–736

Huang ZJ, Kirkwood A, Pizzorusso T, Porciatti V, Morales B, et al. 1999. BDNF regulates the maturation of inhibition and the critical period of plasticity in mouse visual cortex. *Cell* 98:739–55

Hubel DH, Wiesel TN. 1970. The period of susceptibility to the physiological effects of unilateral eye closure in kittens. *J. Physiol. (Lond.)* 206:419–36

Huberman AD, Stellwagen D, Chapman B.

2002. Decoupling eye-specific segregation from lamination in the lateral geniculate nucleus. *J. Neurosci.* 22:9419–29

Huh GS, Boulanger LM, Du H, Riquelme PA, Brotz TM, et al. 2000. Functional requirement for class I MHC in CNS development and plasticity. *Science* 290:2155–59

Huntley GW. 1997. Differential effects of abnormal tactile experience on shaping representation patterns in developing an adult motor cortex. *J. Neurosci.* 17:9220–32

Huttenlocher PR, Dabholkar AS. 1997. Regional differences in synaptogenesis in human cerebral cortex. *J. Comp. Neurol.* 387:167–78

Ichise T, Kano M, Hashimoto K, Yanagihara D, Nakao K, et al. 2000. mGluR1 in cerebellar Purkinje cells essential for long-term depression, synapse elimination, and motor coordination. *Science* 288:1832–35

Imamura K, Kasamatsu T. 1991. Ocular dominance plasticity restored by NA infusion to aplastic visual cortex of anesthetized and paralyzed kittens. *Exp. Brain Res.* 87:309–18

Imamura K, Kasamatsu T, Shirokawa T, Ohashi T. 1999. Restoration of ocular dominance plasticity mediated by adenosine 3′,5′-monophosphate in adult visual cortex. *Proc. R. Soc. London Ser. B* 266:1507–16

Iwai Y, Fagiolini M, Obata K, Hensch TK. 2003. Rapid critical period induction by tonic inhibition in mouse visual cortex. *J. Neurosci.* 23:6695–702

Iwasato T, Datwani A, Wolf AM, Nishiyama H, Taguchi Y, et al. 2000. Cortex-restricted disruption of NMDAR1 impairs neuronal patterns in the barrel cortex. *Nature* 406:726–31

Iwasato T, Erzurumlu RS, Huerta PT, Chen DF, Sasaoka T, et al. 1997. NMDA receptor-dependent refinement of somatotopic maps. *Neuron* 19:1201–10

Jarvis ED, Scharff C, Grossman MR, Ramos JA, Nottebohm F. 1998. For whom the bird sings: context-dependent gene expression. *Neuron* 21:775–88

Jiang B, Akaneya Y, Hata Y, Tsumoto T. 2003. Long-term depression is not induced by low-frequency stimulation in rat visual cortex in vivo: a possible preventing role of endogenous brain-derived neurotrophic factor. *J. Neurosci.* 23:3761–70

Jones E. 2000. Cortical and subcortical contributions to activity-dependent plasticity in primate somatosensory cortex. *Annu. Rev. Neurosci.* 23:1–37

Jouvet-Mounier D, Astic L, Lacote D. 1970. Ontogenesis of the states of sleep in rat, cat, and guinea pig during the first postnatal month. *Dev. Psychobiol.* 2:216–39

Kaba H, Hayashi Y, Higuchi T, Nakanishi S. 1994. Induction of an olfactory memory by the activation of a metabotropic glutamate receptor. *Science* 265:262–64

Kakizawa S, Yamasaki M, Watanabe M, Kano M. 2001. Critical period for activity-dependent synapse elimination in developing cerebellum. *J. Neurosci.* 20:4954–61

Kano M, Hashimoto K, Chen C, Abeliovich A, Aiba A, et al. 1995. Impaired synapse elimination during cerebellar development in PKC gamma mutant mice. *Cell* 83:1223–31

Karni A, Sagi D. 1991. Where practice make perfect in texture discrimination: evidence for primary visual cortex plasticity. *Proc. Natl. Acad. Sci. USA* . 88:4966–70

Kasthuri N, Lichtman JW. 2003. The role of neuronal identity in synaptic competition. *Nature* 424:426–30

Kendrick KM, Guevara-Guzman R, Zorrilla J, Hinton MR, Broad KD, et al. 1997. Formation of olfactory memories mediated by nitric oxide. *Nature* 388:670–74

Kilgard MP, Merzenich MM. 1998. Cortical map reorganization enabled by nucleus basalis activity. *Science* 279:1714–18

Kim G, Kandler K. 2003. Elimination and strengthening of glycinergic/GABAergic connections during tonotopic map formation. *Nat. Neurosci.* 6: 282–90

Kim KH, Relkin NR, Lee KM, Hirsch J. 2002. Distinct cortical areas associated with native and second languages. *Nature* 388:171–74

Kiser PJ, Cooper NG, Mower GD. 1998. Expression of two forms of glutamic acid decarboxylase (GAD67 and GAD65) during postnatal development of rat somatosensory barrel cortex. *J. Comp. Neurol.* 402:62–74

Klausberger T, Roberts JD, Somogyi P. 2002. Cell type- and input-specific differences in the number and subtypes of synaptic GABA(A) receptors in the hippocampus. *J. Neurosci.* 22:2513–21

Knudsen EI. 1998. Capacity for plasticity in the adult owl auditory system expanded by juvenile experience. *Science* 279:1531–33

Knudsen EI, Zheng W, DeBello WM. 2000. Traces of learning in the auditory localization pathway. *Proc. Natl. Acad. Sci. USA* 97:11815–20

Kobatake E, Wang G, Tanaka K. 1998. Effects of shape-discrimination training on the selectivity of inferotemporal cells in adult monkeys. *J. Neurophysiol.* 80:324–30

Kuhl PK. 2000. A new view of language acquisition. *Proc. Natl. Acad. Sci. USA* 97:11850–57

Kuhl PK, Tsao FM, Liu HM. 2003. Foreign-language experience in infancy: effects of short-term exposure and social interaction on phonetic learning. *Proc. Natl. Acad. Sci. USA* 100:9096–101

Lasiter PS, Kachele DL. 1990. Effects of early postnatal receptor damage on development of gustatory recipient zones within the nucleus of the solitary tract. *Dev. Brain Res.* 55:57–71

Lendvai B, Stern EA, Chen B, Svoboda K. 2000. Experience-dependent plasticity of dendritic spines in the developing rat barrel cortex in vivo. *Nature* 404:876–81

Lenneberg EH. 1967. *Biological Foundations of Language*. New York: Wiley

Leonard LB. 1998. *Children with Specific Language Impairment*. Cambridge, MA: MIT Press

Lichtman JW, Colman H. 2000. Synapse elimination and indelible memory. *Neuron* 25:269–78

Linkenhoker BA, Knudsen EI. 2002. Incremental training increases the plasticity of the auditory space map in adult barn owls. *Nature* 419:293–96

Liu D, Diorio J, Day JC, Francis DD, Meaney MJ. 2000. Maternal care, hippocampal synaptogenesis and cognitive development in rats. *Nat. Neurosci.* 3:799–806

Livingston FS, White SA, Mooney R. 2000. Slow NMDA-EPSCs at synapses critical for song development are not required for song learning in zebra finches. *Nat. Neurosci.* 3:482–88

Lorenz KZ. 1958. The evolution of behavior. *Sci. Am.* 199:67–74

Lu HC, Gonzalez E, Crair MC. 2001. Barrel cortex critical period plasticity is independent of changes in NMDA receptor subunit composition. *Neuron* 32:619–34

Lu HC, She WC, Plas DT, Neumann PE, Janz R, Crair MC. 2003. Adenylyl cyclase I regulates AMPA receptor trafficking during mouse cortical 'barrel' map development. *Nat. Neurosci.* 6:939–47

Majewska A, Sur M. 2003. Motility of dendritic spines in visual cortex in vivo: changes during the critical period and effects of visual deprivation. *Proc. Natl. Acad. Sci. USA* 100:16024–29

Maquet P, Smith C, Stickgold R, eds. 2003. *Sleep and Brain Plasticity*. New York: Oxford Univ. Press

Martin SJ, Grimwood PD, Morris RG. 2000. Synaptic plasticity and memory: an evaluation of the hypothesis. *Annu. Rev. Neurosci.* 23:649–711

Maruniak JA, Taylor JA, Henegar JR, Williams MB. 1989. Unilateral naris closure in adult mice: atrophy of the deprived-side olfactory bulbs. *Brain Res. Dev. Brain Res.* 47:27–33

Mataga N, Fujishima S, Condie BG, Hensch TK. 2001. Experience-dependent plasticity of mouse visual cortex in the absence of the neuronal activity-dependent marker *egr-1/zif268*. *J. Neurosci.* 21:9724–32

Mataga N, Nagai N, Hensch TK. 2002. Permissive proteolytic activity for visual cortical plasticity. *Proc. Natl. Acad. Sci. USA* 99:7717–21

Maurer D, Lewis TL, Brent HP, Levin AV. 1999. Rapid improvement in the acuity of infants after visual input. *Science* 286:108–10

McDonald BJ, Amato A, Connolly CN, Benke D, Moss SJ, Smart TG. 1998. Adjacent

phosphorylation sites on GABAA receptor beta subunits determine regulation by cAMP-dependent protein kinase. *Nat. Neurosci.* 1:23–28
Meaney MJ. 2001. Maternal care, gene expression, and the transmission of individual differences in stress reactivity across generations. *Annu. Rev. Neurosci.* 24:1161–92
Mello C, Nottebohm F, Clayton D. 1995. Repeated exposure to one song leads to a rapid and persistent decline in an immediate early gene's response to that song in zebra finch telencephalon. *J. Neurosci.* 15:6919–25
Merzenich MM, Jenkins WM, Johnston P, Schreiner C, Miller SL, Tallal P. 1996. Temporal processing deficits of language-learning impaired children ameliorated by training. *Science* 271:77–81
Miller KD. 1996. Synaptic economics: competition and cooperation in synaptic plasticity. *Neuron* 17:371–74
Miyamoto H, Katagiri H, Hensch TK. 2003. Experience-dependent slow-wave sleep development. *Nat. Neurosci.* 6:553–54
Mooney R. 1999. Sensitive periods and circuits for learned birdsong. *Curr. Opin. Neurobiol.* 9:121–27
Mooney R, Rao M. 1994. Waiting periods versus early innervation: the development of axonal connections in the zebra finch song system. *J. Neurosci.* 14:6532–43
Morales B, Choi SY, Kirkwood A. 2002. Dark rearing alters the development of GABAergic transmission in visual cortex. *J. Neurosci.* 22:8084–90
Mower GD. 1991. The effect of dark rearing on the time course of the critical period in cat visual cortex. *Dev. Brain Res.* 58:151–58
Muir-Robinson G, Hwang BJ, Feller MB. 2002. Retinogeniculate axons undergo eye-specific segregation in the absence of eye-specific layers. *J. Neurosci.* 22:5259–64
Muller CM, Best J. 1989. Ocular dominance plasticity in adult cat visual cortex after transplantation of cultured astrocytes. *Nature* 342:427–30
Nase G, Weishaupt J, Stern P, Singer W, Monyer H. 1999. Genetic and epigenetic regulation of NMDA receptor expression in the rat visual cortex. *Eur. J. Neurosci.* 11:4320–26
Newport EL, Bavalier D, Neville HJ. 2001. Critical thinking about critical periods: perspectives on a critical period for language acquisition. In *Language, Brain and Cognitive Development: Essays in Honor of Jacques Mehler*, ed. E Dupoux, pp. 481–502. Cambridge, MA: MIT Press
O'Leary DDM, Ruff NL, Dyck RH. 1994. Development, critical period plasticity, and adult reorganizations of mammalian somatosensory system. *Curr. Opin. Neurobiol.* 4:535–44
Pantev C, Oostenveld R, Engelien A, Ross B, Roberts LE, Hoke M. 1998. Increased auditory cortical representation in musicians. *Nature* 392:811–14
Pascalis O, de Haan M, Nelson CA. 2002. Is face processing species-specific during the first year of life? *Science* 296:1321–23
Pena M, Maki A, Kovacic D, Dehaene-Lambertz G, Koizumi H, et al. 2003. Sounds and silence: an optical topography study of language recognition at birth. *Proc. Natl. Acad. Sci. USA* 100:11702–5
Penn AA, Riquelme PA, Feller MB, Shatz CJ. 1998. Competition in retinogeniculate patterning driven by spontaneous activity. *Science* 279:2108–12
Pham TA, Rubenstein JL, Silva AJ, Storm DR, Stryker MP. 2001. The CRE/CREB pathway is transiently expressed in thalamic circuit development and contributes to refinement of retinogeniculate axons. *Neuron* 31:409–20
Pizzorusso T, Medini P, Berardi N, Chierzi S, Fawcett JW, et al. 2002. Reactivation of ocular dominance plasticity in the adult visual cortex. *Science* 298:1248–51
Polley DB, Chen-Bee CH, Frostig RD. 1999. Two directions of plasticity in the sensory-deprived adult cortex. *Neuron* 24:623–37
Prusky GT, Douglas RM. 2003. Developmental plasticity of mouse visual acuity. *Eur. J. Neurosci.* 17:167–73
Ramoa AS, Paradiso MA, Freeman RD. 1988. Blockade of intracortical inhibition in kitten striate cortex: effects on receptive field

properties and associated loss of ocular dominance plasticity. *Exp. Brain Res.* 73:285–96

Recanzone G, Merzenich M, Jenkins W, Grajski K, Dinse H. 1992. Topographic reorganization of the hand representation in cortical area 3b owl monkeys trained in a frequency-discrimination task. *J. Neurophysiol.* 67:1031–56

Rema V, Armstrong-James M, Ebner FF. 2003. Experience-dependent plasticity is impaired in adult rat barrel cortex after whiskers are unused in early postnatal life. *J. Neurosci.* 23:358–66

Renger JJ, Hartman KN, Tsuchimoto Y, Yokoi M, Nakanishi S, et al. 2002. Experience-dependent plasticity without long-term depression by type 2 metabotropic glutamate receptors in developing visual cortex. *Proc. Natl. Acad. Sci. USA.* 99:1041–46

Riddle DR, Lo DC, Katz LC. 1995. NT-4-mediated rescue of lateral geniculate neurons from effects of monocular deprivation. *Nature* 378:189–91

Rochefort C, Gheusi G, Vincent JD, Lledo PM. 2002. Enriched odor exposure increases the number of newborn neurons in the adult olfactory bulb and improves odor memory. *J. Neurosci.* 22:2679–89

Roffwarg H, Muzio JN, Dement WC. 1966. Ontogenetic development of the human sleep-dream cycle. *Science* 604–19

Rossi FM, Pizzorusso T, Porciatti V, Marubio LM, Maffei L, et al. 2001. Requirement of the nicotinic acetylcholine receptor beta 2 subunit for the anatomical and functional development of the visual system. *Proc. Natl. Acad. Sci. USA* 98:6453–58

Ruthazer ES, Akerman CJ, Cline HT. 2003. Control of axon branch dynamics by correlated activity in vivo. *Science* 301:66–70

Ruthazer ES, Gillespie DC, Dawson TM, Snyder SH, Stryker MP. 1996. Inhibition of nitric oxide synthase does not prevent ocular dominance plasticity in kitten visual cortex. *J. Physiol.* 494:519–27

Sadato N, Okada T, Honda M, Yonekura Y. 2002. Critical period for cross-modal plasticity in blind humans: a functional MRI study. *Neuroimage* 16:389–400

Sakaguchi H. 1996. Sex differences in the developmental changes of GABAergic neurons in zebra finch song control nuclei. *Exp. Brain Res.* 108:62–68

Sanes JR, Lichtman JW. 1999a. Development of the vertebrate neuromuscular junction. *Annu. Rev. Neurosci.* 22:389–442

Sanes JR, Lichtman JW. 1999b. Can molecules explain long-term potentiation? *Nat. Neurosci.* 2:597–604

Sawtell NB, Frenkel MY, Philpot BD, Nakazawa K, Tonegawa S, et al. 2003. NMDA receptor-dependent ocular dominance plasticity in adult visual cortex. *Neuron* 38:977–85. Erratum. 2003. *Neuron* 39(4):727

Scheich H. 1987. Neural correlates of auditory filial imprinting. *J. Comp. Physiol. A* 161:605–19

Schlaug G, Jancke L, Huang Y, Steinmetz H. 1995. In vivo evidence of structural brain asymmetry in musicians. *Science* 267:699–701

Schoop VM, Gardziella S, Muller CM. 1997. Critical period-dependent reduction of the permissiveness of cat visual cortex tissue for neuronal adhesion and neurite growth. *Eur. J. Neurosci.* 9:1911–22

Schwab ME, Caroni P. 1988. Oligodendrocytes and CNS myelin are nonpermissive substrates for neurite growth and fibroblast spreading in vitro. *J. Neurosci.* 8:2381–93

Scott JP. 1962. Critical periods in behavioral development. *Science* 138:949–58

Shaw C, Aoki C, Wilkinson M, Prusky G, Cynader M. 1987. Benzodiazepine ([3H]flunitrazepam) binding in cat visual cortex: ontogenesis of normal characteristics and the effects of dark rearing. *Brain Res.* 465:67–76

Shaw C, Cynader M. 1984. Disruption of cortical activity prevents ocular dominance changes in monocularly deprived kittens. *Nature* 308:731–34

Shimegi S, Fischer QS, Yang Y, Sato H, Daw NW. 2004. Blockade of cyclic

AMP-dependent protein kinase does not prevent the reverse ocular dominance shift in kitten visual cortex. *J. Neurophysiol.* 90:4027–32

Shingo T, Gregg C, Enwere E, Fujikawa H, Hassam R, et al. 2003. Pregnancy-stimulated neurogenesis in the adult female forebrain mediated by prolactin. *Science* 299:117–20

Sibille E, Pavlides C, Benke D, Toth M. 2000. Genetic inactivation of the Serotonin(1A) receptor in mice results in downregulation of major GABA(A) receptor alpha subunits, reduction of GABA(A) receptor binding, and benzodiazepine-resistant anxiety. *J. Neurosci.* 20:2758–65

Sieghart W. 1995. Structure and pharmacology of γ-aminobutyric acid$_A$ receptor subtypes. *Pharmacol. Rev* . 47:181–234

Simons DJ, Land PW. 1994. Neonatal whisker trimming produces greater effects in nondeprived than deprived thalamic barreloids. *J. Neurophysiol.* 72:1434–37

Smetters DK, Hahm J, Sur M. 1994. An N-methyl-D-aspartate receptor antagonist does not prevent eye-specific segregation in the ferret retinogeniculate pathway. *Brain Res.* 658:168–78

Soghomonian JJ, Martin DL. 1998. Two isoforms of glutamate decarboxylase: Why? *Trends Pharmacol.* 19:500–5

Somogyi P, Tamas G, Lujan R, Buhl EH. 1998. Salient features of synaptic organisation in the cerebral cortex. *Brain Res. Rev.* 26:113–35

Song S, Miller KD, Abbott LF. 2000. Competitive Hebbian learning through spike-timing-dependent synaptic plasticity. *Nat. Neurosci.* 3:919–26

Stellwagen D, Shatz CJ. 2002. An instructive role for retinal waves in the development of retinogeniculate connectivity. *Neuron* 33:357–67

Stern EA, Maravall M, Svoboda K. 2001. Rapid development and plasticity of layer 2/3 maps in rat barrel cortex in vivo. *Neuron* 31:305–15

Stockard CR. 1921. Developmental rate an structural expression: an experimental study of twins, 'double monsters' and single deformities, and the interaction among embryonic organs during their origin and development. *Am. J. Anat.* 28:115–277

Taha S, Stryker MP. 2002. Rapid ocular dominance plasticity requires cortical but not geniculate protein synthesis. *Neuron* 34:425–36

Tang YP, Shimizu E, Dube GR, Rampon C, Kerchner GA, et al. 1999. Genetic enhancement of learning and memory in mice. *Nature* 401:63–69

Tchernichovski O, Lints T, Mitra PP, Nottebohm F. 1999. Vocal imitation in zebra finches is inversely related to model abundance. *Proc. Natl. Acad. Sci. USA* 96:12901–4

Temple E, Deutsch GK, Poldrack RA, Miller SL, Tallal P, et al. 2003. Neural deficits in children with dyslexia ameliorated by behavioral remediation: evidence from functional MRI. *Proc. Natl. Acad. Sci. USA* 100:2860–65

Thompson WJ. 1985. Activity and synapse elimination at the neuromuscular junction. *Cell Mol. Neurobiol.* 5:167–82

Tononi G, Cirelli C. 2001. Modulation of brain gene expression during sleep and wakefulness: a review of recent findings. *Neuropsychopharmacology* 25:S28–35

Trachtenberg JT, Stryker MP. 2001. Rapid anatomical plasticity of horizontal connections in the developing visual cortex. *J. Neurosci.* 21:3476–82

Trepel C, Racine RJ. 2000. GABAergic modulation of neocortical long-term potentiation in the freely moving rat. *Synapse* 35:120–28

Upton AL, Salichon N, Lebrand C, Ravary A, Blakely R, et al. 1999. Excess of serotonin (5-HT) alters the segregation of ispilateral and contralateral retinal projections in monoamine oxidase A knock-out mice: possible role of 5-HT uptake in retinal ganglion cells during development. *J. Neurosci.* 19:7007–24

Van der Loos H, Woolsey TA. 1973. Somatosensory cortex: structural alterations following early injury to sense organs. *Science* 179:395–98

Videen TO, Daw NW, Collins RC. 1986. Penicillin-induced epileptiform activity does not prevent ocular dominance shifts in monocularly deprived kittens. *Brain Res.* 371:1–8

Wallhausser-Franke E, Nixdorf-Bergweiler BE, DeVoogd TJ. 1995. Song isolation is associated with maintaining high spine frequencies on zebra finch LMAN neurons. *Neurobiol. Learn. Mem.* 64:25–35

Walsh MK, Lichtman JW. 2003. In vivo time-lapse imaging of synaptic takeover associated with naturally occurring synapse elimination. *Neuron* 37:67–73

Wartenburger I, Heekeren HR, Abutalebi J, Cappa SF, Villringer A, Perani D. 2003. Early setting of grammatical processing in the bilingual brain. *Neuron* 37:159–70

Welker WI. 1964. Analysis of sniffing of the albino rat. *Behavior* 22:223–44

White SA. 2001. Learning to communicate. *Curr. Opin. Neurobiol.* 11:510–20

Wiesel TN. 1982. Postnatal development of the visual cortex and the influence of environment. *Nature* 299:583–91

Wiesel TN, Hubel DH. 1963. Single-cell responses in striate cortex of kittens deprived of vision in one eye. *J. Neurophysiol.* 26:1003–17

Winkler I, Kushnerenko E, Horvath J, Ceponiene R, Fellman V, et al. 2003. Newborn infants can organize the auditory world. *Proc. Natl. Acad. Sci. USA* 100:11812–15

Wong RO. 1999. Retinal waves and visual system development. *Annu. Rev. Neurosci.* 22: 29–47

Wong WT, Myhr KL, Miller ED, Wong RO. 2000. Developmental changes in the neurotransmitter regulation of correlated spontaneous retinal activity. *J. Neurosci.* 20:351–60

Yamazaki K, Beauchamp GK, Kupniewski D, Bard J, Thomas L, et al. 1988. Familial imprinting determines H-2 selective mating preferences. *Science* 240:1331–32

Yang T, Maunsell JHR. 2004. The effect of perceptual learning on neuronal responses in monkey visual area V4. *J. Neurosci.* 24: 1617–26

Yokoi M, Mori K, Nakanishi S. 1995. Refinement of odor molecule tuning by dendrodendritic synaptic inhibition in the olfactory bulb. *Proc. Natl. Acad. Sci. USA* 92:3371–75

Zheng W, Knudsen EI. 1999. Functional selection of adaptive auditory space map by GABAA-mediated inhibition. *Science* 284: 962–65

Annu. Rev. Neurosci. 2004. 27:581–609
doi: 10.1146/annurev.neuro.27.070203.144238

CEREBELLUM-DEPENDENT LEARNING: The Role of Multiple Plasticity Mechanisms

Edward S. Boyden,* Akira Katoh,* and Jennifer L. Raymond
Department of Neurobiology, Stanford University, Stanford, California 94305; email: boyden@stanford.edu, akato@stanford.edu, jenr@stanford.edu
These authors contributed equally to this work.

Key Words memory, coding, motor systems, consolidation, VOR, synaptic plasticity, LTD

■ **Abstract** The cerebellum is an evolutionarily conserved structure critical for motor learning in vertebrates. The model that has influenced much of the work in the field for the past 30 years suggests that motor learning is mediated by a single plasticity mechanism in the cerebellum: long-term depression (LTD) of parallel fiber synapses onto Purkinje cells. However, recent studies of simple behaviors such as the vestibulo-ocular reflex (VOR) indicate that multiple plasticity mechanisms contribute to cerebellum-dependent learning. Multiple plasticity mechanisms may provide the flexibility required to store memories over different timescales, regulate the dynamics of movement, and allow bidirectional changes in movement amplitude. These plasticity mechanisms must act in combination with appropriate information-coding strategies to equip motor-learning systems with the ability to express learning in correct contexts. Studies of the patterns of generalization of motor learning in the VOR provide insight about the coding of information in neurons at sites of plasticity. These principles emerging from studies of the VOR are consistent with results concerning more complex behaviors and thus may reflect general principles of cerebellar function.

INTRODUCTION

Motor learning is the process of improving the smoothness and accuracy of movements. It is obviously necessary for complicated movements such as playing the piano and climbing trees, but it is also important for calibrating simple movements like reflexes, as parameters of the body and environment change over time. The cerebellum is critical for motor learning. As a result of the universal need for properly calibrated movement, it is not surprising that the cerebellum is widely conserved in vertebrates. Thus the basic architecture of the cerebellum, renowned for its beauty, is also generally useful. In this review we describe the recent progress made in understanding the function of the cerebellum in encoding motor memories.

We begin by describing how the anatomy of the cerebellum inspired a model of motor learning that has influenced much of the research in the field for the last 30 years. We then review the evidence for and against this model from studies of a simple behavior, the vestibulo-ocular reflex (VOR). This model and the main competing model each propose a single plasticity mechanism to explain motor learning, but neither model can account for all of the data regarding motor learning in the VOR. One resolution of these two competing hypotheses is that more than one plasticity mechanism may contribute to cerebellum-dependent motor learning. Recent work suggests that multiple plasticity mechanisms regulate the amplitude, dynamics, and consolidation of learned movements. The properties of motor learning also are influenced by the information-coding strategies used at the sites of plasticity. Studies of the generalization of learning to stimuli and contexts different from those present during learning are revealing constraints on these coding strategies. Analysis of the interaction of plasticity and coding in the VOR circuit is beginning to unveil how the cerebellum and related structures can support motor learning that is reliable yet flexible.

A MODEL OF LEARNING INSPIRED BY THE UNIQUE ARCHITECTURE OF THE CEREBELLUM

The architecture of the cerebellum is often described as crystalline because it is composed of repeated modules, each containing the same few cell types connected in the same manner. Different regions of the cerebellum receive different inputs and project to different targets, yet the uniformity of the cerebellar architecture suggests that these modules process the signals they receive in similar ways. Thus one can be optimistic that studying the operation of the cerebellum during one learning task may reveal general principles that characterize the function of the cerebellum in many tasks.

An influential model of the cerebellum proposes that its general function is to act as a pattern classification device that can be taught to generate an appropriate output in response to an arbitrary input (Figure 1) (Albus 1971, Marr 1969). Mossy fibers provide sensory and motor inputs to the cerebellum. Purkinje cells provide the sole output of the cerebellar cortex and drive specific movements. The mossy fiber inputs are connected to the Purkinje cell outputs by a disynaptic pathway through cerebellar granule cells, as well as by pathways through inhibitory interneurons. Over a hundred thousand granule cell axons, known as parallel fibers, synapse onto a single Purkinje cell. The Marr-Albus model proposes that changes in the strengths of parallel fiber–Purkinje cell synapses could store stimulus-response associations by linking inputs with appropriate motor outputs.

The Marr-Albus model was inspired by some unusual and striking features of the architecture of the cerebellum. One striking feature is the enormous number of granule cells, which make up roughly half the neurons in the brain. According to the Marr-Albus model, the large number of granule cells enables this population of neurons to perform pattern separation. Similar patterns of mossy fiber activity would be sparsely reencoded into largely nonoverlapping populations of

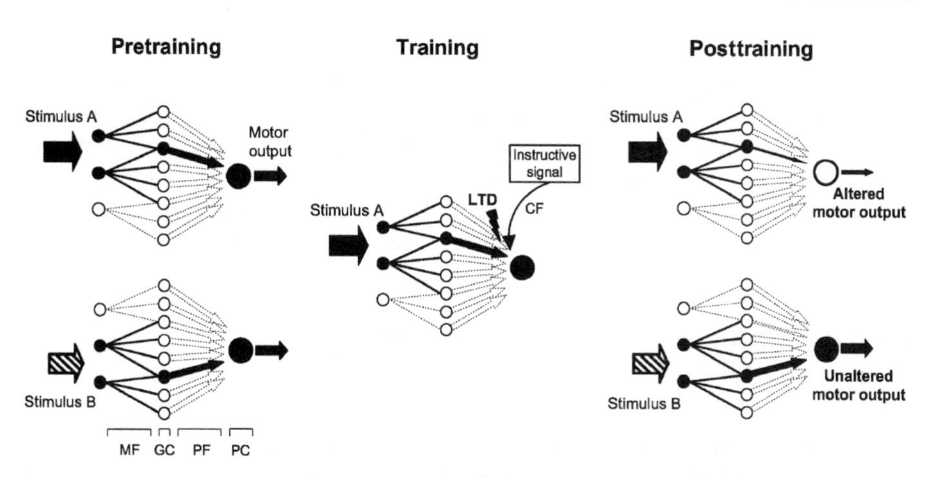

Figure 1 The Marr-Albus model of motor learning. In this model, sparse reencoding in the cerebellum enables learning to create very precise stimulus-response mappings. Granule cells (GC) spike only when sufficient mossy fiber (MF) input is present, causing overlapping mossy fiber input patterns to be reencoded in nonoverlapping populations of granule cells. Plasticity controlled by climbing fibers (CF) weakens the strength of parallel fiber (PF)–Purkinje cell (PC) synapses (via LTD, indicated by the lightning bolt). LTD alters the efficacy of stimulus A firing the Purkinje cell and thus produces altered motor output. Because stimulus B activates different parallel fibers than stimulus A activates, the motor response to stimulus B is unaltered by training. Open circles and dotted lines indicate inactive neurons and synapses. Filled circles and solid lines indicate active neurons and synapses. The thin parallel fiber arrow indicates a synapse weakened by LTD.

granule cell activity, thereby allowing sensory events encoded by similar mossy fiber patterns to be associated with different motor outputs.

Another striking feature of the cerebellum is the climbing fiber input to Purkinje cells. Each Purkinje cell receives input from just one climbing fiber axon, which originates in the inferior olive. The climbing fiber forms a very strong synapse onto the Purkinje cell, with each presynaptic spike triggering a postsynaptic spike. Despite the powerful connection, the climbing fiber makes only a small contribution to the total spike output of the rapidly firing Purkinje cell because the climbing fiber fires at a much lower rate. Therefore it was postulated that the climbing fiber may serve a special function other than ordinary signal transmission. According to the Marr-Albus model, the climbing fiber provides the instructive signal that regulates the strength of parallel fiber–Purkinje cell synapses and thereby guides the encoding of new stimulus-response associations. Marr and Albus differed as to the nature of the instructive signal: Marr believed it to be a positive reinforcer, strengthening parallel fiber synapses when the Purkinje cell output was correct, whereas Albus believed it to be an error signal, weakening synapses when the output was incorrect. Consistent with the latter idea, electrical stimulation of climbing fibers induces a decrease in synaptic strength, called cerebellar LTD, in parallel fibers that are active simultaneously (Ito et al. 1982b). The Albus model proposed that this single plasticity mechanism would mediate motor learning by restricting the expression of a movement coded for by a specific Purkinje cell to specific contexts in which that motor response was appropriate. For many years, the elegance of the Marr-Albus theory has focused attention on cerebellar LTD as a key candidate for the plasticity mechanism mediating motor learning.

MOTOR LEARNING IN A SIMPLE BEHAVIOR: THE VESTIBULO-OCULAR REFLEX

Although motor learning is capable of achieving Olympic feats of skill, much has been learned from studies of simple behaviors. In this review we focus on recent advances in understanding motor learning from studies of the VOR. The VOR has many properties that make it amenable to experimentation, including easily controlled sensory inputs, quantifiable motor outputs, a well-characterized circuit anatomy, and the capacity to be studied in a variety of species.

The VOR is a reflex eye movement that stabilizes images on the retina during head movement by producing an eye movement in the direction opposite to head movement. The VOR elicits eye movements in response to both horizontal and vertical head rotations, as well as head translations. Although the function of the VOR is to support clear vision, the VOR is measured in the dark to isolate the eye movements driven by vestibular stimuli from eye movements driven by visual stimuli. The performance of the VOR is characterized by the gain, which is defined as the ratio between eye and head velocities. If the gain of the VOR is poorly calibrated, then head movements result in image motion on the retina, resulting in blurred vision. Under such conditions, motor learning adjusts the gain of the VOR to produce more accurate eye motion. Such adjustments are needed throughout life, as neurons and muscles develop, weaken, and die—or for humans, when a new pair of eyeglasses changes the magnification of the visual field. In the laboratory, we induce motor learning in the VOR by pairing image motion with head motion. Depending on the relative direction of head motion and image motion, the gain of the VOR can be adaptively increased or decreased. An increase in VOR gain is induced by image motion in the direction opposite that of the head (gain-up stimulus), and a decrease in VOR gain is induced by image motion in the same direction as the head (gain-down stimulus) (Figure 2).

The main neural circuit for the VOR is simple (Figure 3): Vestibular nuclei in the brainstem receive signals related to head movement from the vestibular nerve and project to oculomotor nuclei, which contain motoneurons that drive eye muscle activity. The flocculus and ventral paraflocculus of the cerebellum form an inhibitory side loop in the VOR circuit. Mossy fibers provide vestibular input to the cerebellum, and Purkinje cells inhibit VOR interneurons in the vestibular nucleus. For the VOR, the vestibular nuclei serve the role that the deep cerebellar

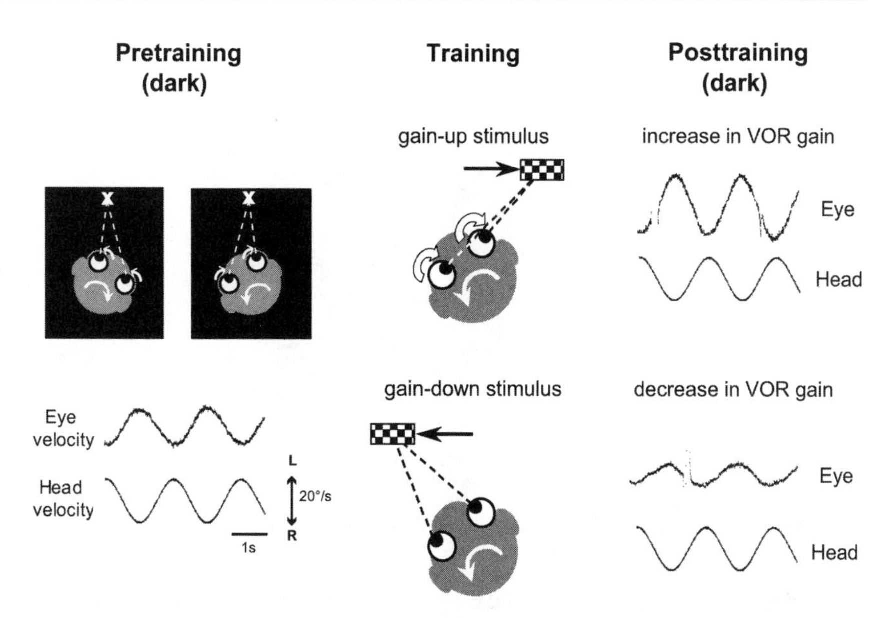

Figure 2 Motor learning in the VOR. Before learning, eyes move with the same speed, but in the opposite direction, as the head, keeping the eyes stationary in world coordinates. Focal points of the eyes are indicated by an *x*. An increase in VOR gain is produced by training with image motion in the direction opposite that of the head (gain-up stimulus). A decrease in VOR gain is produced by training with image motion in the same direction as the head (gain-down stimulus). After each training session, the VOR is remeasured in the dark with the same head movement stimulus used in the pretraining measurements. The data shown are representative traces acquired from monkeys after 2 h of training.

nuclei serve for other cerebellum-dependent behaviors. Thus, when we refer to the contribution of the cerebellum to the VOR, we mean the cerebellar cortex.

ROLE OF THE CEREBELLUM IN MOTOR LEARNING IN THE VOR: TWO INFLUENTIAL MODELS

Lesion and recording studies demonstrate the necessity of the cerebellum for motor learning in the VOR. Surgical lesions, pharmacological inactivation, and genetic disruption of the cerebellum all abolish the ability to adaptively modify VOR gain (Ito et al. 1982a, Koekkoek et al. 1997, Lisberger et al. 1984, Luebke & Robinson 1994, McElligott et al. 1998, Michnovicz & Bennett 1987, Nagao 1983, Rambold et al. 2002, Robinson 1976, Van Alphen et al. 2002), and Purkinje cells exhibit altered responses during the performance of the VOR after motor learning (Hirata & Highstein 2001, Lisberger et al. 1994a, Miles et al. 1980, Nagao 1989,

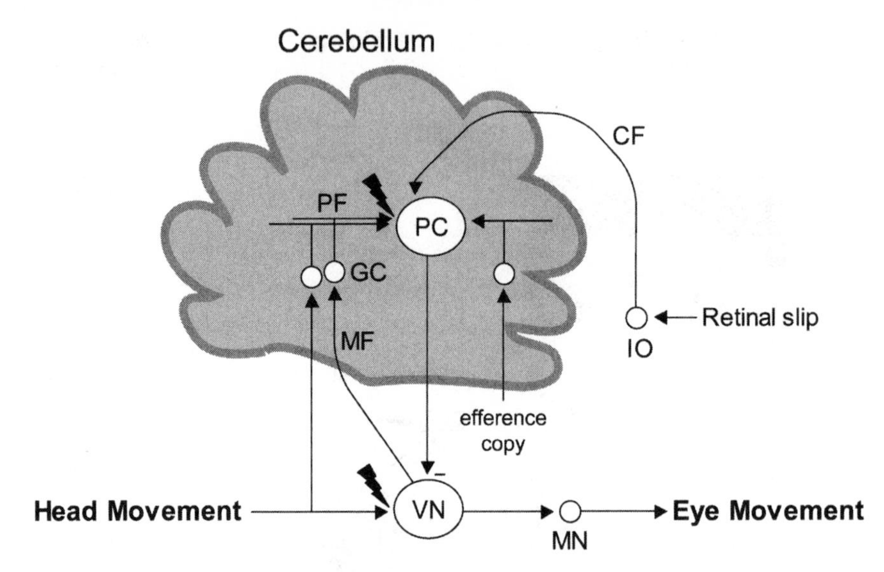

Figure 3 Proposed sites of plasticity in the circuit for the VOR (indicated by lightning bolts). CF, climbing fibers; GC, granule cells; IO, inferior olive; MF, mossy fibers; MN, oculomotor nuclei; PC, Purkinje cells; PF, parallel fibers; VN, vestibular nuclei.

Watanabe 1984). On the basis of what is known about the contribution of Purkinje cells to eye movements, the responses of Purkinje cells during the VOR change in the correct direction to contribute to the altered eye movement response to head movement. This finding suggests that the cerebellum contributes to the expression of the altered VOR gain.

Despite general agreement that the cerebellum is important for motor learning in the VOR, there is disagreement over the specific role it plays; two long-standing hypotheses provide two very different accounts. Ito proposed an implementation of the Marr-Albus hypothesis (Ito 1972, 1982), in which the role of the cerebellum is to store the motor memory for the learned change in VOR gain. More specifically, Ito proposed that during training, the pairing of visual image motion and head motion causes coincident visually driven climbing fiber activity and vestibularly driven parallel fiber activity, which results in LTD of the active parallel fibers. The climbing fiber, which reports retinal slip and therefore the error in the function of the VOR, acts as the instructive signal guiding plasticity. The altered weights of the vestibular parallel fiber–Purkinje cell synapses would cause an altered eye movement response to the vestibular stimulus and thus could encode the motor memory for the altered VOR gain. An alternative model was proposed by Miles & Lisberger (1981). They proposed that the role of the cerebellum was not to store the motor memory but rather to compute the instructive signal guiding the induction of plasticity. According to this model, Purkinje cells convey the instructive signal to the vestibular nucleus, where it triggers heterosynaptic changes in the connections between vestibular afferents and neurons in the vestibular nucleus. The Miles-Lisberger model attributes the altered responses of Purkinje cells to altered input to the cerebellum from mossy fibers carrying an efference copy of the altered eye movement command created in the vestibular nucleus. Thus the Purkinje cells could exhibit learning-related changes in their responses during performance of the VOR, even if no synaptic changes occur within the cerebellum. In contrast, the Marr-Albus-Ito model attributes altered Purkinje cell responses, recorded during performance of the VOR, to synaptic changes onto Purkinje cells themselves.

Thus two long-standing hypotheses propose (*a*) different sites of plasticity for motor learning, (*b*) different instructive signals guiding plasticity at these sites, and (*c*) different explanations for the changes recorded in Purkinje cells during performance of the VOR. Experimental tests of these three predictions provide support for and against each of these two hypotheses.

Evidence for a Role of Plasticity in the Vestibular Nuclei

Support for plasticity in the vestibular nucleus comes from experiments that evaluate whether the changes in Purkinje cells are due to plasticity in the vestibular pathways through the cerebellum or are simply a reflection of altered efference copy resulting from plasticity elsewhere in the circuit. This issue cannot be resolved by recording from Purkinje cells during the performance of the VOR because both vestibular and efference-copy pathways are activated. Therefore, several investigators recorded from Purkinje cells during behavioral paradigms designed to independently assess the signals carried in these two pathways. One way to isolate the input to Purkinje cells carried by vestibular pathways is to rotate the animal while it holds its eyes still (VOR cancellation). Another way is to measure the input from efference-copy pathways during pursuit eye movements and to subtract this from the activity recorded during performance of the VOR. Measured in both of these ways, the sensitivity of Purkinje cells to vestibular input changes in the direction opposite to that predicted by the action of cerebellar LTD, and in the direction opposite that required to account for the change in VOR gain, given the known connectivity of Purkinje cells with their downstream targets (Hirata & Highstein 2001, Lisberger et al. 1994a, Miles et al. 1980). This result is difficult to reconcile with the idea that plasticity of the vestibular inputs to Purkinje cells is the primary mediator of the learned change in the gain of the VOR. Although some investigators have raised concerns about the assumptions associated with comparing neural responses across these different behavioral conditions (Ito 1993, Tabata et al. 2002), further comparisons between neural signals recorded in a single behavioral condition circumvent these concerns and provide sound evidence for extracerebellar plasticity. In particular, the responses of vestibular nucleus neurons during VOR cancellation exhibit learning-related changes that cannot be accounted for by the learning-related changes in the responses of Purkinje cells during VOR cancellation (Lisberger et al. 1994b). Because the changes recorded in the vestibular nucleus neurons cannot be accounted for by the input they receive from Purkinje cells, the changes in the vestibular nucleus must result from plasticity at a site outside the cerebellum.

The latencies of learned responses are also consistent with plasticity in the vestibular nuclei (reviewed in detail in du Lac et al. 1995). Briefly, eye movements that depend only on the direct pathway through the vestibular nuclei should be executed with a shorter latency than eye movements that depend on the side loop through the cerebellum. The short latency of the first modified component of the VOR suggests that plasticity could occur in the direct pathway through the vestibular nuclei. However, in the broad distribution of Purkinje cells in the cerebellum, some cells respond with latencies short enough that they could contribute to the earliest modified component of the VOR. Furthermore, the pathway through the cerebellum could contribute to the modification of the VOR at longer latencies after onset of head movement.

A number of studies have used lesions or inactivation of the cerebellum after training to evaluate whether memory storage occurs outside the cerebellum. Investigators have obtained a variety of outcomes in such studies (Luebke & Robinson 1994, McElligott et al. 1998, Nagao & Kitazawa 2003, Partsalis et al. 1995, Pastor et al. 1994, Robinson 1976), with most finding that lesions did not completely eliminate the expression of learned changes in VOR gain. This retention of learned changes after lesions of the cerebellum would indicate a site of memory storage outside the cerebellum, if the lesion could be shown to be complete, which is difficult to do. In addition, several of these studies reported at least a partial loss of learned changes, which could reflect a site of memory storage in the cerebellum but could also simply reflect a requirement for the cerebellum for the expression of changes stored elsewhere in the VOR circuit.

Anatomical studies of another form of cerebellum-dependent learning also indicate a role for a site downstream of Purkinje cells in the storage of cerebellum-dependent memories. After a classical-conditioning paradigm in which rats were trained to blink in response to a tone (by pairing a tone with an airpuff), excitatory synapse number increased in the deep cerebellar nucleus (Kleim et al. 2002). No changes were found, however, in animals that experienced unpaired tones and airpuffs, which suggests that the anatomical changes observed could be attributed to learning. To the extent that changes in synapse number reflect plasticity, these results corroborate a contribution of the deep cerebellar nuclei to memory storage. In the cerebellar cortex, Purkinje cell morphology also changed after eyeblink conditioning, but these changes also were observed in the unpaired controls and therefore may be induced simply by activity in the cerebellar circuitry, rather than by learning (Anderson et al. 1999).

Evidence for a Role of Cerebellar LTD

In the Marr-Albus-Ito model, changes in VOR gain result from LTD of parallel fibers carrying vestibular signals. In recent years several groups have disrupted molecular pathways required for cerebellar LTD and reported learning impairments. The results of these studies are consistent with a role of cerebellar LTD in motor learning in the VOR, but in each of these studies the learning impairment could potentially be explained by disruption of plasticity mechanisms other than LTD or disruption of signal processing in the cerebellum, resulting in a functional cerebellar lesion. Although no individual result is conclusive, these studies collectively motivate further investigation of how LTD may contribute to motor learning.

Nitric oxide (NO) activity contributes to the induction of cerebellar LTD (Daniel et al. 1993, Lev-Ram et al. 1997, Shibuki & Okada 1991). Blocking cerebellar NO activity by applying scavengers or inhibitors prevents motor learning in the VOR in monkeys, rabbits, and goldfish (Li et al. 1995, Nagao & Ito 1991). Thus a NO-dependent process is important for changes in VOR gain. This NO-dependent process could be cerebellar LTD, but NO has additional actions on the VOR circuit that could potentially mediate its effects on learning. NO is involved in a form of parallel fiber–Purkinje cell long-term potentiation (LTP) (Lev-Ram et al. 2002), and it regulates the intrinsic spiking of Purkinje cells (Smith & Otis 2003), modulates synapses onto granule cells (Wall 2003), and alters processing in the vestibular nuclei (Moreno-Lopez et al. 2002).

Another manipulation that disrupts LTD targets the $\delta 2$ subunit of the glutamate receptor (GluRδ2), which is expressed selectively in Purkinje cells. Mice lacking GluRδ2 show neither cerebellar LTD nor motor learning in the VOR (Katoh et al. 2001). However, these mice have alterations in their basal VOR, so the impairment in motor learning may reflect a performance deficit rather than a learning deficit per se.

Another molecule important for cerebellar LTD is protein kinase C (PKC) (Crepel & Krupa 1988, Linden & Connor 1991). Transgenic mice expressing an inhibitor of protein kinase C (PKCI), under control of the Purkinje cell–specific promoter L7, showed impaired LTD in cultured Purkinje cells. These mice also exhibited no motor learning in the VOR with one hour of training (De Zeeuw et al. 1998). The impairment of LTD may have been responsible for the learning impairment. However, in Purkinje cells, PKC is a multifunctional enzyme that regulates potassium conductances and plasticity of the climbing fiber–Purkinje cell synapse, in addition to plasticity of the parallel fiber–Purkinje cell synapse (Hansel & Linden 2000, Widmer et al. 2003). Thus the transgene may disrupt normal signaling through the Purkinje cells during motor learning and thereby corrupt the instructive signals sent by Purkinje cells to downstream sites like the vestibular nuclei. These animals have normal resting levels of Purkinje cell activity (Goossens et al. 2001), but Purkinje cells have not been recorded during exposure to stimuli that induce motor learning to determine whether the signals they transmit to the vestibular nuclei are abnormal. Fortunately, enough is known about normal signaling in Purkinje cells during the induction and expression of motor learning that in vivo recording studies should readily resolve this confusion.

Consistent with these results from the VOR, studies using other cerebellum-dependent learning tasks have reported that pharmacological or genetic disruption of molecules required for cerebellar LTD impairs motor learning (e.g., Aiba et al. 1994; Chapman et al. 1992; Katoh et al. 2000; Kishimoto et al. 2001a,b; Koekkoek et al. 2003; Nagao & Kitazawa 2000; Shibuki et al. 1996; Shutoh et al. 2002, 2003; Yanagihara & Kondo 1996).

Studies of the instructive signals controlling induction of plasticity also are consistent with a role of cerebellar LTD in motor learning in the VOR. In the Miles-Lisberger model the neural instructive signal is carried by Purkinje cells, which control the induction of plasticity in the vestibular nucleus. Alternatively, in the Marr-Albus-Ito model, the neural instructive signal controlling induction of motor learning is carried by climbing fibers. The climbing fibers can trigger LTD in parallel fibers. Climbing fiber inputs to the flocculus and ventral paraflocculus fire in response to image motion on the retina and thus can indicate errors in VOR performance when the reflex fails to stabilize images on the retina (Ghelarducci et al. 1975, Graf et al. 1988, Simpson & Alley 1974, Stone & Lisberger 1990). Destruction of the inferior olive or the nucleus of the optic tract (which provides input to the olive) alters visual tracking performance and abolishes the ability to change the VOR gain (Haddad et al. 1980, Ito & Miyashita 1975, Tempia et al. 1991, Yakushin et al. 2000b). An analysis of the two candidate instructive signals during motor learning in the VOR induced with high-frequency stimuli also suggests a role for climbing fibers in the induction of motor learning (Raymond & Lisberger 1998). For a neuron to provide an instructive signal for the VOR, its patterns of spiking must discriminate stimuli that increase and decrease VOR gain. During training with stimuli typically used to induce learning (low frequencies of head and visual stimulus oscillation), both climbing fiber and Purkinje cell responses discriminate the gain-up and gain-down stimuli (Simpson & Alley 1974, Watanabe 1984). Therefore, either population of neurons could, in theory, trigger the different changes required under these two conditions, and hence these findings are compatible with both the Marr-Albus-Ito and Miles-Lisberger models. However, during the induction of learning with high-frequency training stimuli, only climbing fibers, not Purkinje cells, discriminate the stimuli that increase or decrease VOR gain (Raymond & Lisberger 1998). This finding indicates that Purkinje cells cannot trigger the different changes induced by the gain-up and gain-down stimuli at high frequencies. Therefore, at least at high frequencies, the climbing fiber is the best candidate for the instructive signal guiding motor learning in the VOR.

MULTIPLE PLASTICITY MECHANISMS CONTRIBUTE TO MOTOR LEARNING IN THE VOR

We have reviewed the support for both the Marr-Albus-Ito and the Miles-Lisberger models, and in brief, neither model can account for all of the experimental data accumulated to date. The single biggest challenge to the Marr-Albus-Ito hypothesis is the observation that after motor learning is induced in the VOR, the vestibular sensitivity of Purkinje cells changes in the wrong direction to account for changes in the vestibular nucleus and to mediate the change in VOR gain. In addition, recordings of climbing fiber responses during exposure to a broad range of stimuli that induce motor learning suggest that the climbing fibers cannot serve as the sole instructive signal (Ke & Raymond 2001, 2002). A major challenge to the Miles-Lisberger hypothesis is the observation that adaptive changes in VOR gain can occur under some conditions in which the Purkinje cells fail to discriminate stimuli that increase and decrease VOR gain and therefore provide no useful instructive signal to guide plasticity (Raymond & Lisberger 1998). Also, the plasticity mechanism predicted by the Miles-Lisberger model has not been observed experimentally. Inputs to the vestibular nuclei undergo both potentiation and depression (Caria et al. 1996, 2001; Racine et al. 1986), but whether Purkinje cell spiking, the key instructive signal in the Miles-Lisberger model, can control the induction of plasticity at these synapses is not clear (Babalian & Vidal 2000).

One possible resolution of the conflicting data for and against these two models is that the plasticity mechanisms proposed by both models contribute to motor learning in the VOR. Recent studies using in vitro preparations have described many plasticity mechanisms in the cerebellum and related circuits (for review, see Hansel et al. 2001). In vivo evidence that multiple plasticity mechanisms contribute to motor learning is accumulating as well. In the next few sections we consider three properties of motor learning in the VOR that seem to be supported by multiple plasticity mechanisms.

Regulation of Movement Dynamics

The contribution of multiple plasticity mechanisms to motor learning in the VOR was first proposed to explain why the vestibular sensitivity of Purkinje cells would change in the wrong direction to mediate the observed change in VOR gain. The VOR circuit contains a positive feedback loop: Purkinje cell activity can drive eye movements, and signals related to eye-movement commands then feed back to drive more Purkinje cell activity. One model postulates that the VOR circuit avoids the instability associated with feedback by using plasticity mechanisms distributed between the cerebellum and vestibular nuclei (Lisberger & Sejnowski 1992). According to this model, the change in VOR gain is due to plasticity in the vestibular nuclei. These nuclear changes alone would cause unstable eye movements, such as continuously accelerating eye movements in response to a constant-velocity head movement. However, additional changes in the cerebellar cortex could restore stability. The observed changes in the vestibular sensitivity of Purkinje cells are in the wrong direction to mediate the change in VOR gain, but they are in the right direction to maintain the proper time dynamics of movement (Hirata & Highstein 2001, Lisberger 1994, Lisberger et al. 1994a).

The cerebellar cortex regulates the timing of other learned movements as well (see Buonomano & Mauk 2004, in this volume). Although a posttraining lesion of the deep cerebellar nucleus abolishes conditioned eyeblink responses completely, a lesion of the cerebellar cortex primarily disrupts the timing of learned responses

(McCormick & Thompson 1984, Perrett et al. 1993). This finding suggests that the cerebellar cortex stores timing-related information, whereas the deep cerebellar nucleus is more important for storing the amplitude of the response. Accordingly, in human patients, disruption of the cerebellum results in deficits in the timing of movement (e.g., Ivry et al. 1988, Spencer et al. 2003).

Consolidation of Memories

Multiple plasticity mechanisms may be used to maintain motor memories over different timescales. Many memories undergo consolidation after learning, making them less labile. Consolidation is a transformation in the way a memory is encoded, from using a short-term plasticity mechanism to using one that is distinctly long-lasting. Much work on consolidation has focused on the declarative memory system, but motor memories can undergo consolidation as well.

In humans learning to perform reaching movements in an altered force field, several hours must pass before the motor memory for one force field becomes resistant to erasure by the learning of a new force field (Brashers-Krug et al. 1996, Shadmehr & Brashers-Krug 1997). For this task, execution of movements immediately after learning results in frontal cortex activation, but in the hours after training is complete, the cerebellum becomes active during task performance (Shadmehr & Holcomb 1997). Thus the role of the cerebellum may change with time after acquisition of learning, which is consistent with the general notion that consolidation can involve redistribution of information between different parts of the brain, from short-term areas to long-term areas (Marr 1971).

Application of the information redistribution hypothesis to motor learning in the VOR has led to the idea that the motor memory for a learned change in VOR gain is transferred from the cerebellar cortex to the vestibular nuclei during long periods of training (Galiana 1986, Nagao & Kitazawa 2003, Peterson et al. 1991). This idea was inspired in part by the variable results obtained when various groups lesioned or inactivated the cerebellum after inducing motor learning in the VOR (Luebke & Robinson 1994, McElligott et al. 1998, Nagao & Kitazawa 2003, Partsalis et al. 1995, Pastor et al. 1994, Robinson 1976). Studies that reported small effects of lesions on the expression of previously acquired changes in VOR gain generally used longer training paradigms than did those that reported large effects. These results are consistent with the storage of motor memories initially depending on the cerebellum and with the storage of longer-term memories depending more upon other structures. However, the correlation was not perfect, and these studies were done with several different species and methods. A more systematic exploration of this hypothesis is currently underway (Kassardjian et al. 2003).

Studies of other cerebellum-dependent tasks provide additional evidence that multiple plasticity mechanisms support motor learning over different timescales. Inactivation of the cerebellum disrupted the expression of short-term, but not long-term, learned changes in visually driven eye movements (Shutoh & Nagao 2003). Similarly, inactivation of the cerebellar cortex with infusions of muscimol abolished the retention of classically conditioned eyeblink responses when performed after the first through fourth training sessions, but it had no effect when infused after the ninth through twelfth training sessions (Attwell et al. 2002). This finding supports the idea that different mechanisms are used to maintain cerebellum-dependent memories induced with brief, versus extended, amounts of training.

Bidirectional Changes in Movement Amplitude

The gain of the VOR can be adaptively increased or decreased. The Marr-Albus-Ito model attributes both an increase and a decrease in VOR gain to a single synaptic plasticity mechanism by suggesting that cerebellar LTD operates independently on parallel fibers that are active for head movements in different directions (Ito 1982). Specifically, the model proposed that LTD of parallel fibers firing during ipsiversive head turns would cause an increase in VOR gain, whereas LTD of parallel fibers firing during contraversive head turns would cause a decrease in VOR gain.

The Marr-Albus-Ito model predicts that increases and decreases in VOR gain would have similar properties stemming from their shared plasticity mechanism. In contrast, increases and decreases in VOR gain are different in several regards, which suggests they depend on different plasticity mechanisms. An early study found that increases in VOR gain passively decayed more rapidly than did decreases in VOR gain (Miles & Eighmy 1980). Furthermore, increases in VOR gain can be actively reversed more readily than can decreases in VOR gain (Boyden & Raymond 2003). Finally, increases in gain generalize less than decreases when measured in a context different from that used during training (see below). The distinct properties of increases and decreases in VOR gain suggest they are mediated by different plasticity mechanisms. Other movements may also depend on different plasticity mechanisms to implement bidirectional changes in movement amplitude. For example, increases and decreases in the gain of saccadic eye movements exhibit different behavioral properties (Robinson et al. 2003, Straube et al. 1997).

Further evidence for the idea that increases and decreases in VOR gain depend on different cellular mechanisms comes from the finding that some pharmacological and genetic manipulations differentially affect increases and decreases in VOR gain. Inhibiting NO or NMDA receptor activity impairs the ability of goldfish to increase, but not decrease, their VOR gain (Carter & McElligott 1995, Li et al. 1995). Mice lacking CaMKIV, a molecule necessary for long-lasting cerebellar LTD, have significantly impaired long-term memory for increases, but not decreases, in VOR gain (Boyden et al. 2003). In addition, a close examination of the published data for mutant mice expressing the PKCI transgene seems to suggest that they are more impaired for increases than decreases in VOR gain with long-term training (Van Alphen & De Zeeuw 2002). Together these pharmacological and molecular genetic studies indicate that increases and decreases in VOR gain depend on different plasticity mechanisms, and they raise the specific possibility that increases in VOR gain, more than decreases in VOR gain, depend on cerebellar LTD.

If increases in VOR gain depend more on cerebellar LTD, decreases in gain may depend more on cerebellar LTP. Implementing bidirectional behavioral changes with a single plasticity mechanism such as cerebellar LTD could result, over time, in a state where the available plasticity had been consumed and no further learning was possible (Lisberger 1996, Sejnowski 1977). On the other hand, a model with inverse plasticity mechanisms operating at each synapse could allow for reversal of prior learning. In vitro, parallel fiber synapses exhibit not only LTD but also two different forms of LTP. Both forms of LTP are induced when parallel fibers are active in the absence of climbing fiber activity (Lev-Ram et al. 2002, Sakurai 1987, Salin et al. 1996). Thus the learning rule for induction of LTP is complementary to that for LTD: In the terminology of Boolean logic, LTD is induced when parallel fiber activity *AND* climbing fiber activity occur simultaneously, whereas LTP is induced when parallel fiber activity *AND NOT* climbing fiber activity, occur.

Bidirectional plasticity at the synaptic level could explain the capacity for bidirectional modifications at the behavioral level. In the VOR, an asymmetry exists in the reversal of increases and decreases in VOR gain: Increases in gain are readily reversed by subsequent gain-down training, but decreases in gain are harder to reverse by gain-up training (Boyden & Raymond 2003). This finding suggests that increases and decreases in VOR gain depend on different plasticity mechanisms that reverse each other with unequal efficacy. One specific model that can explain this asymmetric reversal invokes the asymmetric localization of LTP and LTD expression at the parallel fiber–Purkinje cell synapse. LTD is postsynaptically expressed. The two forms of LTP are expressed at different parts of the synapse: One is expressed presynaptically (Salin et al. 1996), and the other is expressed postsynaptically (Lev-Ram et al. 2002). This asymmetric localization at the anatomical level confers asymmetric reversal of plasticity at the physiological level (Bear & Linden 2000, Lev-Ram et al. 2003). If LTD contributes to an increase in VOR gain and the two forms of LTP contribute to a decrease in VOR gain, then this differential contribution could account for the asymmetric reversal seen at the behavioral level: The postsynaptic LTD contributing to an increase in gain would be fully reversed by the postsynaptic LTP contributing to a decrease in gain, but the presynaptic form of LTP contributing to a decrease in gain would not be reversed by the postsynaptic form of LTD contributing to an increase in gain (Boyden & Raymond 2003).

The plausibility of a role of cerebellar LTP in motor learning is supported by several additional results. A study of the strength of parallel fiber–Purkinje cell synapses in rat cerebellar slices estimates that only 7% of the connections are actually capable of driving postsynaptic responses (Isope & Barbour 2002). Thus in the basal state, there may be little additional room for LTD to navigate, and the greatest capacity for circuit modification could lie in potentiating the large number of silent parallel fiber synapses. In support of this view, a recent study in vivo found that stimulation of parallel fibers in zone C3 of the cerebellum resulted in expansion of Purkinje cell cutaneous receptive fields, which the authors attributed to the action of LTP on parallel fiber–Purkinje cell synapses (Jorntell & Ekerot 2002). Contraction of Purkinje cell receptive fields became possible only after prior expansion had occurred, which suggests that one function of LTD is to reverse previously induced LTP. Evidence for a role of LTP in cerebellum-dependent learning also comes from the observation of an increase in the number of parallel fiber–Purkinje cell synapses after long-term training with an obstacle course (Anderson et al. 1996, Black et al. 1990). Presumably an increase in synapse number would be associated with LTP rather than LTD. Thus LTP may play roles both in the reversal of old memories and in the encoding of new ones.

The evidence described above suggests that increases and decreases in VOR gain depend on different plasticity mechanisms. One difference between increases and decreases in the gain of the VOR and other movements may be their dependence on LTD and LTP at parallel fiber–Purkinje cell synapses, but plasticity mechanisms at other sites may contribute differentially, as well.

INTERACTION BETWEEN PLASTICITY AND CODING: SPECIFICITY AND GENERALIZATION OF LEARNING

Learning is useful only if it is expressed in appropriate contexts. Learning must generalize to situations slightly different from those present during training because, otherwise, natural situational variation could restrict learning from ever being expressed. Too much generalization, however, is not desirable: A learned response could be maladaptive if expressed in an inappropriate context. In the VOR, learning is sometimes expressed only in very specific contexts, and sometimes it generalizes to contexts other than that present during learning. The patterns of specificity and generalization provide insight into the encoding of information at the sites of plasticity.

Specificity

One of the best-explored examples of the context dependency of motor learning in the VOR is the specificity of the adapted gain to the head tilt present during training. If during training the gain of the VOR is adaptively modified with the head tilted at a particular angle relative to gravity, much smaller gain changes are expressed when the VOR is measured with the same rotational stimulus but with the head tilted at a different angle (Baker et al. 1987). Thus, varying the angle of head tilt must cause different sets of neurons to be activated at the site of plasticity in response to the same rotational head movement. Indeed, the evidence suggests that varying the tilt angle of the head can cause the activation of completely nonoverlapping populations of neurons in response to the same rotational stimulus. Gain-up training of the rotational VOR with the head tilted 90° to the right induced an increase in gain that was not at all expressed when the head was tilted 90° to the left (Yakushin et al. 2000a). This finding suggests a sparse, combinatorial encoding of information from the two types of vestibular end organs: the semicircular canals, which measure head rotation, and the otolith organs, which measure linear acceleration and head tilt. At the level of the primary afferents, these signals are in separate pathways. However, at the site of plasticity, the canal stimulus must activate totally different populations of neurons, depending on the otolith signal present, because no learning was expressed when the head was tilted in the direction opposite that present during training. The experiment was done in such a way that the canal afferents would be activated in the very same way regardless of whether the head was tilted 90° to the left or to the right. Also, the otolith signal was static in each condition, not varying with head rotation. Thus, at the site of plasticity the context provided by the static otolith input must gate the responses of neurons to their dynamic canal inputs. Then, during learning, plasticity would specifically alter synapses of the neurons activated in the context present during training.

Where might such a sparse combinatorial code exist in the VOR circuit? Many vestibular nucleus neurons are modulated by both rotation signals from the semicircular canals and dynamic head-tilt signals from otoliths (Angelaki et al. 1993, Baker et al. 1984, Bush et al. 1993, Endo et al. 1995, Kubo et al. 1977, Ono et al. 2000, Uchino et al. 2000). However, whether static head-tilt signals can gate rotational signals in the vestibular nuclei to a degree sufficient to account for the behavioral observations is unclear. If this interaction between dynamic canal and static otolith signals is not found in the vestibular nucleus, then the sparse reencoding may occur in the cerebellum. According to the Marr-Albus model, even if mossy fiber inputs carry the same head-rotation signals (Figure 4*A*, common stimulus), granule cells may fire only for specific combinations of head-rotation and static-tilt contexts (Figure 4*A*, Context A, Context B), with no overlap between the sets of granule cells that fire for different combinations. Then, the induction of plasticity of synapses from granule cells active only during rotations with leftward head tilt would not affect the flow of information through granule cells active only during rotations with rightward head tilt, even though their mossy fiber inputs could have identical head-rotation responses.

Another example of the stimulus specificity of motor learning in the VOR concerns the expression of learning when measured at head-rotation frequencies different from the training frequency. If training is induced with visual and vestibular stimuli at a single frequency, the observed change in VOR gain is much smaller when measured at head-movement frequencies different from the training frequency (Godaux et al. 1983, Iwashita et al. 2001, Lisberger et al. 1983, Powell et al. 1991, Raymond & Lisberger 1996). This finding suggests that the circuit for the VOR contains parallel, frequency-tuned, signal-processing channels, which are individually modifiable during training.

Such sparse encoding of head-movement frequency is not seen in early sensory afferent and vestibular nucleus neurons. Many of these neurons respond to head movements at a broad range of frequencies. Responses of some individual neurons

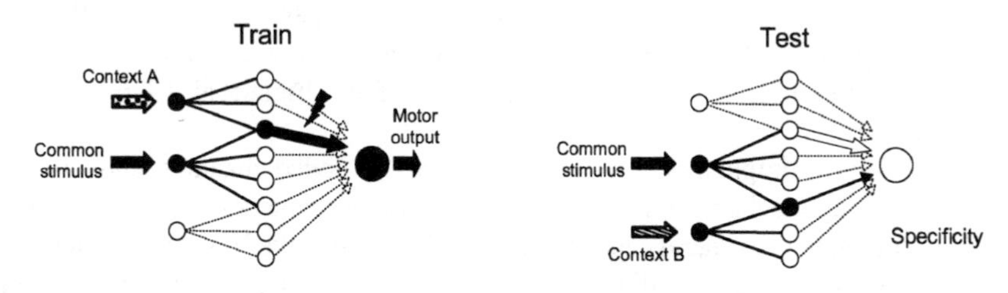

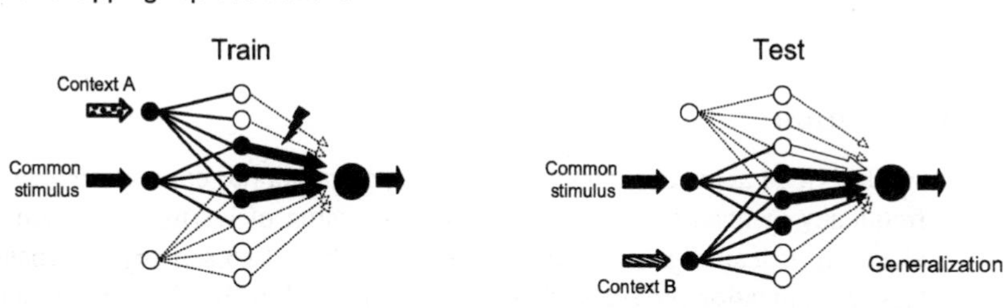

Figure 4 General population coding schemes for achieving specificity and generalization in motor learning. The circuit is divided into three layers: an input layer, an association layer, and an output layer. (*A*) Sparse reencoding enables learning to link particular inputs to particular outputs, as in Figure 1. LTP (lightning bolt) strengthens the synapses active during training with the common stimulus and Context A, causing a learned motor output to be elicited when that combination is present. The common stimulus does not elicit the motor output in the presence of Context B because the neurons activated in the association layer were not active or modified during training. (*B*) Less-sparse representations have more overlap in the association layer, and the common stimulus can elicit the learned motor output in contexts different from those used during training. Open circles and arrows and dotted lines indicate inactive neurons and synapses. Filled circles and arrows and solid lines indicate active neurons and synapses. Thick arrows indicate synapses strengthened by LTP. Circuits using LTD instead of LTP would function in a similar manner.

vary with head-rotation frequency, with some acting as high-pass filters and others acting as low-pass filters (Angelaki & Dickman 2000, Buettner et al. 1978, Jones & Milsum 1971, Schneider & Anderson 1976). These neurons could support some degree of frequency selectivity, but they do not seem to have sufficient selectivity to explain the behavioral data. The cerebellum could act as an adaptive linear filter (Fujita 1982), which could have bandwidths that result in more tightly tuned frequency channels than found in early vestibular pathways.

Another demonstration of independently modifiable channels with different dynamic properties is a component of the VOR that is expressed specifically during fast accelerations (Clendaniel et al. 2002, Minor et al. 1999). Learned changes in this component occur only when training is done with fast accelerations. In addition, if training is performed with stimuli of a particular velocity, the observed change in VOR gain is much smaller when measured with head-movement velocities different from that used during training (Fukushima et al. 1996, Iwashita et al. 2001).

Thus a simple model with parallel, sparsely coded channels, and with a single plasticity mechanism that alters a subset of these channels, can go a long way in explaining the general capacity of motor learning in the VOR to exhibit specificity for the particular stimuli present during training. Additional observations suggest, however, that less-sparse codes and multiple plasticity mechanisms may play an important role as well.

Generalization

As described above, a change in stimulus parameters or context, such as rotation frequency or head tilt, can prevent the expression of learning. But some generalization of learning can be observed across these two sensory dimensions, and this generalization provides additional insight into how head tilt and frequency are coded in the circuit. For generalization of learning to occur, there must be overlap between the populations of neurons activated by the different stimuli or contexts (Figure 4*B*). In such a network, plasticity affecting the association between one stimulus and a motor output could change the association between another stimulus and the motor output, since some of the same neurons would be part of the representations of both stimuli at the site of plasticity. For example, training the rotational VOR with the head tilted 90° to the left induced learning that was partially expressed when the rotational VOR was measured with the head vertical (Yakushin et al. 2003). Hence there is some generalization of learning to different head tilts, which limits the sparseness of the representation of otolith and canal inputs at the site of plasticity. Some of the neurons at the site of plasticity, which are activated and undergo plasticity during rotations with the head tilted 90° to the left, also must respond to a rotation when the head is upright. Thus observing the degree to which learning generalizes can provide insight about coding at the site of plasticity.

The situation is slightly more complex for frequency selectivity because the amount of generalization depends on the frequency of the training stimulus. When training is done with 0.5 Hz stimuli, there is little expression of learning when the VOR gain is measured with 5 Hz head rotation. When training is done with 5 Hz stimuli, however, learning is expressed when the VOR gain is measured with 0.5 Hz head rotation (Raymond & Lisberger 1996). This asymmetric pattern of generalization could potentially result from a coding scheme in which neurons active during low-frequency head rotation are a subset of those active at high frequencies. Consider a simple coding scheme wherein most of the neurons active at 0.5 Hz are also active at 5 Hz, but a relatively small fraction of the neurons active at 5 Hz are active at 0.5 Hz. Then plasticity induced in the neurons active during training at 5 Hz would affect the VOR gain measured at 0.5 Hz because most of the neurons mediating low-frequency head movements would have been activated and altered by the training. However, plasticity induced in the neurons active during 0.5 Hz training would have little effect on the VOR gain measured at 5 Hz because the neurons activated and altered during training would make up only a small fraction of those mediating the VOR response at 5 Hz. Thus, asymmetric generalization of learning could, in theory, result from a single plasticity mechanism acting on a single population of synapses.

However, another possibility is that the different patterns of generalization observed with high- and low-frequency training result from their dependence on different plasticity mechanisms, which operate on separate representations of head movement in the circuit for the VOR. One representation would be sparse like the network shown in Figure 4*A* (intermediate-layer neurons tightly tuned for frequency), and one would be less sparse like that in Figure 4*B* (intermediate-layer neurons responding to a broad range of frequencies). If low-frequency training produced changes in the sparse network, there would be little or no generalization when tested at high frequencies, and if high-frequency training produced changes in the less-sparse network, there would be generalization of learning when tested at low frequencies. One proposal is that the cerebellar cortex could contain the sparse network, and the deep cerebellar nuclei (or for the VOR, the vestibular nuclei) could contain the less-sparse network (Thach et al. 1992). This model would predict that changes induced with low-frequency rotation, which do not generalize, would be localized to the cortex, whereas changes induced with high-frequency rotation, which generalize to lower frequencies, would be localized to the brainstem. However, studies of the instructive signals available to guide motor learning would predict the opposite. At low frequencies, the instructive signals present in the Purkinje cells could potentially drive appropriate changes in the vestibular nuclei, whereas at high frequencies, the best candidate for the instructive signal is the climbing fiber (Raymond & Lisberger 1998). Another challenge to the idea that the cerebellar cortex is responsible for the specificity of learning comes from eyeblink conditioning. With the output of the cerebellar cortex inactivated, the expression of the conditioned response was no less specific for the stimuli used during training (Ohyama et al. 2003).

Even if different plasticity mechanisms are responsble for specific and generalized changes, they need not be located in separate parts of the circuit. For the case of frequency generalization in the VOR, the high- and low-frequency information-processing channels could be carried in the same neurons but by different subcellular signaling pathways. The observation that frequency-tuned coding channels seem to have distinct pharmacological properties supports this idea. Administration of NMDA receptor antagonists reduces low-frequency, but not high-frequency, VOR gain (Priesol et al. 2000). Thus NMDA receptors may be selectively present in the low-frequency coding channels, and they could endow the low-frequency channels with plasticity mechanisms different from those found in the high-frequency channels.

Refinement of Generalization Across Multiple Sensory Dimensions

The extent to which learning generalizes often depends on the precise training paradigm. Additional training can prevent the generalization of learning that would otherwise occur. For example, changes in the rotational VOR gain induced when the head is tilted 90° to one side result in learning expressed when the rotational VOR is measured with the head in the upright position. However, this generalization is not observed if periods of gain-up training with the head tilted 90° to the right are alternated with periods of gain-down training with the head tilted 90° to the left (Yakushin et al. 2003). Appropriate changes in the rotational VOR are expressed when the head is tilted to each side, but no changes are observed when the VOR is measured with the head upright. The changes in the VOR circuit induced by the gain-up and gain-down training, which normally would have each been expressed when measured with the head upright, must somehow cancel each other. This cancellation could occur either by summation (masking) or by reversal of the effects of each training protocol on the neurons activated with the head upright. One hint about whether masking or reversal occurs comes from the observation that gain-down training with the head tilted 90° results in changes expressed in the head-upright position that are significantly larger than those induced by gain-up training with the head tilted 90° (Yakushin et al. 2003). If pure masking occurred, then the alternating-training protocol should result in a net decrease in gain when measured with the head upright. However, no change in gain was observed. Therefore, the changes induced by gain-up and gain-down training did not simply sum linearly when measured in the head-upright position, but rather interacted via a nonlinear process, perhaps involving reversal of plasticity at the cellular level.

This example shows that for a given context, generalization can be actively refined during the training process. To successfully stabilize images on the retina across varying conditions, the circuit for the VOR must coordinate the specificity or generalization of learning across many dimensions of sensory input, including both vestibular stimulus parameters and context cues. For example, the vertical VOR can be adapted to different gains during upward, versus downward, head turns (Hirata et al. 2002). Subjects can be trained to have different horizontal VOR gains when their eyes are angled upward versus downward (Shelhamer et al. 1992). Learning is also specific for the vergence angle held by the eyes during training (Lewis et al. 2003). Motor learning in the translational VOR is expressed only in one eye, if the other is not required to move in response to the training stimulus (Zhou et al. 2003). Despite the large number of dimensions that can provide contexts for specifically expressed learning, not all contexts are equally potent for regulating the expression of learning. For example, the vertical canal-mediated VOR can express different gains for rightward, versus leftward, head tilts, but head tilt that is pitched forward or backward is a less-effective contextual cue (Shelhamer et al. 2002). In the real world, learning is regulated by all of these dimensions simultaneously, and so the underlying coding and plasticity mechanisms in the VOR must be of sufficient richness to enable regulation of learning by multiple dimensions of sensory input.

Sparse, combinatorial coding of multiple dimensions may allow the expression of learned output only when the right combination of stimuli and contexts is present.

Investigators are exploring the patterns of specificity and generalization for other forms of motor learning, as well (a few examples include Collewijn & Grootendorst 1979; Kahlon & Lisberger 1999; Kramer et al. 1995, 1998; Lisberger et al. 1981; Ohyama et al. 2003; and Park et al. 2003). As behavioral observations are combined with studies of how information is encoded in the relevant neural circuits, these studies will constrain hypotheses concerning the sites of plasticity storing motor memories.

Generalization of Increases Versus Decreases in VOR Gain

Increases in VOR gain generalize less than decreases in VOR gain across a number of sensory dimensions. For example, decreasing VOR gain by training with only one eye viewing generalizes to a decreased VOR gain in the other eye, but increasing the gain in one exposed eye does not result in transfer to the other eye (McElligott & Wilson 2001). Gain-down training generalizes across head rotation velocity and frequency more than gain-up training generalizes (Hirata et al. 2002, Iwashita et al. 2001, Kimpo & Raymond 2002, Raymond & Lisberger 1996). Finally, gain-down training of the rotational VOR generalizes more across head tilt angles than gain-up training generalizes (Yakushin et al. 2000a, Yakushin et al. 2003).

These results provide an important new insight about the plasticity mechanisms for increasing and decreasing VOR gain. To account for the above results, the gain changes induced in opposite directions must not simply involve opposite changes in the strength of the same synapses. Rather, different sets of synapses must be modified for increases and decreases in VOR gain. Even though the same vestibular neurons are activated in each case, the visual and other instructive signals must change different subsets of synapses for increases and decreases in gain. One possibility is that synapses at different sites in the circuit could be modified by training paradigms that increase and decrease VOR gain. Neurons at the site storing decreases in gain could encode for head rotation regardless of context or particular stimulus features (like the network in Figure 4*B*), whereas neurons at the site storing increases in gain could be strongly modulated by context (like the network in Figure 4*A*). One specific proposal is that the cerebellar cortex contributes more to gain-up learning, whereas the vestibular nucleus contributes more to gain-down learning (Li et al. 1995). On the other hand, the difference in generalization of increases and decreases in VOR gain could result from the properties of different plasticity mechanisms at a single site. For the model where LTD mediates an increase in gain and LTP mediates a decrease in gain, the patterns of generalization observed at the behavioral level suggest that LTP could spread to nonactivated synapses more than LTD spreads.

Thus the patterns of generalization support the idea that increases and decreases in VOR gain must use different plasticity mechanisms, and these mechanisms must affect different sets of synapses. Whether gain-up and gain-down training are inducing changes at different anatomical sites in the circuit, or just affecting different subpopulations of synapses at the same site, remains to be determined.

CONCLUSION

During the last several decades, attempts to discriminate between the Marr-Albus-Ito and Miles-Lisberger models have dominated research on motor learning in the VOR. Each model can explain some of the data regarding motor learning in the VOR with a single plasticity mechanism. However, several lines of evidence indicate that multiple plasticity mechanisms contribute to the regulation of this simple behavior. Multiple mechanisms may refine the time dynamics of learned responses, enable storage of motor memories over different timescales, and enable bidirectional alteration of movement amplitude.

These multiple plasticity mechanisms must act in combination with appropriate information-coding schemes to equip motor-learning systems with the ability to express learning in correct contexts. Studies of the generalization patterns of motor learning in the VOR provide insights about the coding of information in neurons at the site of plasticity. For a simple system like the VOR, it should be possible to map particular aspects of learning onto specific plasticity mechanisms, thus illuminating how movements can be precisely regulated in a complex world.

ACKNOWLEDGMENTS

We thank Jenelle Jindal, Mike Ke, Rhea Kimpo, and Joyce Liao for helpful comments on the manuscript. E.S.B. is supported by the Fannie and John Hertz Foundation, A.K. is supported by the Human Frontier Science Program and the Zaffaroni Innovation Fund, and J.L.R. is supported by a McKnight Scholar Award and NIH R01 DC04154.

The *Annual Review of Neuroscience* is online at http://neuro.annualreviews.org

LITERATURE CITED

Aiba A, Kano M, Chen C, Stanton ME, Fox GD, et al. 1994. Deficient cerebellar long-term depression and impaired motor learning in mGluR1 mutant mice. *Cell* 79:377–88

Albus J. 1971. A theory of cerebellar function. *Math. Biosci.* 10:25–61

Anderson BJ, Alcantara AA, Greenough WT. 1996. Motor-skill learning: changes in synaptic organization of the rat cerebellar cortex. *Neurobiol. Learn. Mem.* 66:221–29

Anderson BJ, Relucio K, Haglund K, Logan C, Knowlton B, et al. 1999. Effects of paired and unpaired eye-blink conditioning on Purkinje cell morphology. *Learn. Mem.* 6:128–37

Angelaki DE, Bush GA, Perachio AA. 1993. Two-dimensional spatiotemporal coding of linear acceleration in vestibular nuclei neurons. *J. Neurosci.* 13:1403–17

Angelaki DE, Dickman JD. 2000. Spatiotemporal processing of linear acceleration:

primary afferent and central vestibular neuron responses. *J. Neurophysiol.* 84:2113–32
Attwell PJ, Cooke SF, Yeo CH. 2002. Cerebellar function in consolidation of a motor memory. *Neuron* 34:1011–20
Babalian AL, Vidal PP. 2000. Floccular modulation of vestibuloocular pathways and cerebellum-related plasticity: an in vitro whole brain study. *J. Neurophysiol.* 84:2514–28
Baker J, Goldberg J, Hermann G, Peterson B. 1984. Spatial and temporal response properties of secondary neurons that receive convergent input in vestibular nuclei of alert cats. *Brain Res.* 294:138–43
Baker J, Wickland C, Peterson B. 1987. Dependence of cat vestibulo-ocular reflex direction adaptation on animal orientation during adaptation and rotation in darkness. *Brain Res.* 408:339–43
Bear MF, Linden DJ. 2000. The mechanisms and meaning of long-term synaptic depression. In *The Synapse*, ed. WM Cowan, K Davies, pp. 455–517. Baltimore, MD: Johns Hopkins Univ. Press
Black JE, Isaacs KR, Anderson BJ, Alcantara AA, Greenough WT. 1990. Learning causes synaptogenesis, whereas motor activity causes angiogenesis, in cerebellar cortex of adult rats. *Proc. Natl. Acad. Sci. USA* 87:5568–72
Boyden ES, Chatila T, Raymond JL. 2003. Motor memories in the vestibulo-ocular reflex of CaMKIV knockout mice. *Soc. Neurosci.* 73.18 (Abstr.)
Boyden ES, Raymond JL. 2003. Active reversal of motor memories reveals rules governing memory encoding. *Neuron* 39:1031–42
Brashers-Krug T, Shadmehr R, Bizzi E. 1996. Consolidation in human motor memory. *Nature* 382:252–55
Buettner UW, Buttner U, Henn V. 1978. Transfer characteristics of neurons in vestibular nuclei of the alert monkey. *J. Neurophysiol.* 41:1614–28
Buonomano D, Mauk M. 2004. The neural basis of temporal processing. *Annu. Rev. Neurosci.* 27:307–40
Bush GA, Perachio AA, Angelaki DE. 1993. Encoding of head acceleration in vestibular neurons. I. Spatiotemporal response properties to linear acceleration. *J. Neurophysiol.* 69:2039–55
Caria MA, Melis F, Podda MV, Solinas A, Deriu F. 1996. Does long-term potentiation occur in guinea-pig Deiters' nucleus? *Neuroreport* 7:2303–7
Caria MA, Melis F, Solinas A, Tavera C, Mameli O. 2001. Frequency-dependent LTP/LTD in guinea pig Deiters' nucleus. *Neuroreport* 12:2353–58
Carter TL, McElligott JG. 1995. Cerebellar NMDA receptor antagonism by AP5 blocks vestibulo-ocular reflex adaptation in the goldfish. *Soc. Neurosci.* 115.5 (Abstr.)
Chapman PF, Atkins CM, Allen MT, Haley JE, Steinmetz JE. 1992. Inhibition of nitric oxide synthesis impairs two different forms of learning. *Neuroreport* 3:567–70
Clendaniel RA, Lasker DM, Minor LB. 2002. Differential adaptation of the linear and nonlinear components of the horizontal vestibuloocular reflex in squirrel monkeys. *J. Neurophysiol.* 88:3534–40
Collewijn H, Grootendorst AF. 1979. Adaptation of optokinetic and vestibulo-ocular reflexes to modified visual input in the rabbit. In *Reflex Control of Posture and Movement*, ed. R Granit, O Pompeiano, pp. 772–81. Amsterdam: Elsevier
Crepel F, Krupa M. 1988. Activation of protein kinase C induces a long-term depression of glutamate sensitivity of cerebellar Purkinje cells. An in vitro study. *Brain Res.* 458:397–401
Daniel H, Hemart N, Jaillard D, Crepel F. 1993. Long-term depression requires nitric oxide and guanosine 3′:5′ cyclic monophosphate production in rat cerebellar Purkinje cells. *Eur. J. Neurosci.* 5:1079–82
De Zeeuw CI, Hansel C, Bian F, Koekkoek SK, van Alphen AM, et al. 1998. Expression of a protein kinase C inhibitor in Purkinje cells blocks cerebellar LTD and adaptation of the vestibulo-ocular reflex. *Neuron* 20:495–508
du Lac S, Raymond JL, Sejnowski TJ,

Lisberger SG. 1995. Learning and memory in the vestibulo-ocular reflex. *Annu. Rev. Neurosci.* 18:409–41

Endo K, Thomson DB, Wilson VJ, Yamaguchi T, Yates BJ. 1995. Vertical vestibular input to and projections from the caudal parts of the vestibular nuclei of the decerebrate cat. *J. Neurophysiol.* 74:428–36

Fujita M. 1982. Adaptive filter model of the cerebellum. *Biol. Cybern.* 45:195–206

Fukushima K, Fukushima J, Chin S, Tsunekawa H, Kaneko CR. 1996. Cross axis vestibulo-ocular reflex induced by pursuit training in alert monkeys. *Neurosci. Res.* 25:255–65

Galiana HL. 1986. A new approach to understanding adaptive visual-vestibular interactions in the central nervous system. *J. Neurophysiol.* 55:349–74

Ghelarducci B, Ito M, Yagi N. 1975. Impulse discharges from flocculus Purkinje cells of alert rabbits during visual stimulation combined with horizontal head rotation. *Brain Res.* 87:66–72

Godaux E, Halleux J, Gobert C. 1983. Adaptive change of the vestibulo-ocular reflex in the cat: the effects of a long-term frequency-selective procedure. *Exp. Brain Res.* 49:28–34

Goossens J, Daniel H, Rancillac A, van der Steen J, Oberdick J, et al. 2001. Expression of protein kinase C inhibitor blocks cerebellar long-term depression without affecting Purkinje cell excitability in alert mice. *J. Neurosci.* 21:5813–23

Graf W, Simpson JI, Leonard CS. 1988. Spatial organization of visual messages of the rabbit's cerebellar flocculus. II. Complex and simple spike responses of Purkinje cells. *J. Neurophysiol.* 60:2091–121

Haddad GM, Demer JL, Robinson DA. 1980. The effect of lesions of the dorsal cap of the inferior olive on the vestibulo-ocular and optokinetic systems of the cat. *Brain Res.* 185:265–75

Hansel C, Linden DJ. 2000. Long-term depression of the cerebellar climbing fiber–Purkinje neuron synapse. *Neuron* 26:473–82

Hansel C, Linden DJ, D'Angelo E. 2001. Beyond parallel fiber LTD: the diversity of synaptic and non-synaptic plasticity in the cerebellum. *Nat. Neurosci.* 4:467–75

Hirata Y, Highstein SM. 2001. Acute adaptation of the vestibuloocular reflex: signal processing by floccular and ventral parafloccular Purkinje cells. *J. Neurophysiol.* 85:2267–88

Hirata Y, Lockard JM, Highstein SM. 2002. Capacity of vertical VOR adaptation in squirrel monkey. *J. Neurophysiol.* 88:3194–207

Isope P, Barbour B. 2002. Properties of unitary granule cell→Purkinje cell synapses in adult rat cerebellar slices. *J. Neurosci.* 22:9668–78

Ito M. 1993. Cerebellar flocculus hypothesis. *Nature* 363:24–25

Ito M. 1972. Neural design of the cerebellar motor control system. *Brain Res.* 40:81–84

Ito M. 1982. Cerebellar control of the vestibulo-ocular reflex—around the flocculus hypothesis. *Annu. Rev. Neurosci.* 5:275–96

Ito M, Jastreboff PJ, Miyashita Y. 1982a. Specific effects of unilateral lesions in the flocculus upon eye movements in albino rabbits. *Exp. Brain Res.* 45:233–42

Ito M, Miyashita Y. 1975. The effects of chronic destruction of the inferior olive upon visual modification of the horizontal vestibulo-ocular reflex of rabbits. *Proc. Jpn. Acad.* 51:716–20

Ito M, Sakurai M, Tongroach P. 1982b. Climbing fibre induced depression of both mossy fibre responsiveness and glutamate sensitivity of cerebellar Purkinje cells. *J. Physiol.* 324:113–34

Ivry RB, Keele SW, Diener HC. 1988. Dissociation of the lateral and medial cerebellum in movement timing and movement execution. *Exp. Brain Res.* 73:167–80

Iwashita M, Kanai R, Funabiki K, Matsuda K, Hirano T. 2001. Dynamic properties, interactions and adaptive modifications of vestibulo-ocular reflex and optokinetic response in mice. *Neurosci. Res.* 39:299–311

Jones GM, Milsum JH. 1971. Frequency-response analysis of central vestibular unit activity resulting from rotational stimulation of the semicircular canals. *J. Physiol.* 219:191–215

Jorntell H, Ekerot CF. 2002. Reciprocal bidirectional plasticity of parallel fiber receptive fields in cerebellar Purkinje cells and their afferent interneurons. *Neuron* 34:797–806

Kahlon M, Lisberger SG. 1999. Vector averaging occurs downstream from learning in smooth pursuit eye movements of monkeys. *J. Neurosci.* 19:9039–53

Kassardjian CD, Tan YF, Broussard DM. 2003. Effects of partial reversible inactivation of the floccular complex on motor memory in the vestibulo-ocular reflex of the cat. *Soc. Neurosci.* 882.9 (Abstr.)

Katoh A, Kitazawa H, Itohara S, Nagao S. 2000. Inhibition of nitric oxide synthesis and gene knockout of neuronal nitric oxide synthase impaired adaptation of mouse optokinetic response eye movements. *Learn. Mem.* 7:220–26

Katoh A, Yoshida T, Hirano T. 2001. Abnormal reflex eye movements in mutant mice deficient in either the glutamate receptor delta2 subunit or Purkinje cells. *Soc. Neurosci.* 294.7 (Abstr.)

Ke MC, Raymond JL. 2001. Contribution of background vs. target motion to the induction of motor learning in the VOR. *Soc. Neurosci.* 403.25 (Abstr.)

Ke MC, Raymond JL. 2002. Effects of conflicting target and background motion on Purkinje cell activity during motor learning in the VOR. *Soc. Neurosci.* 766.15 (Abstr.)

Kimpo RR, Raymond JL. 2002. Adaptation of the VOR in wild type and PKCgamma knockout mice. *Soc. Neurosci.* 766.7 (Abstr.)

Kishimoto Y, Hirono M, Sugiyama T, Kawahara S, Nakao K, et al. 2001a. Impaired delay but normal trace eyeblink conditioning in PLCbeta4 mutant mice. *Neuroreport* 12:2919–22

Kishimoto Y, Kawahara S, Fujimichi R, Mori H, Mishina M, Kirino Y. 2001b. Impairment of eyeblink conditioning in GluRdelta2-mutant mice depends on the temporal overlap between conditioned and unconditioned stimuli. *Eur. J. Neurosci.* 14:1515–21

Kleim JA, Freeman JH Jr, Bruneau R, Nolan BC, Cooper NR, et al. 2002. Synapse formation is associated with memory storage in the cerebellum. *Proc. Natl. Acad. Sci. USA* 99:13228–31

Koekkoek SK, Hulscher HC, Dortland BR, Hensbroek RA, Elgersma Y, et al. 2003. Cerebellar LTD and learning-dependent timing of conditioned eyelid responses. *Science* 301:1736–39

Koekkoek SK, van Alphen AM, van den Burg J, Grosveld F, Galjart N, De Zeeuw CI. 1997. Gain adaptation and phase dynamics of compensatory eye movements in mice. *Genes Funct.* 1:175–90

Kramer PD, Shelhamer M, Peng GC, Zee DS. 1998. Context-specific short-term adaptation of the phase of the vestibulo-ocular reflex. *Exp. Brain Res.* 120:184–92

Kramer PD, Shelhamer M, Zee DS. 1995. Short-term adaptation of the phase of the vestibulo-ocular reflex (VOR) in normal human subjects. *Exp. Brain Res.* 106:318–26

Kubo T, Matsunaga T, Matano S. 1977. Convergence of ampullar and macular inputs on vestibular nuclei unit of the rat. *Acta Otolaryngol.* 84:166–77

Lev-Ram V, Jiang T, Wood J, Lawrence DS, Tsien RY. 1997. Synergies and coincidence requirements between NO, cGMP, and Ca^{2+} in the induction of cerebellar long-term depression. *Neuron* 18:1025–38

Lev-Ram V, Mehta SB, Kleinfeld D, Tsien RY. 2003. Reversing cerebellar long-term depression. *Proc. Natl. Acad. Sci. USA* 100:15989–93

Lev-Ram V, Wong ST, Storm DR, Tsien RY. 2002. A new form of cerebellar long-term potentiation is postsynaptic and depends on nitric oxide but not cAMP. *Proc. Natl. Acad. Sci. USA* 99:8389–93

Lewis RF, Clendaniel RA, Zee DS. 2003. Vergence-dependent adaptation of the vestibulo-ocular reflex. *Exp. Brain Res.* 152:335–40

Li J, Smith SS, McElligott JG. 1995. Cerebellar nitric oxide is necessary for vestibulo-ocular reflex adaptation, a sensorimotor model of learning. *J. Neurophysiol.* 74:489–94

Linden DJ, Connor JA. 1991. Participation of

postsynaptic PKC in cerebellar long-term depression in culture. *Science* 254:1656–59
Lisberger SG. 1994. Neural basis for motor learning in the vestibuloocular reflex of primates. III. Computational and behavioral analysis of the sites of learning. *J. Neurophysiol.* 72:974–98
Lisberger SG. 1996. Motor learning and memory in the vestibulo-ocular reflex: the dark side. *Ann. NY Acad. Sci.* 781:525–31
Lisberger SG, Miles FA, Optican LM. 1983. Frequency-selective adaptation: evidence for channels in the vestibulo-ocular reflex? *J. Neurosci.* 3:1234–44
Lisberger SG, Miles FA, Optican LM, Eighmy BB. 1981. Optokinetic response in monkey: underlying mechanisms and their sensitivity to long-term adaptive changes in vestibuloocular reflex. *J. Neurophysiol.* 45:869–90
Lisberger SG, Miles FA, Zee DS. 1984. Signals used to compute errors in monkey vestibuloocular reflex: possible role of flocculus. *J. Neurophysiol.* 52:1140–53
Lisberger SG, Pavelko TA, Bronte-Stewart HM, Stone LS. 1994a. Neural basis for motor learning in the vestibuloocular reflex of primates. II. Changes in the responses of horizontal gaze velocity Purkinje cells in the cerebellar flocculus and ventral paraflocculus. *J. Neurophysiol.* 72:954–73
Lisberger SG, Pavelko TA, Broussard DM. 1994b. Neural basis for motor learning in the vestibuloocular reflex of primates. I. Changes in the responses of brain stem neurons. *J. Neurophysiol.* 72:928–53
Lisberger SG, Sejnowski TJ. 1992. Motor learning in a recurrent network model based on the vestibulo-ocular reflex. *Nature* 360:159–61
Luebke AE, Robinson DA. 1994. Gain changes of the cat's vestibulo-ocular reflex after flocculus deactivation. *Exp. Brain Res.* 98:379–90
Marr D. 1969. A theory of cerebellar cortex. *J. Physiol.* 202:437–70
Marr D. 1971. Simple memory: a theory for archicortex. *Philos. Trans. R. Soc. Lond. B Biol. Sci.* 262:23–81
McCormick DA, Thompson RF. 1984. Cerebellum: essential involvement in the classically conditioned eyelid response. *Science* 223:296–99
McElligott JG, Beeton P, Polk J. 1998. Effect of cerebellar inactivation by lidocaine microdialysis on the vestibuloocular reflex in goldfish. *J. Neurophysiol.* 79:1286–94
McElligott JG, Wilson A. 2001. Monocular vestibulo-ocular reflex (VOR) adaptation in the goldfish. *Soc. Neurosci.* 403.1 (Abstr.)
Michnovicz JJ, Bennett MV. 1987. Effects of rapid cerebellectomy on adaptive gain control of the vestibulo-ocular reflex in alert goldfish. *Exp. Brain Res.* 66:287–94
Miles FA, Braitman DJ, Dow BM. 1980. Long-term adaptive changes in primate vestibuloocular reflex. IV. Electrophysiological observations in flocculus of adapted monkeys. *J. Neurophysiol.* 43:1477–93
Miles FA, Eighmy BB. 1980. Long-term adaptive changes in primate vestibuloocular reflex. I. Behavioral observations. *J. Neurophysiol.* 43:1406–25
Miles FA, Lisberger SG. 1981. Plasticity in the vestibulo-ocular reflex: a new hypothesis. *Annu. Rev. Neurosci.* 4:273–99
Minor LB, Lasker DM, Backous DD, Hullar TE. 1999. Horizontal vestibuloocular reflex evoked by high-acceleration rotations in the squirrel monkey. I. Normal responses. *J. Neurophysiol.* 82:1254–70
Moreno-Lopez B, Escudero M, Estrada C. 2002. Nitric oxide facilitates GABAergic neurotransmission in the cat oculomotor system: a physiological mechanism in eye movement control. *J. Physiol.* 540:295–306
Nagao S. 1983. Effects of vestibulocerebellar lesions upon dynamic characteristics and adaptation of vestibulo-ocular and optokinetic responses in pigmented rabbits. *Exp. Brain. Res.* 53:36–46
Nagao S. 1989. Behavior of floccular Purkinje cells correlated with adaptation of vestibulo-ocular reflex in pigmented rabbits. *Exp. Brain. Res.* 77:531–40
Nagao S, Ito M. 1991. Subdural application of hemoglobin to the cerebellum blocks

vestibuloocular reflex adaptation. *Neuroreport* 2:193–96

Nagao S, Kitazawa H. 2000. Subdural applications of NO scavenger or NO blocker to the cerebellum depress the adaptation of monkey post-saccadic smooth pursuit eye movements. *Neuroreport* 11:131–34

Nagao S, Kitazawa H. 2003. Effects of reversible shutdown of the monkey flocculus on the retention of adaptation of the horizontal vestibulo-ocular reflex. *Neuroscience* 118:563–70

Ohyama T, Nores WL, Mauk MD. 2003. Stimulus generalization of conditioned eyelid responses produced without cerebellar cortex: implications for plasticity in the cerebellar nuclei. *Learn. Mem.* 10:346–54

Ono S, Kushiro K, Zakir M, Meng H, Sato H, Uchino Y. 2000. Properties of utricular and saccular nerve-activated vestibulocerebellar neurons in cats. *Exp. Brain. Res.* 134:1–8

Park J, Schalg-Rey M, Schlag J. 2003. How does saccadic adaptation affect antisaccades? *Soc. Neurosci.* 441.9 (Abstr.)

Partsalis AM, Zhang Y, Highstein SM. 1995. Dorsal Y group in the squirrel monkey. II. Contribution of the cerebellar flocculus to neuronal responses in normal and adapted animals. *J. Neurophysiol.* 73:632–50

Pastor AM, de la Cruz RR, Baker R. 1994. Cerebellar role in adaptation of the goldfish vestibuloocular reflex. *J. Neurophysiol.* 72:1383–94

Perrett SP, Ruiz BP, Mauk MD. 1993. Cerebellar cortex lesions disrupt learning-dependent timing of conditioned eyelid responses. *J. Neurosci.* 13:1708–18

Peterson BW, Baker JF, Houk JC. 1991. A model of adaptive control of vestibuloocular reflex based on properties of cross-axis adaptation. *Ann. NY Acad. Sci.* 627:319–37

Powell KD, Quinn KJ, Rude SA, Peterson BW, Baker JF. 1991. Frequency dependence of cat vestibulo-ocular reflex direction adaptation: single frequency and multifrequency rotations. *Brain Res.* 550:137–41

Priesol AJ, Jones GE, Tomlinson RD, Broussard DM. 2000. Frequency-dependent effects of glutamate antagonists on the vestibulo-ocular reflex of the cat. *Brain Res.* 857:252–64

Racine RJ, Wilson DA, Gingell R, Sunderland D. 1986. Long-term potentiation in the interpositus and vestibular nuclei in the rat. *Exp. Brain Res.* 63:158–62

Rambold H, Churchland A, Selig Y, Jasmin L, Lisberger SG. 2002. Partial ablations of the flocculus and ventral paraflocculus in monkeys cause linked deficits in smooth pursuit eye movements and adaptive modification of the VOR. *J. Neurophysiol.* 87:912–24

Raymond JL, Lisberger SG. 1996. Behavioral analysis of signals that guide learned changes in the amplitude and dynamics of the vestibulo-ocular reflex. *J. Neurosci.* 16:7791–802

Raymond JL, Lisberger SG. 1998. Neural learning rules for the vestibulo-ocular reflex. *J. Neurosci.* 18:9112–29

Robinson DA. 1976. Adaptive gain control of vestibuloocular reflex by the cerebellum. *J. Neurophysiol.* 39:954–69

Robinson FR, Noto CT, Bevans SE. 2003. Effect of visual error size on saccade adaptation in monkey. *J. Neurophysiol.* 90:1235–44

Sakurai M. 1987. Synaptic modification of parallel fibre-Purkinje cell transmission in in vitro guinea-pig cerebellar slices. *J. Physiol.* 394:463–80

Salin PA, Malenka RC, Nicoll RA. 1996. Cyclic AMP mediates a presynaptic form of LTP at cerebellar parallel fiber synapses. *Neuron* 16:797–803

Schneider LW, Anderson DJ. 1976. Transfer characteristics of first and second order lateral canal vestibular neurons in gerbil. *Brain Res.* 112:61–76

Sejnowski TJ. 1977. Statistical constraints on synaptic plasticity. *J. Theor. Biol.* 69:385–89

Shadmehr R, Brashers-Krug T. 1997. Functional stages in the formation of human long-term motor memory. *J. Neurosci.* 17:409–19

Shadmehr R, Holcomb HH. 1997. Neural correlates of motor memory consolidation. *Science* 277:821–25

Shelhamer M, Peng GC, Ramat S, Patel V. 2002. Context-specific adaptation of the gain of the oculomotor response to lateral translation using roll and pitch head tilts as contexts. *Exp. Brain Res.* 146:388–93

Shelhamer M, Robinson DA, Tan HS. 1992. Context-specific adaptation of the gain of the vestibulo-ocular reflex in humans. *J. Vestib. Res.* 2:89–96

Shibuki K, Gomi H, Chen L, Bao S, Kim JJ, et al. 1996. Deficient cerebellar long-term depression, impaired eyeblink conditioning, and normal motor coordination in GFAP mutant mice. *Neuron* 16:587–99

Shibuki K, Okada D. 1991. Endogenous nitric oxide release required for long-term synaptic depression in the cerebellum. *Nature* 349:326–28

Shutoh F, Katoh A, Kitazawa H, Aiba A, Itohara S, Nagao S. 2002. Loss of adaptability of horizontal optokinetic response eye movements in mGluR1 knockout mice. *Neurosci. Res.* 42:141–45

Shutoh F, Katoh A, Ohki M, Itohara S, Tonegawa S, Nagao S. 2003. Role of protein kinase C family in the cerebellum-dependent adaptive learning of horizontal optokinetic response eye movements in mice. *Eur. J. Neurosci.* 18:134–42

Shutoh F, Nagao S. 2003. Long-term optokinetic training and adaptation of mouse ocular reflex eye movements. *Soc. Neurosci.* 274.6 (Abstr.)

Simpson JI, Alley KE. 1974. Visual climbing fiber input to rabbit vestibulo-cerebellum: a source of direction-specific information. *Brain Res.* 82:302–8

Smith SL, Otis TS. 2003. Persistent changes in spontaneous firing of Purkinje neurons triggered by the nitric oxide signaling cascade. *J. Neurosci.* 23:367–72

Spencer RM, Zelaznik HN, Diedrichsen J, Ivry RB. 2003. Disrupted timing of discontinuous but not continuous movements by cerebellar lesions. *Science* 300:1437–39

Stone LS, Lisberger SG. 1990. Visual responses of Purkinje cells in the cerebellar flocculus during smooth-pursuit eye movements in monkeys. II. Complex spikes. *J. Neurophysiol.* 63:1262–75

Straube A, Fuchs AF, Usher S, Robinson FR. 1997. Characteristics of saccadic gain adaptation in rhesus macaques. *J. Neurophysiol.* 77:874–95

Tabata H, Yamamoto K, Kawato M. 2002. Computational study on monkey VOR adaptation and smooth pursuit based on the parallel control-pathway theory. *J. Neurophysiol.* 87:2176–89

Tempia F, Dieringer N, Strata P. 1991. Adaptation and habituation of the vestibulo-ocular reflex in intact and inferior olive-lesioned rats. *Exp. Brain Res.* 86:568–78

Thach WT, Goodkin HP, Keating JG. 1992. The cerebellum and the adaptive coordination of movement. *Annu. Rev. Neurosci.* 15:403–42

Uchino Y, Sato H, Kushiro K, Zakir MM, Isu N. 2000. Canal and otolith inputs to single vestibular neurons in cats. *Arch. Ital. Biol.* 138:3–13

Van Alphen AM, De Zeeuw CI. 2002. Cerebellar LTD facilitates but is not essential for long-term adaptation of the vestibulo-ocular reflex. *Eur. J. Neurosci.* 16:486–90

Van Alphen AM, Schepers T, Luo C, De Zeeuw CI. 2002. Motor performance and motor learning in Lurcher mice. *Ann. NY Acad. Sci.* 978:413–24

Wall MJ. 2003. Endogenous nitric oxide modulates GABAergic transmission to granule cells in adult rat cerebellum. *Eur. J. Neurosci.* 18:869–78

Watanabe E. 1984. Neuronal events correlated with long-term adaptation of the horizontal vestibulo-ocular reflex in the primate flocculus. *Brain Res.* 297:169–74

Widmer HA, Rowe IC, Shipston MJ. 2003. Conditional protein phosphorylation regulates BK channel activity in rat cerebellar Purkinje neurones. *J. Physiol.* 552:379–91

Yakushin SB, Raphan T, Cohen B. 2000a. Context-specific adaptation of the vertical vestibuloocular reflex with regard to gravity. *J. Neurophysiol.* 84:3067–71

Yakushin SB, Raphan T, Cohen B. 2003. Gravity-specific adaptation of the angular vestibuloocular reflex: dependence on head orientation with regard to gravity. *J. Neurophysiol.* 89:571–86

Yakushin SB, Reisine H, Buttner-Ennever J, Raphan T, Cohen B. 2000b. Functions of the nucleus of the optic tract (NOT). I. Adaptation of the gain of the horizontal vestibulo-ocular reflex. *Exp. Brain Res.* 131:416–32

Yanagihara D, Kondo I. 1996. Nitric oxide plays a key role in adaptive control of locomotion in cat. *Proc. Natl. Acad. Sci. USA* 93:13292–97

Zhou W, Weldon P, Tang B, King WM. 2003. Rapid motor learning in the translational vestibulo-ocular reflex. *J. Neurosci.* 23:4288–98

Annu. Rev. Neurosci. 2004. 27:611–47
doi: 10.1146/annurev.neuro.26.041002.131039

First published online as a Review in Advance on March 24, 2004

ATTENTIONAL MODULATION OF VISUAL PROCESSING

John H. Reynolds[1] and Leonardo Chelazzi[2]
[1]Systems Neurobiology Laboratory, The Salk Institute for Biological Studies, La Jolla, California 92037-1099; email: reynolds@salk.edu
[2]Department of Neurological and Vision Sciences, Section of Physiology, University of Verona, Strada Le Grazie 8, I-37134 Verona, Italy; email: leonardo.chelazzi@univr.it

Key Words macaque, visual cortex, contrast, feedback, limited capacity

■ **Abstract** Single-unit recording studies in the macaque have carefully documented the modulatory effects of attention on the response properties of visual cortical neurons. Attention produces qualitatively different effects on firing rate, depending on whether a stimulus appears alone or accompanied by distracters. Studies of contrast gain control in anesthetized mammals have found parallel patterns of results when the luminance contrast of a stimulus increases. This finding suggests that attention has co-opted the circuits that mediate contrast gain control and that it operates by increasing the effective contrast of the attended stimulus. Consistent with this idea, microstimulation of the frontal eye fields, one of several areas that control the allocation of spatial attention, induces spatially local increases in sensitivity both at the behavioral level and among neurons in area V4, where endogenously generated attention increases contrast sensitivity. Studies in the slice have begun to explain how modulatory signals might cause such increases in sensitivity.

INTRODUCTION

The central role of attention in perception has been known since the dawn of experimental psychology (James 1890). The advent of new techniques for imaging the human brain complement earlier studies of brain-lesioned patients, enabling neuroscientists to map out the set of areas that mediate the allocation of attention in the human (for recent reviews, see Corbetta & Shulman 2002, Yantis & Serences 2003) and to examine how feedback from these areas alters neural activity in the visual cortices (reviewed by Chun & Marois 2002, Kastner & Ungerleider 2000). The development of techniques to record the activity of neurons in awake behaving animals has enabled researchers to probe the biological underpinnings of attention. In this review, we outline recent progress in understanding the circuits within the visual cortex that are modulated by attentional feedback.

Single-unit recording studies in the monkey have provided detailed, quantitative descriptions of how attention alters visual cortical neuron responses. When attention is directed to a location inside the receptive field (RF), the neuron's contrast-response threshold is reduced, enabling it to respond to stimuli that would otherwise be too faint to elicit a response. For stimuli presented at intermediate contrasts, spatial attention increases the firing rate by a multiplicative factor that is independent of the neuron's tuning for such properties as orientation and direction of motion. This scaling of the response enables neurons to discriminate more reliably the features of the attended stimulus. For stimuli presented at contrasts that exceed the neuron's contrast saturation point, attention has little or no effect on firing rate. Attention has qualitatively different effects when it is directed to one of two stimuli appearing simultaneously inside the receptive field. Both increases and decreases are observed, depending on the neuron's selectivity for the two stimuli. When one of the two stimuli is placed in the receptive field and the other in the surround, attending to the extrareceptive field stimulus can reduce the firing rate. All these phenomena can be accounted for by models developed to explain contrast-dependent modulations of neuronal response if one assumes that attending to a spatial location increases the effective contrast of stimuli appearing there. These attention-dependent modulations of responses in the visual cortex occur as a result of feedback from areas like the lateral intraparietal area (LIP), where elevated responses are associated with increased contrast sensitivity at the behavioral level, and the frontal eye fields (FEF), where microstimulation causes spatially localized increases in sensitivity both at the behavioral level and in visual cortical neurons. Recent studies in the slice have begun to characterize how modulatory signals can change neuronal responsiveness. A similar explanation may account for the effects of feature-based attention, if one assumes that attending to a feature increases sensitivity of neurons selective for that feature. Thus, attention-dependent improvements in our ability to detect faint stimuli, to discriminate stimulus features, and to select a stimulus from among distracters can all be understood as reflecting the operation of a relatively simple set of neural mechanisms.

SPATIAL ATTENTION: FACILITATION AND SELECTION

Psychophysical studies, event-related-potential studies, and brain-imaging studies of spatial attention in humans carefully document the phenomenon of attentional facilitation, the improved processing of a single stimulus appearing alone at an attended location (Posner et al. 1980). Observers can better detect faint stimuli appearing at an attended location (Bashinski & Bacharach 1980, Handy et al. 1996, Hawkins et al. 1990, Muller & Humphreys 1991) and can better discriminate properties of the attended stimulus, such as its orientation (Downing 1988, Lee et al. 1999). The effect of attending to a faint stimulus can be described as an enhancement of signal strength, as measured by the contrast increment that would be required to equate accuracy in discriminating features of stimuli appearing at an attended location versus an unattended location (Carrasco et al. 2000, Lu et al. 2000, Lu & Dosher 1998), and there is recent evidence that attention increases perceived stimulus contrast (Carrasco et al. 2004). This signal enhancement is reflected in greater stimulus-evoked neuronal activity, as measured by scalp potentials (e.g., Luck et al. 1994, reviewed by Hillyard & Anllo-Vento 1998), and brain imaging (e.g., Brefczynski & DeYoe 1999, Heinze et al. 1994; see Pessoa et al. 2003, Yantis & Serences 2003 for recent reviews).

Consistent with these observations, single-unit recording studies in monkeys trained to perform attention-demanding tasks have found that spatial attention often enhances neuronal responses evoked by a single stimulus appearing within the receptive field, an effect observed in neurons throughout the visual system (Ito & Gilbert 1999, McAdams & Maunsell 1999a, Motter 1993, Mountcastle et al. 1987, Roelfsema & Spekreijse 2001, Spitzer et al. 1988, Treue & Maunsell 1996). An example of this attention-dependent response facilitation is illustrated in Figure 1, which shows data recorded by Reynolds et al. (2000). The dashed line in

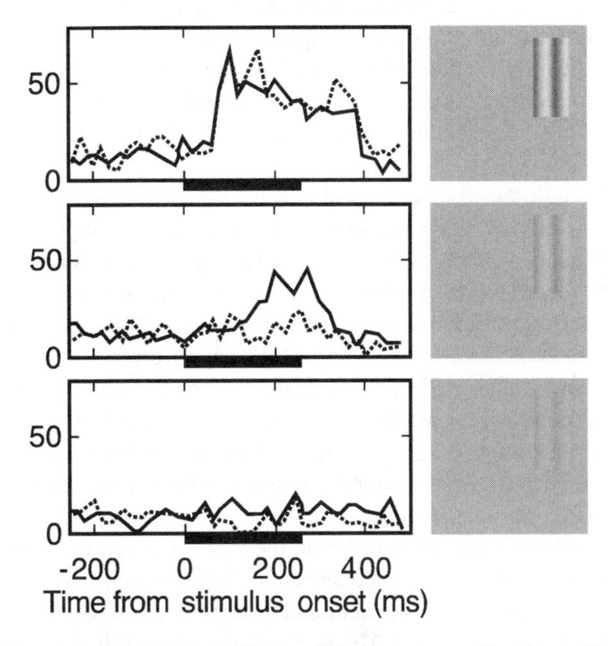

Figure 1 Responses of an example area V4 neuron as a function of attention and stimulus contrast. The contrast of the stimulus in the receptive field varied from 5% (*bottom panel*) to 10% (*middle panel*) to 80% (*upper panel*). On any given trial, attention was directed to either the location of the stimulus inside the receptive field (*solid line*) or a location far away from the receptive field (*dotted line*). The animal's task was to detect a target grating at the attended location. Attention reduced the threshold level of contrast required to elicit a response without causing a measurable change in response at saturation contrast (80%). Adapted from Reynolds et al. (2000).

each panel shows the response elicited by a stimulus that appeared in the receptive field of a V4 neuron while the monkey attended away from the receptive field to detect a target and earn a juice reward. The stimulus appeared at one of three levels of luminance contrast, two of which (5%, bottom panel, and 10%, middle panel) were too faint to elicit a response. That is, they were both below the neuron's contrast-response threshold. The third contrast (80%, top panel) was above the level of contrast at which the neuronal response saturated. The solid line shows the response, under identical sensory conditions, when spatial attention was directed to the stimulus.

Attention had no measurable effect on the response that was elicited at 5% contrast, which was well below the neuron's contrast-response threshold. The 10% contrast stimulus, which was just below the neuron's contrast-response threshold, and thus did not elicit a response when attention was directed away from the receptive field, elicited a clear response when attention was directed to its location in the receptive field. The average response elicited by the 10% contrast stimulus peaked at ~35 spikes per second, well above the baseline activity of the neuron. Thus, with attention directed away from the receptive field, the threshold level of contrast required to elicit a response was above 10%, but when attention was directed to the location of the stimulus, the contrast-response threshold was reduced to a value between 5% and 10%. Attention had no measurable effect on the neuronal response elicited by the stimulus when it was presented above saturating contrast.

Although the psychophysical and neurophysiological data clearly show that attention can facilitate processing of single stimuli appearing against a blank background, a more ecologically relevant purpose is served by attentional selection, the selection of behaviorally relevant stimuli from among competing distracters (Duncan & Humphreys 1989, Palmer et al. 2000, Treisman & Gelade 1980, Verghese 2001, Wolfe et al. 1989). Like any information-processing system, the visual cortex is limited in the quantity of information it can process at each moment in time. A typical visual scene contains a great deal more information than we can process in a single glimpse. Therefore, neural mechanisms must be in place to ensure that behaviorally relevant information will be selected to guide behavior.

Recordings from neurons within the extrastriate cortex have revealed a direct neural correlate of attentional selection. When multiple stimuli appear within a neuron's receptive field, the neuronal response tends to be driven preferentially by the task-relevant stimulus. The first single-unit recording study to document this finding was conducted by Moran & Desimone (1985). Two stimuli appeared within the receptive field: one that was of the neuron's preferred color and orientation, and another that was of a nonpreferred color and orientation. The monkey performed a task that required it to attend to one of the stimuli to report its identity and earn a juice reward. Moran & Desimone found that the neuron's response to the pair was stronger on trials when the monkey attended to the preferred stimulus, as compared to trials when the monkey attended to the nonpreferred stimulus.

A replication of this observation is illustrated in Figure 2, which shows data gathered by Chelazzi et al. (2001). Each line in Panel *B* shows the response averaged across a population of 76 neurons recorded in area V4 of monkeys performing a visual search task. As illustrated in Panel *A* on each trial, a stimulus appeared at fixation and then disappeared for 1500 ms, whereupon the same stimulus that had appeared at fixation (the target) reappeared at a random location in the receptive field, sometimes accompanied by another stimulus, and the monkey had to make a saccade to the target stimulus to earn a juice reward. The two stimuli were selected according to the response selectivity of the neuron. Thus, one (the preferred stimulus) was chosen to be of a shape and color that elicited a strong response, and the other (the poor stimulus) was chosen to elicit a low-firing-rate response.

The upper line in Panel *B* shows the population average response on trials when the preferred stimulus appeared alone as a target. The lowest line shows the average response when the poor stimulus appeared alone as the target. The two middle lines show the response when the pair of stimuli appeared together within the receptive field on trials when the preferred stimulus was the target (upper middle line) and on trials when the poor stimulus was the target (lower middle line). The initial response to the pair was not strongly dependent on which stimulus was the target, but after 150–160 ms, the pair response bifurcated, either increasing or decreasing, depending on whether the target was preferred or poor.

This observation that the attended stimulus exerts preferential control over the neuronal response has been replicated and extended both in the ventral stream areas studied by Moran & Desimone (1985) (Chelazzi et al. 1993, 1998, 2001; Luck et al. 1997; Motter 1993; Reynolds et al. 1999; Reynolds & Desimone 2003; Sheinberg & Logothetis 2001) and in the dorsal stream (Recanzone & Wurtz 2000, Treue & Martínez-Trujillo 1999, Treue & Maunsell 1996; however, see Seidemann & Newsome 1999).

Several of the above studies compared the response elicited by a pair of stimuli in the receptive field when the monkey attended either away from the receptive field or to one of the two stimuli inside the receptive field. These studies have found that attending to the more preferred stimulus increases the response to the pair, but attending to the poor stimulus often reduces the response elicited by the pair (Chelazzi et al. 1998, 2001; Luck et al. 1997; Martínez-Trujillo & Treue 2002; Reynolds et al. 1999; Reynolds & Desimone 2003; Treue & Martínez-Trujillo 1999; Treue & Maunsell 1996). This finding has lent support to models of attention in which inhibition plays a role in selection (Desimone & Duncan 1995, Ferrera & Lisberger 1995, Grossberg & Raizada 2000, Itti & Koch 2000, Lee et al. 1999, Niebur & Koch 1994). These models have accounted for a variety of observations concerning topics as varied as the interplay between attentional selection and oculomotor control, the role of visual salience in guiding attentional selection, and the role of working memory in guiding attention during search. Given the broad explanatory power of such competitive selection models, it is of interest to consider what has been learned from studies in anesthetized animals about the role of response suppression in the visual cortex.

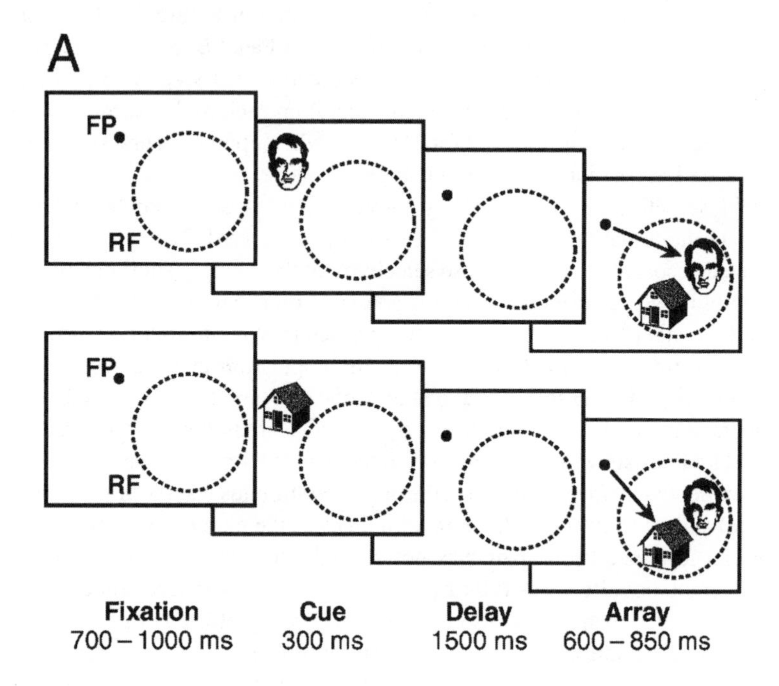
A
FP
RF
FP
RF
Fixation
700 – 1000 ms
Cue
300 ms
Delay
1500 ms
Array
600 – 850 ms

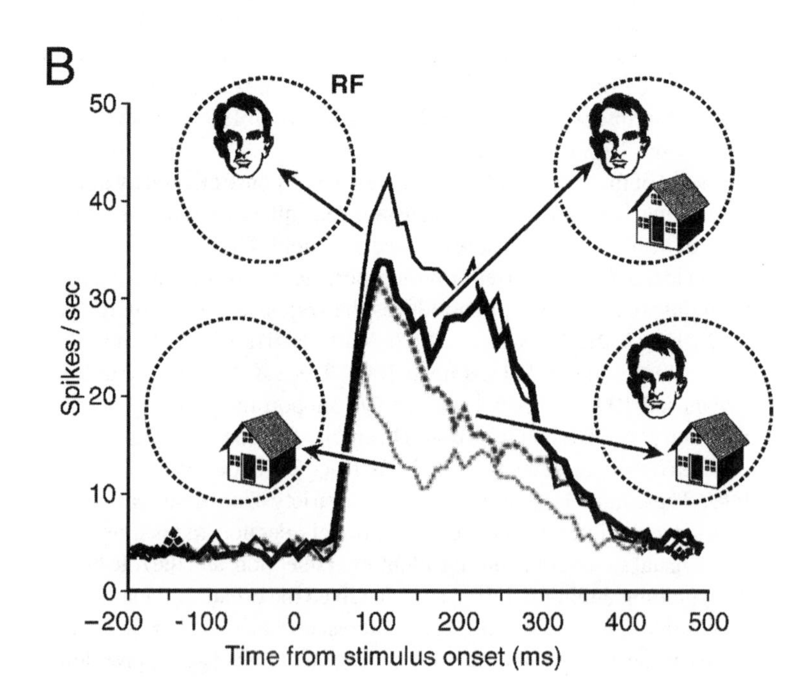
B
RF
Spikes / sec
0
10
20
30
40
50
−200
-100
0
100
200
300
400
500
Time from stimulus onset (ms)

CONTRAST-DEPENDENT RESPONSE MODULATIONS IN ANESTHETIZED ANIMALS PROVIDE CLUES ABOUT THE NEURAL MECHANISMS OF ATTENTION

The role of contrast in modulating the visual response properties of neurons has been extensively documented, and the models developed to account for these modulations rely critically on response suppression. As we will see, the circuitry underlying these modulations can be used to account for the attention-dependent response modulations, mentioned above. Here we describe four ways in which contrast modulates neuronal responses in the anesthetized mammal: Two occur when a single stimulus appears alone within the neuronal receptive field, and two occur when two stimuli appear together, either both within the classical receptive field or one inside the receptive field and the other in the receptive field surround. We then describe models developed to account for these response properties. Finally, we describe a model of attention that is mathematically related to these earlier models and demonstrate that it can therefore account for these contrast-dependent response modulations.

The first phenomenon is that cortical neuronal responses typically saturate as contrast increases, and this saturation firing rate is stimulus dependent. This finding is illustrated in Figure 3*A*, which is adapted from a study by Sclar & Freeman (1982). Each line shows the response of a complex cell recorded in cat area 17

←

Figure 2 Responses of V4 neurons during a visual search task. (*A*) Task. The monkey initiated a trial by foveating a spot at the center of the computer monitor ("FP"), after which a sample stimulus appeared nearby (here, either a face on the trial illustrated above or a house on the trial illustrated below). The sample then disappeared for 1500 ms, at which time the search array appeared, and the monkey had to make a saccade to the stimulus that matched the sample (the "target"). (*B*) Population histogram from 76 V4 neurons recorded during this task. For each neuron studied, the stimuli were selected from a pool of possible stimuli such that one of them (the preferred stimulus—here, the face) elicited a strong response and the other (the poor stimulus—here, the house) elicited a weak response. Responses are time locked to stimulus onset. From bottom left in a clockwise direction, insets illustrate conditions where the poor stimulus alone was presented inside the RF, the preferred stimulus alone was presented, the pair was presented with the preferred stimulus as the target, and finally, the pair was presented with the poor stimulus as the target. In all conditions, the animal made a saccade to the target stimulus, which appeared either alone or as an element of the pair. The initial part of the visual response to the pair falls between the response evoked by the preferred and poor stimuli. At 150–160 ms after display onset, the pair response increased or decreased, depending on which stimulus was the target. By the time of the onset of the saccade, 70–80 ms later, the pair response was driven almost entirely by the attended stimulus. Adapted from Chelazzi et al. (2001).

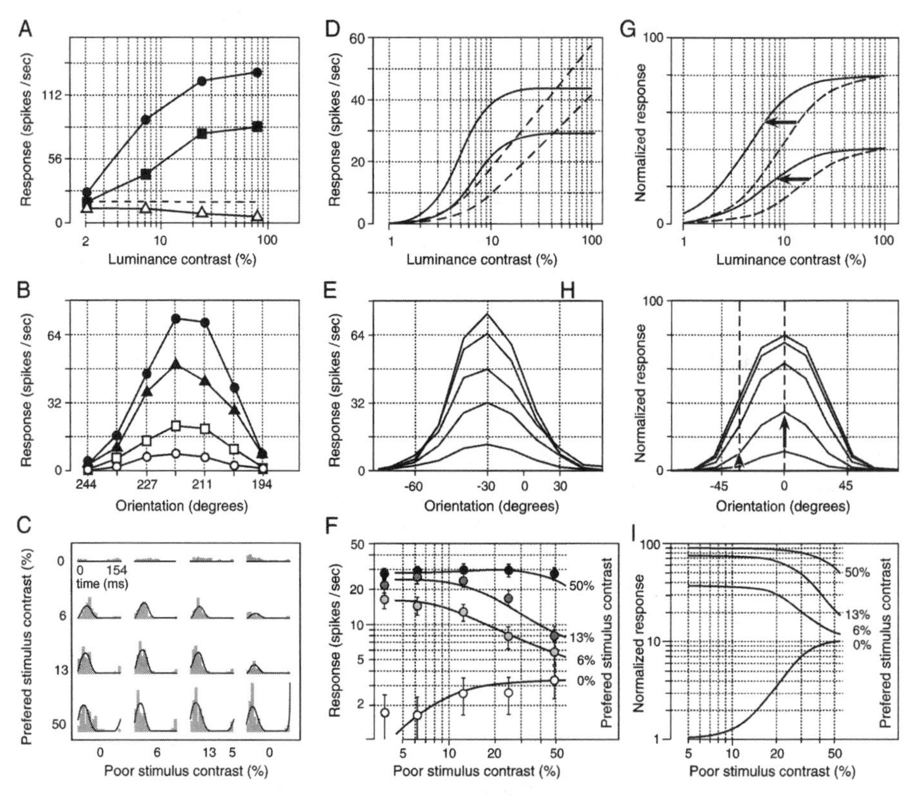

Figure 3 Contrast-dependent response modulations. (*A*) Contrast-response functions for a stimulus of the neuron's preferred orientation (*upper line*), a poor but excitatory orientation (*middle line*), and the null orientation (*bottom line*). Adapted from Sclar & Freeman (1982). (*B*) Orientation tuning curves of a second neuron, measured using a stimulus that varied in contrast from 10% (*empty circles*) to 80% (*filled circles*). Adapted from Sclar & Freeman (1982). (*C*) Responses of neuron recorded in area V1 of the anesthetized macaque. Two spatially superimposed gratings appeared within the receptive field. One grating was of the optimal orientation for the neuron, whereas the second grating was of a suboptimal orientation. Each subpanel illustrates responses obtained with a particular combination of contrast for the two stimuli. The preferred grating varied from 0% contrast (*top row*) to 50% contrast (*bottom row*), and the contrast of the poor grating increased from 0% (*left column*) to 50% (*right column*). Adapted from Carandini et al. (1997). Panels *D–I* show the capacity of several models to account for these contrast-dependent response modulations. See text for details.

when a single grating was presented in the receptive field at different levels of luminance contrast, yielding the neuron's contrast-response function. Each line is the contrast-response function derived using a grating presented at one of three orientations: the cell's optimal orientation (top line), a suboptimal but excitatory orientation (middle line), and the cell's null orientation (bottom line), which was slightly suppressive. The neuron did not respond to either excitatory stimulus presented below a threshold level of luminance contrast. As is typical of cortical neurons, above this threshold, the response increased over a range of contrasts that comprise the dynamic range of the contrast-response function, before reaching a saturation response. The response to the optimally oriented grating (upper line, black circles) saturated at ~130 spikes per second, well above the rate at which a suboptimally oriented grating saturated (black squares, 84 spikes per second).

The second phenomenon is that increasing the contrast of an oriented stimulus characteristically results in a multiplicative increase in the orientation tuning curve, as illustrated in Figure 3*B*. Each curve in this figure (also adapted from Sclar & Freeman 1982) shows a simple cell's response to gratings presented across a range of orientations. The most shallow orientation tuning curve (open circles) was derived using gratings of 10% luminance contrast, and each successively higher curve corresponds to a doubling of contrast. Note that this multiplicative effect of contrast on the orientation tuning curve follows from the observation that the contrast-response functions derived for any two orientations can be related to each other by a gain factor, as is approximately the case for the two contrast-response functions appearing in Figure 3*A*.

The above observations involve increases in response when a stimulus increases in contrast. The effect of increasing contrast is qualitatively different when two stimuli appear within the receptive field, where increasing the contrast of one of them can result in increases or decreases in response, depending on the neuron's selectivity for the two stimuli. This finding is illustrated in Figure 3*C*, which shows data recorded by Carandini et al. (1997) from a neuron in area V1 of the anesthetized macaque, when two spatially superimposed gratings, differing in orientation, appeared simultaneously within the receptive field (see also Bonds 1989, DeAngelis et al. 1992, Morrone et al. 1982). Each subpanel shows the response evoked by a particular combination of contrasts. One grating was of the neuron's preferred orientation, and a second grating was of a suboptimal orientation that nonetheless elicited an excitatory response when presented alone. The contrast of the preferred grating varied from 0% contrast (top row) to 50% contrast (bottom row), and the contrast of the poor grating increased from 0% (left column) to 50% (right column). Because 0% contrast corresponds to the absence of a stimulus, the upper left subpanel shows the neuron's spontaneous activity. Responses elicited by the preferred and poor stimulus presented alone appear in the left column and top row, respectively. The poor stimulus elicited a small but measurable excitatory response that increased with contrast (top row), and yet when paired with the preferred stimulus, it had a suppressive effect. Note, for example, that the response elicited by the 13% contrast preferred stimulus alone (left column, third row) was strongly reduced by the addition of the poor stimulus at 50% contrast (right panel, third row). Figures 3*D–I* show model fits to the data in Figures 3*A–C* and will be described below.

The above data were collected when preferred and poor stimuli were spatially superimposed. Similar competitive interactions among superimposed stimuli have been found in extrastriate visual cortex (Qian & Andersen 1994, Snowden et al. 1991). In the attention studies outlined below, the stimuli appeared at separate locations in the receptive field. It is therefore important to note that contrast-dependent suppressive effects are also observed with nonsuperimposed stimuli. This finding is illustrated in Figure 4, which shows data recorded by Reynolds & Desimone (2003) from a macaque V4 neuron. The first column shows trial-by-trial spike records that illustrate the response of the neuron when a stimulus of the neuron's null orientation appeared alone in the receptive field at luminance contrasts ranging from 5% (bottom panel) to 80% (top panel). The right column shows the response elicited by a stimulus of the neuron's preferred orientation, which was presented at a fixed contrast (40%) at a separate location in the receptive field. The panels are repeated for comparison. The center column shows the response that was elicited by the pair, as a function of poor stimulus contrast. The 5% contrast poor stimulus (bottom panel, center) had no measurable effect on the neuronal response, but at higher levels of contrast it became increasingly suppressive and at 80% contrast (top panel, center) it almost entirely suppressed the response. These observations confirm earlier reports showing that a poor stimulus suppresses the response elicited by a nonsuperimposed preferred stimulus (Miller et al. 1993, Recanzone et al. 1997, Rolls & Tovee 1995; however, see Gawne & Martin 2002). See also Britten & Hauer (1999) and Hauer & Britten (2002) for additional evidence of response normalization in area MT with nonsuperimposed stimuli.

Another form of contrast-dependent response modulation is observed when the second stimulus appears in the surround of the receptive field (Blakemore & Tobin 1972, DeAngelis et al. 1994, Knierim & van Essen 1992, Levitt et al. 1996, Maffei & Fiorentini 1976; see Fitzpatrick 2000 for a review). Suppressive surround effects are commonly attributed to intracortical lateral inhibition. Unlike the contrast-dependent interactions observed when two stimuli appear within the classical receptive field, the maximal suppressive effect of a surround stimulus is usually observed when the stimulus in the surround and the stimulus in the center are both of the neuron's preferred orientation (Cavanaugh et al. 2002a,b;

→

Figure 4 Increasing the contrast of a poor stimulus at one location suppresses the response elicited by a fixed contrast preferred stimulus at a second location in the receptive field of a V4 neuron. The contrast of the poor stimulus, illustrated in the first column, ranged from 5% to 80%. As indicated in the first column of raster plots, this stimulus did not elicit a clear response at any contrast. The right column shows the response elicited by the preferred stimulus, which was fixed in contrast (panels repeated down the column, for comparison). The middle column of raster plots shows the response to the pair. At low contrast (*bottom panel*), the poor stimulus had no measurable effect on the response to the preferred stimulus, but as poor stimulus contrast increased (moving up the column), it became increasingly suppressive, almost entirely suppressing the response at high contrast (*top panel*). Adapted from Reynolds & Desimone (2003).

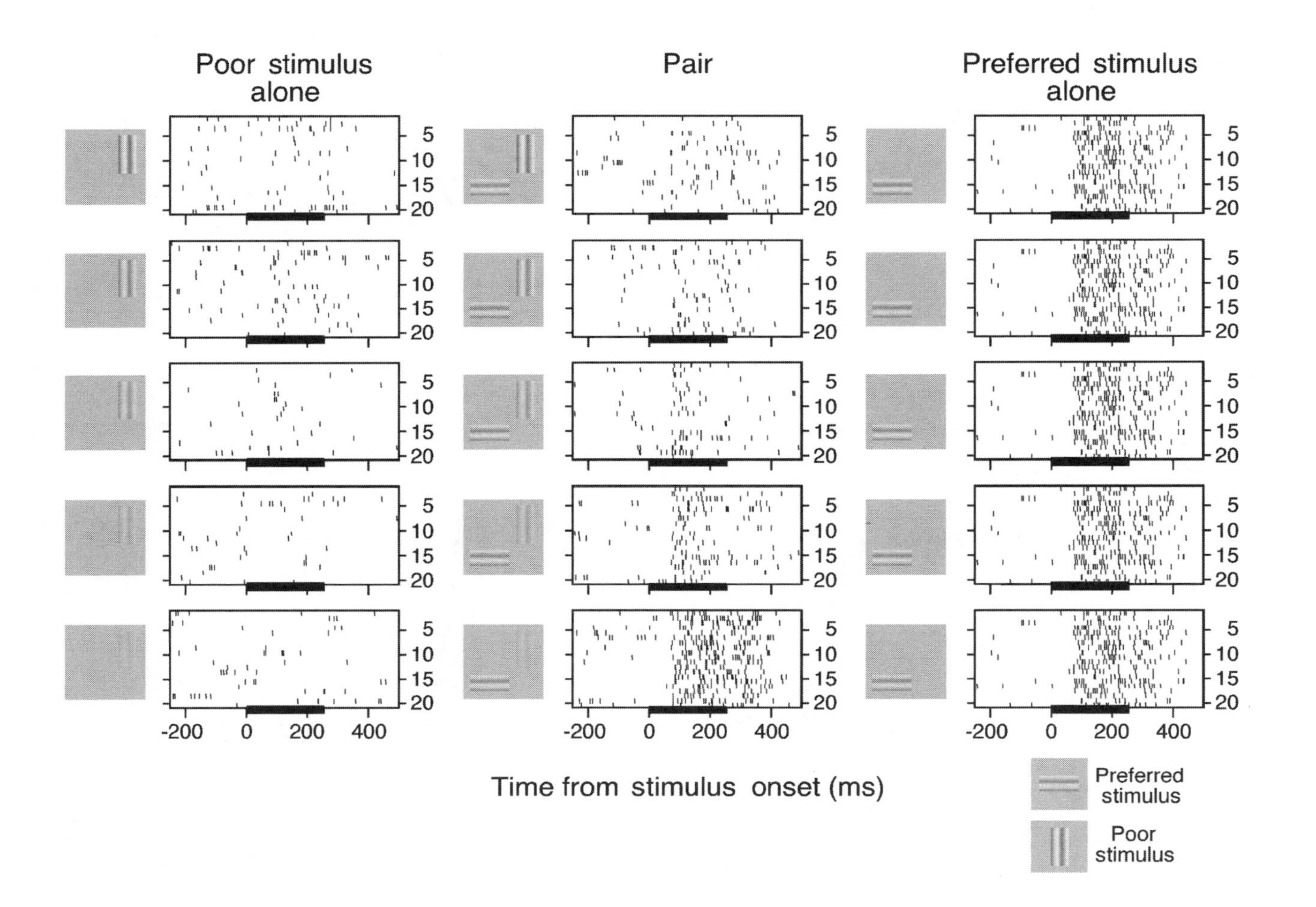
Poor stimulus alone
Pair
Preferred stimulus alone
5
10
15
20
-200
0
200
400
Time from stimulus onset (ms)
Preferred stimulus
Poor stimulus

DeAngelis et al. 1994). Evidence exists that such surround effects rely on divisive inhibition from the receptive field surround tuned for the neuron's preferred orientation (Cavanaugh et al. 2002a,b; Muller et al. 2003; Webb et al. 2003). The effect of varying the contrast of a stimulus in the surround is illustrated in the top two panels of Figure 5, which show responses of two neurons recorded by

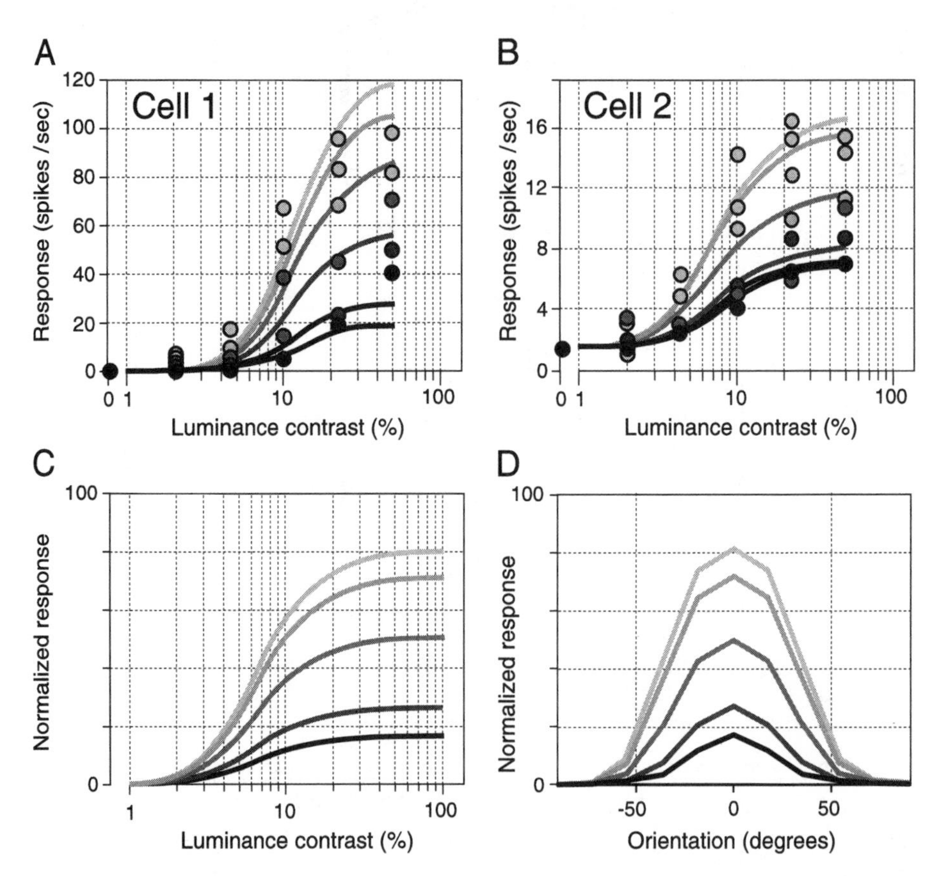

Figure 5 Effect of increasing the contrast of a stimulus in the receptive field surround. Panels *A* and *B* show the responses of two neurons recorded from area V1 of the anesthetized macaque. In this experiment, a circular grating of the neuron's preferred orientation appeared in the receptive field center, in the presence of a second grating in the surround. The circles in each panel indicate responses to the pair of gratings, with the shading of each circle indicating the contrast of the surround grating. Darker shading corresponds to higher contrast. The horizontal axis indicates the contrast of the stimulus in the center of the receptive field. The lines represent model fits to the experimental data. Adapted from Cavanaugh et al. (2002a). (*C*) If the model by Reynolds et al. (1999) is augmented to include an inhibitory surround, it also exhibits this contrast-dependent suppression, which results in a divisive reduction in the orientation tuning curve for the center stimulus (*panel D*).

Cavanaugh et al. (2002a) in area V1 of the anesthetized macaque. In their experiment, a circular grating of the neuron's preferred orientation appeared in the receptive field center, in the presence of a second grating in the surround. The circles in each panel indicate responses to the pair of gratings, with the shading of each circle indicating the contrast of the surround grating. Darker shading corresponds to higher contrast. The horizontal axis indicates the contrast of the stimulus in the center of the receptive field. The lines represent model fits and are discussed below. The effect of increasing the contrast of the surround stimulus is to suppress the response elicited by the preferred stimulus in the center, and Cavanaugh et al. found that this suppression is best described as a multiplicative reduction in the response throughout the contrast-response function.

THESE CONTRAST-DEPENDENT RESPONSE PROPERTIES CAN BE EXPLAINED BY EXISTING MODELS OF THE LOCAL CIRCUIT THAT RELY ON SUPPRESSION

The response properties described above are among the most thoroughly studied in visual neurophysiology, and they have driven the development of increasingly detailed models of the underlying cortical circuit (Albrecht & Geisler 1991, Carandini & Heeger 1994, Carandini et al. 1997, Grossberg 1973, Heeger 1992, McLaughlin et al. 2000, Murphy & Miller 2003, Somers et al. 1995, Sperling & Sondhi 1968, Troyer et al. 1998; for recent reviews, see Ferster & Miller 2000, Geisler & Albrecht 2000, Shapley et al. 2003). These models differ in important respects, such as the relative importance of feedforward inhibition, the role of shunting inhibition, the degree to which inhibition is tuned, and the importance of mutually excitatory interactions between similarly tuned neurons. Which model provides the best approximation to the true microcircuit (which itself may differ in some respects across cell types, according to laminar distribution, across brain areas, and across species) is not yet known. Although these differences are very important, it is beyond the scope of this review to describe them in detail. We therefore simply illustrate fits derived from three of the above models to document their capacity to account for the contrast-dependent response modulations illustrated in Figures 3 and 5.

However, to fix ideas, we begin by outlining briefly the key assumptions of one model, the normalization model, which is described in a recent paper by Carandini et al. (1997). The model is closely related to models proposed by Heeger (1992) and Carandini & Heeger (1994). It has some features in common with models proposed by Albrecht & Geisler (1991), Grossberg (1973), and Sperling & Sondhi (1968). It provides a simple account for the contrast-dependent modulations illustrated in Figures 3*A–C* and, when suitably extended to include a receptive field surround (Cavanaugh et al. 2002a,b), can also account for the contrast-dependent surround effects appearing in Figure 5. The model achieves orientation selectivity as a result of tuned excitatory input, which is stronger for a preferred orientation stimulus than for a nonpreferred stimulus. This excitatory drive increases with contrast, as does a feature nonselective shunting inhibitory drive. Shunting inhibition refers to a synaptically activated conductance with a reversal potential near the resting potential of a neuron. Activating shunting inhibition decreases the input resistance of the cell, which diminishes potential changes induced by excitatory inputs. This action has a divisive effect on subthreshold excitatory postsynaptic potential amplitudes, and the model assumes that this reduction has a divisive effect on firing rate (however, see Holt & Koch 1997). When two stimuli appear together, the effect of varying the contrast of either stimulus depends on the relative contributions of excitatory and divisive inhibitory drive. When a poor stimulus (with proportionally more inhibitory drive) is presented with a preferred stimulus, the additional inhibitory input results in a suppressed response. This response suppression can be magnified by increasing the contrast of the poor stimulus or can be diminished by increasing the contrast of the preferred stimulus.

The capacity of this and two other models to account for contrast-dependent response modulations is documented in Figures 3 and 5. Figure 3*D* illustrates the contrast-response functions that emerge from a model proposed by Somers et al. (1998), which is an extension of an earlier model introduced by Somers et al. (1995). Like the normalization model, this model relies on inhibition to account for orientation-specific saturation responses. However, it differs from the normalization model in several respects, including the fact that it does not depend on shunting inhibition and that it relies on recurrent cortical excitation to sharpen its weakly orientation-tuned excitatory input. Dashed lines in Figure 3*D* show model responses to gratings of the optimal orientation (upper dashed) and a suboptimal but excitatory orientation (lower dashed), when intracortical connections were silenced and only thalamocoritcal inputs were active. The solid lines show model responses to the same two stimuli when intracortical connections were included. The effect of the intracortical connections was to cause the model neuronal response to saturate, while maintaining selectivity for the two stimuli across contrast.

Figure 3*E* illustrates the contrast-invariant tuning that emerges from a model introduced by Troyer et al. (1998). This model incorporates a number of nonlinearities to explain the contrast-dependent gain changes in orientation tuning, including small contrast-dependent conductance changes, spike-rate adaptation currents, and synaptic depression. Each curve indicates the model response across orientation, at different levels of contrast, ranging from 2.5% (lowest curve) to 50% contrast (highest curve). At 5% contrast and above, the orientation tuning curves are approximately related to each other by a multiplicative factor.

Figure 3*F* demonstrates the ability of the model by Carandini et al. (1997, described above) to account for the contrast-dependent suppressive effect of adding a poor stimulus to a preferred stimulus. The circles indicate the mean firing rates taken from the data in Figure 3*C*, with shading indicating the luminance contrast of the preferred orientation grating, ranging from 0% (white) to 50% (black). The contrast of the poor orientation grating is indicated on the horizontal axis. The lines indicate the model's best fit to the data.

Note that the response elicited by the poor stimulus increases with contrast, indicating that although it was poor, it was nonetheless excitatory. The model accounts for the fact that, despite this excitatory effect when presented alone, increasing the contrast of the poor stimulus suppressed the response elicited by the simultaneously presented preferred stimulus. Note, for example, that the response elicited by the 13% contrast preferred stimulus alone (~22 spikes per second, dark gray data point, left) was strongly reduced by the addition of the poor stimulus at 50% contrast (~8 spikes per second, dark gray data point, right). The model also accounts for the fact that increasing the contrast of the preferred grating had the opposite effect: It increased the response to the pair. The highest contrast preferred stimulus (black circles) was virtually immune to this suppressive effect.

Cavanaugh et al. (2002a,b) have extended the model of Heeger & Carandini (1994) to include a divisive inhibitory surround, and with this addition, the model can account for changes in the response to a stimulus in the receptive field center when a second stimulus, in the surround, is varied in contrast. The model's ability to account for this contrast-dependent, center-surround modulation is documented in the upper two panels of Figure 5. The lines indicate model fits to these data, with darker lines corresponding to higher surround contrast. The model fits capture the effect of increasing the contrast of the surround stimulus, which was to suppress the response elicited by the preferred stimulus in the center.

A LINKING HYPOTHESIS: DIRECTING SPATIAL ATTENTION TO A STIMULUS INCREASES ITS EFFECTIVE CONTRAST

How could these local circuit models account for (*a*) the response facilitation that is often observed when attention is directed to a single stimulus appearing alone within the receptive field, and (*b*) the observation that when two stimuli appear within a neuron's receptive field, the neuronal response is dominated by the stimulus that is relevant to current behavior? First, consider the effect of elevating the contrast of a single stimulus, as illustrated in Figures 3*A* and 3*B*. Elevating the contrast of a just-subthreshold stimulus will push it above threshold, thereby eliciting a response. For stimuli falling within the dynamic range of the contrast-response function, elevating contrast leads to a more robust response. Thus, if attention operates by increasing the effective contrast of a stimulus, this elevation of effective contrast would account for the elevations in response found when attention is directed to a single stimulus—the idea of attentional facilitation, described above.

Second, consider how changes in contrast modulate the response when two stimuli—a preferred stimulus and a poor stimulus—appear together within the receptive field. As illustrated in Figures 3*C*, 3*F*, and 4, the poor stimulus suppresses the response elicited by the preferred stimulus, and the magnitude of this suppression depends on the relative contrasts of the two stimuli. At low contrast, the poor stimulus has little or no suppressive effect, but as the contrast of the poor stimulus increases, it drives the response downward. This suppression is diminished if the preferred stimulus is presented at high contrast. Thus, increasing the contrast of one of the two stimuli causes a change in the neuronal response similar to that observed when attention is directed to it: It causes the stimulus to dominate the neuronal response, just as attention causes the neuron to selectively process information about the stimulus that is relevant to the animal's current behavioral goals. An appealing linking hypothesis is that attention operates by multiplying the effective contrast of the behaviorally relevant stimulus, a result that could be achieved by increasing the neuron's contrast sensitivity for the attended stimulus.

This idea is incorporated into a model of attention developed to provide an account of how the behaviorally relevant stimulus gains preferential control over neuronal responses in the visual cortex. This model, described in detail elsewhere (Reynolds et al. 1999), was conceived as a way of formalizing the biased competition model of Desimone & Duncan (1995). Because it operates by multiplying the effective contrast of the attended stimulus, we refer to it as the contrast gain model of attention. It is a functional model in that it is intended to characterize the operations performed by the neural circuit without committing to specific biophysical or biochemical mechanisms. However, it is mathematically related to models that have been used to explain the contrast-dependent effects described above. Therefore, this model can account for the same set of contrast-dependent phenomena, as documented in Figures 3*G–I* and 5*C*.

Figure 3*G* shows the model contrast-response functions for an optimal (upper dashed line) and a suboptimal but excitatory stimulus (lower dashed line). Orientation selectivity arises from differences in the strength of excitatory input across orientation, so here, the optimally oriented stimulus activated greater excitatory input than did the suboptimal stimulus. Inhibitory input was untuned for orientation. Attention leads to increases in the strength of excitatory and inhibitory inputs activated by the attended stimulus (Reynolds et al. 1999), as would occur when increasing the contrast of the stimulus. The effect of this change is to shift the model contrast-response function to the left, as indicated by the arrows. Figure 3*H*, which was obtained using the same set of parameters that yielded Figure 3*G*, documents the ability of the model to exhibit multiplicative increases in the orientation tuning curve with increasing contrast. The vertical lines indicate the orientations whose contrast-response functions are illustrated in Figure 3*G*. Because attention yields a shift in effective contrast, its influence on the tuning curve is the same as an increase in contrast: to cause a multiplicative increase in the tuning curve. This finding is illustrated by the upward arrows, which show the increases in response that result from a leftward shift in the contrast-response function for the two orientations whose contrast-response functions are illustrated in Figure 3*G*.

Figure 3*I* shows the model behavior when the preferred stimulus from Figure 3*G* appears together with a nonpreferred but excitatory stimulus also in the receptive field, at various levels of contrast. As is the case experimentally (Figures 3*C*, 3*F*, 4), the model accounts for the finding that elevating the contrast of the poor stimulus will increase its ability to suppress the response to a fixed-contrast preferred stimulus. Thus, the model accounts for the finding that when two stimuli appear in the receptive field, attending to the more preferred stimulus will cause an elevation in response, and attending to the poor stimulus will lead to a reduction in response.

The model, as originally described by Reynolds et al. (1999), does not specify the geometry of the receptive field, but if we take the lead of Cavanaugh et al. (2002a,b) and assume that the inhibitory kernel extends beyond the excitatory center of the receptive field, the model can then provide a qualitative account for the contrast dependence of the inhibitory surround. This idea is illustrated in Figure 5*C*, which shows the response of the model using the same parameters that were used to derive Figures 3*G–I* but with the addition of a purely inhibitory input increasing monotonically with surround-stimulus contrast. Figure 5*D* shows the model output as a function of the orientation of the center stimulus. As the contrast of the surround stimulus increases, this divisively reduces the response evoked by the center stimulus at each level of contrast. As we summarize below, this has relevance to single-unit recording studies in which monkeys attended to a stimulus in the surround.

ATTENTION-DEPENDENT RESPONSE MODULATIONS MIRROR THE EFFECTS OF A MULTIPLICATIVE INCREASE IN STIMULUS CONTRAST

A number of recent single-unit recording and lesion studies of attention in the macaque have likened attention to increasing visual salience (Bisley & Goldberg 2003; De Weerd et al. 1999; Gottlieb et al. 1998; Martínez-Trujillo & Treue 2002; McAdams & Maunsell 1999a; Reynolds et al. 1999, 2000; Reynolds & Desimone 2003; Treue 2003). Reynolds et al. (2000) directly tested the idea that spatial attention causes a multiplicative increase in the effective contrast of a stimulus. If it does, then as illustrated in Figure 3*G*, attention should cause a leftward shift in the contrast-response function. Such a leftward shift would predict (*a*) that the threshold level of contrast required to elicit a neuronal response should decrease, (*b*) that the largest increases in firing rate should be observed for stimuli that are within (or just below) the upward sloping part of the contrast-response function (the dynamic range of the cell's contrast-response function), and (*c*) that attention should have little or no effect on the firing rate elicited by a stimulus above the dynamic range.

To test these predictions, luminance-modulated gratings were presented within the receptive fields of V4 neurons as monkeys performed a task that required them either to attend to the location of the gratings, or else, on separate trials, to attend to another location far from the receptive field. The monkey's task was to detect a target grating that could appear at an unpredictable time at the cued location. The luminance contrast of each target was selected at random, so to perform the task reliably, the monkey had to continually attend to the location of the upcoming

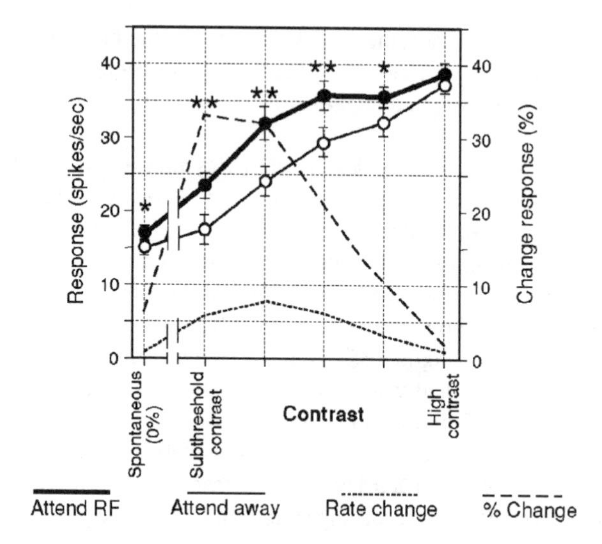

Figure 6 Attention-dependent increases in contrast sensitivity. Each line shows the average responses of a population of area V4 neurons. The monkey attended either to the location of the receptive field stimulus (*thick line, black circles*) or to a location far away from the receptive field (*thin line, white circles*). Luminance-modulated gratings were presented inside the receptive field at five different values of contrast selected to span the dynamic range of each neuron. The monkey's task was to detect a target grating that could appear at an unpredictable time at the cued location. The dashed and dotted lines show, respectively, percent and absolute difference in firing rate across the two attention conditions, as a function of contrast. Adapted from Reynolds et al. (2000).

target. Each grating was presented at a contrast drawn at random from a set of five contrasts that were selected to span the dynamic range of the neuron. Consistent with the predictions of the contrast-gain model, attention caused a reduction in neurons' contrast-response thresholds, caused the strongest increases in response for stimuli within or below the dynamic range of the neuron's contrast-response function, and caused only minimal changes in response for stimuli that were at saturation contrast. This was the case for the neuron illustrated in Figure 1, which showed no change in firing rate with attention at saturating contrast (80%), but which exhibited a clear reduction in response threshold with attention. Figure 6 shows the contrast-response function, averaged across the population. The thin solid line shows the response when attention was directed away from the receptive field, and the thick solid line shows the response to the identical stimuli, when they were attended. The dashed and dotted lines show percent and absolute difference in firing rate across the two attention conditions, as a function of contrast. At zero contrast (no stimulus present), there was a slight elevation in spontaneous activity, consistent with previous reports (e.g., Luck et al. 1997). For stimuli chosen to be below each neuron's contrast-response threshold (subthreshold contrast), there was a clear and significant response. The largest increases in firing rate were observed over the contrasts that were chosen to span the dynamic range of each neuron's contrast-response function. There was no significant effect of attention at the highest contrast tested, which was chosen to be at or slightly above saturating contrast. Similar results were found for both preferred and poor stimuli. In both cases, attention shifted the contrast-response function to the left, consistent with the prediction illustrated in Figure 3*G*. A detailed analysis of the data derived from this experiment found that for a cell to detect an unattended stimulus as reliably as it could detect an attended stimulus, the unattended stimulus would have to be half again as high in contrast as the attended stimulus. That is, under the conditions of this experiment, attention was worth a 51% increase in contrast, in terms of improving stimulus detectability. Although this value would be expected to change as a function of task difficulty (Spitzer et al. 1988) and brain area (Cook & Maunsell 2002, Luck et al. 1997) it is corroborated by other studies that have also quantified spatial attention in units of luminance contrast, including studies conducted in MT by Martínez-Trujillo & Treue (2002) and in V4 by Reynolds & Desimone (2003), who estimated that attending to a stimulus was equivalent to increasing its luminance contrast by 50% and 56%, respectively.

Motter (1993) observed a pattern of responses consistent with the proposal that attention increases the effective contrast of a stimulus. He recorded neuronal responses in V1, V2, and V4, using bars of intermediate levels of contrast. For some neurons, responses were stronger when attention was directed to the stimulus in the receptive field, in contrast to when attention was directed to another stimulus outside the receptive field. Notably, this difference in response was greatest for stimuli presented near the peak of the neuron's orientation tuning curve, as would be expected if attention caused a multiplicative increase in the tuning curve. McAdams & Maunsell (1999a) carefully quantified attention-dependent changes in the orientation tuning curve and found that spatial attention does cause a multiplicative increase in the orientation tuning curve, without otherwise altering its shape. In their experiment, they held contrast constant and varied the orientation of a grating appearing alone in the receptive field. On some trials, monkeys attended to the stimulus in the receptive field to report whether two successive gratings differed in orientation. On other trials, they attended to stimuli appearing at a location across the vertical meridian to report whether they differed in color. This enabled McAdams & Maunsell to map out the neuron's orientation tuning curve under identical sensory conditions and to measure how it changed with attention. As illustrated in Figure 7, attending to the receptive field caused a multiplicative increase in the neuron's orientation tuning curve. A related study demonstrated that this increase in the gain of the orientation tuning curve enabled neuronal signals to better distinguish the orientation of the stimulus (McAdams & Maunsell 1999b).

A third model property is that when two stimuli appear in the receptive field, attending to one of them will cause either an increase or a decrease in response, depending on the cell's relative preference for the two stimuli (see Figure 3*I*), with the magnitude of these changes growing in proportion to the neuron's selectivity

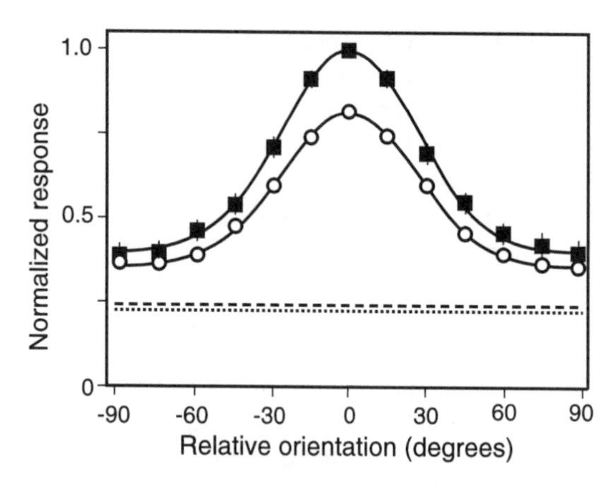

Figure 7 Attention increases tuning curves by a gain factor. Average normalized orientation tuning curves computed across a population of area V4 neurons while the monkey attended either to the location of a grating stimulus inside the receptive field (*filled squares*) or to a location in the opposite hemifield (*empty circles*). The upper curve is approximately a multiplicative version of the lower curve. Adapted from McAdams & Maunsell (1999a).

for the two stimuli. As mentioned earlier, this pattern has been observed in several single-unit recording studies of attention (Chelazzi et al. 1998, 2001; Luck et al. 1997; Martínez-Trujillo & Treue 2002; Reynolds et al. 1999; Reynolds & Desimone 2003; Treue & Martínez-Trujillo 1999; Treue & Maunsell 1996). Figure 8, adapted from a study conducted by Treue & Martínez-Trujillo, illustrates attention-dependent increases and decreases in firing rate. Two patterns of dots appeared within the classical receptive field of an MT neuron. One of them ("pattern A" in Figure 8*A*) moved in the null direction for the neuron, and the other ("pattern B") moved in one of twelve directions of motion, selected at random on each trial. The monkey either attended to the fixation point to detect a change in its luminance or, on separate trials, attended to one of the two patterns of dots to detect a change in the direction or speed of its motion. The neuronal response when attention was directed outside the receptive field is indicated by the middle curve in Figure 8*B*, which shows responses averaged over 56 neurons (sensory response), aligned on the each neuron's preferred direction of motion. Because pattern A moved in each neuron's null direction, pattern B was the more preferred stimulus over a range of contrasts. Over this range, attending to pattern B elevated the response, with the magnitude of this increase growing in proportion to the neuron's selectivity for the two stimuli. When attention was directed to the null stimulus, the response was reduced, again, with changes growing in proportion to selectivity.

A final prediction that follows from the idea that attention is equivalent to an increase in contrast is illustrated in Figure 5*D*. As noted above, increasing

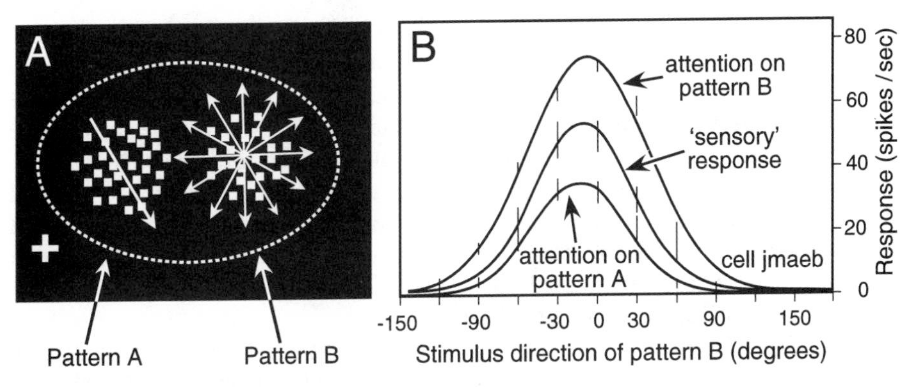

Figure 8 Attention to a poor stimulus decreases the response to a pair of stimuli in MT, and attention to a preferred stimulus elevates the response. (*A*) Stimuli. Two patterns of dots appeared simultaneously in the receptive field of an MT neuron. One pattern (pattern A) moved in the cell's null direction. The other pattern (pattern B) moved in a direction that varied from trial to trial. (*B*) Responses averaged across a population of MT neurons. The center line shows the tuning curve to the pair, as a function of the orientation of Pattern B, when the monkey attended to the fixation point to detect a change in luminance (the sensory response). When attention was directed to the null stimulus (pattern A) the response was reduced (*bottom line*). When attention was directed to pattern B, the response was increased. These increases and decreases reflect a combination of both spatial and feature-based attention, which were found to combine additively. Adapted from Treue & Martínez-Trujillo (1999).

the contrast of a stimulus in the surround causes suppressive effects that are well characterized as a multiplicative reduction in the neuronal response, and this effect has been modeled using a variant of the divisive normalization model described above. If attention is equivalent to an increase in contrast, then, as illustrated in Figure 5*D*, the effect of directing attention toward a stimulus in the receptive field surround should be to reduce the gain of the orientation tuning curve derived from a stimulus in the receptive field center.

This prediction has not been tested directly, but several studies provide provisional support for the proposal that attention modulates center-surround interactions. Ito & Gilbert (1999) measured the effect of directing attention to bars appearing within the receptive fields of neurons in the primary visual cortex of the macaque. They found that whereas attention had no effect when the bar appeared alone in the receptive field, it had a pronounced effect when colinear bars appeared in the receptive field surround, which suggests that attention modulated the sensory interactions between center and collinear surround stimuli. Connor et al. (1996, 1997) found that V4 responses evoked by a stimulus in the RF changed when attention was directed to one of several stimuli in the surround. Responses were often stronger when attention was directed to a stimulus near the location of the probe, possibly reflecting either the falloff of facilitation with distance from the attended location, an increased likelihood of a shift of attention to probes appearing near the attended stimulus, or a shift in the receptive field. They also found a much larger change in the overall responsiveness of the neuron, depending on which surround stimulus was attended. Connor and colleagues did not measure center-surround interactions, but it seems plausible that these changes in response may have resulted from modulation of spatially asymmetric center-surround modulations, as have been found in cat primary visual cortex (Xiao et al. 1995) and in macaque area MT (Walker et al. 1999). Connor and colleagues concluded that such position-dependent changes in responsiveness could help encode the position of stimuli with respect to the locus of attention, which could be useful for object recognition (Salinas & Abbott 1997).

Motter (1993) provided what is perhaps the most direct evidence that attention modulates center-surround interactions. He examined attention-dependent changes in the orientation tuning curve. As noted above, Motter found that some neurons showed an increase in response when attention was directed to a single stimulus within the receptive field. He also found, however, that many neurons were modulated by attention only when multiple stimuli appeared in the visual field. He did not explicitly measure the suppressive surrounds of the neurons he studied. However, the pattern of suppressive effects he observed when attention was directed to stimuli outside the receptive field is consistent with the pattern that would be predicted if attention increased the effective contrast of stimuli in the surround. He examined whether the attention effect reflected an increase in response when attention was directed to the center stimulus or a decrease in response when attention was directed to one of the stimuli outside the receptive field. He measured orientation tuning curves when the center stimulus appeared alone in the visual field during passive fixation and compared these curves with tuning curves measured when the monkey attended either to the stimulus in the receptive field or to one of the extrareceptive field stimuli. For about 50% of the cells modulated by attention under these conditions, attention to the stimulus in the surround caused a reduction in response, especially at the peak of the orientation tuning curve. This attention effect is illustrated in Figure 9, which shows the response for one such neuron. The upper dashed line shows the orientation tuning curve mapped out during fixation. The lower dashed line shows the tuning curve that was mapped out when the monkey attended to the stimulus in the receptive field. There is no clear difference. However, when the monkey attended to the stimuli that appeared outside the classical receptive field, including those in the surround, the peak responses were reduced in magnitude.

SOURCES OF SIGNALS THAT MODULATE RESPONSES IN VISUAL CORTEX

The experiments described above demonstrate that spatial attention causes changes in the neuronal response that mirror the effects of increasing the effective contrast of the attended stimulus. Research from many laboratories using a variety of different

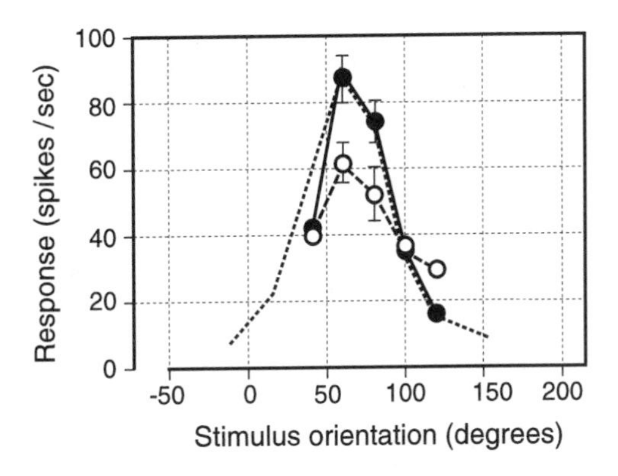

Figure 9 The orientation tuning curve of a V2 neuron was suppressed when the monkey attended to stimuli appearing outside the receptive field. The upper dashed line shows the tuning curve measured when the stimulus in the receptive field appeared alone, during passive fixation. The solid lines show the tuning curve measured when additional stimuli were present outside the receptive field, and the monkey attended to the stimulus in the receptive field (*black circles*). The lower dashed line (*white circles*) shows the tuning curve measured when the monkey attended to the stimuli outside the field. Attending to the stimuli outside the receptive field reduced the response when the stimulus in the receptive field was presented at the neuron's preferred orientation.

techniques have identified potential sources for the feedback signals that modulate visual cortical responses during spatially directed attention (for a recent review, see Corbetta & Shulman (2002). These sources include the pulvinar (reviewed by Sherman (2001), parts of the parietal cortex (Bisley & Goldberg 2003, Colby & Goldberg 1999, Gottlieb et al. 1998, Mountcastle et al. 1987, Steinmetz & Constantinidis 1995), the frontal eye field (FEF; reviewed by Schall 1995), and the superior colliculus (Basso & Wurtz 1998, Wurtz & Goldberg 1972). Space constraints preclude us from detailing the important advances made over the past decade in understanding the roles of these structures in guiding spatial attention. We therefore mention briefly two experiments that have directly examined the role of one of these structures (FEF) in modulating contrast sensitivity at the behavioral and neuronal levels.

Investigators have long known that FEF plays a role in the control of saccadic eye movements (Robinson & Fuchs 1969) and in the selection of stimuli during visual search (for a review, see Thompson et al. 2001). Sensitivity to stimuli is increased at the location targeted by an impending saccade (Chelazzi et al. 1995, Hoffman & Subramaniam 1995, Moore et al. 1998, Shepherd et al. 1986). FEF has direct anatomical projections to visual areas that are modulated by spatial attention, including areas V2, V3, V4, MT, MST, TE, and TEO, as well as to other potential sources of top-down attentional control, such as area LIP (Stanton et al. 1995).

To establish a causal link between attention-dependent increases in contrast sensitivity and FEF activity, Moore & Fallah (2004) measured changes in contrast

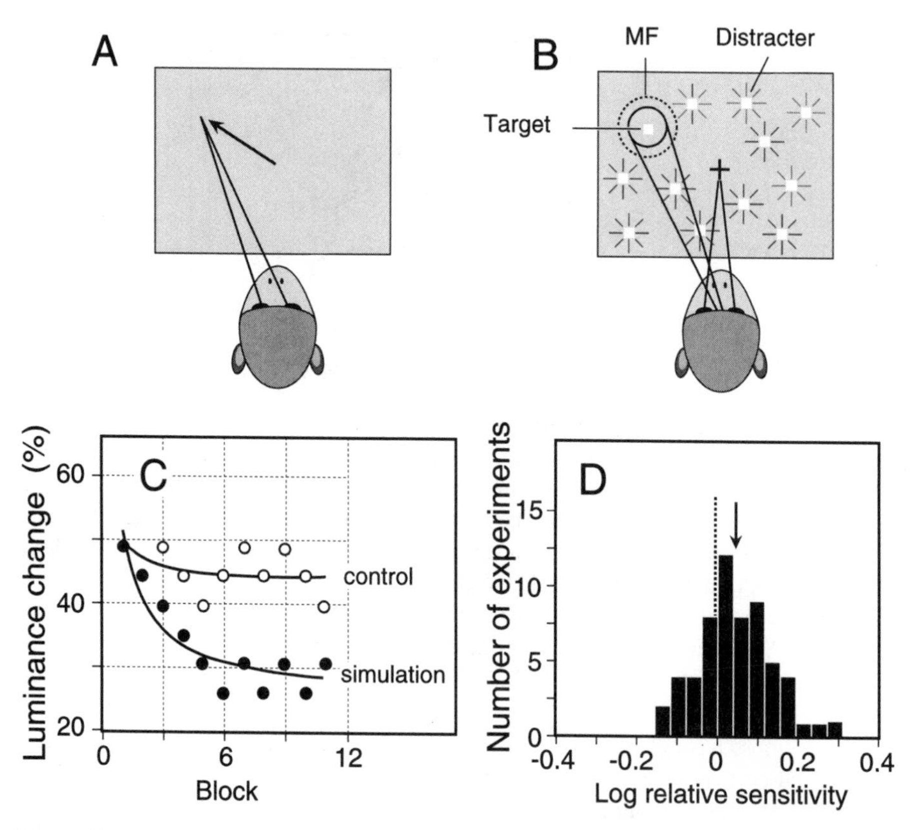

Figure 10 Microstimulation of FEF increases contrast sensitivity. (*A*) Microstimulating a site in FEF at high current moves the eyes from fixation to a location referred to as the movement field. (*B*) During the attention task, the monkey maintained fixation while attending to a target within the movement field (MF) to detect a slight change in luminance. Distracters appeared throughout the visual field. On randomly selected trials, current was injected into FEF at levels too low to elicit an eye movement. (*C*) A staircase procedure was used to determine threshold luminance changes with (*black circles*) and without (*white circles*) FEF stimulation. Thresholds were reduced by FEF stimulation. (*D*) For each session, the change in sensitivity was computed by taking the ratio of threshold luminance changes with and without FEF stimulation. Across sessions, contrast sensitivity increased, as indicated by the rightward shift in the histogram. Adapted from Moore & Fallah (2001).

detection thresholds following electrical stimulation of FEF neurons. As illustrated in Figure 10*A*, electrical stimulation of FEF neurons causes the eyes to move from the fixation point to a particular location, referred to as the movement field of the neurons at the stimulation site. After determining the movement field, Moore & Fallah had monkeys perform a task, in which they reported a brief change in the luminance of a target stimulus, to earn a juice reward. The target appeared either inside or outside the movement field. Distracter stimuli appeared randomly at locations throughout the visual field, which increased task difficulty (Figure 10*B*). On a randomly selected subset of trials, Moore & Fallah injected current calibrated to be too weak to elicit an eye movement. They used a staircase procedure to determine the minimum luminance change required for the monkey to achieve a threshold level of performance, on trials with and without stimulation. Stimulation reduced the level of luminance contrast required to reliably detect the change. This finding is illustrated for an example session in Figure 10*C*, which shows the contrasts generated by the staircase procedure used to determine threshold. For this session, stimulation reduced the threshold luminance change from 44% to 28%. The effect of microstimulation was spatially specific: The monkey only benefited from stimulation when the target appeared within the movement field. Across sessions, stimulation improved sensitivity, as indicated by the rightward shift in the distribution of changes in sensitivity appearing in Figure 10*D*. These results thus show that stimulation of FEF improves contrast sensitivity at the movement field location, in much the same way that spatial attention improves contrast sensitivity in V4 neurons (Reynolds et al. 2000).

Moore & Armstrong (2003) reasoned that FEF stimulation might, therefore, cause increased responsiveness in V4 neurons. To test this hypothesis, they identified the site in FEF whose movement field overlapped with the receptive fields of a set of neurons in area V4 and measured the effect of FEF microstimulation on neuronal responses in V4 (see Figure 11*A*). They found that FEF stimulation caused the neuronal response to increase. This increase in response with FEF stimulation is illustrated in Figure 11*B*, which shows responses of a single V4 neuron with and without stimulation. The time course of stimulus presentation (RF stim) and current injection (FEF stim) are illustrated at the top of the panel. The neuronal responses in the two conditions are indicated at the bottom of the figure. Stimulation was injected 500 ms after the appearance of the stimulus, at which point the stimulus-evoked response had begun to diminish in strength. The average response on microstimulation trials (*gray*) was clearly elevated following electrical stimulation, relative to nonstimulation trials. This increase in response did not simply reflect antidromic activation directly from FEF, as there was no increase in baseline activity when FEF stimulation occurred in the absence of a visual stimulus in the receptive field. Moore & Armstrong (2003) did not vary luminance contrast. However, consistent with an increase in effective contrast, they found that stimulation caused a greater increase in response when a preferred stimulus appeared in the receptive field, in contrast to when a poor stimulus appeared in the RF. This finding is illustrated in Figure 11*C*. The left two bars show the mean increase in response caused by stimulation when a nonpreferred (np) or a preferred (p) stimulus was present in the receptive field.

In a final set of conditions, Moore & Armstrong found that FEF microstimulation appeared to filter out the suppressive influence of distracter stimuli appearing outside the receptive field. They placed a second visual stimulus outside

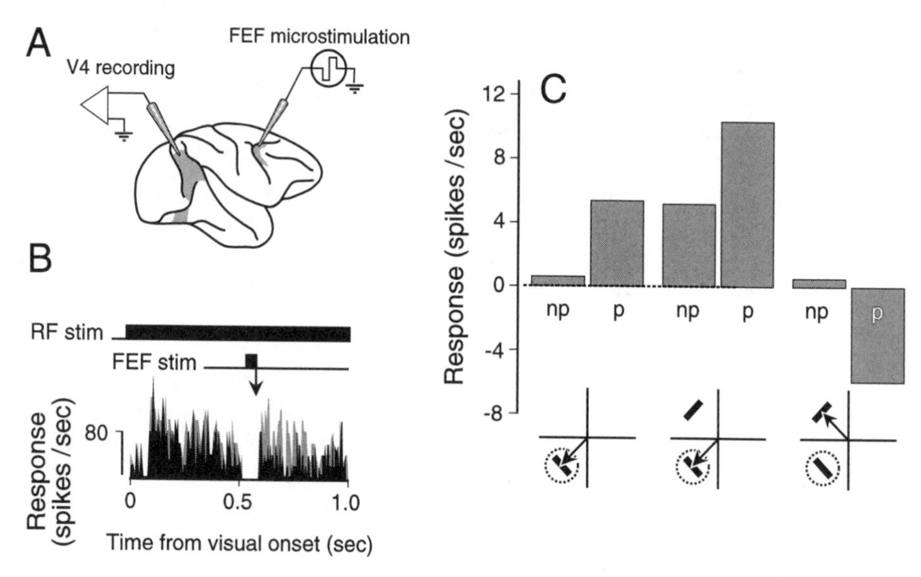

Figure 11 FEF microstimulation increases neuronal responsiveness in V4. (*A*) Subthreshold microstimulating current was injected into FEF while neuronal activity was recorded from V4. (*B*) The visual stimulus appeared for one second (RF stim). And 500 ms after the onset of the visual stimulus, FEF was electrically stimulated for 50 ms (FEF stim). The response of a single V4 neuron with (*gray*) and without (*black*) FEF microstimulation appears below. The apparent gap in response reflects the brief period during which V4 recording was paused while current was injected into FEF. Following this interruption, recording continued. The V4 neuron had elevated responses on trials when FEF was electrically stimulated, as compared to trials without FEF stimulation. (*C*) Mean effect of FEF stimulation with various stimulus configurations. See text for details. Adapted from Moore & Armstrong (2003).

the receptive field and examined the effect of stimulating the portions of FEF whose movement fields corresponded to the RF. They found that the addition of the distracter outside the classical receptive field often led to a reduction of the response elicited by the stimulus in the center, in particular when the center stimulus was a preferred stimulus for the cell, consistent with divisive surround inhibition. Then, when they stimulated FEF, the neuronal response increased. This increase was more than twice the increase observed in the absence of an extra-RF stimulus, consistent with the proposal that stimulation modulated center-surround interactions. In a final condition, they tested whether they could magnify the suppressive effect of the extra receptive field stimulus. They moved their electrode to the FEF location corresponding to the suppressive distracter. Microstimulation resulted in a marked suppression of the response elicited by the preferred stimulus in the RF (rightmost bar). Stimulation at the same location when the distracter was absent had no effect on the response of the recorded V4 neuron. These stimulation findings are strikingly consistent with Motter's (1993) observation that directing spatial attention to an extra-RF stimulus often leads to a decrease in firing rate elicited by a preferred stimulus in the center.

These studies thus establish that FEF microstimulation has effects at the behavioral and at the neuronal level which mimic the effect of spatial attention. At the behavioral level, FEF stimulation causes an increase in contrast sensitivity. At the neuronal level, it increases the response elicited by a stimulus in the RF, an effect that is stronger for a stimulus that is of the neuron's preferred orientation. Stimulation of an FEF site whose movement field corresponds to a stimulus outside the RF magnifies the suppressive influence of the extra receptive field stimulus.

MECHANISMS OF RESPONSE ENHANCEMENT

The experiments described above indicate that spatial attention involves feedback signals from areas, including the FEF, which modulate the responses of visual cortical neurons in ways that mimic increases in the contrast of the attended stimulus. Recent intracellular recording studies using the dynamic clamp technique provide insight into how such a change in effective contrast could take place, by characterizing how changes in the variance of excitatory and inhibitory inputs to a neuron can change its sensitivity to excitatory input. The dynamic clamp technique makes it possible to simulate different patterns of conductance changes that result from the activity of a network of neurons that synapse on a recorded neuron, and to measure their influence on the neuron's response to injected current. Chance et al. (2002) used this technique to record the responses of rat cortical neurons to a steady injected current at different amplitudes. They could change the neuronal response gain by introducing a noisy barrage of excitation and inhibition. Excitation and inhibition were balanced to avoid directly changing the average membrane potential of the neuron. Fellous et al. (2003) extended this finding by varying excitatory and inhibitory modulatory inputs separately (see also Shu et al. 2003 for related findings). Their experiments show that an increase in the variance of either inhibitory or excitatory synaptic inputs can increase neuronal gain. The most obvious way to increase the variance of inputs, and thus to improve neuronal sensitivity, would be to increase the degree of correlation between the neurons that provide input to the cell. Thus, these observations provide a potential biophysical mechanism by which increases in response synchronization among neuronal afferents could cause the increases in neuronal responsiveness observed with attention (for further discussion of this idea, see Crick & Koch 1990, Salinas & Sejnowski 2001, Niebur et al. 2002).

Two recent studies of attention lend support to this proposal by documenting measurable changes in synchrony with attention. Steinmetz et al. (2000) recorded responses in monkey somatosensory cortex and found that the degree to which neurons fired synchronously was higher during a tactile discrimination task than during a visual discrimination task. Fries et al. (2001) found evidence of an increase in high frequency synchronization among macaque V4 neurons when attention was directed to the stimulus within their overlapping receptive fields. Although additional experiments are needed, these studies, taken together with the dynamic clamp studies described above, suggest that response synchronization of afferent neurons may mediate the changes in response sensitivity and response gain observed with attention.

FEATURE-BASED AND OBJECT-BASED ATTENTION

Although we have primarily focused on studies of spatial attention, attention also can be directed to nonspatial features. Selectively enhancing sensitivity to a nonspatial feature may play a role in guiding attention during visual search (Wolfe et al. 1989). Although we know less about the mechanisms by which feature-based attention modulates visual responses, some similarities exist between the two types of attention that suggest they may depend on related mechanisms. Brain-imaging studies have found that directing attention to a particular feature, such as motion, causes increases in neuronal activity in areas selective for the attended feature (e.g., Beauchamp et al. 1997, Corbetta et al. 1991, O'Craven et al. 1997, Saenz et al. 2002, Saenz et al. 2003). Consistent with this finding, single-unit recording studies in the monkey have found feature-selective elevations of neuronal activity (Chelazzi et al. 1993, 1998; Haenny et al. 1988; Haenny & Schiller 1988; Maunsell et al. 1991; Motter 1994a,b; Treue & Martínez-Trujillo 1999). Chelazzi et al. (1993, 1998) recorded responses of inferotemporal cortex neurons in monkeys as they performed the task illustrated in Figure 2. They found that, unlike V4, inferotemporal neurons selective for features of the search target exhibited elevated levels of activity during the blank interval, prior to the appearance of the search array. This finding is consistent with earlier studies of delay activity in inferotemporal cortex (e.g., Fuster & Jervey 1981). As in V4 (see Figure 2) inferotemporal neurons elicited stronger responses to the search array on trials when their preferred stimulus was the target.

Feature-based attention modulates the response elicited by a single stimulus appearing in the receptive field of a V4 neuron (Motter 1994a,b). The task used in this experiment is illustrated in Figure 12*A*. At the beginning of each trial, the fixation point indicated the task-relevant color for that trial. Then, an array of oriented bars appeared in the visual field, half of which were of the task-relevant color, and half of which were of another color. After a delay, all but two bars, one of each color, disappeared, and the monkey had to indicate the orientation of the remaining bar that matched the color of the fixation point. Motter found that the response elicited by the bar in the receptive field during the array presentation was higher when the bar was of the task-relevant color. As illustrated in Figures 12*B–D*, this elevation occurred for both a preferred and a nonpreferred but excitatory stimulus, but it was more pronounced for the preferred stimulus. Figure 12*B* shows the population mean response when a bar of the neuron's preferred color and orientation appeared in the receptive field, and either matched (upper line, M) or did not match (lower line, NM) the cued color. Figure 12*C* shows comparable

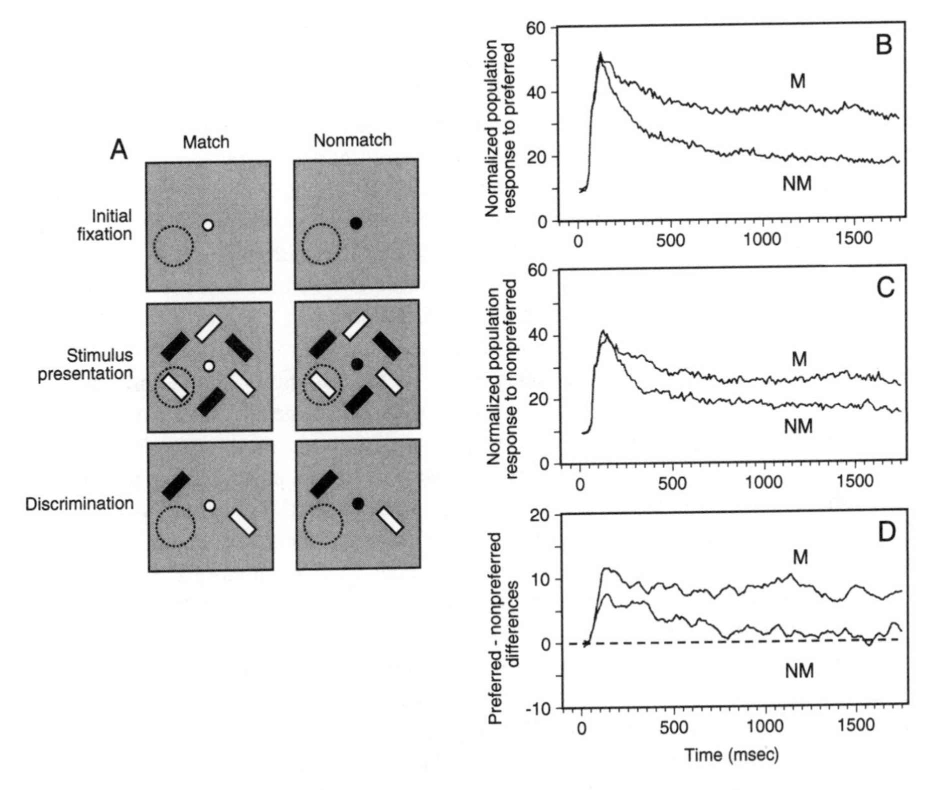

Figure 12 Feature-based attention modulates V4 neurons. (*A*) Task. Each trial began with a fixation point whose color indicated the color that was task relevant and thus had to be attended in that trial ("initial fixation"). An array of oriented bars appeared (stimulus presentation), a subset of which were of the task-relevant color. All but one bar of each color then disappeared, and the monkey reported the orientation of the task-relevant colored bar to earn a juice reward. (*B*) Population response elicited by the bar of the preferred color and orientation, when the color of the bar matched the task-relevant color (M) or did not match the task-relevant color (NM). Responses were much stronger when the bar was task relevant and was therefore attended. (*C*) The response elicited by stimuli that were of a nonpreferred color or orientation was also elevated when their color was attended. (*D*) Neurons maintained their response selectivity regardless of attention condition, as indicated by positive differences between the preferred and the poor stimulus responses.

responses when a nonpreferred stimulus appeared in the receptive field. As was true of the preferred stimulus, the response was higher when the monkey attended to the color of the stimulus in the receptive field. Neurons maintained their selectivity for the preferred and poor stimuli, regardless of which color was attended. This finding is illustrated in Figure 12*D*, which shows differences in the response elicited by the preferred stimulus and the poor stimulus on matching trials (upper line, M) versus nonmatching trials (lower line, NM).

Feature-based attention modulates responses throughout the visual field, even when spatial attention is directed to a particular location (Treue & Martínez-Trujillo 1999). In Treue & Martínez-Trujillo's task, a pattern of dots appeared within the receptive field of an MT neuron, moving in the neuron's preferred direction of motion, and a second pattern of dots appeared at a location in the opposite visual hemifield, moving either in the same or the opposite direction. The monkey performed a task requiring it to detect brief changes in the motion of this second stimulus. Treue & Martínez-Trujíllo found that the neuron's response to the preferred stimulus in its receptive field was stronger when the monkey attended to a contralateral stimulus moving in the same, as compared to the opposite, direction. They concluded that, like spatial attention, this feature-based attention effect is multiplicative. Thus, these experiments raise the possibility that feature-based attention may operate on principles similar to those that govern spatial attention: by increasing the sensitivity of neurons to stimuli that contain the attended feature.

In addition to spatial and feature-based attention, psychophysical, imaging, and event-related-potential studies have demonstrated that attention can select whole objects. When an observer makes a judgment about one feature of an object (e.g., its color) simultaneous judgments about other features of the same object (e.g., its orientation and motion) are made efficiently and do not interfere with the first judgment (e.g., Blaser et al. 2000, Duncan 1984, Mitchell et al. 2003, Reynolds et al. 2003, Valdes-Sosa et al. 1998). This finding is considered evidence that directing attention to one feature of an object causes all of the object's features to be selected together. Consistent with this interpretation, O'Craven et al. (1999) found that discriminating one feature of an object results in increased cerebral blood flow in cortical areas that respond to the task-irrelevant features of the attended object, but not in areas that respond to features of an unattended overlapping object. Schoenfeld et al. (2003) found similar results using a combination of methods [fMRI, event-related potentials (ERPs), event-related magnetic fields (ERFs)] that enabled them to measure the time course over which attention to a task-relevant feature of an object facilitates processing of a task-irrelevant feature of the same object. In their task, observers viewed two spatially superimposed patterns of dots that moved in opposite directions, yielding the percept of two overlapping transparent surfaces. They attended to one of the surfaces to detect a change in speed. Occasionally, one of the surfaces changed color, a change that was task irrelevant. Schoenfeld and colleagues identified a region in the fusiform gyrus that showed elevated BOLD responses when this color change occurred. This area was more active when the color change took place on the attended surface than when it took place on the unattended surface. By carefully comparing the time courses of ERP/ERFs when the color change occurred on either the attended or the unattended surface, they concluded that attention to one feature of an object (here, motion) causes enhancement of other features of the same object (here, color) with a delay of 40–60 ms.

The mechanisms by which attention to one feature spreads to other features of the same object while avoiding features of unattended objects are even less well understood than are the mechanisms that modulate feature-based attention.

However, the psychophysical evidence for object-based attention is compelling, and the above imaging/ERP/ERF studies provide additional evidence that task-irrelevant features of objects are selected, even when stimuli are spatially superimposed and could not, therefore, be selected by a purely spatial attention mechanism. It will be of interest in the coming years to see whether relatively simple mechanisms like those that subserve spatial attention can be identified as neural correlates of feature-based and object-based attention.

ACKNOWLEDGMENTS

We are grateful to Matteo Carandini, Charles Connor, Jean-Marc Fellous, Ken Miller, Brad Motter, Tony Movshon, Terry Sejnowski, and David van Essen for very helpful discussions. We thank E.J. Chichilnisky, Mazyar Fallah, Garth Fowler, Greg Horwitz, Jude Mitchell, and Gene Stoner for providing critical commentary on the manuscript, and Jamie Simon for help with figures. J.R. is supported by grants from the National Eye Institute and The McKnight Endowment Fund for Neuroscience. L.C. is supported by a grant from the Human Frontier Science Program (HFSP).

The *Annual Review of Neuroscience* is online at http://neuro.annualreviews.org

LITERATURE CITED

Albrecht DG, Geisler WS. 1991. Motion selectivity and the contrast-response function of simple cells in the visual cortex. *Vis. Neurosci.* 7:531–46

Bashinski HS, Bacharach VR. 1980. Enhancement of perceptual sensitivity as the result of selectively attending to spatial locations. *Percept. Psychophys.* 28:241–48

Basso MA, Wurtz RH. 1998. Modulation of neuronal activity in superior colliculus by changes in target probability. *J. Neurosci.* 18:7519–34

Beauchamp MS, Cox RW, DeYoe EA. 1997. Graded effects of spatial and featural attention on human area MT and associated motion processing areas. *J. Neurophysiol.* 78:516–20

Bisley JW, Goldberg ME. 2003. Neuronal activity in the lateral intraparietal area and spatial attention. *Science* 299:81–86

Blakemore C, Tobin EA. 1972. Lateral inhibition between orientation detectors in the cat's visual cortex. *Exp. Brain Res.* 15:439–40

Blaser E, Pylyshyn ZW, Holcombe AO. 2000. Tracking an object through feature space. *Nature* 408:196–99

Bonds AB. 1989. Role of inhibition in the specification of orientation selectivity of cells in the cat striate cortex. *Vis. Neurosci.* 2:41–55

Brefczynski JA, DeYoe EA. 1999. A physiological correlate of the 'spotlight' of visual attention. *Nat. Neurosci.* 2:370–74

Britten KH, Hauer HW 1999. Spatial summation in the receptive fields of MT neurons. *J. Neurosci.* 19(12):5074–84

Carandini M, Heeger DJ. 1994. Summation and division by neurons in primate visual cortex. *Science* 264:1333–36

Carandini M, Heeger DJ, Movshon JA. 1997. Linearity and normalization in simple cells of the macaque primary visual cortex. *J. Neurosci.* 17:8621–44

Carrasco M, Penpeci-Talgar C, Eckstein M. 2000. Spatial covert attention increases contrast sensitivity across the CSF: support for signal enhancement. *Vision Res.* 40:1203–15

Carrasco M, Ling S, Read S. 2004. Attention alters appearance. *Nat. Neurosci.* In press

Cavanaugh JR, Bair W, Movshon JA. 2002a. Nature and interaction of signals from the receptive field center and surround in macaque V1 neurons. *J. Neurophysiol.* 88:2530–46

Cavanaugh JR, Bair W, Movshon JA. 2002b. Selectivity and spatial distribution of signals from the receptive field surround in macaque V1 neurons. *J. Neurophysiol.* 88:2547–56

Chance FS, Abbott LF, Reyes AD. 2002. Gain modulation from background synaptic input. *Neuron* 35:773–82

Chelazzi L, Biscaldi M, Corbetta M, Peru A, Tassinari G, Berlucchi G. 1995. Oculomotor activity and visual spatial attention. *Behav. Brain Res.* 71:81–88

Chelazzi L, Duncan J, Miller EK, Desimone R. 1998. Responses of neurons in inferior temporal cortex during memory-guided visual search. *J. Neurophysiol.* 80:2918–40

Chelazzi L, Miller EK, Duncan J, Desimone R. 1993. A neural basis for visual search in inferior temporal cortex. *Nature* 363:345–47

Chelazzi L, Miller EK, Duncan J, Desimone R. 2001. Responses of neurons in macaque area V4 during memory-guided visual search. *Cereb. Cortex* 11:761–72

Chun MM, Marois R. 2002. The dark side of visual attention. *Curr. Opin. Neurobiol.* 12:184–89

Colby CL, Goldberg ME. 1999. Space and attention in parietal cortex. *Annu. Rev. Neurosci.* 22:319–49

Connor CE, Gallant JL, Preddie DC, van Essen DC. 1996. Responses in area V4 depend on the spatial relationship between stimulus and attention. *J. Neurophysiol.* 75:1306–8

Connor CE, Preddie DC, Gallant JL, van Essen DC. 1997. Spatial attention effects in macaque area V4. *J. Neurosci.* 17:3201–14

Cook EP, Maunsell JH. 2002. Attentional modulation of behavioral performance and neuronal responses in middle temporal and ventral intraparietal areas of macaque monkey. *J. Neurosci.* 22:1994–2004

Corbetta M, Miezin FM, Dobmeyer S, Shulman GL, Petersen SE. 1991. Selective and divided attention during visual discriminations of shape, color, and speed: functional anatomy by positron emission tomography. *J. Neurosci.* 11:2383–402

Corbetta M, Shulman GL. 2002. Control of goal-directed and stimulus-driven attention in the brain. *Nat. Rev. Neurosci.* 3:201–15

Crick F, Koch C. 1990. Some reflections on visual awareness. *Cold Spring Harb. Symp. Quant. Biol.* 55:953–62

De Weerd P, Peralta MR 3rd, Desimone R, Ungerleider LG. 1999. Loss of attentional stimulus selection after extrastriate cortical lesions in macaques. *Nat. Neurosci.* 2:753–58

DeAngelis GC, Freeman RD, Ohzawa I. 1994. Length and width tuning of neurons in the cat's primary visual cortex. *J. Neurophysiol.* 71:347–74

DeAngelis GC, Robson JG, Ohzawa I, Freeman RD. 1992. Organization of suppression in receptive fields of neurons in cat visual cortex. *J. Neurophysiol.* 68:144–63

Desimone R, Duncan J. 1995. Neural mechanisms of selective visual attention. *Annu. Rev. Neurosci.* 18:193–222

Downing CJ. 1988. Expectancy and visual-spatial attention: effects on perceptual quality. *J. Exp. Psychol. Hum. Percept. Perform.* 14:188–202

Duncan J. 1984. Selective attention and the organization of visual information. *J. Exp. Psychol. Gen.* 113:501–17

Duncan J, Humphreys GW. 1989. Visual search and stimulus similarity. *Psychol. Rev.* 96:433–58

Fellous J, Rudolph M, Destexhe, Sejnowski T. 2003. Synaptic background noise controls the input/output characteristics of single cells in an in vitro model of in vivo activity. *Neuroscience* 122(3):811–29

Ferrera VP, Lisberger SG. 1995. Attention and target selection for smooth pursuit eye movements. *J. Neurosci.* 15:7472–84

Ferster D, Miller KD. 2000. Neural mechanisms of orientation selectivity in the visual cortex. *Annu. Rev. Neurosci.* 23:441–71

Fitzpatrick D. 2000. Seeing beyond the receptive field in primary visual cortex. *Curr. Opin. Neurobiol.* 10:438–43

Fries P, Reynolds JH, Rorie AE, Desimone R. 2001. Modulation of oscillatory neuronal synchronization by selective visual attention. *Science* 291:1560–63

Fuster JM, Jervey JP. 1981. Inferotemporal neurons distinguish and retain behaviorally relevant features of visual stimuli. *Science* 212:952–55

Gawne TJ, Martin JM. 2002. Responses of primate visual cortical V4 neurons to simultaneously presented stimuli. *J. Neurophysiol.* 88:1128–35

Geisler W, Albrecht D. 2000. *Spatial Vision.* New York: Academic

Gottlieb JP, Kusunoki M, Goldberg ME. 1998. The representation of visual salience in monkey parietal cortex. *Nature* 391:481–84

Grossberg S. 1973. Contour enhancement, short-term memory, and constancies in reverberating neural networks. *Stud. App. Math* 52:217–57

Grossberg S, Raizada RD. 2000. Contrast-sensitive perceptual grouping and object-based attention in the laminar circuits of primary visual cortex. *Vision. Res.* 40:1413–32

Haenny PE, Maunsell JH, Schiller PH. 1988. State dependent activity in monkey visual cortex. II. Retinal and extraretinal factors in V4. *Exp. Brain Res.* 69:245–59

Haenny PE, Schiller PH. 1988. State dependent activity in monkey visual cortex. I. Single cell activity in V1 and V4 on visual tasks. *Exp. Brain Res.* 69:225–44

Handy TC, Kingstone A, Mangun GR. 1996. Spatial distribution of visual attention: perceptual sensitivity and response latency. *Percept. Psychophys.* 58:613–27

Hauer HW, Britten KH. 2002. Contrast dependence of response normalization in area MT of the rhesus macaque. *J. Neurophysiol.* 88:3398–408

Hawkins HL, Hillyard SA, Luck SJ, Mouloua M, Downing CJ, Woodward DP. 1990. Visual attention modulates signal detectability. *J. Exp. Psychol. Hum. Percept. Perform.* 16:802–11

Heeger DJ. 1992. Normalization of cell responses in cat striate cortex. *Vis. Neurosci.* 9:181–97

Heinze HJ, Mangun GR, Burchert W, Hinrichs H, Scholz M, et al. 1994. Combined spatial and temporal imaging of brain activity during visual selective attention in humans. *Nature* 372:543–46

Hillyard SA, Anllo-Vento L. 1998. Event-related brain potentials in the study of visual selective attention. *Proc. Natl. Acad. Sci. USA* 95:781–87

Hoffman JE, Subramaniam B. 1995. The role of visual attention in saccadic eye movements. *Percept. Psychophys.* 57:787–95

Holt GR, Koch C. 1997. Shunting inhibition does not have a divisive effect on firing rates. *Neural Comput.* 9:1001–13

Ito M, Gilbert CD. 1999. Attention modulates contextual influences in the primary visual cortex of alert monkeys. *Neuron* 22:593–604

Itti L, Koch C. 2000. A saliency-based search mechanism for overt and covert shifts of visual attention. *Vision Res.* 40:1489–506

James W. 1890. *The Principles of Psychology.* New York: Holt

Kastner S, Ungerleider LG. 2000. Mechanisms of visual attention in the human cortex. *Annu. Rev. Neurosci.* 23:315–41

Knierim JJ, van Essen DC. 1992. Neuronal responses to static texture patterns in area V1 of the alert macaque monkey. *J. Neurophysiol.* 67:961–80

Lee DK, Itti L, Koch C, Braun J. 1999. Attention activates winner-take-all competition among visual filters. *Nat. Neurosci.* 2:375–81

Levitt JB, Lund JS, Yoshioka T. 1996. Anatomical substrates for early stages in cortical processing of visual information in the macaque monkey. *Behav. Brain Res.* 76:5–19

Lu ZL, Dosher BA. 1998. External noise distinguishes attention mechanisms. *Vision. Res.* 38:1183–98

Lu ZL, Liu CQ, Dosher BA. 2000. Attention mechanisms for multi-location first- and second-order motion perception. *Vision Res.* 40(2):173–86

Luck SJ, Chelazzi L, Hillyard SA, Desimone R. 1997. Neural mechanisms of spatial selective attention in areas V1, V2, and V4 of macaque visual cortex. *J. Neurophysiol.* 77:24–42

Luck SJ, Hillyard SA, Mouloua M, Woldorff MG, Clark VP, Hawkins HL. 1994. Effects of spatial cuing on luminance detectability: psychophysical and electrophysiological evidence for early selection. *J. Exp. Psychol. Hum. Percept. Perform.* 20:887–904

Maffei L, Fiorentini A. 1976. The unresponsive regions of visual cortical receptive fields. *Vision Res.* 16:1131–39

Martínez-Trujillo J, Treue S. 2002. Attentional modulation strength in cortical area MT depends on stimulus contrast. *Neuron* 35:365–70

Maunsell JH, Sclar G, Nealey TA, DePriest DD. 1991. Extraretinal representations in area V4 in the macaque monkey. *Vis. Neurosci.* 7:561–73

McAdams CJ, Maunsell JH. 1999a. Effects of attention on orientation-tuning functions of single neurons in macaque cortical area V4. *J. Neurosci.* 19:431–41

McAdams CJ, Maunsell JH. 1999b. Effects of attention on the reliability of individual neurons in monkey visual cortex. *Neuron* 23:765–73

McLaughlin D, Shapley R, Shelley M, Wielaard DJ. 2000. A neuronal network model of macaque primary visual cortex (V1): orientation selectivity and dynamics in the input layer 4Calpha. *Proc. Natl. Acad. Sci. USA* 97:8087–92

Miller EK, Gochin PM, Gross CG. 1993. Suppression of visual responses of neurons in inferior temporal cortex of the awake macaque by addition of a second stimulus. *Brain Res.* 616:25–29

Mitchell JF, Stoner GR, Fallah M, Reynolds JH. 2003. Attentional selection of superimposed surfaces cannot be explained by modulation of the gain of color channels. *Vision. Res.* 43:1323–28

Moore T, Armstrong KM. 2003. Selective gating of visual signals by microstimulation of frontal cortex. *Nature* 421:370–73

Moore T, Fallah M. 2001. Control of eye movements and spatial attention. *Proc. Natl. Acad. Sci. USA* 98:1273–76

Moore T, Fallah M. 2004. Microstimulation of the frontal eye field and its effects on covert spatial attention. *J. Neurophysiol.* 91(1):152–62

Moore T, Tolias AS, Schiller PH. 1998. Visual representations during saccadic eye movements. *Proc. Natl. Acad. Sci. USA* 95:8981–84

Moran J, Desimone R. 1985. Selective attention gates visual processing in the extrastriate cortex. *Science* 229:782–84

Morrone MC, Burr DC, Maffei L. 1982. Functional implications of cross-orientation inhibition of cortical visual cells. I. Neurophysiological evidence. *Proc. R. Soc. Lond. B Biol. Sci.* 216:335–54

Motter BC. 1993. Focal attention produces spatially selective processing in visual cortical areas V1, V2, and V4 in the presence of competing stimuli. *J. Neurophysiol.* 70:909–19

Motter BC. 1994a. Neural correlates of attentive selection for color or luminance in extrastriate area V4. *J. Neurosci.* 14:2178–89

Motter BC. 1994b. Neural correlates of feature selective memory and pop-out in extrastriate area V4. *J. Neurosci.* 14:2190–99

Mountcastle VB, Motter BC, Steinmetz MA, Sestokas AK. 1987. Common and differential effects of attentive fixation on the excitability of parietal and prestriate (V4) cortical visual neurons in the macaque monkey. *J. Neurosci.* 7:2239–55

Muller HJ, Humphreys GW. 1991. Luminance-increment detection: capacity-limited or not? *J. Exp. Psychol. Hum. Percept. Perform.* 17:107–24

Muller JR, Metha AB, Krauskopf J, Lennie P. 2003. Local signals from beyond the receptive fields of striate cortical neurons. *J. Neurophysiol.* 90:822–31

Murphy B, Miller KD. 2003. Multiplicative gain changes are induced by excitation or inhibition alone. *J. Neurosci.* 23(31):10040–51

Niebur E, Hsiao SS, Johnson KO. 2002. Synchrony: a neuronal mechanism for attentional selection? *Curr. Opin. Neurobiol.* 12:190–94

Niebur E, Koch C. 1994. A model for the neuronal implementation of selective visual attention based on temporal correlation among neurons. *J. Comput. Neurosci.* 1:141–58

O'Craven KM, Downing PE, Kanwisher N. 1999. fMRI evidence for objects as the units of attentional selection. *Nature* 401:584–87

O'Craven KM, Rosen BR, Kwong KK, Treisman A, Savoy RL. 1997. Voluntary attention modulates fMRI activity in human MT-MST. *Neuron* 18:591–98

Palmer J, Verghese P, Pavel M. 2000. The psychophysics of visual search. *Vision Res.* 40(10–12):1227–68

Pessoa L, Kastner S, Ungerleider LG. 2003. Neuroimaging studies of attention: from modulation of sensory processing to top-down control. *J. Neurosci.* 23:3990–98

Posner MI, Snyder CR, Davidson BJ. 1980. Attention and the detection of signals. *J. Exp. Psychol.* 109:160–74

Qian N, Andersen RA. 1994. Transparent motion perception as detection of unbalanced motion signals. II. Physiology. *J. Neurosci.* 14:7367–80

Recanzone GH, Wurtz RH. 2000. Effects of attention on MT and MST neuronal activity during pursuit initiation. *J. Neurophysiol.* 83:777–90

Recanzone GH, Wurtz RH, Schwarz U. 1997. Responses of MT and MST neurons to one and two moving objects in the receptive field. *J. Neurophysiol.* 78:2904–15

Reynolds JH, Chelazzi L, Desimone R. 1999. Competitive mechanisms subserve attention in macaque areas V2 and V4. *J. Neurosci.* 19:1736–53

Reynolds JH, Desimone R. 2003. Interacting roles of attention and visual salience in V4. *Neuron* 37:853–63

Reynolds JH, Pasternak T, Desimone R. 2000. Attention increases sensitivity of V4 neurons. *Neuron* 26:703–14

Reynolds JH, Alborzian S, Stoner GR. 2003. Exogenously cued attention triggers competitive selection of surfaces. *Vision Res.* 43(1):59–66

Robinson DA, Fuchs AF. 1969. Eye movements evoked by stimulation of frontal eye fields. *J. Neurophysiol.* 32:637–48

Roelfsema PR, Spekreijse H. 2001. The representation of erroneously perceived stimuli in the primary visual cortex. *Neuron* 31:853–63

Rolls ET, Tovee MJ. 1995. The responses of single neurons in the temporal visual cortical areas of the macaque when more than one stimulus is present in the receptive field. *Exp. Brain Res.* 103:409–20

Saenz M, Buracas GT, Boynton GM. 2002. Global effects of feature-based attention in human visual cortex. *Nat. Neurosci.* 5:631–32

Saenz M, Buracas GT, Boynton GM. 2003. Global feature-based attention for motion and color. *Vision Res.* 43:629–37

Salinas E, Abbott LF. 1997. Invariant visual responses from attentional gain fields. *J. Neurophysiol.* 77(6):3267–72

Salinas E, Sejnowski TJ. 2001. Correlated neuronal activity and the flow of neural information. *Nat. Rev. Neurosci.* 2:539–50

Schall JD. 1995. Neural basis of saccade target selection. *Rev. Neurosci.* 6:63–85

Schoenfeld MA, Tempelmann C, Martínez A, Hopf JM, Sattler C, et al. 2003. Dynamics of feature binding during object-selective attention. *Proc. Natl. Acad. Sci. USA* 100:11806–11

Sclar G, Freeman RD. 1982. Orientation selectivity in the cat's striate cortex is invariant with stimulus contrast. *Exp. Brain Res.* 46:457–61

Seidemann E, Newsome WT. 1999. Effect of spatial attention on the responses of area MT neurons. *J. Neurophysiol.* 81:1783–94

Shapley R, Hawken M, Ringach DL. 2003. Dynamics of orientation selectivity in the primary visual cortex and the importance of cortical inhibition. *Neuron* 38:689–99

Sheinberg DL, Logothetis NK. 2001. Noticing familiar objects in real world scenes: the role of temporal cortical neurons in natural vision. *J. Neurosci.* 21:1340–50

Shepherd M, Findlay JM, Hockey RJ. 1986. The relationship between eye movements and spatial attention. *Q. J. Exp. Psychol. A* 38:475–91

Sherman SM. 2001. Thalamic relay functions. *Prog. Brain Res.* 134:51–69

Shu Y, Hasenstaub A, Badoual M, Bal T, McCormick DA. 2003. Barrages of synaptic activity control the gain and sensitivity of cortical neurons. *J. Neurosci.* 23(32):10388–401

Snowden RJ, Treue S, Erickson RG, Andersen RA. 1991. The response of area MT and V1 neurons to transparent motion. *J. Neurosci.* 11(9):2768–85

Somers DC, Nelson SB, Sur M. 1995. An emergent model of orientation selectivity in cat visual cortical simple cells. *J. Neurosci.* 15:5448–65

Somers DC, Todorov EV, Siapas AG, Toth LJ, Kim DS, Sur M. 1998. A local circuit approach to understanding integration of long-range inputs in primary visual cortex. *Cereb. Cortex* 8:204–17

Sperling G, Sondhi MM. 1968. Model for visual luminance discrimination and flicker detection. *J. Opt. Soc. Am.* 58:1133–45

Spitzer H, Desimone R, Moran J. 1988. Increased attention enhances both behavioral and neuronal performance. *Science* 240:338–40

Stanton GB, Bruce CJ, Goldberg ME. 1995. Topography of projections to posterior cortical areas from the macaque frontal eye fields. *J. Comp. Neurol.* 353:291–305

Steinmetz MA, Constantinidis C. 1995. Neurophysiological evidence for a role of posterior parietal cortex in redirecting visual attention. *Cereb. Cortex* 5:448–56

Steinmetz PN, Roy A, Fitzgerald PJ, Hsiao SS, Johnson KO, Niebur E. 2000. Attention modulates synchronized neuronal firing in primate somatosensory cortex. *Nature* 404:187–90

Thompson K, Bichot NP, Schall JD. 2001. From attention to action in frontal cortex. In *In Visual Attention and Cortical Circuits*, ed. J Braun, C Koch, JD Davis, pp. 137–56. Cambridge, MA: MIT Press

Treisman AM, Gelade G. 1980. A feature-integration theory of attention. *Cogn. Psychol.* 12:97–136

Treue S. 2003. Visual attention: the where, what, how and why of saliency. *Curr. Opin. Neurobiol.* 13:428–32

Treue S, Martínez-Trujillo JC. 1999. Feature-based attention influences motion processing gain in macaque visual cortex. *Nature* 399:575–79

Treue S, Maunsell JH. 1996. Attentional modulation of visual motion processing in cortical areas MT and MST. *Nature* 382:539–41

Troyer TW, Krukowski AE, Priebe NJ, Miller KD. 1998. Contrast-invariant orientation tuning in cat visual cortex: thalamocortical input tuning and correlation-based intracortical connectivity. *J. Neurosci.* 18:5908–27

Valdes-Sosa M, Bobes MA, Rodriguez V, Pinilla T. 1998. Switching attention without shifting the spotlight object-based attentional modulation of brain potentials. *J. Cogn. Neurosci.* 10:137–51

Verghese P. 2001. Visual search and attention: a signal detection theory approach. *Neuron* 31(4):523–35

Walker GA, Ohzawa I, Freeman RD. 1999. Asymmetric suppression outside the classical receptive field of the visual cortex. *J. Neurosci.* 19(23):10536–53

Webb BS, Tinsley CJ, Barraclough NE, Parker A, Derrington AM. 2003. Gain control from beyond the classical receptive field in primate primary visual cortex. *Vis. Neurosci.* 20:221–30

Wolfe JM, Cave KR, Franzel SL. 1989. Guided search: an alternative to the feature integration model for visual search. *J. Exp. Psychol. Hum. Percept. Perform.* 15:419–33

Wurtz RH, Goldberg ME. 1972. The primate superior colliculus and the shift of visual attention. *Invest. Ophthalmol.* 11:441–50

Xiao DK, Raiguel S, Marcar V, Koenderink J, Orban GA. 1995. Spatial heterogeneity of inhibitory surrounds in the middle temporal visual area. *Proc. Natl. Acad. Sci. USA* 92(24):11303–6

Yantis S, Serences JT. 2003. Cortical mechanisms of space-based and object-based attentional control. *Curr. Opin. Neurobiol.* 13: 187–93

Annu. Rev. Neurosci. 2004. 27:649–77
doi: 10.1146/annurev.neuro.27.070203.144220

THE HUMAN VISUAL CORTEX

Kalanit Grill-Spector[1] and Rafael Malach[2]
[1]*Department of Psychology and Neuroscience, Stanford University, Stanford, California 94305-2130; email: kalanit.grill-spector@stanford.edu*
[2]*Department of Neurobiology, Weizmann Institute of Science, Rehovot, Israel; email: rafi.malach@weizmann.ac.il*

Key Words functional magnetic resonance imaging (fMRI), visual perception, object and face recognition, retinotopic mapping

■ **Abstract** The discovery and analysis of cortical visual areas is a major accomplishment of visual neuroscience. In the past decade the use of noninvasive functional imaging, particularly functional magnetic resonance imaging (fMRI), has dramatically increased our detailed knowledge of the functional organization of the human visual cortex and its relation to visual perception. The fMRI method offers a major advantage over other techniques applied in neuroscience by providing a large-scale neuroanatomical perspective that stems from its ability to image the entire brain essentially at once. This bird's eye view has the potential to reveal large-scale principles within the very complex plethora of visual areas. Thus, it could arrange the entire constellation of human visual areas in a unified functional organizational framework. Here we review recent findings and methods employed to uncover the functional properties of the human visual cortex focusing on two themes: functional specialization and hierarchical processing.

INTRODUCTION

In the past decade the use of noninvasive functional imaging, particularly functional magnetic resonance imaging (fMRI), has dramatically increased our detailed knowledge of the functional organization of the human visual cortex. Thus, more than a dozen putative human visual areas have been described using fMRI (Tootell 1996, 2003). The discovery and analysis of cortical visual areas are major accomplishments in visual neuroscience. The number, locations, and functional roles of these areas are important topics for continuing experimental studies of the human brain.

In trying to account for the multiplicity of visual areas, two main principles have been suggested: hierarchical processing and functional specialization. Hierarchical processing proposes that visual perception is achieved via a gradual stagewise process in which information is first represented in a localized and simple form and, through a sequence of processes, is transformed into more

0147-006X/04/0721-0649$14.00

abstract, holistic, and even multimodal representations (DeYoe & Van Essen 1988). The second principle, functional specialization, proposes that specialized neural pathways exist that process information about different aspects of the visual scene. In particular the visual system may consist of parallel hierarchical sequences, or processing streams, that are specialized for a particular functional task. The dorsal stream, also referred to as the "where" (Mishkin et al. 1983) or "action" (Goodale et al. 1991) stream, has been associated with spatial localization (or visually guided action) and the ventral "what" stream (Mishkin et al. 1983), which are involved in object and form recognition and are the best-known of the processing streams.

Here we review recent findings and methods employed to uncover the functional properties of the human visual cortex, focusing on these two themes: functional specialization and hierarchical processing.

MAPPING THE HUMAN VISUAL CORTEX

In the macaque, visual cortical areas have been distinguished by four main criteria (Felleman & Van Essen 1991, Tootell et al. 2003): (*a*) retinotopy, (*b*) global functional properties, (*c*) histology, and (*d*) intercortical connections. In humans, most visual cortical areas have been revealed by functional MRI, using retinotopic and global functional criteria. Several studies explored histological differences between human brain areas on postmortem specimens, but these have been largely restricted to early visual areas (Clarke 1994a,b; Horton et al. 1990; Horton & Hedley-Whyte 1984; Tootell & Taylor 1995). More recently, anatomical MRI and diffusion tensor imaging (DTI) have revealed histological and connectional distinctions between areas as well (Conturo et al. 1999, Hagmann et al. 2003). We begin by reviewing methods applied most commonly to delineate visual areas: retinotopy and functional specialization.

PRINCIPLES OF RETINOTOPIC MAPPING

Visual field topography is used to identify and map visual areas in animals and humans (DeYoe et al. 1996; Engel et al. 1994, 1997b; Sereno et al. 1995, 2001; Wandell 1999a). Mapping from the retina to the primary visual cortex is topographic in that nearby regions on the retina project to nearby cortical regions. In the cortex, neighboring positions in the visual field are represented by groups of neurons adjacent but laterally displaced within the cortical gray matter. Mapping between the retina and the cortex can be described best as a log-polar transformation, in which standard axes in the retina are transformed into polar axes in the cortex: eccentricity (distance from fovea) and polar angle (angle from horizontal axis; see Figure 1). The logarithmic component of the transformation accounts for the magnification of central representations in the cortex

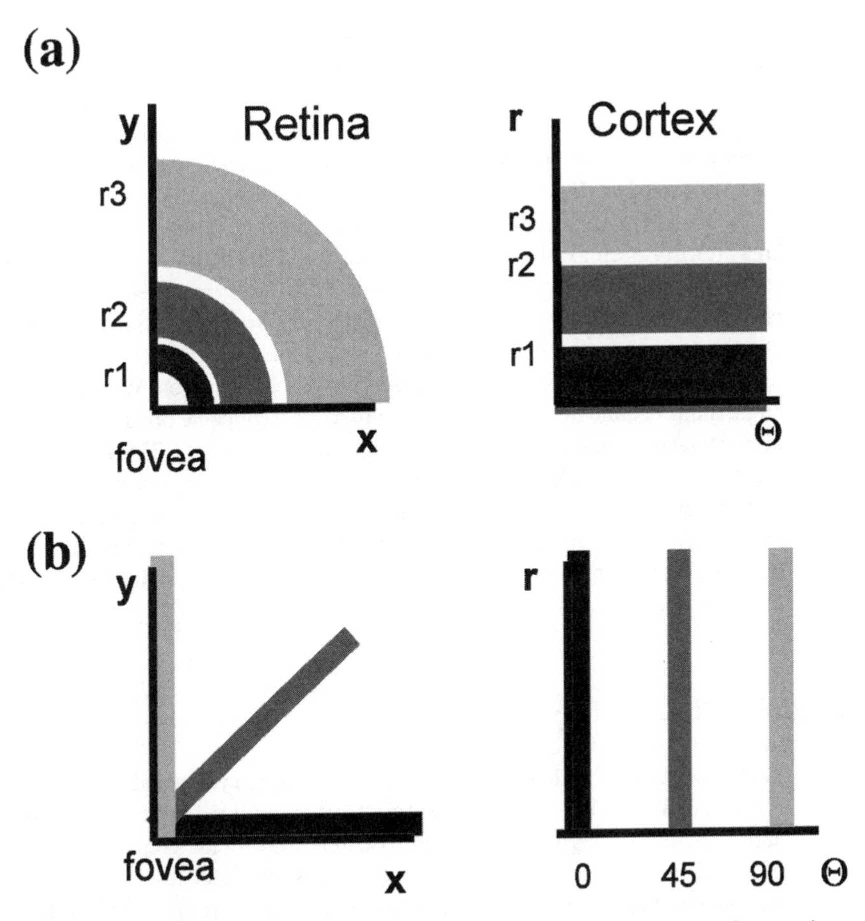

Figure 1 Principles of retinotopic mapping. Log-polar transformation from retinotopic coordinataes to cortical coordinates. (*A*) Eccentricity: distance from the fovea. (*B*) Phase: polar angle from the horizontal.

(Duncan & Boynton 2003, Schwartz et al. 1985). Thus, this transformation from retina to cortex preserves the qualitative spatial relations but distorts quantitative ones.

Topographical cortical representations are revealed when subjects fixate and visual stimuli are presented at selected locations. Mapping the phase (angle) component of the retinotopic map reveals multiple horizontal and vertical meridian representations arranged in approximately parallel bands along the cortical surface (Figure 2, see color insert). These vertical and horizontal meridian representations alternate and define the borders between mirror-symmetric retinotopic areas. Perpendicular to these bands lie iso-eccentricity bands, which constitute an eccentricity gradient (Figure 3, see color insert). However, the representation of the fovea is greatly expanded compared to the representation of the periphery. In humans the foveal representation of low-level retinotopic areas converges in the occipital pole, in the confluent fovea (black arrows in Figure 3). Flanking the confluent fovea there are bands of parafoveal and peripheral representations present both ventrally and dorsally.

By using stimuli that are more optimal for activating mid- and high-level visual areas (Grill-Spector et al. 1998a,b; Hasson et al. 2003a,b; Levy et al. 2001), we revealed consistent retinotopic maps in what is classically regarded "nonretinotopic cortex." Specifically, orderly central and peripheral representations can be found all across visual cortex, extending (albeit with decreased precision) even into high-level regions presumably engaged in face and place perception. Whereas the polar angle maps decline faster than eccentricity maps, replicable polar representations also extend more anteriorly and laterally than initially thought. This finding implies that to subdivide visual cortex into areas that are retinotopic versus nonretinotopic may be oversimplistic. This distinction now appears to be a continuum, not a dichotomy. We are hopeful that these consistent retinotopic representations will be used, in addition to functional specialization, to systematically delineate high-level visual areas.

EARLY VISUAL AREAS

Each retinotopic visual area contains a complete eccentricity and polar angle map (DeYoe et al. 1996, Sereno et al. 1995). On the unfolded cortex, phase lines lie approximately orthogonal to eccentricity bands and provide a one-to-one mapping of visual space in cortical coordinates (Figure 4, see color insert). The details of the anatomic locations of different retionotopic visual areas have been described in detail recently (Wandell 1999a). Here we note that on the unfolded cortical hemisphere, early retinotopic areas V1, V2, V3, V4/V8, and V3a are arranged as parallel, mirror-symmetric bands.

In the dorsal stream, beyond area V3a investigators have identified three additional visual retinotopic areas. However, a consensus has yet to emerge regarding their exact parcelation. Area V7 contains an additional hemifield representation anterior to V3a (Press et al. 2001). V7 horizontal and vertical meridians are rotated compared to V3a meridians (Figure 4). This area may have a separate fovea located along the intraparietal sulcus (IPS) (Press et al. 2001). Others (Press et al. 2001, Smith et al. 1998) have defined another hemifield representation lateral and inferior to V3a, termed V3b. This area shares a parafoveal representation with V3a. Recently, Sereno and colleagues (Sereno et al. 2001) reported yet another hemifield representation more anteriorly along the intraparietal sulcus, which shows robust topographic mapping of a remembered target angle. Sereno and colleagues suggested that this region may correspond to the lateral intraparietal (LIP) area in macaque monkeys, and they named it hLIP.

FUNCTIONAL SUBSYSTEMS: COLOR, MOTION, DEPTH, AND FORM PROCESSING IN THE HUMAN VISUAL CORTEX

A further possible relation between anatomical structure and physiological maps is the suggestion that there are separate neural pathways for processing information about different visual properties such as motion, depth, color, and shape. Physiological and anatomical studies in the monkey reveal such segregation that begins already at the retinal level of neural pathways (Amir et al. 1993, DeYoe & Van Essen 1988, Shipp & Zeki 1985), and they show some functional compartmentalization throughout visual cortex. However, other studies suggest that this compartmentalization is too simplistic (Malach 1994, Schiller 1996). In the next section we review the current knowledge about processing that leads to the perception of motion, depth, color, and form in the human brain.

The Human "Color Center" and Areas V4/V8

In ventral cortex the border between V3v and V4v is defined by an upper vertical meridian. Currently there is vigorous debate about the functional organization and naming conventions beyond V4v (Bartels & Zeki 2000, Hadjikhani et al. 1998, Tootell et al. 2003, Wade et al. 2002, Wandell 1999b, Zeki et al. 1991) and specifically what constitutes the anterior border of V4. The question is whether V4 ends prior to the lower-field representation (Figure 4), leaving area V4v without its corresponding V4d (Sereno et al. 1995), or whether the lower visual field representation lateral to V4v constitutes V4d, thus, forming a complete hemifield representation, complementary to the dorsal V3a representation (Kastner et al. 2001, McKeefry & Zeki 1997, Wade et al. 2002). In contrast, Tootell and colleagues suggest that on the ventral aspect there is a quarter field representation, V4v, and that the neighboring cortex belongs to a separate area termed V8, consisting of a hemifield representation (Hadjikhani et al. 1998), which is rotated relative to V4v.

The debate about the precise definition of V4 is related to the controversy about the location of a color center in the human brain. Clinical studies reveal that color vision loss (achromatopsia) is correlated with damage in ventral occipitotemporal cortex (Damasio 1980, Pearlman 1979, Zeki 1990), suggesting the existence of a color center in the human brain. Indeed, neuroimaging studies reveal regions in ventral visual cortex in the vicinity of V4v that respond more strongly to colored patterns compared to luminance-defined patterns. These regions are referred to as V4 (Bartels & Zeki 2000, Lueck et al. 1989, McKeefry & Zeki 1997), V8 (Hadjikhani et al. 1998, Tootell & Hadjikhani 2000), and sometimes VO (Wandell 1999b). More anteriorly, investigators found additional areas that were activated by attention to color (Bartels & Zeki 2000, Beauchamp et al. 1999).

The controversy of the functional definition of human area V4 illustrates the need for additional dimensions such as histology (Clarke et al. 1994a,b; Tootell & Taylor 1995) and fiber connectivity (Conturo et al. 1999, Hagmann et al. 2003,

Horton & Hocking 1996) that should be supplemented in the attempts to define areas in the human visual cortex systematically.

Whether or not V4 or V8 is principally a color computation center is an open question. Engel et al. (1997a) measured the responses of area V1 to calibrated color stimuli and showed stronger activation per unit of cone contrast for opponent color stimuli than for pure luminance stimuli. On the basis of a definition that uses the criterion, "responds better to color than luminance," it seems that even striate cortex could be a color area. So too could be the central retina. In studies of the effects of color adaptation, an essential part of color constancy, Wade & Wandell (2002) found that much of the computation can be explained by gain changes in retinal photoreceptors. Thus, until the computational basis of human color vision is understood more clearly, it may be more productive to speak of a color-processing stream that begins in the retina and passes through V1, V2, and higher areas including the V4/V8 complex. Each area contributes to color perception (and to other facets of vision), and their precise functions await a more complete characterization.

Motion Perception

Motion processing encompasses different kinds of information such as the derivation of the speed and direction of a moving target, the motion boundaries associated with an object, or judgment of motion direction from optic flow signals. Along the dorsal stream many regions are activated more strongly when subjects view moving versus stationary stimuli. Included are areas MT, MST, V3a, and even low-level areas such as V1 and V2. However, converging evidence suggests that some aspects of motion processing are localized in more specialized regions in the human brain.

A central motion selective focus in the human brain is a region called hMT+, located at the temporo-parietal occipital junction (Figure 4) (Tootell et al. 1995b, Watson et al. 1993). This region is a likely homologue to the macaque motion-sensitive area called MT/V5 (Heeger et al. 2000, Rees et al. 2000). Human MT+ is selectively activated by moving versus stationary stimuli and exhibits high contrast sensitivity. This area contains an orderly eccentricity organization within a hemifield representation (Dukelow et al. 2001, Huk et al. 2002). Recently, several elegant studies used the perceptual illusion of the motion aftereffect to show that hMT+ probably contains direction-selective neural populations (Huk et al. 2001, Tootell et al. 1995a), similar to findings in macaques (Britten et al. 1992, Maunsell & Newsome 1987, Salzman et al. 1992). Furthermore, comparison of coherent versus incoherent motion of light points (in which dots move independently) reveals a significant change in activation within hMT+ that increases linearly with the coherence of motion but shows little change in early visual areas (Rees et al. 2000). Similarly, hMT+ adapts to patterned motion, in contrast to lower visual areas, which adapt to component motion (Huk & Heeger 2002). These results are consistent with the idea of local motion processing in early visual areas and global motion processing in later areas (see also Castelo-Branco et al. 2002). Finally, activation of this area is enhanced when subjects attend to or track motion (Buchel et al. 1998, Culham et al. 1998).

Thus, converging evidence shows that human MT+ response properties parallel both the response properties of single neurons (Rees et al. 2000) and perception (Muckli et al. 2002). This inference is further supported by clinical studies that reveal that akinetopsia (the failure to perceive motion) is associated with lesions in the vicinity of hMT+.

In addition to hMT+, other cortical sites are activated by various types of coherent motion stimuli. Some evidence supports the idea that within the hMT+ complex of areas some subregions show selectivity to particular radial and circular motion patterns (rather than simple translation) and thus may be involved in calculating optical flow (Morrone et al. 2000). Orban and colleagues have suggested the existence of yet another specialized motion-related area, the kinetic-occipital area (KO) (Dupont et al. 1997, Van Oostende et al. 1997), which specializes in processing kinetic boundaries created by discontinuities in motion direction (but see Zeki et al. 2003).

Another intriguing aspect of motion processing is related to the perception of biological motion, the motion of people and animals. One of the striking features of motion perception is that humans can differentiate biological motion, such as the types of motion (running, jumping) and even gender of the actor, from extremely impoverished visual displays in which only a dozen light-points are attached to joints of a person moving in an otherwise dark room. Increasing evidence supports an area specialized for perceiving biological motion. This area is located within a small region on the ventral bank of the occipital extent of the superior-temporal sulcus (STS), located lateral and anterior to human MT/MST, and is selectively activated during viewing of light-point-defined moving figures (Grossman et al. 2000, Grossman & Blake 2002, Vaina et al. 2001), but not by the random movement or inverted motion of the same dots that composed the light-point figures. This region is also activated by other types of biological motion such as movies of people walking (Pelphrey et al. 2003) or of hand, eye, or mouth movements (Puce & Perrett 2003). Particularly interesting is the recent association of these regions with the cortical "mirror" system, which integrates both sensory processing and production of motor movements (see Rizzolatti & Craighero 2004, in this volume).

Depth Perception

Compared to the work on motion and form processing there are fewer studies concerning the processing of depth, surfaces (Stanley & Rubin 2003), and three-dimensional structure (Kourtzi et al. 2003a, Moore & Engel 2001). Like the processing of visual motion, the analysis of depth can involve both low-level cues, such as disparity, derived from the retinal images, and high-level inferred attributes, such as the surfaces corresponding to retinal points with different disparities. In humans, experiments using random dot stereograms have shown a stream of areas that respond to these three-dimensional cues: V1, V2, V3, VP, V3a, and hMT+. Of these areas V3a is the region most sensitive to changes in the disparity range, although all the regions activated by random dot stereograms show a correlation between the amplitude of the fMRI signal and disparity (Backus et al. 2001). In contrast, other high-level areas such as the lateral occipital complex (LOC) respond to random dot stereograms only when these stimuli are used to define object form (Gilaie-Dotan et al. 2002, Kourtzi & Kanwisher 2001a), but these regions do not seem to carry disparity or depth information per se.

OBJECT-SELECTIVE AREAS IN THE HUMAN BRAIN

One of the greatest mysteries in vision research is how humans recognize visually presented objects with high accuracy and speed (Thorpe et al. 1996), despite drastic changes in the appearance of objects caused by changes in the viewing conditions. Interest in how human object recognition works is heightened by the fact that efforts to duplicate this ability in machines have not met extraordinary success. Object-selective regions (Figure 5, see color insert) respond more strongly when subjects view pictures of objects than textures, visual noise, scrambled objects, or scrambled Fourier phase information (which maintains the spatial frequency spectrum) (Grill-Spector et al. 1998b, Malach et al. 1995). These regions compose a large constellation of areas in both the ventral and dorsal visual pathways that lie anterior and lateral to early retinotopic cortex (Figure 5). We discuss the organization of these areas in terms of three main subdivisions: lateral occipital complex (LOC), ventral occipito-temporal (VOT) regions, and dorsal foci.

A central complex of object-selective activation (including faces) occurs in the lateral occipital cortex. These regions show a greater fMRI response to images of familiar and unfamiliar objects, compared with nonobject controls (Grill-Spector et al. 2001, Kanwisher et al. 1996, Malach et al. 1995). The LOC can be divided into at least two putative subdivisions: a dorsal region [LO (lateral occipital)] and a more ventral region [pFus (posterior fusiform)], along the posterior fusiform gyrus, which may show some overlap with the ventrally located VOT.

Ventrally, a set of regions highly selective to object, face (Kanwisher et al. 1997), and place images (Epstein & Kanwisher 1998) have been described in a stretch of cortex bounded by the fusiform gyrus laterally and the parahippocampal gyrus medially (Ishai et al. 1999, Levy et al. 2001, Malach et al. 2002). Here we refer to these more anterior and ventral regions as ventral occipito-temporal or VOT (Figure 6, see color insert). Within the VOT, regions that prefer faces (compared to objects and places) manifest a foveal bias, and regions that respond more strongly to houses and scenes show a peripheral field bias (Hasson et al. 2002, 2003a; Levy et al. 2001, 2004; Malach et al. 2002). Kanwisher and colleagues labeled ventral face-selective regions in the fusiform gyrus as FFA [fusiform face area (Kanwisher et al. 1997)] and place-selective regions as PPA [parahippocampal place area (Epstein & Kanwisher 1998)].

Although object-related activation is commonly associated with ventral-stream areas and object recognition, object and shape information also is necessary in a variety of dorsal stream–related tasks, particularly in the context of object manipulation (Culham et al. 2003, Goodale et al. 1991). Thus, it is not surprising that several regions that respond more strongly to objects than nonobject controls have been found dorsally both in the parietal lobe and in the dorsal aspect of the LOC (Figure 6). Although these regions require more rigorous mapping, there are at least two dorsal foci that activate more strongly when subjects view pictures of objects than when viewing faces or scenes: One focus is located inferior and lateral to V3a (Grill-Spector 2003b, Hasson et al. 2003a), and another focus is located anterior to V3a, partially overlapping V7 (Grill-Spector 2003b). Finally, a region that activates more strongly when subjects look at buildings and scenes (compared to objects and faces) and overlaps with peripheral representations is found dorsally along the transverse occipital sulcus (TOS) (Hasson et al. 2003a). The functional role and involvement of dorsal object-selective regions in perceptual tasks are still poorly understood.

REPRESENTATION OF OBJECTS IN HIGH-LEVEL VISUAL AREAS

Many studies have shown that the activation in object-selective cortex exhibits perceptual invariance; object-selective regions in the ventral stream are activated when subjects view objects defined by luminance (Grill-Spector et al. 1998a), texture (Grill-Spector et al. 1998a, Kastner et al. 2000), motion (Grill-Spector et al. 1998a, Kriegeskorte et al. 2003), or stereo cues (Gilaie-Dotan et al. 2002, Kourtzi & Kanwisher 2001a) but not when subjects view textures, stationary dot patterns, coherently moving dots, or gratings defined by either motion or stereo. Other studies have shown that activation is independent of object format [gray scale or line drawing (Ishai et al. 2000)] and that these regions are also activated when subjects perceived simple shapes created via illusory contours (Kourtzi & Kanwisher 2000, Mendola 1999, Stanley & Rubin 2003). Several recent studies indicate that object-selective regions in the ventral stream represent shapes rather than contours or object features (Andrews et al. 2002, Kourtzi & Kanwisher 2001b, Lerner et al. 2002). Using the well-known Rubin vase-face illusion, Hasson and colleagues (Hasson et al. 2001) and Andrews and colleagues (Andrews et al. 2002) demonstrated that face-selective regions in LO and VOT do not represent local contours or local features, since in both areas the activity was significantly correlated with the percept of a face and uncorrelated with local image features. Interestingly, some regions within LO even show cross-modality convergence, activating more strongly to objects than textures for both visually and haptically sensed objects (Amedi et al. 2001, James et al. 2002b).

Furthermore, to be effective, the object-recognition system must be invariant to external viewing conditions that affect objects' appearance but not their identity.

We have previously shown, using the adaptation paradigm (Grill-Spector et al. 1999, 2001), that regions in the OTS and fusiform gyrus (but not LO) indeed show size and position invariance (Grill-Spector et al. 1999). Interestingly, while at the behavioral level, subjects exhibit viewpoint invariance and invariance to the illumination of objects; at the level of VOT we found that object representations are more sensitive to the viewing angle and direction of illumination (Grill-Spector et al. 1999). Two recent studies (James et al. 2002a, Vuilleumier et al. 2002b) show some degree of viewpoint invariance for the representation of objects in VOT when using small rotation angles.

Taken together, these results demonstrate convergence of a wide range of invariances to visual cues and object transformation within the same cortical regions, which supports the role of occipito-temporal object areas in processing object shape.

THE ROLE OF OBJECT-SELECTIVE REGIONS IN OBJECT RECOGNITION

The strong activation of occipito-temporal object-selective regions when subjects view pictures of objects and have a large degree of perceptual constancy does not by itself prove that this region is the locus in the brain that performs object recognition. Demonstrating that a specific cortical region responds to visual objects is necessary but not sufficient for determining that it is the site of object recognition. Activation during object viewing could be due to other processes, such as visual attention, arousal, figure-ground segmentation, and surface extraction. However, several recent studies show that the activation in ventral (but not dorsal) object-selective areas is correlated to object perception rather than to the low-level features in the visual stimulus.

Experiments in which researchers manipulated a stimulus such that it degraded recognition performance demonstrated the correlation between perception and brain activation in lateral and ventral object areas. Examples include using backward masking and shortening the duration for which pictures of objects were shown (Grill-Spector et al. 2000, see also Figure 11), decreasing the amount of the object that was visible by presenting it behind occluding binds (James et al. 2000), reducing objects' contrast (Avidan et al. 2002a), or presenting letter and objects embedded in noise (Kleinschmidt et al. 2002, Malach et al. 1995). Together, these studies demonstrate that the strength of activation in object-selective regions is lower compared to recognition, and the level of activation increases nonlinearly at recognition threshold.

Other experiments used bistable phenomena such as binocular rivalry (Tong et al. 1998), the Rubin face-vase illusion (Andrews et al. 2002, Hasson et al. 2001), and ambiguous figures to track changes in brain activation that correlate with different perceptual states. The advantage of bistable phenomena is that the percept changes while the physical stimulus remains unchanged. In these studies when subjects perceived a face there was higher activation in face-selective regions in the fusiform gyrus compared to states in which the stimulus was not perceived as a face.

A different approach is to show pictures close to threshold and compare brain activation in trials in which subjects recognized objects to activation in trials in which objects were present but not recognized (Bar et al. 2001; Grill-Spector 2003a, 2004). These experiments revealed that activation in object and face selective areas is correlated with subjects' perceptual reports rather than with the physical presence of the stimulus (Grill-Spector 2003a). Thus, activation in object-selective cortex was highest when a target face (or object) was identified (identification hit), intermediate when it was detected but not identified (detection hit), and lowest when a target was present but not detected (detection miss) (see Figure 7). Importantly, we also showed that false identifications were correlated with a higher signal than identification misses (Grill-Spector 2003a, 2004). In contrast, lower-level visual areas were activated robustly, but their levels of activation were not correlated with subjects' percepts (Figure 7). In sum, activity in early and intermediate visual areas (e.g., V1 and V4, respectively) correlates with the physical properties of the retinal image, but activity in higher-level areas (e.g., LO and FFA) correlates with subjects' percepts in tasks requiring face and object recognition.

PARCELLATION OF OBJECT-SELECTIVE REGIONS TO FUNCTIONAL SUBDIVISIONS

The parcellation of higher-level regions into visual areas becomes difficult because these regions lack the precise retinotopy of early visual areas, and further, different types of objects activate somewhat different regions within this constellation of regions (Figure 6 and 8, see color insert). Several research groups noted regions that show strong activations to specific object categories such as faces (Allison et al. 1994, Ishai et al. 1999, Kanwisher et al. 1997, Levy et al. 2001, Puce et al. 1995), animals (Chao et al. 1999, Martin et al. 1996), body parts (Downing et al. 2001, Grossman & Blake 2002), tools (Beauchamp et al. 2002, Chao et al. 1999, Martin et al. 1996), places or houses (Aguirre et al. 1998, Epstein & Kanwisher 1998, Ishai et al. 1999), and letter strings (Cohen et al. 2000, Hasson et al. 2002). These results are summarized in Figure 8. Importantly, the object-selective patterns of activation across occipito-temporal cortex (e.g., faces and houses) are replicable within and across subjects (Levy et al. 2001, Malach et al. 2002). Still unanswered is whether regions that show preference for an object category should be treated as an independent module for processing that category (see below).

At a higher spatial resolution, possibly at a columnar resolution, there may be additional modular representations at a smaller spatial scale, which can be revealed only with higher-resolution imaging or through indirect methods such as fMR-adaptation (Avidan et al. 2002b, Grill-Spector & Malach 2001).

The parcellation of object cortex to functional subdivisions is still unclear. The debate is centered on which functional criteria should be used to partition this

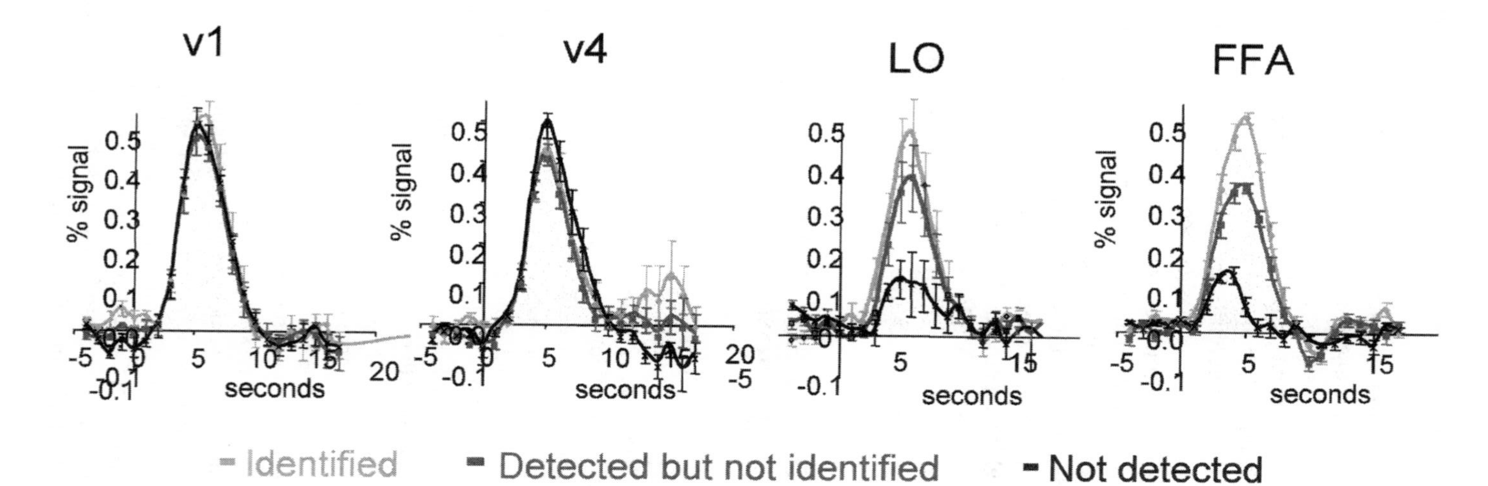

Figure 7 Activation along the ventral visual pathway in a task in which subjects' ability to detect and identify faces was monitored. Shown are time courses of activation in an event-related experiment in which subjects were briefly presented for 33 ms with a displayed picture that was subsequently masked by a texture pattern. Subjects were required to report when they identified a target face, detected a face, or did not detect a face. Here we present the activation pattern for trials in which the target face was presented as a function of subjects' responses. Signal is calculated relative to a fixation baseline (details in Grill-Spector 2003a). Early visual areas are significantly active during this task, but their activation is not correlated with face detection or identification success. In contrast, activation in regions in LO that respond strongly to faces compared to objects and the fusiform face area (FFA) is strongly correlated with subjects' ability to detect and identify faces.

large cortical expanse. At one end of the spectrum is the hypothesis that the whole constellation of regions anterior and lateral to classic retinotopic cortex is one distributed system for object recognition (Haxby et al. 2001, Ishai et al. 1999). At the other end is the hypothesis, proposed by Kanwisher and colleagues, that the constellation of regions in the ventral stream consists of a few domain-specific modules (Kanwisher 2000, Spiridon & Kanwisher 2002) for faces (Kanwisher et al. 1997), body parts (Downing et al. 2001), and places (Epstein & Kanwisher 1998), and the rest is a general purpose object-recognition system. A third hypothesis suggests that occipito-temporal regions should be subdivided on the basis of the semantic attributes of object categories (Chao et al. 1999 Beauchamp et al. 2002). A fourth hypothesis, held by Tarr & Gauthier (2000) suggests that object representations are clustered according to the type of processing required rather than according to their visual attributes. Specifically, Tarr & Gauthier suggest that face-selective regions are engaged in fine-grained discrimination between objects of the same category that is automated by expertise. According to this view, activity in these regions reflects expertise in recognizing certain object categories and is not restricted to faces. Finally, another hypothesis, proposed by Malach and colleagues (Hasson et al. 2003a, Levy et al. 2001, Malach et al. 2002), argues that cortical topography, in particular eccentricity mapping, is the underlying principle of the organization of ventral and dorsal stream object areas. These authors proposed that correlation between regions that prefer faces and central representation and regions that prefer buildings and peripheral representations reflects the basic visual specialization in which different spatial resolutions (cortical magnification factors) are related to different eccentricity bands (which continues into higher-order areas).

These different (but not necessarily contradictory) views emphasize criteria to differentiate between regions involved in object, face, and place perception. Here we propose an alternative framework to examine the human object recognition system as a hierarchical system. The underlying idea implies that object recognition is implemented in the brain through a series of processing stages, in which more global and invariant representations emerge up the hierarchy of the processing stream. The hierarchy is implemented through a gradual transition, from local representations that are closely tied to the retinal image to abstract representations that are closely linked to perception. In the next sections we consider the process of object recognition in the context of a hierarchical processing stream, emphasizing the different types of processing in lower- versus higher-level visual areas in the ventral stream.

FUNCTIONAL MANIFESTATIONS OF HIERARCHICAL PROCESSING IN THE HUMAN VENTRAL STREAM CORTEX

According to hierarchical framework, areas along both the occipito-temporal and occipito-parietal pathways are organized hierarchically, such that low-level inputs are transformed into more abstract representations through successive stages of processing. Virtually all visual-processing tasks activate V1 and V2. However, as one proceeds from one area to the next the neuronal response properties become increasingly complex. In monkeys, along the occipito-temporal pathway for example, many V1 cells function as local spatio-temporal filters, responsive to oriented bars (Hubel & Wiesel 1968), V2 cells respond to illusory contours of figures (Peterhans & von der Heydt 1991), some V4 cells respond only if a stimulus has a specific color or pattern (Gallant et al. 1993, Schein & Desimone 1990), and in occipito-temporal regions cells respond selectively to particular shapes (Booth & Rolls 1998, Desimone et al. 1984, Fujita et al. 1992, Kobatake & Tanaka 1994, Logothetis et al. 1995, Sigala & Logothetis 2002, Tanaka 1993, Wachsmuth et al. 1994).

We examined hierarchical processing along a sequence of ventral visual areas as a function of three criteria: (*a*) retinotopy, (*b*) motion sensitivity, and (*c*) object selectivity (Figure 9). Early retinotopic areas V1, V2, and V3 show a high degree of retinotopy (large print in Figure 9) but a low degree of specificity to either motion or form. Intermediate visual areas such as V3a and hV4 exhibit a lesser degree of retinotopic specificity (intermediate print) and, to some extent, higher stronger responses to objects (versus textures) and moving (versus stationary) low-contrast gratings. Higher-level areas along the processing stream show coarser retinotopy (small print in Figure 9) and a higher degree of specialization. For example, MT shows a strong preference for moving versus stationary stimuli but does not exhibit object selectivity. In contrast LO responds more strongly to objects compared to scenes and textures but has little response for moving versus stationary gratings.

This analysis demonstrates the ability to quantify and assess with fMRI a processing stream along several dimensions of visual computations. These analyses can be extended easily to incorporate additional attributes of visual processing such as color or stereopsis to reveal the hierarchy of computations and representations relevant to these domains.

Hierarchical Processing Along the Ventral Pathway: Local-Image Processing in Early Retinotopic Areas but Not in LOC

Several studies show that activity in early retinotopic areas is correlated with local, low-level aspects of the stimuli and perception of these low-level features. For example, increasing the level of contrast of a stimulus monotonically increases the strength of the fMRI signal in V1 (Avidan et al. 2002a, Boynton et al. 1999). Importantly, Boynton et al. (1999) found that contrast-response functions in early retinotopic areas were consistent with psychophysically measured contrast increment thresholds, which suggests that these early visual areas are involved in contrast detection. Indeed a more recent study (Ress & Heeger 2003) indicates that the activation in early retinotopic cortex predicts the ability of a subject to detect low-contrast patterns embedded in noise. However, as one proceeds along the sequence of visual areas in the ventral stream, higher-level areas along the hierarchy show a more invariant tuning to contrast. Avidan et al. (2002a) showed

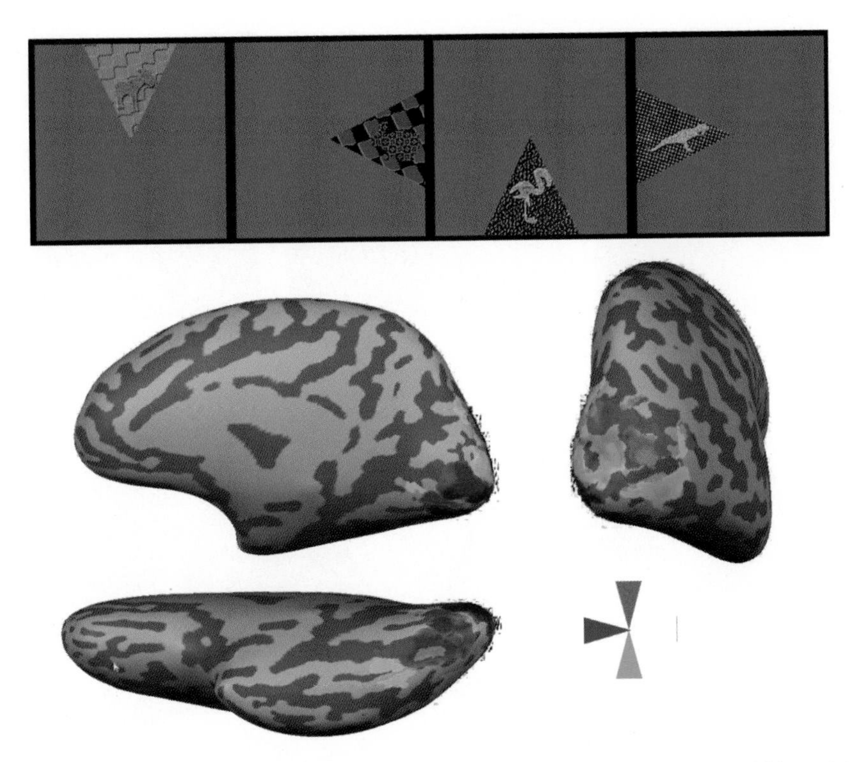

Figure 2 Vertical and horizontal representations in a representative subject. *Top*: Examples of stimuli used in the meridian mapping experiment. Colored objects were displayed on a textured background within each wedge. Stimuli moved back and forth within the wedge during the experiment at a rate of 4 hz. Stimulus radius: 8° *Bottom*: Visual meridians displayed on the inflated right hemisphere of subject EU, displayed from the medial, lateral, and ventral views. Stronger colors indicate higher statistical significance: $10^{-15} < P < 10^{-3}$: Red, blue, and green indicate visual meridians (see icon), purple indicates regions that respond both to the upper and horizontal representations, and turquoise indicates regions that respond both to the horizontal and lower meridian representations.

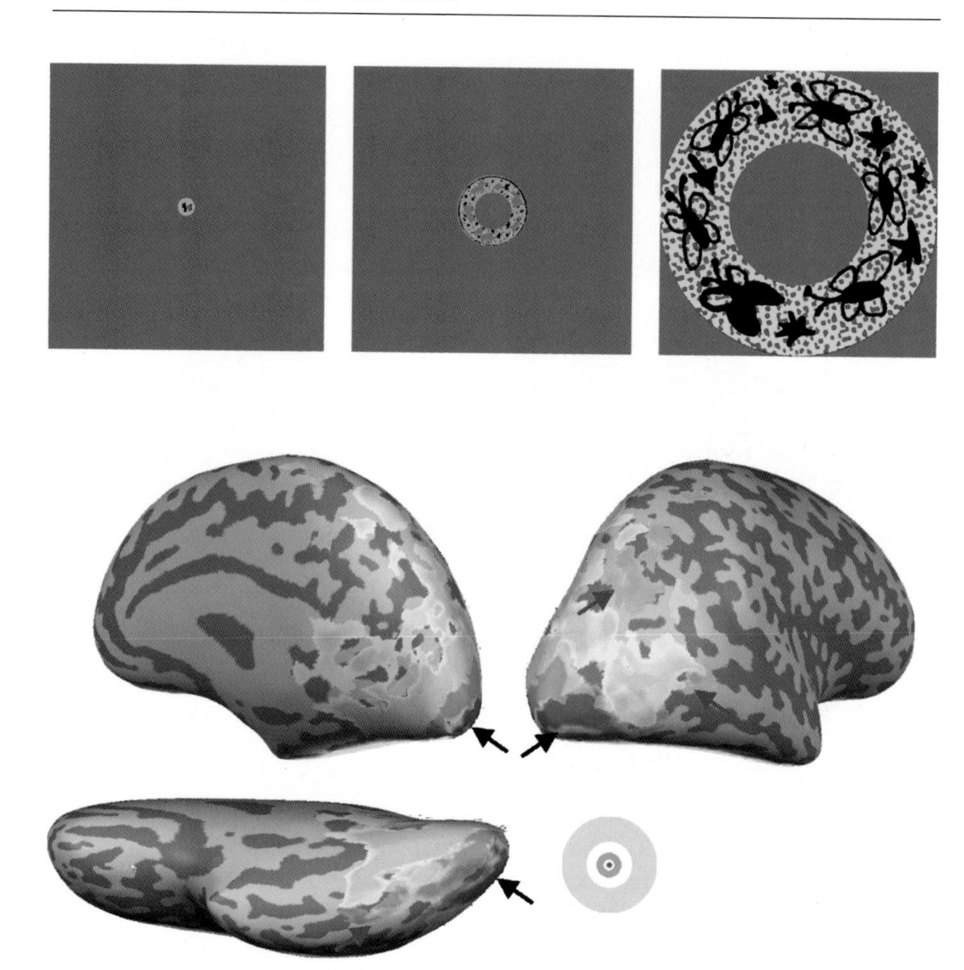

Figure 3 Eccentricity maps in a representative subject. *Top*: Examples of stimuli used in the eccentricity experiment. Stimuli consisted of colored pictures containing objects and textures. Stimuli rotated within the rings at a rate of 2 hz. Stimuli radius: 1.5°, 3–6°, 12–24°. *Bottom*: Eccentricity map of subject EB on the inflated right hemisphere from the medial, lateral, and ventral views. Strong colors indicate high statistical significance, faded colors indicate lower statistical significance: $10^{-15} < P < 10^{-3}$. Pink: central representations (up to 1.5°). Light purple: regions that respond both to central and mid representations. Light blue: parafoveal representations (3–6°). Light green: regions that respond both to mid and peripheral representations. Yellow: peripheral representations: 12–24°. Black arrows indicate the location of the confluent fovea. Red arrows indicate additional fovea in the parietal, lateral temporal, and ventral temporal lobes.

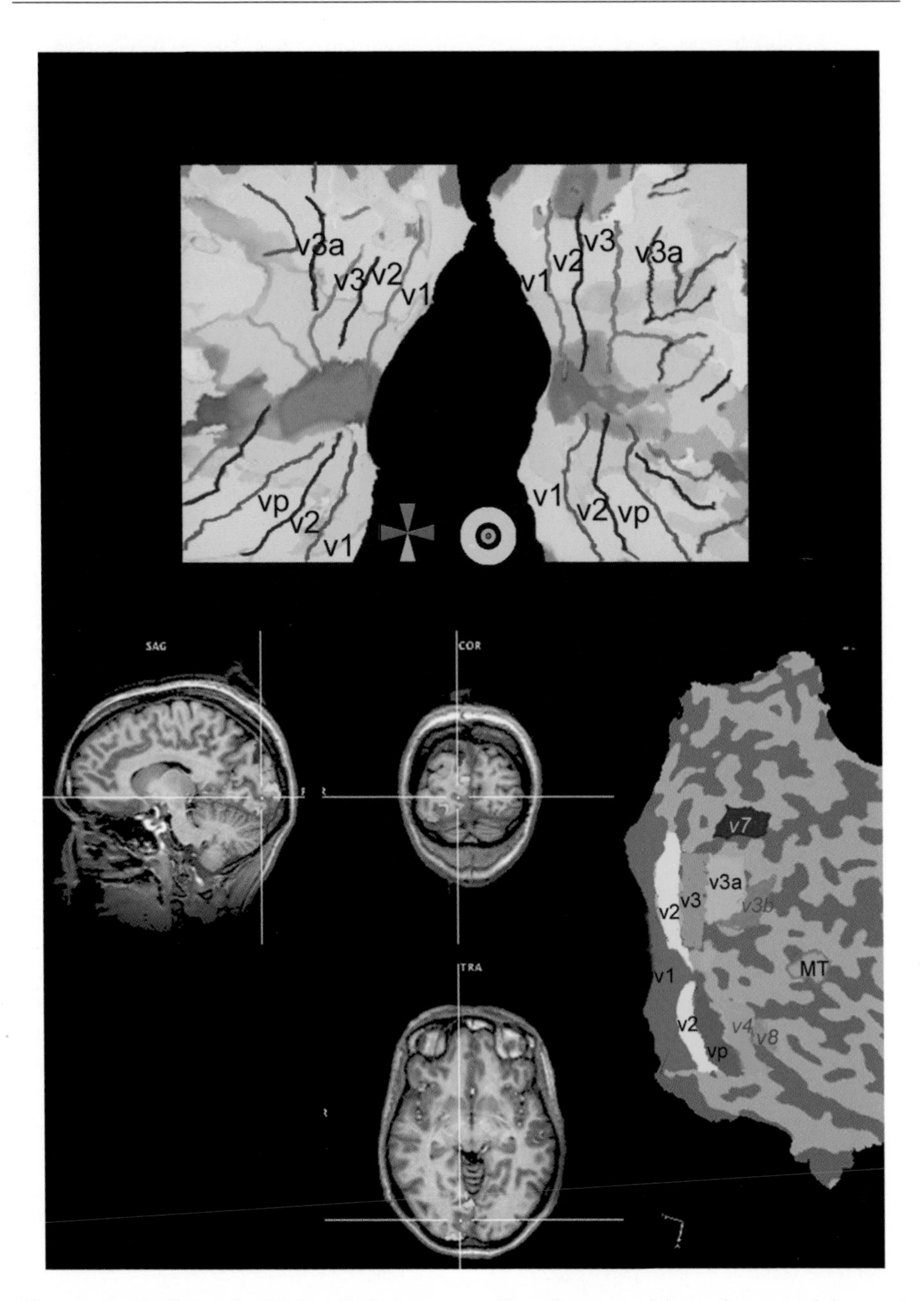

Figure 4 Early and mid-level visual areas. *Top*: Superposition of eccentricity and polar angle maps. Yellow, blue, and pink bands indicate eccentricity maps; lines indicate centers of upper, lower, and horizontal representations (see icons). Note that meridian lines cross all eccentricities orthogonally. *Bottom*: Visual areas on a flattened representation and on the brain volume. Visual area names under consensus are denoted in black, and areas currently under debate are marked in blue italics. All maps in this figure are of subject JHT.

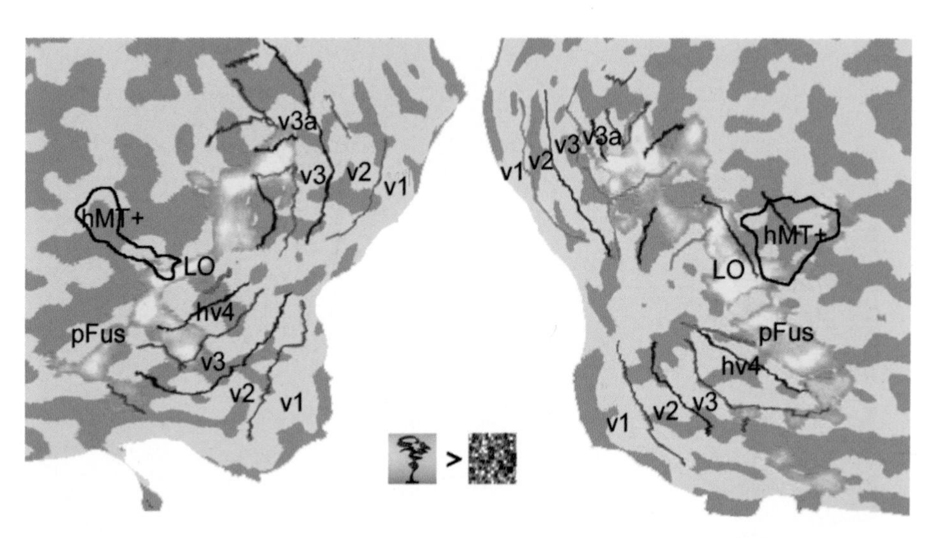

Figure 5 Object-selective regions that respond more strongly to objects than to scrambled objects are displayed on a flattened cortical map (subject EB). Yellow and orange indicate statistical significance: $P < 10^{-12} < P < 10^{-6}$. Colored lines indicate borders of retinotopic visual areas. Area hMT+ was defined as regions in the posterior bank of the inferotemporal sulcus that respond more strongly to moving versus stationary low-contrast gratings (with $P < 10^{-6}$). LO, lateral occipital; pFus, posterior fusiform.

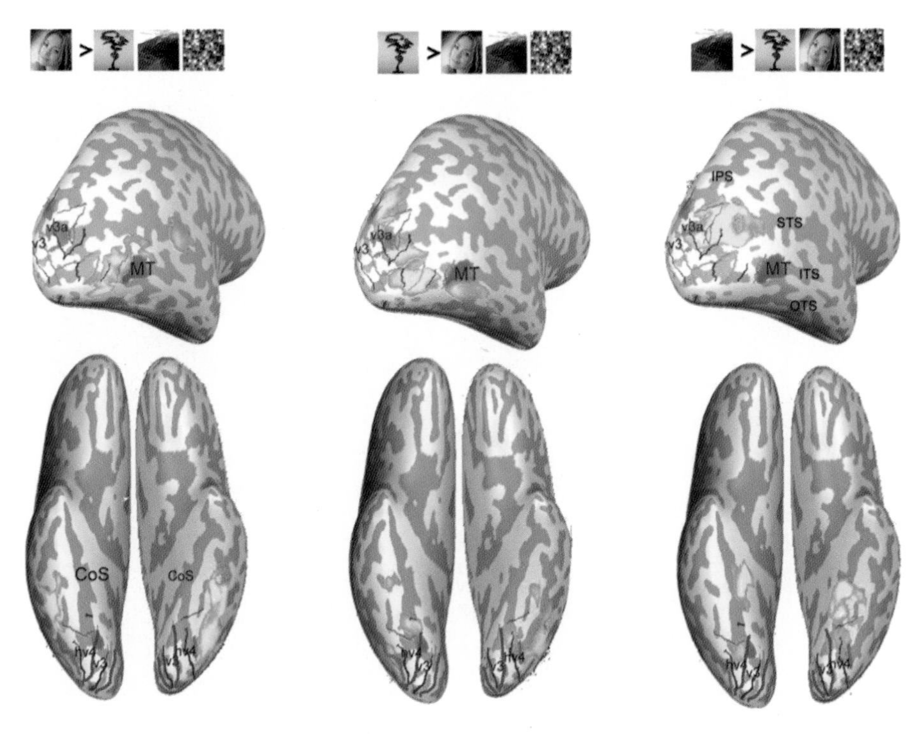

Figure 6 Face-, object-, and place-selective regions in the human brain displayed on an inflated surface representation of subject JHT. Icons indicate the comparison done in the statistical tests. *Left*: areas responding more strongly to faces than objects, places, or textures. *Center*: areas responding more strongly to objects than faces, places, or textures. *Right*: areas responding more strongly to places (scenes) than faces, objects, or textures. Yellow and orange indicate statistical significance: $P < 10^{-12} < P < 10^{-6}$. Colored lines indicate borders of retinotopic visual areas. Blue indicates area hMT+, defined as a region in the posterior bank of the inferotemporal sulcus that responds more strongly to moving versus stationary low-contrast gratings (with $P < 10^{-6}$). Abbreviations: STS, superior temporal sulcus; ITS, inferior temporal sulcus; OTS, occipital-temporal sulcus; CoS, collateral sulcus.

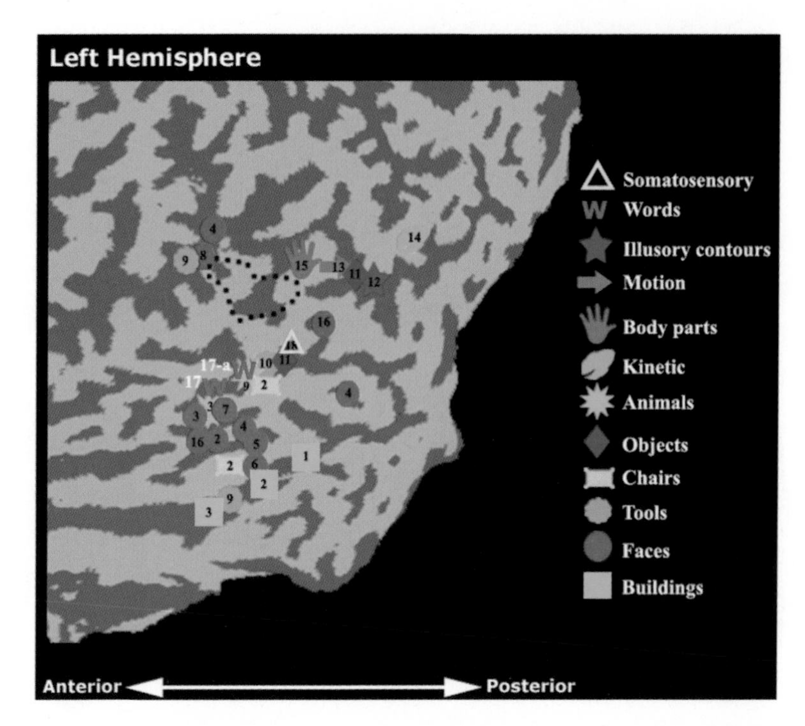

Figure 8 Areas that respond to selected categories: meta-analysis of Talairach coordinates across 18 different studies. Summary of cortical regions showing preferential activations to the following object categories: buildings, faces, tools, chairs, animals, common objects, kinetic movement, body parts, moving objects, illusory contours, words, and haptically defined objects. The numbers appearing on the category icons represent the study number as listed below: **1**, Epstein & Kanwisher 1998; **2**, Ishai et al. 1999; **3**, Maguire et al. 2001; **4**, Hoffman & Haxby 2000; **5**, George et al. 1999; **6**, Kanwisher et al. 1997; **7**, Halgren et al. 1999; **8**, Puce et al. 1998; **9**, Chao et al. 1999; **10**, Beauchamp et al. 2002; **11**, Grill-Spector et al. 1999; **12**, Mendola et al. 1999; **13**, Kourtzi et al. 2001; **14**, Van Oostende et al. 1997; **15**, Downing et al. 2001; **16**, Puce et al. 1996; **17**, Hasson et al. 2003a; **18**, Amedi et al. 2001; **19**, Haxby et al. 1999. The region pertaining to body parts (region 15) is estimated based on the MNI coordinate system. The word-related activation is based on an average activation from six different studies as reported in Hasson et al. 2002. Note the substantial consistency of the various object-related activations across the different studies. Used with permission from Hasson et al. 2003a.

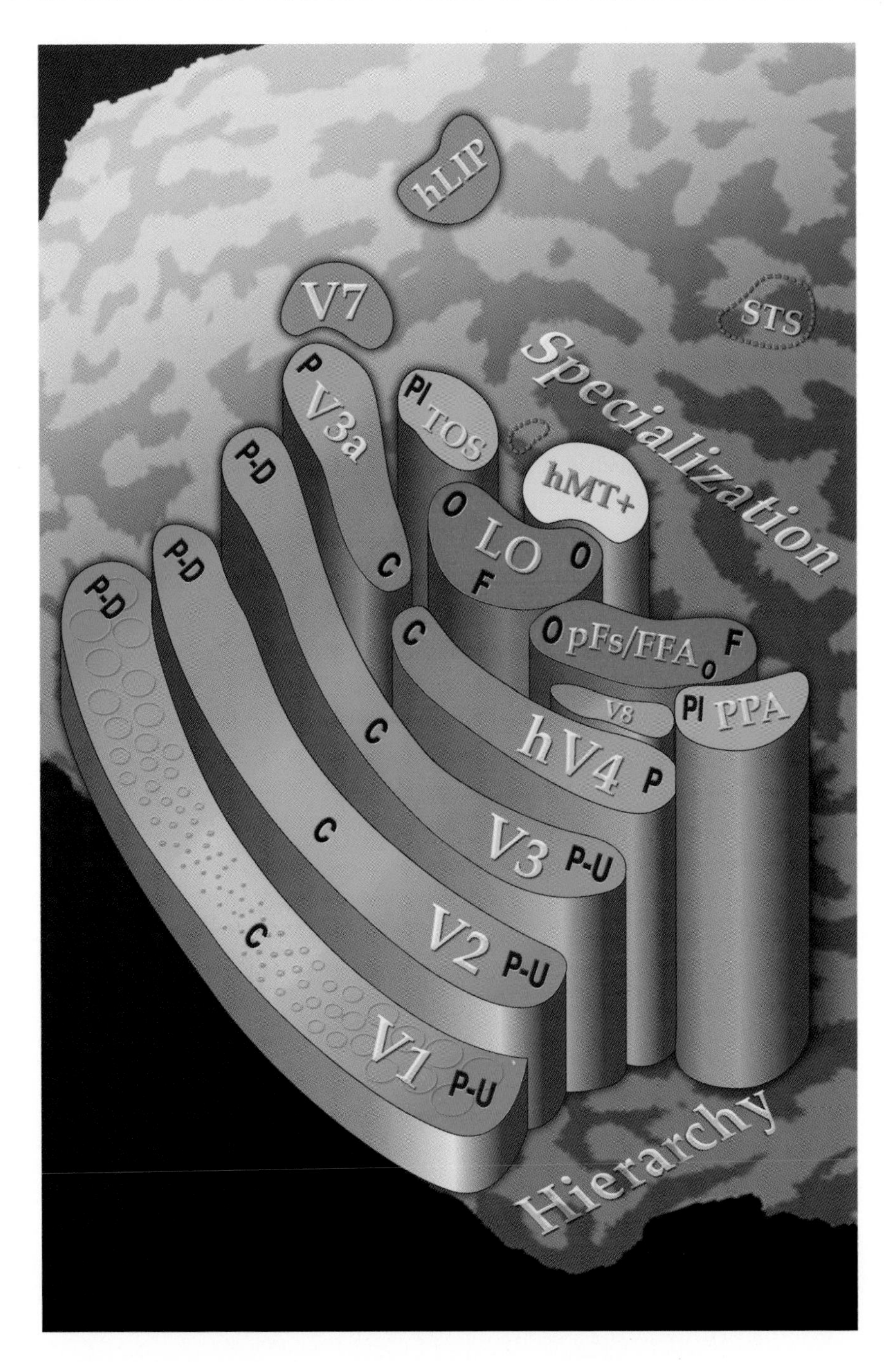

See legend on next page

Figure 12 Schematic atlas of human visual cortex. The schematic layout of human visual areas is presented on an unfolded right hemisphere, illustrating the orthogonal axes of hierarchy and specialization (*labeled along the axis direction*). The visual areas are arranged in a staircase fashion to illustrate the hierarchical sequence of increased abstraction leading from primary visual cortex to high-order visual areas. The orthogonal, specialization axis is illustrated through the color scale. The specialization is manifested in early cortex as a transition from central (*C*) to peripheral (*P*) visual-field representations, associated with high and low magnification factors, respectively. In higher-level cortex, the specialization is manifested as a transition from regions that respond preferentially to objects/faces (*O*, *F*), and are related to central-biased, high-magnification representations, to regions that respond more strongly to places, buildings, and scenes (*B*) and are related to peripheral-biased, lower-magnification representations. B, buildings/scenes; C, central; P-U, P-D, peripheral representation of upper and lower visual fields, respectively; Pl, place; F, faces; O, objects.

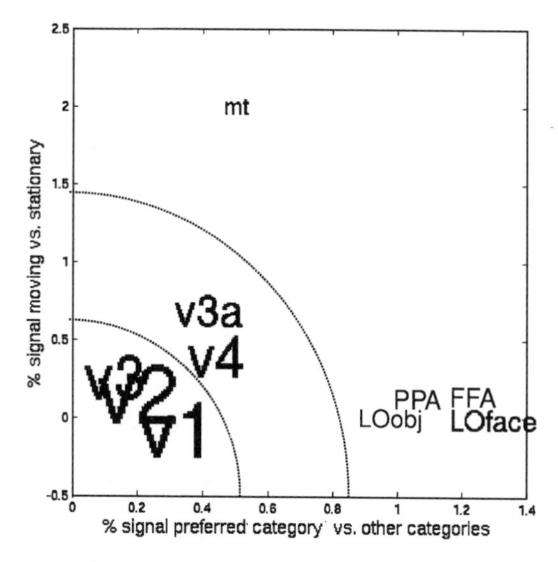

Figure 9 Hierarchical processing in the visual system. Visual areas are plotted as points in a space defined by motion selectivity and category preference averaged across five subjects. The font size illustrates the third dimension of this space: retinotopy (see below). X-axis denotes the object selectivity defined as the average signal (preferred object category) to average signal (nonpreferred categories). Stimuli included unfamiliar objects, faces, and scenes. Y-axis denotes signal (moving low-contrast gratings) to % signal (stationary low-contrast gratings). Retinotopy (denoted by font size) as the difference % signal (preferred eccentricity) to % signal (nonpreferred eccentricity bands) (see Figure 3). In retinotopic areas the measurements were performed on the periphery band. For the FFA and PPA, the measurements were done on the preferred eccentricity band. Dashed arcs indicate putative stages of hierarchical processing.

(Figure 10*A*) that sensitivity to contrast changes decreases from V1 to LO. Specifically, compared to V1, LO showed a flatter, more invariant contrast tuning curve. V4 showed an intermediate profile, exhibiting less contrast sensitivity than V1 but higher contrast sensitivity than LO (Figure 10*A*).

Another indication for hierarchical processing comes from measurements of response to object size along the ventral stream (Figure 10*B*). Again, V1 shows the sharpest sensitivity as peripheral regions in V1 do not respond to intermediate or parafoveal stimuli. V4 shows an intermediate level of tuning, as peripheral regions in V4 do show some response to parafoveal stimuli. Finally, LO shows the flattest (more invariant) curve. Importantly, peripheral regions in LO are significantly activated (but not maximally) by parafoveal stimuli. Recently, several labs have estimated the average receptive field size across retinotopic regions along the ventral and dorsal streams (Kastner et al. 2001, Press et al. 2001, Smith et al. 2001). These studies suggest that the average receptive fields are smallest in V1.

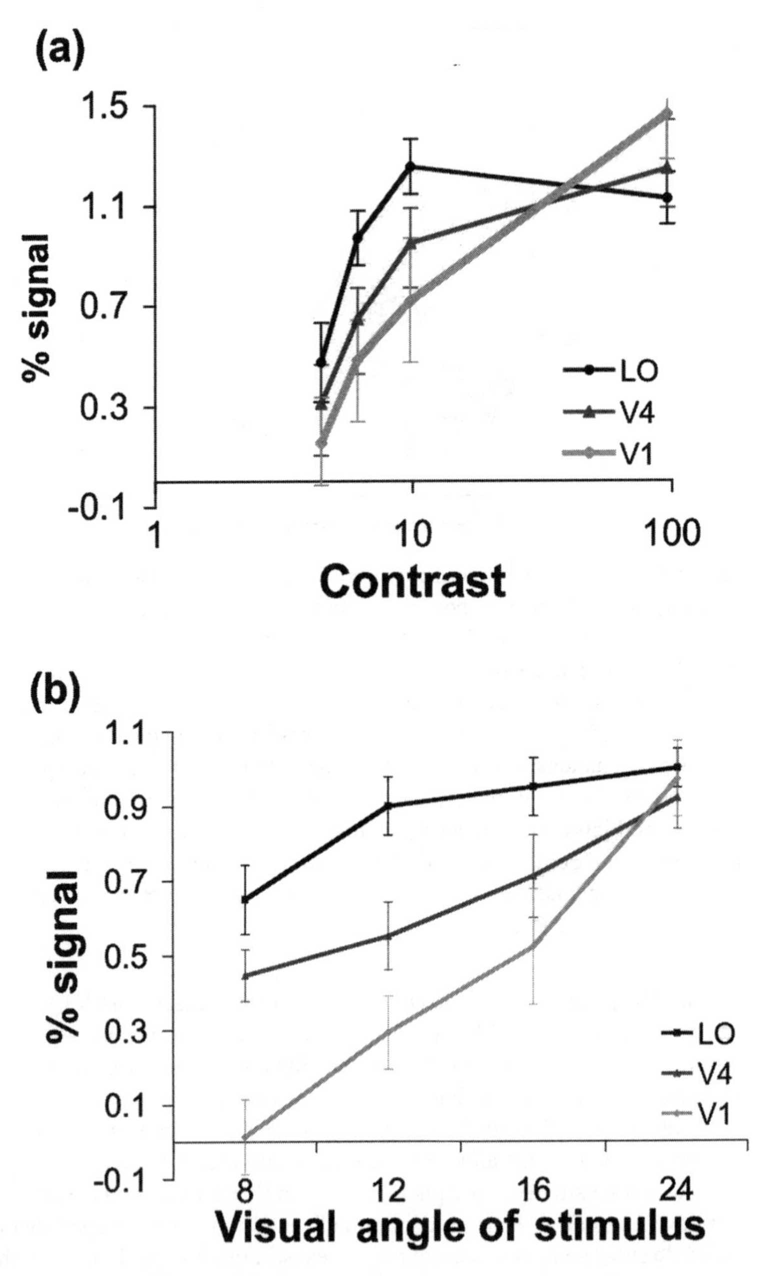

Figure 10 Sensitivity to object contrast and size in retinotopic cortex but not in LOC. (*A*) Response of visual areas as a function of the contrast of line drawings of face images (Avidan et al. 2002a). (*B*) Response to object of different sizes measured on the periphery band of several visual areas.

They are larger in V2, larger again in V3/VP, and largest of all in areas V3a and V4.

From Local Low-Level Representations to Abstract Representations

Hierarchical processing is also evident in the transition from the local, low-level functionality of area V1 to the more holistic object responses in LO and VOT (Figure 11*A*). Gradual scrambling experiments in which object images are broken into an increasingly larger number of fragments reveal areas along the hierarchical sequence that are increasingly more sensitive to image scrambling (Figure 11*A*) (Grill-Spector et al. 1998b, Lerner et al. 1999). The decrease in activation to scrambled objects in object- and face-selective areas may be related to both receptive field size and stimulus selectivity. Specifically, the object (or face) fragment falling in receptive fields within this area needs to contain a large fraction of the object (or face) to produce activation equivalent to activation levels elicited by whole, unscrambled objects.

The question of how the unified perception of a global shape (or good Gestalt) emerges through hierarchical processing has been addressed recently by Kourtzi and colleagues (Altmann et al. 2003, Kourtzi et al. 2003b). Their experiments measured, using fMRI, the recovery from adaptation to patterns containing randomly oriented lines (Grill-Spector & Malach 2001) after the adapting pattern was changed into a different random pattern or into a pattern containing a colinear contour. They observed that both higher (LO and VOT) and lower visual areas (V1 and V2) were involved in the processing of global shapes, but this processing was related to receptive field size. These results suggest that a global gestalt perception may involve multiple brain areas processing information at several spatial scales.

Additional aspects that emerge as correlated with cortical hierarchy are as follows: (*a*) susceptibility to repetition-suppression of repeated object images (adaptation) that occurs in higher-level, object-related areas but less so in early retinotopic regions (Avidan et al. 2002b, Grill-Spector & Malach 2001) (the phenomenon of adaptation indicates that reponses to objects in object-selective cortex also depends on prior exposure to these stimuli); and (*b*) increased temporal nonlinearities that suggest increasing levels of persistent activity in high-order visual areas (Ferber et al. 2003, Mukamel et al. 2003). Temporal nonlinearities were also evident in experiments that used a backward masking paradigm in which object and face stimuli were presented briefly and were followed by a noise pattern (Grill-Spector et al. 2000). Activation in V1 did not vary much with stimulus exposure. However, activation in both LO and VOT drastically decreased with shortened exposure durations (Figure 11*B*). Importantly, the function relating the amplitude of the fMRI signal from LO and VOT was not correlated with exposure duration per se but rather with the psychometric curve of accuracy at an object-recognition task.

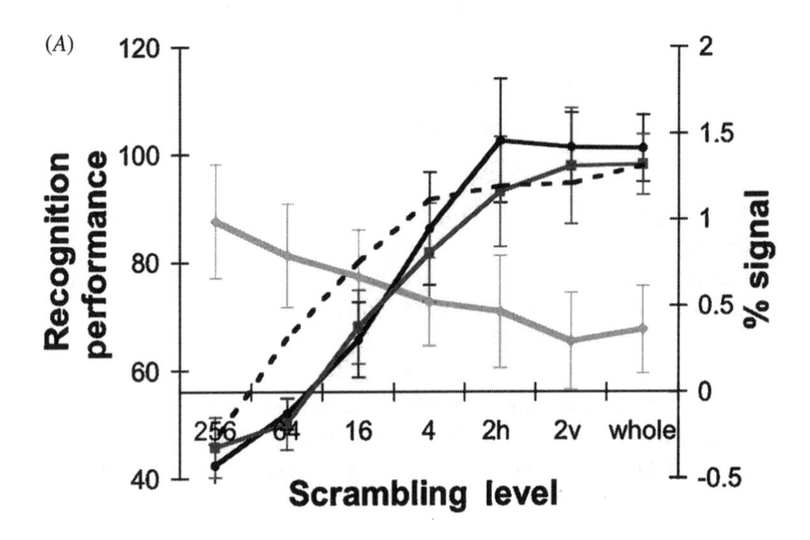

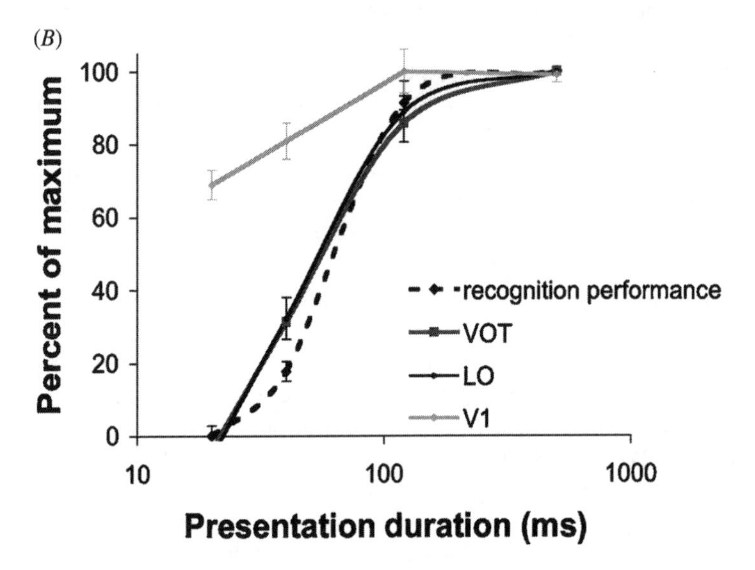

Figure 11 Sensitivity to parameters that affect face and object perception in LO and posterior fusiform (*A*). Scrambling images into smaller parts reduces the signal in LO and posterior fusiform but not in V1 (Lerner et al. 2001). 2v, scrambling the image into two parts along the vertical axis; 2h, scrambling the image into two parts along the horizontal axis; whole, original, unscrambled face images. (*B*) Reducing the exposure duration of object and face images reduces the signal in LO and posterior fusiform in a nonlinear fashion that is correlated with subjects' ability to recognize the objects (Grill-Spector et al. 2000).

TOP-DOWN PROCESSING

We focus on many findings concerning feed-forward hierarchical processing. However, both behavioral studies (Driver et al. 2001) and human neuroimaging studies reveal powerful top-down modulations on visual perception and activity patterns in visual cortex (for a review see Kanwisher & Wojciulik 2000). For example, attention and expectation alone without any visual stimulus can produce brain activation in early visual cortex (Chawla et al. 1999, Kastner et al. 1999, Ress et al. 2000). Directing subjects' attention to different locations in the visual field enhances activation in attended locations and inhibits activations in unattended locations (Brefczynski & DeYoe 1999, Macaluso et al. 2000, Tootell & Hadjikhani 2000) with a specificity comparable to retinotopic mapping. Furthermore, attention to different attributes of the visual stimulus, such as color, motion, or form, focally enhances activation in regions involved in processing these attributes. Thus, attending and tracking motion modulates areas engaged in processing motion (Culham et al. 1998, Lavie & Driver 1996, O'Craven et al. 1999), attending to objects and faces enhances activation in object- and face-selective regions, respectively (Avidan et al. 2003, O'Craven et al. 1999, Wojciulik et al. 1998), and attending to color enhances activation along regions involved in color perception (Beauchamp et al. 1999).

These results reveal that in addition to bottom-up processing of the visual stimulus, driven by the input, information feeds back from high-level cortical regions in the parietal and prefrontal cortex to early processing stations. It is still unclear whether the top-down modulations in early visual areas are limited to spatial location, i.e., attentional spotlight, or whether they constitute more complex computations (Galuske et al. 2002, Hochstein & Ahissar 2002, Lamme & Roelfsema 2000, Pascual-Leone & Walsh 2001).

Other top-down mechanisms that may interact and modulate activity in visual areas include spatial coordinate remapping (Merriam et al. 2003), emotional processing (Armony & Dolan 2002, Hendler et al. 2003, Vuilleumier et al. 2002a), and even stereotypical processing (Golby et al. 2001). Top-down effects measured by fMRI are stronger than those reported in single-unit recordings in behaving monkeys. The source of this discrepancy is unclear but may be related to somewhat different aspects of neuronal signals measured by BOLD imaging (Logothetis 2000), compared to single-unit recording employed in physiology.

LARGE-SCALE RELATIONSHIP BETWEEN HIERARCHICAL AND SPECIALIZED PROCESSING

Figure 12 (see color insert) shows a large-scale summary of known visual areas, painted on the unfolded surface of one hemisphere. Close inspection of such an atlas reveals that if the cortex is unfolded by introducing a cut along the calcarine sulcus, thus "opening" area V1 by splitting its horizontal meridian, an intriguing pattern emerges: The two organizing principles, hierarchical processing and functional specialization, are neatly translated into two orthogonal axes laid upon the unfolded human visual cortex. On the unfolded flat map (Figure 12) the hierarchical progression is arranged along the back-to-front axis (which corresponds to postero-medial to anterior-lateral axis on the real hemisphere), whereas the functional specialization axis is situated orthogonally to the hierarchy axis along the dorso-ventral direction (for axis orientations see hierarchy and specialization labels in Figure 12). Thus, an imaginary line ascending along the hierarchy axis will start from lower area V1 posteriorly and will then ascend in retinotopic cortex going through areas V2, V3, and V3a dorsally and V2, V3, and V4 ventrally, finally reaching into the object-related areas.

Along the axis orthogonal to the hierarchy, i.e. the specialization axis (see label in Figure 12), the visual areas can be examined under the global framework of eccentricity/cortical magnification. Thus, tracing a dorso-ventral imaginary line in early visual cortex (areas V1–V3), eccentricity begins with regions specializing in peripheral vision (peripheral lower-field representation), travels through regions specializing in central vision, at the center of the visual system, and continues ventrally into peripheral representations (this time, upper visual field peripheral representations). A parallel line placed at a higher level of the hierarchy, i.e., more anteriorly along intermediate visual areas (e.g., V3a) will transverse the periphery-center-periphery specialization sequence, but it will now lose the separate upper versus lower visual field representations. Finally, at yet more anterior bands, in high-level, object-selective cortex, one finds a sequence of object-category specializations, which roughly (with some exceptions) parallels the periphery-center-periphery sequence. Regions that respond strongly to places (compared to objects and faces, e.g., the TOS or PPA) are associated with peripheral representations, and regions that respond more strongly to faces (than objects or places, e.g., FFA) are associated with central representations. Thus an imagery line transversing from doral to ventral the TOS, LO, FFA, and PPA will encounter the same periphery-center-periphery map revealed in early visual cortex.

To summarize this point, the center/periphery map provides a global organization principle of the entire visual cortex, cutting across the borders between cortical areas. Surprisingly, this center-periphery specialization extends into high-level, occipito-temporal cortex. Here, the center-periphery specialization translates into object-category specialization. Thus, faces (and words) overlap with central representations, and buildings and scenes overlap with peripheral representations (Hasson et al. 2002, 2003a; Levy et al. 2001, Malach et al. 2002). Overall, the two orthogonal axes of hierarchy and center-periphery specialization could provide an overall framework or grid to define, in a fairly orderly manner, the entire constellation of human visual areas.

FUTURE DIRECTIONS

In less than a decade of research, fMRI has provided extensive information and insight about the functional organization of the human visual cortex and correlations between brain activations and visual percepts. However, we only partially understand how basic visual processes such as color or depth perception are implemented in the human brain, let alone higher-level functions such as object recognition. Further, the relationship between these findings under constrained lab conditions and free viewing in a more natural ecological vision is only beginning to be addressed (Hasson et al. 2004).

Although fMRI is a robust mapping tool and more than a dozen visual areas have been identified, there is consensus about the functional definition of only a handful of visual areas: V1, V2, V3, and probably hMT+. The functional definition of visual areas beyond V3 is currently under vigorous debate. Thus, even the aspect of mapping human visual areas remains a field in which we expect many new findings in the near future. Our recent reports that retinotopic maps extend well beyond "early retinotopic cortex" will be instrumental in deliniating high-level visual areas using similar principles, which were applied to early visual cortex.

Another important issue, which is only beginning to be addressed, is long-range developmental and plasticity effects in the human visual system, such as the role of visual experience in instructing the development of cortical circuitry. So far research in this field has focused mainly on pathological conditions (e.g., Amedi et al. 2003, Lerner et al. 2003).

Despite these limitations our ability to investigate the human brain has dramatically increased in the last several years. We are hopeful that the understanding of visual perception and its implementation in the human brain will further increase in the next decade when researchers will successfuly combine multiple methods, such as fMRI, DTI, VEP (visual evoked potentials), MEG, psychophysics, and computational modeling, to study the computations involved in visual perception. This explosive growth in our knowledge about the human visual cortex offers the exciting hope that broader issues, such as the relationship between conscious awareness and patterns of brain activation, which have long been central to the human quest for knowledge, finally may be addressed productively.

ACKNOWLEDGMENTS

We thank Galia Avidan, Yulia Lerner, and Uri Hasson for contributing data displayed in the figures. We also thank Rainer Goebel for his help with applications of Brainvoyager relevant to data analysis procedures. Finally, we thank Galia Avidan and Uri Hasson for their comments on drafts of this manuscript.

The *Annual Review of Neuroscience* is online at http://neuro.annualreviews.org

LITERATURE CITED

Aguirre GK, Zarahn E, D'Esposito M. 1998. An area within human ventral cortex sensitive to "building" stimuli: evidence and implications. *Neuron* 21:373–83

Allison T, McCarthy G, Nobre A, Puce A, Belger A. 1994. Human extrastriate visual cortex and the perception of faces, words, numbers, and colors. *Cereb. Cortex* 4:544–54

Altmann CF, Bulthoff HH, Kourtzi Z. 2003. Perceptual organization of local elements into global shapes in the human visual cortex. *Curr. Biol.* 13:342–49

Amedi A, Malach R, Hendler T, Peled S, Zohary E. 2001. Visuo-haptic object-related activation in the ventral visual pathway. *Nat. Neurosci.* 4:324–30

Amedi A, Raz N, Pianka P, Malach R, Zohary E. 2003. Early 'visual' cortex activation correlates with superior verbal memory performance in the blind. *Nat. Neurosci.* 6:758–66

Amir Y, Harel M, Malach R. 1993. Cortical hierarchy reflected in the organization of intrinsic connections in macaque monkey visual cortex. *J. Comp. Neurol.* 334:19–46

Andrews TJ, Schluppeck D, Homfray D, Matthews P, Blakemore C. 2002. Activity in the fusiform gyrus predicts conscious perception of Rubin's vase-face illusion. *Neuroimage* 17:890–901

Armony JL, Dolan RJ. 2002. Modulation of spatial attention by fear-conditioned stimuli: an event-related fMRI study. *Neuropsychologia* 40:817–26

Avidan G, Harel M, Hendler T, Ben-Bashat D, Zohary E, Malach R. 2002a. Contrast sensitivity in human visual areas and its relationship to object recognition. *J. Neurophysiol.* 87:3102–16

Avidan G, Hasson U, Hendler T, Zohary E, Malach R. 2002b. Analysis of the neuronal selectivity underlying low fMRI signals. *Curr. Biol.* 12:964–72

Avidan G, Levy I, Hendler T, Zohary E, Malach R. 2003. Spatial vs. object specific attention in high-order visual areas. *Neuroimage* 19:308–18

Backus BT, Fleet DJ, Parker AJ, Heeger DJ. 2001. Human cortical activity correlates with stereoscopic depth perception. *J. Neurophysiol.* 86:2054–68

Bar M, Tootell RB, Schacter DL, Greve DN, Fischl B, et al. 2001. Cortical mechanisms specific to explicit visual object recognition. *Neuron* 29:529–35

Bartels A, Zeki S. 2000. The architecture of the colour centre in the human visual brain: new results and a review. *Eur. J. Neurosci.* 12:172–93

Beauchamp MS, Haxby JV, Jennings JE, DeYoe EA. 1999. An fMRI version of the Farnsworth-Munsell 100-Hue test reveals multiple color-selective areas in human ventral occipitotemporal cortex. *Cereb. Cortex* 9:257–63

Beauchamp MS, Lee KE, Haxby JV, Martin A. 2002. Parallel visual motion processing streams for manipulable objects and human movements. *Neuron* 34:149–59

Booth MC, Rolls ET. 1998. View-invariant representations of familiar objects by neurons in the inferior temporal visual cortex. *Cereb. Cortex* 8:510–23

Boynton GM, Demb JB, Glover GH, Heeger DJ. 1999. Neuronal basis of contrast discrimination. *Vision Res.* 39:257–69

Brefczynski JA, DeYoe EA. 1999. A physiological correlate of the 'spotlight' of visual attention. *Nat. Neurosci.* 2:370–74

Britten KH, Shadlen MN, Newsome WT, Movshon JA. 1992. The analysis of visual motion: a comparison of neuronal and psychophysical performance. *J. Neurosci.* 12:4745–65

Buchel C, Josephs O, Rees G, Turner R, Frith CD, Friston KJ. 1998. The functional anatomy of attention to visual motion. A functional MRI study. *Brain* 121:1281–94

Castelo-Branco M, Formisano E, Backes W, Zanella F, Neuenschwander S, et al. 2002. Activity patterns in human motion-sensitive areas depend on the interpretation of global motion. *Proc. Natl. Acad. Sci.USA* 99:13914–19

Chao LL, Haxby JV, Martin A. 1999. Attribute-based neural substrates in temporal cortex for perceiving and knowing about objects. *Nat. Neurosci.* 2:913–19

Chawla D, Rees G, Friston KJ. 1999. The physiological basis of attentional modulation in extrastriate visual areas. *Nat. Neurosci.* 2:671–76

Clarke S. 1994a. Association and intrinsic connections of human extrastriate visual cortex. *Proc. R. Soc. Lond. B Biol. Sci.* 257:87–92

Clarke S. 1994b. Modular organization of human extrastriate visual cortex: evidence from cytochrome oxidase pattern in normal and macular degeneration cases. *Eur. J. Neurosci.* 6:725–36

Cohen L, Dehaene S, Naccache L, Lehericy S, Dehaene-Lambertz G, et al. 2000. The visual word form area: spatial and temporal characterization of an initial stage of reading in normal subjects and posterior split-brain patients. *Brain* 123:291–307

Conturo TE, Lori NF, Cull TS, Akbudak E, Snyder AZ, et al. 1999. Tracking neuronal fiber pathways in the living human brain. *Proc. Natl. Acad. Sci. USA* 96:10422–27

Culham JC, Brandt SA, Cavanagh P, Kanwisher NG, Dale AM, Tootell RB. 1998. Cortical fMRI activation produced by attentive tracking of moving targets. *J. Neurophysiol.* 80:2657–70

Culham JC, Danckert SL, Souza JF, Gati JS, Menon RS, Goodale MA. 2003. Visually guided grasping produces fMRI activation in dorsal but not ventral stream brain areas. *Exp. Brain. Res.* In press

Dale AM, Fischl B, Sereno MI. 1999. Cortical surface-cased analysis I: segmentation and surface reconstruction. *NeuroImage* 9:179–94

Damasio A, Yamada T, Damasio H, Corbett J, McKee J. 1980. Central achromatopsia: behavioral, anatomic, and physiologic aspects. *Neurology* 30:1064–71

Desimone R, Albright TD, Gross CG, Bruce C. 1984. Stimulus-selective properties of inferior temporal neurons in the macaque. *J. Neurosci.* 4:2051–62

DeYoe EA, Carman GJ, Bandettini P, Glickman S, Wieser J, et al. 1996. Mapping striate and extrastriate visual areas in human cerebral cortex. *Proc. Natl. Acad. Sci. USA* 93:2382–86

DeYoe EA, Van Essen DC. 1988. Concurrent processing streams in monkey visual cortex. *Trends Neurosci.* 11:219–26

Downing PE, Jiang Y, Shuman M, Kanwisher N. 2001. A cortical area selective for visual processing of the human body. *Science* 293:2470–73

Driver J, Davis G, Russell C, Turatto M, Freeman E. 2001. Segmentation, attention and phenomenal visual objects. *Cognition* 80:61–95

Dukelow SP, DeSouza JF, Culham JC, van Den Berg AV, Menon RS, Vilis T. 2001. Distinguishing subregions of the human mt+ complex using visual fields and pursuit eye movements. *J. Neurophysiol.* 86:1991–2000

Duncan RO, Boynton GM. 2003. Cortical magnification within human primary visual cortex correlates with acuity thresholds. *Neuron* 38:659–71

Dupont P, De Bruyn B, Vandenberghe R, Rosier AM, Michiels J, et al. 1997. The kinetic occipital region in human visual cortex. *Cereb. Cortex* 7:283–92

Engel S, Zhang X, Wandell B. 1997a. Colour tuning in human visual cortex measured with functional magnetic resonance imaging. *Nature* 388:68–71

Engel SA, Glover GH, Wandell BA. 1997b. Retinotopic organization in human visual cortex and the spatial precision of functional MRI. *Cereb. Cortex* 7:181–92

Engel SA, Rumelhart DE, Wandell BA, Lee AT, Glover GH, et al. 1994. fMRI of human visual cortex. *Nature* 369:525

Epstein R, Kanwisher N. 1998. A cortical representation of the local visual environment. *Nature* 392:598–601

Felleman DJ, Van Essen DC. 1991. Distributed hierarchical processing in the primate cerebral cortex. *Cereb. Cortex* 1:1–47

Ferber S, Humphrey GK, Vilis T. 2003. The lateral occipital complex subserves the perceptual persistence of motion-defined groupings. *Cereb. Cortex* 13:716–21

Fischl B, Salat DH, Busa E, Albert M, Dieterich M, et al. 2002. Whole brain segmentation: automated labeling of neuroanatomical structures in the human brain. *Neuron* 33:341–55

Fischl B, Sereno MI, Dale AM. 1999. Cortical surface-based analysis II: inflation, flattening, a surface-based coordinate system. *NeuroImage* 9:195–207

Fujita I, Tanaka K, Ito M, Cheng K. 1992. Columns for visual features of objects in monkey inferotemporal cortex. *Nature* 360:343–46

Gallant JL, Braun J, Van Essen DC. 1993. Selectivity for polar, hyperbolic, and Cartesian gratings in macaque visual cortex. *Science* 259:100–3

Galuske RA, Schmidt KE, Goebel R, Lomber SG, Payne BR. 2002. The role of feedback in shaping neural representations in cat visual cortex. *Proc. Natl. Acad. Sci.USA* 99:17083–88

George N, Dolan RJ, Fink GR, Baylis GC, Russell C, Driver J. 1999. Contrast polarity and face recognition in the human fusiform gyrus. *Nat. Neurosci.* 2:574–80

Giese MA, Poggio T. 2003. Neural mechanisms for the recognition of biological movements. *Nat. Rev. Neurosci.* 4:179–92

Gilaie-Dotan S, Ullman S, Kushnir T, Malach R. 2002. Shape-selective stereo processing in human object-related visual areas. *Hum. Brain Mapp.* 15:67–79

Golby AJ, Gabrieli JD, Chiao JY, Eberhardt JL. 2001. Differential responses in the fusiform region to same-race and other-race faces. *Nat. Neurosci.* 4:845–50

Goodale MA, Milner AD, Jakobson LS, Carey DP. 1991. A neurological dissociation between perceiving objects and grasping them. *Nature* 349:154–56

Grill-Spector K. 2003a. The functional organization of the ventral visual pathway and its relationship to object recognition. In *Attention and Performance XX. Functional Brain Imaging of Visual Cognition*, ed. N Kanwisher, J Duncan, pp. 169–93. London: Oxford Univ. Press

Grill-Spector K. 2003b. The neural basis of object perception. *Curr. Opin. Neurobiol.* 13: 159–66

Grill-Spector K, Knouf N, Kanwisher N. 2004. The Fusiform Face Area subserves face perception, not generic within-category identification. *Nat. Neurosci.* 7(5):In press

Grill-Spector K, Kourtzi Z, Kanwisher N. 2001. The lateral occipital complex and its role in object recognition. *Vision Res.* 41:1409–22

Grill-Spector K, Kushnir T, Edelman S, Avidan G, Itzchak Y, Malach R. 1999. Differential processing of objects under various viewing conditions in the human lateral occipital complex. *Neuron* 24:187–203

Grill-Spector K, Kushnir T, Edelman S, Itzchak Y, Malach R. 1998a. Cue-invariant activation in object-related areas of the human occipital lobe. *Neuron* 21:191–202

Grill-Spector K, Kushnir T, Hendler T, Edelman S, Itzchak Y, Malach R. 1998b. A sequence of object-processing stages revealed by fMRI in the human occipital lobe. *Hum. Brain Mapp.* 6:316–28

Grill-Spector K, Kushnir T, Hendler T, Malach R. 2000. The dynamics of object-selective activation correlate with recognition performance in humans. *Nat. Neurosci.* 3:837–43

Grill-Spector K, Malach R. 2001. fMR-adaptation: a tool for studying the functional properties of human cortical neurons. *Acta Psychol. (Amst.)* 107:293–321

Grossman E, Donnelly M, Price R, Pickens D, Morgan V, et al. 2000. Brain areas involved in perception of biological motion. *J. Cogn. Neurosci.* 12:711–20

Grossman ED, Blake R. 2002. Brain areas active during visual perception of biological motion. *Neuron* 35:1167–75

Hadjikhani N, Liu AK, Dale AM, Cavanagh P, Tootell RB. 1998. Retinotopy and color sensitivity in human visual cortical area V8. *Nat. Neurosci.* 1:235–41

Hagmann P, Thiran JP, Jonasson L, Vandergheynst P, Clarke S, et al. 2003. DTI mapping of human brain connectivity: statistical fibre tracking and virtual dissection. *Neuroimage* 19:545–54

Halgren E, Dale AM, Sereno MI, Tootell RB, Marinkovic K, Rosen BR. 1999. Location of human face-selective cortex with respect to retinotopic areas. *Hum. Brain Mapp.* 7:29–37

Hasson U, Harel M, Levy I, Malach R. 2003a. Large-scale mirror-symmetry organization of human occipito-temporal object areas. *Neuron* 37:1027–41

Hasson U, Hendler T, Ben Bashat D, Malach R. 2001. Vase or face? A neural correlate of shape-selective grouping processes in the human brain. *J. Cogn. Neurosci.* 13:744–53

Hasson U, Levy I, Behrmann M, Hendler T, Malach R. 2002. Eccentricity bias as an organizing principle for human high-order object areas. *Neuron* 34:479–90

Hasson U, Nir Y, Levy I, Fuhrmann G, Malach R. 2004. Intersubject synchronization of cortical activity during natural vision. *Science.* 303:1634–40

Hasson U, Nir Y, Levy I, Malach R. 2003b. *Theatre of the mind: organization of human object areas revealed under natural viewing conditions.* Presented at Human Brain Mapping, New York

Haxby JV, Gobbini MI, Furey ML, Ishai A, Schouten JL, Pietrini P. 2001. Distributed and overlapping representations of faces and objects in ventral temporal cortex. *Science* 293:2425–30

Haxby JV, Hoffman EA, Gobbini MI. 2000. The distributed human neural system for face perception. *Trends Cogn. Sci.* 4:223–33

Haxby JV, Ungerleider LG, Clark VP, Schouten JL, Hoffman EA, Martin A. 1999. The effect of face inversion on activity in human neural systems for face and object perception [see comments]. *Neuron* 22:189–99

Heeger DJ, Huk AC, Geisler WS, Albrecht DG. 2000. Spikes versus BOLD: What does neuroimaging tell us about neuronal activity? *Nat. Neurosci.* 3:631–33

Hendler T, Rotshtein P, Yeshurun Y, Weizmann T, Kahn I, et al. 2003. Sensing the invisible: differential sensitivity of visual cortex and amygdala to traumatic context. *Neuroimage* 19:587–600

Hochstein S, Ahissar M. 2002. View from the top: hierarchies and reverse hierarchies in the visual system. *Neuron* 36:791–804

Horton JC, Dagi LR, McCrane EP, de Monasterio FM. 1990. Arrangement of ocular dominance columns in human visual cortex. *Arch. Ophthalmol.* 108:1025–31

Horton JC, Hedley-Whyte ET. 1984. Mapping of cytochrome oxidase patches and ocular dominance columns in human visual cortex. *Philos. Trans. R. Soc. London Ser. B Biol. Sci.* 304:255–72

Horton JC, Hocking DR. 1996. Pattern of ocular dominance columns in human striate cortex in strabismic amblyopia. *Vis. Neurosci.* 13:787–95

Hubel DH, Wiesel TN. 1968. Receptive fields and functional architecture of monkey striate cortex. *J. Physiol.* 195:215–43

Huk AC, Dougherty RF, Heeger DJ. 2002. Retinotopy and functional subdivision of human areas MT and MST. *J. Neurosci.* 22:7195–205

Huk AC, Heeger DJ. 2002. Pattern-motion responses in human visual cortex. *Nat. Neurosci.* 5:72–75

Huk AC, Ress D, Heeger DJ. 2001. Neuronal basis of the motion aftereffect reconsidered. *Neuron* 32:161–72

Ishai A, Ungerleider LG, Martin A, Haxby JV. 2000. The representation of objects in the human occipital and temporal cortex. *J. Cogn. Neurosci.* 12:35–51

Ishai A, Ungerleider LG, Martin A, Schouten JL, Haxby JV. 1999. Distributed representation of objects in the human ventral visual pathway. *Proc. Natl. Acad. Sci. USA* 96:9379–84

James TW, Humphrey GK, Gati JS, Menon RS, Goodale MA. 2000. The effects of visual object priming on brain activation before and after recognition. *Curr. Biol.* 10:1017–24

James TW, Humphrey GK, Gati JS, Menon RS, Goodale MA. 2002a. Differential effects of viewpoint on object-driven activation in dorsal and ventral streams. *Neuron* 35:793–801

James TW, Humphrey GK, Gati JS, Servos P, Menon RS, Goodale MA. 2002b. Haptic study of three-dimensional objects activates extrastriate visual areas. *Neuropsychologia* 40:1706–14

Kanwisher N. 2000. Domain specificity in face perception. *Nat. Neurosci.* 3:759–63

Kanwisher N, Chun MM, McDermott J, Ledden PJ. 1996. Functional imaging of human visual recognition. *Brain Res. Cogn. Brain Res.* 5:55–67

Kanwisher N, McDermott J, Chun MM. 1997. The fusiform face area: a module in human extrastriate cortex specialized for face perception. *J. Neurosci.* 17:4302–11

Kanwisher N, Wojciulik E. 2000. Visual attention: insights from brain imaging. *Nat. Rev. Neurosci.* 1:91–100

Kastner S, De Weerd P, Pinsk MA, Elizondo MI, Desimone R, Ungerleider LG. 2001. Modulation of sensory suppression: implications for receptive field sizes in the human visual cortex. *J. Neurophysiol.* 86:1398–411

Kastner S, De Weerd P, Ungerleider LG. 2000. Texture segregation in the human visual cortex: a functional MRI study. *J. Neurophysiol.* 83:2453–57

Kastner S, Pinsk MA, De Weerd P, Desimone R, Ungerleider LG. 1999. Increased activity in human visual cortex during directed attention in the absence of visual stimulation. *Neuron* 22:751–61

Kleinschmidt A, Buchel C, Hutton C, Friston KJ, Frackowiak RS. 2002. The neural structures expressing perceptual hysteresis in visual letter recognition. *Neuron* 34:659–66

Kobatake E, Tanaka K. 1994. Neuronal selectivities to complex object features in the ventral visual pathway of the macaque cerebral cortex. *J. Neurophysiol.* 71:856–67

Kourtzi Z, Erb M, Grodd W, Bulthoff HH. 2003a. Representation of the perceived 3-D object shape in the human lateral occipital complex. *Cereb. Cortex* 13:911–20

Kourtzi Z, Kanwisher N. 2000. Cortical regions involved in perceiving object shape. *J. Neurosci.* 20:3310–18

Kourtzi Z, Kanwisher N. 2001. Representation of perceived object shape by the human lateral occipital complex. *Science* 293:1506–9

Kourtzi Z, Tolias AS, Altmann CF, Augath M, Logothetis NK. 2003b. Integration of local features into global shapes: monkey and human FMRI studies. *Neuron* 37:333–46

Kriegeskorte N, Sorger B, Naumer M, Schwarzbach J, van den Boogert E, et al. 2003. Human cortical object recognition from a visual motion flowfield. *J. Neurosci.* 23:1451–63

Lamme VA, Roelfsema PR. 2000. The distinct modes of vision offered by feedforward and recurrent processing. *Trends Neurosci.* 23:571–79

Lavie N, Driver J. 1996. On the spatial extent of attention in object-based visual selection. *Percept. Psychophys.* 58:1238–51

Lerner Y, Grill-Spector K, Kushnir T, Malach R. 1999. Holistic versus part-based representation in the human brain. *Neurosci. Lett.* S26

Lerner Y, Hendler T, Ben-Bashat D, Harel M, Malach R. 2001. A hierarchical axis of object processing stages in the human visual cortex. *Cereb. Cortex* 11:287–97

Lerner Y, Hendler T, Malach R. 2002. Object-completion effects in the human lateral occipital complex. *Cereb. Cortex* 12:163–77

Lerner Y, Pianka P, Azmon B, Leiba H, Stolovitch C, et al. 2003. Area-specific amblyopic effects in human occipitotemporal object representations. *Neuron* 5:1023–29

Levy I, Hasson U, Avidan G, Hendler T, Malach R. 2001. Center-periphery organization of human object areas. *Nat. Neurosci.* 4:533–39

Levy I, Hasson U, Harel M, Malach R. 2004. Functional analysis of the periphery effect in human building-related areas. *Hum. Brain Mapp.* In press

Liu J, Harris A, Kanwisher N. 2002. Stages of processing in face perception: an MEG study. *Nat. Neurosci.* 5:910–16

Logothetis N. 2000. Can current fMRI techniques reveal the micro-architecture of cortex? *Nat. Neurosci.* 3:413–14 taf/DynaPage.taf?file = /neuro/journal/v3/n5/full/nn0500_413.html, taf/DynaPage.taf?file=/neuro/journal/v3/n5/abs/nn_413.html

Logothetis NK, Pauls J, Poggio T. 1995. Shape representation in the inferior temporal cortex of monkeys. *Curr. Biol.* 5:552–63

Lueck CJ, Zeki S, Friston KJ, Deiber MP, Cope P, et al. 1989. The colour centre in the cerebral cortex of man. *Nature* 340:386–89

Macaluso E, Frith CD, Driver J. 2000. Modulation of human visual cortex by crossmodal spatial attention. *Science* 289:1206–8

Maguire EA, Frith CD, Cipolotti L. 2001. Distinct neural systems for the encoding and recognition of topography and faces. *Neuroimage* 13:743–50

Malach R. 1994. Cortical columns as devices for maximizing neuronal diversity. *Trends Neurosci.* 17:101–4

Malach R, Levy I, Hasson U. 2002. The topography of high-order human object areas. *Trends Cogn. Sci.* 6:176–84

Malach R, Reppas JB, Benson RR, Kwong KK, Jiang H, et al. 1995. Object-related activity revealed by functional magnetic resonance imaging in human occipital cortex. *Proc. Natl. Acad. Sci. USA* 92:8135–39

Martin A, Wiggs CL, Ungerleider LG, Haxby JV. 1996. Neural correlates of category-specific knowledge. *Nature* 379:649–52

Maunsell JH, Newsome WT. 1987. Visual processing in monkey extrastriate cortex. *Annu. Rev. Neurosci.* 10:363–401

McKeefry DJ, Zeki S. 1997. The position and topography of the human colour centre as revealed by functional magnetic resonance imaging. *Brain* 120:2229–42

Mendola JD, Dale AM, Fischl B, Liu AK, Tootell RB. 1999. The representation of illusory and real contours in human cortical visual areas revealed by functional magnetic resonance imaging. *J. Neurosci.* 19:8560–72

Merriam EP, Genovese CR, Colby CL. 2003. Spatial updating in human parietal cortex. *Neuron* 39:361–73

Mishkin M, Ungerleider LG, Macko KA. 1983. Object vision and spatial vision: two cortical pathways. *Trends Neurosci.* 6:414–17

Moore C, Engel SA. 2001. Neural response to perception of volume in the lateral occipital complex. *Neuron* 29:277–86

Morrone MC, Tosetti M, Montanaro D, Fiorentini A, Cioni G, Burr DC. 2000. A cortical area that responds specifically to optic flow, revealed by fMRI. *Nat. Neurosci.* 3:1322–28

Muckli L, Kriegeskorte N, Lanfermann H, Zanella FE, Singer W, Goebel R. 2002. Apparent motion: event-related functional magnetic resonance imaging of perceptual switches and States. *J. Neurosci.* 22:RC219

Mukamel R, Harel M, Hendler T, Malach RS. 2003. *Enhanced temporal non-linearities in human object-related occipito-temporal cortex.* Presented at Soc. Neurosci., New Orleans

O'Craven KM, Downing PE, Kanwisher N. 1999. fMRI evidence for objects as the units of attentional selection. *Nature* 401:584–87

Pascual-Leone A, Walsh V. 2001. Fast back-projections from the motion to the primary visual area necessary for visual awareness. *Science* 292:510–12

Pearlman AL, Birch J, Meadows JC. 1979. Cerebral color blindness: an acquired defect in hue discrimination. *Ann. Neurol.* 5:253–61

Pelphrey KA, Mitchell TV, McKeown MJ, Goldstein J, Allison T, McCarthy G. 2003. Brain activity evoked by the perception of human walking: controlling for meaningful coherent motion. *J. Neurosci.* 23:6819–25

Peterhans E, von der Heydt R. 1991. Subjective contours–bridging the gap between psychophysics and physiology. *Trends Neurosci.* 14:112–19

Press WA, Brewer AA, Dougherty RF, Wade AR, Wandell BA. 2001. Visual areas and spatial summation in human visual cortex. *Vision Res.* 41:1321–32

Puce A, Allison T, Asgari M, Gore JC, McCarthy G. 1996. Differential sensitivity of human visual cortex to faces, letterstrings, and textures: a functional magnetic resonance imaging study. *J. Neurosci.* 16:5205–15

Puce A, Allison T, Gore JC, McCarthy G. 1995. Face-sensitive regions in human extrastriate cortex studied by functional MRI. *J. Neurophysiol.* 74:1192–99

Puce A, Perrett D. 2003. Electrophysiology and brain imaging of biological motion. *Philos. Trans. R. Soc. London Ser. B Biol. Sci.* 358:435–45

Rees G, Friston K, Koch C. 2000. A direct quantitative relationship between the functional properties of human and macaque V5. *Nat. Neurosci.* 3:716–23

Ress D, Backus BT, Heeger DJ. 2000. Activity in primary visual cortex predicts performance in a visual detection task. *Nat. Neurosci.* 3:940–45

Ress D, Heeger DJ. 2003. Neuronal correlates of perception in early visual cortex. *Nat. Neurosci.* 6:414–20

Rizzolatti G, Craighero L. 2004. The mirror-neuron system. *Annu. Rev. Neurosci.* 27:169–92

Salzman CD, Murasugi CM, Britten KH, Newsome WT. 1992. Microstimulation in visual area MT: effects on direction discrimination performance. *J. Neurosci.* 12:2331–55

Schein SJ, Desimone R. 1990. Spectral properties of V4 neurons in the macaque. *J. Neurosci.* 10:3369–89

Schiller PH. 1996. On the specificity of neurons and visual areas. *Behav. Brain Res.* 76:21–35

Schwartz E, Tootell RB, Silverman MS, Switkes E, De Valois RL. 1985. On the mathematical structure of the visuotopic mapping of macaque striate cortex. *Science* 227:1065–66

Sereno MI, Dale AM, Reppas JB, Kwong KK, Belliveau JW, et al. 1995. Borders of multiple visual areas in humans revealed by functional magnetic resonance imaging. *Science* 268:889–93

Sereno MI, Pitzalis S, Martinez A. 2001. Mapping of contralateral space in retinotopic coordinates by a parietal cortical area in humans. *Science* 294:1350–54

Shipp S, Zeki S. 1985. Segregation of pathways leading from area V2 to areas V4 and V5 of macaque monkey visual cortex. *Nature* 315:322–25

Sigala N, Logothetis NK. 2002. Visual categorization shapes feature selectivity in the primate temporal cortex. *Nature* 415:318–20

Smith AT, Greenlee MW, Singh KD, Kraemer FM, Hennig J. 1998. The processing of first- and second-order motion in human visual cortex assessed by functional magnetic resonance imaging (fMRI). *J. Neurosci.* 18:3816–30

Smith AT, Singh KD, Williams AL, Greenlee MW. 2001. Estimating receptive field size from fMRI data in human striate and extrastriate visual cortex. *Cereb. Cortex* 11:1182–90

Spiridon M, Kanwisher N. 2002. How distributed is visual category information in human occipito-temporal cortex? an fMRI study. *Neuron* 35:1157–65

Stanley DA, Rubin N. 2003. fMRI activation in response to illusory contours and salient regions in the human lateral occipital complex. *Neuron* 37:323–31

Tanaka K. 1993. Neuronal mechanisms of object recognition. *Science* 262:685–88

Tarr MJ, Gauthier I. 2000. FFA: a flexible fusiform area for subordinate-level visual processing automatized by expertise. *Nat. Neurosci.* 3:764–69

Thorpe S, Fize D, Marlot C. 1996. Speed of processing in the human visual system. *Nature* 381:520–22

Tong F, Nakayama K, Vaughan JT, Kanwisher N. 1998. Binocular rivalry and visual awareness in human extrastriate cortex. *Neuron* 21:753–59

Tootell RB, Dale AM, Sereno MI, Malach R. 1996. New images from human visual cortex. *Trends Neurosci.* 19:481–89

Tootell RB, Hadjikhani N. 2000. Attention—brains at work! *Nat. Neurosci.* 3:206–8

Tootell RB, Reppas JB, Dale AM, Look RB, Sereno MI, et al. 1995a. Visual motion aftereffect in human cortical area MT revealed by functional magnetic resonance imaging. *Nature* 375:139–41

Tootell RB, Reppas JB, Kwong KK, Malach R, Born RT, et al. 1995b. Functional analysis of human MT and related visual cortical areas using magnetic resonance imaging. *J. Neurosci.* 15:3215–30

Tootell RB, Taylor JB. 1995. Anatomical evidence for MT and additional cortical visual areas in humans. *Cereb. Cortex* 5:39–55

Tootell RB, Tsao D, Vanduffel W. 2003. Neuroimaging weighs in: humans meet macaques in "primate" visual cortex. *J. Neurosci.* 23:3981–89

Vaina LM, Solomon J, Chowdhury S, Sinha P, Belliveau JW. 2001. Functional neuroanatomy of biological motion perception in humans. *Proc. Natl. Acad. Sci. USA* 98: 11656–61

Van Essen DC, Lewis JW, Drury HA, Hadjikhani N, Tootell RB, et al. 2001. Mapping visual cortex in monkeys and humans using surface-based atlases. *Vision Res.* 41:1359–78

Van Oostende S, Sunaert S, Van Hecke P, Marchal G, Orban GA. 1997. The kinetic occipital (KO) region in man: an fMRI study. *Cereb. Cortex* 7:690–701

Vuilleumier P, Armony J, Clarke K, Husain M, Driver J, Dolan R. 2002a. Neural response to emotional faces with and without awareness: event-related fMRI in a parietal patient with visual extinction and spatial neglect. *Neuropsychologia* 40:2156

Vuilleumier P, Henson RN, Driver J, Dolan RJ. 2002b. Multiple levels of visual object constancy revealed by event-related fMRI of repetition priming. *Nat. Neurosci.* 5:491–99

Wachsmuth E, Oram MW, Perrett DI. 1994. Recognition of objects and their component parts: responses of single units in the temporal cortex of the macaque. *Cereb. Cortex* 4:509–22

Wade AR, Brewer AA, Rieger JW, Wandell BA. 2002. Functional measurements of human ventral occipital cortex: retinotopy and colour. *Philos. Trans. R. Soc. London Ser. B Biol. Sci.* 357:963–73

Wade AR, Wandell BA. 2002. Chromatic light adaptation measured using functional magnetic resonance imaging. *J. Neurosci.* 22:8148–57

Wandell B. 1999a. Computational neuroimaging: color representations and processing. In *New Cognitive Neuroscience*, ed. M Gazzaniga, pp. 291–304. Cambridge, MA: MIT Press

Wandell BA. 1999b. Computational neuroimaging of human visual cortex. *Annu. Rev. Neurosci.* 22:145–73

Watson JD, Myers R, Frackowiak RS, Hajnal JV, Woods RP, et al. 1993. Area V5 of the human brain: evidence from a combined study using positron emission tomography and magnetic resonance imaging. *Cereb. Cortex* 3:79–94

Wojciulik E, Kanwisher N, Driver J. 1998. Covert visual attention modulates face-specific activity in the human fusiform gyrus: fMRI study. *J. Neurophysiol.* 79:1574–78

Zeki S. 1990. A century of cerebral achromatopsia. *Brain* 113(Pt. 6):1721–77

Zeki S, Perry RJ, Bartels A. 2003. The processing of kinetic contours in the brain. *Cereb. Cortex* 13:189–202

Zeki S, Watson JD, Lueck CJ, Friston KJ, Kennard C, Frackowiak RS. 1991. A direct demonstration of functional specialization in human visual cortex. *J. Neurosci.* 11:641–49

Annu. Rev. Neurosci. 2004. 27:679–96
doi: 10.1146/annurev.neuro.27.070203.144343

VISUAL MOTOR COMPUTATIONS IN INSECTS

Mandyam V. Srinivasan and Shaowu Zhang
Center for Visual Science, Research School of Biological Sciences, Australian National University, P.O. Box 475, Canberra, A.C.T. 2601, Australia; email: M.Srinivasan@anu.edu.au, swzhang@rsbs.anu.edu.au

Key Words fly, bee, vision, navigation, behavior

■ **Abstract** With their relatively simple nervous systems and purpose-designed behaviors and reflexes, insects are an excellent organism in which to investigate how visual information is acquired and processed to guide locomotion and navigation. Flies maintain a straight course and monitor their motion through the environment by sensing the patterns of optic flow induced in the eyes. Bees negotiate narrow gaps by balancing the speeds of the images in their two eyes, and they control flight speed by holding constant the average image velocity as seen with their two eyes. Bees achieve a smooth landing on a horizontal surface by holding the image velocity of the surface constant during approach, thus ensuring that flight speed is automatically close to zero at touchdown. Foraging bees estimate the distance that they have traveled to reach a food source by integrating the optic flow experienced en route; this integration gives them a visually driven "odometer." Insects have also evolved sophisticated visuomotor mechanisms for pursuing prey or mates and possibly for concealing their own motion while shadowing objects of interest.

INTRODUCTION

A glance at a fly evading a rapidly descending hand or orchestrating a flawless landing on the rim of a teacup would convince even the most sceptical observer that this insect possesses exquisite visuomotor control, despite its small brain and relatively simple nervous system.

Most insects have compound eyes consisting of a large number of facets, or ommatidia. Each ommatidium includes a lens that focuses light onto a small group of six to nine photoreceptors, and each lens has a visual field that is typically a few degrees in width. The visual axes of neighboring ommatidia are separated by a few degrees, so that the two compound eyes (each containing a few thousand ommatidia in a fly or honeybee, for example) together provide a near-panoramic view of the environment (Wehner 1981). The optics of insect eyes are quite different from those of our own so-called simple eyes, each of which has a single lens and a retina containing approximately one hundred million photoreceptors. In contrast to the composition of insects' eyes, this arrangement endows humans with restricted peripheral vision, but the strong overlap between the visual fields of the two eyes enables good quality stereo vision.

The visual systems of insects differ from those of humans in ways that have more profound consequences for vision and behavior. Unlike vertebrates, insects have immobile eyes with fixed-focus optics. They cannot infer the distance of an object from the extent to which the directions of gaze must converge to view the object, nor can they judge distance by monitoring the refractive power that is required to bring the object's image into focus on the retina. Compared with human eyes, the eyes of insects are positioned much closer together, and they possess inferior spatial acuity. Even if an insect had the neural apparatus required for binocular stereopsis, such a mechanism would be relatively imprecise and restricted to ranges of a few centimeters (Collett & Harkness 1982, Horridge 1987, Rossell 1983, Srinivasan 1993). Not surprisingly, therefore, insects have evolved alternative visual strategies for guiding behavior and locomotion in their three-dimensional world. Many of these strategies use cues derived from the image motion that insects experience as they move in their environment. Vision in insects is a very active process in which perception and action are tightly coupled.

This review outlines a few examples of visuomotor control in insects and highlights the underlying principles. It is by no means exhaustive, but it does provide references to additional topics and more complete accounts.

STABILIZING FLIGHT

For insects, vision provides an important sensory input for flight stabilization. If an insect flying along a straight line is blown to the left by a gust of wind, the image on its frontal retina moves to the right. This causes the flight motor system to generate a counteractive yaw torque that brings the insect back on course (Reichardt 1969). Similar control mechanisms act to stabilize pitch and roll (e.g., Srinivasan 1977). This optomotor response (Reichardt 1969) is an excellent experimental paradigm with which to probe the neural mechanisms underlying motion detection. Largely through studies of the optomotor response, we now know that flies sense the direction of image movement by correlating the intensity variations registered by neighboring ommatidia in the compound eye (Reichardt 1969). Thus, the front end of the movement-detecting pathway consists of an array of elementary movement detectors (EMDs) that perform these correlations. Different sets of EMDs are used to detect motion in various directions by correlating signals from ommatidia that are appropriately positioned relative to each other. During the past 30 years, researchers have discovered several motion-sensitive neurons with large visual fields, each responding preferentially to motion in a specific direction (Hausen 1993, Hausen & Egelhaaf 1989) or to the fly's rotation around a specific axis (Krapp 2000, Krapp & Hengstenberg 1996). These neurons derive their sensitivity and selectivity by pooling signals from EMDs that have the appropriate directional selectivity in different regions of the compound eye. They are likely to play an important role in stabilizing flight and providing the insect with a visually kinesthetic sense. The properties of these motion-sensitive neurons have been reviewed extensively (e.g., Egelhaaf & Borst 1993, Hausen 1993) and we do not elaborate here.

It is generally supposed that motion-sensitive neurons play an important role in the insect's "autopilot" mechanism by detecting deviations from the intended course and generating corrective turning commands. However, the precise means by which course stabilization is achieved remains elusive. Motion-sensitive neurons possess large visual fields, each typically covering most of one eye. Therefore, steering a straight course can only be achieved by balancing the responses of two neurons, each sensitive to front-to-back motion in one eye. Such a scheme works well only when the insect is flying in a symmetrically structured environment. It does not work when the insect flies along a cliff, for example, because the eye that faces the cliff experiences substantially greater image motion than does the contralateral eye. The only way to steer a straight course in an asymmetrical world (which is more often the rule than the exception) is to sense and compensate for image motion in only a small patch of the visual field that faces the direction along which the insect wishes to fly—the frontal visual field, for example, if the objective is to fly straight ahead. Behavioral evidence suggests that hoverflies adopt just such a strategy (Collett 1980). When flying straight ahead, they minimize image motion within a small visual field that looks in the forward direction. When flying obliquely (as hoverflies often do), they minimize image motion within a small visual field that looks in the appropriate, oblique direction (Collett 1980). However, the neural basis of such steering, which requires visuomotor control via an array of motion-sensitive neurons with small visual fields, remains undiscovered.

In flies, course control and stabilization are also aided by the halteres, small hind-wings that oscillate in antiphase with the main wings and act like miniature gyroscopes to provide information on the body's rotation (Dickinson 1999, Nalbach 1993, Nalbach & Hengstenberg 1994). The halteres sense and compensate for rapid rotations, whereas the optomotor reflexes deal with the slower turns (Hengstenberg 1993, Sherman & Dickinson 2003). Long-term course control is aided by a celestial compass (Wehner 1997) and by the use of prominent landmarks or beacons in the environment (Collett & Zeil 1998).

Visual stabilization of roll and pitch are aided by the ocelli, three single-lens eyes situated on top of the head. Each ocellus has a relatively large visual field, more than 40° in width. The two laterally directed ocelli stabilize roll by monitoring the position of the horizon on either side. The medial ocellus stabilizes pitch by monitoring the elevation of the horizon in the frontal field (Stange 1981, Stange & Howard 1979, Stange et al. 2002, Wilson 1978). The neural pathways mediating these reflexes remain to be investigated.

NEGOTIATING NARROW GAPS

When a bee flies through a hole in a window, it tends to fly through the center, balancing the distances to the left and right boundaries of the opening. How does it gauge and balance the distances to the two rims, given that it does not possess stereo vision?

One possibility is that the bee does not measure distances at all, but simply balances the speeds of image motion experienced by its eyes while flying through the opening. To investigate this possibility, Kirchner & Srinivasan (1989) trained bees to enter an apparatus that offered a reward of sugar solution at the end of a tunnel. Each side wall had a pattern consisting of a vertical black-and-white grating (Figure 1). The grating on one wall could be moved horizontally either toward or away from the reward at any speed. After the bees had received several rewards with the gratings stationary, they were filmed from above as they flew along the tunnel. When both gratings were stationary, the bees tended to fly along the midline of the tunnel, i.e., equidistant from the two walls (Figure 1*a*). But when one of the gratings was moved at a constant speed in the direction of the bees' flight—thereby reducing the speed of retinal image motion on one eye relative to the other—the bees' trajectories shifted toward the side of the moving grating (Figure 1*b*). When the grating moved in a direction opposite to that of the bees' flight—thereby increasing the speed of retinal image motion on one eye relative to the other—the bees' trajectories shifted away from the side of the moving grating (Figure 1*c*). When the walls were stationary, the bees maintained equidistance by balancing the speeds of the retinal images experienced by its two eyes. A lower image speed on one eye caused the bee to move closer to the wall seen by that eye. A higher image speed, on the other hand, had the opposite effect. Experiments in which the contrasts and the periods of the gratings on the two sides were varied revealed that this centering response is rather robust to variations in these parameters: Bees continued to fly through the middle of the tunnel even when the contrasts of the gratings on the two sides were substantially different or when their periods varied by a factor of as much as four (Srinivasan et al. 1991). These findings suggest that the bee's visual system is capable of computing the speed of the image of a grating independently of its contrast and spatial-frequency content (Srinivasan et al. 1991).

The neural basis of this capacity remains to be discovered, although there is now some evidence that certain neurons in the visual pathways of the fly (Dror et al. 2001) and honeybee (Ibbotson 2001) encode image speed independently of the spatial texture of the image. The movement-detecting subsystem that mediates the centering response appears to be different, qualitatively as well as quantitatively, from the subsystem that mediates the optomotor response (Srinivasan et al. 1993, Srinivasan & Zhang 1997).

CONTROLLING FLIGHT SPEED

Do insects control the speed of their flight, and if so, how? Work by David (1982) and by Srinivasan et al. (1996) suggests that flight speed is controlled by monitoring the velocity of the image of the environment.

David (1982) observed fruit flies flying upstream in a wind tunnel, lured by the scent of fermenting banana. The inside wall of the cylindrical wind tunnel was lined with a helical black-and-white striped pattern so that rotation of the cylinder

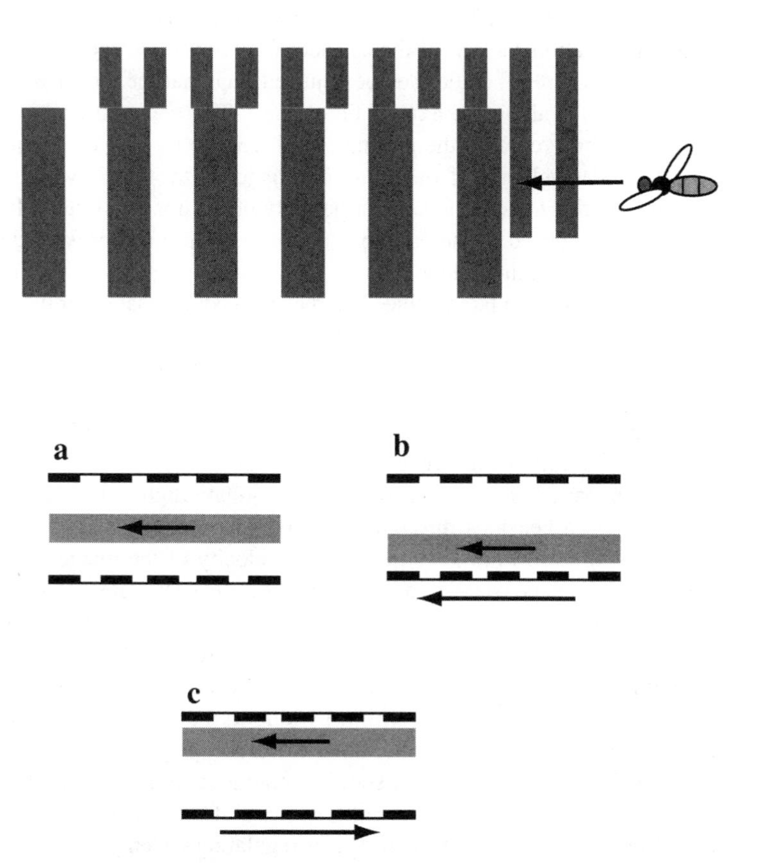

Figure 1 Experiment investigating the centering response of bees. Bees are trained to fly through a tunnel 40 cm long, 12 cm wide, and 20 cm high to collect a reward placed at the far end. The flanking walls of the tunnel are lined with vertical black-and-white gratings of period 5 cm. The flight trajectories of bees, as recorded by a video camera positioned above the tunnel, are shown (*a–c*). In each panel, the shaded area represents the mean and standard deviation of the positions of the flight trajectories, analyzed from recordings of several hundred flights. The dark bars represent the black stripes of the patterns on the walls. The small arrow indicates the direction of bee flight, and the large arrow represents the direction of pattern movement, if any. When the patterns on the walls are stationary, bees tend to fly close to the midline of the tunnel (*a*). When the pattern on one of the walls is in motion, however, bees tend to fly closer to that wall if the pattern moves in the same direction as the bee (*b*) and farther away from that wall if the pattern moves in the opposite direction (*c*). These results indicate that bees balance the distances to the walls of the tunnel by balancing the speeds of image motion that are experienced by their eyes. Adapted from Srinivasan et al. 1991.

around its axis produced apparent movement of the stripes toward the front or the back. The rotational speed of the cylinder (and hence the speed of the backward motion of the pattern) could then be adjusted such that the fly was stationary (i.e., it did not move along the axis of the tunnel). The apparent backward speed of the pattern then revealed the ground speed that the fly was maintaining, as well as the angular velocity of the pattern's image on the flies' eyes. In this setup, fruit flies tended to hold the angular velocity of the image constant. Increasing or decreasing the speed of the pattern caused the fly to move backward or forward, respectively, along the tunnel at a rate such that the angular velocity of the image on the eye stayed at a fixed value. The flies also compensated for headwind in the tunnel, increasing or decreasing their thrust to maintain the same ground speed (as indicated by the angular velocity of image motion on the eye). Experiments in which the angular period of the stripes was varied revealed that the flies were measuring (and holding constant) the angular velocity of the image on the eye, irrespective of the spatial structure of the image.

Bees appear to use a similar strategy to regulate flight speed (Srinivasan et al. 1996). When a bee flies through a tapered tunnel, it decreases its flight speed as the tunnel narrows to keep the angular velocity of the image of the walls, as seen by the eye, constant at approximately 320°/s (Figure 2, see color insert). This suggests that the bee controls flight speed by monitoring and regulating the angular velocity of the environment's image as represented on the eye; that is, if the tunnel's width is doubled, the bee flies twice as fast. On the other hand, a bee flying through a tunnel of uniform width does not change its speed when the spatial period of the stripes lining the walls is abruptly changed (Srinivasan et al. 1996). This indicates that flight speed is regulated by a visual motion-detecting mechanism that measures the angular velocity of the image independently of its spatial structure. In this respect, the speed-regulating system is similar to the system that mediates the centering response described above. Controlling flight speed by regulating image speed allows the insect to automatically slow down to a safer speed when negotiating a narrow passage or a cluttered environment.

COLLISION AVOIDANCE

When an object or surface is approached head-on, its image expands in the eye of the observer (Figure 3*a*, see color insert). This visual cue is used by cruising insects to avoid imminent collisions with surfaces or obstacles. Tammero & Dickinson (2002) filmed and analyzed the trajectories of fruit flies flying in a cylindrical arena lined with a random visual texture. The analysis revealed that the insect consistently turned away from whichever eye experienced greater image expansion, when this expansion exceeded a certain threshold. This reaction ensures that the insect turns away from objects that are dangerously close. Thus, certain neural mechanisms in the insect visual pathway are tuned to detect local image expansion. Such neurons have been found in flies (Borst 1991) and locusts (Gabbiani et al. 2001, Gray et al. 2001, Judge & Rind 1997), and they may be involved in visual sensing to control landing as well as avoid obstacles. However, at least in the fruit fly, the two behaviors may involve partly different neural pathways (Tammero & Dickinson 2002) as one proceeds downstream from the sensory neurons toward the motor command neurons.

ORCHESTRATING SMOOTH LANDINGS

When a flying insect approaches an object to land on it, the insect needs to monitor the expanding image precisely (Figure 3*a*) so that flight speed may be reduced and the forelegs extended in time. Mistakes could lead to unpleasant consequences. Researchers have studied the landing response, using rotating spirals or moving gratings to simulate visual expansion and measuring the strength of the landing response as the probability of foreleg extension (e.g., Braitenberg & Taddei-Ferretti 1966, Borst & Bahde 1988, Eckert & Hamdorf 1980, Tammero & Dickinson 2002). These studies suggest that the strength of the landing response depends on the spatial-frequency content and contrast of the pattern as well as the speed and duration of the pattern's expansion. According to the model proposed by Borst & Bahde (1988; also supported by recent evidence from Tammero & Dickinson 2002), the landing response is triggered when the time-accumulated output of an expansion-detecting system, based on the correlation model (Reichardt 1969), exceeds a preset threshold. Flies may determine when to initiate a landing by computing the time required to contact the object or surface on which they are about to land (Wagner 1982). When an insect approaches a planar surface in a direction normal to the plane (Figure 3*b*), the projected time to contact the surface, τ, is given by the expression $\frac{\theta}{\dot{\theta}}$, if the insect continues to fly toward the surface at a constant velocity (Lee 1976). Here θ is the direction of a visual feature X on the surface, relative to the direction of flight (Figure 3*b*), and $\dot{\theta}$ is the rate of change (increase) of this angle. Computing time to contact in this way is advantageous because it does not require knowledge (or measurement) of the animal's speed or of its distance from the surface: It only requires measurement of θ and $\dot{\theta}$. Flies commence deceleration approximately 90 ms prior to contact (Wagner 1982), suggesting that their nervous systems compute the projected time to contact.

These two strategies provide useful information on when a landing process should commence but not on what the actual landing strategy should be. For example, how rapidly should the insect decelerate, once the landing process is initiated? This question was addressed by Srinivasan et al. (2000b), who filmed bees as they landed on a horizontal surface. On horizontal surfaces, bees usually perform grazing landings in which trajectories are inclined to the surface at an angle that is considerably smaller than 45°. A perpendicular surface approach generates strong looming (image expansion) cues that could be used to control the deceleration of flight. But looming cues are weak when a bee performs a grazing landing, where the motion of the image of the surface is dominated by a strong, front-to-back translatory component in the ventral visual field of the eye.

To investigate how bees execute grazing landings, Srinivasan et al. (1996, 2000b) trained bees to collect a reward of sugar water on a visually textured, horizontal surface. The reward was then removed, and the landings that the bees made on the surface in search of the food were filmed in three dimensions. Analysis of the landing trajectories revealed that the forward speed of the bee decreases steadily as the bee approaches the surface (Figure 4). In fact, the speed of flight is approximately proportional to the height above the surface, indicating that the bee is holding the angular velocity of the surface's image approximately constant as the surface is approached. Analysis of 26 landing trajectories revealed an average image angular velocity of $500°/s \pm 270°/s$, where the variation reflects the fact that each animal maintained a different (but constant) image velocity (Srinivasan et al. 2000b).

Holding the image velocity of the ground constant during landing may be a simple way of decreasing the flight speed progressively (and automatically) and ensuring that its value is close to zero at touchdown. The advantage of such a strategy is that the control is achieved by a simple process and without explicit knowledge of flight speed or distance from the surface.

Detailed analysis of the landing trajectories revealed two characteristic properties (Srinivasan et al. 2000b). First, the instantaneous horizontal flight speed is proportional to the instantaneous height above the surface (as described above). Thus the angular velocity of the image is being held constant as the ground is approached. Second, the instantaneous speed of descent is proportional to the instantaneous horizontal speed. This shows that both the horizontal and the vertical components of flight speed decrease as the ground is approached, and they reach zero simultaneously at touchdown, thus ensuring a safe landing. A mathematical model of the landing process incorporating the above two properties predicts that during landing (*a*) the height should decrease exponentially as a function of time and (*b*) the horizontal distance traveled should also be an exponential function of time. Measurement of the time courses of height and horizontal travel distance in actual landings shows that the data are in excellent agreement with the predictions of the model (Figure 5).

The value of the image angular velocity that is maintained during landing ($\sim 500°/s \pm 270°/s$) is not significantly different from that observed during cruising flight (320°/s) (see Controlling Flight Speed, above). It is possible, therefore, that cruising and landing are controlled by the same, or very similar, visuomotor mechanisms. The only difference between cruising and landing is that during landing the approach is directed toward a surface, causing the bee to slow down automatically as the surface is approached. In principle, the strategy discussed here could also be used to control landing in a head-on approach toward a surface, if the landing insect holds the rate of image expansion constant as the surface is approached.

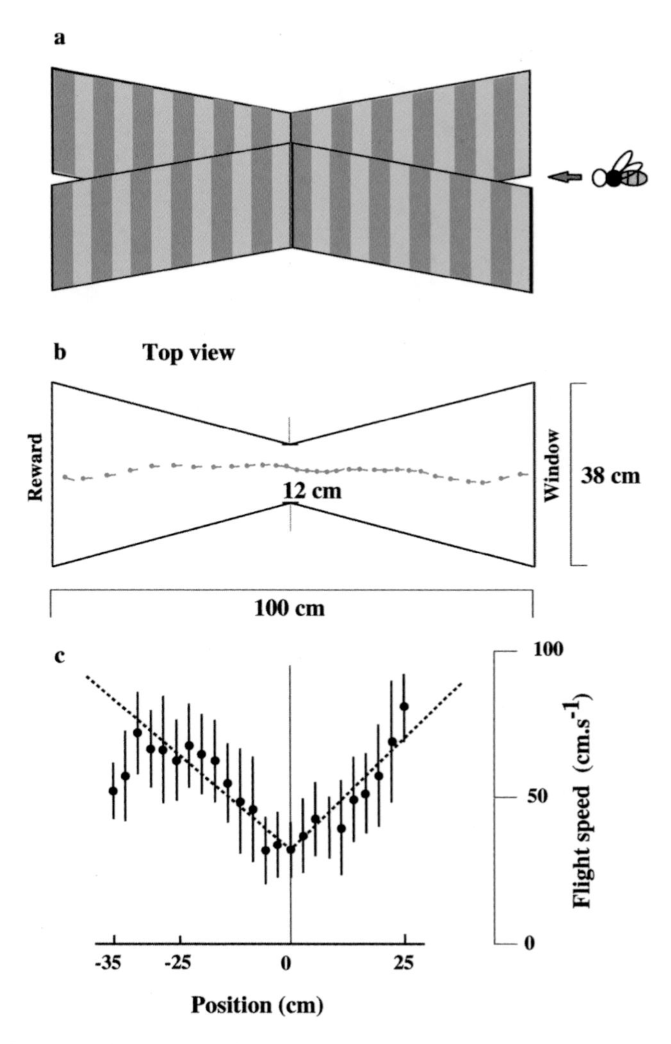

Figure 2 Experiment investigating visual control of flight speed. (*a*) Bees are trained to fly through a tapered tunnel to collect a reward placed at the far end. The walls of the tunnel are lined with vertical black-and-white gratings of period 6 cm. (*b*) A typical flight trajectory, as filmed from above by a video camera, where the bee's position and orientation are shown every 50 msec. (*c*) Mean and standard deviation of flight speeds measured at various locations along the tunnel (data from 18 flights). The dashed line represents the theoretically expected flight speed profile if the bees hold the angular velocity of the images of the walls constant at 320°/s as they fly through the tunnel. The data indicate that bees control flight speed by holding constant the angular velocity of the image of the environment. Adapted from Srinivasan et al. 1996.

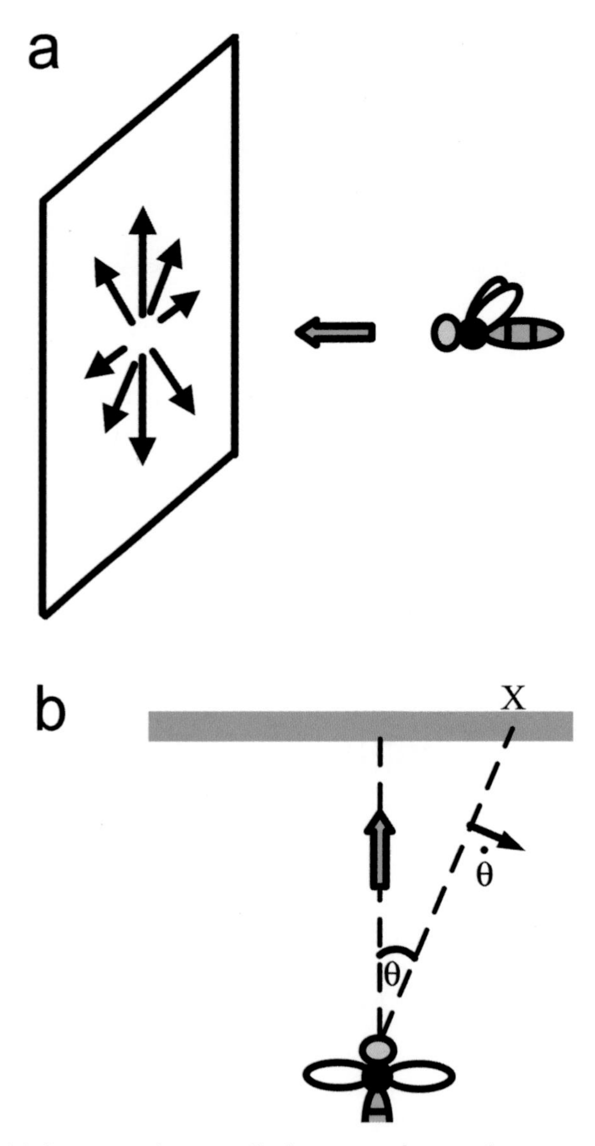

Figure 3 (*a*) An approach perpendicular to a surface produces expansion of the image. (*b*) The time to contact the surface, if the insect were to continue to fly toward the surface at the same speed, is given by the ratio $\frac{\theta}{\dot{\theta}}$, where θ is the direction of a feature X on the surface relative to the direction of flight and $\dot{\theta}$ is the rate of increase of this angle.

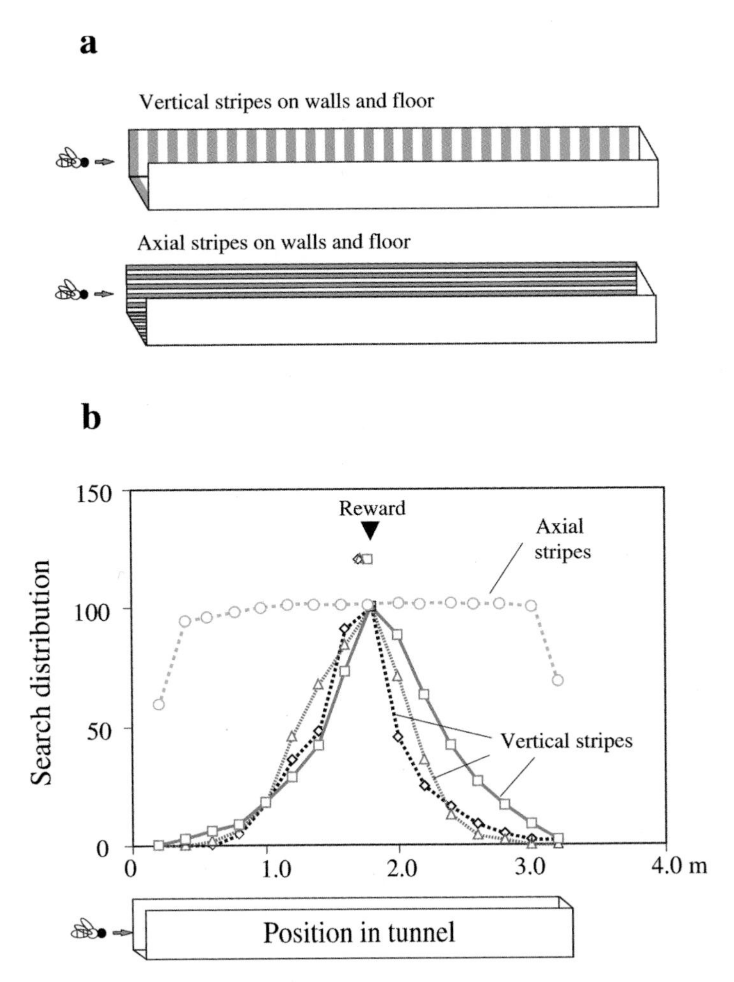

Figure 6 (*a*) Experiment investigating how honeybees gauge distance flown to a food source. Bees were trained to find a food reward placed at a distance of 1.7 m from the entrance of a 3.2-m-long tunnel of width 22 cm and height 20 cm. The tunnel was lined with vertical black-and-white gratings of period 4 cm. (*b*) When the trained bees were tested in a fresh tunnel with the reward absent, they searched at the former location of the feeder, as shown by the bell-shaped search distributions. This is true irrespective of whether the period of the grating was 4 cm (as in the training) (*blue squares*), 8 cm (*red triangles*), or 2 cm (*black diamonds*). The inverted triangle shows the former location of the reward, and the symbols below it depict the mean values of the search distributions in each case. Bees lose their ability to estimate the distance of the feeder when image-motion cues are removed by lining the tunnel with axial (rather than vertical) stripes (*circles*). These experiments and others (Srinivasan et al. 1997) demonstrate that distance flown is estimated visually, by integrating over time the image velocity that is experienced during the flight, and the honeybee's odometer measures image velocity independently of image structure. Adapted from Srinivasan et al. 1997.

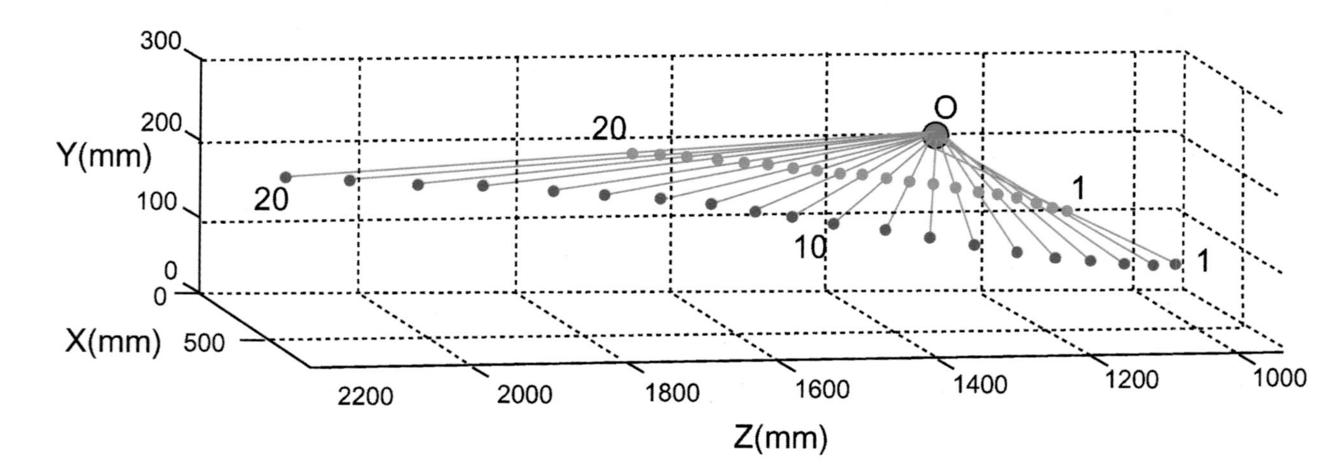

Figure 8 Illustration of motion camouflage in dragonflies. The dots correspond to the locations of the shadower (*blue dots*) and the "shadowee" (*red dots*) in successive video frames. The shadower moves so that, from the point of view of the shadowee, it appears to be a stationary object positioned at O. By moving in this way, the shadower emulates a stationary object located at the intersection point. In other words, the shadower produces the same trajectory in the retina of the shadowee as would a stationary object positioned at O. By camouflaging its motion in this way, the shadower is able to track and approach the shadowee without giving itself away as a live, moving entity. From Mizutani et al. 2003.

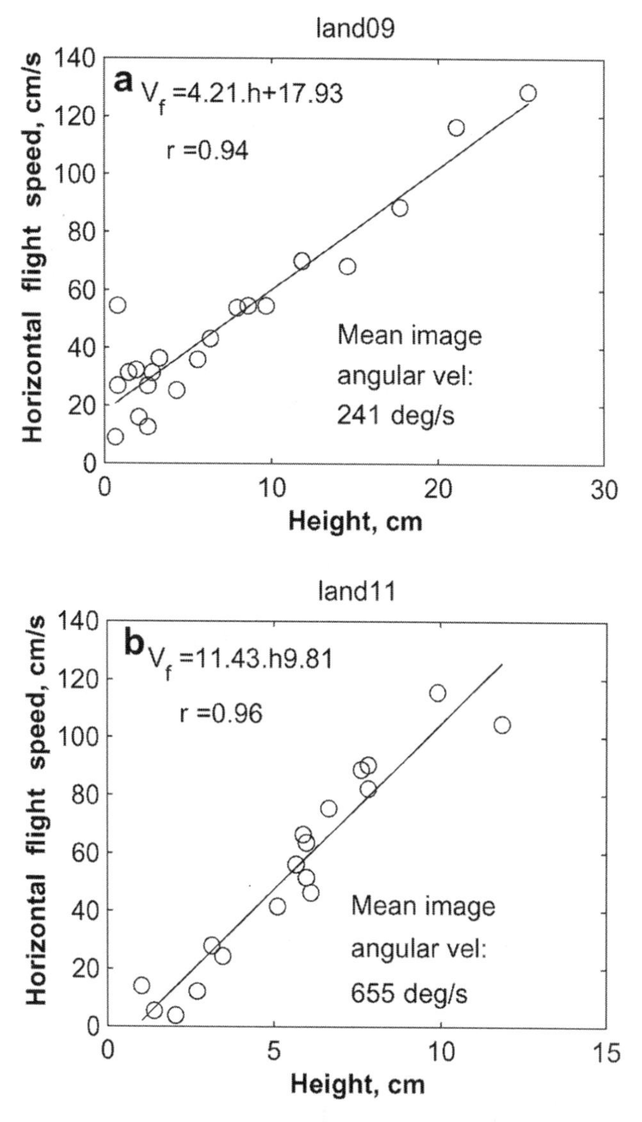

Figure 4 Experiment investigating how bees make a grazing landing on a horizontal surface. The graphs show the approximately linear relationship between horizontal flight speed (V_f) and height (h) for two bees (a, b). The landing bee holds the angular velocity of the image of the ground constant at either 241°/s (a) or 655°/s (b), as calculated from the slopes of the linear regression lines. Also shown are the values of the correlation coefficient (r). Holding the angular velocity of the image of the ground constant during the approach ensures that the landing speed is zero at touchdown. From Srinivasan et al. 2000b.

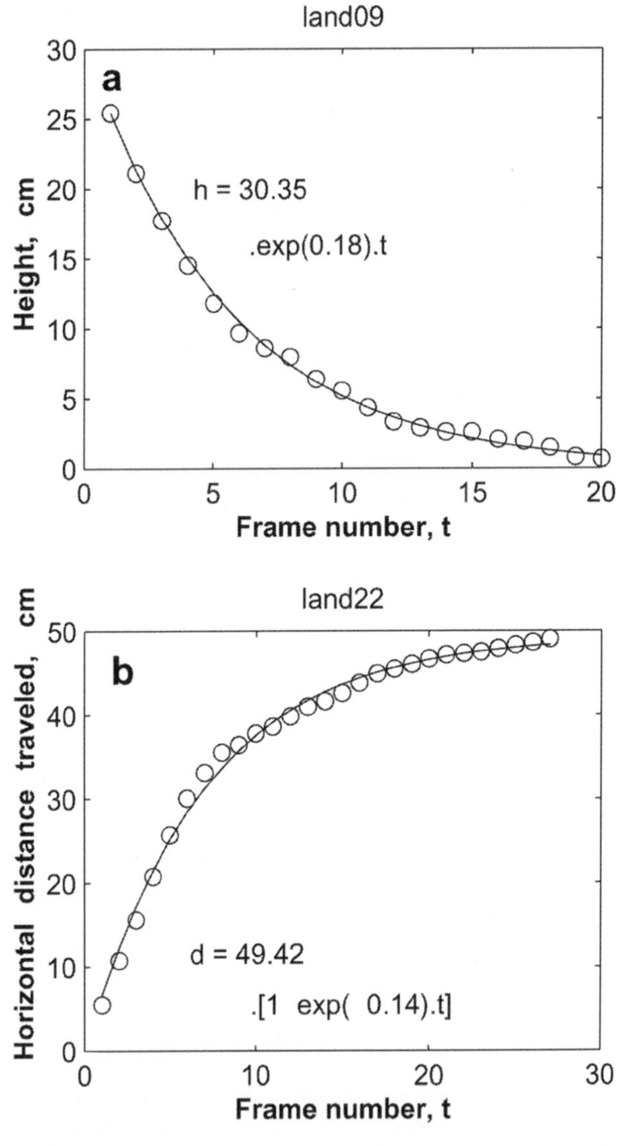

Figure 5 Analysis of honeybee landing trajectories. Variation of height (h) versus time and variation of horizontal distance traveled (d) versus time for two bees (a, b). In each panel, the circles represent experimental measurements, and the curve indicates the theoretical prediction. From Srinivasan et al. 2000b.

ESTIMATING DISTANCE FLOWN

Insects are not merely moment-to-moment navigators concerned only with avoiding collisions or making smooth landings. Honeybees, for example, can navigate accurately and repeatedly to a food source several kilometers away from their nest. Indeed, bees communicate to their nestmates the distance and direction in which to fly to reach the food, through the famous waggle dance (von Frisch 1993). But the cues bees utilize to gauge the distance flown to reach a goal have been a subject of controversy. Early studies of the waggle dance suggested that distance traveled is measured in terms of the total energy expended during flight (reviewed in von Frisch 1993). However, researchers have recently questioned this energy hypothesis, suggesting that an important cue is the extent to which the image of the environment moves in the bee's eye as it flies to the target (Esch & Burns 1995, 1996; Esch et al. 2001; Schöne 1996; Si et al. 2003; Srinivasan et al. 1996, 1997, 2000a). In other words, the honeybee's "odometer" is driven by a visual, rather than an energy-based, signal. Here we describe recent work that led to this new insight.

A few years ago, Esch & Burns (1995, 1996) investigated distance measurement by enticing honeybees to find food at a feeder placed 70 meters away from a hive in an open field. They recorded the distance as signaled by the bees when they danced to recruit other nestmates to visit the feeder. When the feeder was 70 meters away, the bees signaled the correct distance. But when the feeder was raised above the ground by attaching it to a helium balloon, the bees signaled a progressively shorter distance as the height of the balloon was increased, despite the fact that the balloon was now farther away from the hive. Esch and Burns explained this finding by proposing that the bees were gauging distance flown in terms of image motion in relation to the ground, rather than through the energy required to reach the feeder. The higher the balloon was, the lower the total amount of image motion that was experienced by the bee en route to the feeder.

This hypothesis was examined by Srinivasan et al. (1996, 1997), who investigated under controlled laboratory conditions the cues by which bees estimate and learn distances flown. Bees were trained to enter a 3.2-m-long tunnel and collect a reward of sugar solution at a feeder placed in the tunnel at a fixed distance from the entrance. The walls and floor of the tunnel were lined with black-and-white gratings perpendicular to the tunnel's axis (Figure 6*a*, see color insert). During training, the position and orientation of the tunnel were changed frequently to prevent the bees from using any external landmarks to gauge their location relative to the tunnel entrance. The bees were then tested by recording their searching behavior in an identical fresh tunnel that carried no reward and was devoid of any scent cues. In the tests, the bees showed a clear ability to search for the reward at the correct distance (see Figure 6*b*).

How did the bees gauge the distance they had flown in the tunnel? Tests were carried out to examine the participation of a variety of potential cues, including energy consumption, time of flight, airspeed integration, and inertial navigation (Srinivasan et al. 1997). Results show that the bees estimated distance flown by integrating over time the image motion of the walls as registered by their eyes while they flew through the tunnel. In one experiment (Srinivasan et al. 1997) bees were trained and tested in conditions where image motion was eliminated or reduced by using axially oriented stripes on the walls and floor of the tunnel. The bees showed no ability to gauge distance traveled: In these tests, they searched uniformly over the entire length of the tunnel, showing no tendency to stop or turn at the former location of the reward. Trained bees tended to search for the feeder at the same location in the tunnel, even if the period of the gratings lining the walls and floor was varied in the tests. These findings reveal that the odometer integrates image motion robustly and reliably over a fourfold variation in the spatial period of the grating (see Figure 6*b*).

How far do bees "think" they have flown when they return from one of these tunnels? Srinivasan et al. (2000a) and Esch et al. (2001) trained bees to fly directly from a hive into a short, narrow tunnel that was placed close to the hive entrance. The tunnel was 6 m long and 11 cm wide, and its walls and floor were lined with a random visual texture. A feeder was placed 6 m into the tunnel. The dances of bees returning from this feeder were filmed and analyzed. These bees signaled a flight distance of approximately 200 m, despite having flown only a small fraction of this distance. Thus the bees overestimated the distance they had flown in the tunnel because the proximity of the walls and floor of the tunnel greatly magnified the optic flow that they experienced in comparison with what normally occurs when foraging outdoors. This experiment reinforces the conclusion that image motion is the dominant cue that bees use to gauge how far they have traveled. The motion-detecting system that underlies distance estimation, as measured by the waggle dance, seems to be rather robust to variations in the spatial texture and contrast of the environment (Si et al. 2003).

In another experiment, Ugolini (1987) transported wasps from their nests to various sites, then released them, and observed their homing trajectories. He found that the wasps headed accurately toward their homes when they had been taken to the release site in a transparent container—and could thus observe their passage through the environment—but not when they were transported in an opaque container. These findings suggest that wasps, like bees, infer the direction and distance of their travel by observing the apparent motion of the surrounding visual panorama.

What are the consequences of monitoring travel by using visual, rather than energy-based, cues? Unlike an energy-based mechanism, a visually driven odometer is not affected by wind or by the load of nectar carried. Furthermore, it would provide a distance reading that is independent of the speed at which the bee flies to its destination, because the reading would depend only on the total amount of image motion that is registered by the eye. But, as discussed above, a visual odometer works accurately only if the bee follows the same route each time it flies to its destination (or if a follower bee adheres to the same route described by the dancing scout bee). This is because the total amount of image motion that is experienced during the trip depends not only on the distance flown, but also on how visually "tight" or "open" the environment is. Indeed, the dances of bees from a given colony exhibit substantially different distance-calibration curves when the bees are made to forage in different environments (Esch et al. 2001). The strong waggle dances of bees returning from a short, narrow tunnel illustrate this point even more dramatically. However, the unavoidable dependence of the dance on the environment may not be a problem in many natural situations because bees flying repeatedly to an attractive food source tend to remain faithful to the route they have discovered (e.g., Collett 1996). Because the dance indicates the direction of the food source as well as its distance, there is a reasonably good chance that the new recruits, which fly in the same direction as the scout that initially discovered the source, will experience the same environment and therefore fly the same distance. At present, it is not clear whether bees use optic flow information from the ventral as well as the lateral fields of view for odometry (the data on this are presently equivocal: see Si et al. 2003, Srinivasan et al. 1997). If ventral flow is important, bees need to fly at a consistent height, or to account for the height of flight in the computation, to estimate distances reproducibly. Whether bees use either of these strategies remains to be investigated.

What are the neural mechanisms by which the distance signal is computed? Where in the insect's brain is the odometer located? Currently, we have absolutely no idea. This is an intriguing area for future research.

CHASING BEHAVIOR

In addition to navigating safely in their world, insects need to interact with other living creatures to survive and propagate their genes. They must capture prey, chase after potential mates, ward off territorial intruders, and escape from predators. Although much remains to be learned about these behaviors, it is clear that vision plays a significant role in all of them.

Male houseflies, for example, will chase conspecifics with dazzling agility: Females are chased for the purpose of mating, and males are driven away in bouts of territoriality. In a pioneering study, Land & Collett (1974) filmed and analyzed these chases to formulate a model that describes the underlying visual guidance system. The combination of target and pursuer is modeled as a feedback control servomechanism in which the pursuer attempts to keep the target in his frontal (straight-ahead) field of view (see Figure 7). The dynamics of chasing behavior in the male housefly can be characterized accurately by a servomechanism that employs a proportional-derivative controller. Land & Collett (1974) found that with certain values of model parameters, the trajectories of real chases were well reproduced. Recently, it has been shown that an additional control loop, driven by the visual angle that the target subtends in the pursuer's eye, controls the forward speed of the pursuer and ensures that he does not fall too far behind the target (Boeddeker & Egelhaaf 2003, Boeddeker et al. 2003).

The dynamics of the system that mediates chasing behavior are quite different from those of the system that mediates the optomotor response (Srinivasan &

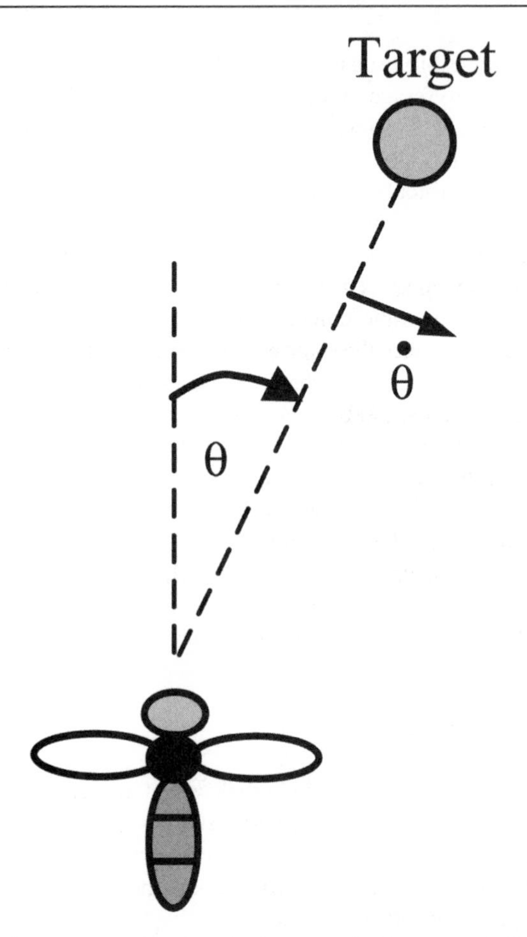

Figure 7 Visual parameters involved in the control of chasing. θ, the bearing of the target relative to the pursuer, is the error angle that the pursuer seeks to reduce to zero to achieve perfect tracking. The pursuer's turning rate, ω, is proportional to two visual parameters, the error angle, θ, and the rate of change of this error angle, $\dot{\theta}$. In other words, $\omega(t+d) = k_1\theta(t) + k_2\dot{\theta}(t)$, where k_1 and k_2 are constants of proportionality and d is a time delay that accounts for the delay between the visual stimulus and the motor response (the turning rate). With values of $k_1 = 20\ \mathrm{s}^{-1}$, $k_2 = 0.7$, and $d = 0.02$ s, the model reproduces well the trajectories of real chases (Land & Collett 1974).

Bernard 1977). Male flies carry "chasing" neurons that respond selectively to small, rapidly moving targets (Gilbert & Strausfeld 1991, Hausen & Strausfeld 1980). These neurons are distinct, both anatomically and physiologically, from the large-field optomotor neurons discussed above, and they are small or absent in female flies. Researchers will likely find chasing neurons of this kind in the males of many species of flying insects that rely on vision, rather than olfaction, to track down their mates.

MOTION CAMOUFLAGE

Observation of chasing behavior in insects provides compelling evidence that these creatures are very adept at detecting other moving objects in the environment. Given this exquisite sensitivity to movement, are there instances in which an insect may try to conceal its own movement to "sneak up" on another insect? Examples of such behavior have been documented in two insect species to date. Hoverflies (Srinivasan & Davey 1995) and dragonflies (Mizutani et al. 2003) occasionally seem to shadow conspecifics by moving in such a way that they appear to be stationary. Figure 8 (see color insert) shows an example of a dragonfly shadowing another dragonfly. Exactly how the shadower computes and executes these stealthy trajectories remains a mystery, although some possibilities are suggested by Srinivasan & Davey (1995).

CONCLUDING REMARKS

Insects are prime subjects in which to study how visual information is exploited to guide a variety of important behavioral tasks. These tasks range from short-term, moment-to-moment guidance, such as that used to stabilize attitude or avoid collisions, to long-term guidance, such as that used to navigate to a food source several kilometers away and return home safely. Flying insects exploit cues derived from image motion to stabilize flight, regulate flight speed, negotiate narrow gaps, infer the ranges to objects, avoid obstacles, orchestrate smooth landings, and monitor distances traveled.

Several motion-sensitive pathways exist in the insect visual system, each with a distinct set of properties and geared to a specific visual function. It seems unlikely, however, that all these systems (and other as yet undiscovered ones) operate continuously. The optomotor system, for instance, has to be switched off, or its corrective commands ignored, when the insect makes a voluntary turn or chases a target (Heisenberg & Wolf 1993, Kirschfeld 1997, Srinivasan & Bernard 1997). It is also impossible to land on a surface without first disabling the collision avoidance system.

Major current challenges are to discover the conditions under which individual subsystems are called into play or ignored, to understand the ways in which these subsystems interact to coordinate flight, and to uncover the neural mechanisms that underlie these visual capacities. Another thrust that many laboratories are now pursuing is to translate these elegant principles into novel algorithms for the guidance of autonomously navigating vehicles (Ayers et al. 2002, Srinivasan & Venkatesh 1997, Srinivasan et al. 1999).

ACKNOWLEDGMENTS

Some work described in this review was supported by the Australia-Germany Collaborative Research Scheme, DP0208683 from the Australian Research

Council, RG 84/97 from the Human Frontiers in Science Program, N00014-99-1-0506 from the U.S. Defense Advanced Research Projects Agency and the Office of Naval Research, and the Deutsche Forschungsgemeinschaft (SFB 554 and GK 200).

The *Annual Review of Neuroscience* is online at http://neuro.annualreviews.org

LITERATURE CITED

Ayers A, Davis JL, Rudolph A. 2002. *Neurotechnology for Biomimetic Robots*. Cambridge, MA: MIT Press. 636 pp.

Boeddeker N, Egelhaaf M. 2003. Steering a virtual blowfly: simulation of visual pursuit. *Proc. R. Soc. London Ser. B* 270:1971–78

Boeddeker N, Kern R, Egelhaaf M. 2003. Chasing a dummy target: smooth pursuit and velocity control in male blowflies. *Proc. R. Soc. London Ser. B* 270:393–99

Borst A. 1991. Fly visual interneurons responsive to image expansion. *Zoologische Jahrbücher (Physiologie)* 95:305–13

Borst A, Bahde S. 1988. Visual information processing in the fly's landing system. *J. Comp. Physiol. A* 163:167–73

Braitenberg V, Taddei-Ferretti C. 1966. Landing reaction of *Musca domestica*. *Naturwissen* 53:155–56

Collett TS. 1980. Some operating rules for the optomotor system of a hoverfly during voluntary flight. *J. Comp. Physiol.* 138:271–82

Collett TS. 1996. Insect navigation *en route* to the goal: multiple strategies for the use of landmarks. *J. Exp. Biol.* 199:227–35

Collett TS, Harkness LIK. 1982. Depth vision in animals. In *Analysis of Visual Behavior*, ed. DJ Ingle, MA Goodale, RJW Mansfield, pp. 111–76. Cambridge, MA: MIT Press

Collett TS, Zeil J. 1998. Places and landmarks: an arthropod perspective. In *Spatial Representation in Animals*, ed. S. Healy, pp. 18–53. Oxford: Oxford Univ. Press. 188 pp.

David CT. 1982. Compensation for height in the control of groundspeed by *Drosophila* in a new, "Barber's Pole" wind tunnel. *J. Comp. Physiol.* 147:485–93

Dickinson MH. 1999. Haltere-mediated equilibrium reflexes of the fruit fly, *Drosophila melanogaster*. *Philos. Trans. R. Soc. London Ser. B* 353:903–916

Dror RO, O'Carroll DC, Laughlin SB. 2001. Accuracy of velocity estimation by Reichardt correlators. *J. Opt. Soc. Am. A* 18:241–52

Eckert H, Hamdorf K. 1980. Excitatory and inhibitory response components in the landing response of the blowfly, *Calliphora erythrocephala*. *J. Comp. Physiol.* 138:253–64

Egelhaaf M, Borst A. 1993. Movement detection in arthropods. See Miles & Wallman 1993, pp. 53–77

Esch H, Burns JE. 1995. Honeybees use optic flow to measure the distance of a food source. *Naturwissen* 82:38–40

Esch H, Burns JE. 1996. Distance estimation by foraging honeybees. *J. Exp. Biol.* 199:155–62

Esch H, Zhang SW, Srinivasan MV, Tautz J. 2001. Honeybee dances communicate distances measured by optic flow. *Nature* 411:581–83

Gabbiani F, Mo C, Laurent G. 2001. Invariance of angular threshold computation in a wide-field, looming-sensitive neuron. *J. Neurosci.* 21:314–29

Gilbert C, Strausfeld NJ. 1991. The functional organization of male-specific visual neurons in flies. *J. Comp. Physiol. A* 169:395–411

Gray JR, Lee JK, Robertson RM. 2001. Activity of descending contralateral movement detector neurons and collision avoidance behavior in response to head-on visual stimuli in locusts. *J. Comp. Physiol. A* 187:115–29

Hausen K. 1993. The decoding of retinal image flow in insects. See Miles & Wallman 1993, pp. 203–35

Hausen K, Egelhaaf M. 1989. Neural mechanisms of visual course control in insects. In *Facets of Vision*, ed. DG Stavenga, RC Hardie, pp. 391–424. Berlin/Heidelberg: Springer-Verlag

Hausen K, Strausfeld NJ. 1980. Sexually dimorphic interneuron arrangements in the fly visual system. *Proc. R. Soc. Lond. B* 208:57–71

Heisenberg M, Wolf R. 1993. The sensory-motor link in motion-dependent flight control of flies. See Miles & Wallman 1993, pp. 265–83

Hengstenberg R. 1993. Multisensory control in insect oculomotor systems. See Miles & Wallman 1993, pp. 285–98

Horridge GA. 1987. The evolution of visual processing and the construction of seeing systems. *Proc. R. Soc. London Ser. B* 230: 279–92

Ibbotson MR. 2001. Evidence for velocity-tuned motion-sensitive descending neurons in the honeybee. *Proc. R. Soc. London Ser. B* 268:2195–201

Judge SJ, Rind FC. 1997. The locust DCMD, a movement-detecting neurone tightly tuned to collision trajectories. *J. Exp. Biol.* 200:2209–16

Kirchner WH, Srinivasan MV. 1989. Freely flying honeybees use image motion to estimate object distance. *Naturwissen* 76:281–82

Kirschfeld K. 1997. Course control and tracking: orientation through image stabilization. See Lehrer 1997, pp. 67–93

Krapp HG. 2000. Neuronal matched filters for optic flow processing in the visual system of flying insects. In *Neuronal Processing of Optic Flow*, ed. M Lappe, pp. 93–120. San Diego: Academic. 321 pp.

Krapp HG, Hengstenberg R. 1996. Estimation of self-motion by optic flow processing in single visual interneurons. *Nature* 384:463–66

Land MF, Collett TS. 1974. Chasing behaviour of houseflies (*Fannia canicularis*). *J. Comp. Physiol.* 89:331–57

Lee DN. 1976. A theory of visual control based on information about time-to-collision. *Perception* 5:437–59

Lehrer M, ed. 1997. *Orientation and Communication in Arthropods*. Basel: Birkhäuser Verlag. 395 pp.

Miles FA, Wallman J, eds. 1993. *Visual Motion and its Role in the Stabilization of Gaze*. Amsterdam: Elsevier. 417 pp.

Mizutani A, Chahl JS, Srinivasan MV. 2003. Motion camouflage in dragonflies. *Nature* 423:604

Nalbach G. 1993. The halteres of the blowfly *Calliphora*. 1. Kinematics and dynamics. *J. Comp. Physiol. A* 163:293–300

Nalbach G, Hengstenberg R. 1994. The halteres of the blowfly *Calliphora*. 2. Three-dimensional organization of compensatory reactions to real and simulated rotations. *J. Comp. Physiol. A* 175:695–708

Reichardt W. 1969. Movement perception in insects. In *Processing of Optical Data by Organisms and by Machines*, ed. W Reichardt, pp. 465–93. New York: Academic. 614 pp.

Rossell S. 1983. Binocular stereopsis in an insect. *Nature* 302:821–22

Schöne H. 1996. Optokinetic speed control and estimation of travel distance in walking honeybees. *J. Comp. Physiol. A* 179:587–92

Sherman A, Dickinson MH. 2003. A comparison of visual and haltere-mediated equilibrium reflexes in the fruit fly *Drosophila melanogaster*. *J. Exp. Biol.* 206:295–302

Si A, Srinivasan MV, Zhang SW. 2003. Honeybee navigation: properties of the visually driven 'odometer.' *J. Exp. Biol.* 206:1265–73

Srinivasan MV. 1977. A visually-evoked roll response in the housefly: open-loop and closed-loop studies. *J. Comp. Physiol.* 119:1–14

Srinivasan MV. 1993. How insects infer range from visual motion. See Miles & Wallman 1993, pp. 139–56

Srinivasan MV, Bernard GD. 1977. The pursuit response of the housefly and its interaction with the optomotor response. *J. Comp. Physiol.* 115:101–17

Srinivasan MV, Chahl JS, Weber K, Venkatesh

S, Nagle MG, Zhang SW. 1999. Robot navigation inspired by principles of insect vision. *Robotics and Autonomous Systems* 26:203–16

Srinivasan MV, Davey M. 1995. Strategies for active camouflage of motion. *Proc. R. Soc. Lond. B.* 259:19–25

Srinivasan MV, Lehrer M, Kirchner W, Zhang SW. 1991. Range perception through apparent image speed in freely-flying honeybees. *Vis. Neurosci.* 6:519–35

Srinivasan MV, Venkatesh S, eds. 1997. *From Living Eyes to Seeing Machines*. Oxford: Oxford Univ. Press. 271 pp.

Srinivasan MV, Zhang SW. 1997. Visual control of honeybee flight. See Lehrer 1997, pp. 95–113

Srinivasan MV, Zhang SW, Altwein M, Tautz J. 2000a. Honeybee navigation: nature and calibration of the 'odometer'. *Science* 287:851–53

Srinivasan MV, Zhang SW, Bidwell N. 1997. Visually mediated odometry in honeybees. *J. Exp. Biol.* 200:2513–22

Srinivasan MV, Zhang SW, Chahl JS, Barth E, Venkatesh S. 2000b. How honeybees make grazing landings on flat surfaces. *Biol. Cybernetics* 83:171–83

Srinivasan MV, Zhang SW, Chandrashekara K. 1993. Evidence for two distinct movement-detecting mechanisms in insect vision. *Naturwissen* 80:38–41

Srinivasan MV, Zhang SW, Lehrer M, Collett TS. 1996. Honeybee navigation *en route* to the goal: visual flight control and odometry. *J. Exp. Biol.* 199:237–44

Stange G. 1981. The ocellar component of flight equilibrium control in dragonflies. *J. Comp. Physiol.* 141:335–47

Stange G, Howard J. 1979. An ocellar dorsal light response in a dragonfly. *J. Exp. Biol.* 83: 351–55

Stange G, Stowe S, Chahl J, Massaro A. 2002. Anisotropic imaging in the dragonfly median ocellus: a matched filter for horizon detection. *J. Comp. Physiol. A* 188:455–67

Tammero LF, Dickinson MH. 2002. Collision-avoidance and landing responses are mediated by separate pathways in the fruit fly, *Drosophila melanogaster*. *J. Exp. Biol.* 205: 2785–98

Ugolini A. 1987. Visual information acquired during displacement and initial orientation in *Polistes gallicus*. *Anim. Behav.* 35:590–95

von Frisch K 1993. *The Dance Language and Orientation of Bees*. Cambridge, MA: Harvard Univ. Press. 566 pp.

Wagner H. 1982. Flow-field variables trigger landing in flies. *Nature* 297:147–48

Wehner R. 1981. Spatial vision in insects. In *Handbook of Sensory Physiology*, ed. H Autrum, 7/6C:287–616. Berlin/Heidelberg: Springer-Verlag

Wehner R. 1997. The ant's celestial compass system: spectral and polarization channels. See Lehrer 1997, pp. 145–85

Wilson M. 1978. The functional organization of locust ocelli. *J. Comp. Physiol.* 124:297–316

Annu. Rev. Neurosci. 2004. 27:697–722
doi: 10.1146/annurev.neuro.27.070203.144148

How The Brain Processes Social Information: Searching for the Social Brain*

Thomas R. Insel[1] and Russell D. Fernald[2]
[1]*National Institute of Mental Health, Bethesda, Maryland 20892; email: insel@mail.nih.gov*
[2]*Neuroscience Program, Stanford University, Stanford, California 94305; email: russ@psych.stanford.edu*

Key Words imprinting, vomeronasal, oxytocin, parental care, affiliation

■ **Abstract** Because information about gender, kin, and social status are essential for reproduction and survival, it seems likely that specialized neural mechanisms have evolved to process social information. This review describes recent studies of four aspects of social information processing: (*a*) perception of social signals via the vomeronasal system, (*b*) formation of social memory via long-term filial imprinting and short-term recognition, (*c*) motivation for parental behavior and pair bonding, and (*d*) the neural consequences of social experience. Results from these studies and some recent functional imaging studies in human subjects begin to define the circuitry of a "social brain." Such neurodevelopmental disorders as autism and schizophrenia are characterized by abnormal social cognition and corresponding deficits in social behavior; thus social neuroscience offers an important opportunity for translational research with an impact on public health.

INTRODUCTION

During the past decade a new field of research, social neuroscience, has emerged as molecular and cellular methods as well as neuroimaging tools have been used to investigate social behavior and social cognition. Social neuroscience has tackled problems as diverse as the neural basis of dominance, the molecular mechanisms of monogamy, and the organization of a "social brain." The emergence of social neuroscience can be traced to three developments. First and perhaps most surprisingly, studies of certain social interactions such as reproductive behaviors or parental care have revealed some simple, yet robust, molecular and cellular mechanisms (Insel & Young 2001, Pfaff et al. 2002). One might expect that these ostensibly complicated behaviors would be the least likely to be reduced to simple neural mechanisms. However, social behavior is essential for reproduction, and therefore, the neural and hormonal processes subserving these behaviors are likely to be highly conserved. Second, there has been increasing recognition of the role abnormal social behavior plays in such human disorders as schizophrenia and autism (Lord et al. 2000). Studies in nonhuman animals may identify neural substrates of normal social behavior that could aid our understanding and treatment of abnormal human social behavior. Third, there is mounting evidence that social isolation and social separation are serious risk factors for medical disorders that may rival well-known traditional risk factors such as smoking and obesity (House et al. 1988). The powerful effect of loneliness on health begs the question of how social interaction protects against illness.

Although social neuroscience has emerged recently, its conceptual underpinnings reach back nearly a century. von Uexküll (1921) suggested that every species experiences life differently, living as it does in its own "Umwelt," or unique perceptual world. Lorenz (1935) expanded von Uexküll's idea of a perceptual world to include not only physical surroundings but also a social system. His landmark article "Companions as Factors in the Bird's Environment" suggests that an animal's perceptual world must include important information about the behavior of other individuals and even the group as a whole. Because successful social behavior requires recognition of key social interactions in their appropriate context, individual responses in social situations must have been important in shaping the species phenotype. However, in many species, little is known either about how such social perception occurs or about how it leads to the behavioral, physiological, cellular, and molecular changes needed for social behavior. Moreover, various epochs of an animal's life pose different requirements for successful and effective social behavior because distinct behavioral patterns may be important at different times. For example, affiliation, meaning contact with a conspecific, can be manifest as attachment during infancy, as maternal care postpartum, or as pair-bonding behavior in reproductive adults.

CONSTRAINTS ON ANALYZING SOCIALLY RELEVANT NEURAL SYSTEMS

Searching for specific neural substrates of social behavior sets important constraints on experimental methods. It is crucial that behavioral studies designed to understand the neural bases of specific behavioral patterns use realistic social situations where animals interact as they would in their natural habitat. Consequently, there are several important caveats for the most successful development of social neuroscience.

First, there is a tendency to use simple behavioral assays to probe complex behavioral patterns. Such assays are proxies for the real events and, although convenient for experimental purposes, can lead to over- or misinterpretation of results. For example, even though "approach" is often taken as a proxy for affiliative behavior, one animal may approach another for many reasons depending critically on individual status, social context, and physiological state of the individuals involved. Therefore, using approach as a measure of affiliation can distort the value of experimental measurements (see Lederhendler & Shulkin 2000 for a more complete discussion of this issue).

Second, many studies make the tacit assumption that typical laboratory housing is appropriate for animals intended for use in the analysis of complex social behaviors. However, studies of rodents have shown the profound effects housing has on brain structures (Rosenzweig & Bennett 1996, van Praag et al. 2000) such that behavioral and genetic manipulations can be obscured by rearing conditions (Henderson 1970). Ethological information should guide decisions about the environmental factors needed to successfully mimic a natural situation. Wiedenmayer (1997) showed, for example, that gerbils develop stereotypies when their environment does not contain shelters that are appropriately based on their natural behavior. Perseveration of stereotypic behavior in captive animals can easily occur and may reflect stress-induced sensitization of dopamine systems (Cabib et al. 2000). These concerns are particularly important in the early experience of animals. Rearing animals in isolation can selectively disrupt higher-order cognitive function as well as sensory filtering (Hall 1998, Robbins et al. 1996). Thus, even relatively subtle changes in rearing can produce enhanced fearfulness in the animals in ways that could seriously confound experimental outcomes (Wurbel 2001).

Third, the failure to use ethologically relevant tasks can compromise the results of studies in which genetic or environmental challenges are used (Gerlai & Clayton 1999). Testing animals in a context irrelevant to the natural behavior of the animal can produce anomalous results for a variety of reasons. For example, the widely used Morris water maze, a spatial learning task, was developed for a rat species that inhabits wetlands (Morris 1981), although now it is used primarily to test mice that evolved to live in burrows in dry regions such as forests and grasslands. Tests that are not matched to natural behavior may subject the animal to significant stress, confounding what is intended to be a cognitive task with behavioral and endocrine responses to a threatening environment. Ideally a combination of field and laboratory studies might be used to structure experiments. Field observations can identify the capabilities of animals in their natural habitat, revealing processes that have been shaped by natural selection. These insights could then be used to frame laboratory procedures useful in controlled analyses of particular behaviors or cognitive tasks. When possible, several complementary tests should be used to assure that the overall outcome reflects the intended assessment.

Finally, growing evidence indicates that the social context can change neural pathways in individual animals. Whether we examine reproductive behavior (Fernald 2002), the size of a litter (Hofer et al. 1993), a companion for birdsong (Hessler & Doupe 1999), or simply group housing with an inevitable dominance hierarchy, it is important to recognize that social context matters. It is a challenge to discover not only how behavior is controlled via physiological processes, but also how social context influences physiological, cellular, and molecular events in the central nervous system.

SENSORY SPECIALIZATIONS FOR SOCIAL PERCEPTION

As noted above, different species have distinct sensory windows into the world and some sensory systems appear to have evolved especially for social behavior. This claim raises the more general issue of how to distinguish between the functioning of a neural system that is dedicated to processing social information and the functioning of a generic sensory system adapted for multimodal processing of complex stimuli such as social interactions.

Perhaps the best-known sensory system specialized for social behavior is that used for detecting pheromones, the compounds used for intraspecies communication (reviewed in Dulac & Torello 2003). All higher eukaryotes show a remarkable convergence toward two distinct olfactory systems. The main olfactory system, which detects volatile odorants that are inhaled via airways, is used to detect food, predators, and prey. Evolved to detect smells that cannot be predicted, it contains a sensory array able to detect a large number of odorants. In contrast, the accessory olfactory system detects a limited set of pheromones that are actively pumped into the interior of the vomeronasal organ where they are sensed by neurons that project to the accessory olfactory bulb (AOB). This system detects and recognizes species-specific olfactory signals that carry information about the sex, reproductive state, and location of possible mates as well as information about territory and social status that regulates various social behaviors. In contrast to the main olfactory system, which faces a large and unknown universe of odors, the accessory olfactory system has a limited and predictable set of signals to detect.

Progress in understanding the olfactory and pheromonal systems has been rapid, with many surprises. The mammalian main olfactory receptors are G protein–coupled receptors (GPCRs) with a conserved seven-transmembrane structure, which facilitated their discovery (Buck & Axel 1991), but it was unexpected that there would be more than 1000 genes encoding olfactory receptors in mammals. Olfactory receptors, (Mombaerts et al. 1996), the signaling cascade (Firestein 2001), and a remarkable spatial encoding of olfactory signals extending from the glomeruli to second-order neurons in the cortex (Zou et al. 2001) have all been well described. More important for social interactions is detection of semiochemicals or pheromones produced by conspecifics, although the issue of human pheromones is hotly contested in some quarters (Meredith 2001). Pheromonal detection via the AOB relies on receptors in the vomeronasal organ (VNO) that are evolutionarily unrelated to those in the primary olfactory system. VNO sensory neurons express receptor genes from three independent supergene families, V1r, V2r, and V3r, arrayed in segregated populations on the VNO neuroepithelium, and they express several immune complex genes including a multigene family, H2-Mv, that represents nonclassical class I members of the major histocompatibility complex (Ishii et al. 2003, Loconto et al. 2003). Cells in the VNO do not express the main components of the signaling cascade used for transducing activation of olfactory receptors (such as G_{olf}), but they do express *trp2*, a cation channel of the transient receptor potential family. However, the VNO in humans is vestigial, disappearing before birth. All members of the human VNO gene family are pseudo genes except for one, and the ligand for this receptor is not known.

Histochemical and optical mapping suggest that there are two or more anatomical subdivisions of the AOB along the antero-posterior axis. The central representations of pheromone receptors differ from the precise spatial representations of the main olfactory system. Instead of the convergence of multiple neurons expressing a single olfactory receptor onto a single glomerulus as seen in the main olfactory bulb, there is a more diffuse topographic projection of pheromone receptors onto multiple glomeruli in the AOB with convergence in the mitral cell, second-order neurons (Del Punta et al. 2002). The significance of this difference between main and accessory bulb organization is not clear, although in both systems there is a high degree of anatomic specificity at this early level of sensory processing. Mitral cell projections also vary between main and accessory systems in the rodent brain, the former represented in the primary olfactory cortex and the latter distributed in the bed nuclei of the accessory olfactory tract and stria terminalis as well as in the vomeronasal amygdala, including aspects of the posteromedial cortical and medial nuclei. Projections from the vomeronasal amygdala are largely to the neuroendocrine hypothalamus, including the medial preoptic area (MPOA) and ventromedial nucleus.

How does the VNO-AOB pathway perceive social signals? Neurophysiological recordings in anesthetized mice and in VNO slices have revealed several interesting aspects of the segregation of information between the VNO and main olfactory bulb (Dulac 2000). Some odorants not known to be pheromones can stimulate neurons in the VNO (Sam et al. 2001, Trinh & Storm 2003), which leads to the speculation that volatile chemicals associated with food or other important environmental signals activate both systems. There is clearly a topography of response. Recording from VNO slices, Leinders-Zufall et al. (2000) found neurons exquisitely sensitive (e.g., 10^{-11} M) to putative pheromone signals in specific regions of the apical VNO. Holy et al. (2000) recorded from the excised VNO sensory system using an electrode array and found neurons sensitive to male or female mouse urine. These in vitro approaches are not able to exploit the fully functioning system, but they offer a view of some capabilities of the system. In an exciting in vivo approach, Luo et al. (2003) recorded from single neurons in the AOB of male mice engaged in natural behaviors. They observed that neuronal firing was modulated by physical contact with male and female anesthetized conspecifics. Their data suggest that pheromone sampling may require sniffing as a prerequisite for pheromonal signaling. Moreover, individual neurons were activated selectively by specific combinations of the sex and strain of conspecifics and failed to respond to an artificial mouse. Furthermore, the intact animals showed no response to chemicals used to stimulate the VNO in anesthetized mice and slices. Presumably, animals need to seek pheromonal sources and actively sample those of interest. An example of the neural recording and associated behavior can be viewed on the Web (http://www.sciencemag.org/content/vol299/issue5610/images/data/1196/DC1/ 1082133S1.mov).

Targeted mutagenesis has also revealed aspects of social perception via the VNO. Stowers et al. (2002) generated a knockout of the *trp2* gene, encoding the cation channel expressed exclusively in neurons of the VNO. In *trp2* knockout animals, copulation was unaffected, but the males apparently could not distinguish between sexes. Rather than attacking male intruders, as seen in wild-type mice, mutant males attempted to copulate with other males. This suggests that the VNO may be essential for gender discrimination, although the receptor family for this behavior is not clear because *trp2* is expressed throughout the VNO. Male mice with a null mutation of the $\beta 2m$ gene, expressed only in neurons that express the the V2R receptor family, do not show copulation with other males, but they lack aggression (Loconto et al. 2003). Thus, different classes of VNO receptors may be linked to specific behavioral responses.

Progress in understanding how the VNO detects social information has been remarkable. Future studies recording simultaneous behavioral and neural events should allow a sophisticated analysis of how the VNO-AOB processes social signals. Salient from the data produced thus far is that pheromonal signaling is different from main olfactory signaling: In pheromonal signaling, the animal appears to seek the signal to be detected, volatile stimuli may not be readily sensed by this system, and receptors in the VNO appear to be more finely tuned and more sensitive than the receptors in the main olfactory epithelium. Although there is a high degree of spatial and molecular organization in the accessory system, natural stimuli represent complex mixtures of pheromones that activate diverse areas in the brain. Rather than serving simply as releasers or activators of behavior or neuroendocrine responses, pheromonal signals may shape diverse sensory systems converging on the hypothalamus (Dulac & Torello 2003).

SOCIAL LEARNING: FROM PERCEPTION TO MEMORY

How does an individual make sense of social information? Social recognition can be considered at several levels: kin, status, gender, and individual. Here we describe two forms of learning about individual identity: imprinting, which is apparently permanently stored; and social recognition in adults, which appears to be short term. Studies in these two areas are beginning to identify some molecules and cells important for social recognition. Some systems or circuits for social recognition are best defined in fMRI studies of humans (described below).

Imprinting: Formation of Long-Term Social Preferences

How do young animals come to "know" their parents, siblings, and appropriate sexual partners? For infants of many species, learning about conspecifics generally and parents and siblings more specifically is achieved via specialized learning processes early in life. This learning is critically important for survival and reproduction. For this reason, it offers an unusual chance to understand how the nervous system evolved to support specialized learning for a social purpose.

Lorenz (1935) first described this specialized form of social learning. He observed that precocial birds (ducklings, goslings, and chicks) follow and become attached or socially bonded to the first moving object they encounter within hours after hatching, usually the mother. Lorenz (1935) discovered that if greylag geese were reared by him from the time of hatching, they would treat him like a parental bird, and upon reaching sexual maturity, they courted him in preference to conspecifics. Lorenz called this process imprinting after the German word "prägung" (printing) because he proposed that the important sensory object met by the newborn bird is stamped immediately and irreversibly onto its nervous system. He also recognized that there was a short, critical period following hatching during which the chick was sensitive to learning. We now distinguish two forms of imprinting: filial (identifying parental and species phenotypes) and sexual (identifying potential future sexual partners).

Imprinting has been extensively studied in the laboratory, in part because its features are a direct challenge to conventional ideas about learning (Hess 1972). Imprinting is fast, requires few trials, has an obligatory sensitive period, and is irreversible in natural situations—all contrary to classical rules of animal learning. As such, imprinting resembles conditioned taste aversion and fear conditioning, two other rapid and enduring forms of learning in adults. But unlike these other forms of single-trial learning, imprinting occurs within a developmentally restricted time window, providing a rich but largely unexplored area of investigation for linking neural changes to social experience. In a series of studies, Horn & McCabe (1984) proposed two distinct processes: (*a*) emergence of filial behavior toward a stimulus without prior exposure to that stimulus, which they called a predisposition, and (*b*) acquiring a preference for a stimulus through exposure to it. The claim that these are dissociable parts of imprinting is based on manipulations such as drug administration and lesions that affect the two phases differentially (Davies et al. 1985).

Predisposition to approach stimuli resembling conspecifics is independent of experience and occurs during a sensitive period (Bolhuis et al. 1985, Johnson et al. 1989b). This predisposition, at least for chicks, appears to depend on the complex structural or configural properties of the stimulus (Johnson & Horn 1988). These data suggest that the chick has inherited some kind of perceptual template that predisposes it to prefer the right class of objects for its attention. Evidence suggests this predisposition is then shaped by the subsequent experience of the animal (Hogan 1988). It is not clear how this evolutionarily essential sensory template is encoded in the brain or how animals match visual experience to such a template.

The second aspect of imprinting, acquiring a preference, has been extensively studied. On the basis of original ethological observations, Bolhuis & Honey (1998) found that the more complex and realistic the stimulus is (e.g., sound, motion, structure), the stronger the imprinting process is, which, in turn, is thought to contribute to formation of an integrated representation of the imprinting object. Are there particular sites in the brain responsible for imprinting? Horn and colleagues (for a review see Horn 1985) have described the importance of a telencephalic midline region, the intermediate and medial hyperstriatum ventrale (IMHV), for imprinting in the chick. In a series of lesion studies, chicks that had their IMHV surgically removed on both sides could no longer retain imprinting and could not recognize the imprinting stimuli. However, lesioned chicks could learn externally rewarded stimuli. In addition, there is increased metabolic activity, *N*-methyl-D-aspartate (NMDA) receptor binding, and c-*fos* gene expression in IMHV during imprinting (McCabe & Horn 1994). Morphological studies have shown that imprinting is correlated with an increase in the length of the postsynaptic density of spine synapses in the IMHV but only in the left hemisphere. Significantly, there does not seem to be an increase in synapse number in the IMHV during imprinting. In a related series of experiments, Bock & Braun (1999) described changes that accompany auditory imprinting in the chick. Here the relevant regions are the mediorostral neostriatum/hyperstriatum ventrale and the dorso-caudal neostriatum. However, in contrast to the increase in the postsynaptic density described with visual imprinting, Bock & Braun (1999) note that auditory imprinting is associated with a loss of spines in these two regions.

In both visual and auditory imprinting, NMDA receptors appear important for experience-dependent plasticity. NMDA receptor antagonists block visual (McCabe & Horn 1991) and auditory (Bock & Braun 1999) imprinting and, in the latter case, prevent the learning-associated loss of spines specifically in the two regions identified as critical for learning. In other models of experience-dependent plasticity associated with sensitive periods (such as the formation of ocular dominance columns in the visual cortex, the formation of barrel fields in the somatosensory cortex, or song learning in the zebra finch), the end of the sensitive period is associated with developmental changes in NMDA receptor physiology along with decreased expression of the NR2B subunit and increased expression of the NR2A subunit within the NMDA receptor complex (Heinrich et al. 2002). In this context, imprinting may represent a specialized form of developmental learning that uses mechanisms adapted for long-term storage in the service of social recognition. However, in other models of developmental plasticity such as avian song learning, downregulation of the NR2B subunit is not sufficient to close the critical period (Heinrich et al. 2003), and in NR2A knockout mice there is no evidence of an extended critical period (Lu et al. 2001).

Imprinting offers an unusual opportunity to explore a well-defined, genetically modulated period of plasticity during which specific brain regions acquire information essential for survival of the individual. In birds this process is largely visual, in rodents imprinting is olfactory (see Sullivan & Wilson 2003), and in sheep, both visual and pheromonal signals may be critical (Kendrick et al. 1998). Two questions still need to be answered: (*a*) What are the neural substrates of the critical or sensitive period? and (*b*) What are the neural consequences of stimulation during this period? It seems likely that with the appropriate experimental paradigm and a careful delineation of the time course, techniques for profiling gene and protein expression will reveal the neural mechanism for the window of filial imprinting, analogous to the recent studies of avian song learning and ocular dominance column formation. In contrast to the evanescent sensitive period, the neural consequences of imprinting are likely to reflect constitutive changes in gene expression possibly via epigenetic mechanisms. A beautiful example of one such mechanism has been described by Meaney and colleagues (Meaney et al. 1996, Champagne et al. 2003) in their studies of the long-term effects of high versus low maternal grooming of rat pups. Grooming apparently induces an epigenetic demethylation of the promoter of the hippocampal glucocorticoid receptor, exposing the promoter to transcription factors that induce gene expression in response to stress. As a result, pups who receive high levels of grooming have more hippocampal glucocortcoid receptors and, because these receptors serve as a brake on the hypothalamic stress response, these pups remain relatively less stress responsive throughout life.

Social Recognition

If imprinting confers an enduring memory that is important for recognizing parents or avoiding incest, how do we recognize familiar individuals encountered later in the life cycle? In rodents, recognizing conspecifics, unlike imprinting, appears to be a short-lived process. Adult social recognition rests on the observation that in a laboratory cage environment and, presumably, in the wild most rodents will enter into a "meet and greet" ritual when exposed to a novel intruder. In the field, rats live in colonies with a common pheromonal signature spread via grooming. When a resident male is exposed to an intruder male or a sexually receptive female, this ritual quickly evolves into either a threat display or an attempted mating bout, respectively, regulated by pheromone detection. But when a resident male is exposed to a juvenile or an ovariectomized female, the male predictably sniffs and grooms the intruder for at least 2 min (depending on the strain). If the intruder is then removed for 30 min before being placed again with the same resident male, the time for investigation falls by approximately 50%. This decrease in investigation has been assumed to reflect recognition of the intruder because (*a*) a novel intruder receives at least 2 min of investigation, (*b*) increasing intervals of separation between the initial and subsequent exposures to the same intruder results in increasing investigation time, and (*c*) drugs or interventions that impair memory formation increase investigation time on the recognition trial (Winslow & Camacho 1995). In male rats, after 90 min of separation there is little or no decrease in investigation time, presumably reflecting a loss of recognition.

In a series of studies dating back nearly two decades, intraventricular administration of the neuropeptide vasopressin (AVP) has been shown to increase social recognition in male rats (Engelmann et al. 1996). Landgraf and colleagues recently reported that a viral-vector-induced increase in the vasopressin V1a receptor specifically in the lateral septum increased social recognition (Landgraf et al. 2003). AVP in the rat lateral septum is much more abundant in males than in females (De Vries et al. 1992), possibly accounting for a gender difference and androgen dependence of social recognition. However, the full circuitry for AVP's effects in the rat brain remains to be described.

Ferguson et al. (2000) recently described mice with a null mutation of another member of the AVP peptide family, oxytocin, as socially amnestic. In the oxytocin-knockout (OT-KO) mouse most aspects of social behavior, such as sexual and maternal behavior, appear unchanged from those of controls (Nishimori et al. 1996, Winslow et al. 2000, Young et al. 1996). In the social recognition paradigm, the responses of male OT-KO and wild-type mice do not differ in an initial encounter with a novel intruder, each spending approximately 150 s investigating the novel mouse (Ferguson et al. 2000). However, when tested 30 min later, wild-type mice show the expected 50% decrease in investigation, whereas OT-KO mice exhibit no change from the initial trial (Ferguson et al. 2000). It is curious to note that OT-KO and wild-type mice do not differ on several tests of nonsocial memory nor do they differ in tests of olfactory function. Indeed, when tested with either a lemon-scented cotton ball or even a lemon-scented mouse, both OT-KO and wild-type mice appear to recognize the stimulus after 30 min of separation (Ferguson et al. 2002). The deficit in the male OT-KO mouse thus appears to be specific to the social domain (although, see also Tomizawa et al. 2003 for cognitive deficits in female OT-KO mice).

Oxytocin receptors are found throughout the main olfactory bulb, the AOB, as well as several telencephalic nuclei in the mouse brain (Insel et al. 1993). Although earlier pharmacological studies implicated the olfactory bulb as the likely site of action for oxytocin effects on social recognition (Dluzen et al. 1998), when Ferguson et al. (2001) compared regional activation in OT-KO and wild-type mice after a brief social exposure, Fos staining was increased in the main and accessory olfactory systems of both strains, with no differences apparent between OT-KO and wild-type mice. However, in contrast to the wild-type mice, the OT-KO mice failed to activate Fos in the medial nucleus of the amygdala and in downstream projections in the bed nucleus of the stria terminalis (BST) and the MPOA. Oxytocin injected into the medial nucleus of the amygdala (a region rich in oxytocin receptors) appeared to reinstate social recognition in the OT-KO mice, at a dose that was ineffective when given by the intracerebroventricular (icv) route.

Consistent with the role of the V1a receptor on social recognition in rats, mice with a null mutation of the V1a receptor (V1a-KO) also manifest a profound deficit in social memory (Bielsky et al. 2004). At first glance, the OT-KO and V1a-KO mice appear to have a murine equivalent of prosopagnosia, a clinical syndrome in which the ability to recognize faces is lost. However, careful consideration of the ethological significance of the behavior suggests that the experimental proxy used for social recognition in these mouse studies is not equivalent to our sense of individual recognition in humans. The grooming ritual in mice that is used to investigate an intruder includes delivery of a pheromonal signature. Thus, it seems possible that the recognition depends on the test mouse detecting his own familiar pheromone rather than recognizing any individual characteristics of the intruder mouse. Could the behavioral results observed in the OT-KO mouse be explained by a deficit in secreting the pheromone rather than an inability to make a social memory? This interpretation would suggest that activation of the medial amygdala is associated with secretion rather than detection of pheromones. At present, we suggest that this social recognition test be used with controls that are attentive to changes or deficits in grooming or pheromone delivery to determine if any alteration in recognition is in fact a problem in information processing or retrieval. Results from other assays, such as the social transmission of food preference test, may be useful to confirm a deficit in social cognition (Wrenn et al. 2003).

SOCIAL MOTIVATION: FROM RECOGNITION TO ACTION

Social attachment, social affiliation, sex behavior, and parental care are among the most highly motivated social behaviors. The motivation for social interaction, as with other appetitive behaviors, can be quantified with operant techniques. For instance, Everitt (1990) has demonstrated that male rats will bar press for access to estrous females, and Lee et al. (1999) have shown that postpartum females will bar press for access to pups. In a recent confirmation of the importance of maternal motivation, postpartum female rats were found to prefer a cage associated with pups to a cage associated with cocaine (Mattson et al. 2001). The laboratory rat, widely used for studies of maternal care, is also useful for studies of maternal motivation (Numan 1994). Unlike many mammals, female rats show little interest in infants of their own species until just before parturition. Approximately one day prior to delivery they shift from avoiding pups to showing intense interest with avid nest building, retrieval, grooming, and defense of young. These behaviors persist through lactation, then abate with weaning. Rats, therefore, provide an opportunity to study two distinct aspects of maternal care: onset and maintenance. The onset of maternal care, switching from avoidance to intense interest, is of particular interest because of the magnitude of the increase in motivation.

The Onset of Rat Maternal Behavior

Given the abundant evidence that mesolimbic dopamine pathways are important for other forms of highly motivated behaviors, from feeding to psychostimulant self-administration (Kelley & Berridge 2002), it is not surprising that these same pathways have been implicated in appetitive social interactions. Maternal behavior is instructive in this regard because of the number of experiments showing the relationship between dopamine and maternal behavior in rats. For example, exposure to pups increases Fos activation (Lonstein et al. 1998) and dopamine release (Hansen et al. 1993) in the nucleus accumbens of maternal but not of nonmaternal females. Depletion of dopamine in the ventral tegmentum by chemical lesion during pregnancy blocks the development of maternal behavior (Hansen et al. 1991). Similar disruptions of maternal behavior result from systemic administration of the dopamine receptor antagonist haloperidol (Giordano et al. 1990, Stern & Keer

1999) and after acute and chronic administration of cocaine (Johns et al. 1997, Kinsley 1994). Either lesions or injections of dopamine antagonists into the nucleus accumbens inhibit selectively the active components of maternal behavior such as retrieval and pup licking, but not the more reflexive aspects such as nursing (Hansen et al. 1991, Keer & Stern 1999). In one study, 6-hydroxydopamine (6-OHDA) lesions of the nucleus accumbens on day 2 or 3 postpartum in primiparous rats markedly reduced retrieval of pups without reducing nest building, nursing, or maternal aggression (Hansen 1991). Lesioned females preferred food to pups, and these same females, tested later for sex behavior, showed no deficits in either proceptive or receptive behavior. Thus the deficit appeared specific to maternal interest. Curiously, after the lesioned females were separated from their pups for 3–6 h, they began to retrieve them, indicating that these females were capable of retrieval but needed some additional incentive to do so.

These experiments suggest dopamine in the nucleus accumbens is responsible for maternal motivation, consistent with research on feeding and sex. Why does the female retrieve her pups rather than respond to myriad other stimuli in her world, such as food and mates? There is increasing evidence that the neuropeptide oxytocin may be critical for linking pup signals to the mesolimbic dopamine stream involved in motivated behaviors. Several investigators have reported that oxytocin given centrally to estrogen-primed, nulliparous female rats facilitates the onset of maternal behavior (reviewed in Insel 1997). Perhaps even more remarkable, blockade of oxytocin neurotransmission results in a significant inhibition of the onset of maternal behavior but fails to affect maternal behavior once it has been established (reviewed in Insel 1997). These results support the notions that oxytocin is necessary for the transition from maternal avoidance to attachment to pups and that a central increase in oxytocin given under the appropriate gonadal steroid conditions facilitates the onset of maternal care. In a sense, the role of oxytocin in the uterus and mammary tissue for providing the physiologic support of the offspring is matched by its role in the brain for subserving the motivational changes essential for maternal care.

Physiological changes in gonadal steroids during pregnancy increase both the synthesis of the peptide and the number of receptors (Crowley et al. 1995, Insel et al. 1992). The changes in oxytocin receptors are not ubiquitous. Only those regions rich in estrogen receptors (e.g., BST and ventromedial nucleus of the hypothalamus) show increased oxytocin receptor binding, but in these regions the changes may be rapid and profound (up to 300% increases in hypothalamic binding in 72 h) (Johnson et al. 1989a).

Results from site-specific injections of an oxytocin antagonist suggest that this peptide may be particularly important for regulating dopaminergic function either by a direct action on the ventral tegmental area or by afferents in the MPOA or BST (Pedersen et al. 1994). A region including the medial aspect of the MPOA and ventral BST has been studied for more than four decades as a "hot spot" for maternal behavior. This region is rich in estrogen receptors, the onset of maternal behavior is associated with a pronounced increase in local estrogen receptor gene expression, and exogenous administration of estrogen into the MPOA stimulates maternal behavior in nulliparous females (Numan & Insel 2003). Moreover, lesions of the MPOA inhibit maternal behavior, and pup stimuli increase the induction of Fos protein in this region, reflecting increased activity (Stack & Numan 2000). What is the connection between the MPOA/BST and the aforementioned dopamine regulation of maternal motivation? Numan & Smith (1984) showed that unilateral lesions of the MPOA (which do not inhibit maternal behavior) in conjunction with lesions of the ventral tegmental area greatly reduced maternal retrieval. A recent follow-up study is even more compelling (Stack et al. 2002). After exposure to pups for 6 h, postpartum rats with unilateral MPOA lesions showed an increase specifically in the nucleus accumbens shell, relative to females not exposed to pups. The Fos increase was unilateral, limited to the side that receives a projection from MPOA/BST, and was not found in the nucleus accumbens core. Taken together, the current evidence supports a model that pup stimuli processed via olfactory and amygdala pathways ultimately activate estrogen- and oxytocin-sensitive MPOA/BST neurons that in turn project to the mesolimbic dopamine pathway in the ventral tegmental area and/or the nucleus accumbens shell.

Null mutations of several genes, including prolactin-receptor, Fos-B, and the paternally imprinted Peg-1 and Peg-3 genes, have disrupted maternal behavior in mice (reviewed in Leckman & Herman 2001). Surprisingly, the OT-KO mouse shows no deficit in maternal behavior (although these mice fail to lactate). This paradox may be resolved by the recognition that most laboratory strains of mice, unlike rats, do not avoid pups and do not require pregnancy or parturition to exhibit maternal care (Russell & Leng 1998). As noted above, pup-directed behavior in rats transforms at parturition from avoidance to approach. Estrogen and oxytocin in the MPOA/BST appear to be essential for this induction of maternal motivation. In mice, none of these factors appear essential for maternal motivation, and there is not a discrete onset of maternal behavior as seen in rats. Therefore, the various mutations that reduce maternal behavior in mice may be influencing various aspects of maternal care, but we have no evidence at this point that they are reducing maternal motivation per se.

Formation of Partner Preferences

Pair bonding in monogamous species provides another interesting example of social motivation. The prairie vole (*Microtus ochrogaster*) is a mouse-sized rodent that manifests the classic features of monogamy: A breeding pair shares the same nest and territory where they are in frequent contact, males participate in parental care, and intruders of either sex are rejected. Getz et al. (1993) reported from field studies that following the death of one of the pair, a new mate is accepted only ~20% of the time (the rate is approximately the same whether the survivor is male or female). Prairie voles also demonstrate a curious pattern of reproductive development: Offspring remain sexually suppressed as long as they remain within the natal group. For females, puberty occurs not at a specific age but after exposure to a chemosignal in the urine of an unrelated male (Carter et al. 1995). Within 24 h of exposure to this signal, the female becomes sexually receptive. She mates repeatedly with an unrelated male and, in the process, forms a selective and enduring preference or pair bond.

As with parturition in rats, mating in these voles is a transformational event resulting in long-term increases in partner preferences that can be quantified in laboratory tests. The available data are largely analogous to data described for rat maternal behavior. Dopamine is released in the nucleus accumbens with mating, dopamine agonists in this region facilitate partner preference formation, and dopamine D2 antagonists inhibit partner preference formation (Gingrich et al. 2000, Wang et al. 1999). However, abundant evidence indicates that mating activates dopamine in the nucleus accumbens in species that do not form a partner preference (Pfaus et al. 2001), so one might ask whether this change is related to pair bonding. Or more generally, what is mating doing in the monogamous brain to confer a preference for the partner?

Because of the evidence that oxytocin and vasopressin are released with sexual behavior (Witt 1995), prairie voles have been treated with these peptides (in the absence of mating) or with their selective antagonists (prior to mating). Both peptides facilitate partner preference formation, and conversely, antagonists reduce partner preference formation without reducing mating behavior (Insel et al. 2001). As with the studies of rat maternal care, much of the recent interest in this area has focused on identifying the neural circuit necessary for pair bonding. In contrast to closely related nonmonogamous voles (and other nonmonogamous rodents such as rats), prairie voles have a high density of oxytocin and AVP V1a receptors in either the nucleus accumbens and prelimbic cortex or the ventral pallidum, respectively (Insel et al. 2001, Lim et al. 2004). Are these receptors important for pair bonding? An oxytocin antagonist injected directly into the nucleus accumbens or prelimbic cortex blocks pair-bond formation in female prairie voles (Young et al. 2001). Increasing AVP V1a receptors via viral-vector administration directly into the ventral pallidum facilitates partner preference formation (Pitkow et al. 2001), but it remains to be shown that an antagonist injected into this region blocks the behavior (Liu et al. 2001).

Molecular studies of both oxytocin and vasopressin receptors suggest that species differences in distribution may result from hypervariable regions found in the promoters of both genes (Insel & Young 2000). Sequence differences in the V1a promoter alter expression in vitro, and mice with a prairie vole transgene, including this promoter, have a prairie vole–like pattern of V1a receptor distribution and exhibit increased social behavior in response to AVP (Hammock & Young 2002, Young et al. 1999). Whatever the mechanism, monogamous species like prairie voles and marmosets have abundant receptors for either oxytocin or vasopressin in mesolimbic pathways such as the nucleus accumbens and the ventral pallidum (Young et al. 2001). One current hypothesis is that these receptors link social information to reward circuits in the brain, providing a neurobiological mechanism for partner preference formation. In its simplest form, the release

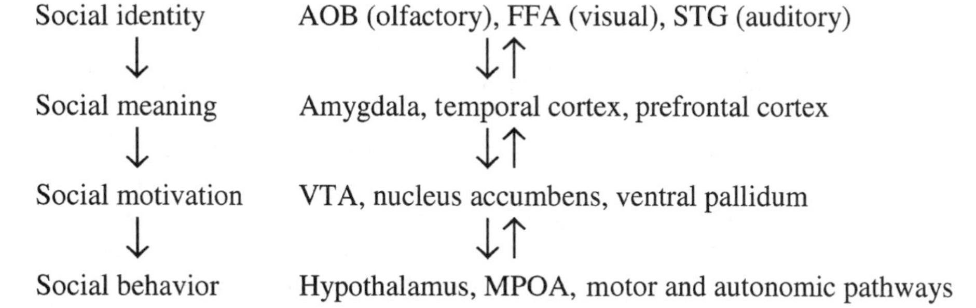

Figure 1 A simplified and highly theoretical model of social information processing in the mammalian brain. Sensory unimodal information is tagged as social in the accessory olfactory bulb (AOB), fusiform area (FFA), or superior temporal gyrus (STG). This signal becomes instantiated as significant or salient in a subsequent multimodal projection to poorly defined fields in the amygdala, temporal cortex, and prefrontal cortex, three regions where emotion, social status, or familiarity may be encoded. Social attachment (maternal behavior, pair bonding, and, potentially, infant attachment) involves recruitment of the mesolimbic dopamine pathway, including the ventral tegmental area, with development of individual preferences. Finally, social behavior involves activation of the neuroendocrine hypothalamus, including the medial preoptic area (MPOA), as well as motor and autonomic centers. The available data, although limited, suggest reciprocal activation between levels (see text for evidence that maternal behavior involves MPOA activation of the nucleus accumbens).

of oxytocin or vasopressin with mating would activate these reward pathways in monogamous species, resulting in a conditioned response or preference just as if the individual had received cocaine or amphetamine (see Figure 1). As noted above, mating activates dopamine release in both monogamous and nonmonogamous species. Therefore, this hypothesis rests on a specific role for oxytocin or vasopressin above and beyond dopamine release in the ventral striatum. Both peptides have been implicated in social recognition, presumably independent of effects on motivation. Recent reports of regional neurogenesis activated by mating in monogamous voles suggest another potential neural correlate of partner preference formation (Fowler et al. 2002).

SOCIAL BEHAVIOR THAT CHANGES THE BRAIN

It seems self-evident that the brain controls behavior, but can behavior also control the brain? Behavior influences specific aspects of brain structure and function in three different time frames. On an evolutionary timescale, the selective forces of the ecological niche of the animal are reflected in body shape, sensory and motor systems, and behavior. Similarly, on a developmental timescale, behavior acts in concert with the environment to establish structural changes in the brain that influence an organism throughout its life. There is now evidence that in real time social behavior also causes changes in the brain of an adult animal. These alterations, caused by behavioral interactions, often are related to reproductive behavior and can be dramatic and reversible. Understanding the mechanisms responsible for such dynamic changes in the nervous systems of adult animals is a major challenge. How does behavior sculpt the brain and how are these changes controlled? To understand this requires a model system in which a complex social system can be manipulated and individuals can be analyzed at the physiological, cellular, and molecular level. A highly social cichlid fish provides one such model system.

In the African cichlid fish, *Haplochromis (Astatotilapia) burtoni*, there are two kinds of adult males: those with territories and those without (Fernald 1977). Territorial (T) males are brightly colored, with a dramatic black stripe through the eye, vertical black bars on the body, a black spot on the tip of the gill cover, and a large red patch just behind it. In contrast, nonterritorial (NT) males are cryptically colored, making them difficult to distinguish from the background and from females that are similarly camouflaged. Whether a male is T or NT depends on the social circumstances.

In their natural habitat, the shallow shorepools and river estuaries of Lake Tanganyika, *H. burtoni* live in a lek-like social system in which T males vigorously defend contiguous territories and solicit females to mate with them. If the female responds to these entreaties, he leads her into his pit where she lays her eggs at the bottom of the pit, collecting them in her mouth almost immediately. After she has laid several eggs, the male swims in front of her, displays the egglike spots on his anal fin (ocelli), and moves his body in a quivering motion (Fernald & Hirata 1977a,b). The male displays his anal fin because the spots may appear to the female to be eggs not yet collected (Wickler 1962). While attempting to "collect" the spots, the female ingests the milt ejected near them by the male, ensuring fertilization. On the other hand, NT males cannot spawn.

The natural behavior of *H. burtoni* reveals the extensive role of visual signals in social interactions and how much the social scene governs the behavior of individual animals. Each behavioral act influences the next. During the behavior, a great deal of information is exchanged between individuals. What does the animal attend to and what are the consequences?

Juvenile males raised with adults show suppressed gonadal maturation relative to those reared without adults (Davis & Fernald 1990). As well as having smaller testes, these animals have smaller gonadotropin-releasing hormone (GnRH)–containing neurons in the preoptic area, an area in the ventral telencephalon adjacent to the hypothalamus. These neurons project to the pituitary (Bushnik & Fernald 1995) where they release GnRH. The somata of GnRH neurons in T males are eight times larger than those in NT males, an effect that depends solely on social conditions. Because GnRH is the main signaling peptide that regulates reproductive maturity, the social control of maturation acts by changing structures in the brain.

Social status determines the physiology of the reproductive state, even in adult fish. Changing males from T to NT or vice versa has dramatic consequences. NT males who become T males have GnRH soma sizes similar to those of T males, whereas T males who become NT males have soma sizes comparable to those of NT males (Francis et al. 1993). The gonad sizes change accordingly. The same result has been shown for GnRH mRNAs using in situ hybridization with GnRH specific probes (White et al. 1995). Thus a change in social status alters brain structures essential for reproduction.

The socially induced GnRH-neuron size changes are remarkably asymmetric (White et al. 2002). Males ascending (NT→T) achieve large GnRH cell sizes in less than a week, whereas males descending (T→NT) have GnRH cell sizes that shrink slowly during a three-week period. This makes ecological sense because the chance to establish a territory may soon arise again, making the maintenance of an active reproductive physiology for a few weeks a reasonable adaptive strategy. Correspondingly, a newly ascended T male should mature sexually as quickly as possible because he may lose his territory sooner rather than later. Social status clearly determines both soma size of GnRH neurons in the preoptic area and relative gonad size, and these effects are reversible. The relatively large testes and GnRH neurons characteristic of T males are a consequence of their social dominance, and when this dominance advantage is lost, both neurons and testes shrink. Exactly how social information is transformed into changes in the brain remains unknown. There is, however, some evidence that visual information may be used to signal the state of individual animals.

Animals that have lost a territory (T→NT) grow more slowly and even shrink (Hofmann et al. 1999). Behavioral stress may play a role. Fox et al. (1997) showed that in *H. burtoni*, status switches in both directions can be accompanied by elevated levels of the major stress hormone cortisol, with the T→NT change showing the most pronounced increase. NT→T fish with increased cortisol levels usually did not maintain territoriality. Descending fish consistently showed high levels of cortisol, which may be elevated by losing a territory, thereby causing the downregulation of somatic growth. Taken together, these data show that social status can directly regulate neuron size, changing reproductive status as well as regulating growth. The complexity of the social interactions suggests that subtle signals from social encounters cause changes in the social brain of these cichlids.

SOCIAL SYSTEMS IN THE HUMAN BRAIN

Is there a social brain? Social perception in primates is largely visual, although auditory, somatosensory, and olfactory cues contribute to identifying kin, gender, and familiar individuals. Face perception has been the primary focus for much of human social neuroscience during the past few years, growing out of earlier neurophysiologic and recent neuroimaging studies that demonstrated cells or fields in the monkey temporal cortex respond to faces (Tsao et al. 2003). As fMRI studies in humans have revealed the categorical nature of cortical processing of a range of visual stimuli, perhaps it is not surprising that the same technique would identify regions activated by faces. The specificity of regional activation for face processing remains unclear. For instance, the fusiform area in the occipital-temporal junction has been variously described as critical for face recognition or for expertise in processing categories with multiple elements, including birds for ornithologists, houses for realtors, or faces for most of us (Gauthier et al. 2000, Kanwisher et al. 1997). Lesions in this region have been associated with deficits in face recognition (prosopagnosia), and more recently, significant reductions in gray matter volume in this region have been reported in patients with chronic schizophrenia who also have difficulty with face recognition (Onitsuka et al. 2003).

An interesting approach to identifying the circuitry for social information has used fMRI to investigate regional brain activation in subjects watching animated vignettes of simple geometric shapes interacting either in a "social," "mechanical," or "random" fashion (Castelli et al. 2002, Martin & Weisberg 2004). These studies focus not on social objects such as faces but on how the brain responds while attributing social interaction to abstract images. These studies have identified a "social" circuit comprising the lateral segment of the fusiform gyrus, the superior temporal sulcus, the amygdala, and the ventromedial prefrontal cortex. Much of this circuit was recognized for social perception from studies in nonhuman primates (Brothers 1990) as well as social cognition in humans (reviewed in Adolphs 2001). For instance, previous evidence from lesion studies as well as functional imaging implicates the amygdala in the recognition of social emotions (guilt, arrogance) as well as perception of fear (Adolphs et al. 2002, but see Amaral et al. 2003). The ventromedial prefrontal cortex is strongly connected to the amygdala (Steffanaci & Amaral 2002) and has been linked to subjective pleasantness (Kringelbach et al. 2003), social judgment (Bechara et al. 1997), and processing of social vocalizations in nonhuman primates (Romanski & Goldman-Rakic 2002). What about extending this analysis from social recognition to social motivation? The first fMRI studies of love and loss in humans implicate the striatum, the medial insula, and the anterior cingulate cortex in romantic attachment (Bartels & Zeki 2000), as well as the anterior cingulate and the right ventral prefrontal cortex in the response to social exclusion (Eisenberger et al. 2003). The activation of the ventral striatum with social motivation in humans is generally consistent with the results from rodent studies presented above.

The identification of a social circuit in the human brain may prove important for identifying the neuropathology of autism (Lord et al. 2000). This neurodevelopmental disorder is defined by deficits in reciprocal social behavior and language as well as the presence of stereotypic behaviors. Children with autism appear to lack social motivation, as measured by eye contact and interest in looking at faces (Klin et al. 2002). Although there are no gross pathognomonic abnormalities in the autistic brain, fMRI studies have shown that people with autism do not activate the fusiform gyrus when presented with faces (Schultz et al. 2000). This could indicate simply a lack of attending to, or expertise with, faces, or the absence of activation in this region may indicate a critical breakdown in the ability to process faces. Some individuals exhibit remarkable expertise with nonsocial categories of information, often in the form of savant skills such as calendar counting and idiosyncratic recall. Thus these individuals are capable of expertise, but apparently not in the social domain.

Studies of social information processing in nonhuman animals will likely point to where to look and what to look for in the brains of children with neurodevelopmental disorders such as autism and schizophrenia. Clinical studies report that children growing up with social deprivation exhibit autistic-like behavior and enduring deficits in attachment (O'Connor et al. 2003). As we understand the principles by which social experience supports normal brain development in animal studies, we may glimpse the process that fails for children with neurodevelopmental disorders who are exposed to healthy social environments yet seem unable to process this information for normal brain development. Much of the past decade has been devoted to searching for the genes, cells, and systems important for normal social behavior in animals. Currently, there is a broad search using linkage and association studies for genes associated with autism and schizophrenia. In the next decade these two approaches may converge by linking the genes that contribute to these clinical syndromes to the pathways that mediate social information, as is already happening with Fragile X (Brown et al. 2001) and Rett Syndrome (Shahbazian & Zoghbi 2002). Thus, we will likely borrow from discoveries in humans to design experiments in mice, just as we have been trying to build a clinical neuroscience based on research in rodents and other animals during this past decade.

CONCLUSION

This review focuses on a few examples from the emerging field of social neuroscience to ask how the brain makes sense of the social world. At the molecular level vomeronasal signals appear critical for perceiving social signals, and the nonapeptides such as oxytocin and vasopressin appear to be important for linking social signals to cognition and behavior. A central assumption in this approach is that the mechanisms for social learning and social motivation are built on well-known, generic neural systems for learning and motivation. It seems likely, although still unproven, that social memory requires many of the molecular steps involved in other forms of learning. Similarly, social preference formation, whether for offspring or a sexual partner, relies heavily on mesolimbic dopamine systems that confer the hedonic value of a wide range of stimuli. Unique to social learning (such as imprinting) and social motivation is (*a*) the rapid, apparently hard-wired nature of acquisition; (*b*) the strength of the response; and (*c*) the ostensible reliance on selective neuropeptides for linking perception to learning and motivation. Less clear is the relevance of these observations to the primate brain, where visual processing trumps vomeronasal signals and cortical networks may override the neuropeptide signals from the hypothalamus. Nevertheless, the search for the molecular and cellular markers of the social brain should provide important insights into the mechanisms for autism, schizophrenia, and other vexing human disorders.

ACKNOWLEDGMENTS

This review was supported in part by a Javits award (NIH-NS 34,950) and NIH-EY-05,051 to R.D.F. The authors appreciate the suggestions from Izja Lederhendler and David Crews, who reviewed this manuscript.

The *Annual Review of Neuroscience* is online at http://neuro.annualreviews.org

LITERATURE CITED

Adolphs R. 2001. The neurobiology of social cognition. *Curr. Opin. Neurobiol.* 11:231–39

Adolphs R, Tranel D, Damasio A. 2002. The human amygdala in social judgment. *Nature* 393:470–74

Amaral DG, Capitanio JP, Jourdain M, Mason WA, Mendoza SP, Prather M. 2003. The amygdala: Is it an essential component of the neural network of social cognition? *Neuropsychologia* 41:517–22

Bartels A, Zeki S. 2000. The neural basis of romantic love. *NeuroReport* 11:3829–34

Bechara A, Damasio H, Tranel D, Damasio A. 1997. Deciding advantageously before knowing the advantageous strategy. *Science* 275:1293–95

Bielsky I, Hu SB, Szegda KL, Westphal H, Young LJ. 2004. Profound impairment in social recognition and reduction in anxiety-like behavior in vasopressin v1a receptor knockout mice. *Neuropsychopharmacology* 29:483–93

Bock J, Braun K. 1999. Filial imprinting in domestic chicks is associated with spine pruning in the associative area, dorsocaudal neostriatum. *Eur. J. Neurosci.* 11:2566–70

Bolhuis JJ, Honey RC. 1998. Imprinting, learning and development: from behaviour to brain and back. *Trends Neurosci.* 21:306–11

Bolhuis JJ, Johnson MH, Horn G. 1985. Effects of early experience on the development of filial preferences in the domestic chick. *Dev. Psychobiol.* 18:299–308

Brothers L. 1990. The social brain: a project for integrating primate behavior and neurophysiology in a new domain. *Concepts Neurosci.* 1:27–51

Brown V, Jin P, Ceman S, Darnell J, O'Donnell W, et al. 2001. Microarray identification of FMRP-associated brain mRNAs and altered mRNA translational profiles in fragile X syndrome. *Cell* 107:477–87

Buck L, Axel R. 1991. A novel multigene family may encode odorant receptors: a molecular basis for odor recognition. *Cell* 65:175–87

Bushnik TL, Fernald RD. 1995. The population of GnRH-containing neurons showing socially mediated size changes project to the pituitary in a teleost, *Haplochromis burtoni*. *Brain Behav. Evol.* 46:371–77

Cabib S, Orsini C, Le Moal M, Piazza Pier V. 2000. Abolition and reversal of strain differences in behavioral responses to drugs of abuse after a brief experience. *Science* 289:463–65

Carter C, DeVries A, Getz L. 1995. Physiological substrates of mammalian monogamy: the prairie vole model. *Neurosci. Biobehav. Rev.* 19:303–14

Castelli F, Frith C, Happe F, Frith U. 2002. Autism, Asperger syndrome and brain mechanisms for the attribution of mental states to animated shapes. *Brain* 125:1839–49

Champagne FA, Francis DD, Mar A, Meaney MJ. 2003. Variations in maternal care in the rat as a mediating influence for the effects of environment on development. *Physiol. Behav.* 79:359–71

Crowley RS, Insel TR, O'Keefe JA, Kim NB, Amico JA. 1995. Increased accumulation of oxytocin messenger ribonucleic acid in the hypothalamus of the female rat: induction by long-term estradiol and progesterone administration and subsequent progesterone withdrawal. *Endocrinology* 136:224–31

Davies DC, Horn G, McCabe BJ. 1985. Noradrenaline and learning: effects of the noradrenergic neurotoxin DSP4 on imprinting in the domestic chick. *Behav. Neurosci.* 99: 652–60

Davis MR, Fernald RD. 1990. Social control of neuronal soma size. *J. Neurobiol.* 21:1180–88

Del Punta K, Puche A, Adams NC, Rodruiguez I, Mombaerts P. 2002. A divergent pattern of sensory axonal projections is rendered convergent by second-order neurons in the accessory olfactory bulb. *Neuron* 35:1057–66

De Vries GJ, Crenshaw BJ, al-Shamma HA. 1992. Gonadal steroid modulation of vasopressin pathways. *Ann. NY Acad. Sci.* 652:387–96

Dluzen DE, Muraoka S, Engelmann M, Landgraf R. 1998. The effects of infusion of arginine vasopressin, oxytocin, or their antagonists into the olfactory bulb upon social recognition responses in male rats. *Peptides* 19:999–1005

Dulac C. 2000. Sensory coding of pheromone signals in mammals. *Curr. Opin. Neurobiol.* 10:511–18

Dulac C, Torello AT. 2003. Molecular detection of pheromone signals in mammals: from genes to behaviour. *Nat. Rev. Neurosci.* 4:551–62

Eisenberger NI, Lieberman MD, Williams KD. 2003. Does rejection hurt? An fMRI study of social exclusion. *Science* 302:290–92

Engelmann M, Wotjak CT, Neumann I, Ludwig M, Landgraf R. 1996. Behavioral consequences of intracerebral vasopressin and oxytocin: focus on learning and memory. *Neurosci. Biobehav. Rev.* 20:341–58

Everitt B. 1990. Sexual motivation: a neural and behavioral analysis of the mechanisms underlying appetitive and copulatory responses of male rats. *Neurosci. Biobehav. Rev.* 14:217–32

Ferguson JN, Aldag JM, Insel TR, Young LJ. 2001. Oxytocin in the medial amygdala is essential for social recognition in the mouse. *J. Neurosci.* 21:8278–85

Ferguson JN, Young LJ, Hearn E, Insel TR, Winslow J. 2000. Social amnesia in mice lacking the oxytocin gene. *Nat. Genet.* 25: 284–88

Ferguson JN, Young LJ, Insel TR. 2002. The neuroendocrine basis of social recognition. *Front. Neuroendocrinol.* 23:200–24

Fernald RD. 1977. Quantitative behavioral observations of haplochromis-burtoni under semi-natural conditions. *Anim. Behav.* 25:643–53

Fernald RD. 2002. Social regulation of the brain: sex, size and status. *Novartis Found. Symp.* 244:169–84; discussion 84–6, 203–6, 53–7

Fernald RD, Hirata NR. 1977a. Field study of haplochromis-burtoni habitats and cohabitant. *Environ. Biol. Fishes* 2:299–308

Fernald RD, Hirata NR. 1977b. Field study of haplochromis-burtoni quantitative behavioral observations. *Anim. Behav.* 25:964–75

Firestein S. 2001. How the olfactory system makes sense of scents. *Nature* 413:211–18

Fowler C, Liu Y, Ouimet C, Wang Z. 2002. The effects of social environment on adult neurogenesis in the female prairie vole. *J. Neurobiol.* 51:115–28

Fox HE, White SA, Kao MH, Fernald RD. 1997. Stress and dominance in a social fish. *J. Neurosci.* 17:6463–69

Francis RC, Soma K, Fernald RD. 1993. Social regulation of the brain-pituitary-gonadal axis. *Proc. Natl. Acad. Sci. USA* 90:7794–98

Gauthier I, Skudlarski P, Gore J, Anderson A. 2000. Expertise for cars and birds recruits brain areas involved in face recognition. *Nat. Neurosci.* 3:191–97

Gerlai R, Clayton NS. 1999. Analysing hippocampal function in transgenic mice: an ethological perspective. *Trends Neurosci.* 22:47–51

Getz LL, McGuire B, Pizzuto T, Hoffman JE, Frase B. 1993. Social organization of the prairie vole (*Microtus ochrogaster*). *J. Mammal.* 74:44–58

Gingrich B, Liu Y, Cascio C, Wang Z, Insel TR. 2000. Dopamine D2 receptors in the nucleus accumbens are important for social attachment in female prairie voles. *Behav. Neurosci.* 114:173–83

Giordano AL, Johnson AE, Rosenblatt JS. 1990. Haloperidol-induced disruption of retrieval behavior and reversal with apomorphine in lactating rats. *Physiol. Behav.* 48:211–14

Hall FS. 1998. Social deprivation of neonatal, adolescent, and adult rats has distinct neurochemical and behavioral consequences. *Crit. Rev. Neurobiol.* 12:129–62

Hammock E, Young LJ. 2002. Variations in the vasopressin V1a promoter and expression: implications for inter- and intraspecific variations in social behaviour. *Eur. J. Neurosci.* 16:399–402

Hansen S, Bergvall AH, Nyiredi S. 1993. Interaction with pups enhances dopamine release in the ventral striatum of maternal rats: a microdialysis study. *Pharmacol. Biochem. Behav.* 45:673–76

Hansen S, Harthon C, Wallin E, Lofberg L, Svensson K. 1991. The effects of 6-OHDA-induced dopamine depletions in the ventral or dorsal striatum on maternal and sexual behavior in the female rat. *Pharmacol. Biochem. Behav.* 39:71–77

Heinrich JE, Singh TD, Nordeen KW, Nordeen EJ. 2003. NR2B downregulation in a forebrain region required for avian vocal learning is not sufficient to close the sensitive period for song learning. *Neurobiol. Learn. Mem.* 79:99–108

Heinrich JE, Singh TD, Sohrabji F, Nordeen KW, Nordeen EJ. 2002. Developmental and hormonal regulation of NR2A mRNA in forebrain regions controlling avian vocal learning. *J. Neurobiol.* 51:149–59

Henderson ND. 1970. Genetic influences on the behavior of mice can be obscured by laboratory rearing. *J. Comp. Physiol. Psychol.* 72:505–11

Hess EH. 1972. Imprinting in a natural laboratory. *Sci. Am.* 227:24–31

Hessler NA, Doupe AJ. 1999. Social context modulates singing-related neural activity in the songbird forebrain. *Nat. Neurosci.* 2:209–11

Hofer MA, Brunelli SA, Shair HN. 1993. The effects of 24-hr maternal separation and of litter size reduction on the isolation-distress response of 12-day-old rat pups. *Dev. Psychobiol.* 26:483–97

Hofmann HA, Benson ME, Fernald RD. 1999. Social status regulates growth rate: consequences for life-history strategies. *Proc. Natl. Acad. Sci. USA* 96:14171–76

Hogan JA. 1988. Cause and function in the development of behavior systems. In *Developmental Psychobiology and Behavioral Ecology*, ed. EM Blass, pp. 63–106. New York: Plenum

Holy TE, Dulac C, Meister M. 2000. Responses of vomeronasal neurons to natural stimuli. *Science* 289:1569–72

Horn G. 1985. *Memory, Imprinting and the Brain: An Inquiry into Mechanisms*. Oxford: Clarendon. 400 pp.

Horn G, McCabe BJ. 1984. Predispositions and preferences: effects on imprinting of lesions to the chick brain. *Anim. Behav.* 32:288–92

House J, Landis K, Umberson D. 1988. Social relationships and health. *Science* 241:540–45

Insel TR. 1997. A neurobiological basis of social attachment. *Am. J. Psychiatry* 154:726–35

Insel TR, Gingrich B, Young LJ. 2001. Oxytocin: Who needs it? In *The Maternal Brain: Neurobiological and Neuroendocrine Adaptation and Disorders in Pregnancy and Postpartum*, ed. JA Russell, A Douglas, R Windle, C Ingram, pp. 59–67. Amsterdam: Elsevier

Insel TR, Winslow JT, Witt DM. 1992. Homologous regulation of brain oxytocin receptors. *Endocrinology* 130:2602–8

Insel TR, Young LJ. 2000. Neuropeptides and

the evolution of social behavior. *Curr. Opin. Neurobiol.* 10:784–89

Insel TR, Young LJ. 2001. The neurobiology of attachment. *Nat. Rev. Neurosci.* 2:129–36

Insel TR, Young LJ, Witt DM, Crews D. 1993. Gonadal steroids have paradoxical effects on brain oxytocin receptors. *J. Neuroendocrinol.* 5:619–28

Ishii T, Hirota J, Mombaerts P. 2003. Combinatorial coexpression of neural and immune multigene families in mouse vomeronasal sensory neurons. *Curr. Biol.* 13:394–400

Johns JM, Noonan LR, Zimmerman LI, Li L, Pedersen CA. 1997. Effects of short- and long-term withdrawal from gestational cocaine treatment on maternal behavior and aggression in Sprague-Dawley rats. *Dev. Neurosci.* 19:368–74

Johnson AE, Ball GF, Coirini H, Harbaugh CR, McEwen BS, Insel TR. 1989a. Time course of the estradiol-dependent induction of oxytocin receptor binding in the ventromedial hypothalamic nucleus of the rat. *Endocrinology* 125:1414–19

Johnson MH, Davies DC, Horn G. 1989b. A sensitive period for the development of a predisposition in dark-reared chicks. *Anim. Behav.* 37:1044–46

Johnson MH, Horn G. 1988. Development of filial preferences in dark-reared chicks. *Anim. Behav.* 36:675–83

Kanwisher N, McDermott J, Chun M. 1997. The fusiform face area: a module in human extrastriate cortex specialized for face perception. *J. Neurosci.* 17:4302–11

Keer SE, Stern JM. 1999. Dopamine receptor blockade in the nucleus accumbens inhibits maternal retrieval and licking, but enhances nursing behavior in lactating rats. *Physiol. Behav.* 67:659–69

Kelley AE, Berridge KC. 2002. The neuroscience of natural rewards: relevance to addictive drugs. *J. Neurosci.* 22:3306–11

Kendrick KM, Hinton MR, Atkins K, Haupt MA, Skinner JD. 1998. Mothers determine sexual preferences. *Nature* 395:229–30

Kinsley CH. 1994. Developmental psychobiological influences on rodent parental behavior. *Neurosci. Biobehav. Rev.* 18:269–80

Klin A, Jones W, Schultz RT, Volkmar F, Cohen DJ. 2002. Visual fixation patterns during viewing of naturalistic social situations as predictors of social competence in individuals with autism. *Arch. Gen. Psychiatry* 59:809–16

Kringelbach M, O'Doherty J, Rolls E, Andrews C. 2003. Activation of the human orbitofrontal cortex to a liquid food stimulus is correlated with its subjective pleasantness. *Cereb. Cortex* 13:1064–71

Landgraf R, Frank E, Aldag J, Neumann I, Sharer C, et al. 2003. Viral vector mediated gene transfer of the vole V1a vasopressin receptor in the rat septum: improved social discrimination and active social behaviour. *Eur. J. Neurosci.* 18:403–11

Leckman JF, Herman AE. 2001. Maternal behavior and developmental psychopathology. *Biol. Psychiatry* 51:27–43

Lederhendler I, Schulkin J. 2000. Behavioral neuroscience: challenges for the era of molecular biology. *Trends Neurosci.* 23:451–54

Lee A, Li M, Watchus J, Fleming AS. 1999. Neuroanatomical basis of maternal memory in postpartum rate: selective role for the nucleus accumbens. *Behav. Neurosci.* 113:523–38

Leinders-Zufall T, Lane AP, Puche AC, Ma W, Novotny MV, et al. 2000. Ultrasensitive pheromone detection by mammalian vomeronasal neurons. *Nature* 405:792–96

Lim MM, Murphy AZ, Young LJ. 2004. Ventral striatopallidal oxytocin and vasopressin V1a receptors in the monogamous prairie vole (Microtus ochrogaster). *J. Comp. Neurol.* 468:555–70

Liu Y, Curtis J, Wang Z. 2001. Vasopressin in the lateral septum regulates pair bond formation in male prairie voles. *Behav. Neurosci.* 115:910–19

Loconto J, Papes F, Chang E, Stowers L, Jones EP, et al. 2003. Functional expression of murine V2R pheromone receptors involves selective association with the M10 and M1

families of MHC class 1b molecules. *Cell* 112:607–18
Lonstein JS, Simmons DA, Swann JM, Stern JM. 1998. Forebrain expression of c-fos due to active maternal behaviour in lactating rats. *Neuroscience* 82:267–81
Lord C, Cook E, Leventhal B, Amaral D. 2000. Autism spectrum disorders. *Neuron* 28:355–63
Lorenz KZ. 1935. Der Kumpan in der Umwelt des Vogels. *J. Ornithol.* 83:137–215
Luo M, Fee MS, Katz LC. 2003. Encoding pheromonal signals in the accessory olfactory bulb of behaving mice. *Science* 299: 1196–201
Lu HC, Gonzalez E, Crair MC. 2001. Barrel cortex critical period plasticity is independent of changes in NMDA receptor subunit composition. *Neuron* 32:619–34
Martin A, Weisberg J. 2003. Neural foundations of understanding social and mechanical concepts. *Cogn. Neuropsychol.* 20:575–87
Mattson BJ, Williams S, Rosenblatt JS, Morrell JI. 2001. Comparison of two positive reinforcing stimuli: pups and cocaine throughout the postpartum period. *Behav. Neurosci.* 115:683–94
McCabe BJ, Horn G. 1991. Synaptic transmission and recognition memory—time course of changes in N-methyl-D-aspartate receptors after imprinting. *Behav. Neurosci.* 105:289–94
McCabe BJ, Horn G. 1994. Learning-related changes in Fos-like immunoreactivity in the chick forebrain after imprinting. *Proc. Natl. Acad. Sci. USA* 91:11417–21
Meaney MJ, Diorio J, Francis D, Widdowson J, LaPlante P, et al. 1996. Early environmental regulation of forebrain glucocorticoid receptor gene expression: implications for adrenocortical responses to stress. *Dev. Neurosci.* 18:49–72
Meredith M. 2001. Human vomeronasal organ function: a critical review of best and worst cases. *Chem. Senses* 26:433–45
Mombaerts P, Wang F, Dulac C, Chao SK, Nemes A, et al. 1996. Visualizing an olfactory sensory map. *Cell* 87:675–86
Morris RGM. 1981. Spatial localization does not require the presence of local cues. *Learn. Motiv.* 12:239–60
Nishimori K, Young LJ, Guo Q, Wang Z, Insel TR, Matzuk M. 1996. Oxytocin is required for nursing but is not essential for parturition or reproductive behavior. *Proc. Natl. Acad. Sci. USA* 93:777–83
Numan M. 1994. Maternal behavior. In *The Physiology of Reproduction*, ed. E Knobil, J Neill, pp. 221–302. New York: Raven
Numan M, Insel TR. 2003. *The Neurobiology of Parental Behavior*. New York: Springer-Verlag
Numan M, Smith HG. 1984. Maternal behavior in rats: evidence for the involvement of preoptic projections to the ventral tegmental area. *Behav. Neurosci.* 98:712–27
O'Connor TG, Marvin RS, Rutter M, Olrick JT, Britner PA, Engl. Rom. Adopt. Study Team. 2003. Child-parent attachment following early institutional deprivation. *Dev. Psychopathol.* 15:19–38
Onitsuka T, Shenton M, Kasai K, Nestor P, Toner S, et al. 2003. Fusiform gyrus reduction and facial recognition in chronic schizophrenia. *Arch. Gen. Psychiatry* 60: 349–55
Pedersen CA, Caldwell JO, Walker C, Ayers G, Mason GA. 1994. Oxytocin activates the postpartum onset of rat maternal behavior in the ventral tegmental and medial preoptic areas. *Behav. Neurosci.* 108:1163–71
Pfaff DW, Frohlich J, Morgan M. 2002. Hormonal and genetic influences on arousal—sexual and otherwise. *Trends Neurosci.* 25: 45–50
Pfaus JG, Kippin TE, Centeno S. 2001. Conditioning and sexual behavior: a review. *Horm. Behav.* 40:291–321
Pitkow L, Sharer C, Ren X, Insel TR, Terwilliger E, Young LJ. 2001. Facilitation of affiliation and pair-bond formation by vasopressin receptor gene transfer into the ventral forebrain of a monogamous vole. *J. Neurosci.* 21:7392–96
Robbins TW, Jones GH, Wilkinson LS. 1996. Behavioural and neurochemical effects of

early social deprivation in the rat. *J. Psychopharmacol.* 10:39–47

Romanski L, Goldman-Rakic P. 2002. An auditory domain in primate prefrontal cortex. *Nat. Neurosci.* 5:15–16

Rosenzweig MR, Bennett EL. 1996. Psychobiology of plasticity: effects of training and experience on brain and behavior. *Behav. Brain Res.* 78:57–65

Russell JA, Leng G. 1998. Sex, parturition and motherhood without oxytocin? *J. Endocrinol.* 157:343–59

Sam M, Vora S, Malnic B, Ma W, Novotny MV, Buck LB. 2001. Odorants may arouse instinctive behaviours. *Nature* 412:142

Schultz RT, Gauthier I, Klin A, Fulbright RK, Anderson AW, et al. 2000. Abnormal ventral temporal cortical activity during face discrimination among individuals with autism and Asperger syndrome. *Arch. Gen. Psychiatry* 57:331–40

Shahbazian M, Zoghbi H. 2002. Rett syndrome and MeCP2: linking epigenetics and neuronal function. *Am. J. Hum. Genet.* 71:1259–72

Stack EC, Balakrishnan R, Numan MJ, Numan M. 2002. A functional neuroanatomical investigation of the role of the medial preoptic area in the neural circuits regulating maternal behavior. *Behav. Brain Res.* 131:17–36

Stack EC, Numan M. 2000. The temporal course of expression of c-Fos and Fos B within the medial preoptic area and other brain regions of postpartum female rats during prolonged mother-young interactions. *Behav. Neurosci.* 114:609–22

Steffanaci L, Amaral D. 2002. Some observations on cortical inputs to the macaque monkey amygdala: an anterograde tracing study. *J. Comput. Neurol.* 451:301–23

Stern JM, Keer SE. 1999. Maternal motivation of lactating rats is disrupted by low dosages of haloperidol. *Behav. Brain Res.* 99:231–39

Stowers L, Holy TE, Meister M, Dulac C, Koentges G. 2002. Loss of sex discrimination and male-male aggression in mice deficient for TRP2. *Science* 295:1493–500

Sullivan RM, Wilson DA. 2003. Molecular biology of early olfactory memory. *Learn. Mem.* 10:1–4

Tomizawa K, Iga N, Lu Y, Moriwaki A, Matsushita M, et al. 2003. Oxytocin improves long-lasting spatial memory during motherhood through MAP kinase cascade. *Nat. Neurosci.* 6:384–90

Trinh K, Storm DR. 2003. Vomeronasal organ detects odorants in absence of signaling through main olfactory epithelium. *Nat. Neurosci.* 6:519–25

Tsao D, Freiwald W, Knutsen T, Mandeville J, Tootell R. 2003. Faces and objects in macaque cerebral cortex. *Nat. Neurosci.* 6:989–95

van Praag H, Kempermann G, Gage FH. 2000. Neural consequences of environmental enrichment. *Nat. Rev. Neurosci.* 1:191–98

von Uexküll J. 1921. *Umwelt und Innenwelt der Tiere*. Berlin: Springer

Wang Z, Yu GZ, Cascio C, Liu Y, Gingrich B, Insel TR. 1999. Dopamine D2 receptor-mediated regulation of partner preferences in female prairie voles: a mechanism for pair bonding. *Behav. Neurosci.* 113:602–11

White SA, Kasten TL, Bond CT, Adelman JP, Fernald RD. 1995. Three gonadotropin-releasing hormone genes in one organism suggest novel roles for an ancient peptide. *Proc. Natl. Acad. Sci. USA* 92:8363–67

White SA, Nguyen T, Fernald RD. 2002. Social regulation of gonadotropin-releasing hormone. *J. Exp. Biol.* 205:2567–81

Wiedenmayer C. 1997. Causation of the ontogenetic development of stereotypic digging in gerbils. *Anim. Behav.* 53:461–70

Winslow JT, Camacho F. 1995. Cholinergic modulation of a decrement in social investigation following repeated contacts between mice. *Psychopharmacology* 121:164–72

Winslow JT, Hearn EF, Ferguson J, Young LJ, Matzuk MM, Insel TR. 2000. Infant vocalization, adult aggression, and fear behavior of an oxytocin null mutant mouse. *Horm. Behav.* 37:145–55

Witt DM. 1995. Oxytocin and rodent sociosexual responses: from behavior to gene expression. *Neurosci. Biobehav. Rev.* 19:315–24

Wrenn C, Harris A, Saavedra M, Crawley J. 2003. Social transmission of food preference in mice: methodology and application to galanin-overexpressing transgenic mice. *Behav. Neurosci.* 117:21–31

Wurbel H. 2001. Ideal homes? Housing effects on rodent brain and behaviour. *Trends Neurosci.* 24:207–11

Young LJ, Gingrich B, Lim MM, Insel TR. 2001. Cellular mechanisms of social attachment. *Horm. Behav.* 40:133–39

Young LJ, Huot B, Nilsen R, Wang Z, Insel TR. 1996. Species differences in central oxytocin receptor gene expression: comparative analysis of promoter sequences. *J. Neuroendocrinol.* 8:777–83

Young LJ, Nilsen R, Waymire K, MacGregor G, Insel TR. 1999. Increased affiliative response to vasopressin in mice expressing the V1a receptor from a monogamous vole. *Nature* 400:766–68

Zou Z, Horowitz LF, Montmayeur JP, Snapper S, Buck LB. 2001. Genetic tracing reveals a stereotyped sensory map in the olfactory cortex. *Nature* 414:173–79

Annu. Rev. Neurosci. 2004. 27:723–49
doi: 10.1146/annurev.neuro.27.070203.144244

First published online as a Review in Advance on April 2, 2004

UNRAVELING THE MECHANISMS INVOLVED IN MOTOR NEURON DEGENERATION IN ALS

Lucie I. Bruijn,[1] Timothy M. Miller,[2] and Don W. Cleveland[2]
[1]The ALS Association, Guilford, Connecticut 06437; email: lbruijn@snet.net
[2]Departments of Medicine and Neurosciences and the Ludwig Institute for Cancer Research, University of California at San Diego, La Jolla, California 92093; email: dcleveland@ucsd.edu; timiller@ucsd.edu

Key Words amyotrophic lateral sclerosis, neurodegenerative, SOD1, superoxide dismutase, Lou Gehrig's disease

■ **Abstract** Although Charcot described amyotrophic lateral sclerosis (ALS) more than 130 years ago, the mechanism underlying the characteristic selective degeneration and death of motor neurons in this common adult motor neuron disease has remained a mystery. There is no effective remedy for this progressive, fatal disorder. Modern genetics has now identified mutations in one gene [Cu/Zn superoxide dismutase (SOD1)] as a primary cause and implicated others [encoding neurofilaments, cytoplasmic dynein and its processivity factor dynactin, and vascular endothelial growth factor (VEGF)] as contributors to, or causes of, motor neuron diseases. These insights have enabled development of model systems to test hypotheses of disease mechanism and potential therapies. Along with errors in the handling of synaptic glutamate and the potential excitotoxic response this provokes, these model systems highlight the involvement of nonneuronal cells in disease progression and provide new therapeutic strategies.

INTRODUCTION

Amyotrophic lateral sclerosis (ALS), commonly known in the United States as Lou Gehrig's disease, is the most common adult motor neuron disease. Described in 1869 by the great French neurobiologist and physician Jean-Martin Charcot, the disease's primary hallmark is the selective dysfunction and death of the neurons in the motor pathways. This leads to spasticity, hyperreflexia (upper motor neurons), generalized weakness, muscle atrophy, and paralysis (lower motor neurons) (Mulder et al. 1986). Failure of the respiratory muscles is generally the fatal event, occurring within one to five years of onset. Selectivity of killing occurs even among motor neurons: The neurons that control the bladder (i.e., Onuf's nucleus in the sacral cord) and eye movements are relatively spared.

Although approximately 5–10% of ALS cases are inherited (familial), the majority of cases have no genetic component (sporadic). The typical age of onset for both forms is between 50 and 60 years. The lifetime risk is approximately 1 in 2000. This corresponds to ~30,000 affected individuals in the United States and ~5000 in the United Kingdom. Motor neuron loss is accompanied by reactive gliosis (Leigh & Swash 1991), intracytoplasmic neurofilament abnormalities, and axonal spheroids (Carpenter 1968, Gonatas et al. 1992, Hirano et al. 1984, Leigh et al. 1991). In end-stage disease, there is significant loss of large myelinated fibers in the corticospinal tracts and ventral roots as well as evidence of Wallerian degeneration and atrophy of the myelinated fibers (Delisle & Carpenter 1984).

The causes for most cases of ALS are unknown and the clinical course is highly variable, suggesting that multiple factors underly the disease mechanism. This review focuses on the proposed mechanisms involved in ALS, why motor neurons are a key target in the disease, and how other cell types may contribute to disease.

The Genetics of Motor Neuron Disease

ALS is a member of a group of heterogeneous disorders that affect the survival and function of upper or lower motor neurons. For some of these disorders, a genetic link has been determined (Table 1). Approximately 10% of classical ALS is familial. Among the familial cases, approximately 20% are caused by dominantly inherited mutations in the protein Cu/Zn superoxide dismutase (SOD1) (Rosen et al. 1993). The striking pathological and clinical similarity between familial and sporadic disease has sparked enthusiasm that the animal models based on mutant SOD1 might provide insight into mechanisms of both sporadic and familial disease. However, to date, there is no direct evidence validating this assumption.

Using modern gene mapping methods, researchers have identified a new gene linked to a rare, recessively inherited juvenile or infantile onset form that progresses slowly (Hadano et al. 2001, Yang et al. 2001). The gene, localized to chromosome 2, encodes a 184-kD protein (named ALS2 or alsin) with three putative guanine-nucleotide-exchange factor (GEF) domains. Small GTP-binding proteins of the Ras superfamily act as molecular switches in signal transduction, affecting cytoskeletal dynamics, intracellular trafficking, and other important biological processes. GEFs catalyze the dissociation of the tightly bound GDP from the small G protein in response to upstream signals. Although widely expressed, the ALS2 protein is enriched in nervous tissue, where it is peripherally bound to the cytoplasmic face of endosomal membranes, an association that requires the amino-terminal RCC1-like GEF domain (Yamanaka et al. 2003). The G protein(s) upon which the ALS2 GEFs act has not been identified, although an initial report has shown that ALS2 can act in vitro as an exchange factor for Rab5a (Otomo et al. 2003), which functions in endosomal trafficking. All of the disease-causing mutants are highly unstable (Yamanaka et al. 2003): This has led to the conclusion that early-onset motor neuron disease is caused by loss of activity of one or more of the GEF domains of this endosomal GEF.

TABLE 1 Genetics of ALS-related diseases

Disease	Inheritance	Linkage	Gene/protein	Onset	Features
Sporadic ALS	None	None	Largely unknown, ?NF-H, ?EAAT2	Adult	>90% of ALS
Familial ALS					
ALS	Dominant	21q22.1	SOD1	Adult	20% of inherited cases; more than 90 mutants known, all but one of which are dominant (Rosen et al. 1993)
ALS	Dominant	16	?	Adult	Typical adult onset, familial ALS without SOD1 mutations (Abalkhail et al. 2003, Ruddy et al. 2003, Sapp et al. 2003)
ALS	Dominant	18	?	Adult	Typical adult onset, famililal ALS without SOD1 mutations (Hand et al. 2002, Sapp et al. 2003)
ALS	Dominant	20	?	Adult	Typical adult onset, familial ALS without SOD1 mutations (Sapp et al. 2003)
ALS with frontotemporal dementia	Dominant	9q21-22		Adult	Dementia (Hosler et al. 2000)
ALS with dementia, Parkinsonism	Dominant	17q21	Tau	Adult	Dementia > Parkinsonism, amotrophy (Wilhermsen 1997)
ALS	X-linked	Xp11-Xq12	?	Adult	
Juvenile-type 1	Recessive	15q15-22	?	Adolescence	Slowly progressing (Hentati et al. 1998)
Juvenile-type 3	Recessive	2q33	ALS2	Adolescence	Slowly progressing; mutant gene product appears to be a guanine exchange factor (GEF) (Hadano et al. 2001, Yang et al. 2001)
Juvenile	Dominant	9q34 (ref 84)	?	Before 25	Slowly progressing (Blair et al. 2000)

Efforts to find new genes linked to the remainder of familial ALS cases continue. The identification of three separate families with linkage to chromosome 16 (Abalkhail et al. 2003, Ruddy et al. 2003, Sapp et al. 2003) has significantly narrowed the region of interest, and rapid identification of the gene involved is anticipated. In addition, efforts to identify disease genes for chromosomes 18 and 20 are under way (Table 1). Researchers have just begun to identify other genetic contributors, including one dominant locus on mouse chromosome 13 that can sharply slow initiation of SOD1 mutant mediated disease (Kunst et al. 2000).

Mechanisms of Selective Motor Neuron Death

There are many animal models of motor neuron degeneration, ranging from chronic aluminum toxicity in rodents to spinal muscular atrophy in cows. Because each of these models may provide new, important information about the motor neuron, they may be relevant to ALS. However, given space constraints, the disease mechanism part of this review focuses almost exclusively on rodent models of mutations in SOD1. This bias is based on the phenotypic and pathologic correlation between sporadic and familial human SOD1 ALS cases and the close resemblance of the mouse model to human disease (discussed below).

A TOXIC PROPERTY FROM CU/ZN SOD1 MUTATIONS

The best-known function of SOD1, a homodimer of an ubiquitously expressed 153–amino acid polypeptide, is to convert superoxide, a toxic by-product of mitochondrial oxidative phosphorylation, to water or hydrogen peroxide. Catalysis by SOD1 is mediated in two asymmetric steps by an essential copper atom, which is alternately reduced and oxidized by superoxide. The disease-causing mutations are scattered throughout the primary and three-dimensional structure of the protein. More than 100 mutations are known (Andersen 2000, Andersen et al. 2001, Gaudette et al. 2000), and all but one mutation, $SOD1^{D90A}$ (aspartate substituted to alanine at reside 90), cause dominantly inherited disease. An updated list of the mutations can be found at the online database for ALS genetics, http://www.alsod.org.

How mutant SOD1 leads to motor neuron degeneration remains unclear. However, it is well established that SOD1-mediated toxicity in ALS is not due to loss of function but instead to a gain of one or more toxic properties that are independent of the levels of SOD1 activity. The main arguments against the importance of loss of dismutase function are that (*a*) SOD1 null mice do not develop motor neuron disease (Reaume et al. 1996) and (*b*) removal of the normal SOD1 genes in mice that develop motor neuron disease from expressing a dismutase inactive mutant ($SOD1^{G85R}$) does not affect onset or survival (Bruijn et al. 1998). In addition, levels of SOD1 activity do not correlate with disease in mice or humans; in fact, some mutant enzymes retain full dismutase activity (Borchelt et al. 1994, Bowling et al. 1995). Finally, chronic increase in the levels of wild-type SOD1 (and dismutase activity) either has no effect on disease (Bruijn et al. 1998) or accelerates it (Jaarsma et al. 2000).

Of the more than 100 mutations in humans, 3 ($SOD1^{G85R}$, $SOD1^{G37R}$, and $SOD1^{G93A}$) have been extensively characterized in transgenic mouse models of ALS (Bruijn & Cleveland 1996, Gurney 1994, Ripps et al. 1995, Wong et al. 1995). In these mice, the mutant human protein is ubiquitously expressed (under control of the human or mouse SOD1 gene promoter) at levels equal to or several fold higher than the level of endogenous SOD1. Unlike the variable pattern of weakness in humans, weakness typically starts in the hind limbs in mice between 3 and 12 months of age, depending on both the mutation and the level to which it is expressed. Hind limb weakness coincides with increased astrogliosis, activation of microglia, and loss of spinal cord motor neurons. Pathology in these mice closely mimics many aspects of the human disease.

A feature common to all examples of SOD1 mutant-mediated disease in mice is prominent ubiquitin-positive, intracellular aggregates of SOD1 in motor neurons and astroctyes (Figure 1, see color insert). Vacuolar pathology that at least partially represents damaged mitochondria is seen in motor neurons of mice or rats expressing high levels of $SOD1^{G93A}$ (Dal Canto & Gurney 1994, Jaarsma et al. 2001, Howland et al. 2002) and $SOD1^{G37R}$ (Wong et al. 1995), but it is not seen in disease from other mutants (Bruijn et al. 1997b, Nagai et al. 2001). This may be due to the higher (by as much as 20-fold) accumulated levels of the catalytically active $SOD1^{G93A}$ and $SOD1^{G37R}$ mutants, as compared with the inactive $SOD1^{G85R}$, which are required to provoke disease in mice. This difference is even more provocative for the frameshift mutation $SOD1^{G127X}$ truncated in the last 26 residues of SOD1. Despite accumulation to only $\sim$1% the level of endogenous wild-type SOD1 in mice (i.e., $\sim$1/400 the level required to cause disease from the dismutase-active mutants in mice), this highly toxic mutant provokes disease accompanied by prominent aggregates (Jonsson et al. 2004).

SOD1-MEDIATED TOXICITY IS NONCELL AUTONOMOUS: ARE MOTOR NEURONS DIRECT TARGETS OF DISEASE?

The obvious loss of motor neurons in the spinal cord initially focused attention on how mutant SOD1 may act within motor neurons to provoke neuronal degeneration and death. However, as in almost all prominent examples of inherited human neurodegenerative diseases, the mutant gene products are expressed widely. In the case of SOD1, expression is ubiquitous, raising the possibility that the toxic cascade may be achieved wholly or in part by mutant SOD1 action in the adjacent nonneuronal cells. In the first set of experiments to address this question, mutant SOD1 was expressed selectively in astrocytes (Gong et al. 2000) or in neurons (Pramatarova et al. 2001, Lino et al. 2002). Although there was astrocytosis in the mice with astrocyte-specific expression, none of the mice in these three sets developed motor neuron disease. Because high mutant SOD1 levels were sustained in the mice accumulating mutant SOD1 only in the astrocytes, mutant action solely within astroctyes appears insufficient to cause disease. In the case of neuron-specific expression [using either neurofilament (Pramatarova et al. 2001) or neural-specific enolase promoters (Lino et al. 2002)], no clear outcome emerged. However, the levels of mutant SOD1 may have been too low to yield disease. Further efforts, however, made it clear that toxicity to motor neurons from SOD1 mutants is noncell autonomous, that is, it requires mutant damage not just within motor neurons but also to nonneuronal cells. Clement et al. (2003) demonstrated this with three sets of chimeric mice including both normal and SOD1 mutant–expressing cells. Mutant motor neurons that chronically expressed $SOD1^{G37R}$ or $SOD1^{G93A}$ mutants at levels that cause early-onset disease when expressed ubiquitously could escape degeneration and death if surrounded by a sufficient number of normal nonneuronal cells (Figure 2, see color insert). In some cases, a relatively small minority (5–20%) of normal cells eliminated disease, as well as degeneration and death of mutant motor neurons. Mice with these cells survived to ages at least twice those of the longest lived parental mice that expressed either mutant ubiquitously. Moreover, even within the same animal, the proportions of wild-type cells frequently differed on the two sides of the spinal cord. In two chimeras, all spinal motor neurons were mutant expressing. Although both animals developed disease, in each case twice as many of these mutant-expressing motor neurons survived on the side of the spinal cord with the higher number of wild-type, nonneuronal cells. Perhaps just as important, normal motor neurons surrounded by mutant-expressing nonneuronal cells acquired intraneuronal ubiquitinated deposits, indicating that the mutant-expressing cells had transfered damage to them. These data clearly demonstrate that the cellular neighborhood does matter: The death of the motor neurons depends, at least in part, on a contribution from surrounding glia and possibly other cell types.

OXIDATIVE DAMAGE AND METAL MISHANDLING AS SUSPECTS IN MUTANT SOD1–MEDIATED DISEASE

Hypotheses for Toxicity from Active-Site Copper-Mediated Aberrant Chemistry

The discovery of mutations in SOD1 immediately prompted hypotheses that toxicity is caused by aberrant chemistry of the active copper and zinc sites of the misfolded enzyme (Beckman et al. 1993). This aberrant chemistry could involve greater access of abnormal substrates to the active site or clumsy handling of the copper and/or zinc ions. Even wild-type SOD1 can exhibit additional enzymatic activities including superoxide reductase or oxidase activities (Liochev & Fridovich 2000) in which either the oxidized or reduced forms use substrates other than superoxide, respectively. Two specific candidate chemistries involve use of inappropriate enzymatic substrates, which produces peroxynitrite and/or hydroxyl radical. Peroxynitrite can nitrate tyrosine residues (Beckman et al. 1994), thus damaging proteins in affected cells. It has been proposed that peroxynitrite can be produced when the enzyme runs catalysis backwards, converting oxygen to superoxide, which then combines with nitric oxide (NO) within the active site (Estevez et al. 1999). In vitro, such chemistry can be done by mutants that bind the catalytic copper, and this backwards catalysis is strongly exacerbated by depletion of zinc from SOD1 (Estevez et al. 1999). Acting as chaperones during the synthesis of metalloenzymes, metallothioneins are important in the regulation of zinc bioavailability within cells (Jacob et al. 1998, Palmiter 1998). Three isoforms exist: MT-I and -II are largely glial; MT-III is neuronal. Metallothionein expression is markedly upregulated in the spinal cord of ALS patients and transgenic mice (Blaauwgeers et al. 1996, Gong & Elliott 2000, Nagano et al. 2001). Consistent with zinc depletion as a factor in toxicity, $SOD1^{G93A}$ mice deficient in either MT-I and MT-II (both expressed in glia) or MT-III (accmulated in neurons) exhibit significant accelerations in disease onset (Nagano et al. 2001), although it is not clear that acceleration in either case is due to altered zinc binding to SOD1.

The other proposed substrate is hydrogen peroxide, the normal end product of the oxidized form of the enzyme (SOD1-Cu^{2+}). Use of peroxide by the reduced form of SOD1 (SOD-Cu^{1+}) may produce the highly reactive hydroxyl radical. Wiedau-Pazos et al. (1996) reported a two- to fourfold increase in the use of hydrogen peroxide by two dismutase-active mutants relative to wild-type SOD1 in vitro. Despite their initial attractiveness, neither hypothesis is likely to represent an underlying toxicity common to the ALS-causing mutants. Evidence against these hypotheses includes the following:

1. Although both of the proposed oxidative mechanisms require active-site copper, a mutant missing all four copper-coordinating histidines still causes progressive motor neuron disease (Wang et al. 2003).
2. Although increased levels of free 3-nitrotyrosine have been reported in the spinal cords of human patients (Beal et al. 1997) and in mouse models of ALS (Bruijn et al. 1997a), there is no evidence of increased levels of nitrotyrosine bound to proteins in ALS patients or in most mutant SOD1 mice (Bruijn et al. 1997a). Furthermore, using mass spectrometry, Williamson et al. (2000) detected no increased nitration in neurofilaments isolated from transgenic mice expressing dismutase-active or -inactive mutants.
3. Although increases in hydroxyl radicals, measured by high-performance liquid chromatography, have been reported (Andrus et al. 1998, Hall et al.

1998) in tissue samples from transgenic mice expressing the catalytically active mutant $SOD1^{G93A}$, these increases have not been detected in other mouse models (Bruijn et al. 1997a).

4. Because NO is critical for the formation of peroxynitrate, alteration in NO synthesis would be expected to substantially alter the disease course if peroxynitrite were an important contributor to disease. However, reduction of neuronal nitric oxide synthase (nNOS) either pharmacologically or by genetic manipulation did not result in a change in disease progression (Dawson 2000). Similarly, in mice deletion of an inducible NOS (iNOS), localized to astrocytes and microglia, did not change the survival of mutant SOD1 mice (Son et al. 2001).

Toxicity Despite Reduced or Absent Catalytic Copper

The discovery that copper acquisition by SOD1 in yeast requires a specific copper chaperone for SOD1 (CCS) (Culotta et al. 1997) provided the basis for a test of whether the catalytic copper loading is important in disease. Both human wild-type and mutant SOD1 subunits load copper in vivo through the action of a mammalian CCS (Corson et al. 1998, Wong et al. 2000). Although eliminating CCS in transgenic mice expressing three different mutations, $SOD1^{G37R}$, $SOD1^{G85R}$, and $SOD1^{G93A}$, significantly lowered copper loading onto mutant SOD1 in transgenic mice [measured either with in vivo labeling or by in vitro catalytic activity of the active mutants (Subramaniam et al. 2002)], onset and progression of motor neuron disease was not affected. Furthermore, mutations in SOD1 that disrupt some or all the copper-coordinating residues do not eliminate toxicity (Wang et al. 2002, Wang et al. 2003). A variant of the oxidative proposal is that the mutants may handle copper more clumsily, thereby releasing the metal that can catalyze aberrant chemistry. Such mishandling, however, is unlikely to be an important contributor to disease. Even though the CCS is the major delivery source of copper to mutant SOD1, toxicity is the same in the presence or absence of the CCS (Subramaniam et al. 2002).

Hypothesis of Aberrant Catalysis from Surface-Bound Copper

A final proposal suggests that copper contributes to toxicity via aberrant chemistry of copper bound not to the active site of SOD1 but rather to one or more residues, including Cys111, on the surface of the subunit (Bush 2002). This idea arose from the observation that SOD1 in vitro can bind copper at its surface (Liu et al. 2000) and that a CCS-independent, ~30% elevation in tissue copper occurred in some SOD1 mouse models of disease (Subramaniam et al. 2002). Although this hypothesis cannot be formally excluded, no in vivo evidence supports such surface-copper binding, and the source of such copper is perplexing given evidence that the average pool of free copper is only a small fraction of a molecule per cell (Rae et al. 1999).

A COMMON FEATURE OF SOD1-MEDIATED TOXICITY IS PROTEIN AGGREGATION: BUT WHY SHOULD AGGREGATES BE TOXIC?

The presence of abnormal protein aggregates or inclusions has been described in many neurodegenerative diseases [amyloid and tau in Alzheimer's disease (Selkoe 2001), α-synuclein in Parkinson's disease (Selkoe 2001), and huntingtin in Huntington's disease (Steffan et al. 2001)]. For ALS, prominent, intracellular, cytoplasmic inclusions in motor neurons and in some cases within the astrocytes surrounding them are found in each of the prominent mouse models of SOD1-mediated disease and in all reported instances of human ALS (Bruijn et al. 1998). In the mouse models, these accumulations are highly immunoreactive for SOD1 (Figure 1). At least some misfolded SOD1 aggregates cannot be readily dissociated and are resistant to strong ionic detergents; some also contain covalent adducts to other components that can be detected biochemically in spinal cord extracts of transgenic SOD1 mice long before (Johnston et al. 2000, Wang et al. 2002), or contemporaneous with (Bruijn et al. 1997b), onset of disease.

An unsolved puzzle is whether these aggregates damage motor neurons (or other cells in the spinal cord), and if so through what mechanism(s). Several possible toxicities of the protein aggregates have been proposed (Figure 3, see color insert), including aberrant chemistry; loss of protein function through coaggregation with the aggregates; depletion of protein folding chaperones; dysfunction of the proteasome overwhelmed with undigestable, misfolded protein; and inhibition of specific organelle function, including mictochondria and peroxisomes, through mutant aggregation onto or within such organelles. Consistent with involvement of the ubiquitin-proteasome system, the aggregates are intensely immunoreactive with antibodies to ubiquitin, a feature common to all SOD1 ALS mouse models (Bruijn et al. 1998, Jonsson et al. 2004, Wang et al. 2003) and human SOD1-mediated familial ALS (Bruijn et al. 1998, Kato et al. 2000).

Partial inhibition of the proteasome is sufficient to provoke large aggregates in cultured nonneuronal cells that express SOD1 mutants. This suggests that proteasome activity is limiting so that decrease in it either prevents the appropriate removal of the mutant protein or compromises the removal of even more important components (Johnston et al. 2000). Mutant SOD1s are selectively degraded by action of the RING finger-type E3 ubiquitin ligase, dorfin, followed by subsequent proteosomal degradation. Like its substrate, dorphin is predominantly localized in inclusion bodies in familial and sporadic ALS (Niwa et al. 2002). The finding that ubiquitin-containing aggregates are a frequent feature of sporadic disease (Mather et al. 1993. Leigh & Swash 1991) could link the mechanisms of familial and sporadic ALS. However, the complete picture must involve other proteins besides SOD1 as the source of the aggregates because SOD1-containing aggregates are not a characteristic feature of sporadic disease (Shibata et al. 1996).

In addition to their effect on the ubiquitin-proteasome system, aggregates affect the functional aspects of protein-folding chaperones, including those normally present and those induced by heat or other stresses. The coimmunoprecipitation of HSP70, HSP40, and $\alpha\beta$-crystallin with mutant SOD1 (Shinder et al. 2001) confirms a likely involvement of such components. Perhaps more important is the finding that cultured motor neurons die when exposed to acute expression of mutant SOD1, but this is ameliorated by contemporaneous expression of high levels of HSP70 (Bruening et al. 1999). Although the conspicuous nature of large aggregates in affected tissue makes them attractive candidates for causing disease, whether these aggregates are integrally involved in the disease process or are beneficial in sequestering toxic by-products is unknown.

A CASCADE OF CASPASES MEDIATES MOTOR NEURON DEATH

Although the primary toxicities of the familial ALS–linked mutations of SOD1 remain unresolved, the final event in the death cascade has been partially clarified. Activation of caspase-1, one of the early events in the mechanism of toxicity of SOD1 mutants, occurs months prior to neuronal death and phenotypic disease onset (Pasinelli et al. 2000, Vukosavic et al. 2000). A central feature in cell death mediated by mutant SOD1 is the activation of caspase-3, one of the major cysteine-aspartate proteases responsible for degradation of many key cellular constituents in apoptotic cell death. Caspase-3 activation occurs in motor neurons (Li et al. 2000b, Pasinelli et al. 2000, Vukosavic et al. 2000) and astrocytes (Pasinelli et al. 2000) contemporaneous with the first stages of motor neuron death in all three of the best-studied mouse models. For SOD1^{G93A}, release of cytochrome c from mitochondria is followed by activation of caspase-9 (Guegan et al. 2001), which may be the effector for the subsequent activation of caspases-3 and -7. In vitro, this temporal cascade of caspase activation occurs within the same neuronal cell (Pasinelli et al. 2000), although this has not been firmly established in mice.

A common step toward toxicity of mutant SOD1 is a sequential activation of at least two caspases, which act more slowly here than they do during the apoptotic death processes of development. Moreover, apparent inhibition of one or more caspases in this cascade is beneficial: Despite a short half-life once in aqueous solution, long-term intrathecal administration of the pan-caspase inhibitor (N-benzylocarbonyl-Val-Ala-Asp-fluoromethylketone or zVAD-fmk) prolongs the life of SOD1^{G93A} mice by approximately 27 days (Li et al. 2000). Further evidence that proteins of the cell death pathways are important for SOD1-mediated neuron death includes the demonstration that increasing expression of the anti-apopotic factor Bcl2 slows disease onset and survival of SOD1^{G93A} mice by three to four weeks (Kostic et al. 1997).

The best-case scenario for modulation of the final apoptotic cell death pathway is to find key mediators of programmed cell death that affect motor, but not other, neurons, thereby providing the basis for specificity. The demonstration that cultured embryonic motor neurons, but not other neuronal populations, are sensitive to killing by Fas ligand may provide such an example (Raoul et al. 2002). Activation of Fas receptors is an important pathway in the immune system, where it eliminates virus-infected cells, cancerous cells, or mature T cells at the end of an immune response. Whether Fas activation is important for the survival of adult motor neurons threatened with abnormal SOD1 proteins remains to be determined.

ARE MITOCHONDRIA TARGETS FOR MUTANT SOD1–MEDIATED DAMAGE?

The presence of what appear to be vacuolated mitochondrial remnants within spinal motor neurons very early in the lines of mice accumulating the highest levels of $SOD1^{G93A}$ and $SOD1^{G37R}$ provided an initial suggestion that mitchondria may be primary targets for SOD1 mutant–mediated damage (Dal Canto & Gurney 1994, Kong & Xu 1998, Wong et al. 1995). For $SOD1^{G93A}$, Mattiazzi et al. (2002) later reported an ~25% decrement in some enzymatic activities of the respiratory chain in isolated mitochondria, although this was neither seen before disease onset nor selective to mitochondria from spinal cord. Decreased activity of mitochondrial cytochrome oxidase has been reported for spinal cord extracts from end-stage sporadic human disease. Such enzymatic losses could arise from action of the mutant SOD1 within mitochondria: Using electron microscopy (Higgins et al. 2002, Jaarsma et al. 2001) and its apparent resistance to protease digestion with intact mitochondria isolated from brain (Mattiazzi et al. 2002), mutant $SOD1^{G93A}$ has been reported to be present in spinal cord mitochondria, presumably within the intermembrane space. Partial deficiency of manganese superoxide dismutase 2 (a mitochondrial enzyme localized to the mitochondrial matrix) exacerbates disease in transgenic SOD1 mice (Andreassen et al. 2000), providing further evidence of mitochondria's role in ALS. Moreover, mis-targeting mutant SOD1 to the mitocondrial matrix of cultured cells is toxic (Takeuchi et al. 2002), although there is no evidence that mutant SOD1 ever enters the matrix without addition of a specific targeting sequence. Administration of creatine (Klivenyi et al. 1999), which may enhance energy storage capacity and inhibit the opening of the mitochondrial transition pore, and minocycline (Van Den Bosch et al. 2002, Zhu et al. 2002), a tetracycline derivative believed to inhibit microglial activation and block release of cytochrome c from the mitochondria (Kriz et al. 2002), slows disease by at least two to four weeks in mice expressing dismutase-active mutants $SOD1^{G93A}$ and $SOD1^{G37R}$.

Despite these findings, the evidence that mitochondria are important targets for damage common to SOD1 mutants with different biochemical characters remains contradictory. Mitochondrial pathology has not been found in other rodent models that develop motor neuron disease from expression of lower levels of the same mutants or in any of the models that develop motor neuron disease from expression of mutants without dismutase activity (Ripps et al. 1995, Bruijn et al. 1997a, Wang et al. 2002, Jonsson et al. 2004, Wang et al. 2003). Moreover, although import of mutant SOD1 into mitochondria requires the CCS in yeast (Field et al. 2003), toxicity from SOD1 mutants is unaffected by the absence of the CCS (Subramanium et al. 2002).

NEUROFILAMENTS: A COMPONENT OF SELECTIVITY OF MOTOR NEURON TOXICITY IN ALS

Neurofilaments, the most abundant structural proteins in many types of mature motor neurons, have long been thought to confer some selective vulnerability to the neurons at risk in disease. Neurofilament assembly in axons is essential for establishing proper axonal diameters (Lee & Cleveland 1996, Garcia et al. 2003). Only the largest caliber, neurofilament-rich, lower motor neurons are at risk either in sporadic ALS (Kawamura et al. 1981) or SOD1-mediated disease in mice (Bruijn et al. 1997b). The discovery that accumulation and abnormal assembly of neurofilaments are common pathological hallmarks in several neurodegenerative diseases, including sporadic (Carpenter 1968, Chou & Fakadej 1971, Hirano 1991) and familial ALS (Hirano et al. 1984), infantile spinal muscular atrophy, and hereditary sensory-motor neuropathy, first suggested that aberrant accumulation of neurofilaments may contribute to disease onset or progression. Multiple genetic manipulations involving either overexpression or deletion of various neurofilament subunits in mice expressing mutant SOD1 have confirmed the importance of neurofilaments in the ALS model. Reduction of axonal neurofilaments and contemporaneous increase either of assembled filaments or the neurofilament-M (NF-M) and neurofilament–H (NF-H) subunits within the perikarya slow the onset of SOD1-mediated disease (Supplemental Table 1: Follow the Supplemental Material link from the Annual Reviews home page at http://www.annualreviews.org). Increased expression of NF-H, which traps most neurofilaments within the neuronal cell bodies, has produced the most robust ameliorization of disease in SOD1 mutant mice thus far, extending life span by as much as six months in one SOD1^{G37R} line (Couillard-Despres et al. 1998).

The results of these manipulations are counterintuitive. Perturbing the normal axonal and perikaryal architecture is expected to add insult to the already present SOD1-mediated injury. Two explanations for why this is not the case have been proposed: First, excess neurofilaments function as a buffer or sink for some other deleterious process. For example, increased perikaryal content of the NF-M and NF-H subunits, which carry multiphosphorylation domains that can be substrates for the prominent neuronal cyclin-dependent kinase 5 (CDK5), may serve as a buffer following disregulation of this kinase. Indeed, the CDK5 neuronal activator p35 is cleaved by proteolysis to p25, producing a mislocalized, constitutively activated form of CDK5 within the cell bodies of SOD1^{G37R} mice (Nguyen et al. 2001b). Second, reducing the axonal burden of neurofilaments (for example, by deletion of the NF-L subunit that is required for filament assembly) may reduce the burden on axonal transport, thereby moderating the damage of mutant SOD1 (Williamson et al. 1998).

Are neurofilament mutations relevant to human disease? At least two studies suggest that they are. By examining the repetitive tail domain of NF-H, a set of small in-frame deletions or insertions has been identified in 1% of more than 1300 ALS patients (Al-Chalabi et al. 1999, Tomkins et al. 1998, Figlewicz

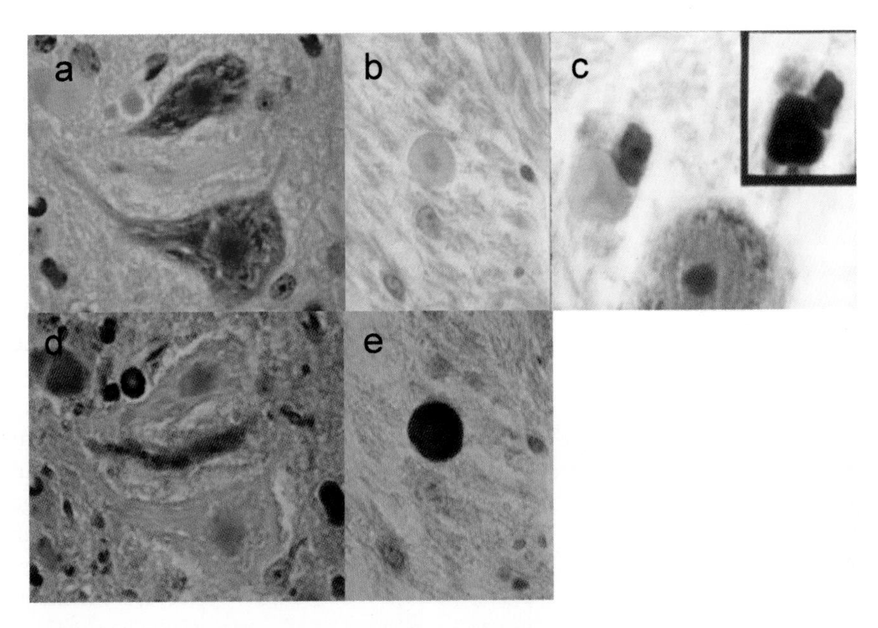

Figure 1 SOD1 reactive inclusions in both motor neurons and astrocytes. (*A*) Motor neuron from an $SOD1^{G85R}$ ALS mouse. (*B*) Motor neuron from an ALS patient with SOD1 frameshift mutation at position 126. (*C*) Astrocyte from a $SOD1^{G85R}$ mouse. Top row (*A*, *B*, *C*) is stained with hematoxylin and eosin. Bottom row (*D*, *E*) and the insert in *C* represent the same sections stained with an antibody to SOD1. These inclusions are also immunoreactive with antibodies to ubiquitin (not shown). Modified with permission from Bruijn et al. 1997b, 1998.

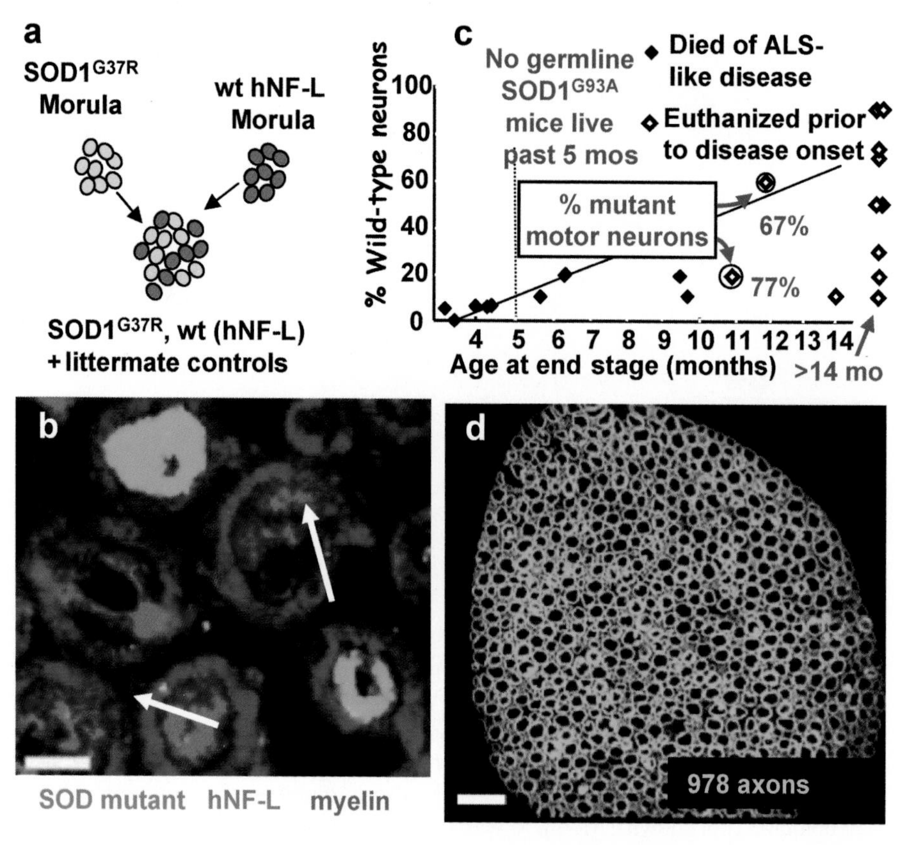

Figure 2 Noncell autonomous toxicity of ALS-causing SOD1 mutants. (*A*) Morula aggregation to produce chimeric mice in which the wild-type neurons were marked by a trace level of human neurofilament-L (NF-L), and mutant neurons and nonneurons were marked by mutant human SOD1. (*B–C*) Even though 30% of mutant motor neurons expressed SOD1^{G37R}, none were killed in one chimeric animal even six months after all mice that expressed the mutant systemically had died from motor neuron loss. (*B*) Immunofluorescent localization of mutant SOD1 (*green*), all axons [with an antibody specific for human NF-L (*red*) and myelin (*blue*)] in a lumbar motor root. (*C*) Robust extension of the life span of chimeric mice with a high proportion of mutant neurons. Chimeras were constructed similar to the scheme in (*A*) except using SOD1^{G93A} mutant morulas. (*D*) There were no signs of degeneration or axon loss, with 978 axons present (normal animals have 927, +/– 99, axons in this root). Scale bar is 40 microns. Reproduced with permission from Clement et al. (2003).

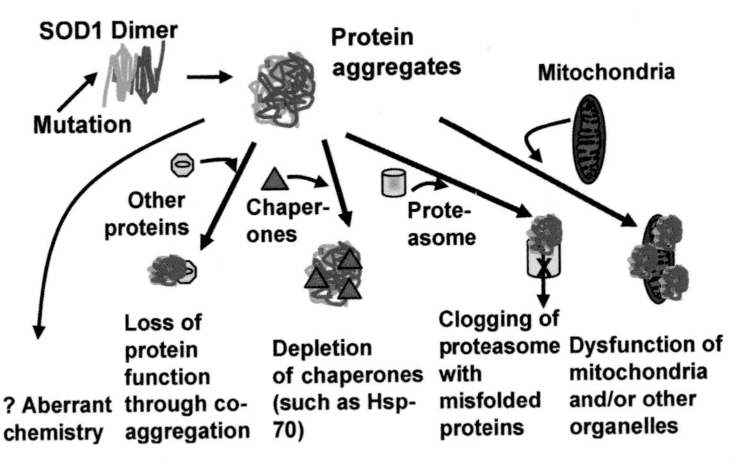

Figure 3 Putative toxicities of protein aggregates. Aberrant chemistry, loss of protein function, depletion of chaperones, loss of proteasome function, and dysfunction of mitochondria are all putative toxicities of protein aggregates.

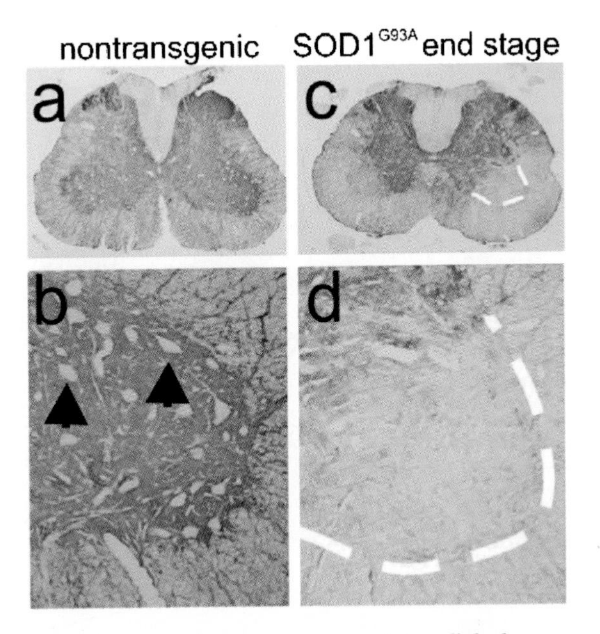

Figure 4 Excitoxicity in ALS. Selective loss of the glial glutamate transporter EAAT2 in the anterior horn during disease within an ALS-linked SOD1 mutant. (*A*) Nearly ubiquitous expression (*brown*) of the glutamate transporter EAAT2 in the gray matter of the spinal cord in nontransgenic animals. (*B*) Higher magnification view, showing EAAT2 staining surrounding the motor neurons (*arrows*). (*C–D*) Striking loss of EAAT2 staining in the anterior horn of end-stage $SOD1^{G93A}$ rats. Reproduced with permission from Howland et al. (2002).

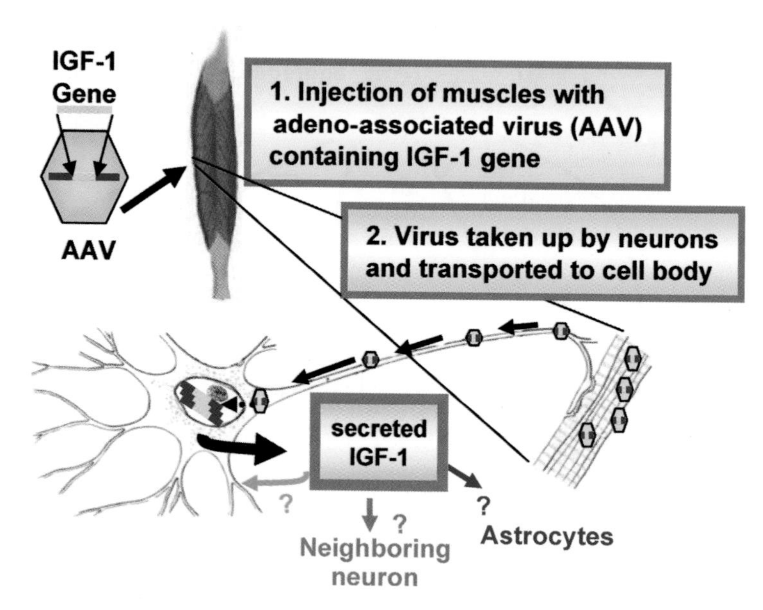

Figure 5 Viral delivery of neurotrophins to motor neurons slows ALS-like disease in an SOD1 mutant model, presumably by forcing the local production of neurotrophins by motor neurons within the spinal cord.

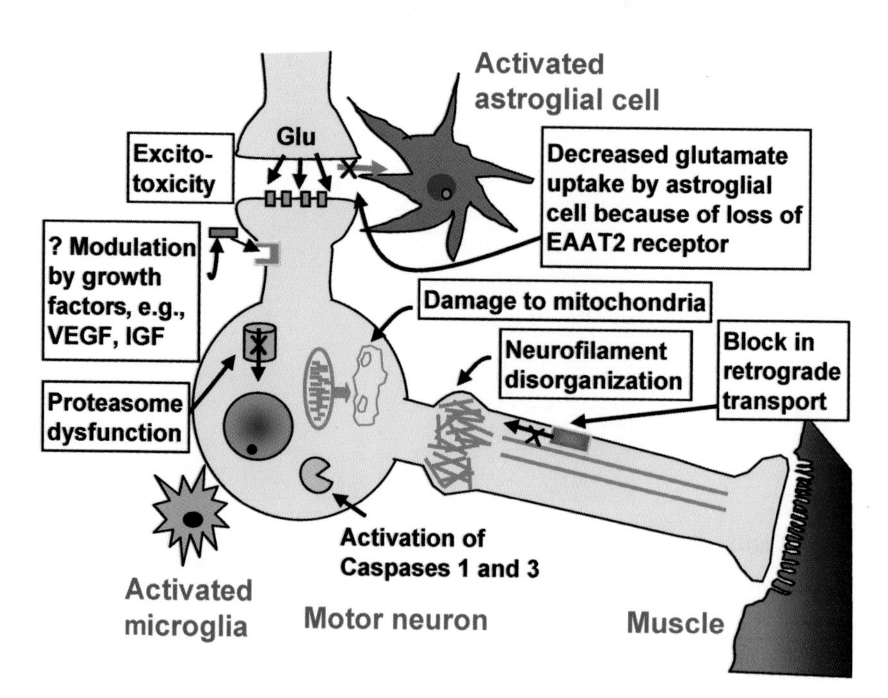

Figure 6 Convergence of multiple pathways that may damage the motor neuron.

et al. 1994). Almost all these mutations appear in sporadic cases, implicating neurofilaments as important risk factors for sporadic disease. Furthermore, investigators (Mersiyanova et al. 2000, De Jonghe et al. 2001) found that a dominant mutation in NF-L is a primary cause of the motor neuropathy Charcot-Marie-Tooth disease (Type II). Taken together, these data suggest that neurofilament content and organization are important contributors and probable risk factors for disease.

PERIPHERIN AND MOTOR NEURON DEGENERATION

Another candidate contributor to motor neuron disease is peripherin, an intermediate protein expressed in spinal motor neurons, peripheral sensory neurons, and autonomic nerves. Corbo & Hays (1992) found peripherin with neurofilament proteins in the majority of axonal inclusions in motor neurons of ALS patients. Increasing expression of the major peripherin isoform (peripherin 58) in motor neurons of transgenic mice leads to late-onset motor neuron disease accompanied by disruption of neurofilament assembly and organization (Beaulieu et al. 1999). In humans, an even more toxic form, peripherin 61, is encoded by an unexpected splice variant in which a part of what is normally an intron provides an additional 3 kD of coding sequence. Expression of peripherin 61 in primary motor neurons is toxic, even at modest levels (Robertson et al. 2003). It is detectable in the lumbar spinal cord of sporadic ALS cases, but not in nondiseased controls (Robertson et al. 2003). Although induced in motor neurons of mutant SOD1 mice, peripherin's role in ALS has been called into question because neither elimination of all isoforms by gene deletion nor overexpression of it in an ALS model had any effect on survival in SOD1 mutant mouse models (Lariviere et al. 2003).

EXCITOTOXICITY

Glutamate-mediated excitotoxicity from repetitive firing and/or elevation of intracellular calcium by calcium-permeable glutamate receptors has long been implicated in neuronal death. In motor neurons, the bulk of the glutamate is actively cleared from the synapse by the glial glutamate transporter, EAAT2, presumably helping to prevent excitotoxicity (Rothstein et al. 1996, Tanaka et al. 1997). Evidence for abnormal glutamate handling in ALS first arose from the discovery that the cerebrospinal fluid of ALS patients had increased glutamate levels (Rothstein et al. 1990, Rothstein et al. 1991, Shaw et al. 1995), a finding now reported in 40% of sporadic ALS patients. Direct measurement of functional glutamate transport in ALS revealed a marked diminution in the affected brain regions, which was the result of pronounced loss of the astroglial EAAT2 protein (Rothstein et al. 1995). Although evidence for EAAT2 loss through RNA mis-splicing in astrocyes (Lin et al. 1998) has not been confirmed in subsequent studies, lowering EAAT2 levels with an antisense oligonucleotide has shown that loss of transport activity directly induces neuronal death (Rothstein et al. 1996). In a patient with sporadic ALS, a mutation in EAAT2 involving a substitution of the putative N-linked glycosylation site asparagine 206 by a serine residue (N206S) was identified (Trotti et al. 2001). Further in vitro studies show that this mutation causes aberrant targeting to the membrane, which decreases glutamate uptake (Trotti et al. 2001), indicating that this may be an important risk factor in the disease. Finally, compared with motor neurons that are spared, the spinal motor neurons at risk are relatively reduced in intracellular calcium-binding components such as calbindin D28k and parvalbumin (Alexianu et al. 1994, Ince et al. 1993), consistent with an excitotoxic component to pathogenesis.

Excitotoxicity has provided one of the few mechanistic links between sporadic and SOD1 mutant–mediated ALS. SOD1 mutants in rodent disease models induce a focal loss (Figure 4, see color insert) of EAAT2 selectively from the astrocytes within the portion of the spinal cord containing the motor neuron cell bodies (Howland et al. 2002). Thus, glutamate excitotoxicity is likely to be an important contributor to neuronal death. Indeed, the only FDA-approved therapy in ALS, Riluzole, functions by decreasing glutamate toxicity.

DYNEIN, DYNACTIN, AND RETROGRADE AXONAL TRANSPORT

A feature that distinguishes motor neurons from other cells is their extreme asymmetry (up to one meter long) and large volume (up to 5000 times that of a typical cell). This size places an enormous metabolic load on the more normally sized cell body whose job is to synthesize the components for this large cell. Components must be transported first into the extended axon. Then via both fast and slow transport, components move toward the synapse. Finally, fast, cytoplasmic dynein–mediated retrograde returns components, including multivesicular bodies and trophic factors such as nerve growth factor, back to the cell body. The finding that SOD1 mutants impair slow axonal transport months prior to disease onset (Williamson et al. 1998) led to the conclusion that diminished transport correlated with the development of motor neuron disease.

Although cytoplasmic dynein has many cellular roles, which include positioning the endoplasmic reticulum and Golgi as well as in assembly of the mitotic spindle, in neurons it is the only known motor for retrograde transport. Two dominant point mutations in it cause a progressive motor neuron disorder in mice (Hafezparast et al. 2003). Similarly, disruption in postnatal motor neurons of the dynactin complex, an activator of cytoplasmic dynein that makes it more processive (King & Schroer 2000), inhibits retrograde axonal transport (LaMonte et al. 2002), provoking a late-onset, progressive motor neuron disease. This is also true in human disease: A dominant point mutation in the p150 subunit of dynactin is the proximal cause of a lower motor neuron disorder that begins with vocal cord paralysis (Puls et al. 2003).

MICROGLIA AND INFLAMMATION

Microglia are the resident immune cells of the central nervous system. They resemble peripheral tissue macrophages and are the primary mediators of neuroinflammation (Kreutzberg 1996). In the healthy adult brain, microglia exist as "resting" microglia, characterized by a small cell body and fine, ramified processes and minimal expression of surface antigens. Upon injury to the central nervous system, these cells become rapidly activated and exert their effects on neurons and macroglia (astrocytes and oligodendrocytes) through the release of cytotoxic and inflammatory substances such as oxygen radicals, NO, glutamate, cytokines, and prostaglandins (Kreutzberg 1996, Hanisch 2002). Microglial involvement has been described in many acute and chronic neurological diseases (Kreutzberg 1996, McGeer & McGeer 1999), although little is known about the role of microglia in ALS. In ALS tissues, there is strong activation and proliferation of microglia in regions of motor neuron loss (Kawamata et al. 1992, Ince et al. 1996). Recent studies showed that expression of proinflammatory mediators [TNF-alpha, Interleukin-1B, and cyclooxygenase 2 (COX-2)] is an early event in mouse models of ALS (Alexianu et al. 2001, Elliott 2001, Nguyen et al. 2001a, Almer et al. 2002, Hensley et al. 2002). Additional evidence for microglial involvement in ALS is provided by pharmocological studies aimed at delaying ALS progression. Minocycline, an antibiotic that also blocks microglial activation (Yrjanheikki et al. 1999), slows disease in ALS mice (Kriz et al. 2002, Van Den Bosch et al. 2002, Zhu et al. 2002). Similarly, inihibition of a key enzyme in prostaglandin synthesis (COX-2) with celecoxib prolongs survival by 25% (Drachman et al. 2002). Both minocylcine and celecoxib are currently being studied in human clinical trials.

GROWTH FACTORS

Vascular Endothelial Growth Factor

A completely unexpected contributor to motor neuron survival emerged from the discovery that ALS-like symptoms and neuropathology can be produced in mice bearing a targeted deletion that eliminates the ability of the vascular endothelial cell growth factor (VEGF) gene to respond to tissue hypoxia (Oosthuyse et al. 2001). VEGF has long been recognized as a crucial factor in controlling the growth and permeability of blood vessels. Hypoxia-induced expression of VEGF through transcription factors that respond to low oxygen tension is crucial for maintaining or restoring the vascular perfusion of normal tissues and for triggering the growth of blood vessels to supply the extraordinary metabolic demands of tumors. Targeted deletion of the hypoxia-response element in the VEGF gene resulted in mice with a normal baseline expression of VEGF but with a pronounced deficit in the ability to induce VEGF in response to hypoxia. Motor neuron deficits first appeared in mice between five and seven months of age and gradually progressed. All the classic features of ALS were observed: accumulation of neurofilaments in spinal cord and brainstem motor neurons, degeneration of motor axons, and the characteristic denervation-induced muscle atrophy.

In a large European study to determine whether alterations in the VEGF gene may be linked to human ALS (Lambrechts et al. 2003), three single nucleotide polymorphisms in the promoter region of the VEGF gene were identified, implicating VEGF as a risk factor in the disease. However, no variations were seen in the hypoxia-response element. The promoter variants in the VEGF gene in these patients coincided with reduced levels of plasma VEGF. In addition, expression of VEGF carrying these single base-pair changes in a culture system lowered the levels of VEGF and resulted in cell death. Decreasing the levels of VEGF in transgenic mice expressing $SOD1^{G93A}$ decreased the age at onset of disease and decreased their life span.

Although these data strongly implicate VEGF as a risk factor for ALS, further validation is required to confirm that VEGF, and not other more important genes in close association with variations in VEGF, is the key gene involved. A subpopulation in England showed no association with these variations (Lambrechts et al. 2003), suggesting that genetic background, the presence of additional modifiers, environment, or lifestyle may also be important. Further studies to determine whether VEGF has a direct function as a neurotrophic factor for motor neurons or a more indirect function in motor neuron survival are under way.

Neurotrophic Factors

Neurotrophic factors selectively regulate the growth and survival of certain populations of neurons in the central and peripheral nervous systems. They have an essential role in neuronal development and in the maintenance of differentiated neurons. Because astrocytes and microglia are important sources of neurotrophic factors, damage to those cells as occurs pathologically in ALS implicates that loss of trophic support is one of the underlying factors causing motor neuron degeneration. Anand et al. (1995) detected decreased levels of ciliary neurotrophic factors (CNTF) in the postmortem tissue of ALS patients, indicating that trophic support may be impaired. Earlier onset of disease occurred in a patient with an SOD1 mutation and a homozygous mutation in the CNTF gene, supporting the idea that neurotrophic factors may play a role in ALS (Giess et al. 2002). However, the discovery that some individuals who lack the CNTF gene do not develop motor neuron disease argues against a strong association between ALS and CNTF (Takahashi et al. 1994). Furthermore, a recent study demonstrated that lack of CNTF does not affect age of onset, clinical presentation, rate of progression, or disease duration in sporadic or familial ($SOD1^{D90A}$) ALS (Al-Chalabi et al. 2003). Other trophic factors that may be involved in ALS pathogenesis include brain-derived neurotrophic factor, glial cell line–derived neurotrophic factor (GDNF), and insulin-like growth factor-I (IGF-1), as they all support the survival of motor neurons in vivo and in vitro (Elliott & Snider 1996, Oppenheim 1996).

Despite the suggestion that trophic factors may be important in ALS, human trials with neurotrophic factors have been disappointing and studies in the mouse models have been relatively unimpressive (Supplemental Table 1). Previous trials based on neurotrophic factors are problematic because the neurotrophin may not have been delivered effectively to the target neurons. In a recent study, Kaspar et al. (2003) used a gene therapy approach to deliver trophic factors directly to neurons, as outlined in Figure 5 (see color insert). Adeno-associated virus (AAV) expressing GDNF or IGF-1 was injected into the hindlimb and intercostal muscles of SOD1^{G93A} mice. IGF-1-containing AAV, which is retrogradely transported back to the neurons of the spinal cord, robustly prolonged survival, although GDNF delivered in a similar manner did not. It is unknown whether the IGF-1, which is probably produced and secreted by the motor neurons, had an effect on neurons producing the IGF-1, neighboring neurons, astrocytes, or all of the above.

MODELING TOXICITY IN ALS

In considering what lessons have now been learned about the mechanisms underlying ALS, the evidence at first glance seems to support a discouraging series of divergent possibilities (Figure 6, see color insert). However, the timing and selectivity for motor neuron killing may arise from the unfortunate convergence of a series of factors—all of which are necessary to place motor neurons at risk—rather than from a single alternative (Figure 6).

For familial ALS, the disease results from an acquired toxicity of mutant SOD1 rather than from any loss of function of the protein. The acquired toxic property affects both neurons and glia, and SOD1 expression in either cell type alone is insufficient to result in disease. The exact nature of this toxicity is uncertain, but in neurons it likely disrupts several basic cellular functions including protein breakdown by the ubiquitin-proteasome system, slow anterograde transport, fast retrograde axonal transport, calcium homeostasis, mitochondrial function, and maintenance of the cytoskeletal architecture. Aggregates are clearly identifiable but thus far have not been directly linked to any of these disruptions in cellular function and could, in theory, be protective rather than pathogenic. Although these changes are occurring in the motor neurons, there is concurrent damage to the astrocytes and activation of microglial cells. The striking loss of glutamate transporters in the astroglial cells likely results in increased levels of glutamate and causes an excitotoxic stress to the struggling motor neurons. Activated microglial cells induce many inflammatory cytokines that may contribute to the death of motor neurons. In the moribund motor neuron, activated caspase-3 deals the final blow.

What about the 98% of disease that does not arise from SOD1 mutations? The convergence model, including a common set of risk factors, may prove correct here too. The key difference is the substitution of an additional combination of initiating genetic or environmental modifiers in lieu of mutant SOD1. Environment can contribute to ALS, as demonstrated by several clusters of disease, e.g., the 50-fold increased risk for ALS in Guam in the 1950s (Mulder & Kurland 1987). More recently, a twofold increased risk of ALS has been recognized among veterans of the Gulf War (Haley 2003, Horner et al. 2003). All the individual events described above likely contribute to the selective loss of motor neurons, thus each represents a target for therapy. Yet despite, or perhaps because of, the multiple possible targets, ALS therapy in both mice and humans has thus far yielded disappointing results.

CHALLENGES: DEVELOPING THERAPIES FOR ALS

ALS is a complex disease with multiple causes, making the discovery of effective pharmacologic therapies challenging. Despite the impressive list of therapies attempted in both mouse and humans (Supplemental Table 1), Riluzole is currently the only FDA-approved compound that may slow disease progression and extend survival, although its effect on both is generously described as modest. Nonpharmacologic therapies such as maintaining nutrition and attending to respiratory function have more significant effects than does any currently available medication, but none of these therapeutic efforts significantly slows disease progression. Many neurologists recommend vitamin supplements and antioxidants, and many discuss with patients medications such as minocycline and celecoxib that have shown some efficacy in mouse models. Clinical trials for several of these compounds are ongoing. As in other neurodegenerative diseases, ALS therapies are advancing in three overlapping areas: (*a*) small molecules including "off-the-shelf" compounds; (*b*) delivering protein, DNA, or RNA; and (*c*) novel gene therapy approaches including viral vectors and stem cells.

Therapies that could be quickly translated to clinical practice may be found among the group of small molecules already available, some of which have been approved by the FDA for other uses. The ALS community has recently embarked on such an approach: In a joint effort by the ALS Association and the National Institute of Neurologic Disorders and Stroke, and involving many different laboratories, researchers have screened 1040 compounds; the results have not been published.

In addition to traditional pharmacologic molecules, proteins, DNA, or RNA may affect disease. In previous ALS trials, growth factors were delivered subcutaneously or directly into the cerebral spinal fluid (Supplemental Table 1). None of these trials were particularly effective. As discussed above, one of the major problems may have been failure to deliver the protein to the target tissue. A recent viral-mediated approach may have solved this problem. When injected into muscles, AAV expressing IGF-1 is retrogradely transported to the spinal cord and prolongs survival in ALS mice. A clinical trial using a similar approach in humans is anticipated. Other novel strategies include the development of antisense and small interfering RNA that block the synthesis of SOD1 (Ding et al. 2003).

Gene therapy and stem cell approaches are being actively and enthusiastically considered for ALS. Delivery of proteins by viral vectors (as discussed above) represents one such gene therapy approach. Stem cells could be engineered to produce a growth factor and then injected or placed within the central nervous system. Both mouse and human embryonic stem cells, when cultured in vitro to produce embroid bodies, are able to differentiate into various types of brain cells (Anderson et al. 2001). The ability to push these cells toward a motor neuron lineage was recently demonstrated (Wichterle et al. 2002). Remarkably, when motor neurons that were created in vitro were introduced into a lesioned chick spinal cord, some stem cell–derived motor neurons extended axons and innervated muscle (Wichterle et al. 2002). Other sources of stem cells such as bone marrow (Brazelton et al. 2000) and muscle (Goodell et al. 2001) are being investigated, but it remains controversial whether these stem cells will provide an abundant source of neurons.

Neural cell replacement therapies are based on the idea that neurological function lost to injury or neurodegeneration can be improved by introducing new cells that can form appropriate connections and replace the function of lost neurons. In ALS, it is hard to imagine that many transplanted motor neurons would form appropriate connections with target muscles, where motor axons need to extend distances up to a meter in length to reach the target. An alternative strategy to the Herculean task of replacing the long motor neurons would be to replace nonneurons because normal nonneuronal cells can sharply extend survival of motor neurons that express SOD1 (Clement et al. 2003).

Another approach could be to stimulate endogenous stem cells in the brain or spinal cord to generate new neurons. Contrary to earlier belief, neurogenesis does occur in the adult nervous system, particularly in the hippocampus and olfactory bulb (Alvarez-Buylla 1992, Alvarez-Buylla & Garcia-Verdugo 2002). Studies to understand the molecular determinants and cues to stimulate endogenous stem cells are under way (Gage 2002, Clarke et al. 2000, Magavi & Macklis 2001). Although promising, we are only beginning to learn the potentials and challenges of these cells, especially for use in neurodegenerative diseases such as ALS.

Given the convergence of multiple pathways leading to disease, various therapies targeting different processes may be the most effective. Although the exact combination of therapies is not yet known, new insights into disease mechanisms and anticipated discoveries of new genes responsible for ALS foster hopes that therapies to significantly slow this disease are within reach.

The *Annual Review of Neuroscience* is online at http://neuro.annualreviews.org

LITERATURE CITED

Abalkhail H, Mitchell J, Habgood J, Orrell R, de Belleroche J. 2003. A new familial amyotrophic lateral sclerosis locus on chromosome 16q12.1-16q12.2. *Am. J. Hum. Genet.* 73:383–89

Al-Chalabi A, Andersen PM, Nilsson P, Chioza B, Andersson JL, et al. 1999. Deletions of the heavy neurofilament subunit tail in amyotrophic lateral sclerosis. *Hum. Mol. Genet.* 8:157–64

Al-Chalabi A, Scheffler MD, Smith BN, Parton MJ, Cudkowicz ME, et al. 2003. Ciliary neurotrophic factor genotype does not influence clinical phenotype in amyotrophic lateral sclerosis. *Ann. Neurol.* 54:130–34

Alexianu ME, Ho BK, Mohamed AH, La Bella

V, Smith RG, Appel SH. 1994. The role of calcium-binding proteins in selective motoneuron vulnerability in amyotrophic lateral sclerosis. *Ann. Neurol.* 36:846–58

Alexianu ME, Kozovska M, Appel SH. 2001. Immune reactivity in a mouse model of familial ALS correlates with disease progression. *Neurology* 57:1282–89

Almer G, Teismann P, Stevic Z, Halaschek-Wiener J, Deecke L, et al. 2002. Increased levels of the pro-inflammatory prostaglandin PGE2 in CSF from ALS patients. *Neurology* 58:1277–79

Alvarez-Buylla A. 1992. Neurogenesis and plasticity in the CNS of adult birds. *Exp. Neurol.* 115:110–14

Alvarez-Buylla A, Garcia-Verdugo JM. 2002. Neurogenesis in adult subventricular zone. *J. Neurosci.* 22:629–34

Anand P, Parrett A, Martin J, Zeman S, Foley P, et al. 1995. Regional changes of ciliary neurotrophic factor and nerve growth factor levels in post mortem spinal cord and cerebral cortex from patients with motor disease. *Nat. Med.* 1:168–72

Andersen PM. 2000. Genetic factors in the early diagnosis of ALS. *Amyotroph. Lateral Scler. Other Motor Neuron Disord.* 1 (Suppl. 1):S31–42

Andersen PM, Spitsyn VA, Makarov SV, Nilsson L, Kravchuk OI, et al. 2001. The geographical and ethnic distribution of the D90A CuZn-SOD mutation in the Russian Federation. *Amyotroph. Lateral Scler. Other Motor Neuron Disord.* 2:63–69

Anderson DJ, Gage FH, Weissman IL. 2001. Can stem cells cross lineage boundaries? *Nat. Med.* 7:393–95

Andreassen OA, Ferrante RJ, Klivenyi P, Klein AM, Shinobu LA, et al. 2000. Partial deficiency of manganese superoxide dismutase exacerbates a transgenic mouse model of amyotrophic lateral sclerosis. *Ann. Neurol.* 47:447–55

Andrus PK, Fleck TJ, Gurney ME, Hall ED. 1998. Protein oxidative damage in a transgenic mouse model of familial amyotrophic lateral sclerosis. *J. Neurochem.* 71:2041–48

Beal MF, Ferrante RJ, Browne SE, Matthews RT, Kowall NW, Brown RH Jr. 1997. Increased 3-nitrotyrosine in both sporadic and familial amyotrophic lateral sclerosis. *Ann. Neurol.* 42:644–54

Beaulieu JM, Nguyen MD, Julien JP. 1999. Late onset death of motor neurons in mice overexpressing wild-type peripherin. *J. Cell Biol.* 147:531–44

Beckman JS, Carson M, Smith CD, Koppenol WH. 1993. ALS, SOD and peroxynitrite. *Nature* 364:584

Beckman JS, Chen J, Crow JP, Ye YZ. 1994. Reactions of nitric oxide, superoxide and peroxynitrite with superoxide dismutase in neurodegeneration. *Prog. Brain Res.* 103:371–80

Blaauwgeers HG, Anwar Chand M, van den Berg FM, Vianney de Jong JM, Troost D. 1996. Expression of different metallothionein messenger ribonucleic acids in motor cortex, spinal cord and liver from patients with amyotrophic lateral sclerosis. *J. Neurol. Sci.* 142:39–44

Blair IP, Bennett CL, Abel A, Rabin BA, Griffin JW, et al. 2000. A gene for autosomal dominant juvenile amyotrophic lateral sclerosis (ALS4) localizes to a 500-kb interval on chromosome 9q34. *Neurogenetics* 3:1–6

Borchelt DR, Lee MK, Slunt HS, Guarnieri M, Xu ZS, et al. 1994. Superoxide dismutase 1 with mutations linked to familial amyotrophic lateral sclerosis possesses significant activity. *Proc. Natl. Acad. Sci. USA* 91:8292–96

Bowling AC, Barkowski EE, McKenna-Yasek D, Sapp P, Horvitz HR, et al. 1995. Superoxide dismutase concentration and activity in familial amyotrophic lateral sclerosis. *J. Neurochem.* 64:2366–69

Brazelton TR, Rossi FM, Keshet GI, Blau HM. 2000. From marrow to brain: expression of neuronal phenotypes in adult mice. *Science* 290:1775–79

Bruening W, Roy J, Giasson B, Figlewicz DA, Mushynski WE, Durham HD. 1999. Up-regulation of protein chaperones preserves viability of cells expressing toxic

Cu/Zn-superoxide dismutase mutants associated with amyotrophic lateral sclerosis. *J. Neurochem.* 72:693–99

Bruijn LI, Beal MF, Becher MW, Schulz JB, Wong PC, et al. 1997a. Elevated free nitrotyrosine levels, but not protein-bound nitrotyrosine or hydroxyl radicals, throughout amyotrophic lateral sclerosis (ALS)-like disease implicate tyrosine nitration as an aberrant in vivo property of one familial ALS-linked superoxide dismutase 1 mutant. *Proc. Natl. Acad. Sci. USA* 94:7606–11

Bruijn LI, Becher MW, Lee MK, Anderson KL, Jenkins NA, et al. 1997b. ALS-linked SOD1 mutant G85R mediates damage to astrocytes and promotes rapidly progressive disease with SOD1-containing inclusions. *Neuron* 18:327–38

Bruijn LI, Cleveland DW. 1996. Mechanisms of selective motor neuron death in ALS: insights from transgenic mouse models of motor neuron disease. *Neuropathol. Appl. Neurobiol.* 22:373–87

Bruijn LI, Houseweart MK, Kato S, Anderson KL, Anderson SD, et al. 1998. Aggregation and motor neuron toxicity of an ALS-linked SOD1 mutant independent from wild-type SOD1. *Science* 281:1851–54

Bush AI. 2002. Is ALS caused by an altered oxidative activity of mutant superoxide dismutase? *Nat. Neurosci.* 5:919

Carpenter S. 1968. Proximal axonal enlargement in motor neuron disease. *Neurology* 18:841–51

Chou SM, Fakadej AV. 1971. Ultrastructure of chromatolytic motoneurons and anterior spinal roots in a case of Werdnig-Hoffmann disease. *J. Neuropathol. Exp. Neurol.* 30:368–79

Clarke DL, Johansson CB, Wilbertz J, Veress B, Nilsson E, et al. 2000. Generalized potential of adult neural stem cells. *Science* 288:1660–63

Clement AM, Nguyen MD, Roberts EA, Garcia ML, Boillee S, et al. 2003. Wild-type nonneuronal cells extend survival of SOD1 mutant motor neurons in ALS mice. *Science* 302:113–17

Corbo M, Hays AP. 1992. Peripherin and neurofilament protein coexist in spinal spheroids of motor neuron disease. *J. Neuropathol. Exp. Neurol.* 51:531–37

Corson LB, Strain JJ, Culotta VC, Cleveland DW. 1998. Chaperone-facilitated copper binding is a property common to several classes of familial amyotrophic lateral sclerosis-linked superoxide dismutase mutants. *Proc. Natl. Acad. Sci. USA* 95:6361–66

Couillard-Despres S, Zhu Q, Wong PC, Price DL, Cleveland DW, Julien JP. 1998. Protective effect of neurofilament heavy gene overexpression in motor neuron disease induced by mutant superoxide dismutase. *Proc. Natl. Acad. Sci. USA* 95:9626–30

Culotta VC, Klomp LW, Strain J, Casareno RL, Krems B, Gitlin JD. 1997. The copper chaperone for superoxide dismutase. *J. Biol. Chem.* 272:23469–72

Dal Canto MC, Gurney ME. 1994. Development of central nervous system pathology in a murine transgenic model of human amyotrophic lateral sclerosis. *Am. J. Pathol.* 145:1271–79

Dawson TM. 2000. New animal models for Parkinson's disease. *Cell* 101:115–18

De Jonghe P, Mersivanova I, Nelis E, Del Favero J, Martin JJ, et al. 2001. Further evidence that neurofilament light chain gene mutations can cause Charcot-Marie-Tooth disease type 2E. *Ann. Neurol.* 49:245–49

Delisle MB, Carpenter S. 1984. Neurofibrillary axonal swellings and amyotrophic lateral sclerosis. *J. Neurol. Sci.* 63:241–50

Ding H, Schwarz DS, Keene A, Affar el B, Fenton L, et al. 2003. Selective silencing by RNAi of a dominant allele that causes amyotrophic lateral sclerosis. *Aging Cell* 2:209–17

Drachman DB, Frank K, Dykes-Hoberg M, Teismann P, Almer G, et al. 2002. Cyclooxygenase 2 inhibition protects motor neurons and prolongs survival in a transgenic mouse model of ALS. *Ann. Neurol.* 52:771–78

Elliott JL. 2001. Cytokine upregulation in a murine model of familial amyotrophic lateral sclerosis. *Brain Res. Mol. Brain Res.* 95:172–78

Elliott JL, Snider WD. 1996. Motor neuron growth factors. *Neurology* 47:S47–53

Estevez AG, Crow JP, Sampson JB, Reiter C, Zhuang Y, et al. 1999. Induction of nitric oxide-dependent apoptosis in motor neurons by zinc-deficient superoxide dismutase. *Science* 286:2498–500

Field LS, Furukawa Y, O'Halloran TV, Culotta VC. 2003. Factors controlling the uptake of yeast copper/zinc superoxide dismutase into mitochondria. *J. Biol. Chem.* 278:28052–59

Figlewicz DA, Krizus A, Martinoli MG, Meininger V, Dib M, et al. 1994. Variants of the heavy neurofilament subunit are associated with the development of amyotrophic lateral sclerosis. *Hum. Mol. Genet.* 3:1757–61

Gage FH. 2002. Neurogenesis in the adult brain. *J. Neurosci.* 22:612–13

Garcia ML, Lobsiger CS, Shah SB, Deerinck TJ, Crum J, et al. 2003. NF-M is an essential target for the mylein-directed "outside-in" signaling cascade that mediates radial axonal growth. *J. Cell Biol.* 163:1011–20

Gaudette M, Hirano M, Siddique T. 2000. Current status of SOD1 mutations in familial amyotrophic lateral sclerosis. *Amyotroph. Lateral Scler. Other Motor Neuron Disord.* 1:83–89

Giess R, Holtmann B, Braga M, Grimm T, Muller-Myhsok B, et al. 2002. Early onset of severe familial amyotrophic lateral sclerosis with a SOD-1 mutation: potential impact of CNTF as a candidate modifier gene. *Am. J. Hum. Genet.* 70:1277–86

Gonatas NK, Stieber A, Mourelatos Z, Chen Y, Gonatas JO, et al. 1992. Fragmentation of the Golgi apparatus of motor neurons in amyotrophic lateral sclerosis. *Am. J. Pathol.* 140:731–37

Gong YH, Elliott JL. 2000. Metallothionein expression is altered in a transgenic murine model of familial amyotrophic lateral sclerosis. *Exp. Neurol.* 162:27–36

Gong YH, Parsadanian AS, Andreeva A, Snider WD, Elliott JL. 2000. Restricted expression of G86R Cu/Zn superoxide dismutase in astrocytes results in astrocytosis but does not cause motoneuron degeneration. *J. Neurosci.* 20:660–65

Goodell MA, Jackson KA, Majka SM, Mi T, Wang H, et al. 2001. Stem cell plasticity in muscle and bone marrow. *Ann. NY Acad. Sci.* 938:208–18

Guegan C, Vila M, Rosoklija G, Hays AP, Przedborski S. 2001. Recruitment of the mitochondrial-dependent apoptotic pathway in amyotrophic lateral sclerosis. *J. Neurosci.* 21:6569–76

Gurney ME. 1994. Transgenic-mouse model of amyotrophic lateral sclerosis. *N. Engl. J. Med.* 331:1721–22

Hadano S, Hand CK, Osuga H, Yanagisawa Y, Otomo A, et al. 2001. A gene encoding a putative GTPase regulator is mutated in familial amyotrophic lateral sclerosis 2. *Nat. Genet.* 29:166–73

Hafezparast M, Klocke R, Ruhrberg C, Marquardt A, Ahmad-Annuar A, et al. 2003. Mutations in dynein link motor neuron degeneration to defects in retrograde transport. *Science* 300:808–12

Haley RW. 2003. Excess incidence of ALS in young Gulf War veterans. *Neurology* 61: 750–56

Hall ED, Andrus PK, Oostveen JA, Fleck TJ, Gurney ME. 1998. Relationship of oxygen radical-induced lipid peroxidative damage to disease onset and progression in a transgenic model of familial ALS. *J. Neurosci. Res.* 53:66–77

Hand CK, Khoris J, Salachas F, Gros-Louis F, Lopes AA, et al. 2002. A novel locus for familial amyotrophic lateral sclerosis, on chromosome 18q. *Am. J. Hum. Genet.* 70:251–56

Hanisch UK. 2002. Microglia as a source and target of cytokines. *Glia* 40:140–55

Hensley K, Floyd RA, Gordon B, Mou S, Pye QN, et al. 2002. Temporal patterns of cytokine and apoptosis-related gene expression in spinal cords of the G93A-SOD1 mouse model of amyotrophic lateral sclerosis. *J. Neurochem.* 82:365–74

Hentati A, Ouahchi K, Pericak-Vance MA, Nijhawan D, Ahmad A, et al. 1998. Linkage of a commoner form of recessive amyotrophic lateral sclerosis to chromosome 15q15-q22 markers. *Neurogenetics* 2:55–60

Higgins CM, Jung C, Ding H, Xu Z. 2002. Mutant Cu, Zn superoxide dismutase that causes motoneuron degeneration is present in mitochondria in the CNS. *J. Neurosci.* 22:RC215

Hirano A. 1991. Cytopathology of amyotrophic lateral sclerosis. *Adv. Neurol.* 56:91–101

Hirano A, Nakano I, Kurland LT, Mulder DW, Holley PW, Saccomanno G. 1984. Fine structural study of neurofibrillary changes in a family with amyotrophic lateral sclerosis. *J. Neuropathol. Exp. Neurol.* 43:471–80

Horner RD, Kamins KG, Feussner JR, Grambow SC, Hoff-Lindquist J, et al. 2003. Occurrence of amyotrophic lateral sclerosis among Gulf War veterans. *Neurology* 61:742–49

Hosler BA, Siddique T, Sapp PC, Sailor W, Huang MC, et al. 2000. Linkage of familial amyotrophic lateral sclerosis with frontotemporal dementia to chromosome 9q21-q22. *JAMA* 284:1664–69

Howland DS, Liu J, She Y, Goad B, Maragakis NJ, et al. 2002. Focal loss of the glutamate transporter EAAT2 in a transgenic rat model of SOD1 mutant-mediated amyotrophic lateral sclerosis (ALS). *Proc. Natl. Acad. Sci. USA* 99:1604–9

Ince P, Stout N, Shaw P, Slade J, Hunziker W, et al. 1993. Parvalbumin and calbindin D-28k in the human motor system and in motor neuron disease. *Neuropathol. Appl. Neurobiol.* 19:291–99

Ince PG, Shaw PJ, Slade JY, Jones C, Hudgson P. 1996. Familial amyotrophic lateral sclerosis with a mutation in exon 4 of the Cu/Zn superoxide dismutase gene: pathological and immunocytochemical changes. *Acta Neuropathol. (Berlin)* 92:395–403

Jaarsma D, Haasdijk ED, Grashorn JA, Hawkins R, van Duijn W, et al. 2000. Human Cu/Zn superoxide dismutase (SOD1) overexpression in mice causes mitochondrial vacuolization, axonal degeneration, and premature motoneuron death and accelerates motoneuron disease in mice expressing a familial amyotrophic lateral sclerosis mutant SOD1. *Neurobiol. Dis.* 7:623–43

Jaarsma D, Rognoni F, van Duijn W, Verspaget HW, Haasdijk ED, Holstege JC. 2001. CuZn superoxide dismutase (SOD1) accumulates in vacuolated mitochondria in transgenic mice expressing amyotrophic lateral sclerosis-linked SOD1 mutations. *Acta Neuropathol. (Berlin)* 102:293–305

Jacob C, Maret W, Vallee BL. 1998. Control of zinc transfer between thionein, metallothionein, and zinc proteins. *Proc. Natl. Acad. Sci. USA* 95:3489–94

Johnston JA, Dalton MJ, Gurney ME, Kopito RR. 2000. Formation of high molecular weight complexes of mutant Cu, Zn-superoxide dismutase in a mouse model for familial amyotrophic lateral sclerosis. *Proc. Natl. Acad. Sci. USA* 97:12571–76

Jonsson PA, Ernhill K, Andersen PM, Bergemalm D, Brannstrom T, et al. 2004. Minute quantities of misfolded mutant superoxide dismutase-1 cause amyotrophic lateral sclerosis. *Brain* 127:73–88

Kaspar BK, Llado J, Sherkat N, Rothstein JD, Gage FH. 2003. Retrograde viral delivery of IGF-1 prolongs survival in a mouse ALS model. *Science* 301:839–42

Kato S, Takikawa M, Nakashima K, Hirano A, Cleveland DW, et al. 2000. New consensus research on neuropathological aspects of familial amyotrophic lateral sclerosis with superoxide dismutase 1 (SOD1) gene mutations: inclusions containing SOD1 in neurons and astrocytes. *Amyotroph. Lateral Scler. Other Motor Neuron Disord.* 1:163–84

Kawamata T, Akiyama H, Yamada T, McGeer PL. 1992. Immunologic reactions in amyotrophic lateral sclerosis brain and spinal cord tissue. *Am. J. Pathol.* 140:691–707

Kawamura Y, Dyck PJ, Shimono M, Okazaki H, Tateishi J, Doi H. 1981. Morphometric comparison of the vulnerability of peripheral motor and sensory neurons in amyotrophic lateral sclerosis. *J. Neuropathol. Exp. Neurol.* 40:667–75

King SJ, Schroer TA. 2000. Dynactin increases

the processivity of the cytoplasmic dynein motor. *Nat. Cell Biol.* 2:20–24

Klivenyi P, Ferrante RJ, Matthews RT, Bogdanov MB, Klein AM, et al. 1999. Neuroprotective effects of creatine in a transgenic animal model of amyotrophic lateral sclerosis. *Nat. Med.* 5:347–50

Kong J, Xu Z. 1998. Massive mitochondrial degeneration in motor neurons triggers the onset of amyotrophic lateral sclerosis in mice expressing a mutant SOD1. *J. Neurosci.* 18:3241–50

Kostic V, Jackson-Lewis V, de Bilbao F, Dubois-Dauphin M, Przedborski S. 1997. Bcl-2: prolonging life in a transgenic mouse model of familial amyotrophic lateral sclerosis. *Science* 277:559–62

Kreutzberg GW. 1996. Microglia: a sensor for pathological events in the CNS. *Trends Neurosci.* 19:312–18

Kriz J, Nguyen MD, Julien JP. 2002. Minocycline slows disease progression in a mouse model of amyotrophic lateral sclerosis. *Neurobiol. Dis.* 10:268–78

Kunst CB, Messer L, Gordon J, Haines J, Patterson D. 2000. Genetic mapping of a mouse modifier gene that can prevent ALS onset. *Genomics* 70:181–89

Lambrechts D, Storkebaum E, Morimoto M, Del-Favero J, Desmet F, et al. 2003. VEGF is a modifier of amyotrophic lateral sclerosis in mice and humans and protects motoneurons against ischemic death. *Nat. Genet.* 34:383–94

LaMonte BH, Wallace KE, Holloway BA, Shelly SS, Ascano J, et al. 2002. Disruption of dynein/dynactin inhibits axonal transport in motor neurons causing late-onset progressive degeneration. *Neuron* 34:715–27

Lariviere RC, Beaulieu JM, Nguyen MD, Julien JP. 2003. Peripherin is not a contributing factor to motor neuron disease in a mouse model of amyotrophic lateral sclerosis caused by mutant superoxide dismutase. *Neurobiol. Dis.* 13:158–66

Lee MK, Cleveland DW. 1996. Neuronal intermediate filaments. *Annu. Rev. Neurosci.* 19:187–217

Leigh PN, Swash M. 1991. Cytoskeletal pathology in motor neuron diseases. *Adv. Neurol.* 56:115–24

Leigh PN, Whitwell H, Garofalo O, Buller J, Swash M, et al. 1991. Ubiquitin-immunoreactive intraneuronal inclusions in amyotrophic lateral sclerosis. Morphology, distribution, and specificity. *Brain* 114:775–88

Li M, Ona VO, Guegan C, Chen M, Jackson-Lewis V, et al. 2000. Functional role of caspase-1 and caspase-3 in an ALS transgenic mouse model. *Science* 288:335–39

Lin CL, Bristol LA, Jin L, Dykes-Hoberg M, Crawford T, et al. 1998. Aberrant RNA processing in a neurodegenerative disease: the cause for absent EAAT2, a glutamate transporter, in amyotrophic lateral sclerosis. *Neuron* 20:589–602

Lino MM, Schneider C, Caroni P. 2002. Accumulation of SOD1 mutants in postnatal motoneurons does not cause motoneuron pathology or motoneuron disease. *J. Neurosci.* 22:4825–32

Liochev SI, Fridovich I. 2000. Copper- and zinc-containing superoxide dismutase can act as a superoxide reductase and a superoxide oxidase. *J. Biol. Chem.* 275:38482–85

Liu H, Zhu H, Eggers DK, Nersissian AM, Faull KF, et al. 2000. Copper(2+) binding to the surface residue cysteine 111 of His46Arg human copper-zinc superoxide dismutase, a familial amyotrophic lateral sclerosis mutant. *Biochemistry* 39:8125–32

Magavi SS, Macklis JD. 2001. Manipulation of neural precursors in situ: induction of neurogenesis in the neocortex of adult mice. *Neuropsychopharmacology* 25:816–35

Mather K, Martin JE, Swash M, Vowles G, Brown A, Leigh PN. 1993. Histochemical and immunocytochemical study of ubiquitinated neuronal inclusions in amyotrophic lateral sclerosis. *Neuropathol. Appl. Neurobiol.* 19:141–45

Mattiazzi M, D'Aurelio M, Gajewski CD, Martushova K, Kiaei M, et al. 2002. Mutated human SOD1 causes dysfunction of oxidative phosphorylation in mitochondria of transgenic mice. *J. Biol. Chem.* 277:29626–33

McGeer EG, McGeer PL. 1999. Brain inflammation in Alzheimer disease and the therapeutic implications. *Curr. Pharm. Des.* 5:821–36

Mersiyanova IV, Perepelov AV, Polyakov AV, Sitnikov VF, Dadali EL, et al. 2000. A new variant of Charcot-Marie-Tooth disease type 2 is probably the result of a mutation in the neurofilament-light gene. *Am. J. Hum. Genet.* 67:37–46

Mulder DW, Kurland LT. 1987. Motor neuron disease: epidemiologic studies. *Adv. Exp. Med. Biol.* 209:325–32

Mulder DW, Kurland LT, Offord KP, Beard CM. 1986. Familial adult motor neuron disease: amyotrophic lateral sclerosis. *Neurology* 36:511–17

Nagai M, Aoki M, Miyoshi I, Kato M, Pasinelli P, et al. 2001. Rats expressing human cytosolic copper-zinc superoxide dismutase transgenes with amyotrophic lateral sclerosis: associated mutations develop motor neuron disease. *J. Neurosci.* 21:9246–54

Nagano S, Satoh M, Sumi H, Fujimura H, Tohyama C, et al. 2001. Reduction of metallothioneins promotes the disease expression of familial amyotrophic lateral sclerosis mice in a dose-dependent manner. *Eur. J. Neurosci.* 13:1363–70

Nguyen MD, Julien JP, Rivest S. 2001a. Induction of proinflammatory molecules in mice with amyotrophic lateral sclerosis: no requirement for proapoptotic interleukin-1beta in neurodegeneration. *Ann. Neurol.* 50:630–39

Nguyen MD, Lariviere RC, Julien JP. 2001b. Deregulation of Cdk5 in a mouse model of ALS: toxicity alleviated by perikaryal neurofilament inclusions. *Neuron* 30:135–47

Niwa J, Ishigaki S, Hishikawa N, Yamamoto M, Doyu M, et al. 2002. Dorfin ubiquitylates mutant SOD1 and prevents mutant SOD1-mediated neurotoxicity. *J. Biol. Chem.* 277:36793–98

Oosthuyse B, Moons L, Storkebaum E, Beck H, Nuyens D, et al. 2001. Deletion of the hypoxia-response element in the vascular endothelial growth factor promoter causes motor neuron degeneration. *Nat. Genet.* 28:131–38

Oppenheim RW. 1996. Neurotrophic survival molecules for motoneurons: an embarrassment of riches. *Neuron* 17:195–97

Otomo A, Hadano S, Okada T, Mizumura H, Kunita R, et al. 2003. ALS2, a novel guanine nucleotide exchange factor for the small GTPase Rab5, is implicated in endosomal dynamics. *Hum. Mol. Genet.* 12:1671–87

Palmiter RD. 1998. The elusive function of metallothioneins. *Proc. Natl. Acad. Sci. USA* 95:8428–30

Pasinelli P, Houseweart MK, Brown RH, Jr., Cleveland DW. 2000. Caspase-1 and -3 are sequentially activated in motor neuron death in Cu, Zn superoxide dismutase-mediated familial amyotrophic lateral sclerosis. *Proc. Natl. Acad. Sci. USA* 97:13901–6

Pramatarova A, Laganiere J, Roussel J, Brisbois K, Rouleau GA. 2001. Neuron-specific expression of mutant superoxide dismutase 1 in transgenic mice does not lead to motor impairment. *J. Neurosci.* 21:3369–74

Puls I, Jonnakuty C, LaMonte BH, Holzbaur EL, Tokito M, et al. 2003. Mutant dynactin in motor neuron disease. *Nat. Genet.* 33:455–56

Rae TD, Schmidt PJ, Pufahl RA, Culotta VC, O'Halloran TV. 1999. Undetectable intracellular free copper: the requirement of a copper chaperone for superoxide dismutase. *Science* 284:805–8

Raoul C, Estevez AG, Nishimune H, Cleveland DW, deLapeyriere O, et al. 2002. Motoneuron death triggered by a specific pathway downstream of Fas: potentiation by ALS-linked SOD1 mutations. *Neuron* 35:1067–83

Reaume AG, Elliott JL, Hoffman EK, Kowall NW, Ferrante RJ, et al. 1996. Motor neurons in Cu/Zn superoxide dismutase-deficient mice develop normally but exhibit enhanced cell death after axonal injury. *Nat. Genet.* 13:43–47

Ripps ME, Huntley GW, Hof PR, Morrison JH, Gordon JW. 1995. Transgenic mice expressing an altered murine superoxide dismutase gene provide an animal model of amyotrophic lateral sclerosis. *Proc. Natl. Acad. Sci. USA* 92:689–93

Robertson J, Doroudchi MM, Nguyen MD, Durham HD, Strong MJ, et al. 2003. A neurotoxic peripherin splice variant in a mouse model of ALS. *J. Cell Biol.* 160:939–49

Rosen DR, Siddique T, Patterson D, Figlewicz DA, Sapp P, et al. 1993. Mutations in Cu/Zn superoxide dismutase gene are associated with familial amyotrophic lateral sclerosis. *Nature* 362:59–62

Rothstein JD, Dykes-Hoberg M, Pardo CA, Bristol LA, Jin L, et al. 1996. Knockout of glutamate transporters reveals a major role for astroglial transport in excitotoxicity and clearance of glutamate. *Neuron* 16:675–86

Rothstein JD, Kuncl R, Chaudhry V, Clawson L, Cornblath DR, et al. 1991. Excitatory amino acids in amyotrophic lateral sclerosis: an update. *Ann. Neurol.* 30:224–25

Rothstein JD, Tsai G, Kuncl RW, Clawson L, Cornblath DR, et al. 1990. Abnormal excitatory amino acid metabolism in amyotrophic lateral sclerosis. *Ann. Neurol.* 28:18–25

Rothstein JD, Van Kammen M, Levey AI, Martin LJ, Kuncl RW. 1995. Selective loss of glial glutamate transporter GLT-1 in amyotrophic lateral sclerosis. *Ann. Neurol.* 38:73–84

Ruddy DM, Parton MJ, Al-Chalabi A, Lewis CM, Vance C, et al. 2003. Two families with familial amyotrophic lateral sclerosis are linked to a novel locus on chromosome 16q. *Am. J. Hum. Genet.* 73:390–96

Sapp PC, Hosler BA, McKenna-Yasek D, Chin W, Gann A, et al. 2003. Identification of two novel loci for dominantly inherited familial amyotrophic lateral sclerosis. *Am. J. Hum. Genet.* 73:397–403

Selkoe DJ. 2001. Presenilin, Notch, and the genesis and treatment of Alzheimer's disease. *Proc. Natl. Acad. Sci. USA* 98:11039–41

Shaw PJ, Forrest V, Ince PG, Richardson JP, Wastell HJ. 1995. CSF and plasma amino acid levels in motor neuron disease: elevation of CSF glutamate in a subset of patients. *Neurodegeneration* 4:209–16

Shibata N, Asayama K, Hirano A, Kobayashi M. 1996. Immunohistochemical study on superoxide dismutases in spinal cords from autopsied patients with amyotrophic lateral sclerosis. *Dev. Neurosci.* 18:492–98

Shinder GA, Lacourse MC, Minotti S, Durham HD. 2001. Mutant Cu/Zn-superoxide dismutase proteins have altered solubility and interact with heat shock/stress proteins in models of amyotrophic lateral sclerosis. *J. Biol. Chem.* 276:12791–96

Son M, Fathallah-Shaykh HM, Elliott JL. 2001. Survival in a transgenic model of FALS is independent of iNOS expression. *Ann. Neurol.* 50:273

Steffan JS, Bodai L, Pallos J, Poelman M, McCampbell A, et al. 2001. Histone deacetylase inhibitors arrest polyglutamine-dependent neurodegeneration in Drosophila. *Nature* 413:739–43

Subramaniam JR, Lyons WE, Liu J, Bartnikas TB, Rothstein J, et al. 2002. Mutant SOD1 causes motor neuron disease independent of copper chaperone-mediated copper loading. *Nat. Neurosci.* 5:301–7

Takahashi R, Yokoji H, Misawa H, Hayashi M, Hu J, Deguchi T. 1994. A null mutation in the human CNTF gene is not causally related to neurological diseases. *Nat. Genet.* 7:79–84

Takeuchi H, Kobayashi Y, Ishigaki S, Doyu M, Sobue G. 2002. Mitochondrial localization of mutant superoxide dismutase 1 triggers caspase-dependent cell death in a cellular model of familial amyotrophic lateral sclerosis. *J. Biol. Chem.* 277:50966–72

Tanaka K, Watase K, Manabe T, Yamada K, Watanabe M, et al. 1997. Epilepsy and exacerbation of brain injury in mice lacking the glutamate transporter GLT-1. *Science* 276:1699–702

Tomkins J, Usher P, Slade JY, Ince PG, Curtis A, et al. 1998. Novel insertion in the KSP region of the neurofilament heavy gene in amyotrophic lateral sclerosis (ALS). *NeuroReport* 9:3967–70

Trotti D, Aoki M, Pasinelli P, Berger UV, Danbolt NC, et al. 2001. Amyotrophic lateral sclerosis-linked glutamate transporter mutant has impaired glutamate clearance capacity. *J. Biol. Chem.* 276:576–82

Van Den Bosch L, Tilkin P, Lemmens G, Robberecht W. 2002. Minocycline delays disease onset and mortality in a transgenic model of ALS. *NeuroReport* 13:1067–70

Vukosavic S, Stefanis L, Jackson-Lewis V, Guegan C, Romero N, et al. 2000. Delaying caspase activation by Bcl-2: A clue to disease retardation in a transgenic mouse model of amyotrophic lateral sclerosis. *J. Neurosci.* 20:9119–25

Wang J, Slunt H, Gonzales V, Fromholt D, Coonfield M, et al. 2003. Copper-binding-site-null SOD1 causes ALS in transgenic mice: aggregates of non-native SOD1 delineate a common feature. *Hum. Mol. Genet.* 12:2753–64

Wang J, Xu G, Gonzales V, Coonfield M, Fromholt D, et al. 2002. Fibrillar inclusions and motor neuron degeneration in transgenic mice expressing superoxide dismutase 1 with a disrupted copper-binding site. *Neurobiol. Dis.* 10:128–38

Wichterle H, Lieberam I, Porter JA, Jessell TM. 2002. Directed differentiation of embryonic stem cells into motor neurons. *Cell* 110:385–97

Wiedau-Pazos M, Goto JJ, Rabizadeh S, Gralla EB, Roe JA, et al. 1996. Altered reactivity of superoxide dismutase in familial amyotrophic lateral sclerosis. *Science* 271:515–18

Wilhelmsen KC. 1997. Disinhibition-dementia-parkinsonism-amyotrophy complex (DDPAC) is a non-Alzheimer's frontotemporal dementia. *J. Neural Transm. Suppl.* 49:269–75

Williamson TL, Bruijn LI, Zhu Q, Anderson KL, Anderson SD, et al. 1998. Absence of neurofilaments reduces the selective vulnerability of motor neurons and slows disease caused by a familial amyotrophic lateral sclerosis-linked superoxide dismutase 1 mutant. *Proc. Natl. Acad. Sci. USA* 95:9631–36

Williamson TL, Corson LB, Huang L, Burlingame A, Liu J, et al. 2000. Toxicity of ALS-linked SOD1 mutants. *Science* 288:399

Wong PC, Pardo CA, Borchelt DR, Lee MK, Copeland NG, et al. 1995. An adverse property of a familial ALS-linked SOD1 mutation causes motor neuron disease characterized by vacuolar degeneration of mitochondria. *Neuron* 14:1105–16

Wong PC, Waggoner D, Subramaniam JR, Tessarollo L, Bartnikas TB, et al. 2000. Copper chaperone for superoxide dismutase is essential to activate mammalian Cu/Zn superoxide dismutase. *Proc. Natl. Acad. Sci. USA* 97:2886–91

Yamanaka K, Vande Velde C, Bertini E, Boespflug-Tanguy O, Cleveland DW. 2003. Unstable mutants in the peripheral endosomal membrane component ALS2 cause early onset motor neuron disease. *Proc. Natl. Acad. Sci.* 100:16041–46

Yang Y, Hentati A, Deng HX, Dabbagh O, Sasaki T, et al. 2001. The gene encoding alsin, a protein with three guanine-nucleotide exchange factor domains, is mutated in a form of recessive amyotrophic lateral sclerosis. *Nat. Genet.* 29:160–65

Yrjanheikki J, Tikka T, Keinanen R, Goldsteins G, Chan PH, Koistinaho J. 1999. A tetracycline derivative, minocycline, reduces inflammation and protects against focal cerebral ischemia with a wide therapeutic window. *Proc. Natl. Acad. Sci. USA* 96:13496–500

Zhu S, Stavrovskaya IG, Drozda M, Kim BY, Ona V, et al. 2002. Minocycline inhibits cytochrome c release and delays progression of amyotrophic lateral sclerosis in mice. *Nature* 417:74–78

Subject Index

D

W

CUMULATIVE INDEXES

CONTRIBUTING AUTHORS, VOLUMES 18–27

CHAPTER TITLES, VOLUMES 18–27

Circadian and Other Rhythms

Clinical Neuroscience

Computational Approaches

Cytoskeleton and Axonal Transport

Degeneration and Repair

Developmental Neurobiology

Glia, Schwann Cells, and Extracellular Matrix

Higher Cognitive Functions

Hippocampus

Miscellaneous

Molecular Neuroscience

Motor Systems

Neuronal Membranes

Neuronal Plasticity

Neuropeptides

Neuroscience Techniques

Neurotrophins

Olfaction/taste

Pain

Plasticity

Receptors and Receptor Subtypes

Sleep and Sleep Disorders

Somatosensory System

Synapses/Synaptic Transmission

Vision

Visual System